Manual of Structural Design and Engineering Solutions

Manual of Structural Design and Engineering Solutions

Maurice E. Walmer, P.E., A.R.A.

with the assistance of

Stephen L. Baron, P.E.

PRENTICE-HALL, Inc. ENGLEWOOD CLIFFS, N. J.

PRENTICE-HALL INTERNATIONAL, INC., *London*
PRENTICE-HALL OF AUSTRALIA, PTY. LTD., *Sydney*
PRENTICE-HALL OF CANADA, LTD., *Toronto*
PRENTICE-HALL OF INDIA PRIVATE LTD., *New Delhi*
PRENTICE-HALL OF JAPAN, INC., *Tokyo*

Library of Congress Cataloging in Publication Data

Walmer, Maurice E.
Manual of structural design and engineering solutions.

1. Structural design—Problems, exercises, etc.
I. Baron, Stephen L., joint author. II. Title.
TA658.35W34 624'.1771'076 72-10236
ISBN 0-13-555573-6

PRINTED IN THE UNITED STATES OF AMERICA

Dedicated to my daughters
Arloene and Rita Carol

Maurice E. Walmer

How the practicing engineer and architect, the junior engineer, the designer, the draftsman, the inspector, and the engineer in training will use the MANUAL of STRUCTURAL DESIGN and ENGINEERING SOLUTIONS

The ***designer, draftsman, inspector,*** *and* ***engineer in training*** *may use this Manual as a complete, self-guided course in structural engineering. Each section includes a summary of the historical development of the design, the design theories and methods, current practice, available materials, and economic considerations, in addition to the example solutions and reference tables. The authors suggest that the most effective study technique is the TEXT-EXAMPLE-TABLE-EXAMPLE method. First, read a single topic of design in the text. Then, study through an example solution illustrating this topic, referring to each table, formula, graph and pilot diagram as it is referenced in the example. Finally, work through several more related examples. Try to understand each calculation—what is known and what is to be found. Take note of the work format and notations; consulting engineers must use a standardized format for synthesis so that design calculations may easily be checked.*

This Manual has been organized with sections, sub-sections, and paragraphs outlined by a decimal indexing system. Major topics within a section are numbered X.X (for example, 4.1 Concrete), *sub-topics are numbered X.X.X* (4.1.5 Concrete mix design), *and example solutions for a sub-topic may be numbered X.X.X.X* (4.1.5.3 Batch design for strength). By using this decimal *system, the reader can organize his study by topic and subtopic.*

The ***junior engineer*** *may use this Manual to prepare and review for state registration examinations in civil and structural engineering and architecture. The authors recommend the EXAMPLE-TABLE-TEXT method. Work through several example solutions for each topic, gaining familiarity with the tables, graphs and pilot diagrams. Refer back to the text for additional explanation. This Manual has been organized so that reference tables are placed near related examples. Many of the examples* *were taken almost directly from state registration examinations. When the reader gains complete familiarity with the reference tables as well as the example solutions, this Manual will be a welcome companion for open-book examinations.*

The ***practicing engineer*** *and* ***architect*** *may use this Manual to determine the formulas which apply to their design projects, to review the steps to follow for a reliable design solution, and to adopt an accepted, concise format for calculations. Many similar design problems may be solved by substituting the actual numerical values into the design formulas in the appropriate example.*

Pilot Diagrams are included as a guide to the solution of large classes of actual design problems. They provide a framework so that the engineer can move quickly to a complete, error-free design.

The authors have attempted to include all necessary reference tables, charts and illustrations, so that the practicing engineer will have a complete desk reference for structural engineering. The Table of Contents includes page numbers in addition to the decimal index numbers, so that a specific example or table may be easily located.

Maurice E. Walmer, P.E., A.R.A.
Stephen L. Baron, P.E.

ACKNOWLEDGEMENTS

I wish to express my gratitude to the following for their assistance and support: Mr. Stephen L. Baron, P.E., my associate in the preparation of this work for publication; the late Carl Lars Svenson, P.E., of Lubbock, Texas; the late Thur Thelander, Professor at Malmo University, Sweden; Mr. James Ainsworth, P.E., of Southwestern Laboratories, Houston, Texas; Mr. Herbert J. Vallat, of Mississippi Valley Equipment Company, St. Louis, Missouri; Mr. Gerald R. Manning, of Raymond International, Inc., Houston, Texas; and the late Leonard H. Bailey, F.A.I.A., of Oklahoma City, Oklahoma.

Also, I wish to acknowledge the very considerable information supplied over a long span of years by such associations as: The American Iron and Steel Institute, for its publication Steelways, *and for* Steel Pipe News, *published by its Committee of Steel Pipe Producers; the American Institute of Steel Construction, publisher of* Modern Steel Construction, *the Canadian Wood Council, of Ottawa; and the Southern Pine Association, of New Orleans, Louisiana.*

Finally, I wish to thank the following building materials and equipment manufacturers for their assistance: Alpha Portland Cement Company, Easton, Pennsylvania; Aluminum Company of America, Pittsburgh, Pennsylvania; Armco Steel Corporation, Kansas City Missouri; Bayou Gasket and Hose Company, Beaumont, Texas; W. R. Grace and Company, Cambridge, Massachusetts; Hohman and Barnard, Inc., Houston, Texas; Inland Steel Company, Chicago Illinois; International Galvanizers, Inc., Beaumont, Texas; Lamson and Sessions Company, Cleveland, Ohio; Lone Star Cement Company, Houston Texas; Mitsubishi International Corp., Tokyo, Japan; Moncrief-Lenoir Manufacturing Company, Houston, Texas; National Tube Division of U. S. Steel Corp., Pittsburgh, Pennsylvania; Republic Steel Corporation, Manufacturing Division, Youngstown, Ohio; Reynolds Metals Company, Richmond, Virginia; Smith Materials Corporation, Beaumont, Texas; Stran-Steel Corporation; Houston, Texas; United States Steel Corporation, Pittsburgh, Pennsylvania. Unit Structures (Division of Koppers Company), Peshtigo, Wisconsin.

Maurice E. Walmer, P.E., A.R.A.

TABLE OF CONTENTS

Detailed listings of contents will be found at the beginning of each section

Manual of Structural Design and Engineering Solutions

I—MECHANICS OF BEAMS

I—MECHANICS OF BEAMS

1

MECHANICS OF BEAMS

MECHANICS OF BEAMS

Contents

Mechanics 1.1

Mechanics is the objective analysis of the action and effect of forces. *Force* is defined as any cause or action tending to produce or modify motion. In the fabrication of a machine, the construction of a structure or the planning of an aircraft, the designing phase must start with a study of the *external* forces, followed by a study of *internal* forces between individual parts. Another subject: *Strength of Materials* refers to the deformation of the structural parts when forces are applied, and also to the determination of the dimensions and properties of the various parts required to support these forces.

Besides *force*, there are other related factors in mechanics from which numerous compound values are derived. These are *distance* and *time*. *Distance* is measured in linear units: inches, feet and meters. *Time* is measured in hours, minutes and seconds.

Action versus reaction: Force 1.1.1

The English philosopher Sir Isaac Newton (1642–1727) first described the force that acts on an apple as it falls from a tree to the ground. Newton called the force which pulls the apple downward toward the earth, the force of gravity. He also first noted that forces always come in pairs. When one force is applied to a body, there is an equal and opposite force resisting the first force. When the body remains motionless or *static*, the two forces are in equilibrium. Thus, in a state of rest, there are two forces acting on a body. They are referred to as the *Action* and *Reaction*. For practical applications, the engineer must provide structural supports with enough reaction resistance to equal the load forces and thereby prevent any movement.

KINETIC ENERGY

In the design of beams and girders, kinetic forces are given only a minimum consideration. Obviously, in locations where seismic forces occur at frequent intervals, they should be taken into account. Kinetic force is actually a product of energy: the kinetic energy of motion, explosive energy or impact energy. An illustration of a kinetic force is seen in the action of a pile hammer. The impact of the hammer acts upon a pile to push it downward and cause movement instead of equilibrium. Kinetic Energy is measured in foot-pounds. In Section IX, *Pile Driving and Fendering*, information on the control of kinetic energy will be presented.

Action versus reaction: Force, continued 1.1.1.

FORCE UNIT MEASURE

From Newton's observations, the unit measure of force has been established as the earth's gravitational pull on a one pound mass. This unit measure is called, in engineering language the *pound-force*, which may be converted to other force units such as *tons* (2000 pounds) or *kips* (1000 pounds). Mass is the quantity of matter that is contained in a body and will be defined in Section IX. One cannot over-emphasize the importance of using the correct units in the solution of problems in Mechanics.

Moment 1.1.2

The moment of a force about a point is equal to the product of the force multiplied by the perpendicular distance from the point to the line of action of the force. This perpendicular distance is also called the *lever arm* of the force.

When a force representing a wind pressure or load acts upon a body such as a beam, and the action is balanced so that no movement takes place, then all the forces acting on the body are in static equilibrium. In order to solve for equilibrium with two or more forces, we give emphasis to the method of moments, and consider the lever arm of each force. In Section VI, we will use lever and moment calculations to find the Center of Gravity and Moment of Inertia of plane sections.

MOMENT ARM

Consider a moment arm as representing a lever with a weight on one end which would tend to cause a rotating action in a circular arc. From a certain point on a beam, this rotating action could be either clockwise or counter-clockwise. This lever is actually a distance and can be measured in inches or feet, and the weight on the end will be in pounds, kips or tons. Then, multiplying weight in pounds times distance in feet gives a *Foot-Pound* result. To illustrate: Let W be a weight of 150 Pounds, and the moment lever (L) is 15.0 Feet. Then Moment M = WL or M = 150 x 15.0 = 2250 Ft. Lb. For convenience, the results can be written as follows: M= 2250′# or as M = 2250 x 12 = 27,000″# (Inch Pounds). It was previously mentioned that the lever is the arm perpendicular to the force. A simple definition may be made:

A moment of a force or load is the product of force times distance, and *the distance is to be measured perpendicular to the line of action of the force.*

It is interesting to recall the many uses of the lever in our modern world. In some form or other, the lever is employed in such contrivances as the phone dial, radio knobs, the claw hammer, door hinges, the broom, the steering wheel, and a vast number of other things.

Reactions in loaded beams 1.2.1

In the case of beams supported by bearing walls or columns, each support reacts with an upward force at the point of support, and this force is called the *Reaction*. To keep the beam in balance or equilibrium, the upward force (R) must be equal to the loads or downward forces

In theory:

If a body is to be placed in equilibrium, then for every action, there must be an equal and opposite reaction.

Newton's Law of Motion can be stated:

Bodies in equilibrium which remain at rest are classed as *Static*. The word is related to *stationary*, which suggests the lack of movement.

A body at rest will remain so until a force acting upon it causes it to move.

Beam shear 1.2.2

The beam loads act downward, and the forces of the reactions at the supports resisting movement act upward. There is a tendency for these opposite acting forces to shear or cut the beam where the supporting member is located. This action is called vertical shear and is denoted by the capital letter V. The shearing force is usually greater near the supports than at any other points on the beam. At the support, the force of reaction (R) is the same as V, or $R_1 = V_1$.

The vertical shear at any point on a beam can be found thus:

From the reaction at the support, subtract the load or loads between that support and the point of interest on the beam.

The Shear Diagrams illustrate this fact very effectively as well as emphasizing the shear action under the two types of load forces (distributed or concentrated loads).

There is another type of shear stress called *Horizontal Shear*. These two shear types must not be confused and we will give more attention to the distinction in Sections II, III, and IV. Both vertical and horizontal shear are internal unit stresses which must be resisted by the beam material. Horizontal shear stress is illustrated in the design examples included in the Sections on Timber and Steel.

Bending moment 1.2.3

Bending moment (M or BM) may best be defined as the summation (Σ) of moments of the external forces about any point on the beam. It should be noted that the bending moment will vary from point to point, as will be illustrated in the algebraic equations and diagrams given in the beam examples which follow.

The loads on the beam and the reactions at the supports constitute the external forces which produce the shear and bending stress in the beam. When constructing a shear diagram, observe that the bending moment will reach a maximum value on the beam at a point where the shear decreases to zero. Since the magnitude of

Bending moment, continued 1.2.3

the maximum bending moment will determine the selection and design of the beam section, there must be no chance for error in the computation. Absolute certainty is best attained by making additional calculations at points close to the point of maximum bending. A Bending moment is expressed in foot pounds, inch pounds or foot kips (foot kilo-pounds or one thousand foot-pounds). Designations and abbreviations may vary in the examples in this manual; we will use several signs: M = 5280 Ft. Lbs., or M = 5280′#, also M = 5280 x 12 = 63,360″# (inch-lbs.). Remember, the magnitude of a moment is the value of a force multiplied by its lever (perpendicular distance) from its line of action to the point about which the moment is taken).

Consider a concentrated load on an overhanging beam which is cantilevered out from the support. Let load P = 2000 Lbs., and locate load 3.0 feet from its support. The moment at the support will be 2000 x 3.0 = 6,000 Foot Pounds, or 2000 x 3.0 x 12 = 72,000″#. The line of action of the force under a short uniform load is at the center of gravity of the load. The center of gravity of a full uniform load is located at the middle of the span. By studying the examples which include uniform loads, it will be noted that the loads are marked off in one foot units, and each unit has its own center of gravity. Thus, for each unit or part of a uniform load, there is a lever point to use to find the distance from the load to the point of interest in calculating the moment.

POSITIVE OR NEGATIVE BENDING

Loads placed on a simply supported beam will give the beam a tendency to sag. This sag is called deflection or deformation. Under such circumstances, the fibers in the lower half of the beam are in tension stress. These lower fibers are being stretched, while the upper fibers are being compressed. Under these conditions the bending moment is positive, and is designated by a plus sign (+M) or, if no sign is given, it will be understood to be a positive moment. In overhanging beams (cantilevers), the load placement causes tension in the top fibers and compression in the bottom fibers. The bending moment is then designated as negative by a minus sign (−M).

The horizontal plane in the beam which divides the fibers in tension from those in compression is called the beam neutral axis or centroid. Horizontal shear is developed along this centroid, because the fibers in which compressive stress is present will tend to slide in an opposite direction from the fibers which contain the tension stress, causing a shearing action in the horizontal plane. Refer to Section VI where we discuss the method of locating the centroid and determining the horizontal shear.

Beam bending formulas 1.2.4

Mathematicians have developed many beam formulas to assist engineers with rapid solutions in finding the reactions, maximum bending moments and probable deflections under various types of load arrangements. These formulas are listed in many handbooks on concrete, steel and aluminum. These formulas are known as *empirical equations.* Before selecting one of these formulas, make a careful survey of load placement and type of load. Examine the end conditions at the supports. Are one or both ends fixed, as would be the case when concrete beams are formed together with the supporting columns in a monolithic structure? If you are not completely satisfied with the results obtained by formula, the results can be checked by the algebraic equations as will be illustrated in the examples.

Solving beam bending problems 1.3

The established method for solving mechanics of beam problems involves following a proper sequence of several steps or stages. Each of these steps should be performed as a separate item and the algebraic equations should be neatly done in the form given in the examples. Use the sequence of steps in the following manner:

STEP I: Beam Elevation

Draw the beam and load placement to a convenient scale, with the loads properly spaced and dimensioned, and the points for moment distances noted. Designate each load and its value. Always work from the extreme left end in the manner illustrated, and designate the bending moment from the left end thus: $M_{12.0}$ implies that the point of moment taken is 12.0 feet from the left end of beam, not necessarily from the support.

STEP II: Computing the Reactions

Calculate the reaction (R_1) at the left support by taking each individual load and multiplying the load value times its distance from right support (R_2). Add the total sum of these moments, then divide the total by the length of the span (L). *The span is considered the distance between the supports R_1 and R_2.* It is not the length of the beam. Note the reactions on the drawing and identify the concentrated loads as P_1, P_2, etc. The uniform loads are identified by a small letter (w) only when given in pound per lineal unit foot along beam. When the uniform load total is given, the capital letter (W) will represent total load. Careful attention must be given to the many formulas which are listed in handbooks, especially those which express the formulas for simple beams thus:

$R = \frac{wL}{2}$, and $R = \frac{W}{2}$, or $M = \frac{wL^2}{8}$.

STEP III: Computing Vertical Shear to zero point

Previously it was stated that the greatest value of shear (V) occurred at the supports. For example: a simple span beam, the reaction $R_1 = V_1$, and on the other end, $R_2 = V_2$. Except for a beam symmetrically loaded, one of the reactions will have a greater value. To find the point on beam where shear diminishes to zero, or to initiate the construction of a shear diagram, proceed in the following manner:

Solving beam bending problems, continued 1.3

Start at left end of beam and deduct the value of each load from the reaction R_1 until the vertical shear decreases to zero. At this point the shear diagram wire will cross the base line, and the bending moment will be greatest.

STEP IV: Calculating Bending Moments

A bending moment can be calculated at any point on a beam regardless of the type load or load arrangement. To calculate a bending moment, first establish a point on the beam where moments are to be taken. Start by multiplying the force value of R_1 by its distance to the established point, and put the figures in an algebraic form. Take each load to the left of the established point and multiply its value times its distance from the point. Subtract the sum of these moments from the reaction moment. The difference between the moments is the bending moment at this point on the beam. A positive bending moment (+M) will result when the reaction moment is greater than the total load moments. A negative bending moment (−M) will result when the second part of the equation is greater.

Symmetrical loads 1.3.1

Any simple span beam with symmetrical loading can usually have the reactions solved by mere observation and a minimum amount of computation. A beam is said to be symmetrically loaded when equal loads are placed the same distance from each support. The loads may vary in number, but must be equal and spaced uniformly so that the reactions are equal ($R_1 = R_2$). A vertical line drawn through the midspan between supports will show immediately if symmetry exists. R_1 and R_2 will not be equal under other loading conditions; therefore the reactions and bending moments must be computed by the method of moments which is illustrated in the examples.

Continuous beams 1.3.2

When a length of beam material is supported by three or more supports, it is referred to as a continuous beam. Such beams are common for roof and floor construction especially in concrete structures. A continuous beam that supports uniform distributed or equally spaced loads with the same load value over two or more spans provides more rigidity than two simple beams end to end. The theory assumes that the continuous beam has fixed ends at the interior supports, and the overhanging portions absorb a part of the positive bending moment. At the interior supports there will be a negative bending moment.

The degree of end restraint given to a beam depends upon the nature of the support connection. A steel welded joint which connects a beam to a column may compare favorably to a monolithic concrete

Continuous beams, continued 1.3.2

column and beam which has been formed together in one operation. A theoretical conception would be to consider a beam completely fixed at the ends only when the restraint at these points is sufficient to keep the beam horizontal at those points.

MOMENT DISTRIBUTION

A method of successive approximation when applied to continuous single line flexural members is essentially an orderly procedure of distributed moments. It is beyond the scope of this work; Continuity theory is a subject in itself. The theory of the *Three Moment Equation* should be pursued by those who desire to explore this phase of design. The three moment equation expresses a relation between the moments at three consecutive supports in terms of the load and the spans. The basic theory is developed from the elastic curve principles of the calculus.

FIXED END BEAM MOMENT FACTORS

For continuous beams having equal spans and uniformly distributed equal loading over each span, the moment factors are given as coefficients of load(w) and span(L). For practical design in concrete where end restraint is present, the following coefficients are acceptable:

For end spans for positive and negative moment $\pm M = \frac{wL^2}{10}$. Where w = Lineal foot load and total load on span W = wL, the equation becomes $\pm M = \frac{WL}{10}$.

For interior spans: When there are more than four supports, use the coefficient for both positive and negative moments as:

$M = \frac{WL}{12}$ or $\frac{wL^2}{12}$.

For single spans with or without end restraint, use the formula: $M = \frac{WL}{8}$.

Inflection point 1.3.3

Bending Moment Diagrams constructed for simple span beams with overhanging ends will have two points on the base line where the curve crosses the base line. At these points, the bending moment changes from positive to negative, and tension and compressive stress cross over the centroid. The point of the crossing is called the *inflection* point, and at this point the bending moment will equal zero, or M = 0.

In the design of concrete beams, it is important to know the location of the inflection point, because it is there that the reinforcing steel must be bent upward to resist the change in stress. Obviously the position of the inflection point will depend upon the loads, their location and magnitude, together with the amount of restraint at the end of the beam. Continuous concrete beams and girders having uniformly distributed loads are considered safe when the point of inflection is taken as 1/5 the clear span distance between the supports. Using this practice the designer notes this $\frac{L}{5}$ point on drawings to assist fabricators and steel bar placement crew workers. Such arbitrary customs do not imply that the designer is bound to a fixed rule. Should a particular problem seem critical

Inflection point, continued 1.3.3

enough to justify the inflection point being given special attention, then the designer and draftsman should note the exact location on the plans.

In the fabrication of steel rigid framed arches which are discussed in Section VII, the moment diagrams will give considerable emphasis to the inflection points. In some instances the rafter section can be reduced in cross-section area at the inflection point to reduce cost and weight.

Beam diagrams 1.3.4

Many handbooks are available which contain beam load diagrams and the coefficients with formulas for computing reactions, maximum bending moments and deflection. Probably the most widely used of these handbooks is the *Manual of Steel Construction* published by the American Institute of Steel Construction (AISC). This handbook should become a part of each engineer's personal library. Many references will be made to this manual in succeeding sections.

Attention should be given to the variation from the standard symbol nomenclature when consulting each handbook. Some publications will reverse the plus and minus signs which represent positive and negative moments tension or/and compressive stresses.

STRUCTURAL DESIGN

Handbooks are intended to assist designers in the selection of suitable cross-sections which will satisfactorily support the loads with the required factor of safety. The actual design of structural members cannot be left to guesswork. Selection of members must not be attempted until all calculations of the mechanics and load behavior are thoroughly understood and checked for accuracy. Choosing a beam or column from handbook load tables without applying the principles of engineering is a dangerous practice.

Office design system 1.3.5

If each individual performing design work in a large engineering or architectural office employed his own system for computations and design calculations, those who check for errors would become bewildered to the point of helplessness. Similar confusion would exist in the preparations of drawings and specifications if each employee used his own system. To avoid this confusion, a method of procedure has been devised which is simple, accurate and generally accepted.

FIRST DESIGN PHASE

In the application of the principles of Structural Mechanics to size beams to safely support the loads, it is first necessary to examine the load conditions, and

Office design system, continued 1.3.5

the effects produced by the location of the loads on the supporting structural member. This rule is true whether the beam under consideration will be formed of concrete, wood, or steel.

ACCURATE COMPUTATIONS

Guesswork or new assumptions must not enter into our calculations. The results must be correct, and we strongly urge that the users of this manual thoroughly study and understand the examples given. The solutions have been worked out using a tested and established office design system. The algebraic equations and format for solving for reactions, bending moments and shear values may seem to be time consuming. Nevertheless they serve as a guide to accuracy and thorough understanding. More important, the equations may be checked by others and remain a permanent, understandable record for the job files. Every structural member designed for a building by an engineer should be given an identification mark and carefully filed so that it can be reviewed later if necessary.

Strength of materials 1.3.6

Resistance to bending and shear in a loaded beam is dependent upon the strength of its extreme fibers. A beam cross-section might be composed of such material that its fibers would sustain greater tension stress than compressive stress. The reverse is true for concrete beams, where the tensile strength of concrete is negligible and steel reinforcing rods are provided for tensile strength in the area concerned.

Every material substance has several significant characteristics in its composition which give it recognizable form. It is not difficult to recognize a timber product as being different from a length of steel or concrete; however, it may require more knowledge to recognize the different characteristics in similar species. This knowledge comes after laboratory tests have been made on each type of material. In the Steel Section of this Manual you will find concise coverage of laboratory test methods as conducted by testing societies for code authorities. See ASTM laboratory tests in Section II. Since wood, steel and concrete are the basic materials of structural engineers, the qualities of each material have been the subject of many tests. Each composition of steel is tested to ascertain its ability to resist tension, compression, deformation and to measure its ductility. Similar tests are conducted on wood products. The object of such test is to give the engineers a base for design purposes. Specifications for each material will state the unit allowable stress to be used for design purposes.

Earlier in this section, it was stated that mechanics was an analysis of external forces tending to produce motion. The second stage in design is relatively simple, and consists of a method for balancing the external forces by selection of a beam cross-section which will support the loads without exceeding the permissible internal stresses in the beam section. A study of the examples given should be sufficient to

Strength of materials, continued 1.3.6

enable the apprentice to compute the bending moment for practically any beam with similar loading which will be encountered in daily practice. Designing a beam for bending consists of computing the maximum bending moment, and then choosing a beam of material and cross-section which can provide a resisting moment (RM) equal to, or greater than, the bending moment. Written in formula, this can be stated as: Bending Moment = Resisting Moment. Brief formula: M = RM.

Beam sections 1.3.7

In the bending formula $\frac{M}{F} = S$ or $S = \frac{Mc}{I}$, there are certain symbols which refer to the properties of beam cross-sections. These properties are given in the tables for each individual beam cross-section. The origin and manner of calculating each property is given extensive coverage in Section VI. The Resisting Moment of a cross-section is dependent upon the property called Moment of Inertia (I). This is the important property which enters into the formulas for bending and deflection, and is also used in computing the radius of gyration for column formulas. When the beam formula, $M = \frac{F}{c} \times I$ is reduced to obtain the Section Modulus (S), it becomes $\frac{I}{c} = S$, and further simplified, the Resisting Moment = SF. Then for design this equation becomes $S = \frac{M}{F}$ for the minimum value of S. An equal resisting moment is correspondingly obtained as: RM = SF. Illustrations are given in the use of these properties as they are applied to the design examples for steel beams in Section II and for timber beams in Section III.

Graphic solutions in mechanics 1.3.8

When good drafting room facilities are available and proper instruments are used, the structural engineer can solve most beam problems by the Graphic method. In some cases, the inexperienced designer or student may better comprehend the force action when he has the opportunity to study a graphic representation. In this section, the solutions for beam reactions and bending moments in several examples will be solved by the Algebraic method of moments. The Shear and Moment Diagrams are in the true sense a part of the Graphic System. Some comparisons can be made by referring to Section V where an identical beam example is solved by constructing a Ray Diagram and Funicular Polygon, and then by the algebraic method. Although the amount of accuracy attained by the Graphic system of forces will be dependent upon the draftsman, the results are usually comparable to the algebraic method using slide rule solutions.

SIMPLE FRAMED STRUCTURES

Any framed device or rigging formed with component parts so arranged as to sustain

Graphic solutions in mechanics, continued 1.3.8

members in tension and compressive stress is considered a simple structure. Their fabrication is accomplished with joint connections consisting of hinges, bolts, pins or similar mechanical attachments. The external forces from loads are assumed to act in several planes of direction at the joints. In the majority of cases, the forces are acting simultaneously as concurrent forces. Such structures are represented by loading devices as: hoist derricks, jib, cranes, ship rigging and roof trusses. Because the design and resolution of forces in such structures involves the use of trigonometry and force diagrams, the mechanics of such frames is illustrated and explained by the examples in Section V.

Force equilibrium 1.4.1

When a beam supports several loads of different weights, and is balanced on a single support, the resultant of the load forces in the same plane of action is balanced by the support reaction. The beam does not move, and therefore is in a state of equilibrium. This is to state simply, that the sum of the forces in one direction must be equal to the sum of the forces in the opposite direction. Direction is not limited to vertical action planes, but the rule will apply when forces are horizontal or make an angle with the horizontal plane. Newton's third law says: To every action, there must be an equal and opposite reaction to obtain equilibrium.

Moment equilibrium 1.4.2

Text books on the subject of Mechanics state that if a system of forces acting on a body is balanced, then the system is in equilibrium. This means that the system of forces produces no motion in a vertical, horizontal or rotating direction; the sum of moments in one direction equals the sum of moments in the opposite direction. Thus $\Sigma M = 0$. With respect to beams, the law is stated thus: The sum of the moments of the forces tending to cause rotation clockwise, must equal the sum of the moments of forces tending to cause rotation counter-clockwise.

Let this be illustrated by a simple problem with three loads placed on a 22.0 foot beam as shown in elevation. First step for solution is to determine the location for the single support which will be the Center of Gravity of the three loads. Loads P_1, P_2 and P_3 are loads which represent three forces acting Colinear or Parallel to vertical

Moment equilibrium, continued 1.4.2

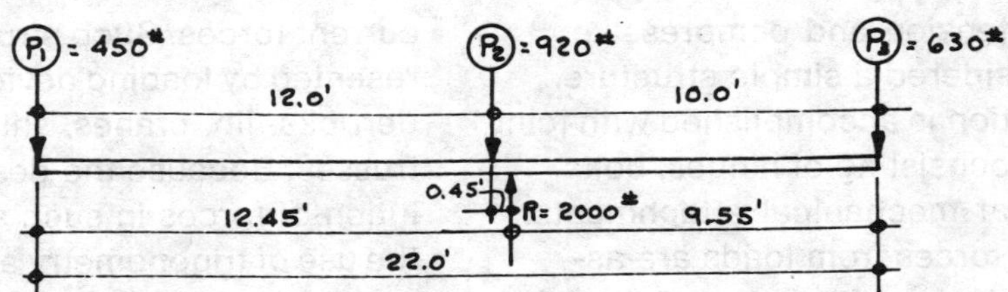

plane of action, therefore the Reaction will be in same plane, but opposite in direction. Total Loads = 2000 Lbs., and is to be supported on one Reaction R. Then R = 2000# and must be located to put the three loads in equilibrium.

Moments can be taken from any point on the beam to find the location for R with respect to that point of taking the moments.

Suppose we take the moments about a point at left end or directly under load P_1. The equation becomes thus:

Distance from left end, $R = \dfrac{(920 \times 12.0) + (630 \times 22.0)}{2000} = 12.45$ Feet.

Moments also may be taken about right end under P_3 as:

Distance from P_3, locate $R = \dfrac{(920 \times 10.0) + (450 \times 22)}{2000} = 9.55$ Feet.

Now that the Center of Gravity of the three loads has been established, the Reaction must be equal to total loads and located at that point for balance. Thus: R = 2000 Lbs.

The law states that the sum of moments at left of R must be equal to the sum of moments to the right of R, and the algebraic sum of moments is zero. Then to prove the law, determine the sum of moments which tend to rotate clockwise about R, and compare it with the sum of moments which tend to rotate counter-clockwise about R. Write the equation thus:

Clockwise: $\Sigma M = 630 \times 9.55 = 6016.50$ Foot Lbs.

Counter-Clockwise: $\Sigma M = (920 \times 0.45) + (450 \times 12.45) = 6016.50$ Ft. Lbs.

When the equation is written in algebraic form, it becomes:

$$\Sigma M = (630 \times 9.55) - \left[(920 \times 0.45) + (450 \times 12.45)\right] = 0$$

Uniform loads 1.4.3

Loads which are distributed over a part of the beam's span are considered uniform loads, as if they were spread along the entire span. Roof and floor live loads are treated as uniform loads. When uniform loads extend over the entire span, their moment lever for maximum bending moment will be taken at the point of their center of gravity which in every case will be at the middle of the span or at the "mid-span." Likewise, any short uniform load distributed over a part of the span must have its center of gravity used as the point for taking moments. When two or more partial uniform loads are placed on a beam, each load length will have a center of gravity and these points are taken for the center of moments. From these centers, moments are taken to determine the reactions at each support and may also be used to find the resultant location or the center of gravity of all the uniform loads combined.

Let the load system be illustrated by drawing a simple span beam which is 20.0 feet between the supports. Start at left end and place a Uniform Load of 400 Pounds per lineal foot. Extend this load to the right a length of 12.0 feet. This becomes a load of 4800 Pounds, and its center of gravity is 6.0 feet from each end of load or a distance of 6.0 feet from Reaction R_1. Now place another Uniform load of 600 Pounds per lineal foot, starting at right end of beam and extend the load for four feet to the left of Reaction R_2. This second load equals 2400 Pounds, and its center of Gravity or moment center is 2.0 feet to left of Reaction of R_2. Total load on beam is 4800 + 2400 or 7200 Pounds. Now use the moment method to compute the Reactions at supports. For Reaction at R_1, use R_2 as the center of Moments, with the moment arms being the distance from R_2 to the center of gravity of each uniform load.

$$R_1 = \frac{(600 \times 4.0 \times 2.0) + (400 \times 12.0 \times 14.0)}{20.0} = 3600 \text{ Lbs.}$$

Solving for Reaction R_2. Take R_1 as center of Moments.

$$R_2 = \frac{(400 \times 12.0 \times 6.0) + (600 \times 4.0 \times 18.0)}{20.0} = 3600 \text{ Lbs.}$$

To find the Center of Gravity or Resultant for both loads, the distance will be from the point of taking the moments, as follows: (Total loads = 7200 Lbs.)

$$\text{From } R_1, \text{ distance} = \frac{(400 \times 12.0 \times 6.0) + (600 \times 4.0 \times 18.0)}{7200} = 10.0 \text{ Feet.}$$

Uniform loads, continued 1.4.3

By drawing a shear diagram in the same plane or line of force action, the location of Maximum Bending will be

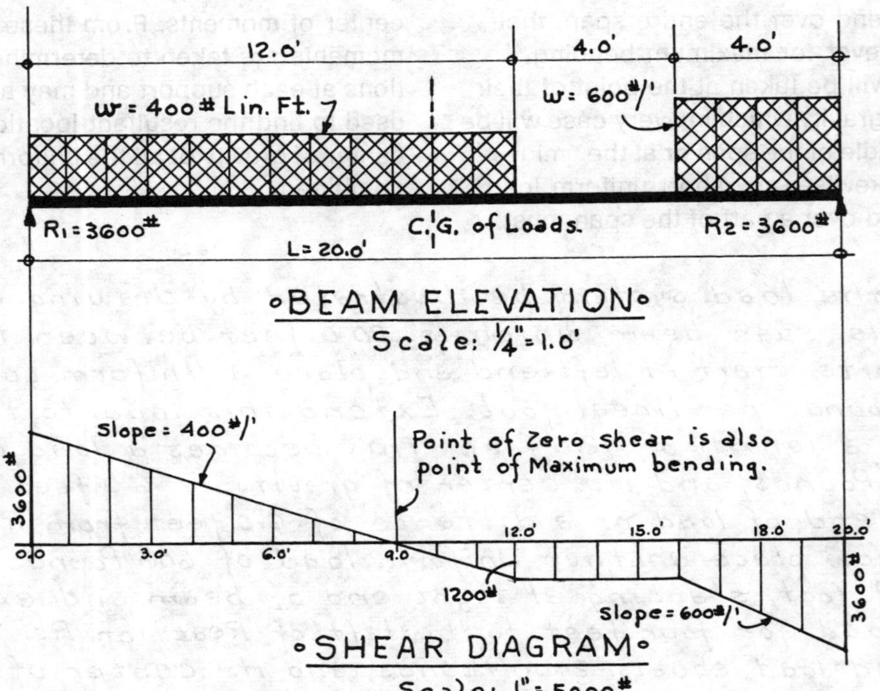

revealed at the point on beam where line crosses zero. It can also be found by reducing Reaction R_1 to zero with the load above, as $3600 \div 400 = 9.0$ Ft.

The Bending Moments are computed at any point on the beam by multiplying the Reaction R_1 by the distance, and deducting all the loads on the left side of moment point times their distance. In the form of an equation, the Bending Moment at 9.0 feet from R_1 would be written thus:

$M_{9.0} = (3600 \times 9.0) - (400 \times 9.0 \times 4.50) =$ 16,200 Foot Pounds, where the figure of 4.50 is the distance from the point of taking moment to the Center of Gravity of 9.0 feet of load. Also note in the shear diagram, that the area to left of zero line forms a triangle. This area is equal to the bending moment or ½ of $3600 \times 9.0 =$ 16,200 Foot Pounds.

Uniform loads, continued 1.4.3

It was previously stated that to produce equilibrium, all clockwise moments must equal the counter-clockwise moments. That is, they must be the same.

Suppose that the right support R_2 is removed. The load forces would produce clockwise rotation about R_1, and if R_1 were removed, the rotation would be about R_2 in Counter-clockwise direction. It was further stated, that the algebraic sum must equal zero. The equation for algebraic summation is written thus:

$$\Sigma M = \frac{\text{Clockwise moments}}{\text{Counter-Clockwise moments}} = 0$$

Now that the resultant point is the center of gravity of all the loads, the moments could be taken from that point. In this case however, if all moments are taken from points R_1 and R_2, they should produce the same result simply because the bending moments at those points is also zero, or should be for simple beam spans. Proceed thus as for bending moment.

$$M_{20.0} = (3600 \times 20.0) - \left[(400 \times 12.0 \times 14.0) + (600 \times 4.0 \times 2.0)\right] = 0$$

$$M_{0.0} = (3600 \times 20.0) - \left[(600 \times 4.0 \times 18.0) + (400 \times 12.0 \times 6.0)\right] = 0$$

The principle of moments for equilibrium is thus expressed by the simple equation, $\Sigma M = 0$.

Beam loading problem introduction 1.5

Many external load arrangements can be placed upon beams and girders which sometimes will seem to give unusual reaction and moment values. This is especially true when loads consist of a combination of uniform loads with a number of concentrated loads. There will follow many examples of such complicated loading arrangements on simple and cantilevered beams, so that the reader can gain familiarity and solve these problems rapidly. The calculations will follow a systematic format, so that the student will become accustomed to presenting his design work in a form which can be filed and later reviewed or rechecked by an independent designer.

To construct the Bending Moment Diagram, the moments are computed at several convenient points on the beam by the algebraic system with equations as given. Study the illustrated examples carefully and practice working through the examples with substituted external force values. The elevation of the beam must be drawn to some convenient scale and the loads applied at the proper locations. Below the elevation drawing, establish a heavily drawn base line of the same length as the beam. From this base line the points are plotted vertically either above or below the base line, depending upon whether the bending moment is positive (+) or negative (−). Positive moments are above the base line, and negative moments below. The distance of these plotted points from the base line when drawn to an accurate scale, will represent the magnitude of the bending moment on the beam at any given point. The vertical lines in the diagram should be drawn for each foot along the length of beam. These lines are called ordinates and they can then be scaled conveniently for the value of each moment.

Certain odd traits will distinguish themselves when drawing the moment curve during the construction of the diagram. Uniform loads will cause a curved profile and should be drawn with a transparent type of ship, railroad or french curve. The points of bending moments as produced from concentrated loads should be connected with straight lines. During the construction of the moment diagram, when a point of moment seems to be out of line, there is an error in the algebraic calculations, and the equation for that point should be rechecked. It is to be kept in mind that the longest vertical line will be the maximum moment, which will later be the governing value for the beam's design. The mechanics of beams, as illustrated in the examples, are the same when applied to beams composed of steel, wood, concrete or aluminum. To select the required beam to sustain the maximum bending moment in equilibrium, the Resisting Moment of the beam cross-section must be equal to, or greater than, the bending moment produced by the load forces. During the process of making the most economical selection for a beam, the subject of "Strength of Materials" is involved. This selection may depend upon several circumstances such as: limited depth; amount of sag or deflection or provision for lateral bracing and support. All these considerations are illustrated in the design examples provided in the sections on Steel, Concrete and Timber.

Mathematical expressions 1.5.1

Analysis = To resolve a problem into its first elements.

Board Measure = (BM) A measured lumber unit being a volume of a board 1″ thick, 12″ wide and 1 foot long. See Section III.

Cancellation = A method for solving equations, by striking out a common factor from the numerator and denominator.

Circle = A plane figure which has a curved circumference where each point is the same distance from center.

Circumference = The curved line which bounds a circle, also a perimeter or periphery.

Cube = The third power of a quantity. As $3^3 = 3 \times 3 \times 3 = 27$.

Cube Root = One of three equal factors. $\sqrt[3]{27} = 3$.

Cubic Measure = A measure of volume with respect to three dimensions: Breadth x Length x Depth.

Diameter = A straight line passing through the center of a circle and connecting the circumference.

Difference = The number or quantity found by subtraction.

Decimal = A system of converting fractions, counting or measurements into units which are powers of ten, hundred or thousand, etc.

Division = A process to determine how many times one number is contained in another of the same kind.

Equate = To solve an equation between two expressions or numbers by reducing to an equal or common standard of comparison.

Equivalent = Equal in value or having the same significant effect.

Evolution = The act of finding the root of a number.

Exponent = The small figure at the upper right of a number to denote the number of times the number is to be taken as a factor. $(125.0^2) = 125.0 \times 125.0 = 15{,}625$.

Factor = One of two or more quantities which, when multiplied together will produce a given quantity.

Gage = Also Gauge. A standard dimension or distance.

Hypotenuse = The longest side of a right triangle and opposite the right angle. Diagonal in rectangle, etc.

Involution = The multiplication of a quantity by itself any number of times or to a given power. Opposite evolution.

Subscript = A reference to an axis, material or stress type as F_t, Sx, Ry, Io, etc.

Standard nomenclature 1.5.2

a = Area cross section above an axis, used is horizontal shear computations. Given in square inches.
A = Area of Section, given in square inches
b = Breadth or width of beam section, in inches.
B = Bending factor used in Steel Columns. $B = A/S$
c = Allowable unit stress (compressive) in timber design
C = Denotes total Compressive value in Concrete and timber.
d = Depth of beam section or distance dimension of lever
D = Total depth - Used in Concrete beams and footings
e = Eccentric distance from load to axis. In feet or inches.
E = Modulus of elasticity of material. Also E = energy in ft. Lbs.
f = Fiber stress. Given in pounds per square inch.
g = Distance from Axis to outer fibers as gage in steel shapes.
h = Height, usually of columns and designated in inches.
H = Height, usually Floor to Floor, Eave height, Columns, in feet.
I = Moment of Inertia of Section, designated as 4 dimension.
J = A design factor used in reinforced concrete
k = A bending factor (kern) for wood column eccentric loading.
l = Length of beam or column, given in inches.
L = Length of spans, beams, etc; given in feet.
M = Bending Moment given in foot or inch pounds. RM = M.
N = Generally denotes a load or dimension ⊥ to surface.
n = A relation of steel to concrete strength ratio. $n = E_s/E_c$
NA = Neutral Axis, also centroid, gravity axis, or center gravity.
p = Percentage of Steel Area to concrete area.
P = A concentrated Point Load on beam, or axial load on Col.
Pe = A concentrated Load which is eccentrically placed.
r = Radius of gyration, a property of a section. $r = \sqrt{I/A}$
R = Reaction of beam loads. Also designates Resistance.
s = Designates spacing rods and joists. Sometimes for stress.
S = The section modulus property of shape, given inches^3.
t = Generally indicates thickness in thin wall plates, etc.
T = Total value of Tension stress as $T = fA$ in pounds.
u = Denotes unit bonding stress between concrete and steel.
v = Unit shear stress $= \frac{V}{A}$ or for horizontal shear, steel stirrups, etc.
V = Total amount Shear, generally at supports, given in pounds, kips.
w = Weight of uniform load per lineal foot on beam.
W = Total Weight of Uniform load on beam. In pounds, kips, etc.

Signs and symbols 1.5.3

Δ	=	Deflection or Distortion in section. Given in inches.
Σ	=	Summation of a group, areas, moments, rod perimeters.
$<$	=	Is less than. As $5.0 < 7.0$.
$>$	=	Is greater than. As $7.0 > 5.0$.
$\perp$	=	Perpendicular to grain, surface or at right angle to plane.
$+$	=	Plus sign, denoting addition. Also positive in moment.
$-$	=	Minus sign, to be subtracted. Also negative moment.
$\pm$	=	Plus or Minus, denoting more or less, or either possible.
$\sqrt{\ }$	=	Square root to be extracted of a number or equation.
$\sqrt[3]{\ }$	=	Cube root to be extracted.
5^2	=	Squared number. (As: $5 \times 5 = 25$)
5^3	=	Cubed number. (As: $5 \times 5 \times 5 = 125$)
□	=	Square (Pounds per square inch = #□".)
⌷	=	Square shaped bar or tube.
○	=	Circle or round in shape.
ϕ	=	Round reinforcing rod or sag rod.
#	=	Pounds or Lbs. Use as PSI is same as: #□".
'#	=	Foot Pounds. Where moment = Pounds times feet.
"#	=	Inch Pounds. Equals foot pounds times 12.
$\times$	=	Multiply, as $5 \times 2 = 10$.
$\div$	=	Divided by, as $6 \div 3 = 2$.
$\frac{A}{b}$	=	Indicates A is divided by b. Or $\frac{6}{3} = 2$.
$\frac{Jdb}{As}$	=	An equation, where: $J \times d \times b$ is divided by As.
( )	=	Parentheses. Used to enclose one distinct equation.
[ ]	=	Brackets. Used to enclose several equations which must be equated and resolved in a single value.
θ	=	Angle under consideration or controlling angle.
$^\circ$	=	Degrees of an angle or circle.
'	=	Minutes of a degree. Also used to indicate feet.
"	=	Seconds of a minute. Also used to denote inches.
%	=	Percent, or a percentage of a quantity or number.
π	=	Pi. A ratio of the circumference to diameter. $= 3.14159+$.
$\frac{\pi}{4}$	=	Pi $\div 4 = 0.7854$ or Diameter squared $\times 0.7854$ = Area of circle.
MWL	=	Mean Water Line. An average elevation for tide conditions.
BM	=	Bending Moment. Also used to denote a Bench Mark.

EXAMPLE: Moment arms on cantilever with concentrated loads 1.6.1

A steel Beam is 15.0 feet long with only 1 supporting Column placed 5.0 feet from Right End. Two Concentrated Loads are placed thus: P_1 of 200 Lbs. is located on extreme left end, and P_2 of 150 Lbs. is located at 6.0 foot to Right of P_1. Extreme right end of Beam is to be anchored to wall of of an abutting structure.

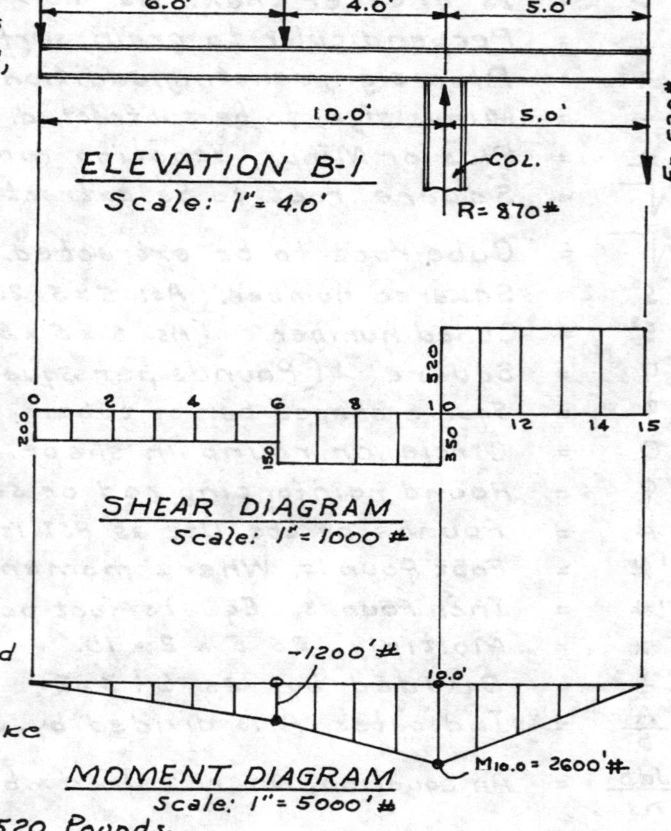

REQUIRED:

To solve for Force at P_3 and Column Reaction to put beam in equilibrium. Draw Shear and Moment Diagrams, and calculate Maximum Bending Moment in Beam.

STEP I:

Sketch Elevation of Beam and place known data. Solving for Force required at P_3, take moments about ℄ Column.

$$F = \frac{(200 \times 10.0) + (150 \times 4.0)}{5.0} = 520 \text{ Pounds.}$$

Total Loads = R = 200 + 150 + 520 = 870 Pounds.

STEP II

Drawing Shear Diagram, Max. Moment will be Negative and $M_{10.0'}$

$M_{10.0'} = (200 \times 10.0) + (150 \times 4.0) = -2600$ Foot Lbs.

Taking Force or P_3 from Support, also negative moment.

$M_{10.0'} = 520 \times 5.0 = -2600$ Foot Lbs.

$M_{6.0'} = 200 \times 6.0 = -1200$ Foot Lbs.

EXAMPLE: Moment arms on cantilever with uniform load 1.6.2

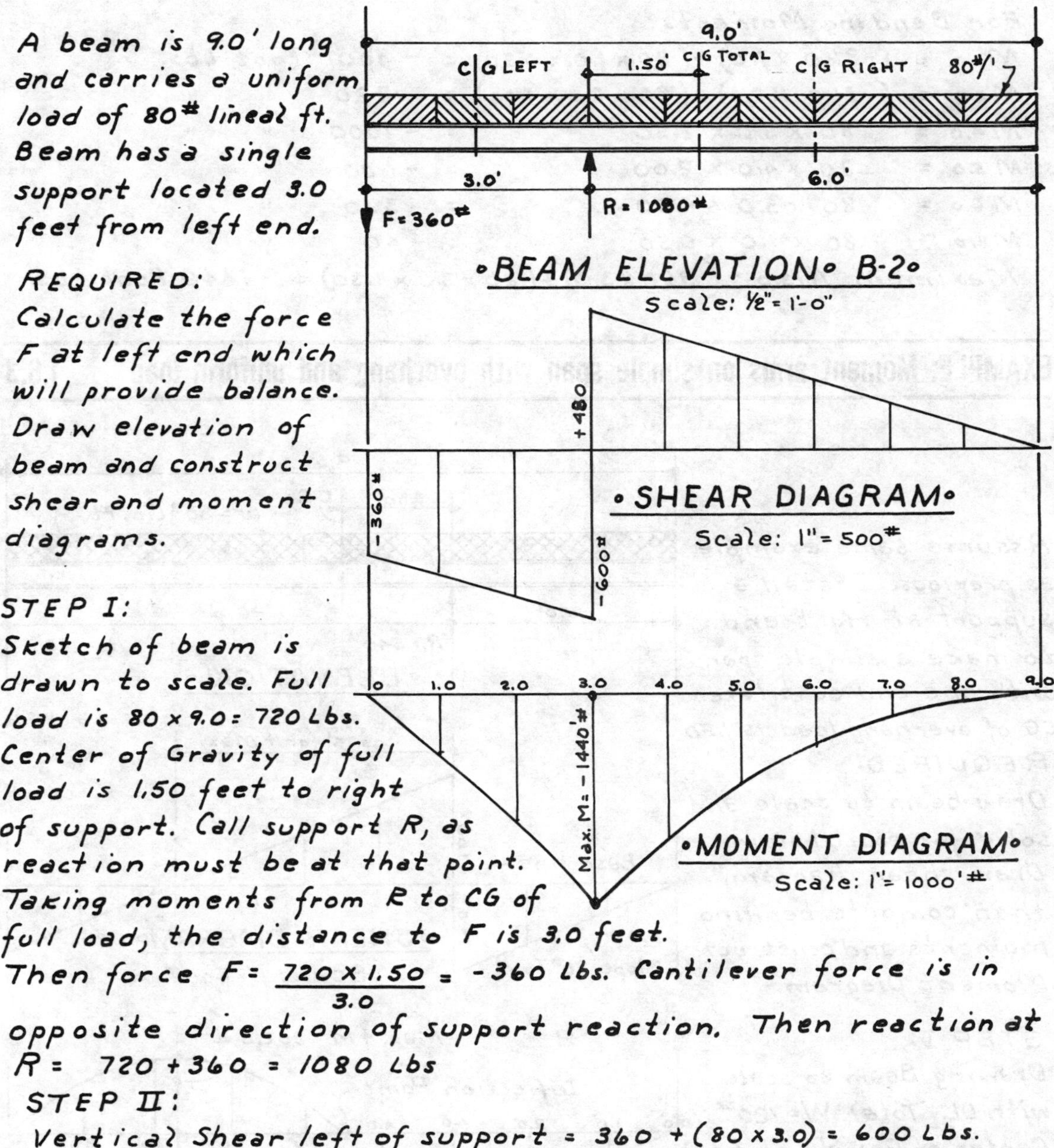

A beam is 9.0' long and carries a uniform load of 80# lineal ft. Beam has a single support located 3.0 feet from left end.

REQUIRED:
Calculate the force F at left end which will provide balance. Draw elevation of beam and construct shear and moment diagrams.

STEP I:
Sketch of beam is drawn to scale. Full load is 80 x 9.0 = 720 Lbs. Center of Gravity of full load is 1.50 feet to right of support. Call support R, as reaction must be at that point. Taking moments from R to CG of full load, the distance to F is 3.0 feet.

Then force $F = \dfrac{720 \times 1.50}{3.0} = -360$ Lbs. Cantilever force is in opposite direction of support reaction. Then reaction at R = 720 + 360 = 1080 Lbs

STEP II:
Vertical Shear left of support = 360 + (80 x 3.0) = 600 Lbs.
Vertical Shear at right of support = 80 x 6.0 = 480 Lbs.

EXAMPLE: Moment arms on cantilever with uniform load, continued 1.6.2

STEP III:

For Bending Moments:

$M_{1.0}$ =	$(-360 \times 1.0)+(80 \times 1.0 \times 0.50)$	= −400	Foot Lbs.
$M_{2.0}$ =	$(-360 \times 2.0)+(80 \times 2.0 \times 1.00)$	= −880	" "
$M_{4.0}$ =	$80 \times 5.0 \times 2.50$	= −1000	" "
$M_{5.0}$ =	$80 \times 4.0 \times 2.00$	= −600	" "
$M_{6.0}$ =	$80 \times 3.0 \times 1.50$	= −360	" "
$M_{8.0}$ =	$80 \times 1.0 \times 0.50$	= −40	" "

Maximum $M_{3.0} = (360 \times 3.0)+(80 \times 3.0 \times 1.50) = -1440$ Foot Lbs.

EXAMPLE: Moment arms on simple span with overhang and uniform load 1.6.3

Assume same example as previous. Install a support at right end to make a simple span with one end cantilever CG of overhang load is 1.50'

REQUIRED:

Draw beam to scale and solve for equilibrium. Draw shear diagram, then compute bending moments and construct Moment Diagram.

STEP I:

Drawing Beam to scale with UL. Total W = 720#. Center of Gravity of total load is 1.50 feet to right of R_1. Solve for R_2 by taking moments about R_1, then reverse for R_1.

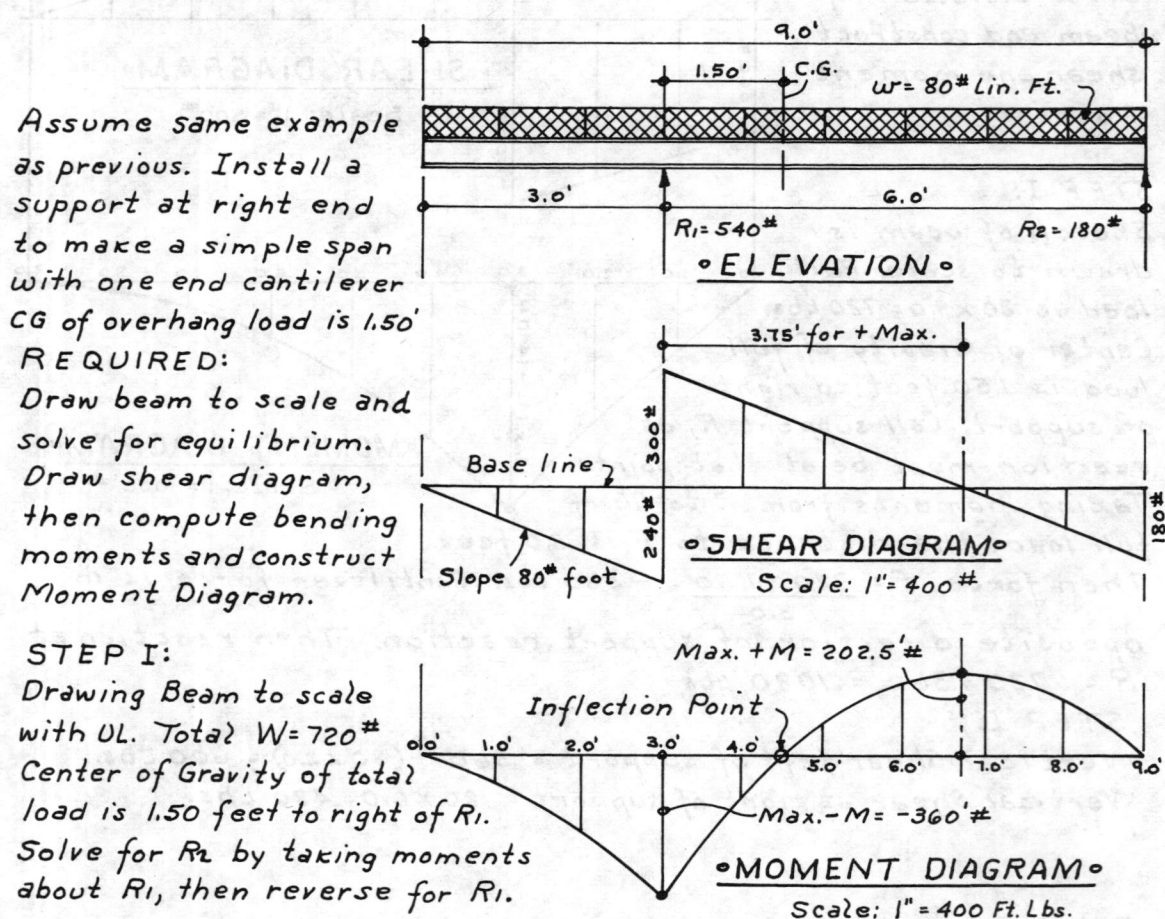

∘ELEVATION∘

∘SHEAR DIAGRAM∘

Scale: 1" = 400#

∘MOMENT DIAGRAM∘

Scale: 1" = 400 Ft. Lbs.

EXAMPLE: Moment arms on simple span, continued 1.6.3

$R_2 = \frac{720 \times 1.50}{6.0} = 180$ Lbs.

$R_1 = \frac{720 \times 4.50}{6.0} = 540$ "

Total Load = 720 Lbs. (Confirms R+R)

STEP II:

Computing Bending Moments. Shear Diagram indicates Max. +M is at a point 6.75 feet from left end of beam.

$M_{1.0'} = 80 \times 1.0 \times 0.50 = -40$ Ft. Lbs.

$M_{2.0'} = 80 \times 2.0 \times 1.0 = -160$ " "

$M_{3.0'} = 80 \times 3.0 \times 1.50 = -360$ " "

$M_{4.0'} = (540 \times 1.0) - (80 \times 4.0 \times 2.0) = -100$ Ft. Lbs.

$M_{5.0'} = (540 \times 2.0) - (80 \times 5.0 \times 2.5) = +80$ " "

$M_{6.0'} = (540 \times 3.0) - (80 \times 6.0 \times 3.0) = +180$ " "

$M_{6.75'} = (540 \times 3.75) - (80 \times 6.75 \times 3.375) = +202.5$ " " Max. +M.

$M_{7.0'} = (540 \times 4.0) - (80 \times 7.0 \times 3.50) = +200$ " "

$M_{8.0'} = (540 \times 5.0) - (80 \times 8.0 \times 4.0) = +140$ " "

$M_{9.0'} = (540 \times 6.0) - (80 \times 9.0 \times 4.5) = 0$ (zero)

EXAMPLE: Inverting beams for alternate solution 1.6.4

A playground See-Saw has a length of 9.0 feet with its fulcrum connected 6.0 feet from right end. A body with a weight of 150 Pounds is placed on short end. Weight of beam to be neglected.

REQUIRED:

(a) Compute load required on long end to put beam in equilibrium

(b) Determine Force or Reaction at Fulcrum support when balanced.

(c) Invert the beam and let end loads become reactions R_1 and R_2, Fulcrum support to become Load P. Assume Reactions unknown.

(d) Provide Shear and Moment Diagrams delineating Positive and Negative Bending Moments.

STEP I:

Sketching Beam Elevation for B-3, then inverting for B-4.

Solving for load P_2, when $P_1 = 150\#$. Take moment about R.

$P_2 = \frac{150 \times 3.0}{6.0} = 75$ Pounds.

EXAMPLE: Inverting beams for alternate solution, continued 1.6.4

Reaction at Fulcrum = Total Loads. $R = 150 \times 75 = 225$ Pounds.

STEP II

For Inverted Beam B-4, Solve for R_1 by taking moments about R_2.

$R_1 = \frac{225 \times 6.0}{9.0} = 150^{\#}$ $R_2 = \frac{225 \times 3.0}{9.0} = 75^{\#}$

STEP III

Bending Moments for B-3 will all be Negative - Stress in top fibers.
Bending Moments for B-4 will all be Positive. - Stress in bottom fibers.
Using Beam B-4 for calculating Bending Moments:

$M_{0.0} = 0$
$M_{1.0'} = 150 \times 1.0 =$ 150 Foot Pounds
$M_{2.0'} = 150 \times 2.0 =$ 300 " "
$M_{3.0'} = 150 \times 3.0 =$ 450 " " (Max. + or −)
$M_{4.0'} = (150 \times 4.0) - (225 \times 1.0) =$ 375 " "
$M_{5.0'} = (150 \times 5.0) - (225 \times 2.0) =$ 300 " "
$M_{6.0'} = (150 \times 6.0) - (225 \times 3.0) =$ 225 " "
$M_{7.0'} = (150 \times 7.0) - (225 \times 4.0) =$ 150 " "
$M_{8.0'} = (150 \times 8.0) - (225 \times 5.0) =$ 75 " "
$M_{9.0'} = (150 \times 9.0) - (225 \times 6.0) =$ 0

STEP IV

Drawing Shear Diagram to serve both Beams B-3 and B-4.
Moment Diagram B-4, drawn above base line for Positive Moments.
Moment Diagram B-3, drawn below base line for Negative Moments.

EXAMPLE: Inverting beams for alternate solution, continued 1.6.4

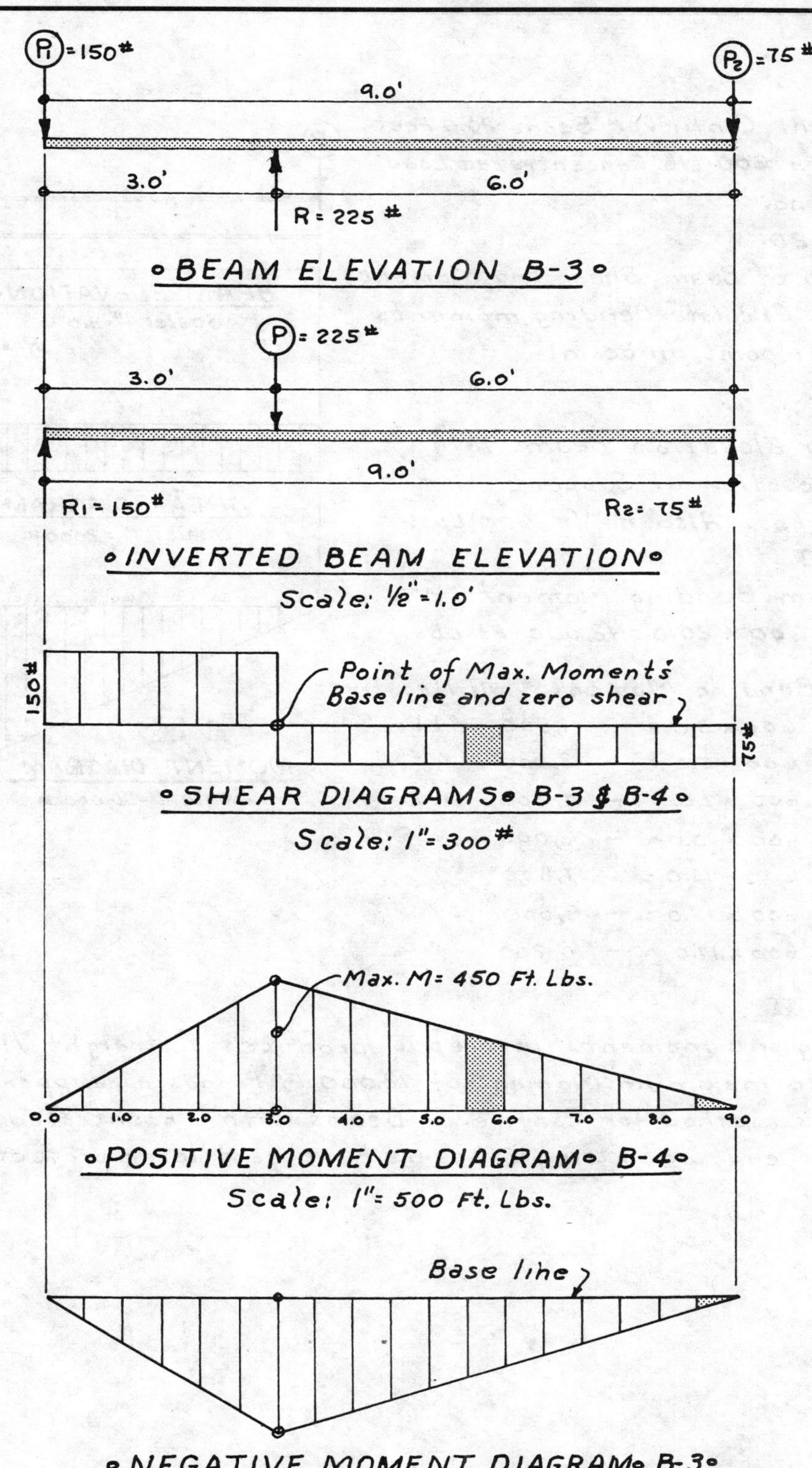

EXAMPLE: Cantilever with concentrated end load 1.7.1

Data Given: Cantilever Beam 20.0 Feet Long with 600 Llb. Concentrated Load on free end.

REQUIRED:

Elevation of Beam, Shear and Moment Diagrams. Calculate Bending moments at several point on beam:

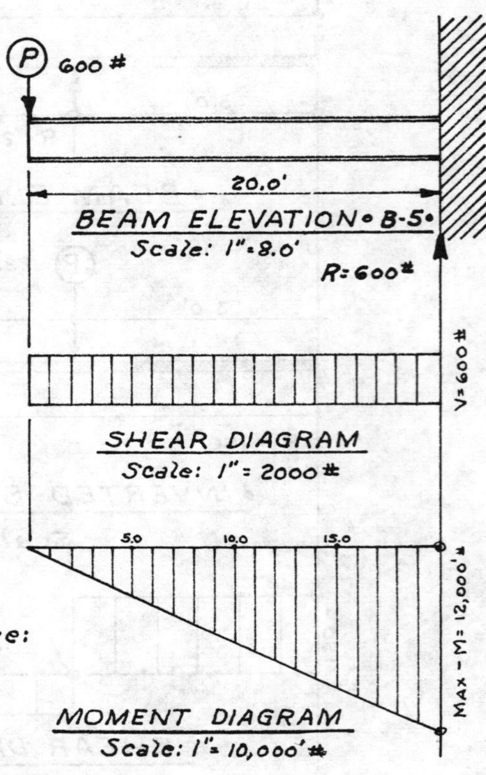

STEP I:

Drawing Elevation Beam. Total Load = Reaction at Support.

W = 600 Lbs. Also R = V = 600 Lbs.

STEP II

Maximum Bending Moment = WL

$M_{20.0}$ = 600 × 20.0 = 12,000 Ft. Lbs.

Other Bending Moments, all Negative:

$M_{3.0'}$ =	600 × 3.0 =	− 1,800	Ft. Lbs.
$M_{5.0'}$ =	600 × 5.0 =	− 3,000	"
$M_{7.0'}$ =	600 × 7.0 =	− 4,200	"
$M_{10.0}$ =	600 × 10.0 =	− 6,000	"
$M_{13.0}$ =	600 × 13.0 =	− 7,800	"
$M_{15.0}$ =	600 × 15.0 =	− 9,000	"
$M_{17.0}$ =	600 × 17.0 =	− 10,200	"

STEP III

Plotting the moments in Step II produces a straight line from zero to maximum Moment of 12,000 Ft. Pounds at support.

Use this method for Cantilever Beams with Concentrated Load at free end, when simple beams have cantilever projections.

EXAMPLE: Cantilever with uniform full-span load 1.7.2

Data Given: Cantilever Beam with 20.0 Foot Span, and Uniform Load of 150 Pounds Lineal Foot for full length.

REQUIRED:
Scale drawing of Beam Elevation, Shear and Moment Diagrams. Calculate Bending moments at several points on beam to plot the Moment Diagram.

STEP I:
The Maximum Moment is at the support. $M_{20.0} = \frac{WL}{2}$

W = 150 x 20.0 = 3000 Lbs. W = R = V

STEP II
The Center of Gravity is the point on span where total load acts, and is equal to L/2 or 20.0/2 = 10.0 feet.
$M_{20.0}$ = 3000 x 10.0 = 30,000 foot Lbs.

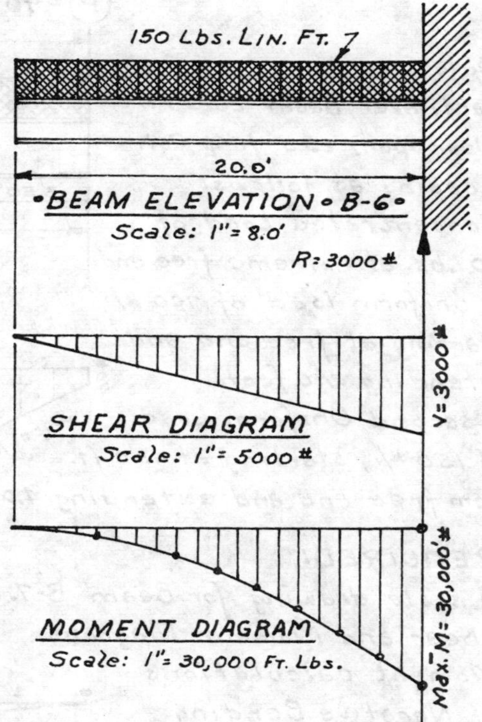

Other Bending Moments at various points on beam are computed as the portion of load to the left of that point acting at their center of gravity, times the distance to the point taken.

$M_{1.0}$ = 150 x 1.0 x 0.50 = − 75 Foot Lbs. CG is ½ foot from point 1.0'
$M_{3.0}$ = 150 x 3.0 x 1.50 = − 675 " "
$M_{6.0}$ = 150 x 6.0 x 3.00 = − 2,700 " "
$M_{10.0}$ = 150 x 10.0 x 5.00 = − 7,500 " "
$M_{12.5}$ = 150 x 12.5 x 6.25 = − 11,718.75 " "
$M_{14.0}$ = 150 x 14.0 x 7.0 = − 14,700 " "
$M_{18.0}$ = 150 x 18.0 x 9.0 = − 24,300 " "
$M_{20.0}$ = 150 x 20.0 x 10.0 = − 30,000 " "

STEP III
Plotting the moment diagram with results from Step II will produce a Curve Line in connecting moment magnitudes from zero at end to maximum at support.

EXAMPLE: Cantilever with two distributed loads 1.7.3

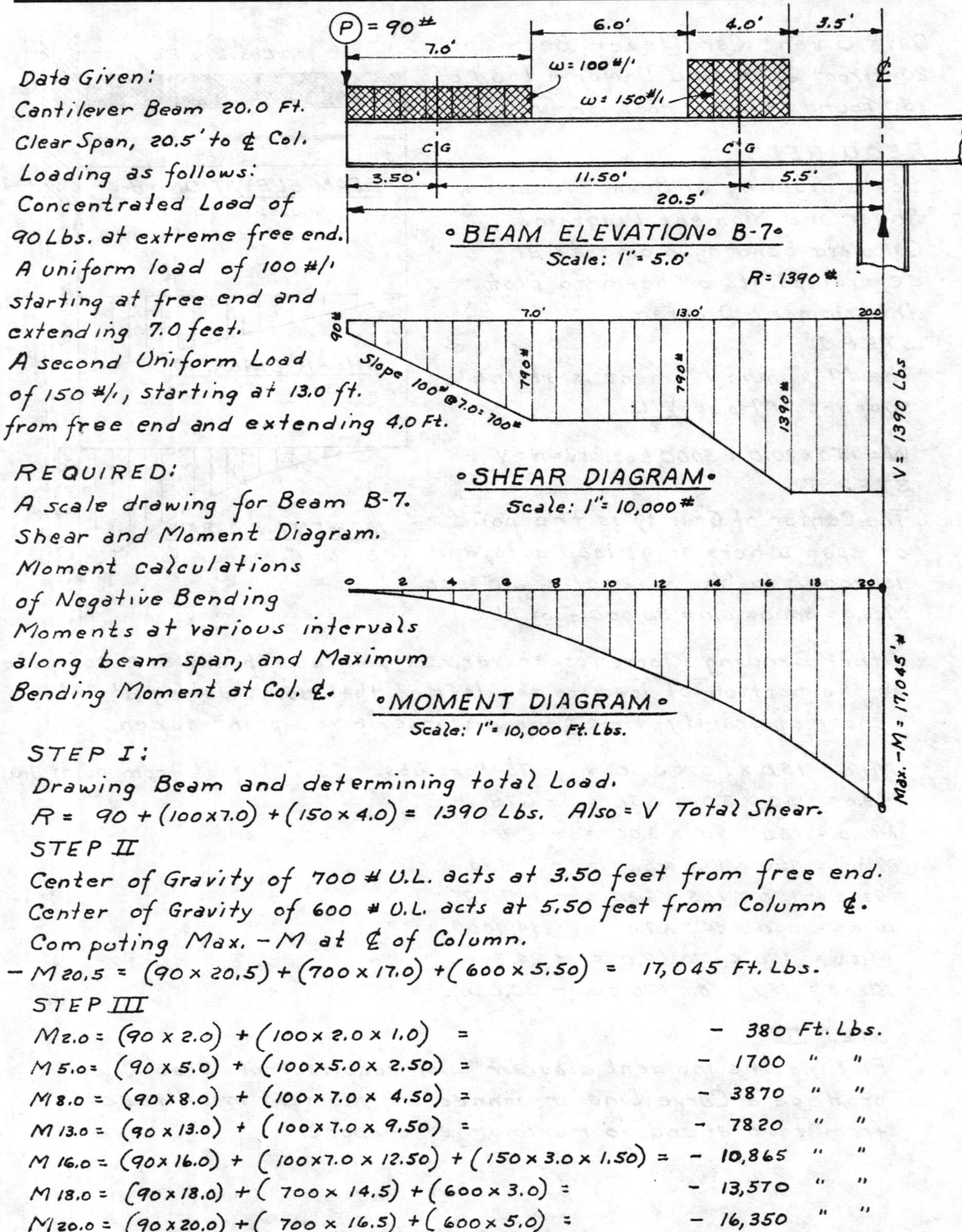

Data Given:
Cantilever Beam 20.0 Ft.
Clear Span, 20.5' to ℄ Col.
Loading as follows:
Concentrated Load of 90 Lbs. at extreme free end.
A uniform load of 100 #/' starting at free end and extending 7.0 feet.
A second Uniform Load of 150 #/', starting at 13.0 ft. from free end and extending 4.0 Ft.

REQUIRED:
A scale drawing for Beam B-7.
Shear and Moment Diagram.
Moment calculations of Negative Bending Moments at various intervals along beam span, and Maximum Bending Moment at Col. ℄.

STEP I:
Drawing Beam and determining total Load.
$R = 90 + (100 \times 7.0) + (150 \times 4.0) = 1390$ Lbs. Also = V Total Shear.

STEP II
Center of Gravity of 700 # U.L. acts at 3.50 feet from free end.
Center of Gravity of 600 # U.L. acts at 5.50 feet from Column ℄.
Computing Max. −M at ℄ of Column.
$-M_{20.5} = (90 \times 20.5) + (700 \times 17.0) + (600 \times 5.50) = 17{,}045$ Ft. Lbs.

STEP III
$M_{2.0} = (90 \times 2.0) + (100 \times 2.0 \times 1.0) = -380$ Ft. Lbs.
$M_{5.0} = (90 \times 5.0) + (100 \times 5.0 \times 2.50) = -1700$ " "
$M_{8.0} = (90 \times 8.0) + (100 \times 7.0 \times 4.50) = -3870$ " "
$M_{13.0} = (90 \times 13.0) + (100 \times 7.0 \times 9.50) = -7820$ " "
$M_{16.0} = (90 \times 16.0) + (100 \times 7.0 \times 12.50) + (150 \times 3.0 \times 1.50) = -10{,}865$ " "
$M_{18.0} = (90 \times 18.0) + (700 \times 14.5) + (600 \times 3.0) = -13{,}570$ " "
$M_{20.0} = (90 \times 20.0) + (700 \times 16.5) + (600 \times 5.0) = -16{,}350$ " "

EXAMPLE: Simple span with uniform full-length load 1.8.1

Simple Beam- Uniform Load Full Length Span L. $\omega = 500^{\#}/_{1}$
$L = 20.0$ FT.

REQUIRED:

(a) Calculate Maximum Moment by Applicable Empirical Formula: $M_x = WL/8$

(b) Draw Elevation Beam, Shear Diagram and Moment Diagram.

(c) Use algebraic method and calculate Bending moments on half span at left.

L = 20.0'
w = 500#/1
R1 = 5000#
R2 = 5000#

ELEVATION BEAM B-9
Scale: 3/16" = 1.0'

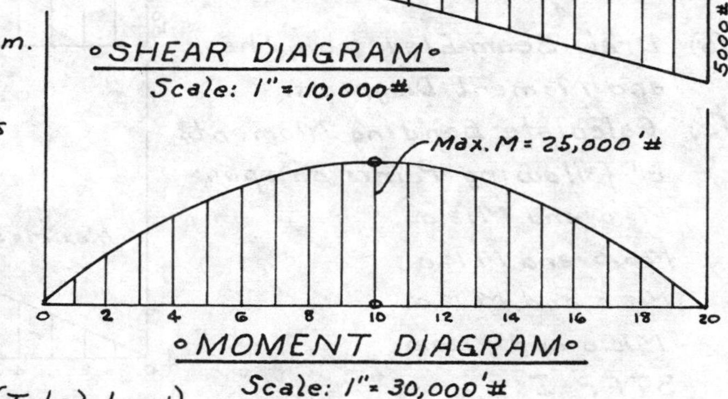

STEP I:

Max. Moment at Midspan and Formula: $M = \frac{\omega L^2}{8}$ or $\frac{WL}{8}$

$\omega = 500\#$ Lin. Ft.

$W = 500 \times 20.0 = 10{,}000$ Lbs. (Total Load)

$M_{10.0} = \frac{10{,}000 \times 20.0}{8} = 25{,}000$ Ft. Lbs. Positive Moment.

STEP II:

Moments at each foot from left end:

$M_{0.0} = (5000 \times 0.0) - 0$

$M_{1.0} = (5000 \times 1.0) - (500 \times 1.0 \times 0.5) = 4{,}750$ Ft. Lbs.

$M_{2.0} = (5000 \times 2.0) - (500 \times 2.0 \times 1.0) = 9{,}000$ "

$M_{3.0} = (5000 \times 3.0) - (500 \times 3.0 \times 1.5) = 12{,}750$ "

$M_{4.0} = (5000 \times 4.0) - (500 \times 4.0 \times 2.0) = 16{,}000$ "

$M_{5.0} = (5000 \times 5.0) - (500 \times 5.0 \times 2.5) = 18{,}750$ "

$M_{6.0} = (5000 \times 6.0) - (500 \times 6.0 \times 3.0) = 21{,}000$ "

$M_{7.0} = (5000 \times 7.0) - (500 \times 7.0 \times 3.5) = 22{,}750$ "

$M_{8.0} = (5000 \times 8.0) - (500 \times 8.0 \times 4.0) = 24{,}000$ "

$M_{9.0} = (5000 \times 9.0) - (500 \times 9.0 \times 4.5) = 24{,}750$ "

$M_{10.0} = (5000 \times 10.0) - (500 \times 10.0 \times 5.0) = 25{,}000$ " (Max. + Mom.)

$M_{11.0} = (5000 \times 11.0) - (500 \times 11.0 \times 5.5) = 24{,}750$ "

$M_{20.0} = (5000 \times 20.0) - (500 \times 20.0 \times 10.0) = 0$

STEP III

Reactions were determined by visual observation. Formula for $R_1 = R_2$ and $R = \frac{\omega L}{2}$.

EXAMPLE: Simple span with concentrated load 1.8.2

Simple Beam with Concentrated Load of 10,000 Lbs. at Mid-Span. L = 20.0 Feet.

REQUIRED:

(a) Calculate Maximum Moment by Empirical Formula:

$$M_x = \frac{PL}{4}.$$

(b) Draw Beam Elevation, Shear, and Moment Diagrams.

(c) Calculate Bending Moments at following Points on span:

$M_{2.0}$ and $M_{18.0}$

$M_{6.0}$ and $M_{14.0}$

$M_{8.0}$ and $M_{12.0}$

$M_{10.0}$ and $M_{20.0}$

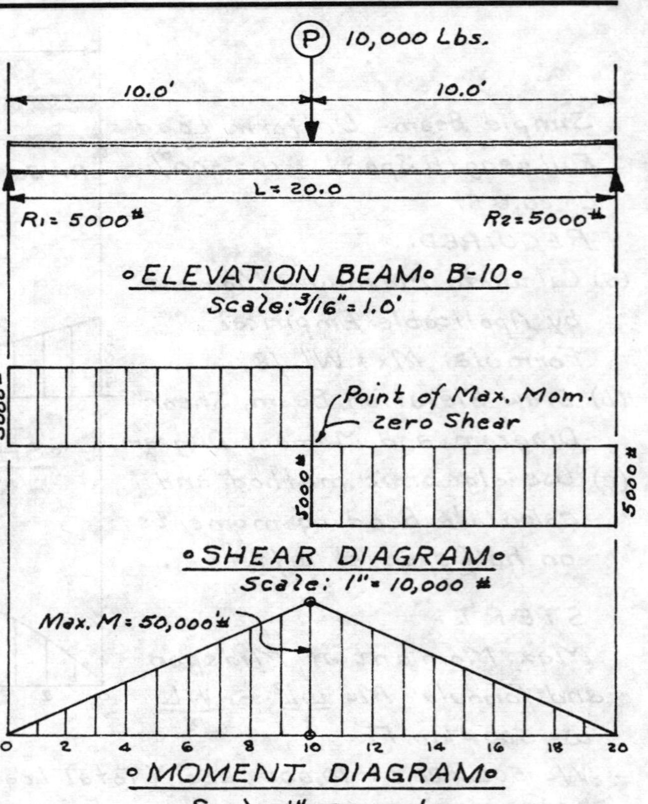

STEP I:

Max. $M_{10.0}$ by Formula:

$$M_x = \frac{10{,}000 \times 20.0}{4} = 50{,}000 \text{ Foot Lbs.}$$

STEP II

Other Bending Moments as required in (c).

$M_{2.0} = 5000 \times 2.0 = 10{,}000$ Ft. Lbs.

$M_{18.0} = (5000 \times 18.0) - (10{,}000 \times 8.0) = 10{,}000$ Ft. Lbs.

$M_{6.0} = 5000 \times 6.0 = 30{,}000$ " "

$M_{14.0} = (5000 \times 14.0) - (10{,}000 \times 4.0) = 30{,}000$ " "

$M_{8.0} = 5000 \times 8.0 = 40{,}000$ " "

$M_{12.0} = (5000 \times 12.0) - (10{,}000 \times 2.0) = 40{,}000$ " "

$M_{10.0} = 5000 \times 10.0 = 50{,}000$ " "

$M_{20.0} = (5000 \times 20.0) - (10{,}000 \times 10.0) = 0$

EXAMPLE: Simple span with concentrated and uniform loads 1.8.3

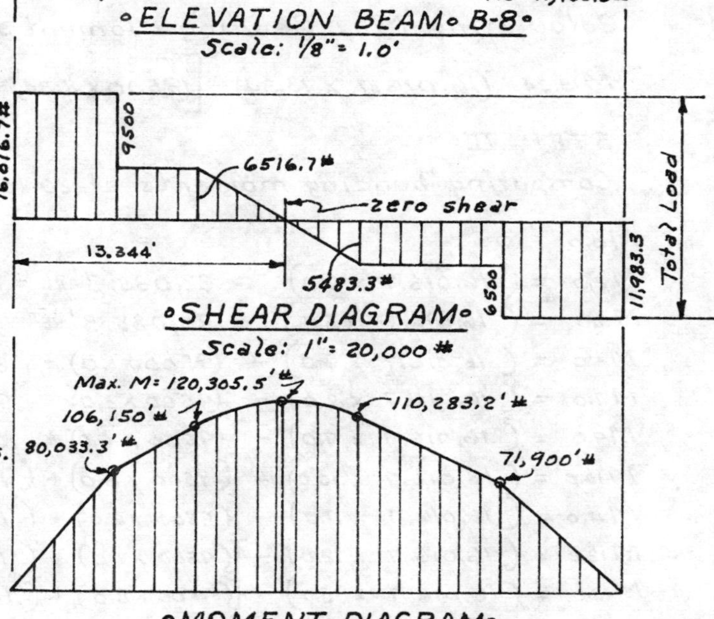

Data given as shown on Beam Elevation. Simple Beam span with end supports.

REQUIRED:

(a) Calculate Reactions R_1 and R_2

(b) Draw Shear Diagram.

(c) Compute Bending Moments at 20 Points on Span.

(d) Draw Moment Diagram and use only the following values for plotting: $M_{5.0'}$; $M_{9.0'}$; $M_{13.0'}$; $M_{17.0'}$ and $M_{24.0}$.

STEP I:

Total Load on Beam:

$9500 + (8.0 \times 1500) + 6500 = 28{,}000$ Lbs.

STEP II

Determine Reactions R_1 and R_2:

Take Moments about R_2 to solve for R_1.

$$R_1 = \frac{(6500 \times 6.0) + (1500 \times 8.0 \times 17.0) + (9500 \times 25.0)}{30.0} = 16{,}016.7 \text{ Lbs.}$$

Take Moments about R_1 to solve for R_2.

$$R_2 = \frac{(9500 \times 5.0) + (1500 \times 8.0 \times 13.0) + (6500 \times 24.0)}{30{,}0} = 11{,}983.3 \text{ Lbs.}$$

Check Total = 28,000.0 Lbs. (OK)

STEP III

Drawing Shear Diagram, start at left end and plot above the base line, the magnitude of R_1. Continue toward right support and reducing under each load in same vertical plane. Maximum Positive moment will be at point on beam where shear is zero and crosses base line. This distance may be scaled for use, or found exact in the next step.

EXAMPLE: Simple span with concentrated and uniform loads, continued 1.8.3

STEP IV

Locating point of zero shear for lever arm producing Max. Bending.

$R_1 - P_1$ = 16,016.7 − 9500 = 6516.7# Uniform Load = 1500 #/1

Length of Uniform Load to equal and absorb = 6516.7/1500 = 4.344 Ft.

Distance from R_1 to point 0 Shear = 5.0 + 4.0 + 4.344 = 13.344 Feet.

STEP V

Calculating Max. Bending Moment at point 13.344 Feet from R_1:

$$M_{13.34} = (16{,}016.7 \times 13.34) - \left[(9500 \times 8.34) + (1500 \times 4.34 \times 2.17)\right] = {}^{+}120{,}305.5 \text{ Ft. Lbs.}$$

STEP VI

Computing bending moments at 20 location points on span, B-8.

$M_{0.0}$ = 0

$M_{2.0}$ = (16,016.7 × 2.0) = 32,033.3 '# = Positive + Moment

$M_{5.0}$ = (16,016.7 × 5.0) = 80,083.3 '#

$M_{6.0}$ = (16,016.7 × 6.0) − (9500 × 1.0) = 86,600.0 '#

$M_{7.0}$ = (16,016.7 × 7.0) − (9500 × 2.0) = 93,116.7 '#

$M_{9.0}$ = (16,016.7 × 9.0) − (9500 × 4.0) = 106,150.0 '#

$M_{10.0}$ = (16,016.7 × 10.0) − [(9500 × 5.0) + (1500 × 1.0 × 0.5)] = 111,916.7 '#

$M_{11.0}$ = (16,016.7 × 11.0) − [(9500 × 6.0) + (1500 × 2.0 × 1.0)] = 116,183.3 '#

$M_{12.0}$ = (16,016.7 × 12.0) − [(9500 × 7.0) + (1500 × 3.0 × 1.5)] = 118,950.0 '#

$M_{13.0}$ = (16,016.7 × 13.0) − [(9500 × 8.0) + (1500 × 4.0 × 2.0)] = 120,216.5 '#

$M_{14.0}$ = (16,016.7 × 14.0) − [(9500 × 9.0) + (1500 × 5.0 × 2.5)] = 119,983.3 '#

$M_{15.0}$ = (16,016.7 × 15.0) − [(9500 × 10.0) + (1500 × 6.0 × 3.0)] = 118,250.0 '#

$M_{16.0}$ = (16,016.7 × 16.0) − [(9500 × 11.0) + (1500 × 7.0 × 3.5)] = 115,016.5 '#

$M_{17.0}$ = (16,016.7 × 17.0) − [(9500 × 12.0) + (1500 × 8.0 × 4.0)] = 110,283.2 '#

$M_{18.0}$ = (16,016.7 × 18.0) − [(9500 × 13.0) + (12,000 × 5.0)] = 104,710.0 '#

$M_{20.0}$ = (16,016.7 × 20.0) − [(9500 × 15.0) + (12,000 × 7.0)] = 93,833.0 '#

$M_{24.0}$ = (16,016.7 × 24.0) − [(9500 × 19.0) + (12,000 × 11.0)] = 71,900.0 '#

$M_{25.0}$ = (16,016.7 × 25.0) − [(9500 × 20.0) + (12,000 × 12.0) + (6500 × 1.0)] = 59,916.5 '#

$M_{28.0}$ = (16,016.7 × 28.0) − [(9500 × 23.0) + (12,000 × 15.0) + (6500 × 4.0)] = 23,966.5 '#

$M_{29.0}$ = (16,016.7 × 29.0) − [(9500 × 24.0) + (12,000 × 16.0) + (6500 × 5.0)] = 11,983.0 '#

$M_{30.0}$ = (16,016.7 × 30.0) − [(9500 × 25.0) + (12,000 × 17.0) + (6500 × 6.0)] = 0

STEP VII

Moment Diagram as plotted shows a curved line where Uniform Load is placed, and a straight line connects points where the Concentrated Loads govern. This a characteristic of the moment diagrams pertaining to the two load types.

EXAMPLE: Simple span with overhang concentrated load 1.8.4

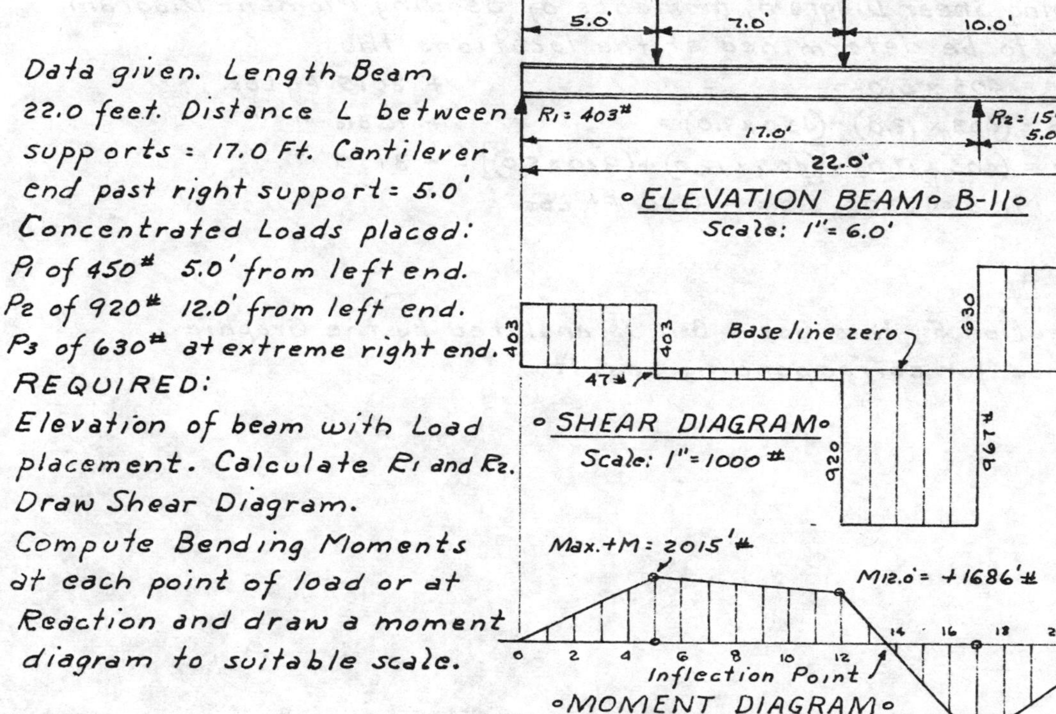

Data given. Length Beam 22.0 feet. Distance L between supports = 17.0 Ft. Cantilever end past right support = 5.0'

Concentrated Loads placed:
P_1 of 450# 5.0' from left end.
P_2 of 920# 12.0' from left end.
P_3 of 630# at extreme right end.

REQUIRED:
Elevation of beam with Load placement. Calculate R_1 and R_2. Draw Shear Diagram.
Compute Bending Moments at each point of load or at Reaction and draw a moment diagram to suitable scale.

STEP I:

Total Loads $P_1 + P_2 + P_3$ = 2000 Lbs:

Take moments about R_1 to solve for R_2.

$$17 R_2 = (630 \times 22.0) + (920 \times 12.0) + (450 \times 5.0) = 27{,}150'\#$$

$$R_2 = \frac{27{,}150'\#}{17.0'} = 1597 \text{ Lbs.}$$

For R_1, take moments about R_2 thus:

Forces P_1 and P_2 action is to rotate beam counter-clockwise about R_2, while force P_3 tends to rotate clockwise about R_2. Then by taking plus moments first:

$$R_1 = \frac{\left[(450 \times 12.0) + (920 \times 5.0)\right] - (630 \times 5.0)}{17.0} = 403 \text{ Lbs.}$$

Check $R_1 + R_2$ to equal Total Load: 1597 + 403 = 2000 Lbs. (OK)

EXAMPLE: Simple span with overhang concentrated load, continued 1.8.4

STEP II

Drawing Shear Diagram, moments of Bending Moment Diagram need to be determined at the locations thus.

$M_{5.0} = 403 \times 5.0 =$ + 2015 Ft. Lbs.

$M_{12.0} = (403 \times 12.0) - (450 \times 7.0) =$ + 1686 " "

$M_{17.0} = (403 \times 17.0) - [(405 \times 12.0) + (920 \times 5.0)] = -3150$ " "

Also $M_{17.0} = 630 \times 5.0 = -3150$ Ft. Lbs.

STEP III

In Section 5, this Beam B-11 is analyzed by the Graphic Method for corresponding results.

EXAMPLE: Simple span with overhang both ends 1.8.5

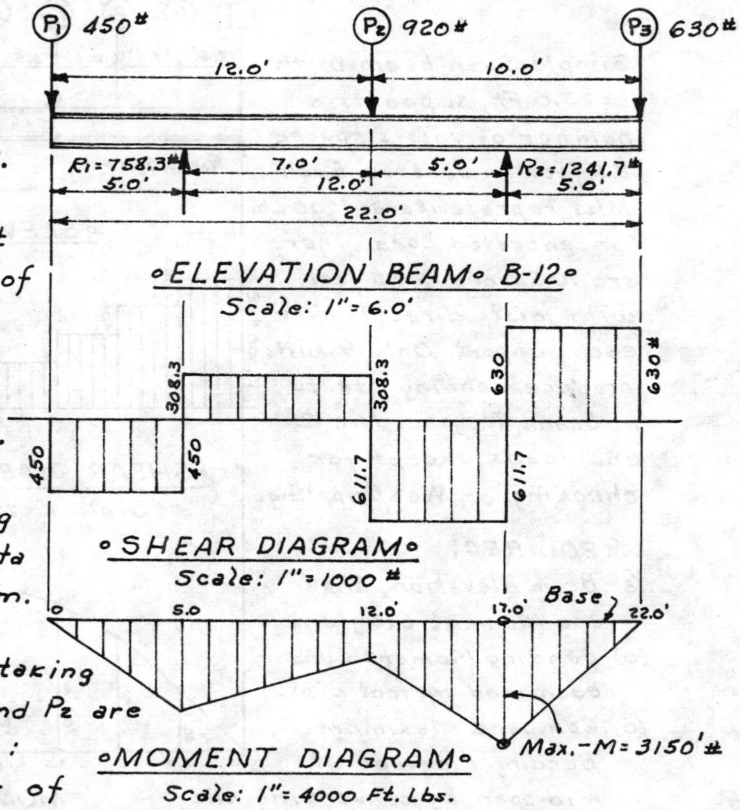

Beam in previous example is to be modified from B-11. Place Load P_1 at extreme left end, and move support R_1 to the right, a distance of 5.0 feet. Span L = 12.0 feet,

REQUIRED:

(a) Calculate Reactions after drawing elevation.

(b) Draw Shear Diagram.

(c) Calculate enough Bending Moments to supply data to draw Moment Diagram.

P1 450# P2 920# P3 630#
12.0' 10.0'
R1 = 758.3# 7.0' 5.0' R2 = 1241.7#
5.0' 12.0' 5.0'
22.0'

°ELEVATION BEAM° B-12°

Scale: 1" = 6.0'

308.3 308.3 630 630
450 450 611.7 611.7

°SHEAR DIAGRAM°

Scale: 1" = 1000#

0 5.0 12.0' 17.0' Base 22.0'

Max. −M = 3150'#

°MOMENT DIAGRAM°

Scale: 1" = 4000 Ft. Lbs.

STEP I:

Solve for reaction R_1 by taking moments about R_2. P_1 and P_2 are counter to P_3. Therefore: Take the greater number of forces in first equation, and deduct the single counter force.

$$R_1 = \frac{[(920 \times 5.0) + (450 \times 17.0)] - (630 \times 5.0)}{12.0} = 758.3 \text{ Lbs.}$$

$$R_2 = \frac{[(920 \times 7.0) + (630 \times 17.0)] - (450 \times 5.0)}{12.0} = 1241.7 \text{ Lbs.}$$

Check by Total Loads: 450 + 920 + 630 = 2000 Lbs.

Also: 758.3 + 1241.7 = 2000 Lbs.

STEP II

$M_{5.0} = 450 \times 5.0 = -2250$ Ft. Lbs.

$M_{12.0} = (758.3 \times 7.0) - (450 \times 12.0) = -91.67$ Ft. Lbs.

$M_{17.0} = (758.3 \times 12.0) - [(920 \times 5.0) + (450 \times 17.0)] = -3150$ Ft. Lbs.

Also $M_{17.0} = 630 \times 5.0 = -3150$ Ft. Lbs.

STEP III

All bending moments are Negative, when plot Moment Diagram below base line. This Problem was submitted to candidates for Engineering Registration by Texas Board in September of 1945. A measure of deception was intended.

EXAMPLE: Simple span with joist loads 1.8.6

Simple Span beam with L = 25.0 Ft., supports a number of Joists spaced 2.5 feet on centers. Each joist represents a 4000 Lb. Concentrated Load. There are 10 Spaces @ 2.5 feet, with joists directly over each support. Only 9 Joists produce bending stress in beam. Neglect the 2 end loads except for checking on Web Crippling.

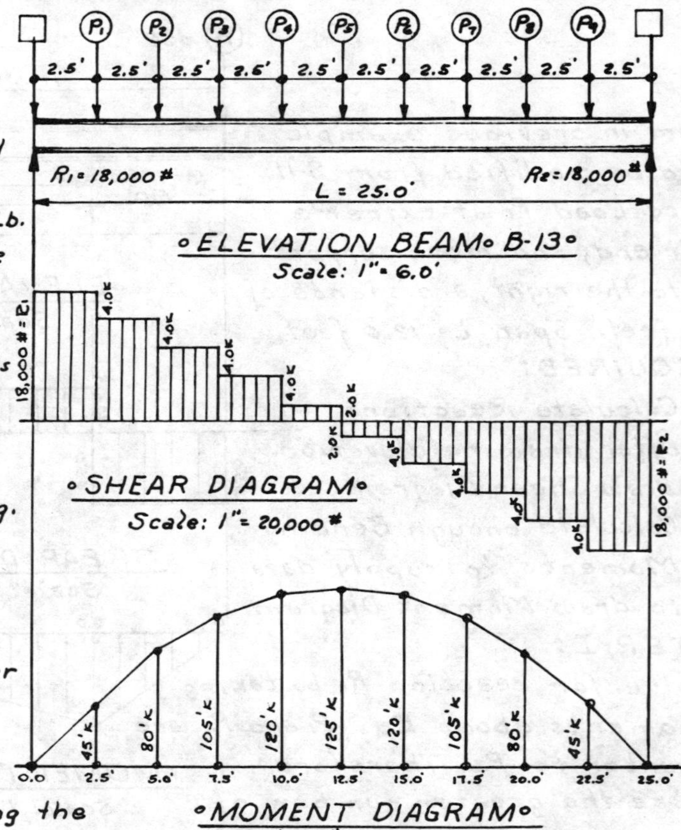

REQUIRED:

(a) Beam Elevation, Shear and Moment Diagram.

(b) Bending Moments under each load in foot kips.

(c) Compare Maximum Bending Moment at mid-span by converting the total Load to equivalent tabular and using formula WL/8 for maximum moment.

STEP I

Drawing Beam Elevation, 9 Loads @ 4000 = 36,000 Lbs.
Beam is symmetrical, therefore $R_1 = R_2$ or 18,000 Lbs. each.

STEP II

Drawing Shear Diagram, shear becomes zero at mid-spam which will be point of Maximum Bending Moment. Reducing and using kips. (1 Kip = 1000 Lbs.) Starting at left end, moments become:

$M_{2.5} = 18.0 \times 2.5 = 45$ Ft. Kips

$M_{5.0} = (18.0 \times 5.0) - (4.0 \times 2.5) = 80$ 'kips.

$M_{7.5} = (18.0 \times 7.5) - \left[(4.0 \times 2.5) + (4.0 \times 5.0)\right] = 105$ Ft. kips

$M_{10.0} = (18.0 \times 10.0) - \left[(4.0 \times 2.5) + (4.0 \times 5.0) + (4.0 \times 7.5)\right] = 120$ Ft. kips

$M_{12.5} = (18.0 \times 12.5) - \left[(4.0 \times 2.5) + (4.0 \times 5.0) + (4.0 \times 7.5) + (4.0 \times 10.0)\right] = 125$ Ft. kips.

Converting to Uniform Load: W = 36,000# (Total Load).

Max. UL Mom. $= \frac{WL}{8}$ — $M = \frac{36{,}000 \times 25.0}{8} = 112{,}500$ Ft. Lbs. (Ans. c)

For Heavy Industrial Floor Loads, use the Greater Moment value.

EXAMPLE: Simple span with joist loads and overhang 1.8.7

6000# 6000 6000 6000 6000 12000
P_1 P_2 P_3 P_4 P_5 P_6
5.0' 5.0' 5.0' 5.0' 5.0'
V= 10,000# 15.0' V= 20,000
25.0'
R_1= 16,000 R_2= 26,000

∘ELEVATION BEAM∘
Scale: 1/8" = 1.0'

4000 4000 12,000 12,000
6000 6000 2000 6000 8000

∘SHEAR DIAGRAM∘
Scale: 1" = 200,000#

0 5 10 15 20 25
−M: 30,000'#
Max.= 60,000'#

∘MOMENT DIAGRAM∘
Scale: 1" = 100,000'#

Given Simple Span with Cantilever at both ends. Length Beam = 25.0 Ft. End overhangs = 5.0 Ft. Clear L = 15.0'
Concentrated Loads as follows:

P_1 = 6000 Lbs. at extreme left end
P_2 = 6000 " 5.0 Ft. to Right of Left end.
P_3 = 6000 " 10.0 ditto
P_4 = 6000 " 15.0 "
P_5 = 6000 " 20.0 "
P_6 = 12,000 " 25.0 "

REQUIRED:
Calculate Reactions based on Loads producing Vertical Shear only.

(a) Draw Beam Elevation, Shear and Moment Diagrams to Convenient Scale.
(b.) Calculate Bending Moment under each Load point and over supports.

STEP I
Loads P_2 and P_5, do not have any effect upon bending and vertical shear stress.
Total Loads = (5 x 6000) + 12,000 = 42,000 Lbs.

STEP II:

$$\text{Reaction } R_1 = \frac{\left[(6000\times20.0)+(6000\times15.0)+(6000\times10.0)+(6000\times5.0)\right]-(12{,}000\times5.0)}{15.0'} = 16{,}000 \text{ Lbs.}$$

Reaction R_1 less P_2 = 16,000 − 6,000 = 10,000 Lbs. equals Shear V at support.

$$R_2 = \frac{\left[(12{,}000\times20.0)+(6000\times15.0)+(6000\times10.0)+(6000\times5.0)\right]-(6000\times5.0)}{15.0'} = 26{,}000 \text{ Lbs.}$$

$V_2 = R_2 - P_5$ = 26,000 − 6,000 = 20,000 Lbs.

STEP III
Calculating Bending Moments. Ignore Loads P_2 and P_5 in computations.

$M_{5.0}$ = 6000 x 5.0 = 30,000 Ft. Lbs (Negative Moment)
$M_{10.0}$ = (10,000 x 5.0) − (6000 x 10.0) = − 10,000 Ft. Lbs.
$M_{15.0}$ = (10,000 x 10.0) − [(6000 x 5.0) + (6000 x 15.0)] = −20,000 Ft. Lbs
$M_{20.0}$ = (10,000 x 15.0) − [(6000 x 5.0) + (6000 x 10.0) + (6000 x 20.0)] = − 60,000 Ft. Lbs.
Also $M_{20.0}$ = 12,000 x 5.0 = −60,000 Ft. Lbs.

DESIGN NOTE:
Loads placed directly over support are in same plane of action and only produce a compressive force between Load and supporting column. Delete such loads in equations for bending moments, but make a notation as shown. V = Shear Load influencing bending, and R = Reaction for Columns and Foundation design.

EXAMPLE I: Simple span with combined loads 1.9.1

Simple Span Beam. L = 21.5 Ft. Supports 200 Lb. Lin. Foot Load entire span, plus Conc. Loads:
P_1 = 12,500 # From left end, 4.0'
P_2 = 14,200 # " " " 10.0'
P_3 = 16,500 # " " " 15.0'

REQUIRED:
(a) Calculate End Reactions.
(b) Draw and Note values on Shear diagram.
(c) Calculate Bending Moments under each load P and Maximum Moment if not under one of P Loads.

P1 12,500#; P2 14,200; P3 16,500
4.0' 6.0' 5.0' 6.5'
w = 200#/'
L = 21.5'
R_1 = 24,900# R_2 = 22,600#

•ELEVATION BEAM• B-15•
Scale: 1" = 6.0'

-800; +24,100; -12,500; +24,900; -1200; +10,400; zero shear; Slope = 200#/'; -3800; 1000; 16,500; 21,300; 1300; 22,600

•SHEAR DIAGRAM•
Scale: 1" = 40,000#

0.0' 4.0' 10.0' 15.0' 21.5
M_{10} = 164,000#

STEP I:
Solving for R_1, by taking moments about R_2. Center of Gravity is point where Uniform Load acts and is at 21.5/2 = 10.75 Ft.
w = 200#/' W = 21.5 × 200 = 4,300# For ULoads $R_1 = R_2$ = 2,150#

$$(CL+UL)\ R_1 = \frac{(16{,}500 \times 6.5) + (14{,}200 \times 11.5) + (12{,}500 \times 17.5) + (4300 \times 10.75)}{21.5'} = 24{,}900 \text{ Lbs.}$$

$$R_2 = \frac{(12{,}500 \times 4.0) + (14{,}200 \times 10.0) + (16{,}500 \times 15.0) + (4300 \times 10.75)}{21.5} = 22{,}600 \text{ Lbs.}$$

Total $R_1 + R_2$ = 47,500 Lbs.

Total Loads: $P_1 + P_2 + P_3 + W$. = 47,500 Lbs.

STEP II:
Drawing Shear Diagram under Beam Elevation, Maximum bending moment will be under P_2 Load, or $M_{10.0'}$

$$M_{4.0} = (24{,}900 \times 4.0) - (200 \times 4.0 \times 2.0) = 98{,}000 \text{ Ft. Lbs.}$$

$$M_{10.0} = (24{,}900 \times 10.0) - [(12{,}500 \times 6.0) + (200 \times 10.0 \times 5.0)] = 164{,}000 \text{ Ft. Lbs.}$$

$$M_{15.0} = (24{,}900 \times 15.0) - [(14{,}200 \times 5.0) + (12{,}500 \times 11.0) + (200 \times 15.0 \times 7.5)] = 142{,}500 \text{ Ft. Lbs.}$$

All moments are Positive (+) Moments. Top beam fibers will be in compressive stress, and lower fibers in Tension stress.

EXAMPLE II: Simple span with combined loads 1.9.2

Simple Supported Beam with Span L = 20.0 Feet. Left Half of span carries a Uniform Load of 1200 Lbs. Lin. Foot. A concentrated Load P = 4000 Lbs., is located 15.0 feet to right of left support.

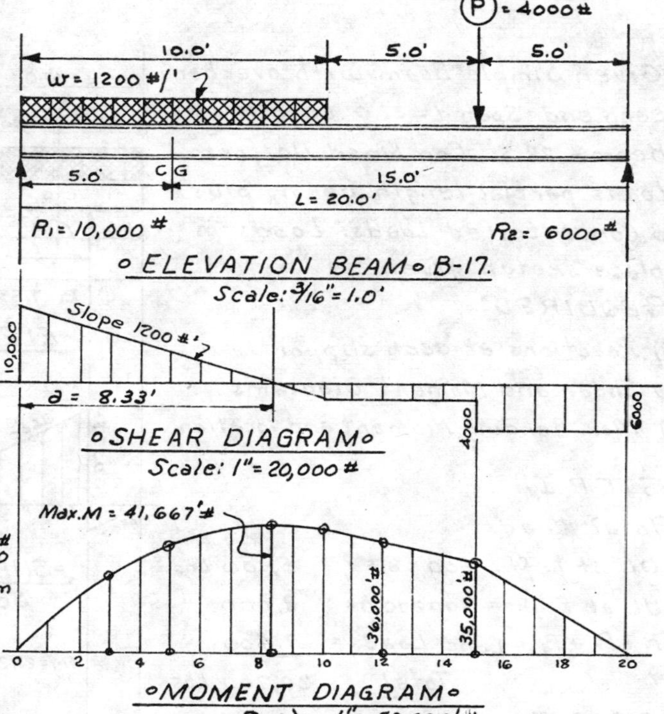

REQUIRED:

(a) Calculate Reactions R_1 and R_2.

(b) Draw Shear Diagram and locate point of no shear.

(c) Compute enough bending moments at points to provide data for an accurate Bending Moment Diagram.

STEP I:

Total Loads: (1200 × 10.0) + 4000 = 16,000#

Center Gravity of 12,000 Lb. Uniform Load acts at point 5.0' from R_1. Taking moments about R_2, to solve for left reaction:

$$R_1 = \frac{(4000 \times 5.0) + (12,000 \times 15.0)}{20.0} = 10,000 \text{ Lbs.}$$

$$R_2 = \frac{(12,000 \times 5.0) + (4000 \times 15.0)}{20.0} = 6,000 \text{ Lbs.}$$

Total Loads = 16,000 Lbs.

STEP II

Point on Beam where shear decreases to zero = $R_1 - (w \times a)$

Point 0 shear = $\frac{10,000}{1200 \times a}$ The $a = \frac{10,000}{1200} = 8.33$ Feet from R_1.

STEP III

Computing Moments to plot Bending Moment Diagram.

$M_{3.0}$ = (10,000 × 3.0) − (1200 × 3.0 × 1.5) = + 25,100 Ft. Lbs.

$M_{5.0}$ = (10,000 × 5.0) − (1200 × 5.0 × 2.5) = + 35,000 Ft. Lbs.

$M_{8.33}$ = (10,000 × 8.33) − (1200 × 8.33 × 4.167) = + 41,667 Ft. Lbs. = Max. B.M.

$M_{10.0}$ = (10,000 × 10.0) − (1200 × 10.0 × 5.0) = + 40,000 Ft. Lbs.

$M_{12.0}$ = (10,000 × 12.0) − (12,000 × 7.0) = + 36,000 Ft. Lbs.

$M_{15.0}$ = (10,000 × 15.0) − (12,000 × 10.0) = + 30,000 Ft. Lbs.

$M_{20.0}$ = (10,000 × 20.0) − [(12,000 × 15.0) + (4000 × 5.0)] = zero (Checks $\Sigma m = 0$)

EXAMPLE I: Simple span with overhangs and combined loads 1.9.3

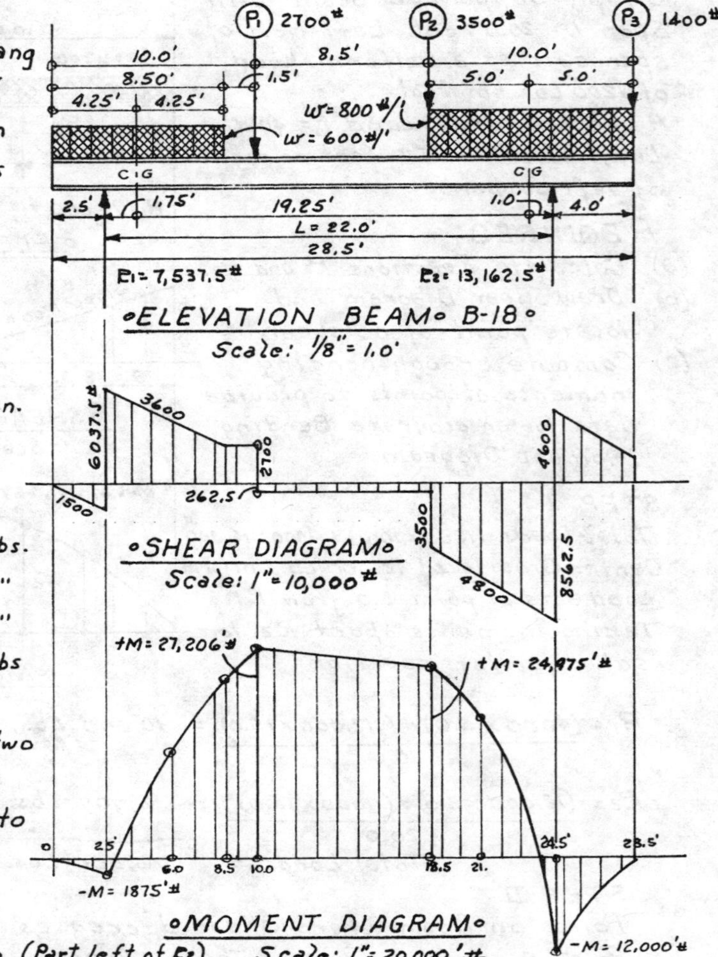

Given Simple Beam with overhang each end. Span L = 22.0 Length of Beam = 28.5'. Combined Uniform Loads partial length beam, plus 3 Concentrated Loads. Loads in place sketch given.

REQUIRED:

(a) Reactions at each support.

(b) Shear and Moment Diagrams

(c) Max. Bending Moment and location.

STEP I:

Total Loads:

UL at Left: 600 × 8.5 =	5,100 Lbs.
UL at Right = 800 × 10.0 =	8,000 "
$P_1 + P_2 + P_3$ = Conc. Loads =	7,600 "
Total =	20,700 Lbs

STEP II:

Locate Center Gravity of the two Uniform Loads and note the distance from each support to action line. Taking moments about R_2 to determine R_1:

22.0 R_1 =	800 × 6.0 × 3.0 =	14,400 (Part left of R_2)
	3500 × 6.0 =	21,000
	2700 × 14.5 =	39,150
	5100 × 20.25 =	103,275

Forces Rotating Left = 177,825 '#

At right side of R_2, Load Forces rotate Clockwise and their moments must be deducted from Moments rotating Counter Clockwise:

UL = 800 × 4.0 × 2.0 = 6400 '#

CL = 1400 × 4.0 = 5600 '#

Forces Rotating R = 12,000 '#

$$\text{Then } R_1 = \frac{177,825 - 12,000}{22.0} = 7,537.5 \text{ Lbs.}$$

EXAMPLE I: Simple span with overhangs and combined loads, continued 1.9.3

STEP III

To solve for R_2, and with R_1 as center of moments, first consider all load forces acting clockwise and to the right of R_1.
Equation will be put in Algebraic Form with clockwise forces taken first.

$$22 R_2 = [(8000 \times 21.0) + (1400 \times 26.0) + (3500 \times 16.0) + (2700 \times 7.5) + (600 \times 6.0 \times 3.0)] = +291,450 \text{ Ft. Lbs.}$$

Minus Moments: $(600 \times 2.50 \times 1.25) = -1875$ Ft. Lbs.

$$R_2 = \frac{291,450 - 1875}{22.0} = 13,162.5 \text{ Lbs.}$$

$R_1 + R_2$ = Total Load check. $7,537.5 + 13,162.5 = 20,700$ Lbs. Checks with Step I.

STEP IV

Calculations for Bending Moments, left to right sequence.

$M_{0.0} = 0$

$M_{1.0'} = 600 \times 1.00 \times 0.50 = -300$ Ft. Lbs.

$M_{2.0'} = 600 \times 2.00 \times 1.00 = -1200$ " "

$M_{2.5'} = 600 \times 2.50 \times 1.25 = -1875$ " "

$M_{5.0'} = (7537.5 \times 2.50) - (600 \times 5.0 \times 2.50) = +11,300$ Foot Lbs.

$M_{7.0'} = (7537.5 \times 4.50) - (600 \times 7.0 \times 3.50) = +19,150$ " "

$M_{8.5'} = (7537.5 \times 6.0) - (600 \times 8.5 \times 4.25) = +23,550$ " "

$M_{10.0'} = (7537.5 \times 7.5) - (5100 \times 5.75) = +27,206$ " " (Max. Pos. Mom.)

$M_{14.0'} = (7537.5 \times 11.5) - [(5100 \times 9.75) + (2700 \times 4.0)] = +26,350$ Ft. Lbs.

$M_{18.5'} = (7537.5 \times 16.0) - [(5100 \times 14.25) + (2700 \times 8.5)] = +24,975$ " "

$M_{20.0'} = (7537.5 \times 17.5) - [(5100 \times 15.75) + (2700 \times 10.0) + (3500 \times 1.5) + (800 \times 1.50 \times .75)] + 18,430'\#$

$M_{22.0'} = (7537.5 \times 19.5) - [(5100 \times 17.75) + (2700 \times 12.0) + (3500 \times 3.5) + (800 \times 3.5 \times 1.75)] + 6,906'\#$

$M_{24.5'} = (7537.5 \times 22.0) - [(5100 \times 20.25) + (2700 \times 14.5) + (3500 \times 6.0) + (800 \times 6.0 \times 3.00)] - 12,000'\#$

Also $M_{24.5} = (1400 \times 4.0) + (800 \times 4.0 \times 2.0) = -12,000$ Ft. Lbs.

EXAMPLE II: Simple span with overhangs and combined loads 1.9.4

Known Data:
Length Beam = 22.0 Feet.
Cantilever of 5.0 Ft. at each end.
Distance between supports = 12.0'
Uniform Load 1000 #/' clear across.
Concentrated Loads as follows:
P_1 = 450 Lbs. Located extreme left end.
P_2 = 920 Lbs. Located 12.0' to right P_1.
P_3 = 630 Lbs Located extreme Right end.

REQUIRED:
(a) Elevation Beam with Loads.
(b) Calculate Reactions and draw shear diagram.
(c) Calculate Maximum Positive and Negative Bending Moments.
(d) Compute other bending moments and draw Moment Diagram.

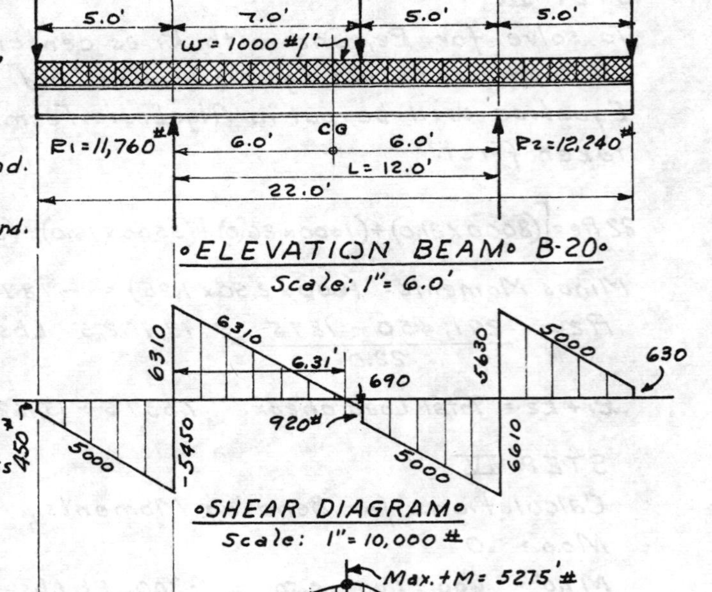

Max. +M = 5275'#
0.0 5.0 8.0 10 12 14 16 18 20 22
+M. 11.31
-M = 14,750'#
-M: 15,650'#
MOMENT DIAGRAM
Scale: 1" = 20,000'#

STEP I:
Determine Total Loads on Beam:

UL. = 22.0 × 1000 =	22,000	Lbs.
P_1 = (Concentrated) =	450	"
P_2 = " =	920	"
P_3 = " =	630	"
Total Loads =	24,000	Lbs.

STEP II
Calculate R_1 by using R_2 as Center of Moments.
Counter-Clockwise forces at left of R_2, rotation taken in equation first.

$$R_1 = \frac{\left[(920 \times 5.0) + (450 \times 17.0) + (1000 \times 17.0 \times 8.5)\right] - \left[(630 \times 5.0) + (1000 \times 5.0 \times 2.5)\right]}{12.0'} =$$

$$R_1 = \frac{(4600 + 7650 + 144{,}500) - (3150 + 12{,}500)}{12.0'} = 11{,}758^{\#} \text{ (Call it 11,760 Lbs)}$$

Calculate R_2 in same manner, Clockwise forces at right of R_1, minus forces acting Counter Clockwise at left of R_1, then divide product by span length L.

$$R_2 = \frac{\left[(920 \times 7.0) + (630 \times 17.0) + (1000 \times 17.0 \times 8.5)\right] - \left[(450 \times 5.0) + (1000 \times 5.0 \times 2.5)\right]}{12.0'} =$$

$$R_1 = \frac{(6440 + 10{,}710 + 144{,}500) - (2{,}250 + 12{,}500)}{12.0'} = 12{,}241.67^{\#} \text{ (Call it 12,240 Lbs.)}$$

Total Reactions must be equal to Loads. 11,760 + 12,240 = 24,000 Lbs. (Checks OK)

STEP III
Point on beam where by deducting loads from amount of Reaction R_1 the amount of shear becomes zero, or nothing.
R_1 = 11,760 Lbs. also equals Total Shear V.
11,760 − (450 + 1000 × 5.0) = 6310# Reducing this amount at rate of 1000 Lbs. Foot = 6310/1000 = 6.31 Feet.

EXAMPLE II: Simple span with overhangs and combined loads, continued 1.9.4

Location on beam for Maximum Positive Moment = 5.0'+6.31' = 11.31 Feet from Left end. Moment identified as $M_{11.31}$.

STEP IV:

Calculate Bending Moments at several points on beam and use scale of 1 inch equals 20,000 Foot Pounds to determine length of ordinates. Location point on beam is to be denoted by subscript thus: $M_{8.0'}$ indicates that Bending Moment is calculated 8.0 feet from extreme left end of the beam. Plot the moments magnitudes on diagram and connect by using celluloid or plastic ship curves.

$M_{2.0} = (450 \times 2.0) + (1000 \times 2.0 \times 1.0)$ − 2,900'#

$M_{5.0} = (450 \times 5.0) + (1000 \times 5.0 \times 2.5)$ − 14,750'#

$M_{6.0} = (11{,}760 \times 1.0) - \left[(450 \times 6.0) + (1000 \times 6.0 \times 3.0)\right] =$ − 8,940'#

$M_{8.0} = (11{,}760 \times 3.0) - \left[(450 \times 8.0) + (1000 \times 8.0 \times 4.0)\right] =$ − 320'#

$M_{11.31} = (11{,}760 \times 6.31) - \left[(450 \times 11.3) + (1000 \times 11.3 \times 5.65)\right] =$ + 5,275'#

$M_{14.0} = (11{,}760 \times 9.0) - \left[(450 \times 14.0) + (1000 \times 14.0 \times 7.0) + (920 \times 2.0)\right] =$ − 315'#

$M_{16.0} = (11{,}760 \times 11.0) - \left[(450 \times 16.0) + (1000 \times 16.0 \times 8.0) + (920 \times 4.0)\right] =$ − 9,535'#

$M_{17.0} = (11{,}760 \times 12.0) - \left[(450 \times 17.0) + (1000 \times 17.0 \times 8.5) + (920 \times 5.0)\right] =$ − 15,650'#

$M_{17.0} = (630 \times 5.0) + (1000 \times 5.0 \times 2.50) =$ − 15,650 Foot Lbs.

EXAMPLE: Simple span with overhangs and uniform load 1.9.5

Given Data: Beam Length = 35.0 Ft. Cantilever both ends. Overhang 5.0' at left end, and 6.0 feet at right end. Clear span between supports = 24.0 Ft. Continuous Uniform Load over whole beam length of 4000 Lbs. Lin. Foot.

REQUIRED:

(a) Compute Reactions R_1 and R_2.

(b) Draw Shear Diagram and compute location of zero shear from end.

(c) Calculate Bending Moments on approximately 2.0 foot intervals to plot a true moment Diagram.

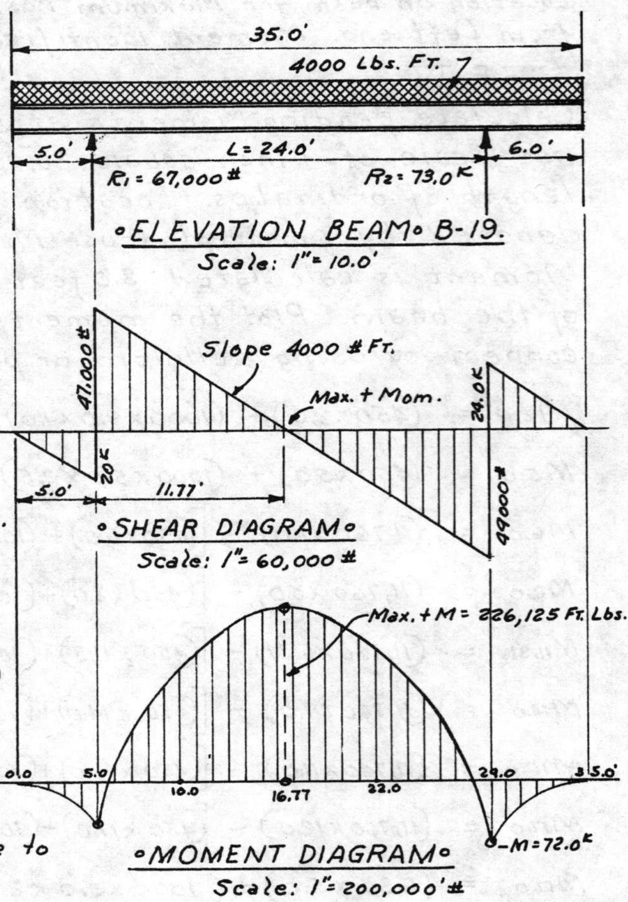

STEP I:

Determine Total Load and Reactions.

$w = 4000\#'$ $W = 4000 \times 35.0 = 140{,}000$ Lbs.

Performing work in Kips:

$+\ 24 R_1 = 4.0 \times 29.0 \times 14.5 = 1{,}682^K$ (left R_2)

$-\ 4.0 \times 6.0 \times 3.0 = 72^K$ (right R_2)

$1{,}610^K$

$R_1 = \frac{1610}{24.0} = 67.0^K$ (Close enough)

Taking Moments Rotating clockwise about R_1 and deducting moments rotating counter-clockwise to solve for R_2.

$24.0\ R_2 = 4.0 \times 30.0 \times 15.0 = +1{,}800^K$

$4.0 \times 5.0 \times 2.5 = -50^K$

$24.0\ R_2 = 1{,}750^K$ $R_2 = 1750/24.0 = 73.0^K$

Totals: $R_1 + R_2 = 67.0 + 73.0 = 140.0^K$ Same as Total Loads.

STEP II

Point of Zero Shear. Shear at R_1 = 67,000 Lbs.

Load Reduction = 4000 Lbs. Foot.

Distance from Left end: 67,000/4000 = 16.75 Ft. (Exact distance = 16.77')

STEP III

Drawing Shear Diagram, the slope lines cross base line at same point as found in Step II or 16.75 feet from left end.

EXAMPLE: Simple span with overhangs and uniform loads, continued 1.9.5

STEP IV

Calculating Bending Moments:

Max. Mom at 16.75 = $(67.0 \times 11.75) - (4.0 \times 16.75 \times 8.375) = +226.125'^{K}$ (Foot Kips)

Max Neg. Moment at R_2: $-M_{29.0} = 4.0 \times 6.0 \times 3.0 = -72.0'^{K}$

Negative Moment at R_1: $-M_{5.0} = 4.0 \times 5.0 \times 2.5 = -50.0'^{K}$

STEP V

Computing Bending Moments between Supports R_1 and R_2, of Beam B-19, for plotting Moment Diagram.

$M_{6.0} = (67.0 \times 1.0) - (4.0 \times 6.0 \times 3.0) = -5.0'^{K}$ (5000 Ft. Lbs.)

$M_{7.0} = (67.0 \times 2.0) - (4.0 \times 7.0 \times 3.5) = +36.0'^{K}$ (Inflection point is rapid)

$M_{9.0} = (67.0 \times 4.0) - (4.0 \times 9.0 \times 4.5) = +106.0'^{K}$

$M_{11.0} = (67.0 \times 6.0) - (4.0 \times 11.0 \times 5.5) = +160.0'^{K}$

$M_{13.0} = (67.0 \times 8.0) - (4.0 \times 13.0 \times 6.5) = +198.0'^{K}$

$M_{15.0} = (67.0 \times 10.0) - (4.0 \times 15.0 \times 7.5) = +220.0'^{K}$

$M_{17.0} = (67.0 \times 12.0) - (4.0 \times 17.0 \times 8.5) = +226.0'^{K}$

$M_{19.0} = (67.0 \times 14.0) - (4.0 \times 19.0 \times 9.5) = +216.0'^{K}$

$M_{21.0} = (67.0 \times 16.0) - (4.0 \times 21.0 \times 10.5) = +190.0'^{K}$

$M_{23.0} = (67.0 \times 18.0) - (4.0 \times 23.0 \times 11.5) = +148.0'^{K}$

$M_{24.0} = (67.0 \times 19.0) - (4.0 \times 24.0 \times 12.0) = +121.0'^{K}$

$M_{27.0} = (67.0 \times 22.0) - (4.0 \times 27.0 \times 13.5) = +6.0'^{K}$

$M_{29.0} = (4.0 \times 6.0 \times 3.0) = -72.0'^{K}$

EXAMPLE: Simple span sloped girder with overhang 1.10.1

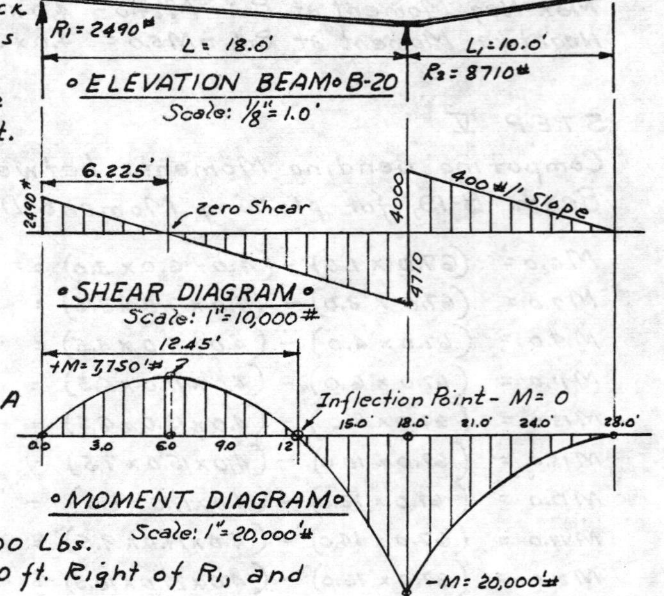

Built up beam of Plate and welded. Length Beam = 28.0' Uniform Load of 400 Lbs. Lin. Foot full length. Right end cantilevers over loading Dock 10.0 Ft. Clear Span between supports is 18.0 ft. Bottom of beam to be sloped with contour to equal the Inertia needs of Bending Moment.

REQUIRED:

(a) Elevation of Beam, Shear and Moment Diagrams.
(b) Calculate Reactions R_1 and R_2.
(c) Compute Moments for Diagram.
(d) Determine where inflection point on beam is located and compute moment at that point. A variable distance of 3 inches is close enough for accuracy.

STEP I:

Total Load = 400 x 28.0 = 11,200 Lbs.
Center of Gravity located 14.0 ft. Right of R_1, and 4.0 feet to left of R_2.

STEP II

With R_2 as center of taking moments, solve for R_1.

$$R_1 = \frac{11,200 \times 4.0'}{18.0'} = 2488.8\# \text{ (Call it 2490 Lbs.)}$$

$$R_2 = \frac{11,200 \times 14.0}{18.0'} = 8711.2\# \text{ (Call it 8710 Lbs.)}$$

Total Reactions = 2490 + 8710 = 11,200 Lbs. Equals Total Load.

STEP III

Drawing Shear Diagram, point of no shear equals point of Maximum Moment. Distance = 2490/400 = 6.225 Ft. (Use 6.25')

STEP IV

Computing Moments for Moment Diagram.

$M_{1.0'} = (2490 \times 1.0) - (400 \times 1.0 \times 0.50) = +2290$ Ft. Lbs.

$M_{3.0'} = (2490 \times 3.0) - (400 \times 3.0 \times 1.50) = +5670$ "

$M_{6.0'} = (2490 \times 6.0) - (400 \times 6.0 \times 3.0) = +7740$ "

$M_{6.25'} = (2490 \times 6.25) - (400 \times 6.25 \times 3.125) = +7750$ "

$M_{9.0'} = (2490 \times 9.0) - (400 \times 9.0 \times 4.50) = +6210$ "

EXAMPLE: Simple span sloped girder with overhang, continued 1.10.1

$M_{12.0}$ = (2490 x 12.0) - (400 x 12.0 x 6.0) = + 1080 "

$M_{14.0}$ = (2490 x 14.0) - (400 x 14.0 x 7.0) = − 4330 "

$M_{18.0}$ = (2490 x 18.0) - (400 x 18.0 x 9.0) = - 20,000 "

$M_{18.0}$ = 400 x 10.0 x 5.0 = − 20,000 Ft. Lbs.

$M_{21.0}$ = 400 x 7.0 x 3.5 = − 9,800 " "

$M_{24.0}$ = 400 x 4.0 x 2.0 = − 3,200 " "

STEP V

Inflection point on beam is the location where the Positive Bending balances the Negative Bending. At the exact location, it is to be assumed that the + moments = the − moments and moment = 0. Under these conditions bending stresses + and − at that point are equal, and a spliced beam should have the splice connection made at such locations.

From observing Moments in Step IV, the bending stress in beam changes between $M_{12.0'}$ and $M_{14.0'}$. However, in drawing the Moment Diagram, the curve appears to intersect the base line at point nearer to 12.50 feet from left end.

Check for actual inflection point by computing $M_{12.50'}$.

$M_{12.5}$ = (2490 x 12.50) − (400 x 12.50 x 6.25) = − 125 Ft. Lbs.

This is very close but a little to the right, and probable point is near to 12.45 feet.

$M_{12.45'}$ (2490 − 12.45) − (400 x 12.45 x 6.225) = 0 (Actual point no moment.)

The proper Engineering term to refer to the conditions on a beam, where Positive bending is countered by an equal amount of Negative bending, is "Contra-Flexure."

EXAMPLE: Simple span with overhanging loads over supports 1.10.2

Length beam = 45.0 Feet.
Cantilever at both ends.
Left end overhang = 5.0 Ft.
Right end overhang = 8.0 Ft.
Length between supports = 32.0 Ft.

Load Criteria:

(1) A 200 Pound Lin. Foot Uniform Load starts at extreme left end and extends 12.0 feet.

(2) A 400 Pound Lineal Foot Uniform Load starts at point 29.0 Feet from left end and extends 16.0 feet, ending at right end beam.

(3) A Concentrated Load of 4000 Lbs. is placed 21.0 feet from left end or 24.0 feet from right extreme end.

REQUIRED:

(a) Calculate Reactions after drawing Beam Elevation.
(b) Draw Shear Diagram.
(c) Calculate Moments and draw bending Moment Diagram.

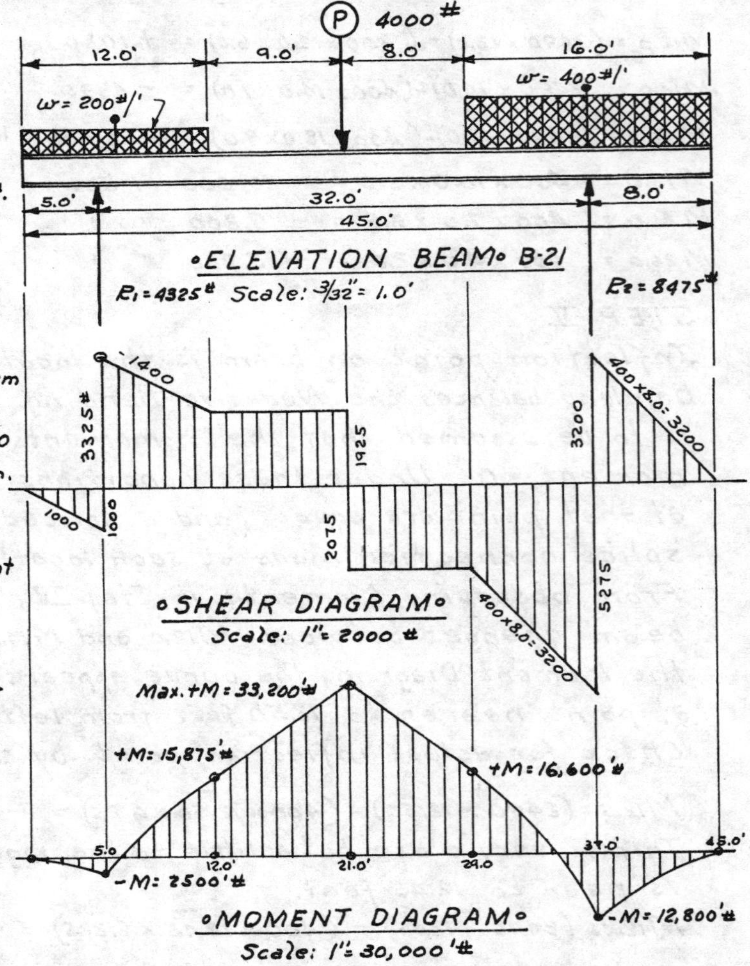

STEP I:
Total Loads on Beam.

Left Uniform Load:	200 × 12.0 =	2,400#	(Center Gravity 1.0' Right of R_1.)
Right Uniform Load:	400 × 16.0 =	6,400#	(Center Gravity is over R_2.)
Concentrated Load P =		4,000#	
Total Load =		12,800#	

STEP II:
Solve for R_1 by using R_2 for center of moments.

$$R_1 = \frac{(2400 \times 31.0) + (4000 \times 16.0) + (6400 \times 0.0)}{32.0'} = \frac{138,400}{32.0} = 4,325 \text{ Lbs.}$$

$$R_2 = \frac{(6400 \times 32.0) + (4000 \times 16.0) + (2400 \times 1.0)}{32.0} = 8,475 \text{ Lbs.}$$

Total Reactions = 12,800 Lbs.

EXAMPLE: Simple span with overhanging loads over supports, continued 1.10.2

STEP III

Cantilever Beams and Overhangs produce Negative Bending.

$-M_{5.0'} = 200 \times 5.0 \times 2.50 = -2500$ Ft. Lbs. Same as Formula $\frac{WL}{2}$

$-M_{37.0'} = 400 \times 8.0 \times 4.0 = -12,800$ Ft. Lbs.

Positive Moments:

$M_{12.0'} = (4325 \times 7.0) - (2400 \times 6.0) = +15,875$ Ft. Lbs.

$M_{21.0'} = (4325 \times 16.0) - (2400 \times 15.0) = +33,200$ " " (Maximum + Mom.)

$M_{29.0} = (4325 \times 24.0) - [(2400 \times 23.0) + (4000 \times 8.0)] = +16,600$ Ft. Lbs.

Checking Algebraic Method for moment over R_2

$M_{37.0} = \overset{+\,Mom}{(4325 \times 32.0)} - \overset{-\,moments}{[(2400 \times 31.0) + (4000 \times 16.0) + (400 \times 8.0 \times 4.0)]} = -12,800'\#$

STEP IV

May check accuracy of Curve on Moment Diagram, by solving to ascertain if moment is very close to zero (0) at point of contra-flexure. Points on curve appear to be at $M_{5.58'}$ and $M_{34.3'}$.

At Left end $M_{5.58} = (\overset{+}{4325} \times 0.58) - (\overset{-}{200} \times 5.58 \times 2.29) = -47'\#$ Close enough.

At. $M_{34.3'} = (4325 \times 29.3) - [(2400 \times 28.3) + (4000 \times 13.3) + (400 \times 5.3 \times 2.65)] = -15.5$ Ft. Lbs.

EXAMPLE: Simple span with overhanging loads over supports; solving uniform and concentrated loads separately 1.10.3

Assume Beam length is 45.0 feet overall, and overhangs Right support (R_2) by 8.0 feet. Overhang of left end is 5.0 feet from Left support (R_1), making span L = 32.0 feet.
A concentrated Load P = 4000 Pounds is located 21.0 feet from left end or 24.0 feet from right end.
A Uniform Load of 200 Lbs. Lineal Foot, starts at left end and extends 12.0 feet on beam. Another Uniform Load of 400 Lbs. Lineal foot starts at Right end and extends to left a length of 16.0 feet on beam.

REQUIRED:
Draw an elevation of the beam for each type of loading. Calculate the Reactions and critical bending moments for each type of load seperately. Leave space below each elevation and construct a shear diagram for each load type. Combine the two bending moments at a point directly under load P, and compare the result with the previous example. Do not draw the moment diagrams.

STEP I:
Drawing elevations of beam with seperate type of loads.

STEP II:
Reactions of Beam with Uniform Loads only: Taking moments about R_1 to solve for R_2.

$$R_2 = \frac{(400 \times 16.0 \times 32.0) + (200 \times 12.0 \times 1.0)}{32.0'} = 6{,}475 \text{ Lbs.}$$

$$R_1 = (200 \times 12.0 \times 31.0) + (400 \times 16.0 \times 0) = 2{,}325 \text{ Lbs.}$$

Total Uniform Loads = $(200 \times 12.0) + (400 \times 16.0) = 8{,}800$ Lbs.
Beam is in equilibrium as $R_1 + R_2$ equal Total loads.

STEP III:
Constructing shear diagram for Uniform loads.
Overhanging load causes moment and starting on left end of beam on base line. Below base line, $200 \times 5.0 =$ 1000 Lbs. Amount shear above Base line = 2325 − 1000 = 1325.#
Slope line crosses base line at point = $\frac{1325}{200}$ = 6.625 feet from support R_1. Amount of shear at point 12.0 feet from left end of beam = $(200 \times 7.0) - 1325 = 75$ Lbs.

STEP IV:

Concentrated Loads:

$R_1 = \frac{4000 \times 16.0}{32.0} = 2000^{\#}$

R_2 = Same as R_1

STEP V:

C.L. Shear Diagram:
Greatest at supports R_1 and R_2.
Maximum Moment will be under P, and will be positive.

STEP VI:

Calculating Moment under load P for concentrated load only.

$+M = 2000 \times 16.0 = 32{,}000'^{\#}$

or may use formula:

$M = \frac{PL}{4}$

$M = \frac{4000 \times 32.0}{4} = 32{,}000'^{\#}$

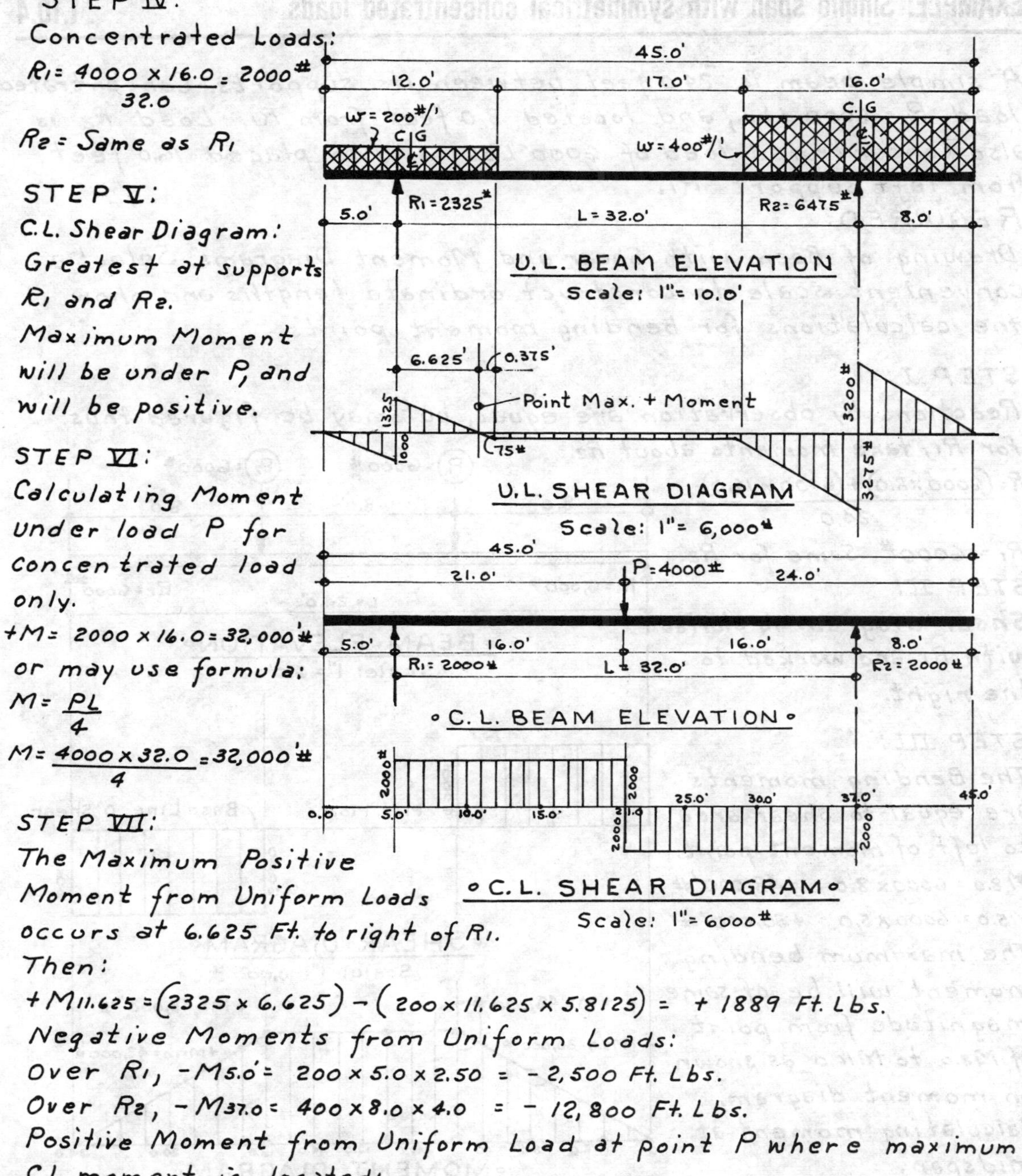

STEP VII:

The Maximum Positive Moment from Uniform Loads occurs at 6.625 Ft. to right of R_1.

Then:

$+M_{11.625} = (2325 \times 6.625) - (200 \times 11.625 \times 5.8125) = +1889$ Ft. Lbs.

Negative Moments from Uniform Loads:

Over R_1, $-M_{5.0'} = 200 \times 5.0 \times 2.50 = -2{,}500$ Ft. Lbs.

Over R_2, $-M_{37.0} = 400 \times 8.0 \times 4.0 = -12{,}800$ Ft. Lbs.

Positive Moment from Uniform Load at point P where maximum CL moment is located:

$M_{21.0'} = (2325 \times 16.0) - (200 \times 12.0 \times 15.0) = +1200$ Ft. Lbs.

STEP VII:

Adding together the Positive Bending for 2 Load types at point P it is: $+M_{max.} = 32{,}000 + 1{,}200 = +33{,}200$ Ft. Lbs.

This checks with results of previous example with Diagram.

EXAMPLE: Simple span with symmetrical concentrated loads 1.10.4

A simple beam is 24.0 feet between its supports. Concentrated load P_1 = 6000 Lbs., and located 8.0 feet from R_1. Load P_2 is also Concentrated load of 6000 Lbs., and is placed 16.0 feet from left support R_1.

REQUIRED:

Drawing of Beam with Shear and Moment Diagrams. Select a convenient scale to construct ordinate lengths and show the calculations for bending moment points.

STEP I:

Reactions by observation are equal, but may be figured thus: For R_1, take moments about R_2.

$$R = \frac{(6000 \times 8.0) + (6000 \times 16.0)}{24.0}$$

R_1 = 6000# Same for R_2.

STEP II:

Shear diagram is started with R_1 and worked to the right.

STEP III:

The Bending moments are equal to shear area to left of moment point.

$M_{8.0} = 6000 \times 8.0 = +48{,}000'\#$

$M_{5.0} = 6000 \times 5.0 = +30{,}000'\#$

The maximum bending moment will be of same magnitude from point of $M_{8.0}$ to $M_{16.0}$ as shown on moment diagram.

Calculating moment at midspan:

$M_{12.0} = (6000 \times 12.0) - (6000 \times 4.0) = +48{,}000'\#$

STEP IV

Check for equilibrium by comparing clockwise moments rotating about R_1, and Counter-Clockwise moments about R_2.

About R_1: $\Sigma M = \left[\frac{(6000 \times 8.0) + (6000 \times 16.0)}{6000 \times 24.0}\right] = 0$

Same equation is written with moments about R_2. $\Sigma M = 0$

EXAMPLE: Overhanging beam with tie down 1.10.5

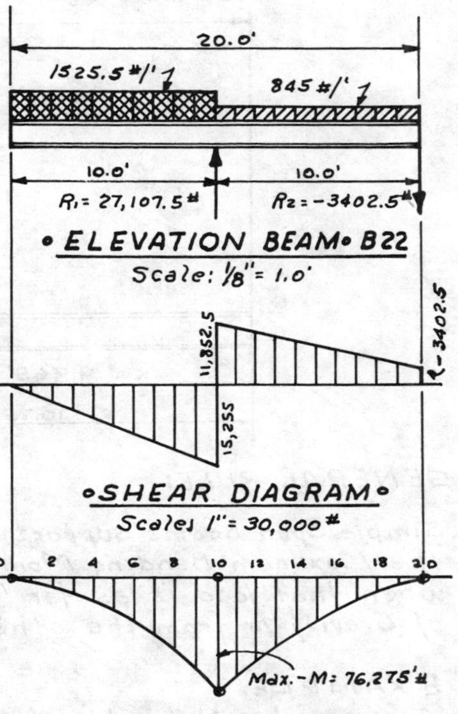

Beam 20.0 Feet long over 3 Supports.
Spacing between supports = 10.0' c-c.
Left half span supports 1525.5 #/' U. Load.
Right Span supports 845#/' U Load.
Loads include dead weight of beam.

REQUIRED:
Remove the left end support and draw load Elevation of Beam B-22. Calculate Reactions at 2 Remaining supports as R_1 and R_2.
Draw Shear Diagram and Calculate the moments for Bending Moment Diagram.

STEP I:
Total Loads: 1525.5 × 10.0 = 15,255 Lbs.
845 × 10.0 = 8,450 "
Total = 23,705 Lbs.

STEP II
Calculating Reaction R_1 by using R_2 as center of Moments: All forces left side of R_2 are rotating counter-Clockwise:

$$R_1 = \frac{(845 \times 10.0 \times 5.0) + (1525.5 \times 10.0 \times 15.0)}{10.0'} = 27,107.5 \text{ Lbs.}$$

$$R_2 = \frac{-\overset{\text{counter}}{(1525.5 \times 10.0 \times 5.0)} + \overset{\text{clockwise}}{(845 \times 10.0 \times 5.0)}}{10.0'} = -3402.5 \text{ Lbs.}$$

Total Loads to agree with Reactions = 27,107.5 - 3402.5 = 23,705 Lbs.
Reaction R_2 acts as required force (or Load) necessary to hold beam in equilibrium and is considered in calculating bending moments: Summation of Moments must equal summation of moments on Right or $\Sigma M = 0$. Thus at R_1, the equation is:
$(1525.5 \times 10.0 \times 5.0) = [(845 \times 10.0 \times 5.0) + (3402.5 \times 10.0)]$ and $\Sigma M = 0$. (checks)

STEP III
Drawing Shear Diagram, Max. Negative Moment is over R_1.

Max. $M_{10.0}$ = 1525.5 × 10.0 × 5.0 = − 76,275 Ft. Lbs.

$M_{15.0} = (27,107.5 \times 8.0) - [(845 \times 5.0 \times 2.5) + (15,255 \times 10.0)] = -27,574.5$ Lbs.

$M_{20.0} = (27,107.5 \times 10.0) - [(845 \times 10.0 \times 5.0) + (15,255 \times 15.0)] = 0$

EXAMPLE: Simple span with moving loads 1.10.6

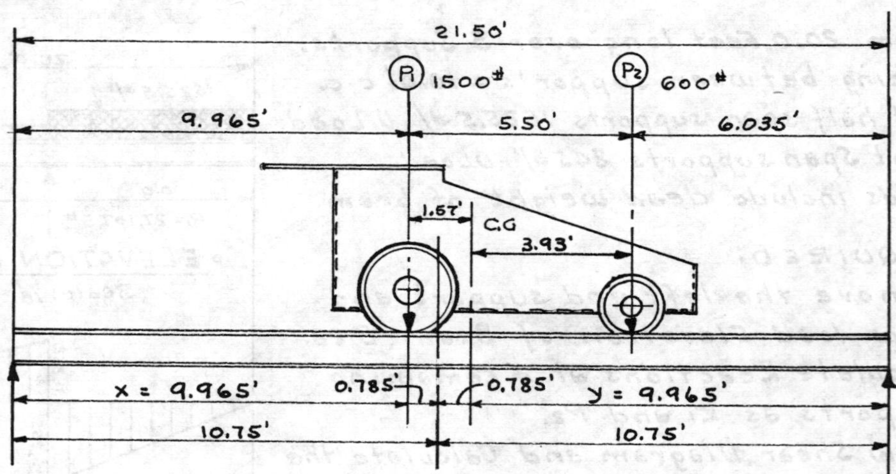

GENERAL RULE:

Simple Span beams supporting Concentrated Moving Loads will have the Maximum Bending Moment produced under one of the Loads, when that load is as far from one support as the loads Center of Gravity is from the other support.

EXAMPLE:

A four wheel slag buggy with load weighs 4200 Pounds, ½ on each track beam. Larger wheels support 1500 Pounds each, and small wheels support 600 Pounds each.
Elevated Spans are supported at 21.50 foot centers, and simple span lengths because of spur tracks. Distance between wheels is 5.50 feet.

REQUIRED:

Determine at which point on beam the cart will produce the Maximum bending moment. Show Center of Gravity of Loads and calculate bending moments and Reactions.

STEP I

Identify large load at $P_1 = 1500^\#$ and $P_2 = 600^\#$ Drawing Elevation of beam and place cart somewhere near midspan. Spacing between P_1 and P_2 = 5.50' Total Loads on beam = 1500 + 600 = 2100 Pounds.

STEP II

Calculating the Center of Gravity of all Loads - Take moments about load wheel P_1.

$$CG \text{ from } P_1 = \frac{(1500 \times 0.0) + (600 \times 5.50)}{2100} = 1.5714' \text{ (Call it 1.57 feet)}$$

CG distance from P_2 = 5.50 - 1.57 = 3.93' Note these on drawing.

EXAMPLE: Simple span with moving loads, continued 1.10.6

STEP III
The Center of Gravity of both loads must be placed the same distance from R_2 as will load P_1 be from R_1. These dimensions are determined as $x = y$ and $\frac{21.50 - 1.57}{2} = 9.965$ feet from each support. Mid-span = $21.50 \times .50 = 10.75$ feet.

STEP IV
All dimensions can now be placed on Elevation of Beam and Reactions computed. Take moments about R_2 to solve for R_1.

$$R_1 = \frac{(600 \times 6.035') + (1500 \times 11.535')}{21.50'} = 973.2 \text{ Pounds}$$

$$R_2 = \frac{(1500 \times 9.965) + (600 \times 15.465)}{21.50} = 1126.8 \text{ Pounds}$$

Total Loads = 2100# and $R_1 + R_2$ = 2100# (Checks)

STEP V
If maximum Moment is under one of the loads, calculate moment under both P_1 and P_2.

Under P_1 - Mom. = $973.2 \times 9.965' = 9{,}698$ Foot Pounds.
Under P_2 - Mom. = $(973.2 \times 15.465) - (1500 \times 5.50) = 6{,}800$ Foot Pounds.

Maximum Moment = 9,698'# when wheel load P_1 is 9.965 feet from left support.

STEP VI
Maximum Vertical Shear will occur when largest wheel load P_1 is near the supports.
Assume P_1 is 1.0 foot to right of R_1

$$R_1 = V = \frac{(1500 \times 20.5) + (600 \times 15.0)}{21.50'} = 1942 \text{ Pounds} \pm$$

EXAMPLE: Continuous beams with uniform loads distribution, 3 spans 1.11.1

A single 1 piece beam extends over 3 equal spans of 15.0 feet. Total length of beam is 45.0 and supports a uniform distributed load of 800 Lbs. Lineal Foot for full length.

REQUIRED:

Draw an elevation of the beam with 3 spans and calculate the Reactions at each support. Calculate Shear and Bending Moments nessary to construct Shear and Moment Diagrams. Use fractions or decimal equivalents to delineate the shear and moment distribution.

STEP I:

Drawing the Beam Elevation below with space to construct the Shear and Moment diagrams in the Vertical action plane.

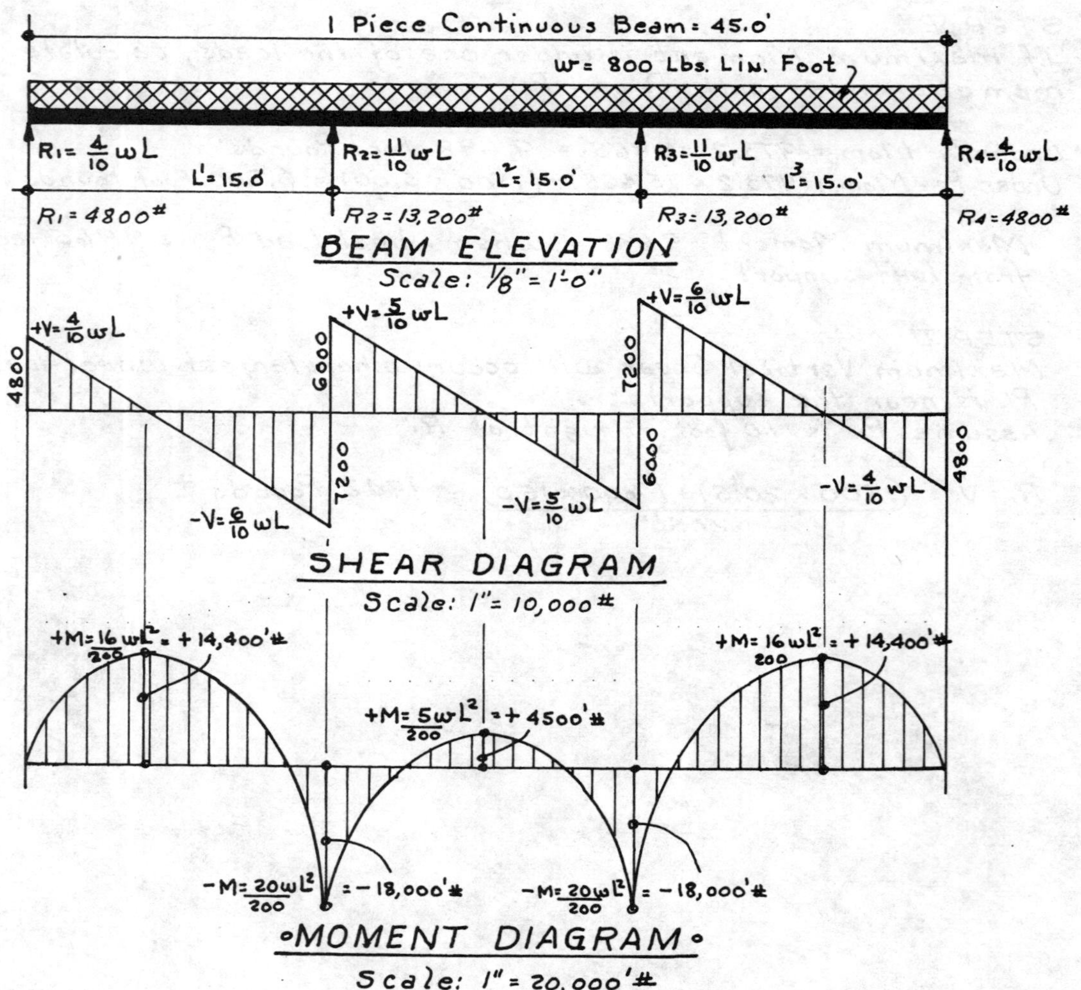

EXAMPLE: Continuous beams with uniform loads, 3 spans, continued 1.11.1

STEP II:

For Reactions R_1 and R_4: $R = \frac{4}{10} wL$ or $R = 0.40 wL$

For Reactions R_2 and R_3: $R = \frac{11}{10} wL$ or $R = 1.10 wL$

R_1 or $R_4 = \frac{4 \times 800 \times 15.0}{10} = 4800$ Lbs.

R_3 or $R_4 = \frac{11 \times 800 \times 15.0}{10} = 13{,}200$ Lbs.

STEP III:

For Vertical Shear: At ends $V_1 = R_1 = 4800$ Lbs. Same at $R_4 = V_4$.

Shear above line will be indicated as +, and below as −.

$+V_2$ and $-V_3 = \frac{5}{10} wL$ or $V = 0.50 wL$.

$-V_2$ and $+V_3 = \frac{6}{10} wL$ or $V = 0.60 wL$.

$+V_2$ and $-V_3 = \frac{5 \times 800 \times 15.0}{10} = 6000$ Lbs.

$-V_2$ and $+V_3 = \frac{6 \times 800 \times 15.0}{10} = 7{,}200$ Lbs.

Shear Diagram may be constructed with data obtained.

STEP IV:

Calculating the Bending Moments:

Distribution for Positive (+) Moment for end spans $= \frac{16}{200} wL^2$

Distribution for Positive (+) Moment at middle span $= \frac{5}{200} wL^2$

Distribution for Negative (−) Moments over supports $= \frac{20}{200} wL^2$

$+M$ for end spans $= \frac{16 \times \overset{4}{\cancel{800}} \times 15.0 \times 15.0}{\cancel{200}} = +14{,}400$ Ft. Lbs

$+M$ for middle span $= \frac{5 \times \overset{4}{\cancel{800}} \times 15.0 \times 15.0}{\cancel{200}} = +4500$ Ft. Lbs.

$-M$ over supports R_2 and R_3: $\frac{20 \times 800 \times 15.0 \times 15.0}{200} = 18{,}000$ Ft. Lbs.

EXAMPLE: Continuous beams with concentrated loads, 3 spans 1.11.2

Single Beam of 1 piece covers 4 supports with 3 Spans of 15.0 feet. Beam supports 3 equal Concentrated Loads, with each load of 4000 Pounds located at its mid-span. Full length of Beam is 45.0 feet.

REQUIRED:
Draw eleavation of full length over 3 spans, place loads and calculate Reactions at each support. Calculate Shear and Bending Moments under each load and over supports, then construct Shear and Moment Diagram. Use fractions or decimals to delineate the distribution of moments.

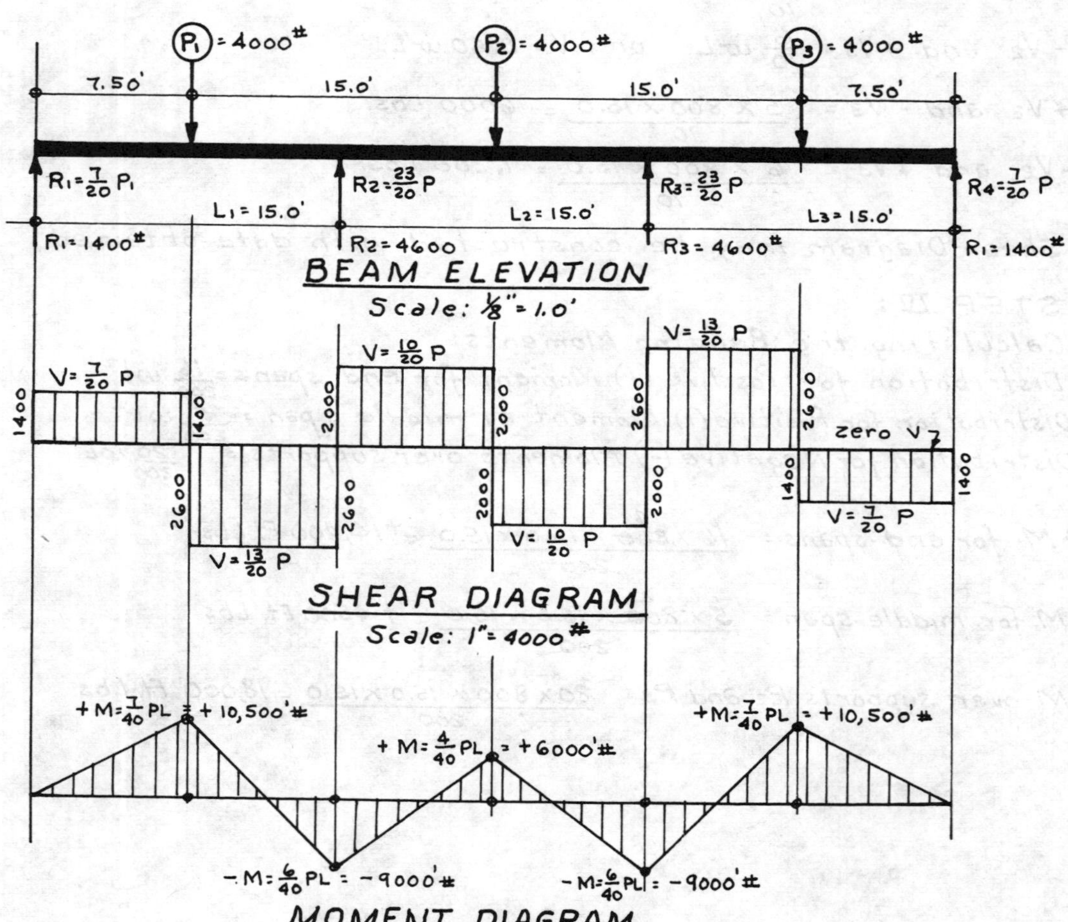

EXAMPLE: Continuous beams with concentrated loads, 3 spans, continued 1.11.2

STEP I:

Drawing Beam Elevation above where Shear and Moment Diagram can be constructed below in same Plane of Action.

STEP II

Load distribution for $R_1 = \frac{7}{20}P$ or $0.350P$

$R_1 = \frac{7 \times 4000}{20} = 1400$ Lbs. R_4 same as R_1

$R_2 = \frac{23 \times 4000}{20} = 4,600$ Lbs. R_3 same as R_2

Total Loads = $3 \times 4000 = 12,000$ Lbs.

Total Reactions = $(4600 \times 2) + (1400 \times 2) = 12,000$ Lbs. (Checks with loads)

STEP III:

To determine vertical Shear:

$V_1 = R_1 = 1400$ Lbs.

At R_2, Shear above $= \frac{10}{20}P = \frac{10 \times 4000}{20} = 2000$ Lbs. Also below at R_3

Below at R_2, Shear $= \frac{13}{20}P = \frac{13 \times 4000}{20} = 2600$ Lbs. Also above at R_3

Shear above and below zero line must total R_2 as $2000 + 2600 = 4600$ Lbs.

Shear Diagram may now be constructed, with scale 1" = 4000 Lbs.

STEP IV:

Moment distribution at $M_{7.50} = \frac{7}{40}PL$, or area in shear diagram left of load P. $+M_{7.5} = 1400 \times 7.50 = 10,500'\#$ or $\frac{7 \times \cancel{4000}^{100} \times 15.0}{\cancel{40}} = 10,500$ Ft. Lbs.

Negative Moment over support R_2:

$-M_{15.0} = \frac{6 \times \cancel{4000}^{100} \times 15.0}{\cancel{40}} = -9,000$ Ft. Lbs. Same over R_3 support

Positive moment under load P_2: $+M_{22.5} = \frac{4}{40}PL$

$+M_{22.5} = \frac{4 \times \cancel{4000}^{100} \times 15.0}{\cancel{40}} = +6000$ Ft. Lbs.

STEP V:

Moments may be pointed off in vertical plane and the moment diagram constructed.

Max. Positive Moment = 10,500'# at center of end spans under the loads P_1 and P_3.

Max. Negative Moment = 9,000'# and over supports R_2 and R_3

EXAMPLE: Continuous beams with combined loads on 2 end spans 1.11.3

A Shop Building 100'-0" Long and 40'-0' wide is to be designed with 2nd Floor for Storage. Combined Live Load with Dead Load for 2nd. Floor = 140 Pounds per Square Foot. Bay Spacing is 20.0 foot Center to Center of Columns.
The Girder supporting 2nd. floor will have a column in Center making 2 Spans of 20.0 feet. At each midspan, a hoist monorail is to support a travelling hoist with 4000 Pound Capacity.
Girder Section is to be continuous on top of center Column and designed for maximum conditions, such as both hoists being fully loaded and directly under girder.

REQUIRED:
Layout a Section of Structure, determine loads, draw shear diagrams. Calculate the maximum bending moments for both Positive and Negative bending, then combine in table for design.

STEP I
Layout Section to 1/8 inch Scale:
Bay Spacing = 20.0' Spans = 20.0' Area Floor for 1 Girder span to support = 20.0 x 20.0 = 400 Sq. Ft. Load W = 400 x 140 = 56,000#
For Uniform Load • w = 56,000/20.0 = 2,800 Pounds Lineal Foot.
For Concentrated Loads at Midspan – P_1 = 4000# and P_2 = 4000#

STEP II
From AISC Manual, Continuous beam diagrams for 2 Spans with Uniform Load are separate from same diagram for beams with Concentrated Loads. Figure reactions and moments seperately and combine later.

STEP III
Determine Reactions and check with Total Combined Loads.

Uniform Load Total =	2800 x 40.0'	=	112,000 Pounds
Concentrated Loads =	4000 x 2	=	8,000 "
Total Loads		=	120,000 "

Call Reaction Supports R_1 – R_2 and R_3 from left to right.

Uniform Load for R_1 = 3/8 wL or R_1 = .375 x 2800 x 20.0 = 21,000#
" " " R_2 = 5/8 wL 2 or R_2 = .625 x 2800 x 20.0 x 2 = 70,000#
Reaction R_3 is same as R_1 = 21,000#
Concentrated Load R_1 = 5/16 P_1 or R_1 = .3125 x 4000 = 1250#
" " R_2 = 11/16 P_1 and P_2 R_2 = .6875 x 4000 x 2 = 5,500#

Checking Reactions – Total UL = 21,000 + 70,000 + 21,000 = 112,000# OK
" " Total CL = 1250 + 5500 + 1250 = 8,000#
TOTAL = 120,000# CKS

TOTAL LOADS:

Uniform Loads: 2800 × 40.0' = 112,000 Pounds

Concentrated Loads: 4000 × 2 = 8,000 "

120,000 Pounds Combined

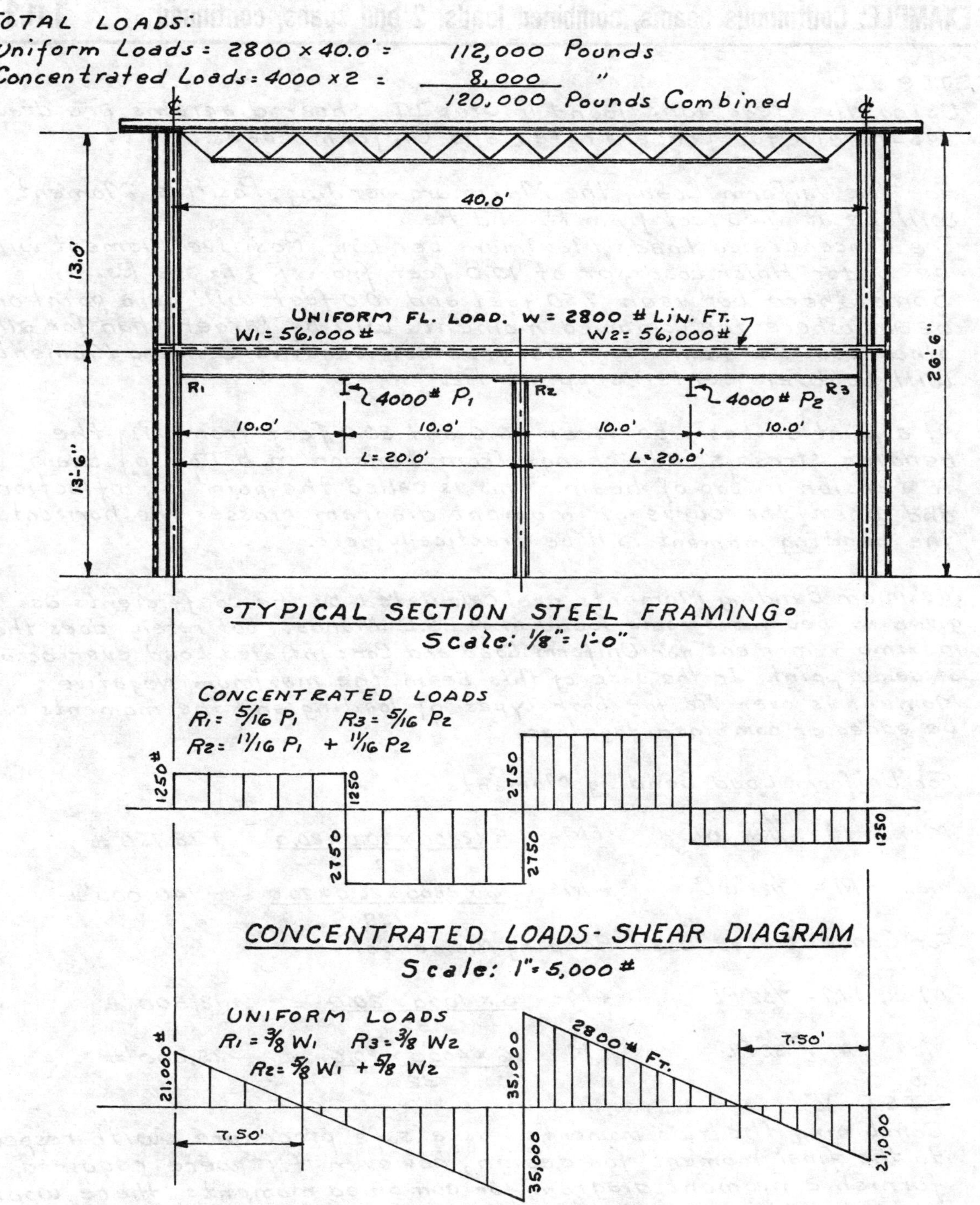

CONCENTRATED LOADS - SHEAR DIAGRAM

Scale: 1" = 5,000#

UNIFORM LOADS - SHEAR DIAGRAM

Scale: 1" = 50,000#

EXAMPLE: Continuous beams, combined loads, 2 end spans, continued 1.11.3

STEP IV

Using the Reactions found in Step III, Shear diagrams are drawn separately for Concentrated and Uniform Loads.

For the Uniform Load, the Maximum bending, Positive Moment will be at 7.50 feet from R_1 and R_3.
The Concentrated Loads, Maximum bending, Positive Moment will be under Hoist Loads, or at 10.0 feet from R_1, R_2 and R_3. Somewhere between 7.50 feet and 10.0 feet will be a point on beam where the Combined moments will be larger than for all other positive moments. The greatest Negative Bending Moment will be over center support R_2.

At a point on beam between 10.0 and 20.0 feet from R_1, the bending stress should change from tension in bottom of beam to tension in top of beam. This is called the point of "inflection", and when the curve on moment diagram crosses the horizontal the bending moment will be practically zero.

Maximum Bending Moments are calculated by the coefficients as given in beam diagrams found in A.I.S.C. Manual, but rarely does the maximum moment for Uniform Load and Concentrated Load ever occur at same point. In the case of this beam, the maximum Negative Moment is over R_2 for both types of loading and the moments can be added or combined together.

For Uniform Load Bending Moments:-

Max. $+M = \frac{9}{128} wL^2$ $\qquad +M = \dfrac{9 \times 2800 \times 20.0 \times 20.0}{128} = +78{,}750$ '#

Max. $-M = \frac{16}{128} wL^2$ $\qquad -M = \dfrac{16 \times 2800 \times 20.0 \times 20.0}{128} = -140{,}000$ '#

For Concentrated Load Bending Moments:

Max. $+M = \frac{5}{32} PL$ $\qquad +M = \dfrac{5 \times 4000 \times 20.0}{32} = +12{,}500$ '#

Max. $-M = \frac{6}{32} PL$ $\qquad -M = \dfrac{6 \times 4000 \times 20.0}{32} = -15{,}000$ '#

STEP V

Combining the two moments is a safe procedure with respect to the final moment for design, however if it were required to furnish a moment diagram for combined moments, there would be some discrepancy in the values because of the different points of inflection for each type of load.

Maximum Positive +M for C.L. is at $M_{10.0} = +12{,}500$ '#
Maximum Positive +M for UL. is at 7.50' from $R_1 = +78{,}750$ '#
Combining the two +Maximum Mom. = 12.500 + 78,750 = + 91,250 '#

P1 = 4000 # P2 = 4000
10.0' 20.0' 10.0'
w = 2800 # LIN. FOOT
L = 20.0' L = 20.0'
R1 = 22,250 #
CL: 1250
UL: 21000
R2 = 75,500 #
UL: 70,000
CL: 5,500
R3 = R1

GIRDER ELEVATION WITH LOADS
Scale: 1/8" = 1.0'

+M = 12,500'# +M = 12,500'#
0 2 4 6 8 10 12 14 16 18 20
-M = 15,000'#

MOMENT DIAGRAM - CONCENTRATED
Scale: 1" = 20,000'#

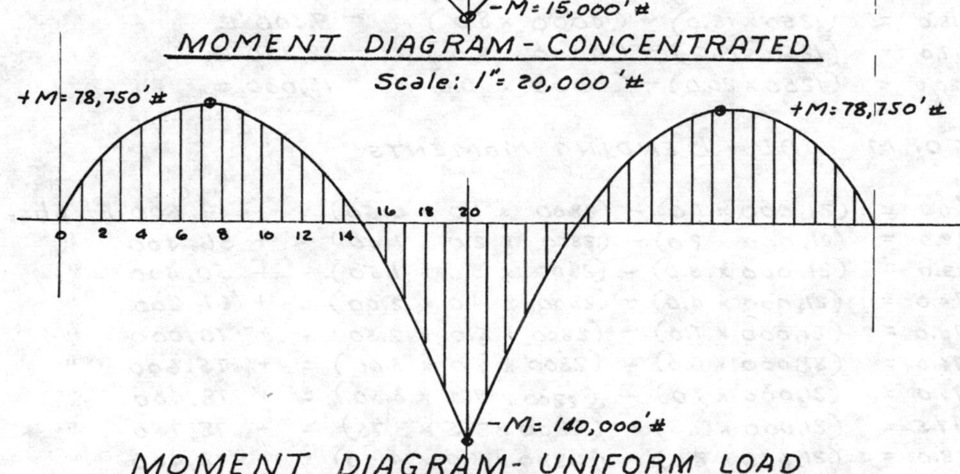

MOMENT DIAGRAM - UNIFORM LOAD
Scale: 1" = 100,000'#

Combining the two - Negative Moments -M, Max. = -140,000 + 15,000 = 155,000'#

Girder Design would be based on larger moment

STEP VI
To prepare a moment diagram with accurate values given with a combination of moments for both types of loads, the values may be figured separately or combined in one equation. The following method is best suited for this purpose. Use the same Reaction values for R1 as found in Step III and work from left to right.

CONCENTRATED LOAD BENDING MOMENTS: + = Pos. - = Neg.

Mom. 0.0' = zero
$M_{1.0'}$ = 1250 × 1.0 = +1250'#
$M_{2.0'}$ = 1250 × 2.0 = +2500'#
$M_{3.0'}$ = 1250 × 3.0 = +3750'#
$M_{4.0'}$ = 1250 × 4.0 = +5000'#
$M_{5.0'}$ = 1250 × 5.0 = +6250'#

EXAMPLE: Continuous beams, combined loads, 2 end spans, continued 1.11.3

Step VI Continued:

M6.0 =	1250 × 6.0 =	+ 7,500 '#	
M7.0 =	1250 × 7.0 =	+ 8,750 '#	
M8.0 =	1250 × 8.0 =	+ 10,000 '#	
M9.0 =	1250 × 9.0 =	+ 11,250 '#	
M10.0 =	1250 × 10.0 =	+ 12,500 '#	*
M11.0 =	(1250 × 11.0) − (4000 × 1.0) =	+ 9,750 '#	
M12.0 =	(1250 × 12.0) − (4000 × 2.0) =	+ 7,000 '#	
M13.0 =	(1250 × 13.0) − (4000 × 3.0) =	+ 4,250 '#	
M14.0 =	(1250 × 14.0) − (4000 × 4.0) =	+ 1,500 '#	
M15.0 =	(1250 × 15.0) − (4000 × 5.0) =	− 1,250 '#	
M16.0 =	(1250 × 16.0) − (4000 × 6.0) =	− 4,000 '#	
M17.0 =	(1250 × 17.0) − (4000 × 7.0) =	− 6,750 '#	
M18.0 =	(1250 × 18.0) − (4000 × 8.0) =	− 9,500 '#	
M19.0 =	(1250 × 19.0) − (4000 × 9.0) =	− 12,250 '#	
M20.0 =	(1250 × 20.0) − (4000 × 10.0) =	− 15,000 '#	*

UNIFORM LOAD - BENDING MOMENTS

M1.0 =	(21,000 × 1.0) − (2800 × 1.0 × 0.50) =	+ 19,600	Ft. Lbs.	
M2.0 =	(21,000 × 2.0) − (2800 × 2.0 × 1.00) =	+ 36,400	"	
M3.0 =	(21,000 × 3.0) − (2800 × 3.0 × 1.50) =	+ 50,400	"	
M4.0 =	(21,000 × 4.0) − (2800 × 4.0 × 2.00) =	+ 61,600	"	
M5.0 =	(21,000 × 5.0) − (2800 × 5.0 × 2.50) =	+ 70,000	"	
M6.0 =	(21,000 × 6.0) − (2800 × 6.0 × 3.00) =	+ 75,600	"	
M7.0 =	(21,000 × 7.0) − (2800 × 7.0 × 3.50) =	+ 78,400	"	
M7.5 =	(21,000 × 7.5) − (2800 × 7.5 × 3.75) =	+ 78,750	"	*
M8.0 =	(21,000 × 8.0) − (2800 × 8.0 × 4.00) =	+ 78,400	"	
M9.0 =	(21,000 × 9.0) − (2800 × 9.0 × 4.50) =	+ 75,600	"	
M10.0 =	(21,000 × 10.0) − (2800 × 10.0 × 5.00) =	+ 70,000	"	
M11.0 =	(21,000 × 11.0) − (2800 × 11.0 × 5.50) =	+ 61,600	"	
M12.0 =	(21,000 × 12.0) − (2800 × 12.0 × 6.00) =	+ 50,400	"	
M13.0 =	(21,000 × 13.0) − (2800 × 13.0 × 6.50) =	+ 36,400	"	
M14.0 =	(21,000 × 14.0) − (2800 × 14.0 × 7.00) =	+ 19,600	"	
M15.0 =	(21,000 × 15.0) − (2800 × 15.0 × 7.50) =	± 0	"	
M16.0 =	(21,000 × 16.0) − (2800 × 16.0 × 8.00) =	− 22,400	"	
M17.0 =	(21,000 × 17.0) − (2800 × 17.0 × 8.50) =	− 47,600	"	
M18.0 =	(21,000 × 18.0) − (2800 × 18.0 × 9.00) =	− 75,600	"	
M19.0 =	(21,000 × 19.0) − (2800 × 19.0 × 9.50) =	− 106,400	"	
M20.0 =	(21,000 × 20.0) − (2800 × 20.0 × 10.0) =	− 140,000	"	*

STEP VII:

Constructing a form for tabulating the above Bending Moments to enable others to check work, follow the columns in the order as listed. Construct diagram with Maximum values in column farthest right.

COMBINED MOMENTS D.L. WITH U.L.			
LOCATION	C.L. MOM. FT. LBS.	U.L. MOM. FT. LBS	TOTAL MOM' #
M1.0'	+ 1,250	+ 19,600	+ 20,850
M2.0	+ 2,500	+ 36,400	+ 38,900
M3.0	+ 3,750	+ 50,400	+ 54,150
M4.0	+ 5,000	+ 61,600	+ 66,600
M5.0	+ 6,250	+ 70,000	+ 76,250
M6.0	+ 7,500	+ 75,600	+ 80,600
M7.0	+ 8,750	+ 78,400	+ 87,150
M8.0	+ 10,000	+ 78,400	+ 88,400
M9.0	+ 11,250	+ 75,600	+ 86,850
M10.0	+ 12,500	+ 70,000	+ 82,500
M11.0	+ 9,750	+ 61,600	+ 71,350
M12.0	+ 7,000	+ 50,400	+ 57,400
M13.0	+ 4,250	+ 36,400	+ 40,650
M14.0	+ 1,500	+ 19,600	+ 21,100
M15.0	− 1,250	-0-	− 1,250
M16.0	− 4,000	− 22,400	− 26,400
M17.0	− 6,750	− 47,600	− 54,350
M18.0	− 9,500	− 75,600	− 85,100
M19.0	− 12,250	− 106,400	− 118,650
M20.0	− 15,000	− 140,000	− 155,000

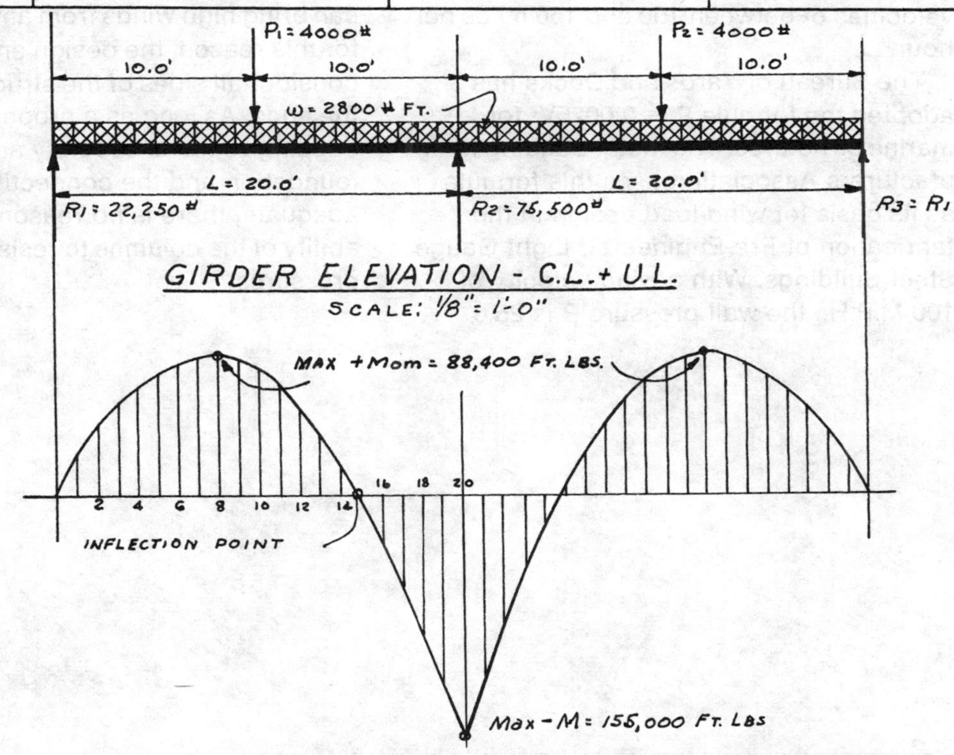

• MOMENT DIAGRAM - WITH COMBINED LOADS •

SCALE: 1" = 100,000 FT. LBS.

Wind pressure against structures 1.12

In the design for wall girts, roof trusses, purlins and columns, and anchor bolts, the wind pressure is a very important part of the design and cannot be neglected. The formula commonly used for converting wind velocity to load pressure per square foot is: $P = 0.004V^2$, where V is the wind velocity given in MPH (Miles Per Hour) and P represents the pressure in pounds per square foot. A 100 MPH wind by formula produces a Wind Load of 40 pounds per square foot. The uncertainty of wind velocities makes it difficult for the designer to establish an accurate estimate of the proper design load. Areas and geographical regions differ in the dangers of hurricane and tornado forces. Large industrial plants such as refineries and chemical works require steel buildings to be designed to sustain periodic wind velocities of between 100 and 150 miles per hour.

The Bureau of Yards and Docks has adopted the formula $P = 0.0025V^2$ for marine structures. The Metal Building Manufacturers Association uses this formula as its basis for wind load design in the fabrication of Pre-Engineered Light Gauge Steel Buildings. With a wind velocity of 100 M.P.H., the wall pressure P is 25.6 P.S.F. (compared with 40 P.S.F. given by the more conservative formula). Referring to the Southern Standard Building Code, 1963 Edition, the wind load requirements are given as 10 P.S.F. for inland regions for structures under 30 feet in height. The same structure for the coastal region, which extends to 125 miles from the coast line, must use 25 P.S.F. for wind loading on walls. The unit pressure load increases as greater heights are exposed until the pressure requirement is given as 50 P.S.F. at 200 feet. See Table 8.8.

WIND FORCE DIRECTION

It is assumed that the wind will apply a force perpendicular to the vertical walls, and will exert a uniform load pressure upon the whole surface of the exposed windward side of the building. Severe storms can bring high winds from any direction and for this reason, the design engineer must consider all sides of the structure including the ends. As long as a properly designed building remains securely anchored to its foundation and the connections remain adequate, there is no reason to doubt the ability of the columns to resist the wind pressures.

Wind load effects 1.12.1

Structures built up of light steel members must transmit the wind load pressure to the supporting columns and their foundations. In general, there are five failure modes which may be a result of high wind loads. These five modes, listed with the most probable first, are as follows:

(a) Tipping or turning over, if the dead load weight of the structure or the strength of the anchor bolts at the column base is not sufficient. The point of overturning is at the base of the column on the leeward side or at R_2.

(b) Collapse of knee brace or connection at top of the column on leeward side. Should wind forces be strong enough to cause the knee brace to buckle, the structure would sway at top of columns and finally collapse.

(c) Failure at end framed wall could be caused by rupture of diagonal tension braces or at girt connections. The effect at end walls would again be by tilting as in (b) above.

(d) Wind load pressure applied on end walls and causing the failure of diagonal tension braces in side walls between column bents. Generally wind bracing in wall bents and ends is accomplished with crossed round rods with end of rods bolted to clips welded to columns. Since the rod bracing is of considerable length, only the rods in tension carry the wind load.

(e) The final method of possible failure would assume that the columns are rigid and framing is adequate to sustain the wind forces without collapse, but the anchor bolts may become reduced in size by rust and fail by shear. In such event, the building could be moved by sliding action or even raised by freak air currents.

To prevent a steel framed building or a rigid arch type structure from collapsing as a result of wind pressure, the design must be one in which the horizontal force is transmitted to the base of the columns. Existing methods accomplish this satisfactorily with two types of design. The portal system or Rigid Frame Arch is becoming more popular in the present era due to economy and new methods of fabrication, sales franchising and erection. The design method for Rigid Frame Structures is covered in Section VII. This type of structure is treated as a truly rigid structure and possesses a certain similarity with other rafter types with respect to the knee bracing: the critical point in all designs. The main difference between the triangular-braced Trussed Arch and the Rigid Frame Arch is the type of stresses involved. Refer to the three illustrations which delineate the Flat Arch, Rigid Arch and Trussed Arch. Compare these three drawings and particularly the moment diagrams for rafters, knee point, and columns. If we consider only the wind load at the left side, the moment diagram for the columns is quite different. In the Rigid Arch, the bending stresses prevail over the axial stresses in both column and rafters. In the Trussed Arch, the axial stresses prevail over the bending stresses. In each case, the knee joint is the critical area.

Examine closely the reactions at base of the columns. The only forces shown are the reactions for a wind load applied to left side. For overturning moment, the arch is assumed to rotate about the column base at R_2. Then the right column must sustain compression and the left column will tend to rise and be in tension. This condition is the same for each type, if one is speaking of wind load. The horizontal reactions H_1

Wind load effects, continued 1.12.1

and H_2 are a result of cantilever shear from wind load acting in a horizontal action line. This horizontal shear is resisted equally by anchor bolts at each column base, or Total $W = H_1 + H_2$, and H_1 is equal to H_2.

The roof load reactions at R_1 and R_2 are equal. The line of action is vertical; therefore the column load is axial in the Flat and Trussed Arch Frames.

At this point in design, the Mechanics of the Rigid Arch differs from the Flat and Trussed Arch. A rigid arch frame is considered to represent a combination of two statically determinate hinged frames. For an ordinary roof beam to support roof loads on the Flat Arch as shown, the moment at midspan is: $M = \frac{WL}{8}$. Horizontal Reactions $H = \frac{WL}{8(h + f)}$

Assuming that the roof beam and columns at top are truly rigid or monolithic, with hinged (free) columns at the base, the bending moment must be computed on the principle theory of statics as: Summation of Moments equals zero. ($\Sigma M = 0$). Then the Moment at Center Line of Roof Beam becomes as Formula: $M = \frac{WL}{8} - Hh$. To calculate the values of H_1 and H_2, the formula are given in Section 7. Our method of design for rigid frame structures is in compliance with the recent specifications of the American Institute of Steel Construction and exceeds any present Code Requirements or the specifications of the metal Building Manufacturers Association.

The purpose of a knee brace for a Trussed Arch is to reduce the danger to the truss end and column connection, which is often limited by other requirements. The load applied across the lower truss chord from wind pressure can be placed at a point on the column where the moment lever is reduced and results in a lesser bending moment. The example to follow can be altered to illustrate the point. Without the knee brace the moment at top of column for wind load would be computed as (H x h) with the theory that H is acting as a concentrated load at the base. Vertical roof loads are not considered in the design of knee braces. Only forces produced from wind pressure are considered.

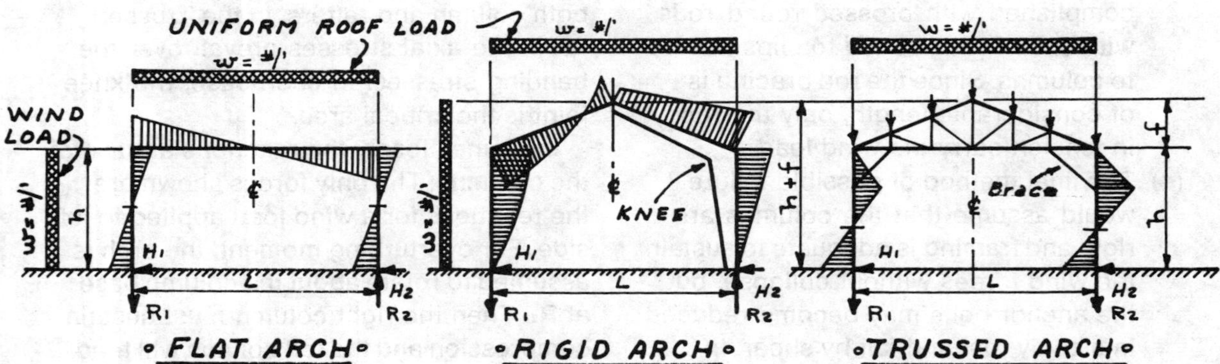

EXAMPLE: Wind pressure on a trussed arch 1.12.2

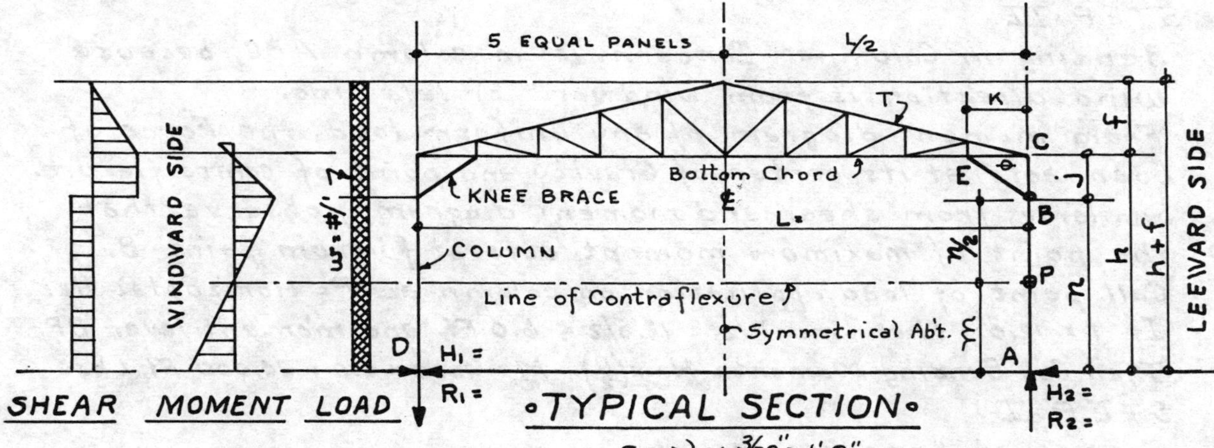

Known Data:

$L = 40.0'$ $h = 15.0'$ $f = 5.0'$ $n = 12.0'$ $J = 3.0'$ $K = 4.0'$

Column bent spacing = 20.0 feet on centers.

Wind pressure on windward side = 20 Lbs. Square Foot.

REQUIRED:

(a) Horizontal and Vertical Reactions from wind pressure.

(b) Bending Moment in Columns at Knee brace, point B.

(c) Force at C, to provide equilibrium with forces at A and B.

(d) Force in knee brace acting concentric on its axis.

STEP I:

Calculate wind load on 1 Bent of 20.0' Height $h+f = 20.0'$

Area of 1 Bent: $20.0 \times 20.0 = 400^{\square}{}'$ Total $W = 400 \times 20 = 8000$ Lbs.

STEP II:

Wind load shared by 2 Columns at Base, Points A and D.

Turn structure 90° clockwise. Treat column ABC as a beam and point B is the fulcrum. H_1 and H_2 = 4000 Lbs each. Then the load on beam is 4000 and support is at B. Bending moment in Column is $H_2 \frac{n}{2}$ = 24,000 Ft. Lbs. Same on other side when wind changes to opposite direction. See moment contraflexure.

STEP III:

Vertical Reactions: Overturning point is at A or R_2, and force is down. Force at D or R_1 is upward. See arrows.

Formula: $RL = \frac{W(h+f)}{2}$ or $R = \frac{W(h+f)}{2L}$, and substituting values:

$$R_1 = \frac{8000 \times (15.0 + 5.0)}{2 \times 40.0} = -2000 \text{ Lbs.} \quad R_2 = +2000 \text{ Lbs.}$$

EXAMPLE: Wind pressure on a trussed arch, continued 1.12.2

STEP IV

Bending in Columns: Concern is on column ABC, because wind direction is from windward or left side.

From moment diagram of any uniform load, the Force of Load acts at its center of Gravity and point of contra-flexure. Similarly, from shear and moment diagrams, observe that the point of maximum moment acts at fulcrum point B.

Call point of load application on column as P= Horizontal Hz.

If n= 12.0', then m = n/2 = 12.0/2 = 6.0 Ft. and moment lever BP.

Then Col. Bending Moment = $Hz \times (\frac{n}{2})$ M = 4000 × 6.0 = 24,000 Ft. Lbs.

STEP V

Forces at Knee Brace:

Assume point B, is fulcrum of beam PBC. Moment arm of PB = n/2 and was 6.0 Ft. J = 3.0 feet, or short end of beam.

Load at P= 4000 Lbs.

Sum of Reactions at points P and C, must equal load at B.

Then h-n = J and moment lever for C and equals 3.0 Feet.

Therefore: $\left[Hz \times (\frac{n}{2})\right] - \left[C \times (h-n)\right] = 0$. And $C = \frac{Hz\, n}{2(h-n)}$

Putting values in formula:

$C = \frac{4000 \times 12.0}{2 \times 3.0} = 8000$ Lbs. Also same as: $C = \frac{P \times \frac{n}{2}}{J}$

Total Reactions at B = P+C RB = 4000 + 8,000 = 12,000 Lbs.

STEP VI

To calculate Axial Force in Knee Brace BE.

Horizontal Force at B = 12,000 Lbs. (from last step V)

Dimension K = 4.0' J = 3.0' Find angle at E.

Tangent $\theta = \frac{3.0}{4.0} = 0.7500$ From Trig. Tables: Angle E = 36° 52'

Stress in B-E = $\frac{K}{\cos\theta}$ K = B = 12,000 Cos. 36°-52' = 0.80021

Force in B-K = $\frac{12,000}{0.80021}$ = 15,000 Lbs.

Checking with Angle knee brace makes with Column Axis.

Angle at B = 89° 60' − 36° 52' = 53° 8' Cosecant = 1.2499 (Call it 1.25)

Force B-K = K Cosec B = 12,000 × 1.25 = 15,000 Lbs. (checks)

DESIGNER'S NOTE:

Section 5 in this manual, explains the graphic method to use for the resolution of forces. The trusses on such structures are generally solved by that method. In Section VII, dealing with Rigid Frames, similar examples for knee design are given.

EXAMPLE: Wind pressure on a hinged arch 1.12.3

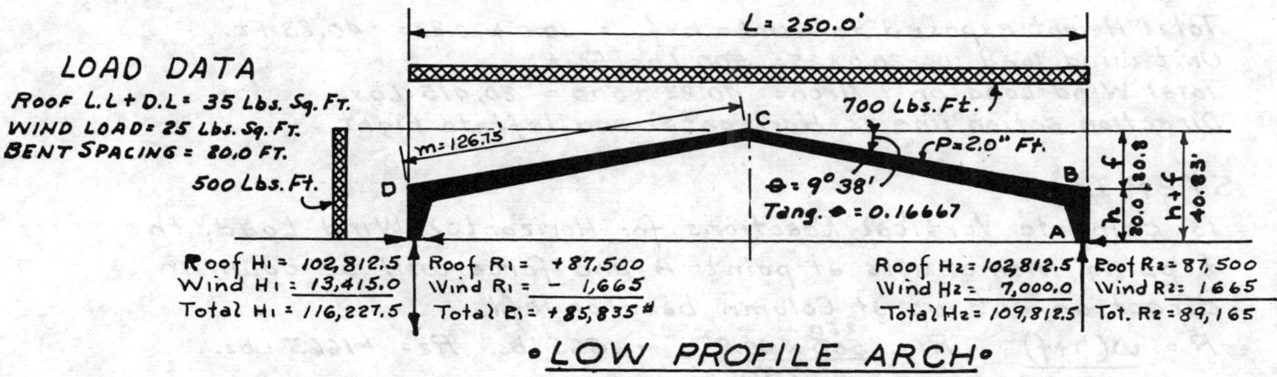

A Low Profile Roof Arch has a span of 250.0 Ft. to Outside Columns.
Eave Height (h) = 20.0 Ft. Roof Pitch (P) = 2.00 inches per foot.
Bent Spacing = 20.0 Ft. Center to Center of Arches.
Dead Load and Live Load at Roof = 35 Lbs. Square Foot.
Wind Pressure Load full height of Arch = 25. Lbs. Square Foot.

REQUIRED:
Using the A.I.S.C. Analysis Theory for Two Hinges Frames as explained in Section 7, determine the mechanics as found by using Charts I and II for values of C_1 and C_6. Refer to Moment Diagrams for applicable Formula, and consult Pilot Diagrams I and III for determining Coefficients Q and K.
Compute the following:

(a) Vertical Reactions R_1 and R_2, resulting from Roof Load.
(b) Vertical Reactions R_1 and R_2, resulting from Wind Load.
(c) Horizontal Reactions H_1 and H_2, resulting from Roof Load.
(d) Horizontal Reactions H_1 and H_2, resulting from Wind Load.
(e) Combine the Reactions and use to compute the Bending Moments at Base of Column R_2, at Knee Joint where Rafter and Column join, and at Roof Ridge on ℄. Call these location points A, B, and C.

STEP I:
Draw Cross Section of Structure with trial elevation:
With 20.0 Ft. Arch spacing, Unit Roof Load $w = 35 \times 20.0 = 700$ Lbs. Ft.
Total Roof Load; $W = 700 \times 250.0 = 175{,}000$ Lbs.
For simple span Arch: $R_1 = R_2$ $R = 175{,}000 \times 0.50 = 87{,}500$ Lbs.

STEP II
To find Wind Load, full height of building must be known.
Roof Pitch = 2.00 inches per foot. $L/2 = 125.0$ Feet
Total Pitch height: $f = \frac{125.0 \times 2.00}{12} = 20.833'$ (20'-10")

EXAMPLE: Wind pressure on a hinged arch, continued 1.12.3

Total Height exposed to Wind = h+f = 20.0 + 20.83 = 40.83 Ft.
Unit wind load w = 20.0 x 25 = 500 Lbs. Foot.
Total Wind Load on 1 Arch = 40.83 x 500 = 20,415 Lbs.
Direction action line is Horizontal and left to right.

STEP III:

To compute Vertical Reactions for Horizontal Wind Load, the tipping moment is at point A and force will be down in direction and up at Column base on left.

$R = \frac{w(h+f)^2}{2L}$ $R_1 = \frac{\overset{250}{\cancel{500}} \times 40.8^2}{\cancel{2 \times 250}} = -1665$ Lbs. $R_2 = +1665$ Lbs.

STEP IV

Horizontal Reactions H_1 and H_2, resulting from Wind Load. Greater portion of Wind Pressure is resisted by Column on (left) windward side and deflection in frame will reduce value H_2. From Section 7, Pilot Diagram III and Chart II for value C_6.

$Q = \frac{f}{h} = \frac{20.83}{20.00} = 1.01$ $K = \frac{h}{m}$ (must solve for m, by Trig.)

If side a = 20.83′ and b = 125.0 Tang. $\theta = \frac{20.83}{125.0} = 0.16667$
From Trig. Tables; Angle $\theta = 9° 36'$
Cos. $\theta = 0.986$ $m = \frac{b}{\text{Cos.}\,\theta}$ m = 125.0/.986 = 126.75 Ft.

K = 20.83/126.75 = 0.1645 From Chart II: $C_6 = 0.70$
$H_2 = C_6 wh$ Horizontal Reaction $H_2 = 0.70 \times 500 \times 20.0 = 7000$ Lbs.
$H_1 = w(h+f) - H_2$ or $H_1 = W - H_2$ $H_1 = 20{,}415 - 7000 = 13{,}415$ Lbs.

STEP V

Horizontal Reactions H_1 and H_2 from Roof Loads: Using Chart I to obtain value of C_1. Q = 1.01 and K = 0.1645 $C_1 = 0.047 \pm$
From Pilot Diagram I: w = 700 # Lin. Ft.
$H_1 = H_2$ $H_2 = \frac{C_1 wL^2}{h}$ $H_2 = \frac{.047 \times 700 \times 250.0 \times 250.0}{20.0'} = 102{,}812.5$ Lbs.

STEP VI

Recapping and Combining Results when all external forces are in effect on Arch:

+ Wind R_2 + Roof R_2 = + 1665 + 87,500 = 89,165 Lbs.
− Wind R_1 + Roof R_1 = − 1665 + 87,500 = 85,835 "
+ Wind H_1 + Roof H_1 = 13,415 + 102,812.5 = 116,227.5 "
+ Wind H_2 + Roof H_2 = 7000 + 102,812.5 = 109,812.5 "

EXAMPLE: Wind pressure on a hinged arch, continued 1.12.3

STEP VII:

Bending Moments: Minus sign – indicates tension.

At B, Roof Load $M = -H_2 h$ $M = 102{,}812.5 \times 20.0' = 2{,}056{,}250'\#$ Outside Col. Flange

At B, Wind Load $M = -H_2 h$ $M = 7000 \times 20.0' = 140{,}000'\#$ – Outside Col Flange.

Total Design Moment for Knee = 2,056,250 + 140,000 = 2,196,250 Ft. Lbs.

Shear Force on Web at Base Plate Column should be designed for basis value of H_1. Shear value can be resisted by long tie rods connecting H_1 to H_2.

Bending Moment in Knee on windward side at Point D.

At D, Wind Moment $M = H_1 h - \dfrac{wh^2}{2}$ $M = (13{,}415 \times 20.0) - \left(\dfrac{500 \times 20.0}{2}\right) = 263{,}300$ Ft. Lbs.

At D. Roof Load $M = H_1 h$ $M = 102{,}812.5 \times 20.0 = 2{,}056{,}250$ Ft. Lbs.

Design Moment for Knee at B and D = 263,000 + 2,056,250 = 2,319,250 Ft. Lbs.

STEP VIII

Bending Moments at Ridge of Roof Rafter, point C.

At C, Wind Load: $M = \left(\dfrac{RL}{2}\right) - \left[H_2 (h+f)\right]$ (Direction R_1 is upward.)

$M = +(1665 \times 125.0) - (7000 \times 40.83) = -77{,}685$ Ft. Lbs. (See Diagram Sect. 7.)

At C, Roof Load; $M = \left(\dfrac{RL}{4}\right) - \left[H - (h+f)\right]$

$M_c = \left(\dfrac{87{,}500 \times 250.0}{4}\right) - \left[102{,}812.5 \times (20.0 + 20.83)\right] = +1{,}270{,}915$ Ft. Lbs. (Positive Moment as Tension is in lower flange of Rafter.)

TABLE: Beam formulas for moment and deflection 1.13.1

BEAM FORMULAS: FOR MAX. BENDING MOMENT & DEFLECTION

BEAM TYPE & SUPPORT ENDS	TOTAL UNIFORM LOAD = W TOTAL CONCENTRATED LOAD = P	MAXIMUM "M" BENDING MOM.	MAXIMUM "Δ" DEFLECTION
CANTILEVER ONE END NOT SUPPORTED	P, L CONCENTRATED LOAD AT UNSUPPORTED END	PL	$\frac{PL^3}{3EI}$
	L UNIFORM DISTRIBUTED OVER WHOLE SPAN	$\frac{WL}{2}$	$\frac{WL^3}{8EI}$
SIMPLE SPANS EACH END FREELY SUPPORTED	P, L CONCENTRATED LOAD AT CENTER OF SPAN	$\frac{PL}{4}$	$\frac{PL^3}{48EI}$
	L UNIFORMLY DISTRIBUTED LOAD OVER WHOLE SPAN	$\frac{WL}{8}$	$\frac{5WL^3}{384EI}$
	L VARYING UNIFORM LOAD FROM ZERO AT ONE END TO MAXIMUM AT OTHER END.	$0.128WL$	$0.013\frac{WL^3}{EI}$
END SPANS ONE END FREE SUPPORTED, OTHER END RIGID SUPPORT	P, L CONCENTRATED LOAD AT CENTER OF END SPAN	$\frac{3PL}{16}$	$0.00932\frac{WL^3}{EI}$
	L UNIFORMLY DISTRIBUTED LOAD OVER WHOLE SPAN	$\frac{WL}{10}$	$0.0054\frac{WL^3}{EI}$
INTERIOR SPANS CONTINUOUS SPANS BOTH SUPPORTS RIGID OR FIXED	P, L CONCENTRATED LOAD AT MID-SPAN	$\frac{PL}{8}$	$\frac{WL^3}{192EI}$
	L UNIFORMLY DISTRIBUTED LOAD OVER WHOLE SPAN	$\frac{WL}{12}$	$\frac{WL^3}{384EI}$

CONVERT SPAN "L" TO INCHES WHEN USED IN DEFLECTION FORMULAS

TABLE: Decimals of an inch 1.13.2

DECIMALS OF AN INCH
For each 64th of an inch
With Millimeter Equivalents

Fraction	1/64ths	Decimal	Millimeters (Approx.)	Fraction	1/64ths	Decimal	Millimeters (Approx.)
...	1	.015625	0.397	...	33	.515625	13.097
1/32	2	.03125	0.794	17/32	34	.53125	13.494
...	3	.046875	1.191	...	35	.546875	13.891
1/16	4	.0625	1.588	9/16	36	.5625	14.288
...	5	.078125	1.984	...	37	.578125	14.684
3/32	6	.09375	2.381	19/32	38	.59375	15.081
...	7	.109375	2.778	...	39	.609375	15.478
1/8	8	.125	3.175	5/8	40	.625	15.875
...	9	.140625	3.572	...	41	.640625	16.272
5/32	10	.15625	3.969	21/32	42	.65625	16.669
...	11	.171875	4.366	...	43	.671875	17.066
3/16	12	.1875	4.763	11/16	44	.6875	17.463
...	13	.203125	5.159	...	45	.703125	17.859
7/32	14	.21875	5.556	23/32	46	.71875	18.256
...	15	.234375	5.953	...	47	.734375	18.653
1/4	16	.250	6.350	3/4	48	.750	19.050
...	17	.265625	6.747	...	49	.765625	19.447
9/32	18	.28125	7.144	25/32	50	.78125	19.844
...	19	.296875	7.541	...	51	.796875	20.241
5/16	20	.3125	7.938	13/16	52	.8125	20.638
...	21	.328125	8.334	...	53	.828125	21.034
11/32	22	.34375	8.731	27/32	54	.84375	21.431
...	23	.359375	9.128	...	55	.859375	21.828
3/8	24	.375	9.525	7/8	56	.875	22.225
...	25	.390625	9.922	...	57	.890625	22.622
13/32	26	.40625	10.319	29/32	58	.90625	23.019
...	27	.421875	10.716	...	59	.921875	23.416
7/16	28	.4375	11.113	15/16	60	.9375	23.813
...	29	.453125	11.509	...	61	.953125	24.209
15/32	30	.46875	11.906	31/32	62	.96875	24.606
...	31	.484375	12.303	...	63	.984375	25.003
1/2	32	.500	12.700	1	64	1.000	25.400

TABLE: Decimals of a foot 1.13.3

DECIMALS OF A FOOT
For each 32nd of an inch

Inch	0	1	2	3	4	5
0	0	.0833	.1667	.2500	.3333	.4167
1/32	.0026	.0859	.1693	.2526	.3359	.4193
1/16	.0052	.0885	.1719	.2552	.3385	.4219
3/32	.0078	.0911	.1745	.2578	.3411	.4245
1/8	.0104	.0938	.1771	.2604	.3438	.4271
5/32	.0130	.0964	.1797	.2630	.3464	.4297
3/16	.0156	.0990	.1823	.2656	.3490	.4323
7/32	.0182	.1016	.1849	.2682	.3516	.4349
1/4	.0208	.1042	.1875	.2708	.3542	.4375
9/32	.0234	.1068	.1901	.2734	.3568	.4401
5/16	.0260	.1094	.1927	.2760	.3594	.4427
11/32	.0286	.1120	.1953	.2786	.3620	.4453
3/8	.0313	.1146	.1979	.2812	.3646	.4479
13/32	.0339	.1172	.2005	.2839	.3672	.4505
7/16	.0365	.1198	.2031	.2865	.3698	.4531
15/32	.0391	.1224	.2057	.2891	.3724	.4557
1/2	.0417	.1250	.2083	.2917	.3750	.4583
17/32	.0443	.1276	.2109	.2943	.3776	.4609
9/16	.0469	.1302	.2135	.2969	.3802	.4635
19/32	.0495	.1328	.2161	.2995	.3828	.4661
5/8	.0521	.1354	.2188	.3021	.3854	.4688
21/32	.0547	.1380	.2214	.3047	.3880	.4714
11/16	.0573	.1406	.2240	.3073	.3906	.4740
23/32	.0599	.1432	.2266	.3099	.3932	.4766
3/4	.0625	.1458	.2292	.3125	.3958	.4792
25/32	.0651	.1484	.2318	.3151	.3984	.4818
13/16	.0677	.1510	.2344	.3177	.4010	.4844
27/32	.0703	.1536	.2370	.3203	.4036	.4870
7/8	.0729	.1563	.2396	.3229	.4063	.4896
29/32	.0755	.1589	.2422	.3255	.4089	.4922
15/16	.0781	.1615	.2448	.3281	.4115	.4948
31/32	.0807	.1641	.2474	.3307	.4141	.4974

TABLE: Decimals of a foot, continued 1.13.3

DECIMALS OF A FOOT
For each 32nd of an inch

Inch	6	7	8	9	10	11
0	.5000	.5833	.6667	.7500	.8333	.9167
1/32	.5026	.5859	.6693	.7526	.8359	.9193
1/16	.5052	.5885	.6719	.7552	.8385	.9219
3/32	.5078	.5911	.6745	.7578	.8411	.9245
1/8	.5104	.5938	.6771	.7604	.8438	.9271
5/32	.5130	.5964	.6797	.7630	.8464	.9297
3/16	.5156	.5990	.6823	.7656	.8490	.9323
7/32	.5182	.6016	.6849	.7682	.8516	.9349
1/4	.5208	.6042	.6875	.7708	.8542	.9375
9/32	.5234	.6068	.6901	.7734	.8568	.9401
5/16	.5260	.6094	.6927	.7760	.8594	.9427
11/32	.5286	.6120	.6953	.7786	.8620	.9453
3/8	.5313	.6146	.6979	.7813	.8646	.9479
13/32	.5339	.6172	.7005	.7839	.8672	.9505
7/16	.5365	.6198	.7031	.7865	.8698	.9531
15/32	.5391	.6224	.7057	.7891	.8724	.9557
1/2	.5417	.6250	.7083	.7917	.8750	.9583
17/32	.5443	.6276	.7109	.7943	.8776	.9609
9/16	.5469	.6302	.7135	.7969	.8802	.9635
19/32	.5495	.6328	.7161	.7995	.8828	.9661
5/8	.5521	.6354	.7188	.8021	.8854	.9688
21/32	.5547	.6380	.7214	.8047	.8880	.9714
11/16	.5573	.6406	.7240	.8073	.8906	.9740
23/32	.5599	.6432	.7266	.8099	.8932	.9766
3/4	.5625	.6458	.7292	.8125	.8958	.9792
25/32	.5651	.6484	.7318	.8151	.8984	.9818
13/16	.5677	.6510	.7344	.8177	.9010	.9844
27/32	.5703	.6536	.7370	.8203	.9036	.9870
7/8	.5729	.6563	.7396	.8229	.9063	.9896
29/32	.5755	.6589	.7422	.8255	.9089	.9922
15/16	.5781	.6615	.7448	.8281	.9115	.9948
31/32	.5807	.6641	.7474	.8307	.9141	.9974

TABLE: Building Code requirements for live loads 1.13.4

BUILDING CODE REQUIREMENTS FOR LIVE LOADS IN POUNDS PER SQUARE FOOT*

Occupancy	Codes								
	Basic Building Code BOCA 1950	Am. Std. Bldg. Code 1945 Nat. Bureau of Stds.	Nat. Bd. of Fire Under-writers 1949	Pacific Coast Bldg. Officials Con-ference 1952	New York 1946	Chicago 1950	Phila-delphia 1949	Detroit	Southern Building Code Congress Southern Std. Bldg. Code 1950
Dwellings, apartment and tenement houses, hotels, club houses, hospitals and places of detention:									
Dwellings, private rooms and apartments	40[30]	40	40	40	40[11]	40	40	40	40[43]
Public corridors, lobbies and dining rooms	100[29]	100	100	100	100	100	100	80	100
School buildings:									
Class rooms and rooms for similar use	60[27]	40	40	40[7]	60[12]	40	50[25]	50[25]	40
Corridors and public parts of the building	100	100	100	100	100	100	100	80	100
Theaters, assembly halls and other places of assemblage:									
Auditoriums with fixed seats	60	60	60	50	75[13]	60	60[26]	80	50
Lobbies, passageways, gymnasiums, grandstands, stages and auditoriums or places of assemblage without fixed seats	100	100	100	100[8]	100	100	100	100[33]	100
Stage floor	150	150	150			150			
Office building:									
Office space	50[2]	80	80	50[2,3]	50[11]	50[21]	60	50[34]	50
Corridors and other public places	100[4]	100	80	100	100[42]	100	100	125[14]	100
Workshops, factories and merchantile establishments:									
Manufacturing—light	120	125	125[2]	75	120	100	120[28]	100[35]	100
Manufacturing—heavy				125	120[41]	100	200[28]	125	
Storage—light	120		125[2]	125	120		120-150[28]	125[36]	150
Storage—heavy	250			250	120[41]		200[28]	150	250
Stores—retail	75[20]	125	125[2]	75	75[15]		100[28]	100[35]	75
Stores—wholesale	120	125	250[2]	100	75		100[28]	100[35]	100
Garages:									
All types of vehicles	175[16]		100[2]	100[9]	175[16]	100[22]	100[4]	150[37]	120
Passenger cars only	75[16]		100[2]	100	75[17]	50[23]	75	80[38]	120
All stairs and fire escapes, except in private residences	100[39]			100	100	100	100	100[39]	100
Roofs (flat)	20-100		20	20[5]	40	25	30	30	20
Sidewalks	250[4]		250	250[4]	300[18]		150[31]	250	250
Wind	Min 20[10]			15-20[1]	0-20[19]	25-35[24]	15-25[32]	20[40]	10-20[44]

Notes:

[1] 15 psf up to 60 ft high, 20 psf over 60 ft.
[2] Or 2000 on any space 2½ feet square.
[3] Where partitions are subject to change, add 20 psf to all other loads.
[4] Or 8000 concentrated.
[5] If area is 200 to 600 sf use 16 psf, over 600 sf, 12 psf; for rise 4 in. per ft use 16 psf under 200 sf, 14 psf for 200-600 sf and 12 psf over 600 sf; for rise 12 in. per ft use 12 psf.
[6] 15 for portions below 40 ft and 30 for portions above 40 ft.
[7] 60 for library reading rooms and 150 for stackrooms.
[8] 150 for armories.
[9] Or concentrated rear wheel of loaded truck in any position.
[10] Increase 0.025 psf for each foot above 100 ft.
[11] Including corridors.
[12] For rooms with fixed seats or, by special permission, other small rooms. 120 for library stackrooms.
[13] 60 for churches.
[14] Including entire first floor.
[15] 100 for entire first floor.
[16] Or 6000 concentrated. Trucking space, 150% max. wheel load; 175 psf on floor construction, 120 psf on beams and girders.
[17] Or 2000 concentrated.
[18] Or 12,000 concentrated for driveways over sidewalks.
[19] 20 psf from top down to 100 ft level, zero below; 30 psf on tanks, stacks and exposed structures.
[20] 100 psf on floor at grade, upper floors 75 psf.
[21] Or 2000 concentrated on any space 3 feet square.
[22] Or 3000 concentrated on any space 4 feet square.
[23] 100 on first floor and alternate of 3000 on area 4 feet square.
[24] 25 for surfaces less than 275 ft high and 35 psf above.
[25] Only school class rooms with fixed seats.
[26] Churches only.
[27] Fixed seats, 60 psf; removable seats, 100 psf.
[28] Every floor beam 4000 concentrated.
[29] Other than residential, 100 psf; hotels and multifamily, 60 psf.
[30] On first floor, 40 psf; upper floors, 30 psf.
[31] Interior courts, sidewalks, etc., not accessible to a driveway.
[32] 15 psf up to 50 ft high, 20 psf from 50 to 200 ft, 25 psf over 200 ft high. Roofs over 30°, 20 psf on windward side, 10 psf on leeward.
[33] 125 for dance halls and drill halls.
[34] Above first floor including corridors.
[35] 125 for first floor.
[36] 150 for first floor.
[37] Or 2500 concentrated on area 6 inches square with such concentrations spaced alternately 2 ft 4 in. and 4 ft 8 in. in one direction and 5 ft and 10 ft in the other direction.
[38] Only structures with clear head room of 8 ft 6 in. or less. Or 1500 concentrated spaced as in 37.
[39] 50 for dwellings and apartments under 3 stories.
[40] For buildings less than 500 ft high.
[41] The minimum for storage or manufacturing is 120 psf, but floors must be designed for any heavier loads contemplated and for any concentrations.
[42] Including entire first floor but not including corridors on floors used for offices.
[43] 30 for one and two family dwellings.
[44] 10 for portions below 40 ft and 20 for portions above 40 ft.

**Courtesy of Concrete Reinforcing Steel Institute.*

TABLE: Recommended live loads for warehouses 1.13.5

CONTENTS OF STORAGE WAREHOUSES

Material	Weight per Cubic Foot of Space, Pounds	Height of Pile, Feet	Weight per Square Foot of Floor, Pounds	Recommended Live Loads, Pounds per Square Foot
Groceries, Wines, Liquors, Etc.				
Beans, in bags	40	8	320	
Canned Goods, in cases	58	6	348	
Coffee, Roasted, in bags	33	8	264	
Coffee, Green, in bags	39	8	312	
Dates, in cases	55	6	330	
Figs, in cases	74	5	370	
Flour, in barrels	40	5	200	
Molasses, in barrels	48	5	240	250 to 300
Rice, in bags	58	6	348	
Sal Soda, in barrels	46	5	230	
Salt, in bags	70	5	350	
Soap Powder, in cases	38	8	304	
Starch, in barrels	25	6	150	
Sugar, in barrels	43	5	215	
Sugar, in cases	51	6	306	
Tea, in chests	25	8	200	
Wines and Liquors, in barrels	38	6	228	
Dry Goods, Cotton, Wool, Etc.				
Burlap, in bales	43	6	258	
Coir Yarn, in bales	33	8	264	
Cotton, in bales, compressed	18	8	144	
Cotton Bleached Goods, in cases	28	8	224	
Cotton Flannel, in cases	12	8	96	
Cotton Sheeting, in cases	23	8	184	
Cotton Yarn, in cases	25	8	200	
Excelsior, compressed	19	8	152	
Hemp, Italian, compressed	22	8	176	
Hemp, Manila, compressed	30	8	240	200 to 250
Jute, compressed	41	8	328	
Linen Damask, in cases	50	5	250	
Linen Goods, in cases	30	8	240	
Linen Towels, in cases	40	6	240	
Sisal, compressed	21	8	168	
Tow, compressed	29	8	232	
Wool, in bales, compressed	48			
Wool, in bales, not compressed	13	8	104	
Wool, Worsted, in cases	27	8	216	

Material	Weight per Cubic Foot of Space, Pounds	Height of Pile, Feet	Weight per Square Foot of Floor, Pounds	Recommended Live Loads, Pounds per Square Foot
Building Materials				
Cement, Natural	59	6	354	300 to 400
Cement, Portland	73	6	438	
Lime and Plaster	53	5	265	
Hardware, Etc.				
Door Checks	45			
Hinges	64			
Locks, in cases, packed	31			
Sash Fasteners	48			
Screws	101			
Sheet Tin, in boxes	278	2	556	300 to 400
Wire Cables, on reels			425	
Wire, Insulated Copper, in coils	63	5	315	
Wire, Galvanized Iron, in coils	74	4½	333	
Wire, Magnet, on spools	75	6	450	
Drugs, Paints, Oil, Etc.				
Alum, Pearl, in barrels	33	6	198	
Bleaching Powder, in hogsheads	31	3½	102	
Blue, Vitriol, in barrels	45	5	226	
Glycerine, in cases	52	6	312	
Linseed Oil, in barrels	36	6	216	
Linseed Oil, in iron drums	45	4	180	
Logwood Extract, in boxes	70	5	350	
Rosin, in barrels	48	6	288	200 to 300
Shellac, Gum	38	6	228	
Soda Ash, in hogsheads	62	2¾	167	
Soda, Caustic, in iron drums	88	3⅓	294	
Soda, Silicate, in barrels	53	6	318	
Sulphuric Acid	60	1⅔	100	
White Lead Paste, in cans	174	3½	610	
White Lead, dry	86	4¾	408	
Red Lead and Litharge, dry	132	3¾	495	
Miscellaneous				
Glass and Chinaware, in crates	40	8	320	
Hides and Leather, in bales	20	8	160	
Hides, in bundles	37	8	296	
Paper, Newspaper and Strawboards	35	6	210	300
Paper, Writing and Calendered	60	6	360	
Rope, in coils	32	6	192	

WEIGHTS OF MATERIAL

Substance	Weight, Pounds per Cubic Foot
Ashlar Masonry	
Granite, syenite, gneiss	165
Limestone, marble	160
Sandstone, bluestone	140
Mortar Rubble Masonry	
Granite, syenite, gneiss	155
Limestone, marble	150
Sandstone, bluestone	130
Dry Rubble Masonry	
Granite, syenite, gneiss	130
Limestone, marble	125
Sandstone, bluestone	110
Brick Masonry	
Pressed Brick	140
Common Brick	120
Soft Brick	100

Substance	Weight, Pounds per Cubic Foot
Concrete Masonry	
Cement, stone, sand	144
Cement, slag, etc.	130
Cement, cinder, etc.	100
Various Building Materials	
Ashes, cinders	40–45
Cement, P'rtl'd, loose	90
Cement, Portland, set	183
Lime, gypsum, loose	53–64
Mortar, set	103
Slags, bank slag	67–72
Slags, bank screenings	98–117
Slags, machine slag	96
Slags, slag sand	49–55
Earth, Etc., Excavated	
Clay, dry	63
Clay, damp, plastic	110
Clay and gravel, dry	100

Substance	Weight, Pounds per Cubic Foot
Earth, Etc., Ex.—Cont'd	
Earth, dry, loose	76
Earth, dry, packed	95
Earth, moist, loose	78
Earth, moist, packed	96
Earth, mud, flowing	108
Earth, mud, packed	115
Riprap, limestone	80–115
Riprap, sandstone	90
Riprap, shale	105
Sand, gravel, dry, loose	90–105
Sand, gravel, dry, packed	100–120
Sand, gravel, dry, wet	118–120
Excavation in Water	
Sand or gravel	60
Sand or gravel and clay	65
Clay	80
River mud	90
Soil	70
Stone riprap	65

Substance	Weight, Pounds per Cubic Foot
Minerals	
Asbestos	153
Barytes	281
Basalt	184
Bauxite	159
Borax	109
Chalk	137
Clay, marl	137
Dolomite	181
Feldspar, orthoclase	159
Gneiss, serpentine	159
Granite, syenite	175
Greenstone, trap	187
Gypsum, alabaster	159
Hornblende	187
Limestone, marble	165
Magnesite	187
Phosphate rock, apatite	200
Porphyry	172
Pumice, natural	40
Quartz, flint	165

TABLE: Weights of materials 1.13.6

WEIGHTS OF MATERIAL—Continued

Substance	Weight, Lb. per Cu. Ft.
Minerals—Continued	
Sandstone, bluestone	147
Shale, slate	175
Soapstone, talc	169
Stone, Quarried, Piled	
Basalt, granite, gneiss	96
Limestone, marble, quartz	95
Sandstone	82
Shale	92
Greenstone, hornblende	107
Bituminous Substances	
Asphaltum	81
Coal, anthracite	97
Coal, bituminous	84
Coal, lignite	78
Coal, peat, turf, dry	47
Coal, charcoal, pine	23
Coal, charcoal, oak	33
Coal, coke	75
Graphite	131
Paraffine	56
Petroleum	54
Petroleum refined	50
Petroleum benzine	46
Petroleum gasoline	42
Pitch	69
Tar, bituminous	75
Coal and Coke, Piled	
Coal, anthracite	47–58
Coal, bituminous, lignite	50–54
Coal, peat, turf	20–26

Substance	Weight, Lb. per Cu. Ft.
Coal and Coke—Cont'd	
Coal, charcoal	10–14
Coal, coke	23–32
Metals, Alloys, Ores	
Aluminum, cast-hammered	165
Aluminum, bronze	481
Brass, cast-rolled	534
Bronze, 7.9 to 14% Sn.	509
Copper, cast-rolled	556
Copper, ore pyrites	262
Gold, cast-hammered	1205
Iron, cast, pig	450
Iron, wrought	485
Iron, steel	490
Iron, speigel-eisen	468
Iron, ferro-silicon	437
Iron, ore, hematite	325
Iron, ore, limonite	237
Iron, ore, magnetite	315
Iron, slag	172
Lead	710
Lead, ore, galena	465
Manganese	475
Manganese ore, pyrolusite	259
Mercury	849
Nickel	565
Nickel monel metal	556
Platinum, cast-hammered	1330
Silver, cast-hammered	656
Tin, cast-hammered	459
Tin, ore, cassiterite	418
Zinc, cast-rolled	440
Zinc, ore, blende	253

Substance	Weight, Lb. per Cu. Ft.
Various Solids	
Cereal, oats, bulk	32
Cereal, barley, bulk	39
Cereal, corn, rye, bulk	48
Cereal, wheat, bulk	48
Hay and Straw, bales	20
Cotton, Flax, Hemp	93
Fats	58
Flour, loose	28
Flour, pressed	47
Glass, common	156
Glass, plate or crown	161
Glass, crystal	184
Leather	59
Paper	58
Potatoes, piled	42
Rubber, caoutchouc	59
Rubber goods	94
Salt, granulated, piled	48
Saltpeter	67
Starch	96
Sulphur	125
Wool	82
Timber, U.S. Seasoned	
Ash, white-red	40
Cedar, white-red	22
Chestnut	41
Cypress	30
Elm, White	45
Fir, Douglas spruce	32
Fir, eastern	25
Hemlock	29
Hickory	49
Locust	46
Maple, hard	43
Maple, white	33
Oak, chestnut	54

Substance	Weight, Lb. per Cu. Ft.
Timber, U.S.—Cont'd	
Oak, live	59
Oak, red, black	41
Oak, white	46
Pine, Oregon	32
Pine, red	30
Pine, white	26
Pine, yellow, long-leaf	44
Pine, yellow, short-leaf	38
Poplar	30
Redwood, California	26
Spruce, white, black	27
Walnut, black	38
Walnut, white	26
Moisture Contents: Seasoned timber 15 to 20% Green timber up to 50%	
Various Liquids	
Alcohol, 100%	49
Acids, muriatic, 40%	75
Acids, nitric, 91%	94
Acids, sulphuric, 87%	112
Lye, Soda, 66%	106
Oils, vegetable	58
Oils, mineral, lubricants	57
Water, 4°C, max. density	62.428
Water, 100°C	59.830
Water, ice	56
Water, snow, fresh fallen	8
Water, sea water	64
Gases, Air=1	
Air, 0°C. 760 mm	.0807

WEIGHTS OF BUILDING MATERIALS

Kind	Weight in Lb. per Sq. Ft.
Floors	
7/8″ Maple finish floor and 7/8″ Spruce under floor on 2″ x 4″ sleepers, 16″ centers, with 2″ dry cinder concrete filling	18
Cinder concrete filling per inch of thickness	7
Cement finish per inch of thickness	12
Asphalt mastic flooring 1½″ thick	18
3″ creosoted wood blocks on ½″ mortar base	21
Solid flat tile on 1″ mortar base	23
Ceilings	
Plaster on tile or concrete	5
Suspended Metal Lath and plaster	10
Roofs	
Five-ply felt and gravel	6
Four-ply felt and gravel	5½
Three-ply ready roofing	1
Cement Tile	16
Slate, ¼″ thick	9½
Sheathing, 1″ thick, Yellow Pine	4
2″ Book Tile	12
3″ Book Tile	20
Skylight with galvanized iron frame, 3/8″ glass	6

Kind	Weight in Lb. per Sq. Ft.: Unplastered	One Side Plastered	Both Sides Plastered
Walls			
9″ Brick Wall	84	89	
13″ Brick Wall	121	126	
18″ Brick Wall	168	173	
22″ Brick Wall	205	210	
26″ Brick Wall	243	248	
4″ Brick, 4″ Tile Backing	60	65	
4″ Brick, 8″ Tile Backing	75	80	
9″ Brick, 4″ Tile Backing	102	107	
8″ Tile	33	38	43
12″ Tile	45	50	55
Partitions			
3″ Clay Tile	17	22	27
4″ Clay Tile	18	23	28
6″ Clay Tile	25	30	35
8″ Clay Tile	31	36	41
10″ Clay Tile	35	40	45
3″ Gypsum Block	10	15	20
4″ Gypsum Block	12	17	22
5″ Gypsum Block	14	19	24
6″ Gypsum Block	16	21	26
2″ Solid Plaster			20
4″ Solid Plaster			32
4″ Hollow Plaster			22

MASONRY

	Weight in Lb. Per Cu. Ft.		Weight in Lb. Per Cu. Ft.
Concrete, cinder	110	Mortar rubble, sandstone	130
Concrete, stone	140 to 150	Mortar rubble, limestone	150
Concrete, reinforced stone	150	Mortar rubble, granite	155
Brick masonry, soft	100	Ashlar sandstone	140
Brick masonry, common	125	Ashlar limestone	160
Brick masonry, pressed	140	Ashlar granite	165

II—STRUCTURAL STEEL DESIGN

II—STRUCTURAL STEEL DESIGN

II

STRUCTURAL STEEL DESIGN

STRUCTURAL STEEL DESIGN

Contents

Steel 2.1

Steel is the foundation of the economy of the great nations of the world. This versatile metal possesses a unique combination of qualities: strength, ductility, ease of fabrication, and economy.

Steel industry 2.1.1

The giant steel industry of today began in the workshops of the ironmasters. Colonial America had many experienced iron workers; the owner of a steel-producing plant was referred to as an Ironmaster. They existed in England and Germany many years before the colonists came to America.

The exact date of the first iron making venture in this country has not been recorded; however, it is known that it took place near Saugus, Massachusetts. In 1720, the master iron worker, Robert Durham, constructed an experimental boat to convey ore and charcoal to his furnace and iron to Philadelphia by way of the Delaware River. The success of this enterprise prompted a group of wealthy Philadelphians to finance Durham in expanding his plant. In 1727, the plant known as Durham's Furnace began to produce large quantities of good quality iron. A large fleet of iron boats was used to carry other cargo to settlements on the Delaware River. On the night of December 25, 1776 General George Washington crossed the Delaware to the Battle of Trenton in a fleet of forty Durham boats. The historical painting of this event by Emanuel Leutz shows the river crossing of General Washington, but the boats bear little resemblance to the boats designed by Robert Durham which were sixty feet in length.

The refining of pig iron from ore requires high temperatures. This heat reduces the initial charge of ore, limestone and fuel coke to molten iron and slag. Powerful air jets are injected into the furnace to raise the temperature; this became known as the "blast furnace." Henry Bessemer (1813–1898) was a British metallurgist who believed that if forced air could fan the fires to raise the temperature, it could also remove the impurities in the fluid pig iron. He developed the process where air is blown through the molten iron and the impurities are burned out. This process produces a softer and more malleable metal: steel. Bessemer could not claim full responsibility for developing the process, he was helped by several others including the prominent Swedish ironmaster, Goransson.

The Bessemer method of producing steel became known as the "pneumatic"

Steel industry, continued 2.1.1

process in America, after the Colonies won independence and chose to avoid many trademarks of English origin. This process was the key to the growth of the greatest steel producing nation on earth. Early America possessed the great area for expansion and the abundant resources of iron ore. With the invention of the steam engine by the Englishman James Watt (1736–1819), the railroads began to open up the West. Men like Andrew Carnegie (1835–1919) and Charles M. Schwab (1862–1939) were the industrial leaders responsible for making this nation the largest steel producer in the world.

European steel production was dominated by a fast-growing steel-producing enterprise in Germany headed by the industrialist Alfred Krupp (1812–1887). The firm was known throughout the world as the Krupp Works. The Krupp family controlled the world's largest cannon and heavy armament plant at Essen. During World War II, the Krupp family was given control of coal mines, imports of scrap iron, and the production of all the steel required for arms. After two great wars in which the Krupp works equipped the German armies, the plants and subsidiaries are no longer controlled by the Krupp family.

Processes for making steel 2.1.2

OPEN HEARTH

Replacing the Bessemer furnaces, the Open Hearth process became the front runner by 1870. This process was developed in France in 1862 by two family shops. They named the process the Siemans-Martin Process. But the "open hearth" name prevailed because it was descriptive of the process. The furnace hearth is open, exposed to the flame of the oxygen blown into the mix. From 1870 to 1950, the open hearth process led in the production of steel in the United States. It is best adapted to handling larger charges of mix and melting down scrap iron.

BASIC OXYGEN FURNACES

In 1956 a pear shaped vessel in a Midwestern plant was tipped to pour its first batch of molten steel—and a new era in steel production technology began. Today, the Basic Oxygen Furnace (BOF) produces more domestic steel than the other two major processes combined. During the year of 1970, the BOF clearly established itself as number one among steel producing furnaces. Growth of the basic oxygen output has come primarily at the expense of the open hearth process which dominated the source in America for more than 60 years. The basic oxygen furnace "pressure-cooks" the steel by forcing a supersonic jet of pure oxygen into the molten iron. It will turn out steel several times faster than an open hearth.

ELECTRIC FURNACE

A few production figures will best illustrate the rapid changes taking place in American steel production. For the first eight months of 1969, the BOF process poured 41.3 percent of the total output. The open hearth furnace poured 44.8 percent. The balance of 13.9 percent was produced by electric furnace. The electric furnace has doubled its production since 1955. The

Processes for making steel, continued 2.1.2

sudden increase in electric furnace steelmaking, which employs the BOF process, is a significant feature of steel industry technological progress and expansion. Why is this so? The electric furnace is better adapted to the making of alloy and stainless steels, and excels in the production of the more common carbon steels. The electrics are also superior for melting down scrap iron and ingots, which is important for conservation of resources. Also, plant locations for electric furnaces are not tied to fuel sources.

In 1970, several steel plants using electric furnaces produced an average of 2800 net tons per day in heats of 325 tons. Most furnaces today are operating with 150,000 KVA transformer capacities. Future plant facilities will be designed for greater KVA rating. Installations will be provided with removable lids, which will permit the combustion chamber to accept several carloads of scrap metal. With this rapid change to electric furnaces, the supply of electrical energy must be enlarged. The nuclear-powered plant will probably meet this challenge when environmental restrictions are satisfied.

Steel markets 2.1.3

America has the largest steel market in the world. It is this market which provides our economic strength. In 1968, the United States used 108 million tons of steel mill products: 90 million tons from domestic sources and 18 million from outside the country. A decade before, imports were less than 3 percent. These imports come from Japan and the countries of western Europe. American producers export very little steel except as finished products.

Metallurgy 2.1.4

Structural engineers and steel designers engaged in the work of selecting members for structures are essentially concerned with the finished material and its properties. A truly competent steel designer should be conversant with the technical terms used in the production of raw and finished, steel. The processes used for producing steel, tempering methods and the use of additive alloys are part of the field of metallurgy.

ASTM MATERIAL SPECIFICATIONS

Virtually every material which is used in construction has been tested and given a specification number by the American Society for Testing Materials. This very effective non-profit society was organized to pool knowledge from scientific, educational and technical sources to develop testing methods and establish specification standards. International in scope, the society reported a membership in 1964 of

Metallurgy, continued 2.1.4

over 12,000 consisting of researchers, educators, material producers, engineers and testing experts. Membership is also open to students and others concerned with the work and aims of the organization.

Although the ASTM may approve and adopt a standard of quality which represents a common viewpoint of concerned parties, the use of such standards is purely voluntary. Neither does the ASTM attempt to prohibit anyone from producing, selling, purchasing or using any product which may not conform to the established standards of the ASTM. The society serves as an aid to industry, governmental agencies, building code officials, the general public, and especially to engineers and architects. A complete index for ASTM standards is issued annually which lists their specification numbers on materials from *adhesives* to *zinc*.

Accepting the ASTM standards and specifications as the fundamental basis for quality, each steel plant producing raw and finished products will maintain its own testing laboratory. A continuous testing program records the production of steel from batches of natural ore to the final testing of the finished, rolled section. In the steel mills, any finished product intended for structural use or for piling, shipbuilding, tubing or bridge work is referred to as the regular "garden variety." Engineers divide this group into medium or mild carbon steels. Other steels with higher physical properties and corrosion resistance (as a result of alloys with nickel, copper, manganese, tungsten, chromium, vanadium and molybdenum) are referred to as the "aristocrats" of steel. From these products are made spring steel, ball bearings, instruments, and stainless steels for utensils and food processing equipment.

BILLET AND BLOOM STEEL

Billet steel is used mainly for making concrete reinforcing bars. *Billet* and *bloom* are synonymous terms used to denote the shape of an ingot of raw steel which is heated and rolled into deformed round or square rods. The deformed ridges on the rods will vary, and serve to identify the producer of the reinforcing rod. Rail steel was used extensively during World War II. However rail steel is brittle; little bending can be done without heating. New Billet steel which is more ductile, is rolled into three grades of ASTM Specification A–15 deformed rods. They are graded as: Structural, Intermediate, and Hard. Also, high strength billets are available, classified by the ASTM Specifications as A–431 and A–432; they are listed in Section III.

Steels to be used for stranded wire rope and pre-stress concrete work are formed from plow steel. These rope and cable sections possess exceptionally high values for tensile stress.

PIG IRON

The crude molten iron which is drawn off the furnace bottom at intervals of from 3 to 5 hours and formed into ingots, is called pig iron. The foundries use this form of raw iron in the making of castings. Foundries return the ingot to a molten state and pour it into a sandy-clay mold form. Products cast in foundries include such items as manhole covers, gratings, cast iron pipe, mooring bits, drains, valve boxes and many others. After being removed from the mold, the surface of the casting rapidly becomes oxidized. This characteristic in cast iron is beneficial; it provides an adherent protective coating which makes cast iron a good choice for locations exposed to weather conditions and underground installations.

Testing 2.1.5

Engineers and designers must be conversant with the test terms used in laboratory reports. Steel producers classify the carbon steels into two groups: *hypo-eutectoid* and *hyper-eutectoid*; the former contains less than approximately 0.9 percent carbon and the latter above 0.9 percent carbon. The instant raw steel solidifies, the product consists of homogeneous austenite, which is to say, all the carbon is in solid solution. The reason for separating the steels into the two groups is done to distinguish their composition. The higher carbon steels offer more difficulty in forging, welding and heat treating than do the low carbon steels.

All steels, regardless of composition, heat treatment, or hardness will, up to a limit, temporarily deform the same amount under the same stress: the modulus of elasticity (E) is almost identical for all steels. The minimum stress which will permanently deform the steel beyond the elastic limit, is a function of composition, heat treatment, and temperature.

In the design of steel girders and beams, the section is selected so that the metal is not stressed beyond its elastic limit, but the deflection must be checked. In service, too great a deflection under the design load may cause permanent distortion. Changes in the steel will not correct the situation, which can only be improved by increasing the section's Moment of Inertia, and thereby lower the unit stress to bring the deflection within limits that are not objectionable.

TENSION TESTS

Tension tests conducted on specimens of steel involve the use of precision laboratory equipment and microscopic measurements. Tension tests determine the force a cross-section will sustain under tensile loads and at what point the load force will cause the specimen to deform without a permanent set.

A steel bar with a measured cross-section and length is clamped into a stretching machine. As the force is increased in increments, measurements are taken on the amount of elongation and reduction in area of the cross-section. Each point of increment force is plotted on a graph. When the specimen will no longer return to its original length and cross-sectional area after releasing the latest force, the yield point and elastic limit will have been found. By continuing to increase the force increments and readings, the ultimate stress and rupture point will be found. Compression tests are conducted by reversing the force direction.

STRESS-STRAIN TEST CURVE

Referring to Curve 2.1.5.1 there are a number of terms which are included in questions usually submitted to applicants in examinations for state registration. These will be investigated in the order that they appear during the test. The following must be defined:

(a) Proportional limit.
(b) Elastic limit.
(c) Proof stress.
(d) Yield stress.
(e) Yield point.
(f) Ultimate stress.

PROPORTIONAL LIMIT

Below the proportional limit, the ratio of unit stress to unit strain is constant. This ratio is the modulus of elasticity (E). This is a property characteristic of the material and should not be confused with the section properties discussed in Section VI. The amount of sag or deflection in a loaded beam is a measure of deformation. It may

Testing, continued 2.1.5

be calculated in most cases by using the proper value of E in formulas for finding the deflection of beams.

The proportional limit in any material is the load per unit area beyond which the increase in deflection ceases to be directly proportional to the increase in stress. In order to construct a stress-strain curve, the load is increased in a regular sequence and data plotted. Note that a straight line results up to the proportional limit. Beyond this point, any additional load increments will result in higher unit stresses and subsequent greater deformation. Beyond this elastic limit, the test specimen will not return to its original measurements when loads are removed.

CURVE: Stress-strain test 2.1.5.1

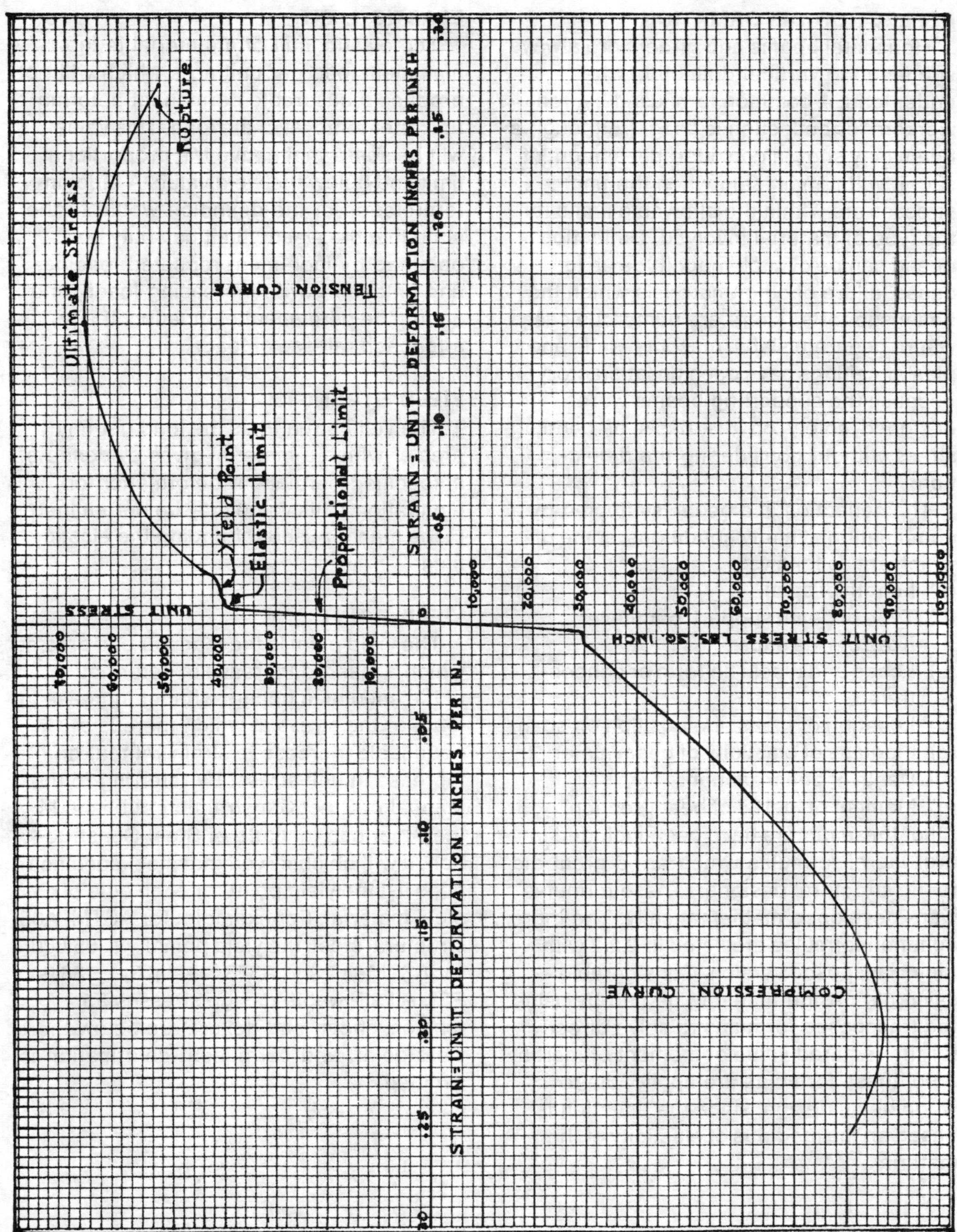

CURVE: Stress-strain test, continued 2.1.5.1

DEFLECTION-DEFORMATION

Horizontal beams and girders are always subjected to vertical deflection when under load. A common term used by ironworkers for deflection is "*sag*." Although a beam may show a considerable amount of sag, it will remain safe when the unit stresses are within the allowable limits for the type of material involved. Deflection or sag can be of such magnitude as to cause failure of other materials which the beam supports. There are numerous cases where concrete slabs, masonry walls and plaster ceilings have been unsatisfactory because the supporting beams were designed solely on the basis of bending stress without considering the deflection.

The deflection in a beam is effected by a combination of many conditions, some of which are as follows:

(a) Length of span.
(b) Type of beam support: cantilever, fixed or free ends.
(c) Loads and types: uniform or concentrated.
(d) Location of loads and position on span.
(e) Modulus of elasticity of beam material.
(f) Depth of beam section.
(g) Moment of Inertia of beam section.

All of these factors have been included in a group of formulas which are a great aid to the structural designer. These formulas were developed mathematically. The more common formulas are listed in 2.2.2 with their transposed version. Considerable saving in design time can be had by using the transposed formulas to solve directly for the required value of moment of inertia. Also, these formulas are applicable to wood beam design, since the equation is based on the modulus of elasticity of the material.

ELASTIC LIMIT

The elastic limit of a material is the maximum load per unit area which will not produce measurable permanent deformation after removal of the load. This value will be somewhat above the proportional limit and below the yield value. A strict interpretation of the definition would further state that the elastic limit can only be obtained by increment repeated loading and unloading with increasing loads, and noting the permanent elongation, if any, after each release of load.

YIELD STRENGTH

The yield strength is the load per unit area at which a material gives evidence of a specified permanent deformation or elongation. This value may be determined in the manner used for finding the elastic limit.

YIELD POINT

The yield point is the load per unit area at which a marked increase in deformation of the specimen occurs without increase in load; or the stress at which there is a noticeable increase in strain without an increase in unit stress. By referring to the Manual of Steel Construction published by the American Institute of Steel Construction, it will be observed that steel specifications list the yield point for each type of steel material. For example: for ASTM–A36 structural steel, the yield point $Fy = 36{,}000$; for ASTM–A7 and A373 Steels, $Fy = 33{,}000$ PSI. All allowable design stresses for tension, shear, and flexure are below the yield point; and permanent deformation is avoided.

CURVE: Stress-strain test, continued 2.1.5.1

PROOF STRESS

A special test required in the specifications for supplying steel for large projects is a certified proof test. Such tests are requested by consulting engineering firms, public building officials and highway officials. The proof stress is a certain load per unit area which a material is capable of sustaining without resulting in a permanent deformation exceeding a stipulated amount per unit of gage length after complete release of the applied load.

ELONGATION-AREA REDUCTION

In certain respects, the testing of a steel specimen can be compared to stretching a rubber band. As the band is elongating, the cross-sectional area is reduced. The elongation under stress and the reduction in the cross-section dimensions are related by a constant ratio called Poisson's Ratio.

The percentage of elongation is the difference in the gage length before any stress is applied and after rupture. This will be expressed as a percentage of the original gage length. Likewise, the percentage reduction of cross-sectional area is the difference between the original cross-section area before stress is applied and the least cross-sectional area at the point of rupture.

ULTIMATE STRESS

Continuing the tensile test and stress-strain curve, the deformation increases until the ultimate stress is reached at 66,000 PSI. After this point the deformation continues with lower stress until rupture. For medium carbon steels, the ultimate stress is between 60,000 and 65,000 PSI.

SAFETY FACTOR

A safety factor is based on the difference between the ultimate stress and the design working stress. Assuming that the ultimate stress (F_u) is 66,000 PSI, and the design allowable stress in bending (F_b) is 22,000 PSI, the safety factor is 3. Safety factors are related to materials ultimate stress. In the AISC Steel Manual Tables, the allowable design stresses apply for static loads, and assume spans are adequately braced for lateral support. Obviously, the higher the factor of safety, the smaller will be the allowable unit design stress. Bridge designers are usually required by state highway officials to use a safety factor of not less than 4. This would limit A36 steel to a design unit bending stress of 16,500 PSI. Structures which are to be equipped with hoists or motorized lifts subjecting the structural members to impact loads should be designed with a safety factor of 6 or possibly even more.

EXAMPLE: Test for modulus of elasticity 2.1.5.2

A square one inch steel bar is 10.0 inches between its upset ends which are firmly secured in a testing device. A stretching force is applied in tension amounting to 20,000 Pounds. This force is sustained over the full 10.0 ins. of rod which was originally 1.0 inch square.
While cross-section is under stress, microscopic measuring instruments reveal the specimen has elongated to a length of 10.006896 inches. After release of 20,000 pound force, the specimen returned to original length.

REQUIRED:
Determine the total and unit deformation from the data supplied, then calculate the modulus of elasticity.

STEP I:
The specimen returned to normal upon release of load and therefore was within the elastic limit.
Total elongation = Total deformation = 0.006896 inches.

STEP II:
Length of specimen = 10 units of 1.0 inch each.
Unit deformation = 0.006896/10.0 = 0.0006896 inches, per In.

STEP III:

$$\text{Modulus of Elasticity} = \frac{\text{Stress per Unit}}{\text{Unit Deformation}} \quad \text{or} \quad E = \frac{S_u}{\Delta_u}$$

The 20,000 Pound Force is applied on each unit.
Then:

$$E = \frac{20{,}000}{0.0006896} = 29{,}000{,}000 \text{ Pound per square inch. (PSI).}$$

HOOKE'S LAW:
The Modulus of Elasticity is defined as being the ratio of unit stress to unit deformation, only so long as it performs within the elastic limit and remains constant.

Allowable design stresses 2.1.6

Structural designers must carry out their calculations within the requirements established by the applicable building code. The Southern Standard Building Code is in complete accord with the allowable stresses as given in the AISC Steel Manual. In the Far West certain calculations must be included due to the frequency of earthquake forces. Large cities in the East and Midwest have building codes which contain many other stipulations at variance with the AISC Manual.

There has been no attempt in the examples in this book to have the allowable stresses comply with any code authority except the AISC. In Section VII, which illustrates the design of rigid frame buildings, the rules and methods proposed by the AISC will be followed. The more liberal and controversial design method presented by the Metal Building Manufacturers Association has not been adopted by design engineers or code authorities. Light gauge materials in pre-engineered buildings reflect the emphasis on economy rather than sturdiness and security. Design unit stresses of 30,000 PSI are used for bending members, and external wind and live loads are assumed to be relatively low.

Allowable design unit stresses should be based upon adequate lateral support. The examples to follow will illustrate the method and formula for stress reduction when lateral support is lacking. Stress is also reduced in some instances when sections are deficient in web stiffening.

TABLE: Working stresses for ASTM steels 2.1.6.1

DESIGN UNIT WORKING STRESSES FOR A.S.T.M. STEELS

YIELD F_y LIMIT ON TEST SPEC. P.S.I.	STEEL SPECIFIED ASTM TYPE DOMESTIC	ALLOWABLE UNIT STRESS — POUNDS PER SQUARE INCH						WELDING ELECTRODE SPECIFIED SERIES MK.
		TENSION IN NET AREA NO VOIDS	SHEAR ON WEB AREA d × t □"	BENDING ON SYMMETRICAL COMPACT SECT.	BENDING ON UNSYMMETRIC SHAPES	BEARING ON MILLED SURFACES	BEARING ON BOLTS-RIVETS NET AREA	
33,000	A·7 & A373	20,000	13,000	22,000 ‡	20,000	30,000	45,000	E60-E70
36,000	A36	22,000	14,500	24,000 ‡	22,000	33,000	48,500	E60-E70
42,000	A242-A440-A441	25,000	17,000	28,000 ‡	25,000	38,000	56,500	E70, LHS
46,000	A242-A440-A441	27,500	18,500	30,500 ‡	27,500	41,500	62,000	E70, LHS
50,000	A242-A440-A441	30,000	20,000	33,000 ‡	30,000	45,000	67,500	E70, LHS

‡ TENSION OR COMPRESSION - BEAM MUST HAVE ADEQUATE BRACING LATERALLY TO USE FULL VALUE F_b. BEAMS WITHOUT LATERAL SUPPORT MUST HAVE UNIT STRESS REDUCED BY FORMULA THUS:

$$F_b = \frac{12,000,000}{\frac{ld}{A_f}}$$

WHERE: l = UNSUPPORTED LENGTH, IN INCHES. d = DEPTH OF SECTION, IN INCHES.
A_f = AREA OF COMPRESSION FLANGE, IN SQUARE INCHES.

Steel design nomenclature 2.2.1

A = Area of a cross-section, plane, etc., given in Sq. Inches (□").
A_b = Area bolt section at root of threads.
A_c = Area Concrete in composite slab design.
A_f = Area of cover plate or flange in compression.
A_g = Area gross, used in columns and composite design.
A_p = Area of plate cover in composite slab design.
A_t = Area in tension, timber or stiffener pairs.
A_s = Area steel, in composite design, pipe and concrete pilings.
A_v = Area of shear web in beam section between fillets.
B = Breadth dimension in base plates, etc., in inches.
B_x = Bending factor with respect to axis x-x. $B_x = A/S_x$.
B_y = Bending factor with respect to axis y-y. $B_y = A/S_y$.
C = Indicates compression, constant, coefficient or dimension.
c = Farthest dimension from centroid axis to extreme fibers.
c_1 = Shortest distance from centroid to outer fibers, in inches.
D = Dimension for total depth, or diameter. In feet or inches.
d = Depth of a section or distance from top to center steel.
bd = Breadth times depth. $A = bd$ in timber sections. In □".
E = Modulus of elasticity, in pounds per square inch. (PSI or #□").
E_s = Modulus of elasticity for steel (29,000,000 PSI)
E_c = Modulus of elasticity for Concrete (See Section IV).
e = Eccentricity or distance of moment lever.
F = Force or Fiber stress, also friction or allowable stress.
F_a = Allowable axial stress in pounds per square inch.
F_b = Allowable bending or flexure unit stress, #□".
F_c = Allowable compressive stress. F_c' = Stress for 28 day Concrete.
F_p = Allowable bearing unit stress for base plates.
F_t = Allowable tension unit stress.
F_y = Minimum yield point of a specified material in PSI.
F_u = Ultimate unit stress of a specified material.
f = Actual unit stress produced by loads, forces, etc., PSI.
f_a = Actual unit axial stress in columns. $f_a = \frac{P}{A}$.
f_b = Actual unit bending stress
f_c = Actual unit compressive stress.
f_v = Actual unit shear stress, beams, bolts, rivets, etc., PSI.
g = Gage dimension in rolled shapes, also gravity (32.174).
H = Height in feet. Used for columns, rigid frames, pile hammers.
h = Height in inches. Also to designate horizontal action plane.

Steel design nomenclature, continued 2.2.1

I_o = Moment of Inertia of a single component, see Section VI.
I_x = Moment of Inertia about x-x axis. Given as: I = "^{4}.
I_y = Moment of Inertia about y-y axis. " " "
j = A design factor for concrete design, see Section IV.
K = A design factor for concrete design, see Section IV.
k = A dimension in base plate design, also a concrete factor.
L = Length designated in feet. Spans, columns, longitude, etc.,
l = Length in inches, unbraced beams, columns, lever arm, etc.
M = Bending moment, force times distance, see Section I.
M_e = Eccentric moment. Moments given in foot-lbs., inch-lbs., etc.
$M_{5.0}$ = Moment at the 5.0 foot distance on a beam, rafter, etc.
M_d = Moment produced by dead loads.
M_L = Moment produced by live and superimposed loads.
M_w = Moment produced by wind pressure.
m = Dimension length for rafters, see rigid frames, Section VII.
N = Denotes a force normal to surface, also as Neutral Axis, NA.
n = Ratio of modulus of elasticity steel and concrete, also used to designate number of hammer blows, see Section IX.
O = Designates a polar point in graphic ray diagram.
P = Concentrated load placed on beam or column, in pounds.
P_e = Concentric load with eccentric moment arm.
P' = An equivalent axial load converted from an eccentric moment by employing the bending factors B_x and B_y.
p = Percentage of steel area in concrete, also weight of a pile.
Q = Ratio of eave height to rafter rise, used in Section VII.
q = A subscript attached to a coefficient in Hiley's formulas.
R = Reaction at a support resulting from loads. Also used to denote radius, and pile's resistance to penetration.
R_1 = Reaction at left support, see Section I.
R_H = Horizontal reaction, used in design stairs, frames, etc.
r = Radius of gyration, given in inches, see Section VI.
r_x = Radius of gyration with respect to axis x-x. In inches.
r_y = Radius of gyration with respect to axis y-y. In inches.
r_o = Occassionally used to denote the least or governing r value.

Steel design nomenclature, continued 2.2.1

S = Section Modulus, a section property given in inches^3.
S_x = Section modulus about major axis x-x.
S_y = Section modulus about minor axis y-y.
S_D = Section modulus value produced by dead loads.
s = Rod spacing, or spacing for rivets, bolts, stirrups, etc.
T = Total force in tension, given in pounds.
t = Thickness dimension, given in inches.
U = Denotes unity, equal to 1.0 or less. Used to compare the ratio of axial compression to bending stress in columns. Also to denote the initial velocity in kinetic energy.
u = Bonding unit stress of steel rods in concrete, in PSI.
V = Vertical shear value in beams, connections, etc. In pounds.
V_H = Horizontal total shear. Reverse to Hv in design of rigid frames and for a horizontal reaction or wind load.
v = Unit vertical shear stress, given in #□', or PSI.
v_h = Unit horizontal shear stress, given in PSI.
W = Total weight of uniform load on beam, joist or deck. In pounds.
W_D = Distributed dead load total on span, given in pounds.
w = Unit load per lineal foot on span, given in pounds.
x = A specified distance to be stated. In inches or feet.
$\bar{y}$ = Used for additional specified distance when x is occupied.
Z = Denotes plastic design section modulus, not used herein.
z = A dimension used for obtaining concrete factor, see Sect. IV.
Δ = Denotes deformation or deflection in inches. (Greek Delta)
Σ_o = Summarized or total quantity or value, taken together.
= Pound symbol, same a lbs. (Pounds per square foot = #□').
□ = Denotes square, applies to shape of rod, tube, section, etc.
' = Dimension in feet, or per foot. Also 1 minute of a degree.
" = Dimension in inches, or inch lbs. = "#. Also 1 second of a minute.
PSF = Pounds per square foot, as #□'.
PSI = Pounds per square inch, as #□".
$\frac{20}{5}$ = Division, 20 divided by 5, same as: $20 \div 5 = 4$.
X = Sign for multiplication, or times. As $5 \times 4 = 20$, or $a \times b = ab$.
+ = Plus sign, add together. Also indicates positive bending as +M, and in graphics, + denotes compressive force.

Steel design nomenclature, continued 2.2.1

− = Minus sign, subtract or deduct. Also indicates a negative bending moment. In graphics it indicates a tension force.

( ) = Parentheses. Used to enclose an equation which is to be reduced to a common value or number.

[ ] = Brackets. Used to enclose a number of equations which are to be reduced to a single equation or common value.

θ = Denotes the angle under consideration, as: Sin θ = 0.0625.

ϕ = Denotes round, a circular reinforcing rod, tube, etc.

⌀ = Denotes square, as a square rod, bar, tube or shape.

% = Percent sign. As 10% of 1500 = 150 or 1500 × 0.10 = 150.

$\sqrt{9.0}$ = Square root sign. Use function of number tables in Sect. V.

3.0^2 = Number 3.0 to be squared, as: 3.0 × 3.0 = 9.0. Applied as l^2, etc.

3.0^3 = Number 3.0 to be cubed, as: 3.0 × 3.0 × 3.0 = 27.0. Applied as l^3, etc.

8° = Denotes 8 degrees. See trigonometry Section V.

± = Plus or minus. Usually preceding a formula. Also used herin after a quantity or value indicating close enough to be accepted as slide rule approximation.

−M = Indicates a negative moment. See Section I on mechanics.

+M = Indicates a positive moment. When plus sign is not used, the moment is assumed positive.

π = Pi. A Greek symbol used for the ratio (3.141592) of the circumference of a circle to its diameter, as: Circ. = πD.

Design formulas with transpositions 2.2.2

SIMPLE SPAN BEAM-UNIFORM LOAD.

$M = \frac{WL}{8}$ or $M = \frac{wL^2}{8}$ $W = \frac{8M}{L}$ or $w = \frac{8M}{L^2}$ $L = \frac{8M}{W}$

$L^2 = \frac{8M}{w}$ and $L = \sqrt{\frac{8M}{w}}$ or $L = \sqrt{L^2}$ $l = L \times 12$

FOR END SPAN OF CONTINUOUS BEAMS:

$M = \frac{WL}{10}$ Transposition is similar to above.

FOR INTERMEDIATE SPANS:

$M = \frac{WL}{12}$ Transposition is similar to simple spans.

PROPERTIES OF PLANE SECTION FORMULAS:

$I = \frac{bd^3}{12}$ $b = \frac{12I}{d^3}$ $d^3 = \frac{12I}{b}$ $d = \sqrt[3]{\frac{12I}{b}}$ $S = \frac{I}{c}$ $I = Sc$ $c = \frac{I}{S}$

$S = \frac{bd^2}{6}$ $b = \frac{6S}{d^2}$ $d^2 = \frac{6S}{b}$ $d = \sqrt{\frac{6S}{b}}$ $RM = SF_b$ $S = \frac{M}{F_b}$ $F_b = \frac{M}{S}$

DEFLECTION FORMULAS

FOR SIMPLE SPAN-UNIFORM LOAD: W = Total load.

$\Delta = \frac{5Wl^3}{384EI}$ $W = \frac{384EI\Delta}{5l^3}$ $I = \frac{5Wl^3}{384E\Delta}$ $l^3 = \frac{384EI\Delta}{5W}$ $l = \sqrt[3]{l^3}$ $L = \frac{l}{12}$

FOR SINGLE CONCENTRATED LOAD-SIMPLE SPAN: P = Load.

$\Delta = \frac{Pl^3}{48EI}$ $P = \frac{48EI\Delta}{l^3}$ $I = \frac{Pl^3}{48E\Delta}$ $l^3 = \frac{48EI\Delta}{P}$ $l = \sqrt[3]{\frac{48EI\Delta}{P}}$

FOR CANTILEVER BEAM WITH UNIFORM LOAD: W = Total load.

$\Delta = \frac{Wl^3}{8EI}$ $W = \frac{8EI\Delta}{l^3}$ $I = \frac{Wl^3}{8E\Delta}$ $l^3 = \frac{8EI\Delta}{W}$ $l = \sqrt[3]{l^3}$

FOR CANTILEVER BEAM WITH CONCENTRATED LOAD:

$\Delta = \frac{Pl^3}{3EI}$ $P = \frac{3EI\Delta}{l^3}$ $I = \frac{Pl^3}{3\Delta E}$ $l^3 = \frac{3EI\Delta}{P}$ $l = \sqrt[3]{l^3}$

TANKS-VESSELS-CYLINDER WALLS

$s = \frac{pr}{t}$ $p = \frac{st}{r}$ $t = \frac{pr}{s}$ $r = \frac{st}{p}$

WHERE:

s = Unit stress, #□" t = Thickness of wall, in inches.

p = Pressure, #□" r = Radius of cylinder, in inches.

Plastic design theory 2.2.3

Plastic design theory, which is comparatively new, is especially applicable to continuous beams or beams with fixed ends. Within the elastic limit, shown on the stress-strain curve, stress is directly proportional to deformation. In the tensile test specimen, the stresses are constant over the whole cross-section. Now, if a beam over continuous spans has the ends fixed, or at each support it is rigidly welded, all fibers in the cross-section correspond to the tensile test condition. Regardless of the fiber distance from centroid, all the fibers are subjected to stress corresponding to the yield limit. At full yield the required moment is denoted as Mp. When the elastic and the plastic section modulus tables are compared, the properties of plastic modulus (Z) will be greater than the elastic modulus (S). Generally, the average increase in plastic design is approximately 12 percent in load capacity when compared with a similar elastic design.

The plastic theory is only used for detailed analysis of a single structure. Plastic design is presently limited to ASTM A7, A373 and A36 steels. Actually, plasticity is assumed in the long-accepted formulas for the fixed and middle spans of continuous beams.

Elastic design theory 2.2.4

A visual examination of a loaded beam will show a measure of sag or deflection between the supports. This is true for all beams; it is more evident in wood beams. This deflection stretches the fibers in the bottom of the beam which are in tension stress. The top fibers are compressed, and therefore in compressive stress. Somewhere between the fibers in tension and those in compression, an area or plane is located where neither stress is present. This plane is called the section's neutral axis, centroid, gravity axis or center of gravity axis.

The term *fibers*, correctly used for wood beams because wood is a fibrous material, has been retained in discussing steel beams even though steel is not a fibrous material. The distance from the neutral axis to the outermost fiber is called dimension c. Calculating the moment of inertia (I) for a rolled shape is equivalent to calculating the value of I for a plane surface. Properties of Sections are presented in Section VI for symmetrical and irregular shapes.

To illustrate how the dimension from the neutral axis to outermost fiber (c) is related to stress in the elastic design theory, refer to the Rolled Shape Tables for angles. Select an unequal leg angle L5 x 3½ x ½. The long leg is 5 inches, and this depth results in the greater value of I. The centroid Y is located 1.66 inches from the corner. The farthest dimension to extreme fibers from axis x–x is found thus: 5.00 − 1.66 = 3.34″ or dimension c. The formula for Section Modulus is: $S = \frac{I}{c}$. Therefore the value of S for this angle is found to check with table as: $S = \frac{9.99}{3.34} = 2.99''^3$ or same as listed in table.

The axis location is a governing factor in the elastic theory. Assuming that the angle was rolled from A36 steel, the allowable unit stress in bending is given as: F_b = 22,000 PSI, and such a section is not considered symmetrical. Used as a beam, the Resisting Moment for the angle is: $RM = SF_b = 2.99 \times 22{,}000 = 65{,}780$ inch-lbs.

It was shown in Section I how external forces on a beam could be resolved into a maximum bending moment. To provide equilibrium, the internal resisting capacity of the beam section must be equal to or exceed the moment produced by external load forces. Simplified by formula, $RM \geqq BM$.

Moment design 2.2.4.1

Before the bending moments computed in Section I can be used for beam design, they must be converted to inch pounds. Since all section properties are in inches, this must not be overlooked. A good means to keep from forgetting this important point is to insert the 12 into the Section Modulus equation. Assume that a beam must sustain a bending moment of 30,000 foot pounds, and the allowable fiber stress is 24,000 pounds per square inch. An equation for solving the value of Section Modulus would be written thus: $S = \frac{M}{F_b}$, and with values: $S = \frac{30{,}000 \times 12}{24{,}000} = 15.0''^3$. To check, the bending moment is: 360,000 inch pounds, and must be matched by the Resisting Moment. $RM = SF_b$ or $15.0 \times 24{,}000 = 360{,}000$ In. lbs.

Solving for the required value of S is the shortest and most accurate method in beam design for bending or flexure. All that is necessary for beam selection is to refer to the Elastic Section Modulus Economy Table. The *lightest weight symmetrical section* which has the required Section Modulus is W12 x 16.5.

TABLES OF STANDARD ROLLED SHAPES

Nominal dimensions, weights, properties for designing and dimensions for detailing for rolled shapes are listed in Tables 2.3.3. In addition there will be found the Elastic Section Modulus Economy Table 2.3.1 and a new Moment of Inertia Economy Table 2.3.2. New shape designations are used and supersede earlier notation for structural shapes. The author is indebted to the United States Steel Corporation for granting permission to reproduce this data from their latest catalog released in May 1971.

BENDING VS. FLEXURE

When a laterally braced beam is elastically designed to resist the bending moment resulting from external loads, the design is said to be based on bending. In such computations, the unit bending stress must fall with the allowable limits for F_b. Deflection or sag in the beam is not considered. If the beam design is restricted to a maximum sag, then the design is said to be based on flexure, and the modulus of elasticity must be used in connection with one of the applicable deflection formulas.

Section I illustrated the methods for determining the maximum bending moments in beams. These moments are given in foot pounds and must be converted into inch pounds before they can be compared to the resisting moment of a section. Simply stated, bending design is based upon the presumption that the resisting moment must be equal to, or exceed, the bending moment. Writing this in formula, it is: $RM \geqq BM$.

ALLOWABLE DESIGN STRESSES

The AISC Steel Manual lists the specifications for each type of steel and the ASTM designation. Since A36 steel is used for structural shapes in greater quantities than other steels, the examples which follow will be based upon A36. To find the resisting moment of a beam cross-section, multiply the value of the section modulus by the allowable unit bending stress. The result will be in inch-pounds. The formula becomes: $RM = SF_b$. When steel beams have adequate lateral bracing for the minor axis (y–y), the allowable stress for A36 steel is $F_b = 24{,}000$ PSI. Note that there are different allowable stresses listed for symmetrical and unsymmetrical cross sections.

Deflection design 2.2.4.2

The deflection of any type of beam is determined by its stiffness or rigidity. In many instances a beam must be rigid to preclude the excessive sag which would damage other materials. A beam which supports a suspended plaster ceiling may deflect under live loads to such an extent that the ceiling plaster might crack. The established rule in designing beams to support plaster ceilings is to limit the deflection to 1/360 of span length in inches. Thus a 30.0 foot simple beam would be limited to a maximum 1.0 inch deflection when fully loaded. In solving for the deflection in a steel or wood beam the dominating factors are the Moment of Inertia (I_O), the modulus of elasticity (E), and the type and position of the load. The deflection for the concentrated loads can be computed by the formula separately from the deflection due to uniformly distributed loads. In a complex loading pattern where several loads of various types are included, it is possible to compute an equivalent load for formula use. The procedure for this approach will be illustrated in Example 2.4.1.6. Where the formula for deflection will be transposed to solve for I. Deflection is given the symbol capital Greek *delta*, Δ.

Vertical shear 2.2.4.3

The maximum shear allowable unit stress (F_v) for ASTM—A36 steel is given as 14,500 PSI, and this is reduced for girder webs according to the ratio of the web thickness to the web height. In the design of structures with longer spans and normal loads, the shear stress can be calculated by the following formula: $f_v = \frac{V}{td}$, where:

V = Total maximum vertical shear or reaction, in pounds.
d = Depth of beam, in inches.
t = Thickness of beam web, in inches.
f_v = Actual unit shearing stress, in PSI. Not to exceed F_v.

Conservative designers, when computing the area to resist vertical shear will refer to the tables of shapes, and consider only the web between the fillet rounds. For example refer to the table and select a wide flange shape W10 x 8 45#. On the dimension sheet, the effective web height is given as T = 7.75 inches. Web thickness is given as ⅜ inches or 0.375″. Area resisting vertical shear (V) is 0.375 x 7.75 = 2.91 Square Inches. The flange areas are ignored when shearing stress is considered. Bridge designers take this conservative method to determine vertical shear area. When the area is found inadequate, they will insert stiffener bars or small angles on each side of web.

The engineering department of a major international oil and chemical producer now requires the shear area between fillets to be investigated for each design. Let us consider why this is important. Excessive shear stresses may be present when the beam has a relatively short span with a large concentrated load near its support. A heavy tank reaction or part of a large hoisting machine could possibly produce a very high shear stress in the web.

Thermal expansion 2.2.4.4

Virtually every type of material is affected by changes in temperature, and steel is no exception. When metal is heated it tends to expand, and the value of the modulus of elasticity decreases in proportion to the rise in temperature. Conversely, the value of E is increased as the metal cools. At room temperature, the modulus of elasticity of steel is 29,000,000 PSI.

From test data taken between room temperature and 200 degrees Fahrenheit, the coefficient for linear expansion for steel is 0.0000065 for each degree. The *coefficient of expansion* is the change in length, per unit of length, per degree change in temperature. To calculate the total change in length of a body for a given change of temperature, multiply the coefficient times the length times the change of temperature in degrees.

Expansion and contraction is a very important factor in the design of bridges, railroad tracks and welded seagoing vessels. Long runs of process piping in refining plants are provided with loops at intervals to absorb the expansion and contraction. Such pipe lines carry heated liquids with temperatures exceeding 240 degrees Fahr. A table of coefficients of thermal expansion (2.2.4.5) is provided for construction materials. Note that the coefficients for hard billet grade steel and concrete are close, which is the reason that they can be used satisfactorily together for reinforcing and composite design.

TABLE: Coefficients of expansion 2.2.4.5

LINEAR COEFFICIENTS OF EXPANSION FOR 1 DEGREE

SUBSTANCE	COEFFICIENT "C" IN INCHES 1° FAHRENHEIT	SUBSTANCE	COEFFICIENT "C" IN INCHES 1° FAHRENHEIT
ALUMINUM, 3s and 4s	0.0000128	PINE, Parellel to grain	0.0000030
BRASS	.0000104	FIR, " " "	.0000021
BRONZE	.0000101	OAK, " " "	.0000027
COPPER	.0000093	MAPLE, " " "	.0000036
SILVER	.0000107	PINE, Normal to grain	.0000190
GOLD	.0000083	FIR, " " "	.0000320
IRON, Cast or Gray	.0000059	OAK, " " "	.0000300
STEEL, Cast	.0000061	MAPLE, " " "	.0000270
STEEL, Hard	.0000073	GYPSUM PLASTER	.0000085
STEEL, Medium	.0000067	BRICK MASONRY	0.0000031
STEEL, Soft	.0000061	CONCRETE	.0000069
LEAD	.0000159	STONE, Ashlar	.0000035
NICKEL	.0000070	GRANITE, Texas	.0000047
PLATINUM-IRIDIUM	.0000045	LIMESTONE, Texas	.0000044
TIN	.0000117	MARBLE, Vermont	.0000056
ZINC, Rolled	.0000173	PLASTER, Cement	.0000092
MONEL METAL, Sheet	.0000080	STONE RUBBLE	.0000035
HARD REINFORCING RODS	.0000073	SANDSTONE	.0000061
GLASS, Plate § Sheet	0.0000047	SLATE	.0000058
PORCELAIN	.0000020	CONCRETE MASONRY	.0000067
GRAPHITE	.0000044	CEMENT, Portland	.0000059

Elastic and thermal deformation 2.2.5

A question frequently asked on tests for state registration concerns the laws of physics which relate to structural design. The modulus of elasticity is also known as Young's modulus, and was derived from the formula that E = Unit stress divided by unit deformation. An English physicist, Robert Hooke, in 1678, was the first to observe that "the deformation of a body is directly proportional to the stress." Hooke reached this conclusion from the results of many experiments with clock springs. Hooke's law will only hold true if the deformation remains within the elastic limit. This can be seen on the Stress-Strain Curve 2.1.5.1. which shows that beyond the elastic limit, the deformation increases more rapidly than the stress.

DEFLECTION FORMULAS

Unit stress is determined by dividing the area into the forces (compression P or tension T). In formula, it is written: $f_a = \frac{P}{A}$ or $f_b = \frac{T}{A}$. These formulas involve two known values. Similarly, two more values are involved when we know the unit deformation and unit length. If any four of the values are known, the fifth and last term is easily determined by the applicable formulas.

$$E = \frac{f}{\Delta u} \quad \text{or} \quad E = \frac{\frac{P}{A}}{\frac{\Delta t}{l}} \quad \text{or} \quad E = \frac{Pl}{A\Delta t} \quad \text{and} \quad \Delta t = \frac{Pl}{AE}.$$

Nomenclature for the above formulas is given as:

E = Modulus of Elasticity (Young's modulus), in PSI.
P = The applied force, P or T in pounds.
A = Cross-section area of member in square inches.
l = Length of member (unit) in inches.
Δu = Unit deformation, in inches.
Δt = Total deformation of member, in inches.

POISSON'S RATIO

It was pointed out in an earlier paragraph that during the elongation by stretching, there is a simultaneous cross-section reduction in a member. Conversely, a compressive force applied to member would tend to enlarge cross-section by bulging. There is a constant ratio of the deformation perpendicular to stress, to the deformation (elongation) parallel to stress. This is called *Poisson's ratio*, expressed as:

$$\frac{\text{Deformation perpendicular to stress}}{\text{Deformation parallel to stress}}$$

For mild steel, the ratio is about 1 to 4 or 1:4 or $\frac{1.00}{4.00}$ = .25 percent of length elongation. For concrete, the ratio is 1:5 or 20 percent.

THERMAL DEFORMATION

Temperature changes cause many failures and considerable damage to heavy construction. Paved highways need

Elastic and thermal deformation, continued 2.2.5

adequate expansion joints to prevent bulging. Long length of continuous steel welded beams may show excessive elongation if erected in colder climates. Structural engineers must take these effects into consideration by providing relief in the form of expansion joints. Masonry walls placed upon steel girders are sensitive to cracking by thermal deformation. The following examples and thermal coefficients table will provide a basis for design in this area.

EXAMPLE: Tie rod deformation 2.2.5.1

A steel tie rod 2.0 inches in diameter connects the column bases together for a 150.0 foot span Rigid Arch. Steel is A36. $E = 29{,}000{,}000$ PSI. $F_y = 36{,}000$ PSI.
Maximum force is horizontal to cause tension in rod and this force is 65,000 Pounds.

REQUIRED:

(a) Determine maximum length of elongation when stress is kept within the limit to allow rod to return to original length upon release of load P.

(b) Calculate the maximum force (P) which would stretch the rod to the elastic limit for return to normal.

(c) With load P = 65,000 Lbs, determine the total length of rod under stress, and what is the unit stress.

STEP I:

Max. Unit elongation, $e = \frac{F_y}{E}$ or $\frac{36{,}000}{29{,}000{,}000} = 0.001241$ " per inch.

$L = 150.0'$ $l = 150.0 \times 12 = 1800$ inches

Max. elongation: $\Delta = \frac{Pl}{EA}$ and also $\Delta = el$. $\Delta = 0.001241 \times 1800 = 2.234''$

Ans. (a).

STEP II:

Max. Δ to stay inside elastic limit is 2.234 inches total length.

Solving for max. force: $P = \frac{AE\Delta}{l}$ $A = 2.0 \times 2.0 \times 0.7854 = 3.1416\ \square''$

Putting values in formula:

Max. $P = \frac{3.1416 \times 29{,}000{,}000 \times 2.234}{1800} = 113{,}068$ Pounds. (Ans. b).

Check by solving for unit stress. $F_y = \frac{P}{A} = \frac{113{,}068}{3.1416} = 36{,}000\ \#\square''$ (OK)

EXAMPLE: Tie rod deformation, continued 2.2.5.1

STEP III:

With P = 65,000 Lbs. $f = \frac{P}{A} = \frac{65,000}{3.1416} = 20,700$ #□" (Part ans. c)

Total deformation $\Delta = \frac{Pl}{EA}$ $\Delta = \frac{65,000 \times 1800}{29,000,000 \times 3.1416} = 1.283$ inches

Total length of rod = 150.0' + 0.1068' = 150.1068' or 150'-1 9/32" (Ans. c)

checking by stress formula; $f = \frac{E\Delta}{l}$. $f = \frac{29,000,000 \times 1.283}{1800} = 20,700$ #□"

EXAMPLE: Suspension rod deformation 2.2.5.2

A monorail supports a travelling hoist with a lifting capacity of 45,000 Pounds. To alleviate monorail sag, 1 3/4 inch diameter rods are spaced at intervals in suspension. When hoist is at the end of its bridge, full load is sustained by suspension rod. Longest rod is 12.0 feet to monoral and steel is A36. Fy = 36,000 PSI.

REQUIRED:

(a) Determine the maximum tension unit stress under full load for suspension rods.

(b) Neglecting any contribution for support from other sources, calculate the stretch deformation per unit inch of rod.

(c) Under maximum load, what will be total length of rod.

STEP I:

Area rod cross section = $0.7854 D^2$ or $A = 0.7854 \times 1.75^2 = 2.40$ □"

Let P = 45,000 Lbs. $f_t = P/A$ or $f_t = \frac{45000}{2.40} = 18,750$ #□" (Ans. a).

Stress is within elastic limit.

Allowable for A36 steel, $F_t = 22,000$ #□" E = 29,000,000 PSI.

STEP II:

Rod length L = 12.0' $l = 12.0 \times 12 = 144$ inches (These are units for Δ)

By formula: $\Delta = \frac{Pl}{AE}$ Total $\Delta = \frac{45,000 \times 144}{2.40 \times 29,000,000} = 0.0931$ inches

For unit deformation, $\Delta = \frac{P}{AE}$ or Unit $\Delta = \frac{0.0931}{144} = 0.0006475$"

STEP III:

Total length rod with sustained load = 144.0931 inches

In fraction conversion: L = 12.0078 Ft. Approximately 3/32 inch.

EXAMPLE: Suspension rod deformation, continued 2.2.5.2

STEP IV:

Check out the work by using a transposed formula where 4 values are known thus: $P = \frac{AE\Delta}{l}$.

Substituting values:

$$P = \frac{2.40 \times 29{,}000{,}000 \times 0.0931}{144} = 45{,}000 \text{ Lbs. (OK)}$$

EXAMPLE: Thermal deformation in bar 2.2.5.3

Assume a steel bar with an exact length of 75.0 feet at zero temperature. Temperature rises to 100° Fahr. Bar cross section is 1.00 x 0.75 with A = 0.75□" Assume bar is fixed at ends and is supported laterally against bending.

REQUIRED:

(a) Determine the elongation for full length if ends not fixed.

(b) With fixed end restraint, what stress is set up by change.

(c) Transpose the formulas to check work, then convert the unit stress into an equivalent force P. Assume E = 30 million.

STEP I:

Let temperature change t = 100° Unit lengths = 75.0 x 12 = 900 inches.

elongation coefficient per unit for each degree: c = 0.0000067"

Total $\Delta = ctl$, or $\Delta = 0.0000067 \times 100 \times 900 = 0.603$ inches

Transposing formulas with values:

$$t = \frac{\Delta}{cl} \quad \text{or} \quad t = \frac{0.603}{0.0000067 \times 900} = 100 \text{ degrees}$$

$$l = \frac{\Delta}{ct} \quad \text{or} \quad l = \frac{0.603}{0.0000067 \times 100} = 900 \text{ inches} \qquad 900/12 = 75.0 \text{ feet.}$$

$$c = \frac{\Delta}{tl} \quad \text{or} \quad c = \frac{0.603}{100 \times 900} = 0.0000067 \text{ inches per degree per in.}$$

STEP II:

Assumed E = 30,000,000 PSI. Area bar = 1.00 x 0.75 = 0.75□"

Deformation $\Delta = 0.603$ inches. To determine unit stress when ends restrained. $f = \frac{P}{A}$ also $f = \frac{E\Delta}{l}$. (f has a greater value than P)

$$f = \frac{30{,}000{,}000 \times 0.603}{900} = 20{,}100 \text{ PSI}$$

$$P = \frac{0.75 \times 30{,}000{,}000 \times 0.603}{900} = 15{,}075 \text{ Pounds} \qquad P = \frac{AE\Delta}{l}$$

$$f = \frac{P}{A} \quad \text{or} \quad f = \frac{15{,}075}{0.75} = 20{,}100 \text{ PSI.}$$

EXAMPLE: Thermal deformation in bar, continued 2.2.5.3

$E = \frac{Pl}{A\Delta}$ or $E = \frac{15,075 \times 900}{0.75 \times 0.603} = 30,000,000$ PSI.

$L = \frac{EA\Delta}{12 \times P}$ or $L = \frac{30,000,000 \times 0.75 \times 0.603}{12 \times 15,075} = 75.0$ feet.

$\Delta = \frac{Pl}{EA}$ or $\Delta = \frac{15,075 \times 75.0 \times 12}{30,000,000 \times 0.75} = 0.603$ inches = Total Δ in 75.0'.

$A = \frac{Pl}{E\Delta}$ or $A = \frac{15,075 \times 75.0 \times 12}{30,000,000 \times 0.603} = 0.75$ Sq. Inches.

Deformation per unit of length $= \frac{\Delta}{l} = \frac{0.603}{900} = 0.00063''$ per inch.

EXAMPLE: Thermal deformation in rails 2.2.5.4

Railroad rail is produced from a type of hard brittle metal which is termed "rail-steel." Lengths of 60.0 feet are most often called standard.

Assume a 60.0 rail is laid when air temperature is 72° and Length is exact.

REQUIRED:

Assume the summer sun will heat rail to 120° Fahr., and the winter will drop the rail temperature to −20° Fahr. Compute the amount of contraction and expansion in total length from lowest temperature to highest temperature.

STEP I:

Unit stress is not a consideration here, nor are ends fixed.

From table of linear coefficients for hard steel:

C = 0.0000073 inches elongation or contraction per unit inch for each degree. Temperature change = t = 120 + 20 = 140°.

Expansion change, t = 120 − 72 = 48 degrees.

Contraction change, t = 72 + 20 = 92 degrees.

STEP II:

From formula: Unit $\Delta = ct$, and Total $\Delta = ctl$.

L = 60.0' l = 60.0 × 12 = 720 inches.

Total from +120° to −20°, $\Delta = 0.0000073 \times 140 \times 720 = 0.73584''$ (about ¾")

EXAMPLE: Thermal deformation in rails, continued 2.2.5.4

Elongation above 72°, Δ = 0.0000073 × 48 × 720 = 0.2522880"
Contraction below 72°, Δ = 0.0000073 × 92 × 720 = 0.4835520"
0.7358400"

DESIGN NOTE:
Additional deformation will be present when rail must sustain moving load from car wheels. See preceding paragraph with regard to Poisson's ratio.

EXAMPLE: Loading due to temperature 2.2.5.5

A Steel square tube column consists of a 4x4 □ 14.52 Section welded between 2 steel girders. Column length is exactly 10.0 feet when temperature is 100 degrees Fahr. Column ends are fixed and no sliding movement can take place. Initial load on column is 65,000 Lbs.

REQUIRED:
Assume the change of temperature in the column changes from 100° to 30° Fahr. Determine the final load in the column when the coefficient of linear expansion or contraction is 0.0000065 per degree.

4x4 □ 14.52#
10.0'

STEP I:
Column length must remain at 10.0 length and initial load of 65,000 Lbs. is a compressive load. Column will stretch equal to the contraction due to lower temperature change. Contraction will set up tension force.

STEP II:
E = 29,000,000. C = 0.0000065 and t = 100° - 30° = 70°
Area cross-section = 4.27 Sq. In.
By formula: P = EAct. Substituting values:

P = 29,000,000 × 4.27 × 0.0000065 × 70 = 56,345 Lbs. in tension.
Total Final load at 30°. P = +65,000 - 56,345 = +8,655 Lbs.

Elastic section modulus economy table 2.3.1

S_x in.^{3}	Shape	S_x in.^{3}	Shape	S_x in.^{3}	Shape	S_x in.^{3}	Shape
1280	**W14X730**	**440**	**W36X135**	**212**	**W27X84**	**114**	**W24X55**
1150	**W14X665**	427	W14X264	202	W18X105	112	W14X74
1110	**W36X300**	414	W24X160	202	W14X127	112	W10X100
1040	W14X605	**406**	**W33X130**	199	S24X100	**110**	**W21X55**
1030	**W36X280**	404	W27X145	198	W21X96	108	W18X60
952	**W36X260**	397	W14X246	**197**	**W24X84**	107	W12X79
933	W14X550	382	W14X237	189	W14X119	104	W16X64
894	**W36X245**	380	W30X132	187	S24X90	103	W14X68
840	W14X500	373	W24X145	185	W18X96	103	S18X70
837	**W36X230**	368	W14X228	183	W12X133	99.7	W10X89
813	W33X240	**359**	**W33X118**	**176**	**W24X76**	**98.4**	**W18X55**
758	W14X455	355	W30X124	176	W14X111	97.5	W12X72
742	**W33X220**	353	W14X219	175	S24X79.9	94.4	W16X58
707	W14X426	339	W14X211	169	W21X82	**93.3**	**W21X49**
671	**W33X200**	332	W24X130	166	W16X96	92.2	W14X61
665	**W36X194**	**329**	**W30X116**	164	W14X103	89.4	S18X54.7
657	W14X398	325	W14X202	163	W12X120	89.1	W18X50
651	W30X210	317	W21X142	161	S20X95	88.0	W12X65
622	**W36X182**	310	W14X193	157	W18X85	86.1	W10X77
608	W14X370	**300**	**W30X108**	**153**	**W24X68**	**81.6**	**W21X44**
587	W30X190	300	W27X114	152	S20X85	80.8	W16X50
580	**W36X170**	300	W24X120	151	W21X73	80.1	W10X72
559	W14X342	296	W14X184	151	W16X88	79.0	W18X45
542	**W36X160**	284	W21X127	151	W14X95	78.1	W12X58
530	W30X172	282	W14X176	145	W12X106	77.8	W14X53
512	W14X314	276	W24X110	142	W18X77	75.1	MC18X58
504	**W36X150**	**270**	**W30X99**	**140**	**W21X68**	73.7	W10X66
494	W27X177	267	W27X102	138	W14X87	72.5	W16X45
493	W14X320	267	W14X167	135	W12X99	70.7	W12X53
487	W33X152	263	W12X190	131	W14X84	70.2	W14X48
465	W14X287	253	W14X158	**130**	**W24X61**	69.7	MC18X51.9
448	**W33X141**	252	S24X120	129	W18X70	**68.4**	**W18X40**
446	W27X160	250	W24X100	128	W16X78	67.1	W10X60
		250	W21X112	128	S20X75	64.8	S15X50
		243	**W27X94**	127	W21X62	64.7	W12X50
		240	W14X150	126	W10X112		
		236	S24X105.9	125	W12X92		
		227	W14X142	121	W14X78		
		222	W12X161	118	W18X64		
		221	**W24X94**	118	S20X65.4		
		220	W18X114	116	W16X71		
		216	W14X136	116	W12X85		

The lightest weight shape that will serve is the first shape in **boldface type** whose S_x value exceeds the required elastic section modulus value.

NOTE: TABLES 2.3.1 through 2.3.3.16 are taken from the UNITED STATES STEEL CATALOG of May 1971.

Elastic section modulus economy table, continued 2.3.1

S_x in.³	Shape
64.6	**W16X40**
64.3	MC18X45.8
62.7	W14X43
61.6	MC18X42.7
60.4	W10X54
60.4	W8X67
59.6	S15X42.9
58.2	W12X45
57.9	**W18X35**
56.5	W16X36
54.7	W14X38
54.6	W10X49
53.8	C15X50
52.0	W8X58
51.9	W12X40
50.8	S12X50
49.1	W10X45
48.6	**W14X34**
48.4	MC13X50
47.2	**W16X31**
46.5	C15X40
46.0	W12X36
45.4	S12X40.8
44.9	MC12X50
43.2	W8X48
42.2	W10X39
4.20	C15X33.9
42.0	MC12X45
42.0	MC13X40
41.9	**W14X30**
39.5	W12X31
39.0	MC12X40
38.8	MC13X35
38.3	**W16X26**
38.2	S12X35
36.8	MC13X31.8
36.4	S12X31.8
36.1	MC12X35
35.5	W8X40

S_x in.³	Shape
35.1	**W14X26**
35.0	W10X33
34.2	W12X27
34.2	MC12X37
33.5	M8X40
32.6	M8X37.7
31.8	MC12X32.9
31.5	MC10X41.1
31.1	W8X35
30.8	W10X29
30.6	MC12X30.9
29.4	S10X35
29.1	M8X34.3
28.9	**W14X22**
28.4	M8X32.6
27.8	MC10X33.6
27.4	W8X31
27.0	C12X30
26.5	W10X25
25.3	**W12X22**
25.3	MC10X28.5
24.7	S10X25.4
24.3	W8X28
24.1	C12X25
23.6	MC10X28.3
22.0	MC10X24.9
21.5	**W10X21**
21.5	**C12X20.7**
21.4	MC10X25.3
21.3	**W12X19**
20.8	W8X24
20.7	M6X33.75
20.7	C10X30
19.7	MC10X21.9
19.6	MC9X25.4
18.9	MCX23.9
18.8	**W10X19**
18.2	C10X25

S_x in.³	Shape
17.6	**W12X16.5**
17.0	W8X20
16.7	W6X25
16.2	W10X17
16.2	S8X23
16.0	MC8X22.8
15.8	C10X20
15.7	M6X25
15.4	MC8X21.4
14.8	**W12X14**
14.4	S8X18.4
14.1	W8X17
13.8	W10X15
13.6	MC8X20
13.6	MC7X22.7
13.5	C10X15.3
13.5	C9X20
13.4	W6X20
13.1	MC8X18.7
13.0	M6X20
12.3	MC7X19.1
12.1	S7X20
11.8	W8X15
11.3	C9X15
11.0	C8X18.75
10.8	MC7X17.6
10.6	**C9X13.4**
10.5	**W10X11.5**
10.5	S7X15.3
10.2	W6X16
10.0	W6X15.5
9.9	W5X18.5
9.9	W8X13
9.9	MC6X18
9.6	M5X18.9
9.0	**C8X13.75**
8.8	S6X17.25
8.7	MC6X16.3
8.5	W5X16
8.5	MC6X15.3
8.3	MC6X15.1

S_x in.³	Shape
8.1	**C8X11.5**
7.8	**W8X10**
7.8	C7X14.75
7.4	S6X12.5
7.3	W6X12
6.9	C7X12.25
6.7	M4X16.3
6.2	MC6X12
6.1	S5X14.75
6.1	**C7X9.8**
5.8	C6X13
5.2	M4X13
5.1	**W6X8.5**
5.1	C6X10.5
4.9	S5X10
4.5	MC4X13.8
4.4	**C6X8.2**
3.6	C5X9
3.4	S4X9.5
3.0	**S4X7.7**
3.0	**C5X6.7**
2.3	C4X7.25
2.1	MC3X9
2.0	S3X7.5
1.9	**C4X5.4**
1.8	MC3X7.1
1.7	S3X5.7
1.4	C3X6
1.2	**C3X5**
1.1	**C3X4.1**

The lightest weight shape that will serve is the first shape in **boldface type** whose S_x value exceeds the required elastic section modulus value.

Moment of inertia economy table 2.3.2

I_x in.^{4}	Shape
20300	**W36X300**
18900	**W36X280**
17300	**W36X260**
16100	**W36X245**
15000	**W36X230**
14400	W14X730
13600	W33X240
12500	W14X665
12300	**W33X220**
12100	**W36X194**
11300	**W36X182**
11100	W33X200
10900	W14X605
10500	**W36X170**
9890	W30X210
9760	**W36X160**
9450	W14X550
9030	**W36X150**
8850	W30X190
8250	W14X500
8160	W33X152
7910	W30X172
7820	**W36X135**
7460	W33X141
7220	W14X455
6740	W27X177
6710	**W33X130**
6610	W14X426
6030	W27X160
6010	W14X398

I_x in.^{4}	Shape
5900	**W33X118**
5760	W30X132
5450	W14X370
5430	W27X145
5360	W30X124
5120	W24X160
4930	**W30X116**
4910	W14X342
4570	W24X145
4470	**W30X108**
4400	W14X314
4140	W14X320
4090	W27X114
4020	W24X130
4000	**W30X99**
3910	W14X287
3650	W24X120
3610	W27X102
3530	W14X264
3410	W21X142
3330	W24X110
3270	**W24X94**
3230	W14X246
3080	W14X237
3030	S24X120
3020	W21X127
3000	W24X100
2940	W14X228
2830	**W27X84**
2830	S24X105.9
2800	W14X219
2690	W24X94
2670	W14X211
2620	W21X112
2540	W14X202
2400	W14X193
2390	S24X100

I_x in.^{4}	Shape
2370	**W24X84**
2270	W14X184
2250	S24X90
2150	W14X176
2110	**S24X79.9**
2100	**W24X76**
2100	W21X96
2040	W18X114
2020	W14X167
1900	W14X158
1890	W12X190
1850	W18X105
1820	**W24X68**
1790	W14X150
1760	W21X82
1680	W18X96
1670	W14X142
1610	S20X92
1600	W21X73
1590	W14X136
1540	**W24X61**
1540	W12X161
1520	S20X85
1430	W21X68
1480	W14X127
1440	W18X85
1370	W14X119
1360	W16X96
1340	**W24X55**
1330	W21X72
1290	W18X77
1280	S20X75
1270	W14X111
1220	W16X88
1220	W12X133
1180	S20X65.4
1170	W14X103
1160	W18X70

I_x in.^{4}	Shape
1140	**W21X55**
1070	W12X120
1060	W14X95
1050	W18X64
1050	W16X78
986	W18X60
971	**W21X49**
967	W14X87
941	W16X71
931	W12X106
928	W14X84
926	S18X70
891	W18X55
859	W12X99
851	W14X78
843	**W21X44**
836	W16X64
804	S18X54.7
802	W18X50
797	W14X74
789	W12X92
748	W16X58
724	W14X68
723	W12X85
719	W10X112
706	W18X45
676	MC18X58
663	W12X79
657	W16X50
641	W14X53
627	MC18X51.9
625	W10X100
612	**W18X40**
597	W12X72
584	W16X45
578	MC18X45.8
554	MC18X42.7
542	W14X53
542	W10X89
533	W12X65

The lightest weight shape that will serve is the first shape in **boldface type** whose I_x value exceeds the required moment of inertia value.

Moment of inertia economy table, continued 2.3.2

I_x in.^{4}	Shape
517	**W16X40**
513	**W18X35**
486	S15X50
485	W14X48
476	W12X58
457	W10X77
447	W16X36
447	S15X42.9
429	W14X43
426	W12X53
421	W10X72
404	C15X50
395	W12X50
386	W14X38
382	W10X66
374	**W16X31**
351	W12X45
349	C15X40
344	W10X60
340	W14X34
315	C15X33.9
314	MC13X50
310	W12X40
306	W10X54
305	S12X50
300	**W16X26**
290	W14X30
281	W12X36
273	W10X49
273	MC13X40
272	W8X67
272	S12X40.8
269	MC12X50
252	MC13X35
252	MC12X45
249	W10X45

I_x in.^{4}	Shape
244	**W14X26**
239	W12X31
239	MC13X31.8
234	MC12X40
229	S12X35
227	W8X58
218	S12X31.8
216	MC12X35
210	W10X39
205	MC12X37
204	W12X27
198	**W14X22**
191	MC12X32.9
184	W8X48
183	MC12X30.9
171	W10X33
162	C12X30
158	W10X29
158	MC10X41.1
156	**W12X22**
147	S10X35
146	W8X40
144	C12X25
139	MC10X33.6
136	M8X40
133	W10X25
132	M8X37.7
130	**W12X19**
129	C12X20.7
127	MC10X28.5
126	W8X35
124	S10X25.4
118	MC10X28.3
116	M8X34.3
114	M8X32.6
110	W8X31
110	MC10X24.9
107	W10X21
107	MC10X25.3

I_x in.^{4}	Shape
105	**W12X16.5**
103	C10X30
98.5	MC10X21.9
97.8	W8X28
96.3	W10X19
91.2	C10X25
88.0	**W12X14**
88.0	MC9X25.4
85.0	MC9X23.9
82.5	W8X24
81.9	W10X17
78.9	C10X20
69.4	W8X20
68.9	W10X15
67.4	C10X15.3
64.9	S8X23
64.7	M6X33.75
63.8	MC8X22.8
61.6	MC8X21.4
60.9	C9X20
57.6	S8X18.4
56.6	W8X17
54.5	MC8X20
53.3	W6X25
52.5	MC8X18.7
52.0	**W10X11.5**
51.0	C9X15
48.1	W8X15
47.9	C9X13.4
47.5	MC7X22.7
47.1	M6X25
44.0	C8X18.75
43.2	MC7X19.1
42.4	S7X20
41.5	W6X20
39.6	W8X13
39.0	M6X20
37.6	MC7X17.6
36.7	S7X15.3
36.1	C8X13.75
32.6	**C8X11.5**
31.7	W6X16

I_x in.^{4}	Shape
30.8	**W8X10**
30.1	W6X15.5
29.7	MC6X18
27.2	C7X14.75
26.3	S6X17.25
26.0	MC6X16.3
25.4	W5X18.5
25.4	MC6X15.3
25.0	MC6X15.1
24.2	C7X12.25
24.1	M5X18.9
22.1	S6X12.5
21.7	W6X12
21.3	**C7X9.8**
21.3	W5X16
18.7	MC6X12
17.4	C6X13
15.2	S5X14.75
15.2	C6X10.5
14.8	**W6X8.5**
14.0	M4X16.3
13.1	C6X8.2
12.3	S5X10
10.5	M4X13
8.91	MC4X13.8
8.90	C5X9
7.49	**C5X6.7**
6.79	S4X9.5
6.08	S4X7.7
4.59	C4X7.25
3.85	**C4X5.4**
3.15	MC3X9
2.93	S3X7.5
2.73	MC3X7.1
2.42	S3X5.7
2.07	C3X6
1.85	**C3X5**
1.66	**C3X4.1**

The lightest weight shape that will serve is the first shape in **boldface type** whose I_x value exceeds the required moment of inertia value.

W
Wide Flange Shapes

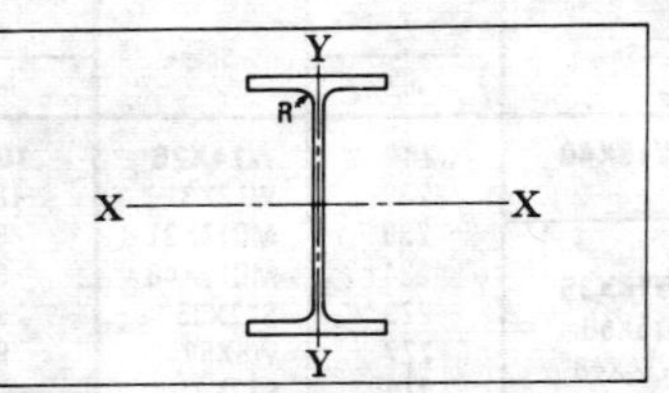

Properties for Designing

Designation and Nominal Size	Weight per Foot	Area of Section	Depth of Section	Flange Width	Flange Thick-ness	Web Thick-ness	Axis X-X *I*	Axis X-X *S*	Axis X-X *r*	Axis Y-Y *I*	Axis Y-Y *S*	Axis Y-Y *r*
In.	Lbs.	In.²	In.	In.	In.	In.	In.⁴	In.³	In.	In.⁴	In.³	In.
W36 36 x 16½ (CB 362)	300	88.3	36.72	16.655	1.680	.945	20300	1110	15.2	1300	156	3.83
	280	82.4	36.50	16.595	1.570	.885	18900	1030	15.1	1200	144	3.81
	260	76.5	36.24	16.551	1.440	.841	17300	952	15.0	1090	132	3.77
	245	72.1	36.06	16.512	1.350	.802	16100	894	15.0	1010	123	3.75
	230	67.7	35.88	16.471	1.260	.761	15000	837	14.9	940	114	3.73
W36 36 x 12 (CB 361)	194	57.2	36.48	12.117	1.260	.770	12100	665	14.6	375	61.9	2.56
	182	53.6	36.32	12.072	1.180	.725	11300	622	14.5	347	57.5	2.55
	170	50.0	36.16	12.027	1.100	.680	10500	580	14.5	320	53.2	2.53
	160	47.1	36.00	12.000	1.020	.653	9760	542	14.4	295	49.1	2.50
	150	44.2	35.84	11.972	.940	.625	9030	504	14.3	270	45.0	2.47
	135	39.8	35.55	11.945	.794	.598	7820	440	14.0	226	37.9	2.39
W33 33 x 15¾ (CB 332)	240	70.6	33.50	15.865	1.400	.830	13600	813	13.9	933	118	3.64
	220	64.8	33.25	15.810	1.275	.775	12300	742	13.8	841	106	3.60
	200	58.9	33.00	15.750	1.150	.715	11100	671	13.7	750	95.2	3.57
W33 33 x 11½ (CB 331)	152	44.8	33.50	11.565	1.055	.635	8160	487	13.5	273	47.2	2.47
	141	41.6	33.31	11.535	.960	.605	7460	448	13.4	246	42.7	2.43
	130	38.3	33.10	11.510	.855	.580	6710	406	13.2	218	37.9	2.38
	118	34.8	32.86	11.484	.738	.554	5900	359	13.0	187	32.5	2.32
W30 30 x 15 (CB 302)	210	61.9	30.38	15.105	1.315	.775	9890	651	12.6	757	100	3.50
	190	56.0	30.12	15.040	1.185	.710	8850	587	12.6	673	89.5	3.47
	172	50.7	29.88	14.985	1.065	.655	7910	530	12.5	598	79.8	3.43
W30 30 x 10½ (CB 301)	132	38.9	30.30	10.551	1.000	.615	5760	380	12.2	196	37.2	2.25
	124	36.5	30.16	10.521	.930	.585	5360	355	12.1	181	34.4	2.23
	116	34.2	30.00	10.500	.850	.564	4930	329	12.0	164	31.3	2.19
	108	31.8	29.82	10.484	.760	.548	4470	300	11.9	146	27.9	2.15
	99	29.1	29.64	10.458	.670	.522	4000	270	11.7	128	24.5	2.10
W27 27 x 14 (CB 272)	177	52.2	27.31	14.090	1.190	.725	6740	494	11.4	556	78.9	3.26
	160	47.1	27.08	14.023	1.075	.658	6030	446	11.3	495	70.6	3.24
	145	42.7	26.88	13.965	.975	.600	5430	404	11.3	443	63.5	3.22
W27 27 x 10 (CB 271)	114	33.6	27.28	10.070	.932	.570	4090	300	11.0	159	31.6	2.18
	102	30.0	27.07	10.018	.827	.518	3610	267	11.0	139	27.7	2.15
	94	27.7	26.91	9.990	.747	.490	3270	243	10.9	124	24.9	2.12
	84	24.8	26.69	9.963	.636	.463	2830	212	10.7	105	21.1	2.06
W24 24 x 14 (CB 243)	160	47.1	24.72	14.091	1.135	.656	5120	414	10.4	530	75.2	3.35
	145	42.7	24.49	14.043	1.020	.608	4570	373	10.3	471	67.1	3.32
	130	38.3	24.25	14.000	.900	.565	4020	332	10.2	412	58.9	3.28

Wide flange shapes (W), continued 2.3.3.1

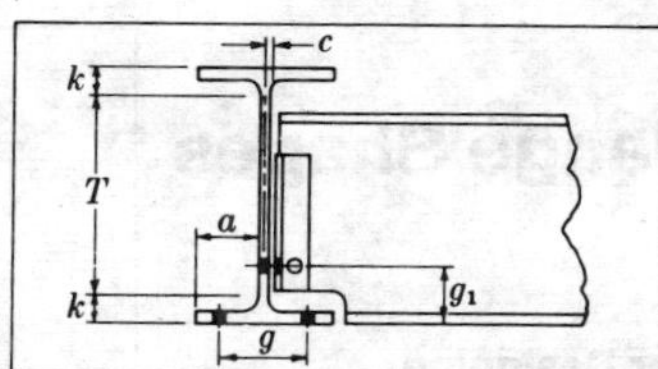

Dimensions for Detailing

Designation and Nominal Size	Weight per Foot	Depth of Section	Flange Width	Flange Thickness	Web Thickness	Web Half Thickness	Distances a	Distances T	Distances k	Distances g_1	Distances c	Usual Gage g	Fillet Radius R
In.	Lbs.	In.	In.	In.	In.	In.	In.	In.	In.	In.	In.	In.	In.
W36	300	36 3/4	16 5/8	1 11/16	15/16	1/2	7 7/8	31 1/8	2 13/16	3 3/4	9/16	5 1/2	
36 x 16 1/2	280	36 1/2	16 5/8	1 9/16	7/8	7/16	7 7/8	31 1/8	2 11/16	3 3/4	1/2	5 1/2	
(CB 362)	260	36 1/4	16 1/2	1 7/16	13/16	7/16	7 7/8	31 1/8	2 9/16	3 1/2	1/2	5 1/2	1.02
	245	36	16 1/2	1 3/8	13/16	3/8	7 7/8	31 1/8	2 7/16	3 1/2	7/16	5 1/2	
	230	35 7/8	16 1/2	1 1/4	3/4	3/8	7 7/8	31 1/8	2 3/8	3 1/2	7/16	5 1/2	
W36	194	36 1/2	12 1/8	1 1/4	3/4	3/8	5 5/8	32 1/8	2 3/16	3 1/2	7/16	5 1/2	
36 x 12	182	36 3/8	12 1/8	1 3/16	3/4	3/8	5 5/8	32 1/8	2 1/8	3 1/4	7/16	5 1/2	
(CB 361)	170	36 1/8	12	1 1/8	11/16	5/16	5 5/8	32 1/8	2	3 1/4	3/8	5 1/2	
	160	36	12	1	5/8	5/16	5 5/8	32 1/8	1 15/16	3 1/4	3/8	5 1/2	.80
	150	35 7/8	12	15/16	5/8	5/16	5 5/8	32 1/8	1 7/8	3	3/8	5 1/2	
	135	35 1/2	12	13/16	5/8	5/16	5 5/8	32 1/8	1 11/16	3	3/8	5 1/2	
W33	240	33 1/2	15 7/8	1 3/8	13/16	7/16	7 1/2	28 5/8	2 7/16	3 1/2	1/2	5 1/2	
33 x 15 3/4	220	33 1/4	15 3/4	1 1/4	3/4	3/8	7 1/2	28 5/8	2 5/16	3 1/2	7/16	5 1/2	.96
(CB 332)	200	33	15 3/4	1 1/8	11/16	3/8	7 1/2	28 5/8	2 3/16	3 1/4	7/16	5 1/2	
W33	152	33 1/2	11 5/8	1 1/16	5/8	5/16	5 1/2	29 3/4	1 7/8	3 1/4	3/8	5 1/2	
33 x 11 1/2	141	33 1/4	11 1/2	15/16	5/8	5/16	5 1/2	29 3/4	1 3/4	3	3/8	5 1/2	.75
(CB 331)	130	33 1/8	11 1/2	7/8	9/16	5/16	5 1/2	29 3/4	1 11/16	3	3/8	5 1/2	
	118	32 7/8	11 1/2	3/4	9/16	1/4	5 1/2	29 3/4	1 9/16	2 3/4	5/16	5 1/2	
W30	210	30 3/8	15 1/8	1 5/16	3/4	3/8	7 1/8	25 3/4	2 5/16	3 1/2	7/16	5 1/2	
30 x 15	190	30 1/8	15	1 3/16	11/16	3/8	7 1/8	25 3/4	2 3/16	3 1/4	7/16	5 1/2	.91
(CB 302)	172	29 7/8	15	1 1/16	5/8	5/16	7 1/8	25 3/4	2 1/16	3 1/4	3/8	5 1/2	
W30	132	30 1/4	10 1/2	1	5/8	5/16	5	26 3/4	1 3/4	3	3/8	5 1/2	
30 x 16 1/2	124	30 1/8	10 1/2	15/16	9/16	5/16	5	26 3/4	1 11/16	3	3/8	5 1/2	
(CB 301)	116	30	10 1/2	7/8	9/16	5/16	5	26 3/4	1 5/8	3	3/8	5 1/2	.70
	108	29 7/8	10 1/2	3/4	9/16	1/4	5	26 3/4	1 9/16	3	5/16	5 1/2	
	99	29 5/8	10 1/2	11/16	1/2	1/4	5	26 3/4	1 7/16	2 3/4	5/16	5 1/2	
W27	177	27 1/4	14 1/8	1 3/16	3/4	3/8	6 5/8	23	2 1/8	3 1/4	7/16	5 1/2	
27 x 14	160	27 1/8	14	1 1/16	11/16	5/16	6 5/8	23	2 1/16	3 1/4	3/8	5 1/2	.86
(CB 272)	145	26 7/8	14	1	5/8	5/16	6 5/8	23	1 15/16	3	3/8	5 1/2	
W27	114	27 1/4	10 1/8	15/16	9/16	5/16	4 3/4	23 7/8	1 11/16	3	3/8	5 1/2	
27 x 10	102	27 1/8	10	13/16	1/2	1/4	4 3/4	23 7/8	1 5/8	3	5/16	5 1/2	
(CB 271)	94	26 7/8	10	3/4	1/2	1/4	4 3/4	23 7/8	1 1/2	2 3/4	5/16	5 1/2	.64
	84	26 3/4	10	5/8	7/16	1/4	4 3/4	23 7/8	1 7/16	2 3/4	5/16	5 1/2	
W24	160	24 3/4	14 1/8	1 1/8	5/8	5/16	6 3/4	20 7/8	1 15/16	3 1/4	3/8	5 1/2	
24 x 14	145	24 1/2	14	1	5/8	5/16	6 3/4	20 7/8	1 13/16	3 1/4	3/8	5 1/2	.70
(CB 243)	130	24 1/4	14	7/8	9/16	5/16	6 3/4	20 7/8	1 11/16	3	3/8	5 1/2	

Wide flange shapes (W), continued 2.3.3.1

W
Wide Flange Shapes

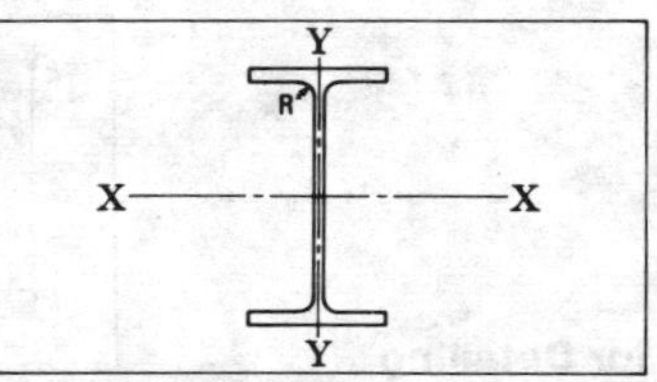

Properties for Designing

Designation and Nominal Size	Weight per Foot	Area of Section	Depth of Section	Flange		Web Thickness	Axis X-X			Axis Y-Y		
				Width	Thickness		*I*	*S*	*r*	*I*	*S*	*r*
In.	Lbs.	In.²	In.	In.	In.	In.	In.⁴	In.³	In.	In.⁴	In.³	In.
W24 24 x 12 (CB 242)	120	35.4	24.31	12.088	.930	.556	3650	300	10.2	274	45.4	2.78
	110	32.5	24.16	12.042	.855	.510	3330	276	10.1	249	41.4	2.77
	100	29.5	24.00	12.000	.775	.468	3000	250	10.1	223	37.2	2.75
W24 24 x 9 (CB 241)	94	27.7	24.29	9.061	.872	.516	2690	221	9.86	108	23.9	1.98
	84	24.7	24.09	9.015	.772	.470	2370	197	9.79	94.5	21.0	1.95
	76	22.4	23.91	8.985	.682	.440	2100	176	9.69	82.6	18.4	1.92
	68	20.0	23.71	8.961	.582	.416	1820	153	9.53	70.0	15.6	1.87
W24 24 x 7 (CBL 24)	61	18.0	23.72	7.023	.591	.419	1540	130	9.25	34.3	9.76	1.38
	55	16.2	23.55	7.000	.503	.396	1340	114	9.10	28.9	8.25	1.34
W21 21 x 13 (CB 213)	142	41.8	21.46	13.132	1.095	.659	3410	317	9.03	414	63.0	3.15
	127	37.4	21.24	13.061	.985	.588	3020	284	8.99	366	56.1	3.13
	112	33.0	21.00	13.000	.865	.527	2620	250	8.92	317	48.8	3.10
W21 21 x 9 (CB 212)	96	28.3	21.14	9.038	.935	.575	2100	198	8.61	115	25.5	2.02
	82	24.2	20.86	8.962	.795	.499	1760	169	8.53	95.6	21.3	1.99
W21 21 x 8¼ (CB 211)	73	21.5	21.24	8.295	.740	.455	1600	151	8.64	70.6	17.0	1.81
	68	20.0	21.13	8.270	.685	.430	1480	140	8.60	64.7	15.7	1.80
	62	18.3	20.99	8.240	.615	.400	1330	127	8.54	57.5	13.9	1.77
	55	16.2	20.80	8.215	.522	.375	1140	110	8.40	48.3	11.8	1.73
W21 21 x 6½ (CBL 21)	49	14.4	20.82	6.520	.532	.368	971	93.3	8.21	24.7	7.57	1.31
	44	13.0	20.66	6.500	.451	.348	843	81.6	8.07	20.7	6.38	1.27
W18 18 x 11¾ (CB 183)	114	33.5	18.48	11.833	.991	.595	2040	220	7.79	274	46.3	2.86
	105	30.9	18.32	11.792	.911	.554	1850	202	7.75	249	42.3	2.84
	96	28.2	18.16	11.750	.831	.512	1680	185	7.70	225	38.3	2.82
W18 18 x 8¾ (CB 182)	85	25.0	18.32	8.838	.911	.526	1440	157	7.57	105	23.8	2.05
	77	22.7	18.16	8.787	.831	.475	1290	142	7.54	94.1	21.4	2.04
	70	20.6	18.00	8.750	.751	.438	1160	129	7.50	84.0	19.2	2.02
	64	18.9	17.87	8.715	.686	.403	1050	118	7.46	75.8	17.4	2.00
W18 18 x 7½ (CB 181)	60	17.7	18.25	7.558	.695	.416	986	108	7.47	50.1	13.3	1.68
	55	16.2	18.12	7.532	.630	.390	891	98.4	7.42	45.0	11.9	1.67
	50	14.7	18.00	7.500	.570	.358	802	89.1	7.38	40.2	10.7	1.65
	45	13.2	17.86	7.477	.499	.335	706	79.0	7.30	34.8	9.32	1.62
W18 18 x 6 (CBL 18)	40	11.8	17.90	6.018	.524	.316	612	68.4	7.21	19.1	6.34	1.27
	35	10.3	17.71	6.000	.429	.298	513	57.9	7.05	15.5	5.16	1.23

Wide flange shapes (W), continued 2.3.3.1

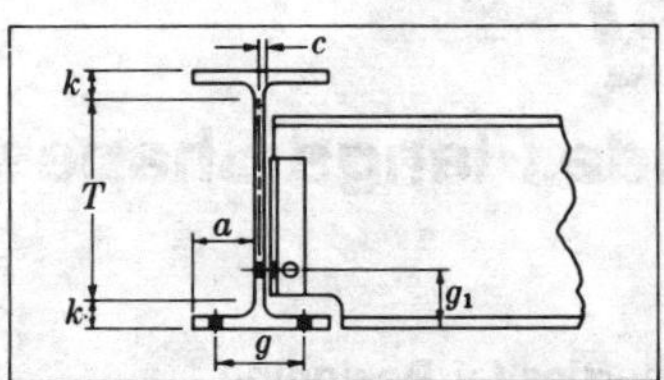

Dimensions for Detailing

Designation and Nominal Size	Weight per Foot	Depth of Section	Flange		Web		Distances					Usual Gage	Fillet Radius
			Width	Thickness	Thickness	Half Thickness	a	T	k	g_1	c	g	R
In.	Lbs.	In.	In.	In.	In.	In.	In.	In.	In.	In.	In.	In.	In.
W24 24 x 12 (CB 242)	120	24¼	12⅛	15/16	9/16	¼	5¾	20⅞	1 11/16	3	5/16	5½	.70
	110	24⅛	12	⅞	½	¼	5¾	20⅞	1⅝	3	5/16	5½	
	100	24	12	¾	7/16	¼	5¾	20⅞	1 9/16	3	5/16	5½	
W24 24 x 9 (CB 241)	94	24¼	9	⅞	½	¼	4¼	21	1⅝	3	5/16	5½	.70
	84	24⅛	9	¾	½	¼	4¼	21	1 9/16	3	5/16	5½	
	76	23⅞	9	11/16	7/16	¼	4¼	21	1 7/16	2¾	5/16	5½	
	68	23¾	9	9/16	7/16	3/16	4¼	21	1⅜	2¾	¼	5½	
W24 24 x 7 (CBL 24)	61	23¾	7	9/16	7/16	3/16	3¼	21	1⅜	2¾	¼	3½	.70
	55	23½	7	½	⅜	3/16	3¼	21	1¼	2¾	¼	3½	
W21 21 x 13 (CB 213)	142	21½	13⅛	1⅛	11/16	5/16	6¼	17¾	1⅞	3	⅜	5½	.65
	127	21¼	13	1	9/16	5/16	6¼	17¾	1¾	3	⅜	5½	
	112	21	13	⅞	½	¼	6¼	17¾	1⅝	2¾	5/16	5½	
W21 21 x 9 (CB 212)	96	21⅛	9	15/16	9/16	5/16	4¼	17¾	1 11/16	3	⅜	5½	.65
	82	20⅞	9	13/16	½	¼	4¼	17¾	1 9/16	2¾	5/16	5½	
W21 21 x 8¼ (CB 211)	73	21¼	8¼	¾	7/16	¼	3⅞	18½	1⅜	2¾	5/16	5½	.54
	68	21⅛	8¼	11/16	7/16	3/16	3⅞	18½	1 5/16	2¾	¼	5½	
	62	21	8¼	⅝	⅜	3/16	3⅞	18½	1¼	2½	¼	5½	
	55	20¾	8¼	½	⅜	3/16	3⅞	18½	1⅛	2½	¼	5½	
W21 4 x 6½ (CBL 21)	49	20⅞	6½	9/16	⅜	3/16	3⅛	18½	1 3/16	2½	¼	3½	.54
	44	20⅝	6½	7/16	⅜	3/16	3⅛	18½	1 1/16	2½	¼	3½	
W18 18 x 11¾ (CB 183)	114	18½	11⅞	1	⅝	5/16	5⅝	15⅛	1 11/16	3	⅜	5½	.60
	105	18⅜	11¾	15/16	9/16	¼	5⅝	15⅛	1⅝	3	5/16	5½	
	96	18⅛	11¾	13/16	½	¼	5⅝	15⅛	1½	2¾	5/16	5½	
W18 18 x 8¾ (CB 182)	85	18⅜	8⅞	15/16	½	¼	4⅛	15⅛	1⅝	3	5/16	5½	.60
	77	18⅛	8¾	13/16	½	¼	4⅛	15⅛	1½	2¾	5/16	5½	
	70	18	8¾	¾	7/16	¼	4⅛	15⅛	1 7/16	2¾	5/16	5½	
	64	17⅞	8¾	11/16	⅜	3/16	4⅛	15⅛	1⅜	2¾	¼	5½	
W18 18 x 7½ (CB 181)	60	18¼	7½	11/16	7/16	3/16	3⅝	15⅞	1 3/16	2¾	¼	3½	.43
	55	18⅛	7½	⅝	⅜	3/16	3⅝	15⅞	1⅛	2¾	¼	3½	
	50	18	7½	9/16	⅜	3/16	3⅝	15⅞	1 1/16	2½	¼	3½	
	45	17⅞	7½	½	5/16	3/16	3⅝	15⅞	1	2½	¼	3½	
W18 18 x 6 (CBL 18)	40	17⅞	6	½	5/16	3/16	2⅞	15¾	1 1/16	2½	¼	3½	.43
	35	17¾	6	7/16	5/16	⅛	2⅞	15¾	1	2½	3/16	3½	

Wide flange shapes (W), continued 2.3.3.1

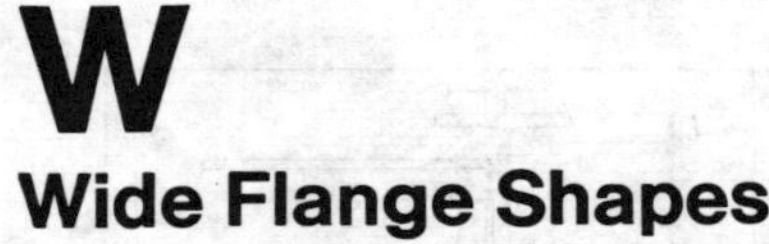

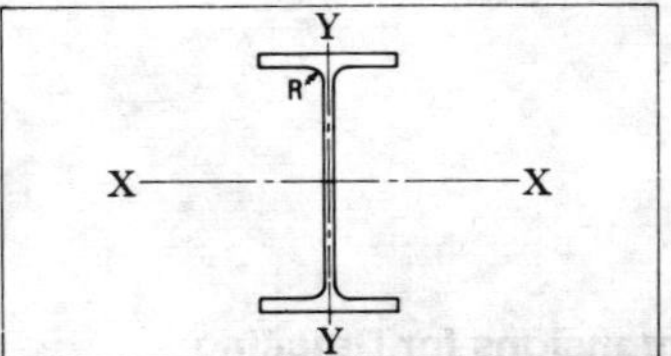

Properties for Designing

Designation and Nominal Size	Weight per Foot	Area of Section	Depth of Section	Flange		Web Thickness	Axis X-X			Axis Y-Y		
				Width	Thickness		I	S	r	I	S	r
In.	Lbs.	In.²	In.	In.	In.	In.	In.⁴	In.³	In.	In.⁴	In.³	In.
W16 16 x 11½ (CB 163)	96	28.2	16.32	11.533	.875	.535	1360	166	6.93	224	38.8	2.82
	88	25.9	16.16	11.502	.795	.504	1220	151	6.87	202	35.1	2.79
W16 16 x 8½ (CB 162)	78	23.0	16.32	8.586	.875	.529	1050	128	6.75	92.5	21.6	2.01
	71	20.9	16.16	8.543	.795	.486	941	116	6.71	82.8	19.4	1.99
	64	18.8	16.00	8.500	.715	.443	836	104	6.66	73.3	17.3	1.97
	58	17.1	15.86	8.464	.645	.407	748	94.4	6.62	65.3	15.4	1.96
W16 16 x 7 (CB 161)	50	14.7	16.25	7.073	.628	.380	657	80.8	6.68	37.1	10.5	1.59
	45	13.3	16.12	7.039	.563	.346	584	72.5	6.64	32.8	9.32	1.57
	40	11.8	16.00	7.000	.503	.307	517	64.6	6.62	28.8	8.23	1.56
	36	10.6	15.85	6.992	.428	.299	447	56.5	6.50	24.4	6.99	1.52
W16 16 x 5½ (CBL 16)	31	9.13	15.84	5.525	.442	.275	374	47.2	6.40	12.5	4.51	1.17
	26	7.67	15.65	5.500	.345	.250	300	38.3	6.25	9.59	3.49	1.12
W14 14 x 16 (CB 146)	730	215	22.44	17.889	4.910	3.069	14400	1280	8.18	4720	527	4.69
	665	196	21.67	17.646	4.522	2.826	12500	1150	7.99	4170	472	4.62
	605	178	20.94	17.418	4.157	2.598	10900	1040	7.81	3680	423	4.55
	550	162	20.26	17.206	3.818	2.386	9450	933	7.64	3260	378	4.49
	500	147	19.63	17.008	3.501	2.188	8250	840	7.49	2880	339	4.43
	455	134	19.05	16.828	3.213	2.008	7220	758	7.35	2560	304	4.37
	426	125	18.69	16.695	3.033	1.875	6610	707	7.26	2360	283	4.34
	398	117	18.31	16.590	2.843	1.770	6010	657	7.17	2170	262	4.31
	370	109	17.94	16.475	2.658	1.655	5450	608	7.08	1990	241	4.27
	342	101	17.56	16.365	2.468	1.545	4910	559	6.99	1810	221	4.24
	314	92.3	17.19	16.235	2.283	1.415	4400	512	6.90	1630	201	4.20
	287	84.4	16.81	16.130	2.093	1.310	3910	465	6.81	1470	182	4.17
	264	77.6	16.50	16.025	1.938	1.205	3530	427	6.74	1330	166	4.14
	246	72.3	16.25	15.945	1.813	1.125	3230	397	6.68	1230	154	4.12
	237	69.7	16.12	15.910	1.748	1.090	3080	382	6.65	1170	148	4.11
	228	67.1	16.00	15.865	1.688	1.045	2940	368	6.62	1120	142	4.10
	219	64.4	15.87	15.825	1.623	1.005	2800	353	6.59	1070	136	4.08
	211	62.1	15.75	15.800	1.563	.980	2670	339	6.56	1030	130	4.07
	202	59.4	15.63	15.750	1.503	.930	2540	325	6.54	980	124	4.06
	193	56.7	15.50	15.710	1.438	.890	2400	310	6.51	930	118	4.05
	184	54.1	15.38	15.660	1.378	.840	2270	296	6.49	883	113	4.04
	176	51.7	15.25	15.640	1.313	.820	2150	282	6.45	838	107	4.02
	167	49.1	15.12	15.600	1.248	.780	2020	267	6.42	790	101	4.01
	158	46.5	15.00	15.550	1.188	.730	1900	253	6.40	745	95.8	4.00
	150	44.1	14.88	15.515	1.128	.695	1790	240	6.37	703	90.6	3.99
	142	41.8	14.75	15.500	1.063	.680	1670	227	6.32	660	85.2	3.97

Wide flange shapes (W), continued 2.3.3.1

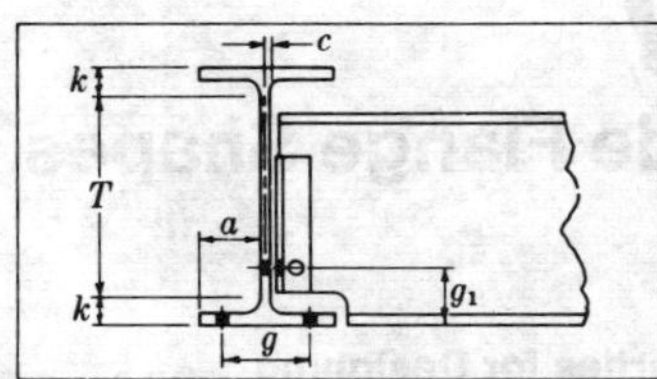

Dimensions for Detailing

Designation and Nominal Size	Weight per Foot	Depth of Section	Flange Width	Flange Thickness	Web Thickness	Web Half Thickness	Distances a	Distances T	Distances k	Distances g_1	Distances c	Usual Gage g	Fillet Radius R
In.	Lbs.	In.	In.	In.	In.	In.	In.	In.	In.	In.	In.	In.	In.
W16 16 x 11½ (CB 163)	96	16⅜	11½	⅞	9/16	¼	5½	13⅛	1⅝	2¾	5/16	5½	.60
	88	16⅛	11½	13/16	½	¼	5½	13⅛	1½	2¾	5/16	5½	
W16 16 x 8½ (CB 162)	78	16⅜	8⅝	⅞	½	¼	4	13⅛	1⅝	2¾	5/16	5½	.60
	71	16⅛	8½	13/16	½	¼	4	13⅛	1½	2¾	5/16	5½	
	64	16	8½	11/16	7/16	¼	4	13⅛	1 7/16	2¾	5/16	5½	
	58	15⅞	8½	⅝	7/16	3/16	4	13⅛	1⅜	2¾	¼	5½	
W16 16 x 7 (CB 161)	50	16¼	7⅛	⅝	⅜	3/16	3⅜	13¾	1¼	2¾	¼	3½	.43
	45	16⅛	7	9/16	⅜	3/16	3⅜	13¾	1 3/16	2½	¼	3½	
	40	16	7	½	5/16	⅛	3⅜	13¾	1⅛	2½	3/16	3½	
	36	15⅞	7	7/16	5/16	⅛	3⅜	13¾	1 1/16	2½	3/16	3½	
W16 16 x 5½ (CBL 16)	31	15⅞	5½	7/16	¼	⅛	2⅝	13¾	1 1/16	2½	3/16	2¾	.43
	26	15⅝	5½	⅜	¼	⅛	2⅝	13¾	15/16	2¼	3/16	2¾	
W14 14 x 16 (CB 146)	730	22½	17⅞	4 15/16	3 1/16	1 9/16	7⅜	11¼	5⅝	7	1⅝	3-7½-3	.60
	665	21⅝	17⅝	4½	2 13/16	1 7/16	7⅜	11¼	5 3/16	6½	1½	3-7½-3	
	605	21	17⅜	4 3/16	2⅝	1 5/16	7⅜	11¼	4⅞	6¼	1⅜	3-7½-3	
	550	20¼	17¼	3 13/16	2⅜	1 3/16	7⅜	11¼	4½	5¾	1¼	3-7½-3	
	500	19⅝	17	3½	2 3/16	1⅛	7⅜	11¼	4 3/16	5½	1 3/16	3-7½-3	
	455	19	16⅞	3 3/16	2	1	7⅜	11¼	3⅞	5¼	1 1/16	3-7½-3	
	426	18¾	16¾	3 1/16	1⅞	15/16	7⅜	11¼	3¾	5	1	3-5½-3	
	398	18¼	16⅝	2 13/16	1¾	⅞	7⅜	11¼	3½	4¾	15/16	3-5½-3	
	370	18	16½	2 11/16	1⅝	13/16	7⅜	11¼	3⅜	4¾	⅞	3-5½-3	
	342	17½	16⅜	2 7/16	1 9/16	¾	7⅜	11¼	3⅛	4½	13/16	3-5½-3	
	314	17¼	16¼	2 5/16	1 7/16	11/16	7⅜	11¼	3	4¼	¾	3-5½-3	
	287	16¾	16⅛	2 1/16	1 5/16	⅝	7⅜	11¼	2¾	4	11/16	3-5½-3	
	264	16½	16	1 15/16	1 3/16	⅝	7⅜	11¼	2⅝	4	11/16	3-5½-3	
	246	16¼	16	1 13/16	1⅛	9/16	7⅜	11¼	2½	3¾	⅝	3-5½-3	
	237	16⅛	15⅞	1¾	1 1/16	9/16	7⅜	11¼	2 7/16	3¾	⅝	3-5½-3	
	228	16	15⅞	1 11/16	1 1/16	½	7⅜	11¼	2⅜	3¾	9/16	3-5½-3	
	219	15⅞	15⅞	1⅝	1	½	7⅜	11¼	2 5/16	3½	9/16	3-5½-3	
	211	15¾	15¾	1 9/16	1	½	7⅜	11¼	2¼	3½	9/16	3-5½-3	
	202	15⅝	15¾	1½	15/16	7/16	7⅜	11¼	2 3/16	3½	½	3-5½-3	
	193	15½	15¾	1 7/16	⅞	7/16	7⅜	11¼	2⅛	3½	½	3-5½-3	
	184	15⅜	15⅝	1⅜	13/16	7/16	7⅜	11¼	2 1/16	3½	½	3-5½-3	
	176	15¼	15⅝	1 5/16	13/16	7/16	7⅜	11¼	2	3¼	½	3-5½-3	
	167	15⅛	15⅝	1¼	¾	⅜	7⅜	11¼	1 15/16	3¼	7/16	3-5½-3	
	158	15	15½	1 3/16	¾	⅜	7⅜	11¼	1⅞	3¼	7/16	3-5½-3	
	150	14⅞	15½	1⅛	11/16	⅜	7⅜	11¼	1 13/16	3¼	7/16	3-5½-3	
	142	14¾	15½	1 1/16	11/16	5/16	7⅜	11¼	1¾	3	⅜	3-5½-3	

Wide flange shapes (W), continued 2.3.3.1

W

Wide Flange Shapes

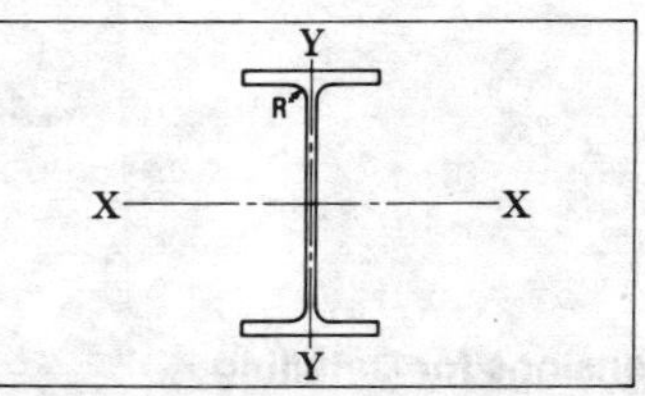

Properties for Designing

Designation and Nominal Size	Weight per Foot	Area of Section	Depth of Section	Flange Width	Flange Thickness	Web Thickness	Axis X-X I	Axis X-X S	Axis X-X r	Axis Y-Y I	Axis Y-Y S	Axis Y-Y r
In.	Lbs.	In.²	In.	In.	In.	In.	In.⁴	In.³	In.	In.⁴	In.³	In.
W14 14 x 16 (CB 146)	320*	94.1	16.81	16.710	2.093	1.890	4140	493	6.63	1640	196	4.17
W14 14 x 14½ (CB 145)	136	40.0	14.75	14.740	1.063	.660	1590	216	6.31	568	77.0	3.77
	127	37.3	14.62	14.690	.998	.610	1480	202	6.29	528	71.8	3.76
	119	35.0	14.50	14.650	.938	.570	1370	189	6.26	492	67.1	3.75
	111	32.7	14.37	14.620	.873	.540	1270	176	6.23	455	62.2	3.73
	103	30.3	14.25	14.575	.813	.495	1170	164	6.21	420	57.6	3.72
	95	27.9	14.12	14.545	.748	.465	1060	151	6.17	384	52.8	3.71
	87	25.6	14.00	14.500	.688	.420	967	138	6.15	350	48.2	3.70
W14 14 x 12 (CB 144)	84	24.7	14.18	12.023	.778	.451	928	131	6.13	225	37.5	3.02
	78	22.9	14.06	12.000	.718	.428	851	121	6.09	207	34.5	3.00
W14 14 x 10 (CB 143)	74	21.8	14.19	10.072	.783	.450	797	112	6.05	133	26.5	2.48
	68	20.0∗	14.06	10.040	.718	.418	724	103	6.02	121	24.1	2.46
	61	17.9	13.91	10.000	.643	.378	641	92.2	5.98	107	21.5	2.45
W14 14 x 8 (CB 142)	53	15.6	13.94	8.062	.658	.370	542	77.8	5.90	57.5	14.3	1.92
	48	14.1	13.81	8.031	.593	.339	485	70.2	5.86	51.3	12.8	1.91
	43	12.6	13.68	8.000	.528	.308	429	62.7	5.82	45.1	11.3	1.89
W14 14 x 6¾ (CB 141)	38	11.2	14.12	6.776	.513	.313	386	54.7	5.88	26.6	7.86	1.54
	34	10.0	14.00	6.750	.453	.287	340	48.6	5.83	23.3	6.89	1.52
	30	8.83	13.86	6.733	.383	.270	290	41.9	5.74	19.5	5.80	1.49
W14 14 x 5 (CBL 14)	26	7.67	13.89	5.025	.418	.255	244	35.1	5.64	8.86	3.53	1.08
	22	6.49	13.72	5.000	.335	.230	198	28.9	5.53	7.00	2.80	1.04
W12 12 x 12 (CB 124)	190	55.9	14.38	12.670	1.736	1.060	1890	263	5.82	590	93.1	3.25
	161	47.4	13.88	12.515	1.486	.905	1540	222	5.70	486	77.7	3.20
	133	39.1	13.38	12.365	1.236	.755	1220	183	5.59	390	63.1	3.16
	120	35.3	13.12	12.320	1.106	.710	1070	163	5.51	345	56.0	3.13
	106	31.2	12.88	12.230	.986	.620	931	145	5.46	301	49.2	3.11
	99	29.1	12.75	12.192	.921	.582	859	135	5.43	278	45.7	3.09
	92	27.1	12.62	12.155	.856	.545	789	125	5.40	256	42.2	3.08
	85	25.0	12.50	12.105	.796	.495	723	116	5.38	235	38.9	3.07
	79	23.2	12.38	12.080	.736	.470	663	107	5.34	216	35.8	3.05
	72	21.2	12.25	12.040	.671	.430	597	97.5	5.31	195	32.4	3.04
	65	19.1	12.12	12.000	.606	.390	533	88.0	5.28	175	29.1	3.02
W12 12 x 10 (CB 123)	58	17.1	12.19	10.014	.641	.359	476	78.1	5.28	107	21.4	2.51
	53	15.6	12.06	10.000	.576	.345	426	70.7	5.23	96.1	19.2	2.48

*Column core section.

Wide flange shapes (W), continued

2.3.3.1

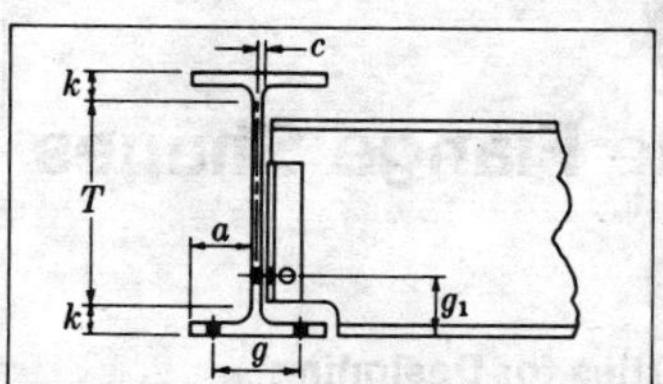

Dimensions for Detailing

Designation and Nominal Size	Weight per Foot	Depth of Section	Flange		Web		Distances					Usual Gage g	Fillet Radius R
			Width	Thickness	Thickness	Half Thickness	a	T	k	g_1	c		
In.	Lbs.	In.	In.	In.	In.	In.	In.	In.	In.	In.	In.	In.	In.
W14 14 x 16 (CB 146)	320*	16¾	16¾	2 1/16	1⅞	15/16	7⅜	11¼	2¾	4	1	3-5½-3	.60
W14 14 x 14½ (CB 145)	136	14¾	14¾	1 1/16	11/16	5/16	7	11¼	1¾	3	⅜	5½	.60
	127	14⅝	14¾	1	⅝	5/16	7	11¼	1 11/16	3	⅜	5½	
	119	14½	14⅝	15/16	9/16	5/16	7	11¼	1⅝	3	⅜	5½	
	111	14⅜	14⅝	⅞	9/16	¼	7	11¼	1 9/16	2¾	5/16	5½	
	103	14¼	14⅝	13/16	½	¼	7	11¼	1½	2¾	5/16	5½	
	95	14⅛	14½	¾	7/16	¼	7	11¼	1 7/16	2¾	5/16	5½	
	87	14	14½	11/16	7/16	3/16	7	11¼	1⅜	2¾	¼	5½	
W14 14 x 12 (CB 144)	84	14⅛	12	¾	7/16	¼	5¾	11¼	1 7/16	2¾	5/16	5½	.60
	78	14	12	11/16	7/16	3/16	5¾	11¼	1⅜	2¾	¼	5½	
W14 14 x 10 (CB 143)	74	14¼	10⅛	13/16	7/16	¼	4¾	11¼	1½	2¾	5/16	5½	.60
	68	14	10	11/16	7/16	3/16	4¾	11¼	1⅜	2¾	¼	5½	
	61	13⅞	10	⅝	⅜	3/16	4¾	11¼	1 5/16	2¾	¼	5½	
W14 14 x 8 (CB 142)	53	14	8	11/16	⅜	3/16	3⅞	11¼	1⅜	2¾	¼	5½	.60
	48	13¾	8	9/16	5/16	3/16	3⅞	11¼	1¼	2½	¼	5½	
	43	13⅝	8	½	5/16	⅛	3⅞	11¼	1 3/16	2½	3/16	5½	
W14 14 x 6¾ (CB 141)	38	14⅛	6¾	½	5/16	3/16	3¼	11⅞	1⅛	2½	¼	3½	.43
	34	14	6¾	7/16	5/16	⅛	3¼	11⅞	1 1/16	2½	3/16	3½	
	30	13⅞	6¾	⅜	¼	⅛	3¼	11⅞	1	2½	3/16	3½	
W14 14 x 5 (CBL 14)	26	13⅞	5	7/16	¼	⅛	2⅜	11⅞	1	2½	3/16	2¾	.43
	22	13¾	5	5/16	¼	⅛	2⅜	11⅞	15/16	2¼	3/16	2¾	
W12 12 x 12 (CB 124)	190	14⅜	12⅝	1¾	1 1/16	½	5¾	9½	2 7/16	3¾	9/16	5½	.60
	161	13⅞	12½	1½	⅞	7/16	5¾	9½	2 3/16	3½	½	5½	
	133	13⅜	12⅜	1¼	¾	⅜	5¾	9½	1 15/16	3¼	7/16	5½	
	120	13⅛	12⅜	1⅛	11/16	⅜	5¾	9½	1 13/16	3	7/16	5½	
	106	12⅞	12¼	1	⅝	5/16	5¾	9½	1 11/16	3	⅜	5½	
	99	12¾	12¼	15/16	9/16	5/16	5¾	9½	1⅝	3	⅜	5½	
	92	12⅝	12⅛	⅞	9/16	¼	5¾	9½	1 9/16	2¾	5/16	5½	
	85	12½	12⅛	13/16	½	¼	5¾	9½	1½	2¾	5/16	5½	
	79	12⅜	12⅛	¾	½	¼	5¾	9½	1 7/16	2¾	5/16	5½	
	72	12¼	12	11/16	7/16	3/16	5¾	9½	1⅜	2¾	¼	5½	
	65	12⅛	12	⅝	⅜	3/16	5¾	9½	1 5/16	2½	¼	5½	
W12 12 x 10 (CB 123)	58	12¼	10	⅝	⅜	3/16	4⅞	9½	1⅜	2¾	¼	5½	.60
	53	12	10	9/16	⅜	3/16	4⅞	9½	1¼	2½	¼	5½	

*Column core section.

Wide flange shapes (W), continued 2.3.3.1

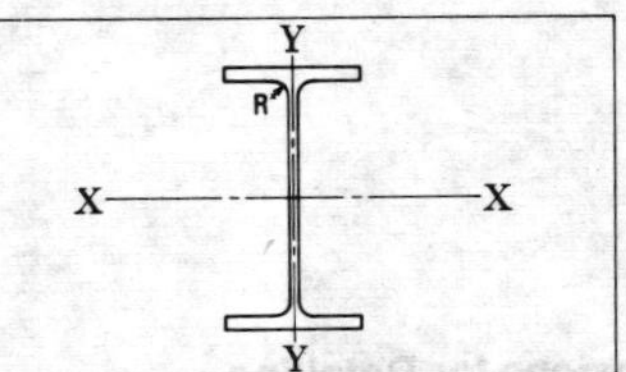

Properties for Designing

Designation and Nominal Size	Weight per Foot	Area of Section	Depth of Section	Flange		Web Thickness	Axis X-X			Axis Y-Y		
				Width	Thickness		I	S	r	I	S	r
In.	Lbs.	In.²	In.	In.	In.	In.	In.⁴	In.³	In.	In.⁴	In.³	In.
W12	50	14.7	12.19	8.077	.641	.371	395	64.7	5.18	56.4	14.0	1.96
12 x 8	45	13.2	12.06	8.042	.576	.336	351	58.2	5.15	50.0	12.4	1.94
(CB 122)	40	11.8	11.94	8.000	.516	.294	310	51.9	5.13	44.1	11.0	1.94
W12	36	10.6	12.24	6.565	.540	.305	281	46.0	5.15	25.5	7.77	1.55
12 x 6½	31	9.13	12.09	6.525	.465	.265	239	39.5	5.12	21.6	6.61	1.54
(CB 121)	27	7.95	11.96	6.497	.400	.237	204	34.2	5.07	18.3	5.63	1.52
W12	22	6.47	12.31	4.030	.424	.260	156	25.3	4.91	4.64	2.31	.847
12 x 4	19	5.59	12.16	4.007	.349	.237	130	21.3	4.82	3.76	1.88	.820
(CBL 12)	16.5	4.87	12.00	4.000	.269	.230	105	17.6	4.65	2.88	1.44	.770
W12 12 x 4 (CBJ 12)	14	4.12	11.91	3.968	.224	.198	88.0	14.8	4.62	2.34	1.18	.754
W10	112	32.9	11.38	10.415	1.248	.755	719	126	4.67	235	45.2	2.67
10 x 10	100	29.4	11.12	10.345	1.118	.685	625	112	4.61	207	39.9	2.65
(CB 103)	89	26.2	10.88	10.275	.998	.615	542	99.7	4.55	181	35.2	2.63
	77	22.7	10.62	10.195	.868	.535	457	86.1	4.49	153	30.1	2.60
	72	21.2	10.50	10.170	.808	.510	421	80.1	4.46	142	27.9	2.59
	66	19.4	10.38	10.117	.748	.457	382	73.7	4.44	129	25.5	2.58
	60	17.7	10.25	10.075	.683	.415	344	67.1	4.41	116	23.1	2.57
	54	15.9	10.12	10.028	.618	.368	306	60.4	4.39	104	20.7	2.56
	49	14.4	10.00	10.000	.558	.340	273	54.6	4.35	93.0	18.6	2.54
W10	45	13.2	10.12	8.022	.618	.350	249	49.1	4.33	53.2	13.3	2.00
10 x 8	39	11.5	9.94	7.990	.528	.318	210	42.2	4.27	44.9	11.2	1.98
(CB 102)	33	9.71	9.75	7.964	.433	.292	171	35.0	4.20	36.5	9.16	1.94
W10	29	8.54	10.22	5.799	.500	.289	158	30.8	4.30	16.3	5.61	1.38
10 x 5¾	25	7.36	10.08	5.762	.430	.252	133	26.5	4.26	13.7	4.76	1.37
(CB 101)	21	6.20	9.90	5.750	.340	.240	107	21.5	4.15	10.8	3.75	1.32
W10	19	5.61	10.25	4.020	.394	.250	96.3	18.8	4.14	4.28	2.13	.874
10 x 4	17	4.99	10.12	4.010	.329	.240	81.9	16.2	4.05	3.55	1.77	.844
(CBL 10)	15	4.41	10.00	4.000	.269	.230	68.9	13.8	3.95	2.88	1.44	.809
W10 10 x 4 (CBJ 10)	11.5	3.39	9.87	3.950	.204	.180	52.0	10.5	3.92	2.10	1.06	.787

Wide flange shapes (W), continued 2.3.3.1

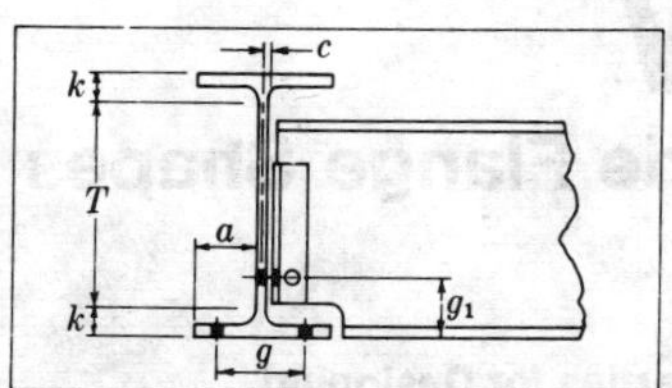

Dimensions for Detailing

Designation and Nominal Size	Weight per Foot	Depth of Section	Flange Width	Flange Thickness	Web Thickness	Web Half Thickness	Distances a	Distances T	Distances k	Distances g_1	Distances c	Usual Gage g	Fillet Radius R
In.	Lbs.	In.	In.	In.	In.	In.	In.	In.	In.	In.	In.	In.	In.
W12 12 x 8 (CB 122)	50	12¼	8⅛	⅝	⅜	3/16	3⅞	9½	1⅜	2¾	¼	5½	.60
	45	12	8	9/16	5/16	3/16	3⅞	9½	1¼	2½	¼	5½	
	40	12	8	½	5/16	⅛	3⅞	9½	1¼	2½	3/16	5½	
W12 12 x 6½ (CB 121)	36	12¼	6⅝	9/16	5/16	⅛	3⅛	10⅛	1 1/16	2½	3/16	3½	.37
	31	12⅛	6½	7/16	¼	⅛	3⅛	10⅛	1	2½	3/16	3½	
	27	12	6½	⅜	¼	⅛	3⅛	10⅛	15/16	2½	3/16	3½	
W12 12 x 4 (CBL 12)	22	12¼	4	7/16	¼	⅛	1⅞	10⅜	15/16	2½	3/16	2¼	.30
	19	12⅛	4	⅜	¼	⅛	1⅞	10⅜	⅞	2¼	3/16	2¼	
	16.5	12	4	¼	¼	⅛	1⅞	10⅜	13/16	2¼	3/16	2¼	
W12 12 x 4 (CBJ 12)	14	11⅞	4	¼	3/16	⅛	1⅞	10⅜	¾	2¼	3/16	2¼	.30
W10 10 x 10 (CB 103)	112	11⅜	10⅜	1¼	¾	⅜	4⅞	7¾	1 13/16	3	7/16	5½	.50
	100	11⅛	10⅜	1⅛	11/16	5/16	4⅞	7¾	1 11/16	3	⅜	5½	
	89	10⅞	10¼	1	⅝	5/16	4⅞	7¾	1 9/16	2¾	⅜	5½	
	77	10⅝	10¼	⅞	9/16	¼	4⅞	7¾	1 7/16	2¾	5/16	5½	
	72	10½	10⅛	13/16	½	¼	4⅞	7¾	1⅜	2¾	5/16	5½	
	66	10⅜	10⅛	¾	7/16	¼	4⅞	7¾	1 5/16	2½	5/16	5½	
	60	10¼	10⅛	11/16	7/16	3/16	4⅞	7¾	1¼	2½	¼	5½	
	54	10⅛	10	⅝	⅜	3/16	4⅞	7¾	1 3/16	2½	¼	5½	
	49	10	10	9/16	5/16	3/16	4⅞	7¾	1⅛	2½	¼	5½	
W10 10 x 8 (CB 102)	45	10⅛	8	⅝	⅜	3/16	3⅞	7¾	1 3/16	2½	¼	5½	.50
	39	10	8	½	5/16	3/16	3⅞	7¾	1⅛	2½	¼	5½	
	33	9¾	8	7/16	5/16	⅛	3⅞	7¾	1	2¼	3/16	5½	
W10 10 x 5¾ (CB 101)	29	10¼	5¾	½	5/16	⅛	2¾	8⅛	1 1/16	2¼	3/16	2¾	.32
	25	10⅛	5¾	7/16	¼	⅛	2¾	8⅛	1	2¼	3/16	2¾	
	21	9⅞	5¾	5/16	¼	⅛	2¾	8⅛	⅞	2¼	3/16	2¾	
W10 10 x 4 (CBL 10)	19	10¼	4	⅜	¼	⅛	1⅞	8⅜	15/16	2¼	3/16	2¼	.30
	17	10⅛	4	5/16	¼	⅛	1⅞	8⅜	⅞	2¼	3/16	2¼	
	15	10	4	¼	¼	⅛	1⅞	8⅜	13/16	2¼	3/16	2¼	
W10 10 x 4 (CBJ 10)	11.5	9⅞	4	3/16	3/16	1/16	1⅞	8⅜	¾	2	⅛	2¼	.30

Wide flange shapes (W), continued 2.3.3.1

W

Wide Flange Shapes

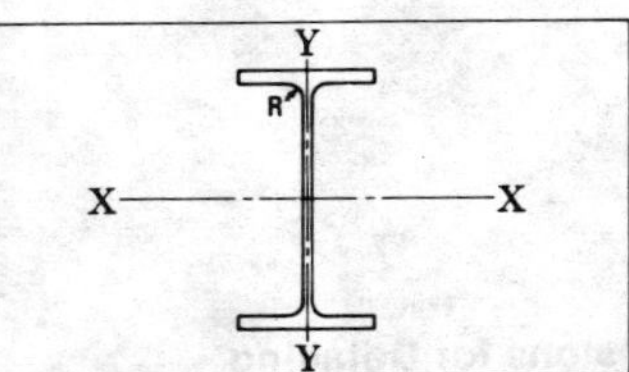

Properties for Designing

Designation and Nominal Size	Weight per Foot	Area of Section	Depth of Section	Flange		Web Thickness	Axis X-X			Axis Y-Y		
				Width	Thickness		I	S	r	I	S	r
In.	Lbs.	In.^{2}	In.	In.	In.	In.	In.^{4}	In.^{3}	In.	In.^{4}	In.^{3}	In.
W8 8 x 8 (CB 83)	67	19.7	9.00	8.287	.933	.575	272	60.4	3.71	88.6	21.4	2.12
	58	17.1	8.75	8.222	.808	.510	227	52.0	3.65	74.9	18.2	2.10
	48	14.1	8.50	8.117	.683	.405	184	43.2	3.61	60.9	15.0	2.08
	40	11.8	8.25	8.077	.558	.365	146	35.5	3.53	49.0	12.1	2.04
	35	10.3	8.12	8.027	.493	.315	126	31.1	3.50	42.5	10.6	2.03
	31	9.12	8.00	8.000	.433	.288	110	27.4	3.47	37.0	9.24	2.01
W8 8 x 6½ (CB 82)	28	8.23	8.06	6.540	.463	.285	97.8	24.3	3.45	21.6	6.61	1.62
	24	7.06	7.93	6.500	.398	.245	82.5	20.8	3.42	18.2	5.61	1.61
W8 8 x 5¼ (CB 81)	20	5.89	8.14	5.268	.378	.248	69.4	17.0	3.43	9.22	3.50	1.25
	17	5.01	8.00	5.250	.308	.230	56.6	14.1	3.36	7.44	2.83	1.22
W8* 8 x 4 (CBL 8)	15	4.43	8.12	4.015	.314	.245	48.1	11.8	3.29	3.40	1.69	.876
	13	3.83	8.00	4.000	.254	.230	39.6	9.90	3.21	2.72	1.36	.842
W8 8 x 4 (CBJ 8)	10	2.96	7.90	3.940	.204	.170	30.8	7.80	3.23	2.08	1.06	.839
W6 6 x 6 (CBS 6)	25	7.35	6.37	6.080	.456	.320	53.3	16.7	2.69	17.1	5.62	1.53
	20	5.88	6.20	6.018	.367	.258	41.5	13.4	2.66	13.3	4.43	1.51
	15.5	4.56	6.00	5.995	.269	.235	30.1	10.0	2.57	9.67	3.23	1.46
W6* 6 x 4 (CBL 6)	16	4.72	6.25	4.030	.404	.260	31.7	10.2	2.59	4.42	2.19	.967
	12	3.54	6.00	4.000	.279	.230	21.7	7.25	2.48	2.98	1.49	.918
W6 6 x 4 (CBJ 6)	8.5	2.51	5.83	3.940	.194	.170	14.8	5.08	2.43	1.98	1.01	.889
W5 5 x 5 (CB 51)	18.5	5.43	5.12	5.025	.420	.265	25.4	9.94	2.16	8.89	3.54	1.28
	16	4.70	5.00	5.000	.360	.240	21.3	8.53	2.13	7.51	3.00	1.26

*These shapes are rolled in the Pittsburgh District with a 3° flange slope. The flange thicknesses shown are the average thicknesses.

Wide flange shapes (W), continued

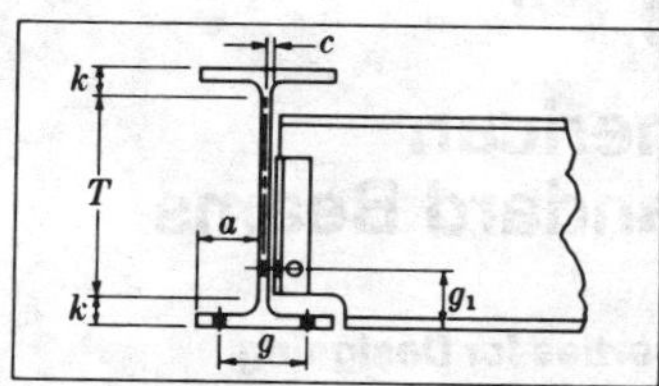

Dimensions for Detailing

Designation and Nominal Size	Weight per Foot	Depth of Section	Flange Width	Flange Thickness	Web Thickness	Web Half Thickness	Distances a	Distances T	Distances k	Distances g_1	Distances c	Usual Gage g	Fillet Radius R
In.	Lbs.	In.	In.	In.	In.	In.	In.	In.	In.	In.	In.	In.	In.
W8 8 x 8 (CB 83)	67	9	8¼	15/16	9/16	5/16	3⅞	6⅛	1 7/16	2¾	⅜	5½	.40
	58	8¾	8¼	13/16	½	¼	3⅞	6⅛	1 5/16	2¾	5/16	5½	
	48	8½	8⅛	11/16	⅜	3/16	3⅞	6⅛	1 3/16	2½	¼	5½	
	40	8¼	8⅛	9/16	⅜	3/16	3⅞	6⅛	1 1/16	2½	¼	5½	
	35	8⅛	8	½	5/16	3/16	3⅞	6⅛	1	2¼	¼	5½	
	31	8	8	7/16	5/16	⅛	3⅞	6⅛	15/16	2¼	3/16	5½	
W8 8 x 6½ (CB 82)	28	8	6½	7/16	5/16	⅛	3⅛	6⅛	15/16	2¼	3/16	3½	.40
	24	7⅞	6½	⅜	¼	⅛	3⅛	6⅛	⅞	2¼	3/16	3½	
W8 8 x 5¼ (CB 81)	20	8⅛	5¼	⅜	¼	⅛	2½	6⅜	⅞	2¼	3/16	2¾	.32
	17	8	5¼	5/16	¼	⅛	2½	6⅜	13/16	2¼	3/16	2¾	
W8* 8 x 4 (CBL 8)	15	8⅛	4	5/16	¼	⅛	1⅞	6½	13/16	2¼	3/16	2¼	.30
	13	8	4	¼	¼	⅛	1⅞	6½	¾	2¼	3/16	2¼	
W8 8 x 4 (CBJ 8)	10	7⅞	4	3/16	3/16	1/16	1⅞	6½	11/16	2	⅛	2¼	.30
W6 6 x 6 (CBS 6)	25	6⅜	6⅛	7/16	5/16	3/16	2⅞	4½	15/16	2¼	¼	3½	.25
	20	6¼	6	⅜	¼	⅛	2⅞	4½	⅞	2¼	3/16	3½	
	15.5	6	6	¼	¼	⅛	2⅞	4½	¾	2¼	3/16	3½	
W6* 6 x 4 (CBL 6)	16	6¼	4	⅜	¼	⅛	1⅞	4½	⅞	2¼	3/16	2¼	.25
	12	6	4	¼	¼	⅛	1⅞	4½	¾	2¼	3/16	2¼	
W6 6 x 4 (CBJ 6)	8.5	5⅞	4	3/16	3/16	1/16	1⅞	4½	11/16	2	⅛	2¼	.25
W5 5 x 5 (CB 51)	18.5	5⅛	5	7/16	¼	⅛	2⅜	3½	13/16	2¼	3/16	2¾	.30
	16	5	5	⅜	¼	⅛	2⅜	3½	¾	2¼	3/16	2¾	

*These shapes are rolled in the Pittsburgh District with a 3° flange slope. The flange thicknesses shown are the average thicknesses.

American Standard beams (S) 2.3.3.2

S

American Standard Beams

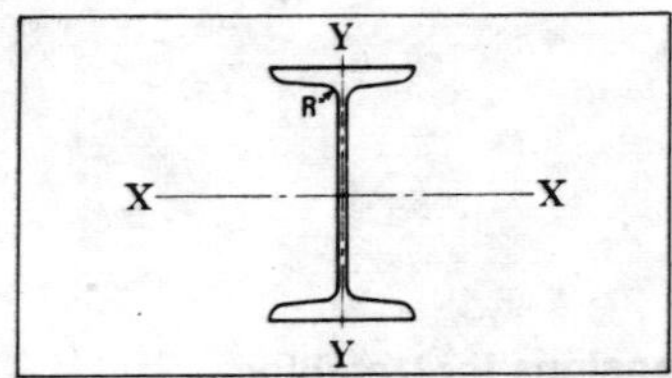

Properties for Designing

Designation and Nominal Size	Weight per Foot	Area of Section	Depth of Beam	Width of Flange	Aver. Flange Thick-ness	Web Thick-ness	Axis X-X			Axis Y-Y		
							I	*S*	*r*	*I*	*S*	*r*
In.	Lbs.	In.^{2}	In.	In.	In.	In.	In.^{4}	In.^{3}	In.	In.^{4}	In.^{3}	In.
S24 24 x 7⅞	120 105.9	35.3 31.1	24.00 24.00	8.048 7.875	1.102 1.102	.798 .625	3030 2830	252 236	9.26 9.53	84.2 78.2	20.9 19.8	1.54 1.58
S24 24 x 7	100 90 79.9	29.4 26.5 23.5	24.00 24.00 24.00	7.247 7.124 7.001	.871 .871 .871	.747 .624 .501	2390 2250 2110	199 187 175	9.01 9.22 9.47	47.8 44.9 42.3	13.2 12.6 12.1	1.27 1.30 1.34
S20 20 x 7	95 85	27.9 25.0	20.00 20.00	7.200 7.053	.916 .916	.800 .653	1610 1520	161 152	7.60 7.79	49.7 46.2	13.8 13.1	1.33 1.36
S20 20 x 6¼	75 65.4	22.1 19.2	20.00 20.00	6.391 6.250	.789 .789	.641 .500	1280 1180	128 118	7.60 7.84	29.6 27.4	9.28 8.77	1.16 1.19
S18 18 x 6	70 54.7	20.6 16.1	18.00 18.00	6.251 6.001	.691 .691	.711 .461	926 804	103 89.4	6.71 7.07	24.1 20.8	7.72 6.94	1.08 1.14
S15 15 x 5½	50 42.9	14.7 12.6	15.00 15.00	5.640 5.501	.622 .622	.550 .411	486 447	64.8 59.6	5.75 5.95	15.7 14.4	5.57 5.23	1.03 1.07
S12 12 x 5¼	50 40.8	14.7 12.0	12.00 12.00	5.477 5.252	.659 .659	.687 .462	305 272	50.8 45.4	4.55 4.77	15.7 13.6	5.74 5.16	1.03 1.06
S12 12 x 5	35 31.8	10.3 9.35	12.00 12.00	5.078 5.000	.544 .544	.428 .350	229 218	38.2 36.4	4.72 4.83	9.87 9.36	3.89 3.74	.980 1.00
S10 10 x 4⅝	35 25.4	10.3 7.46	10.00 10.00	4.944 4.661	.491 .491	.594 .311	147 124	29.4 24.7	3.78 4.07	8.36 6.79	3.38 2.91	.901 .954
S8 8 x 4	23 18.4	6.77 5.41	8.00 8.00	4.171 4.001	.425 .425	.441 .271	64.9 57.6	16.2 14.4	3.10 3.26	4.31 3.73	2.07 1.86	.798 .831
S7 7 x 3⅝	20 15.3	5.88 4.50	7.00 7.00	3.860 3.662	.392 .392	.450 .252	42.4 36.7	12.1 10.5	2.69 2.86	3.17 2.64	1.64 1.44	.734 .766
S6 6 x 3⅜	17.25 12.5	5.07 3.67	6.00 6.00	3.565 3.332	.359 .359	.465 .232	26.3 22.1	8.77 7.37	2.28 2.45	2.31 1.82	1.30 1.09	.675 .705

American Standard beams (S), continued 2.3.3.2

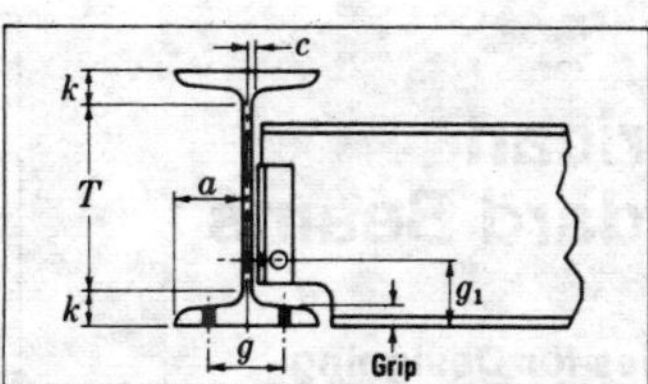

Dimensions for Detailing

Designation and Nominal Size	Weight per Foot	Flange: Width	Flange: Aver. Thickness	Web: Thickness	Web: Half Thickness	Distances: a	Distances: T	Distances: k	Distances: g_1	Distances: c	Usual Gage g	Grip	Max. Flange Fastener	Fillet Radius R
In.	Lbs.	In.	In.	In.	In.	In.	In.	In.	In.	In.	In.	In.	In.	In.
S24 24 x 7 7/8	120	8	1 1/8	13/16	3/8	3 5/8	20	2	3 1/4	7/16	4	1 1/8	1	.60
	105.9	7 7/8	1 1/8	5/8	5/16	3 5/8	20	2	3 1/4	3/8	4	1 1/8	1	
S24 24 x 7	100	7 1/4	7/8	3/4	3/8	3 1/4	20 1/2	1 3/4	3	7/16	4	7/8	1	.60
	90	7 1/8	7/8	5/8	5/16	3 1/4	20 1/2	1 3/4	3	3/8	4	7/8	1	
	79.9	7	7/8	1/2	1/4	3 1/4	20 1/2	1 3/4	3	5/16	4	7/8	1	
S20 20 x 7	95	7 1/4	15/16	13/16	3/8	3 1/4	16 1/4	1 7/8	3	7/16	4	15/16	1	.70
	85	7	15/16	5/8	5/16	3 1/4	16 1/4	1 7/8	3	3/8	4	7/8	1	
S20 20 x 6 1/4	75	6 3/8	13/16	5/8	5/16	2 7/8	16 3/4	1 5/8	3	3/8	3 1/2	13/16	7/8	.60
	65.4	6 1/4	13/16	1/2	1/4	2 7/8	16 3/4	1 5/8	3	5/16	3 1/2	3/4	7/8	
S18 18 x 6	70	6 1/4	11/16	11/16	3/8	2 3/4	15	1 1/2	2 3/4	7/16	3 1/2	11/16	7/8	.56
	54.7	6	11/16	7/16	1/4	2 3/4	15	1 1/2	2 3/4	5/16	3 1/2	11/16	7/8	
S15 15 x 5 1/2	50	5 5/8	5/8	9/16	1/4	2 1/2	12 1/4	1 3/8	2 3/4	5/16	3 1/2	9/16	3/4	.51
	42.9	5 1/2	5/8	7/16	3/16	2 1/2	12 1/4	1 3/8	2 3/4	1/4	3 1/2	9/16	3/4	
S12 12 x 5 1/4	50	5 1/2	11/16	11/16	5/16	2 3/8	9 1/8	1 7/16	2 3/4	3/8	3	11/16	3/4	.56
	40.8	5 1/4	11/16	7/16	1/4	2 3/8	9 1/8	1 7/16	2 3/4	5/16	3	5/8	3/4	
S12 12 x 5	35	5 1/8	9/16	7/16	3/16	2 3/8	9 5/8	1 3/16	2 1/2	1/4	3	1/2	3/4	.45
	31.8	5	9/16	3/8	3/16	2 3/8	9 5/8	1 3/16	2 1/2	1/4	3	1/2	3/4	
S10 10 x 4 5/8	35	5	1/2	5/8	5/16	2 1/8	7 3/4	1 1/8	2 1/2	3/8	2 3/4	1/2	3/4	.41
	25.4	4 5/8	1/2	5/16	1/8	2 1/8	7 3/4	1 1/8	2 1/2	3/16	2 3/4	1/2	3/4	
S8 8 x 4	23	4 1/8	7/16	7/16	1/4	1 7/8	6	1	2 1/2	5/16	2 1/4	7/16	3/4	.37
	18.4	4	7/16	1/4	1/8	1 7/8	6	1	2 1/2	3/16	2 1/4	7/16	3/4	
S7 7 x 3 5/8	20	3 7/8	3/8	7/16	1/4	1 3/4	5 1/4	7/8	2 1/2	5/16	2 1/4	3/8	5/8	.35
	15.3	3 5/8	3/8	1/4	1/8	1 3/4	5 1/4	7/8	2 1/2	3/16	2 1/4	3/8	5/8	
S6 6 x 3 3/8	17.25	3 5/8	3/8	7/16	1/4	1 1/2	4 3/8	13/16	2 1/4	5/16	2	3/8	5/8	.33
	12.5	3 3/8	3/8	1/4	1/8	1 1/2	4 3/8	13/16	2 1/4	3/16	...	5/16	...	

American Standard beams (S), continued 2.3.3.2

S

American Standard Beams

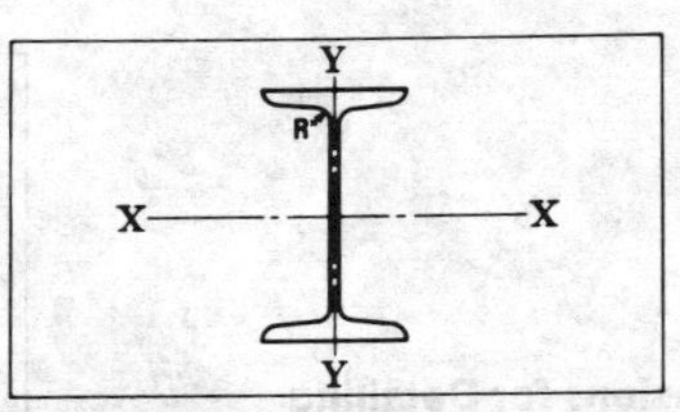

Properties for Designing

Designation and Nominal Size	Weight per Foot	Area of Section	Depth of Beam	Width of Flange	Aver. Flange Thickness	Web Thickness	Axis X-X			Axis Y-Y		
							I	*S*	*r*	*I*	*S*	*r*
In.	Lbs.	In.²	In.	In.	In.	In.	In.⁴	In.³	In.	In.⁴	In.³	In.
S5 5 x 3	14.75	4.34	5.00	3.284	.326	.494	15.2	6.09	1.87	1.67	1.01	.620
	10	2.94	5.00	3.004	.326	.214	12.3	4.92	2.05	1.22	.809	.643
S4 4 x 2⅝	9.5	2.79	4.00	2.796	.293	.326	6.79	3.39	1.56	.903	.646	.569
	7.7	2.26	4.00	2.663	.293	.193	6.08	3.04	1.64	.764	.574	.581
S3 3 x 2⅜	7.5	2.21	3.00	2.509	.260	.349	2.93	1.95	1.15	.586	.468	.516
	5.7	1.67	3.00	2.330	.260	.170	2.52	1.68	1.23	.455	.390	.522

Miscellaneous beams and columns (M) 2.3.3.3

M

Miscellaneous Beam and Column Shapes

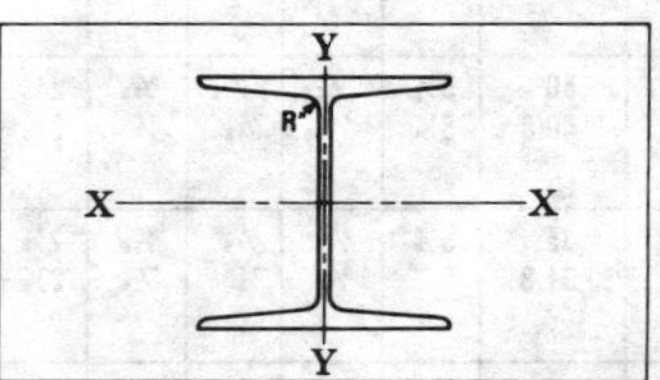

Properties for Designing

Designation and Nominal Size	Weight per Foot	Area of Section	Depth of Beam	Width of Flange	Aver. Flange Thickness	Web Thickness	Axis X-X			Axis Y-Y		
							I	*S*	*r*	*I*	*S*	*r*
In.	Lbs.	In.²	In.	In.	In.	In.	In.⁴	In.³	In.	In.⁴	In.³	In.
M8 8 x 8	40	11.8	8.12	8.088	.521	.463	136	33.5	3.40	41.6	10.3	1.88
	37.6	11.1	8.12	8.002	.521	.377	132	32.6	3.46	40.4	10.1	1.91
	34.3	10.1	8.00	8.003	.459	.378	116	29.1	3.40	34.9	8.73	1.86
	32.6	9.58	8.00	7.940	.459	.315	114	28.4	3.44	34.1	8.58	1.89
M6 6 x 6	25	7.35	6.00	5.942	.480	.317	47.1	15.7	2.53	15.0	5.04	1.43
	20	5.89	6.00	5.938	.379	.250	39.0	13.0	2.57	11.6	3.90	1.40
M5 5 x 5	18.9	5.55	5.00	5.003	.416	.316	24.1	9.63	2.08	7.86	3.14	1.19
M4 4 x 4	16.3	4.80	4.20	3.938	.472	.312	14.0	6.67	1.71	4.44	2.25	.962
	13.0	3.81	4.00	3.940	.371	.254	10.5	5.24	1.66	3.36	1.71	9.39

American Standard beams (S), continued 2.3.3.2

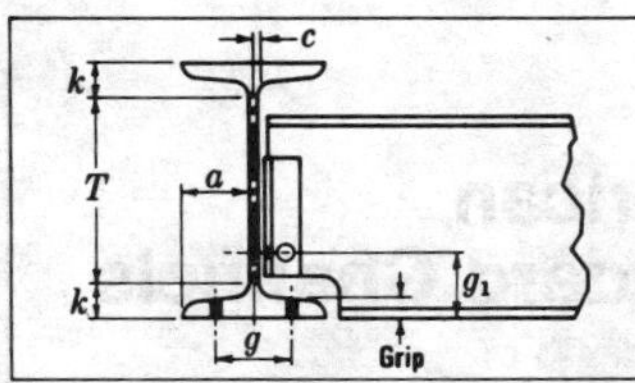

Dimensions for Detailing

Designation and Nominal Size	Weight per Foot	Flange		Web		Distances					Usual Gage g	Grip	Max. Flange Fastener	Fillet Radius R
		Width	Aver. Thickness	Thickness	Half Thickness	a	T	k	g_1	c				
In.	Lbs.	In.	In.	In.	In.	In.	In.	In.	In.	In.	In.	In.	In.	In.
S5 5 x 3	14.75	3¼	5/16	½	¼	1⅜	3½	¾	2¼	5/16	...	5/16	...	.31
	10	3	5/16	3/16	⅛	1⅜	3½	¾	2¼	3/16	...	5/16	...	
S4 4 x 2⅝	9.5	2¾	5/16	5/16	3/16	1¼	2⅝	11/16	2	¼	...	5/16	...	.29
	7.7	2⅝	5/16	3/16	⅛	1¼	2⅝	11/16	2	3/16	...	5/16	...	
S3 3 x 2⅜	7.5	2½	¼	⅜	3/16	1⅛	1¾	⅝	...	¼	...	¼	...	.27
	5.7	2⅜	¼	3/16	1/16	1⅛	1¾	⅝	...	⅛	...	¼	...	

Miscellaneous beams and columns (M), continued 2.3.3.3

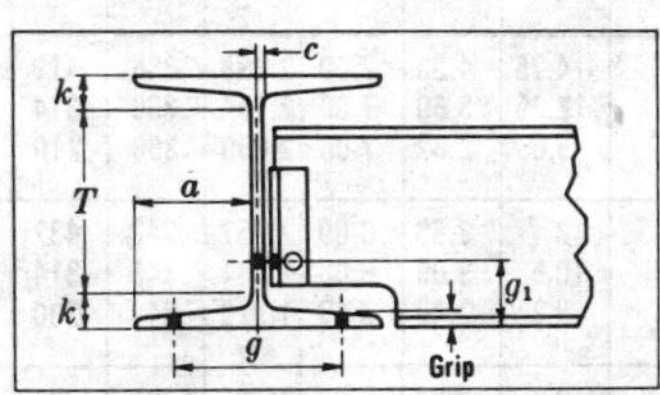

Dimensions for Detailing

Designation and Nominal Size	Weight per Foot	Flange		Web		Distances					Usual Gage g	Grip	Max. Flange Fastener	Fillet Radius R
		Width	Aver. Thickness	Thickness	Half Thickness	a	T	k	g_1	c				
In.	Lbs.	In.	In.	In.	In.	In.	In.	In.	In.	In.	In.	In.	In.	In.
M8 8 x 8	40	8⅛	½	7/16	¼	3⅞	6⅛	1	2½	5/16	5½	½	⅞	.31
	37.7	8	½	⅜	3/16	3¾	6⅛	1	2½	¼	5½	½	⅞	
	34.3	8	7/16	⅜	3/16	3¾	5⅞	1 1/16	2½	¼	5½	7/16	⅞	
	32.6	8	7/16	5/16	3/16	3¾	5⅞	1 1/16	2½	¼	5½	7/16	⅞	
M6 6 x 6	25	6	½	5/16	3/16	2⅞	4⅛	15/16	2½	¼	3½	½	⅞	.31
	20	6	⅜	¼	⅛	2⅞	4⅜	13/16	2¼	3/16	3½	⅜	⅞	
M5 5 x 5	18.9	5	7/16	5/16	3/16	2⅜	3¼	⅞	2½	¼	2¾	7/16	⅞	.31
M4 4 x 4	16.3	4	½	5/16	⅛	1⅞	2⅜	15/16	2	3/16	2¼	½	¾	.31
	13.0	4.1	⅜	¼	⅛	1⅞	2⅜	13/16	2	3/16	2¼	⅜	¾	

American Standard channels (C) 2.3.3.4

C

American Standard Channels

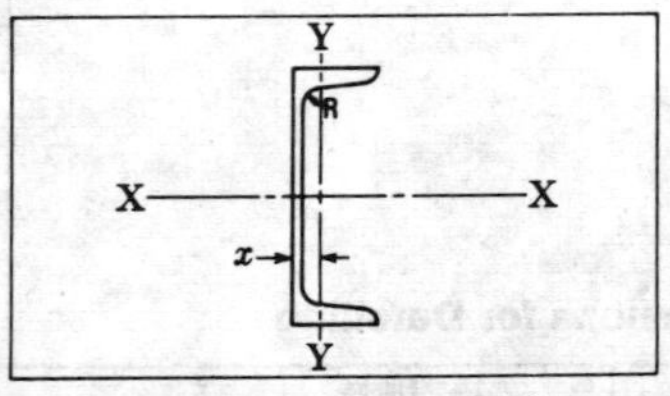

Properties for Designing

Designation and Nominal Size	Weight per Foot	Area of Section	Depth of Channel	Width of Flange	Aver. Flange Thickness	Web Thickness	Axis X-X			Axis Y-Y			
							I	S	r	I	S	r	x
In.	Lbs.	In.^{2}	In.	In.	In.	In.	In.^{4}	In.^{3}	In.	In.^{4}	In.^{3}	In.	In.
C15	50	14.7	15.00	3.716	.650	.716	404.0	53.8	5.24	11.0	3.78	.867	.799
15 x 3⅜	40	11.8	15.00	3.520	.650	.520	349.0	46.5	5.44	9.23	3.36	.886	.778
	33.9	9.96	15.00	3.400	.650	.400	315.0	42.0	5.62	8.13	3.11	.904	.787
C12	30	8.82	12.00	3.170	.501	.510	162.0	27.0	4.29	5.14	2.06	.763	.674
12 x 3	25	7.35	12.00	3.047	.501	.387	144.0	24.1	4.43	4.47	1.88	.780	.674
	20.7	6.09	12.00	2.942	.501	.282	129.0	21.5	4.61	3.88	1.73	.799	.698
C10	30	8.82	10.00	3.033	.436	.673	103.0	20.7	3.42	3.94	1.65	.669	.649
10 x 2⅝	25	7.35	10.00	2.886	.436	.526	91.2	18.2	3.52	3.36	1.48	.676	.617
	20	5.88	10.00	2.739	.436	.379	78.9	15.8	3.66	2.81	1.32	.691	.606
	15.3	4.49	10.00	2.600	.436	.240	67.4	13.5	3.87	2.28	1.16	.713	.634
C9	20	5.88	9.00	2.648	.413	.448	60.9	13.5	3.22	2.42	1.17	.642	.583
9 x 2½	15	4.41	9.00	2.485	.413	.285	51.0	11.3	3.40	1.93	1.01	.661	.586
	13.4	3.94	9.00	2.433	.413	.233	47.9	10.6	3.48	1.76	.962	.668	.601
C8	18.75	5.51	8.00	2.527	.390	.487	44.0	11.0	2.82	1.98	1.01	.599	.565
8 x 2¼	13.75	4.04	8.00	2.343	.390	.303	36.1	9.03	2.99	1.53	.853	.615	.553
	11.5	3.38	8.00	2.260	.390	.220	32.6	8.14	3.11	1.32	.781	.625	.571
C7	14.75	4.33	7.00	2.299	.366	.419	27.2	7.78	2.51	1.38	.779	.564	.532
7 x 2⅛	12.25	3.60	7.00	2.194	.366	.314	24.2	6.93	2.60	1.17	.702	.571	.525
	9.8	2.87	7.00	2.090	.366	.210	21.3	6.08	2.72	.968	.625	.581	.541
C6	13	3.83	6.00	2.157	.343	.437	17.4	5.80	2.13	1.05	.642	.525	.514
6 x 2	10.5	3.09	6.00	2.034	.343	.314	15.2	5.06	2.22	.865	.564	.529	.500
	8.2	2.40	6.00	1.920	.343	.200	13.1	4.38	2.34	.692	.492	.537	.512
C5	9	2.64	5.00	1.885	.320	.325	8.90	3.56	1.83	.632	.449	.489	.478
5 x 1¾	6.7	1.97	5.00	1.750	.320	.190	7.49	3.00	1.95	.478	.378	.493	.484
C4	7.25	2.13	4.00	1.721	.296	.321	4.59	2.29	1.47	.432	.343	.450	.459
4 x 1⅝	5.4	1.59	4.00	1.584	.296	.184	3.85	1.93	1.56	.319	.283	.449	.458
C3	6	1.76	3.00	1.596	.273	.356	2.07	1.38	1.08	.305	.268	.416	.455
3 x 1½	5	1.47	3.00	1.498	.273	.258	1.85	1.24	1.12	.247	.233	.410	.438
	4.1	1.21	3.00	1.410	.273	.170	1.66	1.10	1.17	.197	.202	.404	.437

American Standard channels (C), continued 2.3.3.4

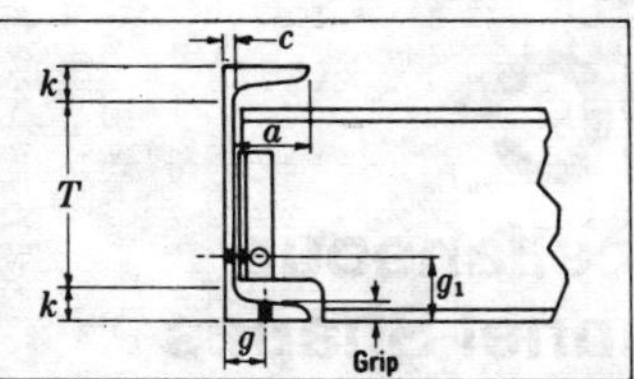

Dimensions for Detailing

Designation and Nominal Size	Weight per Foot	Flange		Web		Distances					Usual Gage g	Grip	Max. Flange Fast-ener	Fillet Radius R
		Width	Aver. Thick-ness	Thick-ness	Half Thick-ness	*a*	*T*	*k*	g_1	*c*				
In.	Lbs.	In.	In.	In.	In.	In.	In.	In.	In.	In.	In.	In.	In.	In.
C15 15 x 3⅜	50	3¾	⅝	11/16	⅜	3	12⅛	1 7/16	2¾	¾	2¼	⅝	1	.50
	40	3½	⅝	½	¼	3	12⅛	1 7/16	2¾	9/16	2	⅝	1	
	33.9	3⅜	⅝	⅜	3/16	3	12⅛	1 7/16	2¾	7/16	2	⅝	1	
C12 12 x 3	30	3⅛	½	½	¼	2⅝	9¾	1⅛	2½	9/16	1¾	½	⅞	.38
	25	3	½	⅜	3/16	2⅝	9¾	1⅛	2½	7/16	1¾	½	⅞	
	20.7	3	½	5/16	⅛	2⅝	9¾	1⅛	2½	⅜	1¾	½	⅞	
C10 10 x 2⅝	30	3	7/16	11/16	5/16	2⅜	8	1	2½	¾	1¾	7/16	¾	.34
	25	2⅞	7/16	½	¼	2⅜	8	1	2½	9/16	1¾	7/16	¾	
	20	2¾	7/16	⅜	3/16	2⅜	8	1	2½	7/16	1½	7/16	¾	
	15.3	2⅝	7/16	¼	⅛	2⅜	8	1	2½	5/16	1½	7/16	¾	
C9 9 x 2½	20	2⅝	7/16	7/16	¼	2¼	7⅛	15/16	2½	½	1½	7/16	¾	.33
	15	2½	7/16	5/16	⅛	2¼	7⅛	15/16	2½	⅜	1⅜	7/16	¾	
	13.4	2⅜	7/16	¼	⅛	2¼	7⅛	15/16	2½	5/16	1⅜	7/16	¾	
C8 8 x 2¼	18.75	2½	⅜	½	¼	2	6⅛	15/16	2½	9/16	1½	⅜	¾	.32
	13.75	2⅜	⅜	5/16	⅛	2	6⅛	15/16	2½	⅜	1⅜	⅜	¾	
	11.5	2¼	⅜	¼	⅛	2	6⅛	15/16	2½	5/16	1⅜	⅜	¾	
C7 7 x 2⅛	14.75	2¼	⅜	7/16	3/16	1⅞	5¼	⅞	2½	½	1¼	⅜	⅝	.31
	12.25	2¼	⅜	5/16	3/16	1⅞	5¼	⅞	2½	⅜	1¼	⅜	⅝	
	9.8	2⅛	⅜	3/16	⅛	1⅞	5¼	⅞	2½	¼	1¼	⅜	⅝	
C6 6 x 2	13	2⅛	5/16	7/16	3/16	1¾	4⅜	13/16	2¼	½	1⅜	5/16	⅝	.30
	10.5	2	5/16	5/16	3/16	1¾	4⅜	13/16	2¼	⅜	1⅛	⅜	⅝	
	8.2	1⅞	5/16	3/16	⅛	1¾	4⅜	13/16	2¼	¼	1⅛	5/16	⅝	
C5 5 x 1¾	9	1⅞	5/16	5/16	3/16	1½	3½	¾	2¼	⅜	1⅛	5/16	⅝	.29
	6.7	1¾	5/16	3/16	⅛	1½	3½	¾	2¼	¼	—	5/16	—	
C4 4 x 1⅝	7.25	1¾	5/16	5/16	3/16	1⅜	2⅝	11/16	2	⅜	1	5/16	⅝	.28
	5.4	1⅝	5/16	3/16	1/16	1⅜	2⅝	11/16	2	¼	—	¼	—	
C3 3 x 1½	6	1⅝	¼	⅜	3/16	1¼	1⅝	11/16	—	7/16	—	5/16	—	.27
	5	1½	¼	¼	⅛	1¼	1⅝	11/16	—	5/16	—	¼	—	
	4.1	1⅜	¼	3/16	1/16	1¼	1⅝	11/16	—	¼	—	¼	—	

Miscellaneous channel shapes (MC) 2.3.3.5

MC
Miscellaneous Channel Shapes

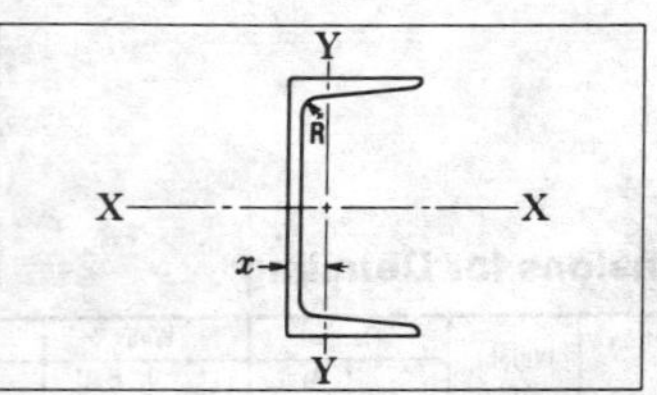

Properties for Designing

Designation and Nominal Size	Weight per Foot	Area of Section	Depth of Channel	Width of Flange	Aver. Flange Thickness	Web Thickness	Axis X-X			Axis Y-Y			
							I	S	r	I	S	r	x
In.	Lbs.	In.²	In.	In.	In.	In.	In.⁴	In.³	In.	In.⁴	In.³	In.	In.
MC18 18 x 4	58	17.1	18.00	4.200	.625	.700	676	75.1	6.29	17.8	5.32	1.02	.862
	51.9	15.3	18.00	4.100	.625	.600	627	69.7	6.41	16.4	5.07	1.04	.858
	45.8	13.5	18.00	4.000	.625	.500	578	64.3	6.56	15.1	4.82	1.06	.866
	42.7	12.6	18.00	3.950	.625	.450	554	61.6	6.64	14.4	4.69	1.07	.877
MC13 13 x 4	50	14.7	13.00	4.412	.610	.787	314	48.4	4.62	16.5	4.79	1.06	.974
	40	11.8	13.00	4.185	.610	.560	273	42.0	4.82	13.7	4.26	1.08	.964
	35	10.3	13.00	4.072	.610	.447	252	38.8	4.95	12.3	3.99	1.10	.980
	31.8	9.35	13.00	4.000	.610	.375	239	36.8	5.06	11.4	3.81	1.11	1.00
MC12 12 x 4	50	14.7	12.00	4.135	.700	.835	269	44.9	4.28	17.4	5.65	1.09	1.05
	45	13.2	12.00	4.012	.700	.712	252	42.0	4.36	15.8	5.33	1.09	1.04
	40	11.8	12.00	3.890	.700	.590	234	39.0	4.46	14.3	5.00	1.10	1.04
	35	10.3	12.00	3.767	.700	.467	216	36.1	4.59	12.7	4.67	1.11	1.05
MC12 12 x 3½	37	10.9	12.00	3.600	.600	.600	205	34.2	4.34	9.81	3.59	.950	.866
	32.9	9.67	12.00	3.500	.600	.500	191	31.8	4.44	8.91	3.39	.960	.867
	30.9	9.07	12.00	3.450	.600	.450	183	30.6	4.50	8.46	3.28	.966	.873
MC10 10 x 4	41.1	12.1	10.00	4.321	.575	.796	158	31.5	3.61	15.8	4.88	1.14	1.09
	33.6	9.87	10.00	4.100	.575	.575	139	27.8	3.75	13.2	4.38	1.16	1.08
	28.5	8.37	10.00	3.950	.575	.425	127	25.3	3.89	11.4	4.02	1.17	1.12
MC10 10 x 3½	28.3	8.32	10.00	3.502	.575	.477	118	23.6	3.77	8.21	3.20	.993	.933
	24.9	7.32	10.00	3.402	.575	.377	110	22.0	3.87	7.32	2.99	1.00	.954
MC10 10 x 3½	25.3	7.43	10.00	3.550	.500	.425	107	21.4	3.79	7.61	2.89	1.01	.918
	21.9	6.43	10.00	3.450	.500	.325	98.5	19.7	3.91	6.74	2.70	1.02	.954
MC9 9 x 3½	25.4	7.47	9.00	3.500	.550	.450	88.0	19.6	3.43	7.65	3.02	1.01	.970
	23.9	7.02	9.00	3.450	.550	.400	85.0	18.9	3.48	7.22	2.93	1.01	.981
MC8 8 x 3½	22.8	6.70	8.00	3.502	.525	.427	63.8	16.0	3.09	7.07	2.84	1.03	1.01
	21.4	6.28	8.00	3.450	.525	.375	61.6	15.4	3.13	6.64	2.74	1.03	1.02
MC8 8 x 3	20	5.88	8.00	3.025	.500	.400	54.5	13.6	3.05	4.47	2.05	.872	.840
	18.7	5.50	8.00	2.978	.500	.353	52.5	13.1	3.09	4.20	1.97	.874	.849
MC7 7 x 3½	22.7	6.67	7.00	3.603	.500	.503	47.5	13.6	2.67	7.29	2.85	1.05	1.04
	19.1	5.61	7.00	3.452	.500	.352	43.2	12.3	2.77	6.11	2.57	1.04	1.08

Miscellaneous channel shapes (MC), continued 2.3.3.5

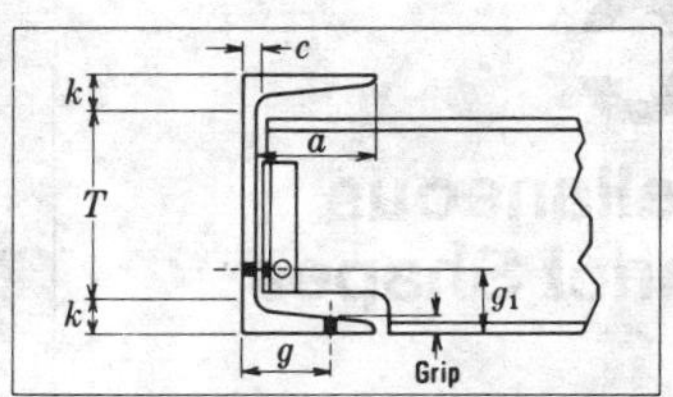

Dimensions for Detailing

Designation and Nominal Size	Weight per Foot	Flange		Web		Distances					Usual Gage g	Grip	Max. Flange Fastener	Fillet Radius R
		Width	Aver. Thickness	Thickness	Half Thickness	a	T	k	g_1	c				
In.	Lbs.	In.	In.	In.	In.	In.	In.	In.	In.	In.	In.	In.	In.	In.
MC18 18 x 4	58	4¼	⅝	11/16	⅜	3½	15¼	1⅜	2½	¾	2½	⅝	1	.62
	51.9	4⅛	⅝	⅝	5/16	3½	15¼	1⅜	2½	11/16	2½	⅝	1	
	45.8	4	⅝	½	¼	3½	15¼	1⅜	2½	9/16	2½	⅝	1	
	42.7	4	⅝	7/16	¼	3½	15¼	1⅜	2½	½	2½	⅝	1	
MC13 13 x 4	50	4⅜	⅝	13/16	⅜	3⅝	10¼	1⅜	2¾	⅞	2½	⅝	1	.48
	40	4⅛	⅝	9/16	¼	3⅝	10¼	1⅜	2¾	⅝	2½	9/16	1	
	35	4⅛	⅝	7/16	¼	3⅝	10¼	1⅜	2¾	½	2½	9/16	1	
	31.8	4	⅝	⅜	3/16	3⅝	10¼	1⅜	2¾	7/16	2½	9/16	1	
MC12 12 x 4	50	4⅛	11/16	13/16	7/16	3¼	9⅜	1 5/16	2½	⅞	2½	11/16	1	.50
	45	4	11/16	11/16	⅜	3¼	9⅜	1 5/16	2½	¾	2½	11/16	1	
	40	3⅞	11/16	9/16	5/16	3¼	9⅜	1 5/16	2½	⅝	2½	11/16	1	
	35	3¾	11/16	7/16	¼	3¼	9⅜	1 5/16	2½	½	2½	11/16	1	
MC12 12 x 3½	37	3⅝	⅝	⅝	5/16	3	9⅜	1 5/16	2½	11/16	2¼	⅝	⅞	.60
	32.9	3½	⅝	½	¼	3	9⅜	1 5/16	2½	9/16	2¼	9/16	⅞	
	30.9	3½	⅝	7/16	¼	3	9⅜	1 5/16	2½	½	2¼	9/16	⅞	
MC10 10 x 4	41.1	4⅜	9/16	13/16	⅜	3½	7½	1¼	2½	⅞	2½	9/16	⅞	.58
	33.6	4⅛	9/16	9/16	5/16	3½	7½	1¼	2½	⅝	2½	9/16	⅞	
	28.5	4	9/16	7/16	3/16	3½	7½	1¼	2½	½	2½	9/16	⅞	
MC10 10 x 3½	28.3	3½	9/16	½	¼	3	7½	1¼	2½	9/16	2	9/16	⅞	.58
	24.9	3⅜	9/16	⅜	3/16	3	7½	1¼	2½	7/16	2	9/16	⅞	
MC10 10 x 3½	25.3	3½	½	7/16	3/16	3⅛	7¾	1⅛	2½	½	2	½	⅞	.50
	21.9	3½	½	5/16	3/16	3⅛	7¾	1⅛	2½	⅜	2	½	⅞	
MC9 9 x 3½	25.4	3½	9/16	7/16	¼	3	6⅝	1 3/16	2½	½	2	9/16	⅞	.55
	23.9	3½	9/16	⅜	3/16	3	6⅝	1 3/16	2½	7/16	2	9/16	⅞	
MC8 8 x 3½	22.8	3½	½	7/16	3/16	3⅛	5⅝	1 3/16	2½	½	2	½	⅞	.52
	21.4	3½	½	⅜	3/16	3⅛	5⅝	1 3/16	2½	7/16	2	½	⅞	
MC8 8 x 3	20	3	½	⅜	3/16	2⅝	5¾	1⅛	2½	7/16	2	½	⅞	.50
	18.7	3	½	⅜	3/16	2⅝	5¾	1⅛	2½	7/16	2	½	⅞	
MC7 7 x 3½	22.7	3⅝	½	½	¼	3⅛	4¾	1⅛	2½	9/16	2	½	⅞	.50
	19.1	3½	½	⅜	3/16	3⅛	4¾	1⅛	2½	7/16	2	½	⅞	

Miscellaneous channel shapes (MC), continued 2.3.3.5

MC
Miscellaneous Channel Shapes

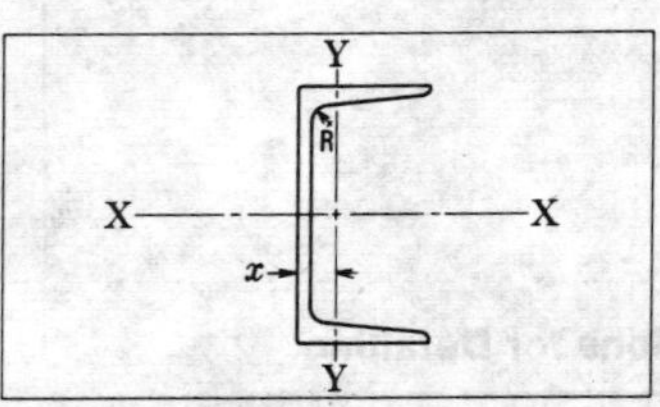

Properties for Designing

Designation and Nominal Size	Weight per Foot	Area of Section	Depth of Channel	Width of Flange	Aver. Flange Thickness	Web Thickness	Axis X-X			Axis Y-Y			
							I	S	r	I	S	r	x
In.	Lbs.	In.²	In.	In.	In.	In.	In.⁴	In.³	In.	In.⁴	In.³	In.	In.
MC7 7 x 3	17.6	5.17	7.00	3.000	.475	.375	37.6	10.8	2.70	4.01	1.89	.881	.873
MC6 6 x 3½	18	5.29	6.00	3.504	.475	.379	29.7	9.91	2.37	5.93	2.48	1.06	1.12
MC6 6 x 3½	15.3	4.50	6.00	3.500	.385	.340	25.4	8.47	2.38	4.97	2.03	1.05	1.05
MC6 6 x 3	16.3 15.1	4.79 4.44	6.00 6.00	3.000 2.941	.475 .475	.375 .316	26.0 25.0	8.68 8.32	2.33 2.37	3.82 3.51	1.84 1.75	.892 .889	.927 .940
MC6 6 x 2½	12	3.53	6.00	2.497	.375	.310	18.7	6.24	2.30	1.87	1.04	.728	.704
MC4 4 x 2½	13.8	4.06	4.00	2.510	.500	.510	8.91	4.46	1.48	2.21	1.34	.738	.858
MCF3* 3 x 1¹⁵⁄₁₆	9 7.1	2.65 2.09	3.00 3.00	2.122 1.938	.351 .351	.497 .312	3.15 2.73	2.10 1.82	1.09 1.14	.967 .712	.677 .561	.604 .583	.694 .669
MC3* 3 x 1¹⁵⁄₁₆	7.1	2.09	3.00	1.938	.351	.312	2.73	1.82	1.14	.712	.561	.583	.669

*MC3 and MCF3 are identical shapes except the MCF3 shapes have flanges that are flared to 3⅛ inch at the toes.

Miscellaneous channel shapes (MC), continued 2.3.3.5

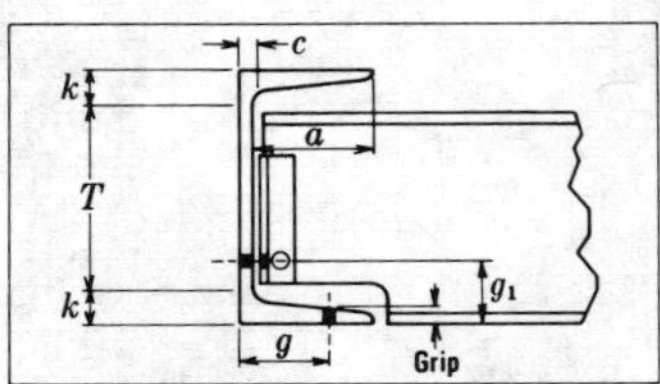

Dimensions for Detailing

Designation and Nominal Size	Weight per Foot	Flange		Web		Distances					Usual Gage	Grip	Max. Flange Fastener	Fillet Radius
		Width	Aver. Thickness	Thickness	Half Thickness	a	T	k	g_1	c	g			R
In.	Lbs.	In.	In.	In.	In.	In.	In.	In.	In.	In.	In.	In.	In.	In.
MC7 7 x 3	17.6	3	½	⅜	3/16	2⅝	4⅞	1 1/16	2½	7/16	1¾	½	¾	.48
MC6 6 x 3½	18	3½	½	⅜	3/16	3⅛	3⅞	1 1/16	2½	7/16	2	½	⅞	.48
MC6 6 x 3½	15.3	3½	⅜	5/16	3/16	3⅛	4¼	⅞	2¼	⅜	2	⅜	⅞	.38
MC6 6 x 3	16.3 15.1	3 3	½ ½	⅜ 5/16	3/16 3/16	2⅝ 2⅝	3⅞ 3⅞	1 1/16 1 1/16	2½ 2½	7/16 ⅜	1¾ 1¾	½ ½	¾ ¾	.48
MC6 6 x 2½	12	2½	⅜	5/16	⅛	2⅛	4⅜	13/16	2¼	⅜	1½	⅜	⅝	.38
MC4 4 x 2½	13.8	2½	½	½	¼	2	2¼	⅞	2	9/16	1½	½	⅝	.28
MCF3 3 x 1 15/16	9 7.1	2⅛ 2	⅜ ⅜	½ 5/16	¼ ⅛	1⅝ 1⅝	1¾ 1¾	⅝ ⅝	... ...	9/16 ⅜	... ...	... ...	... ...	.19
MC3 3 x 1 15/16	7.1	2	⅜	5/16	⅛	1⅝	1¾	⅝	...	⅜	...	...	...	.13

Bar size channels (C) 2.3.3.6

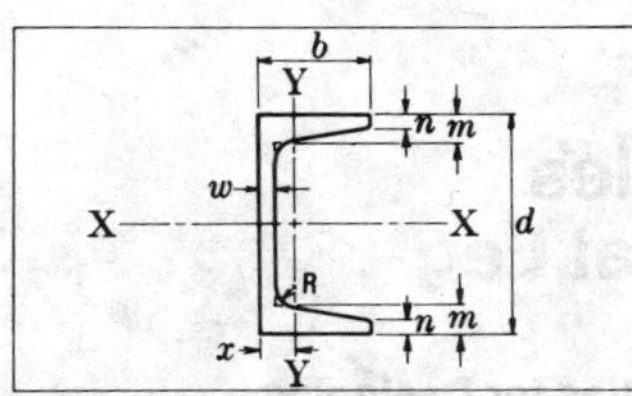

Dimensions and Properties for Designing

Designation and Nominal Size	Weight per Foot	Area of Section	Depth of Section d	Flanges Width b	Flanges Thickness m	Flanges Thickness n	Web Thickness w	Axis X-X I	Axis X-X S	Axis X-X r	Axis Y-Y I	Axis Y-Y S	Axis Y-Y r	Axis Y-Y x	Fillet Radius R
In.	Lbs.	In.²	In.	In.	In.	In.	In.	In.⁴	In.³	In.	In.⁴	In.³	In.	In.	In.
C539															
2½x⅝x3/16	2.27	.668	2½	⅝	21/64	⅛	3/16	.498	.399	.864	.015	.034	.151	.177	*
C597															
2x1x3/16	2.57	.764	2	1	¼	7/32	3/16	.428	.428	.748	.068	.102	.297	.340	9/64
C598															
2x1x⅛	1.78	.528	2	1	3/16	⅛	⅛	.319	.319	.777	.047	.067	.297	.307	7/64
C29															
2x1x3/16	2.32	.683	2	1	3/16	3/16	3/16	.378	.378	.744	.059	.087	.295	.317	3/32
2x1x⅛	1.59	.473	2	1	⅛	⅛	⅛	.279	.279	.768	.044	.062	.304	.295	
C534															
2x⅝x¼	2.18	.641	2	⅝	15/64	9/64	¼	.283	.283	.664	.014	.032	.147	.190	1/16
C73															
2x ⅝ x¼	2.28	.670	2	⅝	5/16	9/64	¼	.300	.300	.669	.015	.035	.150	.198	
2x9/16x3/16	1.86	.545	2	9/16	5/16	9/64	3/16	.258	.258	.688	.011	.028	.140	.174	1/32
2x ½ x⅛	1.43	.420	2	½	5/16	9/64	⅛	.216	.216	.718	.007	.021	.133	.154	
C531															
1¾x½x3/16	1.55	.456	1¾	½	¼	5/32	3/16	.160	.183	.592	.007	.021	.125	.160	*
C639															
1½x1½x3/16	2.72	.801	1½	1½	13/64	3/16	3/16	.274	.366	.585	.175	.188	.467	.570	⅛
C79															
1½x¾x⅛	1.17	.348	1½	¾	⅛	⅛	⅛	.111	.147	.564	.016	.030	.213	.232	3/32
C80															
1½x9/16x3/16	1.44	.433	1½	9/16	¼	⅛	3/16	.113	.151	.518	.009	.023	.144	.180	1/16
1½x½x⅛	1.12	.329	1½	½	¼	⅛	⅛	.096	.128	.540	.006	.018	.136	.161	
C522															
1¼x½x⅛	1.01	.297	1¼	½	¼	⅛	⅛	.060	.096	.450	.006	.017	.138	.171	*
C519															
1⅛x9/16x3/16	1.16	.340	1⅛	9/16	7/32	⅛	3/16	.052	.092	.390	.008	.021	.150	.194	*
C516															
1x½x⅛	.82	.242	1	½	13/64	7/64	⅛	.031	.063	.360	.005	.014	.140	.174	*
C81															
1x⅜x⅛	.68	.199	1	⅜	3/16	7/64	⅛	.024	.048	.346	.002	.008	.100	.128	1/32
C508															
⅞x⅜x⅛	.65	.191	⅞	⅜	7/32	7/64	⅛	.017	.040	.301	.002	.008	.101	.137	*
C38															
¾x⅜x⅛	.56	.164	¾	⅜	3/16	3/32	⅛	.011	.029	.258	.002	.007	.101	.137	1/32

*May be sharp corner to slightly rounded.

Equal leg angles (L) 2.3.3.7

L

Angles Equal Leg

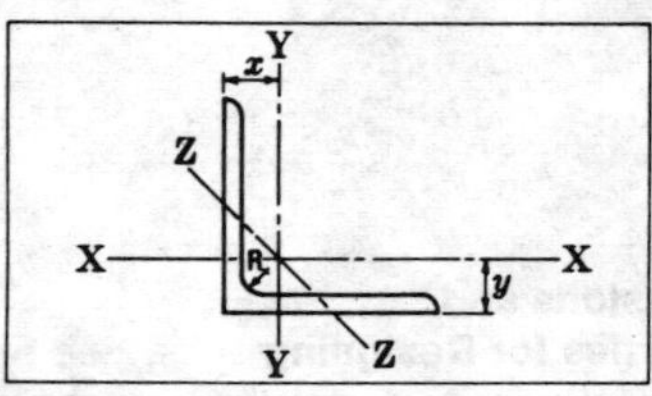

Properties for Designing

Designation and Nominal Size	Thick-ness	Weight per Foot	Area of Section	Axis X-X and Axis Y-Y					Fillet Radius R
				I	S	r	x or y	r_{min}	
In.	In.	Lbs.	In.^{2}	In.^{4}	In.^{3}	In.	In.	In.	In.
L8x8	1⅛	56.9	16.7	98.0	17.5	2.42	2.41	1.56	
	1	51.0	15.0	89.0	15.8	2.44	2.37	1.56	
	⅞	45.0	13.2	79.6	14.0	2.45	2.32	1.57	
	¾	38.9	11.4	69.7	12.2	2.47	2.28	1.58	⅝
	⅝	32.7	9.61	59.4	10.3	2.49	2.23	1.58	
	9/16	29.6	8.68	54.1	9.34	2.50	2.21	1.59	
	½	26.4	7.75	48.6	8.36	2.50	2.19	1.59	
L6x6	1	37.4	11.0	35.5	8.57	1.80	1.86	1.17	
	⅞	33.1	9.73	31.9	7.63	1.81	1.82	1.17	
	¾	28.7	8.44	28.2	6.66	1.83	1.78	1.17	
	⅝	24.2	7.11	24.2	5.66	1.84	1.73	1.18	
	9/16	21.9	6.43	22.1	5.14	1.85	1.71	1.18	½
	½	19.6	5.75	19.9	4.61	1.86	1.68	1.18	
	7/16	17.2	5.06	17.7	4.08	1.87	1.66	1.19	
	⅜	14.9	4.36	15.4	3.53	1.88	1.64	1.19	
	5/16	12.4	3.65	13.0	2.97	1.89	1.62	1.20	
L5x5	⅞	27.2	7.98	17.8	5.17	1.49	1.57	.973	
	¾	23.6	6.94	15.7	4.53	1.51	1.52	.975	
	⅝	20.0	5.86	13.6	3.86	1.52	1.48	.978	
	½	16.2	4.75	11.3	3.16	1.54	1.43	.983	½
	7/16	14.3	4.18	10.0	2.79	1.55	1.41	.986	
	⅜	12.3	3.61	8.74	2.42	1.56	1.39	.990	
	5/16	10.3	3.03	7.42	2.04	1.57	1.37	.994	
L4x4	¾	18.5	5.44	7.67	2.81	1.19	1.27	.778	
	⅝	15.7	4.61	6.66	2.40	1.20	1.23	.779	
	½	12.8	3.75	5.56	1.97	1.22	1.18	.782	
	7/16	11.3	3.31	4.97	1.75	1.23	1.16	.785	⅜
	⅜	9.80	2.86	4.36	1.52	1.23	1.14	.788	
	5/16	8.20	2.40	3.71	1.29	1.24	1.12	.791	
	¼	6.60	1.94	3.04	1.05	1.25	1.09	.795	
L3½x3½	½	11.1	3.25	3.64	1.49	1.06	1.06	.683	
	7/16	9.80	2.87	3.26	1.32	1.07	1.04	.684	
	⅜	8.50	2.48	2.87	1.15	1.07	1.01	.687	⅜
	5/16	7.20	2.09	2.45	.976	1.08	.990	.690	
	¼	5.80	1.69	2.01	.794	1.09	.968	.694	
L3x3	½	9.4	2.75	2.22	1.07	.898	.932	.584	
	7/16	8.3	2.43	1.99	.954	.905	.910	.585	
	⅜	7.2	2.11	1.76	.833	.913	.888	.587	
	5/16	6.1	1.78	1.51	.707	.922	.865	.589	5/16
	¼	4.9	1.44	1.24	.577	.930	.842	.592	
	3/16	3.71	1.09	.962	.441	.939	.820	.596	

Equal leg bar angles (A) 2.3.3.8

Bar Size Angles Equal Legs

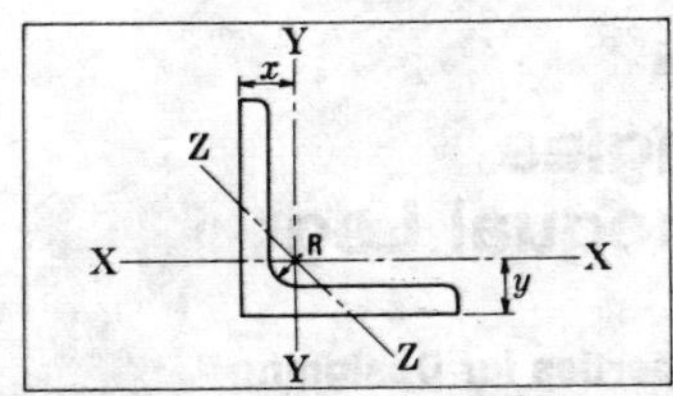

Properties for Designing

Designation and Nominal Size	Thick-ness	Weight per Foot	Area of Section	Axis X-X and Axis Y-Y I	Axis X-X and Axis Y-Y S	Axis X-X and Axis Y-Y r	Axis X-X and Axis Y-Y x or y	Axis Z-Z $r_{min.}$	Fillet Radius R
In.	In.	Lbs.	In.²	In.⁴	In.³	In.	In.	In.	In.
A9 2½ x 2½	½	7.7	2.25	1.23	.724	.739	.806	.487	
	⅜	5.9	1.73	.984	.566	.753	.762	.487	
	5/16	5.0	1.46	.849	.482	.761	.740	.489	5/16*
	¼	4.1	1.19	.703	.394	.769	.717	.491	
	3/16	3.07	.902	.547	.303	.778	.694	.495	
A971 2½ x 2½	⅛	2.07	.609	.378	.207	.788	.671	.499	3/16
A11 2 x 2	⅜	4.7	1.36	.479	.351	.594	.636	.389	
	5/16	3.92	1.15	.416	.300	.601	.614	.390	
	¼	3.19	.938	.348	.247	.609	.592	.391	
	.205	2.65	.778	.294	.207	.615	.575	.393	¼*
	3/16	2.44	.715	.272	.190	.617	.569	.394	
	⅛	1.65	.484	.190	.131	.626	.546	.398	
A12 1¾ x 1¾	5/16	3.39	1.00	.271	.226	.521	.551	.341	
	¼	2.77	.813	.227	.186	.529	.529	.341	
	3/16	2.12	.621	.179	.144	.537	.506	.343	¼*
	⅛	1.44	.422	.126	.099	.546	.484	.347	
A13 1½ x 1½	5/16	2.86	.840	.164	.162	.442	.488	.292	
	¼	2.34	.688	.139	.134	.449	.466	.292	
	3/16	1.80	.527	.110	.104	.457	.444	.293	
	.165	1.59	.468	.099	.093	.460	.436	.294	3/16*
	5/32	1.52	.444	.094	.088	.461	.433	.295	
	⅛	1.23	.359	.078	.072	.465	.421	.296	
A15 1¼ x 1¼	¼	1.92	.563	.077	.091	.369	.403	.243	
	3/16	1.48	.434	.061	.071	.377	.381	.244	3/16*
	⅛	1.01	.297	.044	.049	.385	.359	.246	
A508 1⅛ x 1⅛	⅛	.90	.266	.032	.040	.345	.327	.221	3/32
A16 1 x 1	¼	1.49	.438	.037	.056	.290	.339	.196	
	3/16	1.16	.340	.030	.044	.297	.318	.195	⅛*
	⅛	.80	.234	.022	.031	.304	.296	.196	
A81 ⅞ x ⅞	⅛	.70	.203	.014	.023	.264	.264	.171	⅛*
A17 ¾ x ¾	⅛	.59	.172	.009	.017	.224	.233	.146	⅛*
A513 ⅝ x ⅝	⅛	.48	.141	.005	.011	.185	.201	.122	5/64
A515 ½ x ½	⅛	.38	.109	.002	.007	.145	.170	.098	1/32

*Angles are produced with various size fillet radii depending on where they are rolled. The maximum radii produced are those shown.

Unequal leg angles (L) 2.3.3.9

L

Angles Unequal Leg

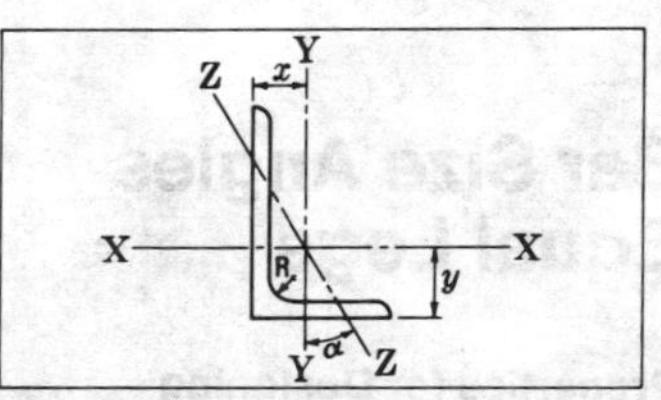

Properties for Designing

Designation and Nominal Size	Thick-ness	Weight per Foot	Area of Section	Axis X-X				Axis Y-Y				Axis Z-Z		Fillet Radius R
				I	S	r	y	I	S	r	x	r_{min}	Tan α	
In.	In.	Lbs.	In.²	In.⁴	In.³	In.	In.	In.⁴	In.³	In.	In.	In.		In.
L8x6	1	44.2	13.0	80.8	15.1	2.49	2.65	38.8	8.92	1.73	1.65	1.28	.543	
	⅞	39.1	11.5	72.3	13.4	2.51	2.61	34.9	7.94	1.74	1.61	1.28	.547	
	¾	33.8	9.94	63.4	11.7	2.53	2.56	30.7	6.92	1.76	1.56	1.29	.551	
	⅝	28.5	8.36	54.1	9.87	2.54	2.52	26.3	5.88	1.77	1.52	1.29	.554	½
	9/16	25.7	7.56	49.3	8.95	2.55	2.50	24.0	5.34	1.78	1.50	1.30	.556	
	½	23.0	6.75	44.3	8.02	2.56	2.47	21.7	4.79	1.79	1.47	1.30	.558	
	7/16	20.2	5.93	39.2	7.07	2.57	2.45	19.3	4.23	1.80	1.45	1.31	.560	
L8x4	1	37.4	11.0	69.6	14.1	2.52	3.05	11.6	3.94	1.03	1.05	.846	.247	
	⅞	33.1	9.73	62.5	12.5	2.53	3.00	10.5	3.51	1.04	.999	.848	.253	
	¾	28.7	8.44	54.9	10.9	2.55	2.95	9.36	3.07	1.05	.953	.852	.258	
	⅝	24.2	7.11	46.9	9.21	2.57	2.91	8.10	2.62	1.07	.906	.857	.262	½
	9/16	21.9	6.43	42.8	8.35	2.58	2.88	7.43	2.38	1.07	.882	.861	.265	
	½	19.6	5.75	38.5	7.49	2.59	2.86	6.74	2.15	1.08	.859	.865	.267	
	7/16	17.2	5.06	34.1	6.60	2.60	2.83	6.02	1.90	1.09	.835	.869	.269	
L7x4	⅞	30.2	8.86	42.9	9.65	2.20	2.55	10.2	3.46	1.07	1.05	.856	.318	
	¾	26.2	7.69	37.8	8.42	2.22	2.51	9.05	3.03	1.09	1.01	.860	.324	
	⅝	22.1	6.48	32.4	7.14	2.24	2.46	7.84	2.58	1.10	.963	.865	.329	
	9/16	20.0	5.87	29.6	6.48	2.24	2.44	7.19	2.35	1.11	.940	.868	.332	½
	½	17.9	5.25	26.7	5.81	2.25	2.42	6.53	2.12	1.11	.917	.872	.335	
	7/16	15.8	4.62	23.7	5.13	2.26	2.39	5.83	1.88	1.12	.893	.876	.337	
	⅜	13.6	3.98	20.6	4.44	2.27	2.37	5.10	1.63	1.13	.870	.880	.340	
L6x4	1	30.6	9.00	30.8	8.02	1.85	2.17	10.8	3.79	1.09	1.17	.857	.414	
	⅞	27.2	7.98	27.7	7.15	1.86	2.12	9.75	3.39	1.11	1.12	.857	.421	
	¾	23.6	6.94	24.5	6.25	1.88	2.08	8.68	2.97	1.12	1.08	.860	.428	
	⅝	20.0	5.86	21.1	5.31	1.90	2.03	7.52	2.54	1.13	1.03	.864	.435	
	9/16	18.1	5.31	19.3	4.83	1.90	2.01	6.91	2.31	1.14	1.01	.866	.438	½
	½	16.2	4.75	17.4	4.33	1.91	1.99	6.27	2.08	1.15	.987	.870	.440	
	7/16	14.3	4.18	15.5	3.83	1.92	1.96	5.60	1.85	1.16	.964	.873	.443	
	⅜	12.3	3.61	13.5	3.32	1.93	1.94	4.90	1.60	1.17	.941	.877	.446	
	5/16	10.3	3.03	11.4	2.79	1.94	1.92	4.18	1.35	1.17	.918	.882	.448	
L6x3½	½	15.3	4.50	16.6	4.24	1.92	2.08	4.25	1.59	.972	.833	.759	.344	
	⅜	11.7	3.42	12.9	3.24	1.94	2.04	3.34	1.23	.980	.787	.767	.350	
	5/16	9.8	2.87	10.9	2.73	1.95	2.01	2.85	1.04	.996	.763	.772	.352	½
	¼	7.9	2.31	8.86	2.21	1.96	1.99	2.34	.847	1.01	.740	.777	.355	
L5x3½	¾	19.8	5.81	13.9	4.28	1.55	1.75	5.55	2.22	.977	.996	.748	.464	
	⅝	16.8	4.92	12.0	3.65	1.56	1.70	4.83	1.90	.991	.951	.751	.472	
	½	13.6	4.00	9.99	2.99	1.58	1.66	4.05	1.56	1.01	.906	.755	.479	
	7/16	12.0	3.53	8.90	2.64	1.59	1.63	3.63	1.39	1.01	.883	.758	.482	7/16
	⅜	10.4	3.05	7.78	2.29	1.60	1.61	3.18	1.21	1.02	.861	.762	.486	
	5/16	8.7	2.56	6.60	1.94	1.61	1.59	2.72	1.02	1.03	.838	.766	.489	
	¼	7.0	2.06	5.39	1.57	1.62	1.56	2.23	.830	1.04	.814	.770	.492	

Unequal leg angles (L), continued 2.3.3.9

L
Angles
Unequal Leg

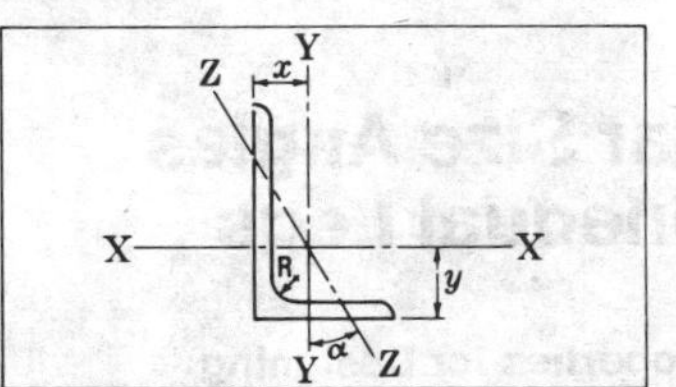

Properties for Designing

Designation and Nominal Size	Thick-ness	Weight per Foot	Area of Section	Axis X-X				Axis Y-Y			Axis Z-Z			Fillet Radius R
				I	S	r	y	I	S	r	x	r_{min}	Tan α	
In.	In.	Lbs.	In.²	In.⁴	In.³	In.	In.	In.⁴	In.³	In.	In.	In.		In.
L5x3	½	12.8	3.75	9.45	2.91	1.59	1.75	2.58	1.15	.829	.750	.648	.357	
	7/16	11.3	3.31	8.43	2.58	1.60	1.73	2.32	1.02	.837	.727	.651	.361	
	⅜	9.8	2.86	7.37	2.24	1.61	1.70	2.04	.888	.845	.704	.654	.364	⅜
	5/16	8.2	2.40	6.26	1.89	1.61	1.68	1.75	.753	.853	.681	.658	.368	
	¼	6.6	1.94	5.11	1.53	1.62	1.66	1.44	.614	.861	.657	.663	.371	
L4x3½	⅝	14.7	4.30	6.37	2.35	1.22	1.29	4.52	1.84	1.03	1.04	.719	.745	
	½	11.9	3.50	5.32	1.94	1.23	1.25	3.79	1.52	1.04	1.00	.722	.750	
	7/16	10.6	3.09	4.76	1.72	1.24	1.23	3.40	1.35	1.05	.978	.724	.753	
	⅜	9.1	2.67	4.18	1.49	1.25	1.21	2.95	1.17	1.06	.955	.727	.755	⅜
	5/16	7.7	2.25	3.56	1.26	1.26	1.18	2.55	.994	1.07	.932	.730	.757	
	¼	6.2	1.81	2.91	1.03	1.27	1.16	2.09	.808	1.07	.909	.734	.759	
L4x3	⅝	13.6	3.98	6.03	2.30	1.23	1.37	2.87	1.35	.849	.871	.637	.534	
	½	11.1	3.25	5.05	1.89	1.25	1.33	2.42	1.12	.864	.827	.639	.543	
	7/16	9.8	2.87	4.52	1.68	1.25	1.30	2.18	.992	.871	.804	.641	.547	
	⅜	8.5	2.48	3.96	1.46	1.26	1.28	1.92	.866	.879	.782	.644	.551	⅜
	5/16	7.2	2.09	3.38	1.23	1.27	1.26	1.65	.734	.887	.759	.647	.554	
	¼	5.8	1.69	2.77	1.00	1.28	1.24	1.36	.599	.896	.736	.651	.558	
L3½x3	½	10.2	3.00	3.45	1.45	1.07	1.13	2.33	1.10	.881	.875	.621	.714	
	7/16	9.1	2.65	3.10	1.29	1.08	1.10	2.09	.975	.889	.853	.622	.718	
	⅜	7.9	2.30	2.72	1.13	1.09	1.08	1.85	.851	.897	.830	.625	.721	⅜
	5/16	6.6	1.93	2.33	.954	1.10	1.06	1.58	.722	.905	.808	.627	.724	
	¼	5.4	1.56	1.91	.776	1.11	1.04	1.30	.589	.914	.785	.631	.727	
L3½x2½	½	9.4	2.75	3.24	1.41	1.09	1.20	1.36	.760	.704	.705	.534	.486	
	7/16	8.3	2.43	2.91	1.26	1.09	1.18	1.23	.677	.711	.682	.535	.491	
	⅜	7.2	2.11	2.56	1.09	1.10	1.16	1.09	.592	.719	.660	.537	.496	5/16
	5/16	6.1	1.78	2.19	.927	1.11	1.14	.939	.504	.727	.637	.540	.501	
	¼	4.9	1.44	1.80	.755	1.12	1.11	.777	.412	.735	.614	.544	.506	
L3x2½	½	8.5	2.50	2.08	1.04	.913	1.00	1.30	.744	.722	.750	.520	.667	
	7/16	7.6	2.21	1.88	.928	.920	.978	1.18	.664	.729	.728	.521	.672	
	⅜	6.6	1.92	1.66	.810	.928	.956	1.04	.581	.736	.706	.522	.676	5/16
	5/16	5.6	1.62	1.42	.688	.937	.933	.898	.494	.744	.683	.525	.680	
	¼	4.5	1.31	1.17	.561	.945	.911	.743	.404	.753	.661	.528	.684	
	3/16	3.39	.996	.907	.430	.954	.888	.577	.310	.761	.638	.533	.688	
L3x2	½	7.7	2.25	1.92	1.00	.924	1.08	.672	.474	.546	.583	.428	.414	
	7/16	6.8	2.00	1.73	.894	.932	1.06	.609	.424	.553	.561	.429	.421	
	⅜	5.9	1.73	1.53	.781	.940	1.04	.543	.371	.559	.539	.430	.428	
	5/16	5.0	1.46	1.32	.664	.948	1.02	.470	.317	.567	.516	.432	.435	5/16
	¼	4.1	1.19	1.09	.542	.957	.993	.392	.260	.574	.493	.435	.440	
	3/16	3.07	.902	.842	.415	.966	.970	.307	.200	.583	.470	.439	.446	

Unequal leg bar angles (A) 2.3.3.10

Bar Size Angles Unequal Legs

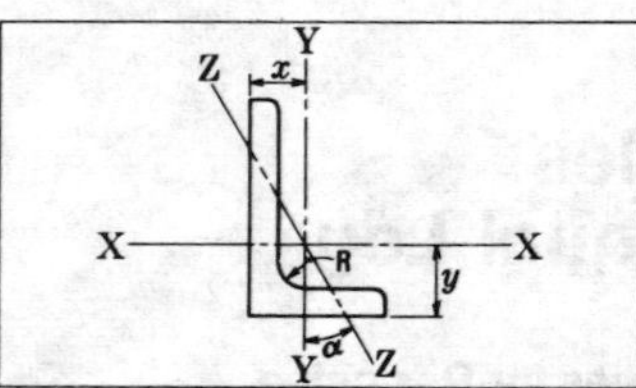

Properties for Designing

Designation and Nominal Size	Thickness	Weight per Foot	Area of Section	Axis X-X				Axis Y-Y				Axis Z-Z		Fillet Radius R
				I	S	r	y	I	S	r	x	$r_{min.}$	Tan α	
In.	In.	Lbs.	In.²	In.⁴	In.³	In.	In.	In.⁴	In.³	In.	In.	In.		In.
A35	3/8	5.3	1.55	.912	.547	.768	.831	.514	.363	.577	.581	.420	.614	
2½ x 2	5/16	4.5	1.31	.788	.466	.776	.809	.446	.310	.584	.559	.422	.620	¼*
	¼	3.62	1.06	.654	.381	.784	.787	.372	.254	.592	.537	.424	.626	
	3/16	2.75	.809	.509	.293	.793	.764	.231	.196	.600	.514	.427	.631	
A48	5/16	3.92	1.15	.711	.444	.785	.898	.191	.174	.408	.398	.322	.349	
2½ x 1½	¼	3.19	.938	.591	.364	.794	.875	.161	.143	.415	.375	.324	.357	3/16*
	3/16	2.44	.715	.461	.279	.803	.852	.127	.111	.422	.352	.327	.364	
A270	3/16	2.28	.668	.344	.229	.718	.745	.124	.110	.431	.370	.326	.440	⅛
2¼ x 1½														
A37	¼	2.77	.813	.316	.236	.623	.663	.151	.139	.432	.413	.320	.543	
2 x 1½	3/16	2.12	.621	.248	.182	.632	.641	.120	.108	.440	.391	.322	.551	3/16*
	⅛	1.44	.422	.173	.125	.641	.618	.085	.075	.448	.368	.326	.558	
A645	5/16	3.12	.918	.353	.278	.620	.731	.104	.117	.337	.356	.268	.367	
2 x 1¼	¼	2.55	.750	.296	.229	.628	.703	.089	.097	.344	.333	.269	.378	¼*
	3/16	1.96	.574	.232	.177	.636	.686	.071	.075	.351	.311	.271	.387	
	⅛	1.33	.391	.163	.122	.645	.663	.050	.052	.359	.237	.274	.396	
A964	3/16	1.81	.527	.214	.170	.638	.738	.037	.048	.263	.238	.213	.258	¼
2 x 1														
A39	¼	.234	.688	.202	.176	.543	.602	.085	.095	.352	.352	.267	.486	
1¾ x 1¼	3/16	1.80	.527	.160	.137	.551	.580	.068	.074	.359	.330	.269	.496	3/16*
	⅛	1.23	.359	.113	.094	.560	.557	.049	.051	.368	.307	.272	.506	
A624	¼	2.13	.625	.130	.130	.456	.500	.081	.093	.361	.375	.260	.667	
1½ x 1¼	3/16	1.64	.480	.104	.104	.464	.478	.065	.073	.368	.353	.261	.676	⅛
A541	⅛	.91	.266	.061	.064	.480	.548	.011	.018	.199	.173	.160	.261	3/32
1½ x ¾														
A40	3/16	1.32	.387	.071	.081	.429	.490	.022	.035	.240	.240	.188	.387	
1⅜ x ⅞	⅛	.91	.266	.051	.056	.438	.467	.016	.025	.247	.217	.190	.401	⅛*
A627	⅛	.70	.203	.020	.030	.312	.332	.009	.017	.216	.207	.160	.543	1/16
1 x ¾														
A42	⅛	.64	.188	.019	.029	.314	.354	.006	.012	.172	.167	.134	.378	⅛*
1 x ⅝														

*Angles are produced with various size fillet radius depending on where they are rolled. The maximum radii produced are those shown.

Tees from W shapes (WT) 2.3.3.11

WT

Structural Tees

Cut from W—Wide Flange Shapes

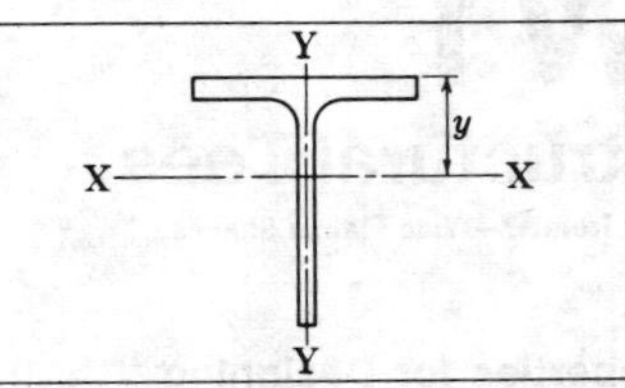

Properties for Designing

Designation	Weight per Foot	Area of Section	Depth of Tee	Flange Width	Flange Thickness	Stem Thickness	Axis X-X I	Axis X-X S	Axis X-X r	Axis X-X y	Axis Y-Y I	Axis Y-Y S	Axis Y-Y r
In.	Lbs.	In.²	In.	In.	In.	In.	In.⁴	In.³	In.	In.	In.⁴	In.³	In.
WT18 From W36	150	44.1	18.360	16.655	1.680	.945	1220	86.0	5.27	4.13	648	77.8	3.83
	140	41.2	18.250	16.595	1.570	.885	1130	80.0	5.25	4.06	599	72.2	3.81
	130	38.2	18.120	16.551	1.440	.841	1060	75.1	5.26	4.05	545	65.9	3.77
	122.5	36.1	18.030	16.512	1.350	.802	995	71.1	5.25	4.03	507	61.4	3.75
	115	33.8	17.940	16.471	1.260	.761	933	67.0	5.25	4.00	470	57.1	3.73
WT18 From W36	97	28.6	18.240	12.117	1.260	.770	905	67.4	5.63	4.81	188	31.0	2.56
	91	26.8	18.160	12.072	1.180	.725	845	63.1	5.61	4.77	174	28.8	2.55
	85	25.0	18.080	12.027	1.100	.680	786	58.8	5.60	4.73	160	26.6	2.53
	80	23.6	18.000	12.000	1.020	.653	742	56.0	5.61	4.75	147	24.6	2.50
	75	22.1	17.920	11.972	.940	.625	698	53.1	5.62	4.78	135	22.5	2.47
	67.5	19.9	17.775	11.945	.794	.598	636	49.5	5.65	4.94	113	18.9	2.39
WT16.5 From W33	120	35.3	16.750	15.865	1.400	.830	823	63.2	4.83	3.73	467	58.8	3.64
	110	32.4	16.625	15.810	1.275	.775	755	58.4	4.83	3.70	421	53.2	3.60
	100	29.4	16.500	15.750	1.150	.715	685	53.3	4.82	3.66	375	47.6	3.57
WT16.5 From W33	76	22.4	16.750	11.565	1.055	.635	592	47.4	5.15	4.26	136	23.6	2.47
	70.5	20.8	16.655	11.535	.960	.605	552	44.7	5.16	4.29	123	21.3	2.43
	65	19.2	16.550	11.510	.855	.580	514	42.2	5.18	4.37	109	18.9	2.38
	59	17.4	16.430	11.484	.738	.554	471	39.4	5.21	4.48	93.4	16.3	2.32
WT15 From W30	105	30.9	15.190	15.105	1.315	.775	579	48.7	4.33	3.31	378	50.1	3.50
	95	28.0	15.060	15.040	1.185	.710	521	44.1	4.31	3.25	336	44.7	3.47
	86	25.4	14.940	14.985	1.065	.655	472	40.2	4.31	3.22	299	39.9	3.43
WT15 From W30	66	19.4	15.150	10.551	1.000	.615	421	37.4	4.65	3.90	98.2	18.6	2.25
	62	18.2	15.080	10.521	.930	.585	395	35.3	4.65	3.89	90.5	17.2	2.23
	58	17.1	15.000	10.500	.850	.564	372	33.6	4.67	3.93	82.2	15.7	2.19
	54	15.9	14.910	10.484	.760	.548	350	32.1	4.69	4.02	73.2	14.0	2.15
	49.5	14.6	14.820	10.458	.670	.522	323	30.1	4.71	4.10	64.1	12.3	2.10
WT13.5 From W27	88.5	26.1	13.655	14.090	1.190	.725	393	36.8	3.88	2.97	278	39.4	3.26
	80	23.6	13.540	14.023	1.075	.658	352	33.1	3.87	2.90	247	35.3	3.24
	72.5	21.4	13.440	13.965	.975	.600	317	29.9	3.85	2.85	222	31.7	3.22
WT13.5 From W27	57	16.8	13.640	10.070	.932	.570	289	28.3	4.15	3.41	79.5	15.8	2.18
	51	15.0	13.535	10.018	.827	.518	258	25.4	4.14	3.38	69.5	13.9	2.15
	47	13.8	13.455	9.990	.747	.490	239	23.8	4.15	3.41	62.2	12.5	2.12
	42	12.4	13.345	9.963	.636	.463	216	22.0	4.18	3.50	52.5	10.5	2.06
WT12 From W24	80	23.6	12.360	14.091	1.135	.656	272	27.6	3.40	2.50	265	37.6	3.35
	72.5	21.4	12.245	14.043	1.020	.608	247	25.2	3.40	2.47	236	33.6	3.32
	65	19.2	12.125	14.000	.900	.565	223	23.1	3.41	2.46	206	29.4	3.28

Tees from W shapes (WT), continued 2.3.3.11

WT

Structural Tees

Cut from W—Wide Flange Shapes

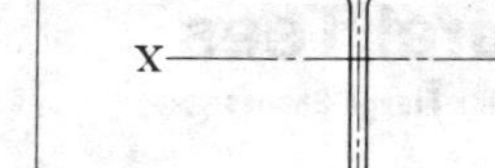

Properties for Designing

Designation	Weight per Foot	Area of Section	Depth of Tee	Flange		Stem Thickness	Axis X-X				Axis Y-Y		
				Width	Thickness		I	S	r	y	I	S	r
In.	Lbs.	In.²	In.	In.	In.	In.	In.⁴	In.³	In.	In.	In.⁴	In.³	In.
WT12	60	17.7	12.155	12.088	.930	.556	215	22.5	3.49	2.62	137	22.7	2.78
From W24	55	16.2	12.080	12.042	.855	.510	195	20.5	3.47	2.57	125	20.7	2.77
	50	14.8	12.000	12.000	.775	.468	177	18.7	3.46	2.53	112	18.6	2.75
WT12	47	13.8	12.145	9.061	.872	.516	186	20.3	3.67	3.00	54.2	12.0	1.98
From W24	42	12.4	12.045	9.015	.772	.470	166	18.3	3.66	2.97	47.2	10.5	1.95
	38	11.2	11.955	8.985	.682	.440	151	16.9	3.68	2.99	41.3	9.20	1.92
	34	10.0	11.855	8.961	.582	.416	137	15.6	3.70	3.07	35.0	7.81	1.87
WT10.5	71	20.9	10.730	13.132	1.095	.659	177	20.8	2.92	2.18	207	31.5	3.15
From W21	63.5	18.7	10.620	13.061	.985	.588	156	18.3	2.89	2.11	183	28.0	3.13
	56	16.5	10.500	13.000	.865	.527	137	16.2	2.88	2.06	159	24.4	3.10
WT10.5	48	14.1	10.570	9.038	.935	.575	137	17.1	3.12	2.54	57.7	12.8	2.02
From W21	41	12.1	10.430	8.962	.795	.499	116	14.6	3.10	2.48	47.8	10.7	1.99
WT10.5	36.5	10.7	10.620	8.295	.740	.455	110	13.8	3.21	2.60	35.3	8.51	1.81
From W21	34	10.0	10.565	8.270	.685	.430	103	12.9	3.20	2.59	32.4	7.83	1.80
	31	9.13	10.495	8.240	.615	.400	93.8	11.9	3.21	2.58	28.7	6.97	1.77
	27.5	8.10	10.400	8.215	.522	.375	84.4	10.9	3.23	2.64	24.2	5.88	1.73
WT9	57	16.8	9.240	11.833	.991	.595	103	13.9	2.48	1.85	137	23.2	2.86
From W18	52.5	15.4	9.160	11.792	.911	.554	94.0	12.8	2.47	1.82	125	21.1	2.84
	48	14.1	9.080	11.750	.831	.512	85.4	11.7	2.46	1.78	112	19.1	2.82
WT9	42.5	12.5	9.160	8.838	.911	.526	84.4	11.9	2.60	2.05	52.5	11.9	2.05
From W18	38.5	11.4	9.080	8.787	.831	.475	75.3	10.6	2.58	1.99	47.1	10.7	2.04
	35	10.3	9.000	8.750	.751	.438	68.2	9.68	2.57	1.96	42.0	9.60	2.02
	32	9.43	8.935	8.715	.686	.403	61.9	8.83	2.56	1.92	37.9	8.70	2.00
WT9	30	8.83	9.125	7.558	.695	.416	64.9	9.32	2.71	2.16	25.1	6.63	1.68
From W18	27.5	8.10	9.060	7.532	.630	.390	59.6	8.64	2.71	2.16	22.5	5.97	1.67
	25	7.36	9.000	7.500	.570	.358	54.0	7.86	2.71	2.13	20.1	5.35	1.65
	22.5	6.62	8.930	7.477	.499	.335	49.0	7.24	2.72	2.16	17.4	4.66	1.62
WT8	48	14.1	8.160	11.533	.875	.535	64.7	9.82	2.14	1.57	112	19.4	2.82
From W16	44	12.9	8.080	11.502	.795	.504	59.5	9.11	2.14	1.55	101	17.5	2.79
WT8	39	11.5	8.160	8.586	.875	.529	60.0	9.45	2.28	1.81	46.3	10.8	2.01
From W16	35.5	10.5	8.080	8.543	.795	.486	54.1	8.57	2.27	1.77	41.4	9.69	1.99
	32	9.41	8.000	8.500	.715	.443	48.3	7.72	2.27	1.73	36.7	8.63	1.97
	29	8.53	7.930	8.464	.645	.407	43.6	7.01	2.26	1.71	32.6	7.71	1.96

Tees from W shapes (WT), continued 2.3.3.11

WT

Structural Tees

Cut from W—Wide Flange Shapes

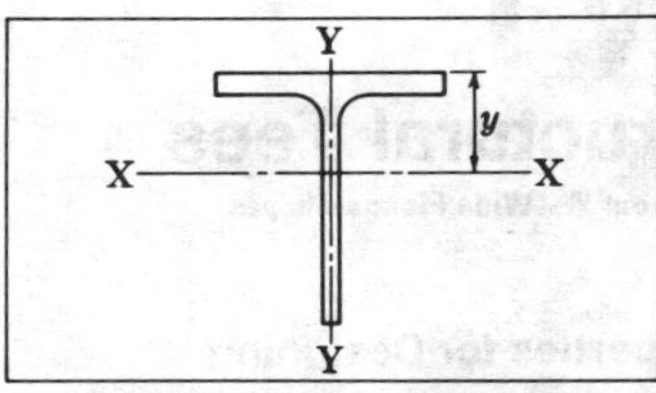

Properties for Designing

Designation	Weight per Foot	Area of Section	Depth of Tee	Flange		Stem Thickness	Axis X-X				Axis Y-Y		
				Width	Thickness		*I*	*S*	*r*	*y*	*I*	*S*	*r*
In.	Lbs.	In.²	In.	In.	In.	In.	In.⁴	In.³	In.	In.	In.⁴	In.³	In.
WT8 From W16	25	7.36	8.125	7.073	.628	.380	42.2	6.77	2.40	1.89	18.6	5.25	1.59
	22.5	6.63	8.060	7.039	.563	.346	37.8	6.10	2.39	1.86	16.4	4.66	1.57
	20	5.89	8.000	7.000	.503	.307	33.2	5.38	2.37	1.82	14.4	4.11	1.56
	18	5.30	7.925	6.992	.428	.299	30.8	5.11	2.41	1.89	12.2	3.49	1.52
WT8 From W16	15.5	4.57	7.920	5.525	.442	.275	27.3	4.62	2.44	2.01	6.23	2.25	1.17
	13	3.84	7.825	5.500	.345	.250	23.3	4.07	2.47	2.08	4.80	1.74	1.12
WT7 From W14	365	107	11.220	17.889	4.910	3.069	740	95.6	2.63	3.47	2360	264	4.69
	332.5	97.8	10.835	17.646	4.522	2.826	623	82.2	2.52	3.25	2080	236	4.62
	302.5	89.0	10.470	17.418	4.157	2.598	525	70.8	2.43	3.05	1840	211	4.55
	275	80.9	10.130	17.206	3.818	2.386	444	61.1	2.34	2.86	1630	189	4.49
	250	73.5	9.815	17.008	3.501	2.188	377	52.8	2.26	2.68	1440	169	4.43
	227.5	66.9	9.525	16.828	3.213	2.008	322	45.9	2.19	2.51	1280	152	4.37
	213	62.6	9.345	16.695	3.033	1.875	288	41.4	2.14	2.40	1180	141	4.34
	199	58.5	9.155	16.590	2.843	1.770	258	37.7	2.10	2.30	1080	131	4.31
	185	54.4	8.970	16.475	2.658	1.655	230	34.0	2.06	2.19	993	121	4.27
	171	50.3	8.780	16.365	2.468	1.545	204	30.5	2.02	2.09	903	110	4.24
	160	47.1	8.405	16.710	2.093	1.890	209	33.3	2.11	2.12	818	97.8	4.17
	157	46.2	8.595	16.235	2.283	1.415	179	27.0	1.97	1.98	816	100	4.20
	143.5	42.2	8.405	16.130	2.093	1.310	157	24.1	1.93	1.87	733	90.9	4.17
	132	38.8	8.250	16.025	1.938	1.205	139	21.5	1.89	1.78	666	83.1	4.14
	123	36.2	8.125	15.945	1.813	1.125	126	19.6	1.86	1.71	613	76.9	4.12
	118.5	34.8	8.060	15.910	1.748	1.090	120	18.7	1.85	1.67	587	73.8	4.11
	114	33.5	8.000	15.865	1.688	1.045	113	17.7	1.84	1.64	562	70.9	4.10
	109.5	32.2	7.935	15.825	1.623	1.005	107	16.9	1.82	1.60	537	67.8	4.08
	105.5	31.0	7.875	15.800	1.563	.980	102	16.2	1.82	1.57	514	65.1	4.07
	101	29.7	7.815	15.750	1.503	.930	95.8	15.2	1.80	1.53	490	62.2	4.06
	96.5	28.4	7.750	15.710	1.438	.890	90.1	14.4	1.78	1.49	465	59.2	4.05
	92	27.0	7.690	15.660	1.378	.840	83.9	13.4	1.76	1.45	441	56.4	4.04
	88	25.9	7.625	15.640	1.313	.820	80.2	12.9	1.76	1.42	419	53.6	4.02
	83.5	24.5	7.560	15.600	1.248	.780	75.0	12.2	1.75	1.39	395	50.7	4.01
	79	23.2	7.500	15.550	1.188	.730	69.3	11.3	1.73	1.34	372	47.9	4.00
	75	22.0	7.440	15.515	1.128	.695	65.0	10.6	1.72	1.31	351	45.3	3.99
	71	20.9	7.375	15.500	1.063	.680	62.1	10.2	1.72	1.29	330	42.6	3.97
WT7 From W14	68	20.0	7.375	14.740	1.063	.660	60.1	9.89	1.73	1.31	284	38.5	3.77
	63.5	18.7	7.310	14.690	.998	.610	54.7	9.05	1.71	1.26	264	35.9	3.76
	59.5	17.5	7.250	14.650	.938	.570	50.4	8.36	1.70	1.22	246	33.6	3.75
	55.5	16.3	7.185	14.620	.873	.540	46.9	7.82	1.69	1.19	227	31.1	3.73
	51.5	15.1	7.125	14.575	.813	.495	42.4	7.10	1.67	1.15	210	28.8	3.72
	47.5	14.0	7.060	14.545	.748	.465	39.1	6.58	1.67	1.12	192	26.4	3.71
	43.5	12.8	7.000	14.500	.688	.420	34.9	5.88	1.65	1.08	175	24.1	3.70

Tees from W shapes (WT), continued 2.3.3.11

WT

Structural Tees

Cut from W—Wide Flange Shapes

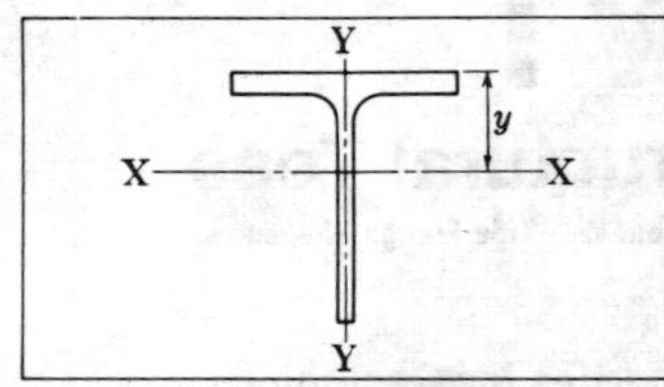

Properties for Designing

Designation	Weight per Foot	Area of Section	Depth of Tee	Flange		Stem Thickness	Axis X-X				Axis Y-Y		
				Width	Thickness		I	S	r	y	I	S	r
In.	Lbs.	In.²	In.	In.	In.	In.	In.⁴	In.³	In.	In.	In.⁴	In.³	In.
WT7	42	12.4	7.090	12.023	.778	.451	37.4	6.36	1.74	1.21	113	18.8	3.02
From W14	39	11.5	7.030	12.000	.718	.428	34.8	5.96	1.74	1.19	103	17.2	3.00
WT7	37	10.9	7.095	10.072	.783	.450	36.1	6.26	1.82	1.32	66.7	13.3	2.48
From W14	34	10.0	7.030	10.040	.718	.418	33.0	5.75	1.82	1.29	60.6	12.1	2.46
	30.5	8.97	6.955	10.000	.643	.378	29.2	5.13	1.80	1.25	53.6	10.7	2.45
WT7	26.5	7.79	6.970	8.062	.658	.370	27.7	4.96	1.88	1.38	28.8	7.14	1.92
From W14	24	7.06	6.905	8.031	.593	.339	24.9	4.49	1.88	1.35	25.6	6.38	1.91
	21.5	6.32	6.840	8.000	.528	.308	22.2	4.02	1.87	1.33	22.6	5.64	1.89
WT7	19	5.59	7.060	6.776	.513	.313	23.5	4.27	2.05	1.55	13.3	3.93	1.54
From W14	17	5.01	7.000	6.750	.453	.287	21.1	3.87	2.05	1.54	11.6	3.44	1.52
	15	4.42	6.930	6.733	.383	.270	19.0	3.56	2.08	1.58	9.76	2.90	1.49
WT7	13	3.83	6.945	5.025	.418	.255	17.2	3.30	2.12	1.72	4.43	1.76	1.08
From W14	11	3.24	6.860	5.000	.335	.230	14.8	2.90	2.13	1.76	3.50	1.40	1.04
WT6	95	27.9	7.190	12.670	1.736	1.060	79.0	14.2	1.68	1.62	295	46.5	3.25
From W12	80.5	23.7	6.940	12.515	1.486	.905	62.6	11.5	1.63	1.47	243	38.9	3.20
	66.5	19.6	6.690	12.365	1.236	.755	48.4	9.04	1.57	1.33	195	31.5	3.16
	60	17.7	6.560	12.320	1.106	.710	43.4	8.22	1.57	1.28	173	28.0	3.13
	53	15.6	6.440	12.230	.986	.620	36.7	7.01	1.53	1.20	150	24.6	3.11
	49.5	14.6	6.375	12.192	.921	.582	33.8	6.48	1.52	1.16	139	22.8	3.09
	46	13.5	6.310	12.155	.856	.545	31.0	5.99	1.51	1.13	128	21.1	3.08
	42.5	12.5	6.250	12.105	.796	.495	27.8	5.38	1.49	1.08	118	19.5	3.07
	39.5	11.6	6.190	12.080	.736	.470	25.8	5.03	1.49	1.06	108	17.9	3.05
	36	10.6	6.125	12.040	.671	.430	23.2	4.54	1.48	1.02	97.6	16.2	3.04
	32.5	9.55	6.060	12.000	.606	.390	20.6	4.06	1.47	.985	87.3	14.6	3.02
WT6	29	8.53	6.095	10.014	.641	.359	19.0	3.75	1.49	1.03	53.7	10.7	2.51
From W12	26.5	7.80	6.030	10.000	.576	.345	17.7	3.54	1.51	1.02	48.0	9.61	2.48
WT6	25	7.36	6.095	8.077	.641	.371	18.7	3.80	1.60	1.17	28.2	6.98	1.96
From W12	22.5	6.62	6.030	8.042	.576	.336	16.6	3.40	1.59	1.13	25.0	6.22	1.94
	20	5.89	5.970	8.000	.516	.294	14.4	2.94	1.56	1.08	22.0	5.51	1.94
WT6	18	5.30	6.120	6.565	.540	.305	15.3	3.14	1.70	1.26	12.7	3.88	1.55
From W12	15.5	4.57	6.045	6.525	.465	.265	13.0	2.69	1.69	1.22	10.8	3.30	1.54
	13.5	3.97	5.980	6.497	.400	.237	11.3	2.37	1.69	1.20	9.15	2.82	1.52

Tees from W shapes (WT), continued 2.3.3.11

WT
Structural Tees
Cut from W—Wide Flange Shapes

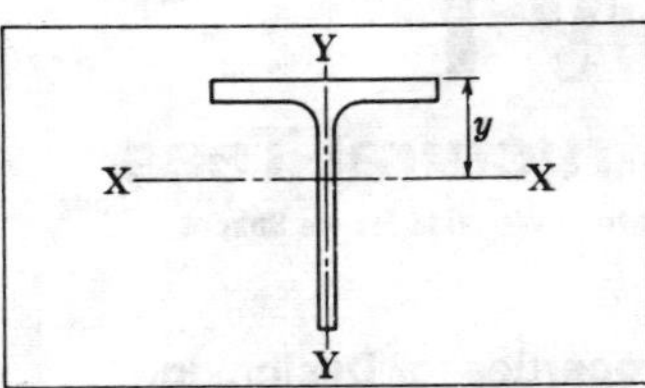

Properties for Designing

Designation	Weight per Foot	Area of Section	Depth of Tee	Flange Width	Flange Thickness	Stem Thickness	Axis X-X I	Axis X-X S	Axis X-X r	Axis X-X y	Axis Y-Y I	Axis Y-Y S	Axis Y-Y r
In.	Lbs.	In.²	In.	In.	In.	In.	In.⁴	In.³	In.	In.	In.⁴	In.³	In.
WT6	11	3.24	6.155	4.030	.424	.260	11.7	2.59	1.90	1.63	2.32	1.15	.847
From W12	9.5	2.80	6.080	4.007	.349	.237	10.2	2.30	1.91	1.65	1.88	.938	.820
	8.25	2.43	6.000	4.000	.269	.230	9.03	2.13	1.93	1.76	1.44	.721	.770
WT6 From W12	7	2.06	5.955	3.968	.224	.198	7.61	1.81	1.92	1.75	1.17	.590	.754
WT5	56	16.5	5.690	10.415	1.248	.755	28.8	6.42	1.32	1.21	118	22.6	2.67
From W10	50	14.7	5.560	10.345	1.118	.685	24.8	5.62	1.30	1.14	103	20.0	2.65
	44.5	13.1	5.440	10.275	.998	.615	21.3	4.88	1.28	1.07	90.3	17.6	2.63
	38.5	11.3	5.310	10.195	.868	.535	17.7	4.10	1.25	.996	76.7	15.1	2.60
	36	10.6	5.250	10.170	.808	.510	16.4	3.83	1.24	.971	70.9	13.9	2.59
	33	9.70	5.190	10.117	.748	.457	14.5	3.39	1.22	.922	64.6	12.8	2.58
	30	8.83	5.125	10.075	.683	.415	12.8	3.03	1.21	.882	58.2	11.6	2.57
	27	7.94	5.060	10.028	.618	.368	11.2	2.64	1.19	.836	52.0	10.4	2.56
	24.5	7.20	5.000	10.000	.558	.340	10.1	2.40	1.18	.809	46.5	9.30	2.54
WT5	22.5	6.62	5.060	8.022	.618	.350	10.3	2.48	1.25	.910	26.6	6.63	2.00
From W10	19.5	5.74	4.970	7.990	.528	.318	8.96	2.19	1.25	.883	22.5	5.62	1.98
	16.5	4.85	4.875	7.964	.433	.292	7.80	1.95	1.27	.875	18.2	4.58	1.94
WT5	14.5	4.27	5.110	5.799	.500	.289	8.39	2.07	1.40	1.05	8.14	2.81	1.38
From W10	12.5	3.68	5.040	5.762	.430	.252	7.13	1.77	1.39	1.01	6.86	2.38	1.37
	10.5	3.10	4.950	5.750	.340	.240	6.32	1.62	1.43	1.06	5.39	1.88	1.32
WT5	9.5	2.81	5.125	4.020	.394	.250	6.70	1.74	1.55	1.28	2.14	1.06	.874
From W10	8.5	2.49	5.060	4.010	.329	.240	6.07	1.62	1.56	1.32	1.77	.885	.844
	7.5	2.20	5.000	4.000	.269	.230	5.46	1.51	1.57	1.37	1.44	.720	.809
WT5 From W10	5.75	1.70	4.935	3.950	.204	.180	4.16	1.16	1.57	1.34	1.05	.532	.787
WT4	33.5	9.85	4.500	8.287	.933	.575	10.9	3.07	1.05	.939	44.3	10.7	2.12
From W8	29	8.53	4.375	8.222	.808	.510	9.12	2.61	1.03	.874	37.5	9.12	2.10
	24	7.06	4.250	8.117	.683	.405	6.92	2.00	.990	.781	30.5	7.51	2.08
	20	5.88	4.125	8.077	.558	.365	5.80	1.71	.993	.740	24.5	6.07	2.04
	17.5	5.15	4.060	8.027	.493	.315	4.88	1.45	.973	.694	21.3	5.30	2.03
	15.5	4.56	4.000	8.000	.433	.288	4.31	1.30	.973	.672	18.5	4.62	2.01
WT4	14	4.11	4.030	6.540	.463	.285	4.22	1.28	1.01	.735	10.8	3.30	1.62
From W8	12	3.53	3.965	6.500	.398	.245	3.53	1.08	1.00	.695	9.12	2.80	1.61
WT4	10	2.95	4.070	5.268	.378	.248	3.67	1.13	1.12	.825	4.61	1.75	1.25
From W8	8.5	2.50	4.000	5.250	.308	.230	3.21	1.02	1.13	.835	3.72	1.42	1.22

Tees from W shapes (WT), continued 2.3.3.11

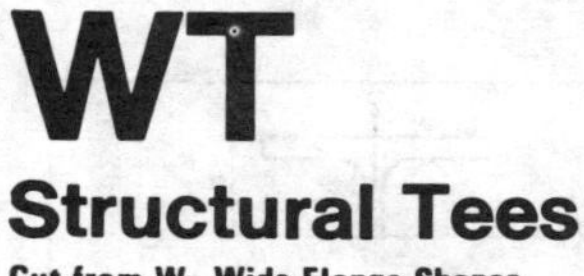

WT

Structural Tees

Cut from W—Wide Flange Shapes

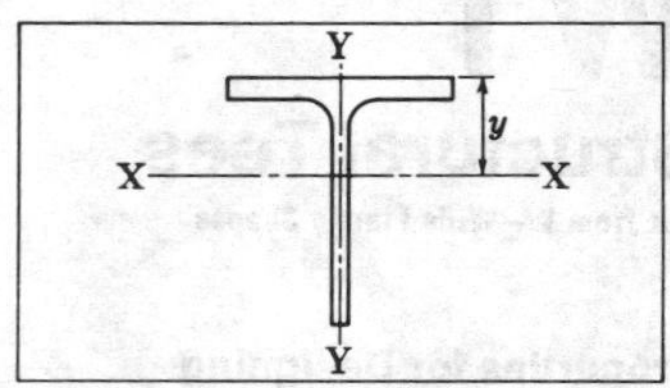

Properties for Designing

Designation	Weight per Foot	Area of Section	Depth of Tee	Flange		Stem Thickness	Axis X-X				Axis Y-Y		
				Width	Thickness		*I*	*S*	*r*	*y*	*I*	*S*	*r*
In.	Lbs.	In.^{2}	In.	In.	In.	In.	In.^{4}	In.^{3}	In.	In.	In.^{4}	In.^{3}	In.
WT4 From W8*	6.5	1.92	4.000	4.000	.254	.230	2.90	.976	1.23	1.03	1.36	.680	.842
WT4 From W8	5	1.48	3.950	3.940	.204	.170	2.15	.719	1.21	.957	1.04	.529	.839
WT3 From W6	12.5	3.67	3.185	6.080	.456	.320	2.27	.883	.787	.609	8.55	2.81	1.53
	10	2.94	3.100	6.018	.367	.258	1.75	.688	.771	.557	6.67	2.22	1.51
	7.75	2.28	3.000	5.995	.269	.235	1.44	.591	.795	.559	4.83	1.61	1.46
WT3 From W6*	6	1.77	3.000	4.000	.279	.230	1.30	.558	.857	.673	1.49	.746	.918
WT3 From W6	4.25	1.25	2.915	3.940	.194	.170	.904	.397	.849	.638	.990	.503	.889

*These shapes are rolled in the Pittsburgh District with a 3° flange slope. The flange thicknesses shown are the average thicknesses.

Tees from S beams (ST) 2.3.3.12

ST

Structural Tees

Cut from S—American Standard Beams

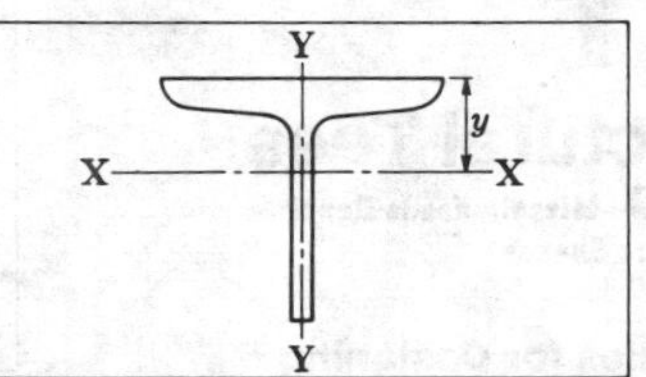

Properties for Designing

Designation	Weight per Foot	Area of Section	Depth of Tee	Flange		Stem Thick-ness	Axis X-X				Axis Y-Y		
				Width	Average Thick-ness		I	S	r	y	I	S	r
In.	Lbs.	In.^{2}	In.	In.	In.	In.	In.^{4}	In.^{3}	In.	In.	In.^{4}	In.^{3}	In.
ST12 From S24	60	17.6	12.000	8.048	1.102	.798	245	28.9	3.72	3.52	42.1	10.5	1.54
	52.95	15.6	12.000	7.875	1.102	.625	205	23.3	3.63	3.19	39.1	9.92	1.58
ST12 From S24	50	14.7	12.000	7.247	.871	.747	215	26.4	3.83	3.84	23.9	6.59	1.27
	45	13.2	12.000	7.124	.871	.624	190	22.6	3.79	3.60	22.5	6.31	1.30
	39.95	11.8	12.000	7.001	.871	.501	163	18.7	3.72	3.30	21.1	6.04	1.34
ST10 From S20	47.5	14.0	10.000	7.200	.916	.800	137	19.7	3.13	3.07	24.8	6.90	1.33
	42.5	12.5	10.000	7.053	.916	.653	118	16.6	3.08	2.85	23.1	6.55	1.36
ST10 From S20	37.5	11.0	10.000	6.391	.789	.641	110	15.9	3.16	3.08	14.8	4.64	1.16
	32.7	9.62	10.000	6.250	.789	.500	92.3	12.8	3.10	2.80	13.7	4.38	1.19
ST9 From S18	35	10.3	9.000	6.251	.691	.711	84.7	14.0	2.87	2.94	12.1	3.86	1.08
	27.35	8.04	9.000	6.001	.691	.461	62.4	9.61	2.79	2.50	10.4	3.47	1.14
ST7.5 From S15	25	7.35	7.500	5.640	.622	.550	40.6	7.73	2.35	2.25	7.85	2.78	1.03
	21.45	6.31	7.500	5.501	.622	.411	33.0	6.00	2.29	2.01	7.19	2.61	1.07
ST6 From S12	25	7.35	6.000	5.477	.659	.687	25.2	6.05	1.85	1.84	7.85	2.87	1.03
	20.4	6.00	6.000	5.252	.659	.462	18.9	4.28	1.78	1.58	6.78	2.58	1.06
ST6 From S12	17.5	5.14	6.000	5.078	.544	.428	17.2	3.95	1.83	1.65	4.93	1.94	.980
	15.9	4.68	6.000	5.000	.544	.350	14.9	3.31	1.78	1.51	4.68	1.87	1.00
ST5 From S10	17.5	5.15	5.000	4.944	.491	.594	12.5	3.63	1.56	1.56	4.18	1.69	.901
	12.7	3.73	5.000	4.661	.491	.311	7.83	2.06	1.45	1.20	3.39	1.46	.954
ST4 From S8	11.5	3.38	4.000	4.171	.425	.441	5.03	1.77	1.22	1.15	2.15	1.03	.798
	9.2	2.70	4.000	4.001	.425	.271	3.51	1.15	1.14	.941	1.86	.932	.831
ST3.5 From S7	10	2.94	3.500	3.860	.392	.450	3.36	1.36	1.07	1.04	1.59	.821	.734
	7.65	2.25	3.500	3.662	.392	.252	2.19	.816	.987	.817	1.32	.720	.766
ST3 From S6	8.625	2.53	3.000	3.565	.359	.465	2.13	1.02	.917	.914	1.15	.648	.675
	6.25	1.83	3.000	3.332	.359	.232	1.27	.552	.833	.691	.911	.547	.705

Tees from M beams (MT) 2.3.3.13

MT

Structural Tees

Cut from M—Miscellaneous Beam and Column Shapes

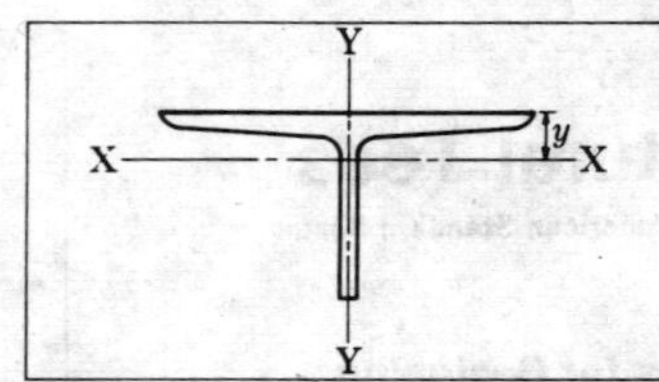

Properties for Designing

Designation	Weight per Foot	Area of Section	Depth of Tee	Flange Width	Flange Average Thickness	Stem Thickness	Axis X-X I	Axis X-X S	Axis X-X r	Axis X-X y	Axis Y-Y I	Axis Y-Y S	Axis Y-Y r
In.	Lbs.	In.²	In.	In.	In.	In.	In.⁴	In.³	In.	In.	In.⁴	In.³	In.
MT3 From M6	12.5	3.67	3.000	5.942	.480	.317	1.88	.774	.715	.572	7.49	2.52	1.43
	10	2.94	3.000	5.938	.379	.250	1.54	.624	.724	.531	5.80	1.95	1.40

Tees (T) 2.3.3.14

T

Tee Shapes

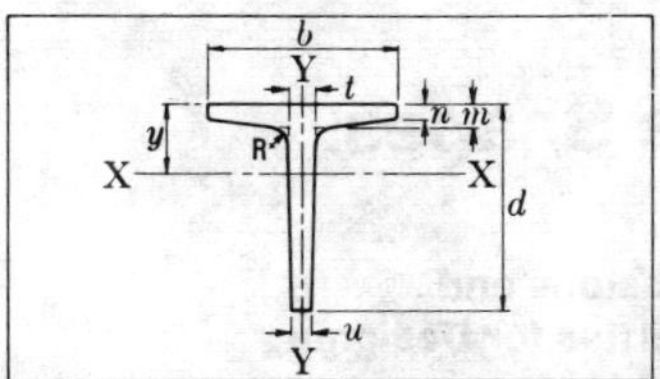

Dimensions and Properties for Designing

Designation and Nominal Size	Weight per Foot	Area of Section	Depth of Section *d*	Flange Width *b*	Flange Thickness *m*	Flange Thickness *n*	Stem Thickness *t*	Stem Thickness *u*	Axis X-X *I*	Axis X-X *S*	Axis X-X *r*	Axis X-X *y*	Axis Y-Y *I*	Axis Y-Y *S*	Axis Y-Y *r*	Fillet Radius R
In.	Lbs.	In.²	In.	In.	In.	In.	In.	In.	In.⁴	In.³	In.	In.	In.⁴	In.³	In.	In.
T4 4 x 4	13.5	3.97	4	4	9/16	1/2	9/16	1/2	6.63	2.33	1.28	1.15	2.82	1.41	.833	1/2
T4 4 x 3	9.2	2.71	3	4	7/16	3/8	7/16	3/8	2.13	.947	.882	.754	2.11	1.06	.880	3/8
T3 3 x 3	7.8	2.31	3	3	7/16	3/8	7/16	3/8	2.05	.957	.942	.860	.902	.601	.625	5/16
T3 3 x 3	6.7	1.98	3	3	3/8	5/16	3/8	5/16	1.76	.811	.942	.833	.754	.503	.617	5/16
T3 3 x 2½	6.1	1.80	2½	3	3/8	5/16	3/8	5/16	.997	.542	.745	.659	.752	.501	.647	1/4

Zees (Z) 2.3.3.15

Z
Zee Shapes

Dimensions and Properties for Designing

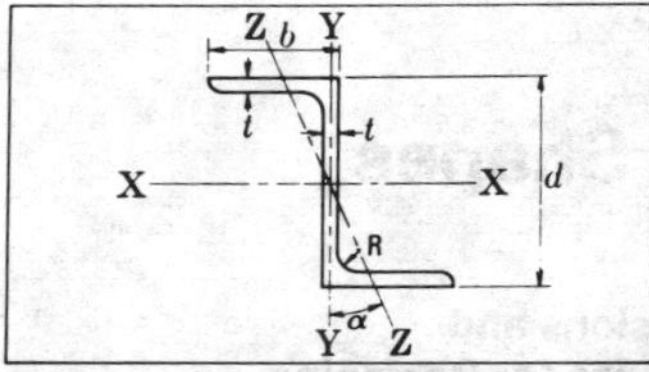

Designation and Nominal Size	Weight per Foot	Area of Section	Depth of Section d	Flange Width b	Thickness t	Axis X-X			Axis Y-Y			Axis Z-Z		Fillet Radius R
						I	S	r	I	S	r	$r_{min.}$	Tan α	
In.	Lbs.	In.^{2}	In.	In.	In.	In.^{4}	In.^{3}	In.	In.^{4}	In.^{3}	In.	In.		In.
Z6	21.1	6.19	6⅛	3⅝	½	34.4	11.2	2.36	12.9	3.81	1.44	.843	.532	
6 x 3½	15.7	4.59	6	3½	⅜	25.3	8.43	2.35	9.11	2.75	1.41	.823	.520	5/16
Z5	17.9	5.25	5	3¼	½	19.2	7.68	1.91	9.05	3.02	1.31	.738	.616	
5 x 3¼	16.4	4.81	5⅛	3⅜	7/16	19.1	7.44	1.99	9.20	2.92	1.38	.768	.626	5/16
	14.0	4.10	5 1/16	3 5/16	⅜	16.2	6.39	1.99	7.65	2.45	1.37	.758	.619	
	11.6	3.40	5	3¼	5/16	13.4	5.34	1.98	6.18	2.00	1.35	.750	.611	
Z4	12.5	3.66	4⅛	3 3/16	⅜	9.63	4.67	1.62	6.77	2.26	1.36	.689	.798	
4 x 3	10.3	3.03	4 1/16	3⅛	5/16	7.94	3.91	1.62	5.46	1.84	1.34	.681	.788	5/16
	8.2	2.41	4	3 1/16	¼	6.28	3.14	1.62	4.23	1.44	1.33	.673	.777	
Z3	12.6	3.69	3	2 11/16	½	4.59	3.06	1.12	4.85	1.99	1.15	.534	1.04	
3 x 2¾	9.8	2.86	3	2 11/16	⅜	3.85	2.57	1.16	3.92	1.57	1.17	.541	1.01	5/16
	6.7	1.97	3	2 11/16	¼	2.87	1.92	1.21	2.81	1.10	1.19	.547	.986	

H piles (HP) 2.3.3.16

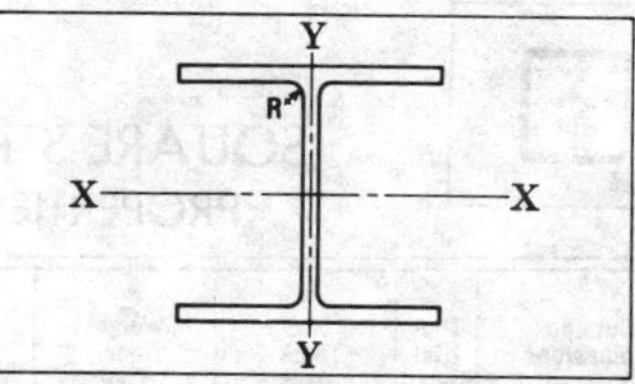

Properties of Sections

Designation and Nominal Size	Weight per Foot	Area of Section	Depth of Section	Flange		Web Thickness	Axis X-X			Axis Y-Y			Fillet Radius R
				Width	Thickness		*I*	*S*	*r*	*I*	*S*	*r*	
In.	Lbs.	In.²	In.	In.	In.	In.	In.⁴	In.³	In.	In.⁴	In.³	In.	In.
HP14 14 x 14½	117	34.4	14.23	14.885	.805	.805	1230	173	5.97	443	59.5	3.59	.60
	102	30.0	14.03	14.784	.704	.704	1050	150	5.93	380	51.3	3.56	
	89	26.2	13.86	14.696	.616	.616	910	131	5.89	326	44.4	3.53	
	73	21.5	13.64	14.586	.506	.506	734	108	5.85	262	35.9	3.49	
HP12 12 x 12	74	21.8	12.12	12.217	.607	.607	566	93.4	5.10	185	30.2	2.91	.60
	53	15.6	11.78	12.046	.436	.436	394	66.9	5.03	127	21.1	2.86	
HP10 10 x 10	57	16.8	10.01	10.224	.564	.564	295	58.8	4.19	101	19.7	2.45	.50
	42	12.4	9.72	10.078	.418	.418	211	43.4	4.13	71.4	14.2	2.40	
HP8 8 x 8	36	10.6	8.03	8.158	.446	.446	120	29.9	3.36	40.4	9.91	1.95	.40
HPS10 10 x 8¼	57	16.8	10.18	8.32	.648	.648	287	56.4	4.14	62.4	15.0	1.93	.50
HPS10 10 x 8	42	12.4	9.85	8.153	.482	.482	205	41.6	4.07	43.7	10.7	1.88	.50
HP8 8 x 8	36	10.6	8.03	8.158	.446	.446	120	29.9	3.36	40.4	9.91	1.95	.40

Square hollow tubing 2.3.3.17

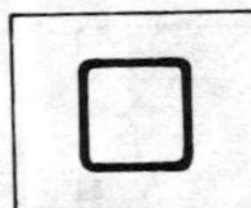

SQUARE STRUCTURAL TUBING
PROPERTIES FOR DESIGNING

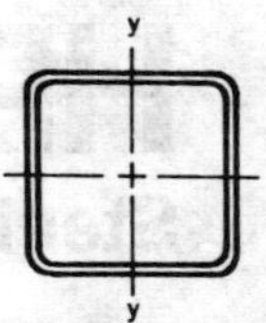

Outside Dimensions	District Rolled	Wall Thickness	Weight per Foot	Area of Metal	I	S	r	Maximum Outside Corner Radius
In.		In.	Lb.	In.²	In.⁴	In.³	In.	In.
1 x 1	F	.095	1.09	.3206	.0420	.0839	.3618	.190
		.133	1.41	.4152	.0484	.0968	.3415	.268
2 x 2	F	.110	2.69	.7914	.4574	.4574	.7603	.268
		.125	3.04	.8934	.5079	.5079	.7540	.268
		.154	3.65	1.0750	.5911	.5911	.7415	.312
		.1875	4.31	1.2688	.6667	.6667	.7249	.375
2½ x 2½	F	.141	4.32	1.2720	1.1498	.9198	.9507	.312
		.1875	5.59	1.6438	1.4211	1.1369	.9298	.375
		.250	7.10	2.0890	1.6849	1.3479	.8981	.500
3 x 3	F	.155	5.78	1.7015	2.2509	1.5006	1.1502	.312
		.1875	6.86	2.0188	2.5977	1.7318	1.1344	.375
		.250	8.80	2.5890	3.1509	2.1006	1.1032	.500
3½ x 3½	F	.156	6.88	2.0240	3.7112	2.1207	1.3541	.312
		.1875	8.14	2.3938	4.2904	2.4517	1.3388	.375
		.250	10.50	3.0890	5.2844	3.0196	1.3079	.500
		.3125	12.69	3.7329	6.0826	3.4758	1.2765	.625
4 x 4	P	.1875	9.31	2.7383	6.4677	3.2338	1.5369	.470
		.250	12.02	3.5354	7.9880	3.9940	1.5031	.625
		.3125	14.52	4.2720	9.2031	4.6016	1.4677	.785
		.375	16.84	4.9543	10.152	5.0760	1.4315	.938
		.500	20.88	6.1416	11.234	5.6169	1.3524	1.250
5 x 5	P	.1875	11.86	3.4883	13.208	5.2831	1.9458	.470
		.250	15.42	4.5354	16.595	6.6380	1.9128	.625
		.3125	18.77	5.5220	19.489	7.7955	1.8786	.785
		.375	21.94	6.4543	21.946	8.7784	1.8440	.938
		.500	27.68	8.1416	25.521	10.208	1.7705	1.250
6 x 6	P	.1875	14.41	4.2383	23.496	7.8322	2.3545	.470
		.250	18.82	5.5354	29.845	9.9482	2.3220	.625
		.3125	23.02	6.7720	35.465	11.822	2.2884	.785
		.375	27.04	7.9543	40.436	13.479	2.2547	.938
		.500	34.48	10.142	48.379	16.126	2.1841	1.250
7 x 7	P	.1875	16.85	4.9577	37.698	10.771	2.7575	.565
		.250	22.04	6.4817	48.052	13.729	2.7228	.750
		.3125	26.99	7.9389	57.306	16.373	2.6867	.940
		.375	31.73	9.3339	65.544	18.727	2.6499	1.125
		.500	40.55	11.927	78.913	22.547	2.5722	1.500
8 x 8	P	.250	25.44	7.4817	73.382	18.346	3.1318	.750
		.3125	31.24	9.1889	88.095	22.024	3.0963	.940
		.375	36.83	10.834	101.46	25.366	3.0603	1.125
		.500	47.35	13.927	124.08	31.021	2.9849	1.500
10 x 10	P	.250	32.23	9.4817	147.89	29.578	3.9494	.750
		.3125	39.74	11.689	179.12	35.824	3.9146	.940
		.375	47.03	13.834	208.21	41.642	3.8795	1.125
		.500	60.95	17.927	259.81	51.962	3.8069	1.500
		.625	73.98	21.761	302.94	60.587	3.7311	1.875

Rectangular hollow tubing

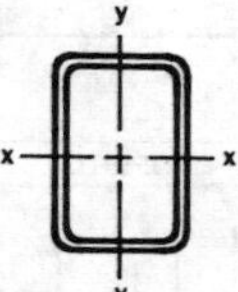

RECTANGULAR STRUCTURAL TUBING
PROPERTIES FOR DESIGNING

Outside Dimensions	District Rolled	Wall Thickness	Weight per Foot	Area of Metal	AXIS X-X I_x	AXIS X-X S_x	AXIS X-X r_x	AXIS Y-Y I_y	AXIS Y-Y S_y	AXIS Y-Y r_y	Maximum Outside Corner Radius
In.		In.	Lb.	$In.^2$	$In.^4$	$In.^3$	In.	$In.^4$	$In.^3$	In.	In.
3 x 2	F	.141	4.32	1.2720	1.4972	.9981	1.0849	.7951	.7951	.7906	.312
		.1875	5.59	1.6438	1.8551	1.2367	1.0623	.9758	.9758	.7704	.375
		.250	7.10	2.0890	2.2030	1.4687	1.0269	1.1466	1.1466	.7409	.500
4 x 2	F	.155	5.78	1.7015	3.3477	1.6738	1.4027	1.1230	1.1230	.8124	.312
		.1875	6.86	2.0188	3.8654	1.9327	1.3837	1.2849	1.2849	.7978	.375
		.250	8.80	2.5890	4.6893	2.3447	1.3458	1.5321	1.5321	.7692	.500
4 x 3	F	.156	6.88	2.0240	4.5198	2.2599	1.4944	2.8949	1.9299	1.1959	.312
		.1875	8.14	2.3938	5.2291	2.6146	1.4780	3.3404	2.2269	1.1813	.375
		.250	10.50	3.0890	6.4498	3.2249	1.4450	4.0988	2.7326	1.1519	.500
		.3125	12.69	3.7329	7.4338	3.7169	1.4112	4.7000	3.1333	1.1221	.625
5 x 3	P	.1875	9.31	2.7383	8.8629	3.5452	1.7991	4.0118	2.6746	1.2104	.470
		.250	12.02	3.5354	10.949	4.3797	1.7598	4.9195	3.2797	1.1796	.625
		.3125	14.52	4.2720	12.612	5.0448	1.7182	5.6255	3.7504	1.1475	.785
		.375	16.84	4.9543	13.907	5.5628	1.6754	6.1552	4.1034	1.1146	.938
		.500	20.88	6.1416	15.355	6.1418	1.5812	6.6839	4.4559	1.0432	1.250
6 x 3	P	.1875	10.58	3.1133	13.991	4.6637	2.1199	4.7545	3.1697	1.2358	.470
		.250	13.72	4.0354	17.438	5.8128	2.0788	5.8675	3.9116	1.2058	.625
		.3125	16.65	4.8970	20.287	6.7622	2.0353	6.7592	4.5061	1.1748	.785
		.375	19.39	5.7043	22.612	7.5373	1.9910	7.4560	4.9706	1.1433	.938
		.500	24.28	7.1416	25.629	8.5431	1.8944	8.2672	5.5115	1.0759	1.250
6 x 4	P	.1875	11.86	3.4883	17.160	5.7198	2.2179	9.1952	4.5976	1.6236	.470
		.250	15.42	4.5354	21.574	7.1913	2.1810	11.509	5.7544	1.5930	.625
		.3125	18.77	5.5220	25.346	8.4487	2.1424	13.463	6.7313	1.5614	.785
		.375	21.94	6.4543	28.553	9.5178	2.1033	15.097	7.5486	1.5294	.938
		.500	27.68	8.1416	33.213	11.071	2.0198	17.400	8.7002	1.4619	1.250
7 x 5	P	.1875	14.41	4.2383	29.380	8.3943	2.6329	17.552	7.0210	2.0350	.470
		.250	18.82	5.5354	37.341	10.669	2.5973	22.241	8.8963	2.0045	.625
		.3125	23.02	6.7720	44.396	12.685	2.5604	26.365	10.546	1.9731	.785
		.375	27.04	7.9543	50.646	14.470	2.5233	29.985	11.994	1.9416	.938
		.500	34.48	10.142	60.642	17.326	2.4453	35.688	14.275	1.8759	1.250
8 x 4	P	.1875	14.41	4.2383	34.828	8.7070	2.8666	11.923	5.9614	1.6772	.470
		.250	18.82	5.5354	44.230	11.058	2.8267	15.030	7.5148	1.6478	.625
		.3125	23.02	6.7720	52.533	13.133	2.7852	17.722	8.8610	1.6177	.785
		.375	27.04	7.9543	59.864	14.966	2.7433	20.042	10.021	1.5874	.938
		.500	34.48	10.142	71.475	17.869	2.6548	23.567	11.784	1.5244	1.250
8 x 6	P	.1875	16.85	4.9577	45.772	11.443	3.0385	29.548	9.8493	2.4413	.565
		.250	22.04	6.4817	58.362	14.590	3.0007	37.608	12.536	2.4088	.750
		.3125	26.99	7.9389	69.617	17.404	2.9613	44.784	14.928	2.3751	.940
		.375	31.73	9.3339	79.643	19.911	2.9211	51.143	17.048	2.3408	1.125
		.500	40.55	11.927	95.916	23.979	2.8358	61.374	20.458	2.2684	1.500
10 x 6	P	.250	25.44	7.4817	100.35	20.070	3.6623	45.879	15.293	2.4763	.750
		.3125	31.24	9.1889	120.45	24.089	3.6205	54.903	18.301	2.4444	.940
		.375	36.83	10.834	138.69	27.739	3.5780	63.026	21.009	2.4119	1.125
		.500	47.35	13.927	169.48	33.896	3.4884	76.541	25.514	2.3443	1.500

Standard pipe sections 2.3.3.19

PIPE

DIMENSIONS							COUPLINGS			PROPERTIES		
Nom. Dia. In.	Outside Dia. In.	Inside Dia. In.	Thick-ness In.	Weight per Foot Lb. Plain Ends	Weight per Foot Lb. Thread & Cplg.	Threads per Inch	Outside Dia. In.	Length In.	Weight Lb.	I In.⁴	A In.²	r In.
STANDARD												
1/8	.405	.269	.068	.24	.25	27	.562	7/8	.03	.001	.072	.12
1/4	.540	.364	.088	.42	.43	18	.685	1	.04	.003	.125	.16
3/8	.675	.493	.091	.57	.57	18	.848	1 1/8	.07	.007	.167	.21
1/2	.840	.622	.109	.85	.85	14	1.024	1 3/8	.12	.017	.250	.26
3/4	1.050	.824	.113	1.13	1.13	14	1.281	1 5/8	.21	.037	.333	.33
1	1.315	1.049	.133	1.68	1.68	11 1/2	1.576	1 7/8	.35	.087	.494	.42
1 1/4	1.660	1.380	.140	2.27	2.28	11 1/2	1.950	2 1/8	.55	.195	.669	.54
1 1/2	1.900	1.610	.145	2.72	2.73	11 1/2	2.218	2 3/8	.76	.310	.799	.62
2	2.375	2.067	.154	3.65	3.68	11 1/2	2.760	2 5/8	1.23	.666	1.075	.79
2 1/2	2.875	2.469	.203	5.79	5.82	8	3.276	2 7/8	1.76	1.530	1.704	.95
3	3.500	3.068	.216	7.58	7.62	8	3.948	3 1/8	2.55	3.017	2.228	1.16
3 1/2	4.000	3.548	.226	9.11	9.20	8	4.591	3 5/8	4.33	4.788	2.680	1.34
4	4.500	4.026	.237	10.79	10.89	8	5.091	3 5/8	5.41	7.233	3.174	1.51
5	5.563	5.047	.258	14.62	14.81	8	6.296	4 1/8	9.16	15.16	4.300	1.88
6	6.625	6.065	.280	18.97	19.19	8	7.358	4 1/8	10.82	28.14	5.581	2.25
8	8.625	8.071	.277	24.70	25.00	8	9.420	4 5/8	15.84	63.35	7.265	2.95
8	8.625	7.981	.322	28.55	28.81	8	9.420	4 5/8	15.84	72.49	8.399	2.94
10	10.750	10.192	.279	31.20	32.00	8	11.721	6 1/8	33.92	125.9	9.178	3.70
10	10.750	10.136	.307	34.24	35.00	8	11.721	6 1/8	33.92	137.4	10.07	3.69
10	10.750	10.020	.365	40.48	41.13	8	11.721	6 1/8	33.92	160.7	11.91	3.67
12	12.750	12.090	.330	43.77	45.00	8	13.958	6 1/8	48.27	248.5	12.88	4.39
12	12.750	12.000	.375	49.56	50.71	8	13.958	6 1/8	48.27	279.3	14.58	4.38
EXTRA STRONG												
1/8	.405	.215	.095	.31	.32	27	.582	1 1/8	.05	.001	.093	.11
1/4	.540	.302	.119	.54	.54	18	.724	1 3/8	.07	.004	.157	.15
3/8	.675	.423	.126	.74	.75	18	.898	1 5/8	.13	.009	.217	.20
1/2	.840	.546	.147	1.09	1.10	14	1.085	1 7/8	.22	.020	.320	.25
3/4	1.050	.742	.154	1.47	1.49	14	1.316	2 1/8	.33	.045	.433	.32
1	1.315	.957	.179	2.17	2.20	11 1/2	1.575	2 3/8	.47	.106	.639	.41
1 1/4	1.660	1.278	.191	3.00	3.05	11 1/2	2.054	2 7/8	1.04	.242	.881	.52
1 1/2	1.900	1.500	.200	3.63	3.69	11 1/2	2.294	2 7/8	1.17	.391	1.068	.61
2	2.375	1.939	.218	5.02	5.13	11 1/2	2.870	3 5/8	2.17	.868	1.477	.77
2 1/2	2.875	2.323	.276	7.66	7.83	8	3.389	4 1/8	3.43	1.924	2.254	.92
3	3.500	2.900	.300	10.25	10.46	8	4.014	4 1/8	4.13	3.894	3.016	1.14
3 1/2	4.000	3.364	.318	12.51	12.82	8	4.628	4 5/8	6.29	6.280	3.678	1.31
4	4.500	3.826	.337	14.98	15.39	8	5.233	4 5/8	8.16	9.610	4.407	1.48
5	5.563	4.813	.375	20.78	21.42	8	6.420	5 1/8	12.87	20.67	6.112	1.84
6	6.625	5.761	.432	28.57	29.33	8	7.482	5 1/8	15.18	40.49	8.405	2.20
8	8.625	7.625	.500	43.39	44.72	8	9.596	6 1/8	26.63	105.7	12.76	2.88
10	10.750	9.750	.500	54.74	56.94	8	11.958	6 5/8	44.16	211.9	16.10	3.63
12	12.750	11.750	.500	65.42	68.02	8	13.958	6 5/8	51.99	361.5	19.24	4.34
DOUBLE-EXTRA STRONG												
1/2	.840	.252	.294	1.71	1.73	14	1.085	1 7/8	.22	.024	.504	.22
3/4	1.050	.434	.308	2.44	2.46	14	1.316	2 1/8	.33	.058	.718	.28
1	1.315	.599	.358	3.66	3.68	11 1/2	1.575	2 3/8	.47	.140	1.076	.36
1 1/4	1.660	.896	.382	5.21	5.27	11 1/2	2.054	2 7/8	1.04	.341	1.534	.47
1 1/2	1.900	1.100	.400	6.41	6.47	11 1/2	2.294	2 7/8	1.17	.568	1.885	.55
2	2.375	1.503	.436	9.03	9.14	11 1/2	2.870	3 5/8	2.17	1.311	2.656	.70
2 1/2	2.875	1.771	.552	13.70	13.87	8	3.389	4 1/8	3.43	2.871	4.028	.84
3	3.500	2.300	.600	18.58	18.79	8	4.014	4 1/8	4.13	5.992	5.466	1.05
3 1/2	4.000	2.728	.636	22.85	23.16	8	4.628	4 5/8	6.29	9.848	6.721	1.21
4	4.500	3.152	.674	27.54	27.95	8	5.233	4 5/8	8.16	15.28	8.101	1.37
5	5.563	4.063	.750	38.55	39.20	8	6.420	5 1/8	12.87	33.64	11.34	1.72
6	6.625	4.897	.864	53.16	53.92	8	7.482	5 1/8	15.18	66.33	15.64	2.06
8	8.625	6.875	.875	72.42	73.76	8	9.596	6 1/8	26.63	162.0	21.30	2.76

LARGE O. D. PIPE

Pipe 14″ and larger is sold by actual O. S. diameter and thickness.
Sizes 14″, 15″, and 16″ are available regularly in thicknesses varying by 1/16″ from 1/4″ to 1″, inclusive.

All pipe is furnished random length unless otherwise ordered, viz: 12 to 22 feet with privilege of furnishing 5 per cent in 6 to 12 feet lengths. Pipe railing is most economically detailed with slip joints and random lengths between couplings.

Courtesy of Bethlehem Steel Co.

EXAMPLE: Beam for concentrated loads 2.4.1.1

Referring to Section I, a simple beam with 25.0 foot span supports 3 Concentrated loads totalling 21,000 Pounds. The maximum bending moment was calculated at 80,240 Foot Pounds. Maximum Reaction at support = 12,520 Lbs.

REQUIRED:
Using A36 Steel with bending and shear allowables, design a beam section to support loads in equilibrium.

STEP I:
Allowables: F_b = 24,000 PSI, and F_v = 14,500 PSI.
Solve for Section Modulus by transposing formula $BM = SF_b$.

$$S = \frac{M}{F_b} \text{ or } S = \frac{80{,}240 \times 12}{24{,}000} = 40.12''^3$$

STEP II:
Refer to Elastic Section Modulus Economy Table:
Select shape W14X30 which gives an $S = 41.9''^3$
The Resisting Moment is SF_b and will exceed the beams bending moment.

STEP III:
Referring to Wide Flange Shape Table: A W14x30 Section is identical to CB141 Section. Depth = 13.86 inches and web thickness t = 0.270 inches.
Shear formula is: $f_v = \frac{V}{dt}$. Maximum shear V is at the support with greatest reaction. Then, R = V.

Actual stress $f_v = \frac{12{,}520}{13.86 \times 0.270} = 3{,}348$ PSI (#□")

Actual unit shear stress is under that allowed and OK.
Accept the W14x30 Section.

EXAMPLE: Beam for uniform load 2.4.1.2

Emergency circumstances have prompted a fabricator to request a substitution for several W12x40 sections which are spaced on 5.0 foot centers with a simple span length of 15.0 feet. Lateral beam support is accomplished by shear connectors in a concrete span. Steel is A36. The proposed section to be substituted is a S12x40.8 also of A36 Steel. All loading is for uniform tabular loads.

REQUIRED:
Determine whether the substitution should be allowed and if allowed, under what conditions.

STEP I:
First determine the design loads from the W12x40 Section:
From tables: $S_x = 51.9$ and $F_b = 24{,}000$ PSI. $RM = SF_b$ also $= M$.

$RM = \frac{51.9 \times 24{,}000}{12} = 103{,}800$ Foot Lbs. $M = \frac{WL}{8}$ and $L = 15.0$ Ft.

Then $W = \frac{8M}{L}$ or $W = \frac{8 \times 103{,}800}{15.0} = 55{,}360$ Pounds.

Load $w = \frac{55{,}360}{15.0} = 3690$ Pounds per Lineal foot on beam.

Spacing is at 5.0' cc. Design Load $= \frac{3690}{5.0} = 732$ Lbs. Square foot.
This load is Dead Load + Live Load.

STEP II:
Checking the substitute section S12x40.8 From tables:
$S_x = 45.4$ and $RM = \frac{45.4 \times 24{,}000}{12} = 90{,}800$ Foot Pounds.

Max. $W = \frac{8M}{L}$ or $W = \frac{8 \times 90{,}800}{15.0} = 48{,}430$ Lbs.

Lineal foot max. load, $w = \frac{48{,}430}{15.0} = 3235$ Lbs.

Design load is 732 Pounds per square foot of floor area.
Then maximum spacing for substitute section S12x40.8 will be: $s = \frac{3235}{732} = 4.42$ Feet.

Substitution may be made with spacings at 4'-5" on centers.

EXAMPLE: Beam flexure design 2.4.1.3

A Cantilever beam projects out 12.0 from its support, and is braced on end for lateral support. Beam is assumed to weigh approximately 50 Lbs. per lineal foot. An added load of 600 Lbs. is to be placed on extreme end at a later date. A36 Steel is specified and maximum sag on end is limited to 1.00 inches when sign is installed.

REQUIRED:

(a) Calculate the beam required to sustain enough rigidity with only its own weight. $E = 29{,}000{,}000$ #□"

(b) Calculate the required section to stay within deflection limit when end load is applied with dead load of beam.

(c) Use bending formula to determine actual bending stress in beam when under conditions (a) and (b).

STEP I:

Conditions involve 2 types of loads. Deflection formulas are different. Resolve the uniform load of 50#/' into an additional concentrated Load called P_2. Then $P = P_1 + P_2$.

Moments used for conversion: $M_{CL} = 600 \times 12.0 = 7200$'#

Uniform Bending Moment of Beam: $M_{UL} = \frac{600 \times 12.0}{2} = 3600$'#

Convert – Total Mom. = 10,800'#

STEP II:

If Concentrated Bending Moment = PL, then $P = \frac{M}{L}$.

$P_1 + P_2 = \frac{10{,}800}{12.0} = 900$ Lbs.

STEP III:

Deflection Formula: $\Delta = \frac{Pl^3}{3EI}$. Solving for: $I = \frac{Pl^3}{3E\Delta}$.

$L = 12.0'$ $l^3 = (12.0 \times 12)^3 = 2{,}985{,}985$ $E = 29{,}000{,}000$ $\Delta = 1.00''$ max.

Then $I_x = \frac{900 \times 2{,}985{,}985}{3 \times 29{,}000{,}000 \times 1.00} = 31.0''^4$ (Required minimum)

STEP IV:

Selecting a Beam Section from Moment of Inertia Economy table: A shape W8x13 has an $I_x = 39.6''^4$ and is acceptable. $S = 9.90''^3$ From $M_{CL} = 7200$'# Then actual bending stress from concentrated load: $f_b = \frac{M}{S}$

$f_b = \frac{7200 \times 12}{9.90} = 9{,}100$ #□"

See investigation in following example with 2 Formulas.

EXAMPLE: Beam deflection with load 2.4.1.4

In the preceding example a W8X13 Symmetrical section was selected for a 12.0 foot long Cantilever beam to support an end load of 600 Pounds. Maximum deflection under load on free end is limited to 1.00 Inch.

REQUIRED:

Calculate the amount of deflection resulting from the dead weight of beam. Use formula for Uniform Load. Calculate the amount of deflection resulting from the load on free end of beam. Use formula for Concentrated Load. Summarize the results for total deflection and compare with result given in previous problem. Slide Rule OK.

STEP I:

Gather all known values thus:

$L = 12.0'$ $l^3 = (12.0 \times 12)^3 = 2,985,985$ $I_x = 39.6''^4$ $E = 29,000,000$

$P = 600$ Lbs. $W = 12.0 \times 13 = 157$ Lbs.

STEP II:

With Uniform Load the formula is: $\Delta = \dfrac{Wl^3}{8EI_x}$

Then:

$$UL.\ \Delta = \frac{157 \times 2,985,985}{8 \times 29,000,000 \times 39.6} = 0.0512''$$

STEP III:

With Concentrated Load on free end. Formula is: $\Delta = \dfrac{Pl^3}{3EI_x}$.

Then:

$$CL\ \Delta = \frac{600 \times 2,985,985}{3 \times 29,000,000 \times 39.6} = 0.522''$$

STEP IV:

Total Deflection from both loads = 0.0512 + 0.522 = 0.5732 inches. From previous example; $\Delta = 1.00$ inch based on a lesser I value of $31.0''^4$. With reduced load beam is stiffer than required.

DESIGNER'S NOTATION:

The Coefficients given in the formulas relate to the type of span and load placement. Minor inconsistencies will appear because the coefficients are given in round numbers for convenience. When calculations are in a critical state, it is recommended that the value of I_x be substantially increased before final analysis as above.

EXAMPLE: Calculating load from deflection 2.4.1.5

A W8x17 beam has a simple clear span of 18.0 feet and is supporting a uniform load of unknown quantity. Steel is A36 and assumed allowable design stresses are as follows: F_b = 24,000 PSI. F_v = 14,500 PSI. E = 29,000,000 PSI. A level instrument was used to measure the sag at the midspan. Deflection is 0.90 inches below elevation at each end support. Lateral bracing is adequate and ends are not rigid or fixed.

REQUIRED:

(a) Calculate the probable uniform load which is causing the deflection of 0.90 inches.
(b) Check the full value of shear allowed at supports.
(c) Determine the actual stress in beam under load.
(d) Calculate the maximum allowable uniform load based upon minimum yield and shear stress.

STEP I:

From Tables of rolled shapes, the properties of an W8x17 are:
$I_x = 56.6''^4$ $S_x = 14.1''^3$ $A = 5.00\square''$ Applicable formula: $\Delta = \dfrac{5Wl^3}{384EI}$.

$L = 18.0'$ $l = 18.0 \times 12 = 216.0''$ $l^3 = 10,077,696$ $\Delta = 0.90$

Transpose formula to solve for $W = \dfrac{384EI\Delta}{5l^3}$. Substituting values:

$$W = \frac{384 \times 29,000,000 \times 56.6 \times 0.90}{5 \times 10,077,696} = 11,350 \text{ Lbs.} \quad \text{(Answer to a).}$$

STEP II:

Depth T of W8x17 = 6.375" Web thickness t = 0.25 Reaction is ½ W = 5675 Lbs. $f_v = \dfrac{V}{tT}$ or $f_v = \dfrac{5675}{0.25 \times 6.375} = 3,440\#\square''$

Maximum Reaction = 14,500 x 0.25 x 6.375 = 23,100 Lbs. and maximum load allowed by shear: W = 23,100 x 2 = 46,200 Lbs. (Ans. b).

STEP III:

Stress in beam under load causing 0.90 inch deflection.

$f_b = \dfrac{M}{S_x}$ and $M = \dfrac{WL}{8}$ $M = \dfrac{11,350 \times 18.0}{8} = 25,540$ Foot Lbs.

$f_b = \dfrac{25,540 \times 12}{14.1} = 21,750\#\square''$ This is within allowable F_b. (Ans. c)

EXAMPLE: Calculating load from deflection, continued 2.4.1.5

STEP IV

Using Max. allowable bending stress $F_b = 24{,}000$ PSI, the load on beam is calculated thus:

Resisting Moment = SF_b or $M = \frac{14.1 \times 24{,}000}{12} = 28{,}200$ Foot Lbs.

$W = \frac{8M}{L}$. $W = \frac{8 \times 28{,}200}{18.0} = 11{,}980$ Lbs.

Using Max. yield stress $F_y = 36{,}000$ PSI to solve for maximum load W. $RM = \frac{14.1 \times 36{,}000}{12} = 42{,}300$ Foot Lbs.

All'd. Yield Load $W = \frac{8 \times 42{,}300}{18.0} = 18{,}800$ Lbs.

Conclusions:

Deflection will be 0.90 inches when load on beam is 11,350 Lbs., or bending stress is 21,750 PSI. It is quite likely that designer used an allowable unit bending stress of 22,000 PSI for calculating Section Modulus thus:

$W = 11{,}350$ Lbs. $F_b = 22{,}000$ PSI. $M = \frac{WL}{8}$ and $S = \frac{M}{F_b}$.

$M = \frac{11{,}350 \times 18.0}{8} = 25{,}500'\#$ $S_x: \frac{25{,}500 \times 12}{22{,}000} = 13.9''^3$

From the Elastic Section Modulus Economy Table, a section W10x15 has an S = 13.8 and next largest is the W8x17 with the S = 14.1''^{3}

EXAMPLE: Equivalent concentrated and uniform loads 2.4.1.6

A simple beam has a span of 10.0 feet between supports and supports 2 Concentrated loads thus: P_1 = 600 Lbs., and located 3.0 feet to right of R_1. P_2 = 600 Lbs., and located 3.0 feet to left of R_2. This is a symmetrically load beam of A-36 Steel.

REQUIRED:

(a) Calculate the maximum bending moment with loads given, then calculate an equivalent uniform tabular load which will produce a corresponding moment value.

(b) Design the beam section by using F_b = 20,000 PSI, and select an economical section. F_v = 14,500 PSI. Neglect lateral bracing $\frac{l}{b}$.

(c) Use the applicable deflection formulas to compute the amount of deflection under each type of loading with the same beam cross section and properties.

STEP I:

Make an elevation of beam with each load type.

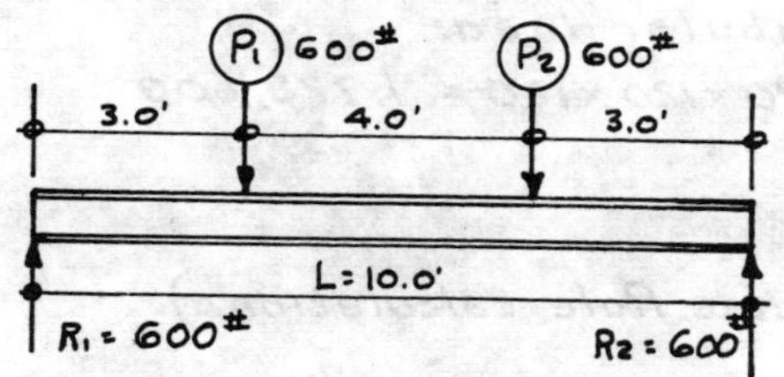

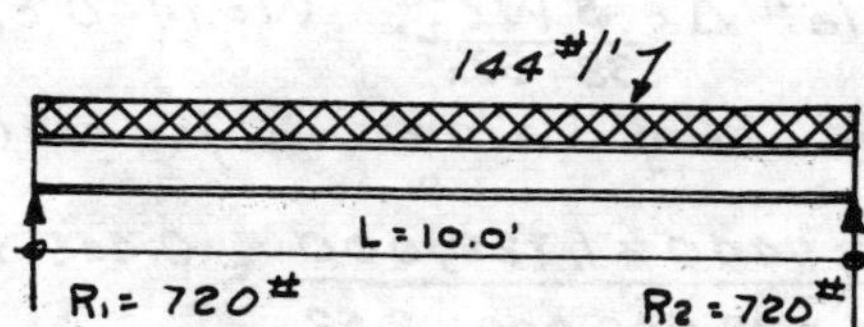

With CL, Max. Moment is under loads:

$M_{3.0}$ = 600 × 3.0 = 1800 Ft. Lbs. For uniform load: $M = \frac{WL}{8}$ or $W = \frac{8M}{L}$

The $W = \frac{8 \times 1800}{10.0} =$ 1440 Pounds. Lineal foot load = 144#

$R_1 = R_2$ for uniform load = 720 Lbs. each.

STEP II:

Section designed with bending stress: F_b = 20,000#□"

$S = \frac{M}{F_b}$ $S_x = \frac{1800 \times 12}{20,000} = 1.08''^3$ From Tables of Rolled Shapes:

Select a S3×5.7# $b = 2\frac{3}{8}''$ $I_x = 2.52''^4$ $S_x = 1.68''^3$ E = 29,000,000 #□"

Shear area = 1.75 × 1.875 = 0.328□" $f_v = \frac{720}{0.328} =$ 2200 PSI. (OK)

STEP III:

Check equivalent tabular load by formula given in AISC Manual.

$W = \frac{8Pa}{L}$, distance a = 3.0 feet. $W = \frac{8 \times 600 \times 3.0}{10.0} =$ 1440# (Checks OK).

EXAMPLE: Equivalent concentrated and uniform loads, continued 2.4.1.6

Equivalent tabular loads produce same moment values and will not produce equal reactions for shear values.

STEP IV:
Computing deflection under 2 concentrated loads on simple span equal in value and symmetrically placed. The formula given in AISC Manual follows: (Max. at center).

$$CL\ \Delta = \frac{Pa}{24EI}(3l^2 - 4a^2).$$ Convert l and a to inches:

$l^2 = (10 \times 12)^2 = 14{,}400$ $a^2 = (3.0 \times 12)^2 = 1296$ Break down equation into 2 parts and substitute values: Deflection Δ equals

$$\frac{600 \times 36.0\left[(3 \times 14{,}400) - (4 \times 1296)\right]}{24 \times 29{,}000{,}000 \times 2.52} = \frac{821{,}145{,}600}{1{,}753{,}920{,}000} = 0.462 \text{ inches.}$$

STEP V:
Deflection with equivalent uniform tabular load:
Formula: $\Delta = \frac{5Wl^3}{384EI_x}$. $W = 1440$ Lbs. $l^3 = 120 \times 120 \times 120 = 1{,}728{,}000$
Substituting values in formula:

$$\Delta = \frac{5 \times 1440 \times 1{,}728{,}000}{384 \times 29{,}000{,}000 \times 2.52} = 0.445 \text{ inches (Slide Rule calculations).}$$

STEP VI:
Same beam section loaded with concentrated loads has produced the greater deflection.
Margin of Δ: $0.462 - 0.445 = 0.027$ inches.

AUTHOR'S NOTE:
The amount of deflection between the two load types in this example are very close due to the small loads and cross section of beam. With greater loads and larger sections, the margin of deflection between the applicable formulas will produce larger margins.

EXAMPLE: End span deflection 2.4.1.7

A single beam section is 20.0 feet long and is supported at both ends and at mid-span. A uniform load of 845 Lbs. per foot extends over full length of beam. Beam section is a S12 x 35, and of A36 steel. The left end support has settled 1/2 inch below the other 2 supports which are still at the same elevation. Beam is connected to each support with adequate anchor bolts.

REQUIRED:

Determine the bending stress in beam due to deflection of 1/2 inch when no other loads have been applied. Give your opinion of the reason for excessive deflection.

STEP I:

First, sketch an elevation of this beam and let deflection be located over support R_1. Total load on 2 spans is 845 x 20.0 = 16,900 Lbs. Mechanics of Beams, Section I states that reactions at R_1 and R_3 should be calculated thus:

$R_1 = 3/8\, wL$, or $R_1 = \dfrac{3 \times 845 \times 10.0}{8} = 3{,}168.75$ Lbs. (Same for R_3)

$R_2 = \frac{10}{8} wL$, or $R_2 = \dfrac{10 \times 845 \times 10.0}{8} = 10{,}562.50$ Lbs. $R_1\text{-}M = \dfrac{16\, w l^2}{128}$

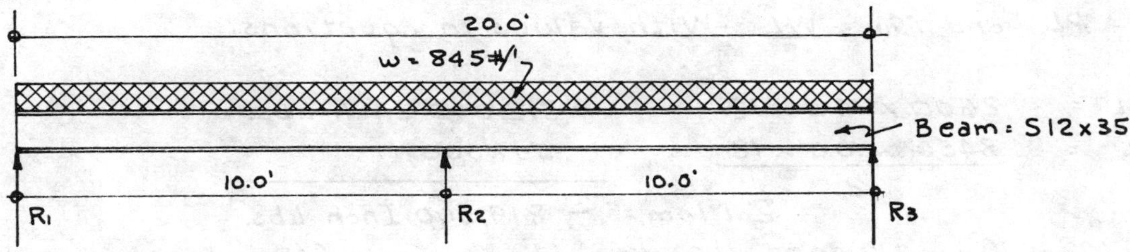

STEP II:

Assume that support at R_1 is removed and left end is a 10.0 foot cantilever beam fixed at R_2. Determine the maximum deflection at free end resulting from applied load. $\Delta = \dfrac{Wl^3}{8EI_x}$.

$L = 10.0'$ $l = 10 \times 12 = 120.0''$ and $l^3 = 1{,}728{,}000$ $E = 29{,}000{,}000$

From Table 2.3.3.2 for S12 x 5 x 35 Section: $I_x = 229.0''^4$ $S_x = 38.2''^3$

Load on Cantilever: W = 845 x 10.0 = 8450 Lbs.

Substituting values in formula: $\Delta = \dfrac{8450 \times 1{,}728{,}000}{8 \times 29{,}000{,}000 \times 229.0} = 0.2748$ In.

EXAMPLE: End span deflection, continued 2.4.1.7

Beam will not deflect 0.5000 inches unless some other force is applied at R_1. Deflection from this force = 0.5000 − 0.2748 = .2252 In.

STEP III:

If elevation at bottom of beam over R_2 and R_3 is at the same level, and support R_1 is of no value, it appears that subsidence is present and R_1 is suspended with bolts. Then this force will be a concentrate type at free end of the cantilever.

To determine this force P, use the deflection formula for Concentrated load on free end of cantilever beam thus:

$P = \frac{3EI\Delta}{l^3}$ which is transposed from $\Delta = \frac{Pl^3}{3EI}$. Therefore

force $P = \frac{3 \times 29{,}000{,}000 \times 229.0 \times 0.2252}{1{,}728{,}000} = 2596.38$ Lbs. (Call it 2600 Lbs.)

STEP IV:

The loads producing bending stress are now known and the bending moments can be computed. Since span between R_1 and R_2 is acting as cantilever beam, the force at R_3 must maintain the 20.0 foot beam in equilibrium.

$M_{CL} = PL$ and $M_{UL} = \frac{WL}{2}$. With values in equations:

$$M_{CL} = 2600 \times 10.0 \times 12 = -312{,}000 \text{ Inch Lbs.}$$

$$M_{UL} = \frac{8450 \times 10.0 \times 12}{2} = -507.000 \text{ " "}$$

$$\Sigma \text{Mom.} = -819{,}000 \text{ Inch Lbs.}$$

$f_b = \frac{M}{S_x}$ or $f_b = \frac{819{,}000}{38.2} = 21{,}440$ Lbs. Sq. Inch. (Negative moment).

DESIGN NOTATION:

Since the bending stress is approximately 60% of the Yield stress F_y = 36,000 PSI, there is reason to assume the left end will raise about 1/4 inch when left end support R_1 anchor bolts are dis-connected. Stress is within the elastic limit. The reactions at R_2 and R_3 in such case will be as follows: Take moments about R_3 to solve for R_2: $R_2 = \frac{845 \times 20.0 \times 10.0}{10.0} = 16{,}900$ Lbs. or total W.

R_1 and $R_3 = 0$

EXAMPLE: Uniform load with lateral bracing 2.4.1.8

A beam with a length of 25.0 feet must support a uniform load of 200 Lbs per foot on total length. Left end is cantilevered 5.0 feet over R_1 support. Clear span between R_1 and R_2 is 20.0 feet. A36 Steel is specified and maximum deflection at cantilever end is restricted to ¼ inches.

REQUIRED:

A design for a beam cross-section which must be investigated for shear, lateral bracing and deflection. For laterally braced beams F_b = 24,000 PSI. E = 29,000,000 and F_v = 14,500 PSI on gross area.

STEP I:

Draw elevation of beam and compute reactions R_1 and R_2.
Maximum + Moment will be at point where shear is zero.
Maximum − Moment will be at point over support R_1.

Total W = 25.0 × 200 = 5000 Lbs.
Cantilever W = 5.00' × 200 = 1000 Lbs.
Center Gravity of total load is located 12.50' to left of R_2 and 7.50' to right of R_1. Loads at R_2 act counter-clockwise.

$R_1 = \frac{5000 \times 12.50}{20.0} = 3125$ Lbs.

$R_2 = \frac{5000 \times 7.50}{20.0} = 1875$ Lbs.

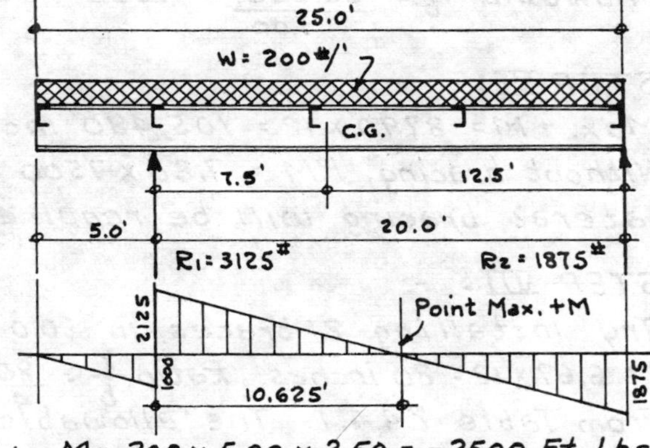

STEP II:

Negative Moment for cantilever: −M = 200 × 5.00 × 2.50 = −2500 Ft. Lbs.
Max. + Moment is at point of zero shear.

Zero shear point = $\frac{3125 - (200 \times 5.00)}{200}$ = 10.625 Feet to right of R_1.

Max. + M = (3125 × 10.625) − (200 × 15.625 × 7.8125) = + 8,790 Foot Lbs.

Required $S_x = \frac{M}{F_b}$ or $S_x = \frac{8{,}790 \times 12}{24{,}000} = 4.40''^3$

Select a W 8×10 Section for further investigation. $S = 7.80''^3$ and $I_x = 30.8''^4$ Flange width b = 4.00" and Web t = 0.170" T = 6.50"

STEP III:

Cantilever requirements for sag of ¼ inch. Formula: $\Delta = \frac{Wl^3}{8EI}$ and transposed $I_x = \frac{Wl^3}{8E\Delta}$. Req. $I_x = \frac{1000 \times 60 \times 60 \times 60}{8 \times 29{,}000{,}000 \times 0.25} = 3.73''^4$

Inertia requirements are sufficient for deflection.

EXAMPLE: Uniform load with lateral bracing, continued 2.4.1.8

STEP IV:
Investigate shear: Max. R_1 = 3125 Lbs. Max. $V = F_v t T$
V = 14,500 × 0.170 × 6.50 = 16,025 Lbs. Exceeds R_1 and stiffeners will not be required.

STEP V:
Lateral bracing when flange b = 4.00 inches and L = 20.0 Feet.
$l = 20.0 \times 12 = 240$ inches. Ratio $\frac{l}{b} = \frac{240}{4} = 60$ Full allowable of F_b can only be used when ratio $\frac{l}{b}$ is not over 15. Bracing will be required if unit stress is over that allowed by formula.

Without lateral bracing $F_b = \dfrac{22,500}{1.00 + \left(\dfrac{l^2}{1800\,b^2}\right)}$

$l^2 = 240 \times 240 = 57,600$

$b^2 = 4.0 \times 4.0 = 16.0$ Then $1.00 + \left(\dfrac{57,600}{1800 \times 16.0}\right) = 3.00$

Allowable $F_b = \dfrac{22,500}{3.00} = 7500$ PSI.

STEP VI:
Max. +M = 8790 × 12 = 105,480 inch Lbs.
Without bracing, RM = 7.80 × 7500 = 58,500 inch Lbs.
Lateral bracing will be required.

STEP VII:
Try installing 2 braces in 20.0 span. $L = \frac{20.0}{3} = 6.67$ Ft. (6'-8").
$l = 6.67 \times 12 = 80$ inches. Ratio $\frac{l}{b} = \frac{80}{4} = 20$
From Table 2.4.4.1 The allowable unit bending stress F_b = 18,410 PSI.
Resisting Moment becomes 7.80 × 18,410 = 143,598 inch pounds.
A section W8×10 will be acceptable only when compression flange is laterally braced every 6'-8" inches along span.

STEP VIII:
Further investigation by using the A.I.S.C. reduced stress formula:
Allowable $F_b = \dfrac{12,000,000}{\left(\dfrac{l d}{A_f}\right)}$. With W8×10 Section. Flange width = 3.94" Flange thickness = 0.204" and d = 7.90"

$\dfrac{l d}{A_f} = \dfrac{240 \times 7.90}{3.94 \times 0.204} = 2359$. Then $F_b = \dfrac{12,000,000}{2359} = 5087$ P.S.I.

Requirements for a beam without lateral bracing: $S = \dfrac{M}{F_b}$ or

$S_x = \dfrac{8790 \times 12}{5087} = 20.73''^3$ Will require a W8×24 which has a flange width of 6½ inches and $S_x = 20.8''^3$ This will probably be more economical due to labor cost for bracing and steel added.

Laterally unbraced beams 2.4.2

Under normal load conditions, the behavior of the top compression flange of a beam is similar to a long column. Excessive span lengths will tend to turn and buckle the flange unless some form of bridging or lateral bracing is provided. Note that the beam load tables given in the AISC Manual assume that the beam has adequate lateral support. Too often, this design requirement is overlooked. Metal decks and concrete slabs may provide adequate lateral bracing but the experienced design engineer will check this assumption. When designing standard S-beams and channels or light gauge Z-sections, make certain that bridging or sag rods are installed within the allowable length ratio.

FORMULAS FOR UNBRACED SPANS

A formula, which applies for all grades of domestic steel, is used to determine the allowable reduced unit stress in the compressive flange, and is written:

$$F_b = \frac{12{,}000{,}000}{\left(\frac{ld}{A_f}\right)}. \text{ Where:}$$

l = Unbraced length, in inches.
d = Depth of beam, in inches.
A_f = Area of the flange, in square inches.

Before this formula was developed, an older set of rules was used, where the flange width (b) was the controlling factor for stress reduction. Depth of beam and area of flange did not enter into the equation. Perhaps it may be a bit premature to abandon these older rules. Many experienced designers continue to use them in transposed form to calculate maximum unsupported length. These older rules are concerned only with the flange width (b) and its ratio to length (l). They are:

(a) For ratios of $\frac{l}{b}$ from 0 to 15, use full allowable stress F_b

(b) For ratios of $\frac{l}{b}$ between 15 and 40, the stress must be reduced by the formula thus: $F_b = \frac{22{,}500}{1.00 + \left(\frac{l^2}{1800\,b^2}\right)}$.

In many wide flange shapes used for normal spans, there will be only slight differences in allowable stress between the old rules and the new formula. A table of unit stresses based on ratios of $\frac{l}{b}$ between 15 and 40 is provided for ready reference (TABLE 2.4.4.1). This table must not be confused with similar tables which present stress reduction values for columns, web shear without stiffeners and web buckling.

Web shear 2.4.3

The unstiffened web of a steel beam may be safe under ordinary bending and vertical shear stresses, but will fail when a short length of the web is highly loaded as an unbraced vertical column. This is especially true when the ends of the beam rest on seat angles, and for short spans with heavy loads near the end supports. This type of failure in the beam is referred to as *web crippling*. Lack of proper lateral support for the top flange at concentrated loads or reactions may cause web crippling in the web plate between the beam fillets (the dimension given in the tables as T). Web crippling occurs most frequently in deep-web welded plate beams and hybrid girders used for bridges, crane booms and knee joints in rigid arches.

Flange buckling 2.4.4

Assume that a rolled-beam compact shape is turned so as to support a load through the minor axis y–y instead of major axis x–x. The most probable mode of collapse would take the form of buckling of the flanges. This event occurred when a large crane picked up a long HP section on its side without a spreader. The flanges on the compression side buckled inward to form a V as the metal bent beyond the yield point.

Flange buckling is not the same as web buckling. Web buckling occurs over the supports and directly under the points of application of concentrated loads. Flange buckling can be avoided by adding a cover plate or channel across the flanges which are in compression.

TABLE: Allowable bending stress without lateral support — 2.4.4.1

ALLOWABLE UNIT DESIGN STRESS

FOR BEAMS AND GIRDERS WITHOUT LATERAL SUPPORTS

$\frac{l}{b}$	UNIT STRESS PSI	$\frac{l}{b}$	UNIT STRESS PSI	$\frac{l}{b}$	UNIT STRESS PSI	$\frac{l}{b}$	UNIT STRESS PSI
15.0	20,000	22.5	17,560	30.0	15,000	37.5	12,630
15.5	19,850	23.0	17,390	30.5	14,830	38.0	12,490
16.0	19,700	23.5	17,220	31.0	14,670	38.5	12,340
16.5	19,540	24.0	17,050	31.5	14,500	39.0	12,200
17.0	19,390	24.5	16,870	32.0	14,340	39.5	12,050
17.5	19,230	25.0	16,700	32.5	14,180	40.0	11,910
18.0	19,070	25.5	16,530	33.0	14,020		
18.5	18,910	26.0	16,360	33.5	13,860		
19.0	18,740	26.5	16,190	34.0	13,700		
19.5	18,580	27.0	16,010	34.5	13,540		
20.0	18,410	27.5	15,840	35.0	13,390		
20.5	18,240	28.0	15,670	35.5	13,230		
21.0	18,070	28.5	15,500	36.0	13,080		
21.5	17,900	29.0	15,340	36.5	12,930		
22.0	17,730	29.5	15,170	37.0	12,780		

FORMULA:

$$F_b = \frac{22,500}{1.0 + \left(\frac{l^2}{1800\, b^2}\right)}$$

l = LENGTH UNBRACED SPAN, IN INCHES
b = WIDTH OF FLANGE OF SECTION, IN INCHES
F_b = ALLOWABLE STRESS REDUCED BY FORMULA

TABLE: Allowable shear stress on web without stiffeners 2.4.4.2

ALLOWABLE UNIT DESIGN STRESS

SHEAR STRESS ON WEB WITHOUT STIFFENERS

$\frac{T}{t}$	F_V = PSI	$\frac{T}{t}$	F_V = PSI	$\frac{T}{t}$	F_V = PSI	$\frac{T}{t}$	F_V = PSI
60	12,000	75	10,110	90	8,470	105	7,110
61	11,870	76	9,900	91	8,370	107	6,950
62	11,730	77	9,870	92	8,270	109	6,790
63	11,600	78	9,760	93	8,180	111	6,640
64	11,470	79	9,640	94	8,080	113	6,490
65	11,340	80	9,530	95	7,990	115	6,350
66	11,210	81	9,420	96	7,890	120	6,000
67	11,090	82	9,310	97	7,800	125	5,680
68	10,960	83	9,200	98	7,710	130	5,380
69	10,840	84	9,090	99	7,620	135	5,100
70	10,710	85	8,980	100	7,530	140	4,840
71	10,590	86	8,880	101	7,450	145	4,590
72	10,470	87	8,780	102	7,360	150	4,360
73	10,340	88	8,670	103	7,280	155	4,150
74	10,220	89	8,570	104	7,190	160	3,950

FORMULA:

$$\frac{V}{A} = F_v = \frac{18{,}000}{1.0 + \left(\frac{T^2}{7200t^2}\right)}$$

V = TOTAL SHEAR OR REACTION, IN POUNDS
A = GROSS AREA OF WEB, IN SQUARE INCHES
F_V = ALLOWABLE UNIT STRESS, PSI
T = UNBRACED WEB DISTANCE, IN INCHES, DETAIL
t = THICKNESS OF WEB, IN INCHES

t T

t T

EXAMPLE: Simple span without lateral support 2.4.4.3

A W12 Wide Flange Beam is to be used on a simple 20.0 Clear Span without any lateral bracing between ends. Bending Moment = 100,000 Ft. Lbs. Steel type = A36.

REQUIRED:

(a) Determine the actual stress in bending and stress in compressive flange of section.

(b) Use formula for laterally unsupported beam to check the allowable stress in top flange.

(c) If results for (a) and (b) are negative, determine where lateral bracing is needed.

STEP I:

From Wide Flange Shape Tables: W12 40# has $S_x = 51.9''^3$

Depth d = 11.94" (call it 12.0") Flange width = 8.00"

Flange thickness = 0.516"

STEP II:

F_b = 24,000 #□" for laterally supported beams.

Actual $f_b = \frac{M}{S}$ or $f_b = \frac{100{,}000 \times 12}{51.9} = 23{,}100$ #□" (OK with brace)

STEP III:

L = 20.0 Ft. l = 20.0 × 12 = 240" d = 12.0 In.

Area flange: $A_f = 8.00 \times 0.516 = 4.128$ Sq. In.

STEP IV:

Allowable stress for this beam: $F_b = \dfrac{12{,}000{,}000}{\dfrac{ld}{A_f}}$

Substituting values in formula:

$$F_b = \frac{12{,}000{,}000}{\dfrac{240 \times 12}{4.128}} = \frac{12{,}000{,}000}{697.3} = 17{,}200 \text{ \#□"}$$

Allowable stress is less than actual stress f_b = 23,100 #□"

Beam will require lateral bracing.

STEP V:

Placing a brace or support at the mid-span, the length is reduced by half, or l = 120"

Then $\frac{120 \times 12}{4.128} = 349.5$ then $F_b = \frac{12.000{,}000}{349.5} = 34{,}400$ #□"

Actual stress will govern design, however control of design was subject to providing lateral support.

EXAMPLE: Bending and flexure design with lateral support 2.4.4.4

A group of simple span beams are spaced 6.0 feet on centers each with a span of 25.0 feet. Deflection must be restricted to not over 1/240 of span l. Uniform load clear across span is 360 Pounds per foot.

REQUIRED:

Steel is to be of A36 with $F_b = 24{,}000$ PSI, and $E = 29{,}000{,}000$.

(a) Design the section for bending, then design for the stiffness required to limit the flexure sag.

(b) Use lateral bracing formulas to determine where bracing should be placed if full stress of 24,000 PSI is to be used

STEP I:

Data given is: $L = 25.0'$ $l = 25.0 \times 12 = 300''$ $\frac{1}{240}l = \frac{300}{240} = 1.25''$ or Max. $\Delta = 1.25''$ $w = 360^{\#'}$ $W = 360 \times 25.0 = 9000^{\#}$

$F_b = 24{,}000^{\#\square''}$ $E = 29{,}000{,}000^{\#\square''}$ $M = \frac{WL}{8}$ For bending design: $S_x = \frac{M}{F_b}$

For flexure design: $\Delta = \frac{5Wl^3}{384EI}$ or $I_x = \frac{5Wl^3}{384E\Delta}$. $l^3 = 27{,}000{,}000$

STEP II:

Bending: $M = \frac{9000 \times 25.0}{8} = 28{,}125'^{\#}$ $S_x = \frac{28{,}125 \times 12}{24{,}000} = 14.06''^3$

Flexure:

$$I_x = \frac{5 \times 9000 \times 27{,}000{,}000}{384 \times 29{,}000{,}000 \times 1.25} = 87.5''^4$$

STEP III:

From Moment of Inertia Economy Table: A W12x14 Beam section has an $I_x = 88.0''^4$ and an $S_x = 14.8''^3$. This section is satisfactory for properties about axis x-x and is in lightweight class formerly listed as a 12x4 B 14#.

Flange is 4.00 inches (b) in width and will not provide support laterally for compression flange.

STEP IV:

Gather physical data on section for 2 Formulas from which to locate points of bracing if required.

$d = 12.0''$ Flange $b = 4.00''$ thickness $= 0.25''$

Area flange $A_f = 4.00 \times 0.25 = 1.00$ Sq. inch. $l = 300$ inches.

EXAMPLE: Bending and flexure design with lateral support, continued 2.4.4.4

Using the older formula for reducing unit stress, the ratio of $\frac{l}{b}$ must be less than 15 in order to use full value of allowable $F_b = 24{,}000$ PSI. In no event shall $\frac{l}{b}$ ratio be over 40.

STEP V

Formula for reduced $F_b = \dfrac{22{,}500}{1.0 + \left(\dfrac{l^2}{1800\,b^2}\right)}$. But if $\frac{l}{b}$ must be not over 15, then when $b = 4.00''$, maximum $l = 15 \times 4.00 =$ 60 inches or 5.0 feet.

Make lateral support consist of 4 Rows of bridging at 5.0 foot spacing, or use rods similar to sag rods, or 5 equal spaces will produce 4 Rows.

STEP VII:

Check the spacing for supports by using the AISC lateral stress formula as: $F_b = \dfrac{12{,}000{,}000}{\left(\dfrac{ld}{A_f}\right)}$.

F_b must be equated to 24,000 PSI

Then: $\dfrac{ld}{A_f} = \dfrac{12{,}000{,}000}{F_b}$ or with values substituted: $\dfrac{12{,}000{,}000}{24{,}000} = 500$ and $= \dfrac{ld}{A_f}$.

Known values:

$d = 12.0''$ $A_f = 1.00\,\square''$ Then Max. $l = \dfrac{500 \times A_f}{d}$.

For full stress of $F_b = 24{,}000\,\#\square''$ $l = \dfrac{500 \times 1.00}{12.0} =$ 41.6 inches.

Requires spacing at approximately 3.5 feet, but could use braces at 5 points as $\dfrac{25.0}{6} = 4.167$ feet.

STEP VIII:

Check results by inserting values in full formula with $l = 42$ inches. $F_b = \dfrac{12{,}000{,}000}{\left(\dfrac{42.0 \times 12.0}{1.00}\right)} = 23{,}800$ PSI.

Actual Stress $f_b = \dfrac{M}{S_x}$ or $f_b = \dfrac{28{,}125 \times 12}{14.8} = 22{,}800\,\#\square''$ (OK)

STEP IX:

With respect to the formula in Step V, the ratio of $\frac{l}{b} = \frac{42.0}{4.0} = 10.5$, and is less than 15 where full stress is OK.

The Table of allowable stresses may be used for a quick reference after $\frac{l}{b}$ ratio is determined.

Web crippling design formula 2.4.5

On the gross area of the webs of beams and girders where the web is not stiffened and T is more than 60 times t, the maximum shear per square inch; $f_v = \frac{V}{A}$. Stress shall not exceed the allowable determined by the formula:

$$F_v = \frac{18{,}000}{1.0 + \left(\frac{T^2}{7200 t^2}\right)}$$. Where:

V = Maximum web shear, in pounds.
A = Gross area of web, in square inches.
T = Distance between flanges or fillets, in inches.
t = Thickness of web, in inches.
F_v = Unit stress for shear, in pounds per square inch.

Table 2.4.4.2 is provided to give stress values for $\frac{T}{t}$ ratios from 60 to 160. When the ratio is less than 60, use a value of F_v = 12,000 pounds per square inch.

The following examples illustrate the method used to design stiffeners to support web. Always use stiffeners in pairs to avoid eccentricity caused by uneven bearing or load placement.

WEB AREA CONSIDERED

Examine the dimensions listed on the cross sections illustrated in Table 2.3.3.1 Wide Flange Shapes. The distance between the fillet rounds is identified as T. The critical area of the web is T times the web thickness (which we shall indicate by a small t). Then A_w = t T. If the end of a beam is supported upon a bearing wall or seat angle, the projected length of end bearing is indicated as N. This also applies to a beam which places a concentrated load on top of another beam.

The reaction at the support is most often where web crippling is of major concern, although other points on the beam may also need attention. The reaction R is usually determined before examining the requirement for stiffeners.

The effective area resisting crippling is the length of bearing N plus k, times the thickness of the web t. The stress allowable is 75% of the specified steel yield point; with A36 steel, the allowable stress is 0.75 x 36,000 = 27,000 PSI.

Using this reasoning, the AISC specifications provide a formula to determine when web stiffening is required: $R = 0.75\,F_y\,[t\,(N + k)]$. When loads are on interior flange of beam, the formula is similar, and is written: $R = 0.75\,F_y[t\,(N + 2k)]$.

Hybrid girders 2.4.6

Welded or riveted girders with abnormal depth are common to bridge structures. Such girders are built up with plate web and cover plates for flanges. Rolled shapes in the form of angles and channels are also used in fabricating these deep sections. In bridge work, impact loads can subject the beam webs to crippling failure, and stiffeners are placed at several intermediate points along the span. Welding has replaced riveting in most hybrid and built-up plate girders. This has simplified the design, as will be seen in the examples given in Section VI for properties of built-up sections. Impact forces can result from moving trucks or trains. Designers substantially reduce the allowable unit stress in such instance, often by as much as 50%.

DESIGN CONSIDERATIONS

The moment of inertia method to design hybrid and built-up plate girders involves the analysis of a number of important requirements as follows:

(a) Provide lateral bracing, or design adequate cover plates.

(b) Check web shear at support for stiffeners and bearing.

(c) Investigate web shear along span for intermediate support.

(d) Calculate bending stress and inertia requirements, and check whether an additional compression cover plate will be required.

(e) Determine stiffener sizes and maximum spacing.

(f) Check horizontal shear for welded joints or rivet spacing. Horizontal shear at the junction of the cover plate and web is described with formula in succeeding paragraphs and examples.

Web stiffeners 2.4.7

END STIFFENERS

The area of the web resisting vertical shear at the supports is calculated as web height times web thickness or T x t. See the detail of cross sections in table for the allowable stresses without stiffeners. When angles are used at the joint of cover plate to web, the T dimension is taken between the angles as shown, otherwise it is taken between the fillets.

Shear stress, determined by the formula $f_v = \frac{V}{Tt}$, may be relatively small, therefore a portion of length along beam will be considered. Assuming a web is 60 inches in depth and ⅜ inches in thickness the area would be: $A_v = 60.0 \times 0.375 = 22.50$ square inches, but if the bearing projection is 10.0 inches, then the area of resisting web is: $22.50 \times 10.0 = 225.0$ square inches. The design theory treats this area as an unbraced column, 0.375″ x 10.0″ and 60.0 inches in height. In stiffener design, the column height is taken as one-half the depth of the web, or $\frac{T}{2}$. This imaginary rectangular column with a section of 0.375″ x 10.0″ is 30.0 inches high and must meet the ratio test for short columns. When this ratio is determined by the $\frac{T}{t}$ method

Webb stiffeners, continued 2.4.7

as given in the table, the allowable stress is calculated by formula. An established rule to follow in end stiffener design is to let the stiffener on one side of the web assume ⅔ of the load reaction.

INTERMEDIATE STIFFENERS

Stiffeners under concentrated loads on beam span are calculated in a manner similar to that used for end stiffeners. The ratio of $\frac{T}{t}$ should not exceed 70 unless web stiffening is added. Another ratio test used to determine the need for web stiffening is the AISC formula: $\frac{8000}{\sqrt{f_v}}$. Where f_v is the actual unit stress, computed as $f_v = \frac{R}{Tt}$, and R = Reaction, T = Web depth, and t = Web thickness. To illustrate this formula for the ratio test:

Assume: R = 128,000 Pounds, T = 60.0 inches and t = 0.375 inches.

By AISC Formula: $Ratio = \frac{8000}{\sqrt{\frac{128,000}{60.0 \times 0.375}}} = \frac{8000}{76.0} = 115$. *Since this ratio is higher and considerably over the allowed ratio of 70, therefore web stiffening is required.*

When stiffeners are designed for welded girders they should extend from cover plate to cover plate. When using angles for stiffeners they should be coped or crimped, or a flat bar filler installed between the angle leg and web of girder, so the stiffener can be welded to the cover plate.

STIFFENER SPACING

In girders which are designed to support uniform or impact loads, the shear stress usually drops as the load moves closer to mid-span. The shear diagrams in Section I show this quite clearly. For better appearance, many designers prefer to continue the minimum stiffener spacing clear across the span.

The AISC provides a spacing formula thus:

$spacing = \left(\frac{270,000\,t}{f_v}\right) \times \sqrt[3]{\frac{f_v t}{T}}$. *Where:* f_v = *Actual unit shear stress as determined in previous paragraph. An older rule which is recommended but has seemingly been abandoned, called for intermediate stiffeners to be placed not over 7.0 feet regardless of calculations.*

Webb stiffeners, continued 2.4.7

To eliminate the possibility of eccentric bearing at the end supports or intermediate points, web stiffeners are installed in pairs: one on each side of the web. Weld joints are in shear, as are the horizontal welds along the joints between the web and the flanges. Stiffeners made from flat bars or small angles should extend the full depth between flanges, and whenever possible the ends should be welded to the flanges. All welding should be skip type, with no weld less than 2 inches in length. The space between welds should never exceed 6 inches. Welds may be staggered on opposite stiffeners.

ECCENTRICITY IN BEARING

It is important that the seat at the ends of beams and girders have uniform bearing to localize the vertical pressure upon each stiffener. A good rule to follow in the design of stiffeners at end supports is to consider one side to sustain ⅔ of the excess reaction force which is not supported by web; then make the pair identical in size. When two or more pairs of stiffeners are used, the ⅔ rule should apply to each pair as a safety measure.

Bearing blocks or sole plates should be required for extra heavy load reactions when the girder is to be set upon a concrete pedestal or plinth. Bearing blocks are cut from thick plate and set level on the concrete before the girder is installed. Bearing blocks with levelling plates will be illustrated and designed in succeeding examples.

EXAMPLE: Web stiffening 2.4.7.1

A wide flange steel W16 x 88 section must sustain a load reaction of 85,000 Pounds. End projected bearing is limited to 4.25 inches. Steel is A36 with F_y = 36,000 PSI. The reaction at support results from concentrated load placed at an intermediate point on this short beam.

REQUIRED:

(a) Determine if stiffeners will be required if end bearing projection cannot be extended beyond 4.25".

(b) Calculate the maximum concentrated load which may be placed at intermediate point without the use of web stiffeners giving protection against crippling of the web between fillets. Let N = 5.25" at load P.

STEP I:

Draw the elevation of this beam at end support and also at an intermediate point. The formula given by AISC for web crippling is thus:

At end bearing: $\text{Max. } R = 0.75\,F_y\left[(N+k)\,t\right]$

At intermediate point on beam; $\text{Max. Load } P = 0.75\,F_y\left[(N+2k)\,t\right]$

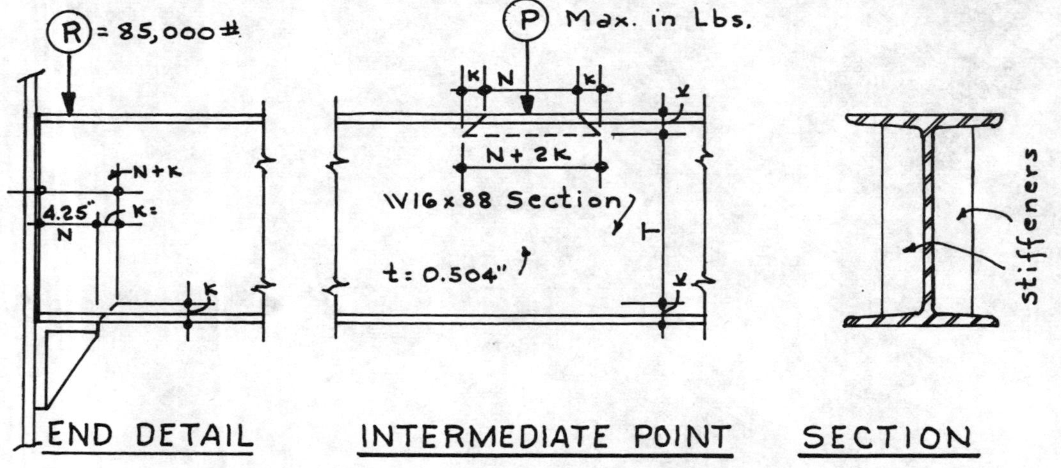

EXAMPLE: Web stiffening, continued 2.4.7.1

STEP II:

From Tables of Shapes: k = 1.50" T = 13.125" t = 0.504"

Substituting in end formula:

Max. R without web stiffening = $0.75 \times 36{,}000 \left[(4.25 + 1.50) \times 0.504\right] =$

Max. R = $27{,}000 \times (5.75 \times 0.504) = 78{,}300$ Lbs.

Web stiffening will be required at end because Reaction is greater than resisting area of web bearing.

STEP III:

At intermediate point under load P = 5.25". Then N+2k equals 5.25 + 3.00 = 8.25" Area under P = (N + 2k) x t, or Resisting Area = 8.25 x 0.504 = 4.158 Sq. Inches.

Allowable web stress = 0.75 Fy = 27,000 PSI

Maximum Load P = 27,000 x 4.158 = 112,265 Pounds without stiffening.

STEP IV:

Design the stiffeners for end web when bearing project is not sufficient.

R = 85,000# and maximum allowed without stiffening web = 78,300#. Then, 85,000 - 78,300 = 6700# which must be sustained by stiffeners.

Area required for 1 pair = $\frac{6700}{27000}$ = 0.248 Sq. inches.

The required area is quite small and no stiffener should be less than a ¼ inch flat bar.

For insurance against eccentric bearing, design one stiffener to take ⅔ of 6700 or 4467 Pounds.

For width of ¼ flat bar: width = $\frac{4467}{0.25 \times 27{,}000}$ = 0.662"

Use 2 minimum FB stiffeners as: ¼ x 1¼. One on each side and weld tight to web and both flanges.

EXAMPLE: Carnegie formula for web stiffening 2.4.7.2

A steel section W16x88 of A36 Steel is resting upon a shelf support angle 4.25 inches. A reaction from loads results in a value, R= 115,000 Pounds.

REQUIRED:
Use the older Carnegie formula to determine whether the web needs stiffening under the 115,000 Lb. Reaction. The formula is written for maximum reaction thus:

$R = F_c\, t\left(a + \frac{d}{4}\right)$ Where: F_c cannot exceed $0.75 F_y$.
t = thickness of web between fillets.
a = bearing depth on support.
d = depth of beam section.

STEP I:
For a W16x88 section: t = 0.504" d = 16.16" and a = 4.25"
A36 Steel, F_y = 36,000 PSI Then F_c = 0.75 x 36,000 = 27,000 PSI

STEP II:
Substituting in formula to determine maximum allowable reaction on bearing length of 4.25 inches:

$$R = 27{,}000 \times 0.504\left(4.25 + \frac{16.16}{4}\right) = 112{,}810 \text{ Pounds.}$$

Web stiffening will be necessary. If end bearing at end could be increased ¼ inch to make a = 4.50, the web would support 116,210 Pounds and stiffening would not be required to support the 115,000 Lb. Reaction.

AUTHOR'S NOTATION:
The formula used in the example above was applied to the identical shape used in the preceding example. Comparing the two formulas, the AISC formula is more conservative because of the difference between $\frac{d}{4}$ and K. Note the allowable stress has remained the same as called for in the older Carnegia formula.

Flange cover plates 2.4.8

For cover plate design, first consideration should be given to the width, because this determines the need for lateral support and locates the bracing points. Tables 2.3.3.1 for rolled sections only extends to wide flange shapes with a maximum depth of thirty-six inches. The designer must calculate the properties for deeper built-up sections. The moment of inertia of the web plate may be taken from Table 6.2.6 for rectangular sections. For the cover plate, a width may be selected for trial after the width has been determined from formula for allowable unit stress as controlled by the ratio of $\frac{l}{b}$. In the examples in Section VI, a format of considerable merit is illustrated for calculating the moment of inertia for plate girders and compound beams. By using this format, the designer can conveniently make adjustments in the cover plate without changing the other components, to arrive at the desired property of I_x.

HORIZONTAL SHEAR AND WELDS

Welds are critical where the cover plate is welded to the web plate. Remember that the top flange is sustaining compressive stress while the lower flange is resisting tension. To determine the horizontal shear at any point on a beam or girder, the following formula may be used:

$$v_h = \frac{Va\bar{y}}{It}$$

Where V = Total shear or reaction at support.
a = Area above horizontal line where shear is to be calculated.
$\bar{y}$ = Distance from neutral axis or x-x.
I = Moment of Inertia of whole section.
t = Thickness of web plate.
v_h = Unit shear stress in pounds per square inch, or actual stress.

For computing the value for weld strength per lineal inch, the value of t is deleted from the formula because a force rather than unit stress is required. Then the formula is re-written: Lineal inch $v_h = \frac{Ra\bar{y}}{I_x}$. R is substituted for V when reaction is used in the formula: An example will follow to illustrate this formula when used for a welded plate girder. Take note of the instructions given for weld lengths and skip welds on each side of the web.

EXAMPLE: Horizontal shear in cover plate welds 2.4.8.1

A fabricated plate girder has a cover plates 12.0 inches wide and ½ inch thick. Web plate is ⅜ inches thick with a depth of 60.0 inches. Welding is to be used to connect web to cover plates. Moment of Inertia = 31,000 in.^{4}

REQUIRED:
Determine the horizontal shear force at joint of web and cover plate when end reaction is assumed to be 130,000 Pounds. Next determine the unit stress at joint in pounds per square inch. Assume that full allowable unit stress F_c is desired, determine lateral bracing points.

STEP I:
Cover plate laid flat is 0.50 and its axis is 0.25 inches.
Axis x-x in web plate is at midpoint or 30.0 inches.
Then distance $\bar{y} = 30.0 + 0.25 = 30.25''$ Area of flange is equal to: $a = 12.0 \times 0.50 = 6.00$ Sq. Inches.

STEP II:
Substitute the knowns in the force formula: $V_h = \dfrac{Ra\bar{y}}{I_x}$

Shear per inch along beam, $V_h = \dfrac{130{,}000 \times 6.0 \times 30.25}{31{,}000} =$

$V_h = \dfrac{23{,}595{,}000}{31{,}000} = 764$ Lbs.

STEP III:
For unit stress the formula is: $f_{vh} = \dfrac{Ra\bar{y}}{I_x t}$ or by putting t into the formula as: 0.375
Then unit stress at web $= \dfrac{764}{0.375} = 2040$ PSI.

STEP IV:
Full allowable unit stress in bending or compression in top flange is permitted only when $\frac{l}{b}$ ratio is 15 or less.

Then with a flange, $b = 12.0''$ $l = 15 \times 12 = 180.0$ inches, or at each 15.0 feet of beam length.

EXAMPLE: Design of hybrid girder 2.4.9

An underpass is to support railroad traffic and span to center of moments is 55.0 feet. Girders are spaced on 8.0 feet centers with adequate provision for any lateral bracing desired.
Each girder is to support a Uniform Load of 3000 Pounds per lineal foot, plus a 65,000 Pound concentrated load placed at mid-span.
End supports consist of a continuous concrete buttress wall with shelf for bearing blocks if required. Impact load has been considered and included in the concentrated load. Depth of steel girder should be restricted to 1/10 of span L.

REQUIRED:
Use the Moment of Inertia method to design this girder. Locate points for lateral support of compression cover plate. Limit the allowable bending design stress to F_b = 20,000 PSI, or as reduced by formulas applicable. For welded fabrication, use a limit of 13,600 PSI for throat shear substituting for fillet rounds. This is to be a complete design for the section with stiffeners, cover plates and welds indicated. Draw a detailed section and elevation, also use suggested form shown in Section VI when computing the properties of cross-section.

STEP I:
This is a simple span type beam 55.0 feet in length for center of reactions. Estimate the weight of steel girder at 210 Pounds per lineal foot.

$w = 3000 + 210 = 3210$ #/' $\quad W = 3210 \times 55.0 = 176{,}550$ Lbs.

Concentrated Load, $P = \quad 65{,}000$ "

Total Load = 241,550 Lbs.

Reactions R_1 and $R_2 = \dfrac{241{,}500}{2} = 120{,}775$ Lbs.

STEP II:
Calculate Bending Moment: Max. at mid span.

For Uniform Load: $\text{Max. } M = \dfrac{WL}{8} = \dfrac{176{,}550 \times 55.0}{8} = 1{,}213{,}780$ '#

For Concentrated Load: $\text{Max. } M = \dfrac{PL}{4} = \dfrac{65{,}000 \times 55.0}{4} = 893{,}750$ '#

EXAMPLE: Design of hybrid girder, continued 2.4.9

Total Maximum Bending moment = 1,213,780 + 893,750 or equal to: $M = 2,107,530$ Foot Lbs.
Bending stress is limited to: $F_b = 20,000$ PSI. Then the least Section Modulus, $S = \frac{M}{F_b}$. $S_x = \frac{2,107,530 \times 12}{20,000} = 1264.50''^3$

STEP III:
Limited depth of Girder = 1/10 of 55.0' or d = 5.50' (66.0 inches.)
For convenience, let depth of Web plate = 60.0 inches.
Web thickness minimum is ¼ inch plate or by rule, it is 1/170 of web depth. Then $t = \frac{60.0}{170} = 0.353''$ Use 3/8 inch plate or $t = 0.375''$
Reaction $R = V$. Then unit shear on web $= \frac{V}{tT}$ and substuting values: $f_v = \frac{120,775}{0.375 \times 60.0} = 5360$ PSI. (OK)

STEP IV:
If full allowable bending stress of $F_b = 20,000$ PSI is to be design basis, the ratio of $\frac{l}{b}$ for cover plate cannot exceed the ratio of 15. To find minimum width of cover plate without bracing, the ratio cannot exceed 40.
Then minimum width to sustain lateral support with some bracing is $= \frac{55.0 \times 12}{40} = 16.5$ inches = b. By referring to Table for $\frac{l}{b}$ stress reduction, it is $F_b = 11,910$ PSI.
Assume that lateral bracing is spaced thus: $\frac{55.0}{4} = 13.75$ Ft.
Then longest unbraced length $l = 13.75 \times 12 = 165.0$ In.
Keep ratio of $\frac{l}{b}$ at 15, the width of cover plate: $b = \frac{165.0}{15} = 11.0''$
Accept this width for the cover plates. 11.00 inches.
This is minimum width and can be enlarged when it becomes necessary to build up for value of I_x later.

STEP V:
Web stiffening at end support is the critical point and the stiffeners under load P will be arbitrarily made same.
Stiffeners represent short columns acting ½ depth of web plate which is 60.0 inches between cover plates.
Maximum axial compressive stress $F_c = 17,000$ PSI.
When all shear stress is removed from web plate the area for stiffeners is: $A = \frac{R}{F_v}$ or $A = \frac{120,775}{17,000} = 7.10$ Sq. inches.

EXAMPLE: Design of hybrid girder, continued 2.4.9

Using 4 Flat bars for stiffeners, allow 1 Pair to assume 2/3 of reaction to compensate for any eccentric bearing on support. Then $A = \dfrac{0.67 \times 120{,}775}{2 \times 17{,}000} = 2.40\ \square''$ (For each flat bar).

From Tables of Rectangular Shapes in Section VI, select a flat bar. ½" x 5.0" with A = 2.50 □" These will extend almost to edge of 11.00 inch cover plate which is desired.
Use 2 Pair stiffeners at end composed of ½ x 5.0 F.B.

STEP VI:
For intermediate stiffeners under load P = 65,000 Lbs.
When ratio of Web depth to Web thickness is over 70, stiffeners are required. In formula, it is: $\frac{T}{t}$ and with T = 60.0" and t = 0.375, the ratio is: $\dfrac{60.0}{0.375} = 160.$ They are required.
For uniform appearance use ½" x 5.0" Flat bars. Install 2 Pair and space to suit width of beam with the R = 65,000 Lbs.

STEP VII:
For intermediate stiffeners along length of Girder span.
The AISC Formula for spacing is: $s = \left(\dfrac{270{,}000 \times t}{f_v}\right) \times \sqrt[3]{\dfrac{f_v t}{T}}$.
In step III, actual stress was: $f_v = 5360$ PSI.
Then substituting values in formula:

$$\text{spacing, } s = \left(\frac{270{,}000 \times 0.375}{5360}\right) \times \sqrt[3]{\frac{5360 \times 0.375}{60.0}} = 18.8 \times 3.2 = 58.24 \text{ In.}$$

Intermediate stiffeners in flat bars should be 1 inch in width for every foot of web depth. Then when depth is 5.0 feet, width of flat bar must be 1" x 5.0' = 5.0 inches.
Use same stiffener in pairs and space not over 4.833 feet.

STEP VIII:
Cross-section must now be designed for properties which will provide a Resisting Moment equal or greater than the Bending Moment as calculated in Step II.
Thus far, no concern has been given to thickness of the cover plates because the AISC formula for reduced stress in lateral bracing was not used. That formula required the value area of cover plate and will be used to check when area is decided.
From Step II, the minimum, $S_x = 1264.5''^3$ If $\frac{d}{2}$ = c dimension,

EXAMPLE: Design of hybrid girder, continued 2.4.9

and $S_x = \frac{I}{c}$, then $I_x = S_x C$. Web plate is 60.0 inches deep, which will make dimension to outer fibers C = 31.0" approx. Cross section will have an I_x value of $1264.5 \times 31.0 = 39,200.0''^4$ Referring to Section VI, Properties of rectangular shape tables: The area of Web: $A = 0.375 \times 60.0 = 22.5$ □" and its moment of inertia $I_x = 6750.0''^4$ The 11.0 inch wide cover plates must make up the rest to arrive at the required I_x.

STEP IX:

Draw the Girder cross-section and identify each component. Proceed as in Section VI. Probably will be necessary to use 2 cover plate where bending moments are larger.

MARK	SIZE	A□"	ʅ"	Aʅ²	I₀	Aʅ² + I₀
1	10.0x0.5	5.00	31.25	4,875.00	0.10	4,875.10
2	11.0x1.00	11.00	30.50	11,332.75	0.92	11,333.67
3	0.50x60.0	30.00	-0-	-0-	9000.00	9,000.00
4	11.0x1.00	11.00	30.50	11,332.75	0.92	11,333.67
5	10.0x0.5	5.00	31.25	4,875.00	0.10	4,875.10
		62.00□"			Σ I_x=	41,417.54"⁴

Section Modulus = $\frac{I_x}{c}$ $S_x = \frac{41,417.54}{31.50} = 1314.0''^3$

Weight less stiffeners: $62.0 \times 3.40 = 210.8$ #/' (close)

Web has been changed to ½ inch thickness in detail. Later it was seen that if the web thickness remained 3/8", the $\Sigma I_x = 39,167.5''^4$ Required $I_x = 39,200''^4$ and is ok with original plate thickness.

SECTION "A"-"A"

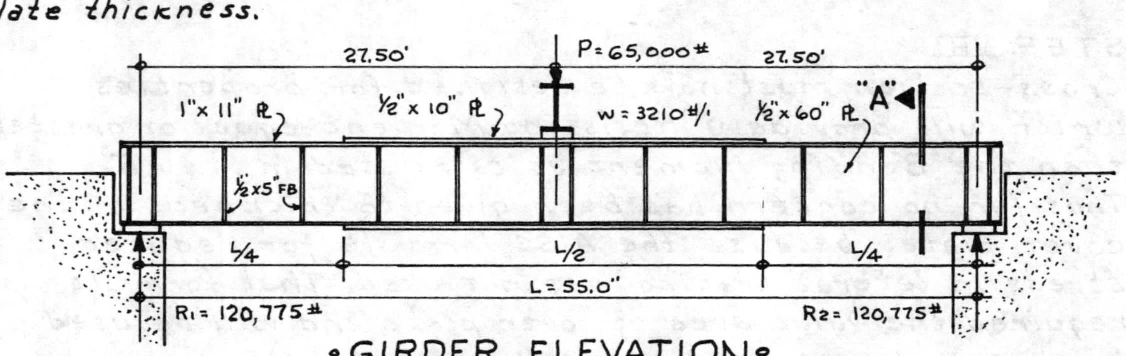

GIRDER ELEVATION

EXAMPLE: Design of hybrid girder, continued 2.4.9

STEP X:

Check unit stress and moment at end of ½" x 10" cover plate.
Dimension at $\frac{L}{4}$ = 55.0 x 0.25 = 13.75 feet from R_1.
Property of I_x at end of top plate = 41,417.14 − (2 x 4875.10)
Without cover plates No. 1 and 5; I = 31,666.95"^{4} and the dimension C = 31.50 − 0.50 = 31.0" $S = \frac{I}{c} = \frac{31,666.95}{31.0} = 1021.5"^3$

Bending Moment at 13.75': $M_{13.75}$ =

$M_{13.75}$ = (120,775 x 13.75) − (3210 x 13.75 x 6.875) = 1,357,200 Foot Lbs.

Actual stress at 13.75: $f_b = \frac{M}{S}$, or $f_b = \frac{1,357,200 \times 12}{1021.5}$ = 15,900 PSI.

Actual unit stress is less than allowable F_b of 20,000 PSI and the cover plates No. 1 and 5 can be cutoff at this point.

STEP XI:

Check horizontal shear at juncture of web plate and cover plate No. 2. The shear force for welds to resist is solved by formula: $V_h = \frac{Va\bar{y}}{I_x}$. Where: $\bar{y}$ = Centroid of both cover plates, and a = area of both cover plates. $\bar{y}$ = 30.0 + 0.75 = 30.75 inches.
a = 5.0 + 11.0 = 16.0□", and I_x = 41,417.54. $V = R$ = 120,775 Lbs.

Then: $V_h = \frac{120,775 \times 16.0 \times 30.75}{41,417.54}$ = 1437 Lbs. per Lineal inch.

With a 3/8" web plate, the unit stress $f_h = \frac{1437}{0.375}$ = 3830 PSI:

STEP XII:

Designing the welds: From tables providing Welding Data: a minimum ¼" fillet weld has a value of 2200 Lbs. per lineal inch. The horizontal shear force per lineal foot is equal to 1437 x 12 = 17,244 Lbs.

Required length of weld per foot = $\frac{17,244}{2200}$ = 7.85 inches.

Use 9 inches per foot, or 3" welds staggered at 4½ inches on centers thus:

3.0" 1.5" 3.0" 1.5" 3.0"
12.0"
¼ 3 − 4½

Weld flat bar stiffeners to web plate with ¼" fillet weld 2 inches long with 6 inch space between welds. At ends of flat bar stiffeners, weld all around to cover plates.

Steel column design 2.5

A column is defined as a structural member which is required to sustain a compressive stress parallel to its length and longitudinal axis. Naval architects refer to a column as a stanchion. The member does not necessarily have to be in a vertical position to be considered or designed as a column. The top chord of a truss and a web strut are designed as columns when the stress is predominantly compressive on the cross-sectional area.

The load which may be safely placed upon a column depends on several factors:

(a) Length of unsupported cross-section.
(b) Area of cross-section.
(c) Load placement; concentric or eccentric.
(d) Type of material; steel, wood, aluminum or concrete.
(e) Profile or shape of member; circular, square, H-shape, rectangle or tubular.
(f) The radius of gyration, and its ratio to the unbraced length (slenderness ratio).
(g) Allowable unit compressive stress which must be compatible with the slenderness ratio stress formula.

Slenderness ratio 2.5.1

In designing steel columns, the least radius of gyration (r) is the first property to be considered in the initial stage. Next, the unbraced length in inches must be determined. Refer to Section VI for an explanation of the least radius of gyration as it applies to differently shaped cross-sections. The slenderness ratio is $\frac{l}{r}$ for a steel section, or $\frac{l}{d}$ for a wood column. The unsupported length (l) is given in inches for both steel and wood.

In the column designs to follow, the property of r about both axis x–x and y–y will be investigated. A column may be sufficiently braced for its minor axis and still remain under the allowable slenderness ratio with regard to its major axis. Building codes limit the slenderness ratio for main compressive members to no greater than 120. For secondary members (braces, struts, and truss webs), the slenderness ratio shall not be greater than 200. For ratios of 60 or less, the members are referred to as short columns and the maximum unit stress F_a = 17,000 P.S.I., is to be used. As the slenderness ratio $\frac{l}{r}$ over 60 is increased, the allowable unit working stress is to be reduced by using the column formulas shown in Tables 2.5.4.3 and 2.5.4.4.

Column design formulas 2.5.1.1

The design unit working stress for a column will depend upon its slenderness ratio, and is reduced by formula as the ratio is increased. The AISC formula is:

For slenderness ratios of $\frac{l}{r}$ between 10 and 120, the unit stress is calculated by the formula:

$$F_a = 17{,}000 - \left(0.485\frac{l^2}{r^2}\right).$$

For $\frac{l}{r}$ ratios between 120 and 200, the formula is:

$$F_a = \frac{18{,}000}{1.0 + \left(\frac{l^2}{18{,}000\,r^2}\right)}.$$

Tables and graphs are provided for convenience in obtaining the allowable stress without evaluating the formulas.

The American Bridge Company employed a very conservative formula which is still used for many bridge structures. There are many other formulas in use by competent engineers, mainly because of preference or reluctance to change.

Straight line formulas are easily used and satisfactory unless code specifications require use of the AISC formula. The American Bridge Formula is:

For ratios of $\frac{l}{r}$ up to 120:

$$\frac{P}{A} = 19{,}000 - 100\frac{l}{r}.$$

For ratios of $\frac{l}{r}$ between 120 and 200:

$$\frac{P}{A} = 13{,}000 - 50\frac{l}{r}.$$

If the unit stress derived by the above formulas is greater than 13,000 PSI, the unit stress is reduced to the maximum allowable of 13,000 PSI.

Designers classify column members according to the slenderness ratio and application. In the examples, a main member is understood to have an $\frac{l}{r}$ ratio below 120. A secondary member is understood to have an $\frac{l}{r}$ ratio between 120 and 200, and is usually a brace or a strut.

Eccentric column loads 2.5.2

Eccentricity in columns means that the center of gravity of the applied load is not plumb with the axes of the column section. The distance from the axes to the load center is an eccentric moment lever which will produce bending in the column. If the load is placed exactly on the intersection of axis x–x and y–y, the load is said to be concentric or axial. An eccentric load is too often overlooked. A small amount of eccentricity in heavy columns may usually be neglected; however, the examples to follow will illustrate the importance of eccentricity.

In the majority of cases, eccentricity is transmitted to the column from the beam connection, or as a result of lateral wind pressure. These types of joints are referred to as moment connectors and will be discussed later in this section under bolt and riveted connectors. In the design of columns which support axial loads and include a bending moment, it is very easy to overlook the fact that the eccentric load is to be added to the axial load.

It will be illustrated in examples how the bending moment can be converted into an equivalent axial load. The final load on the column will be the sum of several terms: Axial load + equivalent load + eccentric load.

Bending factors 2.5.2.1

The effect of the eccentricity of the load can be expressed in terms of an equivalent axial load by using the bending factors B_x and B_y. This equivalent load is added to the actual load as an additional axial load, and should give the same maximum stress as would be computed using the bending moment. Bending factors are indicated as B_x or B_y, depending on the axis taken. This property is found by dividing the section area by the section modulus. In formula form, it is: $B_x = \frac{A}{S_x}$, or $B_y = \frac{A}{S_y}$. All that is necessary to convert an eccentric bending moment into an equivalent axial load is to multiply the bending moment by the bending factor which applies to the proper axis.

To illustrate the convenience of the bending factor, assume a steel section W8x17. From the tables the following properties are taken: A = 5.0 Sq. Inches. Sx = 14.1 and Sy = 2.60. Then $B_x = \frac{5.00}{14.1} = 0.355$ *and* $B_y = \frac{5.00}{2.60} = 1.92$

If the eccentric bending moment were 12,000 inch pounds and applied to the major axis x-x, the equivalent axial load would equal: 12,000 × 0.355 = 4260 Pounds. The actual bending unit stress is therefore: $f_b = \frac{M}{S_x}$ *or* $f_b = \frac{12{,}000}{14.1} = 8520$ *PSI.*

When columns are subjected to both axial and eccentric loads, the AISC specification requires that the quantity $\frac{f_a}{F_a} + \frac{f_b}{F_b}$ shall not exceed unity (1.0).

Ratio of Radius of Gyration 2.5.3

The radius of gyration provides a convenient method to design columns using the load tables provided in the AISC Manual of Steel Construction. Remember that the column loads shown in the tables are based on the *least radius of gyration* and pertain to axial loads only.

Consider the radius of gyration about the major axis, r_x. When this value is used in the column formula instead of the lower value of r_y, the results will return a lower slenderness ratio, and the allowable stress is increased. This simply means that the column could be made of greater length if one only considered the major axis value of r_x. Assuming that a column has been designed on the basis of the minor axis and least r_y value; it is a good check to determine the maximum column length permitted about the major axis x–x. The allowable stress found from the column formula is used, with the same slenderness ratio $\frac{l}{r}$. Let us illustrate by taking a sample from the AISC Manual:

A W14 x 43 Column section will safely support a maximum Axial load of 140,000 Pounds on an unbraced length of 18.0 feet. This load was based upon the least r_y. Now— at what length will the section perform the same service when based upon the greater value of r_x?

$r_y = 1.89$ and $r_x = 5.82$. The ratio $= \frac{r_x}{r_y}$ or $\frac{5.82}{1.89} = 3.08$ and

Area A = 12.65 Sq. inches.

The column will perform the same service about axis x-x without lateral bracing as: $18.0 \times 3.08 = 55.44$ feet.

This can be verified by solving for slenderness ratio for each axis thus:

$L_y = 18.0$ feet, or $l_y = 18.0 \times 12 = 216$ inches. Then $\frac{l_y}{r_y} = \frac{216}{1.89} = 114.2$

$L_x = 55.44$ feet, or $l_x = 55.44 \times 12 = 665.3$ inches. Then $\frac{l_x}{r_x} = \frac{665.3}{5.82} = 114.2$

This illustrates a simple check which should be employed before making a final selection for the column section.

Steel pipe columns 2.5.4

Standard steel pipe and large tubing is produced from material with the same characteristics as A36 steel, which permits the column formulas for rolled shapes to be used. In the design of round or square hollow tubes for columns, the radius of gyration is equal at any point on the exposed surface. Square tubing has an advantage over steel pipe in that the connections for fabrication are made with greater accuracy. Pipe sizes given in tables are the nominal size for the diameter and do not indicate the inside or outside dimension. Architects prefer to specify pipe columns as standard, extra strong, or double extra strong. Engineers use the schedule method of specifying wall thickness: schedule 40, schedule 60, etc. Seamless steel tubing is usually identified by referring to the outside diameter. Very large pipe may be specified from the Pipe Pile Tables in Section IX.

COMPARABLE PROPERTIES OF STEEL PIPE r"

NOMINAL DIAMETER	CLASSIFICATION	OUTSIDE DIAMETER	INSIDE DIAMETER	WALL THICK'N'S.	WEIGHT LIN. FT.	AREA □"	RADIUS GYRATION
8"	STD. STRUCTURAL	8.625"	7.981"	0.322"	28.55	8.39	2.940
8"	EXTRA STRONG	8.625"	7.625"	0.500"	43.39	12.76	2.878
8"	DOUBLE XTRA STR'G.	8.625"	6.875"	0.875"	72.42	21.30	2.757
8"	SCHEDULE 40	8.625"	7.981"	0.322"	28.55	8.39	2.940
8"	SCHEDULE 80	8.625"	7.625"	0.500"	43.39	12.76	2.878
8"	SCHEDULE 120	8.625"	7.177"	0.719"	60.69	17.85	2.392

Concrete-filled pipe columns 2.5.4.1

Steel pipe columns filled with solid concrete are referred to as composite columns. They are not frequently used unless they are an extension of a pile above ground. There is a danger that voids may occur when pipes of 8 inches or less are used for heavy loads. Japanese engineers design their heavy industrial buildings with concrete filled steel pipe columns which are extensions of pipe piles driven to sustain loads of 200 tons or more.

Tests conducted at Lehigh University indicate that composite columns will support *ultimate loads* equal to 85 percent of the 28 day compressive strength of the concrete plus the yield stress of the area of the steel pipe. Written as a formula Ultimate axial load $P_u = (0.85\ F_c') + (F_y A_s)$. To arrive at a working or design formula, it is derived thus:

Safe axial load $P_a = (0.25\ F_c' A_c) + (F_a A_s)$.

Where:

P_a = Safe allowable axial load in pounds.

F_c' = Ultimate strength of concrete at age of 28 days in PSI.

A_c = Area cross section of concrete inside pipe in square inches.

F_a = Permissible steel column stress as obtained from column formulas for $\frac{l}{r}$ ratio, in PSI.

A_s = Area of steel in pipe cross-section, in square inches.

Designers will find the pile tables in Section IX a great aid in the calculations for concrete-filled steel columns. Remember that piles are supported laterally by the soil, and the slenderness ratio is not considered. When piles are laterally unbraced, and long lengths are submerged in water or air, the slenderness ratio is considered as if it were a column.

Rectangular tube columns 2.5.4.2

Architects often lay out room arrangements with the supporting columns concealed within the interior walls. Steel producers began in 1963 to supply a variety of rectangular and square hollow tubes suitable for this type of work. There is no loss of support strength, as will be illustrated for two types which can be concealed in a 2 x 6 wood stud wall.

From the AISC Manual with Column Load Tables, select a W 8 x 17 section with flange width of 5¼ inch. With an unbraced length of 11.0 feet, the safe maximum load is listed as 58,000 pounds about minor axis y–y. The ratio of $\frac{r_x}{r_y} = 2.90$. This column would support the same load about its major axis x–x for the length L = 11.0 x 2.90 = 31.9 feet.

To compare a rectangular hollow tube to the above results, again refer to the load tables. Choose an 8 x 5 tube (with a lower weight of 14.41 pounds). The table lists the maximum load at 11.0 feet as 66,000 pounds. The ratio of $\frac{r_x}{r_y} = 1.71$. Maximum length about axis x–x with the same load is L = 11.0 x 1.71 = 18.8 feet.

Another advantage for the rectangular tube over flanged or pipe column is the case of installation for wood grounds or plate glass on the exterior walls.

TABLE: Allowable stress in MAIN columns 2.5.4.3

l = UNBRACED LENGTH, IN INCHES

r = RADIUS OF GYRATION, IN INCHES

SHORT AND MAIN COLUMNS $\frac{l}{r}$ UP TO 120

AISC STRESS FORMULA:

$$F_a = 17{,}000 - \left(0.485 \frac{l^2}{r^2}\right)$$

AXIAL LOADS ON STEEL COLUMNS ALLOWABLE UNIT COMPRESSIVE STRESSES

$\frac{l}{r}$	F_a = #□"	$\frac{l}{r}$	F_a = #□"	$\frac{l}{r}$	F_a = #□"	$\frac{l}{r}$	F_a = #□"	$\frac{l}{r}$	F_a = #□"
1	17,000	26	16,675	51	15,740	76	14,200	101	12,050
2	17,000	27	16,656	52	15,690	77	14,120	102	11,950
3	17,000	28	16,623	53	15,640	78	14,050	103	11,860
4	16,990	29	16,590	54	15,590	79	13,970	104	11,750
5	16,990	30	16,564	55	15,530	80	13,900	105	11,650
6	16,980	31	16,531	56	15,480	81	13,820	106	11,550
7	16,980	32	16,500	57	15,420	82	13,740	107	11,450
8	16,970	33	16,470	58	15,370	83	13,660	108	11,340
9	16,960	34	16,444	59	15,310	84	13,580	109	11,240
10	16,950	35	16,412	60	15,250	85	13,500	110	11,130
11	16,940	36	16,373	61	15,200	86	13,410	111	11,020
12	16,930	37	16,345	62	15,140	87	13,333	112	10,920
13	16,920	38	16,303	63	15,080	88	13,240	113	10,810
14	16,910	39	16,265	64	15,010	89	13,160	114	10,700
15	16,875	40	16,220	65	14,955	90	13,070	115	10,590
16	16,880	41	16,190	66	14,890	91	12,980	116	10,470
17	16,860	42	16,140	67	14,820	92	12,910	117	10,360
18	16,850	43	16,105	68	14,760	93	12,815	118	10,250
19	16,833	44	16,060	69	14,690	94	12,722	119	10,130
20	16,811	45	16,020	70	14,620	95	12,620	120	10,020
21	16,792	46	15,970	71	14,560	96	12,530		
22	16,776	47	15,930	72	14,490	97	12,440		
23	16,744	48	15,880	73	14,420	98	12,340		
24	16,721	49	15,840	74	14,340	99	12,250		
25	16,700	50	15,790	75	14,270	100	12,150		

TABLE: Allowable stress in SECONDARY members 2.5.4.4

MAIN COLUMNS: $\frac{l}{r}$ = 60 TO 120

SECONDARY MEMBERS: $\frac{l}{r}$ = 120 TO 200

SHORT COLUMNS: $\frac{l}{r}$ = 1 TO 60

FORMULA: $$F_a = \frac{18,000}{1.0 + \left(\frac{l^2}{18,000\, r^2}\right)}$$

AXIAL LOADS ON STEEL COLUMNS
ALLOWABLE UNIT COMPRESSIVE STRESSES

$\frac{l}{r}$	F_a #□"	$\frac{l}{r}$	F_a #□"	$\frac{l}{r}$	F_a #□"	$\frac{l}{r}$	F_a #□"	$\frac{l}{r}$	F_a #□"
60	15,000	89	12,500	118	10,150	147	8,180	176	6,615
61	14,916	90	12,414	119	10,075	148	8,119	177	6,568
62	14,832	91	12,328	120	10,000	149	8,060	178	6,521
63	14,748	92	12,243	121	9,926	150	8,000	179	6,475
64	14,663	93	12,158	122	9,853	151	7,941	180	6,429
65	14,578	94	12,075	123	9,780	152	7,882	181	6,383
66	14,493	95	11,990	124	9,708	153	7,824	182	6,338
67	14,407	96	11,905	125	9,636	154	7,767	183	6,293
68	14,321	97	11,820	126	9,564	155	7,710	184	6,248
69	14,235	98	11,737	127	9,493	156	7,653	185	6,204
70	14,148	99	11,655	128	9,423	157	7,597	186	6,160
71	14,062	100	11,570	129	9,353	158	7,541	187	6,117
72	13,975	101	11,489	130	9,284	159	7,486	188	6,074
73	13,888	102	11,407	131	9,215	160	7,431	189	6,031
74	13,800	103	11,329	132	9,146	161	7,377	190	5,989
75	13,715	104	11,244	133	9,078	162	7,323	191	5,947
76	13,627	105	11,163	134	9,011	163	7,269	192	5,906
77	13,540	106	11,082	135	8,944	164	7,217	193	5,846
78	13,453	107	11,000	136	8,878	165	7,164	194	5,824
79	13,366	108	10,922	137	8,812	166	7,112	195	5,783
80	13,280	109	10,843	138	8,746	167	7,061	196	5,743
81	13,192	110	10,765	139	8,681	168	7,009	197	5,703
82	13,105	111	10,686	140	8,617	169	6,959	198	5,664
83	13,108	112	10,608	141	8,553	170	6,910	199	5,624
84	12,931	113	10,530	142	8,490	171	6,858	200	5,586
85	12,844	114	10,453	143	8,427	172	6,809		
86	12,758	115	10,376	144	8,364	173	6,760		
87	12,672	116	10,300	145	8,302	174	6,711		
88	12,585	117	10,225	146	8,241	175	6,663		

CURVE: Allowable stress in steel columns 2.5.4.5

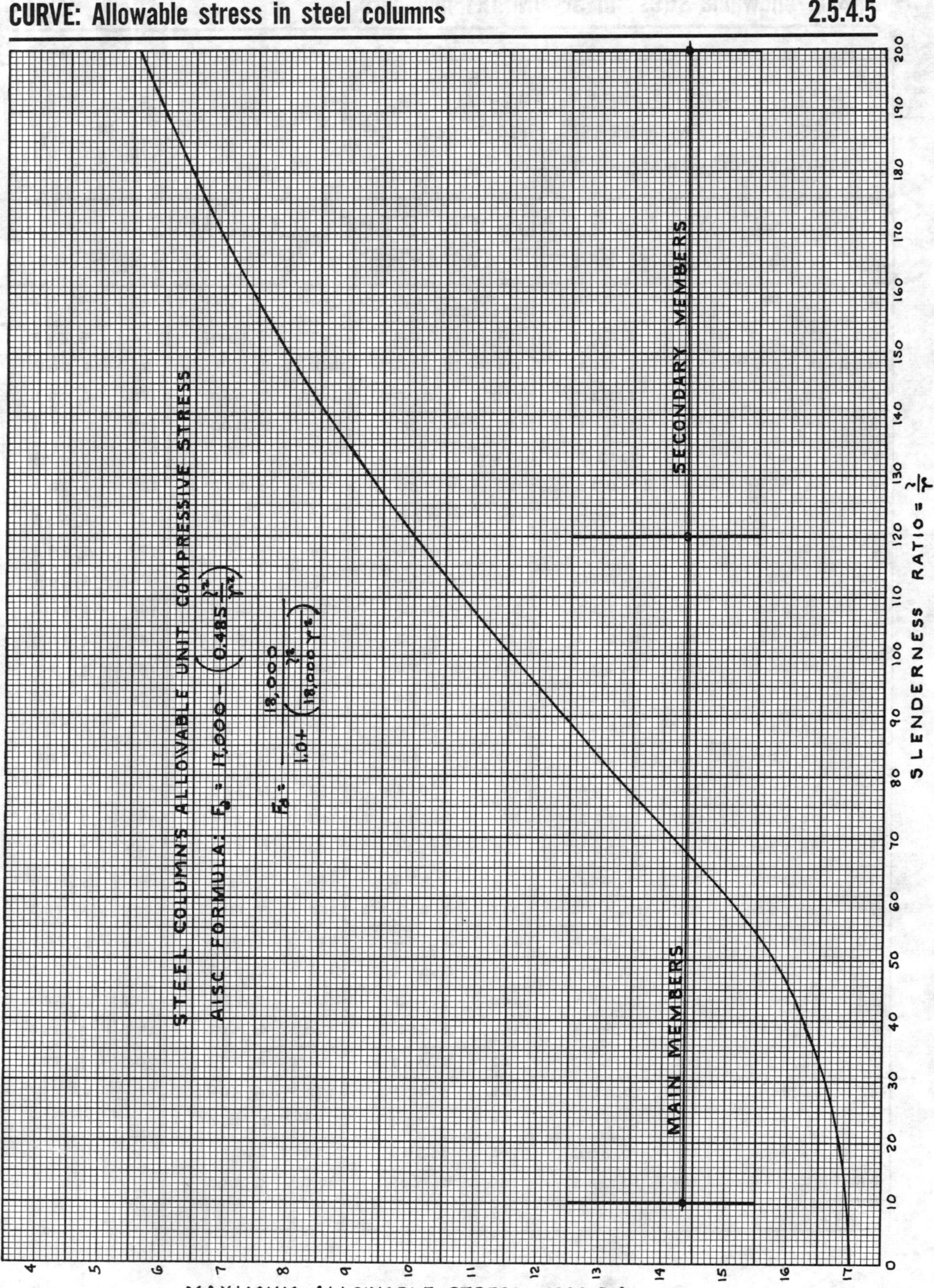

CURVE: Straight line column formulas 2.5.4.6

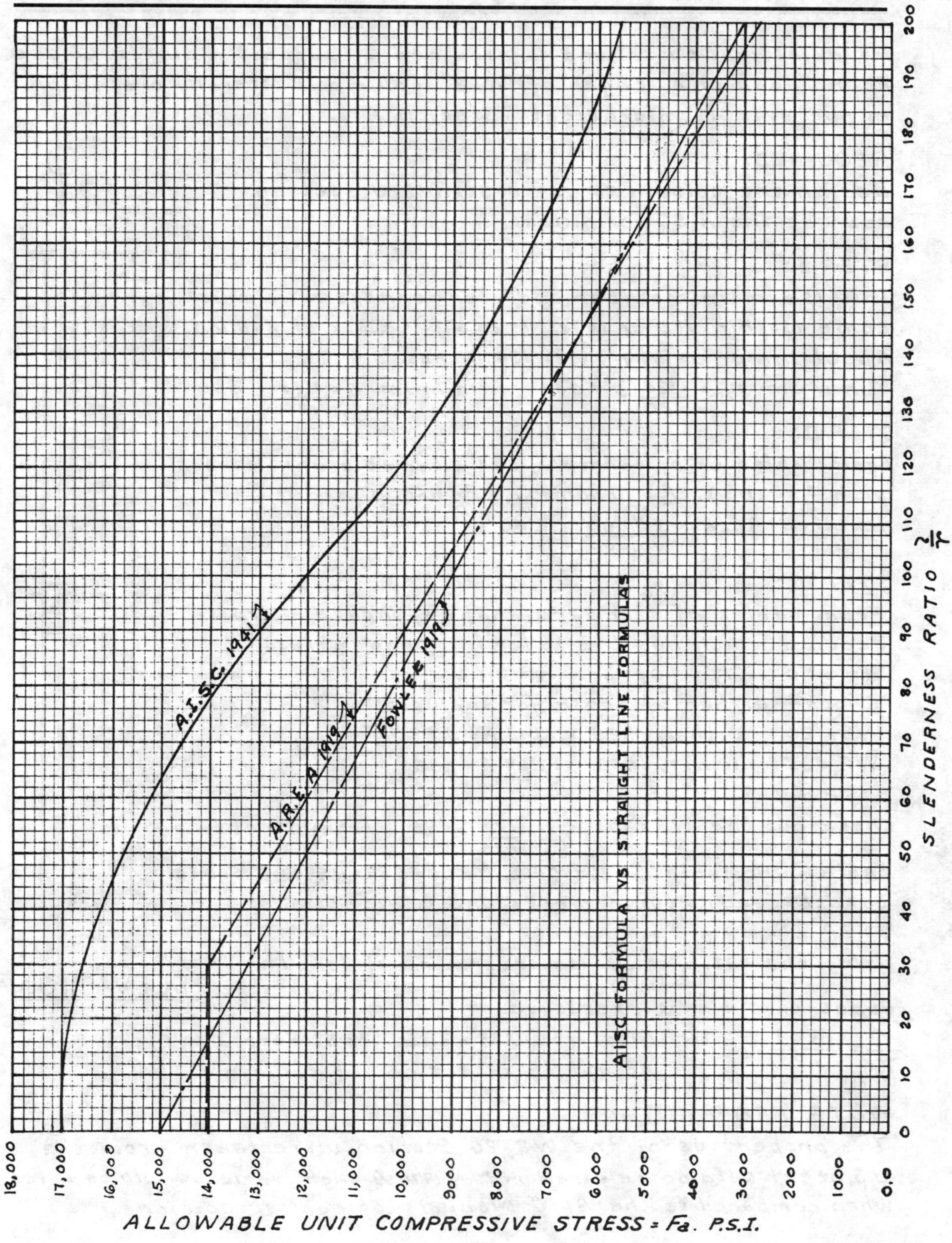

AISC FORMULA VS STRAIGHT LINE FORMULAS

A.I.S.C. 1941

A.R.E.A 1919

FOWLER 1919

SLENDERNESS RATIO $\frac{l}{r}$

0 10 20 30 40 50 60 70 80 90 100 110 120 130 140 150 160 170 180 190 200

ALLOWABLE UNIT COMPRESSIVE STRESS = F_a. P.S.I.

18,000 17,000 16,000 15,000 14,000 13,000 12,000 11,000 10,000 9000 8000 7000 6000 5000 4000 3000 2000 1000 0.0

EXAMPLE: Axial load on column 2.5.5.1

An axial loaded column is required to support a single 65,000 Pound load on a length of 20'-8". Column is laterally braced on minor axis at mid-height. Steel is A36.

REQUIRED:
Design the column to use a standard section and use the AISC Formulas:

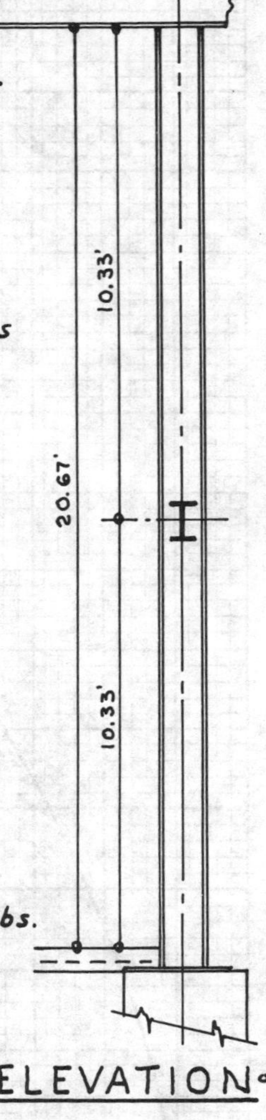

STEP I:
This is a main member and slenderness ratio must come within 120 or less. $L = 20.67$, $l = 20.67 \times 12 = 248$ in.
About minor axis y-y, $l = 124$ inches.
Minimum $r_y = \frac{124}{120} = 1.034$ Minimum $r_x = \frac{248}{120} = 2.06$

STEP II:
From tables or by formula, the allowable unit stress with $\frac{l}{r} = 120$ is: $F_a = 10{,}000$ PSI. $P = 65000$ Lbs. $A = \frac{P}{F_a}$
Required Area $= \frac{65{,}000}{10{,}000} = 6.50$ □"

STEP III:
From tables of standard shapes, locate a cross-section which has the following properties:
$A = 6.50$ □" $r_y = 1.03$ and $r_x = 2.06$
Choose either of these for analysis:
W8 x 20 with $r_y = 1.25$ $r_x = 3.43$ $A = 5.89$ □"
W6 x 25 with $r_y = 1.53$ $r_x = 2.69$ $A = 7.35$ □"

STEP IV:
Check out the W8x20: $\frac{l}{r_y} = \frac{124}{1.25} = 95$ From stress tables, allowable $F_a = 11{,}990$ PSI. Max. $P = 5.89 \times 11{,}990 = 70{,}500$ Lbs.
Load exceeds requirements and is acceptable.

STEP V:
Using the ratio of $\frac{r_x}{r_y}$ to determine if length about axis x-x is excessive.

$\frac{r_x}{r_y} = \frac{3.43}{1.25} = 2.76$ Max. $L_x = 10.33 \times 2.76 = 28.52$ Feet. OK
Accept this section for column.

NOTE FROM AUTHOR:
The properties of the W8x20 Section were taken from the U.S. Steel catalog released in May 1971. A slight variance will be noted when compared to the AISC Manuals of earlier editions.

EXAMPLE: Minimum slenderness ratio for axial load 2.5.5.2

A steel A36 column is to support the end of a pipe rack in a refinery complex. The unsupported length of column is 12'-8" and axial load is 48,750 Pounds.
Plant Engineering requirement stipulate that all main columns shall have their slenderness ratio limited to 85.

REQUIRED:
Use the AISC Column formula to design a suitable section and limit $\frac{l}{r_y}$ to 85 or less.

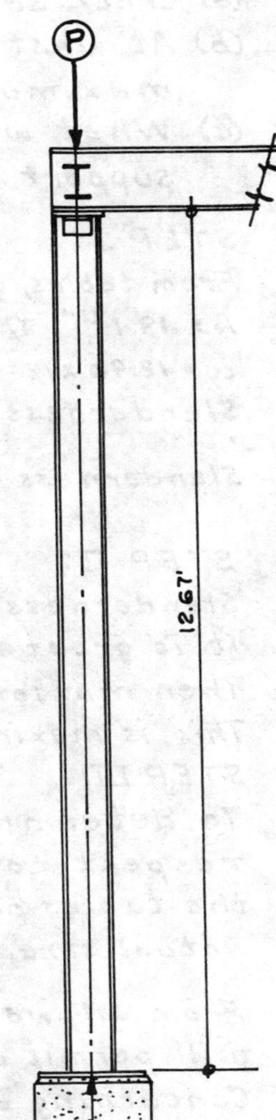

STEP I:
Least radius of gyration permitted:
$L = 12.67'$ $l = 12.67 \times 12 = 152$ inches.
Minimum $r_y = \frac{152}{85} = 1.79$

STEP II:
Determine allowable stress by formula for F_a.

$$F_a = \frac{18{,}000}{1.0 + \left(\frac{l^2}{18000 \times r^2}\right)} \qquad l^2 = 23{,}104 \quad r^2 = 3.20$$

$18{,}000 \times 3.20 = 57{,}600$ $\frac{23{,}104}{57{,}600} = 0.405$

Then: $F_a = \frac{18{,}000}{1.0 + 0.405} = 12{,}840$ PSI.

STEP III:
Required Area: $A = \frac{P}{F_a}$ $A = \frac{48{,}750}{12{,}840} = 3.80$ □"

From tables of Sections:
r_y must be not less than 1.79 and A not less than above.

Accept a Section W8 x 31. This an 8x8 WF or CB83 with an A = 9.12 □" and $r_y = 2.01$"

EXAMPLE: Maximum axial load on columns 2.5.5.3

A W12x65 Section represents a column 23'-9" about its major axis x-x and supports an axial load of 300,000 Pounds. This column is braced about axis y-y at a point 10.85 feet from top. Steel is A36 with Fy = 36000 PSI.

REQUIRED:

(a) Check section to determine if this is a safe load about axis x-x.

(b) At what length on axis x-x will section sustain the greatest maximum Load of 300,000 Lbs.

(c) What will be the maximum safe load the column will safely support with existing lateral support.

STEP I:

From tables, gather the properties of a section W12x65.

$A = 19.1 \square''$ $r_x = 5.28$ $r_y = 3.02$ $L_x = 23.75'$ and $L_y = 12.90'$

$l_y = 12.90 \times 12 = 155$ inches. $l_x = 23.75 \times 12 = 285$ inches.

Slenderness ratio about x-x; $\frac{l_x}{r_x} = \frac{285}{5.28} = 54$

Slenderness ratio about y-y; $\frac{l_y}{r_y} = \frac{155}{3.02} = 51.3$

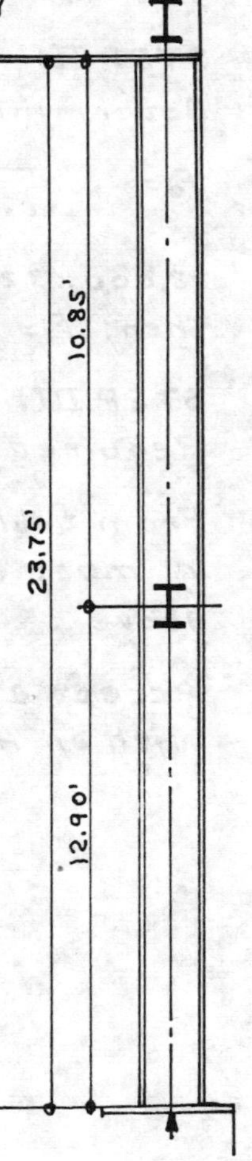

STEP II:

Slenderness ratio about axis x-x will govern since it is greater. From tables: Max. $F_a = 15,590$ #□"

Then maximum load $P = F_a A$ or $P = 15,590 \times 19.1 = 297,770$#

This is maximum load under existing conditions (Ans. c)

STEP III:

To determine maximum length of column with respect to axis x-x under 300,000 Lbs. load, use the tables or formula to compare unit stresses.

Actual stress, $f_a = \frac{P}{A}$ or $f_a = \frac{300,000}{19.1} = 15,720$ PSI.

From allowable unit stress tables, an $\frac{l}{r}$ ratio of 51 will permit an F_a of 15,740 PSI. Then $\frac{l}{r}$ must equal 51.

Concerning axis x-x: l = ratio × r_x and:

$L_x = \frac{51 \times 5.28}{12} = 22.44$ Feet.

STEP IV:

Summarizing the results for requirements:

Answer to (a): No. Maximum P = 297,770 Lbs.

" " (b): Max. L = 22.44 Feet for 300,000 Lbs.

" " (c): Same as (a) = 297,770 Lbs.

EXAMPLE: Maximum length with axial load 2.5.5.4

The AISC Manual column load tables list a permissable axial load of 125,000 Pounds for a W8x40 section with an unsupported length of 20.0 feet. This load is based upon the slenderness ratio by using the minor axis y-y.

REQUIRED:
Analyze the section and determine the following.

(a) Assume column is provided with lateral support at top (20.0 feet), and same load of 125,000 Pounds remains, what amount of extension can be made to support load on axis x-x property.

(b) Compute the slenderness ratio about each axis to confirm the results found in (a).

(c) Compare the actual stress f_a under load with the allowable F_a as obtained by the formula for stress allowed.

STEP I:
Gather the required data for use:
$L = 20.0'$ $l = 20.0 \times 12 = 240$ inches. $P = 125{,}000$ Lbs. From tables for W8x40 Section: $A = 11.80\,\square''$ $r_x = 3.53$ $r_y = 2.04$

Ratio of $\frac{r_x}{r_y} = \frac{3.53}{2.04} = 1.73$ AISC table refers to A36 Steel.

STEP II:
With axis y-y braced at 20.0' $l = 240''$

Slenderness ratio $= \frac{l}{r_y} = \frac{240}{2.04} = 117.5$

Maximum unsupported length for axis x-x:
$L = 20.0 \times 1.73 = 34.6$ Feet or a 14.6 foot extension. (Answer a)

STEP III:
Slenderness ratio about axis y-y = 117.5

About axis x-x: $\frac{l_x}{r_x} = \frac{34.6 \times 12}{3.53} = 117.5$ checks (Answer b)

STEP IV:
Actual stress: $f_a = \frac{P}{A}$ or $f_a = \frac{125{,}000}{11.80} = 10{,}580$ PSI

STEP V
By formula: $F_a = \dfrac{18{,}000}{1.0 + \left(\dfrac{l^2}{18{,}000\, r^2}\right)}$ Using l_y and r_y in formula.

$l_y^2 = 240^2 = 57{,}600$ $r_y^2 = 2.04 \times 2.04 = 4.16$ and $18000 \times 4.16 = 74{,}880$

Then: $\frac{57{,}600}{74{,}880} = 0.757$ and $F_a = \frac{18{,}000}{1.0 + 0.757} = 10{,}250$ PSI.

Load given in AISC Load table is slightly greater. (Answer c)
By example: Max. $P = 10{,}250 \times 11.80 = 120{,}950$ Lbs. Close enough.

EXAMPLE: Column design by formula 2.5.5.5

This problem was submitted to applicants for registration by Texas Board of Architectural Examiners at Austin, Texas in December 1944. Value on examination = 15% on Structural.

REQUIRED:
Select a reliable column formula and apply it to design an economical steel column 21'-6" long, unsupported, and to support a 230,000 load.

STEP I:
Load assumed to be concentric and a standard H Section desired. $L = 21.5'$ $l = 21.5 \times 12 = 258$ inches.
This is a main member and slenderness cannot exceed 120
Least radius of gyration will be about axis y-y.
Then $r_y = \frac{258}{120} = 2.16$

STEP II:
Selecting the American Bridge Company Column Formula:
$\frac{P}{A}$ = Allowable $F_a = 19{,}000 - 100\frac{l}{r} = 7{,}000 \#\square''$ (Too conservative)
The AISC Formula: $F_a = \dfrac{18{,}000}{1.0 + \left(\dfrac{258^2}{18{,}000 \times 2.16^2}\right)} = 10{,}000$ PSI. (Use this)

STEP III:
$P = 230{,}000\#$ $F_a = 10{,}000\#\square''$ Required $A: \frac{230{,}000}{10{,}000} = 23.0\square''$
This appears to require a W10x10-77# Section.

STEP IV:
Investigate a lighter section because r_y is greater.
Try a 10x10-72# with $A = 21.2\square''$ and $r_y = 2.59$

$F_a = \dfrac{18{,}000}{1.0 + \left(\dfrac{258^2}{18{,}000 \times 2.59^2}\right)} = 11{,}570$ PSI $P = 21.2 \times 11{,}570 = 245{,}000$ Lbs

Accept the best of these sections:
A W10x72 or a W12x65 (Answer)

EXAMPLE: Axial load on concrete-filled pipe 2.5.5.6

A steel pipe column with an outside diameter of 16.0 inches and a wall thickness of 0.375 inches is to be filled concrete. Length of unsupported column is 20.0 feet and steel is A36 type. Strength of concrete at age of 28 days is to test $F_c' = 4000$ PSI.

REQUIRED:

(a) Use maximum ultimate compressive strengths of steel and concrete to determine <u>Ultimate Axial Load on Column.</u>

(b) Reduce stress by applicable formulas for steel and concrete to design a safe working load.

(c) From the results of a and b, determine the safety factor. Loads may be converted to tons if found convenient to solution.

16.0" OD

t = 0.375"

STEP I:

This cross-section can be solved similar to a composite pile as illustrated in Section IX: Also, load may be checked by tables.

Area of Steel in Pipe: $A_g = \pi R^2$ or Area gross = $D^2 0.7854$.

Gross $A_g = 16.0 \times 16.0 \times 0.7854 = 201.06$ □" Inside diameter = 16.0 − (0.375×2)

ID = 15.25 inches. Concrete area, $A_c = 15.25^2 \times 0.7854 = 182.66$ □"

Steel area, $A_s = A_g - A_c$: $A_s = 201.06 - 182.66 = 18.40$ □"

STEP II:

The ultimate compressive stress in steel is: $F_u = 60{,}000$ P.S.I, and the yield stress will apply here. $F_y = 36{,}000$ PSI. $F_c' = 4000$ PSI.

Max. Load on steel area = 18.40 × 36,000 = 662,400 Lbs.

Max. Load on Concrete area = 182.66 × 4,000 = 730,640 Lbs.

Max. Load Ult. = 1,393,040 Lbs. (696.52 Tons)

STEP III:

Formula for design allowable concrete stress is: $P_c = 0.25 F_c' A_c$.

Allowable load on Core: $P_c = 0.25 \times 4000 \times 182.66 =$ 182,660 Lbs.

Allowable stress or steel column is based on $\frac{l}{r}$ ratio.

Radius of gyration for 16.0 OD Pipe is given in tables placed in Pile Section IX, and is thus: $r = 5.53$ Then $\frac{l}{r} = \frac{20.0 \times 12}{5.53} = 43.5$

This is a main member where $\frac{l}{r}$ cannot exceed 120.

STEP IV:

From the allowable stress tables with slenderness ratio of 43.5 the max. $F_c = 16{,}083$ PSI. (Interpolated 43 and 44).

Safe load for steel, $P_s = 18.40 \times 16{,}083 =$ 295,927 Lbs.

EXAMPLE: Axial load on concrete-filled pipe, continued 2.5.5.6

Formula for safe total load on composite column is written:
$P = (A_c F_c) + (A_s F_s)$. Where $F_c = 0.25\ F_c'$

Total safe load on column: P = 182,660 + 295,927 = 478,587 Lbs.
In tons: $P = \frac{478{,}587}{2000} = 239.29$ Tons.

STEP V:
Calculating the Safety Factor:

Ultimate Load from Step II: P_U = 696.52 Tons.
Safe Load calculated in Step IV: = 239.29 Tons.
Safety factor = $\frac{696.52}{239.29} = 2.92$ This is a high safety factor in the general sense and is due to the higher concrete mix. Under normal conditions a concrete with $F_c' = 3000$ PSI, will serve satisfactory.

STEP VI:
To calculate the volume of concrete in column core:
Determine volume for 1 lineal foot of column.
Cross section area core = A_c = 182.66□" At a depth of 1.0 foot, the volume is 182.66 x 12 = 2192.0 Cubic Inches.
One (1) Cubic foot = 1726.0 Cubic Inches: Then: $\frac{2192.0}{1726.0} = 1.268$ Cu. Ft.
L = 20.0 feet. Core volume = 1.268 x 20.0 = 25.36 cubic feet.
There are 27 cubic feet in 1 yard of concrete, and plain concrete without steel reinforcing weighs 144 Pounds per Cubic Foot.
Volume Core = $\frac{25.36}{27.0} = 0.938$ Cubic yards in each column.

STEP VI:
For total weight of 1 Column.
Area steel cross section = 18.40□" Wt. per foot = 18.40 x 3.4 = 62.56#
Area core cross-section = 182.66□" Volume per foot = 1.268 cu. Ft.
Weight of 1 cubic foot concrete: 144 x 1.268 = 182.59 Lbs.
Combined weights per lineal foot: 62.56 + 182.59 = 245.15 Lbs.
Total weight of 20.0 foot Column = 245.15 x 20.0 = 4,903 Lbs.

EXAMPLE: Column with three axial loads 2.5.5.7

A Column has an unbraced length about axis x-x of 20.0 feet. Longest length unbraced axis y-y is 10.33 feet. Load are all to be considered axial loads as shown on illustration.
Loads are as follows: $P_1 = 27{,}000$ Lbs.
$P_2 = 5680$ Lbs., and $P_3 = 30{,}320$ Lbs. Steel is A36.

REQUIRED:
Calculate for requirement about both axes x-x and y-y. Use AISC allowable formula for F_a, or take stress from table.

STEP I:
Solve for 10.33 foot length first. $l = 10.33 \times 12 = 124.0$ inches.
This is a main column member and $\frac{l}{r}$ cannot exceed 120.
Total Loads: $27{,}000 + 5680 + 30{,}320 = 63{,}000$ Lbs.

STEP II:
Column is unbraced about axis x-x for 20.0 feet and $l = 240.0$ inches.
Minimum r_x about axis x-x $= \frac{240.0}{120} = 2.00''$ and about axis y-y, minimum $r_y = \frac{124.0}{120} = 1.03''$
From tables: Allowable $F_a = 10{,}000$ PSI.
Area required for section below $P_3 = \frac{P_1 + P_2 + P_3}{F_a}$.
$A = \frac{63{,}000}{10{,}000} = 6.30\,\square''$

STEP III:
Search through tables to find a cross-section with sufficient r values:
For a trial section, choose a W6 x 20, which has these properties: $A = 5.88\,\square''$ $r_x = 2.66$ and $r_y = 1.51$ For slenderness ratios and Allowable F_a.
Ratio on y-y $= \frac{124.0}{1.51} = 82$
Ratio on x-x $= \frac{240.0}{2.66} = 90.3$ Will govern allowable F_a.
From stress tables: At ratio 91, $F_a = 12{,}328$ PSI.
Then max. axial load $= AF_a$ or $P_T = 5.88 \times 12{,}328 = 72{,}600$ Lbs.
This load exceeds $P_1 + P_2 + P_3$ and is therefore acceptable.

Only loads $P_1 + P_2$ are on shorter column length and the top section is safe for length above load P_3.

EXAMPLE: Truss chord angles in compression 2.5.5.8

Two unequal leg angles 4 x 3 x 3/8 are to be used for a truss chord. Angles are placed back to back with 1/2 inch gusset plate between legs.

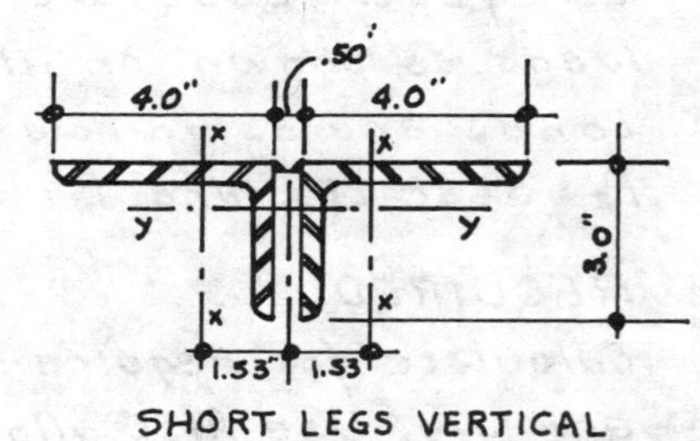

SHORT LEGS VERTICAL

REQUIRED:

With unsupported length of 12.0 feet, determine whether long legs should be vertical or horizontal to give the greatest force in compression. Use the AISC Column formula for the allowable stress F_a.

STEP I:

With short legs vertical, the neutral axis is on ℄ of gusset plate. In drawn section axis x-x is vertical and is 1.53 inches from truss centroid.

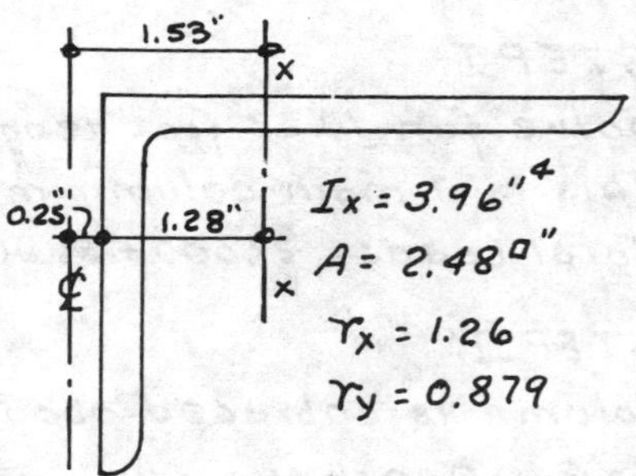

Area $2L^s = 2.48 \times 2 = 4.96^{\square}''$ Lever $l = 1.53''$

$I_x = 3.96''^4$ About ℄: $I = Al^2 + I_o$ and $r = \sqrt{\frac{I}{A}}$

$I_{℄} = (4.96 \times 1.53^2) + (3.96 \times 2) = 19.52''^4$

$$r_{℄} = \sqrt{\frac{19.52}{4.96}} = 1.98''$$

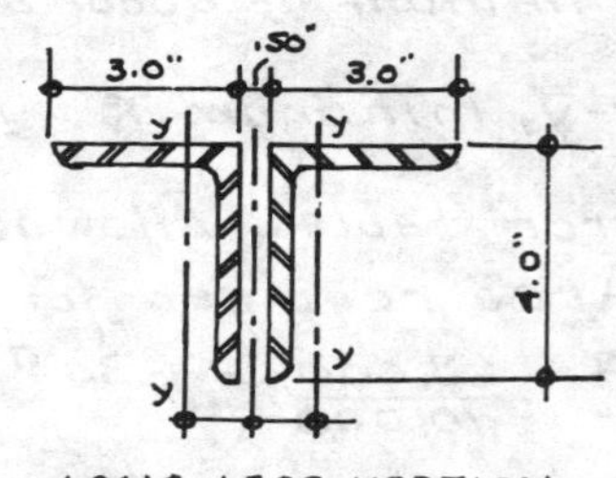

LONG LEGS VERTICAL

STEP II:

With long legs vertical the moment arm $l = 1.03''$ and $I_y = 1.92''$ $A = 4.96^{\square}''$ for $2L^s$.

$$I_{℄} = (4.96 \times 1.03^2) + (1.92 \times 2) = 9.20''^4$$

$$r_{℄} = \sqrt{\frac{9.20}{4.96}} = 1.36'' \text{ (Will govern design).}$$

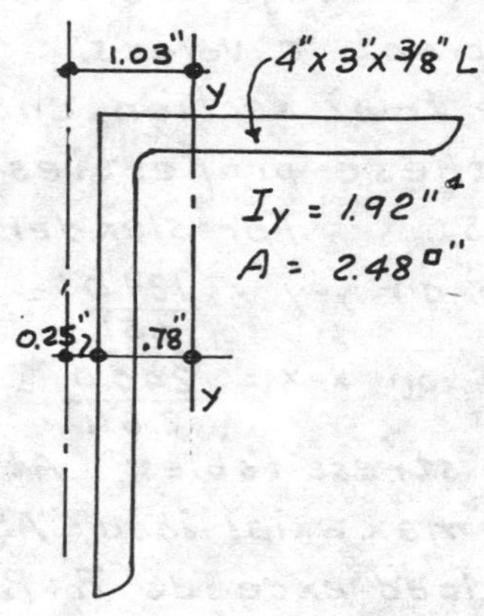

STEP III:

Length of Truss Chord = 12.0 feet.

$l = 12.0 \times 12 = 144.0$ inches. Slenderness ratio $\frac{l}{r} = \frac{144.0}{1.36} = 116$. This is a secondary member and ratio can be over 120.

From tables of allowable compressive stresses, max. $F_a = 10{,}300$ PSI. $A = 4.96^{\square}''$

Maximum force $= P = 4.96 \times 10{,}300 = 51{,}100$ Lbs.

EXAMPLE: Eccentric load using bending factor 2.5.6.1

A Column with an unbraced length of 16.0 feet supports an axial load of 35,000 Pounds in addition to an eccentric load of 12,000 Pounds. Fabricator desires to substitute a 8×6½ WF 24 Lb. section for the column shown on plans. If substitute section is accepted, the eccentric distance (e) will be 6.0 inches from ℄ of section. A36 Steel is specified.

REQUIRED:

Assuming that eccentricity is about major axis x-x, check out the substitute section by using the bending factor B_x for converting bending moment into an equivalent axial load. Check the stress unity and make a recommendation.

STEP I:

From Tables: Properties of 8×6½ WF 24 section are:

$A = 7.06^{\square''}$ $S_x = 20.8''^3$ $r_x = 3.42$ $r_y = 1.61$ $l = 16.0 \times 12 = 192$ inches.

$B_x = \frac{A}{S_x}$ or $B_x = \frac{7.06}{20.8} = 0.339$ $e = 6.00$ inches.

Column must support: Axial load + Eccentric load + Equivalent load.

Axial $P_1 = 35{,}000$ Lbs. $P_2 = 12{,}000$ Lbs.

STEP II:

Moment $= P_2 e$ or $M = 12{,}000 \times 6.00 = 72{,}000$ Inch Lbs.

Equivalent Load $= MB_x$ or $P_3 = 72{,}000 \times 0.339 = 24{,}400$ Lbs.

Total Loads $= 35{,}000 + 12{,}000 + 24{,}400 = 71{,}400$ Pounds axial.

STEP III:

Slenderness ratio: $\frac{l}{r_y} = \frac{192.0}{1.61} = 119$. This is less than 120 which is limit for a main column.

From tables for allowable compressive stresses: $F_a = 10{,}075$ PSI.

Maximum Axial Load $P = F_a A$ or $P = 10{,}075 \times 7.06 = 71{,}130$ Lbs.

This is very close but must be within unity.

STEP IV:

Actual axial stress, $f_a = \frac{P}{A}$ or $f_a = \frac{71{,}130}{7.06} = 10{,}075$ and also $F_a = 10{,}075$ PSI.

Bending stress $= \frac{M}{S_x}$ or $f_b = \frac{72{,}000}{20.8} = 3{,}460$ PSI.

Allowable bending stress for compressive flange is determined by ratio of $\frac{l}{b}$, where b = width of flange. $b = 6.50$ inches.

EXAMPLE: Eccentric load using bending factor, continued 2.5.6.1

STEP V:

Ratio of $\frac{l}{b} = \frac{192.0}{6.50} = 29.5$ Use formula: $F_b = \frac{12,000,000}{\left(\frac{ld}{A_f}\right)}$

depth of section, d = 8.00 inches. flange t = 3/8" (0.375).

Bottom portion of equation: $\frac{ld}{A_f} = \frac{192.0 \times 8.00}{6.50 \times 0.375} = 631$

Then $F_b = \frac{12,000,000}{631} = 19,000$ PSI.

The Alternate formula: $F_b = \frac{22,500}{1.0 + \left(\frac{l^2}{1800\, b^2}\right)}$ in tables gives a lesser $F_b = 15,170$ PSI.

STEP VI:

For unity or less than 1.00. $u = \frac{f_a}{F_a} + \frac{f_b}{F_b}$

$u = \frac{10,075}{10,075} + \frac{3,460}{19,000} = 1.0 + 0.182 = 1.182$

Unity is close enough to accept the substitute under certain conditions, however, with a little more area in section, the unity ratio would be reduced.

STEP VII:

Try using a heavier section as: 8 X 6½ WF 28. Properties are:

A = 8.23 □" b = 6.54" t = 0.463 $S_x = 24.3''^3$

Total Axial loads, $P_1 + P_2 + P_3 = 71,400$ Lbs. $f_a = \frac{71,400}{8.23} = 8,675$ PSI.

Actual bending: $f_b = \frac{M}{S_x} = \frac{72,000}{24.3} = 2,962$ PSI.

Let allowables remain same as in step VI:

Unity, $u = \frac{8675}{10,075} + \frac{2962}{19,000} = 1.016$ OK.

Recommend a section 4 pounds per foot heavier or; 8 X 6½ WF 27.

EXAMPLE: Eccentric plus axial load 2.5.6.2

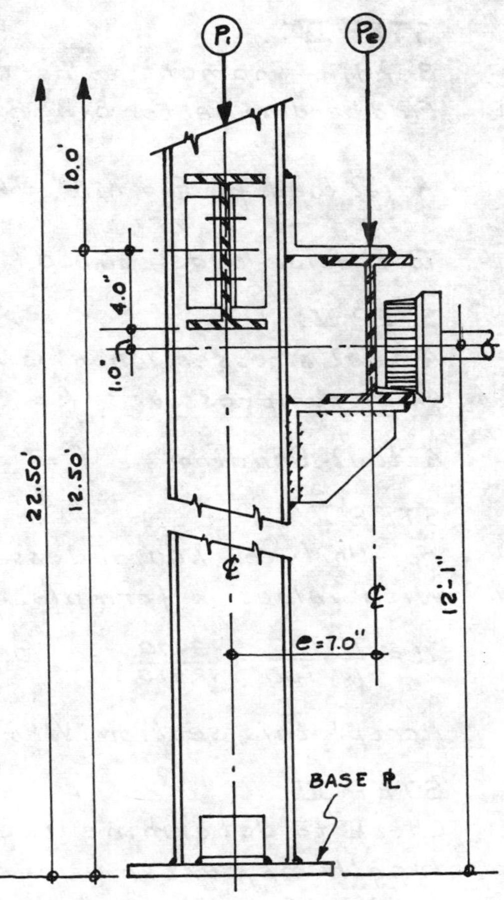

The illustration at right represents a travelling hoist supported on a W8 x 17 Monorail attached to column. Maximum reaction from hoist at column is 10 tons. Axial load on column above hoist is set at 60,000 Lbs. Eccentric distance from ℄ of monorail to ℄ of column is 7.0 Inches.

REQUIRED:
Design the column and restrict the slenderness ratio to 100 or less on axis y-y. Make certain the stress ratio of compressive stress to the bending stress is close to unity.

STEP I:
Max. L on axis x-x is 22.5 Ft. and load is 60,000 Lbs. $l_x = 22.5 \times 12 = 270.0$ in.
$l_y = 12.50 \times 12 = 150.0$ inches. Max. $\frac{l}{r} = 100.$
Least $r_y = \frac{150.0}{100} = 1.50$ $r_x = \frac{270.0}{100} = 2.70$
From tables for allowables, $F_a = 11{,}500$ PSI.

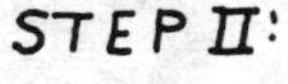
STEP II:
Eccentric bending moment = $20{,}000 \times 7.0" = 140{,}000$ "#.
Column flange assumed at 8.00 inches to use ratio for $\frac{l}{b}$.
Ratio = $\frac{150.0}{8.00} = 18.75$ From tables: Allowable $F_b = 16{,}975$ PSI.
The required $S_x = \frac{M}{F_b}$ or $S_x = \frac{140{,}000}{16{,}975} = 8.26''^3$

STEP III:
The selected cross-section must have these minimum or greater properties: $d = 8.00''$ $S_x = 8.26''^3$ $r_x = 2.70$ $r_y = 1.50$ and $A = 11.80^{\square''}$
Select for trial a section W8 x 40: $A = 11.76^{\square''}$, $b = 8.07''$, $r_x = 3.53$, $r_y = 2.04$, and $S_x = 35.5''^3$ Calculating for ratios to obtain allowable stresses.
$\frac{l}{r_x} = \frac{270.0}{3.53} = 76.5$ $\frac{l}{r_y} = \frac{150.0}{2.04} = 73.6$ and $\frac{l}{b} = \frac{150.0}{807} = 18.6$
From tables: Max. $F_a = 14{,}160$ PSI. Max. $F_b = 18{,}910$ PSI.

EXAMPLE: Eccentric plus axial load, continued 2.5.6.2

STEP IV:

Bending moment to be converted to an equivalent axial load.

Find bending factor Bx: $Bx = \frac{A}{Sx}$ or $Bx = \frac{11.76}{35.5} = 0.331$

Axial load for Pe = MBx. Then Pe = 140,000 × 0.331 = 46,340 Lbs.

Total axial loads: 20,000 + 60,000 + 46,340 = 126,340 Lbs. (Equals P.)

STEP V:

Actual stresses when using section W8×40.

Axial compressive: $f_a = \frac{126,340}{11.8} = 10,730$ PSI.

Actual bending: $f_b = \frac{140,000}{35.5} = 3,940$ PSI.

STEP V:

For Unity of 1.00 or less: $u = \frac{f_a}{F_a} + \frac{f_b}{F_b}$.

With values in formula:

$$u = \frac{10,730}{14,160} + \frac{3940}{18,910} = 0.758 + 0.198 = 0.956 \quad (OK)$$

Accept this section W8×40 for column.

STEP VI.

Check to determine the maximum length about axis x-x the length may extend under conditions.

Ratio $\frac{r_x}{r_y} = \frac{3.53}{2.04} = 1.73$ Length on x-x = 1.73 × 12.50 = 21.63 feet (OK).

DESIGNER'S NOTATION:

By referring to load tables in AISC Manual, an 8×8 WF 40 Section will safely support an axial load of 125,000 Lbs., on an unbraced length of 20.0 feet. This compares favorably with the total of 3 loads obtained in Step IV and serves as a check for this example.

EXAMPLE: Design for axial plus eccentric load 2.5.6.3

A steel column must safely support a cantilever canopy of 8.0' over a walk. Canopy is attached to a 21.50 foot column at 10.0' above base. At end of canopy an electric sign produces a load of 1580 Lbs. Uniform load on canopy is 300 Lbs. perfoot. At top of column, roof load produces an axial load of 9000 Lbs. Axis y-y is braced at 10.0' from base. Axis x-x is unbraced the full length of 21.50 foot column. Steel is A36.

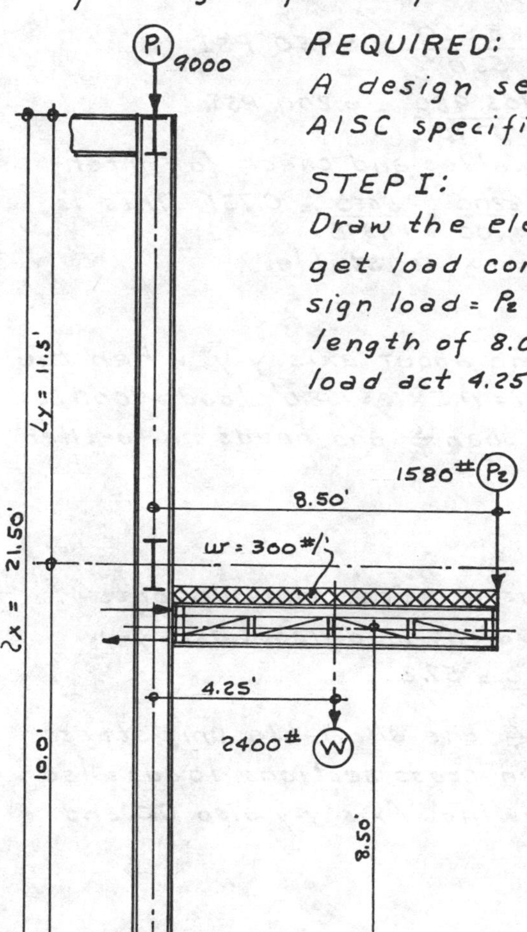

REQUIRED:
A design section for column to meet code and AISC specifications.

STEP I:
Draw the elevation of column with canopy and get load conditions clear. Let $P_1 = 9000$ Lbs., and sign load = P_2 of 1580 Lbs. Uniform load is over the length of 8.0 feet and is, $W = 8.0 \times 300 = 2400$ Lbs. The load act 4.25 feet from assumed column ℄.

STEP II:
Eccentric loads bending moments:
For P_2: $M = 1580 \times 8.50 \times 12 =$ 161,160 In. Lbs.
For W: $M = 2400 \times 4.25 \times 12 =$ 122,400 In. Lbs.
Total Bending $M =$ 283,560 "#

STEP III:
A trial section must be selected and analyzed for requirements. For trial select a W8x58 section. The properties will be gathered thus:
$A = 17.10\,\square''$ $b = 8.22''$ $S_x = 52.0$ $r_x = 3.65$ $r_y = 2.10$
$l_x = 21.5 \times 12 = 258.0''$ $B_x = \frac{A}{S_x} = 0.328$

Converting bending moments into an equivalent axial load = $M \times B_x$.

Then $P_e = 283,560 \times 0.328 =$ 93,000 Lbs. Total loads = $P_1 + P_2 + W + P_e$.
Total axial loads on axis x-x = $P = 9000 + 1580 + 2400 + 93,000 = 105,980^{\#}$

EXAMPLE: Design for axial plus eccentric load, continued 2.5.6.3

STEP IV:

Continue to examine conditions about axis x-x and let y-y come later. Ratio of $\frac{l}{r_x} = \frac{258.0}{3.65} = 70.6$ From tables of unit stress allowables: $F_a = 14,100$ PSI. Longest length where axis x-x is unbraced laterally is 11.50 Feet. $l = 11.50 \times 12 = 138.0''$ $b = 8.22''$ Ratio of $\frac{l}{b} = \frac{138.0}{8.22} = 16.8$ From table: Max. $F_b = 19,390$ PSI.

Actual bending stress: $f_b = \frac{M}{S_x}$ or $f_b = \frac{283,560}{52.0} = 5,450$ PSI.

Actual Axial stress: $f_a = \frac{P}{A}$ or $f_a = \frac{105,980}{17.10} = 6,200$ PSI.

Actual stresses are less than allowables and check for interaction by formula: $U = \frac{f_a}{F_a} + \frac{f_b}{F_b}$. $U = \frac{6200}{14,100} + \frac{5450}{19,390} = 0.731$ This is less than unity of 1.0 and axis x-x is stable.

STEP V:

Check unsupported length of column about axis y-y when the $r_y = 2.10$ $A = 17.10\,\square''$ Above canopy, $l = 11.5 \times 12 = 138.0''$ Load = 9000 Lbs. Ratio $\frac{l}{r_y} = \frac{138.0}{2.10} = 65.7$ This is less than $\frac{l}{r_x}$ and needs no further investigation.

STEP VI:

Unsupported length below canopy is 10.0 feet and supports 3 Loads $P_1 + P_2 + W$. There is no equivalent load about axis y-y. $l = 10.0 \times 12 = 120.0''$ $r_y = 2.10$ $\frac{l}{r_y} = \frac{120.0}{2.10} = 57.2$

With lower slenderness ratios of $\frac{l}{r}$ the allowable unit stress F_c is greater, and with same area in cross sections, loads also become greater than given on drawing. Axis y-y also OK and section W8 x 58 will be acceptable.

Base and bearing plates 2.6

Columns, beams and girders which are supported on concrete or masonry walls must be provided with bearing plates to distribute the load over an area of the supporting material, which has a lower compression value. A *bearing plate* is placed under the end of a beam; a column is supported upon a *base plate*. Column base plates and beam bearing plates perform the same function; however their design will require the use of different formula. The size of the plate will be governed by the load and the area of the bearing material over which this load must be uniformly distributed. In virtually every design the plate must extend beyond the column section or beam flange. Bearing plates under beams are usually restricted in width to the thickness of the masonry wall. Projecting a base or bearing plate beyond the main section induces bending in the plate. This projection will be considered as a cantilever for design purposes.

All base and bearing plates are the design responsibility of the structural engineer. They should be attached at the fabricating plant, with anchor bolt holes punched. Anchor bolts for column base plates should be set in place by using a drilled wood template, spaced in accordance with the anchor bolt setting plan provided by the shop fabricator. True alignment of columns is required for plumb and accurate steel erection. The correct location of anchor bolts is a basic requirement for accurate erection.

Base plate design 2.6.1

Column base plates must be capable of uniformly distributing the load so that the bearing pressure under the plate is less than the allowable bearing pressure, indicated by the symbol F_p. Table 2.6.1.1 can be used to determine the maximum allowable compressive stress in bearing for concrete, masonry, and stone.

When Wide Flange or H sections are used for columns, the load P is assumed to be distributed on the base plate over a square or rectangular area. This effective area of load distribution is calculated as 95 percent of the column depth and 80 percent of the column breadth. Consider a column of section W 10 x 45. The dimension for one side of the effective load area rectangle is 0.95 x 10 inches or 9.50 inches. The other side is the effective flange width, 0.80 x 8 inches or 6.40 inches. The effective area of load distribution on the plate is a rectangle 6.40 x 9.50 inches. When making the calculations for the rectangle on the plate, remember to calculate the dimension *b* parallel with the flanges as 0.80 times the flange width, and the dimension *d* parallel with the web as 0.95 times the section depth.

ALLOWABLE WORKING STRESS

In the design of base and bearing plates, the stress notation may become confused, and subscripts are recommended. With medium steel A36, the allowable bending stress is F_b = 27,000 PSI. Indicate the *allowable bearing pressure* as F_p. The *actual* stress between the plate and bearing surface will be indicated as f_p. The allowable unit bearing stress on concrete will be identified as F_c. Full strength of concrete

Base plate design, continued 2.6.1

at 28 days of curing is indicated as F_c'. A column base plate which covers an entire plinth (a pedestal formed of concrete) will have the bearing pressure F_p determined as 25 percent of F_c', or the allowable $F_p = 0.25\ F_c'$.

BASE PLATE DESIGN FORMULAS

When designing a column base plate, it is the plate thickness (t) which is to be determined. The plate coverage is first calculated by taking the *allowable* bearing pressure F_p from the table, and dividing its value into column load P. Thus the required area is $A = \frac{P}{F_p}$. After the area has been found as above, or has been determined for architectural reasons, the *actual* bearing pressure between plate and bearing is $f_p = \frac{P}{A}$. Again, do not confuse the two bearing symbols F_p and f_p.

A base plate may be either square or rectangular in which case identify the side parallel to the column flanges as side B and the other as side C. Then A = B x C. The cantilever projection in the same direction as side B becomes the moment arm for bending and is indicated as n on both sides of the column. The moment arm in the direction of side C is indicated as m. Moment arms are in inches. The AISC gives the allowable unit bending stress for A36 base plates as $F_b = 27{,}000$ PSI.

Designing for thickness of base plate in direction of side B, and parallel to flange, the formula is written:

$$t_b = \sqrt{\frac{3 f_p n^2}{F_b}}$$

For thickness in direction of side C, the formula becomes:

$$t_c = \sqrt{\frac{3 f_p m^2}{F_b}}$$

Design thickness of plate will be governed by the greater dimension from result of either formula. Examples will follow which may be used as a practical guide and illustrate a method for accuracy. Refer to the examples and note the dimensions for m and n.

TABLE: Allowable bearing pressures 2.6.1.1

ALLOWABLE PRESSURE VALUES FOR BEARING PLATE DESIGN

TYPE OF SUPPORTING COMPOSITION	MAXIMUM COMPRESSIVE BEARING F_p = P.S.I.
HARD AGGREGATE CONCRETE - WHEN ENTIRE CONCRETE IS COVERED.	0.250 F_c'
HARD AGGREGATE CONCRETE - PLATE COVERS ONLY 40% CONCRETE AREA.	0.375 F_c'
LIGHT WEIGHT CONCRETE - PLATE COVERS 100% OR WHOLE AREA OF CONCRETE.	0.200 F_c'
LIGHT WEIGHT CONCRETE - PLATE COVERS UP TO 40% OF CONCRETE AREA.	0.300 F_c'
HARD BURNED OR PAVING BRICK - TYPE M-8000 PSI - CEMENT MORTAR.	400
BURNED CLAY BRICK - 4500 PSI. TEST. TYPE M MORTAR.	250
COMMON CLAY BRICK - 2500 TO 4000 PSI. CEMENT MORTAR.	175
CURED LIGHT WEIGHT MASONRY UNITS - VERTICAL CELLS - CEMENT MORTAR.	125
HAYDITE LINTEL MASONRY UNITS - CEMENT MORTAR FILLED.	200
HAYDITE VERTICAL CELL BLOCK UNITS - CELLS FILLED WITH CEMENT MORTAR.	175
HAYDITE VERTICAL CELL BLOCK UNITS - CELLS LEFT VOID.	85
PRESSED BRICK - LAID UP IN NATURAL MORTAR TYPE B.	120
GRANITE - TEXAS AND VERMONT - TYPE M MORTAR.	800
LIMESTONE - INDIANA, LEUDERS OR CEN-TEX, TYPE M MORTAR.	500
MARBLE - SAWN DOMESTIC OR IMPORTED.	500
CAST STONE CONCRETE OR SPLIT STONE IN CEMENT MORTAR.	400
FIELD STONE, ROUGH RUBBLE IN CEMENT MORTAR.	150
TILE, GLAZED VERTICAL CELLS, CEMENT MORTAR.	175
TILE, CLAY LOAD BEARING STRUCTURAL - TYPE M MORTAR.	175
TILE, CLAY HORIZONTAL CELL BACK UP TYPE IN CEMENT MORTAR.	80
BRICK, HARD BAKED, CAVITY WALL WITH 2" WYTHES - CEMENT MORTAR	200
BRICK, COMMON RED TYPE 2000 TO 3500 PSI. LIME MORTAR.	120

NOTE: ALL APPLICABLE BUILDING CODES SHALL BE GIVEN PREFERENCE OVER THE VALUES GIVEN IN THIS TABLE.
F_c' DENOTES THE COMPRESSIVE STRENGTH AT END OF 28 DAY CURING PERIOD.

EXAMPLE: Column base plate design 2.6.1.2

A Steel Column W14×78 of A36 steel carries a load to a Concrete pedestal as: $P = 300{,}000$ Lbs. Pedestal size to be determined by size of base plate. Concrete strength at age of 28 days is: $f'_c = 2000$ PSI. Max. allowable for steel plate bending is: $F_b = 27{,}000$ Lbs. square inch.

REQUIRED:

Design the base plate for size and thickness; then make a drawing of design for draftsman and file records.

STEP I:

From tables for allowable bearing pressure on 2000 pound Concrete when plate is to cover plinth is: $F_p = 0.25\,f'_c$.
Then $F_p = 0.25 \times 2000 = 500$ Lbs. Square Inch.
Plate area: $\frac{P}{F_p}$ or $A = \frac{300{,}000}{500} = 600$ Sq. In. $\sqrt{600} = 24.5$ In. Sq.
Make size of Plate 24.0×26.0 In. Let side B = 24.0" and C = 26.0 inches.
Area = 24.0×26.0 = 624□" Actual bearing $f_p = \frac{300{,}000}{624} = 480$ PSI.

STEP II:

A W14×78 Section has: d = 14.0" and b = 12.0" These dimensions are changed to a rectangular shape thus: 0.95 d = 13.30" and b = 0.80 × 12.0 = 9.60 Inches. Layout plan of base plate thus:

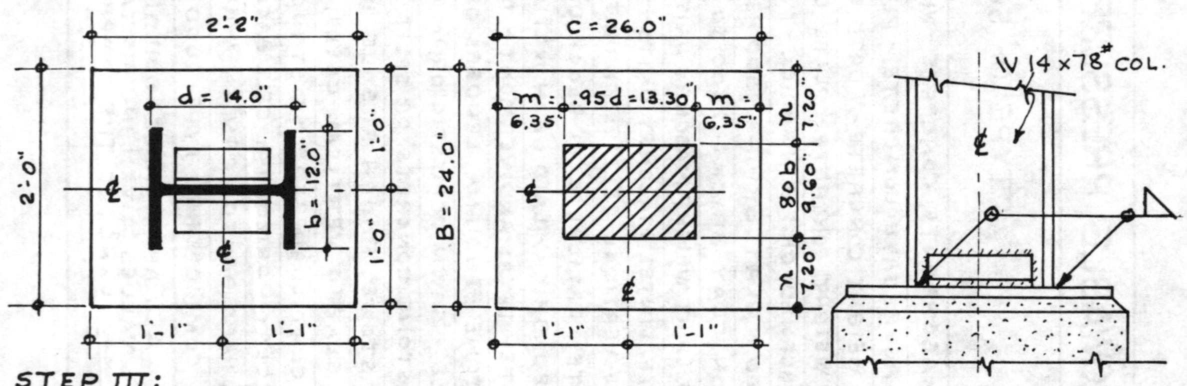

STEP III:

Moment arms: $m = \frac{C-(0.95d)}{2} = 6.35''$ $n = \frac{B-(0.80b)}{2} = 7.20$ Inches

Longer moment lever arm as n will give the greater thickness t.
Formula: $t_b = \sqrt{\frac{3 f_p n^2}{F_b}}$ Substituting values in formula:

$$t_b = \sqrt{\frac{3 \times 480 \times 7.20 \times 7.20}{27{,}000}} \quad \sqrt{2.765} \quad \text{or} \quad 1.663 \text{ Inches.}$$

Accept a Base Plate 24" × 1¾" × 2'-2" for Column.

EXAMPLE: Column base plate analysis 2.6.1.3

The previous example designed a base plate for a W14x78 Column load of 300,000 Pounds to bear on a plinth with a compressive strength $F_c' = 2000$ PSI (28 days). The tables in AISC Manual for column loads and base plates gives the same dimension for plate area with a load of 467,000 Lbs. on concrete with $F_c' = 3000$ PSI, and $F_b = 27{,}000$ PSI.

REQUIRED:
Refer to previous example for moment lever arms m and n and calculate thickness of plate with P= 460,000 Pounds and full coverage on plinth with $F_c' = 3000$ PSI.

STEP I:
From previous example: $B = 24.0''$ $C = 26.0''$ $m = 6.35''$ and $n = 7.20''$ Area plate = BC or $A = 24.0 \times 26.0 = 624\,\square''$
Full coverage on plinth, the allowable $F_p = 0.25 F_c'$
Allowable $F_p = 0.25 \times 3000 = 750\,\#\square''$

STEP II:
Actual bearing pressure $f_p = \frac{P}{A}$ or $f_p = \frac{467{,}000}{750} = 623\,\#\square''$

Formula for $t_b = \sqrt{\frac{3 f_p n^2}{F_b}}$ $n = 7.20''$

Then with actual bearing pressure:

$$t_b = \sqrt{\frac{3 \times 623 \times 7.20^2}{27{,}000}} = 1.895 \text{ inches.}$$

The result obtained with actual bearing pressure will not agree with calculated thickness of plate because the maximum allowable bearing was used in tables. The formula therefore was thus: $t_b = \sqrt{\frac{3 F_p n^2}{F_b}}$
Substitute values in formula:

$$t = \sqrt{\frac{3 \times 750 \times 7.20 \times 7.20}{27{,}000}} = 2.08 \text{ inches.}$$ This thickness

checks with AISC table and is a conservative design.

EXAMPLE: Oversize base plate design 2.6.1.4

An existing footing supports a column in a refinery which is part of a tall TCC unit. Estimated column load is 675,000 Pounds distributed on a pile footing which has a concrete pedestal 32.0 x 34.0 inches. Plans call for the existing TCC unit to be removed and existing footings to support the new structure in same location. The new column to be placed on this pedestal will be a W10 x 100 supporting a lesser load of 350,000 Lbs. Existing concrete when placed was: $F_c' = 3000$ PSI.

REQUIRED:

Make size of base plate 30.0" x 32.0" allowing for 1 inch of grout margin for finish. Use AISC Formula to design thickness of base plate and base the design on actual bearing pressure. Check to determine plate thickness if full allowable F_p was to be used for design. Long side of plate is parallel to web.

STEP I:

A column W10 x 100 has d = 11.12" and b = 10.345" Short side of plate is parallel to flange and is side B. Let this plate be draw as in plan to ascertain accurate arms for m and n.

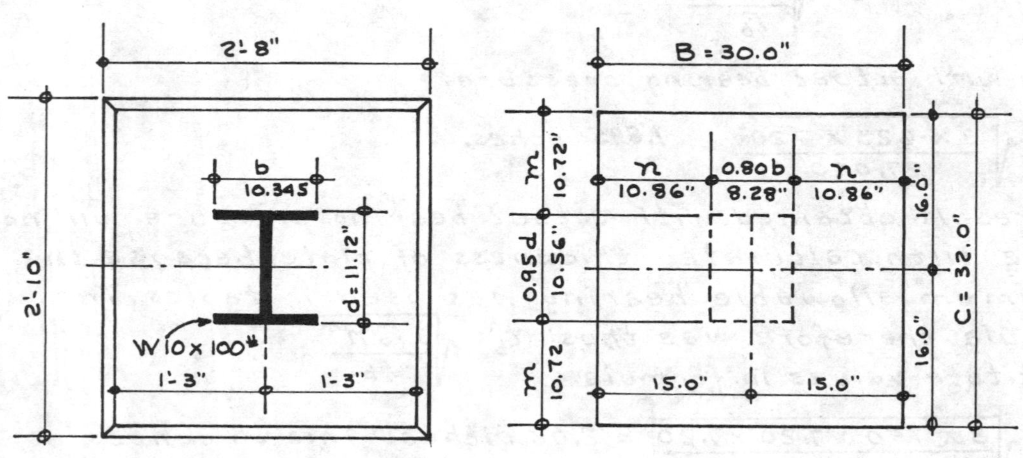

EXAMPLE: Oversize base plate design, continued 2.6.1.4

STEP II

$m = \dfrac{C-(0.95 \times 11.12)}{2} = 10.72''$ $\quad n = \dfrac{B-(0.80 \times 10.345)}{2} = 10.86''$

Area base plate = 30.0 x 32.0 = 960 Square Inches.

$F_p = 0.25 F_c'$ or $F_p = 0.25 \times 3000 = 750$ Psi. P = 675,000 Lbs.

Actual bearing pressure, $f_p = \dfrac{P}{A}$ or $f_p = \dfrac{675,000}{960} = 702$ PSI

STEP III:

Formula for $t_b = \sqrt{\dfrac{3 f_p n^2}{F_b}}$ Longer moment arm is: $n = 10.86''$

With actual stress as: $f_p = 702$ PSI. $F_b = 27,000$ PSI.

$$t_b = \sqrt{\frac{3 \times 702 \times 10.86 \times 10.86}{27,000}} = \sqrt{9.20} = 2.92 \text{ inches}$$

STEP IV:

Checking thickness by using allowable bearing pressure of 3000 PSI Concrete $F_p = 750$ PSI.

$$t_b = \sqrt{\frac{3 \times 750 \times 10.86 \times 10.86}{27,000}} = \sqrt{9.81} = 3.14 \text{ inches}$$

Results are close enough to accept a 3.0 inch thickness of plate.

EXAMPLE: Pipe column base plate 2.6.1.5

The AISC Handbook in its Column Loads on steel pipe 4"φ lists the allowable load as 46,000 Lbs, for an unbraced length of 11.0 feet. The value of r = 1.51 and under the tables for pipe properties, this will be a standard weight, with an outside dimension of 4.50 inches.

REQUIRED:
Design a square base plate for the pipe column under the allowable load. Assume plate will bear upon a covered concrete plinth. Size and area of base plate should be sized to accomodate 4-5/8"φ Anchor bolts at corners. Use a concrete mix for plinths to provide $F'_c = 3000$ PSI at 28 days. Base plate is to be welded to pipe at fabricating shop, therefore the allowed weld is 1/16 inches less than base plate.

STEP I:
Maximum allowable bearing pressure for covered plinth is:
$F_p = F'_c \times 0.25$ or $F_p = 0.25 \times 3000 = 750$ PSI
Required area for plate: $A = \frac{P}{F_p}$ or Min. $A = \frac{46{,}000}{750} = 61.4\square''$
for sides, $\sqrt{61.4} = 7.85''$ Try a size $8.0 \times 8.0 = 64.0\square''$ and will be large enough for anchor bolts. Sides B = C, then moment arms m and n are same. Moment arm: $m = \frac{B-D}{2}$. When B = 8.0" or side of plate, and D = 4.50" outside.
$m = n = \frac{8.00 - 4.50}{2} = 1.75$ inches. With A36 Steel: $F_b = 27{,}000$ PSI

STEP II:
Formula for t: $t = \sqrt{\frac{3 F_p n^2}{F_b}}$. Substituting values in formula:

$$t = \sqrt{\frac{3 \times 750 \times 1.75 \times 1.75}{27{,}000}} = 0.506 \text{ inches.}$$

Accept a base plate as: 8"×½"×8"
Make plinths 9½ inches square.
Details for plan shown at right.

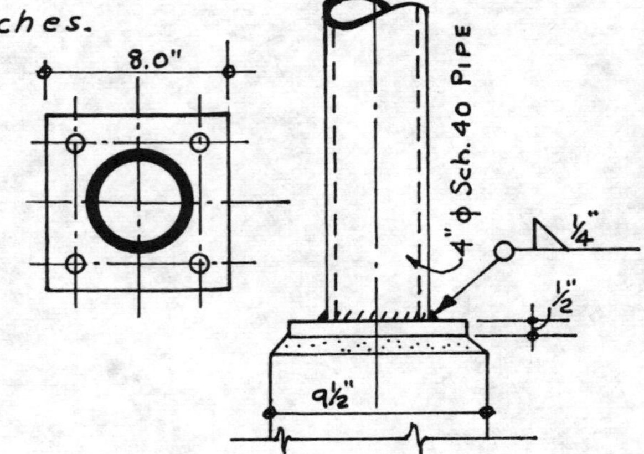

Carnegie formula for bearing plate design 2.6.2.1

The Carnegie formula was popular because the designer has a choice in calculating bearing plate thickness according to the flange location. The size, area, and thickness of a bearing plate depend on the beam reaction, the length and width of bearing, and the allowable stress for steel. The Carnegie formula involves examining two conditions based on two different assumptions.

CONDITION 1:

The first condition assumes that the point of maximum bending moment in the bearing plate occurs at the center of bearing, directly under the beam web.

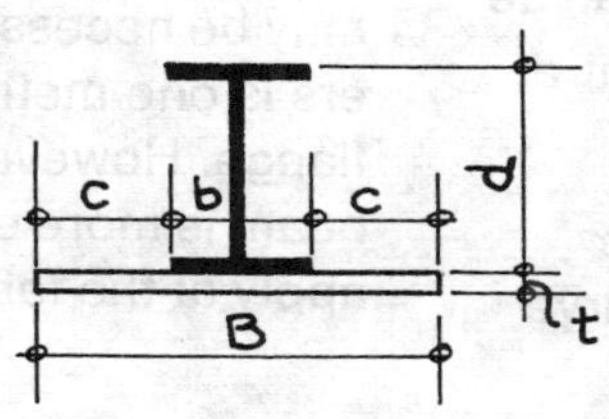

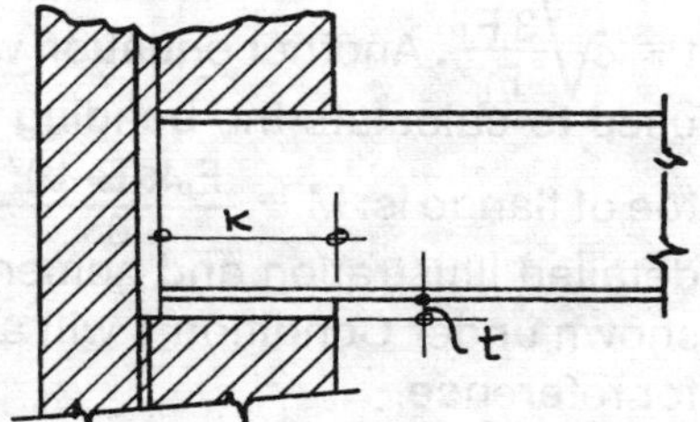

c = Cantilever projection from flange toe. $c = \frac{B-b}{2}$, in inches.
B = Width of bearing plate, in inches.
K = Length of plate in wall projection, in inches.
R = Reaction value on bearing plate in Pounds. $R = BKF_p$.
F_p = Allowable unit bearing pressure in PSI. See table for F_p.
F_b = Allowable unit steel bending stress in PSI.
t = Thickness of plate, in inches.
b = Breadth of beam flange on plate, in inches.
S = Section Modulus of bearing plate when t=d. Inches^3.
M = Bending moment in plate, in inch pounds.

In this condition, the following formulas apply:

$M = \frac{R(B-b)}{8}$ equivalent to $M = \frac{F_p BK(B-b)}{8}$. When $S = \frac{bd^2}{6}$ and $S = \frac{M}{F_b}$, then $M = \frac{F_b k t^2}{6}$ or transposing; $t = \sqrt{\frac{3F_p B(B-b)}{4F_b}}$.

CONDITION 2:

When the plate bending moment under the flange is low because a thick flange is rigidly welded to the bearing plate, the bending moment can be taken at edge of the flange. The cantilever projection of plate is now important, and the bearing reaction at the flange toe is $R = ckF_p$.

Carnegie formula for bearing plate design, continued 2.6.2.1

Dimension $c = \frac{B-b}{2}$. Moment $= \frac{F_p k c^2}{2}$ for cantilever. The equivalent resisting moment must equal the bending moment: $RM = \frac{F_b k t^2}{6}$. From these equations a formula is derived to calculate the value of t. The bending is less than under Condition 1: and the calculated plate thickness will also be less. The formula is written: $t = c\sqrt{\frac{3\,F_p}{F_b}}$. Another equation which can be used to calculate the bending moment at toe of flange is: $M = \frac{F_p k (B-b)^2}{8}$. The detailed illustration and nomenclature shown under Condition 1 will again serve for reference.

BENDING IN FLANGE WITHOUT BEARING PLATE

When the beam end reaction is relatively light or the support bearing material has a high compressive strength, the beam flange may be subjected to excessive bending stress. The area of flange coverage is adequate for distributing the bearing load, but there may be critical bending in the bottom flange, and additional thickness may be necessary. The addition of stiffeners is one method for strengthening the flange. However, a small plate under the beam is more economical and easier to apply at the fabricating plant.

AISC formula for bearing plate design 2.6.2.2

The AISC appears to have modified the Carnegie formulas to include a portion of the flange. When employing the AISC formula, the cantilever is considered to be the distance from the tangent point of the curved fillet to the toe of the flange. The distance from the center line of the web to the tangent point on the flange is given in the tables of beam sections as k. See tables 2.3.3.1. Therefore, the cantilever projection to the toe of the flange is: $C = \frac{b-(2k)}{2}$.

To compute the actual bending stress in the flange without a bearing plate, the formula is $f_b = \frac{3\,F_p c^2}{t^2}$, where t = thickness of flange and F_p = allowable bearing pressure from supporting material.

Maximum allowable bending stress in the flange should not exceed 75 percent of yield point or $F_b = 0.75\,F_y$. With A36 steel, the allowable is $F_b = 0.75 \times 36{,}000 = 27{,}000$ PSI.

The AISC recommended formula for bearing plates is similar to the formula in the paragraph above. However, for bearing plates, the cantilever projection (c) is lengthened to the edge of the bearing plate. Transposing the equation for use in solving for thickness of plate (t) it is re-written: $t^2 = \frac{3\,F_p c^2}{F_b}$ or in final form: $t = \sqrt{\frac{3\,F_p c}{F_b}}$.

EXAMPLE: Beam bearing plate design 2.6.2.3

The end of a W8x17 Steel beam carries a reaction of 18,000 Lbs., to a brick wall of 8.0 inches in thickness. Wall is composed of hard burned brick set in Type M mortar. Under bearing plate the courses are laid up in rowlock fashion and Architect desires that plate cover not less than 8 bricks to preclude possible damage to wall. Limit the narrow side of plate to 7.0 inches.

REQUIRED:
Design the thickness of base plate with the Carnegia formula. Use the allowable bearing pressure F_p only in the event its value is greater than actual bearing f_p. Limit bending stress F_b to 20,000 PSI.

STEP I:
Indicate length of plate as side B which will cover 8 brick or 15.0 inches. Side K = 7.0 inches. Area = 15.0 x 7.0 = 105 □"
Actual bearing = $\frac{R}{A}$ or $f_p = \frac{18,000}{105} = 171$ Lbs. Sq. In.
From Table of allowable bearing pressures for F_p, this wall will sustain a pressure of 250 Lbs. Sq. Inch which is greater than actual bearing f_p.

STEP II:
A section W8x17 has a flange width b = 5.25". Then the cantilever projection is: $C = \frac{B-b}{2} = 4.875"$

The formula becomes:

$$t = \sqrt{\frac{3 F_p B C}{2 F_b}} \quad \text{and} \quad t = \sqrt{\frac{3 \times 250 \times 15.0 \times 4.875}{2 \times 20,000}} = \sqrt{1.371} = 1.17 \text{ inches.}$$

Accept a plate: 7" x 1⅛" x 1'-3"

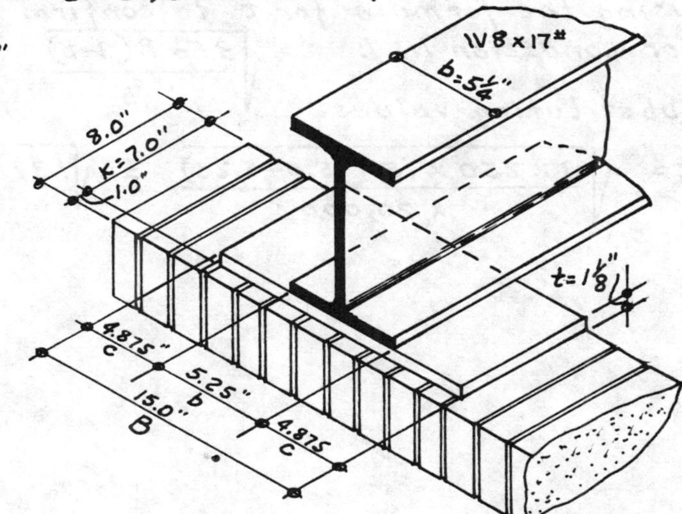

EXAMPLE: Bearing plate design by Carnegie formula 2.6.2.4

A steel W8x17 beam rests upon a masonry wall laid up with hard burned brick and type M mortar. Size of bearing plate is arbitrarily set in office of Architect. Width of plate parallel with flange of beam is 15.0 inches, and projection into wall is 7.0 inches.

REQUIRED:
Refer to the Carnegia formulas and calculate the bearing plate thickness required for condition 1 and 2.

STEP I:
From table for allowable bearing pressures: $F_p = 250$ PSI. Assumed $F_b = 20{,}000$ PSI. Width of plate, $B = 15.0''$ and $K = 7.0''$ Area of plate, $A = BK$ or $15.0 \times 7.0 = 105$ Sq. In.

STEP II:
For condition No. 1, the maximum bending is located under web of beam. Flange width, $b = 5.25$ inches.
Reaction $= F_p BK$ or $R = 250 \times 15.0 \times 7.0 = 26{,}250$ Pounds.

$$M = \frac{R(B-b)}{8} \quad \text{or} \quad M = \frac{26{,}250 \times (15.0 - 5.25)}{8} = 31{,}990 \text{ inch Lbs.}$$

$$S = \frac{M}{F_b}. \quad S = \frac{31{,}990}{20{,}000} = 1.60''^3$$ Also, $S = \frac{bd^2}{6}$. Dimension d in this formula represents thickness t, and $b = K$ or $7.0''$
Transposing for t: $d = \sqrt{\frac{6S}{b}}$, then, $t = \sqrt{\frac{6 \times 1.60}{7.0}} = 1.17$ inches.

STEP III:
Using the formula for t to confirm the result in above for condition No. 1: $t = \sqrt{\frac{3 F_p B (B-b)}{4 F_b}}$
Substituting values:

$$t = \sqrt{\frac{3 \times 250 \times 15\,(15.0 - 5.25)}{4 \times 20{,}000}} = \sqrt{1.371} = 1.17 \text{ inches (checks).}$$

EXAMPLE: Bearing plate design by Carnegie formula, continued 2.6.2.4

STEP IV:

For condition No. 2, bending moment of base plate is calculated at toe of flange. The area of pressure is ck. $k = 7.0''$ and $c = \frac{B-b}{2}$ or $c = \frac{15.0 - 5.25}{2} = 4.875$ inches.

Bearing pressure on cantilever = $F_p ck$.
Pressure = $250 \times 4.875 \times 7.00 = 8531.25$ Lbs. Moment for a cantilever is: $M = \frac{WL}{2}$ or $M = \frac{8531.25 \times 4.875}{2} = 20{,}800$ inch Lbs.

Same results by formula: $M = \frac{F_p k c^2}{2}$

Resisting moment must equal moment.
Then $20{,}800 = \frac{F_b k t^2}{6}$, and t represents d in formula: $S = \frac{bd^2}{6}$

Section Modulus $S = \frac{M}{F_b}$ or $S = \frac{20{,}800}{20{,}000} = 1.04''^3$ $k = b$ or $7.0''$

Then plate $t = \sqrt{\frac{6S}{k}}$ and $t = \sqrt{\frac{6 \times 1.04}{7.0}} = 0.945$ inches.

Confirming the formula previously derived: $t = c\sqrt{\frac{3F_p}{F_b}}$.

$t = 4.875\sqrt{\frac{3 \times 250}{20{,}000}} = 0.945$ inches. (Checks OK).

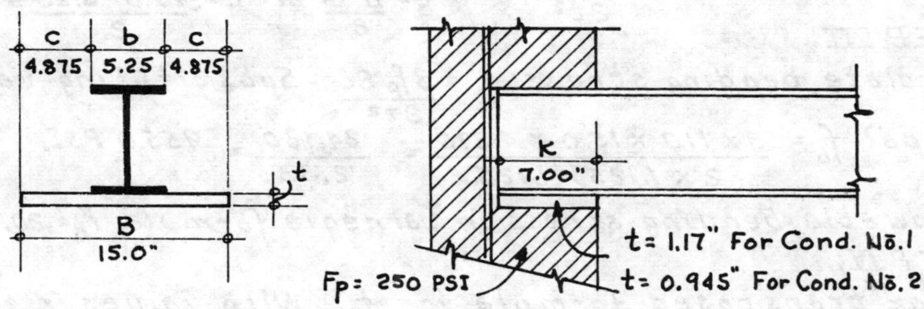

EXAMPLE: Checking pressure under bearing plate 2.6.2.5

A beam section has a flange width of $5\frac{1}{4}$ inches and is welded to a $1\frac{1}{8}$ inch plate. The dimension projecting into wall is 7.0 inches and parallel to wall it is 15.0 in. Reaction from load is 11,500 Lbs. Wall is composed of 8.0 width concrete block units with vertical cells. The mortar is of natural cement mixed with fine sand.

REQUIRED:
Assume the detail of bearing plate in the previous examples and apply the Carnegia formula in the transposed form necessary to check the actual bending stress in plate. Also check the bearing unit pressure under load given.

STEP I:
Collect the data given: $B = 15.0''$ $b = 5.25$ $t = 1.125''$ $K = 7.0''$
$R = 11{,}500$ Lbs. Area bearing = BK and unit bearing on masonary is $\frac{R}{BK}$. Then $f_p = \frac{11{,}500}{15.0 \times 7.0} = 109.5^{\#\square''}$ (Call it $110^{\#\square''}$)
Bearing is below allowable given in table as $F_p = 125$ PSI.

STEP II:
The basic formula is: $t = \sqrt{\frac{3F_p Bc}{2F_b}}$ and $F_p = \frac{2F_b t^2}{3Bc}$ and for stress in plate: $F_b = \frac{3F_p Bc}{2t^2}$. $c = \frac{B-b}{2}$ or $c = \frac{15.0 - 5.25}{2} = 4.875''$

STEP III:
Calculate bending stress: $f_b = \frac{3f_p Bc}{2t^2}$. Substituting values:

Actual $f_b = \frac{3 \times 110 \times 15.0 \times 4.875}{2 \times 1.125 \times 1.125} = \frac{24{,}130}{2.53} = 9550$ PSI.

Allowable bending stress in Carnegia formula. $F_b = 20{,}000$ PSI.

STEP IV:
Check transposed formula for f_p. With values in formula:

$f_p = \frac{2 \times 9550 \times 1.125 \times 1.125}{3 \times 15.0 \times 4.875} = 110$ PSI. checks with Step I.

STEP V:
Use base formula to check other formulas and plate t.

$t = \sqrt{\frac{3 \times 110 \times 15.0 \times 4.875}{2 \times 9550}} = \sqrt{1.260} = 1.125$ inches and checks.

EXAMPLE: Flange bending without bearing plate 2.6.2.6

Given a beam section W8x20 with end reaction of 15,000 Pounds and supported on a concrete wall of 12.0 inch thickness. Compressive strength of concrete at 28 day period is: $F_c' = 3500$ PSI. End of beam projection on wall is limited to 6.0 inches. Steel is A36.

REQUIRED:

Determine whether a bearing plate is necessary and is thickness of flange sufficient to resisting bending moment without stiffening.

STEP I:

From tables of sections: Flange width $b = 5.25$ and the thickness is, 0.375" Dimension at curved fillet is, $k = 0.875$" Projection $K = 6.00$"

Dimensions of cantilever flange: $C = \frac{b-(2k)}{2}$ or

$C = \frac{5.25-(2 \times k)}{2} = 1.75$ inches.

STEP II:

Sketching the condition of beam:

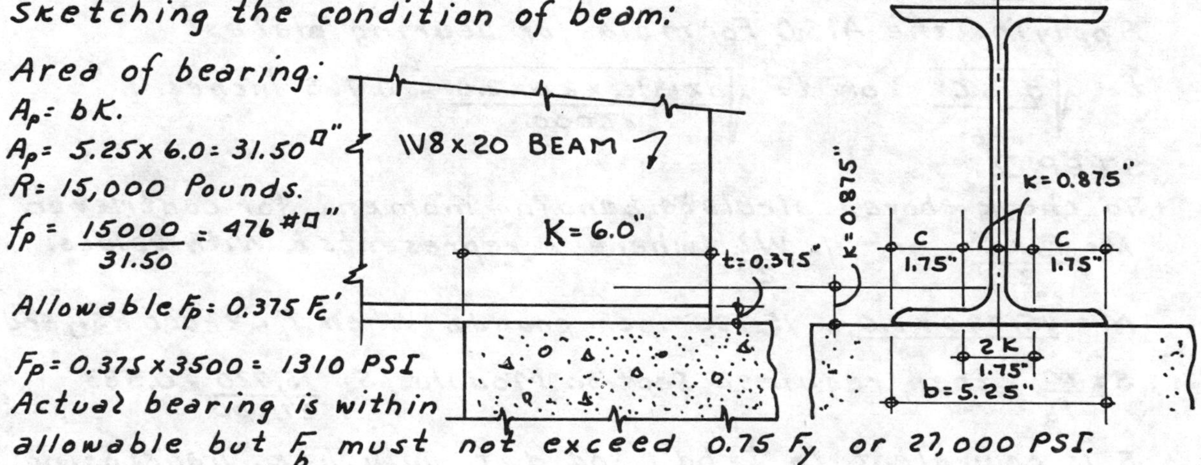

Area of bearing:

$A_p = bK.$

$A_p = 5.25 \times 6.0 = 31.50$ □"

$R = 15,000$ Pounds.

$f_p = \frac{15000}{31.50} = 476$ #□"

Allowable $F_p = 0.375\, F_c'$

$F_p = 0.375 \times 3500 = 1310$ PSI

Actual bearing is within allowable but F_b must not exceed $0.75\, F_y$ or 27,000 PSI.

STEP II:

Effective cantilever portion of flange: $C = \frac{b-(2k)}{2}$, then

$C = \frac{5.25-(2 \times 0.875)}{2} = 1.75$ inches.

Formula for calculated stress: $f_b = \frac{3 f_p C^2}{t^2}$. Substituting values:

$$f_b = \frac{3 \times 476 \times 1.75 \times 1.75}{0.375 \times 0.375} = 31,000 \text{ PSI.}$$

Bending stress is greater than allowable and flange will need to be thicker.

EXAMPLE: Flange bending without bearing plate, continued 2.6.2.6

Calculate the supplementary plate on basis of allowed bearing pressure and add ¼ inch to accomodate welds.
$b = 5.25 + 0.50 = 5.75"$ and $C = 1.75 + 0.25 = 2.0$ inches.

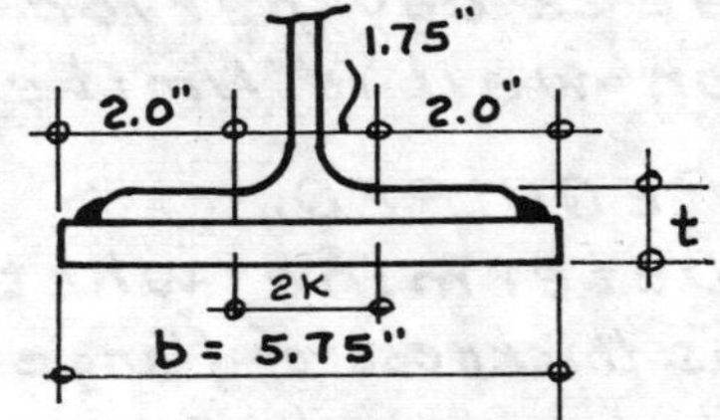

STEP III:

Bearing pressure $R = 1310 \times 5.75 \times 6.0 =$ 45,195 Pounds.
Bearing pressure on cantilever is
$W = 1310 \times 2.0 \times 6.0 = 15,720$ Pounds.

Thickness will include flange and for overall thickness:
$t^2 = \frac{3 F_p C^2}{F_b}$ and with values: $t^2 = \frac{3 \times 1310 \times 2.0 \times 2.0}{27,000} = 0.582$ in.

Then $t = \sqrt{0.582} = 0.765$ in. Deducting flange thickness the thickness of plate $= 0.765 - 0.375 = 0.390$ inches.
A ⅜" plate is close enough (0.375") for acceptance.
t becomes ¾ inches.

STEP IV:
Applying the AISC Formula for bearing plates:

$$t = \sqrt{\frac{3 F_p C^2}{F_b}} \quad \text{or} \quad t = \sqrt{\frac{3 \times 1310 \times 2.0 \times 2.0}{27,000}} = 0.765 \text{ inches.}$$

STEP V:
To check above, calculate bending moment for cantilever.
$M = \frac{F_p K C^2}{2}$ or $M = \frac{W l}{2}$ where l represents t. With values:

$M = \frac{15,720 \times 2.0}{2} = 15,720$ inch pounds. When $F_b = 27,000$ PSI, and
$S = \frac{M}{F_b}$, then required Section Modulus, $S = \frac{15,720}{27,000} = 0.583''^3$

S is equivalent to $S = \frac{bd^2}{6}$ and $d = t$ with $b = K$. Transposing the formula: $t = \sqrt{\frac{6S}{K}}$, or with values substituted:

$$t = \sqrt{\frac{6 \times 0.583}{6.0}} = 0.765 \text{ inches. (checks with step IV)}$$

STEP V:
A formula for bending moment may be derived thus:
$M = \frac{F_b K t^2}{6}$ or $M = \frac{27,000 \times 6.0 \times 0.765 \times 0.765}{6} = 15,720$ inch lbs.

EXAMPLE: Bearing plate design by AISC formula 2.6.2.7

A beam section W8x20 has an end reaction of 26,250 Pounds. Supporting wall consists of a hard burned brick with type M mortar. Allowable bearing pressure as given in Code is 250 PSI. Projection length of beam on wall is limited to 7.0 inches on masonry.

REQUIRED:
Design the size and thickness of bearing plate by using the code requirements which call for the AISC formula. Make a drawing of base plate and check the web crippling to see if stiffeners are necessary. All steel is A36.

STEP I:
Area of plate required is: $A = \frac{R}{F_b}$ or $A: \frac{26,250}{250} = 105$ Sq. In.
Dimension K = 7.0" (limited)
side parallel to wall: $B = \frac{105}{7.0} = 15.0$ inches.
From tables of rolled shapes, dimension for k = 0.875 inches.
Projection for cantilever, $c = \frac{B-(2k)}{2}$
$c = \frac{15.0-(2\times 0.875)}{2} = 6.625$ inches. $F_b = 27,000$ PSI.

STEP II:
Construct drawing of beam and bearing plate with dimensions:

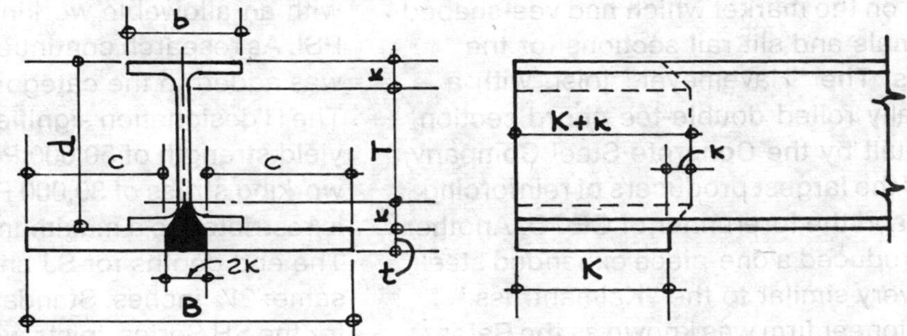

STEP III:
Base formula for thickness of plate t.

$$t = \sqrt{\frac{3F_p c^2}{F_b}} \text{ and } \sqrt{\frac{3\times 250\times 6.625\times 6.625}{27,000}} = 1.115 \text{ inches.}$$

STEP IV:
Investigate web crippling. R = 26,250 # length of web = K+k or 7.0+0.875 = 7.875" Web thickness = 0.25" Allowable $F_v = 0.75\,F_y$.
Allowable $F_v = 0.75 \times 36,000 = 27,000$ PSI
$f_v = \frac{26,250}{7.875\times 0.25} = 8275$ #□" Stiffeners not necessary.

Steel joists 2.7

The first steel web joist was made by using round rods for both the chords and the web, in 1923. The web rod was bent continuously to resemble a simple Warren truss and placed between paired rods which formed the top and bottom chords. Presumably fabrication was by heat welding, with acetylene and oxygen; electric arc welding was not yet in common use. As the demand for the light-weight members increased, fabricators began using small angles for the chords with the bent rod for the web. With little or no engineering data available for design use, claims were made which confused architects. The question centered upon the real strength of the joists to support loads.

One of the early fabricators was the Kalman Steel Company, who produced a popular double-lattice type with diagonal X-bracing in the web. This joist was named the "Kalmantruss." It was fabricated from a single sheet of steel, and the chords were formed by bending up the top and bottom edges. The Gabriel Steel Company placed a joist on the market which had vee-shaped diagonals and slit rail sections for the chords. The "Havemeyer" joist, with a specially rolled double-tee chord section was built by the Concrete Steel Company, one of the largest producers of reinforcing steel, and the forerunner of CECO. Another firm produced a one-piece expanded steel joist, very similar to the "Kalmantruss." This pioneer firm was known as the Bates Expanded Steel Company.

The open-web joist popularity was due to the demand that mechanical features such as steam heating, interior plumbing, and electric lighting be concealed between floors. To save head room, these open-web joists afforded a place to support steel conduits, pipes and vent ducts.

AMERICAN STEEL JOIST INSTITUTE

The early manufacturers of joists were not prepared to furnish engineers and architects with substantial design data. Without an approved set of standards for quality and design, the joist industry was exposed to the hazards of unreliable fabricators who were producing cheap, low quality joists. Building code officials and legitimate engineering and manufacturing firms organized the American Steel Joist Institute (SJI). The first set of standard specifications was approved in 1928, and a load table with joist designations was adopted in 1929.

LONG SPAN AND SHORT SPAN DESIGNATIONS

Designers and architects should take the time to become familiar with the designations of the joist types and series. The first set of joist standards adopted pertained to the short span J Series, restricted to spans of 48 feet or less. These were identified as SJ, and were based upon A36 steel with an allowable working stress of 22,000 PSI. As research continued, the SH Series was added to the category of Short Spans. The H designation signifies a steel with a yield strength of 50,000 PSI and a design working stress of 30,000 PSI. The SH Series is restricted to a maximum span of 48 feet. The end depths for SJ and SH Joists is the same: 2½ inches. Standard specifications for the SH Series Joists were adopted by the Steel Joist Institute in May 1961. The prefix S denotes Short Span. Late in 1961, the Long Span (L) Series standards were revised, and became identified by the designations LA and LH. The LA Series is based upon a design working stress of 22,000 PSI, while the LH Series uses 30,000 PSI. Long Span Series joists are restricted

Steel joists, continued 2.7

to simple spans from 25 feet to 96 feet. The end depth for LA and LH Series is 5 inches. Joist depths run from 18 inches to 48 inches.

LONGSPAN JOISTS: DLJ

A number of joist fabricators produce properly engineered joists for clear spans beyond 96 feet. Some firms use the LH Series designation for spans up to 144 feet. Most fabricators designate these longer spans as DLJ. The end depth is raised to 7½ inches, and joist depths range from 52 inches up to 84 inches. Joists designated as 54LH to 84LH are not a standard of the SJI or AISC, although the basic design will follow the same principles. The long span and deep joists designated as LS are generally built up from A36 steel sections, with an allowable design stress of 22,000 P.S.I.

Steel joist load tables 2.7.1

Using the Standard Load Tables to select the joist series, number and depth is the most convenient method for choosing a joist. These tables are primarily for the use of architects. The joist depth can be found quickly, and will be compatible with span length. Various load tables may show conflicting data. The tables published by some fabricators will list the maximum loads per lineal foot. Other manufacturers will give the whole span maximum load, and spacing will be determined by using a one foot wide strip load. The loads values may separate dead load and live load, or they may be combined. The load tables must be used carefully, and it is a good policy to thoroughly understand one fabricator's design method and use it exclusively.

Tables of standard allowable loads will give two values (separated by heavy or dashed lines) for the load to deflect 1/360 or 1/240 of span length. Consulting engineers in larger offices performing structural design and preparing plans for architects prefer their designers use Tables of Properties and Dimensions. The examples which follow will emphasize this system.

Steel joist design by Resisting Moment 2.7.2

Load and joist-spacing tables will not always provide joist properties such as Moment of Inertia, centroid, section modulus and section area. The property of Inertia I will be required in the deflection formula: $\Delta = \frac{5Wl^3}{384\,EI}$. For short span Series J and H, the Moment of Inertia will be listed in the tabulation of chord and joist dimensions. For long span joists, it may become necessary to compute these properties from available data. An example follows which shows the most accurate method for performing this work.

STRIP LOAD

The strip load design method uses the concept of a load per square foot on a strip one foot wide and extending the full length of the span. The same method is used in concrete slab design. This imaginary strip load is used to calculate a strip bending moment by the simple span formula M = WL/8. Then the strip bending moment is divided into the joist Resisting Moment to determine the maximum spacing. For convenience in checking the work, reduce the Resisting Moment to foot-pounds, and the spacing will be in feet.

An alternative method uses the span formula transposed; the maximum uniform load on a joist is $W = \frac{8\,RM}{L}$. Then the spacing is determined by dividing the strip load into the maximum joist load: Spacing $s = \frac{W}{W_{sl}}$.

CAMBER

The sag in long span joists is partially compensated by camber. The top chord is uniformly sloped upward, starting at the end of the joist and reaching the highest point at the center of the span. For the top chords on a span of 100 feet, the standard camber will exceed four inches. As the effective depth between the top and bottom chords increases, the Moment of Inertia will also increase. Therefore, the Moment of Inertia of a cambered joist will not be given in tables as a single value. It is conservative to calculate the value of I near the end of the joist and to use this value as if the camber did not exist.

Steel joist design by deflection 2.7.3

The Steel Joist Institute approved formula for computing deflection (Δ) in open-web steel joists shows an additional 15 percent deflection over solid-web steel beams. The formula for a simple span with a uniform load is $\Delta = \left(\frac{5Wl^3}{384\,EI}\right) \times (1.15)$. Tests were made at the University of Kansas, Washington University (St. Louis) and Lehigh University to establish this value. Joists are designed for spacing by using this deflection formula.

By transposing, and solving for Maximum Load allowed and spacing, the formula is rewritten and becomes: $W = \frac{384\,EI\,(0.85\Delta)}{5l^3}$

To select a joist by solving for the required value of Inertia, the formula is written: $I = \frac{5Wl^3}{384\,E\,(0.85\Delta)}$

Where:

W = Total Load on Joist in Pounds

l = Length of Span, given in inches

E = Modulus of Elasticity, For Steel: E = 29,000,000 #□"

I = Moment of Inertia of Joist = inches^4

Δ = Deflection of Joist given in inches.

For Joists to support suspended ceilings of plaster, the sag under loads must not exceed 1/360 of span length in inches. For Joists without ceilings or for acoustical tiles suspended from chords, the deflection under full loads shall not exceed 1/240 of clear span.

When the deflection is used in the beam formulas, the factor 0.85Δ should be entered, after deciding on the permissible deflection. The following examples will show this method for joist design.

Joist bridging 2.7.4

The unbraced length of the top chord should have lateral supports placed so that the slenderness ratio l/r does not exceed 200. The AISC and the Steel Joist Institute consider that, within certain limits, the top chords are braced in the lateral direction by the floor slabs or roof deck, *but only when these components actually exist*. Using the recommended spacing for bridging, the compressive stress in top chords at the mid-span, often exceeds the allowable stress found from the slenderness ratio. This ratio can be reduced at the critical point by using a single horizontal member at the top chord only, in addition to the rows of bridging. This situation occurs more often for the Long Span than for the Short Span Series.

Designers, responsible for structural plans should insist upon the installation of diagonal or cross-braced bridging for long spans, and permit no substitute. The installation of all bridging should be complete before applying any construction loads or metal deck. Bridging rows terminating at end walls must be securely fixed and anchored to the wall at both top and bottom chords. Horizontal rod bridging is permitted in the Short Span series when the joists are to remain exposed or support only lightweight suspended ceiling panels of the lay-in type. Any system of joists designed for rigidity or minimum deflection should be installed with cross-braced diagonal bridging. Horizontal bridging if used should be paired with one rod over the other; the rods should not be staggered which is a common but incorrect method. When plans call for diagonal cross-bridging, the substitution of horizontal bridging should not be permitted.

TABLE: Series J and H chord properties 2.7.5.1

COLD FORMED CHORDS							
CHORD PROPERTIES FOR DESIGN - SERIES J & H							
CHORD	A□"	D"	y"	t"	W"	I_o"4	r_o"
2 T	0.464	1.109	0.44	0.109	3.063	0.07	0.40
3 T	0.578	1.125	0.42	0.125	3.438	0.09	0.40
4 T	0.722	1.156	0.43	0.156	3.500	0.12	0.40
5 T	0.855	1.188	0.44	0.188	3.500	0.14	0.40
5 X	0.949	1.188	0.41	0.188	4.000	0.15	0.40
6 T	1.025	1.219	0.44	0.219	3.750	0.17	0.40
6 X	1.066	1.219	0.42	0.219	3.938	0.17	0.40
7 T	1.271	1.219	0.37	0.219	4.875	0.19	0.39
8 T	1.421	1.250	0.39	0.250	4.875	0.22	0.39
2 B	0.351	0.906	0.37	0.094	2.938	0.04	0.33
3 B	0.437	0.922	0.35	0.109	3.188	0.05	0.33
4 B	0.544	0.953	0.37	0.141	3.125	0.06	0.33
5 B	0.654	0.969	0.36	0.156	3.438	0.07	0.33
6 B	0.796	1.000	0.36	0.188	3.563	0.09	0.33
7 B	0.949	1.000	0.32	0.188	4.375	0.10	0.32
8 B	1.066	1.031	0.34	0.219	4.313	0.11	0.33

SERIES J: F_y = 36,000 PSI. F_b = 22,000 PSI.
SERIES H: F_y = 50,000 PSI. F_b = 30,000 PSI.
WEB RODS: F_y = SAME FOR APPLICABLE SERIES.

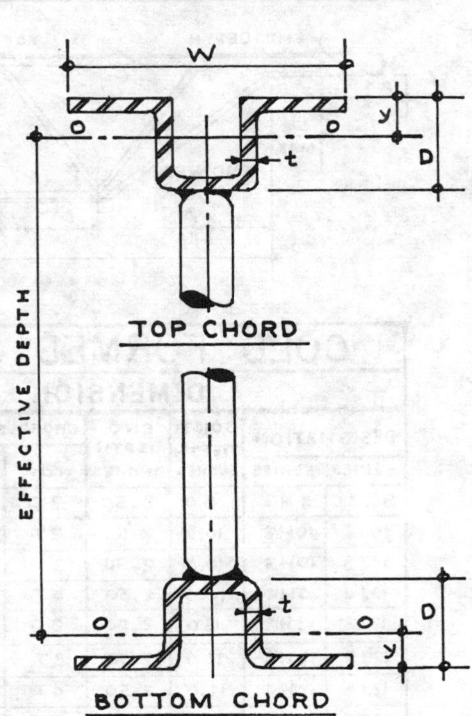

HOT ROLLED CHORDS						
TOP CHORD PROPERTIES FOR DESIGN - SERIES J & H						
SECT.	2 ANGLES ⅂⎾	A□" 2Ls	G"	W"	I_o"4	r_o"
2	1 × 1 × 0.125	0.46	0.30	2.34	0.04	0.30
3	1¼ × 1¼ × 0.125	0.60	0.36	2.84	0.08	0.38
4	1½ × 1½ × 0.125	0.72	0.42	3.34	0.16	0.47
5	1½ × 1½ × 0.188	1.06	0.44	3.47	0.22	0.46
6	1½ × 1½ × 0.188	1.06	0.44	3.47	0.22	0.46
7	1¾ × 1¾ × 0.188	1.24	0.51	3.97	0.36	0.54
8	2 × 2 × 0.188	1.42	0.57	4.53	0.54	0.62

SERIES J: F_y = 36,000 F_b = 22,000 PSI
SERIES H: F_y = 50,000 F_b = 30,000 PSI
WEB RODS: F_y = SAME FOR APPLICABLE SERIES.

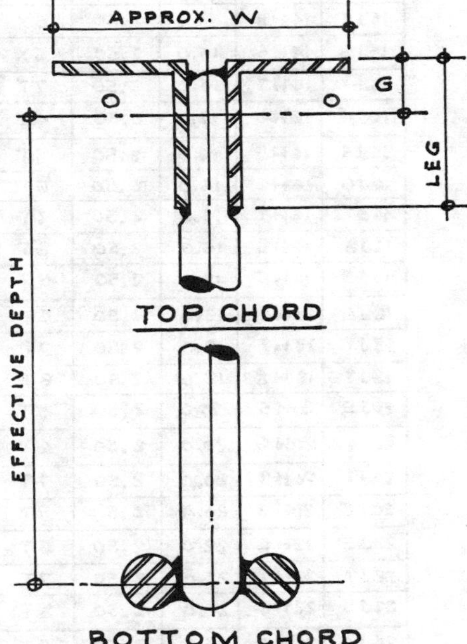

TABLE: Series J and H joist properties 2.7.5.2

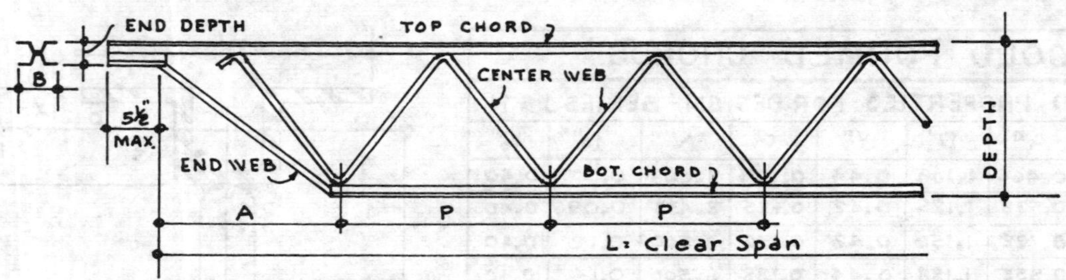

COLD FORMED CHORDS - "J" AND "H" SERIES
DIMENSION DATA AND PROPERTIES

DESIGNATION		JOIST DEPTH	END DEPTH	CHORD SECTION		DIMENSION "A"	WEB BAR DIAMETER		BEARING B	PANEL P	APPROX. WT.	MOMENT OF
SERIES	SERIES	INCHES	INCHES	TOP	BOTTOM	VARIES	END	CENT.	INCHES	INCHES	#/'	INERTIA
8 J2	8 H2	8.0	2.50	2T	2B	20 – 31	0.81	0.59	4.000	24	4.2	10.78
10J2	10H2	10.0	2.50	2T	2B	20 – 31	0.81	0.56	4.000	24	4.2	17.54
10J3	10H3	10.0	2.50	3T	3B	20 – 31	0.81	0.91	4.000	24	5.0	21.78
10J4	10H4	10.0	2.50	4T	4B	20 – 31	0.81	0.63	4.000	24	6.1	27.06
12J2	12H2	12.0	2.50	2T	2B	19 – 30	0.81	0.58	4.000	24	4.5	25.94
12J3	12H3	12.0	2.50	3T	3B	19 – 30	0.81	0.61	4.000	24	5.2	32.18
12J4	12H4	12.0	2.50	4T	4B	19 – 30	0.81	0.64	4.000	24	6.2	40.01
12J5	12H5	12.0	2.50	5X	5B	19 – 30	0.81	0.68	4.375	24	7.1	49.53
12J6	12H6	12.0	2.50	6X	6B	19 – 30	0.81	0.69	5.375	24	8.2	59.29
14J3	14H3	14.0	2.50	3T	3B	19 – 30	0.81	0.63	4.000	24	5.5	44.61
14J4	14H4	14.0	2.50	4T	4B	19 – 30	0.81	0.66	4.000	24	6.5	55.51
14J5	14H5	14.0	2.50	5X	5B	19 – 30	0.81	0.67	4.375	24	7.4	68.67
14J6	14H6	14.0	2.50	6X	6B	19 – 30	0.81	0.70	5.375	24	8.6	82.24
14J7	14H7	14.0	2.50	7T	7B	19 – 30	0.81	0.73	5.375	24	10.0	94.61
16J4	16H4	16.0	2.50	4T	4B	19 – 30	0.81	0.67	4.000	24	6.6	73.55
16J5	16H5	16.0	2.50	5T	5B	19 – 30	0.81	0.70	4.375	24	7.8	86.88
16J6	16H6	16.0	2.50	6T	6B	19 – 30	0.81	0.73	5.375	24	8.6	105.12
16J7	16H7	16.0	2.50	7T	7B	19 – 30	0.81	0.75	5.375	24	10.4	125.24
16J8	16H8	16.0	2.50	8T	8B	19 – 30	0.81	0.77	5.375	24	11.6	144.84
18J5	18H5	18.0	2.50	5T	5B	21 – 32	0.81	0.72	4.375	24	8.0	111.24
18J6	18H6	18.0	2.50	6T	6B	21 – 32	0.81	0.73	5.375	24	9.2	134.63
18J7	18H7	18.0	2.50	7T	7B	21 – 32	0.81	0.77	5.375	24	10.4	160.16
18J8	18H8	18.0	2.50	8T	8B	21 – 32	0.81	0.78	5.375	24	11.6	185.32
20J5	20H5	20.0	2.50	5T	5B	23 – 34	0.81	0.75	4.375	24	8.4	138.61
20J6	20H6	20.0	2.50	6T	6B	23 – 34	0.81	0.77	5.375	24	9.6	167.80
20J7	20H7	20.0	2.50	7T	7B	23 – 34	0.81	0.78	5.375	24	10.7	199.38
20J8	20H8	20.0	2.50	8T	8B	23 – 34	0.81	0.80	5.375	24	12.2	230.77
22J6	22H6	22.0	2.50	6T	6B	27 – 38	0.81	0.72	5.375	24	9.7	204.62
22J7	22H7	22.0	2.50	7T	7B	27 – 38	0.81	0.80	5.375	24	10.7	242.90
22J8	22H8	22.0	2.50	8T	8B	27 – 38	0.81	0.81	5.375	24	12.0	281.23
24J6	24H6	24.0	2.50	6T	5B	29 – 40	0.81	0.81	5.375	24	10.3	245.10
24J7	24H7	24.0	2.50	7T	7B	29 – 40	0.81	0.83	5.375	24	11.5	290.72
24J8	24H8	24.0	2.50	8T	8B	29 – 40	0.81	0.84	5.375	24	12.7	336.67

TABLE: Series J and H joist properties, continued 2.7.5.2

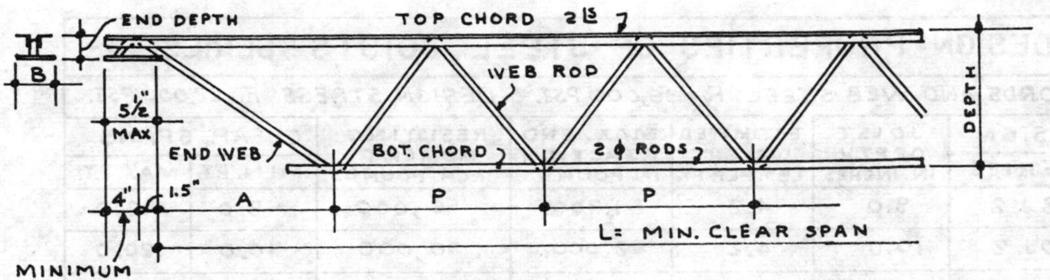

HOT ROLLED CHORDS — "J" AND "H" SERIES

DIMENSION DATA AND PROPERTIES

DESIGNATION		JOIST DEPTH IN.	END DEPTH IN.	CHORD SECT.		DIM. A IN.	WEB RODS DIAMETER		BEARING B INCHES	PANEL P INCHES	APPROX. WT. #/'	MOMENT OF INERTIA
SERIES	SERIES			TOP	BOT. 2-φ		END	WEB				
8J2	8H2	8.0	2.50	2	0.469	7.0	0.469	0.469	4.50	14	4.2	10.95
10J2	10H2	10.0	2.50	2	0.469	9.5	0.469	0.469	4.50	14	4.2	17.57
10J3	10H3	10.0	2.50	3	0.531	9.5	0.531	0.531	4.50	14	5.0	22.37
10J4	10H4	10.0	2.50	4	0.594	9.5	0.531	0.531	4.50	14	6.1	26.72
12J2	12H2	12.0	2.50	2	0.469	12.0	0.531	0.531	4.50	14	4.5	25.76
12J3	12H3	12.0	2.50	3	0.531	12.0	0.594	0.594	4.50	18	5.2	32.89
12J4	12H4	12.0	2.50	4	0.594	12.0	0.594	0.594	4.50	18	6.2	40.25
12J5	12H5	12.0	2.50	5	0.656	12.0	0.594	0.594	4.50	18	7.1	52.48
12J6	12H6	12.0	2.50	6	0.719	12.0	0.656	0.656	4.50	18	8.2	58.22
14J3	14H3	14.0	2.50	3	0.531	14.5	0.594	0.594	4.50	18	5.5	45.45
14J4	14H4	14.0	2.50	4	0.594	14.5	0.594	0.594	4.50	18	6.5	55.73
14J5	14H5	14.0	2.50	5	0.656	14.5	0.656	0.656	4.50	18	7.4	72.63
14J6	14H6	14.0	2.50	6	0.719	14.5	0.656	0.656	4.50	18	8.6	80.79
14J7	14H7	14.0	2.50	7	0.781	14.5	0.656	0.656	4.50	18	10.0	93.23
16J4	16H4	16.0	2.50	4	0.594	17.0	0.656	0.656	5.50	18	6.6	73.72
16J5	16H5	16.0	2.50	5	0.656	17.0	0.656	0.656	5.50	18	7.8	96.31
16J6	16H6	16.0	2.50	6	0.719	17.0	0.719	0.656	5.50	18	8.6	107.05
16J7	16H7	16.0	2.50	7	0.781	17.0	0.719	0.656	5.50	18	10.3	123.75
16J8	16H8	16.0	2.50	8	0.844	17.0	0.719	0.656	5.50	18	11.4	141.62
18J5	18H5	18.0	2.50	5	0.656	19.5	0.719	0.656	5.50	20	8.0	123.22
18J6	18H6	18.0	2.50	6	0.719	19.5	0.719	0.656	5.50	20	9.2	137.00
18J7	18H7	18.0	2.50	7	0.781	19.5	0.719	0.656	5.50	20	10.4	158.60
18J8	18H8	18.0	2.50	8	0.844	19.5	0.719	0.656	5.50	20	11.6	181.64
20J5	20H5	20.0	2.50	5	0.656	22.0	0.781	0.656	5.50	22	8.4	153.41
20J6	20H6	20.0	2.50	6	0.719	22.0	0.781	0.656	5.50	22	9.6	170.66
20J7	20H7	20.0	2.50	7	0.781	22.0	0.781	0.656	5.50	22	10.7	197.78
20J8	20H8	20.0	2.50	8	0.844	22.0	0.781	0.656	5.50	22	12.2	226.83
22J6	22H6	22.0	2.50	6	0.719	24.5	0.844	0.656	5.50	24	9.7	208.02
22J7	22H7	22.0	2.50	7	0.781	24.5	0.844	0.656	5.50	24	10.7	241.28
22J8	22H8	22.0	2.50	8	0.844	24.5	0.844	0.656	5.50	24	12.0	276.95
24J6	24H6	24.0	2.50	6	0.719	27.0	0.844	0.656	5.50	24	10.3	249.07
24J7	24H7	24.0	2.50	7	0.781	27.0	0.844	0.656	5.50	24	11.5	289.12
24J8	24H8	24.0	2.50	8	0.844	27.0	0.844	0.656	5.50	24	12.7	332.08

TABLE: Series J and H design properties 2.7.5.3

DESIGN PROPERTIES OF STEEL JOISTS - SERIES - J -						
CHORDS AND WEB STEEL: $F_y = 36{,}000$ PSI. DESIGN STRESS: $F_t = 22{,}000$ PSI.						
DESIGN SERIES	JOIST DEPTH IN INCHES	ESTIMATED WEIGHT LBS. PER FT.	MAX. END REACTION IN POUNDS	RESISTING MOMENT INCH POUNDS	CLEAR SPANS	
					MIN. FT.	MAX. FT.
8J 2	8.0	4.2	1,900	56,000	8.0	16.0
10J 2	10.0	4.2	2,000	70,000	10.0	20.0
10J 3	10.0	5.0	2,200	89,000	10.0	20.0
10J 4	10.0	6.1	2,400	111,000	10.0	20.0
12J 2	12.0	4.5	2,200	85,000	12.0	24.0
12J 3	12.0	5.2	2,300	108,000	12.0	24.0
12J 4	12.0	6.2	2,500	135,000	12.0	24.0
12J 5	12.0	7.1	2,700	161,000	12.0	24.0
12J 6	12.0	8.2	3,000	196,000	12.0	24.0
14J 3	14.0	5.5	2,400	127,000	14.0	28.0
14J 4	14.0	6.5	2,800	159,000	14.0	28.0
14J 5	14.0	7.4	3,100	190,000	14.0	28.0
14J 6	14.0	8.6	3,400	230,000	14.0	28.0
14J 7	14.0	10.0	3,700	276,000	14.0	28.0
16J 4	16.0	6.6	3,000	173,000	16.0	32.0
16J 5	16.0	7.8	3,300	216,000	16.0	32.0
16J 6	16.0	8.6	3,600	258,000	16.0	32.0
16J 7	16.0	10.3	4,000	310,000	16.0	32.0
16J 8	16.0	11.4	4,300	359,000	18.0	32.0
18J 5	18.0	8.0	3,500	243,000	18.0	36.0
18J 6	18.0	9.2	3,900	293,000	18.0	36.0
18J 7	18.0	10.4	4,200	352,000	18.0	36.0
18J 8	18.0	11.6	4,500	406,000	18.0	36.0
20J 5	20.0	8.4	3,800	265,000	20.0	40.0
20J 6	20.0	9.6	4,100	316,000	20.0	40.0
20J 7	20.0	10.7	4,300	382,000	20.0	40.0
20J 8	20.0	12.2	4,600	455,000	20.0	40.0
22J 6	22.0	9.7	4,200	335,000	22.0	44.0
22J 7	22.0	10.7	4,500	420,000	22.0	44.0
22J 8	22.0	12.0	4,800	493,000	22.0	44.0
24J 6	24.0	10.3	4,400	367,000	24.0	48.0
24J 7	24.0	11.5	4,700	460,000	24.0	48.0
24J 8	24.0	12.7	5,000	540,000	24.0	48.0

TABLE: Series J and H design properties, continued 2.7.5.3

DESIGN PROPERTIES OF STEEL JOISTS – SERIES H						
CHORDS AND WEB STEEL: $F_y = 50{,}000$ PSI. DESIGN STRESS: $F_t = 30{,}000$ PSI.						
DESIGN SERIES	JOIST DEPTH IN INCHES	ESTIMATED WEIGHT LBS. PER FT.	MAX. END REACTION IN POUNDS	RESISTING MOMENT INCH POUNDS	CLEAR SPANS	
					MIN. FT.	MAX. FT.
8 H 2	8.0	4.2	2,000	73,000	8.0	16.0
10 H 2	10.0	4.2	2,200	91,000	10.0	20.0
10 H 3	10.0	5.0	2,500	116,000	10.0	20.0
10 H 4	10.0	6.1	2,800	148,000	10.0	20.0
12 H 2	12.0	4.5	2,400	111,000	12.0	24.0
12 H 3	12.0	5.2	2,800	140,000	12.0	24.0
12 H 4	12.0	6.2	3,200	180,000	12.0	24.0
12 H 5	12.0	7.1	3,600	222,000	12.0	24.0
12 H 6	12.0	8.2	3,900	260,000	12.0	24.0
14 H 3	14.0	5.5	3,200	165,000	14.0	28.0
14 H 4	14.0	6.5	3,500	212,000	14.0	28.0
14 H 5	14.0	7.4	3,800	259,000	14.0	28.0
14 H 6	14.0	8.6	4,200	307,000	14.0	28.0
14 H 7	14.0	10.0	4,600	369,000	14.0	28.0
16 H 4	16.0	6.6	3,800	221,000	16.0	32.0
16 H 5	16.0	7.8	4,300	289,000	16.0	32.0
16 H 6	16.0	8.6	4,600	344,000	16.0	32.0
16 H 7	16.0	10.3	4,900	413,000	16.0	32.0
16 H 8	16.0	11.4	5,200	478,000	18.0	32.0
18 H 5	18.0	8.0	4,500	325,000	18.0	36.0
18 H 6	18.0	9.2	4,800	383,000	18.0	36.0
18 H 7	18.0	10.4	5,200	466,000	18.0	36.0
18 H 8	18.0	11.6	5,400	540,000	18.0	36.0
20 H 5	20.0	8.4	4,800	365,000	20.0	40.0
20 H 6	20.0	9.6	5,100	406,000	20.0	40.0
20 H 7	20.0	10.7	5,400	499,000	20.0	40.0
20 H 8	20.0	12.2	5,600	602,000	20.0	40.0
22 H 6	22.0	9.7	5,400	422,000	22.0	44.0
22 H 7	22.0	10.7	5,600	526,000	22.0	44.0
22 H 8	22.0	12.0	5,800	653,000	22.0	44.0
24 H 6	24.0	10.3	5,600	462,000	24.0	48.0
24 H 7	24.0	11.5	5,800	576,000	24.0	48.0
24 H 8	24.0	12.7	6,000	716,000	24.0	48.0

TABLE: Series LA joist dimensions 2.7.5.4

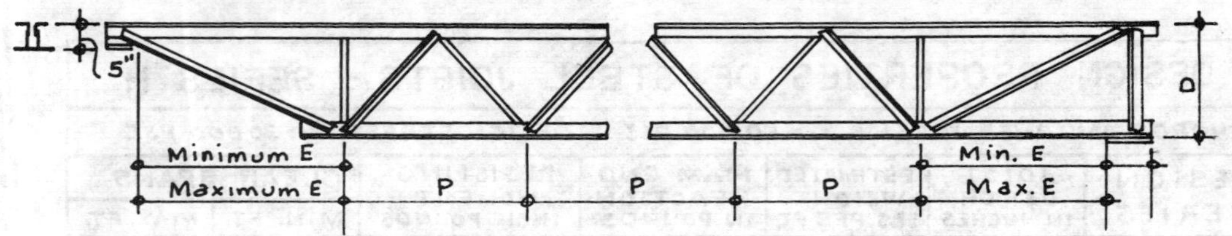

STEEL JOISTS SERIES LA - DESIGN PROPERTIES										
CHORD DESIGN SERIES	TOP CHORDS					BOTTOM CHORDS				END DEPTH INCHES
	CHORD SIZE 2 ANGLES	A□" 2 Ls	G In.	H In.	W In.	CHORD SIZE 2 ANGLES	A□" 2 Ls	G In.	H In.	
LA02	2 x 1½ x 3/16	1.24	0.64	2.0	3.69	1½ x 1½ x 3/16	1.06	0.44	1.5	5.00
LA03	2 x 2 x 3/16	1.42	0.57	2.0	4.69	1½ x 1½ x 3/16	1.06	0.44	1.5	5.00
LA04	2½ x 2 x 3/16	1.62	0.76	2.5	4.69	2 x 1½ x 3/16	1.24	0.64	2.0	5.00
LA05	2 x 2 x ¼	1.88	0.59	2.0	4.69	2 x 2 x 3/16	1.42	0.57	2.0	5.00
LA06	2½ x 2 x ¼	2.12	0.79	2.5	4.69	2½ x 2 x 3/16	1.62	0.76	2.5	5.00
LA07	2½ x 2½ x ¼	2.38	0.72	2.5	5.69	2½ x 2½ x 3/16	1.80	0.69	2.5	5.00
LA08	3 x 2½ x ¼	2.62	0.91	3.0	5.69	2½ x 2 x ¼	2.12	0.79	2.5	5.00
LA09	3 x 3 x ¼	2.88	0.84	3.0	6.69	2½ x 2½ x ¼	2.38	0.72	2.5	5.00
LA10	3½ x 3 x ¼	3.12	1.04	3.5	6.69	3 x 2½ x ¼	2.62	0.91	3.0	5.00
LA11	3 x 3 x 5/16	3.56	0.87	3.0	6.69	3 x 3 x ¼	2.88	0.84	3.0	5.00
LA12	3½ x 3 x 5/16	3.86	1.06	3.5	6.69	3½ x 3 x ¼	3.12	1.04	3.5	5.00
LA13	3½ x 3 x 3/8	4.60	1.08	3.5	6.69	3 x 3 x 5/16	3.56	0.87	3.0	5.00
LA14	4 x 3 x 3/8	4.96	1.28	4.0	6.75	3½ x 3 x 5/16	3.86	1.06	3.5	5.00
LA15	4 x 4 x 3/8	5.72	1.14	4.0	8.81	3½ x 3½ x 5/16	4.18	0.99	3.5	5.00
LA16	4 x 4 x 7/16	6.62	1.16	4.0	8.88	3½ x 3½ x 3/8	4.96	1.01	3.5	5.00
LA17	4 x 4 x ½	7.50	1.18	4.0	8.88	4 x 4 x 3/8	5.72	1.14	4.0	5.00
LA18	5 x 5 x 7/16	8.36	1.41	5.0	10.94	4 x 4 x 7/16	6.62	1.16	4.0	5.00
LA19	5 x 5 x ½	9.50	1.43	5.0	11.00	4 x 4 x ½	7.50	1.18	4.0	5.00

STEEL FOR LA SERIES = F_y = 36,000 PSI. DESIGN STRESS ALLOWABLE = 22,000 PSI.

DEPTH JOIST	INTERIOR P Inches	MINIMUM E Inches	MAXIMUM E Inches
18	30	19.00	34.00
20	33	21.25	37.75
24	39	25.75	45.25
28	45	30.25	52.75
32	51	34.75	60.25
36	57	39.25	67.75
40	63	43.75	75.25
44	69	48.25	82.75
48	75	52.75	90.25

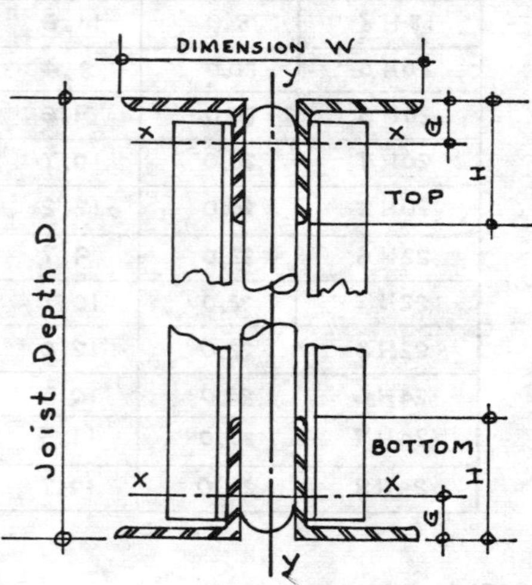

TABLE: Longspan series LH joist dimensions 2.7.5.5

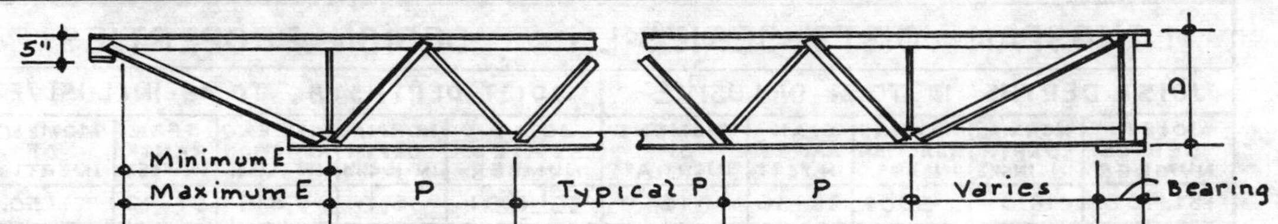

LONGSPAN JOISTS SERIES LH - DESIGN PROPERTIES

CHORD DESIGN SERIES	TOP CHORDS: CHORD SIZE 2 ANGLES	TOP CHORDS: A□" 2 Ls	TOP CHORDS: G In.	TOP CHORDS: H In.	TOP CHORDS: W In.	BOTTOM CHORDS: CHORD SIZE 2 ANGLES	BOTTOM CHORDS: A□" 2 Ls	BOTTOM CHORDS: G In.	BOTTOM CHORDS: H In.	JOIST DEPTH MIN.	JOIST DEPTH MAX.
LH02	2 x 1½ x 3/16	1.24	0.64	2.0	3.69	1½ x 1½ x 3/16	1.06	0.44	1.5	18	20
LH03	2 x 2 x 3/16	1.42	0.57	2.0	4.69	1½ x 1½ x 3/16	1.06	0.44	1.5	18	24
LH04	2 x 1½ x ¼	1.62	0.66	2.0	3.69	2 x 1½ x 3/16	1.24	0.64	2.0	18	24
LH05	2 x 2 x ¼	1.88	0.59	2.0	4.69	2 x 2 x 3/16	1.42	0.57	2.0	18	24
	2 x 2 x ¼	1.88	0.59	2.0	4.69	2 x 1½ x 3/16	1.24	0.64	2.0	28	28
LH06	2½ x 2 x ¼	2.12	0.79	2.5	4.69	2½ x 2 x 3/16	1.62	0.76	2.5	18	28
	2½ x 2 x ¼	2.12	0.79	2.5	4.69	2 x 2 x 3/16	1.42	0.57	2.0	32	32
LH07	2½ x 2½ x ¼	2.38	0.72	2.5	5.69	2½ x 2½ x 3/16	1.80	0.69	2.5	18	28
	2½ x 2½ x ¼	2.38	0.72	2.5	5.69	2½ x 2 x 3/16	1.62	0.76	2.5	32	36
LH08	3 x 2½ x ¼	2.46	0.79	3.0	5.69	2½ x 2 x ¼	2.12	0.79	2.5	18	24
	3 x 2½ x ¼	2.46	0.79	3.0	5.69	2½ x 2½ x 3/16	1.80	0.69	2.5	28	32
	3 x 2½ x ¼	2.46	0.79	3.0	5.69	2½ x 2 x 3/16	1.62	0.76	2.5	36	40
LH09	3 x 2 x 5/16	2.94	1.02	3.0	4.69	2½ x 2½ x ¼	2.38	0.72	2.5	18	32
	3 x 2 x 5/16	2.94	1.02	3.0	4.69	2½ x 2 x ¼	2.12	0.79	2.5	36	40
	3 x 2 x 5/16	2.94	1.02	3.0	4.69	2½ x 2½ x 3/16	1.80	0.69	2.5	44	44
LH10	3 x 2½ x 5/16	3.24	0.93	3.0	5.69	3 x 2½ x ¼	2.62	0.91	3.0	20	28
	3 x 2½ x 5/16	3.24	0.93	3.0	5.69	2½ x 2½ x ¼	2.38	0.72	2.5	32	36
	3 x 2½ x 5/16	3.24	0.93	3.0	5.69	2½ x 2 x ¼	2.12	0.79	2.5	40	48
LH11	3 x 3 x 5/16	3.56	0.87	3.0	6.69	3 x 2 x 5/16	2.94	1.02	3.0	24	28
	3 x 3 x 5/16	3.56	0.87	3.0	6.69	3 x 2½ x ¼	2.62	0.91	3.0	32	36
	3 x 3 x 5/16	3.56	0.87	3.0	6.69	2½ x 2½ x ¼	2.38	0.72	2.5	40	48
LH12	3½ x 3½ x 5/16	4.00	0.95	3.5	7.69	3 x 2½ x 5/16	3.24	0.93	3.0	28	32
	3½ x 3½ x 5/16	4.00	0.95	3.5	7.69	3 x 2 x 5/16	2.94	1.02	3.0	36	40
	3½ x 3½ x 5/16	4.00	0.95	3.5	7.69	3 x 2½ x ¼	2.62	0.91	3.0	44	48
LH13	3½ x 3 x 3/8	4.60	1.08	3.5	6.69	3 x 3 x 5/16	3.56	0.87	3.0	28	40
	3½ x 3 x 3/8	4.60	1.08	3.5	6.69	3 x 2½ x 5/16	3.24	0.93	3.0	44	48
LH14	4 x 3 x 3/8	4.96	1.28	4.0	6.75	3½ x 3 x 5/16	3.86	1.06	3.5	32	40
	4 x 3 x 3/8	4.96	1.28	4.0	6.75	3 x 3 x 5/16	3.56	0.87	3.0	44	48
LH15	4 x 4 x 3/8	5.72	1.14	4.0	8.81	3½ x 3½ x 5/16	4.18	0.99	3.5	32	48
LH16	4 x 4 x 7/16	6.62	1.16	4.0	8.88	3½ x 3½ x 3/8	4.96	1.01	3.5	40	48
LH17	4 x 4 x ½	7.50	1.18	4.0	8.88	4 x 4 x 3/8	5.72	1.14	4.0	44	48

TABLE: Longspan series LH joist properties 2.7.5.6

LONGSPAN JOISTS SERIES LH DESIGN PROPERTIES									
JOIST DEPTHS 18" TO 32" INCLUSIVE					JOIST DEPTHS 36" TO 48" INCLUSIVE				
JOIST SERIES NUMBER	NORMAL DEPTH IN INCH'S	MAX. END REACTION IN LBS.	SPAN RANGE IN FEET	MOMENT OF INERTIA"⁴	JOIST SERIES NUMBER	NORMAL DEPTH IN INCH'S	MAX. END REACTION IN LBS.	SPAN RANGE IN FEET	MOMENT OF INERTIA"⁴
18LH02	18.0	6006	25-36	164.0	36LH07	36.0	8,419	57-72	1150.0
18LH03	18.0	6686	25-36	176.0	36LH08	36.0	9,255	57-72	1183.0
18LH04	18.0	7751	25-36	197.0	36LH09	36.0	11,850	57-72	1444.0
18LH05	18.0	8778	25-36	231.0	36LH10	36.0	13,090	57-72	1600.0
18LH06	18.0	10,382	25-36	251.0	36LH11	36.0	14,272	57-72	1773.0
18LH07	18.0	10,790	25-36	285.0	36LH12	36.0	17,098	57-72	2002.0
18LH08	18.0	11,243	25-36	315.0	36LH13	36.0	20,096	57-72	2335.0
18LH09	18.0	12,017	25-36	341.0	36LH14	36.0	22,146	57-72	2472.0
20LH02	20.0	5672	25-40	205.0	36LH15	36.0	23,326	57-72	2784.0
20LH03	20.0	6018	25-40	220.0	40LH08	40.0	8,339	65-80	1474.0
20LH04	20.0	7366	25-40	247.0	40LH09	40.0	10,900	65-80	1800.0
20LH05	20.0	7905	25-40	288.0	40LH10	40.0	12,049	65-80	1882.0
20LH06	20.0	10,554	25-40	315.0	40LH11	40.0	13,100	65-80	2080.0
20LH07	20.0	11,273	25-40	357.0	40LH12	40.0	15,957	65-80	2499.0
20LH08	20.0	11,653	25-40	396.0	40LH13	40.0	18,813	65-80	2914.0
20LH09	20.0	12,709	25-40	430.0	40LH14	40.0	21,538	65-80	3091.0
20LH10	20.0	13,710	25-40	483.0	40LH15	40.0	24,099	65-80	3477.0
24LH03	24.0	5757	33-48	322.0	40LH16	40.0	26,535	65-80	4074.0
24LH04	24.0	7053	33-48	363.0	44LH09	44.0	10,018	73-88	2000.0
24LH05	24.0	7558	33-48	423.0	44LH10	44.0	11,050	73-88	2295.0
24LH06	24.0	10,167	33-48	465.0	44LH11	44.0	11,970	73-88	2538.0
24LH07	24.0	11,194	33-48	525.0	44LH12	44.0	14,807	73-88	2862.0
24LH08	24.0	11,901	33-48	586.0	44LH13	44.0	17,569	73-88	3360.0
24LH09	24.0	14,006	33-48	641.0	44LH14	44.0	20,221	73-88	3641.0
24LH10	24.0	14,848	33-48	717.0	44LH15	44.0	23,536	73-88	4248.0
24LH11	24.0	15,616	33-48	793.0	44LH16	44.0	27,146	73-88	4977.0
28LH05	28.0	7020	41-56	537.0	44LH17	44.0	29,125	73-88	5657.0
28LH06	28.0	9333	41-56	645.0	48LH10	48.0	10,045	81-96	2749.0
28LH07	28.0	10,520	41-56	727.0	48LH11	48.0	10,861	81-96	3042.0
28LH08	28.0	11,250	41-56	747.0	48LH12	48.0	13,720	81-96	3430.0
28LH09	28.0	13,895	41-56	893.0	48LH13	48.0	16,415	81-96	4029.0
28LH10	28.0	15,187	41-56	997.0	48LH14	48.0	19,395	81-96	4368.0
28LH11	28.0	16,256	41-56	1103.0	48LH15	48.0	22,252	81-96	5095.0
28LH12	28.0	17,873	41-56	1249.0	48LH16	48.0	25,684	81-96	5971.0
28LH13	28.0	18,649	41-56	1370.0	48LH17	48.0	28,828	81-96	6791.0
32LH06	32.0	8393	49-64	800.0					
32LH07	32.0	9411	49-64	900.0					
32LH08	32.0	10,206	49-64	990.0					
32LH09	32.0	12,814	49-64	1188.0					
32LH10	32.0	14,179	49-64	1246.0					
32LH11	32.0	15,520	49-64	1384.0					
32LH12	32.0	18,227	49-64	1659.0					
32LH13	32.0	20,303	49-64	1821.0					
32LH14	32.0	20,937	49-64	1922.0					
32LH15	32.0	21,625	49-64	2168.0					

TABLE: Deep longspan series DLJ joist dimensions 2.7.5.7

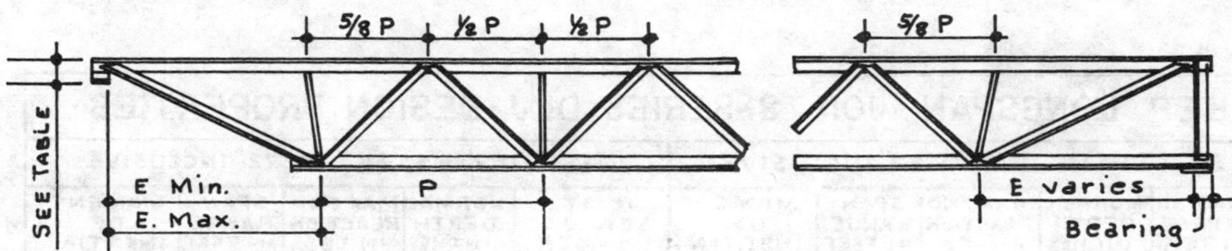

DEEP LONGSPAN JOISTS SERIES DLJ-DESIGN PROPERTIES

CHORD DESIGN SERIES	TOP CHORDS					BOTTOM CHORDS				END DEPTH INCHES
	CHORD SIZE 2 ANGLES	A□" 2 Ls	G In.	H In.	W In.	CHORD SIZE 2 ANGLES	A□" 2 Ls	G In.	H In.	
DLJ 12	3½ x 3 x 5/16	3.86	1.06	3.5	7.00	3 x 2½ x 5/16	3.24	0.93	3.0	5.00
DLJ 13	3½ x 3 x 3/8	4.60	1.08	3.5	7.00	3½ x 3 x 5/16	3.86	1.06	3.5	5.00
DLJ 13	3½ x 3 x 3/8	4.60	1.08	3.5	7.00	3 x 3 x 5/16	3.56	0.87	3.0	5.00
DLJ 14	4 x 3 x 3/8	4.96	1.28	4.0	7.00	3½ x 3½ x 5/16	4.18	0.99	3.5	5.00
DLJ 14	4 x 3 x 3/8	4.96	1.28	4.0	7.00	3½ x 3 x 5/16	3.86	1.06	3.5	5.00
DLJ 15	4 x 4 x 3/8	5.72	1.14	4.0	9.00	3½ x 3 x 3/8	4.60	1.08	3.5	5.00
DLJ 16	4 x 4 x 7/16	6.62	1.16	4.0	9.00	4 x 4 x 3/8	5.72	1.14	4.0	5.00
DLJ 16	4 x 4 x 7/16	6.62	1.16	4.0	9.00	3½ x 3½ x 3/8	4.96	1.01	3.5	5.00
DLJ 17	4 x 4 x ½	7.50	1.18	4.0	9.00	4 x 4 x 7/16	6.62	1.16	4.0	5.00
DLJ 17	4 x 4 x ½	7.50	1.18	4.0	9.00	4 x 4 x 3/8	5.72	1.14	4.0	5.00
DLJ 18	5 x 5 x 7/16	8.36	1.41	5.0	9.00	4 x 4 x ½	7.50	1.18	4.0	7.50
DLJ 18	5 x 5 x 7/16	8.36	1.41	5.0	11.00	4 x 4 x 7/16	6.62	1.16	4.0	7.50
DLJ 19	5 x 5 x ½	9.50	1.43	5.0	11.00	5 x 5 x 7/16	8.36	1.41	5.0	7.50
DLJ 20	6 x 6 x ½	11.50	1.68	6.0	13.25	5 x 5 x ½	9.50	1.43	5.0	7.50
DLJ 21	6 x 6 x ½	11.50	1.68	6.0	13.14	5 x 5 x 7/16	8.36	1.41	5.0	7.50
DLJ 22	6 x 6 x 9/16	12.86	1.71	6.0	13.22	5 x 5 x ½	9.50	1.43	5.0	7.50
DLJ 23	6 x 6 x 5/8	14.22	1.73	6.0	13.39	6 x 6 x 7/16	10.12	1.66	6.0	7.50

STEEL FOR DLJ SERIES = F_y = 36,000 PSI. *DESIGN STRESS ALLOWABLE = 22,000 PSI.*

ALL LONG LEGS VERTICAL IN DESIGN TABLE.

DEPTH INCHES	INTERIOR P INCHES	MINIMUM E INCHES	MAXIMUM E INCHES
52	104.0	48.00	100.00
56	112.0	52.00	108.00
60	120.0	56.00	116.00
64	128.0	60.00	124.00
68	136.0	64.00	132.00
72	144.0	68.00	140.00

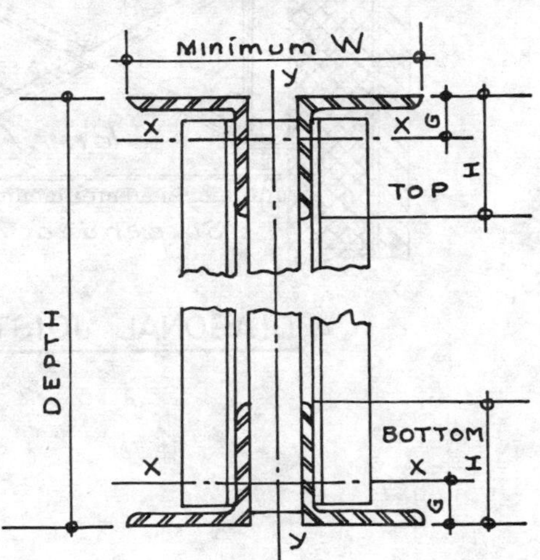

TABLE: Deep longspan series DLJ design properties 2.7.5.8

DEEP LONGSPAN JOISTS SERIES DLJ · DESIGN PROPERTIES									
JOIST DEPTHS 52" TO 62" INCLUSIVE					JOIST DEPTHS 64" TO 72" INCLUSIVE				
JOIST SERIES NUMBER	NORMAL DEPTH INCHES	MAX. END REACTION IN LBS.	SPAN RANGE IN FEET	MOMENT OF INERTIA"4	JOIST SERIES NUMBER	NORMAL DEPTH INCHES	MAX. END REACTION IN LBS.	SPAN RANGE IN FEET	MOMENT OF INERTIA"4
52DLJ12	52.0	12,867	89-104	4,413.0	64DLJ14	64.0	15,345	113-128	8,266.0
52DLJ13	52.0	15,243	89-104	5,228.0	64DLJ15	64.0	18,129	113-128	9,745.0
52DLJ14	52.0	16,943	89-104	5,623.0	64DLJ16	64.0	19,778	113-128	10,856.0
52DLJ15	52.0	18,695	89-104	6,332.0	64DLJ17	64.0	22,733	113-128	12,365.0
52DLJ16	52.0	22,551	89-104	7,598.0	64DLJ18	64.0	27,962	113-128	14,940.0
52DLJ17	52.0	25,465	89-104	8,693.0	64DLJ19	64.0	31,599	113-128	16,676.0
52DLJ18	52.0	29,545	89-104	9,683.0	64DLJ20	64.0	37,282	113-128	19,351.0
56DLJ13	56.0	14,405	97-112	5,871.0	68DLJ15	68.0	17,702	121-136	11,046.0
56DLJ14	56.0	16,554	97-112	6,561.0	68DLJ16	68.0	19,649	121-136	12,304.0
56DLJ15	56.0	18,556	97-112	7,388.0	68DLJ17	68.0	22,630	121-136	14,019.0
56DLJ16	56.0	21,975	97-112	8,867.0	68DLJ18	68.0	26,097	121-136	15,846.0
56DLJ17	56.0	24,856	97-112	10,146.0	68DLJ19	68.0	31,025	121-136	18,923.0
56DLJ18	56.0	29,007	97-112	11,309.0	68DLJ20	68.0	37,108	121-136	21,968.0
56DLJ19	56.0	32,718	97-112	12,609.0	72DLJ16	72.0	19,579	129-144	13,842.0
60DLJ14	60.0	16,167	105-120	7,573.0	72DLJ17	72.0	22,497	129-144	15,776.0
60DLJ15	60.0	18,438	105-120	8,526.0	72DLJ18	72.0	25,933	129-144	17,839.0
60DLJ16	60.0	21,450	105-120	10,235.0	72DLJ19	72.0	30,471	129-144	21,312.0
60DLJ17	60.0	24,250	105-120	11,712.0	72DLJ20	72.0	36,955	127-144	24,752.0
60DLJ18	60.0	28,477	105-120	13,061.0					
60DLJ19	60.0	32,175	105-120	14,571.0					

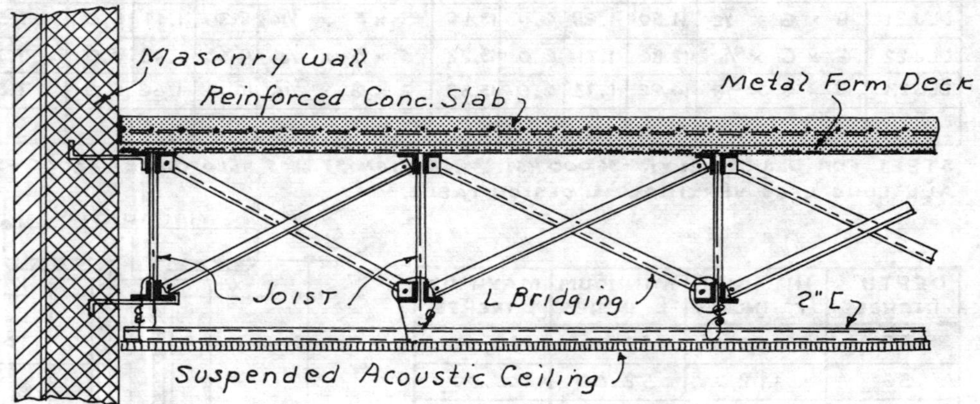

° DIAGONAL JOIST BRIDGING & WALL ANCHORAGE °

ILLUSTRATION: Open web joist accessories 2.7.6.1

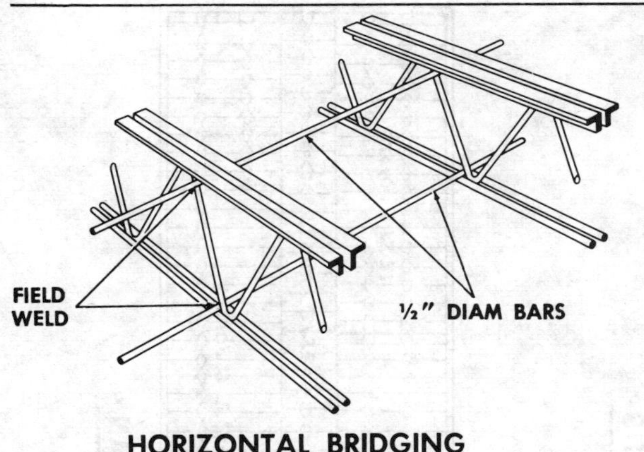

HORIZONTAL BRIDGING

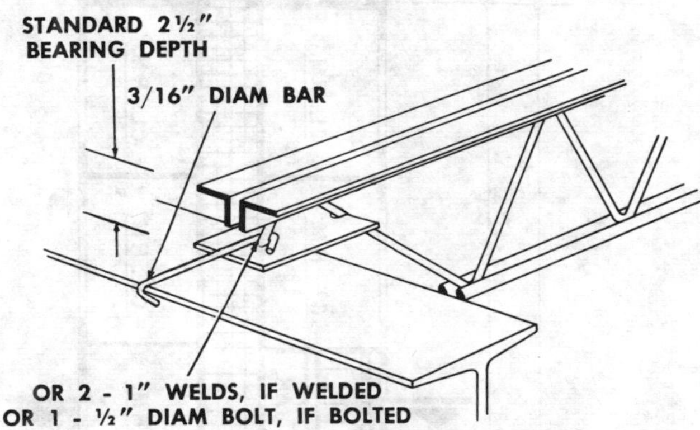

BEAM ANCHOR

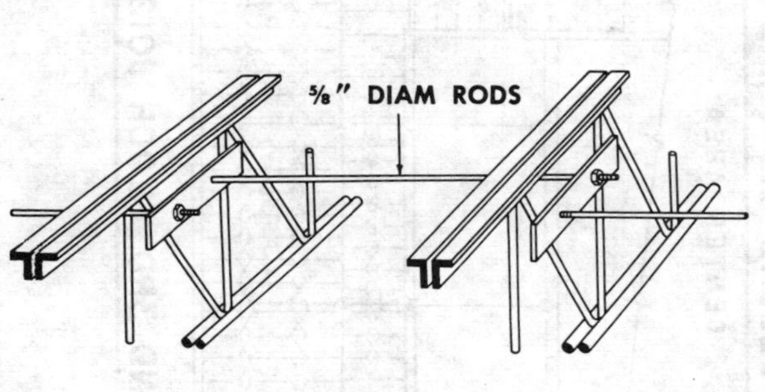

SAG RODS

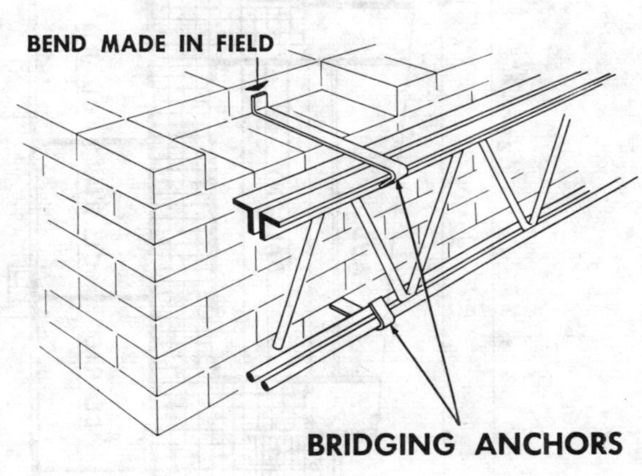

BRIDGING ANCHORS

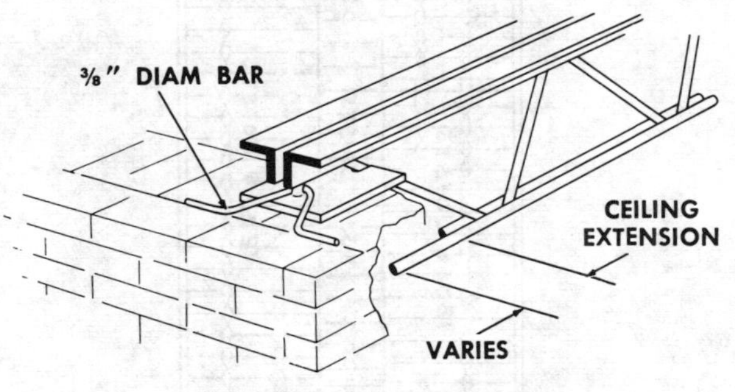

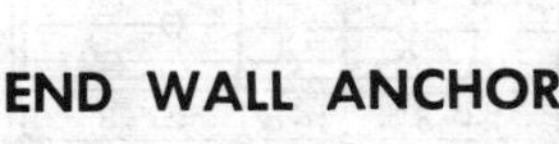

END WALL ANCHOR

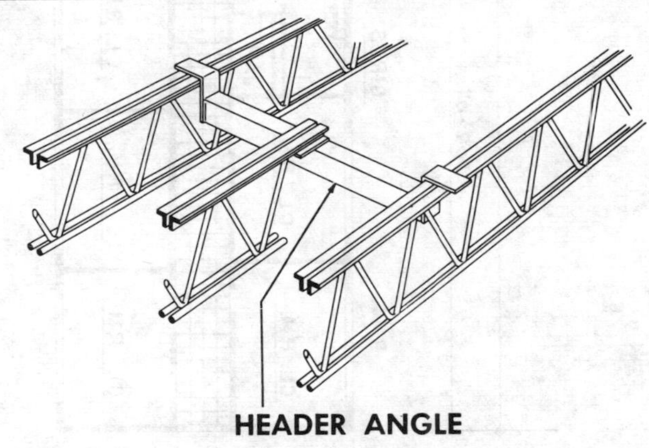

HEADER ANGLE

ILLUSTRATION: Joist plan for school building 2.7.6.2

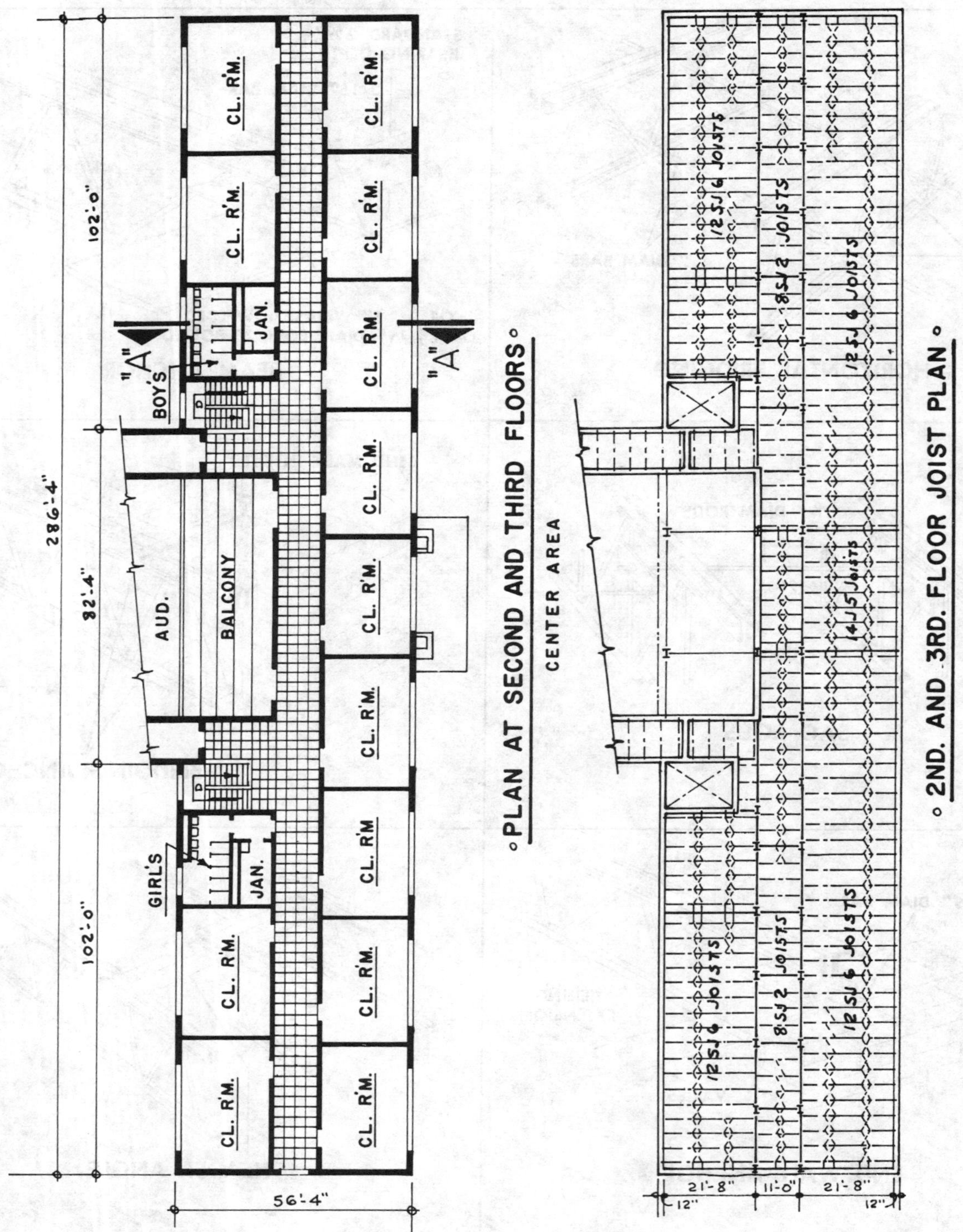

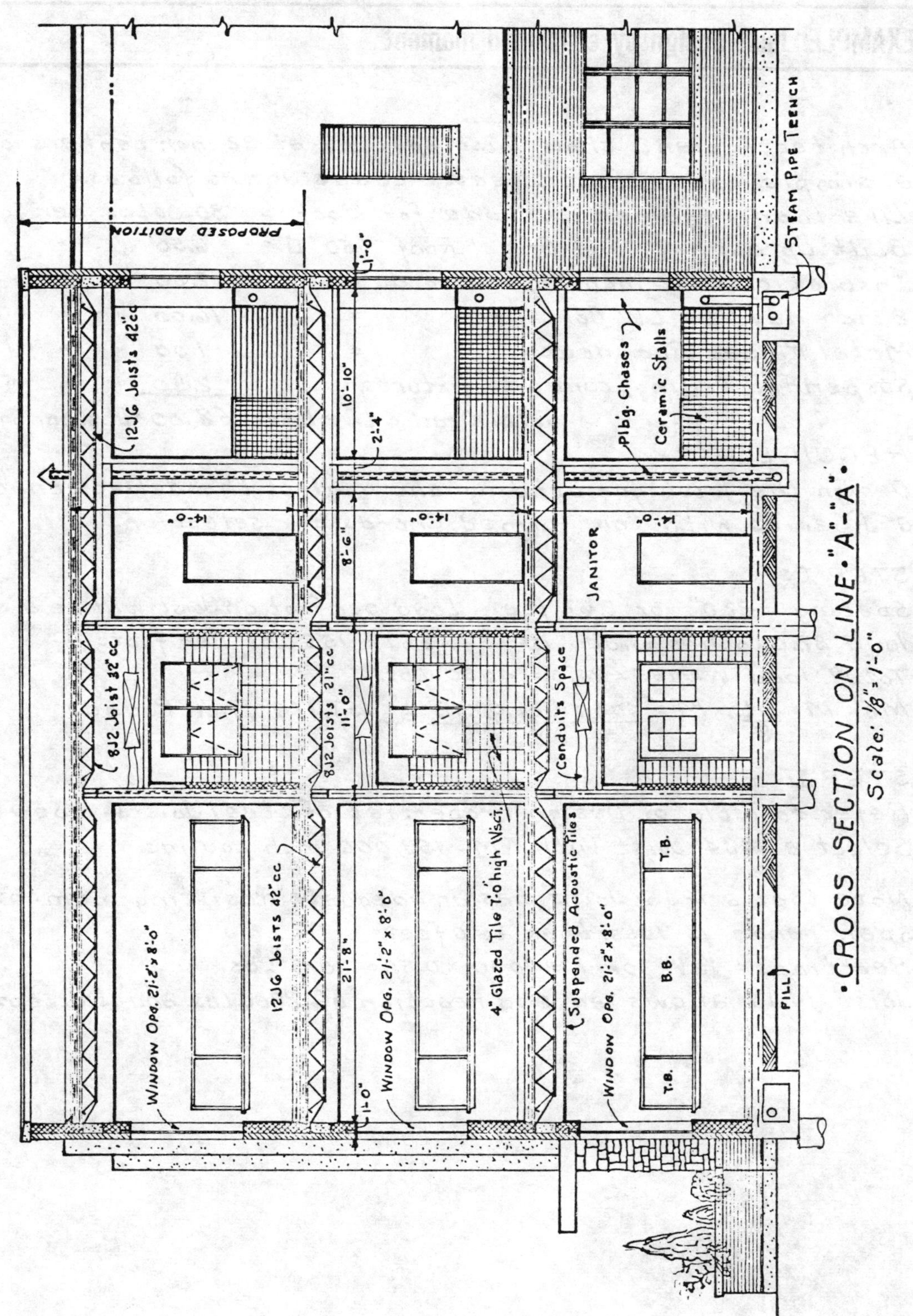

•CROSS SECTION ON LINE "A"-"A"•
Scale: 1/8" = 1'-0"

EXAMPLE: Joist design by strip load moment 2.7.7.1

Architects plans show joist spacing at 32 inch centers on a simple span of 26.0 feet. Loads are as follows:

Live Load required by Code for Roof	=	30.00 Lbs. per Sq. Foot.
Built Up Asphalt and Gravel Roof 650# □	=	6.50 " " " "
Insulation is 1.0 inch Fesco board:	=	2.00 " " " "
2 inch Haydite LW Conc. Deck	=	16.00 " " " "
Metal Ribbed Roof deck	=	1.00 " " " "
Suspended Ceiling, conduits, fixtures	=	2.50 " " " "
Total Dead & Live Load:		58.00 Lbs. per. Sq. Ft.

REQUIRED:

Design the joist for bending and neglect the deflection. Use a J-Series with Cold formed Chords for selection.

STEP I:

Spacing = 32.0" or 2.67 feet. Load per foot on Joist will be a strip load 32.0 inches wide. $\omega = 58 \times 2.67 = 155$ Lbs per foot.

Total load $W = 155 \times 26.0 = 4030$ Lbs.

Max. $M = \frac{WL}{8}$. $M = \frac{4030 \times 26.0 \times 12}{8} = 157{,}170$ Inch Pounds.

STEP II:

Refer to Table of Design Properties of Steel Joists - J Series.

Select a 14J4 Joist with RM = 159,000 inch pounds.

Note that a 12J5 Joist has an adequate Resisting moment but span range is less than 26.0 feet.

Reactions = ½W or $R = 4030 \times 0.50 = 2015$ Lbs.

Joist 14J4 allows an end reaction of 2800 Lbs. and is acceptable.

EXAMPLE: Joist spacing by strip load 2.7.7.2

The design load for a roof is 65.5 Pounds per square foot. Joist span is 22.0 feet and a J-Series joist with Cold Formed Chords is desired. Metal ribbed decking is to be applied and spacing should accomodate even length sheets of 12.0 to 40.0 feet with maximum joist spacing of 4.0 feet on centers.

REQUIRED:

Convert the Resisting Moment of a selected Joist into a 12 inch wide strip load for full length of span. Determine the spacing of joist to meet design load and maximum spacing allowed by deck strength.

STEP I:

From tables select for trial a Cold Formed Chord Joist 14J5. Joist has a resisting moment of 190,000 inch pounds. This figure will represent same as maximum bending moment as RM = M. Reduce RM to Foot Lbs. $M = \frac{190{,}000}{12} = 15{,}830$ Ft. Lbs.

For uniform loads, $M = \frac{wL^2}{8}$ and $w = \frac{8M}{L^2}$. Put values in the latter formula: $w = \frac{8 \times 15{,}830}{22.0 \times 22.0} = 262$ Pounds per lineal foot.

STEP II:

For spacing of joist, divide the lineal foot strip load by the square foot design load thus: $s = \frac{262}{65.5} = 4.00$ ft. centers.

EXAMPLE: Calculating joist resisting moment 2.7.7.3

Select from tables a 16J4 Joist with cold formed chords as follows: Top chord is section 4T and bottom chord is 4B. Also choose a 60 DLJ Series with top chord consisting of 2 angles 4x4x7/16, and bottom chord of 2 angles 4x4x3/8.

REQUIRED:

Calculate the Resisting moment produced by chords when F_b = 22,000 PSI. Take properties for angles from tables applicable to corresponding series. Expect final results for I and resisting moment to vary slightly from values listed in tables.

STEP I:

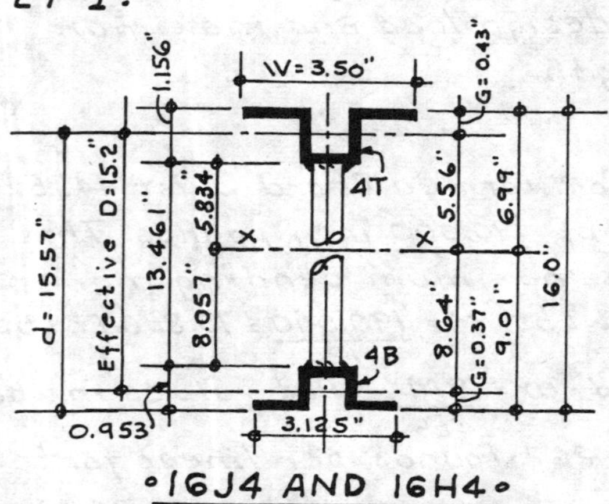

• 16J4 AND 16H4 •

• 60 DLJ 16 •

TOP: 2∟s 4x4x7/16" A = 6.62□"
BOT: 2∟s 4x4x3/8" A = 5.72□"
TOP ANGLES: I_o = 10.0"⁴
BOT ANGLES: I_o = 8.8"⁴

CALCULATIONS FOR 16J4 AND 16H4 JOIST							
M'K.	A□"	d"	Ad	ℓ"	Aℓ²	I_o	Aℓ²+I_o
4T	0.722	15.57	11.242	5.56	22.32	0.12	22.44
4B	0.544	0.37	0.201	8.64	40.61	0.06	40.67
	1.266□"		11.443				73.11"⁴

NA = $\frac{11.443}{1.266}$ = 9.01 Inches from Bot. I_x = 73.11"⁴

$S_x = \frac{I}{C}$ = 8.123" RESISTING MOMENT: $S_x F_b$

RM = 8.123 x 22,000 = 178,700 Inch Pounds. Series J.

STEP II:

CALCULATIONS FOR 60DLJ16 SERIES JOIST							
M'K.	AREA□"	d"	Ad	ℓ"	Aℓ²	I_o	Aℓ²+I_o
TOP ∟s	6.62	58.84	389.52	26.74	4733.5	10.0	4743.5
BOT. ∟s	5.72	1.14	6.52	30.96	5478.0	8.8	5486.8
	12.34		396.04				10,230.3"⁴

$S_x = \frac{10{,}230.3}{32.10}$ = 31.87"³

F_b = 22,000 PSI.

NEUTRAL AXIS = $\frac{396.04}{12.34}$ = 32.10 INCHES FROM BASE.

RESISTING MOMENT = 31.87 x 22,000 = 701,140 INCH POUNDS.

EXAMPLE: Joist deflection design 2.7.7.4

A span is 24.0 feet and design load is 48 Pounds per sq. foot. Ceiling below is suspended from lower chord of Joists and is plaster. Spacing or Joist designation is left for the designer to decide.

REQUIRED:
Design for best joist to meet conditions. Spacing of Joist should not be less than 2.0 feet nor over 4.0 feet. Deflection must be not over 1/360 of span length in inches. Keep in mind that open web joists will deflect 15% more than solid compact sections.

STEP I:
Determine max. deflection: L = 24.0 ℓ = 24.0 x 12 = 288 inches
Max Δ for plaster = $\frac{288}{360}$ = 0.80" When Joist deflects 15% over value obtained by deflection formula for solid beam sections, reduce the maximum at start thus:
For Joist design $\Delta = \frac{0.80}{1.15} = 0.696$ inches as maximum.

STEP II:
Determine a 12 inch wide strip load: 24.0 x 48 = 1152 Lbs.
Choose a spacing of 2.5 feet. W = 1152 x 2.5 = 2880 Lbs.
Formula: $\Delta = \frac{5W\ell^3}{384EI}$. All values may be determined for formula except the Moment of Inertia for joist. Thus the formula becomes: $I = \frac{5W\ell^3}{384E\Delta}$.

STEP III:
Solve for minimum property of I: W = 2880# ℓ = 288"
ℓ^3 = 23,887,872 E = 29,000,000 and Δ = 0.696 inches.
Put values in formula:

$$\text{Min. } I = \frac{5 \times 2880 \times 23{,}887{,}872}{384 \times 29{,}000{,}000 \times 0.696} = 44.5''^4 \text{ (Slide Rule OK)}$$

From Tables: A cold formed chord in Series J and H gives a Moment of Inertia for the 14J3 and 14H3 equal to 44.61"^{3}

STEP IV:
Calculate the bending moment to compare whether Series J or Series should be selected.
$M = \frac{WL}{8}$. $M = \frac{2880 \times 288}{8}$ = 103,750 inch Lbs.

A Series J will be satisfactory as a 14J3 has RM = 127,000"#.

EXAMPLE: Joist spacing by deflection 2.7.7.5

Preliminary plans call for joists to be spaced on 4.0 foot centers and span is 22.0 feet. Combined dead and live loads are 62.5 Pounds per square foot of floor area. Joists are to support a plaster ceiling below which will be suspended from lower chord. An alternate joist is to be substituted if the ceiling is changed to light weight tiles of mineral fibers.

REQUIRED:
Two Joist designations or sizes are to be designed. Solve for joists in J Series. Limit deflection to 1/360 of span when ceiling is plaster and use strip load Resisting Moment to design the alternate joist.

STEP I:
Solve for load on Joist: $L = 22.0$ ft. Spacing $= 4.0$ ft. Area supported on 1 joist $= 22.0 \times 4.0 = 88.0$ □' $W = 88.0 \times 62.5 = 5500$ Lbs.
For Bending: $M = \frac{WL}{8}$. $M = \frac{5500 \times 22.0 \times 12}{8} = 181{,}500$ inch pounds.
The joists Resisting must equal or exceed the moment above. From Table: Design Properties of Steel Joists - Series J, this will require a 12J6 Joist with RM = 196,000"# or could use a 14J5 Joist with Resisting Moment of 190,000 inch pounds.

STEP II:
Design for Joist when plaster ceiling is supported.
Open Web Joists deflect 15% more than compact beam sections.
Formula for deflection is: $\Delta = \left(\frac{5Wl^3}{384IE}\right) \times 1.15$ $\Delta = \frac{l}{360}$ limit
Max. $\Delta = \frac{22.0 \times 12}{360} = 0.733$ inches. If deflection in joist is 15% greater, then joist deflection must be limited to 0.637 in.
Solve for required Joist Moment of Inertia by transposing the formula thus: $I = \frac{5Wl^3}{384E\Delta}$. $l^3 = (22.0 \times 12)^3 = 18{,}399{,}744$
$W = 5500$# $E = 29{,}000{,}000$ and $\Delta = 0.637$ inches max. Put the values in formula: $I = \frac{5 \times 5500 \times 18{,}399{,}744}{384 \times 29{,}000{,}000 \times 0.637} = 71.50''^4$ (Minimum).

From Table of Hot Rolled Chords J and H Series Properties, the best selection is a 14J5 with an $I = 72.63''^4$
Using the 14J5 with Hot Rolled Chords will eliminate the need for an alternate size as required.

EXAMPLE: Calculating forces in joist members 2.7.7.6

A 70.0 foot span joist is required to support a uniform load of 280 Lbs., per lineal foot across entire span. Select a 40LH10 Joist and draw an elevation with the panel dimensions taken from table 2.7.5.4 and member sizes from table 2.7.5.5. Panels P= 63.0" or 5.25 feet.

REQUIRED:
Calculate the forces in chord and web members by using the moment method. Construct a graphic force diagram to check the results. Effective depth of joist is given as 38.28 inches.

STEP I:
The Moment of Inertia of a 40LH10 Joist is listed as: $I = 1882.0''^4$. Check the load by comparing the resisting moment when allowable $F_b = 30{,}000$ PSI. Calculations for location of neutral axis reveal that dimension c is located 24.0 inches from bottom of lower chord.
Then $S = \frac{I}{C}$ or $S = \frac{1882.0}{24.0} = 78.4''^3$ and the resisting moment is: $RM = \frac{78.4 \times 30{,}000}{12} = 196{,}000$ Foot Lbs. $M = \frac{WL}{8}$ or $W = \frac{8M}{L}$.

Then $W = \frac{8 \times 196{,}000}{70.0} = 22{,}400$ Lbs. Maximum allowable load per lineal foot = 22,400/70.0 = 320 Lbs. (ω). Joist will support 280 Lbs. load OK.

STEP II:
Joist elevation is drawn to scale with P= 5.25' and each load point is 5.25 × 280 = 1470 Lbs. W= 70.0 × 280 = 19,600 Lbs. End Reactions are 9,800 Lbs.

STEP III:
Identify each member in joist by using Bow's notations. At bottom chord there are 11 Panels @ 5.25 = 57.75 feet. Each end panel = $\frac{(70.0 - 57.75)}{2} = 6.125$ feet.

STEP IV:
To simplify drawing and shorten calculations convert loads on panel points to kip pounds as: $\frac{1470}{1000} = 1.47^{K\#}$
The effective reaction at supports is $9.31^{K\#}$. Use this value

EXAMPLE: Calculating forces in joist members, continued 2.7.7.6

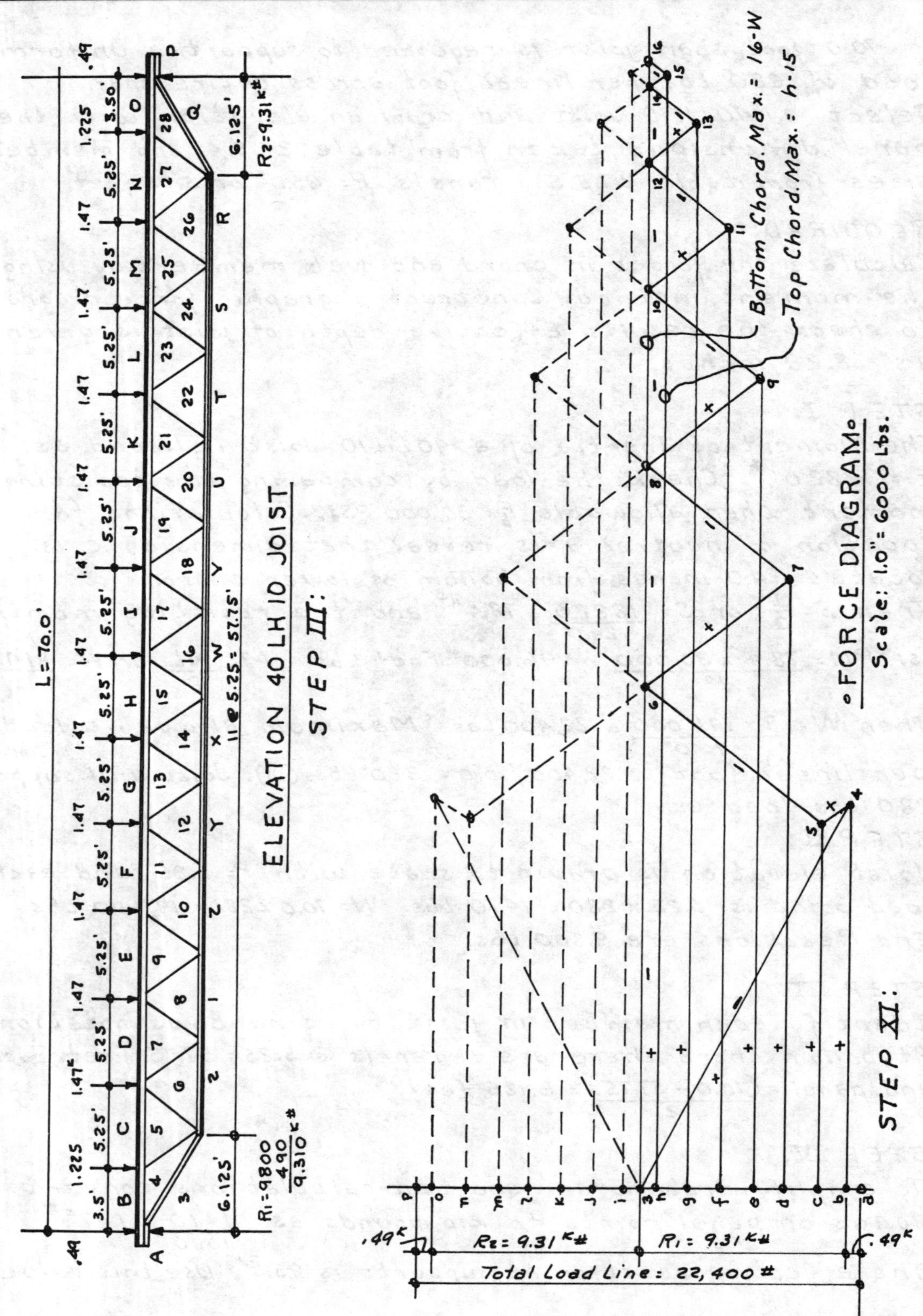

EXAMPLE: Calculating forces in joist members, continued 2.7.7.6

in calculating force in chords and web members. This reaction is 9,310 Lbs., or 9.31 kips.

STEP V:

Calculate the force in top chord by moment method where the effective depth is 38.28 inches. Reduced to suit equation it must be 3.20 feet. Maximum force is in members 15-H and 17-I. Moment distance from center of 15-H is: (5.25×5) + 6.125 or 32.375 feet. First load to left of moment point is 2.625 feet. Equation is written as follows:

$$H\text{-}15 = \frac{(9.31 \times 32.375) - \left[(1.47 \times 2.625) + (1.47 \times 5.875) + (1.47 \times 13.125) + (1.47 \times 18.375) + (1.47 \times 23.625) + (1.225 \times 28.875)\right]}{3.20} \text{ or}$$

$$H\text{-}15 = \frac{(301{,}410) - \left[(3775 + 11{,}740 + 19{,}275 + 27{,}000 + 34{,}760 + 34{,}175)\right]}{3.20} = +\,53{,}340 \text{ Lbs.}$$

Point of moment is middle of D-7 and distance from R_1 = 5.25 + 6.125. First load to left of moment point on top chord is 2.625 ft. Then,

$$D\text{-}7 = \frac{(9.31 \times 11.375) - \left[(1.47 \times 2.625) + (1.225 \times 7.875)\right]}{3.20} = +\,28{,}873 \text{ Lbs.}$$

STEP VI:

Bottom Chord member at midspan will have maximum force value. Midspan is 35.0 feet from R_1, and first load to left is 5.25 feet. D = 3.20'

$$W\text{-}16 = \frac{(9.31 \times 35.0) - \left[(1.47 \times 5.25) + (1.47 \times 10.50) + (1.47 \times 15.75) + (1.47 \times 21.0) + (1.47 \times 26.25) + (1.225 \times 31.5)\right]}{3.20} \text{ or}$$

$$W\text{-}16 = \frac{325{,}850 - (8718 + 15{,}435 + 23{,}152 + 30{,}870 + 38{,}588 + 38{,}587)}{3.20} = -\,53{,}280 \text{ Lbs}$$

STEP VII:

Web members are all at an angle with direction in which the load forces act. Member 3-4 near R_1 has 2 sides known. Side a: 3.20' and b: 6.125'. Tangent of angle at top chord will be angle B and $= \frac{b}{a}$ or Tan B: $\frac{6.125}{3.20} = 1.9140$ From trig tables angle B: 62° 25' and $R_1 = 9.31^k$

Secant of B times vertical equals diagonal side c and is member 3-4. Secant 62° 25' = 2.1596 (Call it 2.16).

Force in members 3-4 and 28-Q = R_1 Secant B.

Web 3-4 and 28-Q = 9,310 × 2.16 = 20,110 Lbs. Tension force.

EXAMPLE: Calculating forces in joist members, continued 2.7.7.6

STEP VIII:

Force in top chord B-4 and web member 4-5.
B-4 is a horizontal member and applies with angle A.
Angle B = 62° 25' then A = 90° − B or A = 27° 35'
Force in member 3-4 represents triangle side c and was calculated to be 20,110 Lbs. When B-4 represents side b, the force is equal to c Cosine A. Cosine of 27° 35' = 0.88634
Force in B-4 and 0-28 = 20,110 × 0.88634 = 17,730 Lbs. (Compressive).

Web member 4-5 and 27-28 are usually made vertical in Joist Series LA and LH. Better load balance is obtaind when slanted as in the case of the DLJ Series. The force action line is vertical and force is 1,225 Lbs. Make a special note to examine this member when drawing the force diagram.
Creating a triangle, side a = 6.125 − 3.50 = 2.625' b = 3.20' Angle A will be at top. Tan. A = $\frac{2.625}{3.20}$ = 0.8203 Angle A = 39° 22'
Secant of A = 1.2935 and Force in 4-5 = 1225 × 1.2935 = 1585 Lbs.

STEP IX:

Top chord C-5 and N-27:
Force in 4-5 is known. Represents side c of triangle and equal to 1585 Lbs. Vertical side b is force equal to 1225 Lbs.
Solving for horizontal side (a) then deduct from force in B-4.
By trig.: a = c Sin. A. Angle A was found to be 39° 22' and the Sine A = 0.63428 and Tan A = 0.8203 Force in B-4 = 17,730 Lbs.
Force in chord C-5 = 17,730 − (1585 × 0.63428) = 16,725 Lbs. (Compressive)

STEP X:

Balance of Web members all have identical angles are the same on each side symmetrical with Joist midspan.
Each panel will be split into two right angles with angle at top becoming angle A. Side b = 3.20 side a = 2.625 and side c = Secant A times b. Vertical forces in direction of side b.
To find angle A: Tan A = $\frac{a}{b}$ or Tan A = $\frac{2.625}{3.20}$ = 0.82031 A = 39° 22'
Side c = Sec. A × b. Beginning at left with Reaction R_1, each load is deducted and equations are written thus: Sec. A = 1.29

Web Member 5-6 = (9310 − 1225) × 1.29 = − 10,430 Lbs.

EXAMPLE: Calculating forces in joist members, continued 2.7.7.6

Forces in Web Members:

Web 6-7 = [9310-(1225+1470)] x 1.29 = + 8,533 Lbs.

Web 7-8 = [9310-(1225+1470)] x 1.29 = - 8,533 "

Web 8-9 = [9310-(1225+1470+1470)] x 1.29 = + 7,025 Lbs

Web 9-10 = Same as 8-9 except tension - 7,025 "

Web 10-11 = [9310-(1225+1470+1470+1470)] x 1.29 = + 6,630 "

Web 11-12 = Same as 10-11 except tension - 6,630 "

Web 12-13 = [9310-(1225+1470+1470+1470+1470)] x 1.29 = + 2,845 Lbs

Web 13-14 = Same as 12-13 except tension - 2,845 "

Web 14-15 = [9310-(1225+1470+1470+1470+1470+1470)] x 1.29 = + 950 Lbs.

Web 15-16 = Same as 14-15 except tension - 950 Lbs

Web 16-17 = Same as 15-16 due to symmetry. - 950 Lbs.

STEP XI:

Drawing Force diagram to check results obtained by method of moments. Deducting the loads in corresponding sequence should also check with equations derived in Step X.

STEP XII:

Investigate top chord for lateral bracing or bridging. From Table of Design Properties for Long Span Series LH: Angles in Top Chord are: 2Ls 3x2½x5/16 with long legs vertical. Radius of gyration shall be effective on axis y-y (vertical). From AISC Manual: Property of these 2Ls on axis y-y with 3/8 inch spacing will be taken. $r_y = 1.14$ and $A = 3.24 \square''$

Ratio of $\frac{l}{r}$ must not exceed 200. Maximum unbraced length equals $\frac{200}{r_y}$ or $l = \frac{200}{1.14} = 175.5$ inches, or $\frac{175.5}{12} = 14.6$ feet.

Use diagonal bridging and space as following by beginning at left end and installing bridge rows at following points. Place first row at 8.75 feet. 2nd. row at 19.25', 3rd. row at 29.75' 4th. row at 40.25', 5th. row at 50.75' and last row at 61.25 feet. Maximum bridging will be spaced at 10.50 feet.

EXAMPLE: Calculating forces in joist members, continued 2.7.7.6

STEP XIII:

Check compressive stress in top chord angles. Maximum bridging and bracing length on axis y-y is 10.50 feet, and on axis x-x the panel length is 5.25 feet.

Area 2Ls = 3.24 □" r_y = 1.14 l = 10.50 x 12 = 126.0 inches.

Slenderness ratio $\frac{l}{r} = \frac{126.0}{1.14} = 101.0$ From Tables of allowable unit stress for Steel Columns F_c = 11,489 PSI. Maximum force in Top Chord was H-15 and calculated in Step V. P = +53,340 Lbs.

Actual $f_c = \frac{P}{A}$ or $f_c = \frac{53,340}{3.24} = 16,450$ PSI. Actual stress at midspan exceeds allowable and can be solved by installing another row of bridging to reduce l. Other values may be checked from from force diagram thus:

Max. P = 11,489 x 3.24 = 37,345 Lbs. Force in Top Chord e-9 scales approximately 36,750 Lbs. Then from this point to midspan add a horizontal angle brace.

STEP XIV:

Stress in bottom chord is considered direct axial tension. Tension T is greatest a W-16 or −53,280 Lbs.

Maximum stress in bottom chord = $\frac{T}{A}$. Lower chord angles are varied in Joist LH-10. Allowable tension in H Series Joists is F_t = 30,000 PSI.

Required Area for Chord: $A = \frac{P}{F_t}$ or $A = \frac{53,280}{30,000} = 1.78$ Sq. In.

Accept the lesser size: 2Ls $2\frac{1}{2} \times 2 \times \frac{1}{4}$ with A = 2.12 □"

EXAMPLE: Joist bridging design 2.7.7.7

The preceding example dealt with the calculations of forces in a long span Joist Series 40LH10. Lateral bracing in top compressive chord was based on a radius of y-y axis. Space between long legs of angles was assumed to be 3/8 inches which would permit flat bars to be used as tension members in web and separate chord Ls. Compression web members can be of other shapes.

Top Chord Angles in Joist 40LH10 consist of 2 - 3x2½x5/16 with long legs vertical and back to back.

REQUIRED:

Calculate the radius of gyration about the vertical axis of the 2Ls 3x2½x5/16 with a space of 3/8 inches (0.375") between the long legs. Assume the slenderness ratio to be 120 and compute the allowable stress by column formula. Give maximum length for bridging when no other lateral support is considered.

STEP I:

Angles are symmetrical about 3/8 inch gusset space. A sketch will be made to delineate the gauge of minor axes.

STEP II:

From tables of Shapes the properties are for 1 Angle.

$A = 1.62 \square''$ $I_y = 0.898''^4$ Gage $x = 0.683''$

$I = Al^2 + I_o$. $l = 0.683 + (0.375 \times 0.50) = 0.8705''$

$l^2 = 0.7578$ $Al^2 = 1.62 \times 0.7578 = 1.2276$

About vertical axis for 2Ls $I_y = Al^2 + I_o$

$I_y = 2 \times (1.2276 + 0.898) = 4.25''^4$ $r_y = \sqrt{\frac{I}{A}}$

$r_y = \sqrt{\frac{4.25}{2 \times 1.62}} = 1.15''$

STEP II:

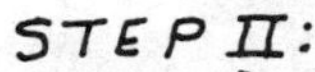

When $\frac{l}{r_y} = 120$ $l = 120 \times 1.15 = 138.0''$ or 11.50 feet max. unsupported lateral length. From allowable compressive stress tables for unbraced columns $F_c = 10{,}000$ PSI.

Maximum force in top chord at maximum bridging length of 11.50 feet is F_cA or $P = 10{,}000 \times (2 \times 1.62) = 32{,}400$ Lbs. (Axial).

Steel ribbed roof deck and slab forms 2.8

Flat metal sheets in several gages, rolled into a corrugated or ribbed profile, have become the most economical material for roof systems and slab forms. Special ribbed sheets are available for roof decks with span lengths up to 40.0 feet. The common roof decks have a depth of 1½ to 3 inches and are classified as type A, B, N and 415B. Steel ribbed or corrugated sheets for slab forms are usually produced with a depth of ½ inch to 1½ inches.

The first use of steel decking was common corrugated iron sheet, which was used for wall siding and roof panels. Manufacturers realized that a steel ribbed system could be made economically feasible for roof structures, and began the necessary tooling to roll sheets with greater depths and various profiles. With greater rib depth and heavy gage steel, the Moment of Inertia can be increased to provide sufficient strength for special design requirements.

The properties of panel types A and B vary only slightly, because each producer of deck material complies with the standards established by the Metal Roof Deck Technical Institute. This organization is the recommended authority for design; building codes have not been updated to include this stipulation. A table of fire ratings is provided to assist designers and architects in the selection of concrete and insulation materials, which will be supported by the decking.

DECK FINISHES

Steel ribbed panels for roof decks and slab forms are available in various finishes. Slab-form deck panels are generally of black, uncoated steel or galvanized. A few producers offer the panels with a sprayed coating which resists the corrosive action of chemicals and concrete admixtures.

Roof deck panels are available in galvanized and coated finishes. If a galvanized finish is specified it should contain not less than a 1.25 ounces of zinc coating per square foot. Galvanizing should be specified to be accomplished by the hot-dipped method and to comply with ASTM Specification A361–59T.

Designers and architects must remember that concrete specifications should forbid the use of calcium chloride or any admixture containing chloride salts when a galvanized deck is used to form the slabs.

Roof panels with coated or painted finishes should receive a chemical bath to remove all grease and scale, then a phosphate coating. The final finish of a rust-inhibitive synthetic enamel primer should be baked for a period of 20 minutes at a minimum temperature of 350°F.

STEEL DECK MATERIAL

Rolled ribbed deck panels for roofs and slabs are produced in commercial grade and high strength steels with yield points of 30,000 to 90,000 PSI. A tough-temper steel with a minimum yield strength of 80,000 to 95,000 PSI is available under ASTM specification A446–60T Grade E. The yield strength varies with the thickness: gages 20 to 24, 80,000 PSI; gage 26, 90,000 PSI; and standard gages in corrugated, 95,000 PSI. Steel decks Type A and B are generally formed from ASTM A245 Grade C and A366 commercial grade, with a yield strength of 33,000 PSI. Galvanized steel conforms to ASTM A–93.

ERECTING DECK PANELS

Designers may select from several methods of fastening the deck to the supports. Spot welding is the most economical and fastest method, but also hazardous and prohibited in plants which produce flam-

Steel ribbed roof deck and slab forms, continued 2.8

mable liquids, acids, solvents and gas. Electric arc welding can be used; each weld should be made through a welding washer. The mechanical clip method is not often used since the system does not have the rigidity necessary for continuous span design. A recent installation was accomplished in a very rapid manner using a new type of bolt. The bolt consists of an integral drill end on a type B self-tapping hex-head bolt. In a single operation using an air-driven nut tightener, the hole is bored and the bolt screwed in place.

Wood plywood or planks should be placed as walkways over the decking while roofing material and insulation panels are stacked and installed. Adequate protection must be provided to avoid disfiguring the ribs. A ribbed deck provides a pleasing and uniform ceiling for commercial and industrial structures.

METAL DECK SPECIFICATIONS

The architect is usually responsible for writing the specifications for building materials. Unfortunately, many design engineers often fail to provide all the necessary design requirements to obtain the proper material. Contractors then may submit bids on deck panels which are not compatible with the design. To eliminate the confusion in bids, the structural drawings should show deck type, gage of metal, minimum yield strength, finish, depth and properties of I or S. With all the design data written into the specifications, the contractor is able to confer with the manufacturer and submit acceptable quotations.

FIRE RATING DESIGN

In most building projects it is the responsibility of the architect to furnish the structural engineer with the composition of the roof structure and the fire-resisting rating required by code or zoning regulations. For steel roof deck construction, the materials must be of an approved type to classify the system as noncombustible. An hourly fire-rated roof structure requires that the design engineer calculate the dead loads, and from these the bending moment and deflection limits.

Insurance rating bureaus throughout the country are now including in their requirements the specifications of Underwriters Laboratories, and Factory Mutual Laboratories to qualify steel roof deck for classification under established fire rates. A series of tests conducted by Underwriters Laboratories and the National Bureau of Standards has been accepted as authority for ratings. See Table 2.8.1.2 for fire-rated conditions.

Steel deck design 2.8.1

Before a metal deck for roof or slab is installed, the supporting joists, purlins or beams must have all the bridging and lateral bracing in place. Deck panels are not considered as a substitute for lateral bracing of joists or purlins. If a deck panel is to be classified as a continuous span, the connections to the supporting joists must be rigid, such as welding or bolting. Panels installed with clips *do not* provide a rigid connection at each support. Since ribbed deck panels are produced in lengths up to forty feet, one sheet may cover from ten to twenty spans. With this beam continuity the Steel Deck Institute and the manufacturers provide design data which may depart from conventional deflection formulas. Catalogues furnished by the producers list various types of decks and provide load tables. Many special types are available; the general design principles remain the same for each type.

Roof panels with deep ribs are generally permanent and supply the main support for the roof materials and live loads. Metal corrugated or ribbed panels used for slab forms are considered as temporary support for wet concrete, workmen and portable equipment. The live load is not considered in the design of forms. Deflection in slab forms is a factor in the design, and shoring of the span is not recommended. A floor slab is trowelled smooth with a level finish over the whole area. When the form shoring is removed while the concrete is curing, before the concrete has reached full strength, the slab will sag. The top surface will then show this sag as depressions between the supports.

DESIGNING WITH LOAD TABLES

Suppliers of metal deck and form materials provide brochures which show deck profiles, section properties, finishes, gage thickness and allowable unit stresses. Design properties will have only slight variances since the I value depends upon depth and profile. The specifications of the metal in panels will range from 40,000 to 90,000 PSI minimum yield strength. Design working stress will vary from 18,000 to 30,000 PSI in bending. After selecting a deck material from a catalog, the designer should list the properties, gage, type and design stress in the specifications. To preclude the chance of obtaining another product with less strength, several comparable producers can be named in specifications, followed by the properties and gages which are acceptable. In addition to the technical data listed in catalogs, load tables will usually be provided. It is wise to investigate the approach used to determine the load. If the allowed load shown in table has been calculated on the basis of bending, then there is a possibility that deflection will be more than the design allows. An example which follows will show how it is possible to confuse bidders who will be furnishing a material produced by several manufacturers.

Deck and form design formulas 2.8.1.1

Steel ribbed deck manufacturers who subscribe to the design criteria of the Metal Deck Technical Institute recommend certain formulas be used for calculating bending moment and deflection. Designers should take note of the fact that the deflection formula is not the conventional equation used for steel girders, beams and joists.

TYPE SPAN	BENDING MOMENT- FT.LBS.	ROOF DECKS DEFLECTION-INCHES	SLAB FORMS DEFLECTION- INCHES
SINGLE	$M = \frac{wL^2}{8}$	$\Delta = \frac{5w\ell^4}{384EI}$	$\Delta = \frac{0.0130w\ell^4}{EI}$
DOUBLE	$M = \frac{wL^2}{8}$	$\Delta = \frac{3w\ell^4}{384EI}$	$\Delta = \frac{0.0054w\ell^4}{EI}$
TRIPLE	$M = \frac{wL^2}{10}$	$\Delta = \frac{3w\ell^4}{384EI}$	$\Delta = \frac{0.0068w\ell^4}{EI}$

w = LOAD PER SQUARE FOOT OR LINEAL FOOT LOAD ON 12" STRIP-LBS.
L = SPAN LENGTH IN FEET. E = 29,000,000 P.S.I.
ℓ = LENGTH OF SPAN IN INCHES. $\ell = 12L$
M = BENDING MOMENT IN FOOT LBS.
Δ = MAXIMUM DEFLECTION AT MIDSPAN, IN INCHES.

TABLE: Fire ratings for roofs and floor slabs 2.8.1.2

ROOF DECKS — DESIGNING FOR FIRE RATINGS

RATED HOURS	TYPE CONSTRUCTION ROOF DECK COMPOSITION	SUSPENDED TYPE CEILING MATERIAL	TEST AUTHORITY DESIGN REFERENCE
1½	MINIMUM 1 INCH RIGID WOOD FIBER INSULATION BOARD	3/4 INCH - CEMENT OR GYP. PLASTER - FIRE RES.	N.B.S. TEST 57 JAN. 15, 1946
2	MINIMUM 1 INCH MINERAL FIBER BOARD INSULATION FESCO, FOAMGLAS, ASBESTOS.	7/8 INCH PLASTER ON METAL LATH - GYP. CEMENT - PERLITE	U. L. R. 3996-3 AUG. 20, 1962
2	MINIMUM 1 INCH INSULATION CEMENT MIXED FIBERS OF WOOD, CANE OR MINERALS	3/4 INCH PERLITE OR GYPSUM PLASTER ON METAL LATHE	N. B. S. TEST 57 JAN. 15, 1946
3	SAME AS TYPE ABOVE	MIN. 1 INCH PLASTER AS ABOVE WITH LATHE	REPORT NO. TR 10235-2FP 2688
3½	MIN. 1¾ INCH INSULATION WITH CEMENT BONDED WOOD, GLASS, CANE, ROCK OR MINERAL FIBERS. PORTLAND CEMENT RIGID TYPE	MIN. 1 INCH PLASTER PERLITE OR GYPSUM ON METAL LATH OR 1 IN. FIRE RES. MINERAL	REPORT NO. TR 10235-2FP 2688
4	MIN. 2 INCHES OF HAYDITE OR LIGHTWEIGHT CONCRETE, OR AN EQUALLY RATED RIGID BOARD INSULATED NON-COMBUSTIBLE.	MIN. 1 INCH PLASTER CEMENT PERLITE OR VERMICULITE ON METAL LATH	N.A.B. TEST 60 JAN. 30, 1949

FLOOR SLABS — DESIGNING FOR FIRE RATINGS

RATED HOURS	SLAB DEPTH - MATERIAL TYPE CONCRETE STRENGTH - FORMS	CONDITION BELOW METAL FORM OR SLAB	TEST AUTHORITY DESIGN REFERENCE
1	4½ INCH REINFORCED 3000 LB. CONCRETE SLAB ON METAL DECK FORM OR WOOD REMOVED	NO CEILING - EXPOSED DECK WITH SUPPORT BEAMS FIREPROOFED	U.L.R. DESIGN NO. 3 NO. 3413-5 MAY 4, 1954
2	5¼ INCH REINFORCED 3000 LB. CONC. SLAB WITH METAL FORM	EXPOSED METAL FORM BEAMS FIREPROOFED	U.L.R. DESIGN NO. 3 NO. 3413-10 AUG. 15, 1956
3	MIN. 4½ INCH REINFORCED 3000 PSI CONCRETE SLAB ON METAL RIBBED FORM.	7/8 INCH MINIMUM PLASTER ON METAL LATH OR EQUALLY RATED CEILING PANELS	U.L.R. DESIGN NO. 5 NO. 3313-4 JULY 1, 1953
3	MIN. 4½ INCH LIGHTWEIGHT OR HAYDITE CONCRETE. REINFORCED. SLAB ON METAL RIBBED FORM.	EXPOSED METAL DECK ON UNDERSIDE WITH FIREPROOFED SUPPORT	U.L.R. DESIGN NO. 7 NO. 3413-9 SEPT. 13. 55
4	MIN. 4½ INCH 3500 P.S.I CONCRETE SLAB. REINFORCED TO SUSTAIN ALL LOADS - PLACED WITH METAL RIBBED FORMS	7/8 INCH MINIMUM OF SPRAYED FIBER INSULATION ON DECK BEAMS FIREPROOFED	U.L.R. DESIGN NO. 10 NO. 3372-2 NO DATE
4	MIN. 4½ INCH REINFORCED SLAB HARD CONCRETE ON METAL FORM RIBBED DECK. CONCRETE SUPPORT OR FIREPROOFED BEAMS	1 INCH CEMENT OR GYPSUM PLASTER ON SUSPENSION TYPE LATHED CEILING	DESIGN NO. 7 U.L.R. 3413-1 MARCH 7, 1952

TABLE: Metal deck properties 2.8.1.3

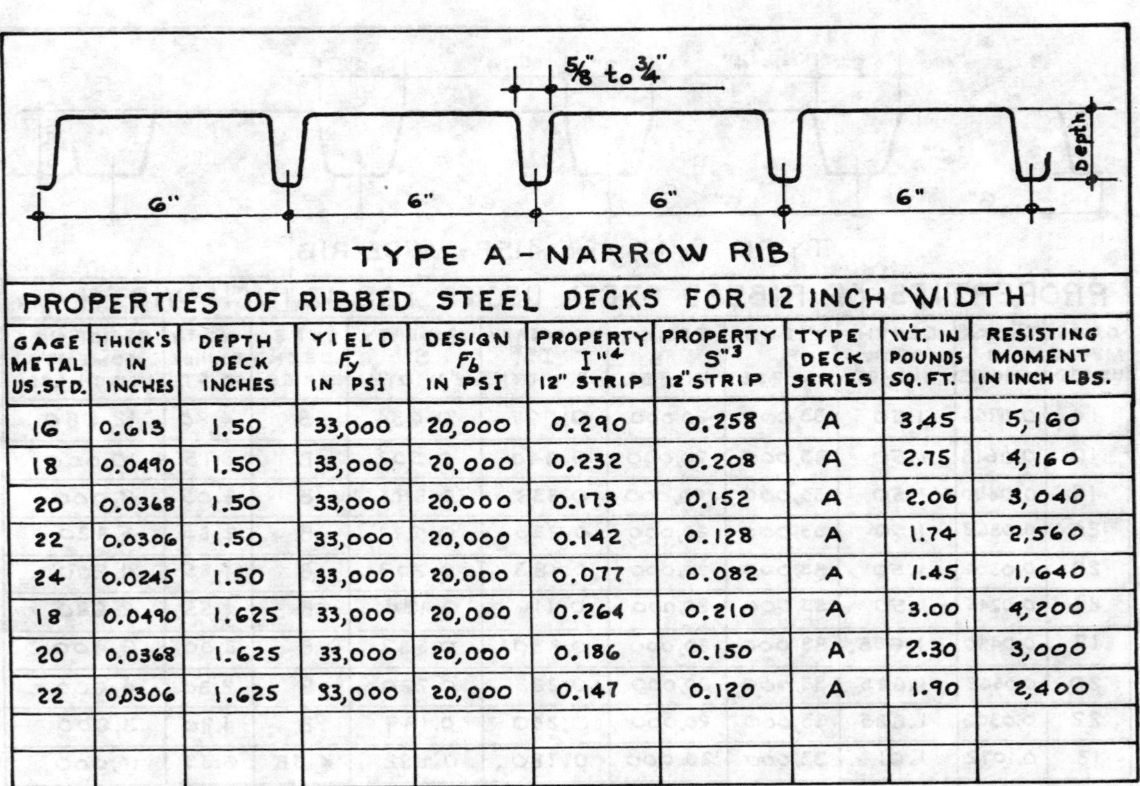

PROPERTIES OF RIBBED STEEL DECKS FOR 12 INCH WIDTH

GAGE METAL US.STD.	THICK'S IN INCHES	DEPTH DECK INCHES	YIELD F_y IN PSI	DESIGN F_b IN PSI	PROPERTY I"4 12" STRIP	PROPERTY S"3 12" STRIP	TYPE DECK SERIES	WT. IN POUNDS SQ. FT.	RESISTING MOMENT IN INCH LBS.
16	0.613	1.50	33,000	20,000	0.290	0.258	A	3.45	5,160
18	0.0490	1.50	33,000	20,000	0.232	0.208	A	2.75	4,160
20	0.0368	1.50	33,000	20,000	0.173	0.152	A	2.06	3,040
22	0.0306	1.50	33,000	20,000	0.142	0.128	A	1.74	2,560
24	0.0245	1.50	33,000	20,000	0.077	0.082	A	1.45	1,640
18	0.0490	1.625	33,000	20,000	0.264	0.210	A	3.00	4,200
20	0.0368	1.625	33,000	20,000	0.186	0.150	A	2.30	3,000
22	0.0306	1.625	33,000	20,000	0.147	0.120	A	1.90	2,400

AVAILABLE IN COATED OR GALVANIZED FINISH AND MINMUM YIELD STEELS OF F_y = 30,000 TO 33,000 PSI. USE F_b = 18,000 OR 20,000 PSI FOR RESISTING MOMENT.

TABLE: Metal deck properties, continued 2.8.1.3

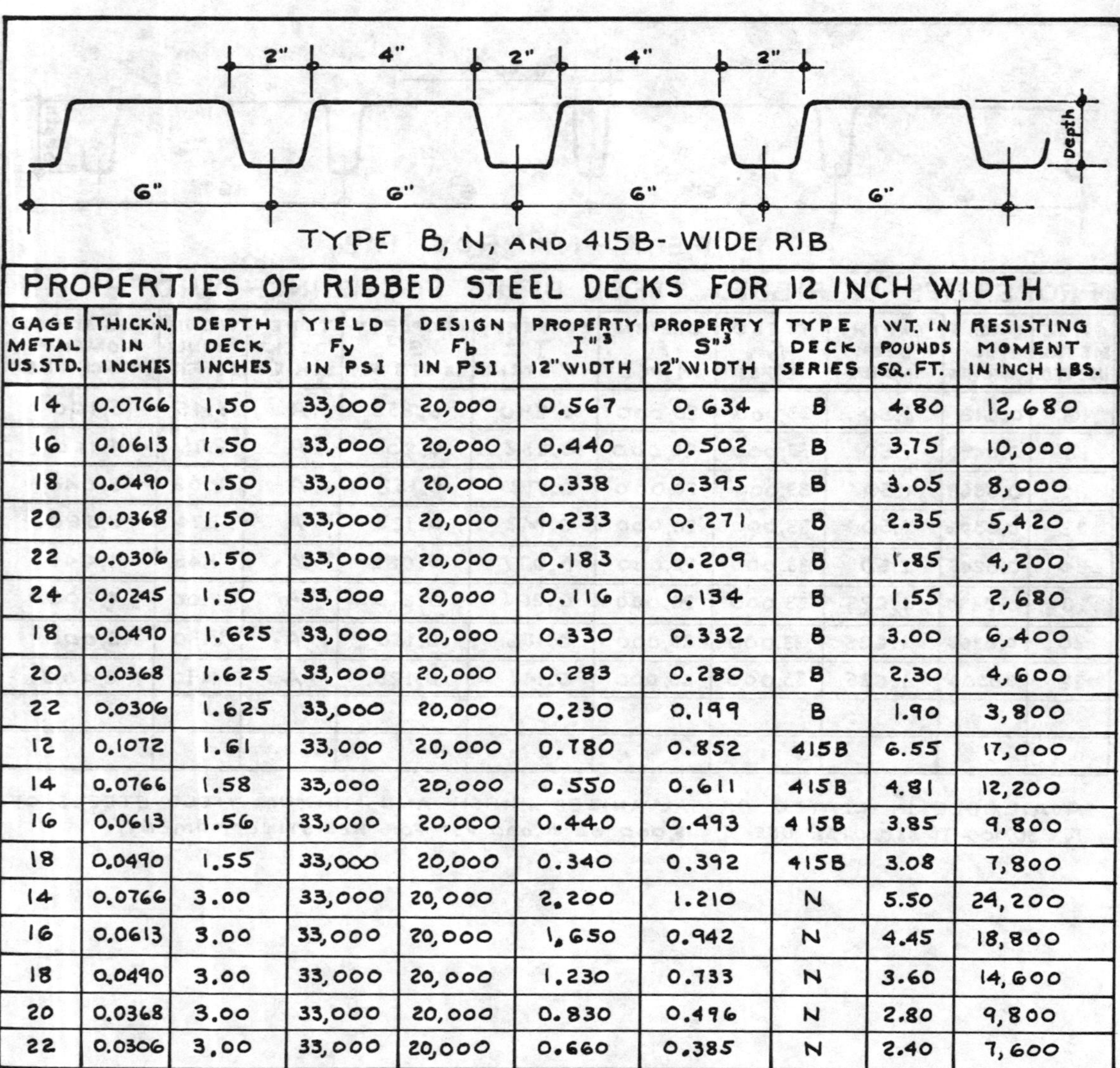

PROPERTIES OF RIBBED STEEL DECKS FOR 12 INCH WIDTH

GAGE METAL US.STD.	THICK'N. IN INCHES	DEPTH DECK INCHES	YIELD Fy IN PSI	DESIGN Fb IN PSI	PROPERTY I"3 12" WIDTH	PROPERTY S"3 12" WIDTH	TYPE DECK SERIES	WT. IN POUNDS SQ. FT.	RESISTING MOMENT IN INCH LBS.
14	0.0766	1.50	33,000	20,000	0.567	0.634	B	4.80	12,680
16	0.0613	1.50	33,000	20,000	0.440	0.502	B	3.75	10,000
18	0.0490	1.50	33,000	20,000	0.338	0.395	B	3.05	8,000
20	0.0368	1.50	33,000	20,000	0.233	0.271	B	2.35	5,420
22	0.0306	1.50	33,000	20,000	0.183	0.209	B	1.85	4,200
24	0.0245	1.50	33,000	20,000	0.116	0.134	B	1.55	2,680
18	0.0490	1.625	33,000	20,000	0.330	0.332	B	3.00	6,400
20	0.0368	1.625	33,000	20,000	0.283	0.280	B	2.30	4,600
22	0.0306	1.625	33,000	20,000	0.230	0.199	B	1.90	3,800
12	0.1072	1.61	33,000	20,000	0.780	0.852	415B	6.55	17,000
14	0.0766	1.58	33,000	20,000	0.550	0.611	415B	4.81	12,200
16	0.0613	1.56	33,000	20,000	0.440	0.493	415B	3.85	9,800
18	0.0490	1.55	33,000	20,000	0.340	0.392	415B	3.08	7,800
14	0.0766	3.00	33,000	20,000	2.200	1.210	N	5.50	24,200
16	0.0613	3.00	33,000	20,000	1.650	0.942	N	4.45	18,800
18	0.0490	3.00	33,000	20,000	1.230	0.733	N	3.60	14,600
20	0.0368	3.00	33,000	20,000	0.830	0.496	N	2.80	9,800
22	0.0306	3.00	33,000	20,000	0.660	0.385	N	2.40	7,600

ALL TYPE B DECKS AVAILABLE IN COATED OR GALVANIZED FINISH.
TYPE 415-B AVAILABLE IN GALVANIZED FINISH ONLY
TYPE N DECK AVAILABLE IN COATED OR GALVANIZED FINISH EXCEPT 14 GAGE WHICH IS AVAILABLE IN GALVANIZED ONLY.

TABLE: Metal slab form properties 2.8.1.4

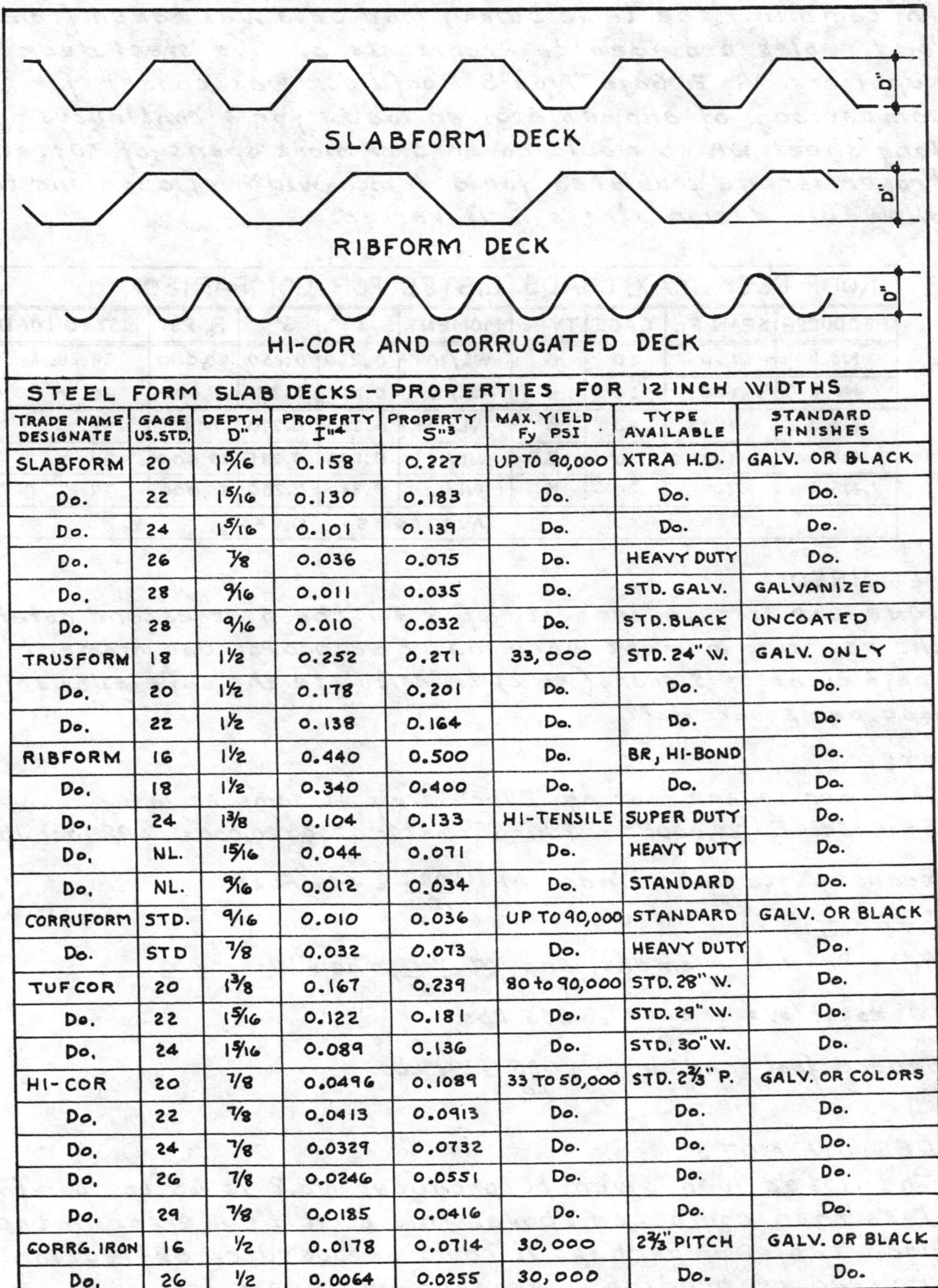

STEEL FORM SLAB DECKS - PROPERTIES FOR 12 INCH WIDTHS

TRADE NAME DESIGNATE	GAGE US.STD.	DEPTH D"	PROPERTY I''^{4}	PROPERTY S''^{3}	MAX. YIELD F_y PSI	TYPE AVAILABLE	STANDARD FINISHES
SLABFORM	20	1 5/16	0.158	0.227	UP TO 90,000	XTRA H.D.	GALV. OR BLACK
Do.	22	1 5/16	0.130	0.183	Do.	Do.	Do.
Do.	24	1 5/16	0.101	0.139	Do.	Do.	Do.
Do.	26	7/8	0.036	0.075	Do.	HEAVY DUTY	Do.
Do.	28	9/16	0.011	0.035	Do.	STD. GALV.	GALVANIZED
Do.	28	9/16	0.010	0.032	Do.	STD. BLACK	UNCOATED
TRUSFORM	18	1 1/2	0.252	0.271	33,000	STD. 24" W.	GALV. ONLY
Do.	20	1 1/2	0.178	0.201	Do.	Do.	Do.
Do.	22	1 1/2	0.138	0.164	Do.	Do.	Do.
RIBFORM	16	1 1/2	0.440	0.500	Do.	BR, HI-BOND	Do.
Do.	18	1 1/2	0.340	0.400	Do.	Do.	Do.
Do.	24	1 3/8	0.104	0.133	HI-TENSILE	SUPER DUTY	Do.
Do.	NL.	15/16	0.044	0.071	Do.	HEAVY DUTY	Do.
Do.	NL.	9/16	0.012	0.034	Do.	STANDARD	Do.
CORRUFORM	STD.	9/16	0.010	0.036	UP TO 90,000	STANDARD	GALV. OR BLACK
Do.	STD	7/8	0.032	0.073	Do.	HEAVY DUTY	Do.
TUFCOR	20	1 3/8	0.167	0.239	80 to 90,000	STD. 28" W.	Do.
Do.	22	1 5/16	0.122	0.181	Do.	STD. 29" W.	Do.
Do.	24	1 5/16	0.089	0.136	Do.	STD. 30" W.	Do.
HI-COR	20	7/8	0.0496	0.1089	33 TO 50,000	STD. 2 2/3" P.	GALV. OR COLORS
Do.	22	7/8	0.0413	0.0913	Do.	Do.	Do.
Do.	24	7/8	0.0329	0.0732	Do.	Do.	Do.
Do.	26	7/8	0.0246	0.0551	Do.	Do.	Do.
Do.	29	7/8	0.0185	0.0416	Do.	Do.	Do.
CORRG. IRON	16	1/2	0.0178	0.0714	30,000	2 2/3" PITCH	GALV. OR BLACK
Do.	26	1/2	0.0064	0.0255	30,000	Do.	Do.

EXAMPLE: Comparing deck load tables 2.8.2

In compiling the table below, load data was taken from load tables provided to Architects by five steel deck suppliers. A 20 Gage Type B Roof Deck was chosen for comparison of allowable given loads for a continuous long sheet which would cover 3 or more spans of 7.0 feet. Properties are tabulated for a 12 inch width of deck and the allowable design stress F_b is varied.

ROOF DECK MAX. LOADS LISTED FOR COMPARISON								
PRODUCER	SPAN FT.	GAGE	TYPE	MOMENT	I''^4	S''^3	F_b PSI	LISTED LOAD
Nō. 1	7.0	20	B	WL/10	0.230	0.250	20,000	89 Lbs. Sq. Ft.
Nō. 2	7.0	20	B	WL/8	0.230	0.251	18,000	26 " " "
Nō. 3	7.0	20	B	WL/10	0.230	0.270	N. L.	92 " " "
Nō. 4	7.0	20	B	WL/10	0.233	0.271	20,000	84 " " "
Nō. 5	7.0	20	B	WL/10	0.230	0.230	20,000	79 " " "
				AVERAGE S =		0.254		

REQUIRED:

Take the average property of S for the 5 decks and obtain the resisting moment when max. F_b = 20,000 PSI. Use moment coefficient of 3 spans (WL/10) to calculate the safe allowable load per square foot.

STEP I:

Load per square foot on 12 inch strip is same as w #/'

$S = 0.254$ $F_b = 20{,}000$ #□" $RM = SF_b$ or $M = 0.254 \times 20{,}000 = 5088$ inch lbs.

Reduce M to foot pounds: $M = \dfrac{5088}{12} = 444$ Ft. Lbs.

STEP II:

RM = Bending Moment, and $M = \dfrac{WL}{10}$, then $W = \dfrac{10 \times M}{L}$

7.0' total $W = \dfrac{10 \times 444}{7.0} = 635$ Lbs.

Square foot load or $w = \dfrac{635}{7.0} = 91$ Lbs.

DESIGN NOTE:

The listed load given by producer Nō. 2 is 65 Lbs. per foot less than computed above. This could be a error in the load table or perhaps it could be based on deflection instead of bending. See following example.

EXAMPLE: Deck load by deflection 2.8.3

Refer to preceding example and load table listing the 5 producers who roll Type B Roof deck. Note Producer # 2 lists a 26 PSF load on 7.0 foot span which is less than the other competitive sections with comparable cross-section properties. It is possible that No. 2 producer has based his listing on deflection limit of 1/250 of span l, or the load could be determined by AISC Formula $\Delta = \frac{5Wl^3}{384EI}$.

REQUIRED:

Examine the No. 2 deck in table 2.8.2 using the S.D.I formula for triple spans. Check loads given to ascertain if tabulated loads are based on deflection or bending.

STEP I:

The S.D.I Formula is: $\Delta = \frac{3wl^4}{384EI}$ and span $l = 84.0$ inches.

$w = 26$ PSF or lineal foot for a strip load 12 inches wide.

Max. deflection: $\Delta = 1/250$ of $84.0 = 0.336$ inches. $E = 29{,}000{,}000$

$I_x = 0.230''^4$ $S_x = 0.251''^3$ and $l^4 = 49{,}787{,}136$.

STEP II:

Transpose formula to solve for load: $w = \frac{384EI\Delta}{3l^4}$, and

with values: $w = \frac{384 \times 29{,}000{,}000 \times 0.230 \times 0.336}{3 \times 49{,}787{,}136} = 5.76$ Lbs. per Ft.

STEP III:

Using AISC formula transposed where $l^3 = 592{,}704$ and total load W is to be found. $\Delta = 0.336$ Then with values:

$W = \frac{384 \times 29{,}000{,}000 \times 0.230 \times 0.336}{5 \times 592{,}704} = 290.39$ Lbs. For unit

load $w = \frac{290.39}{7.0} = 41.5$ Lbs. Foot.

STEP IV:

Determine bending stress when $W = 290.39$ and $\Delta = 0.336$ In.

$M = \frac{WL}{10}$ or $M = \frac{290.39 \times 7.0 \times 12}{10} = 2440$ Inch Lbs. $f_b = \frac{M}{S_x}$

Stress $f_b = \frac{2440}{0.251} = 9{,}722$ Lbs. square inch.

STEP V:

Calculate Max. W when allowable $F_b = 18{,}000$ PSI and $L = 7.0$ Ft.

Resisting Moment $= F_b S_x$ or $M = \frac{18{,}000 \times 0.251}{12} = 376.5$ Ft. Lbs.

If $M = \frac{WL}{10}$, then $W = \frac{10M}{L}$ or $W = \frac{10 \times 376.5}{7.0} = 587.86$ Lbs.

Load unit per foot $= \frac{587.86}{7.0} = 76.84$ Lbs.

DESIGN NOTE:

Obviously the load listed for Product No. 2 is in error and loads given in table are calculated on bending F_b.

EXAMPLE: Fire-rated roof design 2.8.4

A metal ribbed type of roof deck is to be installed on steel joists spaced 4.0 feet on centers. Deck panels will span over three or more supports in continuous series. Roof structure must qualify for a 3 hour fire rating and deflection is limited to 1/240 of span. Code requires flat roofs to be designed with 30 PSF Live Load.

REQUIRED:
Use formulas recommended by Metal Roof Deck Technical Institute for the design and take roof structure data from Fire Rating Tables.

STEP I:
For 3 hour fire rating the weights and material composing roof will be as follows:

(a)	Live Load required by building code	=	30.00	PSF
(b)	Built up 5 Ply Asphalt and gravel roof 650#□	=	6.50	"
(c)	Assumed weight of ribbed deck.	=	6.00	"
(d)	Fire resisting Insulation of 1½" thickness	=	1.50	"
(e)	Ceiling of 1 inch mineral composition	=	6.00	"
	Total Design Load per foot	=	50.00	PSF

For a 12 inch width of deck, strip load $w = 50$ Lbs per foot.

STEP II:
Deck span $L = 4.0'$ $l = 4.0 \times 12 = 48.0''$ Max. $\Delta = \frac{48.0}{240} = 0.20$ inches.

For triple spans, deflection formula for Roof Decks is:
$\Delta = \frac{13.5\, w l^4}{384\, E I}$. Deflection is known, then solve for minimum value of I. Thus: $I = \frac{13.5\, w l^4}{384\, E \Delta}$

STEP III:
Collect values for formula and equation: $E = 29{,}000{,}000$
$w = 50\#'$ $l = 48.0''$ and $l^4 = 5{,}308{,}416$ $\Delta = 0.20''$ Then:

$$I = \frac{13.5 \times \cancel{50} \times \cancel{5{,}308{,}416}}{\underset{7.7}{\cancel{384}} \times \underset{5.46}{\cancel{29{,}000{,}000}} \times 0.20} = \frac{13.50}{8.41} = 1.605''^4 \text{ (Minimum value).}$$

Deck required will be a Type N with 3 inch depth and 16 Gage.

Bending Moment: $M = \frac{WL}{8}$. $W = 50 \times 4.0 = 200$ Lbs. $S''^3 = 0.942$

$M = \frac{200 \times 4.0 \times 12}{8} = 1200$ inch Lbs. $f_b = \frac{1200}{0.942} = 1275$ PSI (below 18,000).

EXAMPLE: Concrete slab form selection 2.8.5

Beam supports for a Concrete slab are spaced on 4.0 foot centers. Concrete slab is reinforced 1 Way to carry a live load of 200 Pounds per square foot. Form deck will be spot welded to beams and span over not less than 3 supports. (2 Spans). During placing of concrete the sag is restricted to 0.20 inches or $\frac{1}{240}$ of span l. Shoring is not permitted.

REQUIRED:
Assume form sheets are 20.0 feet long and will cover 5 spans or 6 supports. Add 15.0 Lbs. per square foot to Concrete dead load for wet mix and workmen to keep sag within limits.

STEP I:
The 200 Lb. Live is to be sustained by slab after concrete has cured. Dead load of wet concrete plus workmens weight will govern design. By referring to 1 Way slab tables in Section IV, a 2½ inch slab weighs 30 PSF and will 212 PSF on simple 4.0 foot span. Design uniform load will total 15 + 30 = 45 Lbs. Sq. Foot. Also strip load w = 45#/'

STEP II:
The formula for deflection in Slab Forms is: $\Delta = \frac{0.0068\, w l^4}{EI}$ Designer may choose a rib section and solve for amount of deflection, or formula may be transposed to solve for minimum value of I when Δ is given.

STEP III:
Solve for minimum moment of Inertia as: $I = \frac{0.0068\, w l^4}{E\Delta}$.
E = 29,000,000 w = 45#/' l = 48" l^4 = 5,308,416 Δ = .20"
Placing values in formula:

$$I = \frac{0.0068 \times 45 \times 5{,}308{,}416}{29{,}000{,}000 \times 0.20} = 0.280''^4 \text{ (Slide rule close enough).}$$

Choose from Tables: 18 Ga. 1½ inch Ribform BR, HiBond with and $I = 0.340''^4$ and $S = 0.400''^3$. This type of deck may be stressed in bending up to its yield point or about 30,000 PSI. Check actual bending stress which should be very low.

$M = \frac{wl^2}{10}$ or $M = \frac{45 \times 4.0 \times 4.0}{10} = 72'^{\#}$ $f_b = \frac{M}{S}$ or $f_b = \frac{72 \times 12}{0.400} = 2160$ PSI.

EXAMPLE: Concrete slab form support spacing 2.8.6

A concrete slab with a 4½ inch depth is to be placed upon a heavy duty corrugated 24 Gage ribbed form and allowed to rust away when concrete curing is accomplished. The supporting beam will consist of precast members and spaced according to form requirements and later design. Form sheet selected is 7/8 inches deep Hi-Core.

REQUIRED:
Determine the maximum span lengths and support beam spacing using the basis of $F_y = 30{,}000$ PSI for bending stress. While concrete is still wet and workmen have slab finished, the maximum deflection should not exceed 1/240 of span length l. Make note if shoring will be required.

STEP I:
The Properties of a 24 Gage High Core steel sheet as taken from tables are as follows: $I = 0.0329''^4$ $S = 0.0732''^3$ Yield point is from 33,000 to 50,000 PSI. For problem $F_b = 30{,}000$ PSI.
The Resisting Moment $= SF_b$. $RM = 0.0732 \times 30{,}000 = 2196$ inch Lbs., or $RM = \frac{2196}{12} = 183$ Foot Pounds.

STEP II:
RM will be equal to continuous span or $M = \frac{wL^2}{10}$ or in feet $L = \sqrt{\frac{10 \times M}{w}}$. Dead weight of Concrete: $w = 4.5 \times 12.5 = 56.25$#/'
Total weight of wet concrete with men workers will be calculated at 70 Lbs. per foot.
Then $L = \sqrt{\frac{10 \times 183}{70}} = 5.10$ ft. (Call it 5.0 feet).

Without working men load and weight of concrete only, the support spacing $L = \sqrt{\frac{10 \times 183}{56.25}} = 5.70$ feet. (Accept for spacing).

STEP III:
$L = 5.0'$ $l = 60.0''$ and $l^4 = 12{,}960{,}000$ Formula: $\Delta = \frac{0.0054\,wl^4}{EI}$
$w = 56.25$#/' $E = 29{,}000{,}000$ $I = 0.0329''^4$
Equation becomes: $\Delta = \frac{0.0054 \times 56.25 \times 12{,}960{,}000}{29{,}000{,}000 \times 0.0329} = 4.10$ inches.

Max. Deflection is 1/240 of l. Max. $\Delta = \frac{5.7 \times 12}{240} = 0.285$ inches.

SHORING UP DECK WILL BE REQUIRED UNTIL CONCRETE SLAB HAS CURED.

Welding 2.9

Welding became a useful and accepted fabrication technique as a result of World War Two, when crash programs in shipbuilding and war supplies accelerated the development of welding methods. Electric arc and gas welding were used before the war, but the lack of skilled welders and of technical controls had prevented wide acceptance. Government financed training schools were established to train workers in the skills necessary for good fusion and finish. It became clear that welded connections could economically replace rivets; welding became acceptable to most fabricators. After the war, welding manpower was abundant due to the war-time training programs. Continued technical improvements by the manufacturers of welding equipment have developed a variety of methods for precision welding with consistent properties.

Welding processes 2.9.1

All welding processes require high temperatures to melt portions of the metals together. This union is called fusion: the blending of two parts into one. Ancient blacksmiths were expert welders. They used the bellows to raise the coals in their forges to white heat, and placed the two iron members into the fire. When the metal also turned a white color, a fine sand flux was sprinkled on the spots to be welded. This flux cleaned the metal. Tongs were used to transfer the heated metal parts to the anvil where they were hammered to press the heated parts together. A series of re-heatings and hammer blows completed the welding operation. To inspect the weld, the smithy filed the surface to see that fusion was complete. Forge welding is classified as pressure welding; a modern counterpart is resistance spot welding, where electric current heats the joint which is fused under high pressure.

Oxy-acetylene welding 2.9.1.1

An extremely hot flame can be obtained with a torch which uses oxygen and acetylene gas, with tips shaped for various types of work. Oxygen is supplied in metal cylinders at about 2000 PSI and acetylene is supplied in a companion cylinder or taken directly from a generator. The acetylene is produced from calcium carbide and water, a process first used to provide a gas lighting system where electric energy was not available in rural communities.

Regulating valves on the welding torch provide the proper flow of oxygen and gas. These regulators keep the feed tip pressure constant regardless of the pressure in the cylinders. Oxy-acetylene welding is accomplished both with and without the addition of extra metal from a filler rod. Heat is applied until the parts melt together. Usually a cleaning flux is applied to the joint before welding. Oxy-acetylene cutting has replaced the large power shears in many plants where thick plate must be cut, shaped and beveled.

Electric arc welding 2.9.1.2

An electric arc is a sustained spark which starts when the electrode is held near the base metal to be welded. Electric arc welding is therefore based upon the idea of bringing the two base metals into contact with the welding metal which will produce an instantaneous fusion. The arc provides the heat to melt and fuse the base metals and the electrode filler metal into a solid uniform mass. The arc produces a temperature of between 6000 and 7000°F. Usually the filler electrode is coated with a flux which cleans the base metal and shields the arc and the hot weld from oxidation.

Hand welders consist of three parts: an electrode or weld stick, a gripping-type stick holder, and an electric cable to which the holder is connected. When the arc is struck and expertly worked, the hot weld deposit will form a stiff fluid mass of tiny round particles which cling to the base metal. This makes overhead welding possible.

The welding power supply must be well regulated and stable. Manufacturers of welding machines make many models which generate low voltages for safety and high amperages for high production and large electrodes. Portable machines for field welding may be purchased with voltage ratings as low as 15 and ampere capacity of 150 or more.

WELDING ELECTRODES

A primary requirement for satisfactory welds is the selection of the proper electrode. The American Welding Society has established a numbering system to classify welding electrodes. The design engineer should become familiar with the designations in Table 2.9.3.5. These designations will be illustrated by selecting one electrode in common use: E6013.

E6013: the capital letter E denotes an Electric arc rod. The first two digits (60) indicate the allowable yield stress in kips (60,000 PSI). The third number will be a 1, 2 or 3 and will denote the following:

1—Can be used in all positions (flat, vertical or overhead).

2—Approved for flat position and horizontal fillet welds.

3—Suitable only for welds made in flat position.

The last digit (3) refers to the coating cover on the rod and the current with which it must be used.

Carbon and hydrogen content are carefully controlled in electrodes, because they have great effect on weld ductility. It should be noted that the specifications for ASTM A36 steel given in the AISC Manual list the approved welding electrodes as series E60 and E70.

In early arc welding, an uncoated or bare metal electrode was used. The molten globules which flow from the rod to the base metal were exposed to the surrounding atmosphere containing oxygen and nitrogen. The molten base metal of the weld was also exposed to these elements. In combination they formed oxides and nitrides in the weld. These impurities tended to weaken and embrittle the steel and reduce its corrosion resistance. Certified electrodes in current use are provided with a heavy coating which, in the heat of the arc, gives off an inert gas which envelopes and shields the arc from the atmosphere. This is referred to as shield welding. The coating on the rod is consumed at a slower rate than the metal core, so that the coating always extends beyond the metal and aids in concentrating the arc stream.

Electric arc welding, continued 2.9.1.2

ELECTRODE PROTECTION

The rod coating of low-hydrogen electrodes must be kept absolutely clean and dry until used. Coated electrodes removed from the hermetically sealed metal container should be used within a four hour period, unless stored in a holding oven. Electrodes exposed to atmosphere where the relative humidity is 75 percent or above for a two hour period should be dried in an electric oven at a temperature of 450°F. Any electrodes which have become wetted by rain or submerged in water should be discarded.

PERSONNEL CERTIFICATION

Steel fabricators of structural members for buildings and vessels realize the importance of good welding and will employ only those who have been certified after passing qualification tests. Many of these tests are conducted by employers, and others are conducted at trade schools in co-operation with local welder union organizations. Proficiency must be demonstrated in welding overhead, vertical, horizontal and flat. The welder must also have a knowledge of symbol and blueprint reading. Architects and engineers should include a clause in their structural specifications which requires all welding to be performed by certified welding personnel in the shop and field. An additional clause should be inserted to restrict the use of a torch to burn corrective holes and provide cutouts in members. Any authority to permit use of cutting torch or burning instrument should receive the approval of the design engineer.

WELD INSPECTION AND TESTS

A non-destructive test (NDT) may be required for critical joints in bridges, building structures, and tanks. Several methods may be used to evaluate the degree of fusion and the strength of the weld. The interior of the weld joint may be evaluated for soundness by X-ray or gamma ray radiography or by ultrasonic probes. The surface of the weld may be evaluated for cracks by dye-penetrant or magnetic particle (Magnaflux) techniques. More information on these tests is available from the Society for Non-Destructive Testing.

TABLE: Welding symbols 2.9.2

Approximately fifteen types of welds may be indicated by using the standard welding symbols. It is estimated that 80 percent of the connections for buildings and bridges will be fillet welds of various size and length. The balance will consist of butt welds, spot welds or other special types. In tank and pressure vessel fabrication, the welds will consist of approximately ninety percent grooved type with the balance of the fillet type. The tables which follow will illustrate the method to be used on drawings to denote the type of weld. This system for indicating welded connections is standard practice in all shop fabricating plants, shipbuilders and pipe fabricators.

TABLE: Welding symbols, continued 2.9.2

STANDARD SYMBOLS FOR WELDED FABRICATION

LOCATION OF WELD	STANDARD FILLET	GROOVE			
		SLOT - PLUG	SLOT - SEAM	SQUARE	VEE
ARROW SIDE					
OTHER SIDE					
BOTH SIDES		NOT USED	NOT USED		
	BEVEL	U	J	FLARE VEE	FLARE BEVEL
ARROW SIDE					
OTHER SIDE					
BOTH SIDES				NOT USED	

SUPPLEMENTARY WELDING SYMBOLS

SOLID DOT FIELD WELD	CIRCLE WELD AROUND	SIDE WELDS STAGGERED	FLUSH FIN. CONTOUR	FIN. SURFACE BUILT UP	INTERMITTENT CHAIN WELD
		3/8 3-6			1/4" 2-4

WELD ALL AROUND IN FIELD INDICATED THUS:

3/8" 3-6 INDICATES 3/8 INCH FILLET WELD 3 INCHES LONG AND ON 6" CENTERS

SAFE ALLOWABLE LOADS FOR FILLET WELDS

SAFE LOAD IN SHEAR PER LINEAL INCH LENGTH OF WELD

SIZE IN.	1/8"	3/16"	1/4"	5/16"	3/8"	1/2"	5/8"	3/4"	7/8"	1.0"
BUILDINGS	1200	1800	2400	3000	3600	4800	6000	7200	8400	9600
BRIDGES	1100	1650	2200	2750	3300	4400	5500	6600	7700	8800

TABLE: Welding symbols, continued 2.9.2

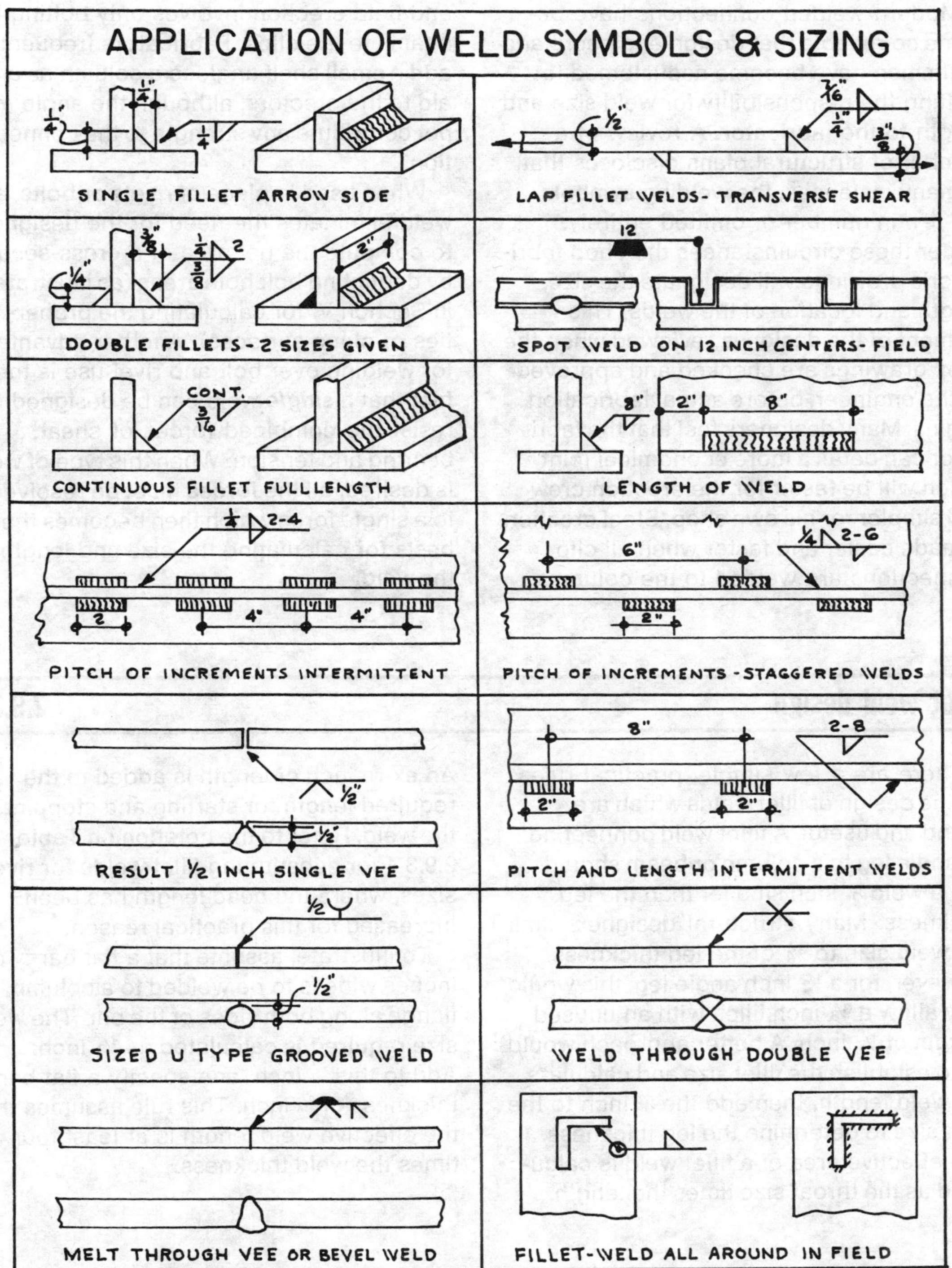

Designing welded connections 2.9.3

Modern welded connections have become so reliable that design engineers and draftsmen have become accustomed to shifting the responsibility for weld size and length to the fabricator. A review of a number of structural plans discloses that, in many instances, the welding symbols are few in number or omitted entirely. Under these circumstances the shop fabricator's drawings will designate the size, length and location of the welds. This element of the design is reviewed when the shop drawings are checked and approved by the engineer, before actual fabrication begins. Many designers feel that the fabricator can detail a more economical joint which will be faster for the erection crew and simpler for his own shop. Steel erection is made easier and faster when all clip connections are welded to the columns, and field erection involves only bolting the beams to the clips. Fabricators frequently add a small shelf angle to a column as an aid to the erectors, although the angle may not contribute any strength to the connection.

When used in place of rivets or bolts, a weld eliminates the need for the designer to compute the net area of a cross-section by deducting bolt hole areas (as illustrated in Section VI for calculating the properties of plane surfaces). Another advantage for welding over bolt and rivet use is the fact that a *single* weld can be designed to resist the combined forces of shear, bearing and tension. When this type of weld is desired, all the forces must be resolved to a single force which then becomes the basis for calculating the size and length of the weld.

Fillet weld design 2.9.3.1

There are a few simple, practical rules for the design of fillet welds which are sound and useful. A fillet weld connecting an angle leg to a column or beam should use a weld 1/16 inch smaller than the leg thickness. Many structural designers limit the weld size to 3/4 of the leg thickness. However, for a 1/2 inch angle leg, this would only allow a 3/8 inch fillet, with an unused margin of 1/8 inch. A better approach would be to establish the fillet size and calculate the weld length, then add the 1/16 inch to the fillet size to determine the leg thickness. The effective area of a fillet weld is calculated as the throat size times the length. An extra inch of length is added to the required length for starting and stopping the weld. Refer to the notation on Table 2.9.3.4 for substituting fillet welds for rivet sizes, where the bead length has been increased for this practical reason.

To illustrate, assume that a flat bar two inches wide is to be welded to a column flange along both sides of the bar. The weld size required is calculated as 1/2 inch. Add to this 1/16 inch, and specify a flat bar thickness of 9/16 inch. This rule assumes that the effective weld length is at least four times the weld thickness.

Shear and tension welds 2.9.3.2

A shear weld is loaded so that the action line of the force is parallel to the length of the weld. A tension weld is loaded so that the action line of the force is perpendicular to the length of the weld. Tension welds are also called transverse welds, and are approximately thirty percent stronger than shear welds. The design examples to follow will illustrate the use of these welds.

THROAT DIMENSIONS

The effective area of a welded connection is calculated as the throat size times the length. Since the common fillet weld is shaped in a 45 degree right triangle, the throat dimension is the distance from the right angle to the center of the hypotenuse or outer side. A fillet weld with ½ inch legs and a 45° slope will have a throat dimension of 0.3535 inches. Using trigonometry, the throat is equal to the leg dimension times Cosine 45°: throat = 0.50 x 0.707 = 0.3535 inches.

Allowable weld stresses 2.9.3.3

Table 2.9.3.4 for safe allowable loads for fillet welds is based upon an allowable design stress of 13,600 PSI. This table assumes that the weld electrodes are of A233 Class E60 submerged series or Class E70 series. Base material is assumed to be A7, A373 or A36 steel.

For steels with greater yield strength such as A242 and A441, the welding electrodes must conform to A233 Class E70 series, and the weld design stress may be increased to 15,800 PSI. This stress is applicable only to submerged-arc or low hydrogen electrode series.

A440 steel is not recommended for welding.

Design engineers must comply with the local building codes, and these may recommend lower allowable stresses than those published by the AISC. Many building codes require the allowable stresses to be in accord with the 1943 directive from the American Welding Society. The allowable stresses would then be as follows:

Welds in shear = 11,300 PSI.

Welds in tension = 13,000 PSI.

Compression welds = 15,000 PSI.

Designers should investigate the requirements for welding in structures which are to be located in areas subjected to periodic seismic forces. Codes are constantly being modified by state and local authorities.

TABLES: Fillet weld size and load

2.9.3.4

SUBSTITUTE FILLET WELDS FOR RIVET SIZES

RIVET DIA. IN INCHES	RIVET SHEAR VALUE IN LBS.	LENGTH OF FILLET WELD TO NEAREST 1/8 INCH.				
		1/4	5/16	3/8	1/2	5/8
1/2	2,950	1 1/2	1 1/4	1 1/8	7/8	3/4
5/8	4,600	2 1/4	1 3/4	1 1/2	1 1/4	1.0
3/4	6,630	3	2 1/2	2 1/8	1 5/8	1 3/8
7/8	9,020	4 1/8	3 3/8	2 7/8	2 1/8	1 3/4
1.0	11,780	5 1/4	4 1/4	3 5/8	2 3/4	2 1/4

RIVET SHEAR VALUES ARE BASED ON STRESS OF 15,000 LBS. SQUARE INCH.
1/4 INCH IS ADDED TO BEAD LENGTH FOR STARTING AND STOPPING THE ARC.

SAFE ALLOWABLE LOADS FOR FILLET WELDS

SAFE LOAD IN SHEAR PER LINEAL INCH LENGTH OF WELD

SIZE IN.	1/8"	3/16"	1/4"	5/16"	3/8"	1/2"	5/8"	3/4"	7/8"	1.0"
BUILDINGS	1200	1800	2400	3000	3600	4800	6000	1200	8400	9600
BRIDGES	1100	1650	2200	2750	3300	4400	5500	6600	7700	8800

TABLES: Properties of weld electrodes and base metals 2.9.3.5

PROPERTIES OF WELD METALS

WELD METAL DESIGNATION	ULTIMATE TENSILE STRENGTH PSI.	YIELD POINT IN LBS. SQ. INCH	ELECTRODE AWS NUMBER
FLEETWELD 5 & 5P	62,000 – 72,000	50,000 – 60,000	E-6010
FLEETWELD 7 & 72	67,000 – 80,000	55,000 – 69,000	E-6012
FLEETWELD 7-MP	67,000 – 75,000	55,000 – 64,000	E-6012
FLEETWELD 35	62,000 – 72,000	50,000 – 60,000	E-6011
FLEETWELD 37	67,000 – 80,000	55,000 – 65,000	E-6013
IMPROVED 47	72,000 – 80,000	60,000 – 70,000	E-6014
JETWELD LH-70	72,000 – 80,000	60,000 – 70,000	E-7018 E-6018
JETWELD LH-90	100,000 – 105,000	88,000 – 93,000	E-9018
JETWELD LH-110	119,000 – 128,000	111,000 – 117,000	E-11018-G
JETWELD 2	62,000 – 72,000	50,000 – 64,000	E-6027
JETWELD 1	72,000 – 90,000	60,000 – 86,000	E-7024 E-6024
JETWELD 2HT	70,000 – 79,000	57,000 – 65,000	E-7020-A1
SHIELD-ARC 85 & 85P	70,000 – 78,000	57,000 – 63,000	E-7010
STAINWELD A5-Cb	90,000 – 100,000	35,000 – 45,000	E-347-15
STAINWELD A7 & A7-Cb	85,000 – 95,000	35,000 – 45,000	E-308-16 E-347-16
STAINWELD B-Cb	85,000 – 95,000	45,000 – 55,000	E-309-Cb-15
STAINWELD C-Cb	85,000 – 95,000	35,000 – 45,000	E-318-15
STAINWELD D	80,000 – 90,000	35,000 – 45,000	E-310-15
ALUMINWELD	17,000 – 22,000	8,000 – 10,000	ALUM.
AERISWELD	20,000 – 40,000	—— - ——	COPPER, BRASS-BR'Z.
GALVWELD	—— - ——	—— - ——	COATING - GAS

PROPERTIES OF BASE METALS

METAL TYPE OR CHARACTER	ULTIMATE TENSILE STRENGTH IN PSI.	YIELD POINT IN LBS. SQ. INCH	ASTM TYPE OR SPECIFICATION
MEDIUM CARBON	55,000 – 65,000	27,500 – 37,500	A7, A36, A373
HI-TENSILE STEEL	90,000	46,000 – 60,000	A242, A440, A441
STAINLESS STEEL	80,000 – 95,000	30,000 – 45,000	18-8
STAINLESS STEEL	80,000 – 100,000	35,000 – 50,000	18-8 MO
STAINLESS STEEL	90,000 – 110,000	40,000 – 60,000	25-12
CAST IRON	15,000 – 25,000	NONE	NONE
ALUMINUM	19,000 – 25,000	9,000 – 20,000	SEE ALLOYS

EXAMPLE: Splice weld design 2.9.4

A steel flat bar 5/8" x 3.0" require an extension and a lap joint is not desired. Tension force in bar is given as 32,500 pounds. Flat bar steel is A36.

REQUIRED:
Design a splice weld which will present the better finished appearance. Try a plain butt weld with ground surface and if found deficient, try a double-vee weld with a raised surface. Make weld square with bar and use a tension allowable stress of $F_t = 13{,}600$ PSI.

STEP I:
For a plain butt weld with surface ground flush the weld throat is same as bar thickness or 0.625 inches.
Area of weld for full width of 3.0 inch bar $= 0.625 \times 3.0 = 1.875^{\square}$"
Tension $T = F_t A$ or Max. $T = 1.875 \times 13{,}600 = 25{,}500$ Lbs.

STEP II:
If diagonal weld were permitted, the length of cut is found thus: $\dfrac{32{,}500}{0.625 \times 13{,}600} = 3.82$ inches. (Call it 4.0 inches).

STEP III:
Calculate the throat size of a Double-Vee weld with bulge.

$$\text{Throat} = \frac{\text{Tension } T}{F_t \text{ Bar width}} \text{ or } \frac{32{,}500}{13{,}600 \times 3.0} = 0.795 \text{ inches.}$$

$$\text{Surface bulge over weld} = \frac{0.795 - 0.625}{2} = 0.085'' \text{ (Approx. } 1/16'')$$

Detail of weld follows:

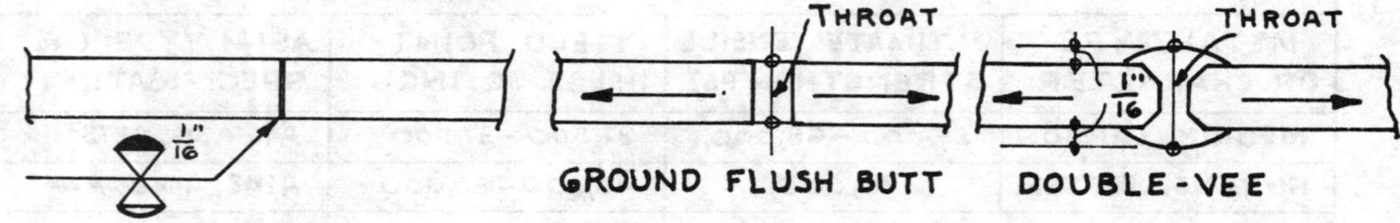

EXAMPLE: Welded clip design 2.9.5

A clip is to be welded all around one leg to the flange of a column with 3/8 inch fillets. Size of angle clip is 3 x 3 x 7/16 and 4.0 inches long. Building Code limits the design unit weld stresses to the following with A7 Steel.
Welds in shear = 11,300 PSI.
Welds in tension = 13,000 PSI.

REQUIRED:
Determine the supporting value of vertical reaction if beam were placed on clip. Separate the values of shear and tension. Draw sketch to illustrate difference.

STEP I:
Legs of fillet welds are 3/8" and Cosine of 45° = 0.707
Throat of weld = 0.625 x 0.707 = 0.265"
For 1 Lineal inch length: A = 0.265 x 1.0 = 0.265 □"

STEP II:
Determine value of each type of weld for 1.0 inch length.

Tension value = 0.265 x 13,000 = 3,445 Pounds.
Shear Value = 0.265 x 11,300 = 2,995 Pounds.

STEP III:
Drawing illustrated conditions:

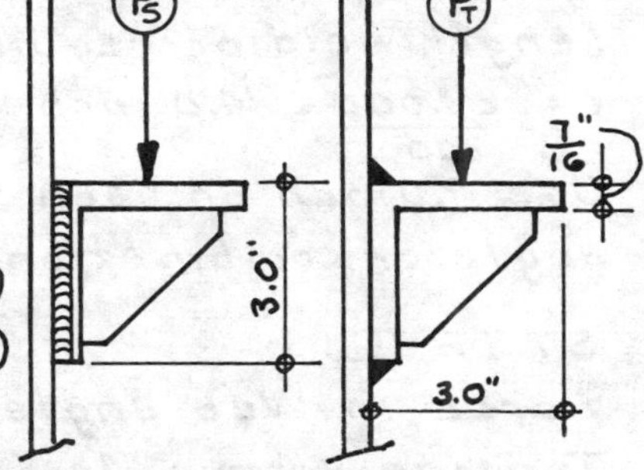

Shear lengths = 6.0 inches.
Tension lengths = 8.0 inches.

Value in Shear = 2,995 x 6.0 = 17,970 Lbs. (P_S)
Value in Tension = 3445 x 8.0 = 27,560 " (P_T)
Total value of Clip = 45,530 Lbs.

STEP IV:
Comparing welded clip value from table of fillet welds: Weld lengths total 14.0" and 3/8" fillet is good for 3600# per inch.

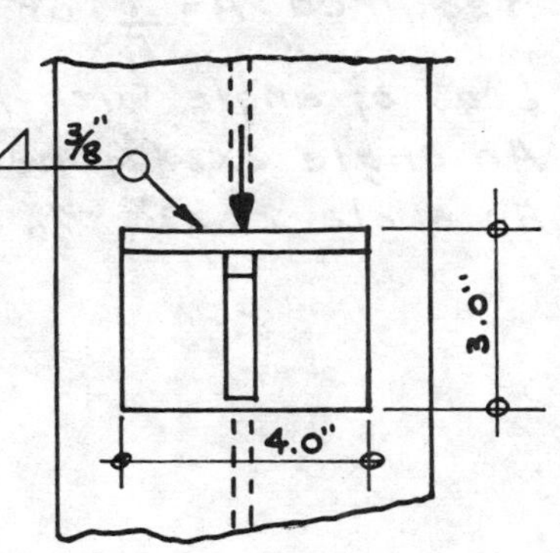

Value of connection = 3600 x 14.0 = 50,400 Lbs.

Since A7 Steel is specified and the yield for A7 is: F_y = 33,000 P.S.I., the results in Step III should be accepted.

EXAMPLE: Welded gusset plate design 2.9.6

The web member of a truss has a tension force of 28,000 Pounds as scaled from a graphic diagram. Lower chord is composed of 2 angles separated with ½ inch gusset plates. Web angle is to be welded to gusset with ¼ inch fillet welds parallel to direction of force. Angle for web member has not been determined, but angle legs should have a 1/16 inch margin over weld leg. Use stresses thus: Shear = 11,300 PSI.

REQUIRED:
Determine the amount of welding required to resist force, then calculate for angle required for web and make choice.

STEP I:
Welding is of shear type when parallel to force.
Weld legs = 0.25" Cosine 45° = 0.707
Throat dimension = 0.25 × 0.707 = 0.177"
Area weld for 1 inch = 0.177
Allowable F_s = 11,300#□"

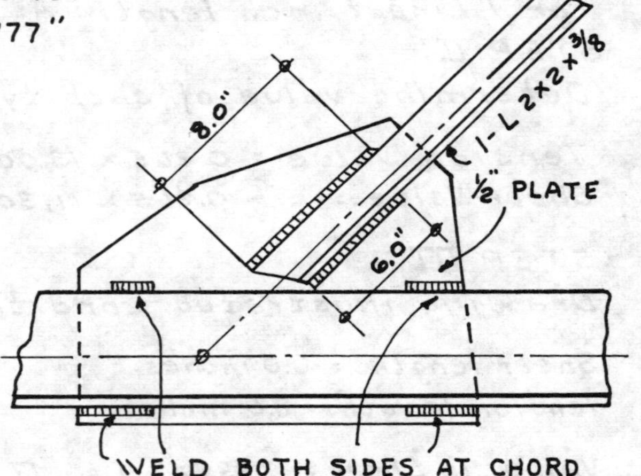

STEP II:
Shear value of 1 lineal inch of weld = 0.177 × 11,300 = 2000 Lbs.
Length welding required:

$$l = \frac{28,000}{2000} = 14.0 \text{ inches.}$$

Use 7.0 inch on each side of angle or combination of 8" and 6".

STEP III:
Force in Web angle = 28,000 Lbs.
Tension allowable for A36 steel is F_t = 22,000 PSI.

$$\text{Required } A = \frac{T}{F_t} \text{ or } A = \frac{28,000}{22,000} = 1.27 \text{ Sq. In.}$$

Legs of angle with 1/16" margin over ¼" weld = 5/16" leg. (Minimum)
An angle 2×2×3/8 has A = 1.36□" Wt. = 4.7#/'
An angle 2½×2½×5/16 has A = 1.46□" with Wt. = 5.0#/'

Accept the L 2×2×3/8 for web.

EXAMPLE: Welded beam to column connection 2.9.7

A steel Column has a flange width of 8.0 inches to which a set of clip angles are to be welded and support a beam with 8.0 inch depth and a 4.0 inch flange. Bottom angles is to serve as shelf and top angle is a 6½ inch long clip cut from a 4x3x7/16 L. Beam will be connected with 2-¾ϕ bolts at top with a horizontal single shear of 30,000 Lbs. Vertical end reaction is 18,000 Lbs.

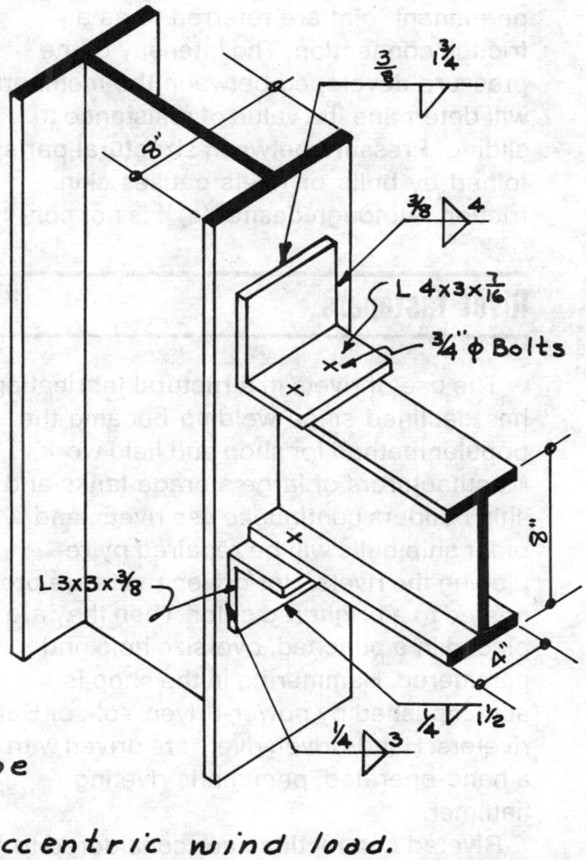

3/8 1¾
8"
3/8 4
L 4x3x 7/16
¾"ϕ Bolts
8"
L3x3x3/8
4"
¼ 1½
¼ 3

REQUIRED:
Design the fillet weld to support beam at top clip connection and bottom clip may be neglected.

STEP I:
This connection appears to be a common moment type as is used in Hi-Rise frames with eccentric wind load. Vertical force = 18,000# and Horizontal force = 30,000 Lbs. or same as bolts. Design for welds will be simplified if both forces are resolved into 1 force in a single direction.

STEP II:
A force diagram will serve to give action line with the tension force if drawn to scale:

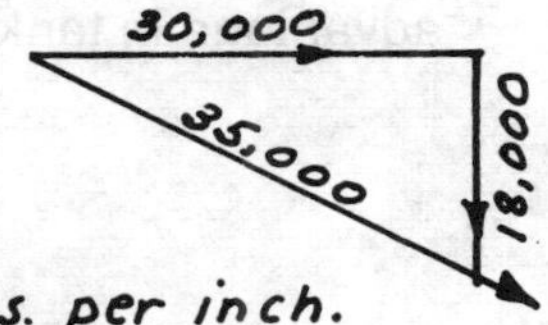

30,000
35,000
18,000

T = 35,000# and bottom clip is in compression.
From Table: A 3/8" fillet weld is good for 3600 Lbs. per inch.
Total weld length required = $\frac{35,000}{3600}$ = 9.72" (Call it 9¾ inches.
Placing angles 4" leg vertical, weld both sides which will give 8 inch length, then weld balance of 1¾ inch at top.
Bottom angle may be of ¼ inch thick legs and weld length can support vertical force of 18,000 Lbs., if desired. Use ¼" weld.
Length of weld on bottom angle = $\frac{18,000}{2400}$ = 7½ inches. Make shelf angle 3x3x¼ L and weld similarly. Beam is not obstructed.

Bolt and rivet fasteners 2.10

Bolts and rivets used to clamp two structural members together in a tight and permanent joint are referred to as a friction connection. The intensity of the pressure developed between the members will determine the value of resistance to sliding. Pressure between structural parts joined by bolts or rivets causes skin friction. Although desirable, it is not considered in the design of riveted and bolted connections.

When a bolt is tightened, the tension stress in the bolt must not exceed the allowed unit stress for the specified bolt. Tests indicate that skin friction will vary from 4000 to 10,000 pounds per fastener with normal tightening of the bolt to 85 to 90 percent of yield.

Rivet fasteners 2.10.1

The use of rivets in structural fabrication has declined since welding became the popular method for shop and field work. Manufacturers of large storage tanks and ship builders continue to use rivets, and older ship hulls will be repaired by replacing the rivets. Hot-driven rivets are preheated to a bright red color. Then they are placed in a punched, oversize hole and hammered. Hammering in the shop is accomplished by power-driven Yolk or Bull riveters. Hand-driven rivets are driven with a hand-operated, pneumatic riveting hammer.

Riveted connections are considered to be of the skin friction type. When the hot rivet is driven, the shank is expanded by the heat. Then, as the rivet cools, the metal contracts, causing a tight clamping between members. Rivets have another advantage in tank and ship work. In driving a hot rivet, the shank tends to bulge inside the hole for a tight fit with full bearing. With storage tanks and pressure vessels, this reduces the possibility of seepage or pressure loss through the hole.

Before rivets are driven to a tight fit, the lapped members must be held together with temporary bolts. In this manner rivets do not loosen as succeeding rivets are driven. Riveting is a very noisy operation. Early fabricating shops were located in rural areas, apart from residential areas and institutional buildings. With the change to welding, it is even possible to make additions to hospital buildings without disturbing the patients.

Design work is reduced when joints are to be welded, since the areas for rivet holes no longer have to be deducted from each cross section.

Bolt fasteners 2.10.2

American standard bolts are manufactured as rough, unfinished bolts or turned, finished bolts commonly referred to as machine bolts. These standard bolts are made from a metal much like plow steel, and must not be confused with high-strength alloy bolts. Rough bolts (like field-driven rivets) are produced directly from rod as received from the rolling mill. Unfinished bolts are used for temporary steel erection and as an aid in rivet installation. Finished bolts are turned on automatic lathes, thus they were given the name machine bolts. Standard bolts in mild steel are also produced with a fluted shank and are classified as ribbed bolts. The deformed flutes on the shank are larger than the hole and must be installed by driving with a hammer. Ribbed bolts will provide a tight fit for bearing, and were originally produced with button heads as a replacement for field driven rivets. The ribbed shank of the earlier bolt became the present deformed shank common to the high-strength bolts.

High strength bolts are manufactured from heat-treated and alloy steels. They are available with various heads and nuts, also with either smooth or deformed shanks. Field erectors refer to high-strength bolts as tempered bolts. Originally when welding supplanted riveting in shop fabrication, the standard bolt had much less strength than the weld when comparisons were made for shear and bearing. A welded connection required an excessive number of standard bolts to equal the value of weld. With the high strength bolt, connections can be designed with fewer bolts so that the advantages of bolting and welding can be efficiently combined.

BOLT IDENTIFICATION MARKS

Manufacturers of high-strength bolts provide a mark on the head and nut to enable field inspectors to identify the bolt. This is important because erection crews may fail to replace an erection bolt with a high-strength bolt. Other crafts will follow the steel erectors and install material which will conceal much of the framing. The changing-out of bolts must be given close scrutiny.

NUTS AND LOCK WASHERS

Various types of nuts and washers are available to secure the bolt and prevent loosening of the nut. Static loads do not usually loosen the nut except when wind and traffic vibrations are present. Bridges are subjected to vibrations and impact loads, and require special attention.

Nut types are produced which have a locking device incorporated into the nut while others are cupped on the contact side and do not require a washer. A nut which is often used with high-strength bolts contains an integral lock and is referred to as a Dardelet self-locking nut. This nut was developed by the Dardelet Threadlock Corporation, with a companion button-head, ribbed-shank bolt. Lock washers, consisting of a split ring of hardened spring steel, maintain a constant pressure on the nut and keep it locked.

Plain cut washers are required under nuts which are not cupped, or if the bolt threads are not cut far enough to allow tightening). Bolt threads which are cut too far will extend into the shear plane and reduce the bolt shear value. There are two alternates for calculating the allowable shear stress when the threads extend into the shear plane. The first alternate provides that the allowable shear stress shall be reduced; the second provides that the shear area of the bolt be calculated on the cross-section at the root of the threads.

Stresses in bolts and rivets 2.10.3

TENSION

When a bolt or rivet is drawn tight, the tension stress extends along the shank from nut to head and stretches the bolt. The thread root has the least cross-sectional area and is the critical point. The tension force in the bolt is calculated by considering the wrench torque and the thread angle. This method is illustrated in the example of the screw-jack in Section V. Bolt suppliers provide tables which give rated values for ultimate load and design stress. These tables should be used with caution when comparing various products.

A moment connection will usually put the bolts in tension as a result of eccentric loading. Here also we must calculate the bolt area at the root of the thread when computing tension value.

Tanks and other structures with lapped joints and bolt or rivet fasteners are designed with the main forces transverse to the shank, which produces shear and bearing stresses. Skin friction between the lapped plates is ignored, and little attention is given to the tension stress in the shank. Tanks and pressure vessels which have accidently been ruptured by internal explosions confirm the theory that shear and bearing are the essential points for design. When designing lap-joints, three conditions must be considered for the design.

(a) Shear stress on the bolt over its cross-sectional area and the allowable shear stress.

(b) Bearing stress on the bolt over its diameter and the thickness of the lapped members.

(c) Area of material removed from the member.

It is not to be expected that the three elements given will produce results equal in value. In each instance, the least value of the three conditions governs the design. These factors will be illustrated in the design examples which follow.

SHEAR AND BEARING

A single lap of two plates or flat bars produces shear and single bearing in the bolt or rivet fastener. If one plate is thicker, the bearing stress on the bolt is governed by the thinner plate. Bearing = Dt F_p, or bolt diameter times plate thickness times allowable bearing stress. Table 2.10.5.4 provides the bearing area for various plate thicknesses and bolt diameters. It is necessary to compare the allowable bearing stress for both the bolt and plate material. The least value will govern the design. Illustrations in the examples will explain single and double shear and single and double bearing.

Gage and pitch 2.10.4.1

In the tables for detailing (2.3.3.1) wide-flange shapes, the dimensions g and g_1 are listed. These are called gage lines (and are also shown for channels and miscellaneous shapes). These gage lines standardize the location of bolts in the web and flange. Most published tables do not give gage lines for angles, however Table 2.10.4.2 lists the gage lines used in many fabricating plants. These gage dimensions are applicable to both bolts and rivets.

The pitch of a bolt or rivet is the center-to-center spacing along the gage line. Minimum pitch should be at least three times the bolt diameter. Standard spacing for ⅞ inch diameter bolts is 3 inches. Excessive pitch wastes clip material, but insufficient pitch may cause hole deformation or splitting which weakens the connection.

TABLE: Angle gages for bolts and rivets 2.10.4.2

ANGLE GAGE DIMENSIONS FOR STANDARD CONNECTIONS

LENGTH OF ANGLE LEG - IN INCHES															
FRACTION	3/4	1	1¼	1⅜	1½	1¾	2	2½	3	3½	4	5	6	7	8
DECIMAL	0.75	1.00	1.25	1.375	1.50	1.75	2.00	2.50	3.00	3.50	4.00	5.00	6.00	7.00	8.00
GAGE g_1	½	⅝	¾	⅞	⅞	1	1⅛	1⅜	1¾	2	2½	3	3½	4	4½
GAGE g_2	---	---	---	---	---	---	---	---	---	---	---	2	2¼	2½	3
GAGE g_3	---	---	---	---	---	---	---	---	---	---	---	1¾	2½	3	3
MAX. BOLT	¼	¼	⅜	⅜	⅜	½	⅝	¾	⅞	⅞	⅞	⅞	⅞	1	1⅛

TABLE IS APPLICABLE TO RIVETED CONNECTIONS:
LIMIT MINIMUM PITCH TO 3 TIMES BOLT DIAMETER.

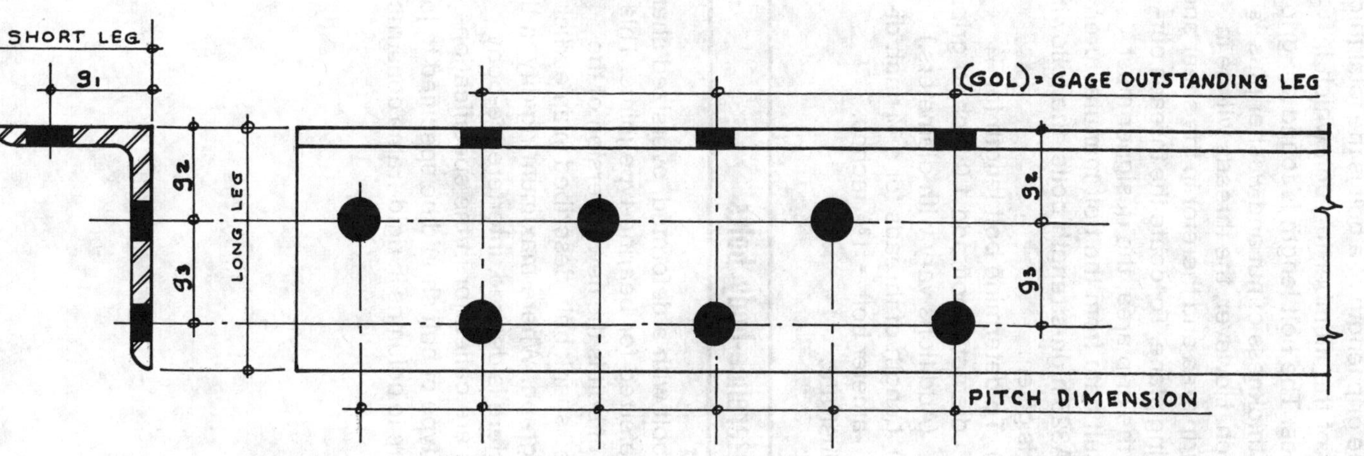

Bolt grip length 2.10.4.3

The grip length of a bolt is the total thickness of the member through which it passes. The bolt length is found as: grip plus thickness of nut and washer plus ⅜ to ½ inch. However, the threads will be in the grip area and therefor in the shear and bearing plane. To locate the threads outside the grip area, the designer must consult data from the bolt manufacturer. For A325 high-strength bolts, the following data is given:

(a) To determine bolt length for ¾ diameter bolt add 1.0 inch to grip. (Additions vary with diameters.)

(b) Length of threads for a ¾ inch diameter bolt = 1⅜ inches.

To illustrate:

Assume two ¾ inch flat bar plates are to be butt spliced with a single ¾ inch cover plate to be bolted to each with ¾ inch diameter A325 Bolts.

Grip is 0.75 + 0.75 = 1.50 inches. Bolt length = 1.50 + 1.0 = 2.50 in. Unthreaded length = 2.50 − 1.375 = 1.125 inches. Length of threads in the shear and bearing plane = 1.50 − 1.125 = 0.375 or ⅜ inch.

To locate the threads outside shear plane, bolt length must be: grip plus threads length T. Then bolt length = 1.50 + 1.375 = 2.875 inches. Use a 3 inch bolt. Table 2.10.5.3 lists the pertinent dimensions for A325 bolts and nuts.

Interference-body bolts 2.10.4.4

A bolt with a deformed, oversized shank is preferred for bearing-type joints. This type of bolt is the newer version of the ribbed shank bolt, described in an earlier paragraph. Where maximum rigidity in the structure is desired, interference-body bolts are called for in the specifications. This type of bolt should be specified to join rafters to columns in rigid-frame construction. (See Section VII.) The purpose of the interference-body bolt is to provide a fastener where bearing stress is the governing factor in the design. These bolts may be purchased to a specified grip length to exclude the threads from the shear plane. Button heads are standard since a wrench is not required to prevent turning while tightening the nut.

Designing bolted connections 2.10.5

Structural failures are often caused by improper connections. Engineers and architects sometimes very carefully select structural members, but then connect them without equal care. This practice is unfortunate but understandable, since member selection is quite simple, while joint connection may require considerable design analysis. Engineers do not always agree on theory or practice for joint design, and older text books did not provide sufficient emphasis on this important phase of design.

ECCENTRICITY

The most common sources of failure in bolted or welded connections are foundation settlement, seismic forces or high velocity wind pressure. A slight settlement in a single column footing will produce bending in the connection of the girder and the column. When bending is present in a connection, tension and compressive forces develop an inflection point. This is referred to as the point of rotation, and is not necessarily related to the centroid of the connected structural member.

An ideal connection for a tension or compression structural member would have the center of gravity of the weld or bolt group line up with the gravity axis of the member and the line of action of the applied load. This ideal is seldom attained because there is usually a certain amount of eccentricity in the standard connections given in many steel handbooks.

WELD AND BOLT GROUPING

The design of connections with eccentric bending is simplified when the fastener groups are placed symmetrically about the centroid of the structural member. This may not always be possible. Should conditions exist where the grouping of welds or bolts causes an unbalanced connection, the following course must be taken.

(a) Calculate the center of gravity axis of each group by the moment method. Moments should be taken from the major axis of the member.

(b) The gravity axis of the fastener group with the greatest distance from major axis of the member will have the greatest stress intensity.

(c) The couple or moment arm is the distance between the action line for tension (T) and the action line for compression (C). These lines of action are through the gravity axis for the outer groups of fasteners.

(d) For wind moment connectors, the design should consider the pivoting point of rotation to be on the gravity axis of the bottom group. Then wind change in the reverse direction will not affect design.

(e) Moment connections should be detailed for design purposes as well as for shop fabrication.

The examples which follow have been solved using the theory of rotation. This assumption is open to debate. Basically, this theory assumes that the fastener farthest from the point of rotation is under greater stress than the inside fasteners. Another theory assumes that the whole group of fasteners acts as a combined group of areas. This theory is not on the conservative side, as will be shown in the example of a bolted wind moment connection. We have followed the more conservative approach.

TABLE: Allowable stresses for standard fasteners 2.10.5.1

ALLOWABLE UNIT STRESSES - STD. FASTENERS

BOLT OR RIVET DESIGNATION	TENSION F_t = P.S.I.	SHEAR F_v = P.S.I.	BEARING F_p = P.S.I.	DBL. BEARING F_p = P.S.I.
RIVETS - FIELD DRIVEN	15,000	10,000	16,000	20,000
RIVETS - SHOP DRIVEN	15,000	13,500	24,000	30,000
BOLTS - STD. ROUGH	15,000	10,000	16,000	20,000
BOLTS - STD. TURNED	15,000	13,500	24,000	30,000
H.S. A325 HOLTS THREADS INCLUDED IN SHEAR PLANE	40,000	15,000	1.35 F_y	
H.S A325 BOLTS THREADS EXCLUDED FROM SHEAR PLANE	40,000	22,000	1.35 F_y	
H.S. BOLTS A354 BC THREADS INCLUDED IN SHEAR PLANE	50,000	20,000	1.35 F_y	
H.S. BOLTS A354 BC THREADS EXCLUDED FROM SHEAR PLANE	50,000	24,000	1.35 F_y	
H.S. BOLTS A490 THREADS INCLUDED IN SHEAR PLANE	60,000	22,000	1.35 F_y	
H.S. BOLTS A490 THREADS EXCLUDED FROM SHEAR PLANE	60,000	24,000	1.35 F_y	

1.35 F_y REFERS TO STEEL ENCLOSED IN BOLTS GRIP. A36 STEEL F_y = 36,000 P.S.I.
FOR A36 STEEL: F_p = 1.35 x 36,000 = 48,500 PSI.

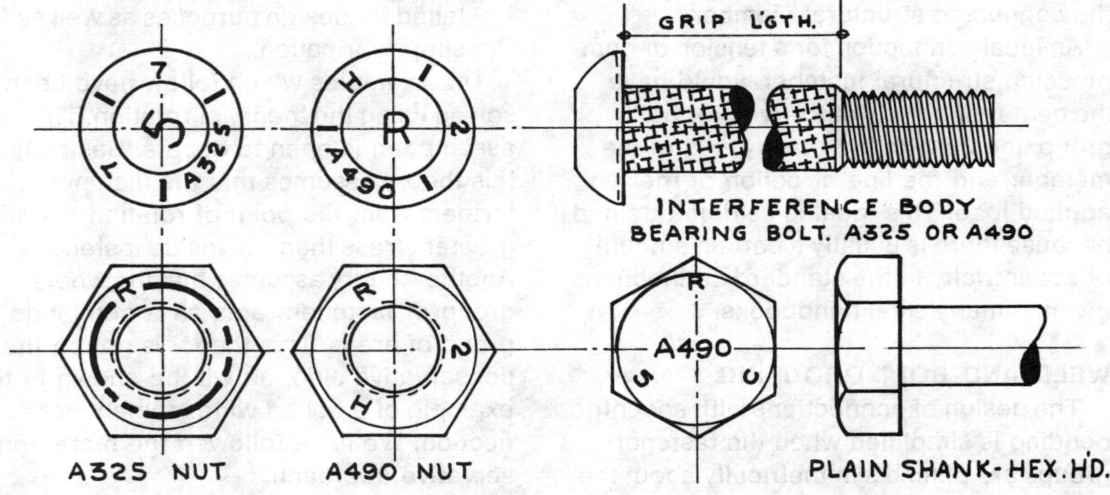

TABLE: Allowable tension and shear for bolts 2.10.5.2

ALLOWABLE TENSION & SHEAR VALUES FOR BOLTS, IN LBS.

TYPE OF STRESS			TENSION			SINGLE SHEAR			DOUBLE SHEAR		
BOLT DIAMETER IN INCHES	NET AREA AT SHANK IN SQ. INCHES	NET AREA AT ROOT IN SQ. INCHES	A-307	A-325	A-490	A-307	A-325	A-490	A-307	A-325	A-490
			F_t = 14 KSI	F_t = 40 KSI	F_t = 60 KSI	F_v = 10 KSI	F_v = 22 KSI	F_v = 32 KSI	F_v = 20 KSI	F_v = 44 KSI	F_v = 64 KSI
1/2	0.1963	0.1420	2750	7850	11,750	1963	4220	6280	3925	8440	12,560
5/8	0.3068	0.226	4300	12,270	18,400	3068	6750	9815	6135	13,500	19,630
3/4	0.4418	0.334	6175	17,670	26,410	4418	9720	14,135	8835	19,440	28,270
7/8	0.6013	0.462	8400	24,050	36,000	6013	13,230	19,240	12025	26,460	38,480
1.0	0.7854	0.606	11,000	31,420	47,125	7854	17,280	25,130	15,700	35,560	50,260
1 1/8	0.9940	0.763	13,940	39,760	59,640	9940	21,865	31,800	19,880	43,730	63,600
1 1/4	1.2272	0.969	17,200	49,100	73,620	12,270	27,060	39,265	24,540	54,120	78,530
1 3/8	1.4894	1.155	20,800	59,400	89,100	14,850	32,660	47,680	29,700	65,320	95,360
1 1/2	1.7671	1.405	24,670	70,800	106,000	17,670	38,875	56,640	35,340	77,750	113,280
1 5/8	2.0739	1.654	29,036	82960	124,440	20,740	45,625	66,365	41,480	91,250	132,730
1 3/4	2.4053	1.900	33,600	96,000	144,600	24,050	53,000	77,120	48,100	106,000	154,240
2.0	3.1416	2.500	44,000	125,600	188,500	34,415	69,125	100,500	62,900	138,250	201,000

VALUES GIVEN IN TABLES ARE BASED UPON CROSS-SECTION AREA OF SHANK.

SHEAR VALUES ARE BASED ON THREADS EXCLUDED FROM SHEAR PLANE.

USE ALLOWABLE UNIT STRESSES FOR REDUCED VALUES WHEN THREADS ENCLOSED IN SHEAR PLANE.

TABLE: Dimensions for high-strength bolts 2.10.5.3

DIMENSION DATA FOR H.S. A325 BOLTS, NUTS & WASHERS

BOLT DIAMETER IN INCHES	LENGTH OF THREADS IN INCHES	STRUCTURAL HEAD BOLT HEIGHT IN INCHES	HEXAGON NUT HEAVY TYPE HEIGHT NUT IN INCHES	WIDTH NUT SHORT SIDE DIMENSION IN INCHES	ADD TO GRIP FOR LENGTH OF BOLT IN INCHES	CIRCULAR HARDENED WASHERS THICKNESS	OUTSIDE DIAMETER WASHER IN INCHES
1/2	1.000	0.31250	0.48438	0.8750	0.6875	0.109	1.375
5/8	1.250	0.39063	0.60938	1.0625	0.8750	0.134	1.500
3/4	1.375	0.46875	0.73438	1.2500	1.0000	0.148	1.750
7/8	1.500	0.54688	0.85938	1.4375	1.1250	0.165	2.000
1.0	1.750	0.60938	0.98438	1.6250	1.2500	0.165	2.250
1 1/8	2.000	0.68750	1.10938	1.8125	1.5000	0.165	2.500
1 1/4	2.000	0.78125	1.21875	2.0000	1.6250	0.165	2.750
1 3/8	2.250	0.84375	1.34375	2.1875	1.7500	0.165	3.000
1 1/2	2.250	0.93750	1.46875	2.3750	1.8750	0.165	3.375

BOLT LENGTHS SHOULD BE ADJUSTED TO NEXT LONGER 1/4 INCH LENGTH INCREMENT WHEN THREADS ARE TO BE EXCLUDED FROM SHEAR PLANE.
THIS TABLE NOT APPLICABLE TO INTERFERENCE-BODY BOLTS, HEADS OR NUTS.

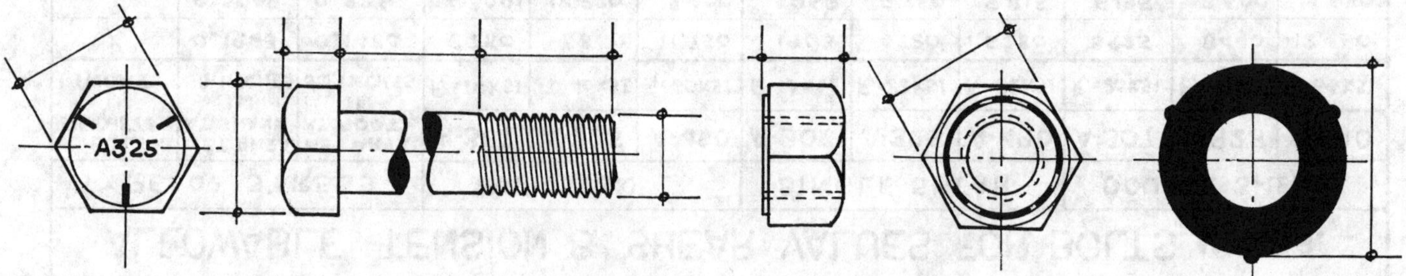

A325

TABLE: Bearing areas for bolts and rivets 2.10.5.4

AREAS OF BOLTS-RIVETS FOR CALCULATING BEARING, IN SQ. IN.

BEARING METAL THICKNESS IN INCHES		DIAMETER OF HOLE - RIVET OR BOLT, IN INCHES													
		0.250	0.375	0.500	0.625	0.750	0.875	1.000	1.125	1.250	1.375	1.500	1.625	1.750	2.000
		1/4	3/8	1/2	5/8	3/4	7/8	1.0	1 1/8	1 1/4	1 3/8	1 1/2	1 5/8	1 3/4	2.0
1/8	0.125	0.031	0.047	0.063	0.078	0.094	0.109	0.125	0.141	0.156	0.172	0.188	0.203	0.219	0.250
3/16	0.1875	0.047	0.052	0.094	0.117	0.141	0.164	0.188	0.211	0.234	0.258	0.281	0.305	0.328	0.375
1/4	0.250	0.063	0.093	0.125	0.156	0.188	0.219	0.250	0.281	0.313	0.344	0.375	0.406	0.438	0.500
5/16	0.3125	0.078	0.117	0.156	0.195	0.234	0.274	0.313	0.352	0.391	0.430	0.469	0.508	0.547	0.625
3/8	0.375	0.094	0.140	0.188	0.234	0.281	0.328	0.375	0.422	0.469	0.516	0.563	0.609	0.656	0.750
7/16	0.4375	0.109	0.164	0.219	0.273	0.328	0.383	0.438	0.492	0.547	0.602	0.656	0.711	0.766	0.875
1/2	0.500	0.125	0.188	0.250	0.313	0.375	0.437	0.500	0.563	0.625	0.688	0.750	0.813	0.875	1.000
9/16	0.5625	0.141	0.211	0.281	0.352	0.422	0.492	0.563	0.633	0.703	0.773	0.844	0.914	0.984	1.125
5/8	0.625	0.156	0.234	0.313	0.391	0.469	0.546	0.625	0.703	0.781	0.859	0.938	1.016	1.094	1.250
11/16	0.6875	0.172	0.258	0.344	0.430	0.516	0.600	0.688	0.773	0.860	0.945	1.030	1.117	1.200	1.375
3/4	0.750	0.188	0.281	0.375	0.469	0.563	0.656	0.750	0.844	0.938	1.031	1.125	1.219	1.313	1.500
13/16	0.8125	0.203	0.304	0.406	0.509	0.609	0.710	0.813	0.914	1.030	1.115	1.218	1.320	1.421	1.625
7/8	0.875	0.219	0.328	0.438	0.547	0.656	0.765	0.875	0.984	1.094	1.203	1.313	1.422	1.531	1.750
15/16	0.9375	0.234	0.351	0.469	0.586	0.703	0.820	0.938	1.055	1.170	1.288	1.405	1.523	1.640	1.875
1.0	1.000	0.250	0.375	0.500	0.625	0.750	0.875	1.000	1.125	1.250	1.375	1.500	1.625	1.750	2.000
1 1/8	1.125	0.281	.422	.563	0.703	.844	.984	1.125	1.266	1.406	1.547	1.688	1.828	1.969	2.250

AREA OF BEARING = BOLT DIAMETER TIMES THICKNESS OF METAL.
VALUE OF BEARING = BEARING AREA TIMES ALLOWABLE BEARING STRESS IN TABLE FOR F_p.

TABLE: Allowable bearing for high strength bolts 2.10.5.5

BEARING VALUES FOR PLAIN SHANK A325 BOLTS, IN LBS.

METAL THICKNESS IN INCHES		DIAMETER OF BOLT, IN INCHES										
		1/2	5/8	3/4	7/8	1.0	1 1/8	1 1/4	1 3/8	1 1/2	1 3/4	2.0
1/8	0.125	3030	3790	4550	5300	6060	6820	7580	8340	9100	10,600	12,125
3/16	0.187	4560	5680	6820	7960	9090	10,230	11,370	12,500	13,620	15,800	18,180
1/4	0.250	6060	7580	9090	10,610	12,130	13,640	15,160	16,675	18,180	21,200	24,260
5/16	0.313	7560	9470	11,370	13,260	15,160	17,050	18,950	20,850	22,750	26,500	30,320
3/8	0.375	9120	11,370	13,640	15,910	18,190	20,460	22,730	25,000	27,300	31,800	36,380
7/16	0.438	10,125	13,260	15,910	18,570	21,220	23,870	26,520	29,200	31,820	37,100	42,440
1/2	0.500	12,125	15,160	18,190	21,220	24,250	27,280	30,310	33,400	36,380	42,300	48,500
9/16	0.563	13,620	17,050	20,460	23,870	27,280	30,690	34,100	37,400	41,000	47,750	54,560
5/8	0.625	15,150	18,900	22,730	26,520	30,310	34,100	37,890	41,600	45,500	53,100	60,620
11/16	0.688	16,670	20,800	25,000	29,180	33,340	37,510	41,680	45,750	50,000	58,300	66,680
3/4	0.750	18,200	22,750	27,300	31,800	36,380	40,920	45,470	50,000	54,600	63,750	72,760
13/16	0.813	19,700	24,600	29,575	34,500	39,410	44,320	49,260	55,750	59,000	69,000	78,820
7/8	0.875	21,230	26,500	31,800	37,100	42,440	47,740	53,050	58,250	63,600	74,250	84,880
15/16	0.938	22,800	28,425	34,100	39,780	45,480	51,150	56,840	62,500	68,200	79,550	90,960
1.0	1.000	24,250	30,310	36,380	42,440	48,500	54,560	60,630	66,690	72,750	85,000	97,000

VALUES IN TABLE ARE IN POUNDS AND CALCULATED FOR CONDITIONS OF ENCLOSED BEARING. NOT APPLICABLE TO FRICTION SHANK TYPE CONNECTORS. VALUES APPLY TO BEARING $F_p = 48{,}500$ PSI AND METAL $F_y = 36{,}000$ PSI.
FRICTION BEARING FASTENERS CONSIDER SHEAR AND TENSION ONLY- SEE OTHER TABLE.

ILLUSTRATION: Typical column-beam connections 2.10.5.6

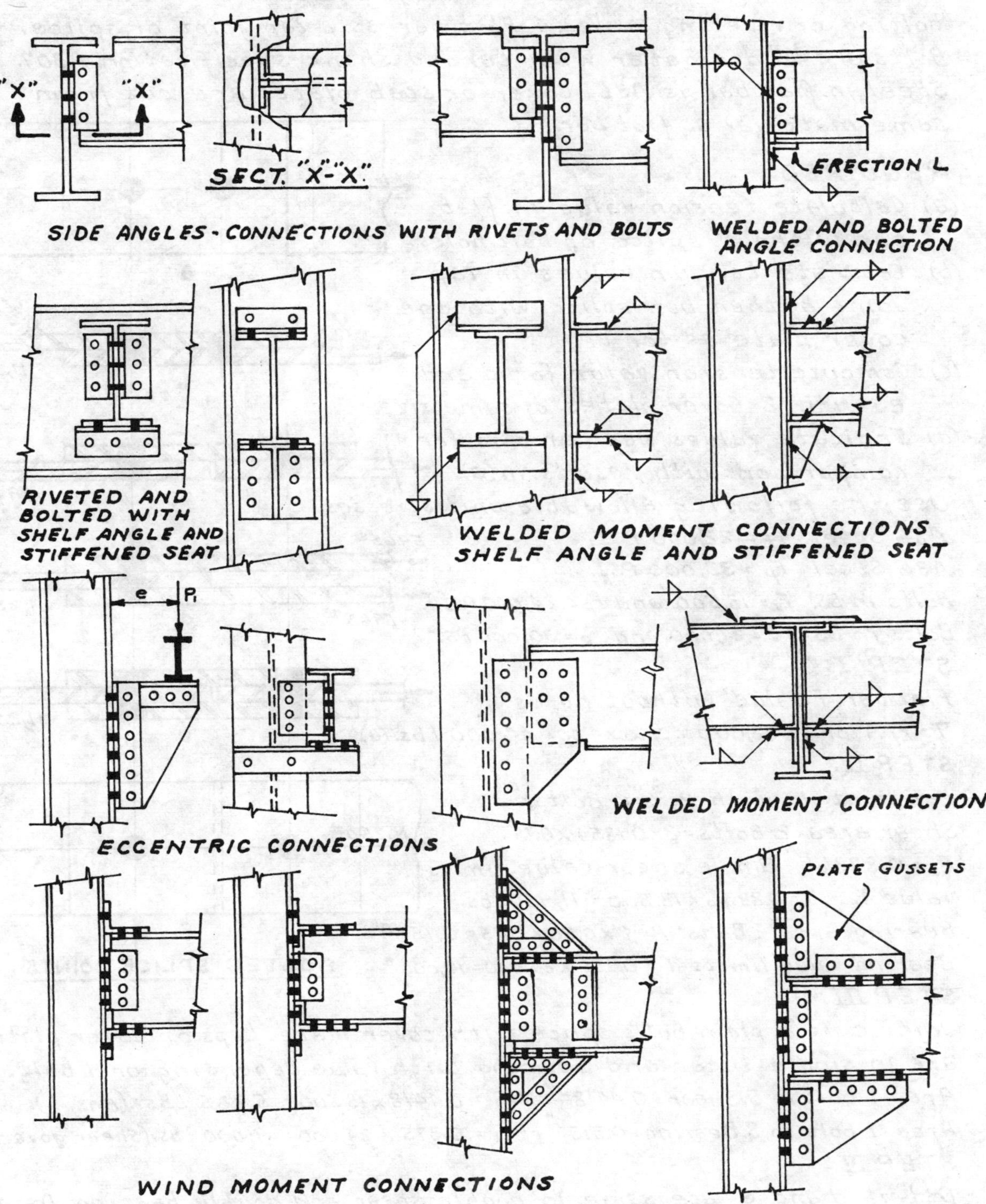

EXAMPLE: Comparing bolted splice joints 2.10.6.1

Illustrations at right delineate common methods used for bolting or riveting a ½ x 4 Flat bar as a lap joint or splice. Bolts are ¾" diameter with turned shank specification A307. Steel in flat bar is A36. Cover or scab plates are cut from same material as flat bar.

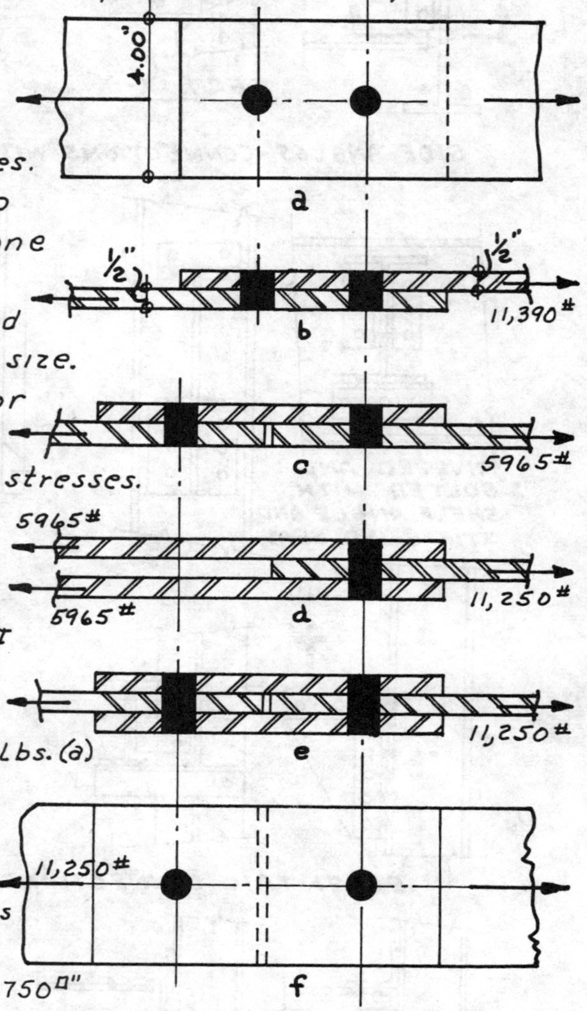

REQUIRED:

(a) Calculate tension value in flat bar without splice or bolt holes.

(b) Calculate tension values in lap Joint b, then butt splice with one cover plate as for c.

(c) Compute tension value for d and e with 2 Cover plates of same size.

(d) Indicate values near arrows for comparison with results in (a).

Use the following allowable unit stresses.

A36 Steel: F_t = 22,000 PSI.

A36 Steel: F_p = 33,000 PSI.

Bolts in SS: F_v = 13,500 and F_p = 24,000 PSI

Bolts in DS: F_v = 27,000 and F_p = 30,000 PSI

STEP I:

Flat bar T value without holes:

$T = F_t A$ or T = 22,000 x 0.50 x 4.00 = 44,000 Lbs. (a)

STEP II:

Lap Joint b with 2-¾"φ Bolts.

Shear area 2 Bolts = 2 x 0.7854 x 0.75²

A_s = 0.8836□" Single shear value limits value T. T = 0.8836 x 13.500 = 11,930 Lbs.

Bearing area 2 Bolts: A = 2 x 0.50 x 0.75 = 0.750□"

Bearing value limits T = 0.750 x 24,000 = 18,000#

BOLTED SPLICE JOINTS

STEP III:

Joint c is a plain butts splice with cover plate. Laps on cover plates are in single shear and bearing, with value depending on 1 bolt.

Area 1 bolt in S. Shear = 0.4418□" V = T = 0.4418 x 13,500 = 5,965 Lbs. (Ans. b).

Area 1 bolt in S. Bearing = 0.375□" P = T = 0.375 x 24,000 = 9,000 Lbs. (Shear governs)

STEP IV:

Detail d and e are same in double shear and double bearing. Detail d shows transfer of force.

1 Bolt in Double Shear = 0.4418 x 27,000 = 11,930 Lbs.

1 Bolt in Double bearing = 0.375 x 30,000 = 11,250 Lbs. (Governs T) (c).

1 Bolt in ½" Plate, bearing = 0.50 x 0.75 x 30,000 = 12,375 Lbs.

EXAMPLE: Calculating bending in angle clip 2.10.6.2

An anchor bolt with 1.0" diameter and type A307 is used to its full allowable tension value of 11,000 Pounds. An angle clip 7x4x½ is proposed to be used as clip for concrete anchorage. Length of angle is 3½ inches. Steel is A36 with F_b = 22,000 PSI. Long leg on angle is shop welded to column on both sides and toe. Total welding is 17½ inches of ⅜" fillets.

REQUIRED:
Calculate the bending moment in short leg and apply the formula to determine if angle leg thickness will be enough to resist tension in bolt. Make a sketch of connection and designate rotation point at column base.

STEP I:
The length of angle clip with a single bolt is considered the same as pitch (p).
From table of angle gages: g_1 = 2½" for 4" leg.
Trial thickness = 0.50"

STEP II:
Formula for leg thickness: $t = \sqrt{\dfrac{3M}{pF_b}}$
To find bending moment:

$$M = \frac{T(g_1 - t)}{2}$$

$$M = 11,000\left(\frac{2.50 - 0.50}{2}\right) = 11,000 \text{ inch lbs.}$$

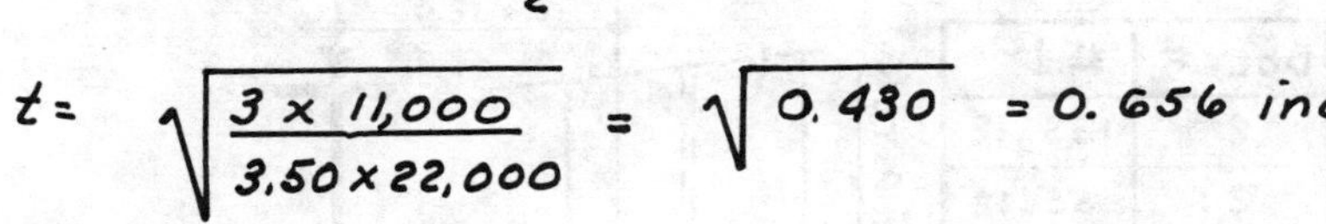

$$t = \sqrt{\frac{3 \times 11,000}{3.50 \times 22,000}} = \sqrt{0.430} = 0.656 \text{ inches.}$$

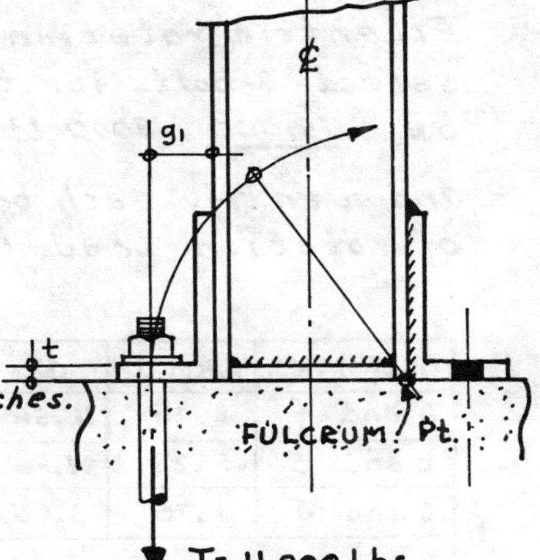

Change clip angle to L 7x4x¾".

EXAMPLE: Eccentric bolted clip design 2.10.6.3

A clip plate with a 3/8 inch thickness is proposed for a bolted connection which must transfer a hoist load of 18,000 Lbs. to flange of steel column. Load is 1'-3" from center line of column. Bolts are to be placed in a single row and consist of standard turned machined type with square heads and nuts.

REQUIRED:

Design the plate connection on column will all bolts of same diameter and length. Draw an elevation of connector and adjust thickness of plate if found necessary.

STEP I:

From allowable stress table for standard finished bolts: F_v = 13,500 PSI and F_p = 24,000 PSI. This is single shear and single bearing friction type connector. Due to eccentric load, there will be a rotating action about the center of gravity of bolt group.

STEP II:

Eccentric rotation will increase shear and bearing therefore select 6 bolts for trial. Vertical shear on each bolt will be $\frac{18,000}{6}$ = 3000 Lbs. Layout the arrangement on sketch and identify each bolt and its moment arm from CG or point of rotation. Load P on each lever = 3000 Lbs.

BOLT M'KS.	ARM d"	d^2	BOLT #	$\#d^2$
a and f	8.75	76.56	2	153.12
b and e	5.25	27.56	2	55.12
c and d	1.75	3.06	2	6.12
				Σ = 214.36

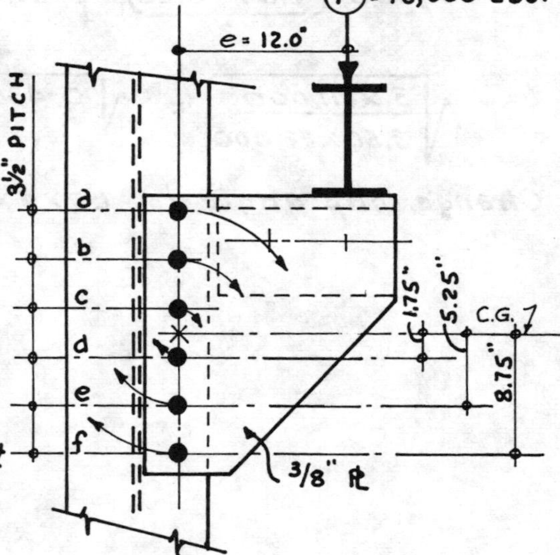

STEP III:

The eccentric bending moment = Pe

M = 12.0 x 18,000 = 216,000 inch lbs.

Summation of $\#d^2$ = 214.36, then the intensity of stress at hub point is equal to $\frac{M}{\Sigma d^2}$ or $\frac{216,000}{214.36}$ = 1008"#

EXAMPLE: Eccentric bolted clip design, continued 2.10.6.3

STEP IV:

The horizontal force on each bolt due to eccentricity is a shear force. Shear for each bolt: V_H = Force x d.

a and f = 1008 x 8.75 = 8816 Lbs.

b and e = 1008 x 5.25 = 5290 "

c and d = 1008 x 1.75 = 1763 "

STEP V:

Now have known force is two directions as vertical shear and horizontal shear. The resultant of these forces may be found by graphic force diagram or trig formula for diagonal as: $c = \sqrt{a^2 + b^2}$. Bolt placed at greatest distance will have larger resultant and govern design.

For bolts: a and f: $R = \sqrt{3000^2 + 8816^2} = 9275$ Lbs (Governs)

b and e: $R = \sqrt{3000^2 + 5290^2} = 6080$ Lbs.

c and d: $R = \sqrt{3000^2 + 1763^2} = 3475$ Lbs.

STEP VI:

Only 1 size of bolt to be used.

Area required for shear: $A = \frac{R}{F_v}$ or $A = \frac{9275}{13{,}500} = 0.6875$ □"

From tables: A ⅞"ϕ Bolt has an area of 0.6013 □" but check for unit stress. $f_v = \frac{9275}{0.6013} = 15{,}400$ #□" (Too high and will require a larger bolt. Dia. 1.0" has A = 0.7854 $f_v = \frac{9275}{0.7854} = 11{,}800$ PSI. (Use 1.0")

Checking pitch: Min. = 3 diameters. Have 3½ inches and OK.

STEP VII:

Plate is ⅜" and in single bearing. Assume column flange is more.

Area of bearing = 0.375 x 1.0 = 0.375 □" $f_p = \frac{9275}{0.375} = 24{,}800$ PSI.

Stress is above the allowable only a small amount and is acceptable. Bearing stress confirms the choice of one inch bolts. A ¾"ϕ Bolt would have been acceptable for shear result as obtained in step VI.

EXAMPLE: Eccentric bolted connector design 2.10.6.4

A standard [10×20 is bolted to flange of a W10×66 wide flange column. Connection uses 4 Bolts 3/4 in diameter. Horizontal gage is 6.0 inches and vertical gage is 5.0 in. Channel supports a hoist monorail load of 12,500 at an eccentric distance of 10.0 inches from column axis y-y.

REQUIRED:

Make a detail of connection and calculate forces involved in rotation. Design for bolt type and determine unit shear and bearing stresses.

STEP I:

Detail is drawn at right with gage dimensions. This joint will revolve about the center of gravity of 4 fasteners. Hub is at center of diagonals. Each bolts distance from hub is the moment arm for bolts.

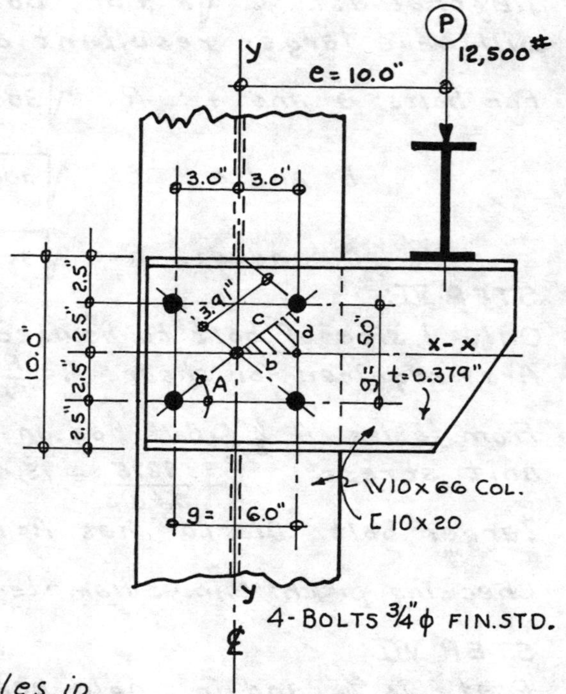

STEP II:

A triangle is formed at each bolt. Let horizontal side = b and vertical size = a. Diagonal side will be side c and arm distance.

$b = 3.00"$ $a = 2.50"$ and $c = \sqrt{a^2 + b^2}$

$c = \sqrt{6.25 + 9.00}$ = 3.91 inches.

Find angle A: $\text{Tan. } A = \frac{a}{b}$ or

$\text{Tan } A = \frac{2.50}{3.00} = 0.83333$ From Trig tables in Section V. Angle A = 39° 48' and Sine A = 0.64011 Cosine A = 0.76828

STEP III:

Vertical shear is resisted by 4 · 3/4" ϕ Bolts

V = 12,500 Lbs. For each bolt $v = \frac{12,500}{4} = 3125$ Lbs.

Additional shear and bearing forces are produced by eccentric lever in vertical and horizontal action planes. Design will be based upon the Resultant of these forces.

EXAMPLE: Eccentric bolted connector design, continued 2.10.6.4

STEP IV:

Eccentric bending moment = Pe. P= 12,500 Lbs. and e= 10.0 in. M = 12,500 x 10.0 = 125,000 inch Lbs. Moment for each bolt is $\frac{125,000}{4}$ = 31,250 inch lbs. Resolve moment into a force acting along diagonal line c. The force formula is derived from the stress formula $f = \frac{M}{n A d^2}$, or $F = \frac{M}{d^2}$. distance d is c or 3.91 inches.

Force in 1 bolt on diagonal = $\frac{31,250}{3.91 \times 3.91}$ = 2050 Lbs.

STEP V:

With force c known, solve for horizontal force b as: b= c Cos. A. Horizontal force in bolt = 2050 x 0.76828 = 1575 Lbs.

Vertical force from rotation when c = 2050#. a= c Sine A, or a = 2050 x 0.64011 = 1315 Lbs.

STEP VI:

Summarize forces to calculate resultant force in one line of action.

Horizontal force: 1575 Lbs. (b)

Two vertical forces = 3125 + 1315 = 4,440 Lbs. (a)

Resultant = $\sqrt{a^2 + b^2}$

$R = \sqrt{1575^2 + 4,440^2}$ = 4700 Lbs. (Design force).

STEP VII:

Cross section of a 3/4" ϕ diameter bolt = $0.7854 \times 0.75^2 = 0.6013$ □"

Actual shear stress: $f_v = \frac{4700}{0.6013}$ = 7840 PSI. Allowed A307: F_v = 13,500 PSI.

Thickness of channel web = 0.379 inches.

Thickness of Column flange = 0.748 inches.

Bearing area bolt on channel web = 0.75 x 0.379 = 0.284 Sq. In.

Actual single bearing stress: $f_p = \frac{4700}{0.284}$ = 16,550 Lbs. Sq. Inch.

Allowable single bearing stress for A307 Finished Bolts = 24,000 PSI.

Allowable bearing for A36 steel on reamed holes or turned and milled surfaces F_p = 33,000 PSI.

Accept 3/4" ϕ Bolts A307 turned shank for connection.

EXAMPLE: Bolted moment connection design 2.10.6.5

An eccentric load P= 22,750 Lbs. is to be supported with a short cantilever W12x45 beam section connected by bolts to a column flange 8.0 inches wide and 0.618 inches thick. Eccentric distance for angle and bolt connection is 18.0 inches. Clip angle at top and bottom is an 8x4x5/8 L and full length of flange. Clips are welded to cantilever beam at shop and need no examination. Either standard turned or high strength A325 Bolts will be accepted.

REQUIRED:

Design and make a detail of the connection. Also check the leg thickness of clip angle to resist the bending stress.

STEP I:

Obviously this is to be a moment connection where bolts are in shear, bearing and tension stress.
Bending moment = Pe or M= 22,750 x18.0 = 409,500 inch pounds.
Assume short leg of angle is connected to column and only one gage line is available. From table: g_1 = 2½ inches.
Flange width of W12x45 = 8.00" and depth = 12.00"
Distance between connecting bolt lines = 12.0 + $2g_1$ = 17.0 inches.

STEP II:

Drawing a side elevation to determine the moment arm for top bolts in tension and calculate required number.
Point of rotation axis is bottom of beam. Lever distance to tension bolts = $d+g_1$ or 14.50 inches. (ℓ).
Total tension in Bolts. $T = \frac{M}{\ell}$ or $T = \frac{409,500}{14.5}$ = 28,300 Lbs.

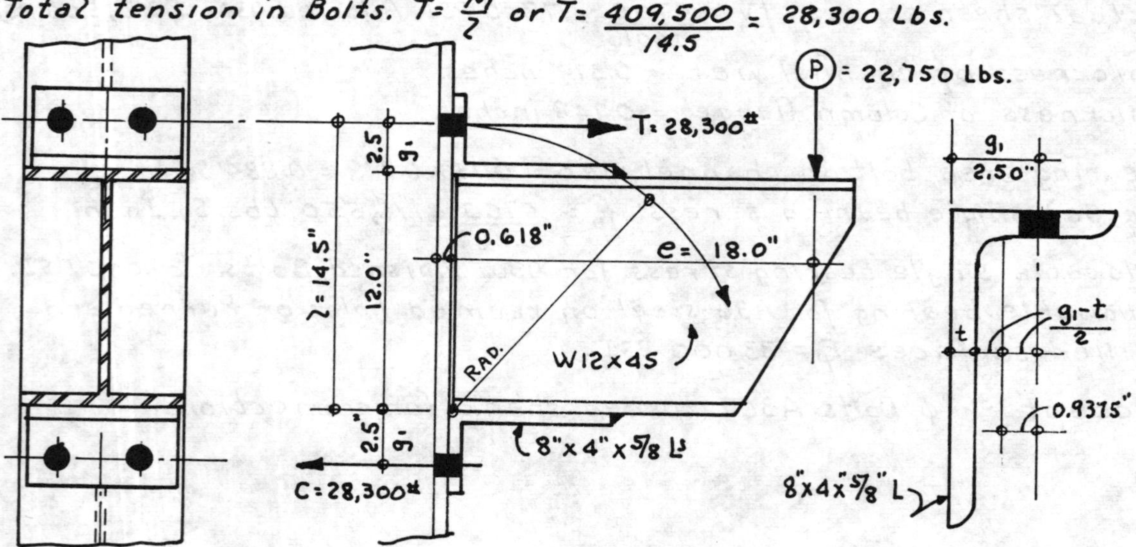

EXAMPLE: Bolted moment connection design, continued 2.10.6.5

Try selecting the least number of bolts which appear to have all requirements for tension, shear and bearing.

For tension of 28,300 Lbs. try 2-Bolts good for 14,150 Lbs. min.
For shear of 22,750 Lbs. try 4-Bolts good for 5,690 Lbs. min.

STEP III:
From tables of bolt values:
A plain High Strength A-325 3/4"ϕ Bolt has these values:
Tension = 17,670 Lbs. (more than required).
S. Shear = 9,720 " " "

STEP IV:
Check a 3/4" diameter bolt for bearing. Allowable F_p = 33,000 PSI when used with A36 steel. Friction type on reamed holes.
Angle leg t = 0.625" D = 0.75" Bearing A = 0.625 x 0.75 = 0.469 □"
Actual bearing, $f_p = \frac{5690}{0.469} = 12{,}140$ PSI. (OK, below F_p).

Check bearing on Column flange when t = 0.618 inches.
A = 0.618 x 0.75 = 0.464 □" $f_p = \frac{5690}{0.464} = 12{,}250$ PSI. OK.
Accept 4-3/4" ϕ Bolts A325 for connection to Column.

STEP V:
To investigate the bending in short leg of clip angle.

From table listing gage dimensions of angles:
For 4.0" leg, g_1 = 2.50 inches. The moment arm $= \frac{g_1 - t}{2}$ or lever $l = \frac{2.50 - 0.625}{2} = 0.9375$ inches. The tension force in a single 3/4"ϕ Bolt = 14,150 Lbs. M = 14,150 x 0.9375 = 13,260 in. Lbs.

Formula for calculating required $t = \sqrt{\frac{3M}{pF_b}}$. Where:
p = pitch or spacing of bolts, and F_b = allowable bending stress for angle steel. p = 4.0 inches.
Then:

$$t = \sqrt{\frac{3 \times 13{,}260}{4.0 \times 24{,}000}} = \sqrt{0.414} = 0.644 \text{ inches.}$$

A 5/8 inch angle is 0.625 inch and close enough to accept as next size is 3/4 inch and much too large.

EXAMPLE: Wind moment bolted connection 2.10.6.6

The bending moment in a beam connection at Column is calculated at 80,000 Foot Pound and reaction from static loading is 72,000 Pounds. Wind from east direction produces tension above beam axis, and when wind direction is from west, the tension is below axis x-x, therefore a joint at column should be made symmetrical about axis x-x.

REQUIRED:

Design the connection to be erected to column with standard A-307 Bolts of ⅞ inch diameter with 3.0" minimum pitch. Use 4x4 clip angles. Make alternate design and details thus:

(a) A shop fabrication on beam and clip angles bolted to flange of column. All shop fabrication to be welded.

(b) Weld clip angles to column flange and brackets to beam. Erection to be with bolts through angle legs and webs of beam and brackets. Change to H.S. Bolts if required.

STEP I:

A preliminary survey must be made to ascertain the number of bolts required to resist the bending moment and to have a moment there must be a lever arm.

From table of bolt properties and values:

A ⅞" φ Bolt A307 has a Tension value of 8400 Pounds. This value is maximum and will have the greater moment arm.

Cross section area = 0.6013 □" and value in single shear = 6013 Lbs.

STEP II:

To find number of bolts for trial, assume a coupling (moment arm) of approximately 15.0 inches which work well with a pitch spacing of 3.00 inches. Force of tension $T = \frac{M}{2}$, but moment must be converted to inch pounds.

$M = 80{,}000 \times 12 = 960{,}000$ inch pounds. $T = \frac{960{,}000}{15.0} = 64000$ Lbs.

Using the average value of bolt to be estimated at 4000 Lbs., the number required $= \frac{64000}{4000} = 16$ Bolts.

STEP III:

To determine thickness of angle leg: Let P = value of 1 Bolt or 8400 Lbs. With A36 Steel, $F_b = 22{,}000$ PSI and pitch $p = 3.00$ inches.

Gage for 4.00" L: $g' = 2.50"$ Assume $t = 0.50"$ Moment arm $= \frac{g_1 - t}{2}$ or

$2 = \frac{2.50 - 0.50}{2} = 1.00"$ Then $M = 8400 \times 1.00" = 8400$ inch lbs.

By Formula: $t = \sqrt{\frac{3M}{PF_b}}$ or $t = \sqrt{\frac{3 \times 8400}{3.0 \times 22{,}000}} = \sqrt{0.382} = 0.620$ inches.

Use 2 L^s 4×4×5/8" for clips.

STEP IV:

Detail sketch can now be constructed for further analysis.

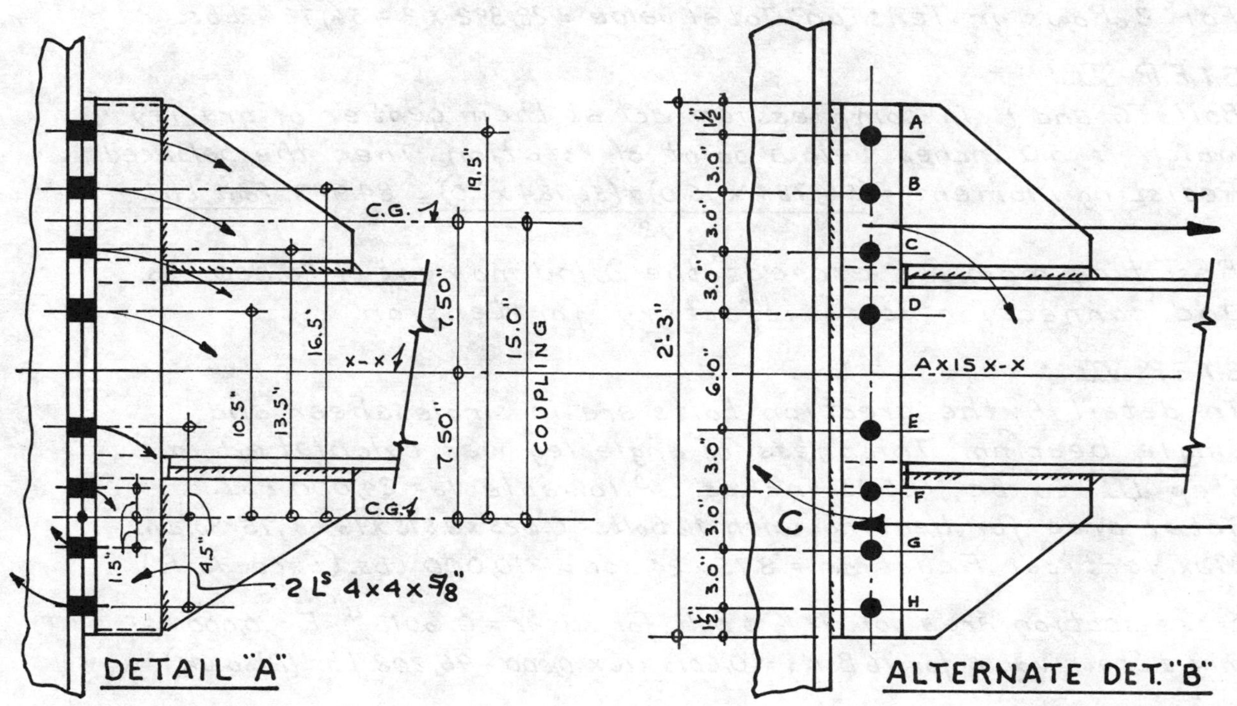

STEP V:

Identify each bolt in Tension row and give bolt "A" the maximum value 8400 Lbs. Compute the resisting values of other bolts by force diagram or by method of moments as used for reactions thus:

Bolt A = $\frac{8400 \times 19.50}{19.5}$ = 8400 Lbs. (1 Bolt)

Bolt B = $\frac{8400 \times 16.50}{19.5}$ = 7060 Lbs.

Bolt C = $\frac{8400 \times 13.50}{19.5}$ = 5820 Lbs.

Bolt D = $\frac{8400 \times 10.50}{19.5}$ = 4525 Lbs.

Bolt E = $\frac{8400 \times 4.50}{19.5}$ = 1940 Lbs.

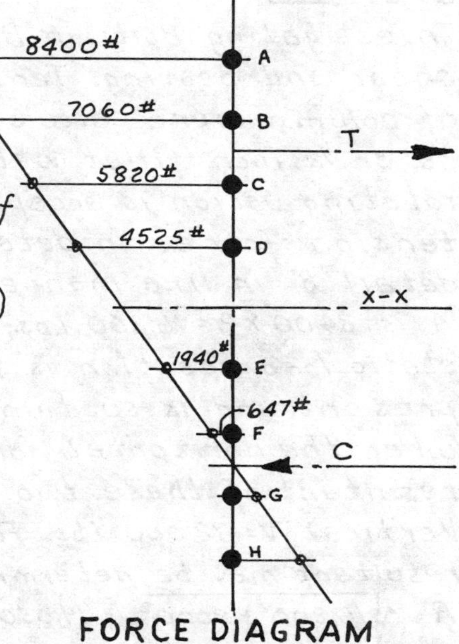

FORCE DIAGRAM

Scale: 1.0" = 5000#

EXAMPLE: Wind moment bolted connection, continued 2.10.6.6

Bolt F = 8400 × 1.50 = 647 Lbs.

Above bolts in tension due to location above rotating point. Bolts below are G and H with compressive values equal to tension. The moment arm is the coupling = 15.0 inches.

For 2 Rows in Tension: Total value = 28,392 × 2 = 56,784 Lbs.

STEP VI:

Bolts G and H in compression act at their center of gravity which is 3.0 inches below point of rotation. Then the balanced

$$\text{Resisting Moment} = \frac{(56,784 \times 15.0) + (56,784 \times 3.0)}{12} = 84,993 \text{ Foot Lbs.}$$

Resisting moment exceeds the wind moment of 80,000 Lbs. and connection is satisfactory for tension.

STEP VII:

In detail A, the erection bolts are in single shear and single bearing. Thickness of angle leg was calculated in Step III to be, 0.625 inches. Allowable F_p = 24,000 PSI.

Total area for bearing with 16 Bolts: 0.625 × 0.875 × 16 = 8.75 Sq. In.

Max. Vertical Reaction = 8.75 × 24,000 = 210,000 Lbs. (Exceeds R)

Cross section area of 1-7/8" ϕ Bolt for shear = 0.6013 □" F_v = 10,000 PSI

Single Shear value for 16 Bolts = 0.6013 × 16 × 10,000 = 96,208 Lbs. (Also OK).

STEP VIII:

Investigating Bolts in Detail B which number 8 in double shear and bearing. Tension values for welding clips to flange of column were calculated in step V and can be made with 3/8 or 1/2 inch fillet welds. When the assumed theory of a rotating action is accepted as the basic method for finding tension forces in detail "A", it should also be applied to detail "B" in like manner. The horizontal force in 2 top bolts A is 8400 × 2 = 16,800 Lbs., or as taken from step V. The total static load reaction is vertical shear and bearing. It is always present and is sustained by the number of fasteners. Then when the horizontal force from wind moment is applied, the resultant of these two forces becomes the basis for design.

Vertical V = 72,000 Lbs. For each bolt $v = \frac{72,000}{8} = 9,000$ Lbs. The resultant may be determined by force diagram or the equation:

$$R = \sqrt{16,800^2 + 9000^2} = 19,050 \text{ Lbs.}$$

Bolts in Detail B are in double shear and double bearing.

Thickness of angle leg = 0.625 inches Bolt diameter = 0.875 in.

Bearing area in 2 Legs = 0.625 x 0.875 x 2 = 1.09 □"

Actual bearing stress, $f_p = \frac{19,050}{1.09} = 17,460$ PSI.

Cross section area of 7/8" ϕ Bolt = 0.7854 x 0.875² = 0.601 □"

Actual shear stress, $f_v = \frac{19,050}{0.601} = \underline{31,750 \text{ PSI.}}$

The allowable bearing stress for A307 Bolts, DS F_v = 20,000 PSI., and for double bearing, F_p = 30,000 PSI.

Allowable shear stress is less than actual and bearing stress is close to limit. <u>Change over to A325 High Strength Bolts.</u>

STEP IX:

To check resultants of other bolts and bolts values with A307 Type

In shear: Max. V = 0.601 x 20,000 = 12,020# Bolt

In bearing: Max \ = 1.090 x 30,000 = 32,700# ✓

Resultants bolts B, C, D, E, F and G

Bolt B: $R = \sqrt{9000^2 + (7060 \times 2)^2} = 16,850$ Lbs.

Bolt C: $R = \sqrt{9000^2 + (5820 \times 2)^2} = 14,700$ Lbs.

Bolt D: $R = \sqrt{9000^2 + (4525 \times 2)^2} = 12,500$ Lbs.

Bolt E: $R = \sqrt{9000^2 + (1940 \times 2)^2} = 9,800$ Lbs.

Bolt F: $R = \sqrt{9000^2 + (647 \times 2)^2} = 9080$ Lbs.

Typical force diagram for all bolts is shown at right.

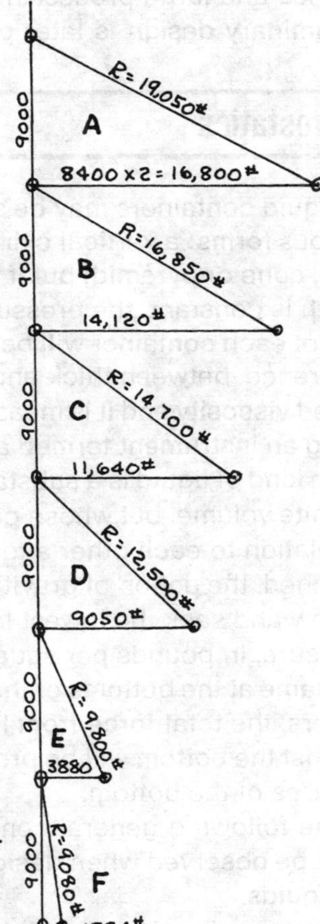

RESULTANT DIAGRAMS
Scale: 1.0" = 10,000#

DESIGN NOTE:-

In detail A as mentioned in step VI, the bolts in compression G and H act in theory only. The actual compression is resisted by end of beam, clip and bracket bearing against column. With a reverse wind direction the Tension T is on the action line C and Compression is at top of connection. Stress in bolt H then is reversed with bolt A.

Thin-walled cylinders 2.11

TANKS

Mechanical engineers are charged with the responsibility for designing pressure vessels and liquid storage tanks, because the design requires a thorough knowledge of hydrostatics. On many occasions however, the structural designer must make preliminary design calculations to be able to establish sizes and loads for supporting structure and foundation requirements. This situation frequently arises with structural design consultants who perform design work and cost estimates for refineries and large processing plants. The preliminary design is later placed with the clients own engineering staff for final chemical and mechanical design.

Cylindrical tanks and pressure vessels are referred to as thin-walled cylinders. In industrial processing plants, many specialized names will be used for the various vessels in a large complex. Tall upright cylindrical vessels may be used as refractor, TCC unit (Thermal Catalytic Cracking), stills, stacks, ethylene condensating unit, and many others. The design of vessels and piping which involves the viscosity of flowing liquids is beyond the scope of this book.

Hydrostatics 2.11.1

Liquid containers may be shaped in various forms: a vertical cylinder, funnel, dish, cone or pyramid; but if the liquid depth is constant, the pressure at the bottom of each container will be the same. The difference between thick and thin liquids is called viscosity and it is measurable by using an instrument termed a viscometer. Each kind of liquid is a substance that has a definite volume, but whose particles move in relation to each other and when not confined, the action of gravity will cause it to flow and seek the lowest level. Since the pressure, in pounds per square inch, is the same at the bottoms of the various containers, the total force from liquid pressure against the bottom will be proportional to the area of the bottom.

The following general considerations must be observed when designing vessels for liquids.

1. Pressure head is the height of the free liquid surface above the bottom (or some other reference point) in feet.
2. In any liquid at rest, the pressure increases directly with the depth.
3. Pressure is exerted with equal intensity in all directions.
4. This pressure acts normal to any surface in contact with the liquid.
5. For curved surfaces, the pressure acts normal to the surface at that point.
6. The total pressure against a submerged plane area is pA: where p = pressure in pounds per square inch, and A = area of the plane.
7. Liquid has a weight per unit volume, or density, given in pounds per cubic foot. The density of fresh water is usually taken as 62.50 pounds per cubic foot. At a depth of 15.0 feet, the pressure in water would be $p = W h / 144$ where W is the density of water in pounds per cubic foot, and h is the depth in feet, and 144 converts from pounds per square foot to pounds per square inch. A cubic foot contains 7.50 gallons.
8. The total force on the bottom of the container is related only to the liquid depth and the area of the bottom, not to the total weight of the liquid in the container. Paradoxically, this total force may be much greater or much less than the total weight of the contents.

Designing for hoop stress 2.11.2

When a thin-wall cylindrical pipe, representative of a steam boiler or a liquid vessel, is subjected to internal pressure, the forces acting to rupture the cylinder are in two directions. The first of these forces called hoop stress, acts tangent to the circumference tending to split the pipe open. The second force acts against the closed ends, tending to stretch the cylinder lengthwise. These forces are resisted by wall thickness in both cases. The stress in wall is resisted by the thickness (t) times the tension stress f_t.

To determine the stress in the cylinder wall, examine a short length of pipe, 3.0 inches long and 4.0 inches inside diameter. Split the pipe into two semi-circles lengthwise. At the plane of the cut, pressure is applied against the plane of each half to cause separation. The area of each plane is Dl, or diameter times length. Assume the internal pressure is p = 25 pounds per square inch. The force from pressure on each half = 4.0 x 3.0 x 25 = 300 pounds. At each wall cut, the wall must sustain one-half this force or 150 pounds. The formula for total force is $R = plD$. Now R is resisted by two pieces of metal of length l, and thickness t, with allowable tension stress F_t. So, $R = plD = 2f_t lt$.

To consider a unit length of only 1 inch, cancel the l from both sides of the equation. Then $pD = 2F_t t$ or: $f_t = \frac{pD}{2t} = \frac{pr}{t}$ since D = 2r where r = radius.
To solve for thickness required with the allowable tension stress F_t: $t = \frac{pD}{2F_t} = \frac{pr}{F_t}$.

Designing for longitudinal stress 2.11.3

The total longitudinal force P will be equal to the pressure times the area of the circle. P = pA. A circular ring of steel is in tension parallel to the axis of the cylinder. This is called longitudinal tension. The end force on the cylinder is $P = p\pi r^2$. The area of the ring in tension is $A = 2\pi rt$. When f_t = unit tension stress, then $P = f_t 2\pi rt = p\pi r^2$. Cancelling, the formula becomes:

$f_t = \frac{pr}{2t}$, or $t = \frac{pr}{2F_t}$.
When the formula for longitudinal stress is compared to the formula for hoop stress, it is found that the hoop stress is twice the longitudinal stress for any given pressure.

HEMISPHERIC ENDS

Cylindrical pressure vessels will have the same longitudinal stress whether the ends are flat plates or hemispheric. Flat end plates require stiffening and bracing at the connection to the cylinder walls so that the internal pressure does not bend the end plate and weaken the weld joint. Hemispheric ends transmit the end pressure to the pipe wall without bending.

EXAMPLE: Calculating wall thickness for pressure pipe 2.11.4

An oil well has been drilled to a depth of 8000 feet and the pressure gauge reading is 4800 PSI. Casing is to be inserted which will have an interior diameter of 3½ inches. After placing well will be capped with screwed fittings.

REQUIRED:

Calculate the required wall thickness for casing and force pressure on end for pipe threads. Use $F_t = 20{,}000$ PSI for the steel allowable unit working stress.

STEP I:

Using the formula for stress in cylinder walls. $f_t = \frac{pr}{t}$ in transposed form thus: $t = \frac{pr}{F_t}$. Substituting values:

$$t = \frac{4800 \times 1.75}{20{,}000} = 0.420 \text{ inches.}$$

STEP II:

For end pressure force, and using formula: $f_t = \frac{pr}{2t}$, to solve for $P = p\pi r^2$ and with values:

$P = 4800 \times 3.1416 \times 1.75 \times 1.75 = 46{,}180$ Pounds.

The longitudinal stress in cylinder walls from end pressure P is obtained by formula: $f_t = \frac{pr}{2t}$. Substituting known values:

$f_t = \frac{4800 \times 1.75}{2 \times 0.420} = 10{,}000$ PSI. (Checks, because it is one half design stress $F_t = 20{,}000$ PSI:

STEP III:

Pressure on end cap is equal to p times area of circle with inside diameter.

Area of small circle $= 0.7854\,D_i^2$ or $A = 9.6211\,\square''$

$P = pA$ or $P = 4800 \times 9.6211 = 46{,}180$ Pounds.

STEP IV:

The area of steel ring is in longitudinal stress due to force P on end cap. Heretofore the formulas neglected the diameter of greater circle and the steel between these circles must resist force P. Area outside Circle $= 0.7854 \times (D+2t)$. Then outside

$D_o = 3.50 + 0.420 + 0.420 = 4.340''$ $A = 0.7854 \times 4.340^2 = 14.793\,\square''$

Net area steel $= 14.7934 - 9.6211 = 5.1723$ Sq. Inches.

Actual stress $= \frac{P}{A}$ or $f_t = \frac{46{,}180}{5.1723} = 8740$ Lbs. Sq. Inch.

To be on safe side, the formula used for $A = 2\pi rt$. Let this be noted, thus $A = 2 \times 3.1416 \times 1.75 \times 0.420 = 4.618\,\square''$ Then result checks with step II where $\frac{P}{A} = \frac{46{,}180}{4.618} = 10{,}000$ Lbs. Sq. Inch.

EXAMPLE: Maximum pressure in welded cylinder 2.11.5

A Steel Pipe has an inside diameter of 15¼ inches with a wall thickness of 3/8 inch. Length of pipe is 7.0 feet between flat end plates which are fillet welded at inside circumference. Pipe wall and end plate material is comparable to A36 Steel which has an ultimate tensile strength of 66,000 PSI and an allowable F_t = 22,000 PSI. Weld allowable shall be based upon a tension weld of 13,500 PSI at root of weld.

REQUIRED:

Areas and Circumferences for large pipes may be obtained from tables in Section IX to calculate the following:

(a) Determine the internal pressure which will be applied to rupture cylinder walls.

(b) Calculate the safe working pressure when pipe with closed ends is fabricated into a steam pressure vessel.

(c) Compute the longitudinal tension force and stress in cylinder walls when pressure is in same amount as found in (b).

(d) Calculate Reaction against end plates and longitudinal stress in cylinder walls. Show details of welds.

STEP I:

Length of vessel has no bearing on stresses, therefore the detail need show only one end thus:

From Tables:

t = 0.375" Area steel ring = 18.41□"

Circumference = 47.91" (weld)

Surface Area end plate = 182.65□"

Design stress F_t = 22,000 PSI.

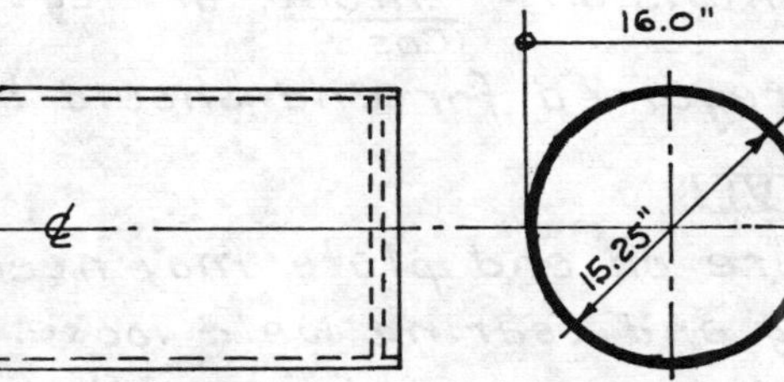

STEP II:

Pipe will probably burst at ultimate tension stress of 66,000 PSI. To find maximum rupture pressure, use formula: $p = \frac{F_t t}{r}$

r = 15.25 x 0.50 = 7.625"

Max. $p = \frac{66,000 \times 0.375}{7.625} = 3245.9$ PSI. (Ans. a)

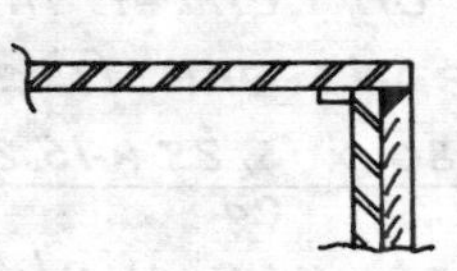

Safe pressure with allowable F_t = 22,000 = 1082 PSI (Ans. b)

For succeeding calculations call p = 1080 Lbs. Sq. Inch.

STEP III:

Longitudinal force on end plate must be resisted by fillet weld and material in steel ring.

EXAMPLE: Maximum pressure in welded cylinder, continued 2.11.5

Let total force of pressure on area of end plate be equal to reaction. $R = pA$ or $R = 1080 \times 182.65 = 197{,}262$ Pounds. Longitudinal press per lineal inch around ring is: $\frac{R}{C}$ where C = circumference or 47.91 inches inside at location of fillet welds. Required weld value $= \frac{197{,}262}{47.91} = 4120^{\#}$

STEP IV:

For longitudinal stress in cylinder walls: Area of steel in ring cross-section = 18.41 square inches. $f_t = \frac{R}{A}$

$f_t = \frac{197{,}262}{18.41} = 10{,}400$ Lbs. Sq. Inch. (Check this by formula and result from step III: Formula for longitudinal stress in cylinder: $f = \frac{pD}{4t}$ or $f_t = \frac{1080 \times 15.25}{4 \times 0.375} = 11{,}000$ PSI (Close enough.)

Also: $f_t = \frac{\text{Weld value}}{\text{plate } t}$ or $f_t = \frac{4120}{0.375} = 11{,}000$ PSI.

STEP V:

Design weld size with given allowable weld stress = 13,500 PSI. Plate in end should be same or thickness greater than leg dimension of weld. Welds are transverse to longitudinal force action line and are tension welds.

Required throat in weld $= \frac{T}{F_t}$ or $\frac{4120}{13{,}500} = 0.305$ inches.

Fillet weld angle = 45° and Cosine 45° = 0.707

Leg dimension $= \frac{\text{throat}}{\text{Cos. } A}$ or leg: $\frac{0.305}{0.707} = 0.432$ inches (7/16 in.)

Plate required for end should be ½ inch thick.

STEP VI:

Pressure on end plate may need to be stiffened to resist bulging and tearing weld loose. Applying a rule of thumb method as employed in similar conditions, assume a 1.0 inch strip on center line with length equal to diameter of 15.25 in. Both ends are fixed and $M = \frac{wl^2}{24}$. Pressure = w or 1080 Lbs. inch.

$M = \frac{1080 \times 15.25 \times 15.25}{24} = 10{,}460$ inch pounds. Because sides are in support, use yield stress to find Section Modulus: $F_y = 36{,}000$ PSI.

$S = \frac{M}{F} = \frac{10{,}460}{36{,}000} = 0.29''^3$ for 1.0 inch width. For rectangle 1.0×0.50 in.

$S = \frac{bd^2}{6}$ or $S = \frac{1.0 \times 0.50 \times 0.50}{6} = 0.0416''^3$. This is less than required and end plates need stiffening. Use ½×1¾ inch flat bars and weld on ends across longest diameter of plate.

EXAMPLE: Designing welded steel pressure vessel 2.11.6

A steel vessel has an overall length of 16.0 feet including hemisphere ends. Vessel is to rest in horizontal position and inside diameter is 70.0 inches. A manhole is to be placed in middle on top side with an inside diameter of 28.0 inches and the cover plate is to be bolted.
Design working pressure is to be a maximum of 200 PSI. Fabrication of cylinder wall and circumference seams at ends is to be fabricated with double vee welds. Allowable unit stress in base metal is not to exceed F_t = 22,000 PSI and allowable unit stress in weld material shall not exceed 12,000 p.s.i. for both shear and tension.

REQUIRED:

Make a drawing of vessel and show side and end elevations. Design for cylinder wall thickness and end thickness. Also design for size of weld throats and in case the weld size requires a greater thickness of wall or end plate, make the adjustment and note the size on drawing. Do not design for manhole or cover plate.

STEP I:

Elevations will be drawn to scale with probable foundation.

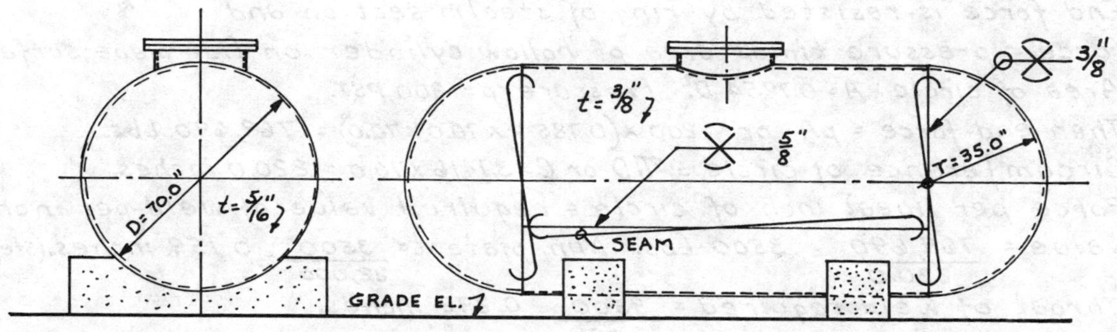

STEP II:

Using the formulas to determine wall and end thickness.

Cylinder wall $t = \frac{pr}{F_t}$. $t = \frac{200 \times 35.0}{22{,}000} = 0.318$ inches

End wall plate $t = \frac{pr}{2F_t}$ or $\frac{pD}{4F_t}$. $t = \frac{200 \times 70.0}{4 \times 22{,}000} = 0.159$ inches.

STEP III:

The thickness results in step II will now be investigated by other means and with calcutations for weld requirements.

EXAMPLE: Designing welded steel pressure vessel, continued 2.11.6

For the force at weld seam per lineal inch of cylinder, take a section of cylinder wall with a length of 1.0 inch, and cut the ring on ℄ into halves. See drawing of top half only.

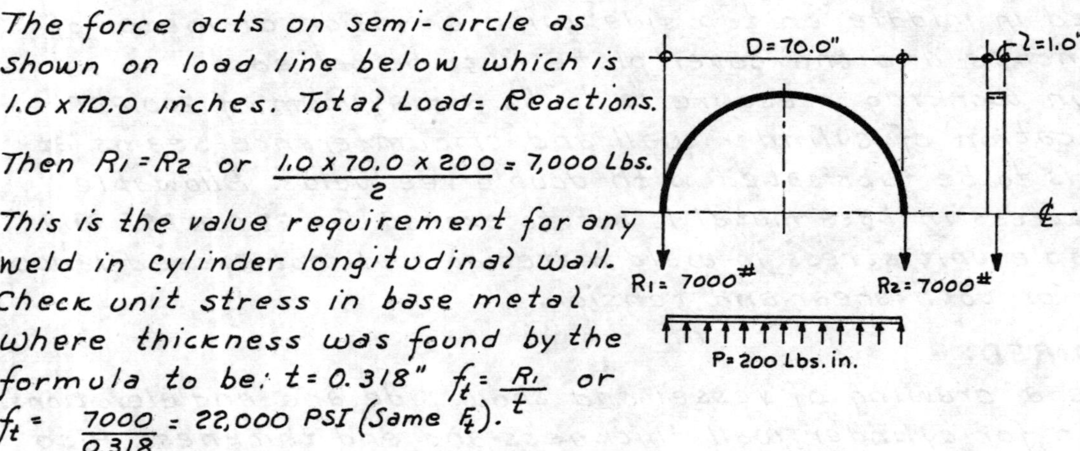

The force acts on semi-circle as shown on load line below which is 1.0 x 70.0 inches. Total Load = Reactions.

Then $R_1 = R_2$ or $\frac{1.0 \times 70.0 \times 200}{2} = 7{,}000$ Lbs.

This is the value requirement for any weld in cylinder longitudinal wall.

Check unit stress in base metal where thickness was found by the formula to be: $t = 0.318"$ $f_t = \frac{R_1}{t}$ or $f_t = \frac{7000}{0.318} = 22{,}000$ PSI (Same F_t).

STEP IV:

To determine throat size for a double-vee when weld unit allowable is: $F_t = 12{,}000$ PSI. Throat $= \frac{R}{F_t}$ or $\frac{7000}{12{,}000} = 0.582$ inches.

Use a 3/8" plate thickness with a 1/8" built up surface both sides.

STEP V:

Cylinder wall — 1/8" — 3/8" — 5/8" Throat — 1/8"

For pressure on end and longitudinal stress.

End force is resisted by ring of steel in section and force = pressure times area of hollow cylinder on flat plane surface.

Area of circle: $A = 0.7854\,D^2$ Pressure $p = 200$ PSI.

Then end force $= pA$ or $200 \times (0.7854 \times 70.0 \times 70.0) = 769{,}690$ Lbs.

Circumference of circle $= \pi D$ or $C = 3.1416 \times 70.0 = 220.0$ inches.

Force per lineal inch of circle = required value of weld per inch.

Value $= \frac{769{,}690}{220.0} = 3500$ Lbs. Min. plate $t = \frac{3500}{22{,}000} = 0.159$ inches. (step II).

Throat of weld required $= \frac{3500}{12{,}000} = 0.292$ inches.

End thickness will require a plate thickness of 5/16" (0.3125).

STEP VI:

Check the actual working stresses in cylinder wall of 0.375" and end wall of 0.3125 in. Transpose the formulas used in step II to do this:

At cylinder walls: $f_t = \frac{pr}{t}$ or $f_t = \frac{200 \times 35.0}{0.375} = 18{,}650$ P.S.I. (OK)

At end walls: $f_t = \frac{pD}{4t}$ or $f_t = \frac{200 \times 70.0}{4 \times 0.3125} = 11{,}150$ P.S.I. (OK).

EXAMPLE: Designing gravity pressure storage tank 2.11.7

A water storage tank is to have a volume capacity between 185,000 and 200,000 Gallons to supply a sprinkling system in a factory building. Height of water head in tank and 10.0 inch pipe outlet must provide a pressure at grade level of 65 PSI. Inside diameter is to be 25.0 feet and have a hemisphere type bottom. All fabrication is to be of steel plate with allowable design stress F_t = 22,000 PSI. Welding electrodes unit allowable for tension and shear shall not exceed 13,500 PSI. In no case shall plate thickness be less than ¼ inch at cylinder or end walls.

REQUIRED:

Determine size and required height of tank. Assume that vessel will be supported with 4 steel columns placed in 90° spacing and vertical cylinder walls will be supported on ring which joins hemisphere bottom to cylinder. Calculate the total weight of steel and liquid when tank is filled to top. Include a 4.0 high cone type cover for tank. Make an elevation drawing and determine reactions for the footings in tons.

STEP I:

Pressure at grade level will depend on height of water head and taken at the highest point. Weight of water = 62.5 Lbs. Cu. Foot., and 1 Cubic foot of liquid = 7.48 Gallons. Pressure required = 65 PSI.
To find required height of water and tank:
The pressure per square inch at bottom of 1 Cubic foot of water = $\frac{62.5 \times 1.0'}{12 \times 12}$ = 0.434 PSI. Transposing equation to find height of head: $H = \frac{P}{0.434}$ or $H = \frac{65}{0.434}$ = 150.0 feet.

STEP II:

Before vertical cylinder height can be computed, the volume of water in hemisphere bottom must be known. The solidity of a full sphere is: $D^3 \times 0.5236$ and a hemispheres volume is ½ of a sphere: Then volume in bottom is: $\frac{25.0^3 \times 0.5236}{2}$ = 4090 cu. ft.
Weight of water in bottom = 4090 × 62.5 = 255,625 Lbs.
Volume of water in bottom = 4090 × 7.48 = 30,590 Gallons.

STEP III:

Required height of cylinder; For volume per each 1.0 foot height is: $0.7854\ D^2$ = V = 0.7854 × 25.0 × 25.0 = 490.8 cu. feet. Volume of water per foot = 490.8 × 7.48 = 3670 Gallons.
Then 44.0 feet of water in cylinder = 3670 × 44.0 = 161,480 Gallons.

EXAMPLE: Designing gravity pressure storage tank, continued 2.11.7

Volume of water in bottom and 44.0 Cylinder:

30,590 + 161,480 = 192,070 Gallons: (Accept capacity as 192,000 Gals.)

Then total weight of water = $\frac{192{,}000 \times 62.5}{7.48}$ = 1,605,600 Lbs.

Reducing to tons:

Weight water = $\frac{1{,}605{,}600}{2000}$ = 802.8 Tons.

STEP IV:

Now have necessary data to draw an elevation of tank and make further calculations for steel sizes and weight of empty tank.

STEP V:

Force against weld will be maximum at juncture of cylinder and hemisphere. Head at ring is 44.0 feet and water pressure acts perpendicular to cylinder walls.

p = 0.434 × 44.0 = 19.10 PSI. Min. plate t = 0.25"

stress $f_t = \frac{pr}{t}$, or $f_t = \frac{19.10 \times 12.50 \times 12}{0.25}$ = 11,460 P.S.I

Value of weld required. Use formula: $P = \frac{pDl}{2}$

Let l = 1.0 inch ring width for weld.

D = 25.0 × 12 = 300 inches   p = 19.10 PSI

Value of 1.0" Lineal weld = $\frac{19.10 \times 300.0 \times 1.0}{2}$ = 2865#

Should be same as equation above transposed: P = pr, or 19.10 × 150.0 = 2865#

Weld stress F_t = 13,500 PSI. Throat = $\frac{2865}{13{,}500}$ = 0.212"

Use a single Vee weld full depth of plate.

STEP VI:

To calculate weight of steel tank in empty condition, the area of surfaces is computed thus:

Surface of sphere: A = 3.1416 D² and the bottom is ½ of a sphere. With values.

Bottom A = $\frac{3.1416 \times 25.0 \times 25.0}{2}$ = 981.75 Sq. Feet.

ID = 25.0'
4.0'
CYL = 44.0'
IDR = 12.5'
COLS. = 106.0'
HEAD = 150.0'

EXAMPLE: Designing gravity pressure storage tank, continued 2.11.7

STEP VII:

Solving for surface area of cone cover. Let D = 26.0 and plate thickness also ¼ inch thick. Slant height of cover = 4.0 feet.

Apron around lid can be a ¼" x 4.0 flat bar with D = 26.0'

Formula to find surface area of a cone is written:

$A = \left(C\,\frac{h}{2}\right) + (0.7854\,D^2)$. Circumference = 3.1416 x 26.0 = 81.68 Sq. Ft.

Slant height $\left(\frac{h}{2}\right) = \frac{4.00}{2} = 2.0$ Then: 81.68 x 2 = 163.36 □' plus area of flat surface: $A = 0.7854 \times 26.0^2 = 530.93$ Sq. Ft.

Area of apron: A = Cd or A = 81.68 x 0.333 = 27.23 Sq. Ft.

Total area of Cone lid = 163.36 + 530.93 + 27.23 = **721.52 Sq. Ft.**

STEP VIII:

Calculating surface area of vertical cylinder.

H = 44.0' D = 25.0 Circumference = πD. C = 3.1416 x 25.0 = 78.54 □'

Cylinder surface area = 78.54 x 44.0 = 3455.75 □'

Total Areas: Bottom, Top and cylinder:

ΣA = 981.75 + 721.52 + 3,455.75 = 5159.02 Square feet.

STEP IX:

Calculated and estimated weights for footing loads:

All plate in tank = ¼" and weight = 10.20 #□' (Table in Section VI)

Empty Tank weight = 5159.02 x 10.20 = 52,622 Pounds.

Weight of water in full tank: Step III = 1,605,600 "

Estimate Columns, bracing, ladder, walkway, railing and 10.0 inch pipe to add another 42,000 Pounds.

Total weight above grade $= \frac{52,622 + 1,605,600 + 42,000}{2000} =$ **850.11 Tons.**

Load for each footing to sustain $= \frac{850.11}{4} = 212.5$ Tons.

EXAMPLE: Wood hooped water tank design 2.11.8

A cypress silo is to be assembled with vertical staves and is to have an inside diameter of 10.0 feet. Height of staves is limited to 16.0 feet. Joints between staves are milled to provide interlocking with splines. Bands around outside circumference are to be provided and consist of 5/8 inch diameter rods with a design allowable $F_t = 16{,}000$ PSI. Completed silo will be tested by filling with water until tightness is assured.

REQUIRED:

Provide drawings for silo and design for hoop spacing. Use a weight of 62.5 Lbs. cubic foot for design basis.

STEP I:

Tank section will drawn to scale and hoop locations added after computations made.

The tension value of a $\frac{5}{8}''\phi$ hoop = AF_t.

Each hoops value: $T = 0.3068 \times 16{,}000 = 4920$ Lbs.

Pressure at lowest point in tank is at the 15.0 foot depth. Head of water is in feet, and pressure at bottom = 6.51 PSI or $0.434 \times 15.0 = 6.51$ PSI.

Another equation: $p = \dfrac{62.50 \times 15.0}{12.0 \times 12.0} = 6.51$ PSI.

STEP II:

The rectangle drawn at right as UVWX will represent a short section of tank with dimension Q unknown. Total force against this rectangle will have two reactions R_1 and R_2. The force against this rectangle is thus: pDQ, and $D = 120.0''$

Although pressure will be higher in intensity at line x-w, it will safe when applied to full height of Q. Hoop will then be located at point $\frac{Q}{2}$.

STEP III:

Total pressure = 2R because other half of circle also has the same R value. For the working area of hoop, the following is derived for equilibrium:

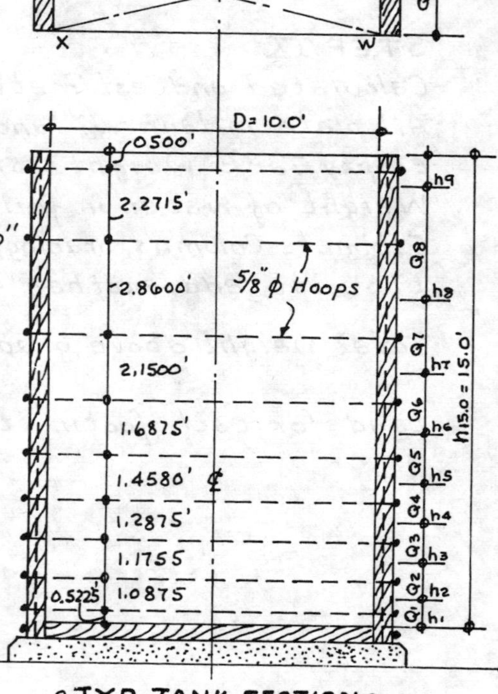

TYP. TANK SECTION

$$2R = \frac{2 \times \pi d^2 \times F_t}{4} \quad \text{and values:}$$

$$2R = \frac{2 \times 3.1416 \times 0.625 \times 0.625 \times 16{,}000}{4} = 9816 \text{ Lbs.}$$

EXAMPLE: Wood hooped water tank design, continued 2.11.8

The force of pressure on line x-w at 15.0' depth and when Q = 1.0". Force = 0.434 × 15.0 × 120 = 781.2 Lbs.

Then $Q_1 = \frac{9816}{781.2} = 12.56$ in. Place hoop h_1 at $\frac{12.56}{2} = 6.28$ inches from bottom. Convert Q_1 to feet to obtain Q_2 point = 1.05'

STEP IV:

The next lowest head of water will be 15.00 - 1.05 = 13.95 ft. The formula can be reduced to save time on other hoops. If 0.434 h = psi. Then 0.434 D = pressure for horizontal.

$0.434\,D = 0.434 \times 120 = 52.08$ and $Q = \frac{9816}{52.08\,h}$

STEP V:

Dimensions for heads (h) and height (Q) will be required for drawings and should be understood by draftsmen.

$h_2 = 15.00 - 1.05 = 13.95'$ $Q_2 = \frac{9816}{52.08 \times 13.950} = 13.55$ In. or 1.130 Feet

$h_3 = 13.95 - 1.130 = 12.820'$ $Q_3 = \frac{9816}{52.08 \times 12.820} = 14.67"$ or 1.221 Feet.

$h_4 = 12.820 - 1.221 = 11.599'$ $Q_4 = \frac{9816}{52.08 \times 11.599} = 16.25"$ or 1.354 Feet.

$h_5 = 11.599 - 1.354 = 10.245'$ $Q_5 = \frac{9816}{52.08 \times 10.245} = 18.38"$ or 1.562 Feet

$h_6 = 10.245 - 1.562 = 8.683'$ $Q_6 = \frac{9816}{52.08 \times 8.683} = 21.76"$ or 1.813 Feet

$h_7 = 8.683 - 1.813 = 6.870'$ $Q_7 = \frac{9816}{52.08 \times 6.870} = 27.44"$ or 2.287 Feet

$h_8 = 6.870 - 2.287 = 4.583'$ $Q_8 = \frac{9816}{52.08 \times 4.583} = 41.20"$ or 3.433 Feet

$h_9 = 4.583 - 3.433 = 1.150'$ $Q_9 = \frac{9816}{52.08 \times 1.150} = 163.75"$ space 6.0" from top.

STEP VI:

Start at bottom and space hoops h_1 to h_2 as follows:
$h_1 = 0.5225'$ $h_2 = 1.0875$ above h_1. etc., $h_3 = 1.1755'$ $h_4 = 1.2875$ $h_5 = 1.4580'$ $h_6 = 1.6875'$ $h_7 = 2.1500'$ $h_8 = 2.8600$ $h_9 = 2.275$ and 0.50 feet from top of tank. Spacing total = 15.0 feet. See Section detail.

STEP VII:

Another common usage formula which may be used to obtain the Q dimension and center of hoop spacing is written thus: $Q = \frac{\pi\, d^2\, Ft}{0.868\, h'D"}$ where: d = hoop diameter in inche and h = head in feet. D = tank diameter in inches.

EXAMPLE: Riveted plate water storage tank design 2.11.9

A farm home with modern plumbing is to be provided with a water system which will deliver a 20 Lb. sq. inch pressure to each fixture during power failure and gravity tank is proposed. Inside tank is 22.0 feet in diameter. Capacity of tank desired is approximately 100,000 gallons. Fabrication shall consist of steel plate with butt joints. Fasteners shall consist of ¾"ϕ rivets with cover plate over joint on inside only. Two lines of rivets may be used in staggered position. Bottom for tank will be fabricated in shop with welded apron. Thickness of plate shall be uniform throughout.

REQUIRED:
Design tank for plate thickness and rivet pitch. Allowable unit stress for steel shall be equivalent of A36 steel. Determine the height for supporting platform based on providing 20PSI to ground when tank is ⅓ full. Assume wt. water at 62.5 Lbs. cu. ft.

STEP I:
For Tank Capacity: Area cross-section = $0.7854\ D^2$
$A = 0.7854 \times 22.0 \times 22.0 = 380.13$ Sq. Ft. 1 Cubic foot has 7.48 Gallons.
1.0 foot depth capacity = $380.13 \times 7.48 = 2843.37$ Gals.
Req'd. $H = \frac{100{,}000}{2843.37} = 35.2'$ (Call it 35.0 feet)

STEP II:
To determine height platform:
⅓ of depth = ⅓ of 35.0 = 11.67 feet.
Pressure = <u>20 PSI.</u> For 1.0 foot head,
pressure = $\frac{62.5}{144} = 0.434$ #□"
$H = \frac{20\#\square''}{0.434} = 46.2$ feet. Height to top of Platform = 46.20 - 11.67 = <u>34.53 feet</u>

STEP III:
Maximum water pressure at bottom of tank: $\frac{62.5 \times 35.0}{144} = 15.19$ PSI.

STEP IV:
Make an elevation of Tank and plan to use 4.0 wide plates except at top.

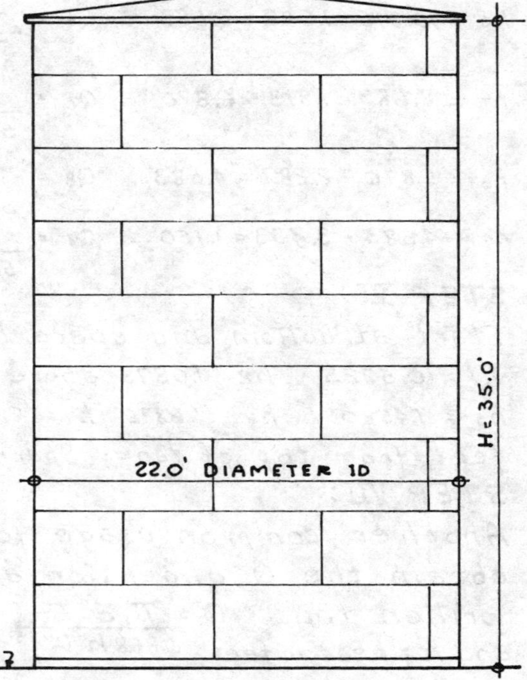

•ELEVATION•

EXAMPLE: Riveted plate water storage tank design, continued 2.11.9

The Reaction at riveted joint is pDQ. Let Q = 1.0 inch in tank height, and convert D to inches. D = 22.0 x 12 = 264 in.
Then R for 1.0 inch ring = 15.19 x 264.0 x 1.0 = 4010 Lbs.
Value of 1-¾" ϕ Rivet in shear = $F_v A$ or 0.4418 x 10,000 = 4,418 Lbs.
Will require 2 Rivets and pitch = $\frac{2 \times 4418}{4010}$ = 2.203"
Pitch is close to maximum of 3d. Max. = 0.75 x 3 = 2.25 in. but let Q = 2.203 in.

STEP V:
Plane of force will be illustrated as DQ:
The force in pitch ring = 4010 x 2.203 = 8834 Lbs.
Area required steel:
Deducting Rivet Hole which is 1/16 inch larger than ¾" ϕ Rivet. Hole D = 0.8125 In.
Width steel effective = 2.203 - 0.8125 = 1.3905"
For tension, area reqd. steel = $\frac{R}{F_t}$. F_t = 20,000#□"
$A = \frac{8834}{20{,}000} = 0.4417$ Sq. In. $t = \frac{A}{Q-\phi}$ or
$t = \frac{0.4417}{1.3905} = 0.318$ inches.

STEP VI:
Shear and Bearing in Rivets: Both are in a single plane and single shear and bearing. Value of shear will be the same as in step IV with value of 8836 Lbs.
Bearing Value of 2-¾" ϕ Rivets in 5/16" Plate.
F_p = 16,000 PSI. A = 2 x 0.75 x 0.3125 = 0.469 □"
Bearing value = 0.469 x 16,000 = 7500 Lbs. (This is less than Force of 8836# and require thicker plate near bottom.
Then, $t = \frac{T}{2 d F_p}$ or $t = \frac{8836}{2 \times 0.75 \times 16{,}000} = 0.368$ inches.
Will require 3/8 in. Plate, t = 0.375 inches.
With this type of tank, plate thickness can be reduced as pressure decreases toward top.
Further calculations will determine the points for transition.

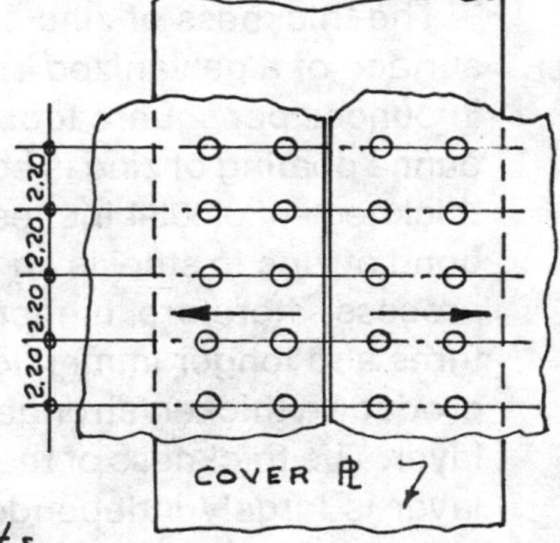

Galvanizing steel members 2.12

When designing structural and fabricated shapes which will be galvanized, the designer should be familiar with the size and depth of the galvanizing bath to be used. This knowledge helps him to size members so they can be completely covered in a single immersion in the zinc bath. This gives the best quality and appearance in the finished coating. Galvanizing plants generally have two standard kettles for daily use, usually less than 40 feet long. Material longer than the kettle, or with a vertical dimension greater than the zinc depth, is subject to a double dip or splicing in the coating.

HOT DIP METHOD

Galvanizing by the hot dip method is accomplished by immersing a thoroughly cleaned member in a bath of molten zinc. Before the steel can be coated properly, it must pass through three stages of cleaning —degreasing, scale and rust removal, and fluxing. Regardless of the cleaning methods, it is essential that the steel be clean.

Grease and paint are removed in a hot alkaline degreasing bath. After rinsing, the metal is descaled by pickling. A pickling vat is a large vessel containing dilute, hot sulfuric acid to which an inhibitor may be added. Mill scale which is deeply imbedded in the steel surface may require grit or sand blasting for removal. The fluxing operation is in two steps after the pickling acid and iron salts have been removed by rinsing. Prefluxing is done by dipping the metal in an aqueous solution of zinc ammonium chloride. This adds a thin salt layer to the steel, and supplements the action of the molten flux, which floats above the hot zinc bath.

The zinc bath is controlled at a temperature between 825° to 860°F. As the member is immersed in the hot zinc bath, bubbling from the interaction of steel, flux and molten zinc is visible. When the bubbling subsides, the material is withdrawn. It retains a continuous coating of zinc.

Coating thickness 2.12.1

The thickness of zinc coverage on the surface of a galvanized article is measured in ounces per square foot of surface. A 2 ounce coating of zinc is equivalent to a thickness of 0.0034 inches or 3.4 mils. The bond of zinc to steel is an alloy diffusion process; therefore, higher bath temperatures and longer immersion time will produce a thicker, stronger alloy bonding layer. The thickness of the pure zinc outer layer is largely independent of the immersion duration. This thickness is generally determined by the speed at which the member is withdrawn from kettle and the extent of drain-off. A rapid withdrawal will carry out more zinc for a heavier coating, but distribution may be less uniform.

The length of time a zinc coating will offer good protection is directly proportional to the zinc coat thickness. This life expectation under several environmental conditions is presented in the graph which follows. In polluted areas, where the sulphur content in the atmosphere is high, the normally protective zinc oxide coating is converted to soluble sulfates. These are washed away by rain, exposing the zinc to further attack, and accelerating the reduction in the coating thickness.

Thickness testing 2.12.2

The weight or thickness of zinc coating on a galvanized article can be determined by either of the following methods.

THE WEIGHT TEST

For articles inspected at the galvanizing plant, the most convenient way to determine the average weight of the coating is by weighing the member after pickling and drying, and then again after galvanizing. The weight of zinc coating per square foot is then determined by dividing the weight gain by the total surface area of a sample piece for each member galvanized. Samples for tests should be of the same shape as the material they represent. In the case of beams, angles and channels, the specimens should be at least three feet long. Test samples must be made from a material with the same composition, and processing must be accomplished in the same time and in the same zinc bath as the material to be galvanized in production.

THE CHEMICAL STRIPPING TEST

This method is used when the material is inspected after galvanizing, and is usually a more accurate test. It is not suitable for the inspection of large or heavy members, unless smaller, representative samples can be substituted. The average weight of the coating is calculated from the weights before and after stripping the sample in a suitable zinc stripping solution. For further details on coating weight determination by stripping, refer to ASTM Specification A90.

Adhesion testing 2.12.3

The ASTM has established two standard methods for testing the adherence of the zinc coating. When these tests are required, they should be performed before the weight and thickness tests.

PIVOT HAMMER TEST

Any type of hand-held hammer test gives widely different results for each inspector. Such tests very often only damage the coating and do not provide any conclusions as to the adhesion quality. Because of this problem, ASTM Specification A123 details a standard hammer test method. In this test, the magnitude and the angle of the blow are controlled. The hammer is similar to a riveting hammer with a chisel-face head having a 90 degree edge. The 12 inch lever handle is mounted on a pivoted base. The weight of head is approximately ½ pound. The hammer assembly is placed on a horizontal surface of galvanized material. The hammer head is raised and allowed to fall freely from a vertical position to strike the coated surface. The patterns resulting from the hammer impacts are illustrated in the ASTM Specification, which gives the interpretation of satisfactory or unsatisfactory adhesion.

PARING TEST

Hammer testing is inherently unsuited to the inspection of light gage metal, which can deflect or deform under the hammer impact. In this case, the paring test is used to determine coating adherence. By cutting into the test specimen with a strong, knife-like instrument which tends to lift a portion of the zinc coating, separation can be visually inspected. If only small bits of coating are removed, the adherence is considered satisfactory. When the openings show a tendency to peel the coating from the steel, the work is unsatisfactory.

Designing and detailing for galvanizing 2.12.4

Closed-end pipe and hollow welded assemblies will not be accepted at the galvanizing plant unless vent holes are included. Should a small amount of water or condensate be contained within the assembly, the pressure of steam when the assembly is immersed in a zinc bath at 850°F could be a great danger. Vent holes must be provided.

Residues from coated welding electrodes are not removed by the pickling solution. It is necessary to clean the welds by sandblasting or grinding before pickling. There is no problem in the galvanizing of articles fabricated with uncoated rods.

Satisfactory galvanizing is impossible if members have been skip welded. The acid pickling bath will penetrate between the joined members, but the zinc will not penetrate the space or seal it closed. Continuous welding has become standard for work to be galvanized.

Nuts and bolts less than 18 inches long are galvanized in perforated buckets placed in the zinc bath. While the zinc is still molten, the bucket is placed in a centrifuge and the excess zinc is spun off. The nuts must first be retapped to fit the oversized threads caused by the zinc coating. Oversize retapping for ⅜ inch and under nuts is 1/64 inches or 0.0156 inch. For ½ inch nuts, the retapping oversize is 1/32 inch (0.0312″). These dimensions apply to American Standard Coarse threads only.

In any large steel assembly fabricated by welding or riveting, there will be a certain amount of thermal distortion set up when the member is removed from the 850°F zinc bath. Most of this distortion occurs at the junction of plates with different thicknesses, which heat and cool at different rates. Another cause for excessive warping is residual internal stresses which have been locked in during fabrication and are relieved during the hot soak in the hot zinc bath.

To minimize warping and distortion, the following precautions should be incorporated in the design and fabrication of members to be galvanized.

(a) Design members with dimensions which will fit bath facilities and can be coated in one dip.

(b) The assembly must be fabricated accurately, so that it is not necessary to force, spring, drift or bend members into position before welding or riveting.

(c) All welds should be relieved of residual stresses by applying a suitable low heat treatment at the fabrication plant.

(d) Parts of an assembled member should be of nearly equal thickness as far as possible.

(e) Sheet and plate should be formed without sharp corners. Square sheet steel bins or tanks should have a liberal radius on edges and corners to minimize stress concentration.

(f) If a uniform finish is required, do not combine old and new steel in the same fabrication.

(g) The fabricating plant should use tags or labels to identify members to be galvanized. Red lead or oil paint must not be used for markings.

(h) Provide vent holes in the base and beam plates on tube columns.

TABLE: ASTM specifications for galvanizing 2.12.5

A.S.T.M. SPECIFICATIONS FOR GALVANIZING	
TITLE OR TYPE OF WORK TO BE COATED	ASTM No.
SPECIFICATION FOR ZINC MATERIAL	B-6
NUTS, BOLTS, WASHERS- HARDWARE, ETC.,	A-153
CHAIN LINK FENCE - WOVEN WIRE FABRICS	A-117
PIPE, WELDED AND SEAMLESS - TUBE STEEL EXTR.	A-120
SHEET STEELS- BINS, FLASHINGS, VENTS, ETC.	A-93
CARBON STEEL ROLLED SHAPES, BARS AND PLATES	A-123
CALL FOR LATEST EDITIONS FOR ALL ASTM SPECIFICATIONS.	

CHART: Life expectancy for galvanized coatings 2.12.6

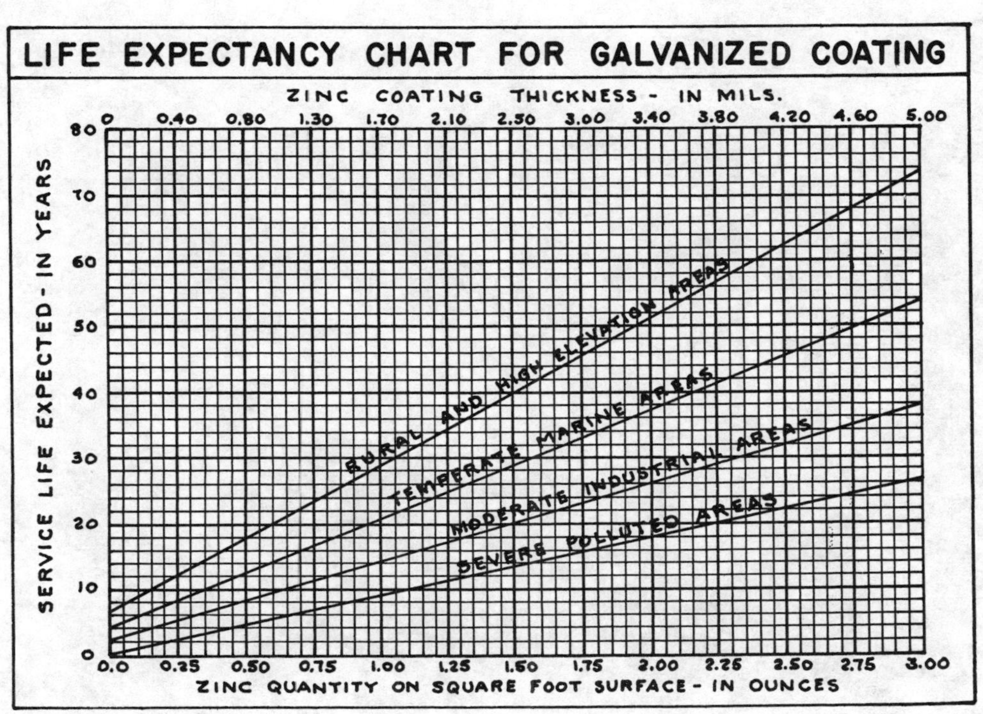

III—TIMBER DESIGN
including Framed Dome Design

III—TIMBER DESIGN
including Framed Dome Design

III

TIMBER DESIGN
including Framed Dome Design

TIMBER DESIGN including Framed Dome Design

Contents

Timber characteristics 3.1

Wood is composed of cellulose, water, resin and lignin: the main binding agent.

A tree which grows to maturity by adding layers of wood cells in concentric rings is called an exogen species. Each year of growth produces new wood outside the previous growth, until the main stem and trunk are large enough to be cut and transported to the saw mill. It is these exogen species which produce the lumber used for structural purposes. In observing the cross section of a tree trunk, we see first the outside bark, then rings of sapwood and, innermost, the heartwood.

Exogen trees include the softwoods or conifers: fir, pine, redwood, and cypress; and the hardwoods or non-conifers: oak, elm, hickory, maple, ash and walnut.

The palms and bamboos are members of the endogen species. The cross-section of this species is made up of bundles of woody fibers running parallel with the stem. They do not show the annual rings of the exogens.

Classification and grading 3.1.1

The sawn products used for general construction are often classed as yard lumber, dimension lumber, structural timbers and shop lumber. Grading rules are applied to each class, and may use such terms as Rough, Surfaced, Kiln Dried, Select, Common, Clear, Sawn Dimension, Framing, Merchantable, Industrial and many others.

Section properties 3.1.2

Rough and dressed sizes:

Specifications for lumber requirements as written by the Architect, will usually call for a size or dimension which is understood to be the nominal size of the rough timber before planing. Prior to 1968, a 2 x 4 in the rough would dress out to an actual size of 1⅝″ x 3⅝″, S4S (surfaced on four sides). Since 1968, the Southern Pine Association has revised the Standard Grading Rules and made some important changes. For the 2 inch dimension, there is in use an alternate dimension of 1½ inch for actual thickness. It is now necessary for the Architect and Designer to specify the original Standard size, or the Alternate size.

Tables of Properties are provided for the actual Standard and Alternate sizes.

TABLE: American Standard timber sizes and properties 3.1.2.1

PROPERTIES FOR DESIGNING

SAWN SIZE	DRESSED SIZE	AREA SECTION	WT. PER FOOT	MOMENT OF INERTIA	SECTION MODULUS	SAWN SIZE	DRESSED SIZE	AREA SECTION	WT. PER. FOOT	MOMENT OF INERTIA	SECTION MODULUS
IN.	INCHES	SQ. IN.	LBS.	$IN.^4$	$IN.^3$	IN.	INCHES	SQ. IN.	LBS.	$IN.^4$	$IN.^3$
2x4	1⅝x3⅝	5.89	1.64	6.45	3.56	10x10	9½x9½	90.3	25.0	679.0	143.0
6	5⅝	9.14	2.54	24.10	8.57	12	11½	109.0	30.3	1204.0	209.0
8	7½	12.2	3.39	57.10	15.3	14	13½	128.0	35.6	1948.0	289.0
10	9½	15.4	4.29	116.0	24.4	16	15½	147.0	40.9	2948.0	380.0
12	11½	18.7	5.19	206.0	35.8	18	17½	166.0	46.1	4243.0	485.0
14	13½	21.9	6.09	333.0	49.4	20	19½	185.0	51.4	5870.0	602.0
16	15½	25.2	7.00	504.0	65.1	22	21½	204.0	56.7	7868.0	732.0
18	17½	28.4	7.90	726.0	82.9	24	23½	223.0	62.0	10,274.0	874.0
3x4	2⅝x3⅝	9.5	2.64	10.4	5.8	12x12	11½x11½	132.0	36.7	1458.0	253.0
6	5⅝	14.8	4.10	38.9	13.8	14	13½	155.0	43.1	2358.0	349.0
8	7½	19.7	5.47	92.3	24.6	16	15½	178.0	49.5	3569.0	460.0
10	9½	24.9	6.93	188.0	39.5	18	17½	201.0	55.9	5136.0	587.0
12	11½	30.2	8.39	333.0	57.9	20	19½	224.0	62.3	7106.0	729.0
14	13½	35.4	9.84	538.0	79.7	22	21½	247.0	68.7	9524.0	886.0
16	15½	40.7	11.30	815.0	105.0	24	23½	270.0	75.0	12,337.0	1058.0
18	17½	45.9	12.80	1172.0	134.0	14x14	13½x13½	182.0	50.6	2768.0	410.0
4x4	3⅝x3⅝	13.1	3.65	14.4	7.9	16	15½	209.0	58.1	4189.0	541.0
6	5⅝	20.4	5.65	53.8	19.1	18	17½	236.0	65.6	6029.0	689.0
8	7½	27.2	7.55	127.0	34.0	20	19½	263.0	73.1	8342.0	856.0
10	9½	34.4	9.57	259.0	54.5	22	21½	290.0	80.6	11,181.0	1040.0
12	11½	41.7	11.60	459.0	79.9	24	23½	317.0	88.1	14,600.0	1243.0
14	13½	48.9	13.60	743.0	110.0	16x16	15½x15½	240.0	66.7	4810.0	621.0
16	15½	56.2	15.60	1125.0	145.0	18	17½	271.0	75.3	6923.0	791.0
18	17½	63.4	17.60	1619.0	185.0	20	19½	302.0	83.9	9578.0	982.0
6x6	5½x5½	30.3	8.40	76.3	27.7	22	21½	333.0	92.5	12,837.0	1194.0
8	7½	41.3	11.40	193.0	51.6	24	23½	364.0	101.0	16,763.0	1427.0
10	9½	52.3	14.5	393.0	82.7	18x18	17½x17½	306.0	85.0	7816.0	893.0
12	11½	63.3	17.5	697.0	121.0	20	19½	341.0	94.8	10,813.0	1109.0
14	13½	74.3	20.6	1128.0	167.0	22	21½	376.0	105.0	14,493.0	1348.0
16	15½	85.3	23.6	1707.0	220.0	24	23½	411.0	114.0	18,926.0	1611.0
18	17½	96.3	26.7	2456.0	281.0	26	25½	446.0	124.0	24,181.0	1897.0
20	19½	107.3	29.8	3398.0	349.0	20x20	19½x19½	380.0	106.0	12,049.0	1236.0
8x8	7½ x 7½	56.3	15.6	264.0	70.3	22	21½	419.0	116.0	16,150.0	1502.0
10	9½	71.3	19.8	536.0	113.0	24	23½	458.0	127.0	21,089.0	1795.0
12	11½	86.3	23.9	951.0	165.0	26	25½	497.0	138.0	26,945.0	2113.0
14	13½	101.3	28.0	1538.0	228.0	28	27½	536.0	149.0	33,795.0	2458.0
16	15½	116.3	32.0	2327.0	300.0	24x24	23½x23½	552.0	153.0	25,415.0	2163.0
18	17½	131.3	36.4	3350.0	383.0	26	25½	599.0	166.0	32,472.0	2547.0
20	19½	146.3	40.6	4634.0	475.0	28	27½	646.0	180.0	40,727.0	2962.0
22	21½	161.3	44.8	6211.0	578.0	30	29½	693.0	193.0	50,275.0	3408.0

TABLE: Southern pine alternate sizes and properties 3.1.2.2

PROPERTIES FOR DESIGNING								
SAWN SIZE	DRESSED SIZE ACTUAL	BOARD MEASURE	AREA SECTION	WEIGHT PER FT.	MOMENT OF INERTIA		SECTION MODULUS	
INCHES	INCHES	LIN. FOOT	SQ. IN.	LBS.	X-X, $IN.^4$	Y-Y, $IN.^4$	X-X, $IN.^3$	Y-Y, $IN.^3$
2 x 2	1 1/2 x 1 1/2	0.33	2.25	0.63	0.42	0.42	0.56	0.56
3	2 9/16	0.50	3.84	1.07	2.10	0.72	1.64	0.96
4	3 9/16	0.67	5.34	1.48	5.65	1.00	3.17	1.36
6	5 1/2	1.00	8.25	2.29	20.80	1.54	7.56	2.06
8	7 1/2	1.33	11.25	3.12	52.73	2.11	14.06	2.81
10	9 1/2	1.67	14.25	3.96	107.17	2.67	22.56	3.56
12	11 1/2	2.00	17.25	4.79	190.11	3.23	33.06	4.31
14	13 1/2	2.33	20.25	5.62	307.55	3.80	45.56	5.06
3 x 3	2 5/8 x 2 5/8	0.75	6.89	1.91	3.96	3.96	3.01	3.01
4	3 5/8	1.00	9.52	2.64	10.42	5.46	5.75	4.16
6	5 1/2	1.50	14.44	4.01	36.40	8.29	13.23	6.31
8	7 1/2	2.00	19.69	5.47	92.30	11.30	24.61	8.61
10	9 1/2	2.50	24.94	6.93	187.57	14.32	39.48	10.91
12	11 1/2	3.00	30.19	8.39	332.71	17.33	57.86	13.20
14	13 1/2	3.50	35.44	9.84	538.24	20.34	79.73	15.50
4 x 4	3 5/8 x 3 5/8	1.33	13.14	3.64	14.39	14.39	7.94	7.94
6	5 1/2	2.00	19.94	5.54	50.26	21.83	18.27	12.04
8	7 1/2	2.67	27.19	7.55	127.45	29.77	33.98	16.42
10	9 1/2	3.33	34.44	9.57	259.02	37.71	54.53	20.80
12	11 1/2	4.00	41.69	11.58	459.13	45.65	79.90	25.18
14	13 1/2	4.67	48.94	13.59	743.28	53.59	110.11	29.56
6 x 6	5 1/2 x 5 1/2	3.00	30.25	8.40	76.25	76.25	27.73	27.73
8	7 1/2	4.00	41.25	11.46	193.35	103.98	51.56	37.81
10	9 1/2	5.00	52.25	14.51	392.96	131.71	82.73	47.89
12	11 1/2	6.00	63.25	17.57	697.07	159.44	121.23	57.98
14	13 1/2	7.00	74.25	20.62	1127.67	187.17	167.06	68.06
8 x 8	7 1/2 x 7 1/2	5.33	56.25	15.62	263.67	263.67	70.31	70.31
10	9 1/2	6.67	71.25	19.79	535.86	333.98	112.81	89.06
12	11 1/2	8.00	86.25	23.96	950.56	404.30	165.31	107.81
14	13 1/2	9.33	101.25	28.12	1537.73	474.61	227.81	126.56
10 x 10	9 1/2 x 9 1/2	8.33	90.25	25.07	678.75	678.75	142.89	142.89
12	11 1/2	10.00	109.25	30.35	1204.03	821.65	209.39	172.98
14	13 1/2	11.67	128.25	35.62	1947.80	964.25	288.56	203.06
12 x 12	11 1/2 x 11 1/2	12.00	132.25	36.74	1457.51	1457.51	253.48	253.48
14	13 1/2	14.00	155.25	43.12	2357.86	1710.98	349.31	297.56
14 x 14	13 1/2 x 13 1/2	16.33	182.25	50.62	2767.92	2767.92	410.06	410.06

Allowable stress 3.1.3

The principle of grading timber for stress rating has, until recently, been based on the visual appearance of the timber and the sound judgment and experience of the lumber grader. Many of the larger mills are now using a principle of mechanical stress-rating, which has been endorsed by the American Lumber Standards Committee, the International Conference of Building Officials, the F.H.A., and the Southern Building Code Congress. Mechanically graded lumber is machine tested at the saw-mill and each piece is rated for fiber stress in bending. Such grading is restricted to lumber 2 inches thick or less. It is necessary for the designer of heavy timber structures to obtain the latest publication of the Standard Grading Rules for Southern Pine and West Coast Douglas Fir in order to obtain accurate information on the allowable working stresses for the larger timber sizes. Since no two pieces of lumber are ever alike in their composition, complete uniformity within a grade is impossible. The stress limitations for rough lumber are the same as those for dressed lumber of like grade.

TABLE: Allowable stress: Douglas Fir 3.1.3.1

VISUALLY GRADED – NORMAL LOADING – IN POUNDS SQ. INCH

SPECIES AND GRADE	EXTREME FIBER IN BENDING F_b & F_t		HORIZONTAL SHEAR F_v		COMPRESSION PERPENDICULAR TO GRAIN F_c ⊥		COMPRESSION PARALLEL TO GRAIN F_c		MODULUS OF ELASTICITY E
DOUGLAS FIR-WESTERN LARCH									
LIGHT FRAMING		MC15		MC15		MC15		MC15	
Dense Select Structural	2050	2300	120	135	455	455	1500	1700	1,760,000
Select Structural	1900	2100	120	135	415	415	1400	1550	1,760,000
1750 F Industrial	1750	2050	120	135	455	455	1400	1600	1,760,000
1500 F Industrial	1500	1750	120	135	390	390	1200	1400	1,760,000
1200 F Industrial	1200	1500	95	110	390	390	1000	1200	1,760,000
JOISTS AND PLANKS									
Dense Select Structural	2050	2300	120	125	455	455	1650	1850	1,760,000
Select Structural	1900	2100	120	125	415	415	1500	1650	1,760,000
Dense Construction	1750	2050	120	125	455	455	1400	1600	1,760,000
Construction	1500	1750	120	125	390	390	1200	1400	1,760,000
Standard	1200	1500	95	110	390	390	1000	1200	1,760,000
BEAMS AND STRINGERS									
Dense Select Structural	2050		120		455		1500		1,760,000
Select Structural	1900		120		415		1400		1,760,000
Dense Construction	1750		120		455		1200		1,760,000
Construction	1500		120		390		1000		1,760,000
POSTS AND TIMBERS									
Dense Select Structural	1900		120		455		1650		1,760,000
Select Structural	1750		120		415		1500		1,760,000
Dense Construction	1500		120		455		1400		1,760,000
Construction	1200		120		390		1200		1,760,000

MC15 DENOTES THE STOCK IS SURFACED AT 15% OR LESS MOISTURE CONTENT

TABLE: Allowable stress: Southern yellow pine 3.1.3.2

IN POUNDS PER SQUARE INCH VISUALLY GRADED

FOR NORMAL LOADING CONDITIONS

SPECIES AND GRADE	EXTREME FIBER IN BENDING F_b OR F_t	HORIZONTAL SHEAR F_v	COMPRESSION PERPENDICULAR TO GRAIN $F_c \perp$	COMPRESSION PARALLEL TO GRAIN F_c	MODULUS OF ELASTICITY E
SOUTHERN YELLOW PINE					
FRAMING					
Dense Select Structural	2400	120	455	1750	1,600,000
Select Str. Long Leaf	2400	120	455	1750	1,600,000
Dense Structural	2000	120	455	1400	1,600,000
Prime Structural Long Leaf	2000	120	455	1400	1,600,000
Dense Structural SE & S	1800	120	455	1300	1,600,000
JOISTS AND PLANKS					
Structural SE&S Longleaf	1800	120	455	1300	1,600,000
Merchantable Struct. L.L.	1800	120	455	1300	1,600,000
Dense No. 1 Structural	1600	120	455	1150	1,600,000
No. 1 Structural Longleaf	1600	120	455	1150	1,600,000
No. 1 Dense Seasoned 2"	1700	150	455	1400	1,600,000
STRINGERS, BEAMS, ETC.					
No. 1 Seasoned Longleaf 2"	1700	150	455	1400	1,600,000
No. 1 Seasoned Shortleaf 2"	1450	125	390	1200	1,600,000
No. 1 Longleaf 3" Thick and Up	1400	140	455	1400	1,600,000
No. 1 Dense Shortleaf 1400	1400	140	455	1400	1,600,000
POSTS AND TIMBERS					
No. 2 Dense Seasoned S.L. 2"	1250	100	455	1025	1,600,000
No. 2 Longleaf Seasoned 2"	1250	100	455	1025	1,600,000
No. 1 1200 f 3" thick & Up.	1200	120	390	1200	1,600,000
No. 2 Seasoned S.L. 2"	1100	85	390	875	1,600,000

Lumber measure 3.1.4

All lumber is sold by the board-measure. The standard unit is called the board-foot. One board-foot of lumber contains 144 cubic inches. Simply illustrated, it is equivalent to a board 12 inches wide, 1 inch thick and 12 inches long.

To calculate the board-feet in any timber, divide the cross-sectional Area by 12 to determine the amount of board feet per foot in length, then by the length of the board. To illustrate:

Compute the board-feet in a 2x4 Section 16.0 feet long.

$$BM = \left(\frac{2 \times 4}{12}\right) \times 16.0 = 10.67 \text{ Bd. Ft.}$$

Always use the rough or nominal size when computing the Board foot measurement. A truck shipment contains 42 pieces of 2x10 YP 16'-0". If the cost is \$155.00 per (MBM) thousand board feet, what is the value of the shipment?

$$42 \text{ Pcs. } 2 \times 10 \text{ YP } 16.0' = 42 \times \left(\frac{2 \times 10}{12}\right) \times 16. = 1120 \text{ Board Ft.}$$

$$\text{Value} = 1120 \times 155.00 = \$173.60$$

Treated lumber 3.1.5

Lumber exposed to the weather will be alternately wet and dry and, therefore subject to decay and dry rot. Timber exposed to salt water will deteriorate from effects of marine life such as the teredo, a boring worm. On land, in damp areas, insects such as ants, beetles and termites will destroy the wood. Wooden marine structures are best protected from decay by pressure treating the timber with creosote or a similar chemical. The American Wood Preservers Association makes available recommended treatments.

Glued laminated members 3.1.6

The length of time required for solid timbers to become properly dried and seasoned can vary from several months to several years. A comparably sized timber can be fabricated by using smaller sections of dried lumber glued together under pressure. These smaller components can be kiln dried rapidly in several days.

When a large beam or girder is fabricated from selected material, and the laminations are held together with approved adhesives, the final product will usually be more satisfactory than a solid member. The lumber must be kiln dried to a moisture content of 8 to 14 percent before placing in the glue clamps. For interior and protected work the laminating adhesives should be casein glue, complying with Federal Specification C–G–456. For exterior or submerged laminated members, a resin glue of phenol, resorcinol or melamine type should be specified, conforming to Military Specifications JAN–A–397, or MIL–A–5534.

The properties of Structural Glued Southern Pine Timber Sections are higher than those allowed for solid sawn members. The following allowable unit stresses are recommended:

Extreme fibers in bending	F_b	=	2400 P.S.I.
Tension parallel to grain	F_t	=	2600 "
Compression parallel to grain	F_c	=	2000 "
Shear parallel to grain	F_v	=	200 "
Compression perpendicular to grain	$F_{c\perp}$	=	385 "
Modulus of elasticity	E	=	1,800,000 "

Laminated sections give the designer a great flexibility in choosing cross-section dimensions and beam length. We include a Table of Properties for laminated sections made up of the commonly-used 1⅝ inch laminations.

TABLE: Laminated section properties 3.1.6.1

PROPERTIES OF SECTIONS
GLUED LAMINATED STRUCTURAL LUMBER

1-5/8 Inch Laminations Only

Arranged in order of ascending section modulus.

Nominal Width in Inches	Number of 1⅝" Laminations*	Net Finished Size* in Inches b x h	Area of Section in Square Inches	Moment of Inertia $I = \frac{bh^3}{12}$	Section Modulus $S = \frac{bh^2}{6}$	Weight in Pounds per Linear Foot at 12% Moisture Content
3	4	2¼ x 6½	14.6	51.5	15.8	3.45
4	4	3¼ x 6½	21.1	74.4	22.9	4.98
3	5	2¼ x 8⅛	18.3	101.	24.8	4.31
5	4	4¼ x 6½	27.6	97.3	29.9	6.52
6	4	5 x 6½	32.5	114.	35.2	7.67
3	6	2¼ x 9¾	21.9	174.	35.7	5.18
4	5	3¼ x 8⅛	26.4	145.	35.8	6.23
6	4	5¼ x 6½	34.1	120.	37.0	8.05
5	5	4¼ x 8⅛	34.5	190.	46.8	8.15
3	7	2¼ x 11⅜	25.6	276.	48.5	6.04
4	6	3¼ x 9¾	31.7	251.	51.5	7.48
6	5	5 x 8⅛	40.6	223.	55.0	9.59
6	5	5¼ x 8⅛	42.7	235.	57.8	10.1
5	6	4¼ x 9¾	41.4	328.	67.3	9.78
4	7	3¼ x 11⅜	37.0	399.	70.1	8.72
8	5	7 x 8⅛	56.9	313.	77.0	13.4
6	6	5 x 9¾	48.8	386.	79.2	11.5
6	6	5¼ x 9¾	51.2	406.	83.2	12.1
4	8	3¼ x 13	42.3	595.	91.6	9.97
5	7	4¼ x 11⅜	48.3	521.	91.7	11.4
6	7	5 x 11⅜	56.9	613.	108.	13.4
8	6	7 x 9¾	68.3	541.	111.	16.1
6	7	5¼ x 11⅜	59.7	644.	113.	14.1
4	9	3¼ x 14⅝	47.5	847.	116.	11.2
5	8	4¼ x 13	55.3	778.	120.	13.0
6	8	5 x 13	65.0	915.	141.	15.3
4	10	3¼ x 16¼	52.8	1,162.	143.	12.5
10	6	9 x 9¾	87.8	695.	143.	20.7
6	8	5¼ x 13	68.3	961.	148.	16.1
8	7	7 x 11⅜	79.6	859.	151.	18.8
5	9	4¼ x 14⅝	62.2	1,108.	152.	14.7
6	9	5 x 14⅝	73.1	1,303.	178.	17.3
5	10	4¼ x 16¼	69.1	1,520.	187.	16.3
6	9	5¼ x 14⅝	76.8	1,369.	187.	18.1
10	7	9 x 11⅜	102.	1,104.	194.	24.2
8	8	7 x 13	91.0	1,282.	197.	21.5
6	10	5 x 16¼	81.3	1,788.	220.	19.2
5	11	4¼ x 17⅞	76.0	2,023.	226.	17.9
6	10	5¼ x 16¼	85.3	1,877.	231.	20.1
12	7	11 x 11⅜	125.	1,349.	237.	29.5

Nominal Width in Inches	Number of 1⅝" Laminations*	Net Finished Size* in Inches b x h	Area of Section in Square Inches	Moment of Inertia $I = \frac{bh^3}{12}$	Section Modulus $S = \frac{bh^2}{6}$	Weight in Pounds per Linear Foot at 12% Moisture Content
8	9	7 x 14⅝	102.	1,825.	250.	24.2
10	8	9 x 13	117.	1,648.	254.	27.6
6	11	5 x 17⅞	89.4	2,380.	266.	21.1
5	12	4¼ x 19½	82.9	2,626.	269.	19.6
6	11	5¼ x 17⅞	93.8	2,499.	280.	22.2
8	10	7 x 16¼	114.	2,503.	308.	26.8
12	8	11 x 13	143.	2,014.	310.	33.8
5	13	4¼ x 21⅛	89.8	3,339.	316.	21.2
6	12	5 x 19½	97.5	3,090.	317.	23.0
10	9	9 x 14⅝	132.	2,346.	321.	31.1
6	12	5¼ x 19½	102.	3,244.	333.	24.2
14	8	12½ x 13	163.	2,289.	352.	38.4
6	13	5 x 21⅛	106.	3,928.	372.	24.9
8	11	7 x 17⅞	125.	3,332.	373.	29.5
6	13	5¼ x 21⅛	111.	4,124.	390.	26.2
12	9	11 x 14⅝	161.	2,867.	392.	38.0
10	10	9 x 16¼	146.	3,218.	396.	34.5
6	14	5 x 22¾	114.	4,906.	431.	26.9
8	12	7 x 19½	137.	4,325.	444.	32.2
14	9	12½ x 14⅝	183.	3,258.	446.	43.1
6	14	5¼ x 22¾	119.	5,151.	453.	28.2
10	11	9 x 17⅞	161.	4,284.	479.	38.0
12	10	11 x 16¼	179.	3,933.	484.	42.2
6	15	5 x 24⅜	122.	6,034	495.	28.8
16	9	14½ x 14⅝	212.	3,780.	517.	50.1
6	15	5¼ x 24⅜	128.	6,336.	520.	30.2
8	13	7 x 21⅛	148.	5,499.	521.	34.9
14	10	12½ x 16¼	203.	4,470.	550.	47.9
6	16	5 x 26	130.	7,323.	563.	30.7
10	12	9 x 19½	176.	5,561.	570.	41.4
12	11	11 x 17⅞	197.	5,235.	586.	46.4
6	16	5¼ x 26	137.	7,690.	592.	32.2
8	14	7 x 22¾	159.	6,868.	604.	37.6
16	10	14½ x 16¼	236.	5,185.	638.	55.6
14	11	12½ x 17⅞	223.	5,949.	666.	52.7
10	13	9 x 21⅛	190.	7,071.	669.	44.9
8	15	7 x 24⅜	171.	8,448.	693.	40.3
12	12	11 x 19½	215.	6,797.	697.	50.6
16	11	14½ x 17⅞	259.	6,901.	772.	61.2
10	14	9 x 22¾	205.	8,831.	776.	48.3

* With glued laminated structural lumber, many additional sizes may be obtained. Greatest economy will result by using standard widths and depths that are multiples of standard board and dimension lumber thicknesses.

TABLE: Laminated section properties, continued 3.1.6.1

PROPERTIES OF SECTIONS
GLUED LAMINATED STRUCTURAL LUMBER

1-5/8 Inch Laminations Only

Arranged in order of ascending section modulus.

Nominal Width in Inches	Number of 1⅝″ Laminations*	Net Finished Size* in Inches b x h	Area of Section in Square Inches	Moment of Inertia $I = \frac{bh^3}{12}$	Section Modulus $S = \frac{bh^2}{6}$	Weight in Pounds per Linear Foot at 12% Moisture Content
8	16	7 x 26	182.	10,250.	789.	43.0
14	12	12½ x 19½	244.	7,724.	792.	57.5
12	13	11 x 21⅛	232.	8,642.	818.	54.8
8	17	7 x 27⅝	193.	12,300.	890.	45.6
10	15	9 x 24⅜	219.	10,860.	891.	51.8
16	12	14½ x 19½	283.	8,960.	919.	66.7
14	13	12½ x 21⅛	264.	9,820.	930.	62.3
12	14	11 x 22¾	250.	10,790.	949.	59.1
8	18	7 x 29¼	205.	14,600.	998.	48.3
10	16	9 x 26	234.	13,180.	1,014.	55.2
14	14	12½ x 22¾	284.	12,270.	1,078.	67.1
16	13	14½ x 21⅛	306.	11,390.	1,078.	72.3
12	15	11 x 24⅜	268.	13,280.	1,089.	63.3
8	19	7 x 30⅞	216.	17,170.	1,112.	51.0
10	17	9 x 27⅝	249.	15,810.	1,145.	58.7
8	20	7 x 32½	228.	20,030.	1,232.	53.7
14	15	12½ x 24⅜	305.	15,090.	1,238.	71.9
12	16	11 x 26	286.	16,110.	1,239.	67.5
16	14	14½ x 22¾	330.	14,230.	1,251.	77.9
10	18	9 x 29¼	263.	18,770.	1,283.	62.1
8	21	7 x 34⅛	239.	23,180.	1,359.	56.4
12	17	11 x 27⅝	304.	19,320.	1,399.	71.7
14	16	12½ x 26	325.	18,310.	1,408.	76.7
10	19	9 x 30⅞	278.	22,070.	1,430.	65.6
16	15	14½ x 24⅜	353.	17,500.	1,436.	83.4
8	22	7 x 35¾	250.	26,650.	1,491.	59.1
12	18	11 x 29¼	322.	22,940.	1,569.	75.9
10.	20	9 x 32½	293.	25,750.	1,584.	69.0
14	17	12½ x 27⅝	345.	21,960.	1,590.	81.5
16	16	14½ x 26	377.	21,240.	1,634.	89.0
10	21	9 x 34⅛	307.	29,800.	1,747.	72.5
12	19	11 x 30⅞	340.	26,980.	1,748.	80.2
14	18	12½ x 29¼	366.	26,070.	1,782.	86.3
16	17	14½ x 27⅝	401.	25,470.	1,844.	94.5
10	22	9 x 35¾	322.	34,270.	1,917.	75.9
12	20	11 x 32½	358.	31,470.	1,936.	84.4
14	19	12½ x 30⅞	386.	30,660.	1,986.	91.1
16	18	14½ x 29¼	424.	30,240.	2,068.	100.
10	23	9 x 37⅜	336.	39,160.	2,095.	79.4
12	21	11 x 34⅛	375.	36,430.	2,135.	88.6
14	20	12½ x 32½	406.	35,760.	2,201.	95.9
10	24	9 x 39	351.	44,490.	2,282.	82.8
16	19	14½ x 30⅞	448.	35,560.	2,304.	106.
12	22	11 x 35¾	393.	41,880.	2,343.	92.8
14	21	12½ x 34⅛	427.	41,390.	2,426.	101.
10	25	9 x 40⅝	366.	50,290.	2,476.	86.3
16	20	14½ x 32½	471.	41,480.	2,553.	111.
12	23	11 x 37⅜	411.	47,860.	2,561.	97.0
14	22	12½ x 35¾	447.	47,590.	2,663.	105.
10	26	9 x 42¼	380.	56,560.	2,678.	89.7
12	24	11 x 39	429.	54,380.	2,789.	101.
16	21	14½ x 34⅛	495.	48,020.	2,814.	117.
10	27	9 x 43⅞	395.	63,350.	2,888.	93.2
14	23	12½ x 37⅜	467.	54,380.	2,910.	110.
12	25	11 x 40⅝	447.	61,460.	3,026.	105.
16	22	14½ x 35¾	518.	55,210.	3,089.	122.
10	28	9 x 45½	410.	70,650.	3,105.	96.6
14	24	12½ x 39	488.	61,790.	3,169.	115.
12	26	11 x 42¼	465.	69,130.	3,273.	110.
16	23	14½ x 37⅜	542.	63,090.	3,376.	128.
14	25	12½ x 40⅝	508.	69,840.	3,438.	120.
12	27	11 x 43⅞	483.	77,420.	3,529.	114.
16	24	14½ x 39	566.	71,680.	3,676.	133.
14	26	12½ x 42¼	528.	78,560.	3,719.	125.
12	28	11 x 45½	501.	86,350.	3,795.	118.
16	25	14½ x 40⅝	589.	81,020.	3,988.	139.
14	27	12½ x 43⅞	548.	87,980.	4,010.	129.
12	29	11 x 47⅛	518.	95,930.	4,071.	122.
14	28	12½ x 45½	569.	98,120.	4,313.	134.
16	26	14½ x 42¼	613.	91,130.	4,314.	145.
12	30	11 x 48¾	536.	106,200.	4,357.	127.
14	29	12½ x 47⅛	589.	109,010.	4,627.	139.
12	31	11 x 50⅜	554.	117,180.	4,652.	131.
16	27	14½ x 43⅞	636.	102,060.	4,652.	150.
14	30	12½ x 48¾	609.	120,680.	4,951.	144.
16	28	14½ x 45½	660.	113,820.	5,003.	156.
14	31	12½ x 50⅜	630.	133,160.	5,287.	149.
16	29	14½ x 47⅛	683.	126,460.	5,367.	161.
16	30	14½ x 48¾	707.	139,990.	5,743.	167.
16	31	14½ x 50⅜	730.	154,470.	6,133.	172.

* With glued laminated structural lumber, many additional sizes may be obtained. Greatest economy will result by using standard widths and depths that are multiples of standard board and dimension lumber thicknesses.

Timber beam design formulas 3.2

The designing of wood beams is similar to the procedure used in steel design. The bending moment (M) is computed, and divided by the allowable stress given in the tables for the selected grade of material.

Then, the flexure formula, $S = \frac{M}{F_b}$, will produce the required Section Modulus. A beam or joist is selected from the Table of Properties which has an equal or greater Section Modulus.

Nomenclature for timber beam design formulas 3.2.1

A = Area cross section, in square inches.
a = Area of laminated section under consideration, in sq. in.
b = Breadth of section, in inches.
c = Distance from neutral axis to extreme fiber, in inches.
C = Total value of Compressive force, in pounds.
D = Diameter of section, in feet or inches.
d = Depth of cross section, in inches. Least column dimension.
E = Modulus of Elasticity of material, in pounds sq. in.
e = Eccentric distance of load or force, in inches or feet.
F = Allowable unit stress, used with subscripts which will indicate shear, bending, tension, compression, etc. P.S.I.
f = Actual unit stress, used with subscripts, same as F.
g = Gross area of section, as A_g, etc., in square inches.
H = Height of column or story, in feet. Also horizontal force.
h = Unsupported height of column, given in inches, also l.
I = Moment of Inertia. Sub-script denotes centroid under consideration, as $I_x''^4$ or $I_y''^4$, etc. Given inches fourth power
k = Coefficient $\frac{S}{A}$, used in design eccentric loaded columns.
L = Length of span, given in feet.
l = Length of span or column, given in inches, = L × 12.
M = Bending moment, given in foot pounds or inch pounds.
n = Number of panels or force normal to surface.
o = A distance point denoting polar dimension.
P = Concentrated load on beam or column, in pounds or tons.
q = Value of horizontal shear for 1 Lineal foot on beam, in lbs.
r = Radius of gyration, used with subscripts for centroid.
R = Reaction at supports or radius of circle.
S = Section Modulus of cross section, in inches to third power.
s = Spacing distance between bolts, holes, connectors, etc.
T = Tension in member, given in pounds, kips or tons.
t = Thickness of gusset plate, washer, etc., in inches
u = Unity ratio between compressive and tension stresses.
V = Total shear at support, given in Lbs. Vertical.
V_H = Total horizontal shear in pounds.
v = Unit shear allowable or actual. Also F_v, F_H, f_v, etc., in lbs.
w = Load or weight per lineal foot on beam, in pounds.
W = Total uniform load or weight on beam, given in pounds.
x = Any distance or dimension from load or support.

Nomenclature for timber beam design formulas, continued 3.2.1

X–X = Major axis or Centroid of section. Denoted by a straight line running through center of gravity.

y–y = Minor axis or Centroid of section. Usually runs perpendicular to axis x-x.

1-1 = Same as X-X axis.

2-2 = Same as y-y axis.

$\bar{y}$ = Dimension from centroid x-x to center of gravity in laminated part of whole section. Used in formula to find horizontal shear at any point in a beam cross-section. Given in inches.

⊥ = Force or shear perpendicular or normal to beam surface or line of action. As indicated in allowable stress tables, $F_{c\perp}$, denotes compression normal to run of wood grain.

Δ = Deflection, deformation or sag in beam under loading. Given in inches.

Σ = Summation of all forces, loads, areas added to determine total loads, total areas, etc.

X = By, a 2X4 etc. Always give least dimension first as bd = 2 inches by 4 inches, etc. Also indicates numbers to be multiplied in equations.

θ = Denotes angle under consideration.

± = Plus or minus quantity.

–M = Negative moment or a member in tension stress.

+M = Positive moment or a compressive stress in member.

P_{1-6} = Summation of Loads as, $P_{1-6} = P_1 + P_2 + P_3 + P_4 + P_5 + P_6$.

NOTE:
In the solution for forces and angles to be used in calculations of Domed hemispheric framed structures, a separate group of symbols is to be used. Another set of symbols is listed preceding the dome example, (See paragraph 3.7.4.1).

Beam bending formulas and transposition 3.2.2

For simple beams with Uniform Load entire span.

$M = \frac{WL}{8}$ or $\frac{wL^2}{8}$ $L = \frac{8M}{W}$ $W = \frac{8M}{L}$ $L = \sqrt{\frac{8M}{w}}$

For Concentrated Load at midspan of simple beams.

$M = \frac{PL}{4}$ $L = \frac{4M}{P}$ $P = \frac{4M}{L}$

For 2 spans, or end spans with Uniform Loads

$M = \frac{WL}{10}$ $L = \frac{10M}{W}$ $W = \frac{10M}{L}$ $L^2 = \frac{10M}{w}$

For interior spans continuous with uniform loading.

$M = \frac{WL}{12}$ $L = \frac{12M}{W}$ $W = \frac{12M}{L}$ $w = \frac{12M}{L^2}$

For Cantilever beam with Uniform Load entire span.

$M = \frac{WL}{2}$ $L = \frac{2M}{W}$ $W = \frac{2M}{L}$ $w = \frac{2M}{L^2}$

For Cantilever beam with Concentrated Load on free end.

$M = PL$ $P = \frac{M}{L}$ $L = \frac{M}{P}$

Shear formulas for rectangular beams 3.2.3

$V = \frac{W}{2}$ or $\frac{P}{2}$ and $V = R_1$ or R_2 See shear diagrams.

Horizontal unit shear stress about centroid x-x.

$v = \frac{3V}{2bd}$ or $\frac{3V}{2A}$ $V = \frac{2bdv}{3}$ $A = \frac{3V}{2v}$ $A = bd$

For horizontal unit shear stress at any point about Centroid.

$V = vbd$ $v = \frac{Va\bar{y}}{I_x b}$ Shear per lineal foot along Centroid is obtained for glue and connectors as: $V = 12bv$

Beam deflection formulas and transposition 3.2.4

For simple span beams with Uniform Load entire span.

$$\Delta = \frac{5W\ell^3}{384EI} \qquad I = \frac{5W\ell^3}{384E\Delta} \qquad W = \frac{384EI\Delta}{5\ell^3} \qquad \ell^3 = \frac{384EI\Delta}{5W}$$

For simple span beams with concentrated load at midspan.

$$\Delta = \frac{P\ell^3}{48EI} \qquad I = \frac{P\ell^3}{48E\Delta} \qquad P = \frac{48EI\Delta}{\ell^3} \qquad \ell^3 = \frac{48EI\Delta}{P}$$

For Cantilever Beam with Uniform Load entire span.

$$\Delta = \frac{W\ell^3}{8EI} \qquad I = \frac{W\ell^3}{8E\Delta} \qquad W = \frac{8EI\Delta}{\ell^3} \qquad \ell^3 = \frac{8EI\Delta}{W}$$

For Cantilever Beam with Concentrated Load at free end.

$$\Delta = \frac{P\ell^3}{3EI} \qquad I = \frac{P\ell^3}{3E\Delta} \qquad P = \frac{3EI\Delta}{\ell^3} \qquad \ell^3 = \frac{3EI\Delta}{P}$$

To reduce ℓ^3 to span in feet: $L = \sqrt[3]{\frac{\ell^3}{12}}$ (Use Tables)

Section property formulas 3.2.5

For square or rectangular shaped sections:

$$A = bd \qquad b = \frac{A}{d} \qquad d = \frac{A}{b} \qquad S = \frac{bd^2}{6} \qquad I = \frac{bd^3}{12} \qquad b = \frac{6S}{d^2} \qquad d = \sqrt{\frac{6S}{b}}$$

$$c = \frac{d}{2} \qquad S = \frac{I}{c} \qquad c = \frac{I}{S} \qquad k = \frac{S}{A} \qquad r = \sqrt{\frac{I}{A}} \qquad A = \frac{I}{r^2} \qquad I = Ar^2$$

For solid Circular shaped section.

$$A = D^2\,0.7854 \quad \text{or} \quad A = \frac{\pi d^2}{4} \qquad c = \frac{d}{2} \qquad I = \frac{\pi R^4}{4} \qquad S = \frac{\pi R^3}{4}$$

$$I = \frac{\pi D^4}{64} \qquad S = \frac{\pi D^3}{32} \qquad r = \frac{D}{4} \text{ or } \frac{R}{2} \qquad c = R$$

EXAMPLE: Joist spacing 3.3.1

A wood floor for a residence is to be designed for a Live Load of 50 Pounds per square foot. Joists proposed are 2X10 S4S, Yellow Pine with allowable F_b = 1000 P.S.I. Clear span is 12.0 feet, and span length for center of moment bearing is 13.0 feet. Dead Loads consist of shiplap sub-floor and T&G finish wood floor.

REQUIRED:
Determine total combined loads on joists and calculate the maximum spacing. Assume dried yellow pine weighs 36 Pounds cubic foot for Dead Load.

STEP I:

Dead Load at 2 inches for shiplap and finish floor =	6.00 Lbs. Sq. Ft.
Joist D.L. Assumed at. =	6.00 " "
Live Design Load =	50.00 " "
Total Design Load =	62.00 " "

STEP II:
A strip load 1.0 foot wide and 12.0 feet long = 744 Lbs.
Simple Span, $M = \frac{WL}{8}$. Strip load $M = \frac{744 \times 13.0}{8} = 1209$ Ft. Lbs.
Required section modulus for strip load: $S = \frac{M}{F_b}$
$S = \frac{1209 \times 12}{1000} = 14.5''^3$

STEP III:
From Table of Properties, a 2X10 Joist has $S = 24.4''^3$
Then spacing is: $s = \frac{24.4}{14.5} = 1.685$ Feet or approx. 20.0" cc.

DESIGNERS NOTE:
An alternate method to solve for spacing is to divide the Joists Resisting moment by the strip load moment as follows:
$RM = SF_b$ For 2X10, $RM = 24.4 \times 1000 = 24{,}400$ Inch Lbs.
Strip Moment = $1209 \times 12 = 14{,}500$ Inch Lbs.
Spacing = $\frac{24{,}400}{14{,}500} = 1.685$ Feet.

EXAMPLE: Laminated beam for special size ratio 3.3.2

Assume Span $L = 12.0$ feet for a laminated wood. Purlin. Desired size must have a depth 2 times greater than b. Load is 600 Lbs. Lineal Foot. Section to be of Southern Yellow Pine, with $F_b = 1200$ PSI.

REQUIRED:

A beam which has a ratio of $b = 1.0$ and $d = 2.0$ Section must support loads and be governed by F_b.

STEP I:

Simplified; $b = ½d$, and $d = 2b$.

Solve for a bending moment and required Section Modulus.

$M = \frac{wL^2}{8}$, $M = \frac{600 \times 12.0 \times 12.0}{8} = 10{,}800$ Ft. Lbs.

$S = \frac{M}{F_b}$, $S = \frac{10{,}800 \times 12}{1200} = 108.0''^3$

STEP II:

The property of S, for a rectangular section $= \frac{bd^2}{6}$

Then: $\frac{bd^2}{6} = 108.0''^3$ as above. b = breadth of section.

Ratio of $S = \frac{b \times (2b)^3}{6}$ or $S = \frac{4b^3}{6}$. When transposed, the formula is written: $b^3 = \frac{6S}{4}$ and $b = \sqrt[3]{\frac{6S}{4}}$

STEP III:

Solve for b^3, with known values placed in formula.

$b^3 = \frac{6 \times 108.0}{4} = 162.0$ Then $b = \sqrt[3]{162.0} = 5.45$ inches.

$d = 2b$ or $d = 2 \times 5.45 = 10.90$ inches

STEP IV:

Check the size by using rectangular formulas for calculating I_o and S_x. The beam is symmetrical about the neutral axis x-x, and $c = \frac{d}{2} = 5.45$ inches.

$I = \frac{bd^3}{12}$ and $S = \frac{I}{c}$.

$I = \frac{5.45 \times 10.90^3}{12} = 588.0''^4$ $S = \frac{588.0}{5.45} = 108.0''^3$ (Checks OK)

Accept a glued section 5.45" x 10.90" S4S net size, however with 1⅝ inch laminations a better section depth should be as: $d = 1.625 \times 7 = 11.375$ inches.

EXAMPLE: Allowable load on laminated beam 3.3.3

A laminated section 8.125" x 11.375" may be glue fabricated of $1\frac{5}{8}$ inch laminations as illustrated. Length of beam = 16.0 Feet. Spacing of beams is 10.0 feet on centers. Material is Southern Yellow Pine. F_b = 1750 PSI.

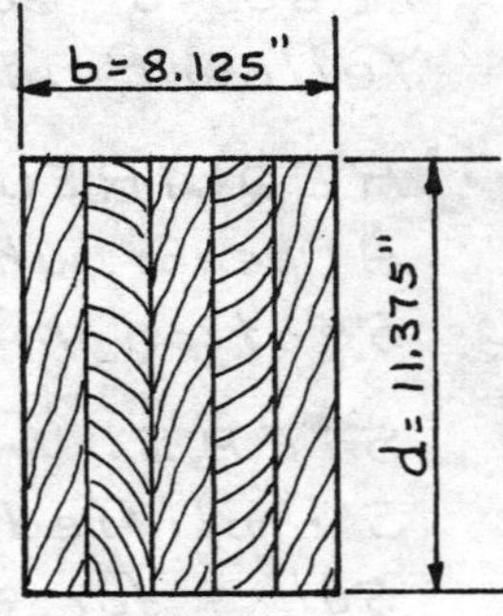

OR

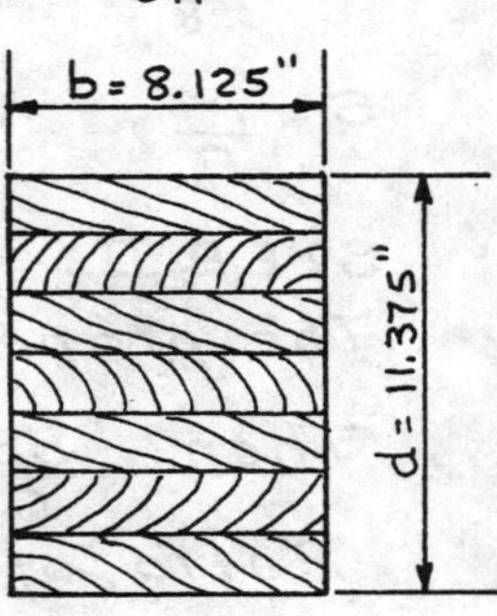

REQUIRED:
Calculate the total dead plus live load per square foot this section will support in bending stress only. Use the property tables for rectangular sections in Section VI to determine value of $S = \frac{I}{c}$.

STEP I:
The value of I_x for a section 1.0" x 11.375" in table is: $I = 122.7''^4$
For full section, $I_x = 122.7 \times 8.125 = 997.0''^4$
Dimension $c = 11.375 \times 0.50 = 5.69$ inches
$S = \frac{I}{c}$ or $S_x = \frac{997.0}{5.69} = 175.0''^3$

STEP II:
Resisting Moment: SF_b. $M = \frac{175 \times 1750}{12} = 25{,}600$ Ft. Lbs.

$W = \frac{8M}{L}$ or Total Loads, $W = \frac{8 \times 25{,}600}{16.0} = 12{,}800$ Lbs.

STEP III:
Area supported by single beam = 16.0 x 10.0 = 160.0 Sq. Feet.
Max. Combined DL+LL = $\frac{12{,}800}{160}$ = 80 Lbs. Sq. Foot.

STEP IV:
Checking for horizontal shear: V = 6,400 Lbs. = (½ of W)

$f_v = \frac{3V}{2bd}$ $A = 8.125 \times 11.375 = 92.42$ Sq. In.

Actual $f_v = \frac{3 \times 6{,}400}{2 \times 92.42} = 104$ PSI.

EXAMPLE: Wood joist size and spacing 3.3.4

Proposed plans require wood joists on simple spans of 11.0 feet, 13.0 feet and 15.0 feet to center of bearing. All joists to be same size, and spacing to accomodate ceiling panels of sheetrock. Spacing may be either 16.0, or 24.0 inches on centers, or both only when load requirements are met. Southern Yellow Pine #2 grade dimension shall be used, with allowable, $F_b = 1100$ PSI. Horizontal shear allowable, $F_v = 85$ PSI.

REQUIRED:

With a Live Load of 45 Lbs. Sq. Foot, and a Dead Load of 15 Lbs. Sq. Foot, design for size of Joists and best spacing.

STEP I:

Square Foot Loads = 45 + 15 = 60 Lbs. Sq. Foot.

Taking longest span of 15.0 feet:

Spacing at 24 inches, $W = 2.0 \times 60 \times 15.0 = 1800$ Lbs. on 1 Joist

Spacing at 16 inches, $W = 1.33 \times 60 \times 15.0 = 1200$ " " 1 "

STEP II:

Calculating Bending Moments for simple spans. $M = \frac{WL}{8}$

At 24.0" spacing: $M = \frac{1800 \times 15.0 \times 12}{8} = 40{,}500$ Inch Lbs.

At 16.0" spacing; $M = \frac{1200 \times 15.0 \times 12}{8} = 27{,}000$ Inch Lbs.

STEP III:

Calculating for Section Modulus or required; $S = \frac{M}{F_b}$

At 24.0", $S = \frac{40{,}500}{1100} = 36.8''^3$

At 16.0", $S = \frac{27{,}000}{1100} = 24.5''^3$

From Tables: A 2X12 has, $S = 35.8''^3$ and a 2X10 has, $S = 24.5''^3$

Either size is close enough to allowable to be used, and choice shall be based on economy.

To cover an area 12.0 x 15.0 feet.

Requires 7-2X12 Joists 16.0' Total Board Feet = 224

Requires 10-2X10 Joists 16.0' Total Board Feet = 267

STEP IV:

Checking shear in 2X12 Joist. Cross Section, A = 18.7 Sq. In.

$V = \frac{W}{2}$ or $V = \frac{1800}{2} = 900$ Lbs. Allowable $v = \frac{3V}{2A}$

Actual shear stress, $v = \frac{3 \times 900}{2 \times 18.7} = 71.2$ PSI. (OK, less than 85 PSI)

Accept 2X12 Joists, and space 24 inches on centers.

EXAMPLE: Floor load analysis 3.3.5

An existing floor structure is using 2X12 #2 S4S, Southern Pine joists spaced 16.0 inches on centers. Clear span is 9.50 feet, and center of moments is 10.0 feet. Dead Loads are given as 11 Lbs. per square foot, and allowable working stress in bending is given as: $F_b = 1000$ PSI. Limit, $v = 95$ PSI.

REQUIRED:

Analyze the floor structure to determine the allowable Live Load per square foot. Assume simple spans.

STEP I:

From Tables, a 2X12 has, $S = 35.82''^3$

Resisting Moment = SF_b. $RM = \frac{35.82 \times 1000}{12} = 2900$ Foot Lbs.

STEP II:

Transpose simple span formula, $M = \frac{WL}{8}$, and solve for load W.

$W = \frac{8M}{L}$ or $W = \frac{8 \times 2900}{10.0} = 2320$ Lbs. (Combined LL plus DL.)

Dead Load = $1.33 \times 10.0 \times 11 = 146.67$ Lbs.

Live Load = $2320 - 146.67 = 2173.33$ Lbs.

STEP III:

Area / Joist supports = $1.33 \times 10.0 = 13.33$ Square feet.

Then $\frac{W_{LL}}{A_F}$ = Load per square foot, maximum allowed.

$L.L = \frac{2173.33}{13.33} = 162.6$ Lbs. Square foot.

STEP IV:

Checking shear with combined DL and LL = 2320 Lbs.

Area 2X12 = 18.70 Sq. In. Actual $v = \frac{3V}{2bd}$ $V = \frac{2320}{2} = 1160$ Lbs.

$v = \frac{3 \times 1160}{2 \times 18.70} = 93$ Lbs. Sq. In. (OK, less than allowable).

DESIGNERS NOTE:

From the given data, it should be noted that the joist end bearing is 3.00 inches, and perpendicular to grain. From the tables, the compressive stress allowable is, 455 PSI for #1 grade. Joist bearing area is $1\frac{5}{8}'' \times 3.0''$ or

$A_b = 1.625 \times 3.0 = 4.875$ Sq. In. Bearing = $\frac{1160}{4.875} = 238$ PSI. (OK)

EXAMPLE: Maximum joist span 3.3.6

Data available: Dead Load plus Live Load = 50 Lbs. Sq. Foot.
Joist = 2X12 S.Y.P #2 Dimension, spaced 18.0 inches on centers.
Simple Spans. $M = \frac{wL^2}{8}$ or $\frac{WL}{8}$. Allowable F_b = 1100 PSI.

REQUIRED:

(a) Calculate the maximum span which joist will support with spacing of 18.0 inch centers, with tension only.

(b) Determine the amount of deflection on above joists when installed under conditions found in answer (a).

(c) By limiting the deflection to 1/360 of span, calculate the required moment of Inertia for joist.

STEP I:

Load on joist per Lineal foot: $w = 1.50 \times 50 = 75$ Lbs.

Table of Properties: For 2X12 Joist, $S = 35.8''^3$ and $I_x = 206.0''^4$

$E = 1,600,000$ F_t = 1100 PSI. Resisting Moment = SF_t

$$M = \frac{35.8 \times 1100}{12} = 3282 \text{ Foot Lbs.}$$

If $M = \frac{wL^2}{8}$ Then, $L^2 = \frac{8M}{w}$ and $L = \sqrt{L^2}$ or $L = \sqrt{\frac{8M}{w}}$.

STEP II:

Substituting values in transposed formula:

$$L = \sqrt{\frac{8 \times 3282}{75}} = 18.7 \text{ Feet.}$$ (Call it 18.0 feet for Answer a).

STEP III:

For deflection on simple spans: $\Delta = \frac{5Wl^3}{384EI}$

$L = 18.0'$ $l = 18.0 \times 12 = 216$ inches

$l^3 = 216 \times 216 \times 216 = 10,077,696$ or take from tables.

$W = 18.0 \times 75 = 1350$ Lbs. Put values in formula:

$$\Delta = \frac{5 \times 1350 \times 10,077,696}{384 \times 1,600,000 \times 206.0} = 0.536 \text{ inches, (Ans. b)}$$

STEP IV:

Limited deflection: $\Delta = \frac{l}{360}$. $\Delta = \frac{216}{360} = 0.60$ inches

To solve for I, transpose formula: $I = \frac{5Wl^3}{384E\Delta}$

$$\text{Then, } I = \frac{5 \times 1350 \times 10,077,696}{384 \times 1,600,000 \times 0.60} = 184.5''^4$$ (Ans. c).

DESIGN NOTE:

All equations in this example were equated by <u>slide rule</u> and any variance within one percent is considered to be close enough for accuracy.

EXAMPLE: Cantilever beam analysis 3.3.7

A West Coast Douglas Fir beam 4.0"x 10.0" nominal size, overhangs its support 8.0 Feet. Beams are spaced 8.0 feet on centers to support an awning. An electric neon sign has been installed at free end of beams, and at last measurement, the deflection at end is down 5/8 inches from horizontal for each beam supporting sign. Allowable F_b = 1500 PSI, and E = 1,760,000.

REQUIRED:

Determine the overstress in beam, and calculate the approximate lineal foot load of sign. Neglect the roof and live loads which may be contributing factors.

STEP I:

Gather all data necessary for solution.

L = 8.0 Ft. From tables: $S_x = 54.5''^3$ $I_x = 259.0''^4$

$l = 8.0 \times 12 = 96$ inches E = 1,760,000 $\Delta = 0.625$ inches

STEP II:

Deflection formula for Cantilever Beam with Concentrated Load P at free end is: $\Delta = \frac{Pl^3}{3EI}$. The unknown = P.

Transposing formula to solve for load, $P = \frac{3EI\Delta}{l^3}$

From tables: l^3 = 884,736

STEP III:

Substituting values in formula:

$$P = \frac{3 \times 1,760,000 \times 259.0 \times 0.625}{884,736} = 965 \text{ Lbs.}$$

Bending, $M = Pl$. $M = 965 \times 96.0 = 92,650$ inch pounds. (Negative)

$f_b = \frac{M}{S}$. Actual stress, $f_b = \frac{92,650}{54.5} = 1,700$ PSI.

Overstress = 1,700 − 1500 = 200 P.S.I.

STEP IV:

Weight of Sign: Spacing of beam supports = 8.0 feet.

Weight per Lineal foot = $\frac{965}{8.0}$ = 121 Lbs.

Deflection ratio = $\frac{96.0}{0.625}$ = 153.5 of span length.

EXAMPLE: Joist spacing by deflection 3.3.8

A floor in a second story structure is to have a plastered ceiling on under-side. Simple spans of 20.0 feet, and 2x12s #2 Pine or Fir are proposed for joists.
Maximum deflection must not exceed 1/360 of span length in inches for plaster.
Combined Loads are 65 Pounds per square foot.
Design stresses are maximum: F_b = 1500 PSI. E = 1,760,000
Horizontal Shear: F_v = 110 PSI.

REQUIRED:

(a) Calculate joist spacing for plaster requirements.
(b) Calculate joist spacing for Acoustical Tile requirements.
(c) Determine actual bending stress in joists for (a) and (b).

STEP I:

Deflection formula for simple spans: $\Delta = \dfrac{5Wl^3}{384EI}$

$L = 20.0'$ $l = 20.0 \times 12 = 240''$ and $l^3 = 13{,}824{,}000$

Max. $\Delta = \dfrac{240}{360} = 0.667''$ for plaster. Max $\Delta = 1.00''$ for Acoustic.

For 2X12, $I_x = 206.0''^4$ and $S_x = 35.8''^3$ $A = 18.7$ Sq. In.

STEP II:

Solving for Max. $W = \dfrac{384EI\Delta}{5l^3}$ Substituting values:

For Plaster, $W = \dfrac{384 \times 1{,}760{,}000 \times 206.0 \times 0.667}{5 \times 13{,}824{,}000} = 1{,}345$ Lbs.

For Acoustic, $W = \dfrac{384 \times 1{,}760{,}000 \times 206.0 \times 1.00}{5 \times 13{,}824{,}000} = 2{,}020$ Lbs.

STEP III:

A strip load 1.0 foot wide = 65 × 20.0 = 1300 Lbs.

Spacing for plaster = $\dfrac{1345}{1300} = 1.035$ Feet. (Call it 12 inches for a.)

spacing for Acoustic = $\dfrac{2020}{1300} = 1.55$ Feet. (Call it 18 inches for b)

STEP IV:

Actual bending stress under loads, $f_b = \dfrac{M}{S}$

For Plaster: $M = \dfrac{1345 \times 20.0 \times 12}{8} = 40{,}350$ In. Lbs. $f_b = \dfrac{40{,}350}{35.8} = 1130$ PSI.

For Acoustic: $M = \dfrac{2020 \times 20.0 \times 12}{8} = 60{,}600$ In. Lbs. $f_b = \dfrac{60{,}600}{35.8} = 1693$ PSI.

Allowable unit stress of F_b = 1500 PSI, is OK for plaster, and is exceeded for Acoustical Tiles. Spacing will have to be based on both deflection and allowable bending stress.

EXAMPLE: Joist spacing by deflection, continued 3.3.8

STEP V:

Re-Calculate spacing for Acoustical Tiles.

Strip Load Moment = $\frac{1300 \times 20.0 \times 12}{8}$ = 39,000 Inch Lbs.

Resistance Moment for a 2X12 = SF_b. RM = 35.8 x 1500 = 53,700 In. Lbs.

spacing = $\frac{RM}{M}$ or spacing = $\frac{53{,}700}{39{,}000}$ = 1.375 Ft. (Call it 16.0 Inches)

STEP VI:

Check for Horizontal shear. Max. F_v = 110 PSI.

For Plaster, W = 1345 Lbs. $V = \frac{W}{2}$, $V = \frac{1345}{2}$ = 673 Lbs.

$f_v = \frac{3V}{2bd}$. $f_v = \frac{3 \times 673}{2 \times 1.625 \times 11.50}$ = 54 PSI. (OK, less than F_v).

STEP VII:

Check shear for Acoustical Tile on spacing found in Step V.

W = 1.33 x 65 x 20.0' = 1740 Lbs. $V = \frac{1740}{2}$ = 870 Lbs.

$f_v = \frac{3 \times 870}{2 \times 18.7}$ = 69.7 PSI. (Also OK, less than 110 PSI Allowed).

DESIGN NOTE:

A kiln dried 2X12 S4S in yellow pine or fir, will have a weight of approximately 5.20 Pounds per lineal foot.

For a 20.0 span, weight = 5.20 x 20.0 = 104 Lbs.

At the start, 5 Pounds was added to dead loads as 60 + 5 = 65 PSF. When spacing came out at 12 inches on centers for plaster, the estimate was equal, however for Acoustic Tile, with spacing at 1.33 feet, the estimate ran over by 1.65 PSF.

EXAMPLE: Wood plank floor 3.3.9

EXAMPLE: (Wood Plank Floor).
A wood deck floor is to consist of 1⅝ inch thick Fir planks which will continue in long lengths over supports. Each plank is tongue and grooved.
Allowable working stresses are as follows: $F_b = 1500$ PSI. $F_v = 130$ PSI.

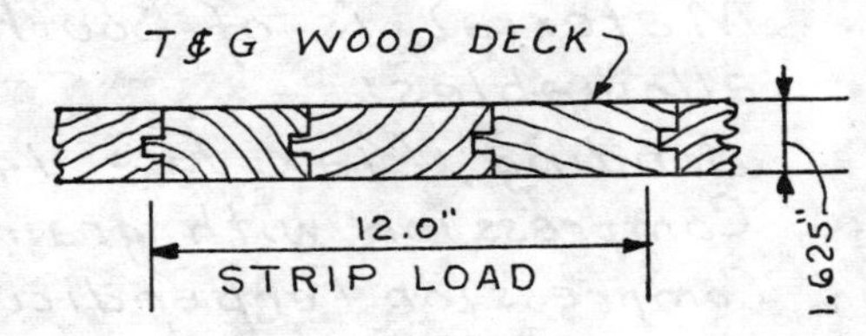

REQUIRED:
With a combined Live Load and Dead Load of 125 Lbs. per Square Foot, determine the maximum span for deck and spacing for supporting beams.

STEP I:
Problem call for solving unknown span L.
Take strip width 12 inches wide at 125 Lbs. Lineal Foot.
$b = 12.0''$ $d = 1.625''$ and $S = \frac{bd^2}{6}$. Also $I = \frac{bd^3}{12}$. $w = 125$ PSF.

$S = \frac{12.0 \times 1.625 \times 1.625}{6} = 5.28''^3$

STEP II:
Resisting Moment = Bending Moment. $M = \frac{wL^2}{10}$ for end spans.

$M = \frac{5.28 \times 1500}{12} = 660$ Foot Lbs.

Transposed formula: $L^2 = \frac{10\,M}{w}$ and $L = \sqrt{L^2}$ or $L = \sqrt{\frac{10M}{w}}$.

$L^2 = \frac{10 \times 660}{125} = 52.75$ and $L = \sqrt{52.75} = 7.26$ Feet. (Call it 7.0' centers).

STEP III:
Checking for shear: $W = 125 \times 7.0 = 875$ Lbs. $V = 875 \times 0.50 = 437.5$ Lbs.
Formula: $f_v = \frac{3V}{2bd}$. Actual $f_v = \frac{3 \times 437.5}{2 \times 12.0 \times 1.625} = 33.6$ PSI. (OK, less than F_v).

Accept support spacing at 7.0 foot centers.

EXAMPLE: Horizontal shear at centroid 3.3.10

A rectangular cross-section with nominal size of 4"x10" S4S, supports a Concentrated Load of 3600 Lbs., at middle of a simple span of 8.0 Feet.
Material is of Southern Pine with following stress allowables:
Bending, Max. F_t = 1400 P.S.I
Compression with grain, F_c = 1500 PSI.
Compression Perpendicular to grain, $F_{c\perp}$ = 350 PSI.
Horizontal Shear, F_{vh} = 130 PSI

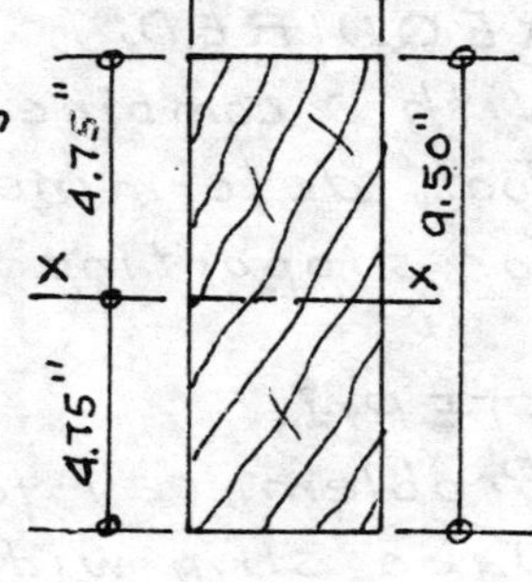

REQUIRED:
(a) Calculate the unit vertical shear maximum at support under load.
(b) Determine the horizontal shear stress at Centroid X-X.
(c) What must be minimum end bearing length on beam to resist compression perpendicular to grain?

STEP I:
Load P = 3600 Lbs. Shear at support = $\frac{P}{2}$ = V.
V = $\frac{3600}{2}$ = 1800 Lbs. Intensity of vertical shear is
$f_v = \frac{V}{bd}$ or $\frac{V}{A}$. In formula: $f_v = \frac{1800}{3.625 \times 9.5}$ = 52.4 PSI. (Ans. a.)

STEP II:
For horizontal shear, formula is: $f_{vh} = \frac{3V}{2bd}$. Then
$f_{hv} = \frac{3 \times 1800}{2 \times 3.625 \times 9.5}$ = 78.5 PSI.' (Ans. b)

STEP III:
V = 1800 Lbs. at support and $F_{c\perp}$ = 350 PSI.
For 1 inch of end projection, area = 3.625 Sq. In.
Bearing with 1 inch projection = 3.625 x 350 = 1269 Lbs.
Minimum projection = $\frac{1800}{1269}$ = 1.415" (Call it 2 inches) Ans. c

STEP IV:
Horizontal tension and compression. Fibers above X-X are in compression, and below X-X, they are in tension.
Max. M = 1800 x 4.0 x 12 = 86,400 inch lbs. $S = \frac{bd^2}{6}$
$S = \frac{3.625 \times 9.5 \times 9.5}{6} = 54.5''^3$ $f_{bct} = \frac{86,400}{54.5}$ = 1580 psi. (Over all'd.)

EXAMPLE: Horizontal shear at any point 3.3.11

A laminated section $3\frac{5}{8}" \times 9\frac{3}{4}"$ is built up of 6 seperate sections $3\frac{5}{8}" \times 1\frac{5}{8}"$ and glued together under pressure. This section is to be used as a simple beam with a span of 8.0 feet, and will support a load of 3600 Lbs. Load is concentrated at mid-span.

REQUIRED:

(a.) Calculate the moment of Inertia of the whole section and then find the horizontal shear at outer glue joint.

(b) Compare the value of shear found with the horizontal shear intensity at the centroid x-x.

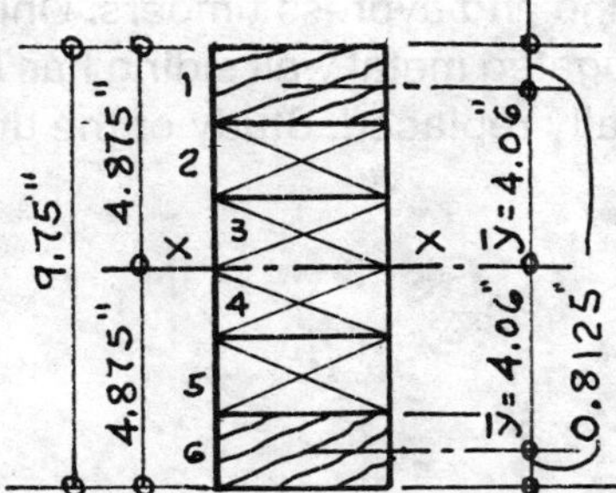

STEP I:

The formula for horizontal shear at any point from centroid is written: $v = \frac{Va\bar{y}}{Ib}$. Where:

$I = \frac{bd^3}{12}$ $I = \frac{3.625 \times 9.75 \times 9.75 \times 9.75}{12} = 280''^4$

Area outer lamination, $a = 3.625 \times 1.625 = 5.89\ \square''$

To center of gravity from x-x, $\bar{y} = 4.06$ inches

Glue joint location $= 1.625 \times 2 = 3.25''$ from centroid x-x.

STEP II:

Total shear at support $= \frac{1}{2}P$. $V = 1800\#$ $b = 3.625''$

Substituting values in formula for unit shear at glue joint 1-2:

$v = \frac{1800 \times 5.89 \times 4.06}{280 \times 3.625} = 42.41$ Lbs. Sq. Inch.

Stress per lineal foot in beam $= 42.41 \times 3.625 \times 12 = 1845$ Lbs.

STEP III:

Unit shear at Centroid x-x: $v = \frac{3V}{2bd}$ and $d = 9.75$ inches.

$v_x = \frac{3 \times 1800}{2 \times 3.625 \times 9.75} = 76.25$ Lbs. sq. inch.

Stress per lineal foot on glue joint x-x:

Horizontal shear: $76.25 \times 3.625 \times 12 = 3320$ Lbs. Lin. Foot.

Timber joints, connectors and fasteners 3.4

An investigation was made of an old, wood-frame, nine-story, abandoned drier elevator formerly used by the Lipton Tea Company of Galveston. This structure was erected prior to 1864 and has many unique timber joints worthy of study. The wrought iron bolts are still intact in the oak, yellow pine and cypress timbers. Only the corrugated metal wall siding has been periodically replaced. Study of the timber joints reveals that each wood component was designed in compliance with sound basic engineering design principles. Look at the details of the important joints and column connections. It is not difficult to see the designer's approach in solving for the interactions of the forces involved. The nomenclature used to identify the joint type is descriptive of its use or place.

ILLUSTRATED DETAILS: Timber joint nomenclature 3.4.1

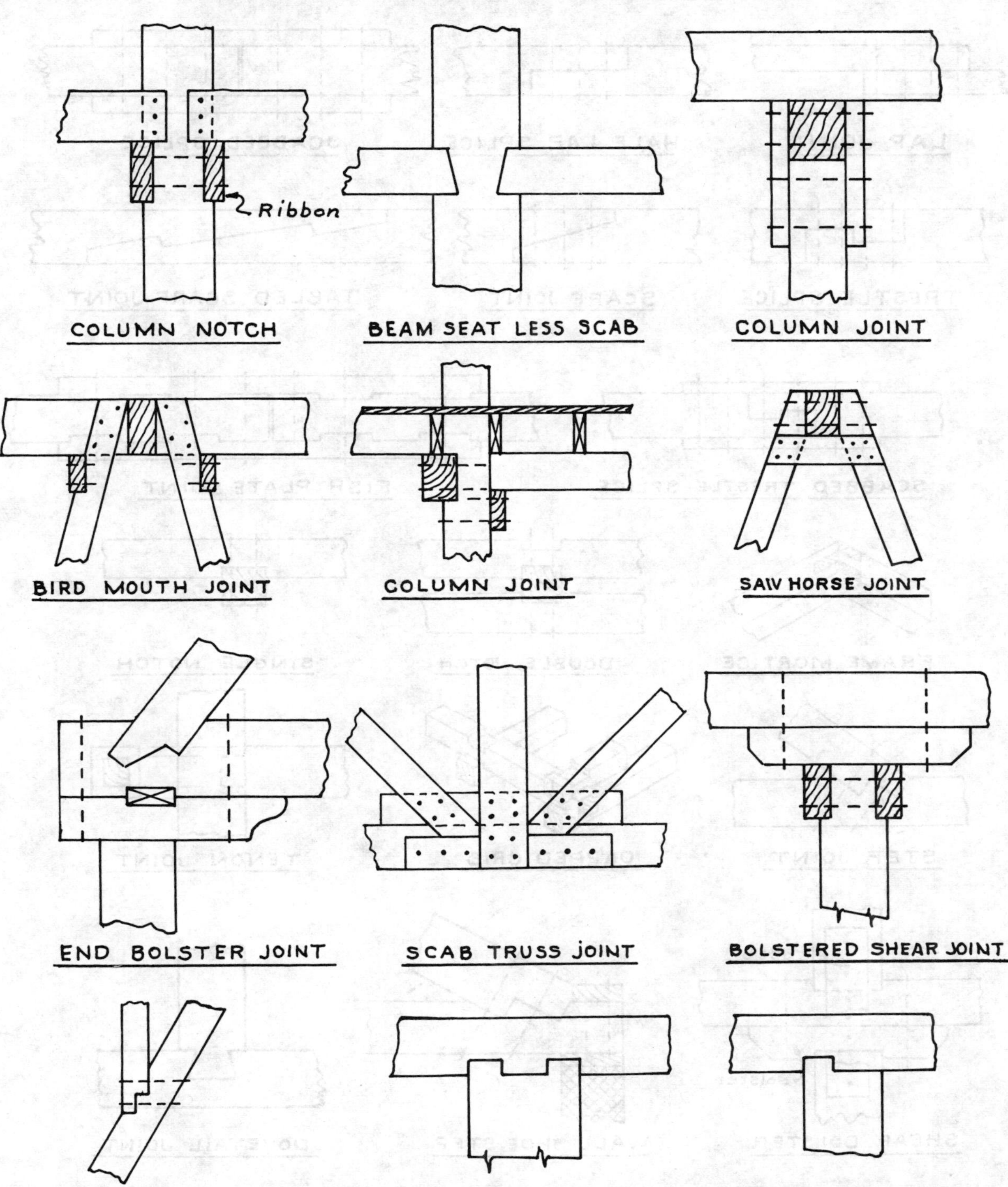

ILLUSTRATED DETAILS: Timber joint nomenclature, continued 3.4.1

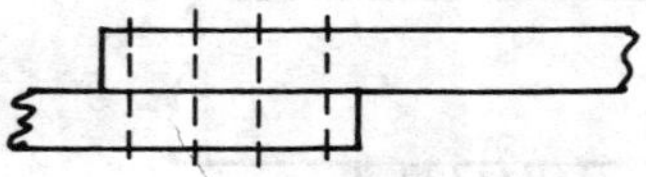

LAP JOINT

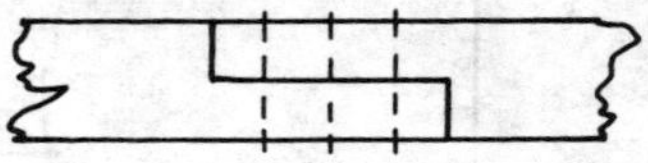

HALF LAP SPLICE

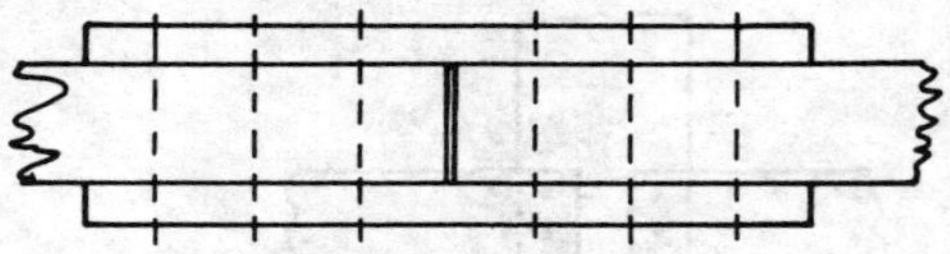

SCABBED SPLICE

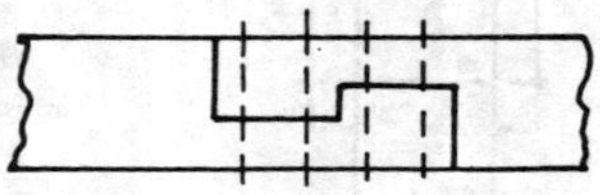

TRESTLE SPLICE

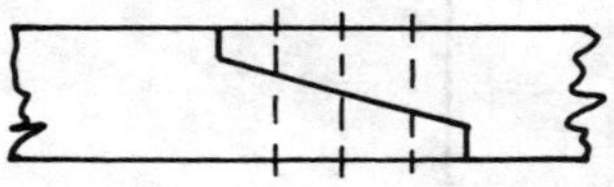

SCARF JOINT

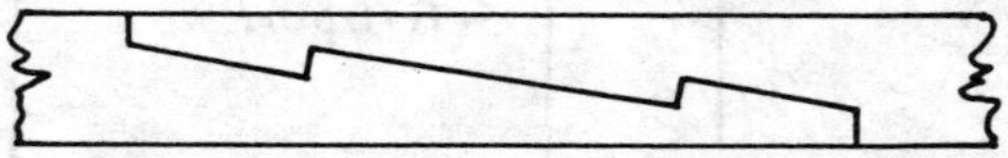

TABLED SCARF JOINT

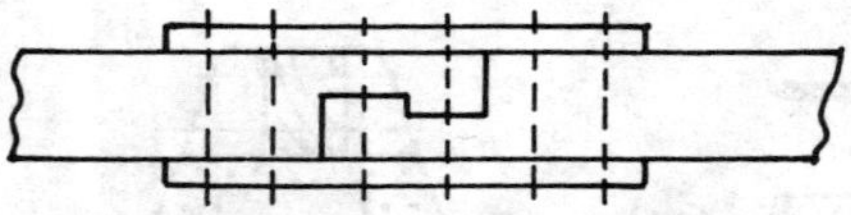

SCABBED TRESTLE SPLICE

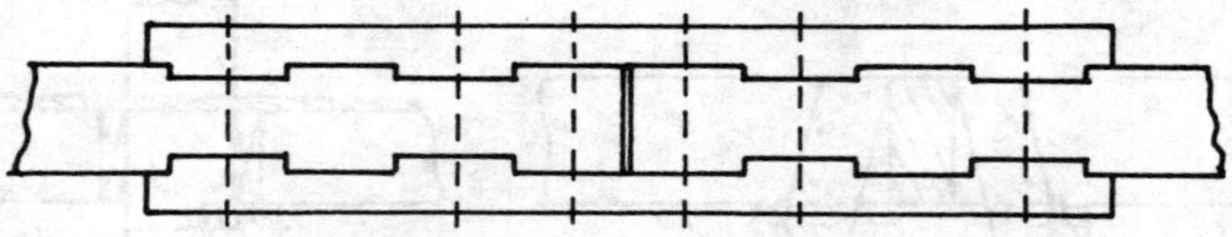

FISH PLATE JOINT

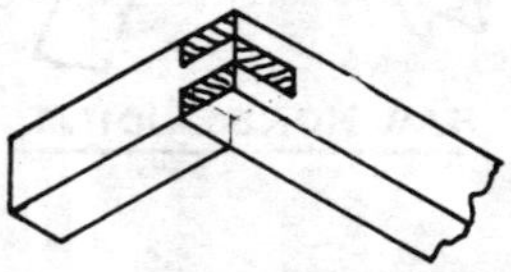

FRAME MORTICE

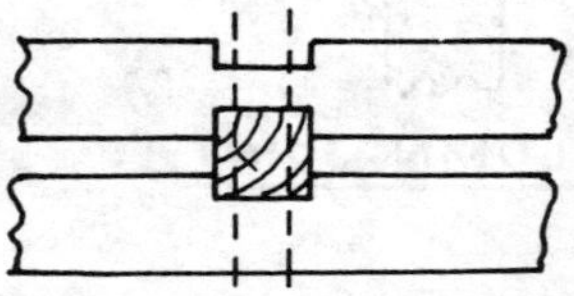

DOUBLE NOTCH

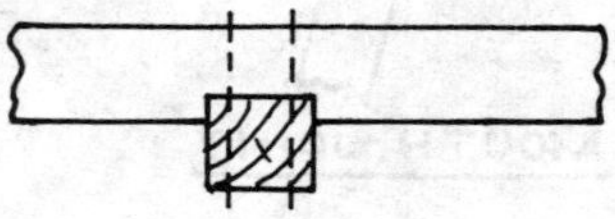

SINGLE NOTCH

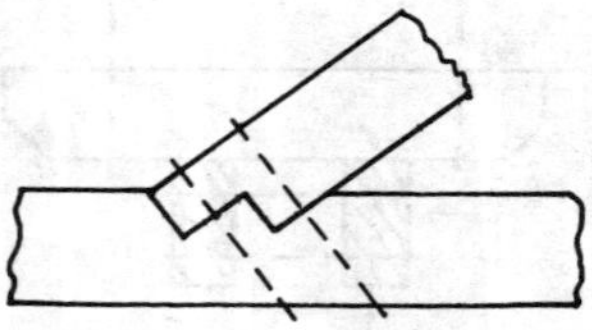

STEP JOINT

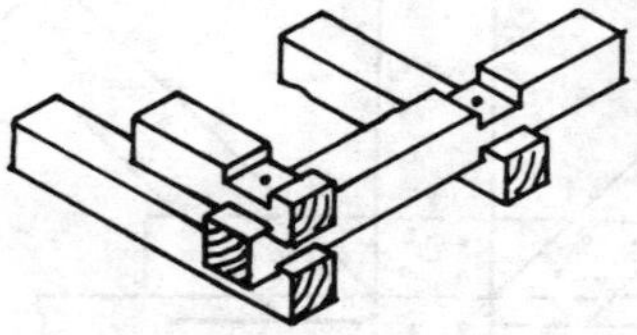

DAPPED CRIB

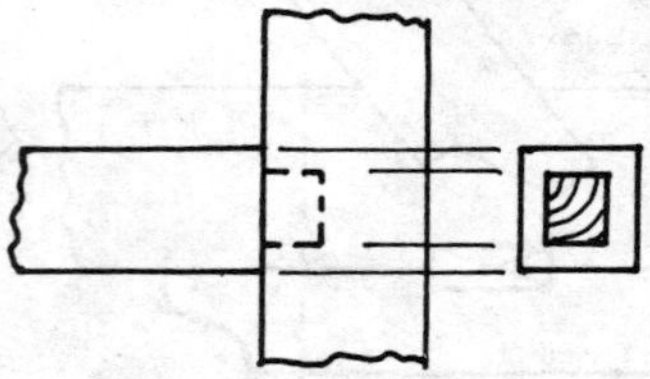

TENON JOINT

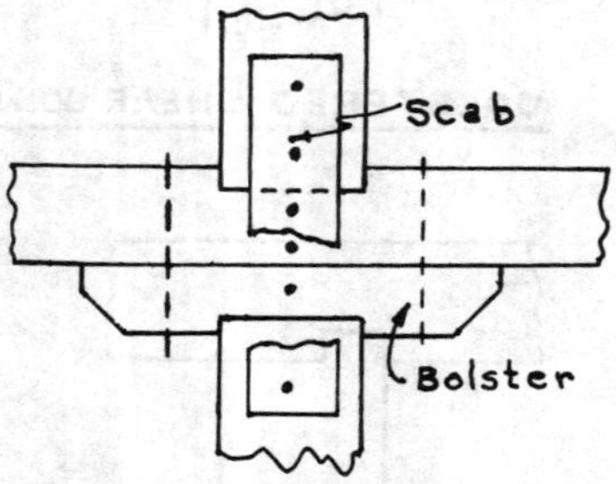

SHEAR BOLSTER

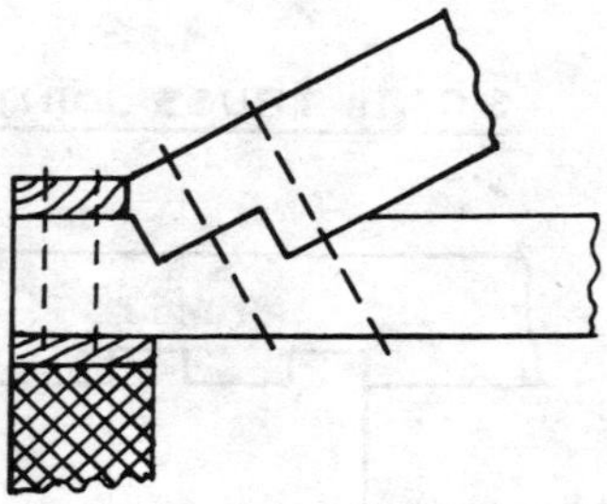

WALL SHOE STEP

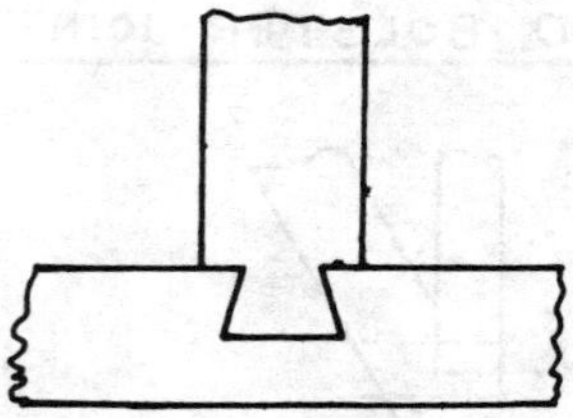

DOVETAIL JOINT

Shear connectors 3.4.2

Modern timber connectors have replaced fish plates and scab joints in the construction of wood joists or large wood trusses. These new connectors resist shear force by use of bolts and split ring claw plate, toothed ring, or alligator connectors, and flange plate connectors.

The values given in the tables for the various types of connectors will show that the resistance to shear stress is greater when the direction of force is in a direction parallel to the grain of the wood. When the force to be resisted is oblique to the wood grain, the value of connector may be obtained by proportion between the two known values based upon a straight line variation for angles between 0 and 90 degrees. An example will provide a formula and give an illustration for the design of loads acting obliquely to the direction of grain.

TIMBER MOISTURE CONTENT

The values given in the Tables are based upon the assumption that the wood is dry and seasoned to a moisture content of less than 15 percent. Green or unseasoned timber will normally contain a moisture content of approximately 28 percent. The table values must be reduced when used in green or damp, unseasoned timber. When designing connections which employ the split ring type, the moisture content in the wood is not an important factor and this reduction is not necessary. The diameter of a split ring type is not fixed as is the case with the other types, and when the ring is installed with its joint spread open, the ring will remain in a tight fit even though the wood will undergo a small amount of shrinking inside the connector ring.

SPLIT RING TYPE

The split ring can be compared to a small length of steel pipe sawn square with center line and then cut to alter the diameter. A circular groove is cut into each wood member, and the ring is inserted between the two pieces of timber. The bolt hole is in the center of ring and bored through the wood. The split ring is tight in the grooves, and the initial shear slippage is taken against the ring, not the bolt.

Only special power-operated tools can be used to cut the ring grooves, and the cut should be slightly larger than the ring. The ring is inserted by spreading and will remain tight.

The split ring connector is the most popular of all the types, and was developed for wood to wood connections.

SHEAR PLATE TYPE

Shear plate connectors are referred to as flange plate types for wood to steel connections. They are also used for demountable wood to wood assemblies. Special power tools are necessary to accomplish satisfactory installation. Shear plate connections consist of a one-piece pressed steel unit which is installed flush with wood surface. Cut grooves similar to split ring installation are required. The bolt passes through the metal plate, or if the connector is of cast steel, a forged, hub will be provided for bolt bearing. This type of connector is particularly well adapted for making splice joints or bolting wood timbers to steel columns or beams. The values given in the tables should be reduced by 20 to 25 percent when used with green or unseasoned timber.

Shear connectors, continued 3.4.2

CLAW PLATE CONNECTORS

This type of connector is intended to serve the same function as the flanged plate type, however the method of installation is different. A precut circular recess is necessary for the hub, and the teeth or claws are forced into the uncut wood by the bolt pressure. The claw plate connectors are installed back to back in timber construction, and one connector will have a hub collar through which the bolt will pass. This member is called the male or hub connector. The female side of the connector fits over the hub which sustains the shear stress and leaves the bolt in tension only. The claw plate type of connector is generally intended to be used for wood to wood connections; however, either the male or female section can be used for wood to steel connections. When used with unseasoned timber, the values listed in table should be reduced approximately ten percent.

TOOTHED RINGS

Toothed rings use a circular shaped ring with saw teeth on each rim which is forced into the wood timbers by bolt pressure. These are also called "alligator" connectors. The installation of toothed rings does not require any power tools. Only the predrilled bolt hole is necessary. The function of the toothed ring is similar to the action of a steel corrugated nail which has been formed into a circle. Bolt holes should be drilled $1/16$ inch oversize so that a high strength bolt can be used for drawing the timbers together, and then replaced by an ordinary bolt after the rings have become embedded in each piece of wood. Correct the values given in the table by a ten to thirty percent reduction when making installations in green or unseasoned wood.

MISCELLANEOUS CONNECTORS

Timber Engineering Company, 1319 Eighteenth Street, N.W., Washington 6, D.C., offers design data for other types of connectors

For securing large timber pile caps to top of wood piles, metal spiked grids may be provided to supply rigidity to cross bracing and give added rigidity against impact energy resulting from ship docking. These spiked grids are fabricated in square and circular shapes for marine and railway structures.

The Bulldog Connector is in some degree similar to the toothed ring shear developer since its teeth must be embedded in the wood by bolt pressure. The thickness of metal used for this type is approximately $1/16$ inch, and they are either square or round in shape. When flanges are installed on this type of connector, they are referred to as "clamping plates." Embedment is accomplished by driving the timbers together with a sledge hammer or jacks.

TABLES: Shear connectors 3.4.2.1

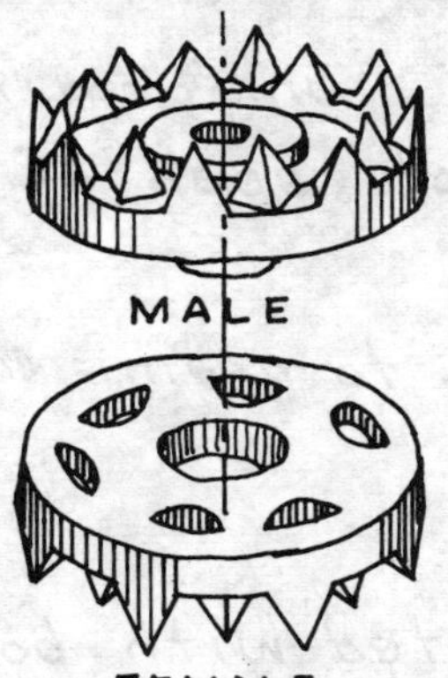
MALE

FEMALE

CLAW PLATES IN PAIRS - VALUES IN LBS.			
CONNECTOR DIAMETER IN INCHES	BOLT DIAMETER IN INCHES	FORCE PARALLEL TO GRAIN	
		WOOD TO WOOD	WOOD TO METAL
2 5/8	1/2	4,200	5,000
3 1/8	5/8	5,200	6,400
4	3/4	7,000	8,600

REDUCE VALUES FOR GREEN TIMBER ACCORDING TO THE RECOMMENDED PERCENTAGES AS GIVEN IN THE DESCRIPTIVE DATA FOR EACH TYPE OF CONNECTOR.

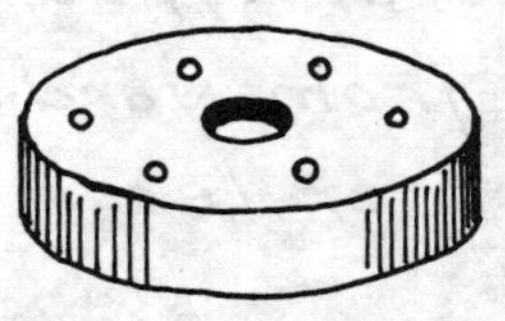
PRESSED STEEL

FLANGE PLATES IN PAIRS - VALUES IN LBS.			
CONNECTOR DIAMETER	BOLT SIZE IN INCHES	CONNECTOR METAL TYPE	PARALLEL TO GRAIN OF WOOD
2 5/8	3/4	PRESSED STEEL	5,500
4	3/4	Do.	7,500
4	7/8	CAST STL. HUB	8,300

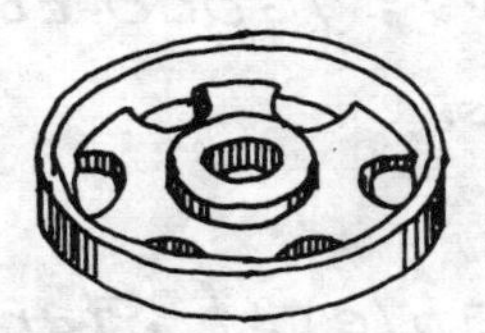
CAST STEEL TYPE

NOTE:
ALL VALUES SHOWN ARE SAFE LOADS ONLY WHEN USED WITH DENSE DRY TIMBER OF NOT MORE THAN 16% MOISTURE CONTENT.

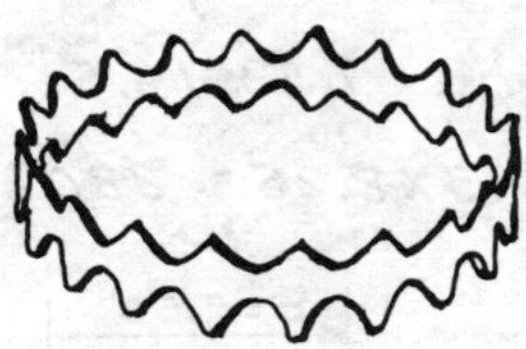

TOOTHED RINGS- SINGLE - VALUE IN LBS.			
CONNECTOR DIAMETER IN INCHES	BOLT DIAMETER IN INCHES	FORCE PARALLEL TO GRAIN	
		BOLT TIGHT FIT	OVERSIZE HOLE
2	1/2	2,400	2,200
2 5/8	5/8	4,200	3,600
3 5/8	3/4	5,800	5,200
4	3/4	6,900	6,300

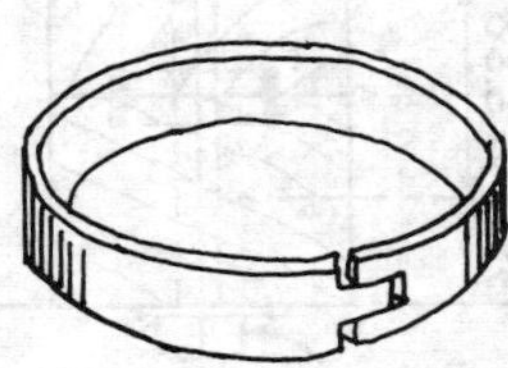

SPLIT RING - SINGLE - VALUE IN LBS.					
RING DIA. INCHES	BOLT SIZE INCHES	RING DEPTH INCHES	TIMBER THICK'N. INCHES	DIRECTION OF GRAIN	
				PARALLEL	PERPENDICULAR
2	1/2	0.75	1.0	2,600	1,550
2	1/2	0.75	1 5/8 +	3,150	1,900
4	3/4	1.00	1 5/8	4,300	2,450
4	3/4	1.00	2.0	4,900	2,850
4	3/4	1.00	2 5/8	6,000	3,450
4	3/4	1.00	3.0+	6,150	3,560

EXAMPLE: Timber shear connector design 3.4.2.2

A timber beam is required to support a uniform load of 1000 Lbs. per lineal foot on a simple span of 16.0 Feet. Allowables are as follows:

F_b = 1600 PSI. F_v = 120 PSI. Compression ⊥ to grain = 400 PSI.

REQUIRED:

Assume that the beam must be fabricated with bolts and timber connectors. Design for several types of connectors and used dimension lumber from yard stock. Limit depth of beam to 18.0 inches or less.

STEP I:

16' × 1000 LB

Neglecting dead load of beam, the load W = 16000 Lbs.

$M = \frac{WL}{8}$. $M = \frac{16{,}000 \times 16.0}{8} = 32{,}000$ Ft. Lbs.

$S = \frac{M}{F_b}$. $S = \frac{32{,}000 \times 12}{1600} = 240''^3$ From table of Standard Timber sizes, an 8X14 has an $S = 228.0''^3$. Then if 3 Pcs. of 6X8 were used, dimensions would be 7.5" X 16.5", and

$S = \frac{7.50 \times 16.5 \times 16.5}{6} = 340.3''^3$ $C = 8.25$ $I_x = 340.3 \times 8.25 = 2807.5''^4$

STEP II:

Drawing Cross-section of Beam using 3 - 6X8s flatwise:

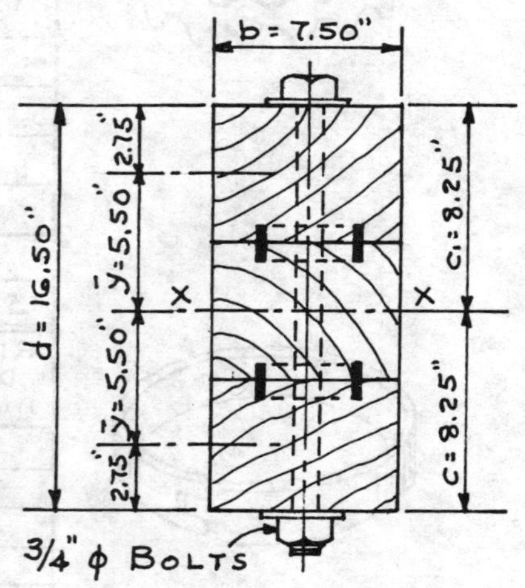

$V = \frac{W}{2}$ $V = \frac{16{,}000}{2} = 8000$ Lbs.

Horizontal shear at Centroid X-X.

$f_v = \frac{3V}{2bd}$. $f_v = \frac{3 \times 8000}{2 \times 7.5 \times 16.5} = 97$ PSI.

EXAMPLE: Timber shear connector design, continued 3.4.2.2

STEP III:

Area of one section, $a = 7.50 \times 5.50 = 41.25^{\square''}$

From x-x to Center of Gravity, $\bar{y} = 5.50''$

Formula for horizontal shear at any point, $f_v = \dfrac{Va\bar{y}}{I_x b}$

Actual unit stress at joint for connector to resist.

$$f_v = \frac{8000 \times 41.25 \times 5.50}{2807.5 \times 7.50} = 86.5 \#^{\square''}$$

STEP IV:

Horizontal Shear force along beam: $q = 12 b f_v$

For connector, shear per foot $= 86.5 \times 7.50 \times 12 = 7785$ Lbs.

STEP V:

Selecting a connector with 4.0 inch diameter to be used with a $\frac{3}{4}''\phi$ Bolt, take values from table of type desired. Using 2 connectors at each location, the spacing becomes:

For Claw Plate Type: $\dfrac{7000 \times 0.50 \times 12}{7785} = 5.40$ Inch spacing.

For Split Ring Type: $\dfrac{12{,}000 \times 0.50 \times 12}{7785} = 9.25$ Inch spacing

For Flanged Connector: $\dfrac{7500 \times 0.50 \times 12}{7785} = 5.75$ Inch spacing.

For Pair Alligator Conns: $\dfrac{6900 \times 0.50 \times 12}{7785} = 5.33$ Inch spacing.

STEP VI:

For length end bearing, $F_{c\perp} = 400$ PSI

Value per lineal inch $= 400 \times 7.50 = 3000$ Lbs.

Minimum length $= \dfrac{8000}{3000} = 2.67$ inches. Use 3 inches min.

Lumber fasteners 3.4.3

One of the unique properties of wood members is the ease with which the pieces can be joined together to provide concrete forms, framing, and other structural projects. Wood fasteners have, over the years, been developed to a high degree of efficiency and many different types are available for various uses. The choice depends upon the intended use and the load to be transferred.

Wood can be joined together with glue or mechanical fasteners. The mechanical fasteners include bolts, nails, dowels, splines, tenons, screws, grip anchors, and studs driven by powder explosives. Only the more common type of wire nail will be taken into consideration in the technical data to follow, since it is the single most used fastener for joining wood members. The tables of allowable loads are based on the National Design Specifications for stress graded lumber, and the results of a variety of tests and experiments conducted by many authorities.

HOLDING POWER OF NAILS

It must be acknowledged that the ability of a nail to hold two pieces of lumber together will depend upon many conditions. The several published tables will show values with considerable variances. Because there is no possible method of testing which can be trusted to provide the absolute value of the holding power of a nail, the published tables must be used with caution. Among the conditions upon which the value of a nailed connection will depend, is the state and character of wood and fasteners. The size, length, penetration, wood species, density of rings, the state of moisture content, side or edge grain joining, and the angle of driving, are only a few of the common circumstances which will affect the holding power of a nail or spike.

Before the common wire nail became the standard type of fastener for dimension lumber, the cut nail with its tapered body was used exclusively by carpenters. This type of nail is still available and requires heavier driving energy. The resistance to pulling offered by the cut nail is initially greater, but the resistance to further withdrawal will decrease very rapidly due to the wedge shape of the body. Tests for resistance to complete withdrawal of the two types have indicated that the wire nail is superior for use in most construction.

LATERAL RESISTANCE OF NAILS

From the average results of many experiments made by several universities, testing laboratories and railroad engineers, the lateral resistance values of a single nail have been compiled into the following table. In referring to the values in table, remember that the values are based upon normal conditions for general work. Average conditions would be considered as being all of the following:

(a) Nail is driven perpendicular to wood surface.
(b) Penetration into secondary member is adequate.
(c) Moisture content in wood does not exceed 16 percent.
(d) Body of nail is bright or galvanized.
(e) Species of wood is compatible with table type shown.

COMMON SIZES AND WEIGHTS

Wire nails as used by the carpenter trade will range in size from sixpenny (6d) to sixtypenny (60d). They are sold by the pound or keg, depending upon needed quantity. The length of common nails runs from 1 inch to 6 inches, and should be noted on drawings. The custom of designating nail by penny, originated back in the

Lumber fasteners, continued 3.4.3

days when nails were sold in the manner of 100 nails for 10 pence, 6 pence, etc. A hundred 10d nails cost 10 pence or 10 pounds equalled 1000 nails. The penny nomenclature has been retained by the producers, although the original meaning is lost because present day suppliers' catalogues will list many different quantities as making a pound in weight. The number of nails in a pound will vary because of the length, gauge and size of head.

DESIGN OF NAILED JOINTS

In the design of lapped joints, the nail values given in tables are predicated on the assumption that the nail is driven perpendicular to the surface of the wood. Toe-nailed joints driven at an angle between 30 and 45 degrees from vertical or perpendicular plane should have the values reduced to 5/6 of the value shown in table.

A formula was developed which can be safely used to determine the lateral load resistance of a single nail used in a lapped joint. Assume P is the safe design lateral load in pounds for one nail. D equals the nail diameter, and the coefficient is 4000. The formula is written as: $P = 4000\ D^2$, and provides a safety factor of 4.75 to 5.5 for the Pine and Fir species.

Using the formula for Oak, Locust, Hedge and Maple, the coefficient is raised and the formula becomes $P = 6000\ D^2$. For work in these hardwood species, the safety factor will range between 4 and 5.

For the softwood species as, White Pine, Ponderosa, Cedar, Spruce, Parana and Redwood, the coefficient is reduced, and the formula is written as: $P = 2500\ D^2$. The safety factor for the softwood species ranges between 3.8 to 4.5.

To illustrate the design formula:
Assume a 16d common wire nail is used to join a 2x8 to a 6x6 Post column. Both woods are of dry yellow pine. A safety factor of 2.5 is desired. Calculate the number of nails required to support the reaction of 1000 Lbs. from the 2x8.

Diameter of 16d nail = 0.165 $D^2 = 0.0272$

$P = 4000 \times 0.0272 = 109\#$ Value with 2.5 SF = $109 \times 2.5 = 272.50$

Required number = $\frac{1000}{272.5} = 3.67$ Use 4 nails.

Nail penetration = $3\frac{1}{2} - 1\frac{5}{8} = 1\frac{7}{8}''$ Allowed 10 diameters = 1.650 Inches.

TABLE: Properties of nails in lateral joints 3.4.3.1

VALUES AND PROPERTIES OF WIRE NAILS IN LATERAL JOINTS

SIZE IN PENNY	LENGTH IN INCHES	WIRE GAUGE NUMBER	DIAMETER IN INCHES	APPROX. NUMBER IN LB	SAFE VALUE S.Y.P. & FIR $P=4000D^2$	ULTIMATE VALUE SYP. & FIR	SAFE VALUE OAK & ELM. $P=6000D^2$	ULTIMATE VALUE OAK & ELM	SAFE VALUE W.P. & CEDAR $P=2500D^2$	ULTIMATE VALUE WP & CEDAR
2d	1	15	0.072	876	21#	83 Lbs.	31#	125 Lbs.	13#	52 Lbs.
3d	1¼	14	0.083	568	28#	108 "	41	160 "	17	67 "
4d	1½	12½	0.102	294	41#	160 "	62	240 "	26	98 "
5d	1¾	12½	0.102	254	42#	168 "	63	250 "	26	98 "
6d	2	11½	0.115	167	53#	240 "	79	310 "	33	124 "
7d	2¼	11½	0.115	161	54#	255 "	80	322 "	34	131 "
8d	2½	10¼	0.124	101	61#	295 "	92	550 "	38	148 "
10d	3	9	0.148	67	88#	325 "	131	685 "	54	200 "
12d	3¼	9	0.148	63	88#	625 "	131	665 "	55	218 "
16d	3½	8	0.165	47	109#	495 "	163	800 "	68	262 "
20d	4	6	0.203	29	165#	750 "	247	1175 "	103	390 "
30d	4½	5	0.220	22	194#	1220 "	290	1390 "	121	453 "
40d	5	4	0.238	17	226#	1200 "	340	1655 "	140	555 "
50d	5½	3	0.259	13	268#	1425 "	404	1925 "	167	660 "
60d	6	2	0.284	10	323#	1780 "	485	1950 "	200	816 "

VALUES IN TABLE ARE BASED ON THE FOLLOWING CONDITIONS:

FOR SIZES 2d TO 6d, THAT ONE PIECE OF TIMBER IS NAILED TO ANOTHER, AND THE NAIL PENETRATION INTO THE SECOND TIMBER IS NOT LESS THAN ½ THE LENGTH OF NAIL.

FOR SIZES 6d TO 60d, THE PENETRATION INTO SECOND TIMBER SHALL BE NOT LESS THAN 10 DIAMETERS OF THE NAIL.

ALL TIMBER SHALL BE SEASONED OR DRIED TO A MOISTURE CONTENT OF NOT OVER 18 PERCENT.

NAILS SHALL HAVE BEEN DRIVEN PERPENDICULAR TO SURFACE OF TIMBERS.

Timber columns 3.5

The use of wood columns in the construction of farm buildings, such as implement sheds and hay barns, is probably as prevalent today as in the past. The Rigid Frame steel building seems not to have made much impact in the farm buildings market due to the absence of loft space for hay storage.

A survey of the past, prior to the Civil War, shows that most of the steel production was located in the eastern part of the country, while the south had a more active forest products industry producing lumber, paper pulp, turpentine and resins. The extensive use of timber for the construction of cotton warehouses, grain elevators and office buildings can be observed by a visit to New Orleans or Galveston. Structures built in the 1860's are still in use, and in a good state of preservation. In fact, with the installation of sprinkling systems acceptable to the fire insurance companies, these fine old office buildings will remain solid and useful for many more years.

Column design 3.5.1

Column design in wood is very similar to the design of steel columns, in that the slenderness ratio will govern the allowable unit stress. Where the least value of the radius of gyration (r) is used in the steel formulas, the least dimension of depth or width (d) is used for wood design. Thus an equation with $\left(\frac{l}{r}\right)$ indicates a steel column, while the ratio of $\left(\frac{l}{d}\right)$ refers to the slenderness of a wood column. With respect to a solid wood circular post such as a wood pile, the diameter is used for the (d) factor.

Most column formulas are written as $\frac{P}{A}$ = allowable stress for the conditions of slenderness ratio. However, in no event must the basic allowable compressive stress for a grade of timber be exceeded, regardless of the results of the slenderness ratio calculation. Under ordinary circumstances, the stress in a column will be parallel to the wood grain and fibers. When a girder, beam or cap is placed on top of column, the bearing stress in the *girder* will be in the nature of compressive stress perpendicular to the grain, and the allowed stress level is much lower. To equalize the bearing force of girder with the axial compressive force of the supporting column, select one of the illustrated column notched joints or bolsters for equal stress distribution.

Limits for long and short columns 3.5.1.1

When the ratio of unsupported length of column (l) in inches, to the least dimension (d) in inches, is 10 or less, the column is classified as a short column or post. For short columns, the full allowable F_c may be used, and it is not necessary to use the column formulas for reducing the allowable unit stress.

Long columns are considered to have an unsupported ratio of $\frac{l}{d}$ which is between 10 and 50. In no event shall a wood column be used which has a slenderness ratio of over 50. See Example 3.5.4.2 for the maximum height of a column.

Moisture conditions 3.5.2

When a thin slat of wood is saturated with water, the slat can be bent to form a complete circle. The same slat in a very dry state would break under the bending forces. The wet wood has become much more flexible because the moisture content has changed the elasticity value. Since the modulus of elasticity (E) in wood timber is governed by the presence of moisture, the design engineer must consider the circumstances and place where the column will be used.

When subjected to heavy loads for long periods, wood columns have a tendency to acquire a permanent set or deformity. This behavior is caused by the material yielding to the proportionate change in the modulus of elasticity. Wood beams and sills have been observed to retain a permanent sag after having been immersed by flood waters or dampened by prolonged roof leaks.

Exposure classifications 3.5.2.1

Lumber producers and design engineers have classified moisture exposure conditions as follows:

(a) Interior use. (Continuously protected to remain dry.)

(b) Semi-protected. (Occasionally wet but rapidly dried in air.)

(c) Moist areas. (Constantly wet, exposed to water, spray, floods, vapors, etc.)

Moisture and the elastic modulus 3.5.2.2

Referring to the Allowable Stress Tables, it will be seen that the Southern Pine Association records the value of E = 1,600,000. The Western Lumber Technical Manual has listed the value of E = 1,700,-000 for its products of Douglas Fir and Western Larch. In each case, the value of E is applicable to wood with a low moisture content—classification (a).

An old directive from the Forest Products Laboratory gives a usable approach to the design conditions for classifications (b) and (c). As a safe basic rule, for structures such as railroad trestles, marine docks, exterior columns, or where the wood is alternately wet and dry, the value of E should be reduced to 1,250,000.

The use of green or unseasoned timber should be used only for wet or constantly moist work. The continual dampness on the green timber serves as a satisfactory seasoning agent for preventing splits, shakes, shrinkage and twists in the timber.

Column design formulas 3.5.3

Design engineering firms have divergent views on what should be the true column formula and slenderness ratio for wood columns. There has never been a standard formula which would meet the safe loading requirements for all building codes, railway engineers associations, and industrial firms. In the many examinations given by the states for registration as a professional engineer or architect, the applicant will be asked to select a "suitable" formula to design a column. The column formula selected should be one of the following:

(a) Southern Pine Association (SPA).
(b) Forest Products Laboratory.
(c) Euler's Formula.
(d) American Railway Engineers Association (AREA)
(e) Winslow's Formula.

Southern Pine Association formula 3.5.3.1

The newest column formula published by the SPA is a replacement of three older formulas.

This formula is written as follows:

$$\frac{P}{A} = \frac{0.30\,E}{\left(\frac{l}{d}\right)^2}, \text{ or further,}$$

$$P = \left[\frac{0.30\,E}{\left(\frac{l}{d}\right)^2}\right] \times A.$$

Where nomenclature represents the following:

P = Total allowable Axial Load, in pounds.
A = Area of cross-section (bd), in square inches.
E = Modulus of Elasticity for wood species and seasoning.
l = Length of column without bracing for support, in inches.
d = Least dimension side of column, in inches.

Forest Products Laboratory formula 3.5.3.2

This formula was developed and published in 1927, but was never given wide acceptance. The formula was written as:

$$\frac{P}{A} = S\left[1.0 - \frac{1}{3}\left(\frac{l}{Kd}\right)^4\right]$$

Nomenclature designations are:

S = Basic unit stress allowable as given in tables for the value of parallel to grain as applicable to a particular species of wood.
K = A coefficient pertinent to wood species, grade, and seasoning condition.

For dry dense Southern Yellow Pine, the coefficient K = 26.1. The same value was used for West Coast Douglas Fir of like grade.

Euler's formula 3.5.3.3

Prominent Design Engineers, such as Euler, Gordon and Rankin, were competing to develop a column formula which could be made applicable to both steel and wood. To a certain extent, their efforts were successful because, in 1929, the Southern Yellow Pine Association and the Forest Products Laboratory of the U. S. Department of Commerce adopted the Euler formula as:

$$\frac{P}{A} = \frac{0.274E}{\left(\frac{l}{d}\right)^2}$$

This formula was considered conservative and was the fore-runner of the present SPA formula. The Eular formula will produce a curve which will run parallel with the SPA curve as shown on the chart.

Winslow's formula 3.5.3.4

Any column formula which uses the simple ratio of $\frac{l}{d}$ as the basis for calculating allowable stress or a reduction of F_c, is called a *straight line formula*. (Remember, the basic maximum allowable stress of F_c parallel to grain can be used for all short columns, or where the slenderness ratio of $\frac{l}{d}$ is not over ten.) When using a straight-line formula to solve for the reduced stress for intermediate lengths (slenderness between 10 and 50), the stress values, when plotted on a graph, will form a straight line. Considerable time and labor can be saved by using the graph, and the results will be close enough for all practical purposes. When plotting the graph, it will be observed that as the ratio of $\frac{l}{d}$ becomes greater, the unit working stress is reduced, which is the main intent of all the column formulas. (See CURVES 3.5.3.7.)

Winslow's formula is a convenient straight-line formula for designing, written:

$$P = C\left(1.0 - \frac{l}{80d}\right).$$

This formula was used in the Chicago area for many years. The symbols are typical of most formulas. In general use, the value of the basic allowable stress for C was listed as 1000 P.S.I. Thus the formula was written as:

$$\frac{P}{A} = 1000\left(1.0 - \frac{l}{80d}\right).$$

In order to adapt the equation to the variety of wood species, another design factor was introduced. These factors were used to raise or lower the value of basic stress C. When using the Winslow formula, use the following factors:

(a) White Pine, Tamarak or Redwood—
C = 1000 PSI.

(b) Douglas Fir or Hemlock—
C = 1.30 X 1000 = 1300 PSI.

(c) Long Leaf Yellow Pine or Oak—
C = 1.40 X 1000 = 1400 PSI.

(d) Short Leaf Southern Yellow Pine—
C = 1.20 X 1000 = 1200 PSI.

(e) Tidewater Cypress, Spruce, Cedar—
C = 0.90 X 1000 = 900 PSI.

The allowable unit stresses also can be read directly from the Design Working Stress Curves (See 3.5.3.7).

American Railway Engineers Association formula 3.5.3.5

Only the coefficient value has been changed to make the AREA formula different from Winslow's formula. This formula is also a straight line type as we can see in the unit stress Curve (3.5.3.7). Nomenclature is identical to the Winslow formula. The formula is:

$$\frac{P}{A} = C\left(1.00 - \frac{l}{60d}\right).$$

Because the Railway Engineers were primarily concerned with the safety factor in the design of wood trestles, they reduced the coefficient as used in the popular Winslow formula, giving more conservative results.

Collective formulas 3.5.3.6

In 1929, another wood column formula was endorsed and published by the Forest Products Laboratory, which appears to bear the trademark of Rankin. This formula required the use of both a coefficient and a constant. It also was accompanied by a published table of values for the constant K. The formula, in original form, is written:

$$\frac{P}{A} = C\left[1.00 - \frac{1}{3}\left(\frac{l}{Kd}\right)^4\right]$$

Where:

C = Allowable basic unit stress (F_c) parallel to grain for the selected species and grade of material as listed in the tables for common woods.

K = The constant which had to be obtained from the table of values for K.

Upon analysis, the constant was found to have been derived as:

$$K = \frac{\pi}{2}\sqrt{\frac{E}{6C}}$$

CURVES: Design working stress for columns 3.5.3.7

WINSLOW'S FORMULA: $\frac{P}{A} = C\left[1.0 - \left(\frac{l}{80d}\right)\right]$ E = 1,600,000 BASIC C = 1400 P.S.I.

AMERICAN RAILWAY ENGINEER'S ASSOCIATION FORMULA: $\frac{P}{A} = C\left[1.0 - \left(\frac{l}{60d}\right)\right]$ BASIC C = 1400 P.S.I.

SOUTHERN PINE ASSOC., FORMULA: $\frac{P}{A} = \frac{0.30E}{\left(\frac{l}{d}\right)^2}$ E = 1,600,000 BASIC C = 1400 P.S.I.

DESIGN WORKING STRESS CURVES FOR SOUTHERN YELLOW PINE AND FIR COLUMNS

WINSLOW FORMULA

AMERICAN RAILWAY ENGINEERS ASSN. FORMULA

SOUTHERN PINE ASSOCIATION FORMULA

SLENDERNESS RATIO = $\frac{l}{d}$

0.0, 5, 10, 15, 20, 25, 30, 35, 40, 45, 50, 55

1400, 1300, 1200, 1100, 1000, 900, 800, 700, 600, 500, 400, 300, 200

ALLOWABLE UNIT STRESS VALUE IN POUNDS PER SQUARE INCH

Column design procedures 3.5.4

The design of wood columns may be accomplished by either of two methods as follows:

(a) By selecting the section directly from the Load Tables as included in 3.5.4.1, or from similar manuals and handbooks.

(b) By assuming a trial section and investigating the allowable reduced stress in accordance with the formula which is selected.

Should the particular column formula be restricted by the Building Code, the ALLOWABLE STRESS CURVES will be helpful in ascertaining the unit stress for the ratio of slenderness, $\frac{l}{d}$. When the Code requirements are not in effect, it is recommended that the Southern Pine Association formula be employed. The advantage of using the SYPA formula is seen in the fact that the entire equation can be resolved with the slide rule when written as follows:

$$P = \left[\frac{0.30E}{\left(\frac{l}{d}\right)^2} \right] \times A.$$

The formula can also be transposed to solve for area of cross section required when axial load P is known. In the event that dimension d was considered as a square section, the area A can be extended to include a rectangular section, without reducing the stress.

TABLE: Column loads 3.5.4.1

COLUMN FORMULA: $\frac{P}{A} = \frac{0.30E}{\left(\frac{l}{d}\right)^2}$

E = 1,600,000

SOUTHERN PINE COLUMNS
SAFE AXIAL LOAD IN POUNDS
DRESSED SIZES

SAWN SIZE INCHES	DRESSED SIZE INCHES	AREA SQ. IN. bd	l/d SLENDER RATIO	LENGTH COLUMN IN FEET	ALLOWABLE COMPRESSION-GRADE PSI			
					1150	1300	1400	1750
4x4	3⅝x3⅝	13.14	13.2	4.0	15,110	17,100	18,400	23,000
			19.9	6.0	15,110	17,100	17,525	17,525
			26.5	8.0	9,900	9,900	9,900	9,900
			33.1	10.0	6,335	6,335	6,335	6,335
			39.7	12.0	4,400	4,400	4,400	4,400
4x6	3⅝x5⅝"	20.39	13.2	4.0	23,450	26,500	28,550	35,680
			19.9	6.0	23,450	26,500	27,200	27,190
			26.5	8.0	15,300	15,300	15,300	15,300
			33.1	10.0	9,850	9,850	9,850	9,850
			39.7	12.0	6,825	6,825	6,825	6,825
6x6	5⅝x5⅝	31.64	17.1	8.0	36,400	41,150	44,300	55,375
			21.3	10.0	36,400	36,825	36,825	36,825
			25.6	12.0	25,500	25,500	25,500	25,500
			29.9	14.0	18,700	18,700	18,700	18,700
6x8	5⅝x7½	42.19	17.1	8.0	48,525	54,850	59,000	73,850
			21.3	10.0	48,525	49,100	49,100	49,100
			25.6	12.0	34,000	34,000	34,000	34,000
			29.9	14.0	25,000	25,000	25,000	25,000
8x8	7½x7½	56.25	12.8	8.0	64,700	73,125	78,750	98,500
			16.0	10.0	64,700	73,125	78,750	98,500
			19.2	12.0	64,700	73,125	78,750	80,575
			22.4	14.0	64,700	59,200	59,200	59,200
			25.6	16.0	45,325	45,325	45,325	45,325
			28.8	18.0	35,800	35,800	35,800	35,800
			32.0	20.0	29,000	29,000	29,000	29,000
8x10	7½x9½	71.25	12.8	8.0	81,950	92,625	100,000	125,000
			16.0	10.0	81,950	92,625	100,000	125,000
			19.2	12.0	81,950	92,625	100,000	102,100
			22.4	14.0	75,000	75,000	75,000	75,000
			25.6	16.0	57,400	57,400	57,400	57,400
			28.8	18.0	45,365	45,365	45,365	45,365
			32.0	20.0	36,750	36,750	36,750	36,750
10x10	9½x9½	90.25	10.1	8.0	104,000	117,325	126,350	158,000
			12.6	10.0	104,000	117,325	126,350	158,000
			15.2	12.0	104,000	117,325	126,350	158,000
			17.7	14.0	104,000	117,325	126,370	152,125
			20.2	16.0	104,000	116,780	116,780	116,780
			22.7	18.0	92,500	92,500	92,500	92,500
			25.3	20.0	74,450	74,450	74,450	74,450
10x12	9½x11½	109.25	10.1	8.0	125,650	142,000	153,000	191,200
			12.6	10.0	125,650	142,000	153,000	191,200
			15.2	12.0	125,650	142,000	153,000	191,200
			17.7	14.0	125,650	142,000	153,000	184,125
			20.2	16.0	125,650	142,000	141,375	141,375
			22.7	18.0	112,000	112,000	112,000	112,000
			25.3	20.0	90,000	90,000	90,000	90,000

TABLE: Column loads, continued 3.5.4.1

COLUMN FORMULA: $\frac{P}{A} = \frac{0.30E}{\left(\frac{l}{d}\right)^2}$

E = 1,600,000 #□"

SOUTHERN PINE COLUMNS
SAFE AXIAL LOAD IN POUNDS
DRESSED SIZES

SAWN SIZE IN INCHES	DRESSED SIZE IN INCHES	AREA SQ. IN. bd	l/d SLENDER RATIO	LENGTH COLUMN IN FEET	ALLOWABLE COMPRESSION GRADE 1150	1300	1400	1750
12 x 12	11½ x 11½	132.25	8.3	8.0	152,100	172,000	185,150	231,440
			10.4	10.0	152,100	172,000	185,150	231,440
			12.5	12.0	152,100	172,000	185,150	231,440
			14.6	14.0	152,100	172,000	185,150	231,440
			16.7	16.0	152,100	172,000	185,150	231,440
			18.8	18.0	152,100	172,000	185,150	197,600
			20.9	20.0	152,100	159,860	159,860	159,860
12 x 14	11½ x 13½	155.25	8.3	8.0	178,550	201,820	217,350	271,690
			10.4	10.0	178,550	201,820	217,350	271,690
			12.5	12.0	178,550	201,820	217,350	271,690
			14.6	14.0	178,550	201,820	217,350	271,690
			16.7	16.0	178,550	201,820	217,350	271,690
			18.8	18.0	178,550	201,820	217,350	231,900
			20.9	20.0	178,550	187,650	187,650	187,650
14 x 14	13½ x 13½	182.25	7.1	8.0	209,600	236,909	255,150	318,950
			8.9	10.0	209,600	236,909	255,150	318,950
			10.7	12.0	209,600	236,909	255,150	318,950
			12.4	14.0	209,600	236,909	255,150	318,950
			14.2	16.0	209,600	236,909	255,150	318,950
			16.0	18.0	209,600	236,909	255,150	318,950
			17.8	20.0	209,600	236,909	255,150	304,700
14 x 16	13½ x 15½	209.25	7.1	8.0	240,650	272,000	293,000	366,200
			8.9	10.0	240,650	272,000	293,000	366,200
			10.7	12.0	240,650	272,000	293,000	366,200
			12.4	14.0	240,650	272,000	293,000	366,200
			14.2	16.0	240,650	272,000	293,000	366,200
			16.0	18.0	240,650	272,000	293,000	366,200
			17.8	20.0	240,650	272,000	293,000	348,700
16 x 16	15½ x 15½	240.25	6.2	8.0	276,300	312,325	336,350	420,450
			7.7	10.0	276,300	312,325	336,350	420,450
			9.3	12.0	276,300	312,325	336,350	420,450
			10.8	14.0	276,300	312,325	336,350	420,450
			12.4	16.0	276,300	312,325	336,350	420,450
			14.0	18.0	276,300	312,325	336,350	420,450
			15.5	20.0	276,300	312,325	336,350	420,450
16 x 18	15½ x 17½	271.25	6.2	8.0	312,000	352,620	379,750	475,000
			7.7	10.0	312,000	352,620	379,750	475,000
			9.3	12.0	312,000	352,620	379,750	475,000
			10.8	14.0	312,000	352,620	379,750	475,000
			12.4	16.0	312,000	352,620	379,750	475,000
			14.0	18.0	312,000	352,620	379,750	475,000
			15.5	20.0	312,000	352,620	379,750	475,000
18 x 18	17½ x 17½	306.25	5.5	8.0	352,200	398,125	428,750	536,000
			6.9	10.0	352,200	398,125	428,750	536,000
			8.2	12.0	352,200	398,125	428,750	536,000
			9.6	14.0	352,200	398,125	428,750	536,000
			11.0	16.0	352,200	398,125	428,750	536,000
			12.3	18.0	352,200	398,125	428,750	536,000
			13.7	20.0	352,200	398,125	428,750	536,000

EXAMPLE: Maximum column height without bracing 3.5.4.2

Building Codes and Designer Engineers have established a maximum height for unsupported columns. Regardless of circumstances. or wood grade and species, no column shall have a slenderness ratio $\left(\frac{l}{d}\right)$ over 50.

REQUIRED:
Refer to Column Tables for net size of sections, and determine the maximum unbraced length, in feet, for the following sections: 4x6; 6x8; 8x10; and 10x12 inches. Convert fractions to decimals for equating values.

STEP I:
Recreate the slenderness ratio formula as: $50 = \frac{l}{d}$.
Let s = slenderness ratio of 50
" l = length of column in inches, and L = length in feet. $\left(\frac{l}{12}\right)$.
" d = least side dimension in inches.

Then: $s = \frac{l}{d} = 50$. Transposing: $l = ds$ and $L = \frac{ds}{12}$.

STEP II
Maximum Length based on actual net sizes.

4x6	Least d = 3.625"	Max. Length = $\frac{3.625 \times 50}{12}$ =	15.10 Feet.
6x8	Least d = 5.50	Max. Length = $\frac{5.50 \times 50}{12}$ =	22.90 "
8x10	Least d = 7.50	Max. Length = $\frac{7.50 \times 50}{12}$ =	31.25 "
10x12	Least d = 9.50	Max. Length = $\frac{9.50 \times 50}{12}$ =	39.58 "

DESIGNERS NOTE:
By referring to Chart for Design Working Stress Curves, (3.5.3.7) the AREA Formula reduces the unit stress to 275 PSI, for limit of 50 ratio.
Using a 4x4, with an area of 13.14 sq. inches, the maximum allowable load, $P = 13.14 \times 275 = 3610$ Lbs.

EXAMPLE: Sizing for axial column load 3.5.4.3

A Southern Yellow Pine Column is to have an unsupported length of 16.0', and sustain an axial load of 52,200 Lbs. Material grade is Dense Structural with E = 1,600,000. This is an Interior Column for continuous dry conditions.

REQUIRED:

Design for size of Column. May be either a square or rectangular section. Use the following column formulas for safe load comparison. Fc = 1200 PSI.

SPA. Formula: $\frac{P}{A} = \frac{0.30\,E}{\left(\frac{l}{d}\right)^2}$, and the Winslow Formula

where: $\frac{P}{A} = C\left[1.00 - \left(\frac{l}{80d}\right)\right]$. $C = 1000 \times 1.20 = 1200$ PSI.

STEP I:

Rather than select a trial section, choose a depth of 7.50 In.

$L = 16.0'$ $l = 16.0 \times 12 = 192''$ $\frac{l}{d} = \frac{192}{7.5} = 25.6$ $P = 52,200$ Lbs.

STEP II:

Solving for Area A: SPA Formula:

$$\frac{P}{A} = \frac{0.30 \times 1,600,000}{25.6 \times 25.6} = \frac{480,000}{655.4} = 733 \text{ Lbs. Sq. In.}$$

Required area, $A = \frac{52,200}{733} = 71.2\,\square''$ $b = \frac{71.2}{7.5} = 9.50$ Inches.

With Southern Pine Association Formula, accept an 8X10 S4S.

STEP III:

Winslow's Formula with values substituted:

$$\frac{P}{A} = 1200\left[1.00 - \left(\frac{192}{80 \times 7.5}\right)\right] = 1200 \times 0.698 = 837.6 \text{ PSI.}$$

Required area, $A = \frac{52,200}{837.6} = 62.4\,\square''$ $b = \frac{62.4}{7.50} = 8.32$ Inches.

Winslow's Formula also requires an 8X10 S4S.

STEP IV:

Checking difference in Allowable Axial Loads.

8X10 S4S Area $A = 7.50 \times 9.50 = 71.25$ Sq. Inches.

For SPA Formula, $P = 52,200$ Lbs.

For Winslow's Formula, $P = 71.25 \times 837.6 = 59,680$ Lbs.

Difference $= 59,680 - 52,200 = 7,480$ Lbs. (About 14 Percent)

EXAMPLE: Maximum allowable column load 3.5.4.4

A 10.0" x 10.0" S4S So. Yel. Pine section 16.0 Feet long is proposed to be used as a column without any bracing. Grade of species is #1 Dense with allowable unit compressive stress parallel to grain, F_c = 1400 PSI. Modulus of Elasticity E = 1,600,000.

REQUIRED:
Use the Southern Yellow Pine Associations Column formula to calculate the allowable axial load. The formula is: $P = \left[\frac{0.30E}{(\frac{l}{d})^2}\right] \times A$. Check Tables for value of E used therein:

STEP I:
Net size of column. d = 9.50" b = 9.50" l = 16.0 x 12 = 192.0"

$\frac{l}{d} = \frac{192}{9.5} = 20.2 \qquad \frac{l^2}{d} = 20.2 \times 20.2 = 408$

Area A = 9.50 x 9.50 = 90.25 Sq. In.

STEP II:
Substitiţing values in Formula: $F_c = \frac{0.30 \times 1,600,000}{408} = 1175$ PSI

Allowable load P = 1175 x 90.25 = 106,045 Lbs.

The tables list this column as capable of supporting a load of 116,780 Lbs. The value of E is greater in the better grades.

STEP III:
To check for value of E, transpose the formula and solve for actual unit stress. $f_c = \frac{116,780}{90.25} = 1294$ PSI.

$E = \frac{f_c \times (\frac{l}{d})^2}{0.30} \qquad E = \frac{1294 \times 408}{0.30} = 1,750,840.$ (Call it 1,751,000)

EXAMPLE: Column axial load check 3.5.4.5

A 10.0" x 12.0" S4S So. Yel. Pine Column, 20.0 Feet long is listed in Column Tables as capable of sustaining an axial load of 90,000 Pounds. This figure was based on the SYP associations column formula. Fc = 1400 PSI.

REQUIRED:
Check the allowable load of 90,000 Pounds by using the Winslow Column Formula as: $\frac{P}{A} = c\left(1.00 - \frac{l}{80d}\right)$.

STEP I:
Winslow's formula does not use the property of E, as is the case with the SYP formula.
Least d = 9.50" b = 11.50" A = 109.25 □" l = 20.0 x 12 = 240.0"
c = 1400 #□"

STEP II:
Solve for values to put in formula.
80d = 80 x 9.50 = 760 Then, $\frac{l}{80d} = \frac{240}{760} = 0.316$

$\frac{P}{A} = 1400 \times (1.00 - 0.316) = 958$ PSI. Same as fc.

STEP III
Winslow load: P = 958 x 109.25 = 104,662 Pounds.
Actual stress from SYP Formula:
$f_c = \frac{P}{A}$ Then $f_c = \frac{90{,}000}{109.25} = 823.3$ PSI.

The Southern Pine Association's Formula is more conservative.

Eccentric column loads 3.5.5

Any column load which does not provide bearing directly through the intersection of the two central axes of the column cross-section, is considered an eccentric load. The placement of such loads upon the column is important, because eccentric loads induce bending stresses which will cause the column to deform with a permanent set. A load which bears directly upon the column center of gravity is referred to as a Concentric or Axial load and identified with the capital letter P. The degree of eccentricity in a column will depend upon the distance of the load from the central axes and the amount of load. Eccentric loads are indicated by the symbol Pe, which has units of foot-pounds or inch-pounds where P = load, and the subscript e= distance of load from central axes, and given in feet or inches.

When calculating the eccentric bending moment, it is simply, M = P x e.

A slight amount of eccentricity in a column can usually be ignored if the columns are plumb. When architectural design requires the columns to be erected on slant angles, the eccentric distance must be measured from a plumb line through the load, horizontally to the center of gravity of the column at its base. This condition is best illustrated when a common wood ladder is placed on an angle against a wall, and the weight of workers acts in a vertical plane to produce bending. A vertical ladder would have no eccentric distance or bending stress. It is important to remember that all of the column formulas assume that the loads are placed concentric to the column axes and the columns are erected plumb with their base.

Eccentric bending factor 3.5.5.1

The bending factor of a section is a property. It is used to convert bending moments caused by eccentric loads to the equivalent of an axial load. Careful attention must be given to the manner in which the bending factors are derived. In timbers of square or rectangular cross-section, the bending factor $K = \frac{S}{A}$ or also as $K = \frac{d}{6}$.

For an 8 x 10 S4S timber section, the area A = 71.3 Sq. inches, and $S_x = 113.0''^3$.

The bending factor, $K_x = \frac{113.0}{71.3} = 1.58''$ or $K_x = \frac{9.50}{6} = 1.58''$ Recall that the eccentric load or moment (P_e) is equal to the load times eccentric distance. To find the equivalent axial load for the eccentric moment, simply divide the moment (M = P_e), by the factor K. The equivalent axial load which would produce the same fiber stress as the eccentric moment, is equal to $\frac{M}{K}$, where the moment and factor are given in like terms (inch pounds and inches). The correct axis with respect to the bending must be kept in mind when calculating the bending factor and equivalent load.

Eccentric load theory 3.5.5.2

Referring to the properties of plane sections and shapes the dimension c is used to indicate the distance from the neutral axis to the outermost fiber of the section. The amount of eccentricity is the distance from the centroid to the point where eccentric load is applied. Let this distance be identified by the small letter a. If P = Load, and I = Moment of Inertia, then the stress from an eccentric load becomes equal to $\frac{Pac}{I}$. Now, the radius of gyration r, is listed in Section VI, on Properties as being a derivation of I and A. Thus, $r = \sqrt{\frac{I}{A}}$, transposed, the formula becomes $I = Ar^2$.

The stress from the eccentric load P must be equated to the axial stress as found by the selected Column Formula.

Assuming the Winslow formula will govern the axial stress, the value of the developed stress due to eccentric loading (P_e) to be equated to the allowable axial stress, can be solved by writing the following equations.

$$\frac{P_e}{A}\left[1.0 + \left(\frac{ac}{r^2}\right)\right] = C\left[1.0 - \left(\frac{l}{80d}\right)\right]$$

An example which follows will illustrate the procedure to be used in designing for eccentric stress, by using formulas which govern axial stress.

Eccentric loads on short columns and piers 3.5.5.3

In the case of short columns and piers, the possibility of buckling due to axial stress may be neglected. This is true only for short posts where the length in inches is not more than ten times the least dimension d. The eccentric distance (a) is the distance from the location of the load to the centroid, along the same axis. Bending moment equals load times eccentric distance. Unit stress is found in the same manner as for a beam, and the formula for stress may be written as follows:

$$f = \frac{P\,a\,c}{I} \quad \text{or} \quad f = \frac{P\,a}{S}$$

The first formula is preferable because it identifies the effective axis.

Bending or compressive stresses are added to the average crushing or compressive stress resulting from the axial load. Likewise, bending tension stresses are subtracted from the average axial compressive stress, and can cause a net tensile stress at one side of the pier. The total stresses are written as follows:

Maximum compression: $f_c = \frac{P}{A} + \frac{P\,a\,c}{I}$

Maximum tension: $f_t = \frac{P}{A} - \frac{P\,a\,c}{I}$

Eccentric loads on brick piers or columns should always have the vertical load action line fall inside the middle third of the pier or column. When this is done, the stresses will always be compressive, without any tensile stress in one side of the pier. It is only when the eccentric action line falls outside of the middle third that tensile stresses are developed. Always limit the eccentric distance to 1/6 of the effective pier width. An example will follow which will illustrate how tensile stresses can result from eccentric load placement.

EXAMPLE: Eccentric column stress 3.5.5.4

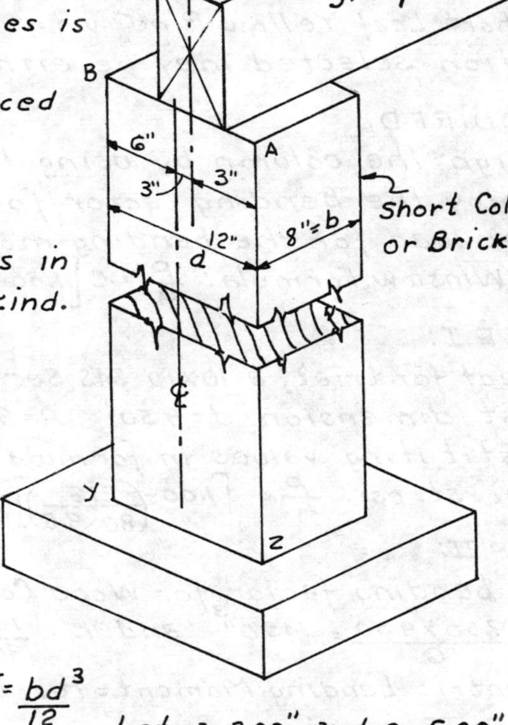

The detail at right will apply to Masonry Piers as well as Columns. P = 16,000 Lbs. or reaction from sills. Eccentric distance of 3.00 inches is outside of middle third. Pier size is net 8X12, with load placed off of axis x-x.

REQUIRED:
Calculate the maximum stress in side AB, and designate stress kind. Calculate the same for side BY.

STEP I:
Assume P is axial on both axes x-x, and y-y.
Area A = 12X8 = 96.0□"

$f_a = \frac{P}{A} = \frac{16{,}000}{96.0} = 166.6$ PSI.

STEP II:
Find value of I on axis x-x. $I = \frac{bd^3}{12}$

$I_x = \frac{8.0 \times 12.0^3}{12} = 1152.0''^4$ Let a = 3.00" and c = 6.00"

Then compressive stress $= \frac{Pac}{I_x}$ $f_e = \frac{16{,}000 \times 3.0 \times 6.0}{1152} = 250.0$ PSI.

Max. Compressive stress in side AZ = 166.6 + 250.0 = 416.6 PSI.

STEP III:
Maximum stress in side BY = 166.6 − 250.0 = −183.4 PSI. (Tension)

DESIGNERS NOTE:
Loading must be placed inside the middle third to assure the maximum stress of remaining compressive. Load placement outside the middle third will set up tension stress as was the case in above example. A good rule to follow is to always limit the eccentric distance to not less than 1/6 of breadth of column or pier.

EXAMPLE: Eccentric column, bending factor solution 3.5.5.5

A Column with an unbraced height of 18.0 feet is required to safely support a load of 40,000 Lbs. The architectural detail calls for the load to be placed 3.0 inches off the Center line, but exactly on the other axis. Column is to be of Short Leaf Yellow Pine with $F_c = 1200$ P.S.I.
Section selected may be either square or rectangular.

REQUIRED:
Design the column by using the Winslow Formula, and employ the Bending Factor for calculating an equivalent axial load for the bending moment.
The Winslow Formula: $\frac{P}{A} = C\left[1.00-\left(\frac{l}{80d}\right)\right]$. $C = 1200$ PSI.

STEP I:
Select for trial, a 10x10 S4S Section. Net size = 9.50"x9.50" with least dimension $d = 9.50"$ $A = 90.25\square"$ $l = 18.0 \times 12 = 216$ inches.
Substituting values in formula for maximum allowable axial stress: $\frac{P}{A} = \left[1.00-\left(\frac{216}{80 \times 9.5}\right)\right] = 860$ PSI.

STEP II:
The bending factor for Wood Columns is: $K = \frac{S}{A}$ and $S = \frac{bd^2}{6}$
$S = \frac{9.50 \times 9.50}{6} = 143.0"^3$ and $K = \frac{143.0}{90.25} = 1.58$
Eccentric bending Moment = Pe. $M_e = 40,000 \times 3.0 = 120,000$ inch lbs.

STEP III:
Equivalent axial load to equal bending $= \frac{M}{K}$, and $Pe = M$.
$P = \frac{120,000}{1.58} = 76,000$ Lbs. Then total loads = 40,000 + 76,000
$P = 116,000$ lbs. Area required $= \frac{116,000}{860} = 135.0$ Sq. Inches.

STEP IV:
Area required is greater than trial section 10x10 S4S.
Retaining the least dimension $d = 9.50"$, try another section as 10x12 S4S. $A = 9.50 \times 11.50 = 109.25\square"$ $S_y = \frac{11.50 \times 9.50 \times 9.50}{6} = 173.0"^3$
$K = \frac{173.0}{109.25} = 1.58$ (No change, there M = same.

STEP V:
Stepping up to a 12X12 S4S. Net size = 11.50"x 11.50." From Table of Sections 3.1.2.1. $A = 132.0\square"$ $S = 253.0"^3$ and $K = \frac{253.0}{1.32} = 1.92$
Equivalent load $= \frac{120,000}{1.92} = 62,500$ Lbs.
Total loads = 62,500 + 40,000 = 102,500 lbs. Actual stress $f_a = \frac{102,500}{132.0} = 777$ #□"
Stress is less and a 12x12 S4S should be first choice

EXAMPLE: Truss chord, axial plus bending stress 3.5.5.6

A compressive chord in a Warren Truss is 12.0 feet long and must sustain an axial compressive force of 14,000 Lbs. In addition to the axial stress, the member must resist the bending moment caused by dead and live loads of roof. The roof loads, w = 600 Lbs. Lineal foot.
Truss material will be of, Dense Structural Southern Pine, with stress values as follows: F_b = 2000 PSI. F_c = 1400 PSI., and E = 1,600,000.

REQUIRED:
The chord appears to be the top of the truss and will be continuous over several panels. Use the Southern Pine Association Formula for axial stress. $\frac{P}{A} = \frac{0.30\,E}{(\frac{l}{d})^2}$. Convert bending moment into an equivalent axial force load by using Bending Factor k.

STEP I:
Calculating Bending Moment in inch pounds. $M = \frac{wL^2}{8}$

$M = \frac{600 \times 12.0 \times 12.0}{8} = 10{,}800'^{\#}$, or M = 129,600 In. Lbs.

Section Modulus for $M = \frac{129{,}600}{2000} = 64.8''^3$

Axial P = 14,000 Lbs.

STEP II:
Referring to Column and Property Tables to get an idea for trial size. An 8X10 has an Area of 71.3 □" and S = 113.0"^{3}

$k_x = \frac{113.0}{71.3} = 1.585$ Then $P_e = \frac{129{,}600}{1.585}$ = 81,750 Lbs. or equivalent axial load. Total axials loads = P + P_e = 95,750 Lbs.

STEP III:
Using column formula: $\frac{P}{A} = \frac{0.30\,E}{(\frac{l}{d})^2}$. l = 12.0 x 12 = 144.0 In.

d = 7.50" $\frac{l}{d} = \frac{144.0}{7.50} = 19.2$

Then $P = \frac{0.30 \times 1{,}600{,}000}{19.2 \times 19.2} \times 71.3$ = 92,750 Lbs.

The results are 3000 Pounds short of requirements, however, the Column Tables list this section as capable of supporting a safe load of 100,000 Lbs. Continuously protected and dry conditions make this possible. Accept the 8x10 section.

EXAMPLE: Column with axial and eccentric load 3.5.5.7

Given data: Column unbraced height = 18.0 Feet
Axial Load = 15,000 Lbs.
Eccentric Load = 6000 Lbs.
Eccentric distance = 5.50 Inches.
Material = S.Y.P. Fc = 1400 PSI. E = 1,600,000

REQUIRED:
Column must be square regardless of size.
Design column with SPA Formula, $\frac{P}{A} = \frac{0.30E}{\left(\frac{l}{d}\right)^2}$. Also use bending factor to find equivalent axial load in lieu of eccentric load.

STEP I:
The Bending moment from eccentric load. M = 6000 X 5.50 =
M = 33,000 Inch Pounds. Bending factor, $k = \frac{d}{6}$

STEP II:
Selecting for trial an 8X8, Actual size = 7.50 x 7.50 inches.
$k = \frac{7.50}{6} = 1.25$ Equivalent load $Pe = \frac{M}{k} = \frac{33,000}{1.25} = 26,400$ Lbs.

STEP III:
Total Loads = 15,000 + 26,400 + 6000 = 47,400 Lbs.
From Column Tables, an 8X8 will support a load of 35,800 Lbs., and an 8X10 is good for 45,365 Lbs.
Square column required, therefore accept a 10X10 Section, which will support a load of 92,500 Lbs.

EXAMPLE: Eccentric loaded column: equating bending to axial stress 3.5.5.8

A 10x10 S4S Timber Column has been proposed to serve as a truss and girder support. Column unsupported height is 18.0 Feet. An Axial load of 5000 Lbs. must be carried by column, plus an additional Eccentric Load of 14,000 Lbs. Eccentric distance is 3.00 inches from one axis only. Continuous dry conditions prevail and allowables are as follows: E = 1,600,000 and F_a = 1400 PSI.

REQUIRED:

Use the following Formula to equate the stresses and determine if Column is acceptable. Slide Rule results OK.

$$\frac{P}{A}\left[1.00+\left(\frac{ac}{r^2}\right)\right]=\frac{P}{A}\left[\frac{0.30\ E}{\left(\frac{l}{d}\right)^2}\right].$$

STEP I:

Taking the second equation for axial stress first:

$L = 18.0'$ $\quad l = 18.0 \times 12 = 216''$ $\quad d = 9.50''$ $\quad E = 1,600,000$

Slenderness ratio $= \frac{l}{d} = \frac{216}{9.50} = 22.7$

Substituting values:

$$\frac{P}{A} = \frac{0.30 \times 1,600,000}{22.7 \times 22.7} = 932 \text{ PSI.}$$

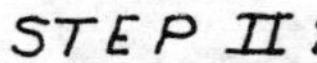

STEP II:

Determine properties of 10x10 for 1st. Equation.

$r = \sqrt{\frac{I}{A}}$ $\quad A = 9.50 \times 9.50 = 90.25$ Sq. In. $\quad I = \frac{bd^3}{12}$ $\quad I = 679.0''^4$

$r = \sqrt{\frac{679.0}{90.25}} = 2.75$ and $r^2 = 2.75 \times 2.75 = 7.56$

c = Distance from axis to extreme fibers = 4.75 Inches.

a = Eccentric distance of 14,000 Load from axis = 3.00 Inches

Then: $1.00 + \left(\frac{3.00 \times 4.75}{7.56}\right) = 2.89$ $\quad$ Axial stress = 932 PSI.

Actual stress $= \frac{932}{2.89} = 323$ PSI.

STEP III:

Max. Eccentric Load: $323 \times 90.25 = 29,150$ Lbs. (Larger than P_e)

Max. Axial Load: $932 \times 90.25 = 94,115$ Lbs.

Column side is acceptable for conditions.

EXAMPLE: Eccentric loaded column: maxiumum allowable load 3.5.5.9

A 10X12 S4S Column of Select Structural West Coast Fir is to be used to support a portion of storage shed. Unbraced length is 12.0 feet. The girder to be supported requires that the load will bear 3.0 inches from major axis.

REQUIRED:

Determine the safe eccentric load column will support when 3.0 inch eccentricity is away from major axis and on line of minor axis. Use the Southern Pine Assocation Formula for axial stress, where: $\frac{P}{A} = \frac{0.30\,E}{\left(\frac{l}{d}\right)^2}$, and $E = 1{,}760{,}000$.

Equalize stress for eccentric load as: $\frac{P}{A}\left[1.0 + \left(\frac{ac}{r^2}\right)\right]$ = allowable stress found in SPA Formula.

STEP I:

For allowable axial stress: $l = 12.0 \times 12 = 144$ inches $d = 9.50''$

With values substituted: $\frac{P}{A} = \frac{0.30 \times 1{,}760{,}000}{\left(\frac{144.0}{9.50}\right)^2} = 2300$ PSI. (This is over base of $F_c = 1550$ PSI.

Use maximum of 1550 #□" for $\frac{P}{A}$. (1st. equation).

STEP II:

Eccentric distance, $a = 3.00''$ Area $A = 109.25$ □"

Centroid distance from short axis = $11.5/2 = 5.75''$

Moment of Inertia along short axis, $I_x = 209.39''^4$ (From Tables).

Then $r_x = \sqrt{\frac{209.39}{109.25}} = 1.915$ $r^2 = 3.67$

STEP III:

To calculate eccentric stress, and equate to axial allowable.

In formula: $1.0 + \left(\frac{3.00 \times 5.75}{3.67}\right) = 5.70$ (2nd. equation).

Allowable, $\frac{P_e}{A} = \frac{1550}{5.70} = 272$ PSI.

STEP IV:

Determine safe Eccentric Load, $P_e = 109.25 \times 272 =$ <u>29,820 Lbs.</u>

DESIGN NOTE: (The importance of eccentricy to design).

With 2.0 inch eccentricity, $P_e = 54{,}000$ Lbs.

With 1.0 inch eccentricity, $P_e = 66{,}000$ Lbs.

A truly axial load on both axes, $P = 109.25 \times 1550 =$ <u>169,335 Lbs.</u>

EXAMPLE: Wood ladder 3.5.5.10

Assume that a painters ladder is to be required to safely support 2 men at frequent intervals. Length of ladder is to be 18.0 foot. The probable and most critical angle for use will call an angle of 60 degrees with the horizontal ground surface.
Place men with weights as follows:
P_1 = 185 Lbs. 5.50' down from top.
P_2 = 160 Lbs. 7.00' up from bottom.

REQUIRED:
Solve for non-coplaner forces resulting from point loads P and P_1. Determine if 2-2x4 stiles will be safe if F_b = 1400 PSI.

STEP I:
Drawing a sketch of ladder in critical position to place loads and identify angles and triangle sides.

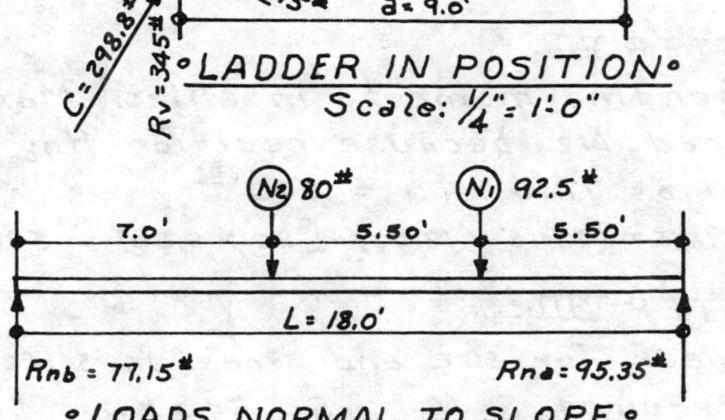

Angle B = 60° and A = 30°
Triangle side c = 18.0 ft.

STEP II:
Refer to Tables in Section V to obtain functions of angles.
For 30° angle: A.
Tan. A = 0.57735 Sin. A = 0.500 CoSin. A = 0.86603 and CoTan. A = 1.7320
For 60° angle: B. Functions are for opposite angle of A:
CoTan. B = 0.57735 CoSin. B = 0.500 Sin. B = 0.86603 and Tan. B = 1.7350

STEP III:
Solve for b = rise on wall. b = c Cos. A = 18.0 × 0.86603 = 15.59 Ft.
Solve for a = distance on ground. a = c Sin. A = 18.0 × 0.5000 = 9.00 Ft.
Note dimensions on drawing.

EXAMPLE: Wood ladder, continued 3.5.5.10

STEP IV:

Total loads = $P_1 + P_2$ = 185 + 160 = 345 Lbs. Act vertical plane as Rv.

STEP V:

Loads P_1 and P_2, act perpendicular to stiles c, or normal to surface of Ladder. Forces are indicated as N_1 and N_2.

Force $N_1 = P_1$ Cos. B. $N_1 = 185 \times 0.5000 = 92.50$ Lbs.

Force $N_2 = P_2$ Cos. B. $N_2 = 160 \times 0.5000 = 80.0$ Lbs.

Neglecting weight of wood ladder, the forces N_1 and N_2 produce the bending moment in stiles.

STEP VI:

Layout ladder as a simple beam of 18.0 foot span, and with 2 Concentrated loads as N_1 and N_2. Reactions will be in same line of action as forces. Solve for Reactions.

$$R_{nb} = \frac{(92.5 \times 5.50) + (80.0 \times 11.0)}{18.0} = 77.15 \text{ Lbs.}$$

$$R_{na} = \frac{(80 \times 7.0) + (92.5 \times 12.50)}{18.0} = 95.35 \text{ Lbs.}$$

Total $N_1 + N_2$ = 172.50 Lbs. Reactions = 172.50 Lbs.

STEP VII:

Bending moments in stiles. Max. Moment will be under load N_2, because reaction R_{nb} is less than N_2.

$M_{7.0'} = 77.15 \times 7.0 = 540'\#$

$M_{12.5} = (77.15 \times 12.50) - (80 \times 5.50) = 524.5'\#$

STEP VIII:

Check for size and stress in stiles. F_b = 1400 PSI

Required $S = \frac{M}{F_b}$ $S = \frac{540 \times 12}{1400} = 4.63''^3$ Applies to both stiles.

Necessary for 1 Stile side = $2.315''^4$ A 2x4 S4S has $S = 3.56''^3$

Stiles for ladder of 2- 2x4 S4S are more than required.

STEP IX:

When inclined, the ladder exerts a horizontal force against wall at top and also a horizontal force at ground level.

At top, Horizontal Reaction, $H_A = R_{na}$ Cos. A

At bottom, Horizontal Reaction, $H_B = R_{nb}$ Cos. A

$H_A = 95.35 \times .86603 = 82.58$ Lbs.

$H_B = 77.15 \times .86603 = 66.80$ "

EXAMPLE: Wood ladder, continued 3.5.5.10

STEP X:

Load P_1 and P_2 develop compressive stress in stiles parallel to grain and incline. Maximum compression will be near bottom under both loads.

Compression $C = (P_1 + P_2) \sin B$. Then compression below the loads, $C = (185 + 160) \times 0.86603 = 298.8$ lbs. Call it 300# for both stiles.

For 1 Stile $C = 150\#$ Area 2x4 = $1.625 \times 3.625 = 5.89$ Sq. Inches

Actual axial stress $= \frac{C}{A}$ or $f_a = \frac{150}{5.89} = 25.5$ PSI.

STEP XI:

Convert eccentric bending moment to an equivalent Axial load by using bending factor K. Factor $K = \frac{S}{A}$

From table 3.1.2.1. $S_x = 3.56''^3$ $K = \frac{3.56}{5.89} = 0.603$

From step VII for bending moment.

Moment in 1 stile $= \frac{540}{2} = 270$ Ft. Lbs.

Equivalent Axial load $P_e = \frac{M}{K}$. $P_e = \frac{270 \times 12}{0.603} = 5360$ Lbs.

Total Axial loads $= C + P_e = 150 + 5360 = 5510$ Lbs.

Actual stress in 1 stile: $f_a = \frac{5510}{5.89} = 935$ PSI.

Accept 2x4s for both Ladder Stiles.

STEP XII:

The external forces of R, N, and C, in this example are subject to checking by drawing a simple Force Polygon, where the scale will represent pounds and not the dimensions. Let 1.0 inch = 100 Lbs. Known are loads P_1 and P_2 which act in vertical direction and total 345 Lbs. Load line will be equal to this value. From bottom of load line draw a line parallel to the action line of N, which is perpendicular to slope of ladder. Now close force polygon by drawing a line from top of load line and in same plane as slope of the ladder. The figure is a right angle triangle and force values will be same in value when scaled.

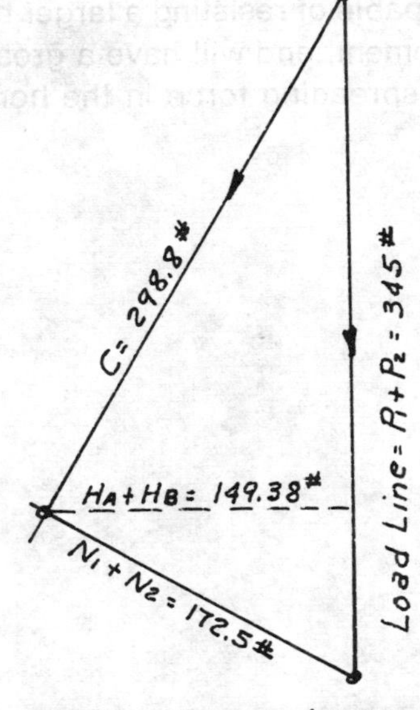

FORCE POLYGON

Laminated segment arches 3.6

Wide, unobstructed spans up to 200 feet are economically feasible with laminated wood segment arches. The ends may be tied or a buttress can be used to support and contain the vertical and horizontal reactions. The buttress is generally formed of reinforced concrete in such shape that the angle of support is the action line of the resultant of the vertical and horizontal reactions. The ends of the segment arches are installed in metal sockets or shoes which are anchored to the buttress foundation. Concrete buttresses are designed in the same way as retaining walls, as explained in Concrete, Section IV. Tapered segment arches can also be placed perpendicular to each other at midspan to from four quadrants of a circle, in which the entire roof structure is supported by four buttresses at the foot of each arch. Such structures are called cross-vaults, and may also be made square with four straight walls. When segment arches are supported upon masonry walls or columns, the ends must be tied together to contain the forces acting horizontally which tend to spread the arch. The horizontal reaction may be computed by using the simple formula: $T = \frac{wL^2}{8H}$. Where w = Load per foot on the arch in pounds, L = clear span in feet, and H = height of arch in feet. We can see from this formula that the flatter the arch (smaller H), the greater the horizontal reaction.

Limiting basic stress 3.6.1

The basic allowable unit stress for laminated wood segment arches is established at a maximum, F_b = 2400 PSI. There are many considerations which must be taken into account in the design. A flat segment with reduced height must be capable of resisting a larger bending moment, and will have a greater amount of spreading force in the horizontal direction. From numerous tests, it was found that the ratio of the height to the radius would best serve to establish an accurate parameter to solve for the unit bending stress. We include a Design Curve to calculate segment arches with limited stress (F_L). The ratio is indicated as $F_{Lb} = \frac{H}{R}$.

DESIGN CURVE for limiting basic stress 3.6.2

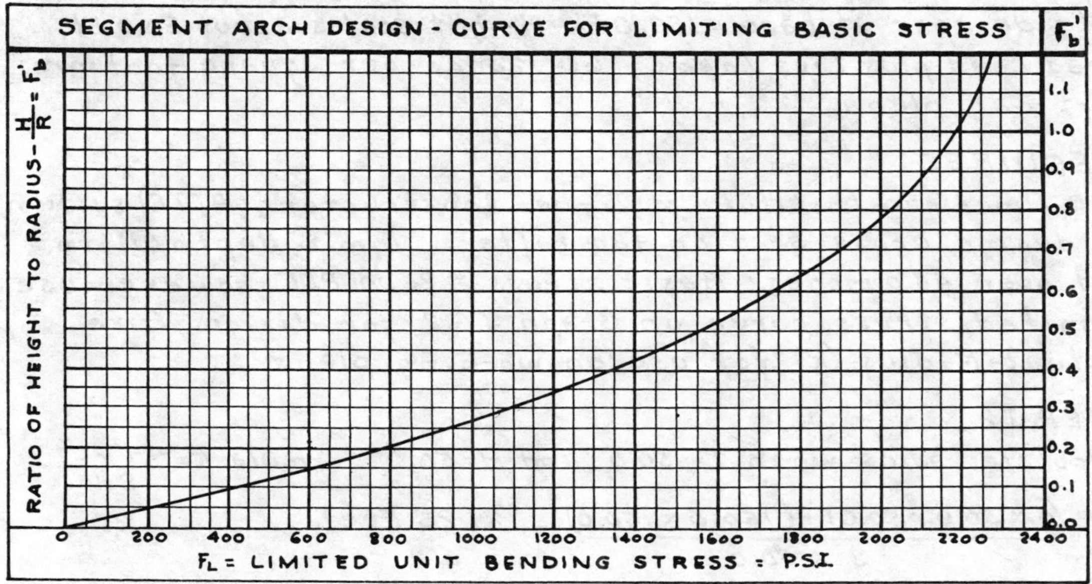

EXAMPLE: Segment roof rafters 3.6.3

Architects preliminary plans call for Segment Roof Rafters to be laminated and spaced in bays 16.0 Ft. CC. Outside span at base = 150.0 Ft. Height scales about 50.0 Ft. Dead Load plus Live Loads = 60 Pounds per square foot on vertical plane.

REQUIRED:

Use American Institute of Timber Constructors (AITC) system to design cross-section for rafters. Limit depth (d) to not over 48.0 inches. Basic stress = 2400 PSI., however use Limited stress curve on Graph 3.6.2. for design. Give actual Radius (R) for use on working plans.

STEP I:

Calculate Radius when L = 150.0' and H = 50.0 Formula $R = \frac{4H^2 + L^2}{8H}$.

$$R = \frac{(4 \times 50.0 \times 50.0) + (150.0 \times 150.0)}{8 \times 50.0} = 81.25 \text{ Feet.}$$

STEP II:

Loading: Spacing of Rafters = 16.0 Ft.

Lineal foot load $w = 16.0 \times 60 = 960$#

Vertical Reactions:

$$R_1 = \frac{960 \times 150.0}{2} = 72{,}000 \text{ Lbs.} = R_2$$

STEP III:

Draw elevation of Segment and solve for Horizontal R_{H1}.

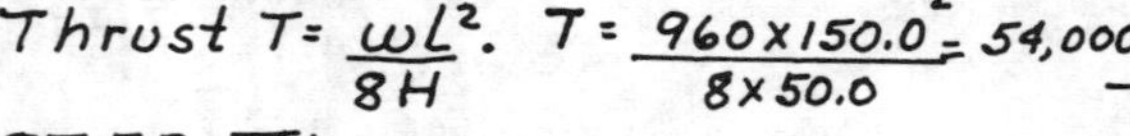

Thrust $T = \frac{wL^2}{8H}$. $T = \frac{960 \times 150.0^2}{8 \times 50.0} = 54{,}000$#

STEP IV:

Calculate Maximum Moment:

AITC Formula: $M = 1.50\,w(H)^2$

$M = 1.50 \times 960 \times 50.0 \times 50.0 = 3{,}600{,}000$ Inch Lbs.

STEP V

Design cross-section with limited stress Formula: $S = \frac{M}{F_L \times F_b'}$.

Ratio Height to Radius: $\frac{H}{R} = \frac{50.0}{81.25} = 0.615$

From Curve: $F_b' = 0.615$ and $F_L = 1775$ P.S.I

$S = \frac{3{,}600{,}000}{0.615 \times 1775} = 3300''^3$ For depth dimension d. $S = \frac{bd^2}{6}$ and

$d = \sqrt{\frac{6S}{b}}$. For trial, assume $b = 9.50''$. $d = \sqrt{\frac{6 \times 3300}{9.50}} = 45.6$ inches.

TABLE: Three hinged buttressed arch sections 3.6.4

BUTTRESSED-THREE HINGED SEGMENT ARCH CROSS-SECTIONS									
L/H	LOAD w IN P.L.F.	CROSS-SECTION FOR SPAN L INDICATED - bd" F_b = 2400 PSI.							
		80.0'	90.0'	100.0'	110.0'	120.0'	130.0'	140.0'	150.0'
3	720	5¼ x 22¾	5¼ x 26	5¼ x 29¼	5¼ x 30⅞	7 x 30⅞	7 x 32½	7 x 35¾	7 x 37⅜
	950	5¼ x 26	5¼ x 29¼	7 x 29¼	7 x 30⅞	7 x 34⅛	7 x 37⅜	7 x 40⅝	9 x 39
	1200	5¼ x 29¼	7 x 29¼	7 x 32½	7 x 34⅛	7 x 37⅜	7 x 40⅝	9 x 40⅝	9 x 42¼
4	720	5¼ x 19½	5¼ x 22¾	5¼ x 24⅜	5¼ x 27⅝	5¼ x 29¼	7 x 29¼	7 x 30⅞	7 x 32½
	950	5¼ x 22¾	5¼ x 24⅜	5¼ x 27⅝	5¼ x 30⅞	7 x 29¼	7 x 32½	7 x 34⅛	7 x 37⅜
	1200	5¼ x 24⅜	5¼ x 27⅝	5¼ x 30⅞	7 x 29¼	7 x 32½	7 x 35¾	7 x 37⅜	7 x 40⅝
5	720	5¼ x 19½	5¼ x 21⅛	5¼ x 22¾	5¼ x 26	5¼ x 27⅝	5¼ x 29¼	7 x 29¼	7 x 30⅞
	950	5¼ x 21⅛	5¼ x 24⅜	5¼ x 26	5¼ x 27⅝	5¼ x 30⅞	7 x 29¼	7 x 32½	7 x 34⅛
	1200	5¼ x 22¾	5¼ x 26	5¼ x 27⅝	5¼ x 30⅞	7 x 30⅝	7 x 32½	7 x 35¾	7 x 37⅜
6	720	5¼ x 17⅞	5¼ x 21⅛	5¼ x 22¾	5¼ x 24⅜	5¼ x 27⅝	5¼ x 29¼	5¼ x 30⅞	7 x 30⅞
	950	5¼ x 21⅛	5¼ x 22¾	5¼ x 24⅜	5¼ x 27⅝	5¼ x 30⅞	7 x 29¼	7 x 30⅞	7 x 34⅛
	1200	5¼ x 22¾	5¼ x 24⅜	5¼ x 27⅝	5¼ x 29¼	7 x 29¼	7 x 32½	7 x 34⅛	7 x 37⅜

3.7 FRAMED DOME DESIGN

History of domes 3.7.1

All written or verbal examinations given for Architectural State Licenses require the candidate to display a knowledge and understanding of architectural form contained in the historical progression of styles from early civilization to the present. A question is certain to be asked relating to the Architectural Renaissance and the elements contributing to its origin. The Dome was a principal factor in the Renaissance movement and, since the function of Domes was contiguous to science and architectural history, it is fitting that the following digest on the subject of Domes should serve both Architect and Engineer. Keep in mind that a fine way to display historical knowledge of a subject would be to write a concise treatment of a particular part of a single representative structure, much as follows:

PRIMITIVE USE OF DOMES

It is a matter of record that primitive man was at first a cave dweller, at least in the mountainous regions. However, during the paleolithic period, which extended 20,000 years, he learned to erect rough huts from available materials. His tools were of wood and stone. The first huts were constructed of light sticks, tied together at the top to form a ribbed cone, with a covering over the ribs of animal hides or mud mixed with leaves and twigs. In this elementary, "Tepee" structure lay the origin of the round hut, the first of the great buildings.

In the frigid north country, the Eskimos discovered that by cutting the packed snow into rectangular blocks and using frozen water for cement, they could build a round hemisphere shaped shelter which would protect them from the elements and wild beasts. These huts, in the language of the Eskimo, became his "Igloo." The entrance opening was very low, and a hole in the top permitted smoke to escape from the burning fat which provided the illumination during long periods of darkness. The Igloo must be considered a dome, and among the first of such structures.

The Stone Age builder in Europe, and especially in Ireland, found that by placing selected stones in a circle for the first row, then by setting each subsequent course a little inward from each row below, that little by little a conical dome-shaped hut developed. Finally the hole at the top became so small that it could be capped with a single stone. One of these primitive stone huts is preserved on the Aran island of Ireland. Others have been restored near Alberobello in Italy.

The Neolithic Period or new stone age saw early man learn to make better tools. He learned that certain animals could be domesticated and used for work, that he could till the soil to produce food, and, most important, he learned that other types of building were possible. The discovery of copper led to the making of bronze and iron. With metal replacing the wood and stone tools, it was inevitable that better weapons would be made. Although

History of domes, continued 3.7.1

the primitive tribes developed metal tools and weapons, concepts of social order and culture did not develop until the Egyptian and Greek periods.

HELLENIC AND HELLENISTIC ARCHITECTURE

Ancient Greek culture came to dominate the world prior to 1000 BC. At that time there came to the Aegean world, wave after wave of new people. Many of these tribes were fine builders, especially the Ionians, Doricans, Spartans, Olympians, and Trojans. Combining the skills of these new peoples with that of the Hellenes and Athenians, great cities were built and great minds began to unravel the secrets of natural science. The colonies of Mycenae and Troy were to become the more powerful in politics. A mass migration of Dorians from the north brought a breed of people who were chiefly military in character. A clash of ideologies was inevitable. Soon wars resulted in the neglect of the social structure, and the government fell into decay.

The conquest of the Persian, Alexander the Great, (356–323 BC) gave Greece a new government structure to replace the former rulers and a new wealth and power which came from shipping and trade. However, these governments soon became so corrupt that in the 4th Century BC, the whole Hellenistic culture and peoples capitulated to the power of Rome.

ROMAN ARCHITECTURE

The Roman Empire was founded by the simple process of conquer first, organize by assimilation, then rule with respect by demanding loyalty only to the Empire, The great and powerful legions from Rome were in command of all the known world. The poet Virgil (70–19 BC) in his tale of the Trojan hero, Aeneas, glorified the potential of Rome to such heights, that symbolically he became the first Roman to be exalted.

All of the architects, sculptors, painters and writers were forcibly brought to Rome to serve their new masters, and the city of Florence became the center of Roman art and culture. Various styles were fused together to produce the architecture of a new civilization. The Etruscans from the north, with their engineering skill, gave the structures a greater artistic scope than had ever been attempted.

Early temples of Roman Architecture were designed with cloistered vaults which rose to great heights. The vault is, in a sense, a square dome without a clear space under the arches. The "Tabularium," as designed and built by the Roman General, Lucius Sulla, (138–78 BC), was a good example of the square dome. The construction of the dome on the Pantheon, in Rome, with its circular ring at top center, was the work of Marcus Agrippa, (63–12 BC) and appears to be the first of such structures to have been analyzed from an engineering standpoint. This hemispheric ribbed dome was first made into a scale model for study. When erected it was 140 feet in diameter. In 76–138 AD, the Emperor Publius Hadrian constructed a new rotunda on the same site. The same structure is still standing and in a very good state of preservation. If one looks carefully, he can observe that the U.S. Capitol Building in Washington contains many similar elements.

MEDIEVAL ARCHITECTURE

The building of many large structures continued in all parts of the known world after the decay and dismemberment of the Roman Empire. This was brought about by the return of tribes to their ancestral

History of domes, continued 3.7.1

regions. Once the migrating peoples from the old Empire had established territorial boundaries and developed their own nations, their building needs for culture and religion were to extend over many centuries. Their design styles showed the influence of the classical Roman buildings, as well as could be executed with the local available wood and stone.

The Santa Sophia, in the Turkish city Byzantium, begun in 532 AD by the Byzantine emperor Flavius Anicius Justinianus (483–565 AD), is perhaps the most elegant decorated dome ever constructed. Designed by Anthemus and Isodorus, the meridian ribs are placed over arches which identify it as a pendentive dome. This edifice has one of the most fabulously decorated interiors in the world.

RENAISSANCE ARCHITECTURE

In the Fourteenth Century, the young architect, Arnolfo di Cambio (1232–1300), designed the Cathedral in Florence, but he did not complete the design. He had undertaken a design concept which surpassed his ability. His attempts to design a vault to cover a circular theater of 138 feet met with failure, and the project was abandoned. Half a century later, the proud city of Florence felt the challenge and held a design competition to complete the Cathedral. They selected Fillippo Brunelleschi (1377–1446). Architect Brunelleschi enlisted the aid of the able sculptor Niccoli di Donatello. Together they made trips to Rome to study the domed Pantheon. Their eventual design called for the construction of a huge octagonal stone dome. Meridian ribs and two thin stone shells provided the structural components for the framework. The shells were installed inside and outside of the ribs which gave lateral support. To cap the dome, a heavy stone lantern with spire rose above the hemisphere to identify the building as a place of worship.

With the completion of the largest cathedral in Italy, it was given the name of "Santa Maria del Fiore." The city of Florence was again the leader in architectural design, and the name of Brunelleschi was acclaimed throughout the empire. The design of this dome is felt to be the greatest single contribution to the Renaissance movement in the 15th. to 18th. century.

The competition for designing the bronze doors for the Florence Cathedral was won by the sculptor, Lorenzo Ghiberti, (1378–1455). Many historians have declared that these classical doors should be given an equal rating with the dome as contributors to the re-birth of classic art.

MODERN DOMES

Looking back, one can see that the Romans used the dome to cover large state buildings and cathedrals. Likewise, to the present day, the dome seems to signify a seat of government or large church. A great number of state capitol buildings, courthouses, and major churches are distinctive only because of their domed tops.

In recent years, the demand for greater comfort in leisure time activities has led to the domed sports arena and gymnasium. The success of the Harris County Domed Stadium (the Astrodome) in Houston, Texas, has started a new "Renaissance" for sport palaces. Although often described as the "Eighth Wonder of the World," this dome is nothing new. With the introduction of new high-strength steels and laminated timbers, this super-sized domed roof concept will be used in many major cities.

Erection of dome framing 3.7.2

Framed domed roof systems are treated as rigid-type shell structures which will not support themselves until all ribs and component members are in place and secure. The field erectors must make extensive preparations before any members are placed in position to ensure safe and orderly erection progress. By building a timber form supported by pipe or wood shoring, the whole dome framework is assembled on the temporary structure. The use of hydraulic or screw jacks permits the framework to be lowered into its final resting position before the shoring is removed. An extra tension ring to contain the horizontal forces in ribs is desirable when the supporting circular wall is composed of a connected series of vertical columns.

Force theory of domes 3.7.3

Such engineers as Schwedler, Loessel, and Schmidt were exponents of the framed dome because they were in a position to thoroughly analyze and appreciate the design of the ancient structures.

The Lamella roof arch with diagonal lattice work was developed in Germany following World War I, and has been used extensively in this country.

Theoreticians agree that framed domes are properly defined as space structures and, further, that all forces act internally within the framework. The forces acting within a dome rib are non-coplanar, and exact analysis is most difficult. The most convenient approach to a safe design is to go from a system of coplanar forces to the forces outside the plane in a series of distinct steps.

Designers of laminated wood framed domes often exclude diagonal ties which are usually used in steel framed domes. The rectangular solid rib and ring sections in wood reduces, to some extent, the need for diagonal ties between those members. Under symetrical loading over the entire hemisphere the diagonals are not stressed. However, with a snow load on only one side, diagonals are stressed and the safer choice would be to design to include diagonal tie members. The same unsymetrical loading can occur in hurricane regions.

Framed rib and ring design 3.7.3.1

Several methods have been advanced for the resolution of forces within the members of the hemispherical dome. An eminent Swedish engineer, Thur Thelander, who designed many important domes in Denmark and Central Europe, used a convenient and rational system developed by Schwedler. Thelander appears to have been proficient with a trigonometric solution, while Schmidt preferred to design with the graphical method. The basic theory of either method of solution is as follows:

(a) The loads are applied at connecting panel points where ribs and rings intersect.

(b) Ribs support the entire load when the whole area is symmetrically loaded.

The stress in the ribs will be compressive.

(c) The diagonals are not stressed when the load is placed symetrically over the whole area.

(d) When all of the dome area above a particular ring belt is fully loaded, the rings above will be in maximum tension stress.

(e) When all of the dome area below a particular ring belt is fully loaded, the rings above will be in compression.

(f) The ribs are in maximum stress when the entire dome is symmetrically loaded.

(g) Point loads shall be carried down the ribs progressively to be applied as floor loads to the supporting column below.

Dome design 3.7.4

The formulas for calculating the stress in dome framing members involve the use of trigonometric functions. These formulas are more understandable when presented in a complete step-by-step example. A preliminary drawing is essential to the design procedure, and should identify angles, locations, types of members, dimensions and load points. This simplifies the solution and makes re-checking by others much more convenient.

Dome design nomenclature 3.7.4.1

P_1 = Vertical point load at first or compression ring and top of meridian rib.

P_{1-4} = Vertical loads as $P_1 + P_2 + P_3 + P_4$.

C_1 = Rib section identification, analogous to side c in triangle.

R_2 = Horizontal Ring identification.

a_2 = Vertical distance between rings, analogous to side a.

b_2 = Horizontal distance between rings, and side b of triangle.

θ_2 = Angle under consideration, function of a_2 and b_2.

r_1 = Dimension of ring between ribs.

b_0 = Radius of Compression ring.

A_0 = Angle of Ribs with ℄ through panel space or center dome.

T_2 = Diagonal ties between Ribs and Ring panels. Stress or length.

L_2 = Wind load per square foot given in pounds.

B = Top angle the tie braces T makes with horizontal plane.

Dome design procedure 3.7.4.2

There are several methods the design engineer can use to shorten his work in calculating panel areas, rib and ring lengths, and loading. In the example to follow we will show that the very accurate dimensions which are necessary for the working drawings at the fabricating plant are not necessary for design work. Precise figuring involves considerable time and effort and does not change calculated areas and stresses enough to justify the extra labor.

After the preliminary plan is adopted, and the number of ribs and rings have been established, the area of panels between ribs and rings is found as follows:

Compute the area enclosed by maximum diameter of tension ring and divide by the number of panels as formed between the ribs. The resulting area multiplied by the square foot load will give the total load to be sustained by a single rib. By calculating the areas enclosed by the other rings, and with similar deductions, the area between ribs and rings can be calculated with sufficient accuracy for practical design purposes.

Accurately drawn sketches are imperative because the many angles and dimensions can be checked for accuracy by the use of a protractor and scale. Graphics may also be used to check the trigonometric solution for forces in members.

Wind and snow loads are customarily added to the vertical live and dead loads. The values selected for these loads should be appropriate for the dome location.

All loads are assumed to be applied at the panel points, as is the case in truss design. The ribs are brought together at the top by abutting them against a horizontal compression ring called the lantern ring. The designer must allow work space for making the erection connections. This factor will determine, in most cases, the diameter of the lantern ring or the number of meridian ribs and panels. The lantern ring, in every case, is under maximum compressive stress and separate calculations may be necessary to design spokes and a hub which can resist this collapsing force. The lowest ring is called the wall ring or, more often, the tension ring because this member is in tension under normal loads. When all ribs are symmetrically loaded, the intermediate rings will be in tension. The lower rings in the high hemispheric dome are an exception to this rule; they may have a small compressive or tensile stress.

The solution of dome framing is best accomplished after an accurate identification system for angles and members has been established. The method recommended for accuracy and efficiency is given thus:

Step I: Prepare a full or half circle plan and elevation to a convenient scale. Select for trial design the number of ribs. Under each rib at tension ring the vertical load reaction must be supported by a wall or column and in such a case, the column spaces may determine the rib spacing.

Dome design procedure, continued 3.7.4.2

Step II: Determine ring spacing. As ribs are sustaining compressive force and are designed as columns, a savings in time and erection can be made when the lengths between rings is close to same length. Refer to Typical Rib Section drawing for illustration.

Step III: Calculate circumference of each ring and the area enclosed. Dimensions should be placed on drawing of a rib section to enable the angles to be determined later with the Trig. functions.

Step IV: Calculate area enclosed by each ring. This may be done accurately by taking values from tables for functions of numbers in Section V.

Step V: Determine area between adjoining rings by deducting area of smaller ring from next larger ring. Divide this area by number of rib spaces, and result will be area of each panel between ribs. The length of ring between ribs is obtained by dividing circumference of each ring by the number of rib spaces.

STEP VI: On drawing of rib section, calculate the Tangent of angle each rib makes with horizontal plane of ring. Two sides are given in each instance as a and b, corresponding with Trigonometric function tables for sines, co-tangents, etc. Designate angle and functions for later reference and for checking by others.
Use nomenclature for indicating angles, ribs, rings and sides of each triangle.

Step VII: Panel point loads are computed for area of each panel bounded by two ribs and two rings. Multiply area panel by square foot combined live, dead, wind and snow load. Indicate each load as P with subscript number of ring.
Make an enlarged drawing of each panel formed between two ribs and denote area in each panel.
Check the total area in panels with area found in area enclosed by tension ring. Deduct area of lantern ring.

Dome design procedure, continued 3.7.4.2

Step VIII: Tabulate each load as shown in following example. Calculate the stress in rib between rings by formula given. All stress will be compressive.

Step IX: Calculate stress in each ring member by using the formula given in example. Note that the angle A where rib abutts lantern ring is one half the angle from center line of panel, therefore the formula contains this function as: 2 Sin. A.

STEP X: Stresses in Tie braces (T).

Under symmetrical dome loading only, the diagonal cross ties are assumed to be without stress. The theory therefore requires that the ribs be only braced for wind or snow loads which tend to develop unsymmetrical loading. Under such conditions, the diagonal ties must resist the load difference for the adjacent rib panel, as would be the case in a one sided snow load. The designer should allow an additional vertical load of from 20 to 30 pounds per square foot for safety when designing the diagonal bracing. In computing the forces in ties resulting from wind or snow, use only the wind or snow load values for figuring the panel loads L_1, L_2, etc. See Step XI to Step XIII in the example. The dimension (a) is the same distance of rise for diagonals as for ribs, however the longer length of the diagonal tie will result in a lesser angle. It is more convenient to work from the top angle B, with the secant of B, becoming the main function. Thus: Stress in $T_1 = L_1 \times \left(\frac{T_1}{a_1}\right)$, or T_1 = Cosecant of the angle A times wind load L_1. Secant of angle B = length of Tie over Rise a. The stresses in diagonal ties may also be found by drawing force diagrams as will shown in the example.

STEP XI: Design of Lantern Ring.

By conducting a simple experiment, it will be

Dome design procedure, continued 3.7.4.2

found that a circular ring will deform into an elliptical shape when external forces are applied on the opposite quadrants of the original ring. This action is the basic theory upon which the top Lantern Ring is to be designed.
The Ring must be truly rigid as to resist any rib forces which are unequal in abuttment and cause bending stress in the ring.
With horizontal forces acting upon two opposite quadrants of the ring, and assuming that the other quadrants will not be subjected to any similar forces, the bending moment in the ring can be found by using the formula: $M = \frac{\omega r^2}{5}$.
Where, ω = Uniform horizontal load as applied on ring per lineal foot of circumference, and r = radius of ring in feet. Since bending produces a corresponding and tension unit stress, the axial stress will be added to the bending stress to obtain maximum stress. The example will illustrate the procedure for the design of the Lantern Ring, or Compression Ring as it is somtimes called.

STEP XII: Final design of Rings.
Certain load applications consisting of wood deck covering, glazing and roofing materials will produce bending in the horizontal rings connecting the ribs. The line of force action will depend upon the location, and in some instances the load bending moment would effect the minor axis of the ring cross-section. The loads which resulted in the panel point load P, is the total of two (2) ring reactions. The bending moment in ring is obtained as $\frac{WL}{8}$. This bending stress is to be added to the stresses previously calculated for each ring designated as R_1, R_2, etc. Total stress in ring is thefore according to the general formula for two kinds of stress action as: $f = \frac{P}{A} \pm \frac{Mc}{I}$.

EXAMPLE: Framed dome in wood or steel: complete design example 3.7.4.3

Architects preliminary plans call for a framed dome of wood or steel with a diameter of 80.0 feet and a height of 29.0 feet. Shape of dome shall be of the hemispherical type with a compressive lantern ring of 9.0 Diameter.

A statue, with a weight of 12,000 Pounds, and an interior chandelier with a weight of 3000 Pounds, are to be supported symmetrically on this lantern ring.

Live loads plus dead and wind loads are estimated to be approximately 60 Pounds per square foot.

REQUIRED:

Draw a plan and elevation of a Half Dome. Show the spacing of Ribs and Belt Rings with all necessary dimensions, angles and identify each member according to nomenclature previously given. Calculate all forces in ribs, belt rings and diagonals, but do not design sizes for sections.

STEP I:

Draw half circle plan to scale and elevation. Determine radius for shell elevation.

Given: $L = 80.0$ Ft. $H = 29.0$ Ft. $\text{Radius} = \dfrac{(4H^2) + L^2}{8H}$

$$R = \frac{(4 \times 29.0 \times 29.0) + (80.0 \times 80.0)}{8 \times 29.0} = 42.086 \text{ Feet.}$$

From Tables of number functions:

Circumference of Lantern Ring = 28.274 Feet.

Circumference of Tension Ring = 251.33 Feet.

STEP II:

Selecting Rib and Belt Ring spacing:

Provide a minimum of 1½ feet at Lantern Ring for making connections and work room for Rib connections.

Choosing 18 Rib spaces, distance between ribs on the tension ring $= \dfrac{231.33}{18} = 13.963$ Feet. (Enough for erection)

At Lantern or Compression Ring, distance $= \dfrac{28.274}{18} = 1.571$ Ft.

Spacing in degrees $= \dfrac{360}{18} = 20°$ Angle $A = 10$ degrees

EXAMPLE: Framed dome in wood or steel, continued 3.7.4.3

STEPS I to V: (Dome example)

Drawing preliminary plan and elevation for spacing of Ribs and Rings. For steel framing compute all frame components as straight members.
Glued laminated wood members can be fabricated to obtain radius curves by sawing or formed.

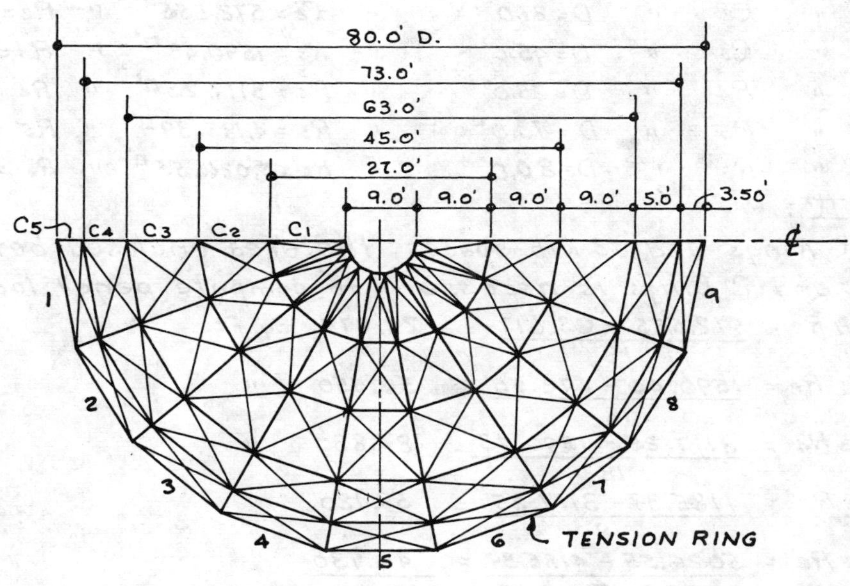

∘HALF PLAN DOME∘

Scale: 1" = 20.0 Ft.

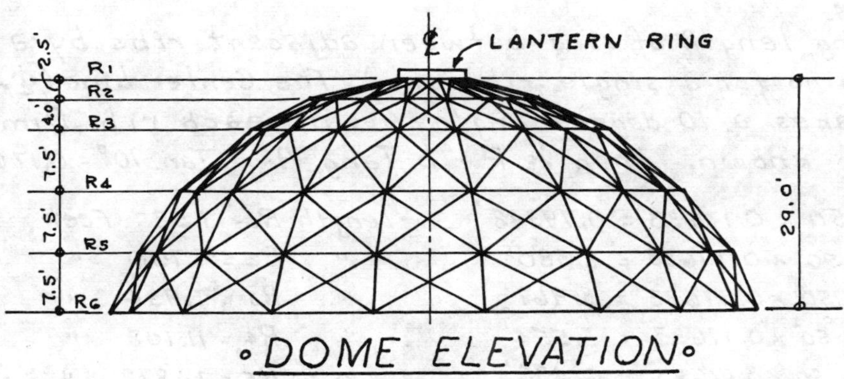

∘DOME ELEVATION∘

EXAMPLE: Framed dome in wood or steel, continued 3.7.4.3

STEP III:

When selecting the spacing for ring belts, the loads will be better distributed if the rib length is kept close to the same dimension. Ribs are more often sustaining compressive stress and the slenderness ratio of $\frac{l}{r}$ or $\frac{l}{d}$, can be kept uniform.

Calculate the Area and Circumference enclosed in each ring. This data can be obtained from Tables of number functions.

Area inside R_1	when	D = 9.0'	Area	R_1 = 63.617 □'	Circ.	R_1 = 28.274'
" "	R_2 "	D = 27.0'	"	R_2 = 572.555 □'	"	R_2 = 84.823'
" "	R_3 "	D = 45.0'	"	R_3 = 1590.43 □'	"	R_3 = 141.37'
" "	R_4 "	D = 63.0'	"	R_4 = 3,117.25 □'	"	R_4 = 197.92'
" "	R_5 "	D = 73.0'	"	R_5 = 4,185.39 □'	"	R_5 = 229.34'
" "	R_6 "	D = 80.0'	"	R_6 = 5,026.55 □'	"	R_6 = 251.33'

STEP IV:

With 6 Rings and 18 Rib spaces, the area enclosed between 2 ribs and 2 Rings is necessary to compute panel load.

Area $C_1 R_2$ = $\frac{572.555 - 63.617}{18}$ = 28.274 Sq. Ft.

Area $C_2 R_3$ = $\frac{1590.43 - 572.55}{18}$ = 56.550 "

Area $C_3 R_4$ = $\frac{3117.25 - 1590.43}{18}$ = 84.830 "

Area $C_4 R_5$ = $\frac{4185.39 - 3117.25}{18}$ = 59.120 "

Area $C_5 R_6$ = $\frac{5026.55 - 4185.39}{18}$ = 46.730 "

Total Areas = 275.504 Sq. Ft.

Total Area panels between 2 Ribs checked thus.

Area between R_6 and R_1 = $\frac{5026.55 - 63.62}{18}$ = 275.72 □' (Close enough).

STEP V:

Calculating length of ring between adjacent ribs by a flat plan drawing for a single rib panel. The Center Line (℄) of panel makes a 10 degree angle A with each rib. Dimension b, is also known. Then ½ R = b Tang. A. Tan. 10° = 0.17633

½ R_1 = 4.50 × 0.17633 = 0.79348'	Length	R_1 = 1.587 Feet
½ R_2 = 13.50 × 0.17633 = 2.380'	"	R_2 = 4.760 "
½ R_3 = 22.50 × 0.17633 = 3.967'	"	R_3 = 7.934 "
½ R_4 = 31.50 × 0.17633 = 5.554'	"	R_4 = 11.108 "
½ R_5 = 36.50 × 0.17633 = 6.436'	"	R_5 = 12.872 "
½ R_6 = 40.00 × 0.17633 = 7.053'	"	R_6 = 14.106 "

EXAMPLE: Framed dome in wood or steel, continued 3.7.4.3

The dimensions of rings may be checked by scaling the plan drawing. Also, by dividing the circumference of each ring circle by number of rib spaces will be close enough.

STEP VI:

The angles which the ribs make with horizontal plane will be necessary to calculate the stresses. By drawing a section through shell with an enlarged scale profile, compute the length of ribs between rings. Identify each rib as side c of triangle, and angle A as θ_2, etc.
2 dimensions are known in each instance.

Tang. $A = \frac{a}{b}$ and $C = \frac{b}{\text{Cos.}\,\theta}$. Locate angle and function in table.

Angle θ_1: $b_1 = 9.0'$ $a_1 = 2.50'$ Tan. $\theta_1 = \frac{2.50}{9.00} = 0.2777$

Angle θ_2: $b_1 = 9.0'$ $a_2 = 4.00'$ Tan. $\theta_2 = \frac{4.00}{9.00} = 0.4444$

Angle θ_3: $b_1 = 9.0'$ $a_3 = 7.50'$ Tan. $\theta_3 = \frac{7.50}{9.00} = 0.8333$

Angle θ_4: $b_4 = 5.00'$ $a_4 = 7.50'$ Tan. $\theta_4 = \frac{7.50}{5.00} = 1.5000$

Angle θ_5: $b_5 = 3.50'$ $a_5 = 7.50'$ Tan. $\theta_5 = \frac{7.50}{3.50} = 2.1428$

From Tables of Trigonometric Functions:

Angle $\theta_1 = 15°30'$ Sin. $\theta_1 = 0.26724$ Cos. $\theta_1 = 0.96363$ Cot. $\theta_1 = 3.6059$
Angle $\theta_2 = 23°57'$ Sin. $\theta_2 = 0.40594$ Cos. $\theta_2 = 0.91390$ Cot. $\theta_2 = 2.2513$
Angle $\theta_3 = 39°48'$ Sin. $\theta_3 = 0.64011$ Cos. $\theta_3 = 0.76828$ Cot. $\theta_3 = 1.2002$
Angle $\theta_4 = 56°20'$ Sin. $\theta_4 = 0.83228$ Cos. $\theta_4 = 0.55436$ Cot. $\theta_4 = 0.66608$
Angle $\theta_5 = 64°59'$ Sin. $\theta_5 = 0.90618$ Cos. $\theta_5 = 0.42288$ Cot. $\theta_5 = 0.46666$

Solve for Rib lengths between rings. Slide rule figures OK.

$C = \frac{a}{\text{Sin.}\,\theta}$ or $\frac{b}{\text{Cos.}\,\theta}$. Other function will be used later.

$C_1 = \frac{9.00}{0.96363} = \frac{2.50}{0.26724} = 9.36$ Feet

$C_2 = \frac{9.00}{0.91390} = \frac{4.00}{0.40594} = 9.85$ Feet

$C_3 = \frac{9.00}{0.76828} = \frac{7.50}{0.64011} = 11.72$ Feet

$C_4 = \frac{5.00}{0.55436} = \frac{7.50}{0.83228} = 9.02$ Feet

$C_5 = \frac{3.50}{0.42288} = \frac{7.50}{0.90618} = 8.28$ Feet

EXAMPLE: Framed dome in wood or steel, continued 3.7.4.3

STEP VI: Shell section for computing angles θ, etc.

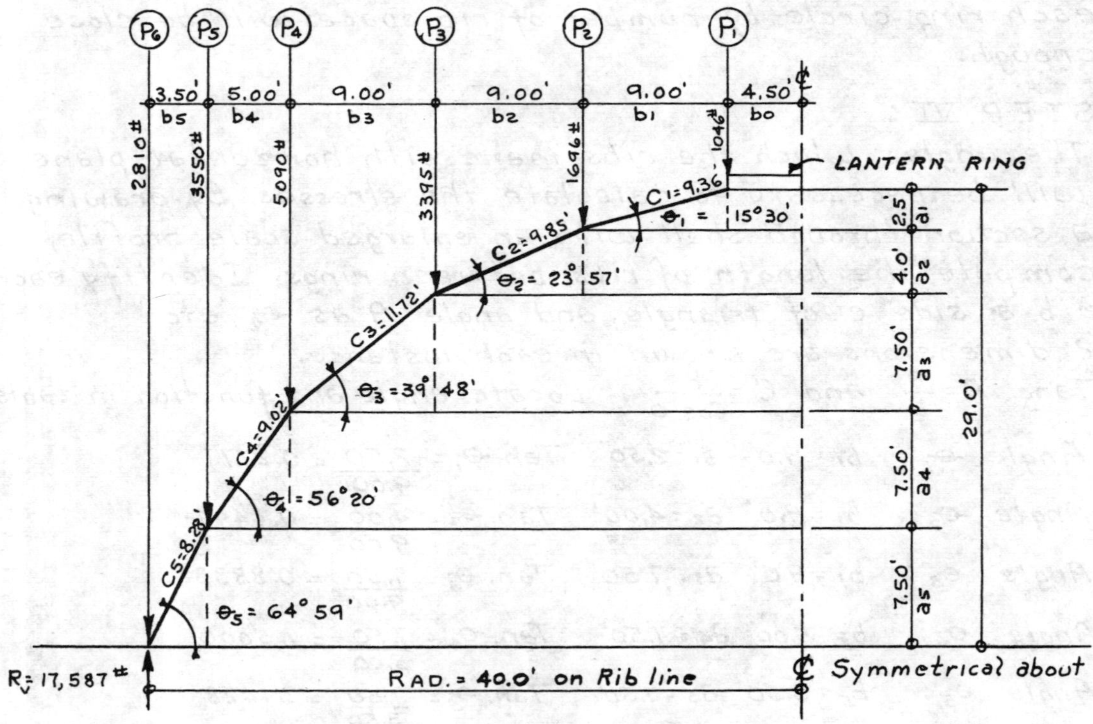

∘TYPICAL RIB SECTION∘

Scale: 1.0" = 10.0 Ft.

STEP V:
For calculating ring lengths, etc.

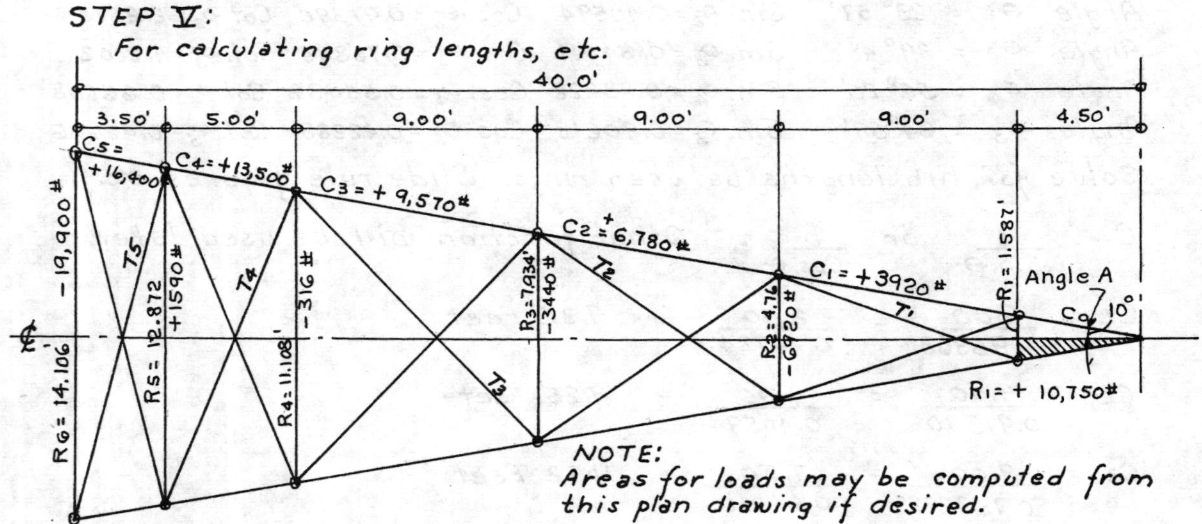

∘PLAN TYPICAL RIB PANEL∘

Scale: 1.0" = 6.0 Ft.

EXAMPLE: Framed dome in wood or steel, continued 3.7.4.3

STEP VII:

Computing loads to be applied at panel points. Denote load as P_1 when applied to rib C_1 and ring R_1, etc. Areas of each panel was calculated in step IV.

Compute area for full circle of Dome to check later.

Flat Area = 5026.55 × 60 =	301,593	Pounds
Weight of Statue at top =	12,000	"
Add for Light Fixture =	3,000	"
Dome Total =	316,593	Pounds.

18 Rib spaces, load under rib $C_5 = \frac{316,593}{18} = 17,588$ Pounds.

Vertical Live Load = 30#□' Wind Load = 20#□' Dead Load = 10#□'

For point load at P_1

$$P_1 = \frac{(63.617 \times 60) + 12,000 + 3,000}{18} = 1,046 \text{ Pounds}$$

$P_2 = 28.27 \times 60 =$	1,696	"
$P_3 = 56.55 \times 60 =$	3,395	"
$P_4 = 84.83 \times 60 =$	5,090	"
$P_5 = 59.12 \times 60 =$	3,550	"
$P_6 = 46.73 \times 60 =$	2,810	"
Total under 1 Rib =	17,587	Pounds. OK

STEP VIII:

Calculating stress in Rib Sections C_1 to C_5. Add loads from above to next rib under. Thus indicate P_{1-4} as denoting $P_1 + P_2 + P_3 + P_4$, etc. By Formula for stress:

$$C_1 = \frac{P_1}{\sin\theta_1} \quad \text{or} \quad C_1 = \frac{1046}{0.26724} = +3,920 \text{ Lbs.}$$

$$C_2 = \frac{P_{1-2}}{\sin\theta_2} \quad \text{or} \quad C_2 = \frac{2,742}{0.40594} = +6,780 \text{ "}$$

$$C_3 = \frac{P_{1-3}}{\sin\theta_3} \quad \text{or} \quad C_3 = \frac{6,137}{0.64011} = +9,570 \text{ "}$$

$$C_4 = \frac{P_{1-4}}{\sin\theta_4} \quad \text{or} \quad C_4 = \frac{11,227}{0.83228} = +13,500 \text{ "}$$

$$C_5 = \frac{P_{1-5}}{\sin\theta_5} \quad \text{or} \quad C_5 = \frac{14,777}{0.90618} = +16,400 \text{ "}$$

All stress in ribs is compressive.

EXAMPLE: Framed dome in wood or steel, continued 3.7.4.3

STEP IX:

Compute stress in each ring belt. Lantern Ring R_1 will be in compression, and Tension or Wall Ring R_6 will be in Tension. Intermediate rings may be in either compression or tension stress, depending upon the location or arrangement of the ring. Let the formula decide the type of stress.

Compression or Lantern Ring Formula: $R_1 = \frac{P_1 \text{ Cot. } \theta_1}{2 \text{ Sine } A}$

Angle $A = 10$ degrees as described is step V.

$$R_1 = \frac{1046 \times 3.6059}{2 \times 0.17365} = +10{,}750 \text{ Lbs. (Force of 1 Rib against Ring).}$$

$$R_2 = \frac{(P_1 \text{ Cot. } \theta_1) - (P_{1\text{-}2} \text{ Cot. } \theta_2)}{2 \text{ Sin. } A} \qquad R_2 = \frac{(1046 \times 3.6059) - (2742 \times 2.2513)}{2 \times 0.17365} = -6920 \text{ Lbs.}$$

$$R_3 = \frac{(P_{1\text{-}2} \text{ Cot. } \theta_2) - (P_{1\text{-}3} \text{ Cot. } \theta_3)}{2 \text{ Sin. } A} \qquad R_3 = \frac{(2742 \times 2.2513) - (6137 \times 1.2002)}{2 \times 0.17365} = -3440 \text{ Lbs.}$$

$$R_4 = \frac{(P_{1\text{-}3} \text{ Cot. } \theta_3) - (P_{1\text{-}4} \text{ Cot. } \theta_4)}{2 \text{ Sin. } A} \qquad R_4 = \frac{(6137 \times 1.2002) - (11{,}227 \times 0.66608)}{2 \times 0.17365} = -316 \text{ Lbs.}$$

$$R_5 = \frac{(P_{1\text{-}4} \text{ Cot. } \theta_4) - (P_{1\text{-}5} \text{ Cot. } \theta_5)}{2 \text{ Sin. } A} \qquad R_5 = \frac{(11{,}227 \times 0.66608) - (14{,}777 \times 0.4666)}{2 \times 0.17365} = +1590 \text{ Lbs.}$$

$$\text{Tension Ring } R_6 = \frac{P_{1\text{-}5} \text{ Cot. } \theta_5}{2 \text{ Sin. } A} \qquad R_6 = \frac{14{,}777 \times 0.4666}{2 \times 0.17365} = -19{,}900 \text{ Lbs.}$$

STEP X:

Length of Diagonal Ties indicated as T_1, T_2, etc. A single panel bounded by 2 Ribs and 2 Rings forms a geometrical plane figure called an, "Isosceles Trapezoid." The parallel top and bottom represent the parallel rings. The sides represent the Ribs C, and are 10 degrees from vertical plumb line as is shown on plan inside lantern ring.

Should a typical panel trapezoid be rigidly assembled and assumed to be hinged at bottom, the sides and diagonal ties will always have the same height of rise with horizontal plane, however angle will change. See illustrated detail.

In step V, the dimension of ring to ℄ of panel (trapezoid) was computed by using the sine of angle A.

Draw to scale the panel formed by 2 Ribs C_1 and Rings R_1 and R_2. Dimensions for all four sides are known. The dimensions for ties T_1 to T_5 are easily solved by making one rib a Right angle of 90 degrees.

EXAMPLE: Framed dome in wood or steel, continued 3.7.4.3

STEP X:
Illustration of Trapezoid.

Create right triangles in each panel with two sides known.

$c = \sqrt{a^2 + b^2}$

c = Tie

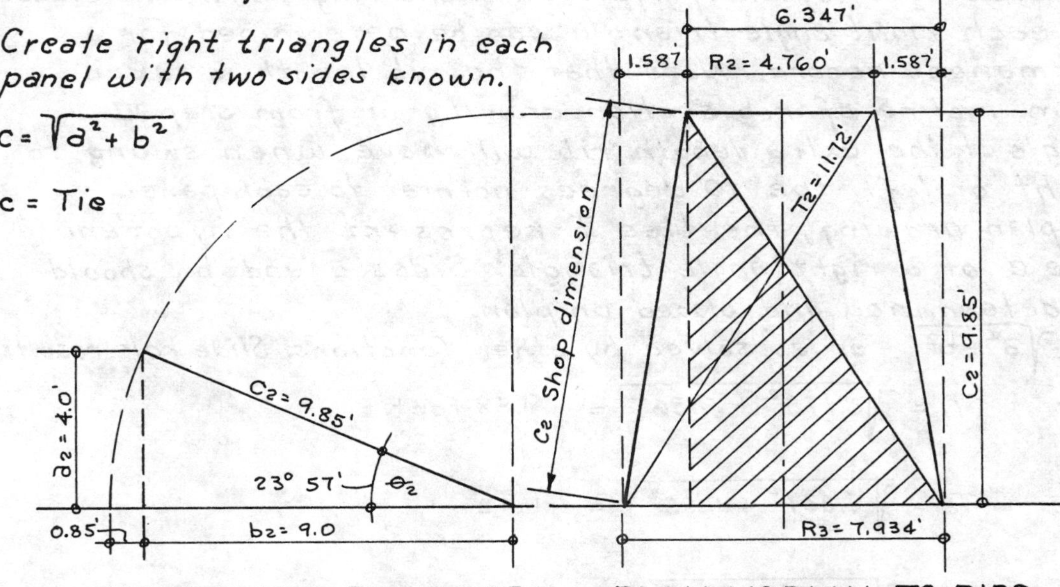

• HORIZONTAL RIB SLOPE • • PLAN NORMAL TO RIBS •

Side a of triangle = $\left(\frac{R_3 - R_2}{2}\right) + R_2$.   a = 6.347 Ft.

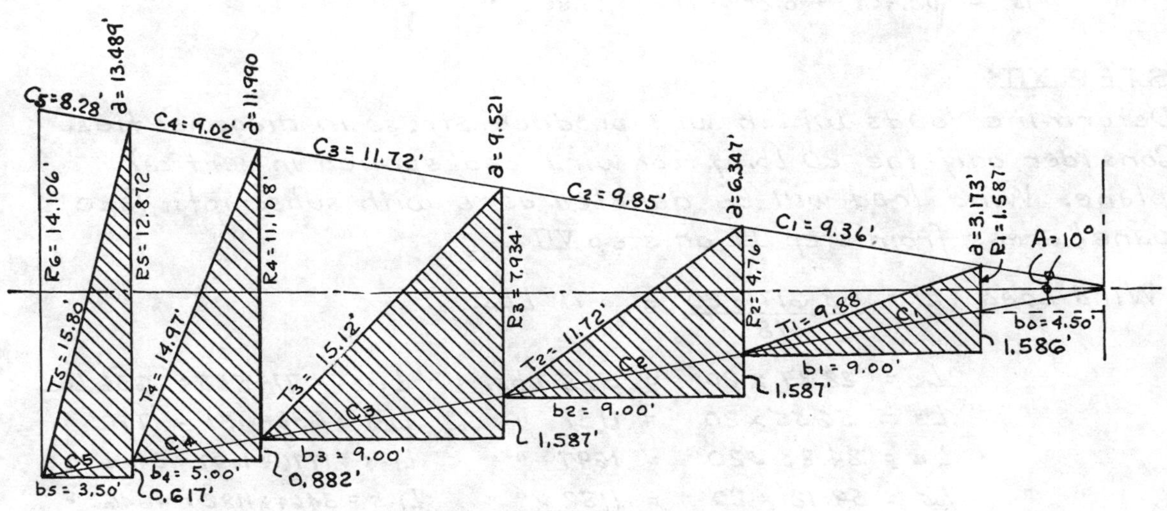

PLAN RIB PANELS ON HORIZONTAL PLANE

EXAMPLE: Framed dome in wood or steel, continued 3.7.4.3

STEP XI:

By drawing a layout of typical rib and ring panel, the sides of each right angle triangle can be determined for a permanent record. Note that the rib length is taken from incline of rib section drawing or from step VI. Neglect the extra length rib will have when swung to right or left the 10 degrees normal to each panel.

In plan drawing, the ties T represent the hypotenuse side c of a right angle triangle. Sides a and b, should be determined and placed on plan.

$T = \sqrt{a^2 + b^2}$ or as solved by other functions. Slide rule results.

$T_1 =$ $T_1 = \sqrt{3.173^2 + 9.36^2} =$ 9.88 Feet.

$T_2 = \sqrt{6.347^2 + 9.85^2} =$ 11.72 "

$T_3 = \sqrt{9.521^2 + 11.72^2} =$ 15.12 "

$T_4 = \sqrt{11.99^2 + 9.02^2} =$ 14.97 "

$T_5 = \sqrt{13.489^2 + 8.28^2} =$ 15.80 "

STEP XII:

Determine loads which will produce stress in diagonal ties. Consider only the 20 Lb. Sq. Foot wind load applied in vertical plane. Wind load will be denoted as L with subscripts. Take panel areas from step IV or step VII.

Wind Load $L_1 = \dfrac{63.617 \times 20}{18} =$ 71 Lbs.

$L_2 = 28.27 \times 20 = 565$ Lbs. $L_{1-2} = 71 + 565 = 636$ Lbs.

$L_3 = 56.55 \times 20 = 1131$ " $L_{1-3} = 636 + 1131 = 1776$ "

$L_4 = 84.83 \times 20 = 1697$ " $L_{1-4} = 1776 + 1697 = 3464$ "

$L_5 = 59.12 \times 20 = 1182$ " $L_{1-5} = 3464 + 1182 = 4646$ "

$L_6 = 46.73 \times 20 = 935$ "

Load L_6 does not influence stress in diagonals.

EXAMPLE: Framed dome in wood or steel, continued 3.7.4.3

STEP XIII:

Computing stresses in diagonal Ties. Wind loads applied at same panel points as shown for P on shell section in step VI drawing. Use the same section drawing to get the rise dimension a_1, a_2, etc., for ties which are the same for ribs.

When length of Tie T_1 = 9.88 Feet, and rise a_1 = 2.50 Feet, a right angle triangle can be formed with 2 sides known. The sides are a_1 and c.

The Secant of B = Cosecant of A, and equals $\frac{c}{a_1}$

$$\text{Sec. } B = \frac{9.88}{2.50} = 3.952$$

Stress in T_1 = L_1 Sec. B.

L₁ = 71#, T_1 = c = 9.88', a_1 = 2.50', A', B, C, b

Stresses in diagonals:

$T_1 = L_1 \times \left(\frac{9.88}{2.50}\right)$	$T_1 = 71 \times 3.952 =$	281 Pounds + or −.
$T_2 = L_{1\text{-}2} \times \left(\frac{11.72}{4.00}\right)$	$T_1 = 636 \times 2.93 =$	1863 "
$T_3 = L_{1\text{-}3} \times \left(\frac{15.12}{7.50}\right)$	$T_3 = 1776 \times 2.02 =$	3588 "
$T_4 = L_{1\text{-}4} \times \left(\frac{14.97}{7.50}\right)$	$T_4 = 3464 \times 1.996 =$	6927 "
$T_5 = L_{1\text{-}5} \times \left(\frac{15.80}{7.50}\right)$	$T_5 = 4646 \times 2.1066 =$	9803 "

The stresses in diagonal ties may be tension or compressive, and when the rings in wood laminated domes are placed between the wood ribs for lateral support, the diagonal ties could be omitted. Connections must be sufficient to provide rigidity in such cases.

STEP XIV:

Recap the stresses in each member for quick reference.

RIBS	RINGS	TIES
C_1 = + 3920 Lbs.	R_1 = + 10,750 Lbs.	T_1 = 281 Lbs.
C_2 = + 6780 "	R_2 = − 6920 "	T_2 = 1863 "
C_3 = + 9570 "	R_3 = − 3440 "	T_3 = 3588 "
C_4 = + 13,500 "	R_4 = − 316 "	T_4 = 6927 "
C_5 = + 16,400 "	R_5 = + 1590 "	T_5 = 9803 "
	R_6 = − 19,900 "	

EXAMPLE: Framed dome in wood or steel, continued 3.7.4.3

STEP XV:

Drawings to check work by scale and graphics

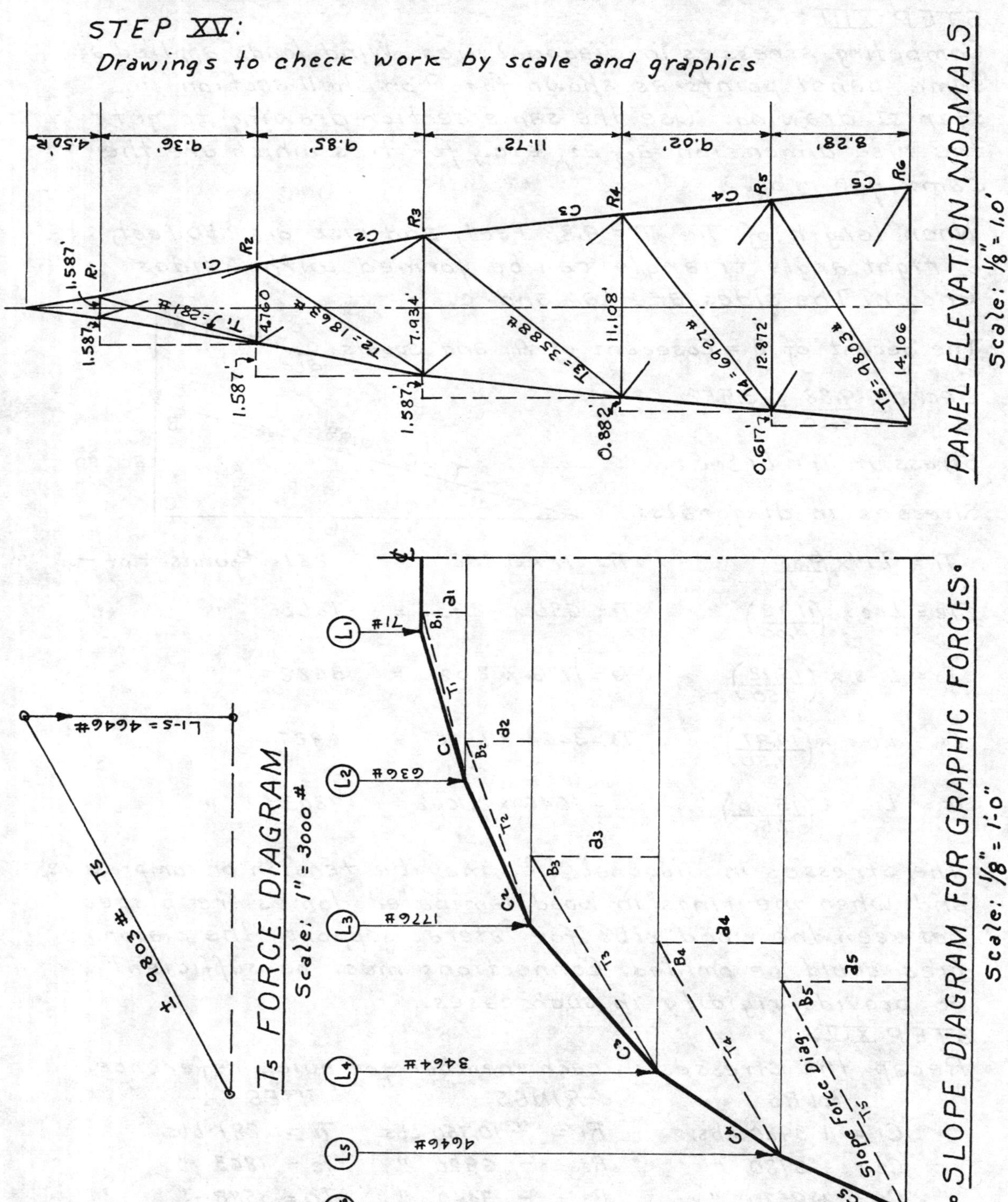

EXAMPLE: Framed dome in wood or steel, continued 3.7.4.3

STEP XVI:

Designing the Compression or Lantern Ring.
Dividing the circle into quadrants, the opposite two quadrants resist the bending in ring.
The formula for bending moment is: $M = \frac{wr^2}{5}$ where,

w = Uniform load from Ribs in Pounds.
r = Radius of Lantern Ring in feet. See Step VI Rib section.

Area Panel acting against ring = 63.17 Sq. Ft.
Vertical Loads = 60 Lbs. Sq. Feet. Load = 63.17 x 60 = 3817 Lbs.

Rise a_1 = 2.50' Horizontal plane = 4.50'

Tangent of Top angle $B = \frac{4.50}{2.50} = 1.80$

Horizontal force on Ring from 1 Rib = 3817 x 1.80 = 6870 Lbs.
Converting to a uniform Load around rib with spaces at 1.587 feet for ribs.

Uniform horizontal Load on Ring, $w = \frac{6870}{1.587} = 4325$# Lin. Ft.

Using formula: Bending $M = \frac{4325 \times 4.50^2}{5} = 17{,}515$ Foot Lbs.

STEP XVII:

Calculate compression in Lantern Ring. This is the additional stress to be added to the compressive stress as resulting from bending moment.

Circumference of Ring = 28.274 Feet.
Length of 2 equal quadrants = ½ of 28.274 = 14.137 Lin. Ft.

Force of Compression = 4325 x 14.137 = 61,145 Lbs. (Axial)

STEP XVIII:

The formula for Maximum Stress = $\frac{Mc}{I} + \frac{P}{A}$ or it may be rewritten as: $\frac{M}{S} + \frac{P}{A}$.

The maximum allowable stress for dense glued Southern Yellow Pine, parallel to grain is F_c = 1400 P.S.I. To choose a trial section, reduce this stress 200#□" to obtain the probable Section Modulus and cross sectional Area.

$S = \frac{17{,}515 \times 12}{1200} = 175.15''^3$ $A = \frac{61{,}145}{700} = 87.35$ Sq. Inches

EXAMPLE: Framed dome in wood or steel, continued 3.7.4.3

STEP XIX:

In selecting for trial a rectangular section, the correct axis subjected to bending must be indicated.

From the tables of Alternate sizes, choose a 10"x 14" S4S as being close to having required properties. A net size of 9"x 13" is given in Laminated tables as a stock size.

Area = 9.0 x 13.0 = 117.0□" $S_y = \frac{db^2}{6}$ $S_y = \frac{13.0 \times 9.0^2}{6} = 175.5''^3$

Max. Stress = $\frac{17{,}515 \times 12}{175.5} + \frac{61{,}145}{117.0}$ = 1522 Lbs. Sq. In. This is over the allowable of 1400 PSI, therefore another trial must be made. Choose a net size of 11.0"x 13.0" and compute properties.

$S_y = \frac{13.0 \times 11.0^2}{6} = 262.0''^3$ A = 13.0 x 11.0 = 143.0□"

Max. $f_c = \frac{17{,}515 \times 12}{262.0} + \frac{61{,}145}{143.0}$ = 1230 PSI. (OK, accept this size).

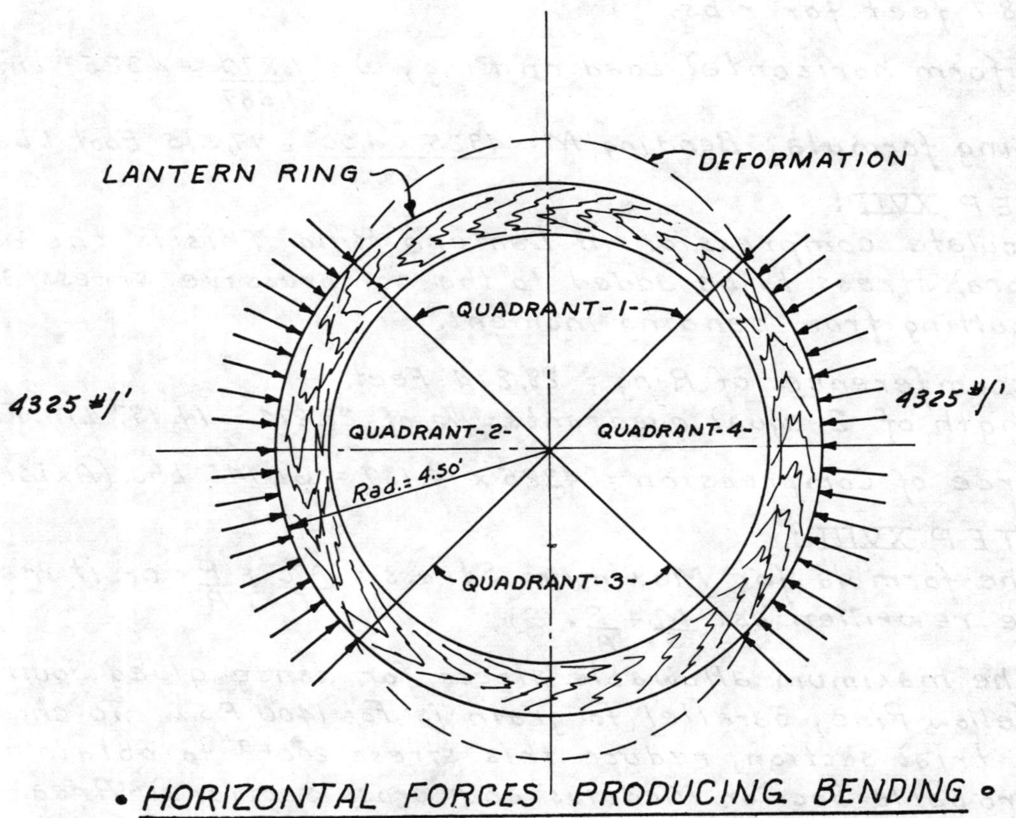

• HORIZONTAL FORCES PRODUCING BENDING •

IV—CONCRETE DESIGN

IV

CONCRETE DESIGN

CONCRETE DESIGN

Contents

PORTLAND CEMENT

There are examples of masonry construction dating back to Roman times in existence today. The builders of these ancient buildings and bridges used a cement made of hydrated lime and ash mixed to the consistency of a paste to fill voids and level stones. After the water evaporated, there was little adhesion; only pozzolan, a volcanic ash, would cling to the stones. This type of cement was used until 1756, when an English engineer, John Smeaton, discovered that if hydraulic lime was added to the mix, it would harden the paste and add a certain amount of adhesion.

In 1796, another Englishman, James Parker, developed a cement by grinding clayey limestone into powder and then burning it to a calcium compound. This was the first natural cement produced by grinding and calcining limestone.

In 1824 an improved cement was patented by Joseph Aspdin of Leeds, England. It was given the name Portland Cement because when it hardened it had a close resemblance to the stone taken from the quarry on Portland Isle. The first Portland cement was made in America in 1872 by David O. Saylor. Portland cement was not as popular as natural cement until, in 1892, the method of burning the cement slag in rotary kilns was developed. The cement industry has since found other sources for raw materials: washed oyster shell from the coast line provides cement plants with raw materials.

REINFORCED CONCRETE

The use of reinforcement in cement dates back to an occasion when a French fisherman, Lambot, observed a small Swedish girl using reeds to reinforce a basket made of clay and leaves. In 1850, Lambot constructed a small boat with a concrete shell and metal reinforcing. Many patents were taken out by Joseph Monier, a Paris gardener who produced garden tubs and flower pots with metal-ribbed reinforcement. Reinforced concrete construction spread rapidly after two German engineers, Wayss and Bauschinger, in 1886, applied the Monier system to the casting of slabs and long rectangular shapes.

In the United States reinforcing appeared in several early structures. A reinforced-wall, concrete house was constructed in 1872 at Port Chester, New York by William E. Ward. In 1896, bridge builder Edwin Thatcher began to construct spans with steel shapes serving as the reinforcing. Thatcher built a distinguished career as a bridge builder. The reinforced arch construction was employed by the eminent Austrian engineer Melan in 1894. The first flat-slab, fully monolithic structure was poured in 1906 in Minneapolis. This building was designed without girders or beams by C. P. Turner, who called it a mushroom floor. The design is still in use today. It employs a dropped panel above a flared capital on a circular column. An example of this design is illustrated in Example 4.9.3 along with the spiral-hooped column, which is an essential part of the modern technique.

Concrete, continued 4.1

MANUFACTURING PROCESS

Portland cements are produced from selected materials under stringent controls. Calcareous materials such as limestone, shell, or marl, and argillaceous clayey materials such as shale, slag and clay are used in its manufacture. The raw ingredients are crushed and pulverized, mixed in the proper proportion for the chemical composition, and placed in rotating kilns where the mixture is calcined at a temperature of 2700°F. During the calcining process, a clinker is formed. This clinker is removed from the kiln and allowed to cool. Then it is pulverized and mixed with a gypsum additive which controls the setting time. This pulverized clinker is ground to a powder so fine that it will pass through a sieve with 40,000 openings per square inch. The finished product is portland cement.

Mixing the cement powder with water forms a paste which will set and become a solid mass. The setting and hardening are the results of a chemical reaction between the cement powder and water. This chemical action is referred to as hydration or curing. Standard Portland cements are manufactured in several types and covered by ASTM Designation C–150.

Types of Portland cement 4.1.1

Under the ASTM specifications there are five types of portland cement:

TYPE I: NORMAL PORTLAND CEMENT
A general purpose cement satisfactory for all uses except when the special properties of the other types are desired for reasons of design or exposure.

TYPE II: MODIFIED PORTLAND CEMENT
A cement which generates a lower heat during a longer hydration period. Compared to Type I, this cement is better for large abutments and dam structures where the heat of hydration may become a problem.

TYPE III: HIGH-EARLY STRENGTH PORTLAND CEMENT
A cement which quickly reaches a working strength which allows forms to be removed earlier or the structure to be put into service sooner. Also, it requires less cold weather protection. This type of cement generates high heat during the early period of hydration and finishing work must be performed rapidly. Results are best with mild air temperatures and the use of Type III should be discouraged in temperatures above 90°F.

TYPE IV: LOW-HEAT PORTLAND CEMENT
A special cement for massive structures where the heat generated by hydration must be kept to a minimum. Strength development is at a slower rate.

TYPE V: SULFATE-RESISTANT PORTLAND CEMENT
A slow hydration cement for use in soils and water with high alkali content. It is

Types of Portland cement, continued 4.1.1

used for structures subject to severe exposure to sulfate fumes. Restrict its use to these conditions.

AIR-ENTRAINING PORTLAND CEMENT

There are three types of air-entraining cement which correspond to Types I, II, and III in ASTM Designation C150. ASTM Designation C175 covers these. By grinding air-entraining materials into the cement during clinker producing operations, the cement develops a resistance to severe frost and salt (from ice removal). Entrained air in fresh concrete is in the form of minute air bubbles, so small that there are billions of them in a cubic foot of concrete. This type of cement is desirable for highway or paving construction which may be subject to freezing and thawing.

SPECIAL CEMENTS

Oil-well cementing requires a cement which can be jetted into place under water around the casings in deep wells. Such cement must harden properly at high temperatures.

White cement is produced from special raw materials for use in pre-cast shapes. It is also used in mixed portions for patching up honeycombs and plaster finishes.

Waterproof cement is produced by grinding water-repellant material into the mix before the clinker is made.

Non-shrinking cement is made by adding metal filings to regular portland cements. It is used for grouting joints in pre-cast panels and pipes passing through masonry walls below water level.

Colored cements are required by building codes to encase electric conduits underground. Regular cement is used and the coloring is added as a powder at the batching plant.

Cement volume measure 4.1.2

A standard cloth or paper bag of portland cement contains one cubic foot and weighs 94 pounds. Large users of cement will have cement shipped in by barrels or in bulk in special rail cars or trucks. A barrel of portland cement contains the equivalent of four bags or 376 pounds. Since modern batching is based on weight proportion, not volume, the mix for a specified strength requires that the ingredients be accurately weighed. The engineer must supply the batching plant with the design weights of materials.

Strength of concrete

The strength of hardened concrete depends upon the ability of the cement paste to bond the aggregate together into a solid mass. The paste must fill all the void spaces. When the mix is placed in forms it is necessary to hand puddle or use a vibrator to completely fill all cavities.

The strength of the adhesive bond resulting from the chemical reaction between cement and water is principally affected by the amount of water used in the mix. Excessive water tends to dilute the mix; the correct amount of water is of great importance if the desired strength is to be achieved. Although wet mixes are easier to work with less labor, they will have a lower strength. Stiff mixes require more labor in placing; however, they are more economical if strength and durability are considered.

Hydration in concrete will continue for several months. Concrete designers usually specify that test specimens be sent to the laboratory to be broken after curing for 7, 14 and 28 days.

SLUMP TEST

The consistency of a concrete batched mix is determined by the slump test in accordance with ASTM Specification C143. This test should be made by the inspector in the field by taking concrete samples from the truck spout or from the fresh concrete as it is placed in the forms. A slump cone is made of 16 gage steel and is 12 inches in length. The larger open end is 8 inches in diameter and the hole at top is 4 inches in diameter. Two handles for lifting are placed on opposite sides 4 inches from the top. In making a slump test the following procedure must be followed:

(a) Dampen the metal cone and place on a flat surface.

(b) Fill the cone from the top by placing three layers of fresh concrete.

(c) Each layer of concrete must be rodded with 25 vertical strokes.

(d) The stroking rod is a smooth ⅝ inch diameter rod with a length of approximately two feet. The end used for stroking is hemispherical in shape with a slight taper extending 1 inch. The rod may be steel, plastic or wood.

(e) All stroking should be uniform. The bottom layer is rodded throughout its depth. The two top layers are rodded with rod penetrating the underlying layer.

(f) Strike-off the top layer or add mix until the cone is full and level.

(g) Lift the cone from mold of mix very carefully in a vertical direction, then place it beside the mold just formed.

(h) Place a straight edge across the top of the cone, extending over the mold.

(i) Measure the difference between the height of the cone and the molded concrete. This measurement is taken at the vertical axis of the mold specimen.

(j) Record the slump test in inches of subsidence of the specimen: Slump equals 12 inches minus height of mold.

A well-proportioned concrete mix with proper water content will gradually slump but retain its original, cohesive form. A poor mix with an excessive water content will slump quickly, aggregates will separate, and the specimen will disintegrate. Slump tests should be conducted continually by field inspectors to control water content.

CYLINDER TEST

Standard compression test cylinders are made in the field in straight cylinder molds. Fill the cylinder and rod each layer as

Strength of concrete, continued 4.1.3

described for the slump test. Take specimen concrete from several locations of the same batch or direct from the truck spout at intervals. Cylinders for test specimens are supplied either by the contractor or the concrete supplier, with test specimens made by the engineer inspector or under his personal supervision.

For specimens with coarse aggregate of sizes 2 inches and under, the cylinder is 6 inches in diameter and 12 inches in height, with one end closed. The cylinder walls are of thick, waxed cardboard or of light sheet metal. The specimen cylinders should be cured under conditions identical to the work which they represent. Cylinder specimens must not be handled or moved until concrete has hardened. Immediately after rodding the top layer in the cylinder, tap the sides lightly to close any voids left by rodding. Fill the cylinder and level the top with a trowel or by placing a glass plate on top and rotating until sealed at the walls. This cap should be left on the specimen until the cylinder is ready to be moved.

FLEXURE TEST

This test which measures the flexural resistance strength of concrete is important to the engineers who design concrete highways and airport runways. Flexure tests are conducted as follows:

A test specimen is made with 2 layers of concrete placed in a wood form which will provide a mold cross-section 6 by 6 inches square. Length is not less than three times the depth of the section plus 6 inches to permit seating in the breaking machine. Specimens for flexure tests are made under conditions similar to other specimens, except that curing must correspond closely with site conditions. It is advisable to set the specimen in damp earth and cover the top with a double layer of wet burlap. Keep the specimen wet or damp for 24 hours, then remove the wood form and place the specimen on the ground. Bank the sides with damp earth, and remove to the laboratory after 7 or 10 days.

RELATING TESTS RESULTS TO DESIGN FACTORS

Flexure tests have only limited use in structural design. This is true because structural concrete design assumes that concrete supplies only compressive strength and contributes no resistance at all to tension. Of course, cracks in concrete caused by shrinkage or temperature change can be annoying to the architects and engineer, but there is little cause for alarm unless the cracks occur below the neutral axis of the cross-section. In a concrete beam with a positive bending moment, the compressive stress is resisted by the concrete above the neutral axis (NA), and the tension stress is resisted by the reinforcing steel placed below the neutral axis. Here are two materials being used which differ widely in their ability to resist the type stresses mentioned. The location of the neutral axis depends on the strength of the concrete above the NA and the value of steel resistance below the NA. Design factors are coefficients obtained from several basic formulas which consider the compressive concrete strength and the allowable tension stress of the steel. Paragraphs which follow will provide additional information on the methods used to determine the location of neutral axis and coefficients for design purposes.

CURVES: Curing temperature vs. strength during aging 4.1.3.1

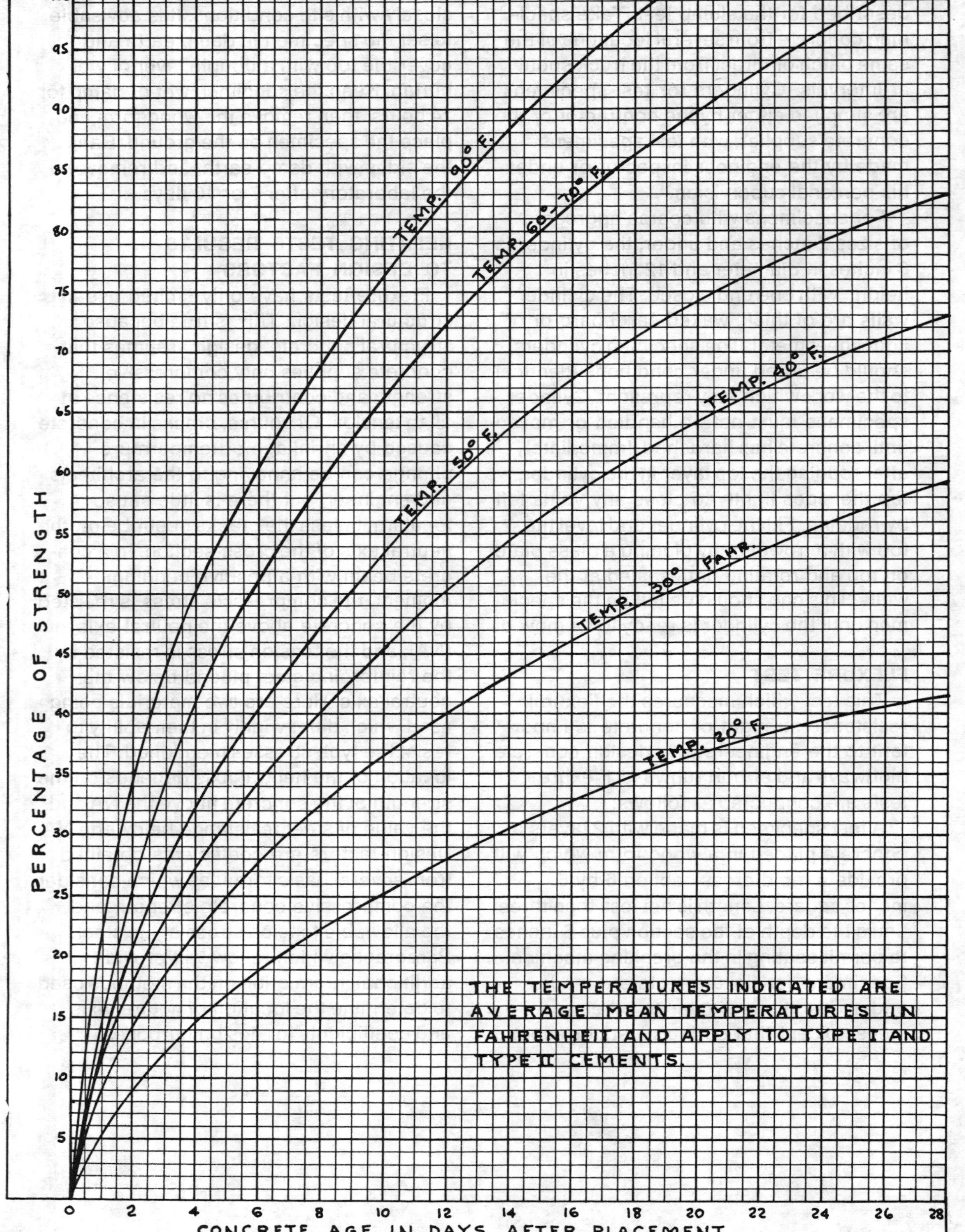

Concrete components 4.1.4

COARSE AGGREGATES

Coarse aggregates used in concrete mixes are inert materials such as gravel, crushed stone, slag or volcanic rock. When high-strength concrete is required, the coarse aggregate is usually commercial gravel supplied by a reliable plant which has adequate facilities for crushing, washing and screening the stones. The usual sizes of coarse aggregate in commercial construction will run from ¾ inch to a maximum of 1½ inches. The designers of a section which requires close rod spacing should specify a maximum size for coarse aggregate. To make certain that the fluid concrete will surround the reinforcing rods, the aggregate should not be larger than ¾ of the clear spacing between rods.

FINE AGGREGATE

The usual fine aggregate for all types of concrete is a good grade of clean, washed sand. Sand can be produced from crushed rock but river bed sand is a superior product. It is the fine aggregate in fluid concrete which supplies the plasticity and workability of the mix. After washing and screening, the rough sand must be graded and mixed for the proper size distribution. Experience has shown that for smooth surfaces where concrete is cast against wood or metal forms, the fine aggregate should contain not less than 15 percent passing a number 50 sieve and at least 4 percent passing a number 100 sieve. These minimums must be met to regulate the water gain and add to the cohesiveness of the mix.

MIXING WATER

Water used for mixing concrete must not contain acids, alkalies, oils or decayed vegetable matter. Specifications should require that mixing water be suitable for drinking.

ADMIXTURES

Proper use of admixtures or additives comes with experience and visual inspection. Additives are placed in the batched mix to improve workability, reduce segregation, accelerate setting, retard setting, entrain air, reduce shrinkage, or harden the concrete surface. When the use of an admixture is being considered, it is well to explore other methods which may be more convenient and economical.

The only approved additive for use as an accelerator is calcium chloride. The amount used should be restricted to less than 2 percent by weight of the cement in the mix. An approved retarding agent is referred to as a plastiment. The amount required to retard setting time is governed by temperature, humidity and wind. In North Africa and the Middle East where air temperatures reach 120°F, 3 to 4 ounces per sack was found adequate for satisfactory results.

CURING AND HARDENING AGENTS

The paragraph on the strength of concrete stated that the concrete strengthens with age. This gain in strength is very rapid in the early stages, but continues more slowly for an indefinite period. Good concrete can only be obtained by proper control. It is as important to control the curing as it is to control the fluid mix. Curing proceeds at a very slow rate when temperatures are below normal, and at freezing temperatures there is virtually no hydration or chemical action. The reverse is true for temperatures above 90°F.

Concrete components, continued 4.1.4

In hot, dry weather, wood forms dry out and must be kept moist by spraying. Concrete can be kept moist by leaving the forms in place. Slab surfaces can be ponded by flooding with water. Continuous sprinkling is most desirable, not alternate flooding and drying out. Burlap or curing quilts are good covers for slabs and wall tops when kept damp or saturated.

In the curing of concrete there is no substitute for water. Curing agents and sealants have a place in the final curing stages, but examine carefully the claims for curing agents. Spraying a curing agent on freshly set concrete is a cheap shortcut which may take advantage of a careless or novice inspector. No reliable manufacturer of curing agents or sealants has ever claimed that their product is a substitute for water. These products are intended to reduce the labor involved in prolonged curing by a spray coating after the hydration action has slowed and the concrete has obtained a high percentage of final strength. Spraying slabs or flat surfaces immediately after trowelling should not be allowed; insist on a few days of water curing.

Concrete floor hardening compounds such as antacoid or lapidolith should be applied after the concrete is at least 28 days old. If concrete floor slabs are to be hardened or made resistent to oils, membrane-curing agents previously mentioned must not be used. They must be applied according to the manufacturer's recommendations. When properly applied, they can increase resistance to abrasion by 40 percent and resistance to water leakage under pressure by 90 percent. The hardening agents penetrate the cellular formation and chemically react with the lime content to produce a dense, flint-like surface. Liquid hardeners are applied with a brush or mop, and must never be sprayed on fresh new concrete even when diluted with water.

EPOXY AGENTS

The bonding of new concrete to old concrete or masonry is possible when the older material is brush-coated with an approved epoxy mix. The epoxy bonding agents permit parts of a structure to be precast in casting beds located near the project. The reinforcing rods in the sections project from the precast member and are encased in the forms for the main structure. The precasting assembly method is often used for struts in concrete docks which join concrete piles. Before placing the new concrete, the ends of the precast struts are brush-coated with the epoxy mix. Speed is important for a satisfactory bond, because the pot life of the epoxy mix is very short.

Only reliable materials should be used. Resin-based types are preferred. Recommendations should be obtained from engineers who have had experience with epoxies. Suppliers' sales literature should be objectively evaluated. Epoxy products are affected by age; reliable manufacturers prefer to ship from factory-fresh stock.

Concrete epoxy for adhesion and bonding consists of a liquid base to which a hardening compound is added just prior to application. Mixing must be in small amounts to be rapidly applied. When new concrete is placed against the sticky, brushed surface a polymerization action begins. The old surface reacts to start an extended hydrating action which will fuse the old into the new. Considerable heat is generated during the fusing action. Epoxy mixed with cement and sand grout is ideal for patching pile caps damaged by hard driving. Patched piles can be driven within three days.

Concrete mix design 4.1.5

CONCRETE TRIAL BATCHES

Two methods are used to design a trial batch mix for laboratory testing. The older of these is the volume mix. Quantities are measured with a shovel or container. It is roughly assumed that a 1:2:4 mix will give a 28 day breaking strength of 2000 PSI. A mix 1:2:4 by volume means that the mix has 1 sack (cubic foot) of cement mixed with 2 cubic feet of sand and 4 cubic feet of gravel. Water is added during mixing for the desired workability; from 5½ to 7½ gallons per sack of cement. The amount of moisture in the sand and gravel will determine the amount of water to use with volume mixes. The coarser the aggregate, the less free water it will carry into the batch.

Designing batches by weight is the modern method for making trial batch designs. The data thus obtained can be passed on to the batching plant operator, who also uses the weight method for mixing in bulk. There is a certain amount of void space in any loose bulk material. The absolute volume of a loose aggregate is therefore the actual total volume of the solid matter in all particles. The absolute volume of a bulk unit volume can be computed when the specific gravity of the material is known. This may be written thus:

$$\text{Absolute volume} = \frac{\text{Unit bulk weight}}{\text{Specific gravity x unit weight of water}}$$

where the bulk unit weight is based on dry surfaces. Before proceeding with instructions for design examples using the absolute weight method, the meaning of voids and specific gravity must be understood. These terms will be explained and by conducting two simple field experiments the properties may be readily obtained.

PERCENTAGE OF VOIDS

The void space in coarse aggregate is the space which must be filled by the cement paste to provide for the adhesion of particles and form a solid material. Imagine an experiment as follows: Take two new concrete test cylinder containers. Fill one cylinder with dry loose gravel level with the top. Fill the other cylinder with clean fresh water and place it above the first cylinder. Using a small, flexible tube, siphon water from the top cylinder to fill the void spaces in the gravel. With 1½ inch gravel aggregate, it will be found that approximately 6 inches of water will be required to fill the voids and bring the water level to the top of the gravel. The percentage of voids is therefore 50 percent, and the percentage solid is 50 percent. The same experiment conducted with the fine dry sand will require approximately 30 percent water to fill the voids, leaving 70 percent for solids.

SPECIFIC GRAVITY

The specific gravity of a substance is found by weighing a volume of a substance in air, then in water, and dividing the weight in air by the loss of weight in water. The specific gravity of coarse, dry, gravel aggregate is determined as follows: Obtain a small-mesh onion sack from a grocer and fill with a volume of dry coarse gravel. Use a suspension scale to weigh the contents in air. Next swing the sack over a water tank and lower into the water until completely submerged. Record the weight while submerged. The weight in water will be less; this weight subtracted from the weight in air will be the loss in water. The weight in air divided by the loss in water will be the specific gravity.

$$\text{Specific gravity} = \frac{\text{Weight of Volume in air}}{\text{Loss in water}}.$$

Concrete mix design, continued 4.1.5

The absolute weight of a substance per cubic foot is equal to: Specific gravity times weight of 1 cubic foot of water. To illustrate, with weight of water = 62.5 pounds per cubic foot: A volume of dry gravel weighs 5.55 pounds in air. Same volume submerged in water weighs 3.46 pounds. The loss in water = 5.55 − 3.46 or 2.09 pounds. Specific gravity = 5.55/2.09 = 2.65. A cubic foot of this gravel will weigh: 2.65 x 62.5 = 165.36 pounds. The specific gravity of dry sand is usually approximately the same as gravel or 2.65. The specific gravity of water is 1.0. The specific gravity of portland cement is 3.10.

Water-cement ratio 4.1.5.1

It may not always be necessary to design trial batches for testing in order to obtain a mix which will give the desired strength and character. But if small aggregates are to be used or an exceptionally high-strength concrete is required for cast-in-place piles, a trial batch may be required. The supplier of concrete which is proportioned by the absolute weight method and mixed in transit will have file records for various strengths. They will know the mix which can be used under given conditions. To be conservative and for estimating purposes, some designers prefer to make small trial batches before starting the project. The supplier will be asked to submit samples of various aggregates to be used in these small batches. The essential factor in making a series of trial batches is to maintain the water-cement ratio constant.

All aggregates contain moisture since they are stored without any protection from rains. Moist sand contains approximately ¼ gallon of water per cubic foot; when very wet, ¾ to 1 gallon of water per cubic foot. Moist gravel contains about ¼ gallon per cubic foot: on 1½ inch gravel after rain exposure from ¼ to ¾ gallons per cubic foot. The amount of mixing water must be reduced to allow for the aggregate moisture. An experienced concrete inspector will make an early morning visit to the batching plant to ascertain the moisture content in the aggregate as it travels up the conveyor to the batching hopper. Water reduction is then calculated and adjustments made during the loading of the first truck.

EXAMPLE: Converting test batch to one yard batch 4.1.5.2

A trial batch is mixed to a desired consistency by using the measured materials in the following proportions:

Cement weight = 6.0 Lbs. Specific gravity = 3.10
Sand aggregate = 9.0 " Specific gravity = 2.65
Gravel aggregate = 19.5 " Specific gravity = 2.65
Mixing water = 0.312 Gallons. Specific gravity = 1.00
With water at 62.5 Lbs. cubic foot- Water wt.: 0.312 x 8⅓ = 2.60 Pounds

REQUIRED:
Design the batch mix for absolute volume and bulk weight. Calculate quantities required for a 1 Yard batch and give cement quantity in sacks or parts therof. Give water content in gallons per yard and aggregates in weight in Lbs. Use 94 Lbs. for each sack cement. Calculate weight of 1 yard batch and check against an equivalent bulk batch weight.

STEP I:
The calculations will be placed in the form as given here, leaving the right column for 1 Yard quantities:

MATERIALS TRIAL MIX	BULK WEIGHT IN LBS. USED	ABSOLUTE VOLUME IN CUBIC FEET	QUANTITIES FOR A 1.0 YARD BATCH
CEMENT	6.0	$\frac{6.0}{3.10 \times 62.50} = 0.0310$	$\frac{6.0 \times 110}{94}$ = 7 SACKS
SAND	9.0	$\frac{9.0}{2.65 \times 62.50} = 0.0545$	9.0 x 110 = 990 Lbs.
GRAVEL	19.50	$\frac{19.50}{2.65 \times 62.50} = 0.1176$	19.50 x 110 = 2145 Lbs.
WATER	2.60	$\frac{2.60}{1.0 \times 62.50} = 0.0416$	$\frac{0.0416 \times 110 \times 62.5}{8.33}$ = 34.2 Gal's.
TOTAL WT. = 37.10 Lbs.		TOTAL = 0.2447	

STEP II:
Required for a 1.0 Yard batch of 27 Cubic Feet = $\frac{27.0}{0.2447}$ = 110 batches.

Weight per cubic foot of wet mix = $\frac{37.10}{0.2447}$ = 151.5 Lbs.

Put required quantities of each material in last column on right.
Total weight for a 1.0 Yard batch = 37.10 x 110 = 4081 Lbs.
A test cylinder 6.0 inches in diameter with 12.0 inch depth has a volume of 0.196 cubic feet, thus the test batch has enough volume to fill cylinder. 1 Cubic Foot contains 7.48 Gallons.

EXAMPLE: Batch design for strength 4.1.5.3

A 28 day Concrete with required $F_c' = 2500^{\#\square"}$ is to be designed with a mix containing 4 sacks cement. Fine aggregate contains 70% solids and Coarse aggregate contains 50% voids.

REQUIRED:

Select proportions as follows for mix trial batch:
Cement = 1.0 volume. Sand = 3.5 volumes and Gravel = 6.0 volumes.
Assume water at 7 Gallons per sack but reduce amount if necessary to make absolute volume come out to 27.0 Cu. Ft.
Specific gravity of cement = 2.65 and aggregates = 2.65. Water is taken as 7.48 Gallons per cubic foot.

STEP I:

Calculate proportions by bulk weight and volumes:

Cement: 4 sk. @ 94# = 376 Lbs. = 4.00 Cubic Ft.
Sand: 3.50 Cubic Ft. sk = 3.50 × 4 = 14.00 " "
Gravel: 6.00 " " " = 6.00 × 4 = 24.00 " "
Water: 7.0 Gal. Sack = $\frac{7.0 \times 4}{7.48}$ = 3.74 " "

Total Bulk Volume = 45.74 Cubic Ft.

STEP II:

Put these volumes in table form to obtain absolute volumes and net weights. These results are for file records.

MATERIAL	BULK VOLUME IN CUBIC FT.	ABSOLUTE VOLUME IN CUBIC FEET	THEORETICAL WEIGHT IN POUNDS
CEMENT	4 sacks = 4.00	$\frac{4.00 \times 94}{3.10 \times 62.5}$ = 1.94	4 × 94 = 376
SAND	3½ cu. Ft. per sack 3.50 × 4 = 14.00	70% Solids 14.00 × 0.70 = 9.80	9.80 × 2.65 × 62.5 = 1620
GRAVEL	6.0 cu. Ft. per sack 6.0 × 4 = 24.00	50% Solids 24.00 × 0.50 = 12.00	12.00 × 2.65 × 62.5 = 1985
WATER	7.0 Gal. per sack $\frac{7.0 \times 4}{7.48}$ = 3.74	100% Solids 3.74	3.74 × 62.5 = 234
	TOTAL BULK: 45.74	TOTAL SOLIDS = 27.48	TOTAL BATCH WT = 4215

STEP III:

Reducing water to get even 27.00 Cubic Ft. solids.
$27.48 - 27.00 = 0.48'^3$ Weight = 0.48 × 62.5 = 30 Lbs.
Reduced volume in gallons for whole batch = 0.48 × 7.48 = 3.59 Gals
Accurate Mixing Water = $\frac{(3.74 - 0.48) \times 7.48}{4}$ = 6.10 Gals. per sack.

EXAMPLE: Batch design for strength 4.1.5.4

Tests reveal that a concrete batch with 5½ sacks of cement per yard properly proportioned with a 1:6 mix will produce a 28 day concrete of 3000 PSI or more. Break the mix in proportions as: 1:2¼:3¾.

REQUIRED:

Interpret the mix as specified volumes and design on absolute volumes. Use 6½ Gallons water per sack and gravel contains 50% voids and sand contains 30% voids. Calculate the absolute weight for 1 yard design, then determine the weight percentage of each material in batch. There are 7.48 Gallons of water in a cubic foot.

STEP I:

Specific Gravity of materials are as follows:

Water = 1.0 Cement = 3.10 Sand and Gravel = 2.65 Cement in sack = 1.0 cubic foot and weighs 94 Lbs.

STEP II:

Working in a design form thus:

TYPE MATERIAL	BULK VOLUME IN CUBIC FEET	ABSOLUTE VOLUME IN CUBIC FEET	NET WEIGHTS IN POUNDS
CEMENT	5½ sacks = 5.50	$\frac{5.50 \times 94}{3.10 \times 62.50} = 2.67$	5.50 × 94 = 517.0
SAND	2.25 Cu.Ft. Per Sack 2.25 × 5.50 = 12.38	30% Voids 70% Solids 12.38 × 0.70 = 8.67	8.67 × 2.65 × 62.5 = 1435.4
GRAVEL	3.75 Cu.Ft. per sack 3.75 × 5.50 = 20.63	50% Voids 50% Solids 20.63 × 0.50 = 10.32	10.32 × 2.65 × 62.5 = 1711.7
WATER	6½ Gallons per sack $\frac{6.50 \times 5.50}{7.48} = 4.78$	4.78 × 1.0 = 4.78	4.78 × 62.50 = 298.7
	TOTAL = 43.29	TOTAL = 26.44	TOTAL = 3,962.8

STEP II:

PERCENTAGE PROPORTIONS BY WEIGHT

CEMENT $\frac{517.0}{3962.8} = 13.0\%$ SAND $\frac{1435.4}{3962.8} = 36.25\%$ GRAVEL $\frac{1711.7}{3962.8} = 43.25\%$ WATER $\frac{298.7}{3962.8} = 7.50\%$

Weight per Cubic Foot of Wet Concrete = $\frac{3962.8}{27} = 147$ Lbs

Water Content = 5.50 × 6.50 = 35.75 Gallons per Yard.

EXAMPLE: Converting truck batch to test batch 4.1.5.5

A concrete supplier has submitted a proposed batch mix which, it is claimed, has recorded 3500 PSI 28 day strength from previous projects. Concrete coarse aggregate must consist of maximum 3/4 inch gravel. Quantities submitted are for a 6 yard truck batch and are as follows:

Cement = 3,102 Lbs. Sand = 7,152 Lbs. Gravel = 12,636 Lbs., and water = 196 Gallons per 6 yd. batch. Wt. water = 1633 Lbs.

REQUIRED:

Design engineer desires to have laboratory break 2 test cylinders of this equivalent mix. A small trial batch will be made with enough absolute volume to fill 2 cylinders. For calculations, design a trial batch of 0.50 cubic feet. Prepare table and compute proportions in bulk, absolute and weight volumes.

STEP I:

A 6.0 Yard batch contains: 27 × 6 = 162 Cubic feet. Then the quantity for a 1.0 Cubic foot batch will used for base.

Total weight of 6.0 yard batch given = 24,523 Lbs., and for 1.0 Cubic foot, weight = $\frac{24,523}{162}$ = 151 Lbs. (About right for 3/4")

Number of sacks cement per yard: $\frac{3102}{6 \times 94}$ = 5½ SK.

Gallons of water per yard = $\frac{1633}{6 \times 8.333}$ = 32.67

STEP II:

To determine proportionate ratios, use weights:

Cement in 1.0 Yard = 5.5 × 94 = 517 Lbs.

Sand in 1.0 Yard = $\frac{7152}{6}$ = 1192 Lbs. Mix ratio = $\frac{1192}{517}$ = 2.31

3/4" Gravel in 1.0 Yard = $\frac{12,636}{6}$ = 2106 Lbs. Mix ratio = $\frac{2106}{517}$ = 4.07

Mix may be described as: 1 : 2⅓ : 4.

STEP III:

To further reduce the volumes for a ½ cubic foot test batch, the divisor for 1.0 yd. batch is 27 × 2 = 54.

Trial batch Cement = $\frac{517}{54}$ = 9.57 Lbs.

EXAMPLE: Converting truck batch to test batch, continued 4.1.5.5

Trial batch Sand = $\frac{1192}{54}$ = 22.10 Lbs.

Trial batch Gravel = $\frac{2106}{54}$ = 39.00 Lbs.

Trial batch water = $\frac{32.6}{54}$ = 0.603 Gallons.

STEP IV:

1 Sack of cement = 94 Lbs., and volume = 1.0 Cubic foot.

1 Cubic foot of water contains 7.48 Gallons and weighs 62.5 Lbs.

Specific gravity of dry cement = 3.10

Specific gravity of dry sand = 2.65 and 30% voids or 70% solids.

Specific gravity of dry 3/4" Gravel = 2.65 and 40% voids or 60% solids

Put data obtained in form in order to determine the trial test batch absolute volume which should be close to $0.50'^3$

MATERIALS TRIAL MIX	BULK WEIGHT IN LBS. TRIAL	ABSOLUTE VOLUME IN CUBIC FEET	QUANTITIES FOR A 1.0 YARD BATCH
CEMENT	9.57	$\frac{9.57}{3.10 \times 62.50}$ 0.0495	$\frac{9.54 \times 54}{94}$ = 5½ Sacks 517 Lbs.
SAND	22.10	$\frac{22.10}{2.65 \times 62.50}$ = 0.1332	22.10 × 54 = 1192 Lbs.
GRAVEL $\frac{3}{4}$"	39.00	$\frac{39.00}{2.65 \times 62.50}$ = 0.2357	39.00 × 54 = 2106 Lbs.
WATER	5.02	$\frac{5.02}{1.0 \times 62.50}$ = 0.0805	$\frac{5.02 \times 54}{8.333}$ = 32.6 Gals. 272 Lbs.
TOTAL BULK WT. = 75.69 Lbs.		TOTAL = 0.4989	TOTAL WT = 4087 Lbs.

STEP V:

Calculate number trial batches to make a yard of concrete. Show figures in right column above for each quantity.

Weight per cubic foot of wet mix = 75.69 × 54 = 4087 Lbs.

Number of trial batches for 1.0 Yard = $\frac{27.0}{0.4989}$ = 54

Check weight with result in Step I: Wt = $\frac{4087}{27.0}$ = 151 Lbs. Cu. Ft.

Values in right hand column also check with results in Step II.

EXAMPLE: Converting volume to weight batch 4.1.5.6

Due to shutdown of batching plants, the General Contractor has arranged to proceed by batching at site of project. Maximum capacity of mixer is 4 Cubic feet and mix will be accomplished by bulk volumes and checked by the total weight of a 4.0 Yard bath. Bulk volume mix is set at proportions of: 1: 3½: 6, with 6 gallons of water per sack.

REQUIRED:
Furnish contractor with information for batching by bulk volumes and total weight for each batch. Use the specific gravity and void percentages thus:
Cement; specific gravity = 3.10 1 Sack = 1.0 Cubic foot.
Fine aggregate; specific gravity = 2.65 Solids = 70 Percent.
Coarse aggregate; specific gravity = 2.65 Solids = 50 Percent.
Water; 62.5 Lbs. Cubic foot. 7.48 Gallons = 1.0 Cubic foot.

STEP I:
Prepare the bulk, absolute volume and weight in the form similar to preceding examples. Absolute volume for a 4 yard batch should be approximately 27 × 4.0 = 108 Cubic feet, and total weight approximately 108 × 150 = 16,200 Lbs.

QUANTITIES FOR 4 YARD BATCH			F'_c = 2500 PSI. 28 Days.	
MATERIAL	1.0 YARD IN BULK IN CUBIC FT.	ABSOLUTE VOL. 1.0 CUBIC YARD	ABSOLUTE VOL. 4 CUBIC YDS.	TOTAL WEIGHT 4.0 YARD BATCH
CEMENT	1 Sack = 1.0 Cu. Ft. 4 Sacks = 4.00	$\frac{4.00}{3.10 \times 62.5}$ = 2.07	4 × 2.07 = 8.28	1 Sack = 94 Lbs. 4 × 4 × 94 = 1504
SAND	3½ Cu. Ft. per sack 4.0 × 3.5 = 14.00	70% Solids 0.70 × 14.00 = 9.80	4 × 9.80 = 39.20	2.65 × 62.5 × 39.2 = 6490
GRAVEL (1½")	6 Cu. Ft. per sack 4.0 × 6 = 24.00	50% Solids 0.50 × 24.00 = 12.00	4 × 12.00 = 48.00	2.65 × 62.5 × 48.0 = 7950
WATER	6 Gals. per sack $\frac{4.0 \times 6}{7.48}$ = 3.20	1.0 × 3.20 = 3.20	4 × 3.20 = 12.80	12.80 × 62.5 = 800
TOTAL BULK VOLUME = 45.20		TOTAL = 27.07	TOTAL VOL. 108.28	TOTAL WT. = 16,744

WET BATCH WT. PER CUBIC FOOT = $\frac{16{,}744}{108.28}$ = 154.5 LBS.

The control operator has accurate weigh scales for aggregate, cement, and water being loaded into each truck. The tables below the scales give weight proportions for various strength mixes.

Batching is controlled by a single operator at an electronic control panel, in radio communication with the truck driver, and in phone communication with the job-site inspector so that slump as placed may be continuously adjusted.

Reinforcing steel 4.2

The reinforcing is that part of the beam composition which provides all of the resistance to bending, and is usually in the form of steel rods with deformed surfaces. In Section II, it was stated that the modulus of elasticity (Es) is nearly constant at 29,000,000 PSI for most types of steel. This figure applies to billet steels, axle steels and rail steels, which are in common use for reinforcing concrete. Cured concrete is strictly a compressive material, and a concrete beam will fail unless steel is embedded in the action plane of the tension force. Reinforcing steel is also used for other purposes such as:

(a) To resist shear stress.
(b) To resist tension due to negative bending moments.
(c) To supplement the compressive strength of concrete when necessary as in columns, piles and girders.
(d) To reduce shrinkage and temperature cracks.
(e) To tie concrete structural members firmly together.

Tables 4.2.3.1, 2, 3 provide data on standard sizes and specifications for steel reinforcing rods, bars and welded wire mesh. Field engineers with the responsibility of inspecting the reinforcing placement in forms before concrete is placed should become familiar with the grade and identification marks as illustrated. Steel rods with deformed rib surfaces come in many patterns (which are the manufacturer's trade mark); however, the ASTM specification A305 for deformation and rib spacing must be followed.

Steel area and depth 4.2.1

The *depth to steel* is the dimension (d) from the top of the beam to the center of gravity of the steel area. In order to satisfy architectural considerations, the structural designer may find it necessary to depart from the usual allowable concrete and steel unit stresses and use higher strength material. The use of higher grades of steel and concrete will permit the cross-section to be reduced in size until there is enough room to accommodate the steel. Two or three layers of rods may be necessary to fulfill the requirements for shear and bond stress. The rods used in groups are not always the same size in each layer, and the center of gravity may shift its location. The effective plane is found by the method of moments which is explained in Section VI. An example will follow to illustrate this procedure.

EXAMPLE: Depth to steel for rod group 4.2.1.1

Assume a rectangular beam is limited to a breadth of 10.0 inches and a total depth of 14.0 inches. Area of tension rods has been determined and addition perimeter surface for bond requires 8.20 Sq. Inches. The following is selected: 2- #6 ϕ Rods and 3 #3 ϕ which give Σ_o = 8.24 Sq. Inches.

REQUIRED:

Determine the depth to steel (d) when the larger #6 Rods are placed above the #3 rods a distance of 2.50 inches. Allow 2.0 inches for rust and fire protection. If steel F_s = 18,000 PSI and J = 0.852, calculate the resisting moment.

STEP I:

A cross-section of beam is provided to locate steel and moment arms. Moments for CG plane will be taken from bottom of beam and put in regular format.

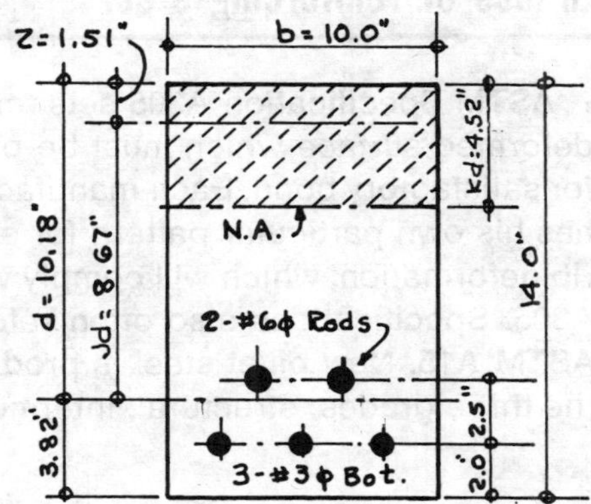

RODS	A□"	d	Ad
2-#6 ϕ	0.88	4.50	3.96
3-#3 ϕ	0.33	2.00	0.66
ΣA_s= 1.21		ΣM= 4.62	

Center of Gravity from beam bottom:

$CG = \frac{4.62}{1.21} = 3.82$ inches.

STEP II:

Depth to steel: d = 14.000 - 3.820 = 10.18 inches

A_s = 1.21 Sq. In. F_s = 18,000 PSI. Resisting Moment = $A_s F_s Jd$.

RM = 1.21 × 18,000 × 0.852 × 10.18 = 189,000 Inch Pounds. (Slide rule).

STEP III:

Beams of this type must be investigated to determine if Compression steel is needed in compressive area above NA. From Coefficient Tables: k = 0.444 and F_c = 1800 PSI. A_c above Neutral Axis = kdb. Compressive A_c = 0.444 × 10.18 × 10.0 = 45.20 Sq. Inches. Average compressive stress in concrete is equal to $\frac{F_c}{2}$, because there is zero stress at NA and maximum at top of beam. Then average stress = $\frac{1800}{2}$ = 900 PSI. C must equal T or be above the force in tension of $F_s A_s$. T = 18,000 × 1.21 = 21,780 Lbs.

C = 45.20 × 900 = 40,680 Lbs. By formula: $C = \frac{F_c kdb}{2}$.

Uses of reinforcing 4.2.2

Engineers assume that the tensile strength of concrete is zero, and design to use steel for this purpose. Plain concrete is a term used to designate concrete without reinforcing, and is applied to wet concrete as well as cured concrete. In the early stages, concrete was reinforced with plain smooth-surfaced rods and twisted wire. The many failures of these early designs led to several conclusions:

(a) When a loaded beam fails under conditions of overloading, it breaks on an angle of 45 degrees.

(b) The failure is caused by insufficient bond or adhesion of the steel to the concrete.

(c) By bending the rods at the ends in a hook pattern, the bond can be improved.

(d) If the rods simply lay in position without adhesion to the concrete, they have no value in resisting tension stress. Keep in mind that failure in beams results in a break on a 45 degree angle. Web reinforcement, in the form of stirrups for shear and diagonal tension, will be based upon this fact.

Grades of reinforcing steel 4.2.3

ASTM Specification A305 sets forth the deformed surface which must be provided for satisfactory bond. Each manufacturer has his own particular pattern for surface rib deformation, which will comply with A305. Specifications also often refer to ASTM A15. New billet steel is produced in the three grades: structural, intermediate, and hard. Specifications should stipulate the grade desired. High-strength billet steels A431 and A432 are more compatible with the higher strength concrete mixes. In such cases a smaller beam cross-section can be designed, which will result in less concrete volume and less form material.

TABLES: Markings, sizes and specifications for reinforcing rods 4.2.3.1

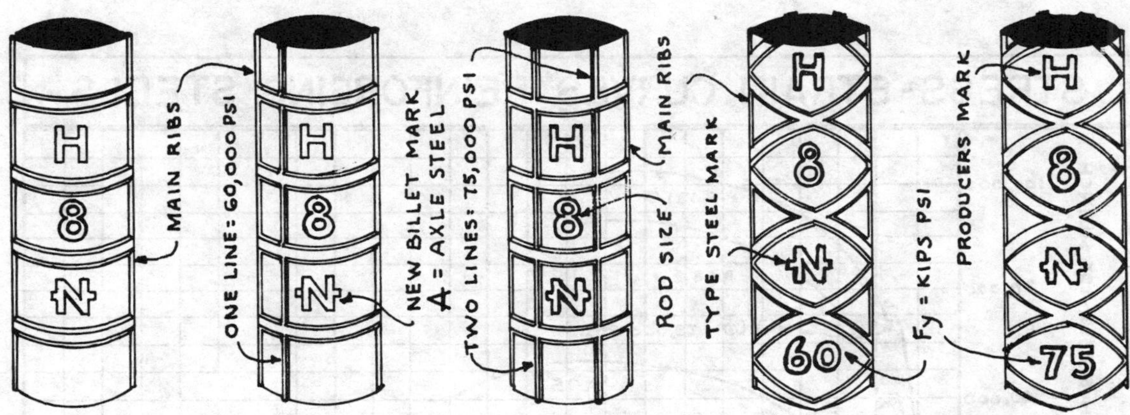

ORDINARY
A15 A408
33,000 - 50,000 PSI
A408 = 14S - 18S

HIGH STRENGTH GRADE STEELS MARKING SYSTEMS

60,000 PSI	75,000 PSI	60,000 PSI	75,000 PSI
A432	A431	A432	A431

PROPERTIES AND DATA - STANDARD REINFORCING RODS

ROD NUMBER	2	3	4	5	6	7	8	9	10	11	14S	18S
FRACTION SIZE	1/4	3/8	1/2	5/8	3/4	7/8	1.0	1.0	1 1/8	1 1/4	1 1/2	2.0
ROD PROFILE	●	●	●	●	●	●	●	■	■	■	■	■
AREA SECT.-SQ. IN.	0.050	0.110	0.20	0.31	0.44	0.60	0.79	1.00	1.27	1.56	2.25	4.00
PERIMETER-SQ. IN.	0.786	1.178	1.571	1.963	2.356	2.749	3.142	3.544	3.990	4.430	5.320	7.090
WEIGHT. LBS. FOOT	0.167	0.38	0.67	1.04	1.50	2.04	2.67	3.40	4.30	5.31	7.65	13.60

SPECIFICATIONS FOR REINFORCING STEELS

ASTM NUMBER	RANGE SIZES	GRADE DESIGNATION	TYPE OF STEEL	ULTIMATE TENSION PSI	MINIMUM YIELD PSI
A15	2 - 11	STRUCTURAL	NEW BILLET	55,000 TO 75,000	33,000
A15	2 - 11	INTERMEDIATE	NEW BILLET	70,000 TO 90,000	40,000
A15	2 - 11	HARD	NEW BILLET	80,000 MINIMUM	50,000
A160	2 - 11	STRUCTURAL	AXLE A	55,000 TO 75,000	33,000
A160	2 - 11	INTERMEDIATE	AXLE A	70,000 TO 90,000	40,000
A160	2 - 11	HARD	AXLE A	80,000	50,000
A16	2 - 11	REGULAR	RAIL I	80,000 MINIMUM	50,000
A61	3 - 11	DEFORMED	RAIL I	90,000 MINIMUM	60,000
A432	3 - 18S	HIGH STR'GTH.	H.S. BILLET N	100,000 MIN.	75,000
A431	3 - 18S	HIGH STR'GTH.	H.S. BILLET N	100,000 MIN.	75,000

GRAPH: Stress-strain curves for reinforced steels 4.2.3.2

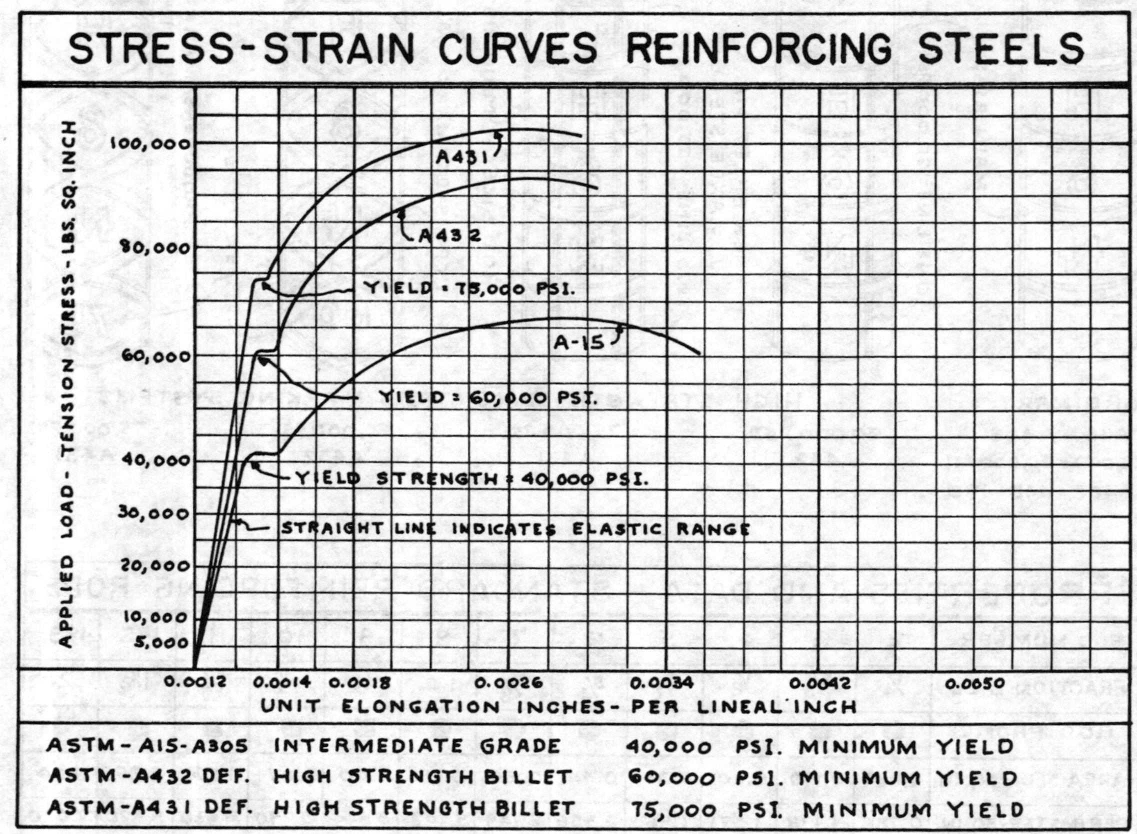

ILLUSTRATION: Reinforcing rod placing accessories 4.2.3.3

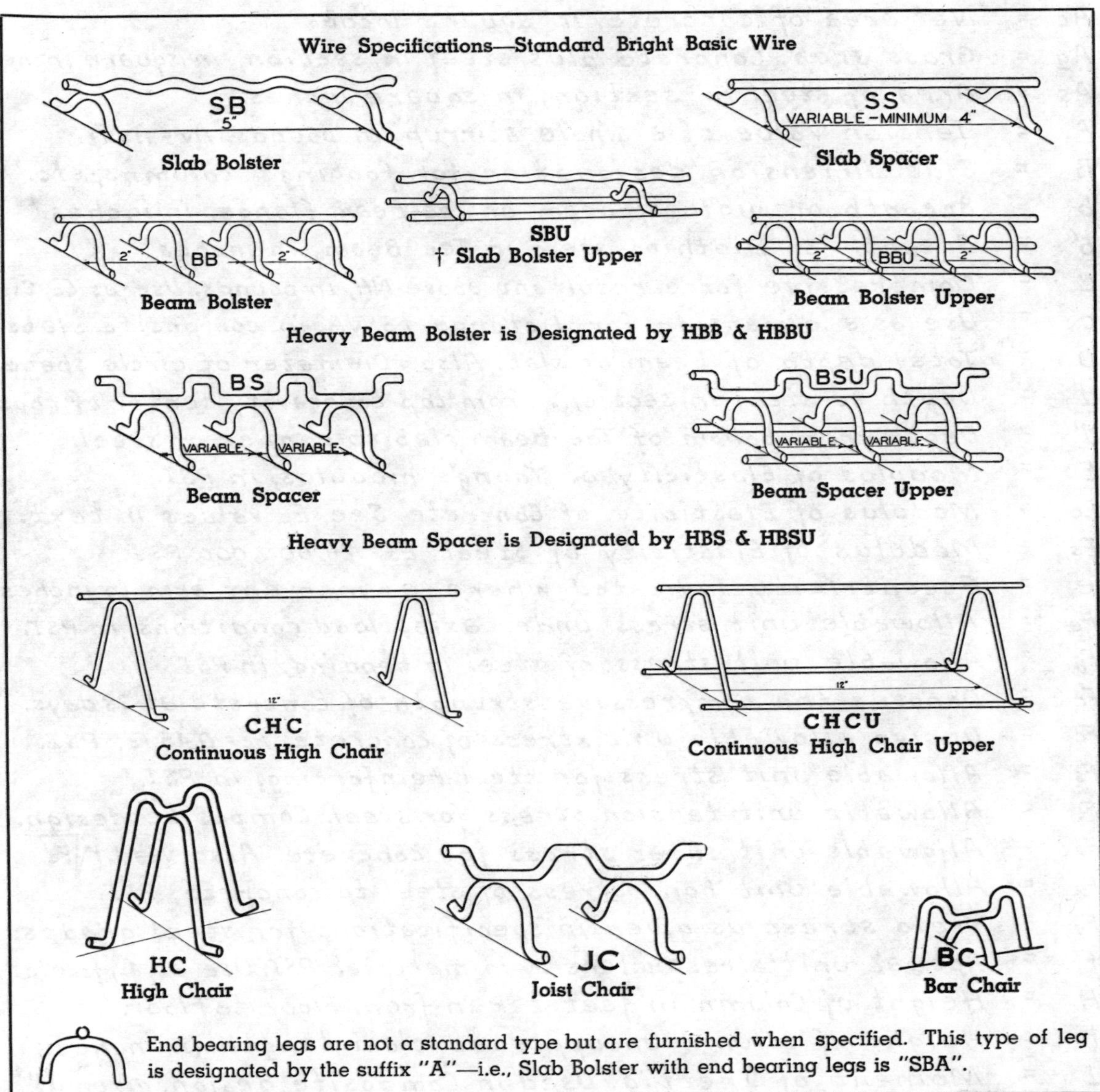

Galvanized legs furnished, when required, for small additional charge

Concrete design nomenclature 4.3.1

a = Distance from beam support which will require stirrups.
Ac = Net area of concrete, in square inches.
Ag = Gross area. Concrete plus steel in section, in square inches.
As = Area of steel in section, in square inches.
Av = Tension value of a single stirrup, in pounds. $A_v = A_s F_t$.
B = Side dimension designation for footings, columns, etc.
b = Breadth or width of beam or Tee-Beam flange, in inches.
b' = Breadth or width of stem in Tee-Beam, in inches
C = Compressive force resultant above NA, in pounds. Use as C_c, C_s.
c = Use as a dimension for distance to NA in composite slabs.
D = Total depth of beam or slab. Also = Diameter of circle shapes.
d = Depth to steel in section. From top to gravity center of rods.
d' = Depth from bottom of Tee-Beam slab to center of steel.
E = Modulus of Elasticity or Young's modulus, in PSI.
Ec = Modulus of Elasticity of Concrete. See E_c values in text.
Es = Modulus of Elasticity of Steel. E_s = 29,000,000 PSI.
e = Eccentricity indicated where e = moment arm in inches.
Fa = Allowable unit stress under axial load conditions, in PSI.
Fb = Allowable unit stress of steel in bending, in PSI.
Fc' = Denotes the compressive strength of concrete at 28 days.
Fc = Design allowable unit stress of concrete. $F_c = 0.45 F_c'$, PSI.
Fs = Allowable unit stress for steel reinforcing, in PSI.
Ft = Allowable unit tension stress for steel. Composite design.
Fv = Allowable unit shear stress for concrete. Also; $v_c = 1.1\sqrt{F_c'}$
Fu = Allowable unit bond stress of steel to concrete, PSI.
Fy = Yield stress as given in specifications for steel grades.
f = Actual unit stress intensity in material PSI. Use as: f_c, f_s, etc.
H = Height of Column in feet taken from Floor to Floor.
h = Height of Column unsupported clear length in inches.
I = Moment of Inertia. Used in Composite design, given I''^4
j = A design factor used to establish moment arm effective.
jd = Effective depth of concrete. Also distance C to T. In inches.
K = A design coefficient to determine b and depth to steel.
k = Ratio of distance, top fibers to NA and to effective depth.
kd = Dimension from top of beam to NA. Compression depth.
L = Length of span given in feet.
l = Length of span given in inches.
M = Indicates Moment or Resisting moment. In foot or inch lbs.
-M = Moment is negative or bending is in upper fibers.

Concrete design nomenclature, continued 4.3.1

+M = Positive bending moment, in foot or inch pounds.
M_x = Moment at specified point on beam or maximum moment.
M_w = Moment at wall resulting from wind pressure.
M_e = Eccentric bending moment or $P_e = M_e$.
M_{DL} = Moment resulting from dead loads only.
M_{LL} = Moment resulting from live loads only.
N = Number of required stirrups or force normal to surface.
NA = Neutral axis, centroid or gravity axis.
n = Ratio of modulus of elasticity, as: $n = \frac{E_s}{E_c}$. See tables.
o = Denotes a polar distance point for ray diagram.
P = Concentrated or axial load on beam or column.
P_e = Concentrated load with eccentric moment lever arm.
PSI = Pounds per square inch, also in examples as #□".
p = Percentage of steel reinforcing in relation to concrete.
R = Reaction at support or resultant of several forces.
r = Radius of gyration or radius of circular column or pipe.
S = Section modulus. Used in composite design with subscripts.
s = Spacing dimension for stirrups. In inches.
T = Tension force value in pounds. $T = A_s F_t$ or $T = C$.
t = Thickness of slab or flange of tee-beam. Also temperature.
t_1 = Thickness of slab or dropped panel in flat slab design.
u = Bond stress on rods, allowable or actual, in pounds sq. Inch.
V = Total vertical shear usually equals R. In pounds.
v = Unit shear stress intensity at support, in PSI. ($v = v_c + v'$).
v_c = Unit shear in Concrete, in PSI.
v_s = Unit shear for steel, in PSI.
v' = Unit shear to be resisted by stirrups, in PSI. ($v' = v - v_c$)
W = Total uniform distributed load on span in Pounds.
w = Uniform load per lineal foot on span, in Pounds.
x = Multiplication sign. Also to designate an unknown length.
y = Designates unknown dimension or moment lever arm.
z = Distance from top fibers to compression center = 1/3 kd.
# = Denotes pounds or rod number. Used in examples.
□ = Square rod in tables. In examples: PSI = #□" or PSF = #□'.
φ = Indicates a round rod, bolt, pipe, etc.
⌽ = Indicates a square reinforcing bar.
Σ = Sum total. Σo = Sum of rod perimeters in sq. In.
Δ = Deflection or deformation. Used in composite design.
± = Plus or Minus. More or less. Approximate value.

Modulus of elasticity for concrete 4.3.2

As concrete ages and gains strength, the modulus of elasticity will increase. It is assumed that at 28 days the required compressive strength is attained, and the modulus of elasticity should be measured at that time. Concrete strength at 28 days is identified by the symbol Fc′, and allowable unit stress is taken as forty-five percent of Fc′. Then Fc = 0.45 Fc′. These symbols must not be confused.

Recall that the deflection formulas derived to calculate the amount of sag in steel and wood beams used the modulus of elasticity (E). The ratio of the modulus of elasticity of steel (Es) to the modulus of elasticity of concrete (Ec) is the essential basis for concrete design. This ratio is expressed as $n = \frac{Es}{Ec}$. The modulus for concrete Ec may be found by the formula: $Ec = 57{,}255\sqrt{Fc'}$.

Fc′ = 2500 PSI at 28 day period.	Ec = 2,870,000 PSI
Fc′ = 3000 " " " " "	Ec = 3,150,000 PSI
Fc′ = 3500 " " " " "	Ec = 3,335,000 PSI
Fc′ = 4000 " " " " "	Ec = 3,625,000 PSI
Fc′ = 5000 " " " " "	Ec = 4,100,000 PSI
Fc′ = 5500 " " " " "	Ec = 4,400,000 PSI
Fc′ = 6000 " " " " "	Ec = 4,550,000 PSI

For all steels the modulus of elasticity Es = 29,000,000 PSI.

The breaking strength of the test cylinders of the selected mix will not be equal. The important thing in the mix design is that no test cylinder may break at a lower strength than specified.

Locating the neutral axis 4.3.3

The properties of rectangular sections explained in Section VI showed that under positive bending moment, the material above the neutral axis sustains compressive stress and the material below the neutral axis is under tensile stress. In a symmetrical cross-section when all of mass is of the same material, the neutral axis is located at one-half the depth of the section.

In a concrete section the concrete below the neutral axis is assumed not to carry any tensile stress; this stress is carried by the steel rods. Since the steel has the greater strength for both types of stress, the neutral axis will be positioned at some point where there is a balanced resisting stress area of equal potential value. A balanced design is achieved when the concrete above the neutral axis and steel below the neutral axis are both stressed to their respective safe working *allowable stress*. The location of the neutral axis depends on three conditions:

(a) The value of n, the ratio of Es to Ec.
(b) The allowable unit stress of the concrete, Fc.
(c) The allowable unit stress of the steel, or Fs.

The neutral axis may be located by formula, although a better understanding may be obtained when it is located by graphics. The dimension from the top of the beam to the neutral axis is denoted kd. By formula $kd = \frac{Fc \times d}{\frac{Fs}{n} + Fc}$. Divide both sides of the equation by the depth to steel (d) to obtain the design factor k. This design factor is used very often in concrete design, because of its importance in balancing the design and deriving the proper lever arm to calculate the resisting moment of a beam.

CENTER OF COMPRESSION ABOVE NEUTRAL AXIS

Often in architectural plans, the size of a beam may be restricted so that the area of concrete above the neutral axis available to resist compression is not sufficient. In this event, the area of concrete must be supplemented with steel. The compressive stresses are resolved into a single force C which acts at a center of gravity of the area in compression. The concrete and steel in compression are stressed in proportion to the ratio of their modulii of elasticity. Thus, if $n = \frac{Es}{Ec} = 10$, then the unit stress on the steel in compression would be ten times the unit stress on the concrete.

The center of gravity of the compression area C may be found as ⅓ altitude of the compression triangle. The height of this imagined triangle is kd, the distance from the top of the beam to the neutral axis, and the base of the triangle is width b. This triangle is illustrated in Example 4.3.3.1. ⅓ of the triangle height is equal to ⅓ of kd.

MOMENT LEVER ARM

When steel is used for both tension and compression the lever arm is used to find the resisting moment of the steel. This lever dimension is the vertical distance from the plane of the compression steel to the plane of the tension steel. It is frequently referred to as the *resisting couple.* As explained in Section I dealing with the mechanics of beams, there are two internal the external forces are the loads plus the dead load of beam. Following the set pattern, the lever is therefore perpendicular to the line of action of the internal forces.

EXAMPLE: Locating the neutral axis in rectangular beam 4.3.3.1

A rectangular concrete beam is 8.0" × 12.0" and depth to steel is 10.0 inches. Concrete design mix is to produce 3000 PSI at age of 28 day period. Allowable unit stress for steel is F_s = 18,000 PSI. E_s = 29,000,000 and E_c = 3,150,000 PSI.

REQUIRED:

Determine the position of Neutral Axis for positive bending moment by formulas, then confirm the location by graphic method of forces. Draw cross-section to convenient scale.

STEP I:

Allowable compressive stress for concrete is $F_c = 0.45 F_c'$ or $F_c = 0.45 \times 3000 = 1350$ PSI. F_s = 18,000 PSI and $n = \frac{E_s}{E_c}$ or

$n = \frac{29{,}000{,}000}{3{,}150{,}000} = 9.2$ $d = 10.0"$

By Formula: $NA = kd$ or $k = \frac{F_c}{\frac{F_s}{n} + F_c} = \frac{1350}{\frac{18{,}000}{9.2} + 1350} = 0.408$ In.

$kd = 0.408 \times 10.0 = 4.08$ inches.

STEP II:

The ratio of E_s to E_c equals: $n = 9.2$ and $\frac{F_s}{n} = \frac{18{,}000}{9.2} = 1967$ #□"

Cross section of beam is drawn to architectural scale and values will be measured with engineers scale.

On graphic section vertical line ab is drawn. Lay off on top of beam line fc equal to F_c or 1350 Lbs. Lay off from point e the ratio of $\frac{F_s}{n}$. This will be de = 1967 Lbs. Connect point c with point d by drawing a straight line cd. Where this line crosses ab will be location of Neutral Axis which is kd when used in formulas. The center of Compression C is 1/3 of kd = 1.36 inches from top of beam.

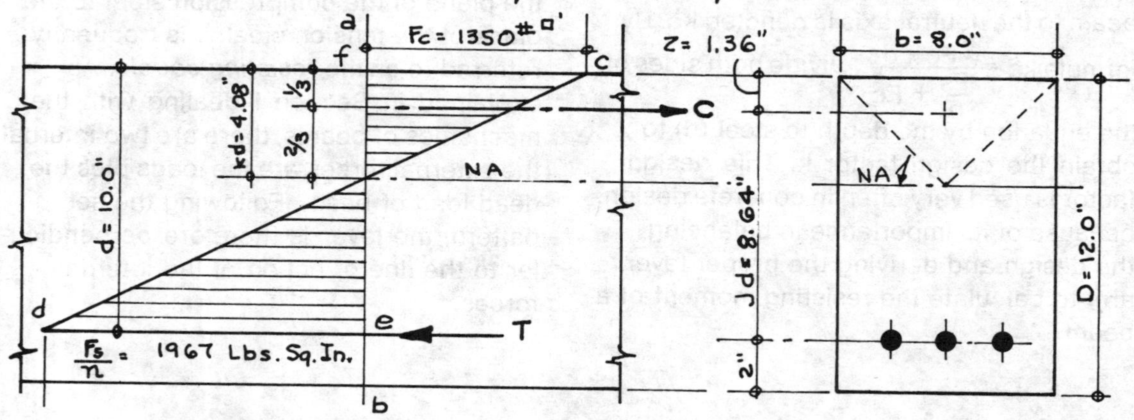

GRAPHIC METHOD
Scale: 1" = 1000 Lbs.

SECTION
Scale: 2" = 1.0 Ft.

EXAMPLE: Locating the neutral axis in tee-beam 4.3.3.2

A T-Beam has an effective flange width of 27.5 inches with a slab thickness $t = 4.00$ inches. Depth to steel $d = 12.0$ inches and stem width $b' = 12.0$ inches. $F_c' = 3000$ PSI, $F_s = 20{,}000$ PSI and $n = 9.2$

REQUIRED:

Sketch a cross section of T-Beam and calculate the location of Neutral Axis. Show this location in graphic form if desired.

STEP I:

Drawing section and making graphic illustration:

Jd is taken as: $Jd = d - \frac{t}{2}$ or $Jd = 13.10 - 2.00 = 9.10$ inches.

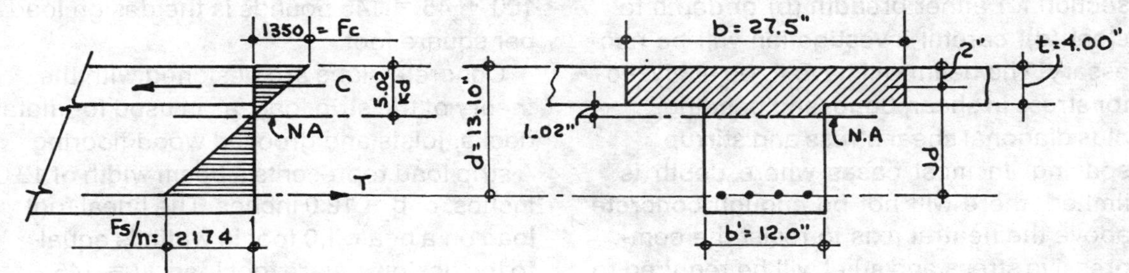

STEP II:

Calculate design factors: For T-Beams $J = d - \frac{t}{2}$ $F_c = 0.45 F_c'$

$$K = \frac{1.00}{1.00 + \left(\frac{F_s}{nF_c}\right)} = \frac{1.00}{1.00 + \left(\frac{20{,}000}{9.2 \times 1350}\right)} = 0.383 \qquad Jd = 13.10 - \left(\frac{4.0}{2}\right) = 11.10 \text{ inches}$$

check factors by solving for F_c. Formula: $F_c = \frac{F_s k}{n(1.00 - k)}$

$$F_c = \frac{20{,}000 \times 0.383}{9.2 \times (1.00 - 0.383)} = 1350 \text{ PSI (OK)}. \qquad kd = 0.383 \times 13.10 = 5.02 \text{ inches.}$$

Location of NA: $kd - t = 5.02 - 4.00 = 1.02$ inches below slab.

FORMULAS FOR TEE BEAMS

$$A_s = \frac{M}{F_s\left(d - \frac{t}{2}\right)} \quad \text{and} \quad f_c = \frac{2M}{bt\left(d - \frac{t}{2}\right)}$$

When the Neutral Axis occurs in the slab flange, use the formulas for design as given for rectangular beams.

When the Neutral Axis occurs in the stem, the amount of concrete area in stem above NA is usually small when compared area contained in flange. If desired, this small area may be neglected.

Effective depth 4.3.4

The term *effective depth* is used to describe the depth of concrete under stress; it corresponds to a moment arm or resisting couple. Effective depth is denoted as Jd. With d equal to depth to steel, J is found as $J = 1.0 - \frac{k}{3}$. In the preceding example, k = 0.408, which gives the value for J as:

$$J = 1.00 - \frac{0.408}{3} = 0.864.$$

The effective depth of the beam is therefore: Jd = 0.864 x 10.0 = 8.64 inches. Dimension z = ⅓ k = 1.36 in.

Beam and slab depth 4.3.5

Beam and girder depth may be fixed by architectural considerations. When the designer must accept a limited cross-section for either breadth (b) or depth to steel (d), careful investigation will be necessary. The beam will have to be analyzed for stress in shear, bond, and bending, plus diagonal shear stress and stirrup spacing. In most cases where depth is limited, there will not be enough concrete above the neutral axis to resist the compressive stress and steel will be required to carry this compressive load.

If the depth is not restricted, a useful formula to find the depth when the breadth has been established is to use the design coefficient K. The formula is: $K = \frac{Fc\, Jk}{2}$ and resisting moment $M = Kbd^2$. Transposing, $d^2 = \frac{M}{Kb}$, or $b = \frac{M}{Kd^2}$.

To illustrate the method for finding the design factor K, the formula requires that three values be known. Assume Fc = 1350 PSI. J = 0.864 and k = 0.408. With these values in the formula:

$$K = \frac{1350 \times 0.864 \times 0.408}{2} = 238.0$$

This checks with value 238.0 given in table 4.3.6.2 for Design Coefficients.

To further illustrate the use of this convenient factor K, assume that a floor slab has a simple span length of 8.0 feet. Live load to be supported is 100 pounds per square foot. General practice dictates that a slab of about 3 to 4 inches should be adequate, and the weight of slab would be about 45 pounds per square foot. Thus: 100 + 45 = 145 pounds is the design load per square foot.

Concrete slabs are designed with the theory of the strip load, as is used for metal decks, joists and grooved wood flooring. A strip load represents a beam width of 12.0 inches, or b = 12.0 inches. The lineal foot load on a beam 1.0 foot in width is equal to the design square foot load, w = 145 pounds per foot. W = wL or 145 x 8.0 = 1160 pounds. Bending moment $M = \frac{WL}{8}$ or $\frac{1160 \times 8.0}{8}$ = 1160 foot-pounds. Converting to inch-pounds: M = 1160 x 12 = 13,920 inch-pounds. Values can now be put in the formula to find the depth to steel (d):

$$d = \sqrt{\frac{M}{Kb}} = \sqrt{\frac{13{,}920}{238.0 \times 12.0}} = \sqrt{4.88} = 2.21$$

inches. To this depth, add a minimum of ¾ inch of concrete for embedment, rust and fire protection, which makes the total depth of the slab 3.0 inches. Note that the value of the design factors J and k vary depending on Fs. *This formula for depth to steel represents only the bending requirements, and further investigation must be made with respect to shear at the supporting ends*. When the depth (d) is known, and beam breadth (b) is required, the formula is transposed: $b = \frac{M}{Kd^2}$.

FORMULAS: for design factors 4.3.6

PRINCIPAL DESIGN COEFFICIENTS

$$k = \frac{n}{n + \frac{F_s}{F_c}} \qquad k = \frac{1.0}{1.0 + \frac{F_s}{nF_c}} \qquad k = \frac{F_c}{\frac{F_s}{n} + F_c} \qquad F_s = n\left(\frac{F_c}{k}\right) - F_c$$

$$J = 1.00 - \left(\frac{k}{3}\right) \qquad J = 1.00 - (0.333k) \qquad F_c = 0.45F_c' \qquad n = \frac{29{,}000{,}000}{57{,}255\sqrt{F_c'}}$$

$$F_c = \frac{F_s k}{n(1.00 - k)} \qquad F_s = \frac{nF_c(1.00 - k)}{k} \qquad K = \frac{F_c J k}{2} \qquad K = pF_s J$$

$$p = \frac{k F_c}{2F_s} \qquad p = \frac{K}{F_s J} \qquad p = \frac{A_s}{bd} \qquad n = \frac{E_s}{E_c}$$

F_c' = Concrete 28 day PSI.

DESIGN FORMULAS FOR RECTANGULAR SECTIONS

$$A_s = \frac{M}{F_s J d} \qquad A_s = pbd \qquad A_c = kdb \qquad Jd = \frac{M}{F_s A_s} \qquad b = \frac{M}{Kd^2} \qquad F_s = \frac{M}{A_s J d}$$

$$f_c = \frac{M}{pJbd^2} \qquad f_c = \frac{2M}{Jkbd^2} \qquad M_c = \frac{f_c k j b d^2}{2} \qquad M = A_s F_s J d \qquad z = \frac{kd}{3}$$

$$M_s = pf_s J d^2 b \qquad b = \frac{A_s}{pd} \qquad bd^2 = \frac{2M}{F_c J k} \qquad d = \sqrt{d^2} \qquad b = \frac{M}{Kd^2}$$

$$f_c = \frac{2pf_s}{k} \qquad d = \sqrt{\frac{M}{Kb}} \qquad bd^2 = \frac{M}{K} \qquad d^2 = \frac{M}{\left(\frac{F_c}{2}\right) J k b} \qquad C = \frac{kdbF_c}{2}$$

FORMULAS FOR SHEAR-BOND AND COMPRESSION

$$C = T \qquad T = A_s f_s J d \qquad C = \frac{f_c k d b}{2} \qquad -M = Kbd^2 \qquad R = V \qquad F_v = 1.1\sqrt{F_c'}$$

$$v = \frac{V}{Jdb} \qquad b = \frac{V}{Jdv} \qquad d = \frac{V}{Jbv} \qquad V' = V - F_v \qquad A_v = V'bs \qquad V = Jdb(v' + v_c)$$

$$u = \frac{V}{\Sigma o J d} \qquad \Sigma o = \frac{V}{Jdu} \qquad Jd = \frac{V}{\Sigma o\, u} \qquad u = \frac{vb}{\Sigma o} \qquad u = 3\sqrt{F_c'}$$

DISTANCE FOR WEB STIRRUPS AND SPACING

$a = \left(\frac{L}{2}\right) \times \left(\frac{V_c}{V'}\right)$. Value 1 Stirrup U Type, $A_v = A_\phi F_s 2$ or $2A_s f_s$.
Value 1 Stirrup W Type, $A_v = A_\phi F_s 4$ or $4A_s f_s$.

Stirrup spacing; $s = \frac{A_v}{v'b}$ Maximum spacing; $s = 0.45d$

Slab rod spacing; $s = \frac{A_\phi \times 12}{A_s}$ Inflection distance $= \frac{L}{5}$

FORMULAS: for bending moments in continuous spans 4.3.6.1

Reinforced concrete structural members are designed to resist bending moment using the theory of continuity. Beams and slabs of uniform length, freely supported or formed monolithically with the columns, and carrying uniform loads should be designed for maximum moments at critical locations as follows:

(a) Beams and slabs of one span: (simple spans).
  1. Maximum positive moment near midspan. .... $M = \frac{WL}{8}$

(b) Beams and slabs with two spans only:
  1. Maximum positive moment near midspans. .... $M = \frac{WL}{10}$
  2. Negative moment over interior support. $M = \frac{WL}{8}$

(c) Beams and slabs continuous over more than two spans:
  1. Maximum positive moment near mid-span and negative moment over supports of interior spans .... $M = \frac{WL}{12}$
  2. Maximum positive moment near center of end spans and negative moment at first interior support. .... $M = \frac{WL}{10}$

(d) Beams and slabs built into masonry walls which give partial end restraint, negative moment at support. .... $M = \frac{WL}{16}$

(e) Beams and slabs of equal spans *formed to act integrally with columns or other restraining supports*, and carrying uniform loads:
  1. Maximum positive moment near the center of interior spans. .... $M = \frac{WL}{16}$
  2. Negative moment at interior supports, except the first. .... $M = \frac{WL}{12}$

TABLE: Concrete design coefficients 4.3.6.2

$$k = \frac{1.0}{1.0 + \left(\frac{F_s}{n F_c}\right)} \qquad J = 1.0 - \frac{k}{3} \qquad K = \frac{F_c J k}{2} \qquad p = \frac{K}{F_s j} \qquad n = \frac{29{,}000{,}000}{57{,}255 \sqrt{F_c'}}$$

$F_c = 0.45 F_c'$ $F_v = 1.1\sqrt{F_c'}$ $u = 3\sqrt{F_c'}$ F_c' = Concrete PSI - Age 28 days.

u value based on deformed rods A15. $n = \frac{E_s}{E_c}$

F_c'	2500	3000	3500	4000	4500	5000	5500	6000
F_c	1125	1350	1575	1800	2025	2250	2475	2700
F_v	55	60	65	70	74	78	82	85
n	10.1	9.2	8.7	8.0	7.47	7.1	6.65	6.50
u	150	165	175	200	225	250	265	270
				F_s = 16,000 PSI				
k	0.415	0.437	0.452	0.474	0.475	0.500	0.506	0.517
K	201.0	252.0	301.3	359.0	410.0	468.0	518.0	580.0
j	0.862	0.854	0.849	0.842	0.842	0.833	0.831	0.828
p	0.0146	0.0184	0.0222	0.0266	0.0301	0.0351	0.0391	0.0436
				F_s = 18,000 PSI				
k	0.387	0.408	0.435	0.444	0.466	0.470	0.477	0.488
K	189.5	238.0	293.6	341.0	400.0	446.0	495.0	552.0
j	0.871	0.864	0.855	0.852	0.845	0.843	0.841	0.837
p	0.0121	0.0153	0.0191	0.0222	0.0263	0.0294	0.0327	0.0367
				F_s = 20,000 PSI				
k	0.362	0.383	0.408	0.419	0.432	0.444	0.452	0.462
K	179.0	226.0	278.0	324.0	375.0	426.0	473.0	516.0
j	0.879	0.872	0.864	0.860	0.856	0.852	0.849	0.846
p	0.0102	0.0129	0.0161	0.0188	0.0290	0.0250	0.0279	0.0305
				F_s = 22,000 PSI				
k	0.341	0.361	0.384	0.396	0.408	0.421	0.428	0.441
K	170.0	214.0	263.0	309.0	357.0	407.0	453.0	506.0
j	0.886	0.880	0.872	0.868	0.864	0.860	0.857	0.853
p	0.0087	0.0111	0.0133	0.0162	0.0188	0.0215	0.0239	0.0269
				F_s = 24,000 PSI				
k	0.321	0.341	0.364	0.375	0.387	0.400	0.407	0.418
K	161.0	204.0	252.0	295.0	341.0	390.0	435.0	486.0
j	0.893	0.886	0.879	0.875	0.871	0.867	0.864	0.861
p	0.0075	0.0096	0.0120	0.0141	0.0163	0.0187	0.0209	0.0235

TABLE: Concrete design coefficients, continued 4.3.6.2

F'_c	2500	3000	3500	4000	4500	5000	5500	6000
F_c	1125	1350	1575	1800	2025	2250	2475	2700
F_v	55	60	65	70	74	78	82	85
n	10.1	9.2	8.7	8.0	7.47	7.1	6.65	6.50
u	150	165	175	200	225	250	265	270
F_s = 27,000 PSI								
k	0.296	0.315	0.337	0.348	0.361	0.372	0.379	0.394
K	150.0	190.0	236.0	277.0	322.0	366.0	412.0	462.0
j	0.901	0.895	0.888	0.884	0.880	0.876	0.874	0.869
p	.0062	0.0079	0.0098	0.0116	0.0135	0.0155	0.0175	0.0194
F_s = 30,000 PSI								
k	0.275	0.293	0.313	0.324	0.336	0.347	0.364	0.369
K	140.0	178.0	220.0	260.0	302.0	346.0	389.0	438.0
j	0.908	0.902	0.896	0.892	0.888	0.884	0.879	0.877
p	0.0052	0.0066	0.0082	0.0097	0.0113	0.0130	0.0146	0.0167
F_s = 33,000 PSI								
k	0.256	0.273	0.294	0.304	0.315	0.320	0.332	0.346
K	132.0	168.0	208.0	246.0	286.0	327.0	363.0	413.0
j	0.915	0.909	0.902	0.899	0.895	0.891	0.889	0.885
p	0.0044	0.0056	0.0070	0.0083	0.0097	0.0111	0.0124	0.0143
F_s = 35,000 PSI								
k	0.245	0.262	0.274	0.292	0.302	0.314	0.320	0.333
K	126.0	161.0	196.0	237.0	276.0	316.0	354.0	400.0
j	0.918	0.913	0.909	0.903	0.899	0.895	0.893	0.889
p	0.00392	0.00505	0.00616	0.00748	0.00767	0.0110	0.0113	0.0129

TABLE: Total areas of steel for rod groups 4.3.6.3

TOTAL RODS IN GROUP	BAR DESIGNATION AND DIAMETER												As
	2	3	4	5	6	7	8	9	10	11	14S	18S	
	¼" ●	⅜" ●	½" ●	⅝" ●	¾" ●	⅞" ●	1" ●	1" ■	1⅛" ■	1¼" ■	1½" ■	2" ■	
1	0.05	0.11	0.20	0.31	0.44	0.60	0.79	1.00	1.27	1.56	2.25	4.00	AREA IN SQUARE INCHES
2	0.10	0.22	0.40	0.62	0.88	1.20	1.57	2.00	2.54	3.12	4.50	8.00	
3	0.15	0.33	0.60	0.93	1.32	1.80	2.37	3.00	3.81	4.68	6.75	12.00	
4	0.20	0.44	0.80	1.24	1.76	2.40	3.16	4.00	5.08	6.24	9.00	16.00	
5	0.25	0.55	1.00	1.55	2.20	3.00	3.95	5.00	6.35	7.80	11.25	20.00	
6	0.29	0.66	1.20	1.86	2.64	3.60	4.74	6.00	7.62	9.36	13.50	24.00	
7	0.33	0.77	1.40	2.17	3.08	4.20	5.53	7.00	8.89	10.92	15.75	28.00	
8	0.38	0.88	1.60	2.48	3.52	4.80	6.32	8.00	10.16	12.48	18.00	32.00	
9	0.43	0.99	1.80	2.79	3.96	5.40	7.11	9.00	11.43	14.04	20.25	36.00	
10	0.48	1.10	2.00	3.10	4.40	6.00	7.90	10.00	12.70	15.60	22.50	40.00	
11	0.54	1.21	2.20	3.41	4.84	6.60	8.69	11.00	13.97	17.16	24.75	44.00	
12	0.59	1.32	2.40	3.72	5.28	7.20	9.48	12.00	15.24	18.72	27.00	48.00	
13	0.64	1.43	2.60	4.03	5.72	7.80	10.27	13.00	16.51	20.28	29.25	52.00	
14	0.69	1.54	2.80	4.34	6.16	8.40	11.06	14.00	17.78	21.84	31.50	54.00	
15	0.74	1.65	3.00	4.65	6.60	9.00	11.85	15.00	19.05	23.40	33.75	60.00	
16	0.79	1.77	3.20	4.96	7.04	9.60	12.64	16.00	20.32	24.96	36.00	64.00	

TABLE: Slab rod spacing for required steel area 4.3.6.4

ROD SIZE	ROD DIA.	SPACING OF RODS PER FOOT OF WIDTH - IN INCHES																					A_s
		2	2½	3	3½	4	4½	5	5½	6	6½	7	7½	8	8½	9	9½	10	10½	11	11½	12	AREA IN SQUARE INCHES
2	¼"	.30	.240	.200	.173	.150	.133	.120	.102	.100	.092	.086	.080	.075	.071	.066	.063	.060	.057	.054	.052	.05	
3	⅜"	.66	.527	.440	.377	.333	.395	.267	.242	.220	.205	.190	.177	.165	.156	.147	.139	.132	.125	.120	.115	.11	
4	½"	1.20	.960	.800	.685	.600	.533	.480	.436	.400	.370	.343	.320	.300	.282	.267	.253	.240	.228	.218	.208	.20	
5	⅝"	1.86	1.49	1.24	1.06	.930	.827	.745	.677	.620	.572	.533	.497	.465	.437	.413	.392	.372	.354	.338	.323	.31	
6	¾"	2.64	2.11	1.76	1.51	1.32	1.17	1.06	.960	.880	.812	.755	.705	.660	.622	.586	.557	.528	.503	.481	.460	.44	
7	⅞"	3.60	2.88	2.40	2.06	1.80	1.60	1.44	1.31	1.20	1.11	1.02	.960	.900	.847	.800	.757	.720	.686	.655	.625	.60	
8	1.0"	4.74	3.79	3.16	2.72	2.37	2.12	1.89	1.73	1.58	1.46	1.35	1.26	1.18	1.11	1.05	1.00	.948	.900	.862	.825	.79	
9	1.0"	6.00	4.80	4.00	3.43	3.00	2.67	2.40	2.18	2.00	1.84	1.73	1.60	1.50	1.41	1.33	1.26	1.20	1.15	1.08	1.04	1.00	
10	1⅛"	7.62	6.10	5.08	4.35	3.81	3.39	3.05	2.77	2.54	2.35	2.18	2.03	1.91	1.79	1.70	1.61	1.52	1.45	1.39	1.32	1.27	

SPACING FORMULA:

$$S = \frac{\text{Area 1 Rod} \times 12 \text{ Inches}}{\text{Required } A_s}$$

AREA OF STEEL FORMULA:

$$A_s = \frac{\text{Area 1 Rod} \times 12 \text{ Inches}}{\text{Spacing } (s)}$$

AREA OF 1-ROD GIVEN IN 12" COLUMN

TABLE: Perimeter summations for rod groups 4.3.6.5

TOTAL RODS IN GROUP	ROD DIAMETER AND BAR DESIGNATION												
	2	3	4	5	6	7	8	9	10	11	14S	18S	Σ_o
	1/4" ●	3/8" ●	1/2" ●	5/8" ●	3/4" ●	7/8" ●	1" ●	1" ■	1 1/8" ■	1 1/4" ■	1 1/2" ■	2" ■	
1	0.786	1.178	1.571	1.963	2.356	2.749	3.142	3.544	3.990	4.430	5.320	7.090	AREA IN SQUARE INCHES
2	1.57	2.35	3.14	3.93	4.71	5.50	6.28	7.09	7.98	8.86	10.64	14.18	
3	2.36	3.53	4.71	5.89	7.07	8.25	9.43	10.63	11.97	13.29	15.96	21.27	
4	3.14	4.71	6.28	7.85	9.42	11.00	12.57	14.18	15.96	17.72	21.28	28.36	
5	3.93	5.89	7.85	9.82	11.78	13.75	15.71	17.72	19.95	22.15	26.60	35.45	
6	4.72	7.07	9.43	11.78	14.14	16.49	18.85	21.26	23.94	26.58	31.92	42.54	
7	5.50	8.25	11.00	13.74	16.49	19.24	21.99	24.81	27.93	31.00	37.24	49.63	
8	6.29	9.42	12.57	15.70	18.85	21.99	25.14	28.35	31.92	35.44	42.56	56.72	
9	7.07	10.60	14.14	17.67	21.20	24.74	28.28	31.90	35.91	39.87	47.88	68.31	
10	7.86	11.78	15.71	19.63	23.56	27.49	31.42	35.44	39.90	44.30	53.20	70.90	
11	8.65	12.96	17.28	21.59	25.92	30.24	34.56	38.98	43.89	48.73	58.52	78.00	
12	9.43	14.14	18.85	23.55	28.27	32.99	37.70	42.53	47.88	53.16	63.84	85.08	
13	10.22	15.31	20.42	25.52	30.63	35.74	40.85	46.07	51.87	57.59	69.16	92.17	
14	11.00	16.49	21.99	27.48	32.98	38.49	44.00	49.62	55.86	62.02	74.48	99.26	
15	11.79	17.67	23.57	29.45	35.34	41.23	47.13	53.16	59.85	66.45	79.80	106.35	
16	12.58	18.85	25.14	31.41	37.70	43.98	50.27	56.70	63.84	70.88	85.12	113.44	

Slab design by moment coefficient table 4.3.7

A coefficient is generally considered to be a constant applying to some formula. A better understanding is possible if it were explained that a coefficient is intended to be a true figure which unites other numerical values.

Table 4.3.8.2 is a guide for the rapid selection of slab depth in one-way reinforced slabs and also the steel rod size and spacing. An example will illustrate the derivation of these coefficients. Note that this table is based on allowable stress in concrete and steel of $Fc' = 3000$ PSI and $Fs = 20{,}000$ PSI. Dead load of concrete is taken as 12 pounds per inch depth per square foot, or 144 pounds per cubic foot. The allowance for reinforcing bar concrete cover varies; for rust and fire protection, the tabulated value is usually adequate. In some texts, live load is referred to as superimposed load, which is taken to include the load from interior walls and other permanent fixtures.

To arrive at the safe live load on a given span, select a moment coefficient suitable to the span end conditions. Square the span length (L) in feet, and divide the result into the moment coefficient. Then deduct the slab dead load. To illustrate:

Assume L = 16.0 feet. For a continuous span, the bending moment is $M = WL/12$. Desired live load is 100 pounds per square foot. Refer to Table 4.3.7.2 and note there is a 20 pound difference on a 6½ inch slab spanning 14.0 or 15.0 feet. From the table, 46000 is the moment coefficient. Then

$$\text{safe load} = \left(\frac{46{,}000}{16.0 \times 16.0}\right) - 78 = 102$$

pounds per square foot and acceptable.

In continuous slabs there is a negative bending moment over the supports which requires tensile reinforcement in the top of the slab. This reinforcement can be provided by bending up alternate bottom rods at a point 1/5 of the span out from the support and extending them over the support to the 1/5 point of the adjacent span. This design for negative bending will be shown by detailed drawings in the slab design examples.

EXAMPLE: Derivation of slab moment coefficients 4.3.7.1

A 1-Way reinforced slab is to have a total depth of 4.0 inches and span lengths are varied throughout the project from 7.0 feet to 14.0 feet. Specifications call for concrete to be 3000 PSI at 28 days, and steel Fs = 20,000 PSI No. 3 φ size rods are to be used spaced 4½" centers. Depth to steel d = 3.25 inches with ¾" protection.

REQUIRED:

Calculate the Resisting Moment of a strip load cross section 12.0 inches wide. Let span L = 1.0' and compute moment coefficient for a simple span, end span and continuous span. Determine whether a safe live load of 100 PSF can be supported on continuous spans of 10.0 feet.

STEP I:

From table of Design Factors, collect the following values:

$F_c = 0.45 \times 3000 = 1350$ PSI. $F_s = 20{,}000$ PSI. $k = 0.383$ $K = 226.0$, $b = 12.0"$

$j = 0.872$ $F_v = 60$ PSI $u = 164$ PSI. $n = 9.2$ (Not all are required)

Area of 1 #3 φ Rod = 0.11 Sq. In. Perimeter #3 φ Rod = 1.18 □" d = 3¼"

Area of steel in strip when spaced 4.50" cc. $A_s = \frac{A\phi \times b}{s}$, or

$A_s = \frac{0.11 \times 12}{4.50} = 0.294$ □"

STEP II:

Resisting Moment = As Fs Jd. Will be converted to foot pounds.

$RM = \frac{0.294 \times 20{,}000 \times 0.872 \times 3.25}{12} = 1390$ Ft. Lbs.

STEP III:

RM = M. L = 1.0' and $M = \frac{WL}{8}$ or $\frac{WL}{10}$ and continuous span, $M = \frac{WL}{12}$

For Simple Span, $\frac{WL}{8}$: Moment Coefficient = 1390 x 8 = 11,120

For End spans, $\frac{WL}{10}$: Moment Coefficient = 1390 x 10 = 13,900

For Continuous span $\frac{WL}{12}$: Moment Coefficient = 1390 x 12 = 16,680

STEP IV:

To determine safe Live Load on Continuous spans of 10.0 Ft.

Dead load slab = 4.0 x 12# = 48 Lbs. Sq. Ft. $L^2 = 10.0 \times 10.0 = 100.0$

Safe LL = $\left(\frac{16{,}680}{100.0}\right) - 48 = 118.8$ Lbs. Sq. Foot. W = 1188 Lbs.

STEP V:

Maximum shear for cross-section: $V = F_v jdb$. In formula:

Max. V = 60 x 0.872 x 3.25 x 12.0 = 2040 Lbs. at supports.

TABLE: Slab moment coefficients and safe loads for one-way slabs 4.3.7.2

CONCRETE 28 DAYS: F'_c = 3000 PSI. F_c = 0.45 F'_c = 1350 PSI j =
F_s = 20,000 PSI. LIVE LOAD PER SQ. FT. = $\frac{\text{Coefficent Mom.}}{L^2}$ − DL.

SLAB D IN.	STEEL d IN.	SIZE ROD #	TOT. As □"	SPACE RODS IN.	DEAD LOAD SLAB	MOMENT COEFF'T FLEX.	M SPAN TYPE	L = LENGTH OF SPAN IN FEET 5.0	6.0	7.0	8.0	9.0	10.0	11.0	12.0	13.0	14.0	15.0
2½	1.75	3	0.19	7	30#	3867	WL/8	125	87	49	30							
						4833	WL/10	163	104	68	45							
						5800	WL/12	202	132	88	60							
3	2.25	3	0.22	6	36#	5750	WL/8	194	124	81	54	35						
						7200	WL/10	252	164	102	70	47						
						8620	WL/12	309	204	140	100	70						
3½	2.75	3	0.24	5½	42#	7660	WL/8	266	171	115	78	53	35	21				
						9600	WL/10	342	224	154	108	76	58	37				
						11500	WL/12	418	278	193	138	100	73	53				
4	3.25	3	0.29	4½	48#	10960	WL/8	389	256	176	124	87	62	42	28	17		
						13700	WL/10	500	332	232	172	127	95	71	47	39		
						16440	WL/12	609	407	267	208	155	116	88	66	48		
4½	3.75	4	0.32	7½	54#	13900	WL/8		332	230	163	118	85	61	42	28	17	
						17400	WL/10		423	296	214	158	120	88	65	48	34	
						20900	WL/12		526	373	273	204	155	119	91	70	52	
5	4.00	4	0.37	6½	60#	17200	WL/8			290	209	152	112	82	60	42	28	17
						21500	WL/10			400	292	218	155	121	96	73	55	40
						25900	WL/12			468	345	260	200	154	120	93	72	55
5½	4.50	4	0.43	5½	66#	22800	WL/8				290	216	162	122	92	69	50	36
						28500	WL/10				379	286	219	170	132	102	79	61
						34200	WL/12				469	357	276	217	172	136	108	86
6	5.00	4	0.48	5	72#	27900	WL/8						207	158	121	93	70	52
						34900	WL/10						277	206	162	127	100	77
						41850	WL/12						346	273	218	176	141	114
6½	5.50	4	0.48	5	78#	30600	WL/8								134	104	78	58
						38300	WL/10								188	148	117	92
						46000	WL/12								242	194	157	137
7	6.00	5	0.57	6½	84#	39,850	WL/8									152	119	93
						49750	WL/10									210	170	137
						59700	WL/12									270	221	182
7½	6.50	5	0.62	6	90#	46800	WL/8										149	118
						58500	WL/10										225	170
						70250	WL/12										268	222
8	7.00	6	0.62	6	96#	50500	WL/8										162	128
						63000	WL/10										226	184
						75600	WL/12										290	240

Expansion and contraction of concrete 4.4

Although steel and concrete are very different in physical properties, the two materials are compatible in structures under stress because they exhibit similar coefficients of thermal expansion. Section II discusses the thermal expansion of steel. The average coefficient of linear expansion of steel (A36) is 0.0000065 per inch for each degree of temperature change. The coefficient for concrete is 0.0000079 per degree Fahrenheit. The difference in these values is so slight that there is little effect on the bond between steel and concrete. For ordinary building structures, the effect of temperature change is usually neglected, but for large structures, such as dams, bridges and dock terminals, provisions must be included to reduce the effect of expansion and contraction.

A reinforced slab or beam, when free to move, will increase in length with a rise in temperature and decrease in length when the temperature drops. Maintaining complete control of concrete temperature during the early days of curing is a requirement for a satisfactory structure.

Slabs more than any other form of concrete are susceptible to expansion or contraction cracks. This is especially true when the flat surface is exposed to the sun's heat during the day. The rapid cooling during the night may drop the surface temperature by 40°F. There are many precautions to prevent cracks in slabs, which can also be applied to other formed members. These precautions are:

(a) The wet concrete must contain a minimum of water: only enough to provide workability.
(b) Slump tests should be made to see that excessive water is not present in the mix. Excessive water is responsible for shrinkage cracks during curing.
(c) Monolithic areas should not be too large.
(d) Immediately after the finished surface is trowelled, the entire surface must be wetted with water and kept wet. This can be done by damming the edges, and flooding or continuously spraying with perforated pipes connected to a small pump.
(e) The designer should include a generous quantity of perpendicular rods for temperature steel.
(f) A curing compound should not be used as a substitute for water curing.
(g) Freshly placed concrete must not be left unattended, if the curing period extends into a holiday or weekend.
(h) The design should make use of expansion and contraction joints with resilient properties between pours. Construction butt joints may be used between expansion joints when work stoppages are unavoidable.

Designing for expansion 4.4.1

Heavy marine structures and concrete turnpikes must be designed for the contraction and expansion of the full weather season. In many regions the ambient temperatures may vary from −30 to 100°F. This amounts to a 130 degree change and must be considered, especially in large structures. The apprentice design engineer or architect may not be greatly concerned with expansion and contraction problems until he must explain an expansion crack to a dissatisfied client.

Thermal expansion in a rigid concrete section will be investigated in Example 4.4.1.1. Note that this section of dock is only a typical part of a structure with total length of 1350 feet. The entire project was constructed in the summer of 1965, with daytime temperatures of 116°F. Since completion the lowest winter temperature recorded was 20°F. Not one of the nine sections has crack problems.

EXAMPLE: Thermal expansion joints in dock 4.4.1.1

Illustrated is a typical section of marine cargo dock with a length of 152'-4". Width each section = 52'-4". Dimensions were taken inside forms before placing concrete. Entire project including piles is concrete. Recorded temperature at time of placing concrete was 70° Fahr. (mean). Structure is exposed to air temperature on all surfaces. Highest recorded temperature since completion was 115° F., and lowest reading was 20° Fahr.

REQUIRED:
Calculate the dimensions for contraction and expansion in the long direction due to temperature changes given. State the requirements and safe width of expansion joint.

STEP I:
From table in Section II for linear coefficients:
Concrete; c = 0.0000079 inches per inch per degree F.
Reinforcing: c = 0.0000067 " " " " "
Dimensions established when mean temperature was 70° F.
Heat expansion change: = t = 100° − 70° = 30°
Contraction change under mean: t = 70° − 20° = 50° (Greater.)
L = 152.33 Ft. ℓ = 152.33 × 12 = 1828 inches. Deformation Δ = ctℓ.

STEP II:
Concrete expansion in 30°: e = 0.0000079 × 30 = 0.000237 in. per inch.
Concrete contraction in 50°: s = 0.0000079 × 50 = 0.000395 " " "
Steel Rod expansion in 30°: e = 0.0000067 × 30 = 0.000201 " " "
Steel Rod contraction in 50°: s = 0.0000067 × 50 = 0.000335 " " "

STEP III:
For contraction and elongation for full length of ℓ:
Concrete elongation = 0.000237 × 1828 = 0.433 inches
Concrete shrinkage = 0.000395 × 1828 0.722 "
steel elongation = 0.000201 × 1828 0.367 "
steel shrinkage = 0.000335 × 1825 0.611 "

STEP IV
Concrete shrinkage will increase steel stress which will be neglected. Contraction is 0.722 − 0.611 = 0.111 greater than steel over whole length of 152'-4" which will not hinder bond.
Total dimension change over 80° = 0.433 + 0.722 = 1.155 inches.
Min. expansion Joint = ½", At lowest temperature the joint will open to: 0.50 + 0.433 = 0.933 inches at end of each section.

EXAMPLE: Thermal expansion joints in dock, continued 4.4.1.1

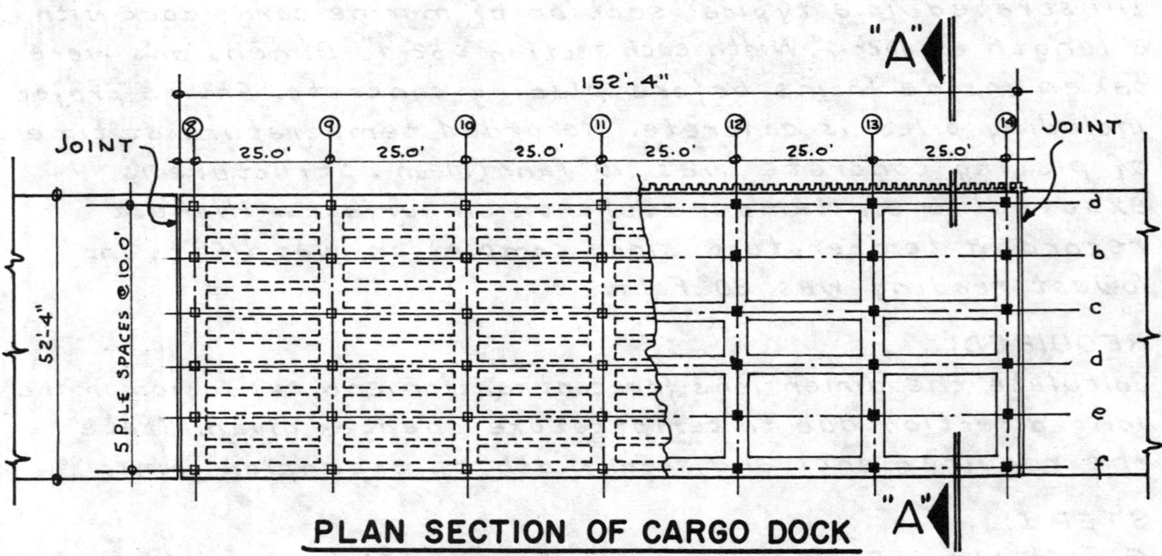

PLAN SECTION OF CARGO DOCK

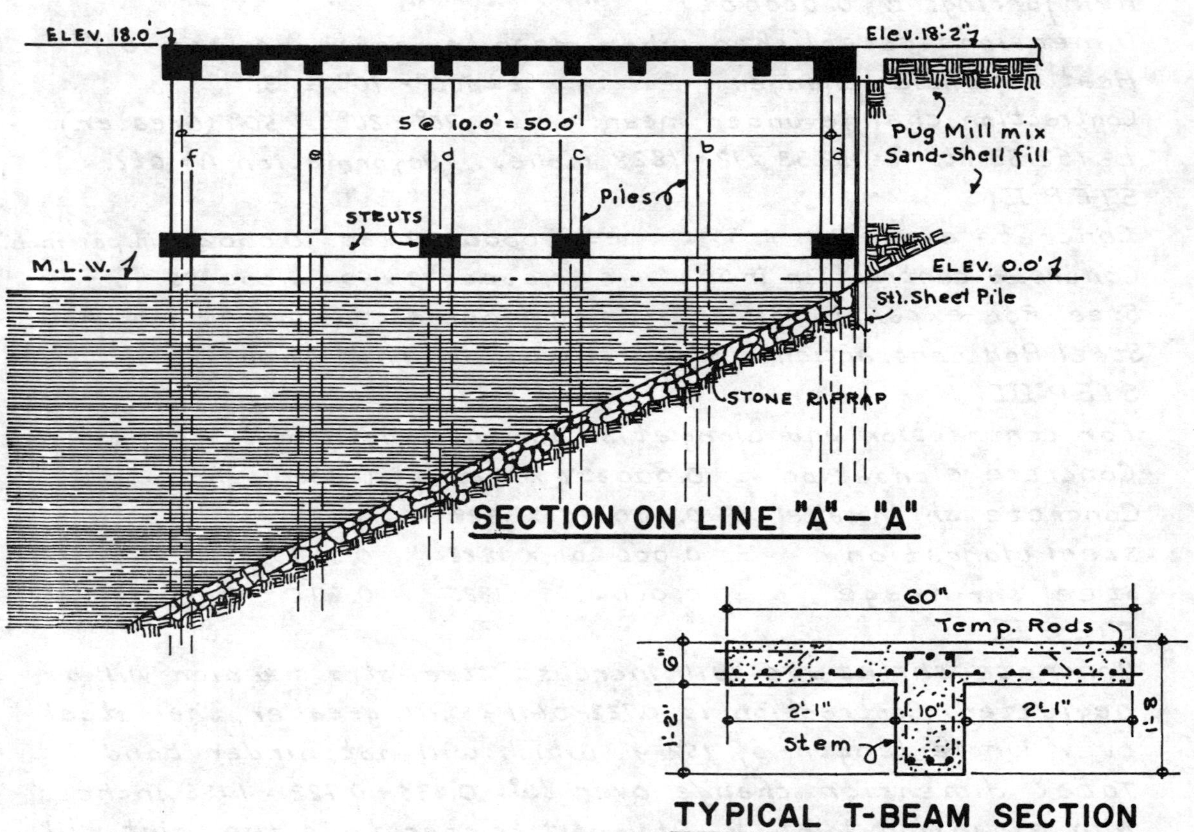

SECTION ON LINE "A"—"A"

TYPICAL T-BEAM SECTION

Internal temperature stress 4.4.2

Heavy industrial structures such as marine wharfs, grain elevators, car parking structures, dams and overhead viaducts are exposed to ambient temperature changes. There is no control of temperature, as there is for buildings enclosed and provided with mechanical heating and cooling. Contraction and expansion therefore must be of greater concern to the designers of exposed structures.

In Example 4.4.1.1 a concrete dock was used to illustrate exposure to seasonal temperature changes. Relief or expansion joints were provided to open in cold weather and close when temperature rises. During this movement the different expansion coefficients of steel and concrete cause an internal stress in the member. The area of steel is required to resist the tension stress from the external loads plus the temperature stress. When the design takes temperature stress into consideration, the stress may rise above the allowable working stress and the yield point and elastic limit become the important limits.

Stress due to temperature change is found by the simple formula: $f_t = Etc$; where E = modulus of elasticity and t = change in temperature in degrees. The coefficient c is taken from Table 2.2.4.5 in Section II, and is the deformation in inches per inch of length for each degree of temperature change.

To illustrate:

ES = 29,000,000, temperature change t = 70° and linear coefficient of expansion for steel c = 0.0000067 per degree. Then the stress in steel is: f_{st} = 29,000,000 x 70 x 0.0000067 = 13,601 PSI.

Elongation per unit of length e = ct. A length of steel with a length l = 500 inches, will have a total deformation: $\Delta = ctl$ or Δ = 0.0000067 x 70 x 500 = 0.2345 inches.

Temperature reinforcement in slabs 4.4.2.1

Shrinkage and temperature rods for stresses perpendicular to the direction of principal bending reinforcement must be provided in one-way slabs for floor and roof. Shrinkage cracks are not as critical a problem in the short dimension as the main rod design which runs the length of the slab. Temperature reinforcement, placed normal to the principal rods, must meet the local Code requirements. If the Code does not call for temperature reinforcement, it should be installed in accordance with the percentage ratio given below. Temperature steel normal to the principal reinforcing should never be spaced over five times the slab depth or more than 12 inches apart. See Example 4.4.2.4 for designing slab temperature rods.

TEMPERATURE REINFORCEMENT FOR CONCRETE SLABS

TYPE OF SLAB AND LOCATION	TYPE STEEL REINFORCEMENT	SPECIFIED MAX. YIELD STRESS PSI	PROPORTIONATE RATIO TO CONCRETE
ROOF	PLAIN RODS NO. 2	33,000 TO 50,000	0.0030 A_c
ROOF	DEFORMED RODS	33,000 TO 60,000	0.0025 A_c
ROOF	W. WIRE MESH FABRIC	60,000	0.0022 A_c
FLOOR	PLAIN RODS	33,000 TO 50,000	0.0025 A_c
FLOOR	DEFORMED RODS	33,000 TO 60,000	0.0020 A_c
FLOOR	W. WIRE MESH FABRIC	60,000	0.0018 A_c
RIBBED AND T-BEAMS	DEFORMED RODS	33,000 TO 50,000	0.0020 A_c

SLABS WITH TOP AND BOTTOM SURFACE EXPOSED TO AMBIENT CHANGE IN TEMPERATURES DURING SEASON SHALL BE INCREASED 25 PERCENT A_s.

TABLE: Properties of wire mesh fabric 4.4.2.2

PROPERTIES WIRE MESH FABRIC- ELECTRICALLY WELDED						
MESH SPACING IN INCHES		STANDARD WIRE GAGE NUMBER		SECTIONAL AREA SQUARE IN. PER FOOT		WEIGHT PER 100 SQ. FEET
LONGITUDINAL	TRANSVERSE	LONGITUDINAL	TRANSVERSE	LONGITUDINAL	TRANSVERSE	
4	12	10	12	0.043	0.009	18.6
4	12	6	6	0.087	0.029	41.6
4	4	4	4	0.120	0.120	85.3
4	4	6	6	0.087	0.087	61.9
4	4	8	8	0.062	0.062	44.1
6	8	12	12	0.017	0.013	11.1
6	12	4	4	0.080	0.040	43.8
6	6	4	4	0.080	0.080	57.8
6	6	5	5	0.067	0.067	48.8
6	6	6	6	0.058	0.058	42.0
6	6	8	8	0.049	0.049	35.7
6	6	9	9	0.035	0.035	25.0
6	6	10	10	0.029	0.029	20.7
2	2	12	12	0.052	0.052	36.8

SIZES IN TABLE ARE GENERAL STOCK ITEMS - OTHER SIZES AVAILABLE

EXAMPLE: Internal temperature stress in beam 4.4.2.3

Refer to Concrete Dock plan and Section through the T-Beam. This beam is assumed to be rigid and restrained against the ends. Let temperature be 90° Fahr., which will soon drop to 20°. T-Beam concrete is 3000 PSI at 28 day period. L = 25.0 Feet. Yield stress for steel = 50,000 PSI.

REQUIRED:

Calculate the internal unit stress in concrete and steel when temperature changes 70 degrees when L = 25.0 at 90° temperature. $E_s = 29{,}000{,}000$ $E_c = 3{,}000{,}000$. Coefficients for each degree of change are: Unit = 1.0 inch.

Concrete; c = 0.0000079 Steel; c = 0.0000067 inches per degree

STEP I

Put length into inches or $l = 25.0 \times 12 = 300$ inches (1 Bent only)

$t = 90° - 20° = 70°$

Total Conc: $\Delta = 0.0000079 \times 70 \times 300 = 0.1659$ inches

Total Steel: $\Delta = 0.0000067 \times 70 \times 300 = 0.1407$ "

STEP II:

Assuming that concrete does not contain steel reinforcing, the stress in concrete = Etc.

$f = 3{,}000{,}000 \times 70 \times 0.0000079 = 1659$ Lbs. Sq. Inch.

Stress is shrinkage or tension and T-Beam will break. The steel must resist all tension stress.

STEP III:

When steel is stressed same as concrete with $\Delta = 0.1659$ inches for full length, then $f_s = \dfrac{E\Delta}{l}$ or $f_s = \dfrac{29{,}000{,}000 \times 0.1659}{300} = 16{,}037$ PSI.

May also be found thus: $f_s = Etc$, or

$f_s = 29{,}000{,}000 \times 70 \times 0.0000079 = 16{,}037$ PSI.

STEP IV:

Stress from temperature change must be added to the unit bending stress produced from dead and live loads. These stresses must not total greater in amount than the yield stress. A15 Hard Billet steel has a yield of $F_y = 50{,}000$ PSI and design allowable maximum is 20,000 PSI. Thus: 16,037 + 20,000 = 36,037 PSI and safely under the yield.

EXAMPLE: Internal temperature stress in temperature steel 4.4.2.4

In the preceding example we investigated temperature stress and deformation in long dimension of Cargo dock. Temperature conditions are the same in short direction which is 52.33 feet and change is 70° Fahr.

REQUIRED:

Refer to table for temperature reinforcement in slabs and obtain the proportionate ratio of temperature steel to area of concrete. Observe Tee-Beam section and assume a strip of slab 12.0 inches wide for typical design. Calculate the As and required spacing. Compute the stress in steel with the 70° temperature change.

STEP I:

Depth of T-Beam slab is 6.00 inches and width assumed = 12.0"
Area concrete: $A_c = 6.0 \times 12.0 = 72.0$ Sq. Inches. Percentage ratio of Ac to As = 0.002 Ac. When all surfaces are exposed the ratio is increased 25%.
$A_s = 0.002 \times 72.0 \times 1.25 = 0.180$ □" for each 12.0 inch strip.

STEP II:

Try using #4 φ Rods where $A\phi = 0.20$ Sq. in.
Spacing $s = \frac{A\phi \times 12}{A_s}$ or $s = \frac{0.20 \times 12}{0.180} = 13.33$ inch centers.
Spacing is too wide since 12 inches is limit allowed by code.

STEP III:

Re-figure spacing by using a smaller rod. A #3 φ has $A\phi = 0.11$ □"
spacing $= \frac{0.11 \times 12}{0.180} = 7.33$ inches. Use 7⅜ inch spacing c-c.

STEP IV:

For temperature stress when change t = 70 degrees. $f_t = E_s t c$.
$E_s = 29{,}000{,}000$ PSI $c = 0.0000067$ inches per inch per degree.
$f_s = 29{,}000{,}000 \times 70 \times 0.0000067 = 13{,}600$ PSI. This is well within the yield range of Fy = 33,000 PSI. Temperature stress is same in all directions.

STEP V:

Total deformation (elongation or shrinkage) = Δ L = 52.33 Ft.
$l = 52.33 \times 12 = 628$ inches. Formula: $\Delta = ctl$ and with values:

$\Delta = 0.0000067 \times 70 \times 628.0 = 0.2945320$ inches (about 9/32 inches).

One and two-way slab design 4.5

A one-way slab is designed as a rectangular beam, supported at each end or extending over several supports as a continuous member. The principal reinforcing is in the lower portion of the slab, and temperature steel for shrinkage is required normal to the main tension rods. Using the strip load design concept, a strip of the slab 12 inches wide is assumed for dimension b. Many one-way slab designs span several supports; then, the maximum moment is at mid-span, and is comparable to the negative moment over the supports. Equal areas of tension steel are required in the bottom for maximum positive moment and are also required in the top for the negative moment. It is possible to accomplish this steel area balance by bending up alternate bottom rods, and adding negative steel of the same rod size between the rods bent up from the bottom. The point of contra-flexure, where the bending moment changes from positive to negative, is taken as 1/5 the clear span between supports. The point for bending the rods is designated on drawings as $\frac{L}{5}$, and fabricators refer to this detail as the inflection bend.

The area of temperature steel required for one-way slabs is given as a fraction of the concrete area in the slab. These ratios are given in Paragraph 4.4.2.1 according to the type and yield stress of the steel. The ratios given are the minimum for average slab construction. The spacing of rods to resist shrinkage and temperature cracking should be limited to 12 inches or less. Although many designers permit spacing up to 18 inches, they do so because they have special confidence that the inspector on the project will be alert to the water content in the mix and the curing conditions. (See slump tests Par. 4.1.3.).

Two-way slabs are designed with the slab panel supported on four sides. The load is transferred in two directions by placing reinforcing rods at right angles to each other. When a slab panel is square, with four sides for support, the load distribution will be equal for each supporting beam. One half of the load will be resisted by the reinforcing steel running in each direction, and temperature steel is not necessary. Usually a two-way slab design will be more economical due to savings in material and labor. Slab panels longer in one dimension than the other are assumed to have the greater load transmitted on the shorter span. When two sets of reinforcing rods are used perpendicular to each other, the rods running in short direction are placed under the longer rods. The effective depth and moment arm is greater for the lower rods. For load distribution on a rectangular panel, a simple formula is used: $W_s = \left(\frac{L}{S} - 0.50\right) wS$, where:

W_s = Total load on short span in pounds.
L = Longer span length, in feet.
S = Shorter span length, in feet.
w = Load per foot on 12 inch width, in pounds.

In designing a two-way slab, follow the same procedure used for one way design. A strip of slab one foot wide is taken for each direction. Loads will be greater for the short direction when the load distribution formula is used. The bending moments will be maximum at midspan, and decrease nearer the supporting beams on all four sides. At a point customarily taken as ¼ of span, the entire computed steel area may be reduced. At the $\frac{L}{4}$ point, increase spacing in each direction up to 100 percent.

One and two-way slab design, continued 4.5

However, the rods within the¼ span dimension must not be spaced over 12 inches, and rod size should not be changed.

One-way and two-way slabs must not be confused with flat slab construction. This type of slab is used without supporting beams or girders. Slabs will be designed with drop panels and supported by circular columns with flared capitals. The reinforcement for flat slabs is never placed in less than two directions, and in some structures the rods may run in four directions. Flat slab, girderless floors were originally a patented design, and were referred to as *mushroom floors*. This floor system will be discussed in Paragraph 4.9.

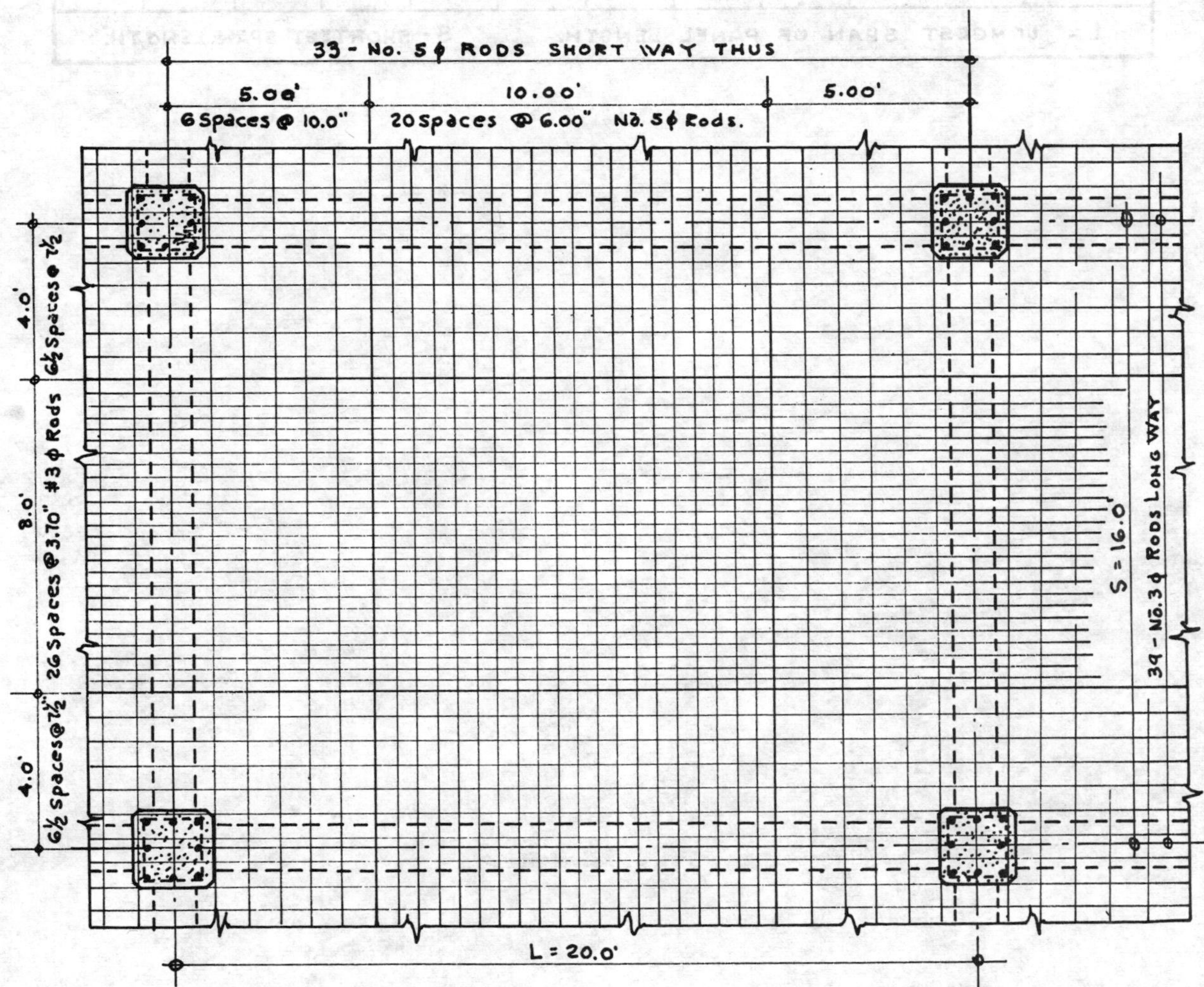

TYPICAL 2-WAY RECTANGULAR SLAB PANEL

TABLE: Load distribution for two-way slabs 4.5.1

LOAD DISTRIBUTION FOR 2-WAY CONCRETE SLABS											
RATIO $\frac{L}{S}$ =	1.00	1.05	1.10	1.15	1.20	1.25	1.30	1.35	1.40	1.45	1.50
SHORT SPAN PORTION	0.50	0.55	0.60	0.65	0.70	0.75	0.80	0.85	0.90	0.95	1.00
LONG SPAN PORTION	0.50	0.45	0.40	0.35	0.30	0.25	0.20	0.15	0.10	0.05	0.00

L = LONGEST SPAN OF PANEL LENGTH. S = SHORTEST SPAN LENGTH.

Rod laps and splices 4.5.2

ROD LAPS AND SPLICING

Splices must be made at construction joints to accommodate field work. Splices must also be provided in long beams and slabs which are continuous over several supports. *No rod splice should ever be made at a point of maximum bending moment*. Rod splicing is accomplished by lapping the rods a specified length, such as 10, 20 or 30 diameters. Lapping uses the bond strength to prevent the tension force in the steel rods from pulling away from the concrete. In some structures, it may be required that the rods be hooked together. For rods in compression, splicing is better accomplished by fastening the rods with u-bolts or a mechanical sleeve device. This method is required for columns in high-rise structures in most codes. Field-welded splicing is an accepted method for joining rods in tension and compression. Using a back-up plate and welding each rod in the same plane of stress will eliminate eccentric bending. Single-vee butt joints with angle back-up are more desirable for welded splices.

In Table 4.3.6.2 design coefficients and allowable bond stresses, it will be noted that the higher strength mixes provide a higher allowable bonding stress. For 2500 PSI concrete, u = 150 PSI, and for 4000 PSI concrete, u = 200 SPI. To check the strength of a lapped splice, for example, examine a #4 rod with an area of 0.20 square inches and a perimeter of 1.57 square inches. If F_t = 18,000 PSI, the tension value of the rod is T = 0.20 x 18,000 = 3600 pounds. With an allowable bond stress u = 150 PSI, the tension value of the bond per lineal inch of rod = 1.57 x 150 = 235.5 pounds. Minimum length required to pull rod from concrete = 3600/235.5 = 15.3 inches. Full embedment is not present if the rods are placed together for this length, and additional length is required. The increased length for lapped splices in contact is generally 20 percent of the above dimension. Thus: 15.3 x 1.20 = 18.3 inches for minimum lap length. Since a #4 rod has a diameter of 0.50 inches, the lap would be noted on drawings as 40 diameters to give a lapped splice of 20 inches.

The American Concrete Institute recommended code states that splices at maximum tensile stress shall be accomplished by welding, lapping or other means, where the computed stress from bar to bar shall not exceed 75 percent of the bond value. The allowable bond stress in this case is computed for A305 bottom rods by the formula: $u = \frac{4.8\sqrt{Fc'}}{D}$ but not greater than 500 PSI. In the formula, D equals rod diameter. In the illustration above, the allowable bond stress would be reduced from 150 to 120 PSI for a ½ inch diameter rod. Further, when Fc′ = 3000 PSI or more, the length of the lap for deformed rods shall be 20, 24 or 30 bar diameters for yield strengths F_y = 50,000 PSI and under, 60,000, and 75,000 PSI, respectively, and not less than 12 inches in any case.

Table 4.5.2.1 is provided to serve as a guide for specifying minimum laps based on the yield strength of the steel. In using this table, it is not difficult to convert from inches to diameters. For instance: A #7 rod is 0.875 inches in diameter, and the table gives a lap length of 21 inches for yield stress in tension of 33,000 PSI. Number of diameters = 21.0/0.875 = 24. When plain rods without deformations are used for reinforcing, the lap length is double that of deformed rods. All rods and bars larger than number 11 are not to be lap spliced, but should be welded, or if enough bulk is available in the girder, they may be hooked tightly together and welded.

TABLE: Lapped splices for steel reinforcement 4.5.2.1

LAPPED SPLICES FOR STEEL REINFORCEMENT

BASED ON A.C.I. 316-63
A.S.T.M DESIGNATIONS
BARS 2 THROUGH 11 = A305
BARS 14S AND 18S - A408

SIZE D	PROFILE	BAR NO.	As SQ.IN.	DIAMETER IN DECIMAL D"	TENSION: TYPE "A" SPLICE			TENSION: TYPE "B" SPLICE			COMPRESSION: F'_c = 3000 PSI +			COMPRESSION: F'_c = UNDER 3000 PSI		
					YIELD STRENGTH OF REINFORCING RODS - F_y = PSI.											
					33,000 TO 40,000	50,000	60,000	33,000 TO 40,000	50,000	60,000	33,000 40,000 50,000	60,000	75,000	33,000 40,000 50,000	60,000	75,000
					MINIMUM BAR LAP - IN INCHES											
1/4	●	2	0.05	0.250	12	15	18	15	18	22	12	12	15	14	16	20
3/8	●	3	0.11	0.375	12	12	14	12	14	17	12	12	12	12	12	15
1/2	●	4	0.20	0.500	12	15	18	15	18	22	12	12	15	14	16	20
5/8	●	5	0.31	0.625	15	19	23	18	23	27	13	15	19	17	20	25
3/4	●	6	0.44	0.750	18	23	27	22	27	33	15	18	23	20	24	30
7/8	●	7	0.60	0.875	21	27	32	26	32	38	18	21	27	24	28	35
1	●	8	0.79	1.000	24	30	36	29	36	44	20	24	30	27	32	40
1	■	9	1.00	1.128	28	34	41	33	41	49	23	28	34	31	37	46
1 1/8	■	10	1.27	1.270	31	39	46	37	46	55	26	31	39	34	41	51
1 1/4	■	11	1.56	1.410	34	43	51	41	51	61	29	34	43	38	46	57
1 1/2	■	14S	2.25	1.693	DO NOT LAP WHEN BAR IN TENSION						34	41	51	46	55	68
2	■	18S	4.00	2.257							36	55	68	61	73	91

TYPE A SPLICE	VALUES APPLY TO CONTACT SPLICES LATERALLY SPACED OVER 12 DIAMETERS APART. AND LOCATED OVER 6 DIAMETERS, OR 6 INCHES FROM OUTSIDE EDGE. ALSO APPLIES TO OTHER CONTACT SPLICES WHEN STIRRUPS ARE AS SPECIFIED IN A.C.I. RECOMMENDED CODE, OR IF CLOSELY SPACED SPIRALS ENCLOSE ENTIRE SPLICE.
TYPE B SPLICE	DIMENSION VALUES APPLY TO CONTACT SPLICES LATERALLY SPACED CLOSER THAN 12 BAR DIAMETERS OR LOCATED LESS THAN 6 DIAMETERS, OR 6 INCHES FROM THE OUTSIDE EDGE, AND EXCEPT WHERE STIRRUPS OR SPIRALS ARE USED IN TYPE A.

EXAMPLE: Designing tension steel and splice length 4.5.2.2

A beam section has a positive bending moment at center of span equal to 220,000 inch pounds. $F_c' = 3000$ PSI and $F_s = 20,000$ PSI. Rods are to be A305-56T deformed with $F_y = 40,000$ PSI, depth to steel = 14.5 inches.

REQUIRED:
Calculate the required area of steel reinforcement, then select 2 suitable round rod for tension. Compute the minimum splice lap with splice rods in contact and fully embedded in concrete. Use formula: $u = \frac{4.8\sqrt{F_c'}}{D}$.

STEP I:
From tables: $J = 0.872$ and $A_s = \frac{M}{F_s J d}$. With values in formula:

$$A_s = \frac{220,000}{20,000 \times 0.872 \times 14.5} = 0.87 \text{ Square inches.}$$

2-#6 ϕ Rods have an $A_s = 0.88^{\square''}$ and 2 Perimeters $\Sigma o = 4.72^{\square''}$

STEP II:
Allowable $u = \frac{4.8\sqrt{3000}}{0.75} = 350$ P.S.I. (D = diameter of 1 Rod)

Tension Force in 2 Rods: $F_s A_s$. $T = 20,000 \times 0.88 = 17,600$ Lbs.

Length of lap $= \frac{T}{\Sigma u}$ or $\frac{17,600}{4.72 \times 350} = 10.65$ inches.

Converting inches to diameters: $\frac{10.65}{0.75} = 14.2$ diameters.

STEP III:
By referring to table for lapped splices, note that about 57% has been added to the result obtained in step II, or minimum length in inches = 18.0
When table value is converted thus: $\frac{18.0}{0.75} = 24$ Diameters.

Lap will sustain a force $T = 18.0 \times 4.72 \times 350 = 29,735$ Lbs.
Unit stress in steel $= \frac{29,735}{0.75} = 39,650$ PSI, and within F_y.

EXAMPLE: Designing one-way slab on simple span 4.5.3

The interior dimensions of a vault are 12.0' x 20.0' and walls are of 16.0 inch masonry. A concrete slab is to be used for top of vault and support a storage room above. Live load desired is 100 Lbs. square foot. Inside of vault will have plaster walls and ceiling. Specifications call for a 1-Way reinforced slab with $F_c' = 3000$ PSI and $F_s = 18{,}000$ PSI.

REQUIRED:
Design the slab to be supported on 12.0 inches of brick wall on 4 sides and rods to run the short direction. Provide the temperature rods in long dimension according to schedule. A plan drawing with section shall accompany the design.

STEP I:
To establish a slab depth, gather the applicable design factors from tables as follows: b = 12.0 inch strip of slab.
$F_c' = 3000$ PSI $F_c = 1350$ PSI $F_s = 18{,}000$ PSI $K = 238.0$ $J = 0.864$ $L = 12.0'$
$u = 165$ PSI and $F_v = 60$ PSI. $M = WL/8$

STEP II:
Assume Dead Load of 70 Lbs. Sq. Foot. Design Load = 100 + 70 = 170 Lbs. Sq. Ft.
$W = 170 \times 12.0 = 2040$ Lbs. $M = \dfrac{2040 \times 12.0 \times 12}{8} = 36{,}720$ Inch Lbs.

$d = \sqrt{\dfrac{M}{Kb}} = \sqrt{\dfrac{36{,}720}{238.0 \times 12.0}} = 3.59$ inches. Call it 3.75 inches and make slab depth $3.75 + 0.75 = 4.50$ inches.

STEP III:
For area of Steel: $A_s = \dfrac{M}{F_s J d}$ or $\dfrac{36{,}720}{18{,}000 \times 0.864 \times 3.75} = 0.630$ □"

Try #5 ϕ Rods: $A\phi = 0.31$ □" $s = \dfrac{0.31 \times 12.0}{0.630} = 5.90"$ Use spacing $5\frac{3}{4}$ to 6 inches.

Temperature steel to run long dimension.
From table ratio = 0.0025 Ac. $A_c = 12.0 \times 4.50 = 54$ Sq. Inches
Temp. $A_s = 0.0025 \times 54 = 0.135$ □" Try using #3 ϕ Rods with $A\phi = 0.11$ □"
Spacing for temp. steel. $s = \dfrac{0.11 \times 12.0}{0.135} = 9.80"$ Use about 10.0" cc.

STEP IV:
Check Shear at wall: W = 2040 Lbs. $R = V = \dfrac{2040}{2} = 1020$ Lbs.

$f_v = \dfrac{V}{Jdb}$ or $f_v = \dfrac{1020}{0.864 \times 3.75 \times 12.0} = 26.2$ PSI (OK below allowable F_v)

EXAMPLE: Designing one-way slab on simple span, continued 4.5.3

STEP V:

Check bond to determine if rods must have ends hooked:

Number of rods in 12.0 inch strip when spacing = 5.90 inches.

$n = \frac{12.0}{s}$ or $n = \frac{12.0}{5.90} = 2.04$

Perimeter of 1-#5φ Rod: 1.963 Sq. In. $\Sigma_o = 2.04 \times 1.963 = 4.00$ Sq. In.

$u = \frac{V}{\Sigma_o\, jd}$ or $u = \frac{1020}{4.00 \times 0.864 \times 3.75} = 78.7^{\#\square''}$ (OK allowed 165 PSI)

Rods will not require hooked ends.

STEP VI:

The design may now be drawn in plan and section through the short dimension. Small adjustments for spacing are permissible.

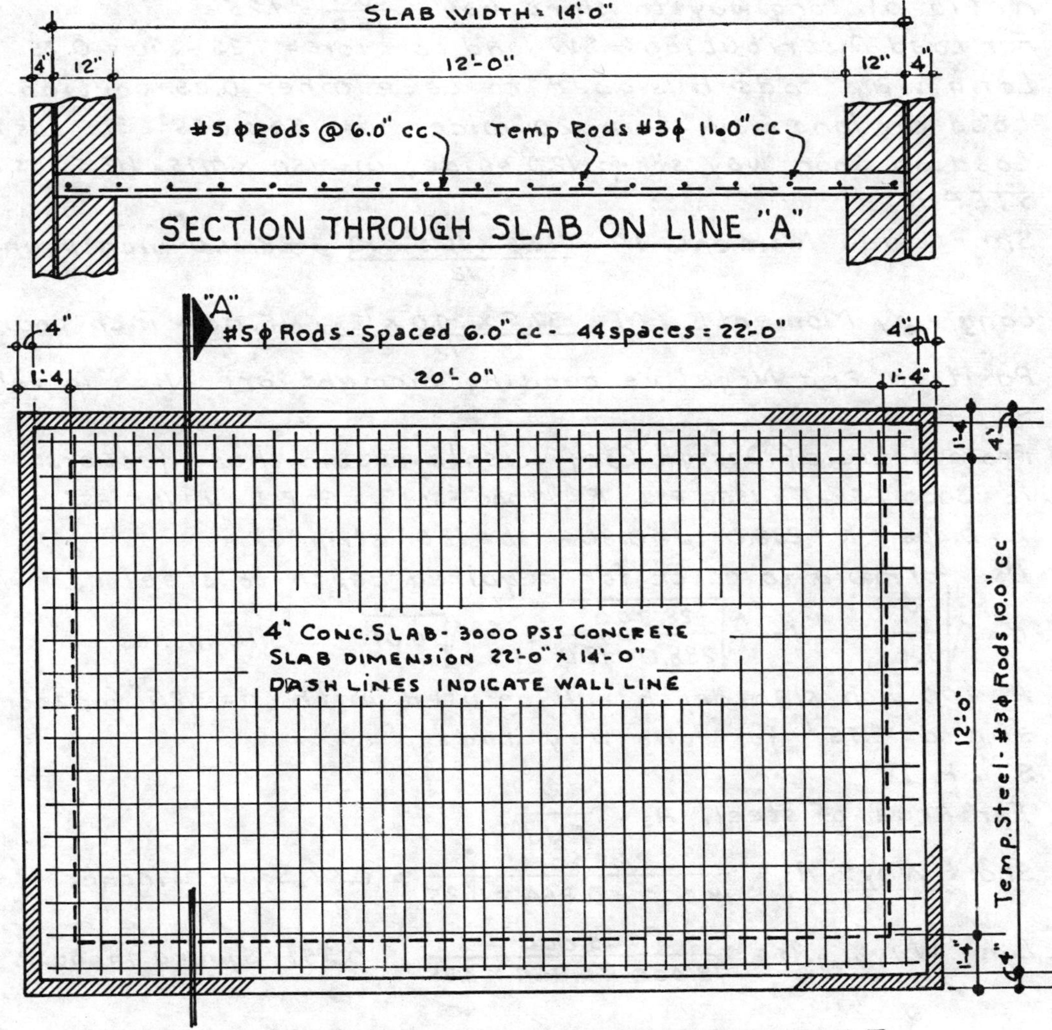

•PLAN OF 1-WAY SLAB OVER STORAGE VAULT•

EXAMPLE: Designing two-way interior slab 4.5.4

A typical interior floor panel slab is 20.0' x 16.0' as shown in plan for this example (Page 4068). Specifications are as follows: $F_c' = 3000$ P.S.I. $F_t = 18{,}000$ PSI. $+M = -M$ for moment or $M = \frac{WL}{12}$.

REQUIRED:

Design slab for 100 PSF Live load and 2-Way reinforcement. Check shear and bond stress to meet allowable values given in Tables for Design Factors. Make a section through floor to illustrate how rods are to bent up for negative moment.

STEP I:

Assume minimum slab depth to be 4.00 inches and Live Load plus Dead Load = 150 Lbs. Sq. Foot and equals w.

Ratio of Long way to short way $= \frac{20.0}{16.0} = 1.25$

For Load Distribution: SW load portion = 1.25 − .50 = 0.75

Long way rods designed to take other 0.25 portion.

Load on long way strip 12.0" wide; $w = 150 \times 0.25 = 37.5$ Lbs. Ft.

Load on Short way strip 12.0" wide; $w = 150 \times 0.75 = 112.5$ Lbs. Ft.

STEP II:

Short way Moment: $M = \frac{112.5 \times 16.0^2 \times 12}{12} = 28{,}800$ inch pounds.

Long way Moment: $M = \frac{37.5 \times 20.0^2 \times 12}{12} = 15{,}000$ inch pounds.

Positive and Negative bending moment are alike in value.

STEP III:

From Table of Design Coefficients obtain these factors:

$F_c' = 3000$ PSI. $F_c = 1350$ PSI $F_s = 18{,}000$ PSI. $F_v = 60$ PSI $u = 165$ PSI.

$k = 0.408$ $K = 238.0$ $J = 0.864$ $b = 12.0"$ strip.

Use formula to check for required depth to steel d;

$$d = \sqrt{\frac{M}{Kb}} \qquad d = \sqrt{\frac{28{,}800}{238.0 \times 12.0}} = \sqrt{10.01} = 3.16 \text{ inches.}$$

A 4.00 inch slab depth will be used with d = 3.25" for short way and d = 2.75" for long way rods.

STEP IV:

For area of steel: $A_s = \frac{M}{F_s J d}$.

Short Way: $A_s = \frac{28{,}800}{18{,}000 \times 0.864 \times 3.25} = 0.57$ Square inches.

Long Way: $A_s = \frac{15{,}000}{18{,}000 \times 0.864 \times 2.75}$ 0.351 Square inches.

EXAMPLE: Designing two-way interior slab, continued 4.5.4

From table of slab rod spacing the As with #5 φ Rods spaced 6½ inches is given as As = 0.572 □". This is for short way. For long way rods which should have a closer spacing try using a No.3 φ rod and calculate spacing. Aφ for #3 = 0.11 □"

Long way: $s = \frac{0.11 \times 12}{0.351} = 3.75$ inches. Use #3 φ spaced 3¾" cc.
For short way, use #5 φ rods spaced 6½" cc.

Alternate rods to be bent up at inflection point as $\frac{L}{5}$ to take care of negative moments. Between the alternates which were bent up, drop in another rod of same size and length equal to: $2 \times \frac{L}{5}$. Section drawing follows.

STEP V:

Check depth for shear intensity at supports.

Short Way: W = 112.5 x 16.0 = 1800 Lbs. $R = \frac{W}{2} = 900$ Lbs. = V.

Long Way: W = 37.5 x 20.0 = 750 Lbs. Use greater value of V.

$f_v = \frac{V}{Jdb}$ or $f_v = \frac{900}{0.864 \times 3.25 \times 12.0} = 26.8$ PSI. (Allowable F_v = 60 PSI)

STEP VI:

Check for bond stress: Allowable u = 165 PSI. $u = \frac{V}{\Sigma_o Jd}$.

Perimeter of #5 φ Rod = 1.963 Sq.In. Spacing = 6.50 In.

Number of rods in effective strip = $\frac{12.0}{6.5} = 1.85$

Sum of Perimeters: $\Sigma_o = 1.85 \times 1.963 = 3.63$ Square inches.

Bond stress intensity: $u = \frac{900}{3.63 \times 0.864 \times 3.25} = 88.3$ PSI. (OK)

DESIGNER'S NOTATION:

The above results giving rod sizes, spacing and slab thickness can now be placed in the custody of draftsmen who will fill in the slab panel plan and make details of sections as shown. Since maximum bending moment is in the area of the mid-spans, it follows that rod spacing can be widened near the supports. This point is taken as ¼ of span or $\frac{L}{4}$. The general rule applied in such cases is to delete the alternate rod and use double the spacing. In no event must spacing be over 12.0 inches. To be on the safe side, adjust the spacing in edge areas to suit the panel dimensions as is shown in the panel plan. Section drawings are drawn through the critical area at center of panel.

EXAMPLE: Designing two-way interior slab, continued 4.5.4

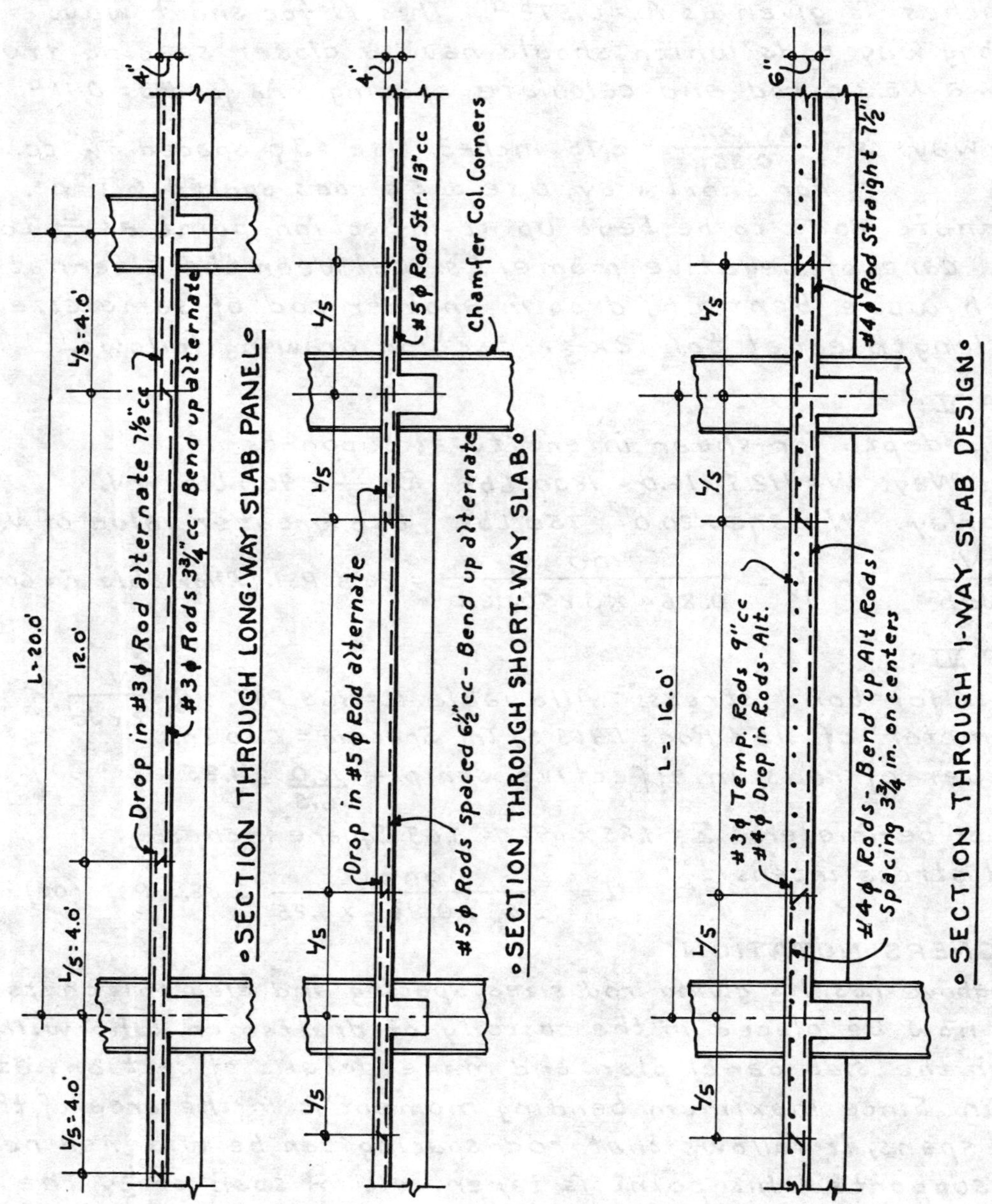

EXAMPLE: Designing one-way slab for continuous span 4.5.5

Refer to previous example of a design for a 2-Way slab as illustrated with short span S = 16.0 Feet. Loading is to remain same or Live Load = 100 P.S.F. Specifications for Concrete and steel are: $F_c' = 3000$ PSI. $F_s = 18000$ PSI.

REQUIRED:
Check thickness of slab and design for a Continuous type span with principal reinforcing running in short direction for a 1-Way slab design. Also design for temperature rods running normal to main reinforcing.

STEP I:
By referring to table for Live Loads on 1-Way reinforced slabs it will be noted that span lengths over 15.0 feet are not given. Also the allowable Fs is over 18,000 PSI. A slab thickness of 6.00 inches is close to requirements and the dead load = 72 Lbs. Sq. Foot. Design Load = 72 + 100 = 172 Lbs. Sq. Ft.

STEP II:
$w = 172^{\#}/_{1}$ and $W = 172 \times 16.0 = 2752$ Lbs. $+M = \frac{WL}{12}$ and $-M = \frac{WL}{10}$.

$$+M = \frac{2752 \times 16.0 \times 12.0}{12} = 44{,}032 \text{ Inch Lbs.}$$

$$-M = \frac{2752 \times 16.0 \times 12.0}{10} = 52{,}835 \text{ Inch Lbs.}$$

STEP III:
Check depth d by formula and gather design coefficients:
$b = 12.0"$ $J = 0.864$ $K = 238.0$ $u = 165$ PSI $F_v = 60$ PSI $F_s = 18{,}000$ PSI

$$d = \sqrt{\frac{M}{Kb}} \text{ or } d = \sqrt{\frac{52{,}835}{238.0 \times 12.0}} \text{ ac } \sqrt{18.5} = 4.32 \text{ inches}$$

With steel in top and bottom d is a coupling and total depth D = 4.32 + 0.75 + 0.75 = 5.82 inches. Use 6.0 inch slab and make moment arm for steel, d = (6.00 − 5.82) + 4.32 = 4.50 Inches.

STEP IV:
Computing for area of steel: $As = \frac{M}{Fs\,Jd}$

Positive in bottom +M: $As: \frac{44{,}032}{18{,}000 \times 0.864 \times 4.50} = 0.630$ Sq. In.

Negative in top −M: $As: \frac{52{,}835}{18{,}000 \times 0.864 \times 4.50} = 0.755$ Sq. In.

EXAMPLE: Designing one-way slab for continuous span, continued 4.5.5

STEP V:

Rod selection and spacing:

Try #5 ϕ Rods for bottom. $A\phi = 0.31\,\square''$ $\quad s = \dfrac{0.31 \times 12.0}{0.630} = 5.90$ in.

A little closer is desirable - try #4 ϕ with $A\phi = 0.20$ Sq. In.

$s = \dfrac{0.20 \times 12.0}{0.630} = 3.80$ inches. Use spacing of $3\frac{3}{4}$ inches.

Bend up alternate rods over supports at $\frac{L}{5}$ span length.

STEP VI:

Negative moment being larger requires the drop-in rod at alternate spaces to be larger.

$-As = 0.755 \quad +As = 0.630$ Then difference is: $0.755 - 0.630 = 0.125\,\square''$

Number of rods in 12.0 in strip $= \dfrac{12.0}{3.75} = 3.2$ Rods.

Additional steel for drop-in rod $= \dfrac{0.125}{3.20} = 0.39\,\square''$

Use alternate rods in top of #5 ϕ with $A\phi = 0.31\,\square''$

STEP VII:

Check for shear at supports: $V = \dfrac{W}{2}$ or $V = \dfrac{2752}{2} = 1{,}376$ Lbs

$f_v = \dfrac{V}{Jdb} = \dfrac{1376}{0.864 \times 4.50 \times 12.0} = 29.5$ PSI (Allowed 60 PSI).

STEP VIII

In step VI there are 3.2 Rods of #4 size in 12.0 strip.

Perimeter of #4 ϕ rod is = 1.571 Sq. Inches.

Summation: $\Sigma_o = 1.571 \times 3.20 = 5.03$ Sq. In.

$u = \dfrac{V}{\Sigma_o Jd}$ or $u = \dfrac{1376}{5.03 \times 0.864 \times 4.50} = 70.3$ PSI (Allowed 165 PSI)

STEP IX

Temperature steel to run normal to main steel:

From Table: Proportionate ratio of steel to concrete is given as; 0.0020 Ac. Maximum spacing not over 12.0 inches.

$Ac = bD$ or $Ac = 12.0 \times 6.0 = 72.0$ Sq. In. $\quad A_{TS} = 72.0 \times 0.0020 = 0.144$ Sq. In.

Try using #3 ϕ Rods with $A\phi = 0.11\,\square''$ Spacing $s = \dfrac{A\phi \times 12.0}{Ats}$

Spacing: $s = \dfrac{0.11 \times 12.0}{0.144} = 0.917''$ Space at 9.0 inches.

DESIGN NOTE:

See sections through slabs given at end of previous example 4.5.4 for typical rod bending.

EXAMPLE: Designing continuous two-way slab on piles 4.5.6

A warehouse floor slab and steel building are to be supported on wood piling due to alluvial soil and subsidence. Live load is specified at 250 Pounds per square foot. Preliminary plans propose a monolithic type slab with haunched beams for pile caps. Panels are laid out to accomodate column bents and are 12.0'x16.0'. Specified concrete is to be 3000 PSI and Steel allowable = 18,000 PSI.

REQUIRED:
Spans obviously will be continuous and slab depth will be determined at edge of haunch. Design slab for an end panel on basis of 2-Way reinforcing. Assume an 8.00 inch slab depth for analysis. Provide drawing of plan and section.

STEP I:
Total D=8.00" Depth to steel short way is d=7.25" and for long dimension d= 6.50." Dead Load = 8.00 x 12.5 = 100 Lbs. Sq. Foot.
Design load = 250 + 100 = 350 Lbs. Sq. Foot.

STEP II:
For 2-Way load distribution on short and long spans:
$L = 16.0'$ $S = 12.0'$ Ratio $= \frac{16.0}{12.0} = 1.33$ Proportion for $S = 1.33 - 0.50 = 0.83$

Proportion of load for long span = 1.00 − 0.83 = 0.17
Load on Short span: $W = 350 \times 12.0 \times 0.83 = 3486$ Lbs. $V = 1743$ Lbs.
Load on Long span: $W = 350 \times 16.0 \times 0.17 = 952$ Lbs. $V = 476$ Lbs.

STEP III:
Collecting data and design coefficients from tables thus:
$F_c' = 3000$ PSI. $F_s = 18{,}000$ PSI. $F_v = 60$ PSI. $u = 165$ PSI
$K = 0.238$ $J = 0.864$ $b = 12.0''$
Check depth for shear using greater value of V: $f_v = \frac{V}{Jdb}$

$$f_v = \frac{1743}{0.864 \times 7.25 \times 12.0} = 23.2 \text{ PSI.}$$ (Depth is adequate for shear)

STEP IV:
Calculating bending moments for end spans as: $M = \frac{WL}{10}$.

For Short span: $M_S = \frac{3486 \times 12.0}{10.0} = 4185$ Foot Lbs.

For Long span: $M_L = \frac{952 \times 16.0}{10} = 1523$ Foot Lbs.

EXAMPLE: Designing continuous two-way slab on piles, continued 4.5.6

STEP V:

Calculate for Area of Steel: $A_s = \dfrac{M}{F_s J d}$ Convert M to In.Lbs.

Short Span: $M = \dfrac{4185 \times 12}{18{,}000 \times 0.864 \times 7.25} = 0.446$ Sq. inches

Long Span: $M = \dfrac{1523 \times 12}{18{,}000 \times 0.864 \times 6.50} = 0.183$ Sq. inches.

Try using #4 ϕ Rods for short way, $A_\phi = 0.20^{\square}$"

Spacing: $s = \dfrac{0.20 \times 12}{0.446} = 5.38$ in. Space about 5⅜" on centers.

Try using #3 ϕ Rods for long way, $A_\phi = 0.11^{\square}$"

spacing: $s = \dfrac{0.11 \times 12}{0.183} = 7.25$" Space about 7¼" on centers.

STEP VI:

Check bond for hooked end requirements:

Short Way: Perimeter of #4 ϕ Rod = 1.571 Sq. In.

Number of rods when spaced 5.375" $n = \dfrac{12}{5.375} = 2.23$ Rods.

$u = \dfrac{V}{\Sigma o J d}$ $\quad u = \dfrac{1743}{2.23 \times 1.571 \times 0.864 \times 7.25} = 79.5$ PSI. (OK all'd. 164 PSI)

STEP VII:

May now draw rod plan spacing with sections through Long and Short dimensions. Bend up alternate rods to resist negative bending. Moment −M is same as +M, thus the drop-in top rods will be the same size. Reinforcing rods for grade and supporting haunches is not shown.

EXAMPLE: Designing continuous two-way slab on piles, continued 4.5.6

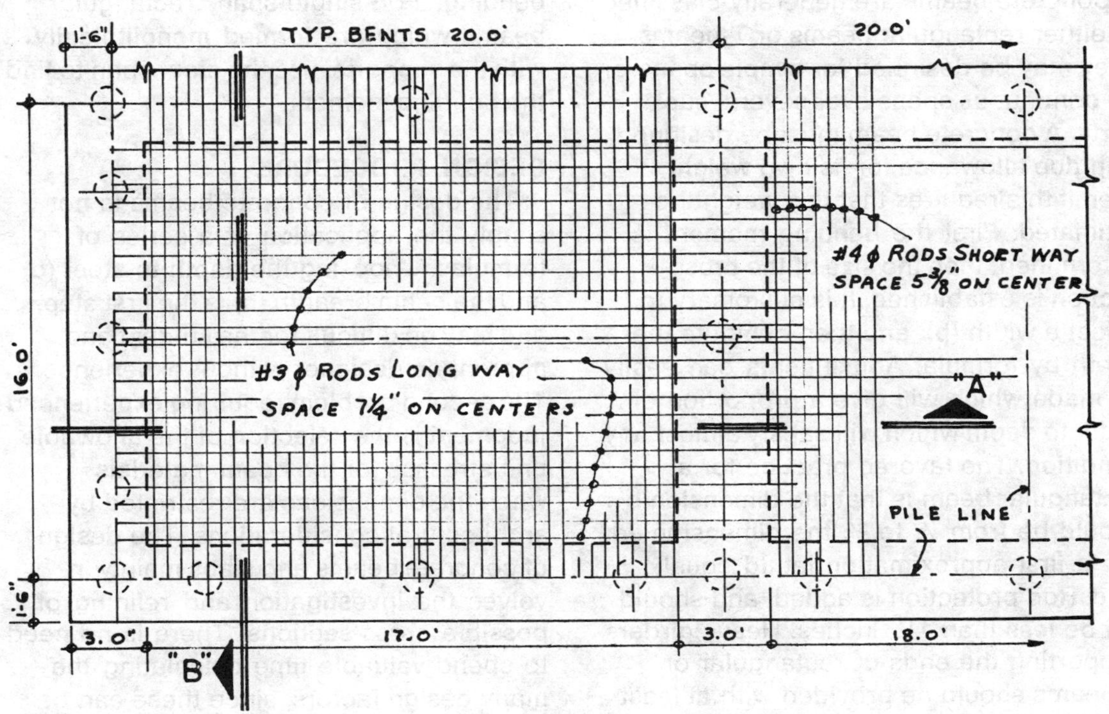

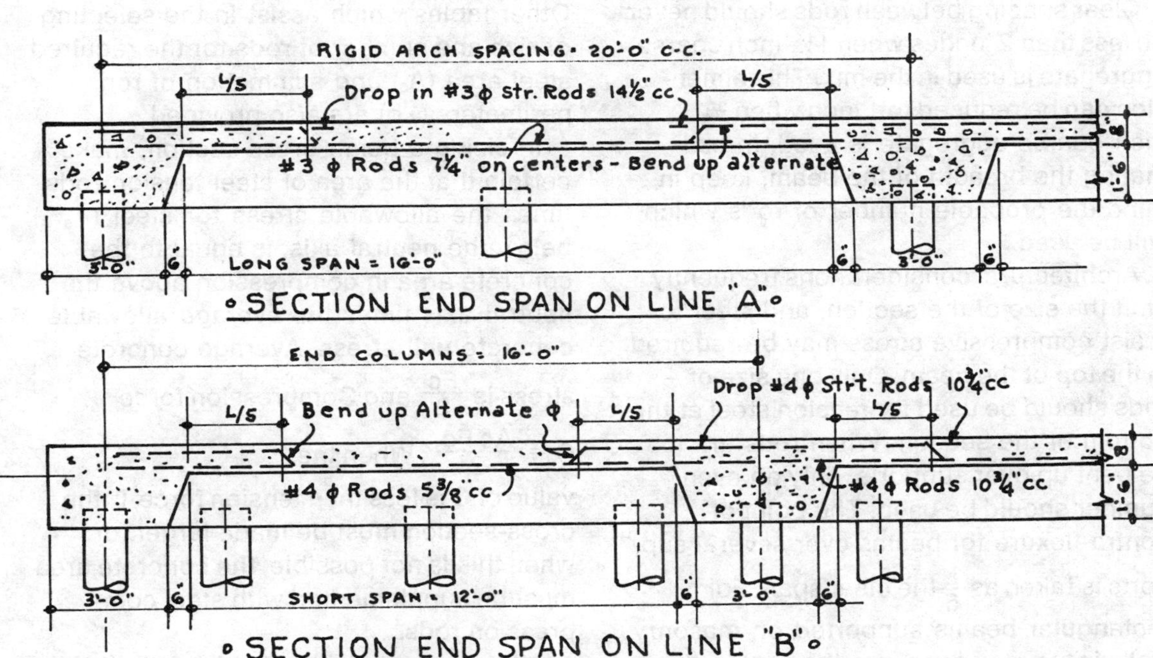

Rectangular concrete beam design 4.6

Concrete beams are generally classified as either rectangular beams or T-beams. They may be designed for simple spans or for continuous spans over several supports. A concrete beam must be designed with due allowance for its own weight, even if this requires that this weight be estimated. First the bending moment is determined; then the size of the cross-section is established. It is customary to select a width (b), and then compute the depth by formula. Adjustments can easily be made which will give a proportion of width to depth which will satisfy almost any condition. The favored practice for a rectangular beam is that the dimension b should be from ½ to ¾ the dimension Jd. For a first approximation let Jd equal ⅞ d. Rod protection is added, and should not be less than 1½ inches. Heavy girders supporting the ends of rectangular or T-Beams should be provided with at least 2 inches of fire protection.

Clear spacing between rods should never be less than 2 inches when 1½ inch coarse aggregate is used in the mix. This dimension can be reduced to 1 inch when ¾ inch coarse aggregate is used. In estimating the breadth of the beam, keep in mind the probable number of rods which will be used.

Architectural considerations frequently limit the size of the section, and steel to resist compressive stress may be required in the top of the beam. Only one size of rods should be used for tension steel at the bottom of the section. When rods are to be bent up over supports, only an even number should be used. The point of contra-flexure for beams over several supports is taken as $\frac{L}{5}$ the clear span. For rectangular beams supported on masonry walls in simple spans, use the center of end bearing as the center of moments for bending. For single-span, rectangular beams which are formed monolithically with the supports, use the clear span to find the bending moment.

DESIGN PROCEDURE

The design of concrete beams is not simply the application of a series of formulas. Choosing the depth to steel (d) and the beam breadth (b) is the first step, and may be difficult for the student and apprentice designer without experience. The second problem requiring experienced judgment is the selection of the allowable unit stresses for the beam materials when the dimensions are restricted by architectural considerations. The design of concrete beams and slabs mainly involves the investigation and refining of possible cross-sections. There is no need to spend valuable time calculating the many design factors, since these can be obtained from the tables in this section. Other tables which assist in the selecting of size and number of rods for the required steel area (A_s) and summation of rod perimeters (Σo) are also provided.

In sizing a beam cross-section, make certain that the area of steel tension rods times the allowable stress for steel (A_sF_s) below the neutral axis, is equal to the concrete area in compression above the neutral axis times the *average* allowable concrete unit stress. Average concrete stress is $\frac{Fc}{2}$ and Compression force $C = \frac{Ac\,Fc}{2}$. When the value of C is less than tension force T, the cross-section must be made larger, or when this is not possible, the concrete area must be supplemented with steel compression rods.

After assuming either the breadth (b) or

Rectangular concrete beam design, continued 4.6

depth to steel (d) dimensions, investigate the section for bending, shear and rod bond. Adjustments are made to meet the needs for each type of stress. Examples which follow illustrate the sequence for calculating the depth dimension, as well as the steps used to investigate the other stresses.

HOOKED-END RODS

The standard hooked-end rod is bent in the shape of a half circle, although many designers approve a 90 degree bend. The radius of the curved end should vary according to the rod size. Generally the radius should be between 3 and 6 inches. Billet steel rods can be bent cold; rail steel rods should be heated red before bending. The hooked end supplements the bond strength to resist tension loads, hence when the bond stress is above the allowable stress, the rods must be hooked. All beams for simple spans, cantilevers, and other conditions which limit the use of fully lapped rods should have the rods hooked. Columns, footing plinths and foundation pads should be designed with hooked end rods, even though the bond stress may appear to be adequate. Steel base and bearing plates supported by concrete or masonry shall have the anchor bolts installed with the embedded end hooked with a right-angle leg of not less than 6 to 8 diameters.

Transformed sections 4.6.1

A simple formula defines the n factor, $n = Es/Ec$. This indicates that the steel is n times stiffer than the concrete, or that it would require n times as much stress to deform the steel as would be required for the same deformation in concrete. This ratio has become the essential factor in the design of transformed sections and the basic theory for composite floor systems. The steel beam with concrete slab acts as a single unit. Transformed section theory assumes that an area of concrete replaces the steel *above the neutral axis*. This transformed area depends on the ratios of the moduli of elasticity. Transformed section theory makes it possible to design composite beam sections with formulas similar to those used for homogeneous rectangular and T-beam sections. Composite beam design will be discussed in 4.8.

Diagonal tension 4.6.2

In the concrete beam the horizontal shear and vertical shear have the same intensity at any point in a beam. If the shear force is resolved in a force diagram, the resultant maximum will act on a 45 degree diagonal line. This shear stress is therefore called *diagonal tension*. The cracks which form when a beam fails lie on this diagonal plane. An example to follow will show how a beam fails near the support where the reaction and shear are greatest.

Theoretically, the steel rods used to resist this shear should cross this plane at right angles, but this is not practical. It would be difficult to place stirrups in an inclined position; they are placed in a vertical position, and their spacing is determined by the shear component forces. Generally, the spacing for stirrups and the length of beam which requires stirrup reinforcing will not work out to a simple multiple. In drafting the working plans to show spacing, it is important to see that the spacing is not stretched beyond the calculations to fill the length of beam. Of course, the draftsman may call for closer spacing; this would be on the side of safety. Maximum spacing of stirrups for diagonal tension is thus limited to 0.45d.

Shear 4.6.3

The allowable unit shear stress in a concrete beam is determined by a formula which is based on the 28 day compressive strength of the concrete. Total shear (V) is determined by the loads and the reactions at the supports. The formula is $V = vJbd$, or $v = \frac{V}{Jdb}$. Note that Jd represents the effective depth of the beam or slab. When the actual shear stress is over the allowable, stirrups are required. Stirrups are referred to as web reinforcement.

Previously it was stated that concrete beams fail by breaking in a diagonal direction of about 45 degrees. This failure angle is caused by the resultant vertical and horizontal stresses known as *diagonal tension*. This angle becomes more vertical toward the center of the span, where if they did occur, they could be imagined to be vertical. Although this diagonal tension is a tensile stress, it is still referred to as shear by most design engineers.

In most beams and girders the greatest shear value is close to the supports and (V) equals the reaction (R). The bending moment is usually very small near the supports and increases to a maximum near midspan. Beams designed for uniform loads across the span have zero shear at midspan; therefore, stirrups to resist shear may be spaced at greater intervals.

The critical point in the design of stirrups is to make sure the spacing does not exceed forty-five percent of the depth to steel. In addition to the vertical stirrups, a number of tension rods can be bent up at points where they no longer are important to resist tension from bending moment. Do not overlook the contribution of the concrete in resisting the vertical shear component of the diagonal tension.

For beams and slabs with no web reinforcement, the Codes recommend that the concrete unit allowable shear stress be limited to $F_v = 1.1\sqrt{F_c'}$. The values for F_v

Shear, continued 4.6.2

are provided in Table 4.3.6.2 for design coefficients. Web reinforcement in stirrup form is only required when the actual shear stress f_v is greater than the allowable F_v. To illustrate with Slab Strip:

Load on strip of 8.0 foot length = 1160 pounds, and reactions are 580 pounds at supports. Then R = V. The concrete strength at 28 days is $F_c' = 3000$ PSI, and the allowable unit shear stress (without web reinforcement) is $F_v = 1.1\sqrt{3000} =$ 60 PSI. Depth to steel d = 2.21 inches, b = 12.0 inches, and J = 0.864. Substituting values in the formula for actual shear stress: f_v or $v = \frac{580}{0.864 \times 2.21 \times 12.0} =$ 25.3 PSI. Actual shear stress is less than allowable, and stirrups are not necessary.

Note the symbols used for actual shear stress: v and f_v. These designations are interchangeable. In the examples which follow, the small v will be used with superscripts and subscripts to assist in distinguishing between the shear value of the concrete and the shear value of the stirrups.

Stirrup design 4.6.3.1

Stirrups for shear resistance can be fabricated from standard reinforcing rod with deformed surfaces. They may be bent to resemble a capital U or a capital W. Rod sizes for building construction will generally vary from #3 to #6 size. In heavy dock work, stirrups may be used up to 1½ inch, square bars. Where there are unusual loads such as impact forces, it may be prudent to provide stirrups across the entire span.

A #4 rod stirrup has a diameter of ½ inch. When bent in the form of a U, it has 2 legs which gives it the value of 2 rods. The value of a #4 U stirrup is computed as: $2As\ F_t$, where As equals the cross-sectional area of the rod and F_t equals the allowable tension stress for steel. This value is denoted as Av. For a #4 U stirrup, Av = $2 \times 0.20 \times 18{,}000 = 7200$ pounds if $F_v =$ 18,000 PSI. A stirrup formed in the shape of a W has four (4) legs, and its value is double a U stirrup. Thus, for a #4 W stirrup, with the same allowable stress Av = $4 \times 0.20 \times 18{,}000 = 14{,}400$ pounds.

When the actual unit shear stress (v_c) exceeds the allowable for the concrete (F_v), the additional amount must be provided by stirrups. This excess shear stress is identified as v'. Then: $v' = v_c - F_v$, in pounds per square inch. This investigation of shear stress to be resisted by stirrups is illustrated in several design examples which follow. Remember the meaning of the different stress symbols; refer to 4.3, Nomenclature for Design Formulas when in doubt.

SPACING OF STIRRUPS

Recall that Av equals the stress value of a stirrup, and v' equals the amount to be carried by the stirrups. The stirrups must be spaced toward midspan, until the concrete alone can carry the shear stress. This is determined by the area of concrete (Ac) and the breadth of the beam (b). $A_v = v'bs$, where s equals the spacing. Transposing, $s = \frac{A_v}{v'b}$. This equation determines the proper spacing.

Stirrup design, continued 4.6.3.1

To illustrate this formula, assume: v_c = 146 PSI, F_v = 55 PSI and Av = 7200 Lbs. b = 12.0 inches. Then the stirrups must take v′ = 146 − 55 = 91 PSI. Spacing s = $\frac{7200}{91 \times 12.0}$ = 6.62 inches, or about 6⅝ inches.

DISTANCE FROM SUPPORT REQUIRING STIRRUPS

Beams supporting uniform distributed loads across their full span have the greatest shear intensity at their supports: V = Reaction. The intensity of the shear stress may be plotted on the shear diagrams. Moving away from the supports the shear stress decreases, until the stress is less than the allowable F_v and can be carried by the concrete. Then there is no need to continue stirrups. This distance (a) is determined by the formula:

$$a = \frac{L}{2} \times \left(\frac{v'}{F_v} \right),$$

where L = length of span, in feet, and a = distance from *each support* where web reinforcing is required. This formula is illustrated in the following examples:

Span length L = 18.0 feet: b = 8.0 inches and d = 20.0 inches. Reaction R = V = 18,000 Lbs. J = 0.864 and allowable F_v = 60 PSI. Stress intensity at support: $v = \frac{V}{Jdb}$.

Then total stress $v = \frac{18{,}000}{0.864 \times 20.0 \times 10.0} = 104$ PSI.

For stirrups: $v' = v - v_c$ or $v' = 104 - 60 = 44$ PSI. These values can now be placed in the formula for distance (a).

$a = \left(\frac{18.0}{2}\right) \times \left(\frac{44}{104}\right) = 9.0 \times 0.422 = 3.80$ feet. The first stirrup should always be placed as close to support as possible and never over ⅓ depth of beam.

Reinforcing bond stress 4.6.4

It is essential that the steel reinforcing rods be securely embedded in the concrete with good adhesion of the steel to the concrete throughout the full length of the span. Rod placement requires careful inspection before placing concrete and constant attention during placing and vibrating. This adhesion of concrete to steel is referred to as bond stress and identified by the small letter u. The allowable stress for this bond in most building codes is:
For ASTM A305 deformed top rods,
allowable $u = \frac{3.4\sqrt{Fc'}}{D}$.
For ASTM A305 tension bottom rods,
allowable $u = \frac{4.8\sqrt{Fc'}}{D}$.
For ASTM A408 large, square top bars,
allowable $u = 2.1\sqrt{Fc'}$.
For ASTM A408 large, square bottom bars,
allowable $u = 3\sqrt{Fc'}$.
In the formula for A305 rods, D is the nominal diameter of the rod in inches. Since these formulas give different results for each size of rod, they are seldom used by designers because they are difficult to remember.

An older, more convenient formula, which is used for allowable bond stress in the Table 4.3.6.2 design coefficients is $u = 3\sqrt{Fc'}$. This formula will provide a conservative stress for all practical uses, and will be used in all design examples. The use of this formula is supported by many experienced designers who subscribe to the theory that the cement paste in the mix has more influence on bond adhesion rather than the size of the rod.

The bond stress is checked after the rod size and number has been established to resist the bending moment. If the bond stress is found to be over the allowable, the rod ends are bent to form hooks. However, it may be desirable to use a greater number of smaller rods which will supply the same cross-sectional area, but offer more surface area for bond. Bond stress is calculated by adding the number of rods times the perimeter of each rod, and is indicated: Σo = summation of perimeters of compression or tension rods. For example, the total area of four #7 rods is written: $\Sigma o = 4 \times 2.75 = 11.0$ square inches.

The equation for bond stress in a slab or beam is: $u = \frac{V}{\Sigma o\, Jd}$. To illustrate, assume the shear at the support is 20,000 pounds, and the effective depth Jd = 16.0 inches. Rods are four #7, with $\Sigma o = 4 \times 2.75 = 11.0$ square inches. Intensity of bond stress $u = \frac{20{,}000}{11.0 \times 16.0} = 113.5$ PSI.

To find the sum of perimeters required, transpose the formula: $\Sigma o = \frac{V}{u\, Jd}$. Table 4.3.6.5 gives the summation of perimeters for various sizes of standard rods used in groups from two to sixteen rods. This table will be found convenient for investigating bond stress, and will save the designer considerable labor when making rod selection.

EXAMPLE: Locating neutral axis and calculating beam bending stress 4.6.5.1

A rectangular beam section consist of the following:
$b = 9.0$ In. $d = 18.0''$ $A_s = 2.50^{\square''}$ $n = 12$

REQUIRED:
Calculate the location for Neutral Axis, then assume that beam has a bending moment of 600,000 Inch Pounds and calculate the unit stresses in concrete and steel under bending load.

STEP I:
The basic formula for a balanced design may be written as: $b\bar{y} \times \frac{\bar{y}}{2} = nA_s(d-\bar{y})$. Where the dimension from top of beam to neutral axis $= \bar{y}$.
Substituting values in formula:

$$9\bar{y} \times \frac{\bar{y}}{2} = 12 \times 2.5(18.0 - \bar{y}) \quad \text{or} \quad 4.5\bar{y} + 4.5\bar{y}^2 = 12 \times 2.5 \times 18.0 = 540$$

Then: $4.5\bar{y} + 4.5\bar{y}^2 = 540$ and $\bar{y}^2 + 4.5\bar{y} = 120$ To complete the square: If $A_s^2 = 2.50^2 = 6.25$ and equation becomes:

$$\bar{y}^2 + 4.5\bar{y} + 6.25 = 120 + 6.25 \quad \text{or} \quad (\bar{y} + 2.5)^2 = 126.25 \quad \text{so,} \quad \bar{y} + 2.5 = 11.27 \text{In.}$$

Therefore $\bar{y} = 11.27 - 2.50$ or 8.77 Inches to NA. (Kd).

STEP II:
To calculate the stresses, the moment lever (Coupling) has to be determined. Depth to steel: $d = 18.0''$ Middle third of Concrete compressive triangle $= \frac{1}{3}\bar{y}$ or $\frac{8.77}{3} = 2.92$ inches.

Moment arm $= d - \frac{\bar{y}}{3}$ or lever $= 18.00 - 2.92 = 15.08$ inches. (Jd).

STEP III:
Calculating stress in concrete and steel with $M = 600{,}000''\#$.
Compression C must balance Tension T, or $C = T$.

Either C or $T = \frac{M}{Jd} = \frac{600{,}000}{15.08} = 39{,}788$ Lbs.

Compression in Concrete: $f_c = \frac{39{,}788}{8.77 \times 9.0} = 504^{\#\square''}$ (same as $\frac{C}{Kdb}$).

Tension in steel: $f_s = \frac{39{,}788}{2.50} = 15{,}915^{\#\square''}$ (same as $\frac{T}{A_s}$).

DESIGNER'S NOTATION:
This example should be compared to Example 4.3.3.1 for locating the Neutral axis if not clearly understood.

EXAMPLE: Evaluating beam bending stress 4.6.5.2

A rectangular beam is 10.0" x 24.0" with depth to steel d=22.0"
Reinforcing consists of 2-#9 ⌀ Rods with As=2.00□" $n=15$

REQUIRED:

Calculate the probable 28 day concrete strength design stress Fc' when Fs=16,000 PSI. Locate the neutral axis, then compute the Resisting Moment of cross section.

STEP I:

The basic formula: $b\bar{y} \times \frac{\bar{y}}{2} = nA_s(d-\bar{y})$. Where $\bar{y}$ = distance to NA.
b = 10.0" d = 22.0" As = 2.00 Sq.In. $n=15$ With values in formula:

$10\bar{y} \times \frac{\bar{y}}{2} = 15 \times 2.00 \times (22.0-\bar{y})$ equals $5\bar{y}^2 + 30\bar{y} = 660$

and $y^2 + 6\bar{y} = 132$ To complete the square, add As^2 or 4.00

Now $\bar{y}^2 + 6\bar{y} + 4.00 = 132 + 4.00$ or, $\bar{y} + 4.00 = \sqrt{136}$

Then $\bar{y} = 11.66 - 2.00 = 9.66$ inches

STEP II

Dimension $\bar{y}$ is equivalent of kd and dimension to center of compression is: $z = \frac{kd}{3}$ and $z = \frac{\bar{y}}{3}$ or $z = \frac{9.66}{3} = 3.22$ inches.

Moment lever couple is distance from C to T:
Moment arm = d-z or 22.0-3.22 = 18.78 inches. (same as Jd).

STEP III:

Resisting Moment: M = As Fs Jd. M = 2.0 x 16,000 x 18.78 = 600,960"#
T = As Fs. T = 2.00 x 16,000 = 32,000 Lbs. C must equal T.
Area concrete $A_c = b\bar{y}$ and $f_c = \frac{C}{A_c}$ Ac = 10.0 x 9.66 = 96.6□"

Then $f_c = \frac{32,000}{96.6} = 332$ PSI.

STEP IV:

Compressive stress in concrete runs from zero at NA to max. at extreme top fibers. Average stress = $\frac{Fc}{2}$ and Fc = 0.45 Fc'.

Average stress = 332#□" Fc = 2 x 332 = 664 PSI. Required 28 day strength of Concrete = $\frac{664}{0.45}$ = 1475 PSI or Fc' = 1500 PSI (Approx.)

STEP V:

A cross-section drawing is made for records. Hatched area above NA indicates concrete area in compression. This area appears to be unusually large which is due to the low grade of concrete and high n value.

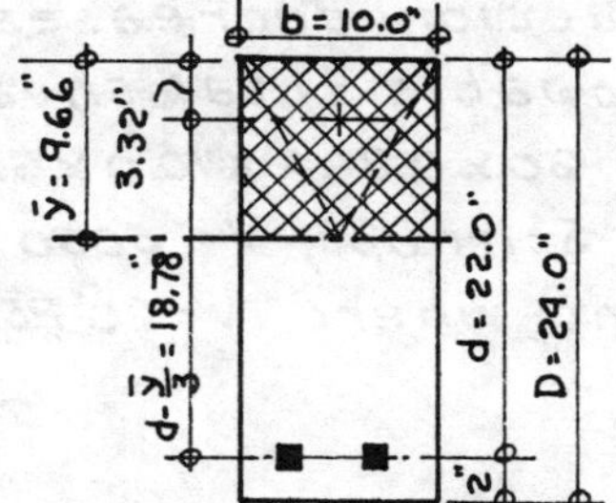

EXAMPLE: Designing tension steel and web stirrups 4.6.5.3

A short rectangular beam is simply supported with a clear span of 10.0 feet and reactions from the uniform load upon beam is 5000 Pounds at each end. Width of beam is 6.50 inches and depth to steel is 10.0 inches. Concrete at 28 day strength $F_c' = 3000$ PSI. $F_s = 18{,}000$ PSI.

REQUIRED:

Design cross section for tension steel and web stirrups. Furnish data to draftsman with a sketch of elevation showing rod placement. Use this sketch to indicate the critical point of failure by diagonal tension.

STEP I:

$R_1 = R_2 = 5000$ Lbs. each. $W = 2 \times 5000 = 10{,}000$ Lbs. $M = WL/8$ and center of moments will be taken as $L = 11.0$ feet.

$$M = \frac{10{,}000 \times 11.0}{8} = 13{,}750 \text{ Ft. Lbs.}$$

STEP II:

Design factors taken from tables:

$F_c = 1350$ PSI $F_v = 60$ PSI $u = 165$ PSI $J = 0.864$ $x = 0.408$

$K = 238.0$ $F_s = 18{,}000$ PSI. $R = V = 5000$ Lbs. $b = 6.50''$ $d = 10.0''$

$$A_s \ \frac{M}{F_s J d} \text{ or } A_s = \frac{13{,}750 \times 12}{18{,}000 \times 0.864 \times 10.0} = 1{,}06 \text{ Sq.In.}$$

Requires 2-#7ϕ Rods giving a steel area $A_s = 1.20$ Sq.In.

STEP III:

For shear and web reinforcement: $V = 5000$ Lbs. $v = \dfrac{V}{Jdb}$

$$v = \frac{5000}{0.864 \times 10.0 \times 6.50} = 89 \text{ PSI. (Includes steel and concrete)}$$

For stirrups: $v' = v - F_v$ or $v' = 89 - 60 = 29$ Lbs. sq. inch.

Length required for stirrups: $a = \dfrac{L}{2} \times \dfrac{v'}{v}$ or

$a = \dfrac{10.0}{2} \times \dfrac{29}{89} = 1.63$ Feet. This distance will be checked by location a point on beam where $v = 60$ PSI allowable.

Reaction decreases 1000 Lbs. per foot as moved to right.

Allowable total shear for concrete only: $V = F_v Jdb$, or

$V = 60 \times 0.864 \times 10.0 \times 6.50 = 3370$ Lbs. $w = 1000$ #/'

For Stirrups, $V = 5000 - 3370 = 1630$ Pounds.

Point where $v = 60$ PSI: $a = \dfrac{1630}{1000} = 1.63$ Feet. (Checks OK).

EXAMPLE: Designing tension steel and web stirrups, continued 4.6.5.3

STEP IV:
Selecting size for stirrups. Maximum spacing, $s = 0.45d$.

Try using #3ϕ Rods. $A_\phi = 0.11\square''$ With 2 Legs, $A_s = 0.22\square''$

spacing $s = \dfrac{A_s F_s}{V' b}$ or $s = \dfrac{0.22 \times 18{,}000}{29 \times 6.50} = 21.0$ inch centers.

This spacing is well over maximum of $0.45 \times 10.0 = 4.50$ inches and smaller rods for stirrups can be used. Try again and use the smallest in size or a #2ϕ Rod. $A_\phi = 0.05\square''$ and 2 Legs: 0.10 Sq.In. Spacing $s = \dfrac{0.10 \times 18{,}000}{29 \times 6.50} = 9.55$ inches.

Rods are still too large but must be used with $s = 4.50''$cc.

STEP V:
Checking bond stress: Perimeter of a #7ϕ tension rod $= 2.75\square''$ and with 2 Rods: $\Sigma_o = 2 \times 2.75 = 5.50$ Sq. In. Allowable $u = 165$ PSI.

$u = \dfrac{V}{\Sigma_o J d}$ or $u = \dfrac{5000}{5.50 \times 0.864 \times 10.0} = 105$ PSI. Bond stress is less than allowable however cantilever and single span beams should have hooked ends to be on safe side.

STEP VI:
In drawing rod placement in elevation two smaller rods will be required to hold stirrups in vertical position. Bottom tension rods will be bent up at points $\frac{L}{5}$ and the 2 lower rods will be set with hooked end. Also, these rods will be lapped and tied to lower tension rods near the point of bending up on 45° angle. Critical point of possible failure is indicated near left support.

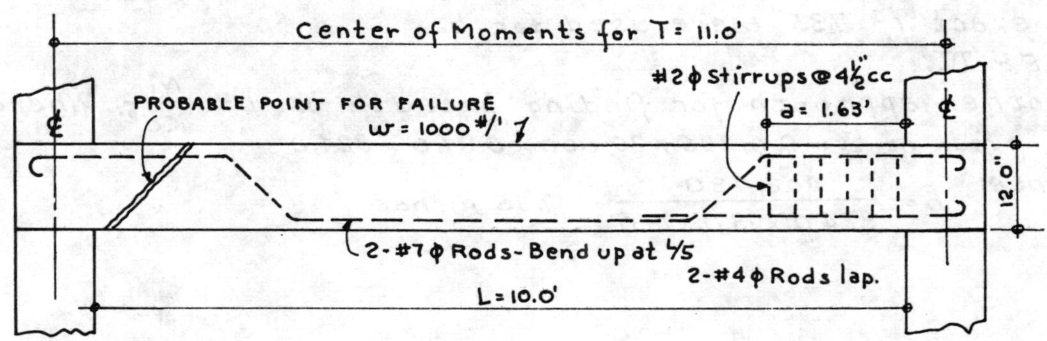

EXAMPLE: Beam design using percentage of steel factor 4.6.5.4

A cross-section has a depth to steel $d = 13.0$ inches and $As = 1.80$ Sq. Inches. A balanced design is desired. $n = 8$. $Fc = 1800$ PSI and $Fs = 20{,}000$ PSI.

REQUIRED:
Calculate the breadth of beam b. Use formulas to solve for necessary design coefficients k, j, p, K and the resisting moment.

STEP I:
Basic coefficient is k. By formula: $k = \dfrac{n}{n + \left(\frac{Fs}{Fc}\right)}$. With values

$$k = \frac{8}{8 + \left(\frac{20{,}000}{1800}\right)} = 0.419$$

Can now solve for $p = \dfrac{Fc\,k}{2\,Fs}$.

$$p = \frac{1800 \times 0.419}{2 \times 20{,}000} = 0.01885$$

Then $b = \dfrac{As}{pd}$ or $b = \dfrac{1.80}{0.01885 \times 13.0} = 7.35$ In.

STEP II:
With fire and rust protection added, the size of beams cross section is approximately 7.50" x 15.0".

$J = 1.00 - \left(\frac{k}{3}\right)$ or $J = 1.00 - \left(\frac{0.419}{3}\right) = 0.860$

Resisting Moment = $As\,Fs\,Jd$. $RM = 1.80 \times 20{,}000 \times 0.860 \times 13.0 = 402{,}480$ "#

STEP III:
$kd = 0.419 \times 13.0 = 5.45$ inches = NA dimension. Concrete area above NA = kdb, or $Ac = 5.45 \times 7.50 = 40.875$ □" Average compressive stress in concrete above NA is $\frac{Fc}{2}$ or $\frac{1800}{2} = 900$ PSI. (Allowable).

$C = 40.875 \times 900 = 36{,}780$ Lbs. $T = As\,Fs$. $T = 1.80 \times 20{,}000 = 36{,}000$ Lbs.

Difference between C and T is less than 1 square inch of concrete above NA and is considered balanced and would be exact if 7.35" were used for b.

STEP IV:
Another approach for finding breadth b: $b = \dfrac{M}{Kd^2}$. Where $K = pFsJ$ or $K = 0.01885 \times 20{,}000 \times 0.860 = 324.0$

Hence: $b = \dfrac{402{,}480}{324.0 \times 13.0 \times 13.0} = 7.35$ inches

EXAMPLE: Complete design of simple span beam 4.6.5.5

A concrete reinforced beam is 12.0 inches wide with a 22.0 inch depth to center of steel. Beam is simply supported on a 16.0 foot clear span and is reinforced for tension with 4-No.7 ϕ Rods. Beam carries a load of 30,000 Lbs., uniformly distributed over entire span. $E_s/E_c = 10.1$

REQUIRED:
Add the weight of beam to load and calculate the following:

(a) Maximum stress in Concrete and Steel.
(b) Distance along beam where web reinforcement ceases to be necessary. Allowable $v_c = 55$ PSI.
(c) Proper spacing from support for #3 ϕ stirrups. Neglect value of bent up bars and use $F_t = 16{,}000$ PSI for stirrups.
(d) Make a 3/8 inch scale elevation of 1/2 beam length and indicate bending and stirrup spacing dimension points.

STEP I:
Former E for steel was given as 30,000,000 PSI when the formula for n was E_s/E_c. Then $E_c = \frac{30{,}000{,}000}{10.0} = 3{,}000{,}000$ PSI.
This E_c called for concrete of 2500 PSI.
Thus: $F_c' = 2500$ PSI. $F_c = 1125$ PSI. $F_v = 55$ PSI. $U = 150$ PSI $F_s = 16{,}000$ PSI.
From tables: $k = 0.415$ $K = 201.0$ $J = 0.862$

STEP II:
Calculate weight of beam at 150 Lbs. Cubic foot. Total depth of beam with steel protection = 22.0 + 2.00 = 24.0" or 2.0 feet.
Beam contains 2.0 Cubic feet per lineal foot. $w = 2.0 \times 150 = 300$ #/'
Total Load = 30,000 + (300 x 16.0) = 34,800 Lbs. = W

STEP III:
Bending Moment: $M = \frac{WL}{8}$ or $M = \frac{34{,}800 \times 16.0}{8} = 69{,}600$ Foot Lbs.

Area Steel: $A_s = \frac{M}{F_s J d}$ A_s = 4-#7 ϕ Rods or $A_s = 2.40$ Sq. In.

Transpose formula to solve for: $f_s = \frac{M}{A_s J d}$. With values in formula:

$$f_s = \frac{69{,}600 \times 12}{2.40 \times 0.862 \times 22.0} = 18{,}350 \text{ PSI} = \text{Answer (a)}$$

Concrete $M = \frac{f_c k J b d^2}{2}$ or $f_c = \frac{2M}{J k b d^2}$

$$f_c = \frac{2 \times 69{,}600 \times 12}{0.862 \times 0.415 \times 12.0 \times 22.0 \times 22.0} = 804 \text{ PSI} \quad \text{Answer (a).}$$

EXAMPLE: Complete design of simple span beam, continued 4.6.5.5

STEP IV:

Web reinforcement: W = 34,800 Lbs. V = 17,400 Lbs. F_v = 55 PSI.

$v = \frac{V}{Jdb}$. Shear intensity $v = \frac{17{,}400}{0.862 \times 22.0 \times 12.0} = 76.5$ PSI.

Shear for stirrups: $v' = 76.5 - 55 = 21.5$ PSI.

Length required on beam: $a = \frac{L}{2} \times \frac{v'}{v}$. $a = \frac{16.0}{2} \times \frac{21.5}{76.5} = 2.25$ Ft. (Ans. b)

STEP V:

Stirrup spacing: $s = \frac{A_s F_s}{v' b}$ Using #3 φ Rods and F_s = 16,000 PSI.

Area #3φ Rod = 0.11 □" with 2 legs $A_s = 2 \times 0.11 = 0.22$ Sq. In.

spacing: $s = \frac{0.22 \times 16{,}000}{21.5 \times 12.0} = 13.6$ inches.

Maximum spacing must not exceed 0.45d. Then $s = 0.45 \times 22.0 = 9.90$ In. (Ans. c)

STEP VI:

Half elevation of Beam with reinforcement is shown. All reinforcing shown is symmetrical about midspan ℄. Bond was not calculated because it is customary to hook ends of rods in simple span beams. Place first stirrup 3.0 inches from support and bend up 2 bottom rods at L/5.

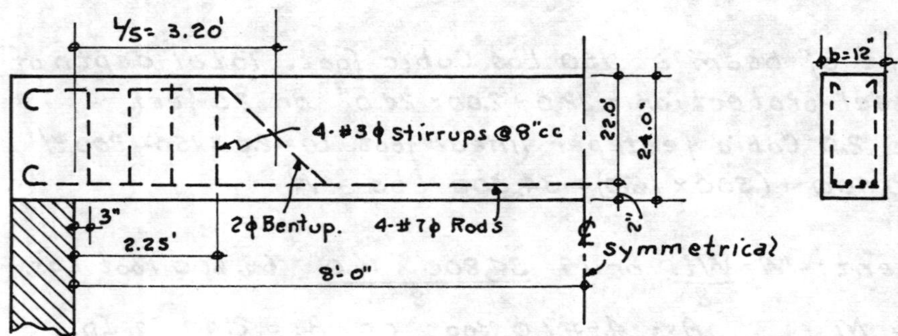

DESIGN NOTE:

Since maximum spacing is 9.90 inches and length required from support is 27 inches, the sketch above is to be drawn on plans. Note that 3 spaces at 8 inches, plus 3 inches at support equals 27 inches. Spacing is uniform and does not exceed the code.

EXAMPLE: Reducing beam section with high-strength materials 4.6.5.6

A rectangular or square beam must support a bending moment of 20,000 foot pounds. Established width is 9.0 inches and depth is limited to a maximum of 11.0 inches. Beam is used as a lintel. Other materials used on project calls for concrete to be $F_c' = 2500$ PSI (28 day) and $F_s = 16{,}000$ PSI. Stirrups are not desired and ends of tension rods shall be hooked.

REQUIRED:

Design a balanced beam using same allowables as given. If the section is oversize use $F_c' = 5000$ PSI and $F_s = 30{,}000$ PSI. Check the shear value maximum for each beam without web stirrups.

STEP I:

For 1st. beam section: $F_c' = 2500$ PSI, $F_c = 1125$ PSI. $F_v = 55$ PSI. $u = 150$ PSI.
Factors for design: $k = 0.415$ $K = 201.0$ $J = 0.862$ $b = 9.0$ in. $M = 20{,}000'\#$.

Formula for $d = \sqrt{\dfrac{M}{Kb}}$ or $d = \sqrt{\dfrac{20{,}000 \times 12}{201.0 \times 9.0}} = 11.62''$ (over 11.0" limit)

With 1½" rod protection: $D = 11.62 + 1.50 = 13.12$ inches.

Shear value without stirrups: $V = F_v Jdb$ or $V = 55 \times 0.862 \times 11.62 \times 9.0 = 4{,}975\#$.

STEP II:

Area tension rods: $A_s = \dfrac{M}{F_s Jd}$ or $A_s = \dfrac{20{,}000 \times 12}{16{,}000 \times 0.862 \times 11.62} = 1.495\,\square''$

Use 2-#8 ϕ Rods with $A_s = 1.57\,\square''$

STEP III:

Designing with High Strength concrete and steel for 2nd. beam:
$F_c' = 5000$ PSI: $F_c = 2250$ PSI., $F_v = 78$ PSI, $u = 250$ PSI. $F_s = 30{,}000$ PSI.
Design factors from Tables: $K = 346.0$ $J = 0.884$ $k = 0.347$

$d = \sqrt{\dfrac{20{,}000 \times 12}{346.0 \times 9.0}} = 8.80$ inches. Then $D = 8.80 + 1.50 = 10.30''$ (Call it 10½ in.).

STEP IV:

Area steel: $A_s = \dfrac{20{,}000 \times 12}{30{,}000 \times 0.884 \times 8.80} = 1.03\,\square''$ Use 2-#7 ϕ Rods with $A_s = 1.20\,\square''$

Shear value at support: $V = 78 \times 0.884 \times 8.80 \times 9.0 = 5{,}450$ Lbs.

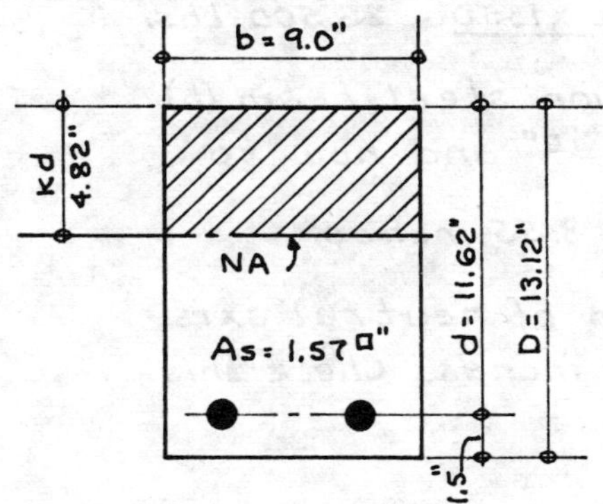

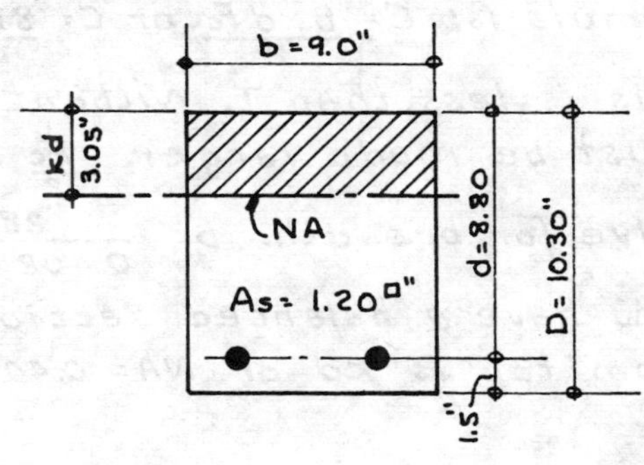

EXAMPLE: Rectangular beam design for bending moment 4.6.5.7

A rectangular beam with a length of 18.0 feet has a bending moment 22,000 Foot Pounds. Specifications for steel and concrete are as follows: F_s = 18,000 PSI and at age of 28 days F_c' = 3000 PSI. This beam is uniformly loaded.

REQUIRED:

(a) Determine the dimensions b and d for beam.
(b) Calculate the required area of steel rods.
(c) Check the tension force T against the compression force C so that compression rods are not required.
(d) Investigate the shear stress to determine if web reinforcement is required.
(e) Examine the bond stress to determine whether ends of rods should be hooked.

Use tables to collect design factors. Slide rule results will be acceptable when sufficiently close.

STEP I:

Gathering the applicable design factors from tables:
F_s = 18,000 PSI. F_c' = 3000 PSI $F_c = 0.45 \times 3000 = 1350$ PSI.
$k = 0.408$ $K = 238.0$ $J = 0.864$ $F_v = 60$ PSI $u = 164$ PSI.

STEP II:

To find the depth d and lever Jd, assume for trial that b = 8.00 In.

$$d = \sqrt{\frac{M}{Kb}} = \sqrt{\frac{22{,}000 \times 12}{238.0 \times 8.00}} = 11.8 \text{ inches. Call it 12.0 inches.}$$

Area steel: $A_s = \dfrac{M}{F_s Jd}$ or $A_s = \dfrac{22{,}000 \times 12}{18{,}000 \times 0.864 \times 12.0} = 1.42$ Sq. In.

From rod table: Try using 2-#8 ϕ with $A_s = 1.57^{\square''}$ $\Sigma o = 6.28^{\square''}$

STEP III:

Tension $T = A_s F_s$ or $T = 1.57 \times 18{,}000 = 28{,}250$ Lbs. Must = C.
Compression average stress from NA to top is $\frac{F_c}{2}$ and formula is: $C = \dfrac{bkdF_c}{2}$ or $C = \dfrac{8.0 \times 0.408 \times 12.0 \times 1350}{2} = 26{,}500$ Lbs.

This is less than T. Without compression steel, beam (b) must be made larger. $\frac{F_c}{2} = \frac{1350}{2} = 675\ \#^{\square''}$ and now to solve for breadth. $b = \dfrac{28{,}250}{0.408 \times 12.0 \times 675} = 8.55$ inches.

Now have a balanced section. Location of neutral axis from top is kd or $NA = 0.408 \times 12.0 = 4.90$ inches. Check this:

EXAMPLE: Rectangular beam design for bending moment, continued 4.6.5.7

Area of concrete in compression above neutral axis is kdb and to check: $C = \frac{kdbFc}{2}$ or $C = \frac{0.408 \times 12.0 \times 8.55'' \times 1350}{2} = 28,250^{\#}$ OK.

STEP IV:

To determine Reaction and value of V at support, assume a simple span beam where $M = \frac{WL}{8}$. Then $W = \frac{8M}{L}$. $L = 18.0$ Ft.

$W = \frac{8 \times 22,000}{18.0} = 9,800$ Lbs.

$R = V = \frac{W}{2}$ or $V = \frac{9,800}{2} = 4900$ Lbs. $f_v = \frac{V}{Jdb}$. Allowable $F_v = 60$ PSI

$f_v = \frac{4900}{0.864 \times 12.0 \times 8.55} = 55.3$ PSI Web stirrups not required.

STEP V:

Investigating bond: Allowable $u = 164$ PSI and $\Sigma o = 6.28$ Sq. In.

Formula for bond stress: $u = \frac{V}{\Sigma o Jd}$. Putting values in formula:

$u = \frac{4900}{6.28 \times 0.864 \times 12.0} = 75.4$ PSI. Rod choice is OK and ends need not be hooked.

To make adequate space for rods and providing rust and fire protection make the size of beam as shown thus. Other information shown for review:

Hatched portion above NA is concrete in compression.
d = Depth to steel.
D = Total depth.
b = Breadth or width of beam.
Kd = Distance to neutral axis NA.
z = Distance to compression center
Jd = Coupling, moment arm and effective depth.
Area of concrete below NA contributes no resistance to tension.
Shear area is Jdb.

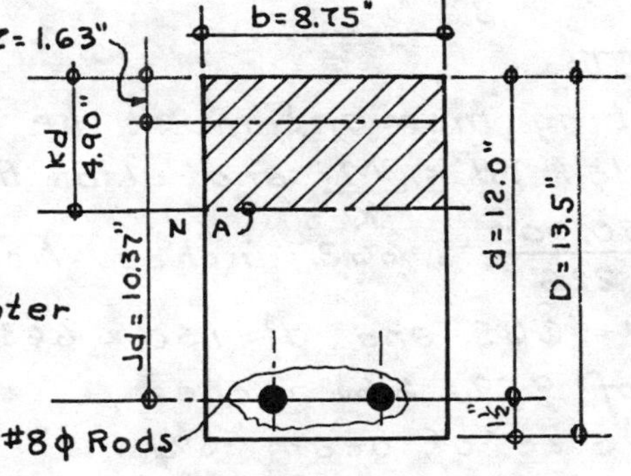

SECTION

EXAMPLE: Rectangular beam design for simple span 4.6.5.8

A concrete rectangular beam is simply supported on a clear span of 10.0 feet and must support a uniform load of 1000 Lbs. per foot including dead weight of beam. Proportion beam section to make width approximately 2/3 depth. $F_c' = 3000$ PSI. $F_s = 20{,}000$ PSI.

REQUIRED:
A beam section design which will result in a balance of compression concrete area above neutral axis to tension value of steel at effective depth Jd. Use the steel percentage (p) given in tables to check As. Make a drawing of beam elevation being supported on 12 inch masonry walls at ends. Use span L for center of moments.

STEP I:
Collecting design factors from tables:
$F_c = 1350$ PSI $F_s = 20{,}000$ PSI. $F_v = 60$ PSI $u = 165$ PSI, $k = 0.383$
$K = 226.0$ $J = 0.872$ $p = 0.0129$

STEP II:
Computing loads and bending moment: $M = WL/8$
$w = 1000$ #/' $L = 10.0'$ $W = 1000 \times 10.0 = 10{,}000$ Lbs. $V = 5000$ Lbs.

$$M = \frac{10{,}000 \times 10.0 \times 12}{8} = 150{,}000 \text{ Inch Lbs.}$$

STEP III:
Resisting moment must be equal to 150,000"# and formula: $bd^2 = \frac{M}{K}$ and also: $bd^2 = \frac{2M}{F_c J k}$. Hence:
$bd^2 = \frac{150{,}000}{226} = 663.7$ inches. Now if $2/3\, d = b$, then by ratio
$d \times d \times d = 663$ and $d^3 = 1.50 \times 663 = 996$ inches. $d = \sqrt[3]{996} = 9.99''$
$b = 2/3$ of $9.99 = 6.66$ inches.
Make size of beam $6\frac{3}{4} \times 10''$ $(6.75'' \times 10.0'')$, to fit forms.

STEP IV:
Area of cross-section $= 6.75 \times 10.0 = 67.50\,\square''$ and $p = 0.0129$
Area of steel should be approx: $A_s = A_c p$ or $A_s = pbd$.
Then: $A_s = 67.50 \times 0.0129 = 0.87$ Sq. In. Also formula: $A_s = \frac{M}{F_s J d}$.

$$A_s = \frac{150{,}000}{20{,}000 \times 0.872 \times 10.0} = 0.86 \text{ Sq. In. (Close enough to use).}$$

Selecting #6 φ Rods: 2 Rods have $A_s = 0.88$ Sq. Inches.

EXAMPLE: Rectangular beam design for simple span, continued 4.6.5.8

STEP V:

Check for Compress moment to equal Tension moment:

$M_c = K b d^2$ or $M_c = 226.0 \times 6.75 \times 10.0 \times 10.0 = 152{,}500$ In. Lbs.

This result would be same as Step II if b = 6.63 in.

Compressive steel is not required.

STEP VI:

Check shear to determine need for stirrups: V = 5000#.

$v = \frac{V}{Jdb}$ or $v = \frac{5000}{0.872 \times 10.0 \times 6.75} = 85$ Lbs. Sq. In. $F_v = 60$ PSI.

Shear for stirrups: $v' = 85 - 60 = 25$ Lbs. Sq. In.

Length from support stirrups required: $a = \frac{L}{2} \times \frac{v'}{v}$.

$a = \frac{10.0}{2} \times \frac{25}{85} = 1.47$ feet. Max. spacing = 0.45 d or 4.50 inches.

Maximimum spacing is close and a stirrup with small rods will be economical. Try using #2ϕ Rods: $A_\phi = 0.05$ □"

Value of 1 stirrup with 2 legs = 2 × 0.05 × 20,000 = 2000 Lbs.

spacing $s = \frac{A_s F_s}{v' b}$. $s = \frac{2000}{25 \times 6.75} = 11.85$ inches. Will have to use maximum spacing. Place first stirrup close to support.

STEP VII:

Checking bond stress: Using 2-#6ϕ Rods. Perimeter #6 = 2.36□"

$\Sigma_o = 2 \times 2.36 = 4.72$ □".

$u = \frac{V}{\Sigma_o J d}$ or $u = \frac{5000}{4.72 \times 0.872 \times 10.0} = 121.5$ #□". This is less than allowable of 165 PSI, however on simple beams it is the custom to be on the side of safety and hook rod ends. Only 2-#6ϕ Rods are used for tension and will not be bent up, therefore add 2 Rods at top to hold stirrups. Since the span is very short these rods will save labor if they are run full length of beam and ends hooked.

STEP VIII:

Drawing of cross-section and beam elevation showing rod arrangement with stirrups is drawn thus:

Note that 2 inches has been added to depth for fire and rust protection. The depth d found in step III is 10.0 inches or depth to steel.

EXAMPLE: Rectangular beam design for simple span, continued 4.6.5.8

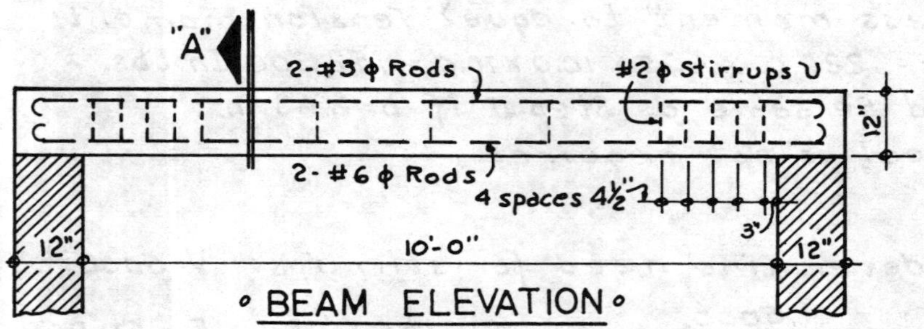

• BEAM ELEVATION •

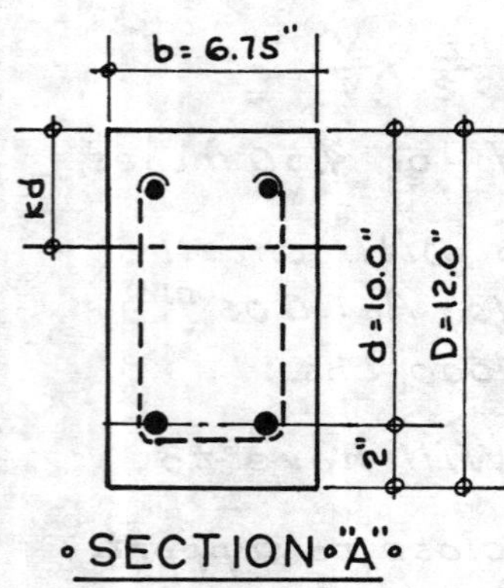

• SECTION • "A" •

STEP IX:

Checking a balanced design C = T.

kd = dimension for Neutral Axis.

Concrete area above NA is in compression and $A_c = kdb$.

$A_c = 0.383 \times 10.0 \times 6.75 = 25.85$ Sq. Inches.

Average stress in Concrete = $F_c/2$

Then value of $C = \frac{kdb\,F_c}{2}$

$C = \frac{25.85 \times 1350}{2} = 17{,}450$ Pounds.

STEP X:

Check for tension force T:

From step IV, $A_s = 0.87$ □" and $F_s = 20{,}000$ PSI.

$T = A_s F_s$, or $T = 0.87 \times 20{,}000 = 17{,}400$ Pounds.

Then C = T and design is a beam in balance.

EXAMPLE: Designing compression steel for continuous beam 4.6.5.9

A number of continuous rectangular beams are laid out with spans of 18.0 feet centers. Size of cross-section is: b = 12.0" depth to steel d = 18.0" and D = 20.0" Load per lineal foot of beam, w = 2500# and include dead weight of beam.
F_s = 18,000 PSI and F_c' = 2500 PSI $n = 10.1$

REQUIRED:
Complete design of beam shall include:
(a) Tension steel required and rod selection.
(b) Compression value of concrete above NA.
(c) Moment arm coupling between center of C and T.
(d) Moment and As for compression if required.
(e) Design of stirrups, selection and spacing.
(f) Check bond stress and lap bent up rods for negative moment.
(d) An elevation drawing for plans with reinforcement.

STEP I:
From tables gather and note applicable design coefficients:
F_c = 1125 PSI. F_s = 18,000 PSI. F_v = 55 PSI U = 125 PSI
$k = 0.387$ $K = 189.5$ $J = 0.871$ $p = 0.0121$ $w = 2500$ #/' $L = 18.0'$

STEP II:
Bending moment and value of tension T: $+M = wL^2/12$

$$+M = \frac{2500 \times 18.0 \times 18.0}{12} = 67{,}500 \text{ Ft. Lbs.}$$

$$+A_s = \frac{M}{F_s J d}. \quad +A_s = \frac{67{,}500 \times 12}{18{,}000 \times 0.871 \times 18.0} = 2.87 \text{ Sq. In. (Use same for } -M).$$

Select from table:
Room for only 4- #8 ϕ Rods. with $A_s = 3.16^{\square}$" $\Sigma_o = 12.57^{\square}$"

STEP III:
Concrete area in compression = $A_c = bkd$. $A_c = 12.0 \times 0.387 \times 18.0 = 83.6^{\square}$"
Compressive value $C = \frac{A_c F_c}{2}$ or $C = \frac{83.6 \times 1125}{2} = 47{,}000$ Pounds.
Tension value $T = A_s F_s$. $T = 3.16 \times 18{,}000 = 57{,}000$ Pounds.
T is greater than C and compression steel is required.

STEP IV:
Compressive Moment: $M_c = Kbd^2$ or CJd.
$M_c = 189.5 \times 12.0 \times 18.0 \times 18.0 =$ 737,000 Inch lbs.
$M_t = 67{,}500 \times 12 =$ 810,000 Inch lbs.
Moment compressive steel must sustain: $M_t - M_c$. Then the Steel $M_{cs} = 810{,}000 - 737{,}000 =$ 73,000 inch lbs.

EXAMPLE: Designing compression steel for continuous beam, continued 4.6.5.9

Compression steel $A = \frac{M}{F_s d'}$. Where d' = coupling arm between C and T.

Center of compression $C = \frac{1}{3}kd$ or $\frac{0.387 \times 18.0}{3} = 2.32$ inches.

Coupling arm d' = 18.0 − 2.32 = 15.68 inches.

STEP V:

$A_{cs} = \frac{73{,}000}{18{,}000 \times 15.68} = 0.259$ Sq. In. Select 3-#3 ϕ Rods with A_s = 0.33□"

STEP VI:

Checking shear for web reinforcement: $v = \frac{V}{Jdb}$.

Total Load W = 2500 × 18.0 = 45,000 Pounds. $V = \frac{W}{2}$

$V = \frac{45{,}000}{2} = 22{,}500$ Lbs.

Total $v = \frac{22{,}500}{0.871 \times 18.0 \times 12.0} = 120$ Lbs. Sq. In. Concrete F_v = 55 PSI.

Will require stirrups. Stirrups must take v' = 120 − 55 = 65 PSI.

Distance required: $a = \frac{L}{2} \times \frac{v'}{v}$. $a = \frac{18.0}{2} \times \frac{65}{120} = 4.88$ Feet.

Max. spacing = 0.45 d, or Mas s = 0.45 × 18.0 = 8.10 inch centers.

Try using No. 4 ϕ Rods. A_s = 0.20□" 2 Legs A_s = 0.40 Sq. In.

spacing $s = \frac{A_s F_s}{v' b}$. $s = \frac{0.40 \times 18{,}000}{65 \times 12.0} = 9.25$ inches on centers.

Maximum spacing must be limited to 8.10 inches. (0.45d).

STEP VII:

Checking bond stress: Allowed u = 125 PSI.

Using 4-#8 ϕ Rods from step II and Σo = 12.57□" $u = \frac{V}{\Sigma o J d}$.

$u = \frac{22{,}500}{12.57 \times 0.871 \times 18.0} = 114$ Lbs. Sq. In. (Below allowable and OK).

STEP VIII:

Preparing sketch of rod placement follows:

In continuous beams lap both bent up rods at top with same rods from other beams and full As is obtained for +M and −M.

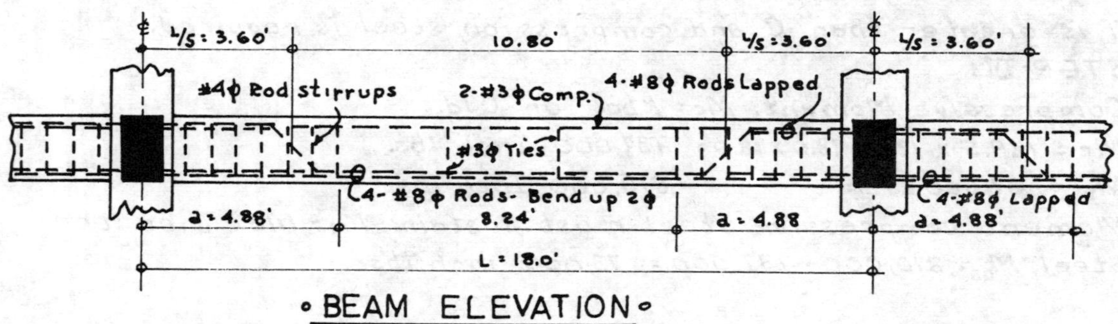

• BEAM ELEVATION •

Concrete T-beam design 4.7

T-beams have a cross-section which consists of a flange and stem. They are formed so that the slab flange and beam stem are poured in a single operation, and this results in a monolithic section. Forming for T-beams uses either inverted steel pans or built-up laminated plywood. Square steel pan forms are available which shape T-beams which extend in both directions. When removed, the underside of the slab has the appearance of a waffle or honeycomb. Frequently the straight type of floor system is referred to as ribbed construction.

The slab flange of the T-beam requires transverse tension rods, but due to the short, continuous span length, only a minimum amount of steel is required. The wide flange provides an excess of compressive concrete, which raises the neutral axis close to the bottom of the slab. The depth of a T-beam is usually controlled by the shear at the supports. One or two tension rods in the stem generally are sufficient to resist bending.

There are established rules for the design of T-beams. These rules govern the effective width of the flange in compression. A beam along a wall with the slab flange on only one side is called an L-beam. The examples to follow will illustrate the design step for both these beams.

The following rules govern the design of T-beams:

CASE 1
When the neutral axis of a T-beam falls in the slab flange area of the cross section, use the design formulas for rectangular beams.

CASE 2
When the neutral axis falls below the slab, in the stem area, the amount of compression usually is very small compared to the amount in flange, and may be neglected.

CASE 3
The formula for computing unit shear in a T-beam is: $v = \frac{V}{b'Jd}$. where b′ is the width of the stem. To determine the distance from the supports where web reinforcing is no longer required, the formula is: $a = \frac{L}{2}\left(\frac{v'}{v}\right)$, where v′ = unit shear to be resisted by stirrups and v = unit shear intensity at supports.

CASE 4
Flange widths shall be limited to the minimum dimension of the following:

(a) Flange width shall not exceed ¼ of clear span.
(b) Overhang on either side of stem shall not exceed eight times slab thickness.
(c) Overhang shall not be greater than ½ the clear space between stems of adjacent T-beams.

CASE 5
L-beams along wall with overhang on one side shall not exceed the minimum dimension:

(a) Width of flange overhang shall not be greater than 1/10 of clear span.
(b) overhang shall not exceed six times slab thickness.
(c) Overhang shall not be greater than ½ the clear space of the adjacent T-beam.

Effective depth in T-beams

In designing for Tension (T) and Compression (C), the value for Jd is given by the equation $Jd = d - \frac{t}{2}$. Therefore, the dimension z is ½ the slab thickness.

When the concrete moment of resistance for compression is to be computed, use the formula: $M_c = C \times \left(d - \frac{t}{2} \right)$,

or transposed: $C = \frac{M}{\left(d - \frac{t}{2} \right)}$.

To find the tension resisting moment for the steel, the formula is repeated with T replacing C: $M_s = T \times \left(d - \frac{t}{2} \right)$,

or transposed: $T = \frac{M}{\left(d - \frac{t}{2} \right)}$.

From these formulas, we derive the equations for finding the area of tension steel (As) and the intensity of the compressive stress (f_c) in the flange area:

$$As = \frac{M}{F_s \left[d - \left(\frac{t}{2} \right) \right]} \text{ and } f_c = \frac{2M}{bt \left[d - \left(\frac{t}{2} \right) \right]}$$

COMPRESSIVE STRESS

In the design for every slab, beam, or girder, investigate the compressive stress in the concrete above the neutral axis. If investigation shows that there is not enough concrete in flange to meet the compressive resistance required, the designer has the following choices:

(a) Increase the thickness of the slab.
(b) Increase the depth of the beam.
(c) Widen the stem (b′).
(d) Provide compression steel in the top of the beam.

In many T-beams, the flange width will be much greater than that required for adequate compression area. This is not important. What is important is that the *effective flange width* (that part of the flange which may be assumed to resist compression) has adequate area.

Wind load moment and negative steel 4.7.2

Wind pressure may produce positive or negative bending at the mid-span of T-beams and L-girders. Section VIII, dealing with wind forces on hi-rise buildings, illustrates the design system used to calculate these forces. Note that in the design of tall structures, these bending moments and shear values are carried by the columns, beams, girders and connectors. The slabs, joists, and smaller T-beams are not considered to carry any portion of the wind forces.

When designing a girder with the static load moment and the wind moment combined in a single moment value, the stress in the steel must not exceed the minimum yield stress of the grade of reinforcing steel specified. (Since maximum wind moments occur only at infrequent intervals, it is not necessary to limit the stress in the steel to the design working stress.) It is also advisable to use the yield stress to size the negative wind bending moment steel area. Any negative steel placed in the upper part of the girder will provide additional resistance to compression for the concrete area above the neutral axis.

Example 4.7.5 will illustrate the shortest method for calculating the required area of steel for static and wind load moments. Designers may be restricted by the applicable building code for this phase of the design.

Percentage of steel factor 4.7.3

Table 4.3.6.2 Design Coefficients gives a value for the percentage of reinforcement. For a balanced design, the formula for the area of steel is $A_s = pdb$. Transposed, the formula is $p = \frac{A_s}{bd}$. Many designers make use of the factor p for design, although this is not recommended because of the opportunities for error when a balanced design is not used. Where conditions permit, use the formula $A_s = \frac{M}{F_s Jd}$. When either dimension b or d is restricted due to architectural considerations, the safest policy is to consider the bending moment as the primary design criteria.

Other formulas may be derived to obtain the percentage of steel:

$$p = \frac{F_c k}{2 F_s} \text{ or } p = \frac{K}{F_s J} \text{ and } p = \frac{M}{F_s J b d^2}.$$

An example will follow which will illustrate the limits for which the factor p will give satisfactory results.

EXAMPLE: T-beam slab design for hi-rise 4.7.4

Architects plans call for a multi-story structure with column spacing 20.0 feet on center in each direction. Floor system is to be designed with T-Beams or formed with steel pans which produce ribbed beams. First floor portion of structure is to support mechanical equipment which will require a 200 Pound per square foot live load. Wind load bending moment and shear will be neglected as discussed in Section VIII for Hi-Rise projects. $F_c = 3000$ PSI. $F_s = 20{,}000$ PSI. A15 deformed rods to be used.

REQUIRED:

Prepare a structural floor plan and layout for T-Beams. Since each floor panel is 20.0' × 20.0' and beams will be in a continuous span arrangement, refer to rules for limited spacing between stems. Design for Moment as: $M = \frac{WL}{12}$.

STEP I:

Case IV (a) states that flange width shall not exceed ¼ span. Spacing of stems would be limited to: 0.25 × 20.0 = 5.0 feet. If slab thickness of 4.0 inches is used, overhang is limited to 8t or 8×4.0 = 32.0." Should stem width b' = 12.0", then width of flange is limited to: b = 32.0 + 32.0 + 12.0 = 76.0 inches

Also, rule states that overhang on each side shall not exceed ½ the clear distance distance between adjacent stems. Then stems are 5.0 foot on centers, and b' = 12.0 inches, space between stems = 5.0' - 1.0' = 4.0'. Flange width = 48.0" + 12.0" = 60.0 In. Appears two rules will govern flange width and b = 60.0 inches.

STEP II:

Loading: LL = 200 PSF Slab DL = 4.00 × 12.5 = 50 PSF. 150# Ft = stem DL.

Sq. Foot Load = 250 PSF. Load on 1 Tee-Beam = 250 × 5.0 = 1250 Lbs. Lineal Ft.

Total load on T-Beam = 1250 + 150 = 1400 Lbs. Lineal Foot.

W = 1400 × 20.0 = 28,000 Lbs. Shear = ½ W or V = 14000 Lbs. at girder.

STEP III:

Positive Bending moment: $M = \frac{WL}{12}$ or $M = \frac{28{,}000 \times 20.0}{12} = 46{,}667$ Foot Lbs.

Convert moment as M = 46,667 × 12 = 560,000 Inch Lbs.

STEP IV:

To determine depth d: Allowable unit shear for concrete when $F_c' = 3000$ is $v_c = 60$ PSI. From tables: J = 0.872 and U = 165 PSI.

By formula: $d = \frac{V}{b'Jv}$ or $d = \frac{14{,}000}{12.0 \times 0.872 \times 60} = 22.4$ inches.

EXAMPLE: T-beam slab design for hi-rise, continued 4.7.4

When 1.50 inches is added to 22.4 the total depth = 23.9 inches and is probably to great to suit architect. Stirrups will reduce the depth.

Stirrups can take 120 #□" of shear as v'. Concrete $v_c = 60$ #□"

Then $V_s = V' - V_c = 120 - 60 = 60$ PSI. Should this condition work, then depth d will be reduced to almost half. Then:

$$d = \frac{14{,}000}{12.0 \times 0.872 \times 120} = 11.15$$ inches. With 1½" rod protection, total depth of beams will be set at 13.0 inches.

In step II, the DL of stem was estimated at 150 Lbs. Foot and $13.0 \times 12 = 156$ Lbs. which is only 6 Pounds per foot over. This can be neglected.

STEP V

Computing for tension steel As:

$Jd = d - t/2$ or $Jd = 11.15 - \frac{4.00}{2} = 9.15$ inches.

$$A_s = \frac{M}{F_s\, Jd} = \frac{560{,}000}{20{,}000 \times 9.15} = 3.06 \text{ Sq. Inches.}$$

When $b' = 12.0$ inches, 4 φ Rods can be used and 4-#8 φ Rods have $A_s = 0.79 \times 4 = 3.16$ Sq. In.

Two of these rods will be bent up over supports at L/5.

STEP VI:

Determine if concrete above NA is adequate to resist compression:

$T = F_s A_s$: or $T = 20{,}000 \times 3.16 = 63{,}200$ Pounds. Location of NA is equal to kd. From tables: $k = 0.383$ and NA: $0.383 \times 11.15 = 4.27$ In.

Neutral Axis falls below bottom of slab about ¼ inches.

When $F_c' = 3000$ PSI, and $F_c = 1350$ PSI. Average compressive stress of concrete is taken as: zero at NA, and 1350 PSI at extreme top fibers. Average unit allowable stress is therefore: $\frac{F_c}{2}$ or $\frac{1350}{2} = 675$ PSI.

Neglect the concrete area between NA and bottom of slab.

The compressive value of a 4.0 inch slab per inch of width, is $4.0 \times 675 = 2700$ Lbs. Width of flange required to resist compression when C = T: $b = \frac{63{,}200}{2700} = 23.40$ inches.

Flange width was taken as 60.0 inch width, therefore area in flange slab is adequate.

EXAMPLE: T-beam slab design for hi-rise, continued 4.7.4

Since compression steel is not required, nor will flange slab need to be thickened, the next step must concern the slab in transverse direction.

STEP VII:

Checking a 1-Way slab with continuous spans of 4.0 feet clear between stems of T-Beam. In step II: $w = 250$ PSF.

$L = 4.0'$ $W = 250 \times 4.0 \times 1.0 = 1000$ Lbs. $M = \frac{WL}{12} = \frac{1000 \times 4.0}{12} = 333.3$ Ft. Lbs.

$As = \frac{M}{Fs\,Jd}$. Let $d = 3.75$ inches. $J = .872$

$As = \frac{333.3 \times 12}{20{,}000 \times 0.872 \times 3.75} = 0.0612$ Sq. In. This area can be filled by using welded wire mesh if desired. Shear $V = \frac{1}{2}W$ and is small in intensity for V_c. Slab appears to be satisfactory.

STEP IX:

Before calculations are made for stirrups and bond stress, a review of the design thus far will be drawn to scale. The plans and steel elevations are yet to be laid out and this sketch is essential to those items.

Note that depth to steel $d = 11.10''$ and $K = 0.383$ Depth to neutral axis is Kd and $11.10 \times 0.383 = 4.25$ inches.

STEP X:

Check for shear and web stirrups: from Step IV:

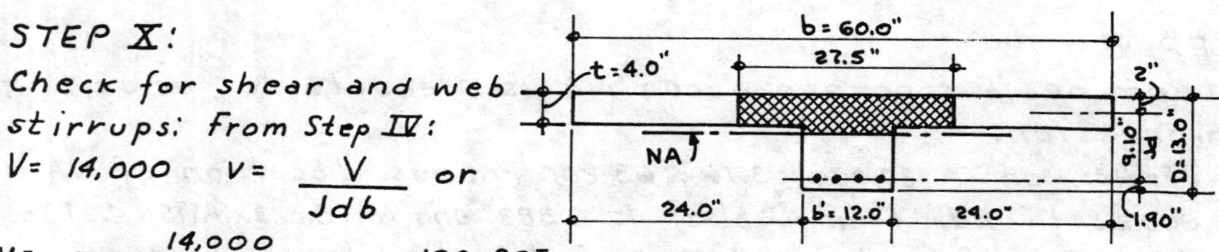

$V = 14{,}000$ $v = \frac{V}{Jdb}$ or

$v = \frac{14{,}000}{0.872 \times 11.10 \times 12.0} = 120$ PSI.

$F_v = 60$ PSI, then stirrups must resist: $v' = v - F_v$ or $v' = 120 - 60 = 60$ PSI.

For Length requiring stirrups: $a = \frac{L}{2} \times \frac{v'}{v_c}$. $a = \frac{20.0}{2} \times \frac{60}{120} = 5.00$ feet.

Maximum spacing: $s = 0.45d$ or $0.45 \times 11.10 = 5.00$ inches on centers.

A #3 ϕ Rod has $A\phi = 0.11$ □" and with 2 Legs $As = 0.22$ □" The value of 1-#3 ϕ stirrup = $AsFs$ or $0.22 \times 20{,}000 = 4400$ Lbs. $s = \frac{AsFs}{v'b}$, or spacing $s = \frac{4400}{60 \times 12.0} = 6.11$ inches. Use max. $s = 5.00$ inches.

STEP XI:

To check bond: $u = \frac{V}{\Sigma o\,Jd}$ Perimeters of 4-#8 ϕ Rods = 12.57 Sq. In.

Then: $u = \frac{14{,}000}{12.57 \times 0.872 \times 11.10} = 115$ PSI. (This is below allowable 165 PSI)

STEP XII:

Plan and sections will now be drawn. T-Beam designed is designated B-2.

EXAMPLE: T-beam slab design for hi-rise, continued 4.7.4

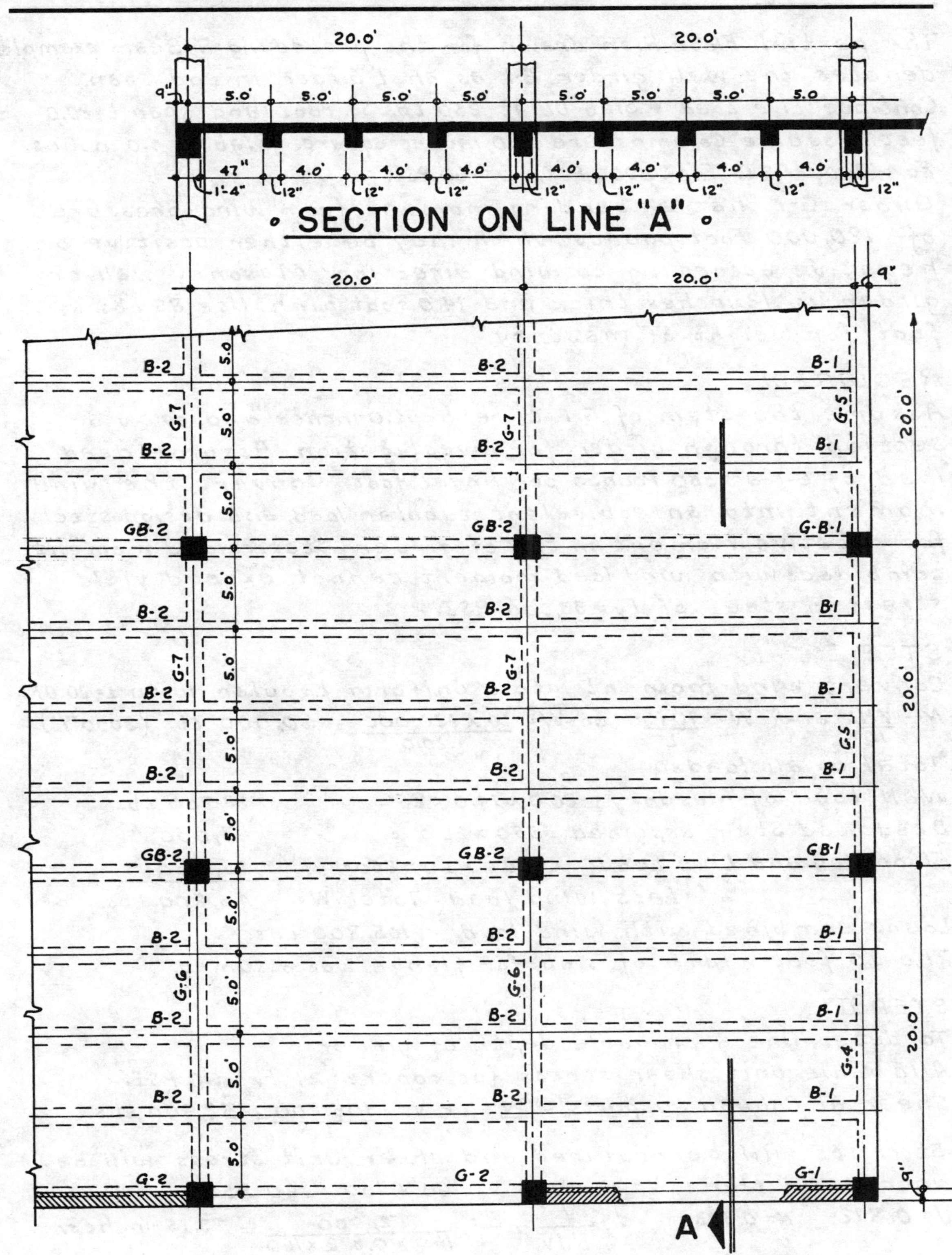

EXAMPLE: End-span L-girder design with wind moment 4.7.5

The partial Floor Plan drawn for the preceding T-Beam example denotes the wall girder G-1 as an L-Girder in end span. Continue Live Load + Slab DL at 250 Lbs. Sq. Foot and span $L = 20.0$ feet. Assume Columns are 18.0 inches square. Slab $t = 4.0$ inches. $F_c' = 3000$ PSI. $F_s = 20{,}000$ PSI. $F_v = 60$ PSI.

Girder G-1 has a bending moment from wind-pressure of 120,000 Foot pounds which may be either positive or negative according to wind direction. Masonry wall on girder is 12 inches thick and 14.0 foot high. Use 85 Lbs. sq. foot for weight of masonry.

REQUIRED:

Assume the stem of G-1 to be $b' = 16.0$ inches and draw a section through girder for investigation. Assume dead load of G-1 as 350 Pounds per lineal foot. Convert the wind moment into an equivalent tabular load and design steel for negative moment in top of girder. Static load moment combined with wind load moment cannot exceed yield stress in steel of $F_y = 33{,}000$ PSI.

STEP I:

Convert wind-moment into a uniform tabular with $L = 20.0$ Ft.

$M = \frac{WL}{10}$ and $W = \frac{10M}{L}$ or $W = \frac{10 \times 120{,}000}{20.0} = 60{,}000$ Lbs. (3000#/').

Total up all loads:

Wall Load of Masonry	$= 20.0 \times 14.0 \times 85 =$	23,800 Lbs.
Dead Load Stem assumed	$= 350 \times 20.0 =$	7,000 "
Floor Slab and Live Load	$= 20.0 \times 3.0 \times 250 =$	15,000 "
	Less wind load - Total W =	45,800 Lbs.

Loads combined with wind load = 105,800 Lbs.

The 3.0 foot width of slab for flange was assumed.

STEP II:

To determine a probable depth of girder:

Allowable unit shear stress for concrete: $F_v = 60$ PSI.

Shear at Column supports $= \frac{W}{2}$ or $V = \frac{105{,}800}{2} = 52{,}900$ Lbs.

Stirrups will be required and shear unit stress will be increased thus: $V = V_c + V_s$. $V = 60 + 100 = 160$ PSI.

$J = 0.872$ $\quad k = 0.383$ $\quad d = \frac{V}{b'Jv}$., $d = \frac{52{,}900}{16.0 \times 0.872 \times 160} = 23.75$ inches.

EXAMPLE: End-span L-girder design with wind moment, continued 4.7.5

Try using a depth to steel as d=23.0", and total D= 25.0 In.
Now $Jd = d-(\frac{t}{2})$ or Jd = 23.0 - 2.00 = 21.0 inches. Slab t = 4.00"

STEP III:

A working drawing of girder section is necessary to find exact flange width (b) and calculate compressive area of concrete above neutral axis.

Flange width allowed by Rules:

(a) Max. 1/10 of span L = 1/10 of 20.0' = 24.0 inches.
(b) Max. 6 times slab thickness t = 6 x 4.0 = 24.0 inches.
(c) Max. 1/2 clear dimension between stems = 1/2 of 47.0" = 23.5 In.
Accept least (c) as governing factor for design.

STEP IV:

Locating Neutral Axis:
kd = NA distance from top.
K = 0.383 d = 23.0" kd = 8.80 In.
NA is below slab 4.8 inches

Compressive area concrete is above NA and total value for $C = \frac{A_c F_c}{2}$.

Area slab overhang
23.5 x 4.00 = 94.0 □"

Area Girder between top and NA = 16.0 x 8.80 = 140.8 □"

Total A_c = 94.0 + 140.8 = 234.8 □"

Compression $C = \frac{234.8 \times 1350}{2} = 158,490$ Lbs.

◦ SECTION GIRDER G-1 ◦

STEP V:

To determine area steel to resist static loads and wind load moment. When static moment and wind moment are combined the stress cannot exceed yield of F_y = 33,000 PSI. Neither must stress exceed F_s = 20,000 PSI when static load only is to be considered. Wind behavior could likely cause the wind momemt to become negative and steel in top of beam for this purpose will designed with stress based upon F_y.

Static Load +Moment: $\frac{WL}{10}$ or $+M = \frac{45,800 \times 20.0}{10} = 91,600$ Foot Lbs.

Wind Load +Moment = −Moment and M: 120,000 Foot Lbs.

Combined Static and Wind Moments: +M: 211,600 Foot Lbs.

EXAMPLE: End-span L-girder design with wind moment, continued 4.7.5

STEP VI:

Areas for required steel: Jd: 21.0 inches. $As = \frac{M}{Fs\,Jd}$

Without Wind: Static $^{+}As = \frac{91{,}600 \times 12}{20{,}000 \times 21.0} = 2.62$ Sq. In.

Steel required when static and wind moments are combined and yield stress Fy = 33,000 PSI is substituted in equation:

$$+As = \frac{211{,}600 \times 12}{33{,}000 \times 21.0} = 3.66 \text{ Square inches. (over } 2.62^{\square''} \text{ and OK)}$$

Required to satisfy wind moment = 3.66 - 2.62 = 1.04 Sq. Inches.

Steel in top of beam for negative wind moment:
Use full moment from wind and Fy.

$$-As = \frac{120{,}000 \times 12}{33{,}000 \times 21.0} = 2.08 \text{ Sq. Inches.}$$

STEP VIII:

Selecting rods for positive and negative bending:
Have room for either 4 or 6 rods. If 4 Rods used, $A\phi = 3.75/4 = 0.94^{\square''}$
This is close to a #9 ϕ with 1.00 Sq. In. Use 4-#9 ϕ bars in bottom and bend up 2 bars at points L/5.
For top negative steel: 2 #9 ϕ Rods give 2.00 Sq. inches and close enough. Use these 2 bars in top and straight with ends hooked into columns. Also use hooked ends on all bars in girder. Thus: Bond stress calculations are not needed.

STEP IX:

Compare value of Tension T to the value of C calculated in step IV, where C = 158,490 Lbs.
Tension T = AsFs. or T = 4.00 × 20,000 = 80,000 Lbs. (C is OK).

STEP X:

In step II size of girder was based on using stirrups and effective depth Jd = 21.0 inches. V = 52,900 Lbs. with wind load.
Maximum spacing = 0.45 d or Max. s = 0.45 × 23.0 = 10.35 inches.

Shear intensity $v = \frac{V}{Jdb}$ or $v = \frac{52{,}900}{21.0 \times 16.0} = 158\ \#^{\square''}$

Fv = 60 PSI and for stirrups, v' = 158 - 60 = 98 Lbs. Sq. In.

Length required: $a = \frac{L}{2} \times \frac{v'}{v}$ or $a = \frac{20.0}{2} \times \frac{98}{158} = 6.20$ feet.

EXAMPLE: End-span L-girder design with wind moment, continued 4.7.5

Try using #6 ϕ Rods.

Value of 1 Stirrup with 2 Legs = 2 x 0.44 x 20,000 = 17,600 Lbs.

$$\text{spacing} = \frac{17{,}600}{98 \times 16.0} = 11.20 \text{ inches.}$$ Will have to use max. of 10.35 in.

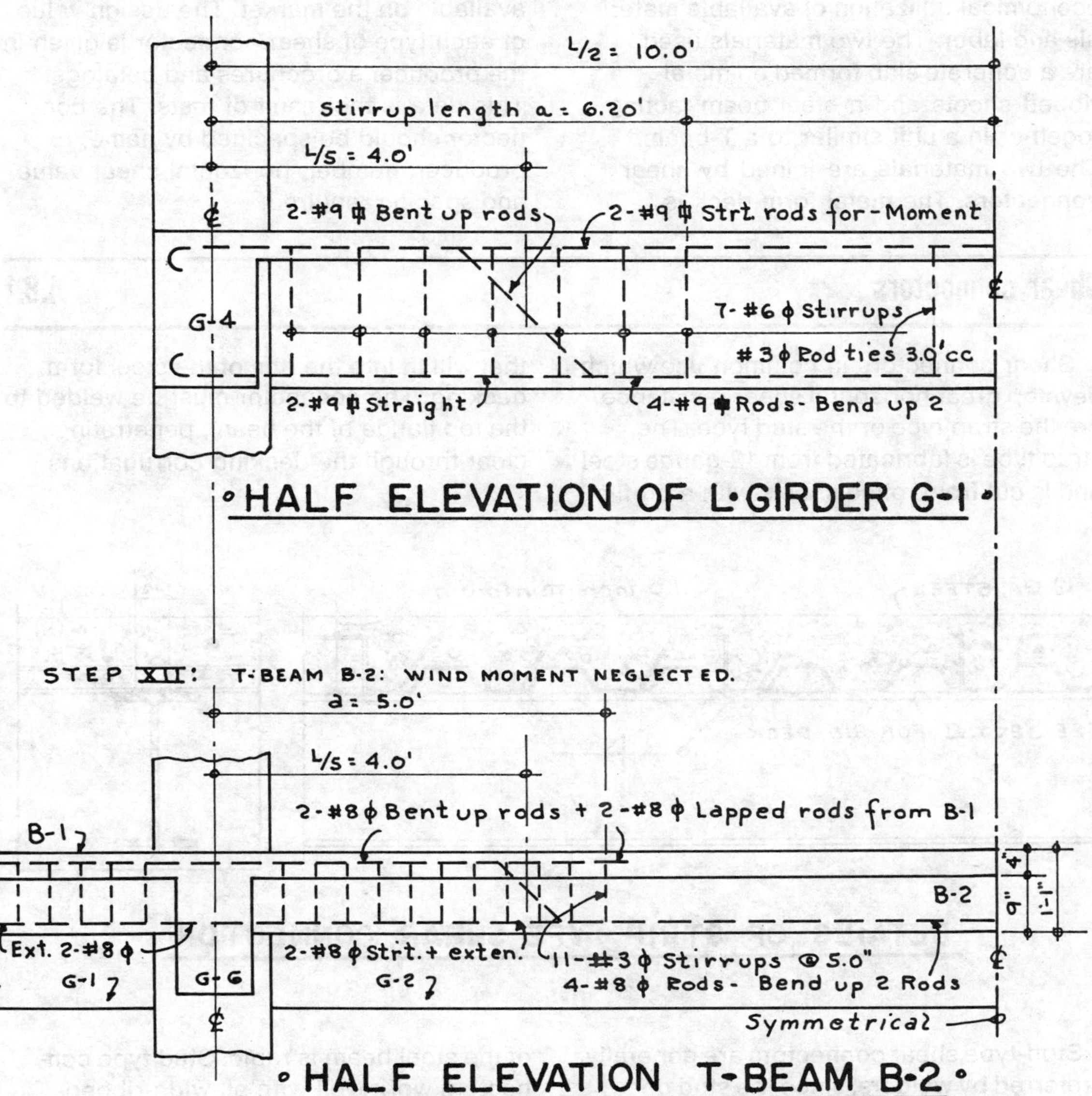

Composite slab design 4.8

Composite floor and roof slab construction is a relatively new technique which employs two materials which are interconnected to act as an integrated unit in resisting the superimposed load. The composite slab system is an example of the building industry's movement toward the economical utilization of available materials and labor. The two materials used are a concrete slab formed on metal ribbed sheets and a steel beam, acting together in a unit similar to a T-beam. The two materials are joined by shear connectors. The metal form deck is not considered to contribute any strength, and serves only for forming. The shear connectors provide the resistance to horizontal shear stress and preserve the firm connection between the slab and the steel beam.

Several types of shear connectors are available on the market. The design value of each type of shear connector is given in the producer's brochures and catalogs. This data is the result of tests. The connector should be specified by name, producer, number, horizontal shear value and spacing required.

Shear connectors 4.8.1

Shear connectors in common use which develop great horizontal shear resistance are the strap type or the stud type. The strap type is fabricated from 12-gauge steel and is cut from rolled sheets with a profile that will fit into the ribs of the steel form decking. The connector must be welded to the top flange of the beam, penetrating clear through the decking corrugations.

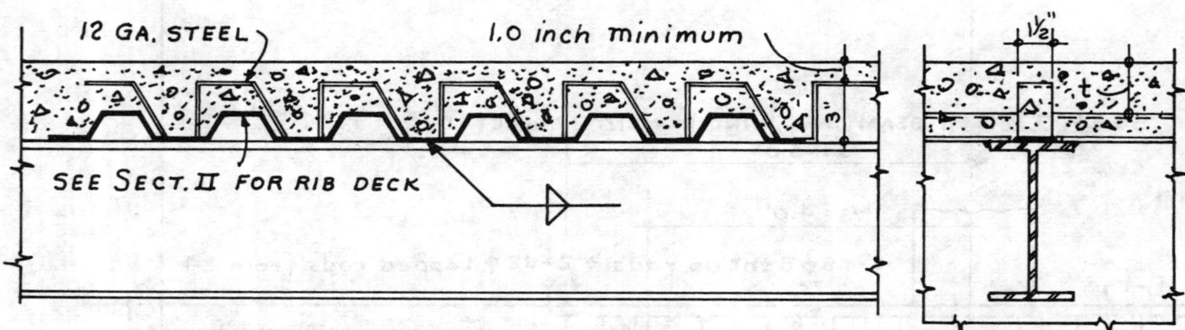

DETAILS OF STRIP TYPE SHEAR CONNECTOR

Stud-type shear connectors are generally preferred by welders, since the stud does not burn away and fusion with the top of the steel beam is rapid. Stud type connectors work well with all wide-ribbed decks, regardless of profile.

Shear connectors, continued 4.8.1

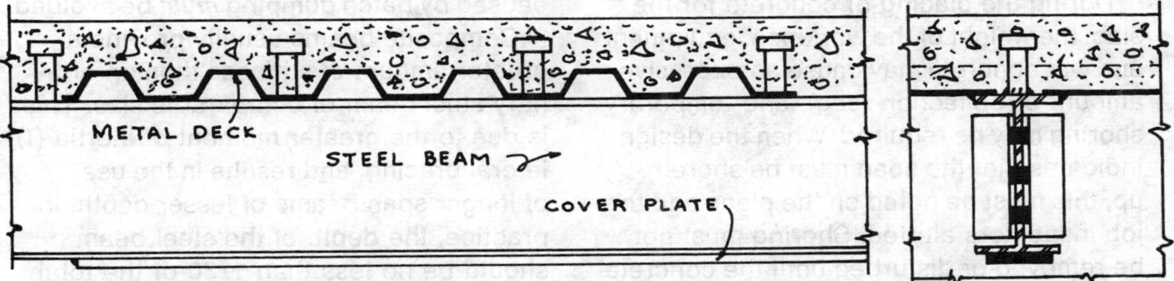

DETAILS OF STUD TYPE SHEAR CONNECTORS

SHEAR CONNECTOR VALUES

The horizontal shear resisting values of stud and strap connectors are based upon a 28 day concrete compressive strength at least 3000 PSI. Stud connector design values are as follows:

diameter	length	shear value
½ inch	2½ inch	5,100 pounds
⅝ ″	2½ ″	8,000 ″
¾ ″	3 ″	11,500 ″

Shear strap connectors of 12-gauge steel have a standard 3 inch height, and have a shear value of 15,000 pounds per foot. Minimum standard width and height is 1½″ x 2½″. The connector should project a minimum of 1 inch above the ribbed deck.

The symbol for connector value is q, and the symbol for number connectors required is N.

Composite section properties 4.8.2

Data and properties for composite sections are supplied in catalogs and steel manuals, and the producers of metal decking also give many tables in their brochures. Some tables provide data based on an effective flange width equal to 16 times slab thickness plus the width of the steel flange. They may also include properties for beams with and without cover plates. Most tables apply to 3000 PSI concrete and A36 steel, and should not be used when the value of n is less than 9.

EFFECTIVE FLANGE WIDTH

The effective width of the concrete flange is the least of the following:

(a) ¼ of beam span in inches, or

$$b_c = \frac{L \times 12}{4}$$

(b) Beam spacing, in inches.

(c) 2 x 8t plus the width of the steel beam flange. $b = 16t + b_s$. The thickness of slab t is taken from the slab top to the top surface of the steel deck, as shown in the examples.

Deflection and shoring 4.8.2.1

During the placing of concrete for the slab, the weight of the workmen, equipment and wet concrete may cause an excessive amount of deflection (sag), and temporary shoring may be required. When the design indicates that the span must be shored up, this must be noted on the plans and the job inspectors alerted. Shoring must not be removed or disturbed until the concrete has attained 70 percent of its strength; otherwise the shear connector bond may be damaged. Satisfactory composite floor systems require careful attention during slab placement and curing. Impact forces caused by batch dumping *must* be avoided.

Composite beams usually provide a greater amount of stiffness than an ordinary steel beam of comparable size. This is due to the greater moment of inertia (I), lateral bracing, and results in the use of longer span beams of lesser depth. In practice, the depth of the steel beam should be no less than 1/30 of the total span, and the composite section depth should be no less than 1/24 of the span. The formulas used to determine deflection will be as used in the design example to follow.

Unshored beam limits 4.8.2.2

The section modulus for lower fibers of the composite section cannot exceed the value given by the following formula:

$$\text{Maximum } S_b = 1.35 + \left[0.35 \left(\frac{M_l}{M_d} \right) \right] \times S_s,$$

where:

S_b = Section modulus of composite section.

M_l = Bending moment for live loads only.

M_d = Bending moment for dead loads only.

S_s = Section modulus for steel beam.

The formula is used in the later design steps to determine if shoring will be required, or it can be applied in the earlier stages to size the section to avoid the need for shoring.

Welding deck to beam 4.8.2.3

The intent of the composite system is to obtain rigidity, and the lateral beam support must not be neglected. Welding heavy gauge steel deck to each beam is a standard requirement. The shear connectors must be welded through the steel deck to the top flange of the steel beam. Deck properties and steel beam section tables are included in Section II. Composite beams permit wider beam spacing, which results in the use of a heavier deck to support the deck span. Select a deck long enough to span a minimum of four supports.

Concrete mix 4.8.2.4

Attention must be given to the size of coarse aggregate used with a particular type of ribbed decking. The narrow-ribbed types require gravel size of ¾ or 1 inch. Common, 1½ inch gravel is satisfactory for the wider-ribbed decks, such as type BH and N. The concrete must be fluid and worked well between the ribs and around the connectors to eliminate voids.

Wind load moment 4.8.2.5

When the composite beam system is used in Hi-Rise projects, the shear and bending moment due to wind pressure must be given careful consideration. A large girder can be designed between the columns to resist these forces, and take the place of the usual beam. An alternate method distributes the wind shear and moment to each composite section on an equal basis. Since steel girders are used to support the ends of the composite beams, the lateral horizontal forces can be resisted by the girder. Lateral bending force in the usual homogenous T-beam is not practical, unless this bending moment is resisted by placing steel in the side of the girder. When all the bending moments are combined with the wind moment, the stress must not exceed the elastic limit or yield stress of the steel.

Building code restrictions 4.8.2.6

Because the composite beam system for floors is a relatively new method, existing codes give little, if any, guide for design. In such a case, the structural designer should consult the local authority before proceeding with the initial design layout. All references should be based on the system as described in the AISC Manual of Steel Construction and the American Concrete Institute Building Code Recommendations for Concrete.

Fire resistance test ratings to submit are as follows:

Design No. C130–3 Underwriters Laboratories results for sprayed-on fiber or cementitious materials: Tests include: U.L. Design 43–3; 50–3; 58–3; 61–3; 62–3; 52–2; 90–2 and 203–2. Suspended Ceiling test reports include the following: U.L. Design 49–3; 68–3; and 73–3. Copies of the above reports and test data are available from sale engineers representing the metal deck producers.

EXAMPLE: Designing composite beam slab 4.8.3

An apartment project has preliminary plans drawn which denote span lengths to be 28.0 feet long and spacing at 7.50 feet. A composite floor system is being considered for economy reasons. A single span area is to be used for storage which will have a 150 PSF Live load. Slabs to be poured on 1½ inch deck and depth of slab including deck is 4 inches. F_c' = 3000 PSI. F_s = 24,000 PSI. Depth of steel beams is restricted to 18 inches. Ceiling to consist of ½ inch lay-in panels with suspension system.

REQUIRED:

Design of beam on basis of simple span. M = WL/8
Limit deflection under Live Load to 1/360 span L.
Limit deflection under Dead Load to 1.0 inch at mid-span.
Locate centroid and calculate properties of section.
Check design to determine if cover plate is required.
Check to determine whether temporary shoring is required.
Make sketches of slab plan and composite beam section which will illustrate the transformed area if n = 9.2

STEP I:

Floor framing plan is an aid for computing loads and will be drawn thus: (See following sheet).

Live Load = 150 PSF. Area supported by 1 beam = 7.50 x 28.0' or Area = 210 Sq. Ft. Live load, W_L = 210 x 150 = 31,500 Lbs.

Dead Loads: Slab = 4.0 x 12 = 48.0 Lbs. Sq. Ft.
Ceiling = 2.0 " " "
Total DL = 50.0 " " "

DL on beam + weight beam assumed at 40#/'
W_D = (210 x 50) + (40 x 28.0) = 11,620 Lbs.

Combined Loads: $W_L + W_D$ = 31,500 + 11,620 = 43,120 Lbs.

STEP II:

Calculate Bending Moments for 3 Conditions:

Maximum DL Mom; $M_D = \frac{11,620 \times 28.0}{8}$ = 40,670 Foot Lbs.

Maximum LL Mom: $M_L = \frac{31,500 \times 28.0}{8}$ = 110,250 Foot Lbs.

Total Combined Moment = 150,920 Foot Lbs.

EXAMPLE: Designing composite beam slab, continued 4.8.3

STEP III:

Calculating Section Modulus for 3 conditions: $S = \frac{M}{F_s}$.

Dead Load: $S_D = \frac{40,670 \times 12}{24,000} = 20.385''^3$

Live Load: $S_L = \frac{110,250 \times 12}{24,000} = 55.125''^3$

Combined $S_D + S_L = 75.510''^3$

STEP IV:

Determine width of effective slab flange for control b.

(a) $b = \frac{1}{4} L$ in. $b = 0.25 \times 28.0 \times 12 = 84.0''$

(b) $b =$ spacing $b = 7.50 \times 12 = 90.0''$

(c) $b = 16t +$ flange $b = (16 \times 2.50) + 7.00 = 47.0''$ (Controls).

($t = 4.00''$ - Deck thickness of 1.50") (Width of flange 7.00" assumed)

STEP V:

Choose a steel beam for further investigation:

Try a 16 WF 36# with $S_s = 82.6''^3$ (without cover plate $A_p = 0$)

Transformed area of concrete to steel equivalent $= \frac{b}{n} t$

width transformed Sect. $= \frac{47.0}{9.2} = 5.11''$ $A_T = 5.11 \times 2.50 = 12.78 \square''$

STEP VI:

Locating position of Centroid in Composite section using format recommended in Property Section VI: Take the moments for NA about base of steel beam: $I_o = \frac{bd^3}{12}$. (Rect.)

SECT.	SIZE	A□"	d"	Ad	ℓ"	Aℓ²	Io	Aℓ² + Io
1	16 WF36	10.59	8.00	84.72	5.84	361.12	446.30	807.42
2	5.11 X 2.50	12.78	18.75	239.63	4.91	308.00	6.65	314.65

A = 23.37□" Ad = 324.35 $I = 1122.07''^4$

Location of NA: Distance from Base Line $= \frac{324.35}{23.37} = 13.84$ inches

Distance $c = 13.84''$ and $c' = 20.00 - 13.84 = 6.16$ inches. Cross section of Composite beam will show these dimensions.

Section Modulus of Composite Beam: $S_b = \frac{I}{C} = \frac{1122.07}{13.84} = 81.07''^3$

Property of S_b exceeds requirements for same property when Dead Load and Live loads are combined.

For transformed area in Compression: $S_t = \frac{1122.07}{6.16} = 182.5''^3$

EXAMPLE: Designing composite beam slab, continued 4.8.3

STEP VII:

Check deflections: Max. Δ under Live Load = $\frac{28.0 \times 12}{360}$ = 0.93 In.
Max. Δ under dead loads limited to 1.00 inch.

$$\Delta = \frac{5Wl^3}{384EI} \qquad \Delta_L = \frac{5 \times 31{,}500 \times 37{,}933{,}000}{384 \times 29{,}000{,}000 \times 1122.07} = 0.48 \text{ inches (ok).}$$

$$\Delta_D = \frac{5 \times 11{,}620 \times 37{,}933{,}000}{384 \times 29{,}000{,}000 \times 1122.07} = 0.177 \text{ inches (ok).}$$

Deflection under combined Dead Load + Live Load is the critical design factor and also limited to 0.93 inches.

$$\Delta_{DL} = \frac{5 \times 43{,}120 \times 37{,}933{,}000}{384 \times 29{,}000{,}000 \times 1122.07} = 0.656 \text{ Inches.}$$

STEP VIII:

Check unit stresses in Concrete and Steel under combined Loads: In step III, the required $S_{DL} = 75.51''^3$ based on 24,000 PSI.

Tension stress in Steel $f_t = \frac{M_{DL}}{S_{DL}}$ or $f_t = \frac{150{,}920 \times 12}{81.07} = 22{,}340$ PSI.

Compressive stress in Concrete area: Allowable $F_c = 1350$ PSI.

$$f_c = \frac{M}{nS_t} \quad \text{or} \quad f_c = \frac{150{,}920 \times 12}{9.2 \times 182.5} = 1080 \text{ PSI. (Ok less than } F_c\text{).}$$

Section is adequate and will not require a cover plate to lower location of centroid.

STEP IX:

To check on whether temporary shoring will be required, the Section Modulus for composite section ($S_b = 81.25$) shall not exceed that obtained by the formula thus:

$$\text{Maximum } S_b = \left[1.35 + \left(0.35 \frac{M_L}{M_D}\right)\right] \times S_s$$

Moment used in this formula will be in foot pounds. S_s = Property of 16WF36.

$$\text{With values: } S_b = \left[1.35 + \left(0.35 \times \frac{110{,}250}{40{,}670}\right)\right] \times 82.6 = 190.0''^3$$

Result of formula in not exceeded and shoring will not be required during construction.

STEP X:

Designing for stud type shear connectors:

The steel beam controls the horizontal shear because the centroid must remain in the steel beam component.
Formula 18 AISC. Spec. 1.11.4 gives: $V_h = \frac{0.85 F'_c A_c}{2}$. This formula is for average compressive shear.

EXAMPLE: Designing composite beam slab, continued 4.8.3

The effective area of Concrete (A_c) is flange times clear thickness of concrete or $A_c = bt$. $F_c' = 3000$ PSI. $b = 47.0$ In. With values substituted in formula:

$$V_h = \frac{0.85 \times 3000 \times 47.0 \times 2.50}{2} = 150{,}000 \text{ Lbs. (Slide rule result).}$$

Selecting a stud shear connector 3/4"ϕ and 3" long, the value of one connector is 11,500 Lbs.

Required number $N = \frac{150{,}000}{11{,}500} = 13.03$

V_h is shear for 1/2 of span L and therefore N is doubled for full span. Space 26 studs at equal spaces and weld through deck to top flange of 16WF36. Deck must also be welded to beam from underside.

STEP XI:

Substituting strap type connectors for studs:
Value of strap 1½" wide, 3" high, 12 Gage and 1.0 foot long is 15,000 Lbs. Feet required $= \frac{150{,}000}{15{,}000} = 10.0$ feet for each side of mid-span. Cut and space 20.0 feet of strap uniformly across 28.0 foot beam length.

STEP XII:

Design for metal form deck: Live Load can be neglected and only dead load of wet concrete plus temporary work personnel will be considered. Deck must extend over 3 spans or more. Spacing beams or deck $L = 7.50$ feet and $M = \frac{wL^2}{12}$. Limit sag in deck to 1/240 span in inches.

$\text{Max.}\Delta = \frac{7.50 \times 12}{240} = 0.375$ inches. Refer to Metal Deck tables in section II and deflection formula for slab forms. The formula for triple span: $\Delta = \frac{0.0068\, w l^4}{EI}$ Transpose this formula and solve for I.

$I = \frac{0.0068\, w l^4}{E\Delta}$.

Dead load concrete	= 48	Lbs. Sq. Ft.
Assume workers	= 17	" " "
Assume deck weight	= 03	" " "
Design Load w	= 68	Lbs. Sq. Ft.

$l = 7.50 \times 12 = 90.0$ inches.

$l^4 = 65{,}610{,}000$ $E = 29{,}000{,}000$ $\Delta = 0.375"$

$$I = \frac{0.0068 \times 68 \times 65{,}610{,}000}{29{,}000{,}000 \times 0.375} = 2.79"^4$$

EXAMPLE: Designing composite beam slab, continued 4.8.3

Properties given in metal deck tables list a 14 Ga. Type N deck as having a 2.20"^{4} value for I, however the depth is 3.0 inches. An 1½ inch depth was used for design and a heavy gauge could be used if middle was shored up with temporary supports. Further calculations for slab steel reinforcing could possibly support a good percentage of dead loads.

STEP XIII:

Designing slab reinforcing steel for a 1-Way floor:

Dead Load slab and deck = 50 Lbs. Sq. Foot.

Live Load on floor = 150 " " "

Design Load = 200 Lbs. Sq. Foot.

Span L = 7.50 Ft. $M = \frac{WL}{12}$. W = 200 x 7.50 = 1500 Lbs. on strip 1.0'

Total depth slab = 4.0" Above deck t = 2.50 inches. depth to steel d = 2.50 - 0.75 = 1.75 inches. Concrete f'_c = 3000 PSI.

$M = \frac{1500 \times 7.50}{12} = 937.5$ Ft. Lbs. F_s = 18,000 PSI J = 0.864

$A_s = \frac{M}{F_s J d} = \frac{937.5 \times 12}{18,000 \times 0.864 \times 1.75} = 0.413$ Sq. Inches.

Select for trial #4 ϕ Rods with A_ϕ = 0.20 □"

$s = \frac{A_\phi \times 12}{A_s}$. Spacing $s = \frac{0.20 \times 12}{0.413} = 5.81$ inches center to center.

This may be too close for 1½ inch coarse aggregate, so try #5 ϕ Rods.

#5 ϕ Rod A_ϕ = 0.31 Sq. In. $s = \frac{0.31 \times 12}{0.413} = 9.00$ inches cc.

STEP XIV:

Slabs will require temperature steel to run parallel with the composite beams. Area concrete in 12.0 inch width at top of deck: A_c = 2.50 x 12 = 30 Sq. In. From table ratio A_s to A_c = 0.0020

A_{ST} = 0.0020 x 30 = 0.060 Sq. Inches per 12 inches of width.

Select #2 ϕ Rods A_ϕ = 0.05 □" spacing $s = \frac{0.05 \times 12}{0.06} = 10.0$ cc.

Accept #2 ϕ for Temperature steel.

AUTHORS NOTE:

The steel section in the composite beam design of this example is oversized and selected for the purpose of design sequence. Students are expected to compare the steel section with the tables provided in AISC Manual, select another section and investigate the results by following the same procedure.

b = 47.0"

5.11"

t = 2.5"

4.0"

6.6

1.50"

NA

7.0"

16 WF 36# SECT.

C = 13.84"

16.0"

TRANSFORMED $A_c = 5.11" \times 2.50" =$

$\frac{b}{n} t$ $A_T = 12.78\ \square''$

SECTION A

NA

3/16

16 WF 36# BEAM

SECTION B

L = 28.0'

16 WF 36

A

A

16 WF 36

#2 φ Temp Rods @ 10.0" cc - Straight

GIRDER

A

16 WF 36

7.50'

7.50'

7.50'

7.50'

30.0'

4" Concrete Slab on 16 Ga. Type B form deck - Shore up between beams

Slab Reinforcing #5 φ Rods @ 7½ cc Straight.

A

16 WF 36

B

All Concrete 3000 PSI AT AGE 28 DAYS

GIRDER

26 Stud Connectors ¾" φ 3.0" Each Beam

A

16 WF 36

TYPICAL PLAN COMPOSITE FLOOR SYSTEM

Flat slab floor systems 4.9

Flat slab construction describes a girderless slab, supported only by columns. Originally the flat slab system was a patented method, originated by Mr. C.A.P. "Cap" Turner, and was called the *"mushroom floor."* The patents have long since expired. The system has been used extensively for heavy industrial buildings and warehouses, because of the clear ceiling advantage. Electrical conduits, process piping and sprinkler systems can be installed under the ceilings with less difficulty than with beam and girder construction. Flat slab systems do not adapt well to interior architectural design unless the ceilings are suspended below the flared capital.

Forming for the slabs, dropped panel and columns is less expensive than for ribbed-beam and girder construction. Steel forms are available which can be moved and reused as the work progresses. The interior support columns incorporate spiral-wrapped hooped steel with vertical reinforcing, and are usually round. Wall and corner columns are rectangular, square or L-shaped. Near the column, the design is critical and more vulnerable to fire hazards, but flat slab resistance to fire damage is greater than beam and girder construction. The flat slab must have at least three continuous bays in each direction to meet the requirements of building codes. Bay panels may be either square or rectangular. Successive span lengths must not differ more than twenty percent of the shorter span.

Flat slab design 4.9.1

A flat slab floor is designed as a two-way reinforced slab. The reinforcement runs parallel to the columns center lines in both directions. The formulas used for moment distribution are empirical and have no mathematical basis. Building codes have adopted these formulas; frequently the designer will find that some codes contain modifications of the formulas given here. An illustration of a typical flat slab plan will show that the two-way slab is designed with two strips of reinforcing. One strip, extending a quarter panel on each side of the column center line, is called the *column strip*. The other strip is of half-panel width, and is known as the *middle strip*. Two sets of strips run at right angles to each other, thus the two-way design term. Always designate the column strip as A and the middle strip B, to correspond to the moment distribution tables which will be prepared.

Each strip of the whole panel is treated as a wide beam, and the bending moment is calculated for the whole panel. After the bending moment is obtained for the entire panel, certain percentages of the total moment are applied to act on strip A and the balance on strip B. The total bending moment is divided up between the two types of strips according to percentages given in Chart 4.9.1.2.

DROPPED HEADS AND CAPITALS

When the columns are spaced evenly with square panel bays, each column supports one whole panel load. In this case, the shear is concentrated around the

Flat slab design, continued 4.9.1

perimeter of the column, and provision must be made to resist this punching shear. This resistance to shear is accommodated by expanding the perimeter of the column with a flared capital. If the shear stress is still over the allowable concrete unit stress, the slab depth must be increased by using a dropped head between the capital and the slab. Punching shear stress should not exceed six percent of the ultimate 28 day concrete strength ($0.06F_c'$). The width of the dropped head in either direction should not be less than 3/2 of the diameter of column capital. The diameter at the top of the column capital should not be less than 1/5 the center-to-center column distance in the long direction of span.

THICKNESS OF DROPPED HEAD AND SLAB

The effective depth of the slab should include an allowance of ¾ inch cover over the rods. When the rods cross, make an allowance of one rod diameter plus 0.3 inch for deformations. Slab thickness should not be less than L/40, and should be sufficient to keep bending and shearing stresses within the limits of the applicable code. The minimum slab thickness is given by the following formula (but never less than 4 inches): $t_2 = 0.375L \sqrt[3]{\frac{2000}{Fc'}}$. For example, if L = 20.0 feet and Fc′ = 3000 PSI, the formula calls for a 6.53 inch minimum slab thickness.

The maximum total thickness at the drop panel (used in computing the negative area of steel for the column strip A) should be 1.5 times the thickness of slab t_2. In this case, the side dimension or diameter of the drop panel should not be less than ½ the span L in any direction.

WHOLE PANEL MOMENT

The total sum of the positive and negative bending moment in either direction of a rectangular panel may be obtained with the following formula:

$M_o = 0.09\,WL\left(1.0 - \frac{2c}{3L}\right)^2$, where:

W = total panel load
L = span length in feet.
c = effective support size.

When column capitals are used, the value of c is the diameter of the flared cone at the bottom of the dropped head or slab (when the dropped head is not used). When the column is without a flared capital, the dimension c is the diameter of the column.

COLUMN SIZES

Without a complete set of tables giving the safe load and diameter of hooped columns, the designer must calculate the diameter c by making a number of "cut and try" investigations. This may be a long, tedious process. A condensed Table 4.9.1.3 for this purpose is provided. In no case should a column supporting a flat slab have a minimum core dimension less than 10 inches, or have a minimum moment of inertia of less than 1000″⁴. (The formula used to calculate the value of I in a round, solid column is $I_c = 0.7854\,R^4$, where R = radius of column.)

TABLE: Moment distribution for two-way slab design 4.9.1.1

MOMENT DISTRIBUTION FOR 2-WAY FLAT SLAB DESIGN

STRIP MARK	PERCENTAGE OF TOTAL PANEL MOMENT		PANEL LOCATION
	WITH DROPPED HEAD	WITHOUT DROPPED HEAD	
A	+ MOMENT = 0.20	+ MOMENT = 0.22	INT. COLUMN
A	− MOMENT = 0.50	− MOMENT = 0.46	DO.
B	+ MOMENT = 0.15	+ MOMENT = 0.16	MIDDLE INT.
B	− MOMENT = 0.15	− MOMENT = 0.16	DO.

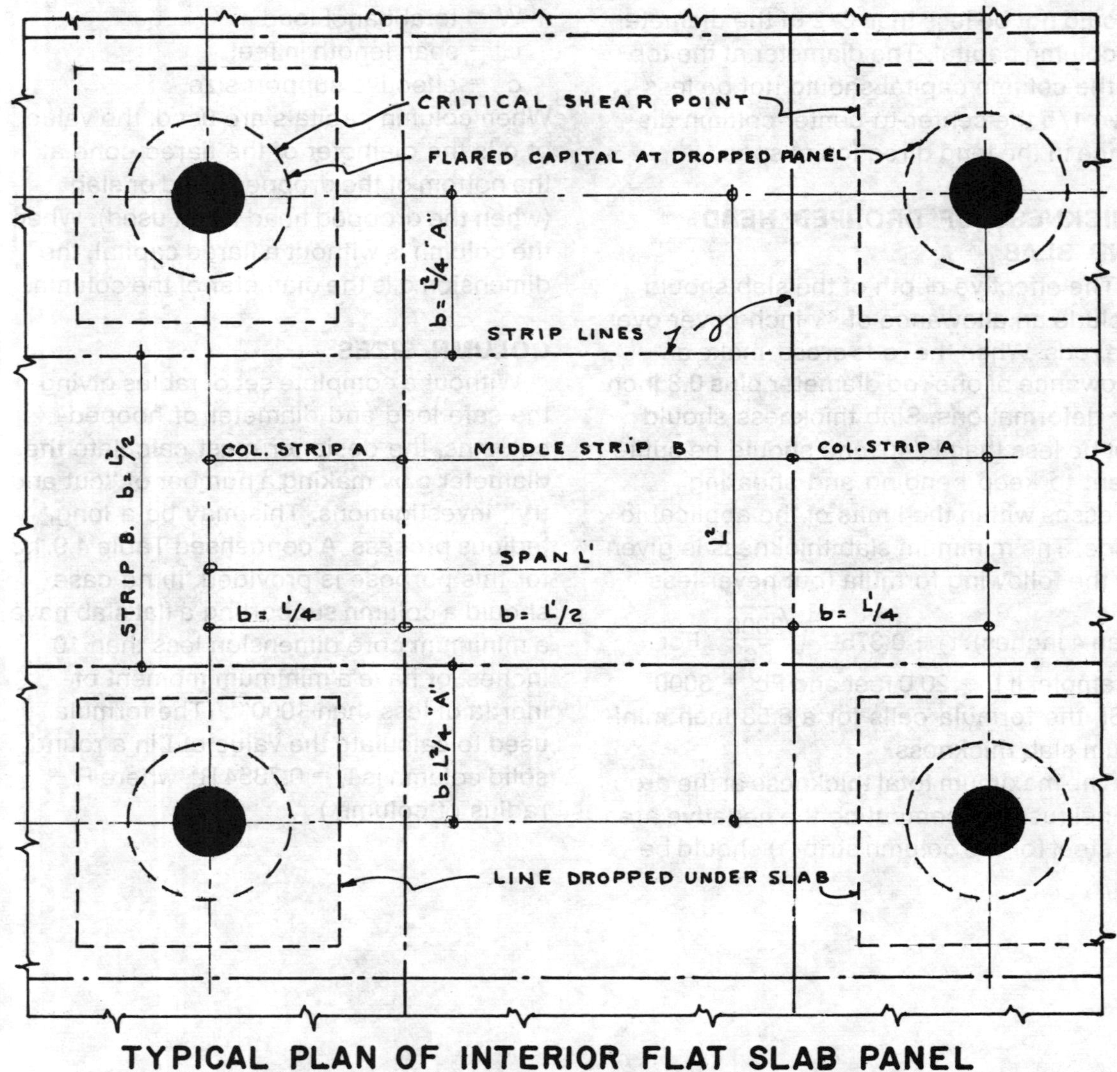

TYPICAL PLAN OF INTERIOR FLAT SLAB PANEL

CHART: Moment distribution for flat slab floor system 4.9.1.2

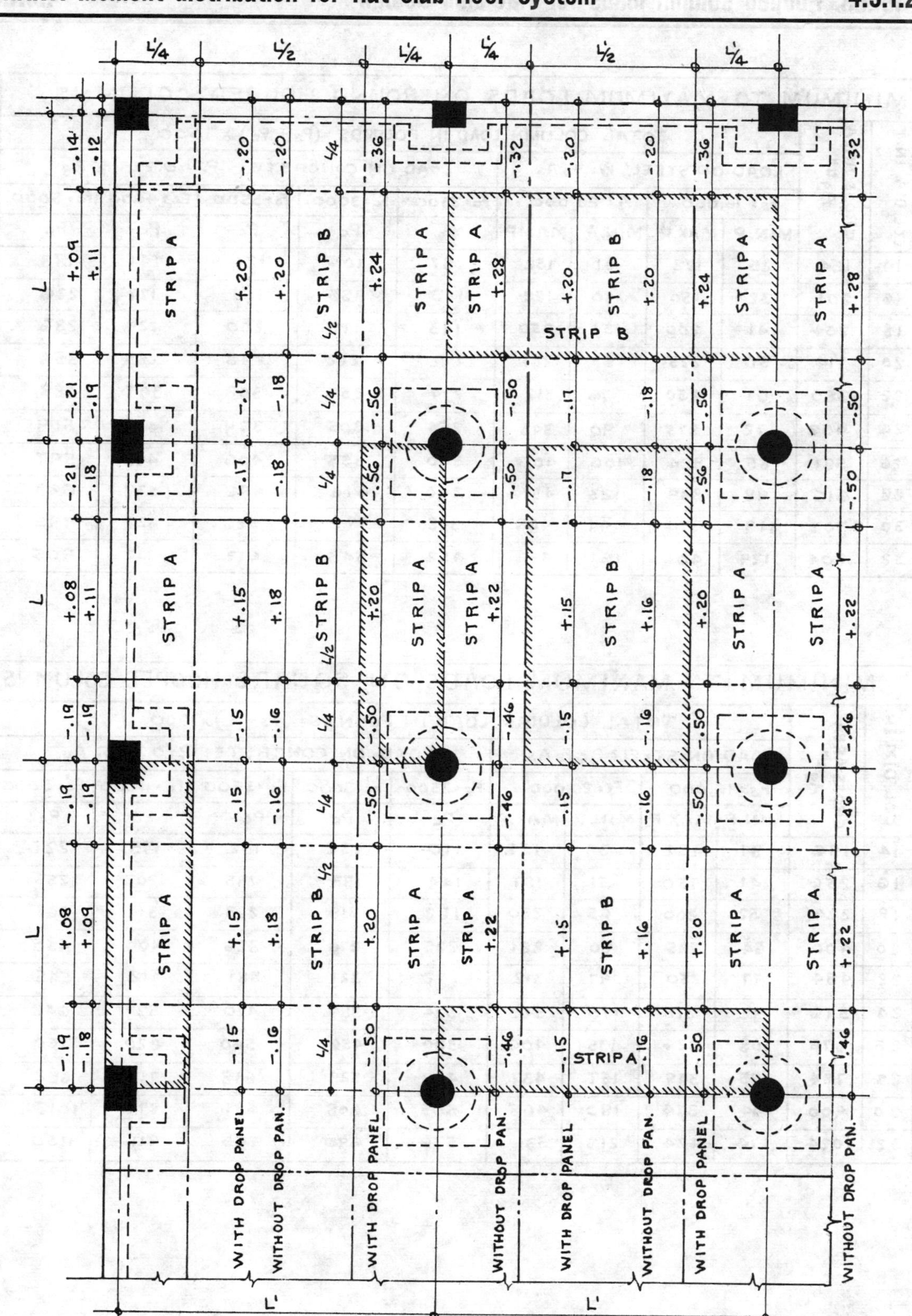

TABLES: Hooped column loads for flat slab design 4.9.1.3

MINIMUM TO MAXIMUM LOADS ON ROUND HOOPED COLUMNS										
COL. DIA. IN.	GROSS AREA A_g □"	TOTAL COLUMN LOAD IN POUNDS = $(P_s + P_c) \times 1000$								
		LOAD ON STEEL: $P_s = F_s A_s$				LOAD ON CONCRETE: $P_c = 0.225 F_c' A_g$				
		$F_s = 16{,}000$		$F_s = 20{,}000$		$F_c' = 2500$	$F_c' = 3000$	$F_c' = 3500$	$F_c' = 4000$	$F_c' = 5000$
		MIN. P	MAX. P	MIN. P	MAX. P	P_c	P_c	P_c	P_c	P_c
14	154	25	122	31	152	87	104	126	141	173
16	201	32	150	40	187	113	136	151	177	226
18	254	41	200	51	250	143	172	200	229	286
20	314	50	225	63	281	177	212	248	283	354
22	380	61	250	76	312	214	257	308	347	428
24	452	72	275	90	343	254	305	364	410	509
26	531	85	324	106	406	300	358	406	485	597
28	616	98	349	123	437	346	416	498	572	693
30	707	113	374	141	468	398	477	526	622	795
32	804	129	400	161	499	452	543	617	717	905

MINIMUM TO MAXIMUM LOADS ON SQUARE HOOPED COLUMNS										
COL. DIM. IN.	GROSS AREA A_g □"	TOTAL COLUMN LOAD IN POUNDS = $(P_s + P_c) \times 1000$								
		LOAD ON STEEL: $P_s = F_s A_s$				LOAD ON CONCRETE: $P_c = 0.225 F_c' A_g$.				
		$F_s = 16{,}000$		$F_s = 20{,}000$		$F_c' = 2500$	$F_c' = 3000$	$F_c' = 3500$	$F_c' = 4000$	$F_c' = 5000$
		MIN. P	MAX. P	MIN. P	MAX. P	P_c	P_c	P_c	P_c	P_c
14	196	31	122	39	152	110	132	152	173	221
16	256	41	150	51	187	144	173	185	202	253
18	324	52	200	65	250	182	219	263	311	365
20	400	64	225	80	281	225	270	326	404	450
22	484	77	250	97	312	272	327	381	412	545
24	576	92	275	115	343	324	389	470	527	648
26	676	108	324	135	406	380	456	560	620	760
28	784	125	349	157	437	440	529	618	715	882
30	900	144	374	180	468	505	608	691	877	1013
32	1024	174	424	218	531	576	690	766	911	1150

EXAMPLE: Calculating slab and drop panel depth 4.9.2

A flat slab panel has a column spacing of 20.0 feet in each direction or L= 20.0 feet. Flared diameter at top is dimension C with a 4.20 foot diameter. The combined Dead load plus Live Load is ω' = 295 Pounds per square foot. Concrete strength F_c' = 3000 PSI.

REQUIRED:

Calculate the total thickness, t_1, in inches for slab without drop panel, or through the drop panel if one is used. Use the following formula for minimum t_1:

$$t_1 = 0.028\,L\left[\left(1.00 - \frac{2C}{3L}\right)\sqrt{\frac{\omega'}{F_c'/2000}}\right] + 1.50$$

Also calculate the total thickness, t_2, in inches for slab with dropped panel, at points beyond the dropped panel. Use the following formula for minimum thickness t_2:

$$t_2 = 0.024\,L\left[\left(1.00 - \frac{2C}{3L}\right)\sqrt{\frac{\omega'}{F_c'/2000}}\right] + 1.00$$

STEP I:

It should be noted that dimensions C and L are given in feet, and t_1 and t_2 are in inches. ω' is given in pounds per square foot. With values substituted in formula:

$$t_1 = 0.028 \times 20.0\left[\left(1.00 - \frac{2 \times 4.20}{3 \times 20.0}\right) \times \sqrt{\frac{295}{3000/2000}}\right] + 1.50 = 8.29 \text{ or}$$

Reducing formula:

$$t_1 = 0.56\left[(1.00 - 0.14) \times \sqrt{197}\right] + 1.50 =$$

$$t_1 = 0.56(0.86 \times 14.1) + 1.50 = 8.29 \text{ inches, minimum.}$$

STEP II:

Formula for t_2 can be reduced since values enclosed in brackets are identical and will be used thus:

$$t_2 = 0.48(0.86 \times 14.1) + 1.00 = 6.82 \text{ inches, minimum.}$$

DESIGN NOTE:

The side dimension of a dropped head panel shall be not less than 0.33 L. In the above example the drop panel will be square and not less than 0.33 x 20.0 = 6.60 Ft.

EXAMPLE: Complete flat slab floor design 4.9.3

Refer to prepared plan of an interior panel plan of a flat slab. Assume plan of column spacing long-way is 24.0 feet. Let short way column spacing be 20.0 feet. Round columns from floor above are 18.0 inches in diameter and carry a load of 80,000 Pounds. Live Load for floor is to be not less than 300 Lbs. Square foot. Slab shall have a dropped head and flared capitals at each column. Concrete $F'_c = 3000$ PSI. $F_s = 16{,}000$ PSI. Use $F_v = 2\sqrt{F'_c}$ for allowable shear.

REQUIRED:

Design slab by empirical method. Continue round columns. Give size and thickness of dropped panel and flared type capital. Designate column strips A, and middle strips B for each direction, then show rods in plan with a section of plan with dimensions necessary for detailing.

STEP I:

Establish a column size to support panel corners and plus Loads on column above. Long way $L_L = 24.0'$ and $L_S = 20.0'$.

Area floor panel = 24.0 x 20.0 = 480 Sq. Ft.

Live Load on panel = 480 x 300 = 144,000 Lbs.

Assume slab $t_2 = 5.0''$ thick. Dead load = 63 Lbs. Sq. Foot.

Panel Dead Load = 480 x 63 = 30,240 Lbs.

Total Load = L.L + DL = 144,000 + 30,240 = 174,240 Lbs. = W.

Column Load: 80,000 + 174,240 = 254,240 Lbs. (Call it 255 kip Lbs.).

STEP II:

1 Column supports equivalent load of whole panel, and from table of Min.-Max. Round column Loads:

A 20.0 inch diameter column can be designed to support a load P thus: $P_s = 225{,}000^{\#}$ and $P_c = 270{,}000^{\#}$. Then when $P_s + P_c$ are totaled: P = 495,000 Lbs. A hooped column is designed with $F_s = 16{,}000$ PSI and $F'_c = 3000$ PSI. Use a 20.0" Column.

STEP III:

To determine size of flared capital:

Minimum size = 1/5 of longest span or, C = 24.0/5 = 4.80' (Use 5.0').

Size of dropped head panel:

Shall be not less than 1/3 longest span or 3/2 x flared capital.

Min. side dimension = 24.0/3 = 8.0 feet.

EXAMPLE: Complete flat slab floor design, continued 4.9.3

Also minimum side dimension = $\frac{3 \times 5.0}{2} = 7.50'$ (Use 8.0' square panel)

STEP IV:

Use formula to determine slab thickness t_2:

$t_2 = 0.375\,L\sqrt[3]{\frac{2000}{F_c}}$ or $t_2 = 0.375 \times 24.0\sqrt[3]{\frac{2000}{3000}} = 7.875$ inches.

Try using $t_2 = 8.00$ inches until depth is checked for shear. The Maximum thickness of dropped panel to be not more than $1.5\,t_2$ when computing for negative steel in Column strip "A." Thus: $t_1 = 1.5 \times 8.00 = 12.0''$ Max. Now hold this figure until later.

STEP V:

Calculating Moment (M_o) on whole panel for moment distribution.

$M_o = 0.09\,WLF\left(1.00 - \frac{2C}{3L}\right)^2$. C = 5.0' or Diameter of Flared Capital.

Revise W to include extra thickness of slab and neglect the dead load of drop panel. Slab DL = 480 x 8.00 x 12.5 = 48,000 Lbs. Then W = 144,000 + 48,000 = 192,000 Lbs. L = 24.0 Ft. for longer Span. $F = 1.15 - \frac{C}{L}$, but not less than 1.00 $F = 1.15 - \frac{5.00}{24.0} = 0.942$ (Delete the value of F from formula for M_o.)

Long Span $M_o = 0.09 \times 192{,}000 \times 24.0\left(1.00 - \frac{2 \times 5.00}{3 \times 24.0}\right)^2 = 307{,}308$ Ft. Lbs.

Short Span $M_o = 0.09 \times 192{,}000 \times 20.0\left(1.00 - \frac{2 \times 5.00}{3 \times 20.0}\right)^2 = 239{,}846$ Ft. Lbs.

STEP VI:

Obtain coefficients from Table with Dropped head panels and use for Moment distribution on Column strips "A" and Middle strips "B."

LONG SPAN:	$-M_A = 0.50 \times 307{,}308 =$	$-$ 153,654	Foot Pounds.
	$+M_A = 0.20 \times 307{,}308 =$	$+$ 61,462	" "
	$-M_B = 0.15 \times 307{,}308 =$	$-$ 46,096	" "
	$+M_B = 0.15 \times 307{,}308 =$	$+$ 46,096	" "
SHORT SPAN:	$-M_A = 0.50 \times 239{,}846 =$	$-$ 119,923	" "
	$+M_A = 0.20 \times 239{,}846 =$	$+$ 47,970	" "
	$-M_B = 0.15 \times 239{,}846 =$	$-$ 35,977	" "
	$+M_B = 0.15 \times 239{,}846 =$	$+$ 35,977	" "

STEP VII:

Check shear at edges where slab and drop panel join:

Use Live Load = 300#□' Dead Load: 8.00 x 12.5 = 100#□' Combined, w = 400#□'

Area whole panel = 24.0 x 20.0 = 480 Square feet. Area DP = 8.0 x 8.0 = 64.0□'

Area slab around drop panel: 480 − 64 = 416 Square feet.

EXAMPLE: Complete flat slab floor design, continued 4.9.3

Load on slab area only = 416 x 400 = 166,400 Lbs. Equals shear V.
Perimeter of dropped panel = 4 x 8.0 x 12 = 384 In. Equals side b"
Depth slab = 8.00" and assume depth to steel d = 7.00 inches
From Tables: J = 0.854 shear intensity $v = \frac{V}{Jdb}$.

$v = \frac{166,400}{0.854 \times 7.00 \times 384} = 72.49$ Lbs. sq. inch. Allowable $F_v = 2\sqrt{F_c'}$

$F_v = 2\sqrt{3000} = 110$ P.S.I. Shear at 8.00" slab depth is OK.

STEP VIII:

Check punching shear around circumference of flared round Capital and solve for depth $d = t_1 + t_2$ when $t_1 = 8.00"$ and t_2 is depth drop panel below slab. Load W = 192,000 Lbs. = V.
Circumference capital = πd or C = 3.1416 x 5.0 x 12 = 188.5 inches and represents shear dimension b. To find depth d, transpose the formula as: $b = \frac{V}{J b F_v}$ and substitute values in formula:

$d = \frac{192,000}{0.854 \times 188.5 \times 110} = 10.84$ inches. With a slab thickness of 8.00 inches, $t_2 = 10.84 - 8.00 = 2.84$ inches. The empirical rules state that dropped panels for negative steel in strip A shall be 1.5 t_2 where thickness t_2 = slab and drop panel and further, the drop panel shall be not less than 4.0 inches. Then $t_1 = 4.00"$
Then (1.50 x 8.00) − 8.00 = 4.00 inches for drop panel thickness below slab.
For designing negative steel use d = 10.0 inches.

STEP IX:

Calculate widths of Strips A and B for each span: $A = \frac{L}{4}$ and $B = \frac{L}{2}$.
Long Way Strip A = 20.0/4 = 5.0 Ft Short Way Strip A = 24.0/4 = 6.0 Ft
Long Way Strip B = 20.0/2 = 10.0 Ft. Short Way Strip B = 24.0/2 = 12.0 Ft.

STEP X:

Determine approximate depth to steel (d) for positive and negative bending moments. Negative d in strip A will equal slab + drop panel less fire cover protection of top rods and ½ diameter of lower rods, less 0.03 for deformations. (Empirical Rule).
Then d for −As = (8.00" + 4.00") − (0.75" + 0.875 + 0.4375 + 0.75) = 9.19 Inches
And d for +As = 8.00 − (0.75 + 0.875) = 6.37 Inches.

Depths above are based on the assumption that #7 ϕ Rods with ⅞ inch diameter will be required and fire protection cover will be ¾ inches. The deformation of 0.03 has been neglected and if used: − d = 9.19 − 0.03 = 9.16 inches

EXAMPLE: Complete flat slab floor design, continued 4.9.3

STEP XI

Calculate the As for Column and Middle strips A and B for Long and Short spans. As is for dimension b or full width of each strip. $F_s = 16{,}000$ PSI. $F_c' = 3000$ PSI and $J = 0.854$ The formula is $As = \frac{M}{F_s J d}$.

LONG WAY STRIPS A and B: L = 24.0 Feet

Column Strip A: $-As = \frac{153{,}654 \times 12}{16{,}000 \times 0.854 \times 9.19} = 14.68$ Sq. In.

Column Strip A: $+As = \frac{61{,}462 \times 12}{16{,}000 \times 0.854 \times 6.37} = 8.47$ Sq. In.

Column Strip B: $-As = \frac{46{,}096 \times 12}{16{,}000 \times 0.854 \times 6.37} = 6.37$ Sq. In.

Column Strip B: $+As = \frac{46{,}096 \times 12}{16{,}000 \times 0.854 \times 6.37} = 6.37$ Sq. In.

SHORT WAY STRIPS A and B: L = 20.0 Feet

Column Strip A: $-As = \frac{119{,}923 \times 12}{16{,}000 \times 0.854 \times 9.19} = 11.46$ Sq. In.

Column Strip A: $+As = \frac{47{,}970 \times 12}{16{,}000 \times 0.854 \times 6.37} = 6.61$ Sq. In.

Column Strip B: $\pm As = \frac{35{,}977 \times 12}{16{,}000 \times 0.854 \times 6.37} = 4.96$ Sq. In.

STEP XII:

Selecting rods and quantity for each strip: Spacing of rods must be wide enough to permit 1½ inch coarse aggregate to pass through crossed rods, but not wider than depth of slab which is 8.00 inches. In Column strips A, the lapped rods from adjoining panels double the steel area, therefore bend up alternate rods and width of strip A becomes L/2.

For Long Way Strip A: b = 120.0 Inches. Req. As = 14.68 Sq. In. Select 19-#8ϕ Rods with As = 14.92 Sq. In. Lay 9 rods straight in bottom and bend up 10 rods for negative bending. With 10 lapped rods bent up from adjacent panel strip there will be 20 rods for negative bending with -As = 15.70 Sq. In. Add for positive bending 2 rods in bottom to give 11 rods and +As = 8.64□"

For Short Way Strip A: b = 144.0 In. Required -As = 11.46 Sq. In. Select 15-#8ϕ Rods with As = 11.78 Sq. In. Bend up 8 rods for negative bending and with lapped rods bent gives an -As = 12.57 Sq. In. If spacing exceeds 8 inches, go to #7ϕ rods. For positive bending the 7 straight rods give +As = 5.50 Sq. In. and requires 2 additional straight rods to give +As = 7.07 Sq. In.

EXAMPLE: Complete flat slab floor design, continued 4.9.3

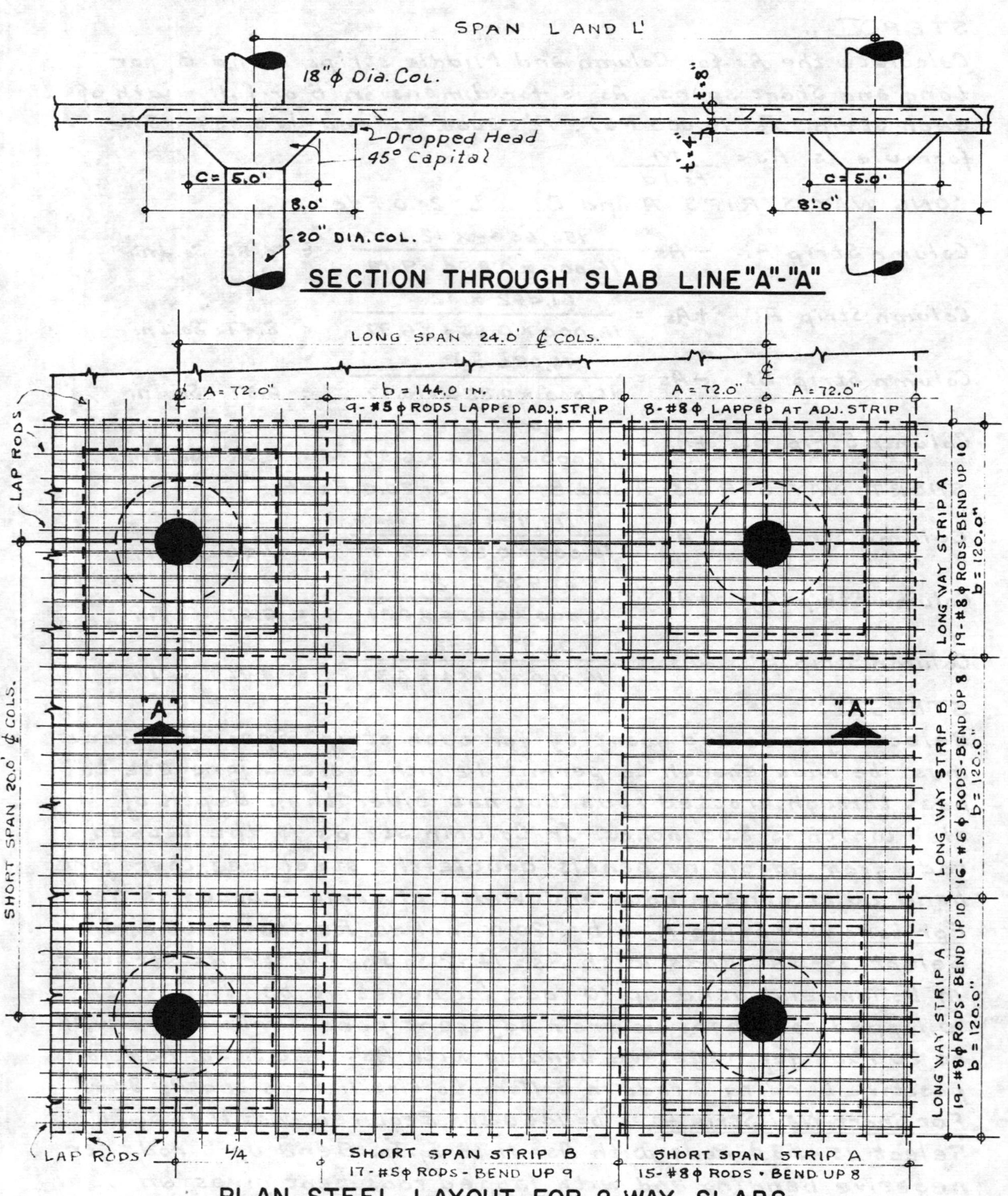

EXAMPLE: Complete flat slab floor design, continued 4.9.3

Short Way Strip B:

Required As = 4.96 Sq. In., for both Positive and Negative bending. Width of short span middle strip b = 144.0 Inches. For trial, select #5 φ rods. Area each rod = 0.3068 Sq. In. Number of rods required = $\frac{4.96}{0.3068}$ = 16+. Thus, use 17 Rods. Spacing of rods must not be greater than depth of slab t or 8.0 In. s = $\frac{144.0}{18}$ = 8.00 inches. Bend up alternate rods and leave 8 rods straight in bottom. The lapped bars from adjacent panels will then provide 17 bars for positive and negative bending as necessary. A #4 φ rod will require 26 rods and may be used with spacing of 5.76 inches with alternate rods bent up for negative bending.

Long Way Strip B:

Required As = 6.36 Sq. In. Width strip b = 120.0 inches. Select an even number of #6 φ Rods. Area #6 φ = 0.4418 Sq. In., and number required = $\frac{6.36}{0.4418}$ = 14.40. Use 16 Rods which give an area steel of 7.07 Sq. In. Use 8 Straight and 8 bent rods and with lapped rods the steel area for positive and negative bending will be 7.07 Square inches. Spacing = $\frac{120.0}{15}$ = 8.00 inches.

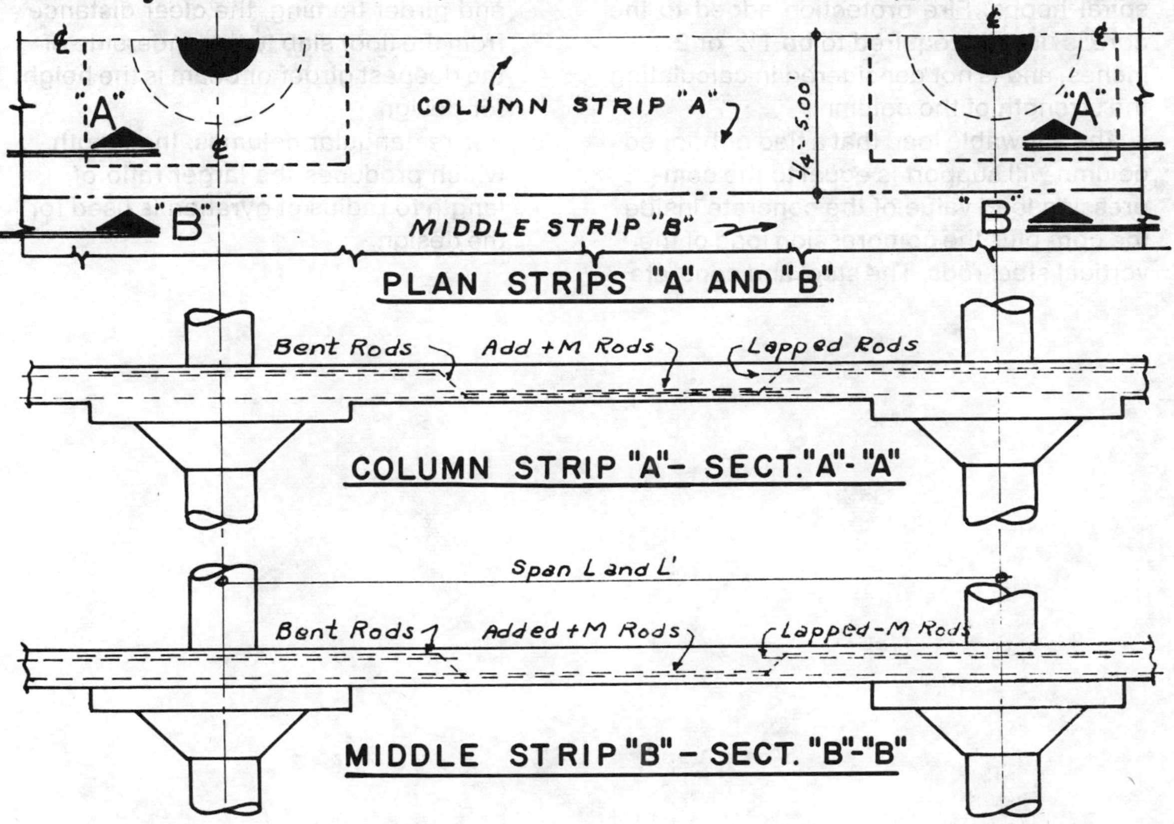

Concrete columns 4.10

Concrete columns are commonly classified by the type of reinforcement employed. Generally, there are three types:

(a) *Tied* columns, in which the vertical (longitudinal) steel rods are tied laterally to prevent buckling.

(b) *Spiral* or *hooped* columns: closely spaced, smaller rods are wrapped around the vertical steel rods in spirals, enclosing a concrete core. Field workers usually refer to this type as a *wrapped column.*

(c) *Composite* columns: a concrete column with a steel column in the center as well as wrapped vertical steel rods. Composite columns are used where space must be conserved or where there is extreme fire hazard.

EFFECTIVE AREA

In all types of concrete columns, only the core area can be considered effective in carrying the compression load. The core is the area encased in the lateral ties or spiral hoops. Fire protection added to the core is usually required to be 1½ or 2 inches, and is not considered in calculating the strength of the column.

The allowable load that a tied or hooped column will support is equal to the compression load value of the concrete inside the core plus the compression load of the vertical steel rods. The steel and concrete act together as a unit, and the stress in each is proportional to the modulus of elasticity, since the unit deformation is the same. By formula, $f_s = nf_c$, where f_c = actual stress in concrete and f_s = actual stress in steel. The ultimate axial load may be represented by the following equation: $P = (A_cF_c) + (A_sF_s)$. The allowable steel and concrete unit stresses will be provided in tables and discussed in examples.

COLUMN LENGTH FOR DESIGN

Building codes usually set a maximum for the ratio of column length to the diameter or least side dimension. To determine the unsupported height, the following rules should be applied:

(a) For columns with flared capitals supporting flat slab construction, the clear dimension from the bottom of the flare to the slab floor below is the height (H) for design.

(b) For columns supporting T-beam or ribbed-slab construction with beam and girder framing, the clear distance from the floor slab to the underside of the deepest girder or beam is the height for design.

(c) For rectangular columns, the length which produces the larger ratio of length to radius of gyration is used for the design.

Code requirements for columns 4.10.1

The minimum size of the concrete column is limited by code requirements. The Southern Standard Building Code, 1965 Edition, follows the American Concrete Institute Code 318–63 which limits the minimum diameter of a round floor or roof column to not less than 10 inches. For a rectangular column, the thickness must be at least 8 inches and the gross area not less than 96 square inches.

Vertical reinforcement in columns must be not less than 0.01 nor more than 0.08 times the gross area of the section. For spiral hooped columns, the vertical reinforcement must not be less than six #5 rods. For tied columns, the minimum reinforcement is four #5 rods. The ratio of spiral reinforcement (p′) must not be less than the value obtained by the equation: $p' = 0.45\left(\frac{A_g}{A_c} - 1.00\right)\frac{F'_c}{F_y}$, where F_y = yield strength of spiral steel up to 60,000 PSI.

For flat slab construction, the least dimension or diameter of the column must not be less than 1/15 of the longest span (center-to-center of columns), and in no event less than 16 inches.

The spiral reinforcement must not be less than ¼ of the vertical reinforcement. Spiral spacing should not be greater than 1/6 the diameter of core, and in no case more than 3 inches. Lateral ties in square or rectangular columns should not be less than ¼ inch in diameter, and should not be spaced over 12 inches for the full column height. In square and rectangular columns with more than 4 vertical reenforcing rods, the codes usually require each rod to be wired to the lateral ties. The arrangement for tying multiple vertical rods is illustrated with cross section drawings in 4.10.3.2.

Reduced load in long columns 4.10.2

A short column has a ratio of length to least dimension of ten or less. The slenderness ratio is the ratio of the unsupported height (h) to the least dimension or diameter (d). When the $\frac{h}{d}$ ratio exceeds ten, the allowable load P′ is reduced by a percentage (rather than using the reduction of allowable unit stress method). To compute the allowable load on a long column, the ACI has provided the following formula: $P' = P\left(1.30 - 0.03\frac{h}{d}\right)$. Where

P′ = allowable load on long column and

P = allowable safe axial load when $\frac{h}{d}$ is less than 10. Hence when $\frac{h}{d} = 10$, the value of P′ = P x 1.00 or 100 percent. Note that the allowable axial load for a short column must first be calculated and then inserted in the equation. TABLE 4.10.3.3 will be convenient to find the allowable percentage of P for various ratios of h/d.

Column design methods 4.10.3

When building codes permit a choice, there are two methods for the design of concrete columns. Both methods should be carefully examined, so that the corresponding notation for loads will not be confusing. The more conservative, and older method, is called *working stress design* or WSD. We will use this method exclusively, because it incorporates a safety factor in all formulas and tables.

The second method is called *ultimate strength design* or USD. Ultimate strength is based on calculating ultimate strength values. A safety factor for the USD design method involves selecting a coefficient based on experience and study of the following variables:

(a) Possible loss of strength in materials over the life of the structure.
(b) Workmanship in controlling and curing mixes.
(c) Maintaining dimensions within design tolerances.
(d) Strict and knowledgeable job supervision.

ALLOWABLE WORKING STRESS—WSD

The allowable unit working design stress for concrete in spiral-hooped columns is $F_c = 0.25\ F_c'$. When the vertical reinforcement is laterally tied, the value of P for the tied column is 85 percent of a corresponding spiral-hooped column. Hence, the allowable stress for a tied column is $F_c = 0.2125\ F_c'$.

Steel reinforcing in spiral-hooped columns is assigned an allowable unit working stress of 40 percent of the yield stress of the particular grade used, but in no event more than 30,000 PSI. The formula is written: $F_s = 0.40\ F_y$. For tied columns, with the 85 percent reduction from the spiral-hooped type, the allowable unit design working stress is: $F_s = 0.34\ F_y$. Values for various concrete strengths and steel grades are given in TABLE 4.10.3.3.

Concrete column design nomenclature 4.10.3.1

- A_t = Transformed area of concrete, in square inches = nA_s.
- A_c = Area of Concrete in core, given in square inches.
- A_g = Area gross in core, includes $A_c + A_s$, in square inches.
- A_s = Area vertical steel reinforcement, in square inches.
- A_{st} = Area of structural steel shape in composite section.
- A_r = Same as A_{st} above when used in formula with F_r.
- C = A coefficient calculated for all columns with same design.
- c = Distance from NA to center of steel in outer rim, in inches.
- b = Dimension for breadth of rectangular core, in inches.
- d = Depth dimension or diameter of core, in inches.
- e = Eccentric dimension from load point to NA, in inches.
- E = Modulus of elasticity - See tables for E_c.
- F_c' = Concrete compressive strength at age 28 days, in PSI.
- F_c = Unit design concrete stress - See tables for percentage PSI.
- F_y = Yield stress of reinforcing grade steel - see tables, PSI.
- F_r = Allowable unit stress in steel shape embedded in composite.
- f_c = Actual compressive stress for concrete under load, PSI.
- f_s = Actual compressive stress intensity in steel rods, PSI.
- H = Height of Column, in feet.
- H' = Column height floor to floor, in feet. See Section VIII.
- h = Long columns unsupported height, in inches.
- I = Moment of Inertia of columns core section. = I''^4
- M = Bending moment resulting from wind or eccentricity.
- N = Usually used for designating number of rods, hoops, etc.
- NA = Neutral axis or centroid of core section, in inches.
- P = Total load to be sustained on column, or allowable, in lbs.
- P' = Reduced load allowed by reason of long columns length ratio.
- P_c = Allowable axial column load sustained on concrete area.
- P_s = Allowable axial column load sustained on steel rod area.
- r = Radius of gyration. For round columns = $\frac{d}{4}$, in inches.
- s = Spacing of spiral hoops or vertical ties, in inches.
- n = Ratio of E_s/E_c. See tables for values of E_c.
- p_g = Ratio of A_s to the gross area A_g, where A_s = Vertical rods.
- p' = Ratio of volume of spiral reinforcing to the core volume of concrete.

Concrete column design formulas 4.10.3.2

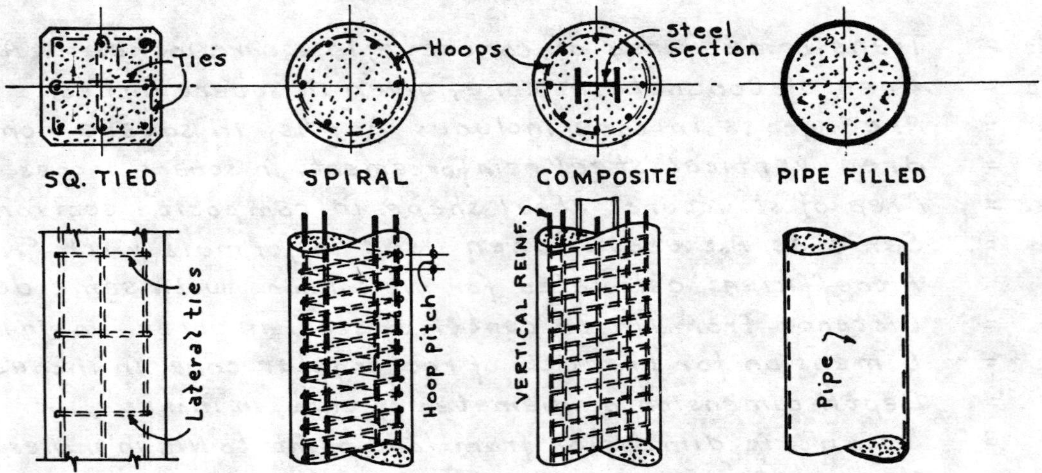

These design formulas apply to the types of concrete column illustrated above. they give the maximum axial load P, for short columns where $\frac{h}{d}=10$ or less.

SPIRAL HOOPED COLUMNS

Max $P = A_g(0.25\,F_c' + F_s\,\rho_g)$, where the ratio of spiral reinforcement must not be less than that given by the formula: $\rho' = 0.45\left(\frac{A_g}{A_C} - 1.00\right)\frac{F_c'}{F_Y}$, and F_Y may not exceed 60,000 PSI,

TIED COLUMNS

The maximum axial load P for a tied column is 85 percent of a spirally hooped column. Revising the formula above for tied columns: Max $P = A_g(0.25\,F_c' + F_s\,\rho_g)\,0.85$. When unit stress tables are used, an alternate formula may be used: Max $P = (0.2125\,F_c'\,A_c) + (0.34\,F_Y\,A_s)$.

COMPOSITE COLUMNS

The maximum axial load P for a composite column with vertical rod reinforcing and a steel structural shape in the core (thoroughly embedded in concrete and enclosed by spiral hoops) is Max $P = (0.225\,A_g\,F_c') + (F_s\,A_s) + (F_r\,A_r)$.

CONCRETE FILLED PIPE COLUMNS

Pipe columns with a concrete core are designed with JC Pile formula, given in Paragraph 9.3.3 of section IX

TABLES: Concrete column design factors 4.10.3.3

LONG COLUMN REDUCTION PERCENTAGES FOR LOAD P'					
$\frac{h}{d}$	$\left(1.30-0.03\frac{h}{d}\right)$	$\frac{h}{d}$	$\left(1.30-0.03\frac{h}{d}\right)$	$\frac{h}{d}$	$\left(1.30-0.03\frac{h}{d}\right)$
10	1.00	20	0.70	30	0.40
11	0.97	21	0.67	31	0.37
12	0.94	22	0.64	32	0.34
13	0.91	23	0.61	33	0.31
14	0.88	24	0.58	34	0.28
15	0.85	25	0.55	35	0.25
16	0.82	26	0.52	36	0.22
17	0.79	27	0.49	37	0.19
18	0.76	28	0.46	38	0.16
19	0.73	29	0.43	39	0.13
WITH ECCENTRICALLY LOADED COLUMNS $\frac{h}{d}$ IS NOT TO EXCEED 20					

ALLOWABLE STEEL DESIGN UNIT STRESSES FOR COLUMNS			
GRADE-TYPE REINFORCING	YIELD F_y=PSI	SPIRAL: $F_s = 0.40 F_y$	TIED: $F_s = 0.34 F_y$.
A15 STRUCTURAL BILLET	33,000	F_s = 13,200 PSI	F_s = 11,220 PSI
A15 INTERMEDIATE BILLET	40,000	F_s = 16,000 "	F_s = 13,600 "
A15 HARD GRADE BILLET	50,000	F_s = 20,000 "	F_s = 17,000 "
A61 RAIL-DEFORMED GRADE	60,000	F_s = 24,000 "	F_s = 20,400 "
A432 HIGH STRENGTH BILLET	60,000	F_s = 24,000 "	F_s = 20,400 "

ALLOWABLE CONCRETE DESIGN UNIT STRESSES FOR COLUMNS		
CONCRETE STRENGTH AT AGE OF 28 DAYS: F'_c.	SPIRAL HOOPED COLUMNS $F_c = 0.25 F'_c$. PSI.	TIED LATERALLY COLUMNS $F_c = 0.2125 F'_c$. PSI.
2000 PSI	500	425
2500 "	625	530
3000 "	750	640
3500 "	875	745
4000 "	1000	850
4500 "	1125	955
5000 "	1250	1065
5500 "	1375	1170
6000 "	1500	1275

Composite column design 4.10.4

Composite columns have a steel section embedded in the core center, with vertical rods and spiral hoops to encase the steel and concrete. The allowable axial load P is given by the formula $P = (0.225\ A_g F_c') + (F_s A_s) + (F_r A_r)$. When an A-36 steel section is used, the maximum allowable unit stress must not exceed $F_r = 18{,}000$ PSI. For A-7 steel maximum $F_r = 16{,}000$ PSI. When hollow tubes or pipe sections are used in the core, the hollow portion must be thoroughly filled with concrete.

The cross-sectional area of the metal shape in the core must not exceed 20 percent of the gross area of the column. Thus, the maximum $A_r = 0.20\ A_g$. Concrete clearance between the hoops and the steel core shape must not be less than 3 inches at any point. When the core shape is a structural H column, the clear space between the hoops and the H-shape may be not less than 2 inches.

Eccentric column loads 4.10.5

A tied or spiral-hooped column with a load placed a distance from the center axis will subject one side of the column to greater stress intensity and reduce the stress on the opposite side. In the general case, columns will have an axial load and an eccentric load. The eccentric load in most Hi-Rise structures is the result of the wind moment, which increases as the calculations proceed to the lower stories. The design of columns with both axial load and bending moment involves the use of the *transformed section*. Such design may be best explained with examples.

When a column is required to sustain compressive stress from an axial load in addition to tension stress from an eccentric load or wind moment, the two kinds of stress must be examined for equal distribution. The formula used to determine the maximum and minimum stress under the above conditions is written:

$f_c = \frac{P_a + P_e}{A_{gt}} \pm \frac{M_e c}{Io}$. A_{gt} = Gross area plus transformed area (Asn). Should there be an axial load plus an eccentric load plus a wind moment, the two moments are added together $M_e + M_c$. This formula may also be written: $f_c = \frac{P}{A} \pm \frac{M}{S}$, where $S = \frac{I}{c}$. The examples to follow will show the practical application.

Spiral hoop spacing 4.10.6

Certain codes may require the spiral hoop spacing (pitch) to be not less than 3 inches, while others may limit the spacing to 1/6 the core diameter. The material for hoops must not be less than #2 rods; when cold drawn wire is used for wrapping the core, the minimum size is limited to #4 Gage United States Standard. Splicing for lapped spirals is generally 1½ turns, but welding the rods is preferred.

The spacing of spiral hoops on columns can be calculated by the following procedure: Calculate the ratio for percentage of steel p′ as given by the formula. Assume for example that the ratio p′ = 0.013 and that the gross area of the 16.0 inch core is A_g = 201.0 square inches. Then the area of hoop steel must be 0.013 x 201.0 = 2.61 square inches. *This is the area of steel required to wrap one foot of column core.* The perimeter of the 16.0 inch core is πd or 3.1416 x 16.0 = 50.26 inches or 4.19 feet. Choosing #3 rod with a cross-sectional area of 0.11 square inches, the length of rod required to wrap one foot = 2.61/0.11 or 23.8 lineal feet. The number of turns required to wrap one foot = 23.8/4.19 or 5.75 turns. The pitch or hoop spacing = 12/5.75 = 2.10 inches center-to-center. This method for calculating pitch is used in the examples.

EXAMPLE: Calculating long column load reduction 4.10.7.1

A round concrete column has a core diameter of 18.0 inches and vertical reinforcement consists of 8-#8φ Rods. This is a tied column with a length of 19.5 feet. Steel reinforcing is $F_y = 30{,}000$ PSI and $F_c' = 2500$ PSI.

REQUIRED:
Calculate the allowable axial load if designed as a short column. If the h/d ratio places column in the long class, determine the safe load.

STEP I:
Gross section area: $A_g = \pi R^2$ or $0.7854 D^2$

$A_g = 3.1416 \times 9.0 \times 9.0 = 254.47$ Sq. inches.

$A_s = 0.7854 \times 1.0 \times 8 = 6.28$ Sq. inches

$A_c = A_g - A_s$ or $A_c = 254.47 - 6.28 = 248.19$ Sq. Inches.

STEP II:
Allowable stresses for tied columns:

$F_s = 0.34 F_y$. or $F_s = 0.34 \times 30{,}000 = 10{,}200$ PSI.

$F_c = 0.25 F_c'$. or $F_c = 0.25 \times 2500 = 625$ PSI.

$P_s = 6.28 \times 10{,}200 = 64{,}056$ Pounds.

$P_c = 248.19 \times 625 = 155{,}119$ Pounds.

Short Column axial load $= 64{,}056 + 155{,}119 = 219{,}175$ Pounds.

STEP III:
Long column class when $\frac{h}{d}$ is greater than 10.

Ratio $= \frac{19.5 \times 12}{18.0} = 13$ Over 10 ratio and load must be reduced.

Reducing P: When $\frac{h}{d}$ is between 10 and 40 use the formula: $P' = P\left(1.30 - 0.03 \frac{h}{d}\right)$. with values substituted:

$P' = 219{,}175 \left[1.30 - (0.03 \times 13)\right] = 219{,}175 \times 0.91 = 199{,}450$ Lbs.

EXAMPLE: Column design for axial load plus wind moment 4.10.7.2

An interior column must support an axial load from floor and column above of 360,000 Lbs. This column will be subjected to a wind moment of 60,000 Foot pounds. Column may be square or rectangular and core dimension is limited in one direction to 14.0 inches. Wind moment will be applied on x-x axis which may be not less tha 14.0 inches. Percentage of steel being used is 0.02. Concrete $F_c' = 4000$ PSI $F_y = 40{,}000$ PSI for Intermediate grade A15 billet steel. $n = 8$. This is a tied column in the short column class where $h/d = 9.7$

REQUIRED:
Assume that previous columns on this project have been designed with same criteria and constant for combined stress is: $C = 1105$ PSI. Design column and draw cross-section of required size.

STEP I:
For axial load: $A_g = \frac{P}{C}$ or $A_g = \frac{360{,}000}{1105} = 326.0$ Sq. Inches

Area Steel $= 0.02 \times 326.0 = 6.52$ Sq. In. Using an even number of rods, try #7 ϕ Rod with $A\phi = 0.60\square''$ $n = 12$ Rods.
With 1 dimension limited to 14.0 inches as b, then side $d = \frac{326.0}{14.0} = 23.3''$ Call it 24.0 inches.

STEP II:
A cross-section drawing is made thus: The wind load moment will be resisted on As parallel with axis x-x. Only the vertical rods in outer plane will be of much use in resisting bending. Area of 8-#7ϕ Rods $= 4.80\square''$ and their distance from x-x is 12.0 inches.

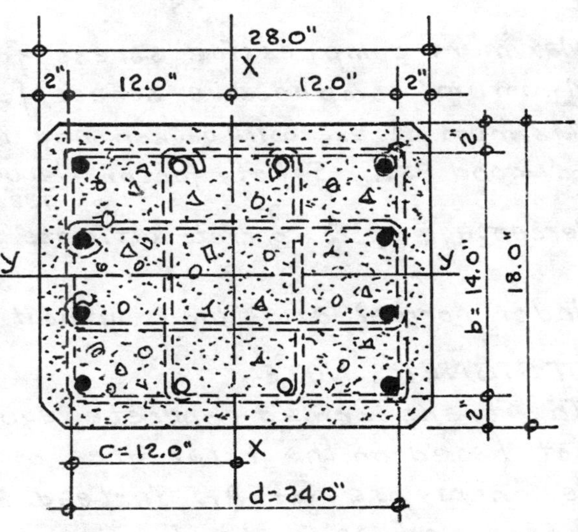

STEP III:
Outer rods will be converted to a transformed concrete area.

EXAMPLE: Column design for axial load plus wind moment, continued 4.10.7.2

Step III Continued:

$n = Es/Ec$ or $n = 8$ and $n - 1.0 = 7$.
Transformed area $A_t = 4.80 \times 7 = 33.50$ Square inches.

STEP IV:
To determine bending resistance of section, the moment of Inertia must be calculated. $I = \frac{bd^3}{12}$.
$b = 14.0"$ $d = 24.0"$ and Area $A_t = 33.50$ □"

$I_g = \frac{14.0 \times 24.0 \times 24.0 \times 24.0}{12} = 16,130''^4$ $I_t = A_t l^2$ and $l = 12.0"$

$I_t = 33.50 \times 12.0 \times 12.0 = 4,825''^4$

$I = \Sigma I_o + A_t l^2 = 20,955''^4$

STEP V:
Refigure the accurate A_g, and <u>Add</u> the transformed area to determine the axial compression stress from P.
$A_g = bd$ or $A_{gt} = (14.0 \times 24.0) + 33.50 = 369.50$ Square inches.

STEP VI:
Caluulate the maximum and minimum unit stress on concrete due to bending action. The formula used is thus: $f_c = \left(\frac{P}{A_{gt}}\right) \pm \left(\frac{Mc}{I_{gt}}\right)$. $c = 12.0"$ $M = 60,000'\#$ $P = 360,000\#$

With values in equation: $f_c = \left(\frac{360,000}{369.50}\right) \pm \left(\frac{60,000 \times 12 \times 12.0}{20,955}\right)$ or

$f_c = 974 \pm 412$ Therefore:

Maximum compressive stress, $f_c = 974 + 412 = 1386$ P.S.I.
Minimum compressive stress, $f_c = 974 - 412 = 562$ PSI.
Maximum stress will govern and is well below the ultimate of $F_c' = 4000$ PSI. Safety factor $= \frac{4000}{1386} = 2.89$

Percentage of F_c' concrete ultimate stress $= \frac{1386}{4000} = 0.347$ (about 35%.)

Under normal axial load without wind $f_c = \frac{360,000}{369.50} = 972$ PSI.

DESIGNER'S NOTE:
The transformed concrete equivalent of steel area was not based on the total area of vertical steel which most designers use for At. Instead, only the outside rods were considered as in step II. This results in an increase in the safety factor which is not obvious.

EXAMPLE: Column design for axial and eccentric loads 4.10.7.3

A wall type tied column is limited to 16.0 inches overall in wall depth and height unsupported is 9'-8." Axial load P_a = 215,000 Lbs. Eccentric load P_e = 88,000 Lbs., with e = 4.50 inches from axis parallel to wall.
Concrete F_c' = 3500 PSI. Use A15 Hard billet steel for vertical reinforcing. Use 1½ percent of A_g for steel area As.

REQUIRED:
Design wall column for size and make a plan of cross-section showing rod placement and application point of eccentric load Pe. Adjust final design to make even dimensions.

STEP I:
With 2.0 inches of rod cover, core dimension is 12.0 inches for one dimension. h = 116 and $\frac{h}{d} = \frac{116}{12} = 9.67$. Less than 10 and is designed as short column.
Total Loads = 215,000 + 88,000 = 303,000 Lbs. Bending moment from eccentric load: Me = 88,000 x 4.50 = 396,000 inch pounds.
STEP II:
From tables for tied columns: Steel F_y = 5000 PSI. F_s = 17,000 PSI F_c = 745 PSI. Ratio n = 8.70
No attempt will be made in this example to refer to a column load table or assume a trial section. Instead of estimating a size we will solve for a coefficient C which will be of use for other columns on same project.
Assume A_g = 400□" F_c = 745#□" then As = 400 x 0.015 = 6.00□"
A_c = 400 - 6.00 = 394□" For calculating a usable coefficient, the same unit allowables must be used.
P_c = 394 x 745 = 293,530# P_s = 6.00 x 17,000 = 102,000 Lbs.
Total P = 293,530 + 102,000 = 395,530 Lbs. Now the average stress with allowables given and steel percentage is $\frac{P}{A}$ and = C.
Then average stress C or $\frac{P}{A_g}$ = 395,530/400 = 989 PSI.

STEP III:
Returning to example from step I: $P_a + P_e$ = 303,000 Lbs. Average stress when As = 0.015 A_g = C or 989 PSI. Then area for core of wall column = 303,000/989 = 306.0 Sq. In. Dimension d = 12.0 inches. and b = 306.0/12 = 25.5 inches. Call it 26.0 inches.
Core size of wall column is 12.0" x 26.0" A_g = 312.0□". Outside dimensions with 2 inch fire protection = 16.0" x 30.0" (Rectangular).

EXAMPLE: Column design for axial and eccentric loads, continued 4.10.7.3

Cross-section of column is now drawn. Adjustments in steel area can be made to meet design requirements. Area core: $A_g = 312.0\,\square''$ With 1½% for steel, $A_s = 0.015 \times 312.0 = 4.68$ Sq. Inches. Select about 8 Rods and place 4 on each side of weak axis y-y. Closest will be 8-#7 φ Rods with $A_s = 4.80\,\square''$
Net $A_c = 312.0 - 4.80 = 307.2$ Sq. In.

STEP IV:
Check load for axial value P:
$P_a = (A_c F_c) + (A_s F_s)$ and substituting:
$P = (307.2 \times 745) + (4.80 \times 17{,}000) = 310{,}465$ Lbs.
This is in excess of step I and OK.

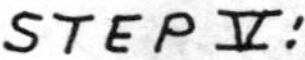

STEP V:
Calculate the area of the transformed section when $n = 8.70$ $n - 1.00 = 7.70$ $A_t = 7.70 \times 4.80 = 37.00\,\square''$
$A_t = (n - 1.00) A_s$ is a formula which allows for rod adjustments.

STEP VI:
Now required is the moment of Inertia about weak axis where eccentric distance of bending moment is 4.50"
For cross-section A_g: $I = \frac{bd^3}{12}$ and $b = 26.0''$ and $d = 12.0''$

$I_g = \frac{26.0 \times 12.0^3}{12} = 3744.0$ For transformed section the lever distance from axis y-y is same as steel rods or 6.0 inches.
$I_t = A l^2$ or $I_t = 37.0 \times 6.0 \times 6.0 = 1332.0$
Total I_o about axis y-y $= 3744.0 + 1332.0 = 5076.0''^4 = I_y$.
Distance to extreme core fiber = 6.00 inches.

STEP VII:
Computing stresses in concrete: $A_g = A_c + A_t$ or $A_g = 349.0\,\square''$
$P_a + P_e = 303{,}000$ Lbs. $M_e = 396{,}000$ Inch Lbs. $S = I/c = 5076/6.0 = 846.0''^3$

Formula for Maximum and Minimum $F_c = \left(\frac{P_a + P_e}{A_g}\right) \pm \left(\frac{Mc}{I_o}\right)$.
Simplified formula with values:

$$\text{Maximum } f_c = \frac{303{,}000}{349.0} + \frac{396{,}000}{846.0} = 867 + 467 = 1334 \text{ PSI}$$

Minimum $f_c = 867 - 467 = 400$ PSI.
Maximum stress fc is about $0.38 F_c'$ and can be reduced if percentage of steel is increased. Refigure from Step III at top.

EXAMPLE: Calculating safe column load and height 4.10.7.4

A concrete column is 14.0" x 16.0" and reinforced with 4-#6 ϕ Rods vertical and tied with $\frac{1}{4}\phi$ ties on 8.0 inch centers. Allowable unit compressive stress on concrete is 530 PSI. $n = 15$. Height of unsupported length of column is 12.0 feet.

REQUIRED:
Calculate the safe load column will support and state the height limit of unsupported length to make cross-section come within limits of a short column. Design of column is intended to meet ACI Code and applicant shall state his recommendation for meeting code. Show cross section.

STEP I:
This examinee will assume that the size given is the core dimensions enclosed in ties. $Ag = 14.0 \times 16.0 = 224.0$ Square. With 4-#6ϕ Rods $As = 1.76$ Square inches. NOTE: ACI specifications require a minimum of 4-Rods and also a minimum of 1 Percent Ag for steel area: Steel area As should be not less than $224.0 \times 0.01 = 2.24$ Sq. Inches. 6-#6ϕ Rods an area: $As = 0.44 \times 6 = 2.64$ □" (This column will be revised to meet ACI.

STEP II:
With revised cross-section:
$Ac = Ag - As$: $Ac = 224.0 - 2.64 = 221.36$ Sq. In. $Fc = 530$ PSI and $n = 15$
$Fs = Fcn$, or $Fs = 530 \times 15 = 7950$ PSI. $H = 12.0'$ and $h = 12.0 \times 12 = 144.0$ In.
$d = 14.0"$ Slenderness ratio $h/d = 144.0/14.0 = 10.4$ or over 10 and over short column classification.
Axial $Pa = AcFc + AsFs$: $Pa = (221.36 \times 530) + (2.64 \times 7950) = 138,310$ Lbs. only as a short column. (Note that 4-#5ϕ Rods is ACI minimum).

STEP III:
Reducing load for column when $h/d = 11$. Formula: $P' = P_a\left(1.30 - 0.03\frac{h}{d}\right)$.
With values in equation:

$$P' = 138,310\left[1.30 - \left(0.03\,\frac{144.0}{14.0}\right)\right] = 136,925 \text{ Pounds.}$$

Drawing of cross-section at right:

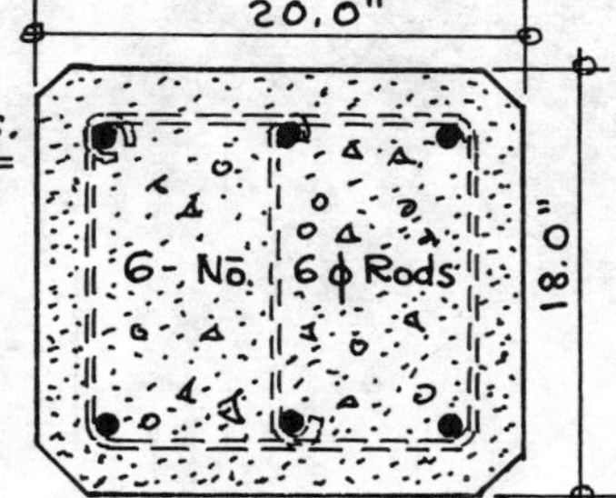

AUTHORS NOTE:
This method for column analysis is a very old and based on factor n, which is all the examinee had available for solution.

EXAMPLE: Calculating safe column load 4.10.7.5

Refer to previous example submitted by a State Board of Architectural examiners and with the following design criteria:
Overall dimensions are 20.0"x 18.0" and core = 16.0"x 14.0".
As = 6-#6 ϕ Rods with area of 2.64 Square inches. Use intermediate grade of A15 billet steel with Fy = 40,000 PSI and Fc' = 2500 PSI.
Ties will remain #2 ϕ Rods and spacing shall not exceed 16 times vertical rod diameters nor more than 48 tie rod diameter.

REQUIRED:
Use the tables for allowable design working stress for concrete and steel. Also use table for long column reduced percentage for P.' Determine maximum spacing of tie rods in this column section. Column height H = 9

STEP I:
Vertical rods are 3/4 inch diameter and As = 2.64□". Tie rods are 1/4 inches in diameter. To check for maximum tie rod spacing: Max. s = 16 × 0.75 = 12.0" or Max. s = 48. × 0.25 = 12.0 inches.
Spacing of 8 inches in previous example is OK.

STEP II:
From tables, allowable stresses are: (Tied columns)
F_s = 13,600 PSI. Fc = 530 PSI. Reduction: 1.30 - 0.03 h/d = 0.97
Maximum Axial load P = (Ac Fc) + (As Fs). Ac = (16.0 × 14.0) - 2.64 = 221.36□"

Axial P = (221.36 × 530) + (2.64 × 13,600) = 153,225 Lbs.
Reduced load P' = 153,225 × 0.97 = 148,628 Lbs.

Modern design methods rely upon better control of concrete mix and curing thus the increase of nearly 15,000 Lbs. over previous example.

EXAMPLE: Computing column combined stress constant 4.10.7.6

A project consists of five floors with slab roof and all columns are spaced equal in each direction running from top to bottom. All interior and wall columns will be of tied type and percentage of steel to gross area of column core is established as 0.03 Ag. ACI sets the minimum at 0.01 and maximum as 0.08 for vertical steel rod reinforcement. Minimum size rod is #5 and not less than 4 rods. $F_c' = 3000$ PSI and $F_y = 50{,}000$ PSI.

REQUIRED:

Assume a core size 18.0" x 18.0" and design for an average combined stress constant as $C = \frac{P}{A_g}$. This constant shall then be used for sizing.

Check the value of constant C by analyzing a column which supports a 650,000 Lb. axial load. Use square section.

STEP I:

$A_g = 18.0 \times 18.0 = 324.0$ Sq. In. $A_s = 0.03 \times 324.0 = 9.72$ Sq. In.

$A_c = 324.0 - 9.72 = 314.28$ □"

From tables tied Columns: $F_c = 745$ PSI and $F_s = 17{,}000$ PSI:

$P_a = (A_s F_s) + (A_c F_c)$, or $P_a = P_s + P_c$. $C = \frac{P_a}{A_g}$. With values in formula:

$P_a = (9.72 \times 17{,}000) + (314.28 \times 745) = 399{,}380$ Pounds.

STEP II:

A combination of 2 materials, each having a different value for unit stress with a constant quantity ratio can be considered as the working unit. Then C = average unit stress to use in determining other sizes of column sections. Therefore,

$$C = \frac{P_a}{A_g} \text{ or } C = \frac{399{,}380}{324} = 1235 \text{ PSI.}$$

STEP III:

Use the value of C to determine the size of a column to safely support a load of 650,000 Lbs. Formula: $A_g = P_a/C$.

Then $A_g = \frac{650{,}000}{1235} = 526$ Sq. Inches. With square Column the dimensions for each side ane $\sqrt{526} = 23.0$ inches.

Area steel = $0.03 \times 526 = 15.78$ Square inches.

$A_c = 526 - 15.78 = 510.22$ □"

EXAMPLE: Computing column combined stress constant, continued 4.10.7.6

STEP III Continued:

Selecting the reinforcing for $A_s = 15.78$ Sq. In. From tables of areas: 16-#9 ∯ Rods have an $A_s = 16.00$ Sq. inches.
Net area concrete $A_c = 526.0 - 16.0 = 510.0$ Sq. In.

STEP IV:
Compute safe load on this column to check the size of cross-section required to support 650,000 Pounds: Same formula as used in step I:

$$P_a = (510.0 \times 745) + (16.0 \times 17{,}000) = 651{,}950 \text{ Lbs.}$$

DESIGN NOTE:

When coefficient C is to be used as the basic value for computing column sizes and area of steel reinforcing, a check column should be analyzed for confirming value. Unit stress allowables cannot be altered nor can the steel percentage be changed.

EXAMPLE: Designing spiral hooped column 4.10.7.7

Load from roof column is 88,000 Pounds and brought down to next floor which has a DL+LL = 118,000 Lbs. Column height (H') floor to floor is 12.0 feet. Code requires 2 inches of fire protection. Percentage of vertical steel to gross core area is set at 2%, or 0.02 Ag. Concrete strength at 28 days is $F_c' = 3000$ PSI. Vertical steel is specified as hard grade A15 billet steel with $F_y = 50{,}000$ PSI. Hoop steel will be either #2 or #3 ϕ Rods A15 with $F_y = 33{,}000$ PSI.

REQUIRED:

Design the column to sustain loads given, plus approximately 2800 Pounds for weight of column. Calculate the punching shear around column perimeter to determine if a flared capital is required when slab thickness is 5.0 inches. Sketch a cross section of design and an interior panel plan of area supported with elevation of column. Floor panel size is 20.0' x 20.0' with combined DL+LL = 295 Lbs. Sq. Ft.

STEP I:

Formula for spiral columns: $P = A_g \left[(0.25 F_c') + (0.40 F_y P_s)\right]$.

The combined stresses are: $(0.25 \times 3000) + (0.40 \times 50{,}000 \times 0.02)$, or combined allowable $F_{cs} = 1150$ PSI. Then the gross core area $= P/F_{cs}$. If we use 12.0 foot height as $h = 12.0 \times 12 = 144''$ and short column ratio of $\frac{h}{d}$ not less tha 10, the minimum core diameter cannot be less than 14.4 inches.

$$A_g = \frac{P}{F_{cs}} \quad \text{or} \quad A_g = \frac{88{,}000 + 118{,}000 + 2800}{1150} = 181.5 \square'' \text{ (Call it } 182.0 \square'')$$

With transposed formula, core diameter $= 2\sqrt{\frac{A_g}{\pi}}$ and with values substituted: $d = 2\sqrt{\frac{182.0}{3.1416}} = 15.24''$ (Make it even inches and call it 16.0 inches). Overall Dia. = 16.0 + 4.00 = 20.0".

STEP II:

Punching shear around perimeter of 20.0 inch diameter column. Circumference = 3.1416 x 20.0 = 62.832 inches. Slab is 5.0 inches thick and floor load = 118,000 Lbs. Assume depth to slab steel = 4.25 inches and J = 0.854 V = 118,000 # and b = 62.8" Allowable $F_v = 2\sqrt{F_c'}$ or $F_v = 2\sqrt{3000} = 110$ PSI.

$$f_v = \frac{V}{Jdb} \quad \text{or} \quad f_v = \frac{118{,}000}{0.854 \times 4.25 \times 62.832} = 518 \text{ Lbs. sq. inch.}$$

Column will require a flared capital and probably a dropped

EXAMPLE: Designing spiral hooped column, continued 4.10.7.7

head also. Refer to design of flat slabs to solve for this condition.

STEP III:
Calculating area of vertical steel and 1 turn hoop length.
Core d = 16.0 inches. Perimeter = 3.1416 x 16.0 = 50.26 inches.
1 Spiral turn = 50.26/12 = 4.19 feet. Percentage of vertical steel = 0.02 Ag. $Ag = 0.7854\,d^2$ or $Ag = 0.7854 \times 16.0 \times 16.0 = 201.0$ □"
As = 0.02 x 201.0 = 4.02 square inches. Use 10-#6 ϕ Rods - As = 4.40□"
Ac = Ag - As: Ac = 201.0 - 4.40 = 196.60 □"

STEP IV:
Calculating for spiral hoop steel when Fy = 33000 PSI and Fc' = 3000 PSI.
Minimum amount of hoop steel allowed by code = 0.01 Ag, also by the formula for percentage which shall be not less than:

$$P_s = 0.45\left[\left(\frac{Ag}{Ac} - 1.00\right) \times \left(\frac{Fc'}{Fy}\right)\right].$$

With values in equation:

$$P_s = 0.45\left[\left(\frac{201.0}{196.6} - 1.00\right) \times \left(\frac{3000}{33{,}000}\right)\right] = 0.001$$

This is also less than 1% minimum, thus the minimum of 0.01 shall be used. Hoop steel A = 0.01 x 201.0 = 2.01 Sq. Inches.

STEP V:
Using #3 ϕ Rods for hoops: Area 1 Rod = 0.11 Sq. inches.
From step III, 1 turn = 4.19 Lineal feet hoop rod.
Length required to wrap 12 inches of column core is equal to $\frac{2.01}{0.11}$ or 18.25 lineal feet. Number of turns require for 1.0 foot of core = 18.25/4.19 = 4.36
Spacing or hoop pitch = 12.0/4.36 = 2.75". 2¾" is less than maximum of 3.0 inches required by code. Also, the maximum is 1/6 of core diameter for center to center hoop spacing which would be equal to 2.66 inches.
Number of turns per foot = 12.0/2.66 = 4½. (Use 2.66" spacing.)

Step VI
Plan of support area for 1 column is drawn with rules applying as given for flat slabs. Elevation will delineate the negative bending theory in column strip A.

EXAMPLE: Designing spiral hooped column, continued 4.10.7.7

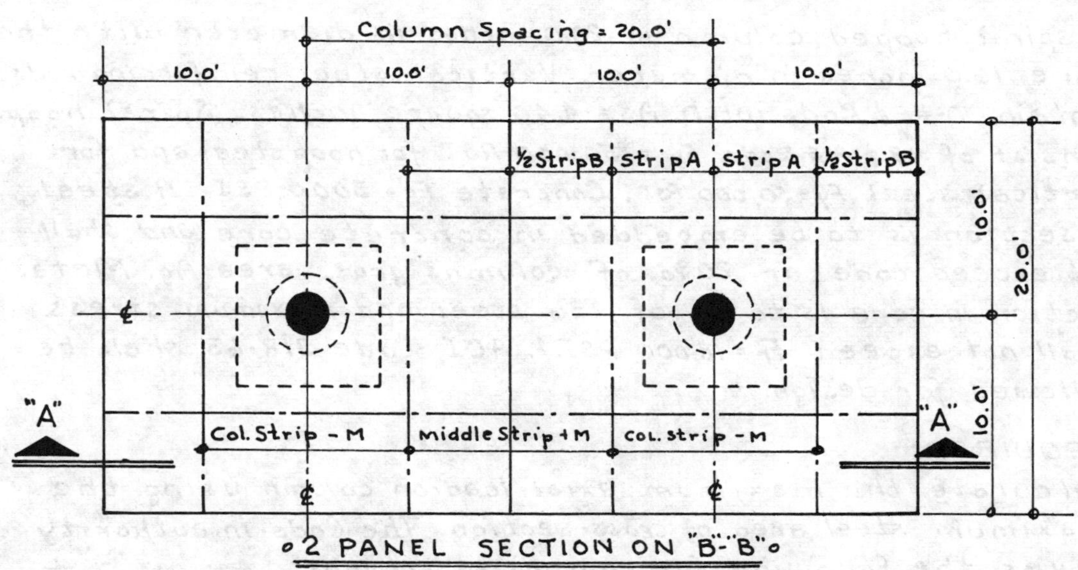

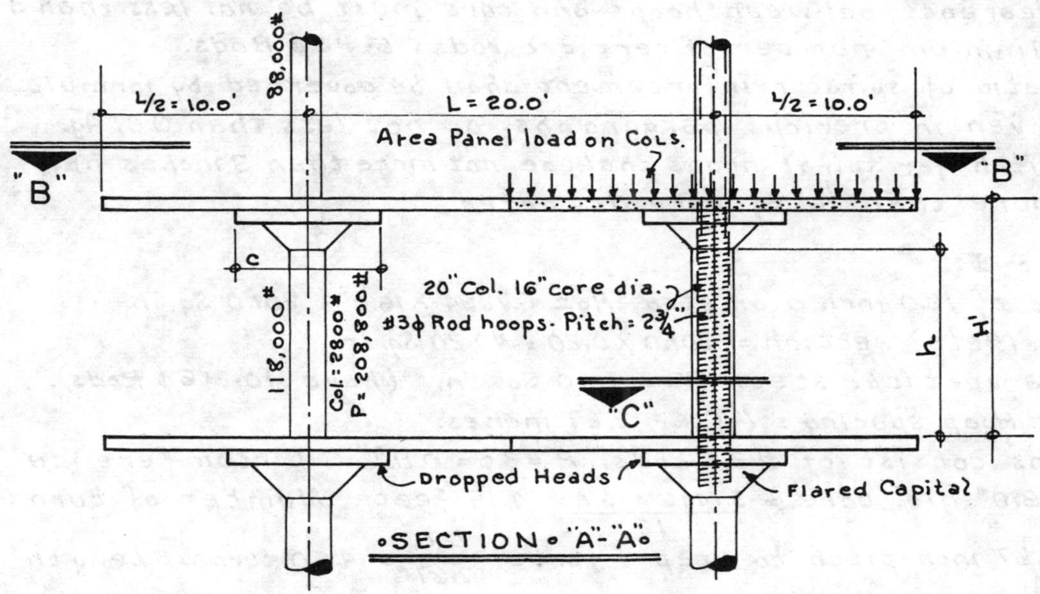

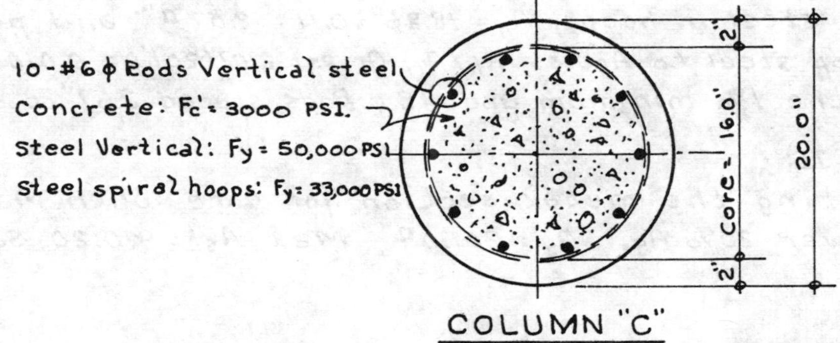

EXAMPLE: Designing composite column 4.10.7.8

A spiral hooped column is 20.0 inches in diameter with the core 16.0 inches in diameter. Vertical steel reinforcing rods contain 10-#6ϕ Rods with As = 4.40 square inches. Spiral hoops consist of #3ϕ A15 Rods. F_y = 50,000 PSI for hoop steel and for vertical steel Fy = 50,000 PSI. Concrete Fc' = 3000 PSI. A steel H section is to be embedded in concrete core and shall not exceed code or 20% of columns gross area Ag. Metal section in core is to be of A36 steel and maximum stress shall not exceed F_r = 18,000 PSI. ACI Code 318-63 shall be followed for design.

REQUIRED:
Calculate the maximum axial load on column using the maximum steel area of cross-section. The Code in authority gives the following for composite columns:

(a) Clearance between hoops and core must be not less than 3 in.
(b) Mininium number of vertical rods: 6-#5ϕ Rods.
(c) Ratio of spiral reinforcement shall be governed by formula given in previous paragraphs, or not less than 0.01 Ag.
(d) Pitch for spiral hoops shall be not more than 3 inches nor more than 1/6 of core diameter.

STEP I:
Core of 16.0 inch diameter. Ag = 0.7854×16.0^2 = 201.0 Sq. In.
Max. Metal section = 201.0 x 0.20 = 40.20 Sq. In.
Area vertical steel As = 4.40 Sq. In. (have 10-#6ϕ Rods).
Max. Hoop spacing = 16.0/6 = 2.67 inches.
Hoops consist of #3ϕ Rods. A #3ϕ = 0.11□" Circumference of 16.0" dia. core = $\frac{3.1416 \times 16.0}{12}$ = 4.19 feet. Number of turns at 2.67 inch pitch to wrap 1 foot = $\frac{12.0}{2.67}$ = 4.50 turns. Length of rod to wrap 1 foot column = 4.19 x 4.50 = 18.86 lineal feet. Area steel in hoops, A_h = 18.86 x 0.11 = 2.07□" and percentage of hoop steel to Ag is: A_h/Ag. Or ps = 2.07/201.0 = 0.0103 This is over the 1% minimum and #3ϕ Rods spaced 2.67" cc are Ok.

STEP II:
Selecting the metal section for core when Max. Ast is not over 20% Ag. Ag = 201.0□" Max. Ast = 40.20 Sq. In.

EXAMPLE: Designing composite column, continued 4.10.7.8

To make certain that code clearance requirements of 3 inch minimum is maintained between metal section and spiral hoops the cross-section will be drawn to scale. The ACI code permits this clearance to be reduced to 2 inches when a symmetrical H type section is used. Steel is A36 grade.

20.0"
2" 16.0" 2"
#3 d Spirals 2¼ Pitch

The largest 8x8 H section listed is a W8x67 with an area $A_{st} = 19.70^{\square}{}''$

STEP III:

Calculating the maximum allowable load: For short column with $\frac{h}{d}$ of 10 or less, the maximum length without bracing = $10 \times 16.0 = 160$ inches or 13'-4".

Formula for Maximum axial load is:

$$P = (0.225\ A_g F_c') + (A_s F_s) + (F_r A_r).$$

$F_c' = 3000$ PSI $A_g = 201.0^{\square}{}''$ (See note below).

$A_s = 4.40^{\square}{}''$ $F_s = 0.40\, F_y$ or $F_s = 0.40 \times 50000 = 20{,}000$ PSI

F_r for A36 Steel 18,000 PSI. $A_r = 19.70^{\square}{}''$

With values substituted in formula:

$$P = (0.225 \times 3000 \times 176.90) + (4.40 \times 20{,}000) + (19.70 \times 18{,}000)$$

$$\text{or } P = 119{,}400 + 88{,}000 + 354{,}600 = 562{,}000 \text{ Lbs.}$$

DESIGN NOTE:

With respect to composite columns with metal section embedded in core, A_g is taken as the area of concrete encasement. In equation above, $A_g = 201.0 - (4.40 + 19.70) = 176.90^{\square}{}''$

Concrete stairs 4.11

The usual concrete stair is a simple slab with long-way reinforcing and temperature steel placed across the width of each tread and riser. Stairs are inclined, and supported at the top and bottom. The concrete which forms the steps does not add to the strength of the run. A single run of stairs should be limited to ten treads or eleven risers. The intermediate support between two flights is called the landing.

Building codes usually require stairs to be designed for a minimum live load of 100 pounds per square foot. The stairs are formed after the main structure has been completed. Dowel rods should extend from the main structure to tie the ends of the stair slab into the floors and landings. Key joints should also be provided to restrict movement at the supports. This type of end joint is also a great aid in the placing of concrete.

STAIR FINISHES

Concrete stairs which are enclosed in a stair well, close to the elevators, are generally regarded by codes as interior fire escapes, and are left as rough stairs without any finish except an abrasive nosing along the edge of each tread. Stairs which are to receive architectural treatment such as terazzo or a colored surface are formed with clearance to allow the finish to be applied later. Such treatment is referred to as the *applied topping*. Concrete stairs in industrial plants are often subjected to rough treatment, which causes damage to the edge of tread. All stairs in industrial plants and warehouses should be provided with steel angle nosings, with anchors embedded in concrete.

RISE AND RUN RATIOS

The *rise* of a flight of stairs is the height from floor to floor or from one landing to the next landing. The *run* of a flight of stairs is the total width of the treads taken horizontally between landings in a single flight. The ratio of riser height to tread width must be properly proportioned for comfort and safety. Many rules have been proposed for calculating the comfortable ratio of rise to run, but it is best to examine the building code. For example the Southern Standard Building Code, which has been adopted by municipalities, gives the following rules for the proportion of stairs, treads and risers:

(a) The sum of two risers and a tread, less its projecting nosing, shall not be less than 24 inches nor more than 25 inches. The height of the riser shall not exceed 7¾ inches. The width of the tread, exclusive of nosing, shall not be less than nine inches, and every tread less than ten inches wide shall have a clear nosing projection of one inch over the immediate level below that tread.

(b) Any one flight of stairs shall have uniform riser height and tread width.

(c) The use of winding treads or spiral stairways is prohibited in stairways serving as required exits.

(d) No flight of stairs shall have a vertical rise of more than 12.0 feet between floors or landings. In theaters or assembly halls this vertical rise is limited to 8.0 feet.

(e) The length and width of landings shall be not less than the width of the stairwell in which they occur.

Concrete stairs, continued 4.11

(f) The width of any stair serving as a means of egress shall not be less than 3.0 feet, and handrails shall not project in more than 3½ inches at each side of the required width.

(g) Stairways serving as fire exits in buildings of four stories or more shall be completely enclosed with walled partitions of not less than 2-hour fire-resistance.

It should be noted that the code fails to mention a minimum height for risers. On occasion, an institutional client caring for senior citizens will stipulate that no riser in any stair flight shall exceed six inches in height. If rule A above is followed, the width of the tread would have to be twelve inches. Use the TABLE 4.11.1 for proportioning riser and tread.

TABLE: Recommended ratios of riser to tread 4.11.1

RECOMMENDED RATIO SIZES OF RISER TO TREAD					
RISER IN INCHES	TREAD IN INCHES	RISER IN INCHES	TREAD IN INCHES	RISER IN INCHES	TREAD IN INCHES
6.000	12.625	6.875	11.000	7.750	9.625
6.125	12.375	7.000	10.750	7.875	9.500
6.250	12.125	7.125	10.625	8.000	9.250
6.375	11.875	7.250	10.375	8.125	9.125
6.500	11.625	7.375	10.250	8.250	9.000
6.625	11.500	7.500	10.000	8.375	8.750
6.750	11.250	7.625	9.875	8.500	8.625
BUILDING CODES LIMIT HEIGHT OF RISERS TO 7.750 INCHES IN ALL COMMERCIAL, PUBLIC AND STORE BLDGS. SUCH RESTRICTIONS DO NOT APPLY TO INDUSTRIAL PLANTS, TANK FARMS OR PROCESSING STRUCTURES.					

Multi-story stair exits 4.11.2

For several decades, architects usually located the main stair fire exits close to the elevators. This arrangement is convenient to occupants who prefer not to wait for the elevator, and is time-saving during elevator break-downs. But, darkened stairs in the multiple flights of a high-rise structure are very dangerous during the frantic rush of an emergency evacuation. Multi-story apartments, dormitories and condominiums are now required by several municipal codes to have the fire exit stairs located at the ends of the main corridors, at a distance from the elevators. Natural light can be introduced in the stair wells by glazed openings.

Architects now tend to plan the stairs in a single shaft, adjacent to the exterior wall of the main structure. Multiple stair flights in a single well will produce horizontal component forces in addition to vertical reactions. The method to calculate these horizontal forces, and those parallel and normal to the inclined stair, is explained in Section V, Trigonometry and Graphic Analysis, in Example 5.1.5.7 illustrating a leaning ladder.

EXAMPLE: Complete multi-story stair design 4.11.3

Architects preliminary plan layout gives floor heights of 12'-10". This is from floor to floor with 1 Landing. The inside dimension of stair well is 9'-2 x 17'-6". Between each short flight an 8 inch glazed tile partion wall will separate the void space and serve as a fume and smoke curtain. Vertical Live load for design = 150 Lbs. square foot. Concrete is 3000 PSI at 28.0 days. Reinforcing shall consist of new billet steel, intermediate grade with Fy = 40,000 PSI.

REQUIRED:
Determine the most comfortable ratio of riser height to tread run and draw a section through the full flight floor to floor. Also show plan layout of treads, risers and landing which will comply with Southern Standard Building Code rules previously given.
Set out the short flight from floor to landing and design the required reinforcing steel. Calculate the concurrent component forces by assuming the flights to be a free body, and note the horizontal force at landing.

STEP I:
Well hole is 9'-2 x 17'-6 and with 8 inch partition the width of stair can be made 4'-3" wide. With a single landing maximum number of treads in each flight is limited to 10 treads or 11 risers.
Total height floor to floor = 12'-10" or 154 inches. With 22 risers, each riser will be 154/22 = 7.0 inches. Using code rule (a) 2 risers plus 1 tread run cannot be less than 24.
Then tread run = 24 - (7.0"+7.0") = 10.0 inches. Landing must be same as stair width or 4'-3".
Run of stair with 10 Treads + landing = (10 x 10.0) + 4'-3 = 12'-7"
Clearance at stair entry = 17'-6" — 12'-7" = 4'-11" (Adequate).

STEP II:
Since entry space is larger, the layout in plan and section the landing will be made 4'-7" and entry will also be 4'-7"
10 Treads plus landing = 8'-4" + 4-7 = 12'-11"
Details will delineate trowelled treads without nosings.
Height each side of landing = 12'-10"/2 = 6'-5" (even riser number).

EXAMPLE: Complete multi-story stair design, continued 4.11.3

STEP III:

Ordinarily a 12.0 inch strip load is used for slab design but in this example the width of stair 4.25 feet can be used and considered as a wide beam 51.0 inches = b. This will be similar to flat slab strips A and B.

Assume for dead load estimate that slab will be 6.0 inches in thickness less treads and risers. Calculate total length of slab on incline. Horizontal dimension b = 8.33 and vertical dimension a = 6.42 − 0.58 = 5.84 feet. Incline c = hypotenuse, and

$c = \sqrt{a^2 + b^2}$. Then $c = \sqrt{5.84^2 + 8.33^2} = 10.16$ Feet. Area of slab is 4.25 x 10.16 = 43.14 Sq. Feet. Now there are 10 triangular treads 4.25 ft. long or 4.25 x 10 = 42.50 feet. Each tread = 7" x 10" = 70.0□"

Volume in treads = $\frac{42.50 \times 12 \times 70.0}{2} = 17{,}850$ Cubic Inches.

Reducing volume to feet: Then, $\frac{17{,}850}{1728} = 10.33$ Cubic Feet

Calculating design load: Wt. Concrete = 150 Lbs. cubic foot.

Dead Load Slab: 43.14 x 0.50 x 150 = 3235.50 Lbs.
Dead Load triangular treads: 10.33 x 150 = 1545.00 "
Total Dead Load = 4780.50 Lbs.

Live load action is vertical and horizontal dimension = 8.33 Ft.
Area = 4.25 x 8.33 = 35.40□' Live load = 35.40 x 150 = 5310 Lbs.
Design Load W = 4780.5 + 5310 = 10,190.5 Lbs. (<u>Call it 10,200 Lbs.</u>)

STEP IV:

Vertical reactions as simple beam = W/2 or 10,200/2 = 5100 Lbs. Force perpendicular to incline is N (Normal to plane). Determine incline angle with horizontal when sides a = 7.0" as riser and b = 10.0" width of tread. $\text{Tan } A = \frac{a}{b} = \frac{7.00}{10.00} = 0.7000$ From Trig. tables in Section V: Angle A = 35 degrees. Other trig functions of 35° are: Sine = 0.5730 Cos = 0.819 Cotan = 1.4281 Angle B = 90° − 35° = 55° Vertical Force W = 10,200 and acts at its center of gravity or mid-span. Force N = W Cos A.

N = 10,200 x 0.819 = 8356 Lbs. (W represents side a of triangle. Parallel to incline = side c, and $c = \frac{a}{\text{Sin } A}$ or $\frac{10{,}200}{0.5730} = 17{,}800$ Lbs.

Horizontal force is side b and H_A = a Cot A. Load W = a.
H_A = 10,200 x 1.4281 = 14,566 Lbs. An elevation representing inclined stair will now be drawn and values checked by a force diagram. When beam is connected at each end as

EXAMPLE: Complete multi-story stair design, continued 4.11.3

show in triangle below, the ends must maintain the beam in equilibrium by the reactions. The force diagram is drawn by starting with known force W=10,000 lbs. and the side a is scaled to 10,200 lbs. Angle A is also known. Set your adjustable triangle to 35°. Draw the incline from top of a. Next close the triangle by drawing the horizontal line from bottom of line a. c= incline and b= horizontal line. Use the same scale to read values for lines c and b. Force N is perpendicular to line c. A line drawn from bottom of a to line c = the force N. (normal to slope). However N acts at the center of gravity as determined by the dash diagonal. The values all check with results obtained by trig. formulas.

DESIGN NOTE:

The flight of stairs above landing has the same force reactions and the horizontal force against landing = 14,566 lbs.

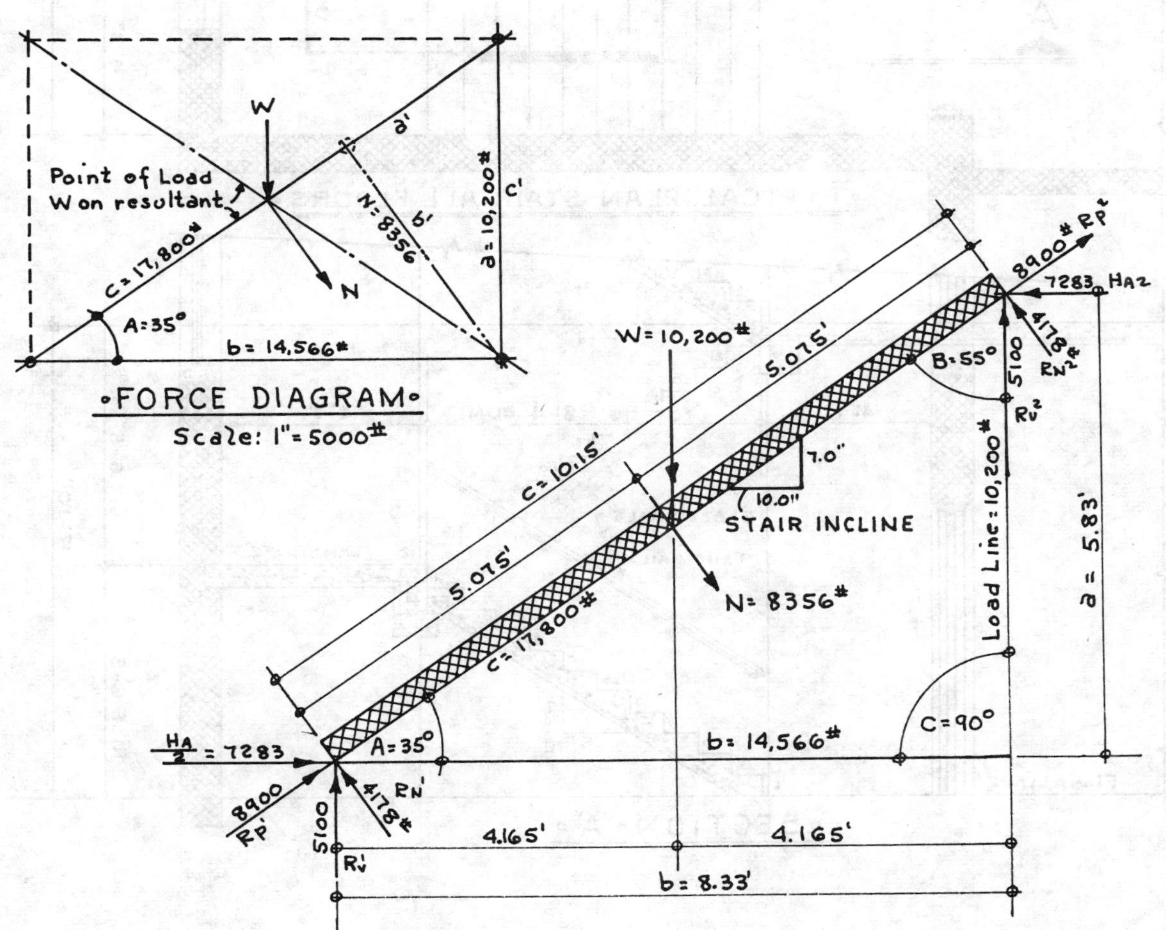

EXAMPLE: Complete multi-story stair design, continued 4.11.3

Plan and section through stair well will now be drawn to scale. A 12 inch masonry wall serves for fire wall enclosure. Partition between serves to prevent smoke or fumes from lower floors from being drawn upward. An additional door can be added to completely seal off each floor.

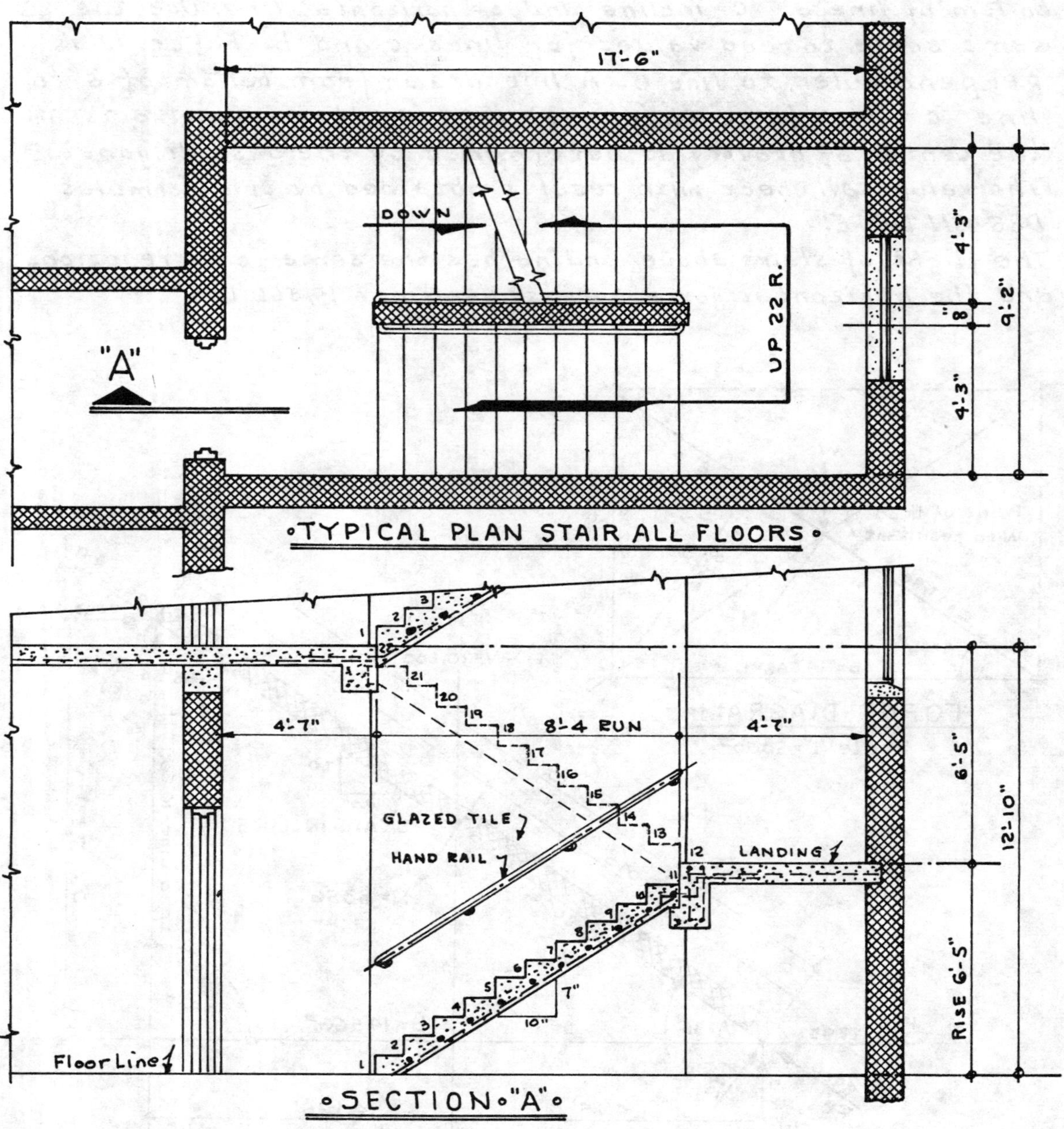

EXAMPLE: Complete multi-story stair design, continued 4.11.3

STEP V:

Calculating bending moment: Bending moment can be taken with vertical load and horizontal span as L, or Normal load N with span equal to incline. If results are not the same in value an error has been made and a re-check is in order.

$M = \frac{WL}{8}$ or $M = \frac{Nc}{8}$. With values substituted:

$M = \frac{10,200 \times 8.33}{8} = 10,620$ Ft. Lbs. $M = \frac{8356 \times 10.15}{8} = 10,620$ Ft. Lbs.

STEP VI:

Check depth of slab by formula: $d = \sqrt{\frac{M}{Kb}}$. Recall that the whole width of stair is a beam. $b = 51.0''$

From tables, when $Fc' = 3000$ PSI and $F_s = \frac{1}{2} F_y$ or $F_s = 20,000$ PSI.

$K = 226.0$ $J = 0.872$ With values in formula.

$d = \sqrt{\frac{10,620 \times 12}{226.0 \times 51.0}} = 3.35''$ Adding 1.0 inch for fire protection, d can be made 4½ inches. Use a 5.0 inch depth and depth to steel d = 4.00 inches. Loading is safe since a slab of 6.0 inch thickness was used for estimating dead load.

STEP VII:

Determine area steel required for 51.0 width:

$As = \frac{M}{F_s J d}$ or $As = \frac{10,620 \times 12}{20,000 \times 0.872 \times 4.0} = 1.83$ Sq. In.

Choosing #4 ϕ Rod with area of 0.20 □"

Number required = 1.83/0.20 = 9.1 Use 10-#4 ϕ Rods and space thus: Deduct 2½ inches from each side of stair making width 51.0 - 5.0 = 46.0 inches. 10 Rods will use 9 spacing and s = 46.0/9. = 5.11 inches. This is less than usual and acceptable.

Add 1-#4 ϕ Rod across width of stair 51.0" long at each weak point where shown in drawing section A.

Foundations and footings 4.12

When the planning for a new structure is initiated, the foundation is usually the first item considered. A safe, permanent base which will support the superstructure is required. To meet these requirements, a complete design procedure must be followed:

(a) The soil bearing investigation must be accurate, and of proven durability.
(b) The materials used for constructing the footings must be proof against decay or deteriorating influences.
(c) The foundation must not be overstressed; in the future loads may be added or the structure may be put to another use.
(d) Future excavation for adjoining structures must be anticipated.

The first step is to make a site investigation, including a nexhaustive study of the soil characteristics on which the footings will rest. The depth to the various strata can be analyzed by laboratory test borings and settlement load tests. In the older cities, it is not uncommon to discover buried foundations which supported buildings long since torn down to an elevation below the grade. Artificial fills such as dredged spoil from canals and river beds are liable to continuous settlement and sliding upon deeper vegetable layers. Uniform bearing must be maintained over the entire building area. When portions of a structure will rest on hard, compressed soils and another section of the building will rest on soil of doubtful stability, the unit design bearing load allowable on the higher strength soil will have to be reduced to equal the allowable bearing on the lower strength soil.

Load bearing tests 4.12.1

Load tests to determine a safe allowable design bearing pressure are conducted on below-grade soil strata to determine the safe loads for continuous spread footings or for independent footings. Building codes require load tests on soil areas which are isolated to verify the proposed bearing value. The procedure used to make the test must be approved by the local authorities. A settlement curve should be prepared for the local building officials.

A satisfactory load bearing test may be conducted as follows: Obtain a steel plate 2.0 feet square. Place it upon the undisturbed soil in the excavation pit at the elevation of the expected footings. Place a large square timber or steel pipe at the exact center of the plate in a vertical position. Any guy wires or stays to this vertical member to prevent tipping should be kept horizontal. Construct a platform to receive loading material upon the vertical member. Load the platform uniformly and carefully, so that all bearing is directed to the center of the plate. Materials for loading may consist of pre-measured weights of bagged sand, steel ingots or a water container.

The first load increment to be placed on the platform should equal the weight of the proposed bearing pressure. This load is allowed to remain 24 hours, after which an instrument reading is taken to determine the initial settlement. Allow the first load increment to remain until further readings show no additional settlement over a 24 hour period. When the last reading shows

Load bearing tests, continued 4.12.1

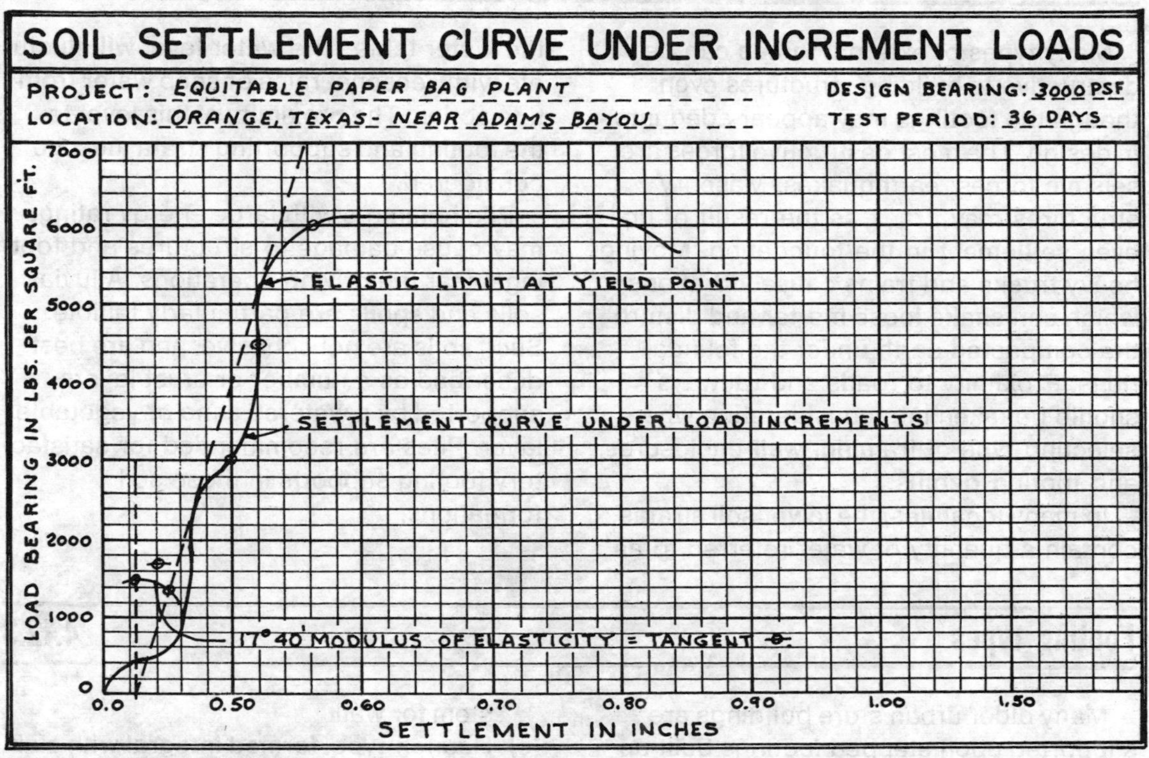

that the plate has stopped settling, another load increment of 50 percent should be added. Take readings at 24 hour intervals and record the rate of settlement. The final load increment on the bearing plate should bring the total bearing pressure on the soil to double the value proposed for design purposes. Readings must be taken at 24 hour intervals until no further settlement can be measured. The settlement curve is constructed as shown.

The proposed safe design bearing is acceptable if the increment of settlement resulting from the 50 percent overload does not exceed 60 percent of the settlement under the 100 percent design load. The settlement under the allowed load should not exceed ½ inch. If, at the proposed safe bearing, these conditions are not satisfied, the allowable bearing value must be reduced accordingly. When the proposed design load produces more than ¾ inches of settlement, of the 50 percent overload increment exceeds 60 percent of the settlement from design bearing, the load test is not satisfactory. As the loading continues to increase beyond the 100 percent design bearing, the test may be extended to find the ultimate load and limit of soil cohesion. The settlement curve shows the result when applied loads are applied beyond the elastic yield point.

Destructive external forces 4.12.2

Many types of external forces can be destructive to building structures even though the footings may appear adequate in design. The most destructive forces are seismic forces: earthquakes. Masonry structures may crack as the result of uneven settlement in the foundation. Moving heavy trucks and trains cause vibrations which can shake loose mortar and disturb the compacted earth under the foundations. Proximity to roads and railways should be taken into consideration when selecting type of framing, wall enclosures, and footing depths.

In many localities, the lower soil stratas contain a quantity of water referred to as the water table. The water level will fluctuate with seasonal rains or seep water from tide action. The proximity of this level to the footing and supporting strata must be considered.

Pile hammers with large energy ratings may cause damage to structures and footings near the driving operations. Alluvial soils and spoils are particularly fallible. Such soils are not cohesive, and are best described as a number or crust layers supported by a watered sand or vegetable layer. Piles are recommended for satisfactory footing supports in these soil formations.

Footing types 4.12.3

Many older urban store buildings are supported upon stepped footings built up from brick and mortar. Stepped footings are constructed with a lower pad which supports several layers called pedestals or plinths. In more modern buildings, concrete, which can sustain higher compressive and shear stress than common clay brick is used, the number of pedestals can be reduced. Footings are formed from concrete into various types which can be described by the forming, location, and shape. The following are several types:

(a) Independent spread footing: square or rectangular.
(b) Continuous-wall spread footing, with stem for wall.
(c) Haunch-type, formed integral with slab.
(d) Connected-type which joins and supports an interior column and a wall column on the same footing. This type may also be referred to as cantilever, strap or combined footing.
(e) Drilled footing, also referred to as bell-bottom, cone-bottom or shaft-type cast-in-place.
(f) Continuous footing: resembles a continuous beam when inverted, or can support several columns.
(g) Pile-supported footing: usually used in connection with types a, b, c, and plain pile caps.

Footing design 4.12.4

Independent footings supporting columns are designed with the area of spread governed by the column load and the allowable safe soil bearing. Consider the soil bearing as a pressure which acts as a uniform upward force. A simple, inverted drawing will serve as an aid for a better examination of the design. Using this method, the column load is represented as a reaction, and the bearing pressure appears as a uniform load. The overhang beyond the face of the column is analyzed as a wide cantilever beam with negative bending moment.

The depth of the footing is usually determined by the safe allowable unit shear stress. For square and rectangular footings with columns and pedestals, the punching shear around the periphery of support should be investigated, and the footing depth designed to keep within the allowable shear stress. Pedestals are used to increase the punching shear perimeter, while also reducing the area of bearing pressure outside the frustum and pedestal.

When the punching shear around the column is excessive for a reasonable footing depth, the designer has two choices:

(a) Provide a pedestal or plinth under the column which will give a greater shear depth and reduce the outer rim area subjected to soil pressure.

(b) Make the footing block under column deeper for more effective shear area.

FOOTING DEPTH

The trial depth to steel in a footing block may be initially calculated by using the beam bending formula for depth: $d = \sqrt{\frac{M}{Kb}}$. After obtaining this result for depth, check the depth required for shear, and add a pedestal if necessary. The formula for punching shear is: $f_v = \frac{V}{Jdb}$, where dimension b is the perimeter of the column, pedestal or frustum. For the effective shear depth of footing pad, there are a few codes which do not use the design factor J. In this case, the unit shear stress formula is: $f_v = \frac{V}{bd}$.

MINIMUM DEPTH TO STEEL

The code requirements for minimum depth to steel from top of footing pad is as follows:

(a) Reinforced footings supported by *soil bearing* shall have a thickness above reinforcement at the edge of the footing not less than 6 inches.

(b) Reinforced footings supported *by piles* shall have a minimum thickness of 14.0 inches above the tops of the piles.

Bending moment in footings 4.12.4.1

Two methods are in common use for calculating the bending moment in the footing pad. Each of these following methods will be illustrated in the examples: (a) On one side of the column or pedestal plane, draw a line clear across footing. The forces acting upon the area *outside* the column cause the bending moment in the slab. The moment arm is taken from face of column to center of gravity of the plane rectangle. (b) In independent or isolated footings, the moment shall be computed in the manner prescribed in (a) when the conditions are thus:

(1) From the face of the column, pedestal or wall, when footings support a concrete column, pedestal or stem.
(2) Half the distance between the edge of the stem and the extended projection of the footing, when the footing supports masonry walls.
(3) From face of column to center of piles when footings are supported by piles.

Shear and diagonal tension 4.12.4.2

The pedestal and footing slab must be investigated for shear in the vicinity of concentrated reactions at the following critical sections:

(a) The footing slab (acting in the manner of a wide beam) is highly loaded in shear along a plane vertical to the slab at the face of the column. Nominal shear stress is computed as: $f_v = \frac{V}{bd}$, where b = length of line along entire footing at face of column, pedestal or wall.
(b) Punching shear is the critical stress, and usually will govern the design depth for slab and pedestal. Action is two-way. When evaluating the diagonal tension (or the potential for a 45° crack), the bottom of the frustum or truncated cone is limited to a distance of ½ depth to steel. This is indicated in examples as d/2. Shear area to sustain punching shear is taken as the periphery of the frustum times the depth. Reaction or total shear (V) is the upward force outside the frustum area.

Allowable shear stress for footings designed without web reinforcement is: $F_v = 2\sqrt{Fc'}$. Shear allowable for footings with web reinforcement is increased to: $F_v = 3\sqrt{Fc'}$. Shear reinforcement such as deformed rods or bars is not considered effective in concrete footings with a total thickness less than 12 inches.

Drilled footings 4.12.5

Under-reamed or straight-shaft footings drilled by motorized equipment are unquestionably the most economical and popular type of footing for modern foundations. Drill depth can be adjusted at the job site to reach safe soil bearing in the desired stratum. Excavating work is reduced and back-filling, which is required for spread-type footings, is not required. Reinforcing steel in drilled footings consists of tied vertical rods, which are assembled on the surface and pushed down into the hole after the concrete is placed and still fluid. These rods should extend down into the under-ream to within approximately six inches of the bottom. Curing of concrete is not a problem since the underground soil is usually damp and temperature is relatively constant. Holes should be filled with concrete immediately after inspection. In no event should holes be left open overnight. With two drill rigs in operation, a large project can have the footings in place in a matter of days. In regions where the soil stratum of the underream is non-cohesive, and there is danger of cave-in or collapse, the concrete trucks must maintain a safe distance from the shaft opening. In such cases, plan for the use of flexible tremmies with funnel tops. Spouting the concrete into the tremmie will place the concrete gently into the frustum and build the pressure upward against the top of cone.

Engineers and architects can save the contractor valuable time on a project by minimizing the number of different diameter footings. Changing drill bits consumes drilling crew time. Also, concrete volume estimates will be simplified. Estimating the volume of concrete for drilled under-reams depends on the type of cone and angle. The volume in a 45° truncated cone is about equal to the bell-bottom or spherical-sector zone. Tables 4.12.9.1 may be used for estimating either type. A 60° cone frustum bottom will contain a larger volume, and Table 4.12.9.2 is provided for figuring concrete requirements in this type. There is less danger of cave-in when the 60 degree frustum is used.

The design of drilled footings is relatively simple. All that is required is the computation for bearing area and hole diameter. Use the following formula for diameter with the slide-rule: $d'' = \sqrt{\frac{A' \times 144}{0.7854}}$. The formula is illustrated below:

Load to footing = 78,500 lbs. Soil bearing = 2500 PSF.

Area on soil = 78,500/2500 = 31.50 sq. ft.

Then diameter $d'' = \sqrt{\frac{31.50 \times 144}{0.7854}} = 76.0$ inches or 6.33 feet.

The examples and tables to follow will furnish formulas and check values for calculating cone volumes.

EXAMPLE: Designing continuous wall footing 4.12.6.1

A masonry wall 12.0 inches thick is to be supported on a continuous concrete footing at a depth 4.0 feet below grade elevation. The stem or grade beam supporting wall is also to be 12.0 inches thick. Loads from floor, roof and dead load of masonry total 18,500 Pounds per lineal foot of wall. Safe design load on soil at 4.0 foot depth is 3750 Lbs. square foot. Reinforcing rods to be new billet steel with $F_s = 18{,}000$ PSI. Concrete will be 2500 PSI at 28 days. Rust protection to be not less than 3.0 inches.

REQUIRED:

A design of footing for width and depth to steel to be determined by bend moment or shear whichever is the greater. Stem of footing will be formed and concrete poured later. Draw cross-section and show key joint with dowels for stem. Run temperature rods long way.

STEP I:

Weight of concrete must be added and estimate this dead load at approximately 8% of load or 1500 Lbs. Foot.
Then Lineal foot load on soil = 18,500 + 1500 = 20,000 Lbs. Foot.
Safe bearing = 3750 PSF. Width Footing = $b = \frac{20{,}000}{3750} = 5.33'$ (5'-4").
Overhang from stem = $\frac{5.33 - 1.00}{2} = 2.167'$

STEP II:

When stem and footing is inverted it becomes a twin cantilever beam on a single support and load acts at center of gravity of clear overhang. Axial bearing is on ℄ of stem. Moment lever = $\frac{2.167}{2} + 0.50 = 1.583$ feet.

STEP III:

Load on overhang = 2.167 x 3750 = 8125 Lbs. Moment for a cantilever beam is $M = WL/2$. With values W becomes a Concentrated load P and Moment arm $L = 1.583'$ Then the

$M = PL$ or $M = 8125 \times 1.583 \times 12 = 154{,}345$ inch pounds.

STEP IV:

Calculating depth to steel d by shear allowable and moment formula: From tables: With $F_c' = 2500$ PSI and $F_s = 18{,}000$ PSI. $K = 189.5$ $J = 0.871$ $n = 10.1$ $F_v = 2\sqrt{F_c'} = 100$ PSI and $u = 150$ PSI

Max. shear at edge of stem = 8125 Lbs. $d = \frac{V}{F_v J b}$ and b = 12.0 in.

$d = \frac{8125}{100 \times 0.871 \times 12.0} = 7.78$ inches. Required for shear.

Required depth by moment formula: $d = \sqrt{\frac{M}{Kb}}$. With values:

$d = \sqrt{\frac{154,345}{189.5 \times 12.0}} = 8.25$ inches. (Call it 9.0 inches and with 3 inches of rust protection, total depth D = 12.0 inches.

STEP V:

With all required dimensions cross-section can now be drawn. The inverted diagram will show the same condition as a twin cantilever beam with uniform load and a single support at center.

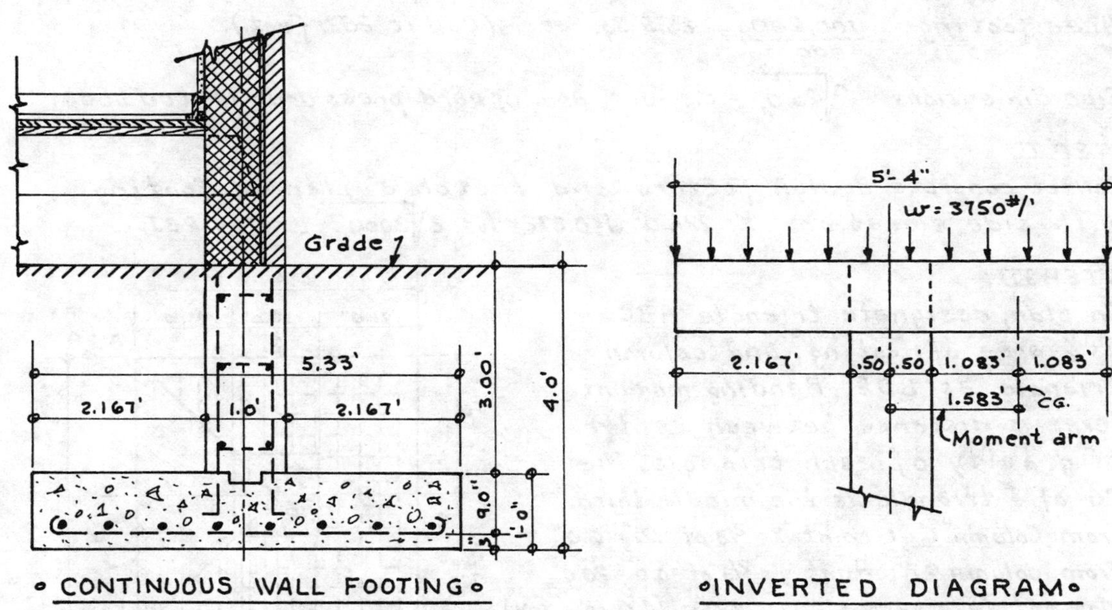

• CONTINUOUS WALL FOOTING •

• INVERTED DIAGRAM •

STEP VI:

For steel requirments perpendicular to stem:

$A_s = \frac{M}{F_s J d}$ or $A_s = \frac{154,345}{18,000 \times 0.871 \times 9.0} = 1.095$ Sq. inches.

This quantity is for each lineal foot parallel to wall.

If #7 ϕ Rods are selected: Area each rod = 0.60 Sq. In.

Space rods to provide 1.095 sq. inches for each foot of footing.

Spacing formula: $s = \frac{\text{Area 1 Rod} \times 12 \text{ inches}}{\text{Area steel required}}$.

Thus: $s = \frac{0.60 \times 12}{1.095} = 6.57$ inches on centers.

EXAMPLE: Designing independent footing 4.12.6.2

A concrete column with outside dimensions 18.0" x 18" is square and supports a load of 94,000 Pounds including lower column DL. Weight. Safe soil bearing = 4000 P.S.F. Concrete strength at 28 day, $F_c' = 3000$ PSI and $F_s = 20{,}000$ PSI. Code requires 4 inches of rust protection due to high salinity soil.

REQUIRED:

A design for a 2-Way reinforced footing without a plinth or pedestal. Footing shall be square. Use $F_v = 1.1\sqrt{F_c'}$

STEP I:

Assume Dead Load weight of footing to be approximately 6600 Pounds or around 7% of Column Load. Then axial load is P = 94,000 + 6600 = 100,600 Lbs. (Total bearing on soil).

$$\text{Area footing} = \frac{100{,}600}{4000} = 25.5 \text{ Sq. Feet. (Call it 25.0 feet.)}$$

Side dimensions = $\sqrt{25.0}$ = 5.0 feet and upward pressure = 100,000 Lbs.

STEP II:

Collect concrete design factors and sketch a plan of footing with side elevation: K = 226.0 J = 0.872 $F_v = 2\sqrt{3000}$ u = 164 PSI.

STEP III:

In plan, designate triangle ABC as 1/4 area of footing, and column triangle as BDE. Bending moment lever is distance between center of gravity of each triangle. The CG of a triangle is the middle third.

From Column ℄1-1 Point x = 2/3 of 9.0" = 6.0"

From column ℄1-1 Point y = 2/3 of 30.0" = 20.0"

Moment lever arm = 20.0 - 6.0 = 14.0 in. (x-y).

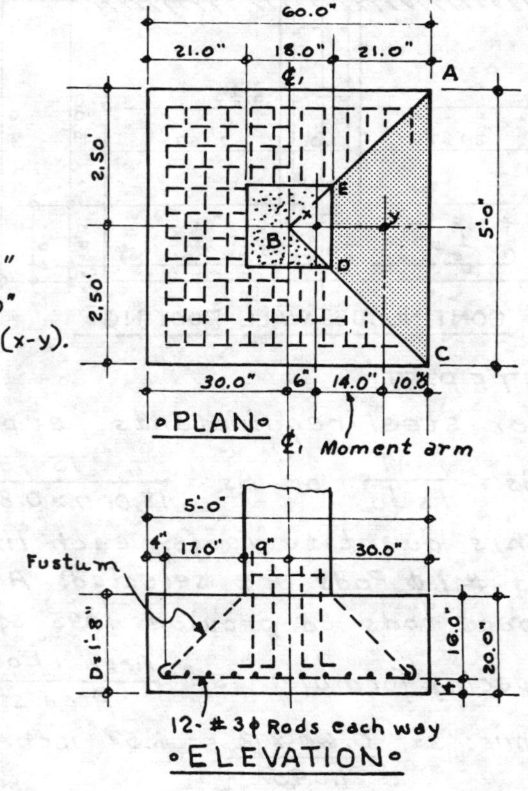

STEP IV:

Computing bending moment: The triangle ABC supports 1/4 of Load, or P = 100,000/4 = 25,000 Lbs. Then M = 25,000 x 14.0 = 350,000 Inch Lbs.

STEP V:

Calculate punching shear stress at column: Periphery of square column = 18.0 x 4 = 72.0 inches

Load P = 100,000 Lbs. Allowable unit shear $F_v = 2\sqrt{3000}$ = 110 PSI

EXAMPLE: Designing independent footing, continued 4.12.6.2

Depth footing required for shear: $d = \frac{V}{F_v J p}$, Where V = Load P and p = periphery. d = depth to steel.

$d = \frac{100{,}000}{110 \times 0.872 \times 72.0} = 14.5$ inches. With 4.0 inches of rust cover

D = 14.5 + 4.0 = 18.5 inches. (Call it 20.0 inches with d = 16.0 in.)

STEP VI:

Calculating area steel: $A_s = \frac{M}{F_s J d}$, and with values:

$A_s = \frac{350{,}000}{20{,}000 \times 0.872 \times 16.0} = 1.26$ Sq. Inches.

Area for rod spacing = 60.0 − (2×4.0) = 52.0 inches. Use an equal number of rods. From tables for total steel areas, 12-#3φ Rods give an A_s = 1.32 □" 12 Rods require 11 spaces, then Spacing s = 52.0/11 = 4.72" (space 4¾" on centers each way.)

STEP VII:

Check bond: Code requires all rods in footings to be placed with hooked ends. Bond will be checked anyhow. $u = \frac{V}{\Sigma o J d}$.

12-#3φ Rods have Σo = 14.14 Sq. In. V = ¼P or 25,000#

With values: $u = \frac{25{,}000}{14.14 \times 0.872 \times 16.0} = 127$ #□" (within allowable).

STEP VIII:

Depth of footing was based on allowable unit shear of 110 PSI. Punching shear has been determined and now diagonal tension will be checked at base of frustum. Width of frustum with steel depth d is 18.0 + (2×16.0) = 50.0 inches square. Area of frustum = $\frac{50.0 \times 50.0}{144} = 17.35$ □' Area whole footing = 25.0 □'

Pressure bearing area outside frustum = 25.0 − 17.35 = 7.65 □'

Shear V outside Cone frustum = 7.65 × 4000 = 30,600 Lbs.

Periphery of cone = 50.0 × 4 = 200 inches. All'd. $F_v = 1.1\sqrt{3000} = 60$ PSI

Diagonal tension stress = $\frac{30{,}600}{200 \times 0.872 \times 16.0} = 10.95$ PSI. This is below 60 PSI allowable.

STEP IX:

Check dead load weight of footing when concrete = 150 #³'

Volume = 5.0 × 5.0 × 1.67 = 41.67 cubic Foot.

Weight = 41.67 × 150 = 6250 Lbs. (This is very close to 6600 Lbs. used for estimate in step I). Percentage of column load for footing = $\frac{6250}{94{,}000} = 0.0665$

EXAMPLE: Designing independent footing 4.12.6.3

Assume the same conditions in the previous example for a 5.0 foot square independent footing. This problem was submitted to applicants for Engineering Registration by a State Board of Examiners in 1950. The previous example illustrates the older method for solution and design.

REQUIRED:
Design footing to comply with ACI Code 318-63 using the same soil bearing, column load, footing area and allowable stress values for concrete and steel. Draw a plan and the elevation. Design may include a pedestal if desired.

STEP I:
Review data from previous example:
Size footing = 5.0'x 5.0' Column size = 18.0" x 18.0" P = 94,000 Lbs.
F_c' = 3000 Ps F_s = 20,000 PSI F_v = 110 PSI Soil bearing = 4000 PSF.
J = 0.872 K = 226.0

STEP II:
Load on soil with weight of footing = 5.0 x 5.0 x 4000 = 100,000 Lbs. Drawing plan of footing the area at right of footing sustaining bend force is rectangle ABCD. Line AD in drawn at right side face of column but is the same on other 3 sides. The moment arm is ½ of dimension AB or 21.0/2 = 10.50 inches.

STEP III:
Bending moment on line AD:
Pressure on rectangle ABCD = 1.75 x 5.0 x 4000 = 35,000 Lbs.
M = 35,000 x 10.50 = 367,500 Inch lbs.

STEP IV:
Determine approximate depth to steel: $d = \sqrt{\frac{M}{Kb}}$. Length of line AD = b. b = 5.0 x 12 = 60.0 inches.

$d = \sqrt{\frac{367{,}500}{226.0 \times 60.0}}$ = 5.21 inches. For minimum depth to steel for footing bearing on soil, d must be not less than 6.0 inches.

STEP V
Investigate shear on line AD:
b = 60.0" Fv = 110 PSI and V = 35,000 Lbs. Solving for $d = \frac{V}{F_v J b}$.

depth $d = \frac{35{,}000}{110 \times 0.872 \times 60.0}$ = 6.08 inches

EXAMPLE: Designing independent footing, continued 4.12.6.3

Check diagonal tension around column when frustum is formed with ½d. Comply with code which requires footings on steel to have a minimum d= 12.0 inches. Thus: d/2= 6.0 inches. Column sides are 18.0" and frustum will be 18.0 + 12.0 = 30.0 inches. Area of column frustum = 2.5×2.5 = 6.25 □' Bearing on total footing = 100,000 Lbs. Bearing outside column frustum = 100,000 − (6.25×4000) = 75,000 Lbs. Perimeter of frustum = 30.0×4 = 120 inches.

STEP VI:

Punching shear around Column frustum: V= 75,000# b= 120.0"
d= d/2 or 6.00" and J= 0.872.

$f_v = \frac{V}{Jdb}$ or $f_v = \frac{75{,}000}{0.872 \times 6.0 \times 120.0} = 120$ PSI. This is a little over the allowable and for safety, a pedestal should be used or footing depth increased.

Choose a pedestal with a dimension 6.0 inches all around the column, and make the height ½ column or 9.0 inches. Then side dimension of pedestal = 18.0 + 12.0 = 30.0 inches. Frustum sides of pedestal will be 30.0 + 12.0 = 42.0 inches. Perimeter = 42.0×4 = 168.0"
Area pedestal frustum = 3.5×3.5 = 12.25 □' Pressure on area outside frustum = 100,000 − (12.25 × 4,000) = 49,000 Lbs.

Stress $f_v = \frac{49{,}000}{0.872 \times 168.0 \times 6.0}$ = 55.75 Lbs. Sq. In. (Less than allowable F_v).

STEP VII:

Calculating area of steel reinforcing: $As = \frac{M}{FsJd}$ and d= 12.0"

$As = \frac{367{,}500}{20{,}000 \times 0.872 \times 12.0} = 1.75$ □" From area tables select even number of rods as: 10-#4ϕ bars. Run rods each way.

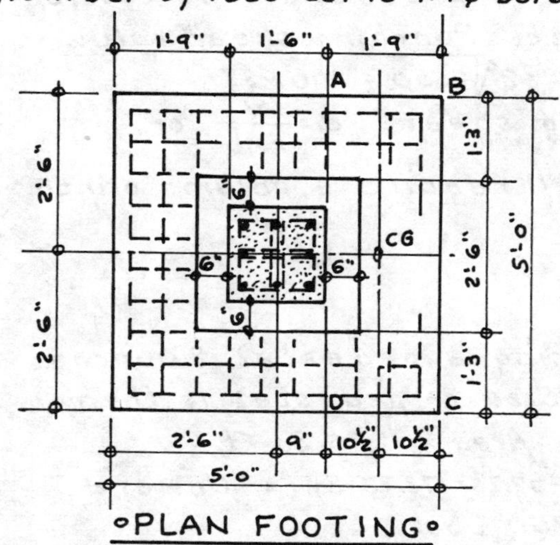

°PLAN FOOTING°

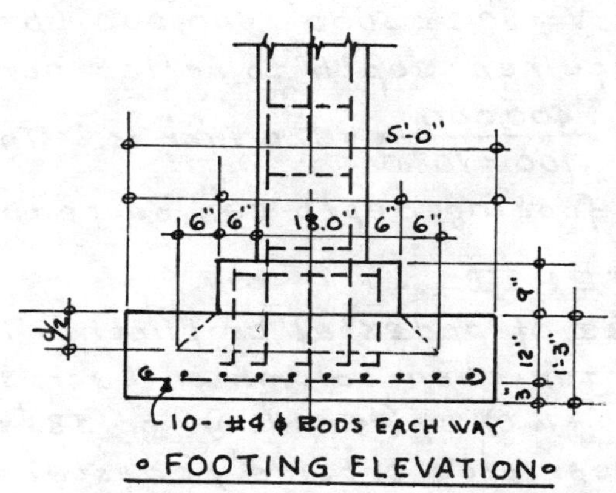

°FOOTING ELEVATION°

EXAMPLE: Designing rectangular spread footing 4.12.6.4

A rectangular wall column is 30.0 x 24.0 inches and brings a load of 400,000 Pounds to footing. Property line parallel to 30.0 inch column side is 3'-6" from ℄ of column and this line must not be crossed, therefore dimension for short way is limited to 6.75 feet. Design allowable for safe soil bearing is 5000 Lbs. square foot.
Concrete $F_c' = 2500$ PSI and $F_s = 18{,}000$ PSI. Design must be based on Code recommended by ACI 318-63.

REQUIRED:
Design a rectangular footing and provide drawings to be placed on plans of structure. The following ACI 318-63 Code uses the following:
Shear $F_v = 2\sqrt{F_c'}$, and diagonal tension is $\frac{d}{2}$ and $1.1\sqrt{F_c'}$
Bond in tension rods: $u = 4.8\sqrt{F_c'}$. Rust cover = 3.0" minimum.

STEP I:
Foot will be rectangular with 2-Way reinforcing. Estimate weight of concrete footing at about 25,000 Lbs. which is 6.250 percent of column load. Load on soil = 425,000 Lbs. and area footing required = 425,000/5000 = 85.0 Sq. feet.
Short dimension = 6.75' Long dimension = 85.0/6.75 = 12.5 feet.

STEP II:
Making a plan drawing of footing pad 6'-9" x 12'-6" and place column on both axes for axial loading.
Area of column = 2.50 x 2.0 = 5.0 Sq. Ft. Periphery around column is: (24.0 + 30.0) x 2 = 108 inches. Upward pressure area outside column = 85.0 - 5.0 = 80.0 Square feet. Punching shear value or V = 80.0 x 5000 = 400,000 Lbs. $F_v = 2\sqrt{2500} = 100$ PSI
Required depth to resist punching shear: $d = \frac{V}{F_v b}$ or

$d = \frac{400{,}000}{100 \times 108.0} = 37.0$ inches. This will require a column plinth so footing depth can be reduced.

STEP III:
Size of pedestal or plinth: Try using a pedestal 8.0 inches larger than column on each side. Size of pedestal is then (30.0 + 16.0) by (24.0 + 16.0) or 3.83' x 3.33' Area = 12.75 Sq. Ft.
Shear area outside pedestal = 85.0 - 12.75 = 72.75 Sq. Ft. Upward force of Shear: V = 72.75 x 5000 = 363,750 Lbs.

EXAMPLE: Designing rectangular spread footing, continued 4.12.6.4

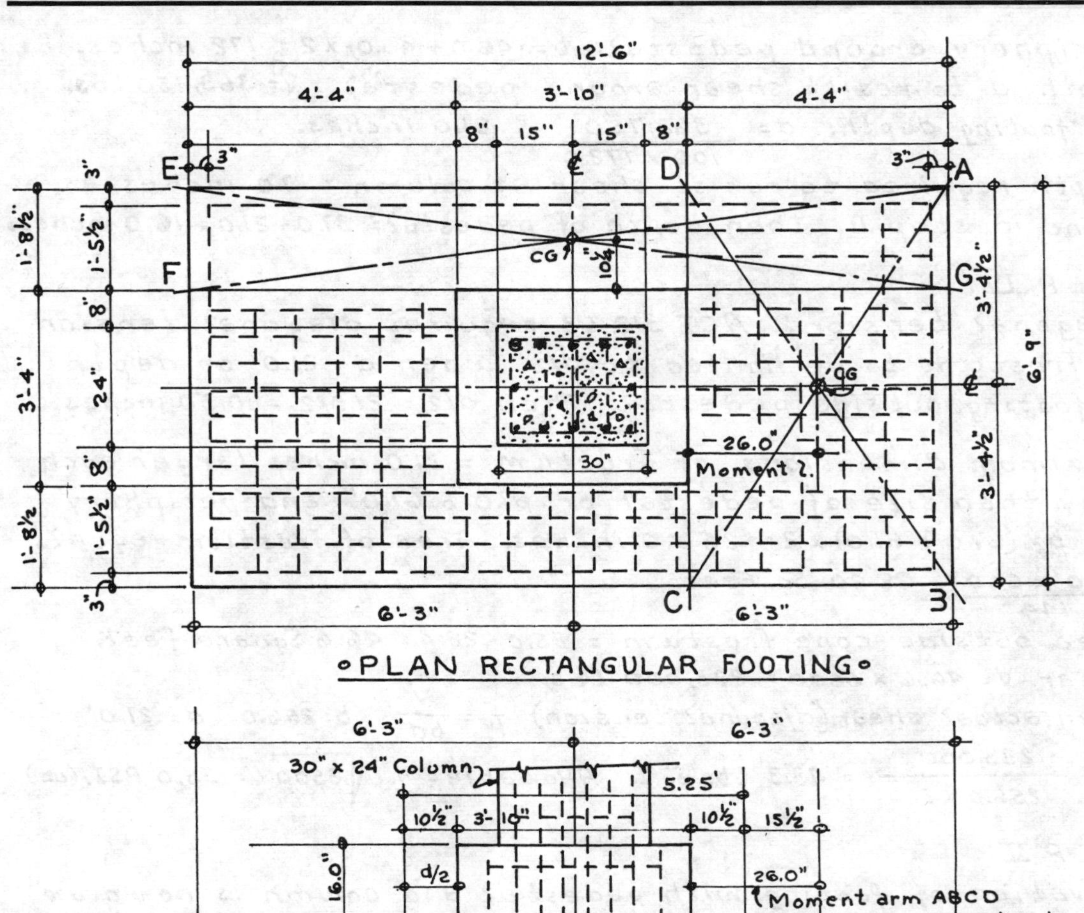

∘PLAN RECTANGULAR FOOTING∘

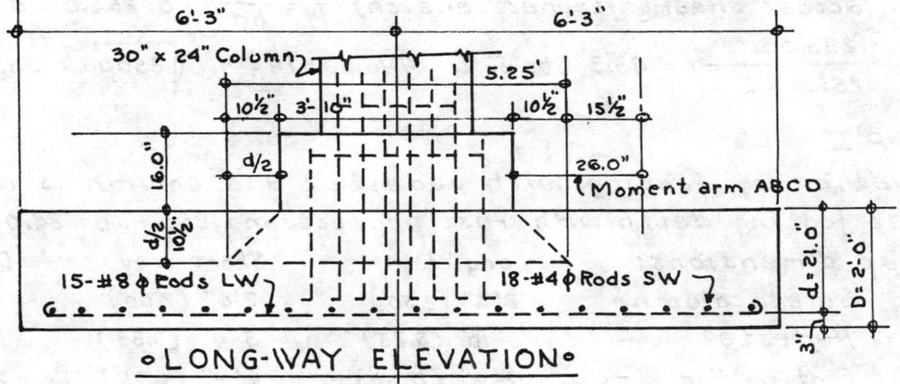

∘LONG-WAY ELEVATION∘

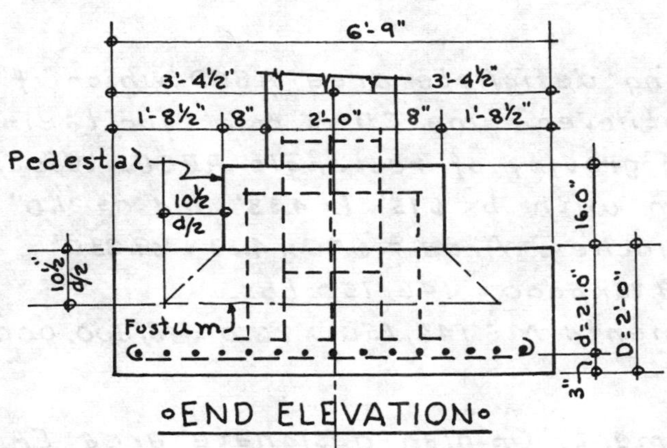

∘END ELEVATION∘

EXAMPLE: Designing rectangular spread footing, continued 4.12.6.4

Periphery around pedestal: $b = (46.0 + 40.0) \times 2 = 172$ inches.
Depth d to resist shear around pedestal: V= 363,750 Lbs.
For footing depth: $d = \frac{363{,}750}{100 \times 172.0} = 21.0$ inches.
Depth required to resist shear at column = 37.0 inches as found in step II. Then depth of pedestal= 37.0 - 21.0 = 16.0 inches.

STEP IV:
Diagonal tension: ACI 318-63 requires diagonal tension of frustum to be limited to d/2 when d = 21.0 or depth of footing outside pedestal. Then d/2 = 21.0/2 = 10.50 inches.

Maximum dimensions of frustum = 21.0 inches larger each way than size of pedestal or 67.0" by 61.0" and periphery is: $b = (67.0 + 61.0) \times 2 = 256.0$ inches. Area of frustum equals $\frac{67.0 \times 61.0}{144} = 28.40$ Sq. Feet.
Area outside cone frustum = 85.0 - 28.4 = 46.6 Square feet.
Shear $V = 46.6 \times 5000 = 233{,}000$ Lbs.
Then actual shear (diagonal tension) $f_v = \frac{V}{bd}$. b: 256.0" d= 21.0"

$$f_v = \frac{233{,}000}{256.0 \times 21.0} = 43.3 \text{ Lbs.Sq.I.} \quad \text{Allowable} = 1.1\sqrt{2500} = 55.0 \text{ PSI. (ok)}$$

STEP V:
Elevation of footing with pedestal and column is now draw
Total footing depth with rust protection = 21.0 + 3.0 = 24.0" or 2.0 feet.

Recap dimensions:	Long Way	Short way	Depth.
Sides column -	2'-6" (2.50')	2'-0" (2.00')	—
Pedestal -	3'-10" (3.83')	3'-4" (3.33')	1'-4" (1.33')
Footing Pad -	12'-6" (12.50')	6'-9" (6.75')	2'-0" (2.00')

STEP VI:
On plan drawing designate area ABCD which if inverted would be a cantilever. Line CD is point for taking moments from center of gravity of rectangle ABCD. Width CD is taken as a wide beam with b= 6.75' L= 4.33' and d= 21.0" Moment arm is ½ L or 26.0 inches. Area = 6.75 x 4.33 = 29.23 □'
Pressure W= 29.23 x 5000 = 146,150 Lbs.
Long Way moment: M= 146,150 x 26.0 = 3,800,000 inch Pounds.

STEP VII:
Short Way moment: On plan, designate area EAGF.

EXAMPLE: Designing rectangular spread footing, continued 4.12.6.4

Dimension AG = 3'-4½" − 1'-8" = 20½ inches or (1.71'). EA = 12.50' or 150.0 In.
Area EAGF = 1.71 x 12.5 = 21.38□' Moment arm = ½ of 20.5 = 10.25 inches.
Upward pressure V = 21.38 x 5000 = 106,900 Lbs.
Short-Way moment: M = 106,900 x 10.25 = 1,095,725 inch pounds.

STEP VIII:
Checking depth to steel by formula, $d = \sqrt{\frac{M}{Kb}}$, when b = 81.0 inches.
Long dimension has greater Moment. K = 189.5 and J = 0.871

$$d = \sqrt{\frac{3,800,000}{189.5 \times 81.0}} = \sqrt{249} = 15.8 \text{ inches.}$$

Let shear depth govern as this depth would require stirrups.

STEP IX:
Designing for area of steel rods: Solve for Long-Way dimension and a beam width of 81.0 inches. $As = \frac{M}{FsJd}$ Use d = 21.0"

$$As = \frac{3,800,000}{18,000 \times 0.871 \times 21.0} = 11.56 \text{ Sq. In.}$$

Short-Way steel area: b = 12.0' or 150.0 inches.

$$As = \frac{1,095,725}{18,000 \times 0.871 \times 21.0} = 3.33 \text{ Sq. In.}$$

STEP X:
Selecting rods for convenient spacing:
Long directing rod width = 81.0" − 6.00" = 75.0 inches.
Try using #8 φ Rods. Aφ = 0.79□" n = 11.56 / 0.79 = 14.65 Rods.
Use 15-#8 φ Rods and spacing = 75.0/14 = 5.36" about 5⅜" cc.
Short-way rods:
Width of beam = 12'-6" or 150.0" spacing width = 150.0 − 6.0 = 144.0 In.

As = 3.33□" Select smaller rod: Try #4 φ Rod with Aφ = 0.20 Sq. In.
Number required: n = 3.33 / 0.20 = 16.65 Use 18-#4 φ Rods.
Spacing s = 144.0/17 = 8.47" or about 8½" centers.
These rods can be added to drawings.

AUTHOR'S NOTE:
The core of column supported by pedestal is 26.0" x 20.0" and area: 520.0 Sq. In. Pedestal must contain an equal or greater area of steel contained in column. Rods extending into the pedestal and footing must be of equal size and larger. Hook column rods and each end of rods found in step X.

EXAMPLE: Designing connected footing 4.12.6.5

A building is to be erected on a downtown plot between two existing masonry buildings where exterior face of wall will be plumb with property line. A connected type of footing (Cantilever) is proposed. Column spacing is on 20.0 foot center line spacing. Interior columns of concrete are 2.0 feet square. Wall Columns are 1.50 x 2.50 feet with wide side running parallel to wall. Safe soil bearing is 3750 Lbs. Sq. Ft. Specifications call for concrete $F_c' = 3000$ PSI. Steel $F_s = 20{,}000$ PSI. Load on interior column $P_1 = 330{,}000$ Lbs., and axial load on wall columm $P_2 = 245{,}000$ Lbs.

REQUIRED:
A design and detailed drawings of footing. A shear diagram under an inverted beam sketch representing footing will serve best for Code authorities.

STEP I:
Drawing an elevation as inverted beam with column loads used for reaction the resultant of the 2 Column loads must be at middle of total footing length. To calculate the resultant or gravity center, take moments about ℄ of interior column at R_1.

$$\text{Dimension } a = \frac{330{,}000 \times 20.0'}{330{,}000 + 245{,}000} = 11.50 \text{ feet from } R_2 \text{ column center.}$$

Right half of footing = 11.50 + 0.75 = 12.25 feet. With resultant middle of footing, length must be 12.25 x 2 = 24.50 feet. Add to left side of interior column: 24.50 − (1.0 + 20.0 + 0.75) = 2.75 feet.

STEP II:
To calculate size of footing when length = 24.50 Feet. Estimate weight of concrete footing as 10% of total column loads.
$P_1 + P_2 = 330{,}000 + 245{,}000 = 575{,}000$ Lbs. Ftg. Wt. = 57,500 Lbs.
Total pressure on soil = 575,000 + 57,500 = 632,500 Lbs. With soil bearing of 3750 PSF, Area Footing = $\frac{632{,}500}{3750}$ = 168.50 Sq. Ft.
Width footing = 168.50 / 24.5 = 6.88 ft.
Use 7.0 foot width for footing.

Equivalent uniform load per foot length of inverted beam or soil bearing = 7.0 x 3750 or $w = 26{,}250$ Lbs per foot.

STEP III:
Calculate value of shear at several points in order to draw

EXAMPLE: Designing connected footing, continued 4.12.6.5

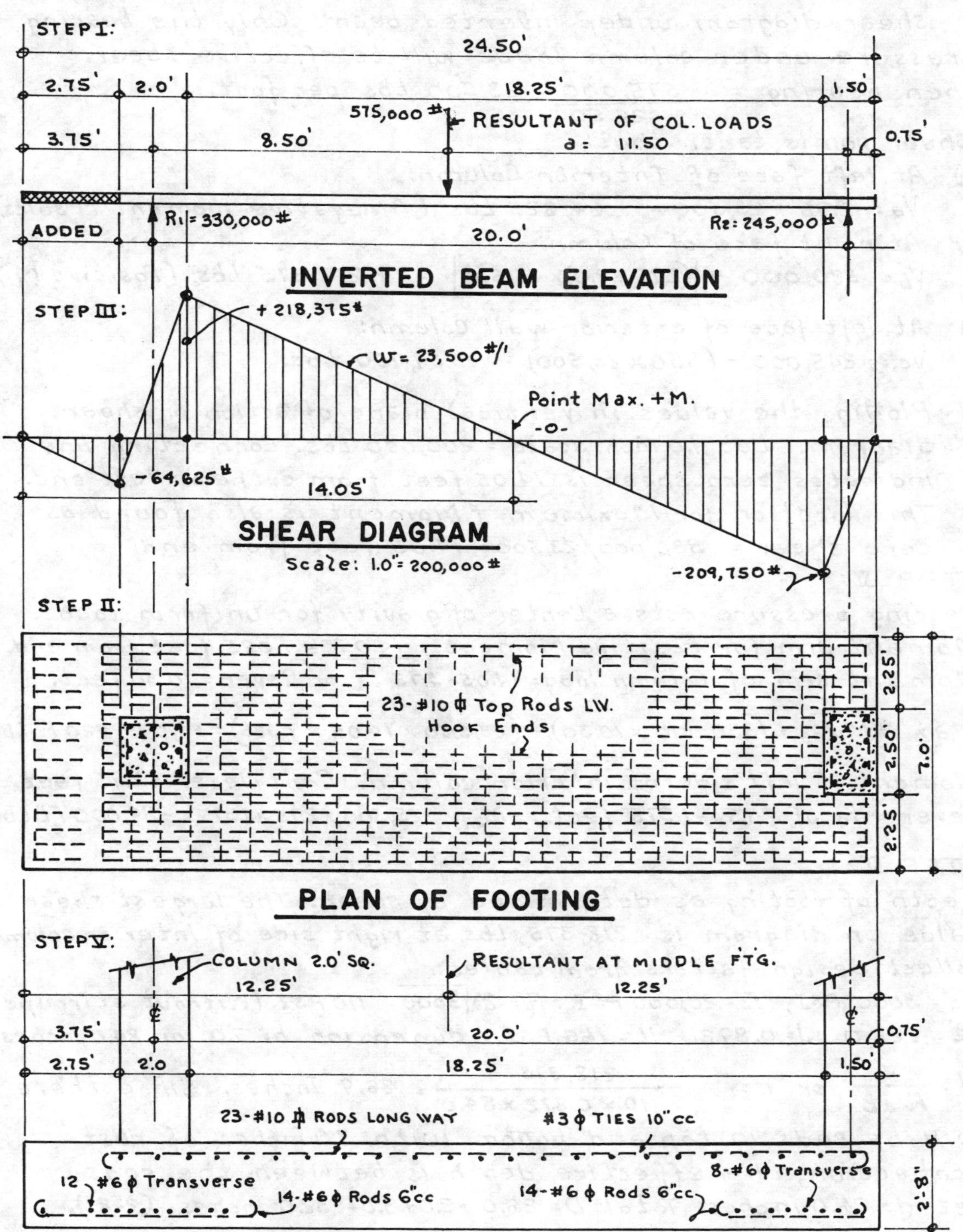

EXAMPLE: Designing connected footing, continued 4.12.6.5

a shear diagram under inverted beam. Only the bearing pressure under column loads will be effective shear.

Then bearing = $\frac{575,000}{24.50}$ = 23,500 Lbs. per foot.

Shear points taken thus:

(a) At left face of Interior Column:
V_a = 2.75' × 23,500 = −64,625 Lbs. (A negative moment results.)

(b) At right face of Column:
V_b = 330,000 − [(2.75 + 2.0) × 23,500] = +218,375 Lbs. (Positive M.)

(c) At left face of exterior wall Column:
V_c = 245,000 − (1.50 × 23,500) = −209,750 Lbs.

Plotting the values in vertical plane of action or shear diagram. Let 1.0 inch scale = 200,000 Lbs. Connecting line indicates zero shear is 14.05 feet from extreme left end. This location for Maximum +Moment is also found as: Zero shear = 330,000/23,500 = 14.05 feet from end.

STEP IV:

Bearing pressure acts a Center of gravity for uniform loads.
Moment arm for Positive Moment: $\frac{14.05}{2}$ = 7.025 feet from end.
Moment arm of Column load: 14.05 − 3.75 or lever = 10.30 feet.

Max. $+M_{14.05}$ = (330,000 × 10.30) − (23,500 × 14.05 × 7.025) = +1,079,520 Ft. Lbs.

Moment at left side of interior column: Cantilever = 2.75 feet.
Lever = ½ of 2.75 = 1.375 feet. $-M_{2.75}$ = 23,500 × 2.75 × 1.375 = −89,000 Ft. Lbs.

STEP V:

Depth of footing as determined by shear: The largest shear value on diagram is 218,375 Lbs. at right side of interior column.
Collect design factors from tables:
F_c' = 3000 PSI, F_s = 20,000 PSI $F_v = 2\sqrt{3000}$ = 110 PSI (Without stirrups)
K = 226.0 J = 0.872 U = 165 PSI. dimension b = 7.0' or 84.0 inches.

$d = \frac{V}{F_v J b}$ or $d = \frac{218,375}{110 \times 0.872 \times 84.0}$ = 26.9 Inches. Since there will be rods in top and bottom with 3.0 inches of rust protection, the effective depth is between the rods.
Let d = 26.0 inches. Total D = 26.0 + 3.0 + 3.0 = 32.0 Inches. (2'-8").

STEP VI:

Check footing depth by formula: $d = \sqrt{\frac{M}{Kb}}$

EXAMPLE: Designing connected footing, continued 4.12.6.5

Using larger moment for top rods:

$$d = \sqrt{\frac{1{,}079{,}520 \times 12}{226.0 \times 84.0}} = 26.2 \text{ inches. (Continue to use 26.0 inches).}$$

STEP VII:

Calculate required steel area from moments: For top rods;

$$A_s = \frac{M}{F_s J d} \text{ or } A_s = \frac{1{,}079{,}520 \times 12}{20{,}000 \times 0.872 \times 26.0} = 28.60 \text{ Square inches.}$$

This is required for full footing width of 7.0 feet.
Width for rods = 84.0 - 6.0 = 78.0 inches.
Try using #9 ϕ rods: 29 Rods give area of 29.0 Sq. inches and 28 equal spaces. S = 78.0/28 = 2.80 inch centers. This is too close as clear space between rods = 2.80 - 1.00 = 1.80 inches and coarse aggregate of 1½ inches may fail to pass between rods.

A larger side rod such as #10 ϕ has a cross-section area of 1.27 □" and is 1⅛" or 1.125 in.
Number required = 28.60/1.27 = 22½ . Use 23 Rods. Spacing is to be 78.0/22 = 3.55 inch centers. Clearance: 3.55 - 1.125 = 2.425 In.
Accept 23-#10 ϕ Rods and hook ends for bond. (Top steel).

STEP VIII:

Longitudinal steel in overhang at left of interior column:
M = -89,000'# Rods to be placed in bottom below column.

$$A_s = \frac{89{,}000 \times 12}{20{,}000 \times 0.872 \times 26.0} = 2.36 \text{ Square inches. Select minimum}$$

size rod as #4 ϕ with Aϕ = 0.20 □" Number = 2.36/0.20 = 11.8
Spacing should be about 6.0 inches, the number required is 78.0/6.0 = 13 spaces and 14 rods. Place same number under each column in long direction and in bottom of footing.

STEP IX:

Calculating steel in short or transverse direction.
Cantilever projection at interior column on each face of column = $\frac{7.0 - 2.0}{2}$ = 2.50 feet. Width of footing = 7.0 feet.
Column load P_1 = 330,000 Lbs. Bearing per foot under column = $\frac{330{,}000}{7.0}$ = 47,150 Lbs. Distribution width of overhang

EXAMPLE: Designing connected footing, continued 4.12.6.5

should be about ½ of footing width or 2 times column width. Use 3.50 feet or ½ of 7.0 foot width.

Moment lever arm for projection = 2.50/2 = 1.25 feet
M = 47,150 × 2.50 × 1.25 = 147,345 Foot Lbs.

$$As = \frac{147{,}345 \times 12}{20{,}000 \times 0.872 \times 26.0} = 3.92 \text{ Sq. In.}$$

Rods work well in this direction when about 4.0 inches on centers. Number required = $\frac{3.50 \times 12}{4.0}$ = 10.5 Use 11 Rods and size = 3.92/11 = 0.356 Sq. Inches for each rod.

Accept #6 ϕ Rods which have Aϕ = 0.44 Sq. Inches.

Use same rods for wall column in same direction.

STEP X:
Temperature steel is not required by code however a number of small tie rods may be used to hold the longer top rods in place. Add #3 ϕ Rods in transverse direction on top rods and space on 10.0 inch centers.

DESIGN NOTE:
Bond stress will not be calculated since the majority of Codes require all ends to be hooked when placed in any type of footing.

EXAMPLE: Designing continuous footing 4.12.6.6

A line of exterior columns are spaced at 14.0 feet on column center line and each column is 2.0 feet square. From ℄ of column row the property line is 2.0 feet and outer edge cannot extend beyond this line.
Each Column brings a 215,000 Lb. axial load to footing. Safe soil bearing allowable is 3750 Lbs. per square foot. Concrete is specified at $F_c' = 3000$ PSI, and $F_s = 20{,}000$ PSI. Footing must be without pedestal

REQUIRED:
A design of typical continuous footing with sufficient drawings for draftsmen to use in preparing plans. Should tie rods or stirrups be used as an aid to re-bar craft in placing rods, they shall be shown on details.

STEP I:
Construct an inverted beam drawing with 2 spans and 3 supports. Let columns represent supports and the loads equal the reactions R_1, R_2 and R_3.
Estimate dead load weight of footing 14.0 feet in length at 13% of Column Load 215,000 Lbs.
Weight of footing = 215,000 × 0.13 = 27,950 Lbs.
Total load on a single 14.0 foot span = 215,000 + 27,950 = 242,950 Lbs.
This load is distributed on 14.0 foot length, and soil pressure is 3750 PSF. Required footing area = $\frac{242{,}950}{3750}$ = 64.8 Sq. Feet.
L = 14.0' then width = 64.8/14.0 = 4.63 feet.
Call the width 4.67' or 4'-8"
Load per foot on beam = 3750 × 4.67 = 17,500 Lbs.

STEP II:
With outside edge of footing 2.0 feet from Column's ℄, the inside edge will be 2.67' or 2'-8" from ℄. A plan drawing of continuous footing can now be drawn.
The continuous inverted beam will now be figured as having a uniform distributed load of 17,500 #/' and will have both positive and negative moments. From Code:

$+M = \frac{wL^2}{12}$ and $-M = \frac{wL^2}{12}$ for continuous footings.

Then $\pm M = \frac{17{,}500 \times 14.0 \times 14.0}{12} = 285{,}833$ Foot Lbs.

EXAMPLE: Designing continuous footing, continued 4.12.6.6

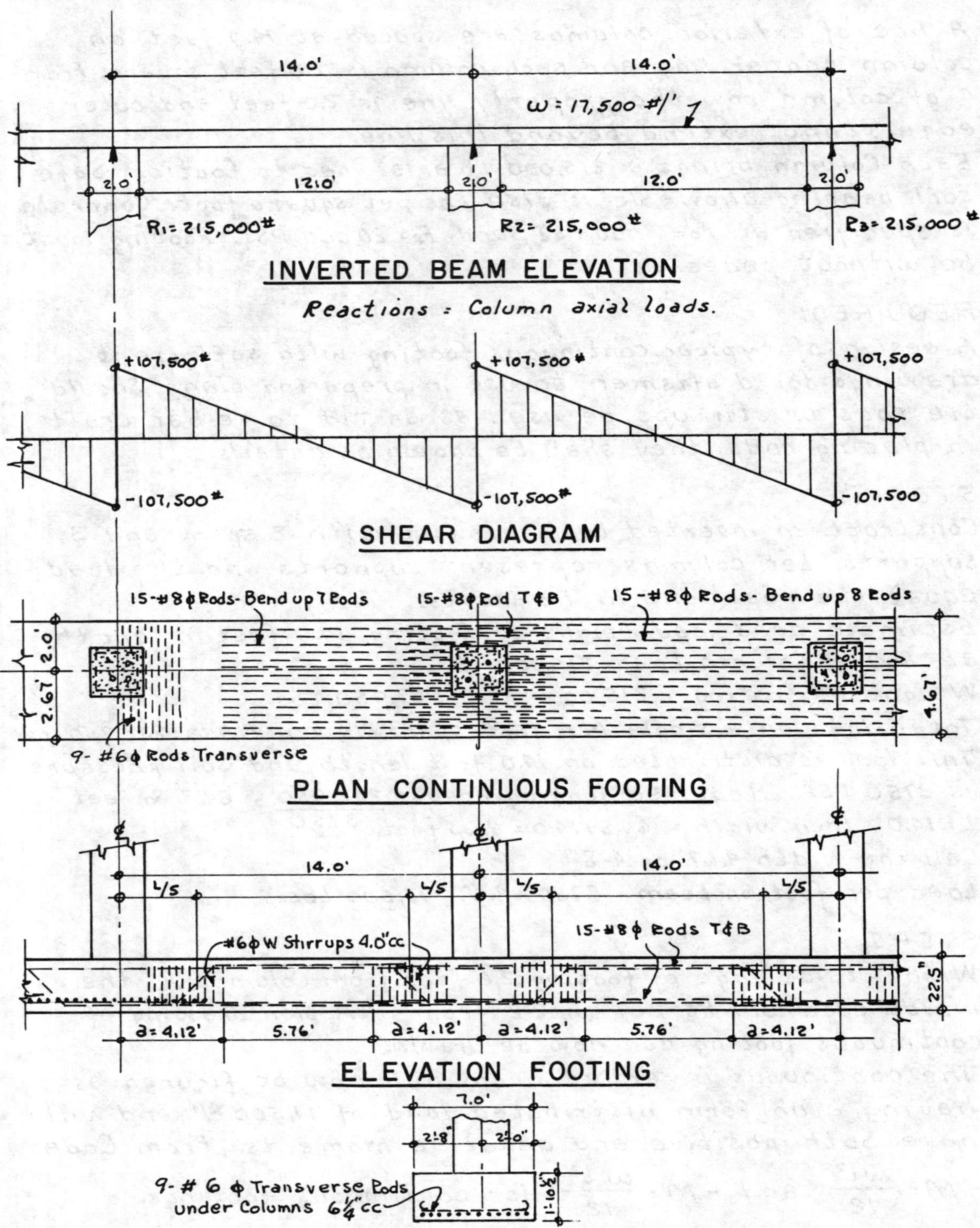

EXAMPLE: Designing continuous footing, continued 4.12.6.6

STEP III:

Calculating depth by Moment formula:
Take design factors from tables: $F_c' = 3000$ PSI. $F_s = 20{,}000$ PSI
$K = 226.0$ $J = 0.872$ $F_v = 2\sqrt{3000} = 110$ PSI.
Width of footing = b or b = 56.0 inches. Ends of rods will not be hooked and bond will be calculated. $u = 165$ PSI.

With values in formula: $d = \sqrt{\dfrac{285{,}833 \times 12}{226.0 \times 56.0}} = 16.46$ inches. Call it $16\frac{1}{2}$" and with 3" rust protection for top and bottom rods the depth D = 16.5 + 3.0 + 3.0 = 22.5 inches. (Decimal = 1.875 Feet).

STEP IV:
Check weight of footing as width and depth are now known.
b = 4.67' D = 1.875' and L = 14.0' Wt. Concrete = 150 # Cubic foot.
Wt. of footing = 4.67 x 1.875 x 14.0 x 150 = 18,400 Lbs. This is 900 Lbs. over estimated weight calculated in step I and need not be changed to influence design.

STEP V:
Calculate steel area for positive and negative bending. The Moments are equal and As will be same for top and bottom. Effective depth = coupling dimension of d = 16.5 inches.

$A_s = \dfrac{M}{F_s J d}$ or $A_s = \dfrac{285{,}833 \times 12}{20{,}000 \times 0.872 \times 16.5} = 11.90$ Square inches.

Set the desired spacing at about 3.50 inch centers. Spacing width will be 56.0 − (3.0 + 3.0) = 50.0 Inches. Number spaces = 50.0/3.50 = 14
Use 15 Rods. Area steel each rod = 11.90/15 = 0.793 Sq. Inches.
Select #8 ϕ Rods which have cross-section area of 0.79 Sq. In.
Bend up alternate rods and run alternate bars straight. Lap each rod top and bottom which gives required Area steel.
Bending location is L/5 or 2.80 feet from Column ℄.

STEP VI:
With footing depth and reinforced rod arrangement solved, the elevation can be drawn.
Check bond to ascertain whether 30 diameters for laps is safe.
Shear value V on 1 side of Column = ½ of 215,000 = 107,500 Lbs.
Perimeter of 1-#8 ϕ Rod = 3.14□" $\Sigma_o = 3.14 \times 15 = 47.10$ Sq. Inches

EXAMPLE: Designing continuous footing, continued 4.12.6.6

Unit bond stress $u = \frac{V}{\Sigma o Jd}$, or $u = \frac{107,500}{47.10 \times 0.872 \times 16.5} = 158.$ PSI.

Bond allowable is 165 PSI and 30 diameter laps will be OK.

STEP VII:

Check shear on diagonal to determine need for web reinforcing. Shear value on 1 side of column = 107,500 Lbs.

Shear formula: $f_v = \frac{V}{Jbd}$, or $f_v = \frac{107,500}{0.872 \times 56 \times 8.25} = 267$ PSI. (v').

Effective depth for diagonal = d/2 or 16.5/2 = 8.25 inches. (See ACI Code 318-63 Section 1207 d). The factor J may be left out of formula but will be used here for safety. Also allowable shear stress with web reinforcement can be increased to $3\sqrt{F'_c}$. This will also be neglected since the design is close to border line. Let F_v or V_c = 110 PSI.

Shear to be resisted by steel stirrups: $v_s = 267 - 110 = 157$ PSI. Distance from face of column stirrups will be required:

$a = \frac{L \times V_s}{2 \times V'}$. Then; $a = \frac{14.0 \times 157}{2 \times 267} = 4.12$ feet.

Choose a #6 ϕ Rod for stirrup trial. Stirrups for wide footings such as this must be formed as a shape "W" and calculated with 4 Legs. For #6 ϕ Stirrup thus, the area of steel = 4 x 0.44 = 1.76 Sq. Inches.

Spacing $= \frac{A_s F_s}{V_s \; b}$. $S = \frac{1.76 \times 20,000}{157 \times 56.0} = 4.00$ inches. Accept 3/4 ϕ size #6 Rods for stirrups.

STEP VIII:

Elevation may drawn and rods designated. Calculate the transverse bending moment under column using the larger overhang and the width same as footing at 56.0 inches.

W = 107,500 Lbs. Extension = 32.0-12.0 = 20.0" Moment arm = ½ of 20.0 = 10.0"

M = 107,500 x 10.0 = 1,075,000 Inch Lbs. d = 16.50 inches.

$A_s = \frac{1,075,000}{20,000 \times 0.872 \times 16.50} = 3.74$ Sq. In. Try using #6 ϕ Rod with area of 0.44 □" Number required = 3.74/0.44 = 8.5 Use 9-#6 ϕ Rods. spacing = 56.0/8 = 7.00 inches on centers.

EXAMPLE: Designing footing on piles 4.12.7.1

A spiral wrapped column core = 17.0 inches in diameter and carries a 175,000 Pound load to footing. Steel in column consists of 8-#6ϕ Vertical rods and core 12 inches in diameter. Code requires piles to be spaced not less than 3.0 feet on centers and with wood piles 3 piles driven to 30 tons each for bearing would support 180,000 Lbs. Other specifications are as follows: Fc' = 3000 PSI and Fs = 20,000 PSI.

REQUIRED:
With 3 Piles under footing the design will be in a triangular figure with each pile having the same length moment arm. Spacing of piles remains at 3.0 feet on centers. Design the footing with 3-Way reinforcing and provide drawings for plan, elevation and reinforcing layout.

STEP I:
The layout in plan is drawn to scale and center of gravity is in center where column load is applied. Drawing the 17.0 inch diameter column in location the moment arm can now be calculated. Angles are 30 degrees as designated for pile 1, 2, and 3. Let pile line 1-2 equal side c of triangle. Angle A is 30° and side a = 18.0 inches. Solving for the vertical side b, when side c = 36.0 inches. Side b = c Cos A. Functions of A (30°) as taken from Section V Trig. tables: Sine = 0.5000 Cos = 0.866 and Tan. = 0.57735. Side b = 0.866 x 36.0 = 31.177 inches. Now triangle is formed by 3 Piles with equal sides and middle third of each side = Center of gravity of triangle 1-2-3. Dimension on side or horizontal b = 1/3 of 31.17" = 10.39 inches. (Call it 10.40 inches.) From ℄ of Column to line 1-2 is side, a = 10.40" and b = 18.0", then $c = \frac{a}{\text{Sin.}A}$ or $c = \frac{10.40}{0.500}$ = 20.80 inches. Deducting the radius dimension of 8.50 inches, the moment arm distance for surface of column to ℄ of pile = 20.80 - 8.50 = 12.30 inches.

STEP II:
Calculating bending moment for each pile driven to 30 tons of bearing. Upward pressure P = 30 x 2000 = 60,000 Lbs.
Moment = 12.30 x 60,000 = 738,000 inch pounds.

STEP III:
Calculate punching shear around circumference of column:
Upward pressure of 3 Piles = 3 x 60,000 = 180,000 Lbs.

EXAMPLE: Designing footing on piles, continued 4.12.7.1

Circumference of 17.0 inch diameter column $= 3.1416 \times 17.0 = 63.41''$
Consult tables for concrete design factors:
$F_c' = 3000$ PSI $F_s = 18{,}000$ PSI. $K = 238.0$ $J = 0.864$ For peripheral shear in footings, ACI 318-63 allows $F_v = 2\sqrt{F_c'}$ and $u = 165$ PSI.

$F_v = 2\sqrt{3000} = 110$ PSI Formula for shear: $f_v = \dfrac{V}{Jdb}$ and to solve for depth to steel without a pedestal, transpose the formula as $d = \dfrac{V}{JbF_v}$. Circumference is substituted for value b.

$$d = \frac{180{,}000}{0.864 \times 63.41 \times 110} = 29.5 \text{ inches.}$$

With 3 inches of rust cover, total footing depth $= 29.5 + 3.0 = 32.5''$ or $2'\text{-}8\tfrac{1}{2}''$.

STEP IV:
Bending Moment to figure steel area: Only 1 Pile will be considered as moment was calculated in step II and rods will run 3-way direction. $A = \dfrac{M}{F_s J d}$. With values placed in formula:

$$A_s = \frac{738{,}000}{18{,}000 \times 0.864 \times 29.5} = 1.63\,\square''$$

This is a small amount of steel and requires further investigation. Explore the possibility of bending on vertical axis through ℄ column. With 2 piles the up-ward pressure = 120,000 Lbs. and moment arm = 10.40" $M = 1{,}248{,}000$ "#:

$$A_s = \frac{1{,}248{,}000}{18{,}000 \times 0.864 \times 29.5} = 2.78 \text{ Sq. In.}$$

9-#5 ϕ Rods give $A_s = 2.79\,\square''$

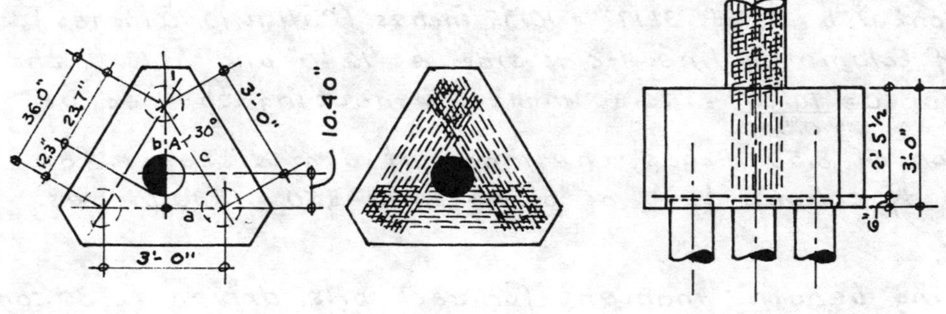

EXAMPLE: Designing footing on 8 pile support 4.12.7.2

A square tied Column 14.0 x 14.0 inches brings an axial load of 215,000 Lbs to an independent footing. Wood piles are to support footing with each pile being driven to a safe load capacity of 15 Tons. Code requires spacing of piles to be not less than 3.0 foot on centers. Concrete at age of 28 days is, $F_c' = 3000$ PSI and steel $F_s = 20{,}000$ PSI.

REQUIRED:
Design of footing with a plan and elevation drawing for plans.

STEP I:
Number of piles to support column load plus weight of concrete in footing: Estimate footing weight between 8 and 10 percent of column load. Then $P = 215{,}000 + 20{,}000 = 235{,}000$ Lbs. Number piles required $= 235{,}000 / 30{,}000 = 7.8$ Use 8 Piles and use pile upward reactions for all calculations. Plan layout will be made on following sheet. Allow 15.0 inches ℄ Pile to edge.

STEP II:
Total reaction of 8 Piles $= 30{,}000 \times 8 = 240{,}000$ Lbs. Number piles in sequence shown. Construct rectangle ABCD with line AB at left face of column. From line AB to ℄ of piles 1 and 6 the moment arm = 29.0 inches, and from AB to ℄ of pile #4, the lever is 11.0 inches.
Moment about line AB $= (2 \times 30{,}000 \times 29.0) + (30{,}000 \times 11.0) = 2{,}070{,}000$ inch pounds.

For piles 1, 2, and 3: Draw line EF across footing at top face of column. Distance from EF to ℄ of 3 top piles = 24.5 inches.
3 Pile reaction $= 3 \times 30{,}000$ Lbs. $M = 90{,}000 \times 24.5 = 2{,}205{,}000$ inch lbs.

STEP III:
The greater moment will control design. Calculate the required depth by shear value along line EF when $V = 90{,}000$ Lbs.
Length $EF = 8.50 \times 12 = 102.0''$ and represents value b.
From tables obtain design factors; For footings, $F_v = 2\sqrt{F_c'}$ or

$F_v = 2\sqrt{3000} = 110$ PSI. $K = 226.0$ $J = 0.872$ Rods must be hooked and bond stress will be neglected. $d = \frac{V}{F_v J b}$. With values in formula: $d = \frac{90{,}000}{110 \times 0.872 \times 102.0} = 9.20$ inches. This is less than required by Code which is a minimum of 12.0" above steel.

EXAMPLE: Designing footing on 8 pile support, continued 4.12.7.2

For punching shear: Neglect reactions of piles #4 and 5 since they will likely be below pedestal and frustum cone. Then $V = 6 \times 30{,}000 = 180{,}000$ Lbs. For diagonal tension in concrete the maximum depth to frustum is taken as $d/2$ or 6.0 In. Required area to resist V when $F_v = 110$ PSI $= \frac{180{,}000}{110} = 1635$ □"

Perimeter of frustum = 1635/6.0 = 273 inches. Side dimensions for square frustum = 273/4 = 68.0 inches.
Side of pedestal = 68.0 − (6.0 + 6.0) = 56.0 inches (4'-8" square).

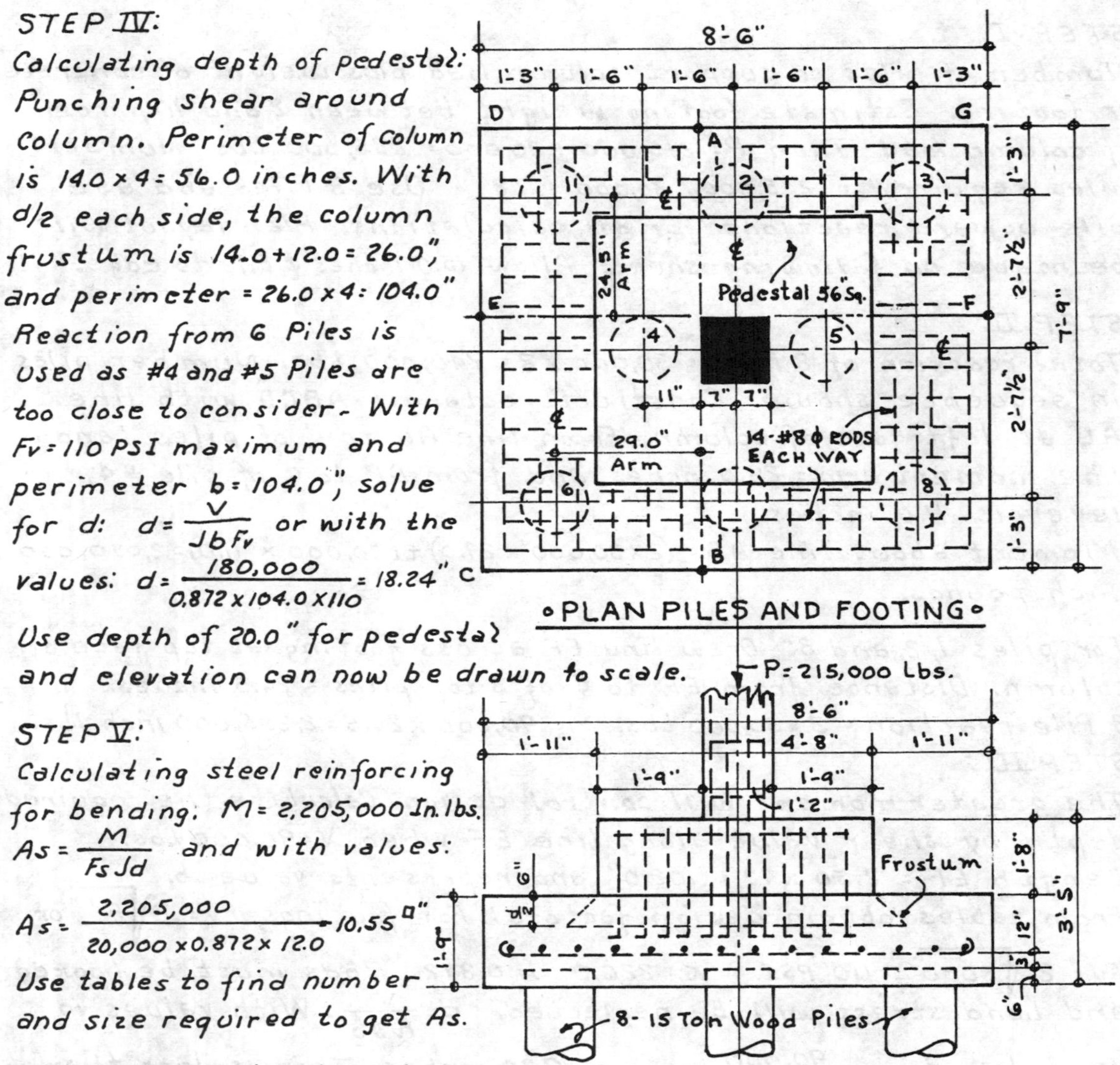

STEP IV:
Calculating depth of pedestal:
Punching shear around column. Perimeter of column is $14.0 \times 4 = 56.0$ inches. With $d/2$ each side, the column frustum is $14.0 + 12.0 = 26.0"$, and perimeter $= 26.0 \times 4 = 104.0"$
Reaction from 6 Piles is used as #4 and #5 Piles are too close to consider. With $F_v = 110$ PSI maximum and perimeter $b = 104.0"$, solve for d: $d = \frac{V}{JbF_v}$ or with the values: $d = \frac{180{,}000}{0.872 \times 104.0 \times 110} = 18.24"$
Use depth of 20.0" for pedestal and elevation can now be drawn to scale.

STEP V:
Calculating steel reinforcing for bending: $M = 2{,}205{,}000$ In. lbs.
$A_s = \frac{M}{F_s J d}$ and with values:

$$A_s = \frac{2{,}205{,}000}{20{,}000 \times 0.872 \times 12.0} = 10.55 \text{ □"}$$

Use tables to find number and size required to get A_s.

Choose 14-#8 ϕ Rods and run rods each way. Use 13 even spaces to get 14 rods. Allow a minimum of 2 to 3 inches for space between rods and pile cut-off. Let pile penetrate footing not less than 6 inches nor more than 9.0 inches.

PILOT DIAGRAMS: Pile cap footings 4.12.7.3

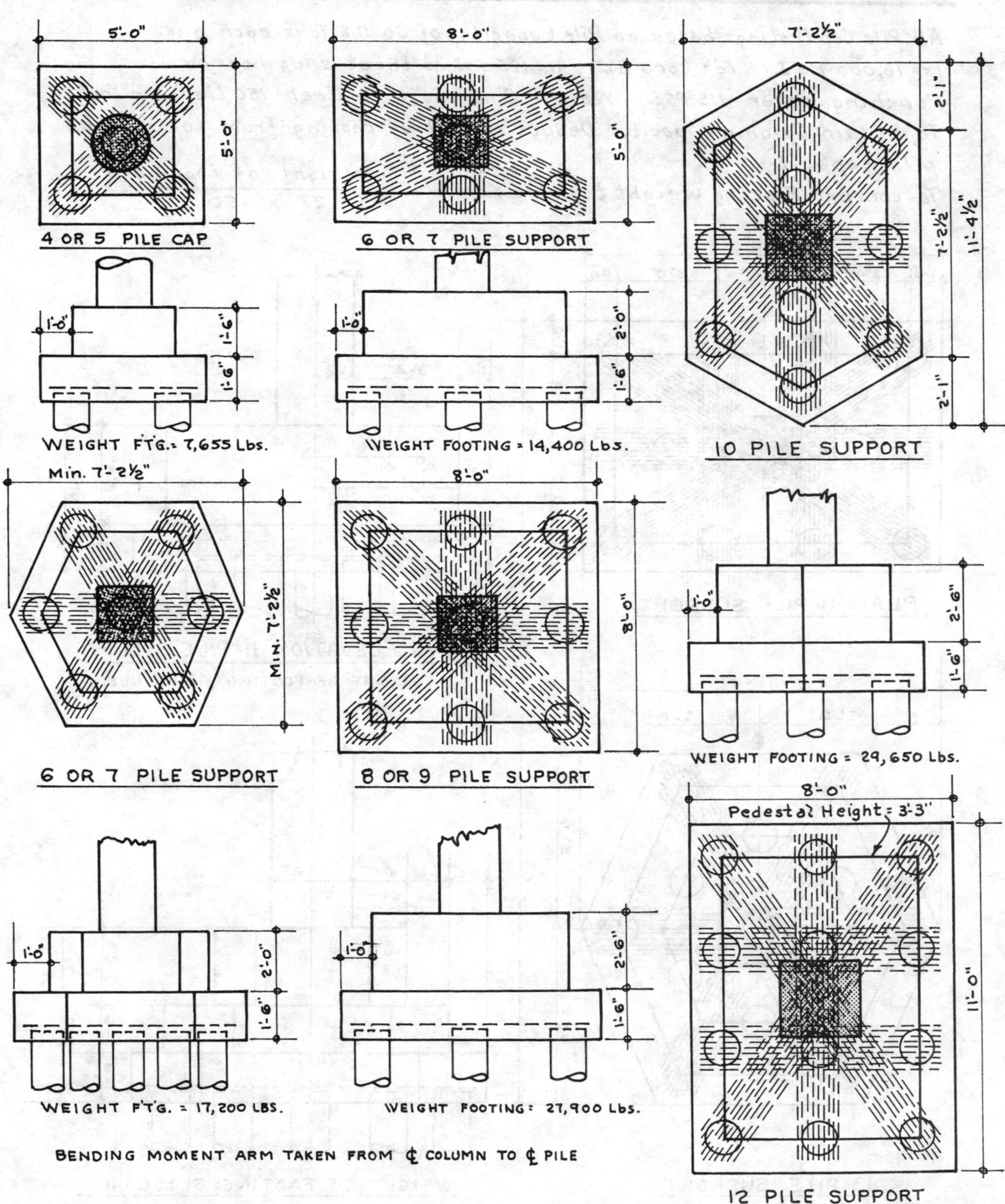

BENDING MOMENT ARM TAKEN FROM ℄ COLUMN TO ℄ PILE

PILOT DIAGRAMS: Pile cap footings, continued 4.12.7.3

All Pile Cap Footings based on Pile Capacity of 30 U.S. Tons each pile.

F_s = 16,000 P.S.I. F_c' = 2500 P.S.I. U = 125 PSI without ends hooked.

Punching shear = 115 PSI. Weight Concrete with steel = 150 Lbs. Cubic Foot.

To obtain Column capacity: Deduct weight of Footing from total group pile capacity.

To convert Footing weight to Cubic Yards = $\frac{\text{Weight of Footing}}{27 \times 150}$.

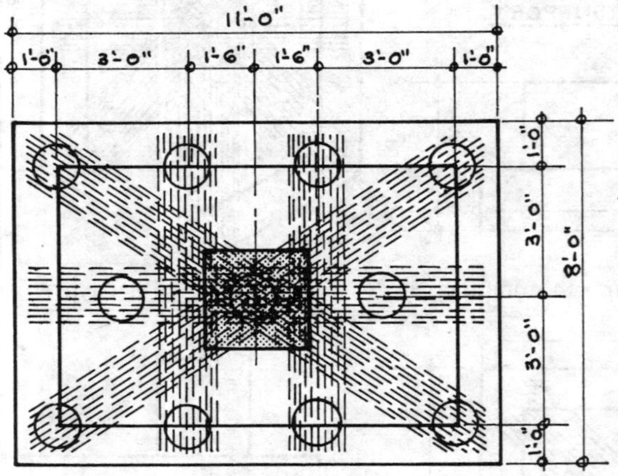

PLAN 11-PILE SUPPORT

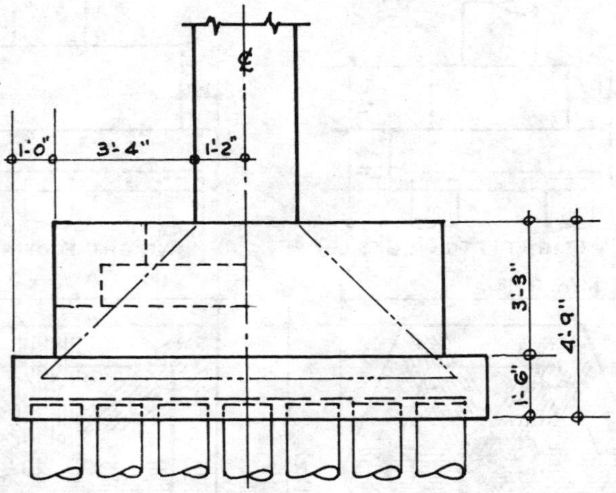

LONG ELEVATION 11-PILE GROUP

WEIGHT OF FOOTING = 40,700 Lbs.

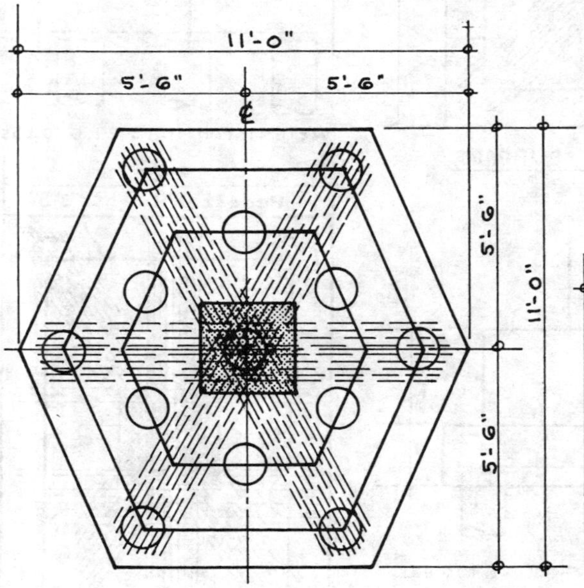

13 PILE SUPPORT

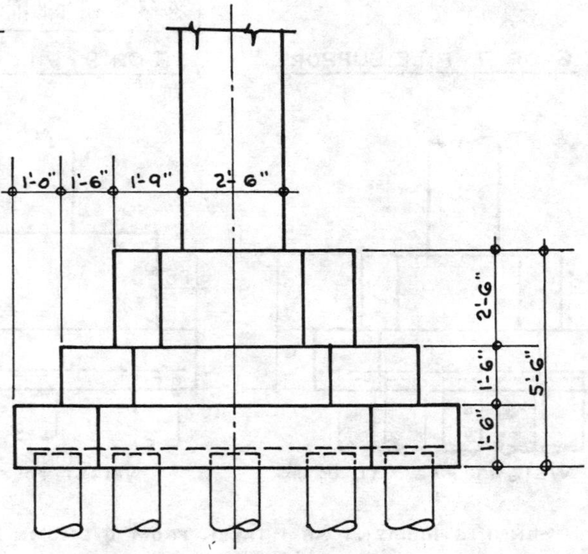

WEIGHT OF FOOTING: 51,150 Lbs.

EXAMPLE: Designing drilled footing 4.12.8.1

A steel column is supported upon a plinth which is to be an integral part of a continuous grade beam. Grade beam is 12.0" x 30.0" and column spacing is 16.0 feet on bay center. Load from column is 35,000 Lbs. Soil bearing for footing at 10.0 foot depth is safe at 4000 Lbs. Square foot. Figure weight of concrete at 150 Lbs. cubic foot.

REQUIRED:
Calculate the diameter for required footing under column. Compute the dead load of grade beam and add to column load. Determine the volume of concrete in footing shaft and cone when footing is drilled to 10.0 feet below grade. Draw section through grade beam and elevation of shaft and cone. Use a cone angle of 60 degrees for volume solution.

STEP I:
Calculating weight of grade beam with 16.0 foot length.
b = 1.0' d = 2.50' L = 16.0' Concrete = 150 # cubic ft.
W = 1.0 x 2.50 x 16.0 x 150 = 6000 Lbs. Column P = 35,000 Lbs.
Estimate weight shaft and cone at 1500 Lbs.
Total weight for bearing = 35,000 + 6,000 + 1500 = 42,500 Lbs.

STEP II:
Calculate area bearing and diameter:

A = 42,500/4000 = 10.625 Square feet. Formula of circle area is $A = \frac{\pi D^2}{4}$ or Circle area = 0.7854 x Square area. Then Diameter of Circle $= \sqrt{\frac{A}{0.7854}}$ Substituting values in this formula: Footing Diameter $D = \sqrt{\frac{10.625}{0.7854}} = 3.70'$ (44.4 inches)

From underream footing tables: A diameter of 44.0 inches gives a bearing area of 10.57 Sq. Feet and acceptable.

For a cone with 60° angle with horizontal, the shaft diameter will be either 16.0 or 18.0 inches. See Table 4.12.9.2.

Volume shaft + 60° cone = 24.54 Cubic feet of concrete
Draw section for next step and calculate volume on 60°.

EXAMPLE: Designing drilled footing, continued 4.12.8.1

STEP III:
Calculate the Volume of Concrete in shaft: Height of cone (frustum) will have to be calculated.
Angle A of 30° and side a = 1.165 feet are known. Side b = height cone.
From Trig. formulas and tables in Section V: Side b = a Cot A.
Cotangent of 30° = 1.7320
Height Cone = 1.165 × 1.7320 = 2.02 feet.
Length of shaft = 9.0 − 2.02 = 6.98 feet but call it 7.0 feet.

STEP IV:
The Volume of a Cylinder by formula is: Vol. = $0.7854\ D^2 \times L$. When this formula is in inches, divide the result by 1728 to get Cubic feet.
Shaft Vol = $0.7854 \times 1.33^2 \times 7.0 = 9.82'^3$

STEP V:
To find the volume of a "Truncated Cone" (Solid).

1.00'
1.50'
2.50'
GRADE
1.00'
6.98'
9.0'
a = 1.165'
d2 = 1.33'
1'-2"
A = 30°
B = 60°
b = 2.02'
d1 = 3.67'

Established RULE:

Square the two diameters and add together. $(d_1^2 + d_2^2)$.
Add to result the sum of the two end diameters. $(d_1 + d_2)$.
Multiply the product by 0.7854 and the result multiplied by height of cone (h) and then divide by 3. The result will be in cubic inches when diameters are taken in inches.
Written into formula, it is thus:

$$\text{Vol.} = \frac{\left[(d_1^2 + d_2^2) + (d_1 + d_2)\right] 0.7854\ H}{3}$$

With values in feet substituted:

$$\text{Vol.} = \frac{\left[(3.67^2 + 1.33^2) + (3.67 + 1.33)\right] \times 0.7854 \times 2.0}{3} = 10.60 \text{ Cubic Feet}$$

Total Volume Shaft + Cone = 10.60 + 9.82 = 20.42 Cubic feet.
In yard Concrete: Volume = $\frac{20.42}{27.0}$ = 0.756 Cubic yards.

Weight of footing = 20.24 × 150 = 3063 Lbs.
Weight of footing in Step I was under estimated. Wt = 7½% Load.

EXAMPLE: Estimating drilled footing volume 4.12.8.2

Architects plans show under-ream type footings on plans with several diameters at bearing bottom. Shaft diameter is estimated to be ⅓ of bottom diameter. Angle of cone frustum will be governed by drill rig bit. There are 80 footings with bearing diameter of 4.0 feet, and shaft with frustum is 6.50 feet below grade.

REQUIRED:

Calculate the concrete volume necessary to fill each type of footing with 45 and 60 degree under-reams. Determine the difference in Concrete Volume based on quantity of 80 footings on project.

STEP I:

If $d_2 = 4.00$ feet, $d_1 = \frac{1}{3}$ of $4.00 = 1.33$ ft.
Take dimension a for 60° cone first.
If $d_1 = 1.33'$, dimension $a = 1.33$ feet and $h = 1.33 \times 1.7320 = 2.31$ feet. Length of shaft above cone: $6.50 - 2.31 = 4.19$ Ft.

STEP II:

Calculating volume in 60° solid cone.

Formula: $Vol. = \dfrac{\left[\left(d_1^2 + d_2^2\right) + \left(d_1 \times d_2\right)\right] 0.7854h}{3}$

$d_1^2 = 1.33 \times 1.33 = 1.77$ $d_2^2 = 4.0 \times 4.0 = 16.0$

$d_1 \times d_2 = 1.33 \times 4.00 = 5.32$

$\Sigma = 1.77 + 16.00 + 5.32 = 23.09$

Volume $= \dfrac{23.09 \times 0.7854 \times 2.31}{3} = 13.97$ Cu. Ft.

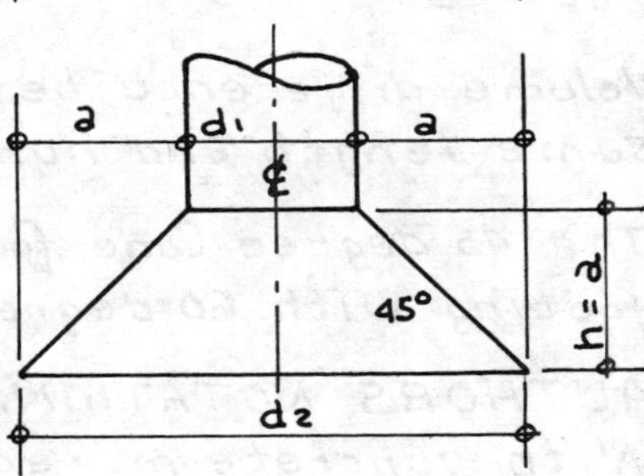

STEP III:

Calculate Volume in shaft per lineal foot since the length with the 45° cone will be longer.

$D_1 = 1.33'$ Area $= 0.7854\,D^2$ Volume $= 0.7854 \times 1.77 \times 1.0 = 1.39$ Cubic Ft.

Volume in shaft for 60° cone $= 4.19 \times 1.39 = 5.82$ cubic feet.

Total Volume in Footing $= 13.97 + 5.82 = 19.79$ Cubic Feet.

Total volume in 80 Footings with 60° Cone. Result in cu. yards.

Volume $= \dfrac{19.79 \times 80}{27} = 58.64$ Yards.

EXAMPLE: Estimating drilled footing volume, continued 4.12.8.2

STEP IV:

Calculating Volume in 45° Cone.
d_1 and d_2 are same as used in Step II.
a = 1.33 feet and will be dimension for h in 45° angle.

Σ = 23.09 Change value of h in equation as used for 60° cone in step II:

$$\text{Volume} = \frac{23.09 \times 0.7854 \times 1.33}{3} = 8.06 \text{ Cubic Feet.}$$

STEP V:
Volume in Shaft over 45° Cone:
Shaft length = 6.50 − 1.33 = 5.17 feet.
Volume per foot was calculated in step III to be 1.39 cu. Ft.
Total Volume in shaft = 5.17 × 1.39 = 7.19 Cubic feet.

Combined Volume Cone + Shaft = 8.06 + 7.19 = 15.25 Cubic Feet.

Volume in 80 Footings with 45° cone = 15.25 × 80 = 1220 Cubic Ft.

Reducing volume to yard of Concrete: $\text{Volume} = \frac{1220}{27} = 45.1$ Yds.

STEP VI:

Volume difference between 60° and 45° cone footings of same length and number = 58.64 − 45.10 = 13.54 Cubic Yards.

The 45 degree Cone footing requires less concrete than a footing with 60 degree cone.

AUTHOR'S NOTATION:
With concrete mixed in transit costing 16.00 per yard, the total cost difference on 80 footings may be a factor in selecting the best bid for drilling the holes.
Cost difference = 13.54 × 16.00 = $216.65 The proposed bid for drilling holes should be carefully examined with respect to the type of under-reaming bit each drilling rig will use.

TABLE: Footing volume: Spherical sector and 45° cone 4.12.9.1

CUBIC FOOT VOLUME IN DRILLED FOOTINGS				
BEARING AREA IN SQ. FEET	BOTTOM DIAMETER IN INCHES	SHAFT DIAMETER IN INCHES	UNDER-REAM VOLUME IN CUBIC FT.	SHAFT VOLUME PER FOOT HEIGHT IN CUBIC FEET
1.39	16	12	0.50	0.785
1.76	18		0.90	0.785
2.18	20		1.20	0.785
2.64	22		1.60	0.785
3.14	24		2.00	0.785
4.00	27		3.00	0.785
4.90	30		4.10	0.785
5.58	32	↓	5.20	0.785
6.30	34	14	5.80	1.07
7.07	36		7.20	1.07
7.88	38		7.50	1.07
8.72	40		8.30	1.07
9.62	42	↓	11.20	1.07
10.57	44	16	13.00	1.40
11.53	46		14.80	1.40
12.56	48	↓	16.70	1.40
13.63	50	18	19.00	1.77
14.75	52		21.20	1.77
15.90	54	↓	23.80	1.77
17.10	56	20	26.40	2.18
18.35	58		29.80	2.18
19.62	60	↓	32.70	2.18
20.93	62	24	35.60	3.14
22.30	64		38.80	3.14
23.80	66		43.50	3.14
25.28	68		48.30	3.14
26.20	70		51.80	3.14
28.30	72	↓	56.00	3.14
29.87	74	27	60.00	4.00
31.50	76		64.70	4.00
33.18	78		69.30	4.00
34.91	80		76.10	4.00
36.67	82	↓	81.80	4.00
38.48	84	30	86.40	4.90
40.34	86		94.10	4.90
42.24	88		99.60	4.90
44.18	90	↓	106.50	4.90
46.16	92	32	112.30	5.60
48.19	94		119.70	5.60
50.27	96	↓	125.00	5.60
54.54	100	36	169.20	7.07
56.75	102		182.70	7.07
63.62	108		217.60	7.07
70.88	114	↓	257.30	7.07
78.54	120	48	301.40	12.60

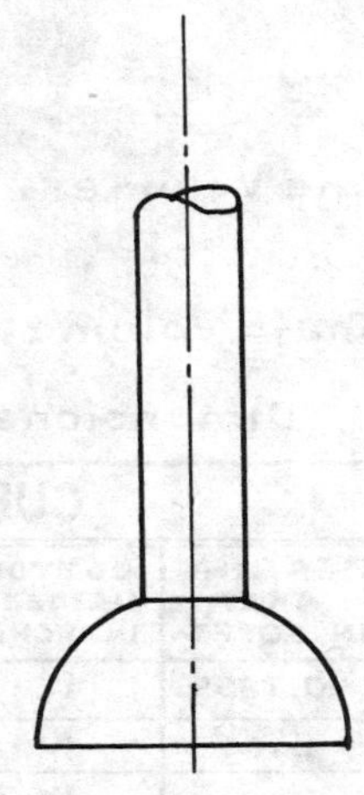

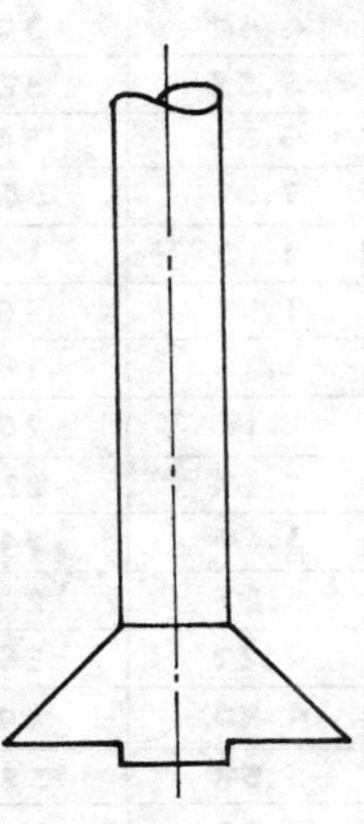

TABLE: Footing volume: 60° cone 4.12.9.2

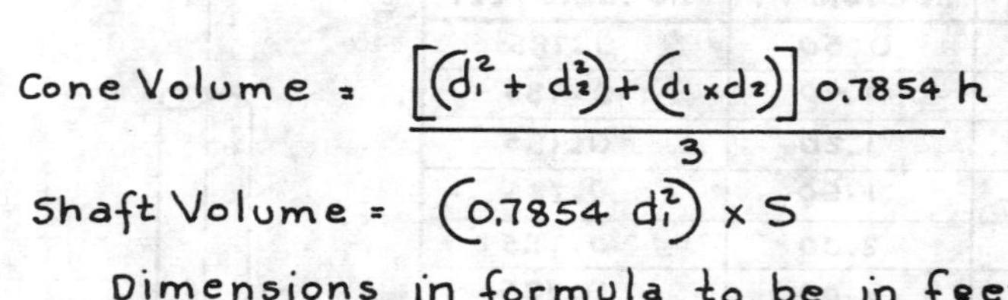

Cone Volume = $\frac{\left[(d_1^2 + d_2^2) + (d_1 \times d_2)\right] 0.7854\,h}{3}$
Shaft Volume = $(0.7854\,d_1^2) \times S$
Dimensions in formula to be in feet.

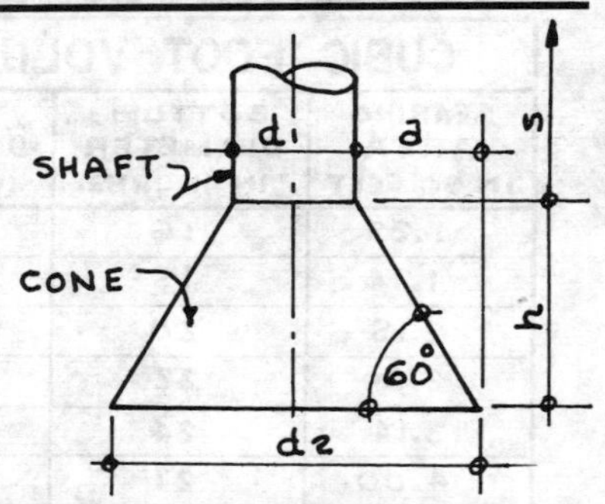

CUBIC FOOT VOLUME IN DRILLED FOOTINGS

BEARING AREA IN SQ. FT.	BOTTOM DIAMETER IN INCHES	SHAFT DIAMETER IN INCHES	DEPTH TO BOTTOM IN FEET									
			1.0	2.0	3.0	4.0	5.0	6.0	7.0	8.0	9.0	10.0
0.7854	12	12	0.78	1.56	2.34	3.12	3.90	4.68	5.47	6.24	7.02	7.80
1.10	14		0.80	1.58	2.36	3.14	3.92	4.70	5.48	6.26	7.04	7.82
1.39	16		0.86	1.64	2.42	3.20	3.98	4.76	5.54	6.32	7.10	7.86
1.76	18		0.97	1.75	2.53	3.31	4.09	4.87	5.65	6.48	7.21	7.99
2.18	20		1.14	1.92	2.70	3.48	4.26	5.04	5.92	6.60	7.38	8.16
2.64	22		1.38	2.16	2.94	3.72	4.50	5.28	6.06	6.84	7.62	8.41
3.14	24		1.68	2.46	3.24	4.02	4.80	5.58	6.38	7.14	7.92	8.70
3.69	26			2.82	3.60	4.38	5.16	5.94	6.72	7.50	8.28	9.26
4.26	28			3.27	4.05	4.83	5.61	6.39	7.17	7.95	8.73	9.97
4.90	30			3.81	4.59	5.37	6.15	6.93	7.71	8.49	9.27	10.05
5.58	32			4.45	5.23	6.01	6.79	7.57	8.35	9.13	9.91	10.69
6.30	34			5.18	5.96	6.74	7.52	8.80	9.08	9.86	10.64	11.42
7.07	36			6.02	6.80	7.58	8.36	9.14	9.92	10.70	11.48	12.26
1.10	14	14	1.06	2.12	3.18	4.24	5.30	6.36	7.42	8.48	9.54	10.60
1.39	16		1.08	2.14	3.20	4.26	5.32	6.38	7.44	8.50	9.56	10.62
1.76	18		1.15	2.21	3.27	4.33	5.39	6.45	7.51	8.57	9.63	10.69
2.18	20		1.27	2.33	3.39	4.45	5.51	6.57	7.63	8.69	9.75	10.81
2.64	22		1.48	2.54	3.60	4.66	5.72	6.78	7.84	8.90	9.96	11.02
3.14	24		1.73	2.79	3.85	4.91	5.97	7.03	8.09	9.15	10.21	11.27
3.68	26		2.06	3.12	4.18	5.24	6.30	7.36	8.42	9.48	10.54	11.60
4.27	28		2.47	3.58	4.59	5.65	6.71	7.77	8.83	9.89	10.95	12.01
4.90	30			4.02	5.08	6.14	7.20	8.26	9.32	10.38	11.44	12.50
5.58	32			4.60	5.66	6.72	7.78	8.94	9.90	10.96	12.62	13.08
6.30	34			5.29	6.35	7.41	8.47	9.53	10.59	11.65	12.71	13.77
7.07	36			6.12	7.18	8.24	9.30	10.36	11.42	12.48	13.54	14.60
7.88	38			7.00	8.06	9.12	10.18	11.24	12.30	13.36	14.42	15.48
8.72	40			8.02	9.08	10.14	11.20	12.26	13.32	14.38	15.44	16.50
9.62	42			9.17	10.23	11.29	12.35	13.41	14.47	15.53	16.59	17.65
1.39	16	16	1.39	2.78	4.17	5.56	6.95	8.34	9.73	11.12	12.51	13.90
1.76	18		1.41	2.80	4.19	5.58	6.97	8.36	9.75	11.14	12.53	13.92
2.18	20		1.49	2.88	4.27	5.66	7.05	8.44	9.83	11.22	12.61	14.00
2.64	22		1.64	3.03	4.42	5.81	7.20	8.59	9.98	11.37	12.76	14.15
3.14	24		1.85	3.24	4.63	6.02	7.41	8.80	10.19	11.58	12.97	14.36
3.68	26		2.13	3.52	4.91	6.30	7.69	9.08	10.47	11.86	13.25	14.64
4.27	28		2.50	3.89	5.28	6.67	8.06	9.45	10.84	12.23	13.62	15.01
4.90	30		2.95	4.34	5.73	7.12	8.51	9.90	11.29	12.68	14.07	15.46
5.58	32			4.87	6.26	7.65	9.04	10.43	11.82	13.21	14.60	15.99
6.30	34			5.51	6.90	8.29	9.68	11.07	12.46	13.85	15.24	16.63

TABLE: Footing volume: 60° cone, continued 4.12.9.2

CUBIC FOOT VOLUME IN DRILLED FOOTINGS												
BEARING AREA IN SQ. FEET	BOTTOM DIAMETER IN INCHES	SHAFT DIAMETER IN INCHES	DEPTH TO BOTTOM - IN FEET									
			1.0	2.0	3.0	4.0	5.0	6.0	7.0	8.0	9.0	10.0
7.07	36	16		6.28	7.67	9.06	10.45	11.84	13.23	14.62	16.01	17.40
7.88	38			7.14	8.53	9.92	11.31	12.70	14.09	15.48	16.87	18.28
8.72	40			8.11	9.50	10.89	12.28	13.67	15.06	16.45	17.84	19.23
9.62	42			9.22	10.61	12.00	13.39	14.78	16.17	17.56	18.95	20.34
10.57	44			10.41	11.80	13.19	14.58	15.97	17.36	18.75	20.14	21.53
11.53	46				13.17	14.56	15.95	17.34	18.73	20.12	21.51	22.90
12.56	48				14.69	16.08	17.47	18.86	20.25	21.64	23.03	24.42
2.18	20	18	1.78	3.54	5.30	7.06	8.82	10.58	12.34	14.10	15.86	17.62
2.64	22		1.88	3.64	5.40	7.16	8.92	10.68	12.44	14.20	15.96	17.72
3.14	24		2.04	3.80	5.56	7.32	9.08	10.84	12.60	14.36	16.12	17.88
3.69	26		2.27	4.03	5.79	7.55	9.31	11.07	12.83	14.59	16.35	18.11
4.27	28		2.58	4.34	6.10	7.86	9.62	11.38	13.14	14.90	16.66	18.42
4.90	30		2.98	4.74	6.50	8.26	10.02	11.78	13.54	15.30	17.06	18.82
5.58	32		3.46	5.22	6.98	8.74	10.50	12.26	14.02	15.78	17.54	19.30
6.30	34			5.81	7.57	9.33	11.09	12.85	14.61	16.37	18.13	19.89
7.07	36			6.51	8.27	10.03	11.79	13.55	15.31	17.07	18.83	20.59
7.88	38			7.32	9.08	10.84	12.60	14.36	16.12	17.88	19.64	21.40
8.72	40			8.22	9.98	11.74	13.50	15.26	17.02	18.78	20.54	22.30
9.62	42			9.32	11.08	12.84	14.60	16.36	18.12	19.88	21.64	23.40
10.57	44			10.46	12.22	13.98	15.74	17.50	19.26	21.02	22.78	24.54
11.53	46			11.78	13.54	15.30	17.06	18.82	20.58	22.34	24.10	25.86
12.56	48				15.01	16.77	18.53	20.29	22.05	23.81	25.57	27.33
13.63	50				16.60	18.36	20.12	21.88	23.64	25.40	27.16	28.92
14.75	52				18.32	20.08	21.84	23.60	25.36	27.12	28.80	30.64
15.90	54				20.31	22.07	23.83	25.59	27.35	29.11	30.87	32.63
2.64	22	20	2.21	4.39	6.57	8.75	10.93	13.11	15.29	17.47	19.65	21.83
3.14	24		2.30	4.48	6.66	8.84	11.02	13.20	15.38	17.56	19.74	21.92
3.69	26		2.47	4.65	6.83	9.01	11.19	13.37	15.55	17.73	19.91	22.09
4.27	28		2.74	4.92	7.10	9.28	11.46	13.64	15.82	18.00	20.18	22.36
4.90	30		3.06	5.24	7.42	9.60	11.78	13.96	16.14	18.32	20.50	22.68
5.58	32		3.51	5.69	7.87	10.05	12.23	14.41	16.59	18.77	20.95	23.13
6.30	34		4.03	6.21	8.39	10.57	12.75	14.93	17.11	19.29	21.47	23.65
7.07	36			6.85	9.03	11.21	13.39	15.57	17.75	19.93	22.11	24.29
7.88	38			7.58	9.76	11.94	14.12	16.30	18.48	20.66	22.84	25.02
8.72	40			8.44	10.62	12.80	14.98	17.16	19.34	21.52	23.70	25.88
9.62	42			9.42	11.60	13.78	15.96	18.14	20.32	22.50	24.68	26.86
10.57	44			10.55	12.73	14.91	17.09	19.27	21.45	23.63	25.81	27.99
11.53	46			11.81	13.99	16.17	18.35	20.53	22.71	24.89	27.07	29.25
12.56	48			13.20	15.38	17.56	19.74	21.92	24.10	26.28	28.46	30.64
13.63	50				16.92	19.10	21.28	23.46	25.64	27.82	30.00	32.18
14.75	52				18.59	20.77	22.95	25.13	27.31	29.49	31.67	33.85
15.90	54				20.28	22.46	24.64	26.82	29.00	31.18	33.36	35.54
17.10	56				22.49	24.67	26.85	29.03	31.21	33.39	35.57	37.75
18.35	58				24.70	26.88	29.06	31.24	33.42	35.60	37.78	39.96
19.62	60				27.01	29.19	31.37	33.55	35.78	37.91	40.09	42.27

TABLE: Footing volume: 60° cone, continued 4.12.9.2

CUBIC FOOT VOLUME IN DRILLED FOOTINGS

BEARING AREA IN SQ. FEET	BOTTOM DIAMETER IN INCHES	SHAFT DIAMETER IN INCHES	DEPTH TO BOTTOM IN FEET									
			1.0	2.0	3.0	4.0	5.0	6.0	7.0	8.0	9.0	10.0
3.14	24	24	3.14	6.28	9.42	12.56	15.70	18.84	21.98	25.12	28.26	31.40
3.68	26		3.18	6.32	9.46	12.60	15.74	18.88	22.02	25.16	28.30	31.44
4.27	28		3.29	6.43	9.57	12.71	15.85	18.99	22.13	25.27	28.41	31.55
4.90	30		3.49	6.63	9.77	12.91	16.05	19.19	22.33	25.47	28.61	31.75
5.58	32		3.78	6.92	10.06	13.20	16.34	19.48	22.62	25.76	28.90	32.04
6.30	34		4.19	7.33	10.47	13.61	16.75	19.89	23.05	26.17	29.31	32.05
7.07	36		4.70	7.84	10.98	14.12	17.26	20.40	23.54	26.68	29.82	32.96
7.88	38		5.24	8.38	11.52	14.66	17.80	20.94	24.08	27.22	30.36	33.50
8.72	40			9.16	12.30	15.44	18.58	21.72	24.86	28.00	31.14	34.28
9.62	42			10.02	13.16	16.30	19.44	22.58	25.72	28.86	32.00	35.14
10.57	44			10.99	14.13	17.27	20.41	23.55	26.69	29.83	32.97	36.11
11.53	46			12.09	15.23	18.37	21.51	24.65	27.79	30.93	34.07	37.21
12.56	48			13.40	16.54	19.68	22.82	25.96	29.10	32.24	35.38	38.52
13.63	50			14.80	17.94	21.08	24.22	27.36	30.50	33.64	36.78	39.92
14.75	52			16.34	19.48	22.62	25.76	28.90	32.64	35.18	38.32	41.46
15.90	54				21.21	24.35	27.49	30.63	33.77	36.90	40.05	43.19
17.10	56				23.05	26.19	29.33	32.47	35.61	38.75	41.99	45.03
18.35	58				25.12	28.26	31.40	34.54	37.68	40.82	43.96	47.10
19.62	60				27.44	30.58	33.72	36.26	40.00	43.14	46.28	49.42
20.93	62				29.85	32.97	36.73	39.27	42.41	45.55	48.69	51.85
22.30	64				32.44	35.58	38.72	41.86	45.00	48.14	51.28	54.42
23.80	66				35.27	38.41	41.55	44.69	47.83	50.47	54.11	57.25
25.28	68					41.35	44.49	47.63	50.77	53.91	57.05	60.19
26.80	70					44.59	47.73	50.87	54.01	57.15	60.29	65.43
28.30	72					48.16	51.30	54.44	57.58	60.72	63.86	67.50
4.27	28	28	4.27	8.54	12.81	17.08	21.35	25.62	29.89	34.16	38.43	42.70
4.90	30		4.31	8.58	12.85	17.12	21.39	25.66	29.93	34.20	38.47	42.74
5.58	32		4.44	8.71	12.98	17.25	21.52	25.79	30.06	34.33	38.60	42.87
6.30	34		4.68	8.95	13.22	17.49	21.76	26.03	30.30	34.57	38.84	43.11
7.07	36		5.03	9.30	13.57	17.84	22.11	26.38	30.65	34.92	39.19	43.46
7.88	38		5.47	9.74	14.01	18.28	22.55	26.82	31.09	35.86	39.63	43.90
8.72	40		6.04	10.31	14.58	18.85	23.12	27.39	31.66	35.93	40.20	44.47
9.62	42		6.73	11.00	15.27	19.54	23.81	28.02	32.35	36.62	40.89	45.16
10.57	44			11.80	16.07	20.34	24.61	29.38	33.15	37.42	41.69	45.96
11.53	46			12.77	17.04	21.31	25.58	29.85	34.12	38.39	42.66	46.93
12.56	48			13.87	14.14	22.41	26.68	30.95	35.22	39.49	43.76	48.03
13.63	50			15.13	19.40	23.67	27.94	32.21	36.48	40.75	45.62	49.29
14.75	52			16.53	20.80	25.07	29.34	33.61	37.88	42.15	46.42	50.69
15.90	54			18.09	22.36	26.63	30.90	35.17	39.44	43.71	47.98	52.25
17.10	56			19.79	24.06	28.33	32.60	36.87	41.14	45.41	49.68	53.95
18.35	58				25.97	30.24	34.51	38.78	43.05	47.32	51.59	55.86
19.62	60				28.12	32.39	36.66	40.95	45.20	49.47	53.74	58.01
20.93	62				30.26	34.53	38.80	43.07	47.34	51.61	55.88	60.15
22.30	64				32.80	37.07	41.34	45.61	49.88	54.15	58.42	62.69
23.75	66				35.47	39.74	44.01	48.28	52.56	56.82	61.09	65.36

TABLE: Footing volume: 60° cone, continued 4.12.9.2

CUBIC FOOT VOLUME IN DRILLED FOOTINGS												
BEARING AREA, IN SQ. FEET	BOTTOM DIAMETER IN INCHES	SHAFT DIAMETER IN INCHES	DEPTH TO BOTTOM - IN FEET									
			1.0	2.0	3.0	4.0	5.0	6.0	7.0	8.0	9.0	10.0
25.25	68	28			38.26	42.53	46.80	51.07	55.34	59.61	63.88	68.15
26.80	70				41.33	45.60	49.87	54.14	58.41	62.68	66.95	71.22
28.30	72					49.02	53.29	57.56	61.83	66.10	70.37	74.64
29.87	74					52.43	56.70	60.97	65.24	69.51	73.78	78.05
31.50	76					56.19	60.46	64.73	69.00	73.27	77.54	81.81
33.18	78					60.18	64.45	68.72	72.99	77.26	81.53	85.90
34.91	80					64.29	68.56	72.83	77.10	81.37	85.64	89.91
36.67	82					68.76	73.03	77.30	81.57	85.24	90.11	94.38
38.48	84					73.50	77.77	82.04	86.31	90.58	94.85	99.12
4.90	30	30	4.90	9.80	14.70	19.60	24.50	29.40	34.30	39.20	44.10	49.00
5.58	32		4.95	9.85	14.75	19.65	24.55	29.45	34.35	39.25	44.15	49.05
6.30	34		5.10	10.00	14.90	19.80	24.70	29.60	34.50	39.40	44.30	49.20
7.07	36		5.45	10.35	15.25	20.15	25.05	29.95	34.85	39.75	44.65	49.55
7.98	38		5.71	10.61	15.51	20.41	25.31	30.21	35.11	40.01	44.91	49.81
8.72	40		6.19	11.09	15.99	20.89	25.79	30.69	35.59	40.49	45.39	50.29
9.62	42		6.80	11.70	16.60	21.50	26.40	31.30	36.20	41.10	46.00	50.90
10.57	44		7.52	12.42	17.32	22.22	27.12	32.02	36.92	41.82	46.72	51.62
11.53	46			13.30	18.20	23.10	28.00	32.90	37.80	42.70	47.60	52.50
12.56	48			14.31	19.21	24.11	29.01	33.91	38.81	43.71	48.61	53.51
13.63	50			15.47	20.37	25.27	30.17	35.07	39.97	44.87	49.77	54.67
14.75	52			16.78	21.68	26.58	31.48	36.38	41.28	46.18	51.08	55.98
15.90	54			18.26	23.16	28.06	32.96	37.86	42.76	47.66	52.56	57.46
17.10	56			19.90	24.80	29.70	34.60	39.50	44.40	49.30	54.20	59.10
18.35	58			21.71	26.61	31.51	36.41	41.31	46.21	51.11	56.01	60.91
19.62	60				28.62	33.52	38.42	43.32	48.22	53.12	58.02	62.92
20.93	62				30.79	35.69	40.59	45.49	50.39	55.29	60.19	65.09
22.30	64				33.08	37.98	42.88	47.78	52.68	57.59	62.48	67.38
23.80	66				35.74	40.64	45.54	50.44	55.34	60.24	65.14	70.04
25.28	68				38.44	43.34	48.29	53.14	58.04	62.94	67.84	72.74
26.80	70				41.43	46.33	51.23	56.13	61.03	65.93	70.83	75.73
28.30	72				44.65	49.55	54.45	59.35	64.25	69.15	74.05	78.95
29.87	74					52.99	57.89	62.79	67.69	72.59	77.49	82.39
31.50	76					56.65	61.55	66.45	71.35	76.25	81.15	86.05
33.18	78					60.57	65.47	70.37	75.27	80.17	85.07	89.97
34.91	80					64.58	69.48	74.38	79.28	84.18	89.08	93.98
35.67	82					68.98	73.88	78.78	83.68	88.58	93.48	98.38
38.48	84					73.63	78.53	83.43	88.33	93.23	98.13	103.03
40.34	86					78.39	83.29	88.19	93.09	97.99	102.89	107.79
42.24	88						88.45	93.35	98.25	103.15	108.05	112.95
44.18	90						94.07	98.97	103.87	108.77	113.67	118.57

Retaining walls 4.13

Retaining walls are erected to retain and contain water, coal, or earth. Similar methods are used to construct basement walls, swimming pools, seawalls, dikes, dams and flood control projects. There are several methods used to stabilize such walls, for adequate support in holding back the pressure applied on one side.

Retaining walls which protect docks and other marine structures are constructed of interlocking steel sheet piles driven at the water's edge. The top of the wall is braced against water pressure with underground tie rods connected to buried anchor blocks or with another sheet pile wall with compacted earth in between. Steel sheet pile are not water-tight, but they will retain fine sand. They are not usually subject to the water undermining problem which can occur with solid concrete walls with inadequate drainage.

Another type of retaining wall which resists horizontal pressure from water and soil is constructed as a solid wall. Without drainage the water pressure under the wall may build up, and the wall may fail by tipping over or sliding horizontally. Basement walls are usually designed to be supported by fixed ends or anchored to slab floors at the bottom and beam structures along the top. Concrete basement walls may be designed similar to one-way and two-way slabs in a vertical position.

Retaining walls which contain water or earth depend upon their weight and sliding resistance for stability. Resistance to sliding is governed by the friction between the wall and the supporting base soil at the bottom. The greater the weight of the wall, the greater the resistance to sliding. The frictional resistance developed between two surfaces in contact depends on the composition and finish of the materials in contact, or the force pressing them together, and on the lubricant, if any, between them. (See Section V, Trigonometry.)

Angle of repose 4.13.1

The angle which a heaped pile of loose material makes with the horizontal is the *angle of repose*. When gravel and sand are stored in large piles, it may be observed that they will each have distinct, but different, slope angles. Even if more material is added the pile will assume the original slope angle. As material is added, gravity pulls the particles down until the slope of the pile decreases and frictional resistance equals the pull of gravity. Then the particles will sleep at the angle of repose. Often, for stone surfaces and certain soils, the angle of repose is known as the *angle of friction*.

Sliding wedge theory 4.13.2

In 1925, the Austrian engineer Dr. Karl von Terzaghi published his theories and experimental methods in soil mechanics and soil behavior under gravity forces. He examined the *sliding wedge* of soil behind a retaining wall. This sliding wedge consists of the retained earth which lies above the angle of repose. This wedge produces a force parallel with the slope angle of repose. The horizontal component of this wedge force acts to cause the wall to slide and creates an overturning moment. The vertical component force acts on the base or foundation of the footing.

The *middle-third theory* requires that the line of action of the result of the sliding wedge force must pass through the middle third of the wall; otherwise the wall will not be stable. This is illustrated in the examples to follow. The designer may make several adjustments in the wall cross-section, to be certain that the resultant is within the middle-third. When this requirement is met, the earth pressure will be applied over the entire footing base.

Cantilever retaining walls 4.13.3

Large tank farms, used near refineries and chemical plants, are given fire and pollution protection by constructing dikes around each tank. These tanks may contain up to a million gallons of gasoline, crude oil, or chemicals. A ruptured tank could devastate the surrounding area if its contents were not confined. Concrete retaining walls of the cantilever type are constructed around each tank for protection. This economical retaining wall consists of three basic elements: a cantilever vertical slab (stem), the toe slab and the heel slab. This type of wall is also well adapted for large outdoor swimming pools where a portion of the pool can remain above ground level. The type can be sloped for landscaping and drainage. In tank farms, the dikes are well landscaped and appear to be soil embankments.

Retaining wall design 4.13.4

The first step in the design of a retaining wall involves assembling all the available information. Draw a preliminary outline of the cross-section. There are no precise rules to proportion the wall, but the design must be refined. After analyzing the preliminary drawing, modify and improve the design until the cross-section is satisfactory. The unit of design is one lineal horizontal foot of wall length. The design examples to follow will employ the Trautwine Theory of the Sliding Wedge and the Rankine Theory of Earth Pressure. There is a certain amount of soil adhesion to the vertical wall surface. This adhesion is not considered in our design examples, since the cohesion of retained materials will differ and cannot be depended upon.

Retaining wall design, continued 4.13.4

Gravity walls depend on their weight alone to resist the sliding thrust and the overturning moment. Cantilever walls are stabilized in addition by the weight of the retained earth. The design, therefore, is basically a cut and try procedure where the mechanics of the vertical and horizontal forces are investigated. The stability of wall must have enough weight and resistance to overcome the forces from earth thrust. A safety factor of 2.0 is adequate for both tipping and sliding. When sliding stability has a safety factor of less than 1.5, a key should be installed under the base footing as an integral part of the foundation. The steel reinforcing in the base is designed following the examples given for the design of spread footings. The cantilever stem is designed as a cantilever beam supported by the base.

Engineering theories on earth pressure are contained in many voluminous works by different authors, and extensive research reveals obvious uncertain features. The different kinds of substances in inert material will give varying angles of repose. Authorities agree the correct angle cannot be accurately determined and the angle of no friction is merely a theoretical term. The pressure exerted against a wall will vary according to the content of water in the retained earth or material. Earth pressure is increased if the backfill is saturated with water. Deep holes near the bottom of the stem should be provided, and adding a layer of loose gravel or slag will be of considerable aid in promoting proper drainage and resistance to sliding. The usual angle of repose for retained earth is taken by Rankines theory to be 33° 40′ with the horizontal plane. This angle is based on a slope of 1.0 foot rise for each 1.50 foot of run. The formulas are based on the trigonometric function of this angle. The sliding wedge design method using the angle of *no friction* will produce satisfactory results. It often compares favorably with Rankines designs.

The illustrated sections 4.13.4.1 given for cantilever and gravity walls to retain earth and water are representative of the usual types. When using Rankines theory of earth pressure and thrust, select the formula corresponding to the wall type under consideration and with or without surcharge. The resultant of the vertical force W and horizontal force P must intersect at middle third through the base at bottom. It is located by graphics. The formulas for soil pressure can be ascertained at toe and heel by the formulas given for p_1 and p_2.

STEEL REINFORCING IN WALLS

The design for reinforcing steel in retaining walls should follow the same methods used for spread footings. Use the formulas given for calculating soil bearing and upward pressure at toe and heel. The vertical wall is designed as a cantilever beam, with its support at base where shear is greatest. Use temperature steel generously and not less than the ratio of 0.0025 Ac. Smaller rod diameters are more effective with closer spacing. The bulky type of gravity wall requires a much longer curing period. Arrangements for adequate water supply should be made well in advance of the concrete placement.

Retaining wall formulas 4.13.4.1

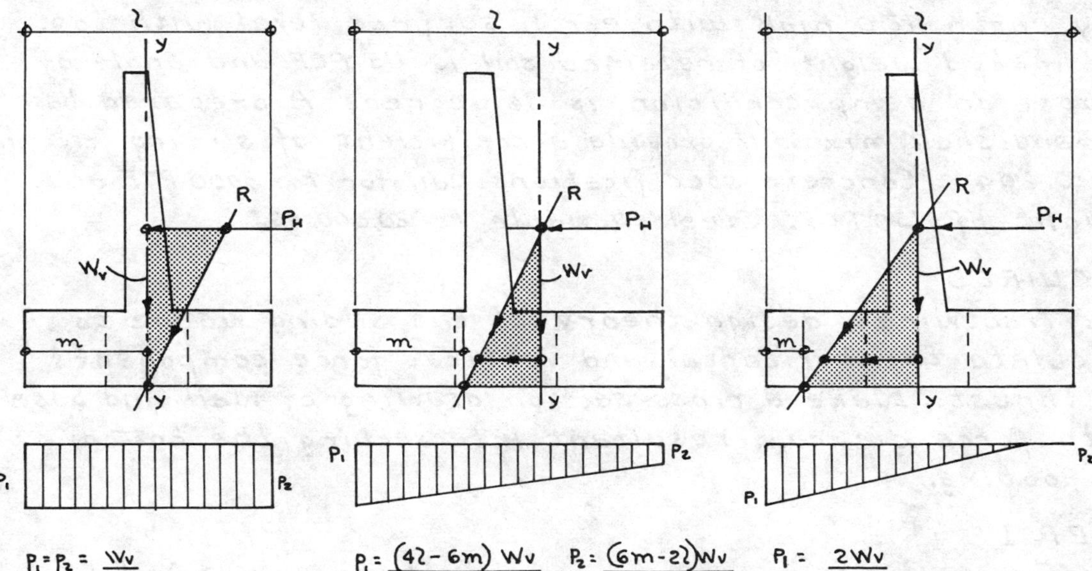

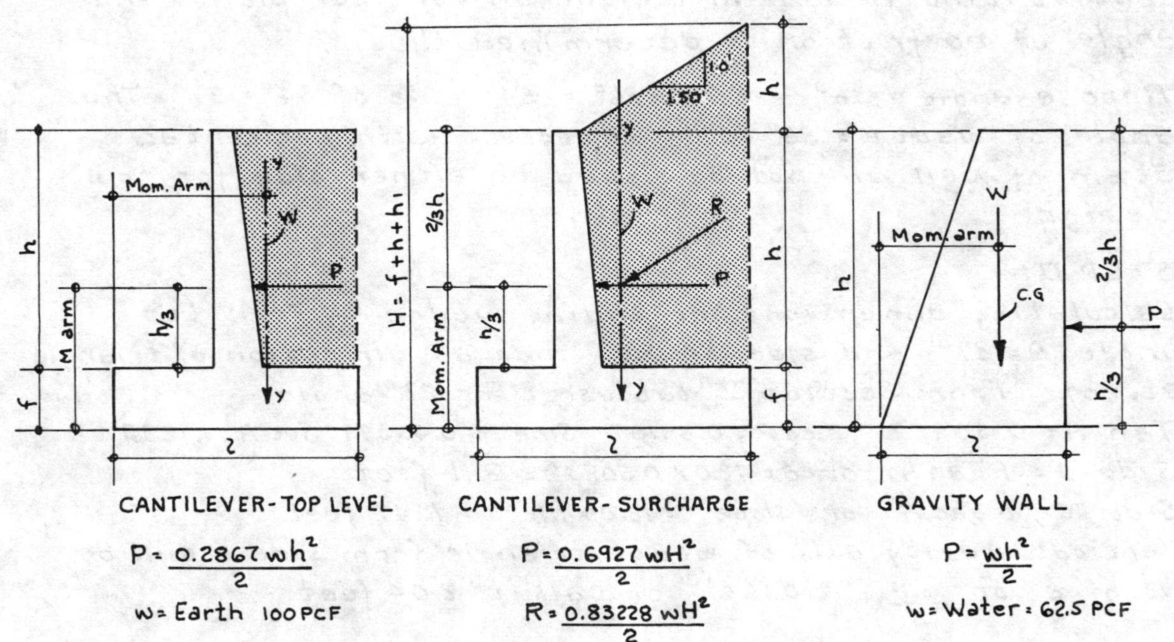

EXAMPLE: Designing cantilever retaining wall 4.13.4.2

A retaining wall is required to support a column of clay earth 12.0 high with earth surface level with top. Estimated weight of retained soil is 115 PCF and angle of repose in damp condition is 36 degrees. A prepared base of sand-shell mix will provide a coefficient of sliding friction of 0.604 Concrete specifications call for $F_c' = 3000$ PSF and weight of 150 PCF. Steel allowable $F_s = 20{,}000$ PSI.

REQUIRED:
Use Trautwine's design theory of the sliding wedge to calculate the horizontal and vertical force components of thrust. Make a cross-section drawing of wall and base with force diagram resultant intersecting the bottom of footing.

STEP I:
Preliminary working drawing will have a 10.0 foot base and retained fill on footing will be assumed to be a contributing factor in weight on vertical plane. The angle of no friction is determined thus:

Repose angle = 36° 90° - 36° = 54° ½ of 54° = 27° Thus angle of wedge = 36° + 27° = 63 degrees with horizontal. Stem of wall will not be sloped on either side for trial design.

STEP II:
Calculating dimensions for sliding wedge by Trig: Let angle A = 27° and side b = 12.0' side a = dimension of triangle at top. From Section V tables: For 27° angle:

Tan A = 0.50952 Cos A = 0.89107 Sine A = 0.45399 Sec. A = 1.1223

Side a = b Tan A or a = 12.0 × 0.50952 = 6.11 feet

Side c = b Sec A or slope c = 12.0 × 1.1223 = 13.47 feet.

Vertical gravity axis of wedge triangle from side b will be 1/3 of a or $\frac{6.11}{3} = 2.0366'$ or call it 2.04 feet.

STEP III:
Calculating weight of concrete and wedge acting in vertical plane on soil bearing under footing. A unit of wall is taken as 1.00 foot in length.

EXAMPLE: Designing cantilever retaining wall, continued 4.13.4.2

Sliding Wedge Triangle: $A = \frac{6.11 \times 12.0}{2} = 36.66^{\square'}$ Wt.: 36.66 x 115 = 4215.90 #

Footing Rectangle: A = 2.50 x 10.0 = 25.00□' Wt: 25.00 x 150 = 3750.00 #

Wall stem Rectangle: A: 1.50 x 12.0 = 18.00□' Wt.: 18.00 x 150 = 2700.00 #

Total combined Areas and Wts. 79.66□' ΣW=10,665.90 #

STEP IV:

Calculate Thrust against wall from weight of earth:
Vertical force = Weight of sliding wedge = 4215.90 Lbs.
Force parallel with slope angle of no friction is Resultant of Vertical and Horizontal:

R = b Sec A or R = 4215.90 x 1.1223 = 4731.5 Lbs.

P = c Sin A or P = 4731.50 x 0.45399 = 2148.0 Lbs. (Horizontal force).

STEP V:

Placing stem to left of center on footing, try 6.0 feet for cantilever projection under wedge. Toe will be A and the point of overturning moment from force P.

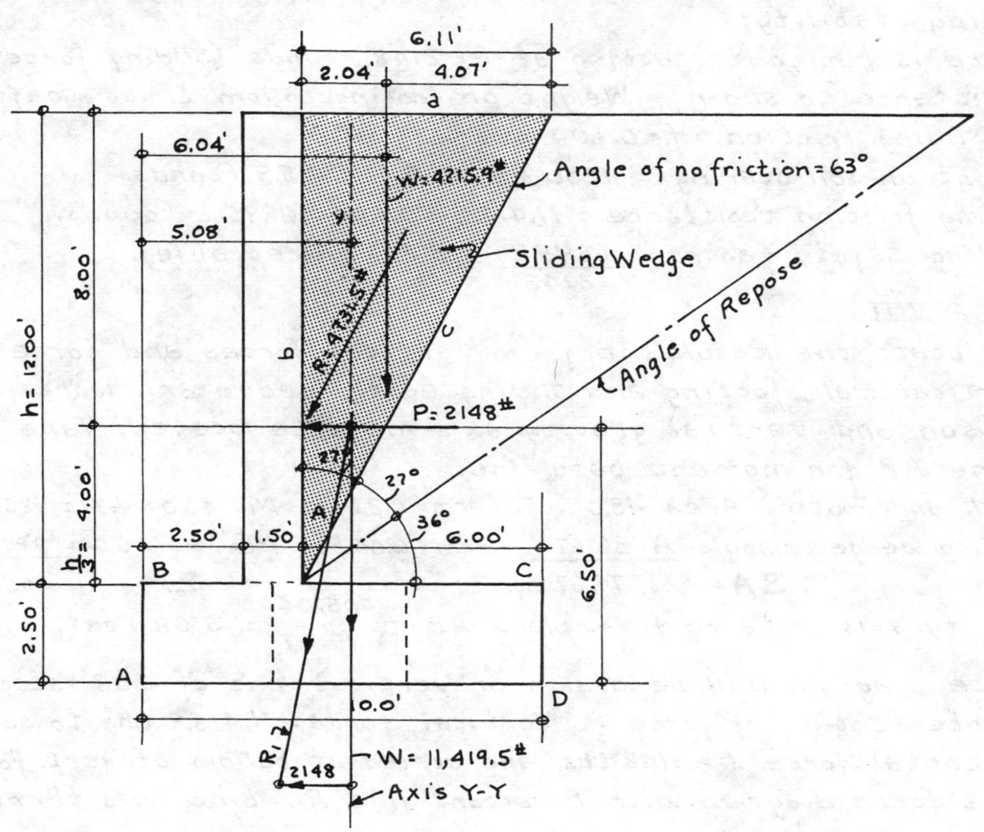

EXAMPLE: Designing cantilever retaining wall, continued 4.13.4.2

Use the moment method to calculate vertical gravity axis of footing and wall stem. Take plane of AB as base line.

Footing Rectangle: A = 2.50 x 10.0 = 25.00 □' ℓ = 5.00' Aℓ = 25.0 x 5.0 = 125.00
Wall stem Rectangle: A = 1.50 x 12.0 = 18.00 □' ℓ = 3.25' Aℓ = 18.0 x 3.25 = 58.50
ΣA = 43.00 □' ΣM = 183.50

Vertical axis distance = $\frac{183.50}{43.0}$ = 4.27 feet.

STEP VI:
Stability of wall: (Tipping)
Overturning moment produced by force P. Point of tipping at A.
lever = 6.50' P = 2148# M = 2148 x 6.50 = 13,962 Foot Lbs.
Resistance to overturning: Moments from total weights.

From Concrete Mass: Wt. = 7200# ℓ = 4.27' RM = 7200 x 4.27 = 30,744 Ft. Lbs.
From Earth wedge: Wt. = 4215.9# ℓ = 6.04' RM = 4215.9 x 6.04 = 25,464 Ft. Lbs.
ΣM = 56,208 Ft. Lbs.

Safety factor against overturning = $\frac{56,208}{13,962}$ = 4.02 OK.

STEP VII:
Sliding stability:
Force is horizontal acting at P = 2148 Pounds. (Sliding force).
Resistance to sliding = Weight on footing bottom times coefficient of sliding friction. f = 0.604
Weight on soil bearing = 7200 + 4215.9 = 11,415.9 Pounds.
Sliding friction Resistance = 11,415.9 x 0.604 = 6895.2 Pounds.
Sliding safety factor = $\frac{6895.2}{2148}$ = 3.21 (Acceptable).

STEP VIII:
To locate the Resultant from vertical forces and force P. Vertical wall, footing and sliding earth wedge are working in unison and vertical gravity axis must be located. Take the plane AB for moment base line.

Wall and Footing Area 43.0 Sq. Ft. arm = 4.27 Ft. M = 43.0 x 4.27 = 183.61
Sliding wedge triangle A = 36.67 □' arm = 6.04 Ft. M = 36.67 x 6.04 = 221.43
ΣA = 79.67 Sq. Ft. ΣM = 405.04

Gravity axis distance from plane AB = $\frac{405.04}{79.67}$ = 5.08 Feet.

Force diagram will be drawn on vertical axis of 5.08' starting at intersection of force P. Vertical force = 11,415.9 Lbs. to scale. Horizontal force P = 2148 Lbs., and scaled at bottom of vert. force. Close force diagram with Resultant line R_1. Line cuts through the middle third and wall is stable.

EXAMPLE: Designing cantilever retaining wall with surcharge 4.13.4.3

A Cantilever retaining wall is proposed to restrain an earth mass which will contain a surcharge graded on the angle of repose or 33° 40' Height of wall to be 10.0 feet. Coefficient of sliding friction is rated to be 0.40 Weight of retained earth is 100 Lbs. cubic foot. Concrete weight = 150 Lbs. cubic foot: Bearing on soil safe at 6000 PSF.

REQUIRED:

The wall proposed is illustrated in the cross-section shown and drawn to scale. Investigate the walls stability for resistance to overturning and sliding. Use Rankine's method for all calculations. Do not design steel reinforcing.

STEP I;

All retained earth is on top of footing with shear greatest at base of stem. Adding dimensions to drawing and marking sections for identifications.

Calculate height surcharge h_1 or side a.

From Trig. tables in Section V

Angle A = 33° 40'

Tan. A = 0.66608

Sine A = 0.55436

Cos. A = 0.83228

Sec. A = 1.2015 Side b = 4.50'

a = b Tan A or height h_1

$h_1 = 4.50 \times 0.66608 = 3.00'$

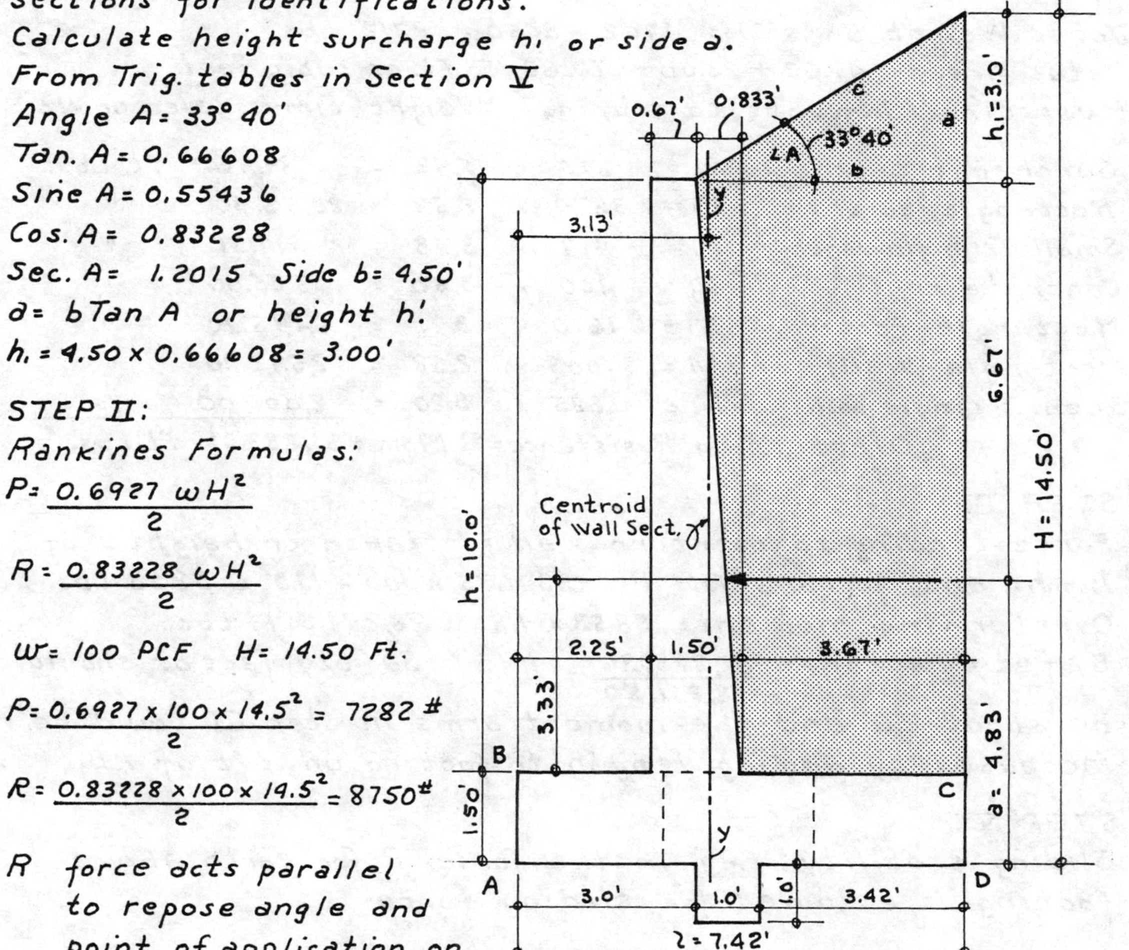

STEP II:

Rankines Formulas:

$$P = \frac{0.6927\ wH^2}{2}$$

$$R = \frac{0.83228\ wH^2}{2}$$

w = 100 PCF H = 14.50 Ft.

$$P = \frac{0.6927 \times 100 \times 14.5^2}{2} = 7282\#$$

$$R = \frac{0.83228 \times 100 \times 14.5^2}{2} = 8750\#$$

R force acts parallel to repose angle and point of application on wall is same as P.

EXAMPLE: Designing cantilever retaining wall with surcharge, continued 4.13.4.3

STEP III:

Calculate the weight on soil at bottom of footing. Resisting moments against overturning at toe or point A, are also computed thus: For Retained earth: ℓ = Moment lever.

Surcharge:	A = 4.50 x 3.0 x 0.50 =	6.75□'	Wt. = 6.75 x 100 = 675#	ℓ = 5.92'
Rectangle:	A = 3.67 x 10.0 =	36.70□'	Wt. = 36.70 x 100 = 3670#	ℓ = 5.59'
Triangle:	A = 0.833 x 10.0 x 0.50 =	4.17□'	Wt. = 4.17 x 100 = 417#	ℓ = 3.48'
		ΣA = 47.62□'	ΣW = 4,762	

For Concrete in Section:

Base Key:	A = 1.00 x 1.00 =	1.00□'	W = 1.00 x 150 = 150#	ℓ = 3.50'
Footing:	A = 1.50 x 7.42 =	11.13□'	W = 11.13 x 150 = 1670#	ℓ = 3.71'
Stem Rect:	A = 0.67 x 10.0 =	6.70□'	W = 6.70 x 150 = 1005#	ℓ = 2.58'
Stem Triangle:	A = 0.833 x 10.0 x .5 =	4.17□'	W = 4.17 x 150 = 625#	ℓ = 3.20
		ΣA = 23.00□'	ΣW = 3450#	

Total Weight on soil = 4762 + 3450 = 8212 Lbs.

Total Areas = 47.62 + 23.00 = 70.62 Sq. Ft. or cubic feet.

Resisting Moments to tipping: Weight times lever or Wℓ

Surcharge earth:	M =	675	x 5.92	=	39.96	Ft. Lbs.
Rectangle earth:	M =	3670	x 5.59	=	20,515.30	"
Small Triangle earth:	M =	417	x 3.48	=	14.51	"
Concrete key:	M =	150	x 3.50	=	525.00	"
Rectangle-Footing:	M =	1670	x 3.71	=	6,195.70	"
Rectangle-Wall:	M =	1005	x 2.58	=	2,592.90	"
Stem Triangle wall:	M =	625	x 3.20	=	2,000.00	"

Overturning Resistance = Σ Mom. = 31,883.36 Ft. Lbs.

STEP IV:

Force tending to overturn wall = P taken on heights of $h + h_1$ and formula is: $P = 0.34635 \times 100 \times 13.0^2 = 5853$ Lbs.

Overturning moment = 5853 x 4.83 = 28,271.5 Ft. Lbs.

Safety Factor = $\frac{31,883.36}{28,271.50} = 1.13$ Safety factor should be about 2.0 and the moment arms in step III could be increased by adding length to footing on left of AB.

STEP V:

Sliding stability: Again the force P for earth above footing is assumed for sliding force.

Sliding force = 5853 Lbs. horizontal. Coefficient of sliding friction given = 0.40 Weight of earth and concrete on bearing area = 8212 Lbs.
Sliding resistance without key = 8212 x 0.40 = 3285 Lbs.
Wall is NOT stable and will slide.
Weight Required for safety factor of 2.0
Then $P = 0.40\,W \times 2$ or $W = \frac{2P}{f}$ $W = \frac{2 \times 5853}{0.40} = 29,265$ Lbs.

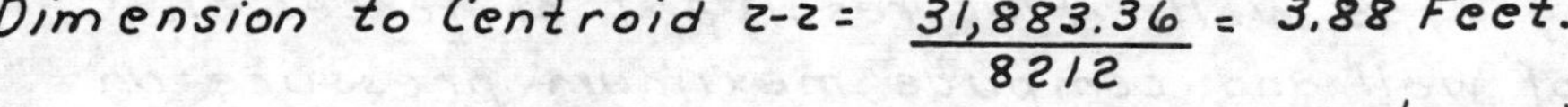

Weight to be added = 29,265 − 8212 = 21,053 Lbs. A safety factor of 1.50 may serve only if a footing bed were prepared to develop a higher coefficient of friction.

STEP VI:
To find point on bottom where the resultant R_1 of forces P and W intersects plane AD: Find the centroid in vertical plane where weight total acts.
In step III: W = 8212 Lbs. and ΣM = 31,883.36 Ft. Lbs. from plane AB.
Dimension to Centroid z-z = $\frac{31,883.36}{8212}$ = 3.88 Feet.

Force thrust comes from $h + h_1$ or 13.0'
$R = \frac{0.83228 \times 100 \times 13.0^2}{2} = 7033$ Lbs. on slope.
Constructing force diagram at right, the resultant R_1 comes out of the middle-third and does not meet code

STEP VII:
Calculate pressure on soil at Toe point A: When R_1 is out of the Middle-Third the formula for p_1 is thus:

$$p_1 = \frac{2W}{3m} \text{ or } p_1 = \frac{2 \times 8212}{3 \times 2.30} = 2380 \text{ \#}/\square'$$

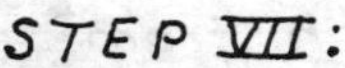

A uniform bearing pressure on soil = $\frac{8212}{7.42}$ = 1107 Lbs. Sq. Ft.

AUTHOR'S NOTE:
The modified cross-section for a stable wall is left for the apprentice or student. After making changes, simply follow the steps outlined until design meets requirements.

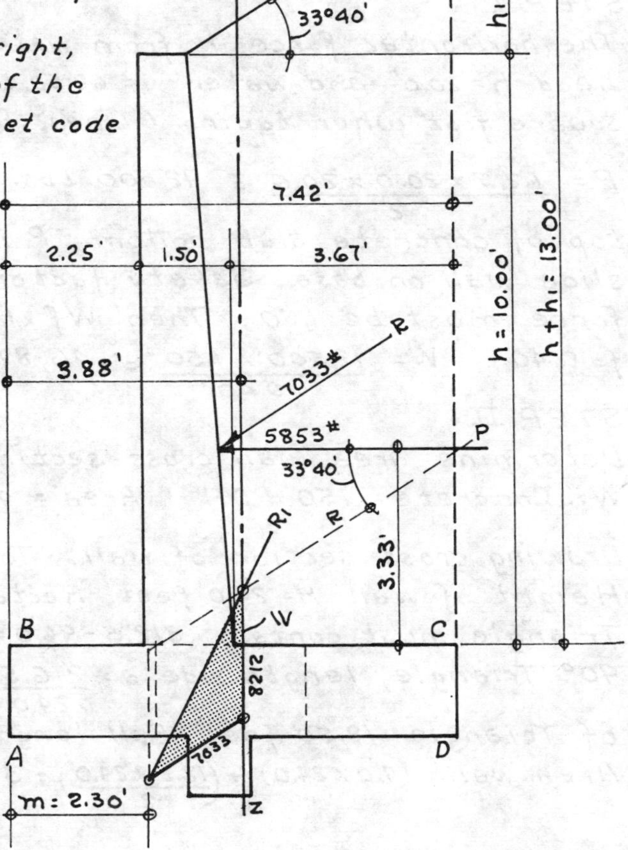

EXAMPLE: Designing gravity reservoir wall 4.13.4.4

A generating plant proposes to construct a 120.0 x 180.0 Foot concrete reservoir to cool water by a spray system. Depth of wall top to bottom is 24.0 feet. Water head of 20.0 feet will be constant. Bottom of reservoir consists of a concrete slab on tamped base. Thickness is 18.0 inches and sloped to center. Safe soil bearing under wall is 2750 PSF and coefficient of sliding friction is 0.40 Safety factor for sliding shall be not less than 1.50 and for overturning the safety factor is to be 2.5 to 3.0 Leave a flat walkway on top of wall for the installation of sprinkling pipe supports of not less than 4.0 feet wide. Water weight = 62.5 PCF and Concrete = 150 Lbs. Cu. Ft.

REQUIRED:
Calculate the mechanics of Vertical and Horizontal forces for the Gravity Type wall. Plot resultant of forces on cross-section of wall and compute maximum pressure on soil bearing at toe and heel.

STEP I:
The horizontal force P from water pressure is: $P = \frac{wh^2}{2}$. head h = 20.0' and water w = 62.50 Lbs. cubic foot, or square foot when taking 1 Lineal foot of wall.

$P = \frac{62.5 \times 20.0 \times 20.0}{2} = 12{,}500$ Lbs. Force P acts a ⅓h from top of concrete slab bottom. P is also the force tending to slide wall on base. Safety factor for resistance to sliding force must be 1.50 Then $Wf \times 1.50 = P$ or $W = \frac{P \times 1.50}{f}$

$f = 0.40$ $W = \frac{12{,}500 \times 1.50}{0.40} = 46{,}875$ Lbs. (Minimum)

STEP II:
Determine area wall cross-section:
Wt. Concrete = 150 PCF Area $= \frac{46{,}875}{150} = 312.5$ Sq. Feet Min.

Drawing cross-section of wall: Top must have min. 4.0 feet. Height of Wall H = 24.0 feet. Rectangle area = 24.0 x 4.0 = 96.0□' Triangle must contain: 312.5 − 96.0 = 216.5□' H = 24.0 Ft. For a 90° Triangle, length side $a = \frac{216.5}{24.0} \times 2 = 18.04$ Feet. Call base of Triangle 18.50 feet. Full length base = 18.50 + 4.0 = 22.50 Ft. Area wall $= (4.0 \times 24.0) + \left(\frac{18.5 \times 24.0}{2}\right) = 318$ Sq. Ft. W = 318 x 150 = 47,700 Lbs.

STEP III:

Check stability for overturning:

Force P acts to tip wall at toe Point A. Moment arm = 9.67 feet.

Tipping M = 12,500 × 9.67' = 120,875 Foot Lbs.

Resistance comes from Weight W, and Moment arm is 14.80 feet.

RM = 47,700 × 14,80 = 705,960 Foot Lbs.

Safety Factor = $\frac{705,960}{120,875}$ = 5.84 (Well above requirements)

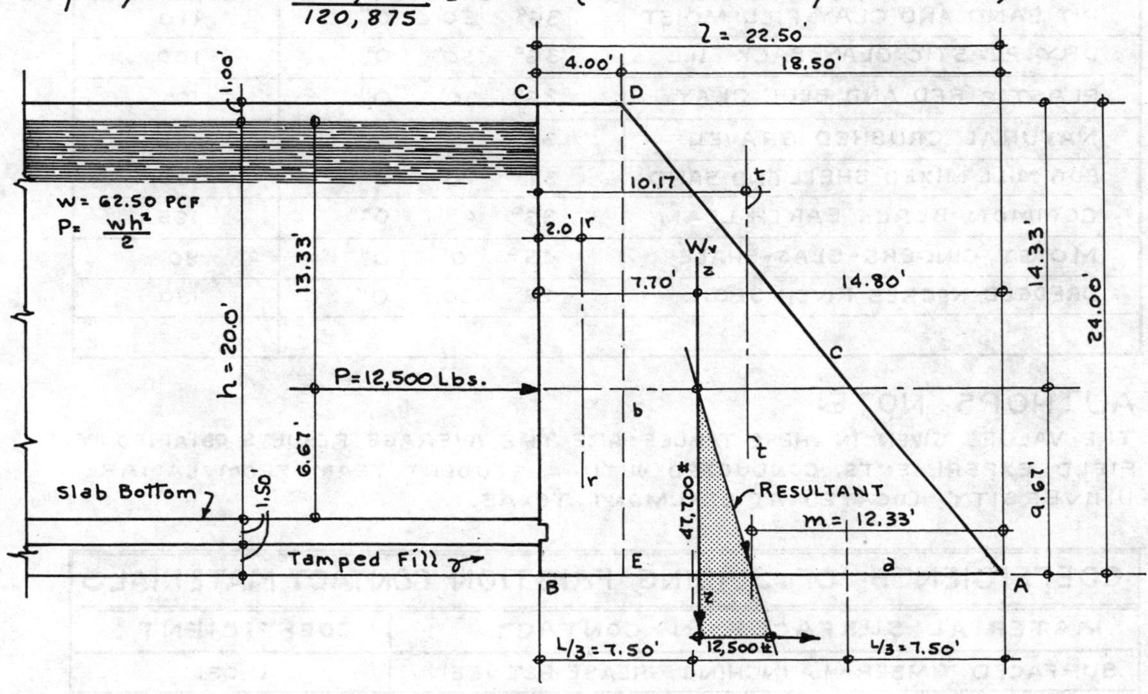

GRAVITY AXIS Z-Z FOR WALL SECTION				
SECTION	SIZE	AREA SQ.FT.	d'	Ad
RECT. BCDE	4.0 × 24.0	96.00	2.00	192.00
TRIA. ADE	18.50 × 24.0	222.00	10.17	2257.74
		ΣA = 318.00□'		ΣM = 2449.74

AXIS Z-Z FROM BC = $\frac{\Sigma M}{\Sigma A}$ Z-Z = $\frac{2449.74}{318.00}$ = 7.70 FEET

WEIGHT WALL = 318.0 × 150 = 47,700 LBS.

STEP IV:

Calculate pressure on soil at Toe (A) and Heel (B): m = 12.33'

Reaction at A = $\frac{(4l - 6m) W_v}{l^2}$ or $p_1 = \left[\frac{(4 \times 22.5) - (6 \times 12.33)}{22.5 \times 22.5}\right] \times 47,700 = 1510$ Lbs.□'

Reaction at B = $\frac{(6m - 2l) W_v}{l^2}$ or $P_2 = \left[\frac{(6 \times 12.33) - (2 \times 22.5)}{22.5 \times 22.5}\right] \times 47,700 = 2730$ Lbs □'

Pressure on soil does not exceed bearing given of 2750 Lbs. sq. foot.

Angle of Resultant = 14° 41' and secant = 1.0338 b = 47.7K R = 47,700 × 1.0338 = 49,312#

Average soil pressure = $\frac{49,312}{22.50}$ = 2192 Lbs. Sq. Foot.

TABLES: Angles of repose and coefficients of friction 4.13.5

ANGLES OF REPOSE AND WEIGHTS OF MATERIALS		
MATERIAL OR SUBSTANCE	ANGLE OF REPOSE	WT. CU. FOOT
MOIST CLEAN WASHED SAND	33° 41' 0"	100
PIT SAND AND CLAY FILL-MOIST	36° 50' 0"	110
DRY PLASTIC CLAY BACK-FILL	36° 50' 0"	100
PLASTIC RED AND BLUE CLAY	26° 30' 0"	105
NATURAL CRUSHED GRAVEL	37° 15' 0"	140
PUG MILL MIXED SHELL AND SAND	37° 20' 0"	130
COMMON BLACK EARTH LOAM	36° 45' 0"	125
MOIST CINDERS-SLAG-SHALE	45° 0' 0"	80
DREDGED NECHES RIVER SPOIL	26° 20' 0"	130

AUTHORS NOTE:

THE VALUES GIVEN IN THESE TABLES ARE THE AVERAGE RESULTS OBTAINED BY FIELD EXPERIMENTS CONDUCTED WITH A STUDENT TEAM FROM LAMAR UNIVERSITY LOCATED AT BEAUMONT, TEXAS.

COEFFICIENTS OF SLIDING FRICTION CONTACT MATERIALS	
MATERIAL SURFACES IN CONTACT	COEFFICIENT
SURFACED TIMBER - LAUNCHING GREASE BETWEEN	0.031
MASONRY UPON CONCRETE SCREEDED SURFACE	0.650
MASONRY UPON WOOD TIMBER SAWN SURFACE	0.600
CONCRETE UPON WET CLAY	0.330
CONCRETE UPON MOIST SAND	0.400
CONCRETE UPON CRUSHED GRAVEL BED	0.625
CONCRETE UPON DRY CLAY	0.500
ASHLAR STONE UPON UNPAINTED STEEL	0.400
SURFACED TIMBER UPON TROWELLED CONCRETE	0.400
BLACK BUTYL RUBBER UPON ABRASIVE STEEL	0.600
LEATHER UPON ABRASIVE STEEL	0.750
CARBON BLACK RUBBER UPON DRY SURFACED WOOD	0.730
LEATHER UPON DRY SURFACED WOOD	0.800
CARBON BLACK RUBBER UPON ROUGH CONCRETE	0.680
CARBON BLACK RUBBER UPON SMOOTH CONCRETE	0.620
CARBON BLACK RUBBER UPON MACADAM SURFACE	0.690

V—TRIGONOMETRY AND GRAPHICS

V—TRIGONOMETRY AND GRAPHICS

V

TRIGONOMETRY AND GRAPHICS

TRIGONOMETRY AND GRAPHICS

Contents

Trigonometry and Graphic Analysis 5.1

There are many problems encountered in structural design which must be solved using trigonometry. Without a working knowledge of this subject, no candidate for state registration can expect to pass the examination. This section is not intended to provide a complete text on this science, but rather to emphasize the principles used in the solutions for dimensions and forces.

Graphic analysis was included in this section because this science provides an alternate and complementary system for the resolution of forces. The authors believe that the junior engineer will be best prepared if he is familiar with both methods and their close relation. Often the solution of a problem can be checked and refined if the work is solved graphically, and minor adjustments made by trigonometric analysis. The graphic method usually saves considerable time and labor, and often checks itself.

The origin of trigonometry 5.1.1

Thales (640–546 BC) was one of the seven sages of ancient Greece. History credits this philosopher and mathematician with the development of trigonometry. Thales used the system to measure the height of the Great Pyramid in Egypt. The Greek astronomer Hipparchus (146–126 BC) used Thales' system in his study of the movements of celestial bodies. The system of trigonometry spread slowly throughout India, Arabia and Egypt, and finally was studied in Rome. The Romans introduced several terms used in trigonometry. The Latin word *sinus* used to describe the hanging fold of a Roman toga, was later reduced to *Sine*, and came to mean the law for solutions. *Secant* is derived from the Latin word *secans*, which meant a cutting or diagonal intersecting line. The Arabic nations within the Roman Empire added the Latin word *tangens* which means to touch, and from this came the word tangent. Trigonometry was originally called *trigon*, a Greek word for triangle. The trigonometer was the original name for an instrument which is now called the transit. The practical use of trigonometry is evident in the Greek and Roman buildings.

Thales' solution 5.1.2

In his experiments, Thales discovered that the shadows from the rays of the sun produced angles with the horizontal plane. The angle was similar, whether the shadow was cast by a tree, building or cliff. He also noted that the apparent movement of the sun constantly changed the shadow angle; to obtain accurate measurements of two angles, he made simultaneous readings.

Crossing the Mediterranean to Egypt, Thales planned to measure the Pyramids. He drove a pole into the ground. The distance from top of the pole to the ground was a known dimension, and became side a in the smaller triangle. The length of both shadows was measured as side b of the similar triangles. The height of the pyramid was side a of the larger triangle.

Assume that the height of driven post is six feet. The shadow cast by the post is eight feet along the ground. The distance from the center of the pyramid to the outer base is 180 feet. From the base of the pyramid to the end of the shadow cast by the pyramid measures 220 feet. Side b of larger triangle equals 180 + 220 = 400 feet.

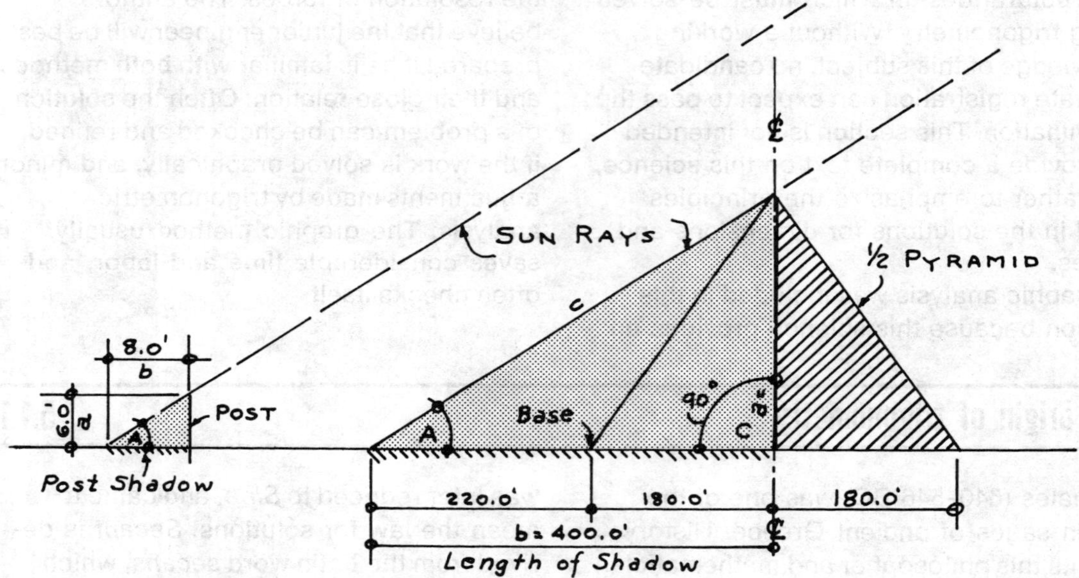

THALES SOLUTION OF A PYRAMID

RATIO: Post a = 6.0 Ft. Shadow side b = 8.00 Ft.

Angles A of small and large triangle are the same.

For small triangle: Tangent $A = \frac{a}{b}$ or $\text{Tan. } A = \frac{6.00}{8.00} = 0.750$

Solving for Side a in large triangle: $a = b \text{ Tan. } A$.

Height of Pyramid: $a = 400.0 \times 0.750 = 300.0$ Feet.

Thales' solution, continued 5.1.2

From Thales' problem, the sides referred to the following:
a = altitude or short side
b = base line along ground
c = connection for sides a and b

The angles are identified as A, B and C where each angle is opposite the side.

Using trigonometric functions 5.1.3

The solution of unknown sides or angles of a right-angle triangle is based on the Law of Sines and Cosines, and the rule that the sum of the three interior angles in a triangle equals 180 degrees. Any triangle can be solved if three parts are known, if at least one known part is a side. In the case of a right triangle, where angle C = 90°, only two additional parts will be needed for solution: two sides, or one side, and another angle A or B.

Formula symbols 5.1.3.1

Standard symbols for angles are used in solving problems in trigonometry. Capital letters A, B and C denote the angles. Angle C always represents the right angle of 90 degrees. Angle A usually refers to the most acute angle. Then Angle B must equal 90° − A. The sides of the triangle are given lower case letters a, b and c, c = hypotenuse, a = altitude, and b = base side. Note that each side is opposite the angle which has the same letter.

Trigonometric functions and formulas 5.1.3.2

Experienced designers realize that the best way to solve triangles of forces or dimensions is the shortest. When angle functions are used in solutions for dimensions or stress, they can be checked by drawing a simple force diagram to scale.

To solve for a Natural Function of any angle in a right triangle, select two known sides and use one of the formulas given in the following table. When one side and an acute angle is known, the formulas can be transposed to solve for the other side.

Trigonometric functions and formulas, continued 5.1.3.2

RIGHT ANGLE TRIANGLES:

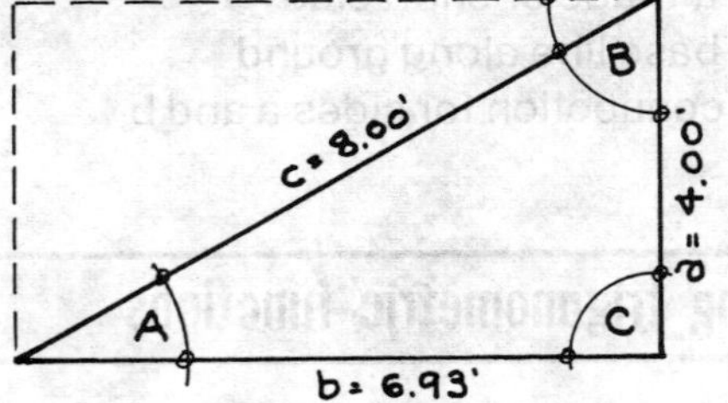

Known data assumed. 2 Dimensions or 2 forces as a = 4.00 feet or a force of 4000 lbs. b = 6.93' or 6930 lbs.

$c = \sqrt{a^2 + b^2}$ or $c = \sqrt{4.00^2 + 6.93^2} = 8.00'$

AREA TRIANGLE = $\frac{ab}{2}$

FORMULAS FOR OBTAINING THE FUNCTION OF ANGLES					
ANGLE A	A = 30°	FUNCTION	ANGLE B	B = 60°	FUNCTION
Sin. A = $\frac{a}{c}$	Sin. = $\frac{4.00}{8.00}$	0.50000	Cos. B = $\frac{a}{c}$	Cos. = $\frac{4.00}{8.00}$	0.50000
Tang. A = $\frac{a}{b}$	Tang. = $\frac{4.00}{6.93}$	0.57720	CoTan. B = $\frac{a}{b}$	CoTan = $\frac{4.00}{6.93}$	0.57220
Cos. A = $\frac{b}{c}$	Cos. = $\frac{6.93}{8.00}$	0.86625	Sine B = $\frac{b}{c}$	Sine = $\frac{6.93}{8.00}$	0.86625
Sec. A = $\frac{c}{b}$	Sec. = $\frac{8.00}{6.93}$	1.15440	CoSec. B = $\frac{c}{b}$	CoSec. = $\frac{8.00}{6.93}$	1.15440
CoSec. A = $\frac{c}{a}$	CoSec. = $\frac{8.00}{4.00}$	2.00000	Sec. B = $\frac{c}{A}$	Sec. = $\frac{8.00}{4.00}$	2.00000
CoTan. A = $\frac{b}{a}$	CoTan = $\frac{6.93}{4.00}$	1.73250	Tan. B = $\frac{b}{a}$	Tan. = $\frac{6.93}{4.00}$	1.73250

TRANSPOSED FORMULA:

Solving for dimensions or forces in sides or planes:

$b = \frac{c}{\text{Sec. A.}}$ or $b = \frac{8.00}{1.15440} = 6.93$ Feet. (6930 Lbs.)

$c = b$ Sec. A. or $c = 6.93 \times 1.15440 = 8.00$ Feet (8000 Lbs.)

$a = c$ Sin. A. or $a = 8.00 \times 0.50000 = 4.00$ Feet (4000 Lbs.)

$b = c$ Cos A. or $b = 8.00 \times 0.86625 = 6.93$ Feet (6930 Lbs.)

$c = \sqrt{a^2 + b^2}$ or $c = \sqrt{4.00^2 + 6.93^2} = 8.00$ Feet (8000 Lbs.)

Branches of trigonometry 5.1.4

Trigonometry as a mathematical subject has been divided into three branches: *plane, spherical* and *analytical*. Plane trigonometry deals with the solution of plane triangles. It is the branch which is used by designers and structural engineers for the solution of angles and lengths of truss members, and the resolution of forces in structural members. The other two branches are used in navigation and plotting the routes for space travel. Only plane trigonometry will be considered in the manual.

Plane trigonometry 5.1.4.1

There are certain axioms which form the basis for plane trigonometry, and these facts should be understood and committed to memory.

(a) Triangles which have the same angles are *similar* triangles.

(b) All three angles of any triangle must add up to 180°.

(c) In a right triangle, the two acute angles will add up to 90°.

(d) Any plane triangle can be divided into right triangles to permit the solution to be reduced to a series of right triangles.

(e) Any angle of less than 90° is termed an *acute* angle.

(f) An angle of 90° is called a *right* angle.

(g) Any angle over 90° is termed an *obtuse* angle.

(h) Any circle contains 360 degrees. A degree (°) contains 60 minutes (′). A minute contains 60 seconds (″). The angle thirty degrees, fifteen minutes, and 30 seconds would be written 30°15′30″.

EXAMPLE: Natural angle functions illustrated 5.1.4.2

The First Quadrant of a Circle is drawn to a radius of 1.00 which is referred to as Unity. There are six (6) natural trigonometric functions in any angle formed within this quadrant. Let the functions be found for an angle A = 30 degrees, then check the results with the tables.

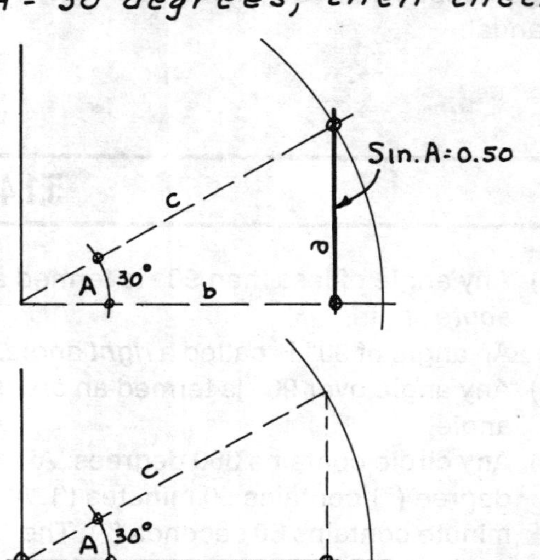

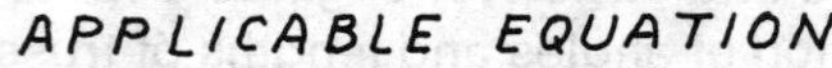

1. $Sine^2 + Cos^2 = Unity$
2. Versed Sine = Unity − Cosine
3. Co-versed Sine = Unity − Sine
4. Tan = Sine ÷ Cosine
5. $Secant^2 = Tan.^2 + Unity$
6. Secant = Unity ÷ Cosine
7. Co-Secant = Unity ÷ Sine
8. $Co\text{-}Sec.^2 = CoTan.^2 + Unity$
9. CoTan. = Cosine ÷ Sine
10. CoTan. = Unity ÷ Tangent

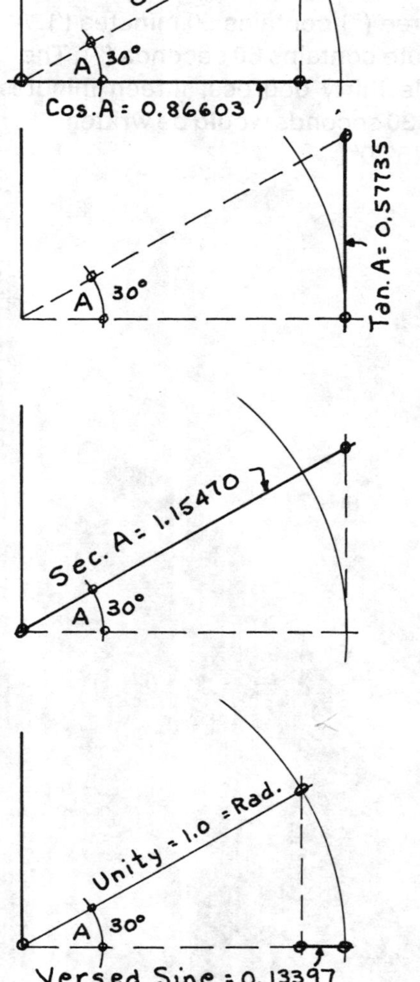

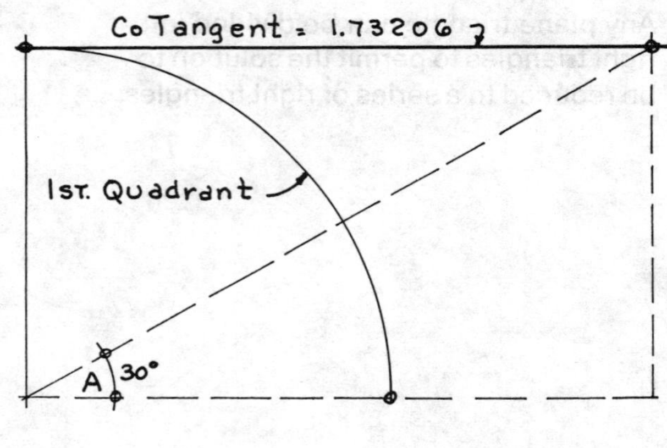

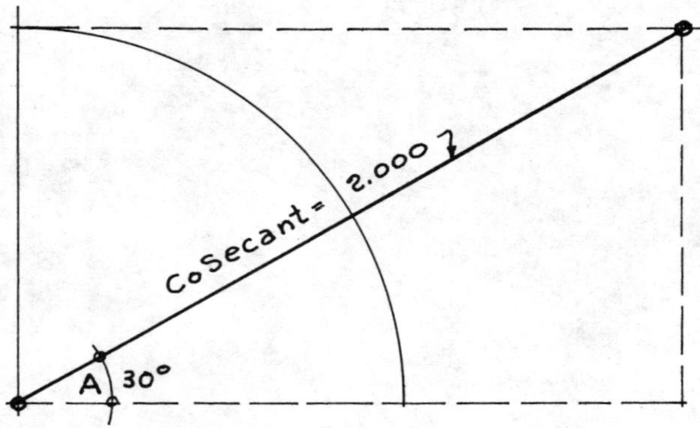

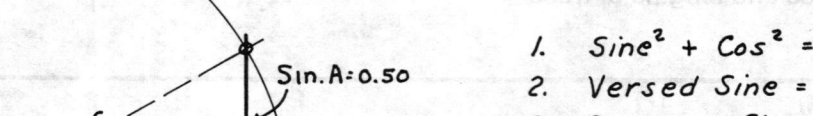

Solving triangles 5.1.5

As already discussed the solution of triangles is based on the Law of Sines and Cosines. Unknown angles and sides can be solved when three parts of the triangle are known. This fact is also applicable when the sides of the triangle represent forces.

There is another method for the solution of triangles which is, in some respects, more convenient and time saving when one has access to a drafting table. A triangle is solved by drawing a plane figure to accurate scale. A protractor is used to measure the angles. Unknown sides may be scaled for length. This method is called the *graphic* method. Again, force and dimension values may be represented by the sides.

A more accurate method for the solution of triangles is the *algebraic* system, which uses equations and natural trigonometric functions of the angles. In the formulation of the algebraic method, there are three basic laws.

PYTHAGOREAN THEOREM:

Ancient Greece produced another philosopher and teacher of mathematics, Pythagoras, who lived during the time of Thales (582–500 B.C.). The Pythagorean Theorem is properly a rule of Geometry, and the formula is always given in the tables of formulas. The Pythagorean Theorem states:

In any right triangle, the square of the hypotenuse is equal to the sum of the squares of the two sides or

$c^2 = a^2 + b^2$ and $c = \sqrt{a^2 + b^2}$ and $b = \sqrt{c^2 - a^2}$.

LAW OF SINES:

In any triangle, the sides are to each other as the sines of the opposite angles.

To simplify the wording, note that angle A is opposite side a, and the others are likewise opposite. Then by formula, the law of sines is written:

$$\frac{a}{\text{Sin A}} = \frac{b}{\text{Sin B}} = \frac{c}{\text{Sin C}}.$$

Also it can be written: $\frac{a}{b} = \frac{\text{Sin A}}{\text{Sin B}}$.

The law holds true when the sides represent forces.

LAW OF COSINES:

In any triangle, the square of any side is equal to the sum of the squares of the other two sides minus twice their product times the cosine of the included angle.

To illustrate the law, let a, b, and c equal the sides of a triangle. The angle opposite side a is angle A. Then to solve for side a, the included angle is A, and the law is put into a formula: $a^2 = (b^2 + c^2) - 2\,bc\text{ Cos A}$ or $a = \sqrt{(b^2 + c^2) - 2bc\text{ Cos A}}$.

Triangles in structural design 5.1.5.1

In structural engineering, the designer will be solving for forces as often as for dimensions. Use the most direct method when solving for a side dimension, angle, or force in a triangle. Select the proper formula, substituting the known values, and solve for the unknown. The graphic method can be used to check the results of the solution by formula.

Using the trig tables 5.1.5.2

Table 5.6.8 gives the natural functions of angles from 0° to 90°, by one minute intervals. More condensed tables may present values for 10 minute intervals. For design purposes, the closest tabulated value will provide sufficient accuracy. For shop detailing considerably more accuracy is required. The angle functions must be evaluated to read degrees, minutes and seconds to insure absolute fits in erection.

When the design requires a value not given in the tables, the method most often used is called *interpolation*: to introduce additional data between the ten minute listings and find intermediate values. The following example will illustrate this method: Should the calculations require a value in seconds, the same procedure would apply; find the values for one minute and divide by sixty for the value of one second.

EXAMPLE: Interpolation in the trig tables 5.1.5.3

A legal paper is presented in a court of law and the angle of a property line is listed as 64 degrees and 32 minutes or written on survey as 64°-32'-0".

REQUIRED:
In order to check the property line dimensions the exact functions must be used. Assuming that the available tables list the functions at 10 minute intervals, solve for functions of Sine, Tangent and Cosine. Check results with tables provided herein. Use interpolation method.

STEP I:
Check Tangent: Angle = 64° 32' Functions listed in Tables are:
Tan. 64° 40° = 2.11233
Tan. 64° 30° = 2.09654
10 Minutes = +0.01579 Difference of 2' = 1/5 of 0.01579 = 0.003158
Then Tan 64° 32' = 2.09654 + 0.003158 = 2.099698 (checks with Tables.)

STEP II:
Check Sine: From tables with 10 minute intervals:
Sine 64° 40' = 0.90383
Sine 64° 30' = 0.90259
10' Interval = + 0.00124 2 minutes = $\frac{0.00124}{5}$ = 0.000248
Sine of 64° 32' = 0.90259 + 0.000248 = 0.902838

STEP III:
Solving for Cosine:
Cos. 64° 40' = 0.42788
Cos. 64° 30' = 0.43051
10' Interval : − 0.00263 2 minutes = 1/5 of 0.00263 = 0.000526
Cosine 60° 32' = 0.43051 − 0.000526 = 0.429984

STEP IV:
Solve for Secant of 64° 32'
Sec. 64° 40' = 2.3371
Sec. 64° 30' = 2.3228
10 Minutes = +0.0143 2 minutes = $\frac{0.0143}{5}$ = 0.00286
Secant 64° 32' = 2.3228 + 0.00286 = 2.32566

EXAMPLE: Right-triangle with two sides known 5.1.5.4

A Right-angle Triangle has these known sides:
Side a = 4.50 feet, and side b = 9.10 feet.

REQUIRED:
Solve for diagonal hypotenuse, or slope side c.
Determine angles A, and B.
Calculate the rise per foot the slope makes with the horizontal plane.

STEP I:
From formula tables: $\text{Tan } A = \frac{a}{b}$, or $\text{Tan } A = \frac{4.50}{9.10} = 0.49450$

Trim Trig. tables; Angle A = 26° 19′ and Sin. A = 0.44333
Angle B: 90° or (89° 60′) − (26° 19′ = 63° 41′ (Given in tables opposite A).

STEP II:

$c = \sqrt{a^2 + b^2}$ or $c = \frac{a}{\text{Sin.} A}$ or $c = b \text{ Secant } A.$

$c = \sqrt{4.50^2 + 9.10^2} = 10.15$ Feet. Also:

$c = \frac{4.50}{0.44333} = 10.150$ Feet. Or $c = 9.10 \times 1.1156 = 10.15$ Feet.

Tables also list secant of 26° 19′ as 1.1156

STEP III:
The rise per foot of a sloped rafter, ramp or incline is equal to the Tangent of Angle A and is 0.49450 feet per lineal foot of run.
Indicate Roof Pitch and slopes thus:

EXAMPLE: Interpolating to find ramp height 5.1.5.5

A truck ramp measures 11.0 Feet on slope. The top angle was measured with with a steel protractor and read as 67° 26'.

REQUIRED:
To find the volume of dirt fill in ramp, solve to find the side dimension thus: Rise = a, and Run = b.

STEP I:
Drawing a Triangle, given angle B = 67° 26'
Given side c = 11.0 Feet.
Angle A = 89° 60' − 67° 26' = 22° 34'

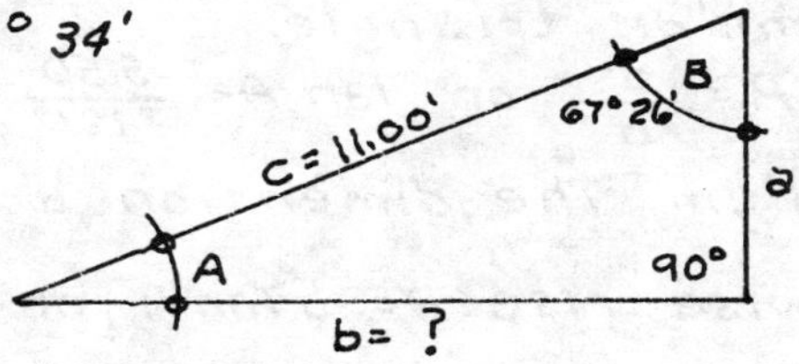

STEP II:
From Trigonometric Formulas:

$b = c \text{ Cos. } A. \quad a = c \text{ Sin. } A$

Turning to Tables of Natural Functions:
Cos. 22° 34' = 0.92343 and Sine = 0.38376

STEP III:
Substituting values, with c = 11.00 Ft.
Side b = 11.00 × 0.92343 = 10.158 Feet.
Side a = 11.00 × 0.38376 = 4.221 Feet.

DESIGNERS NOTE:
Interpolating for 4 minute variance thus:

Sine 22° 40' = 0.38537
Sine 22° 30' = 0.38268
10 = 0.00269

$4' = \frac{0.00269}{2.5} = 0.00108$

Sine 22° 34' = 0.38268 + 0.00108 = 0.38376

Cos. 22° 40' = 0.92276
Cos. 22° 30' = 0.92388
10' = − 0.00112

$4' = \frac{0.00112}{2.5} = 0.00045$

Cos. 22° 34' = 0.92388 − 0.00045 = 0.92343

EXAMPLE: Calculating truss dimensions 5.1.5.6

A Roof Truss has a span $L = 70.0$ feet center to center of columns and is symmetrical about mid-span ℄. The roof pitch is 3.50 inch rise per foot of run.

REQUIRED:

Calculate the height of truss at mid-span, then determine the length of Top Chord. Note on a sketch elevation the angle top chord makes with horizontal bottom chord.

STEP I

Side c = top chord and ½ of span = side b or 35.0 feet.
Pitch = 3.50" per 12.0 inches of b. Let a = 3.50" and b = 12.0" for a similar triangle.

$\text{Tan } A = \frac{a}{b}$ or $\text{Tan } A = \frac{3.50}{12.0} = 0.291667$ and like Thales problem, The dimension a at ℄ is $\frac{3.50 \times 35.0}{12} = 10.208$ Feet.

Likewise rise $a = b \text{ Tan } A$, or $a = 35.0 \times 0.291667 = 10.208$ Feet.

STEP II:

The dimension of Top Chord on slope: $c^2 = a^2 + b^2$ or by usual formula: $c = \sqrt{a^2 + b^2}$. when $a = 10.208'$ and $b = 35.0'$

$c = \sqrt{10.208^2 + 35.0^2} = 36.466$ Feet. Calculating other functions:

$\text{Sec } A = \frac{c}{b}$ and $\text{Cos. } A = \frac{b}{c}$ also $\text{Sin. } A = \frac{a}{c}$.

$\text{Sec. } A = \frac{36.466}{35.0} = 1.04189$ $\text{Cos. } A = \frac{35.0}{36.466} = 0.95980$

$\text{Sine } A = \frac{10.208}{36.466} = 0.27993$

STEP III:

Turning to tables of Natural Trig. Functions: An angle of 16° 16' 0" has a Tangent of 0.29179 and the next lower one minute angle = 0.29147 Now Angle A Tangent is 0.291667. You need 0.291667 − 0.29147 = 0.000197 and by using interpolation given in previous examples, Angle A = 16° 15' and 37 seconds.

STEP IV:

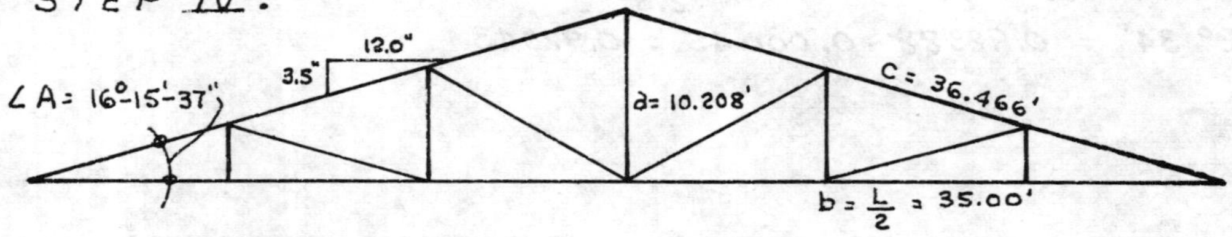

EXAMPLE: Solving right triangles 5.1.5.7

Two 20.0 foot wood ladders are placed against a vertical wall on different angles. Ladder #1 makes an angle of 63° with horizontal plane, while ladder #2 makes an angle of 43° with horizontal.

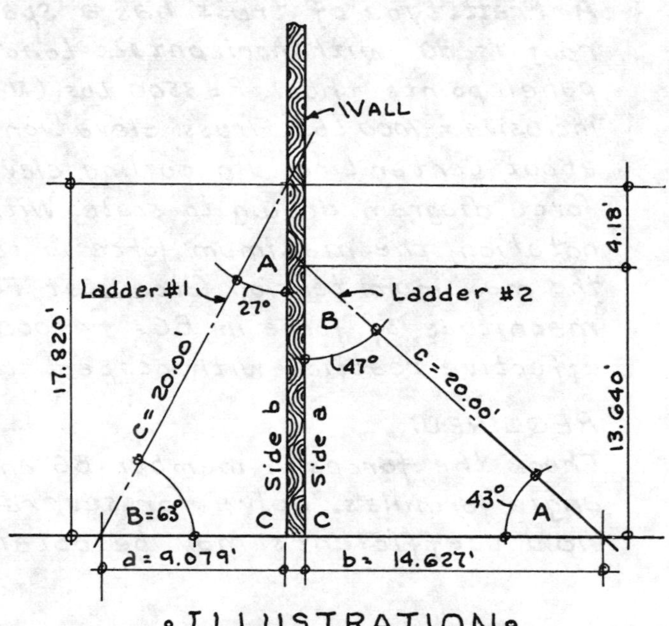

• ILLUSTRATION •

REQUIRED:

Draw an illustration of this problem and determine the horizontal distance of each ladder from wall on level ground.

STEP I:

The known values are the angles and length of ladder which represents side c as 20.0 feet.

Take ladder #1 first with angle of 63 degrees. This will be the angle B in triangle since it contains over 45 degrees. Then angle A will be at top and $A = 90° - 63° = 27°$ From tables of Natural Functions: Tan 27° or Tan A = 0.50952 and Sine A = 0.45399

Now side a = horizontal distance from wall and side b = vertical distance on wall. By formula: $a = c \sin A$, or $a = 20.0 \times 0.45399 = 9.079$ Ft.

Wall dimension $b = c \cos A$ or, $b = \sqrt{c^2 - a^2}$

$$b = \sqrt{20.0^2 - 9.079^2} = 17.82 \text{ Feet}$$

STEP II:

Ladder No. 2: Angle A = 43° because it is the most acute angle. side a is opposite angle A and = vertical dimension on wall. Side b is opposite angle B and equals horizontal dimension.

Functions of 43° angle: Sine A = 0.68200 and Cos. A = 0.73135

Side $b = c \cos A$ or $b = 20.0 \times 0.73135 = 14.627$ feet.

Side $a = c \sin A$ or $a = 20.0 \times 0.68200 = 13.640$ feet

STEP III:

Dimensions are now plotted on illustration

Ladder is placed 4.18 feet on wall above Ladder No. 2

EXAMPLE: Calculating forces in a truss 5.1.5.8

A Pratt type of truss has a span L = 60.0 feet and slope of roof is 30° with horizontal. Loads are placed on top chord panel points thus: P_1 = 3500 Lbs. (Not on truss system). P_2 to P_5 inclusive = 7000 Lbs. Truss elevation and loading is symmetrical about Center Line. An outline elevation is given with a graphic force diagram drawn to scale. With members identified by Bow's notation, the maximum force in top chord is BG and in bottom chord the maximum tension force is at FG. Scaling the stress diagram the magnitude of force in BG = + 49,000 Lbs., and GF = − 42,435 Lbs. The effective reaction with respect to Truss outline = 24,500 Lbs.

REQUIRED:

Check the forces in member BG and FG by using the trigonometric angle formulas. Solve for several web members and show how coefficients may be obtained

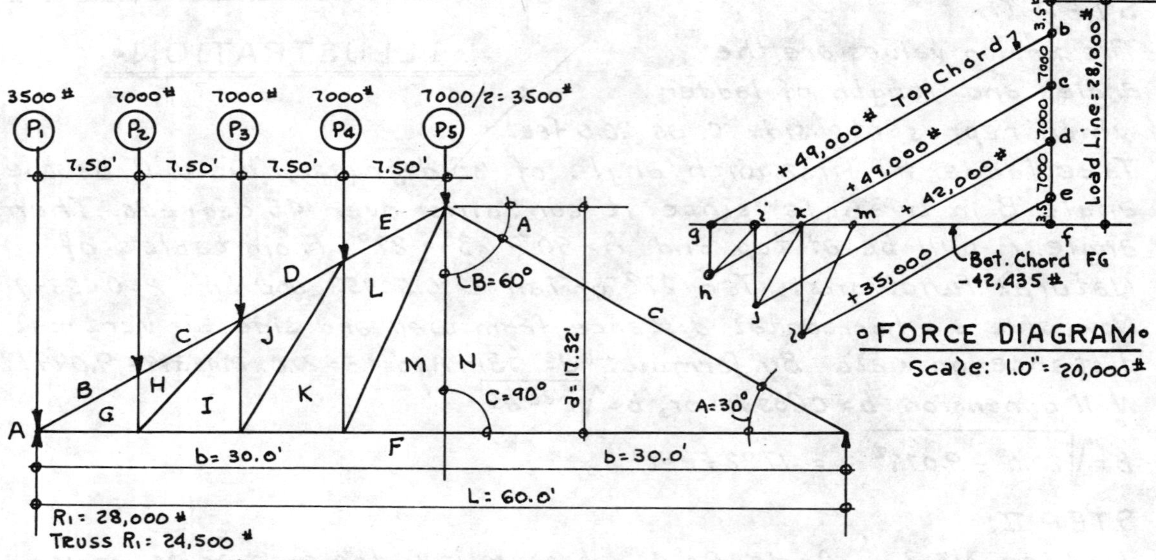

∘ELEVATION OF TRUSS∘
Scale: 1" = 12.0 Ft.

∘FORCE DIAGRAM∘
Scale: 1.0" = 20,000#

STEP I:

Let member BG represent side c of Triangle and force R_1 of 24,500 acts in same plane as GH which is vertical and represents side a. Then angle A = 30° and side a = 24,500 Lbs. Members BG and GH will be in compression and GF is in tension. + = Comp. − = Tension.

STEP II:

From Trig. Tables collect functions for angle A of 30°:

Sin. A = 0.5000 Cos. A = 0.86603 Sec. A = 1.1547 CoSec. A = 2.0000

Member BG or $c = \frac{a}{\text{Sin. } A}$ or $BG = \frac{24{,}500}{0.5000} = 49{,}000$ Lbs. (Checks with dia.)

EXAMPLE: Calculating forces in a truss, continued 5.1.5.8

STEP III:
Consider lower chord member GF to represent side b.
Then by formula: b = C Cos. A. GF = 49,000 × 0.86603 = 42,435 Lbs.

STEP IV:
Load Line in force diagram will aid in illustrating the vertical forces at load points:

C-H = 7000 Lbs.
I-J = 7000 + 3500 = 10,500 Lbs.
K-L = 7000 + 7000 = 14,000 Lbs.
M-N = 24,500 − (7000 + 7000 + 7000 + 3500) = 0 Thus MN is a member for appearance only and is classified as a reduntant.

STEP V:
The Secant of 60° = 2.000 or same as CoSecant of angle A.
Forces in Top Chord members are thus:

B-G = 24,500 × 2.000 = + 49,000 Lbs.
C-H = 24,500 × 2.000 = + 49,000 Lbs.
D-J = (24,500 − 3500) × 2.000 = + 42,000 Lbs.
E-L = (24,500 − 7000) × 2.000 = + 35,000 Lbs.

STEP VI:
In step III, the Cosine A and vertical force R_1 was used to calculate force in bottom chord GF as 42,435 Lbs. Cos. A = 0.86603

I-F = 42,435 − (7000 × 0.86603) = −36,373 Lbs.
K-F = 42,435 − [(7000 + 7000) × 0.86603] = − 30,310 Lbs.
M-F = 42,435 − [(7000 + 7000 + 7000) × 0.86603] = − 24,248 Lbs.

STEP VII:
Coefficient are derived from loads (w = 7000#) and Forces used in above. Coefficients can be used with other values of load W only when truss outline is identical with angle A = 30°

Coefficient B-G = 49,000 ÷ 7000 = + 7.00
Coefficient C-H = 49,000 ÷ 7000 = + 7.00
Coefficient E-L = 35,000 ÷ 7000 = + 5.00
Coefficient G-F = 42,435 ÷ 7000 = − 6.06
Coefficient M-F = 24,248 ÷ 7000 = − 3.46
Coefficient J-K = 12,110 ÷ 7000 = − 1.73

Force in truss members is found as W times coefficient when certain manuals use this method for similar truss designs.

EXAMPLE: Trigonometry in navigation 5.1.5.9

A ship passes Buoy Nō. 1 and heads in the direction of Buoy Nō. 2 which is 55 miles North and 70° East. Continuing from Buoy #2 the course changes to 60° North of East, and Buoy Nō. 3 is 35 miles from Buoy Nō. 2. Total travel from Buoy Nō. 1 is 90 miles.

REQUIRED:

Calculate the distance ship will have traveled if course had been on a straight line from Buoy Nō 1 to Buoy Nō. 3 and Buoy Nō. 2 had been by-passed. Give the compass angle direction ship must take from Buoy Nō. 1.

STEP I:

A base line North to South is laid out and another base line East to West is also drawn. Buoy Nō. 1 is the starting point of travel and will be plotted at vertex of these two base lines N-S and E-W. Plot points of Buoy Nō. 2 and Buoy No. 3 and connect ship's course by a dash line. Instruments or protractors will help in drawing the outline of problem.
From Buoy Nō. 1, Buoy Nō. 2 is on a line 70° East of North 55 miles. Buoy Nō. 3 is 35 miles on a line from Buoy Nō. 2, 60° North of East or same as 30° East of North.

STEP II:

If a line is drawn on course from Buoy Nō. 1 to Buoy Nō. 3 a large triangle is formed by using base lines N-S and E-W. Other triangles within the large triangle are formed by points of Buoy's 1, 2 and 3. In each case, the known side is C, and angle A can be easily determined.
Take first triangle where C = 55 miles and angle A is equal to 90° - 70° or 20 degrees.
Sine 20° (A) = 0.34202 Cos. 20° = 0.93969
Then, side a = C Sin. A or a = 55 x 0.34202 = 18.81 Miles.
And, side b = C Cos. A or b = 55 x 0.93969 = 51.68 Miles.
Plot these dimensions on first triangle.

EXAMPLE: Trigonometry in navigation, continued 5.1.5.9

STEP III:
The second triangle is formed by points of Buoy's No. 1 and 3.

Side c = 35.0 miles. Course of ship is 60° N of E, and angle A = 90° - 60° = 30° where side a is East-West and side b, is North-South. Then, Sin. A = 0.50000 and Cos. A = 0.86603

Then side a = c Sin. A, or a = 35.0 × 0.50000 = 17.50 Miles.
And side b = c Cos. A, or b = 35.0 × 0.86603 = 30.31 Miles.

STEP IV:
The acute angle of larger triangle is not known but sides a and b can be determined thus:

Side b = 51.68 + 17.50 = 69.18 Miles.
Side a = 18.81 + 30.31 = 49.12 Miles.

Tangent $A = \frac{a}{b}$ or Tan. $A = \frac{49.12}{69.18} = 0.71003$

From Tables: Angle A from E-W Base Line = 35° 22′ 26″

Distance c = travel distance on direct course from Buoy No. 1 to Buoy No. 3 and by-passing No. 2 Buoy.

Sine A (35° 22′ 26″) = 0.57892 Cos. A = 0.81538

$c = \frac{a}{\text{Sin. } A}$ or $\frac{b}{\text{Cos. } A}$ or $\sqrt{a^2 + b^2}$

Distance $c = \frac{49.12}{0.57892} = 84.8476$ Miles

STEP V:
Course directions for travel on direct line from Buoy No. 1 to Buoy No. 3: Now 90 degrees = 89° 59′ 60″
Deducting Angle A: 35° 22′ 26″
Angle B is = 54° 37′ 34″

Travel East of North 54° 37′ 34″ for 84.8476 Miles, or
Travel North of East 35° 22′ 26″ for 84.8476 Miles

EXAMPLE: Artillery trigonometry 5.1.5.10

An army supply train is moving at a speed of 60 M.P.H as it passes a depot at point X. Direction of travel is 22° 30' East of North. At point Z and 17° 50' South of East a long range gun is in position with an effective range of 100 Miles radius. Airline distance from point X to point Z is 25 miles. Gun is able to swing a full 360°.

REQUIRED:
Assume that supply train will maintain the 60 MPH speed on a straight course, determine the time period train will be in range of gun after it has passed depot at point X.

STEP I:
Lay out Base Lines to compass as North-South and East-West. Plot depot point X at point of base line intersections, and gun emplacement at point Z. Draw a line 22°30' East of vertical base line N-S which is the track and direction of travel. Establish gun location at Z with a line drawn from X to Z and 17° 50' south of base line E-W. Length of line = 25 miles. At point Z (Gun) draw a circle with 100 mile radius. Where circle intersects railroad track will be maximum range and point R. Connect points Z and R with a dash line and will be side of a triangle = to 100 miles. Now-triangle XZR is not a Right Angle Triangle.

STEP II:
Solve for line X-R by constructing two Right Angle Triangles. The angle R.R track makes with base line EW = 90° - 22°30' = 67° 30'
Then the angle line X-Z makes with R.R track = 67°30' + 17° 50' = 85° 20'
If a line is drawn from point Z to make a 90° angle with R.R. track line X-R, the angle A near point Z will be found as 90° - 85° 20' or A = 4° and 40'. Side c of this small triangle

EXAMPLE: Artillery trigonometry, continued 5.1.5.10

is equal to 25 miles. Thus: All angles A, B, and C can now be known with side dimension c.

Angle C near point y = 90° for two triangles

Angle B = 85° 20' and angle A = 4° 40'

Solving for length of track between points X and Y.

Line X-Y represents side a. a = C Sine A. From the tables:

Sine 4°40' = 0.08136 Then X-Y = 25 × 0.08136 = 2.034 Miles.

STEP III:

The larger triangle remains as RZY. Angle at point Y is 90° however angles near points Z and R are unknown. Side c = 100 miles thus we return to smaller triangle and solve for length YZ which will represent side a of the large triangle. Side YZ = b and b = C Cos A.

Cosine 4° 40' = 0.99668 The large triangle side a = YZ and a = 25 × 0.99668 = 24.917 miles.

STEP IV:

Solve for track length RY which represents side b of large triangle: Two sides are now known. a = 24.917 miles and side c = 100 miles. Angle A is near point R. Then the Sine $A = \frac{a}{c}$ as given in formulas 5.1.3.2.

Sine $A = \frac{24.917}{100.0} = 0.24917$ and from Tables, 5.6.8 the angle is approximately 14° 26' and Cosine A = 0.96844

Distance RY or b = C Cos. A. Track length RY = 100.0 × 0.96844 = 96.844 miles.

STEP V:

Total length of RR track from depot point X to R equals XY + RY. RX = 2.034 + 96.844 = 98.878 Miles.

Train travels at 60 MPH or 1 mile per minute.

Time the train is within effective gun range after passing depot is 98.878 minutes or 1 hour 38 minutes 52.68 seconds.

AUTHOR'S NOTE:

The original problem for this example used Kilometers instead of miles. Actual gun range was assumed to be not over 40 kilometers. Converting to miles, the effective range = 40 × 62.137 or 24.854 miles. 100 Kilometers is equal to 62.137 Miles.

Archimedes Law 5.2.1

The law of buoyancy was discovered by the Greek mathematician and inventor Archimedes of Syracuse (287–212 BC). When Archimedes entered his bath, he noticed that the water rose higher when his body was submerged and lowered again when his body was removed. He noticed that his body displaced liquid, and seemed lighter in weight when submerged.

A solid piece of steel will float in a vat containing mercury or molten lead, because the weight of mercury is 846 pounds per cubic foot, and lead has a weight of 664 pounds per cubic foot. Since steel weighs 490 pounds per cubic foot, it is lighter than either mercury or lead. The steel will submerge only until the weight of mercury displaced is equal to its own weight. The Rule of Displacement may be stated: When a body (barge or ship) is placed in a liquid, the body will be supported in that liquid by a force equal to the weight of the liquid that it displaces.

Further study on buoyancy is given in the design of Steel Vessels, Section II.

Naval architecture 5.2.2

Where structural engineers are primarily concerned with structures which are supported by solid earth, the designers of floating vessels must depend upon water for support. Naval architects are, in fact, developers of water-born structures. It is possible for naval architects to determine the weight of a floating vessel by computing the volume of the submerged portion below the water line. Conversely, when the weight of all the ships parts are totalled, and the weight of men and cargo added, the naval architect will be able to calculate the volume of water which the ship will displace.

Pontoons and rafts 5.2.3

Army engineers use pontoon bridges to transport soldiers, equipment, and heavy weapons across rivers encountered during an advance. This type of pontoon structure can also be used during commercial construction operations. Hydraulic dredging requires pontoon rafts to support the spoil outfall tube. Temporary pontoon supports are used to erect bridge components. A pontoon raft can be used to float a pile driver to the opposite bank of a river project.

Water density: temperature and salinity 5.2.4

A temperature change in any liquid will change the density of the liquid. The addition of salt (sodium chloride) to fresh water will also cause a density change. A ship will float higher in the open sea salt water than in a fresh water port. The naval architect determines the location of the trim lines on the hull of a vessel after considering the density of the water in which the vessel will sail.

The specific gravity of steel is 7.84, the weight per cubic foot equals 489.6 pounds. The specific gravity of 7.84 is based upon distilled water at a temperature of 4 degrees Centigrade. American engineers use the Fahrenheit thermometer for temperature: 4 degrees Centigrade is equal to 39.2 degrees Fahrenheit. At this temperature, water weight is 62.43 pounds per cubic foot. This is the value of a specific gravity of 1.00.

To convert a Centigrade reading to a Fahrenheit reading: $F^\circ = \left(\frac{C^\circ \times 9}{5}\right) + 32^\circ$.

Illustrated: $F^\circ = \left(\frac{4^\circ \times 9}{5}\right) + 32^\circ = 39.2^\circ$.

When water is frozen into ice, the ice weighs 56 pounds per cubic foot. The weight of water at 212° Fahr. or 100°C is 59.83 pounds per cubic foot.

Ocean water has a weight of approximately 64.0 pounds per cubic foot, and a specific gravity of 1.025. This density is generally used for the design basis of displacement for sea going vessels. Fresh water weighs 62.43 pounds per cubic foot; however, for convenience in problems involving gravity tanks or pontoons, the weight is taken as 62.5 pounds per cubic foot or 8.33 pounds per gallon. (The number of gallons in a cubic foot = 7.482.) An unusual characteristic of water is that it expands below and above the point of maximum density, at 39.2°F.

Hydrostatic theorem 5.2.5

Hydrostatic refers to fluids in equilibrium, which are motionless. The basic theorem in hydrostatics is stated: At any point in a liquid, the pressure due to the weight of the liquid is the same in all directions. This pressure is equal to the column head height, and is given in pounds per square inch. To illustrate: Water weighing 62.5 pounds per cubic foot is used to fill a tank having a height of 15.0 feet. The pressure at bottom of tank in all directions is:

$\frac{15.0 \times 62.5}{12 \times 12} = 6.51$ pounds per square inch.

A simple formula can be written:

$$p = \left(\frac{h}{144}\right) \times W, \text{ where}$$

p = Water pressure in pounds per square inch.

h = Height (head) or liquid in vessel in feet.

W = Weight of liquid in pounds per cubic foot.

For another example, if the water head (h) is 10.0 feet, and the weight (W) is 62.5 pounds per cubic foot, then the pressure (p) at this depth would be $p = \frac{10.0}{12 \times 12} \times 62.5 = 4.34$ Pounds per Square Inch.

Simplifying the formula for fresh water, it becomes $p = 0.434\ h$. This figure can be used for gravity water towers, standpipes and similar structures.

Pressure on curved surfaces 5.2.6

The end pressure on a closed-end pipe submerged in a liquid is equal to the end area times the pressure: $P = p\,\pi r^2$. When a circular tube with the ends closed is only partially submerged, as in the case of a floating pipe section the underwater portion will have the cross-sectional shape of a segment of a circle. The chord of the segment will be the water surface. The greatest dimension from chord to the arc of the circle is the depth of the floating pipe in the liquid, and may be calculated as the versed sine. Refer to Example 5.1.4.2 which illustrates this angle function. The versed sine divides the segment into equal parts, and is normally identified as H, the height of chord above the arc.

Circular segment coefficient tables 5.2.6.1

Solutions which require the designer to calculate the length of segment chords, arcs, areas and heights may become very involved and time consuming. By drawing a vertical line through the center of the circle, and drawing lines from the ends of the chord to the center, a sector of the circle will be outlined. To find the area of the sector, determine the whole area of the circle, and deduct the area of the sector. There are two right triangles below the chord. The area of these triangles can be quickly found. The area of the segment enclosed inside the chord and the arc is then found by deducting the area of triangles from the area of the sector. An example will illustrate this method of solving for segment areas.

Mathematicians have developed a coefficient method for solving segment areas when the sector angle, the chord C, and the rise H are known. Table 5.6.6 is self explanatory and includes angles from 1 to 180 degrees.

Another coefficient table is available to shorten the design work, and this is based upon the ratio of rise in the segment to the diameter of the circle. This Table 5.6.7 is more convenient to use, as will be shown in several examples which follow. Another advantage of this table is that the coefficients may be obtained with great accuracy by the method of interpolation. By transposing the formula, other values can readily be solved. To illustrate: Let D = 20.0″ and rise H = 7.00″ Ratio of $\frac{H}{D} = \frac{7.00}{20.0} = 0.350$.
From table: C = 0.24498 Then $A = CD^2$
Area = 0.24498 x 20.0 x 20.0 = 97.99 sq. in.
Transpose when A and D are known. Find coefficient C.

$C = \frac{A}{D^2}$ or $C = \frac{97.99}{400} = 0.24498$ From table
$\frac{H}{D} = 0.350$ Solving for Rise:
H = 0.350 x 20.0 = 7.00 inches

EXAMPLE: Solving circle sectors and segments 5.2.6.2

A circle with a diameter of 18.0 inches, contains an arc Sector of 70 degrees. Connecting the radius lines at intersection of circumference a Segment is produced.

REQUIRED:

(a) Make a drawing of Circle and identify sector and segment.

(b.) Calculate the area in whole circle, then the area of sector, and finally the area of Segment.

(c) Solve for the length of the versed sine in circle sector, and identify the location.

(f) Calculate the length of chord, and arc in Segment.

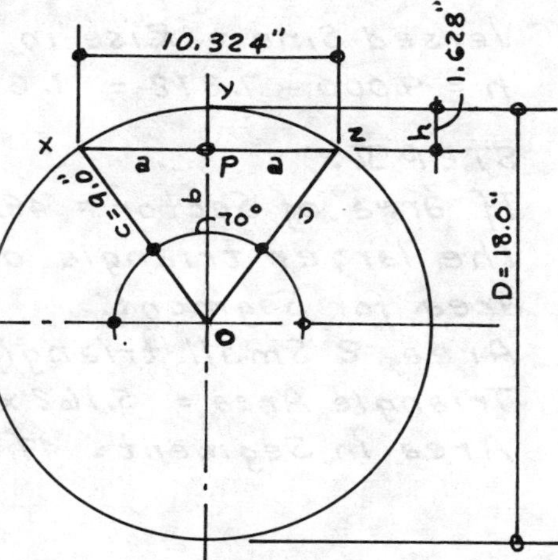

STEP I:

Drawing illustrated. Sector = xyzo.

Segment arc = xyz. Segment Chord = xz.

Radius = ox or oz, and Versed Sine = yp or h.

STEP II:

Area of Complete Circle = $0.7854\ D^2$ = 254.47 Square Inches

Circumference of Circle = 3.1416 D = 56.55 Inches.

Full Circle = 360° and Sector = 70°

Sector Area = $\dfrac{254.47 \times 70}{360}$ = 49.480 Sq. In.

Length of Segment arc is same as Sector arc, or 70°.

Arc xyz = $\dfrac{56.55 \times 70}{360}$ = 10.996 inches

STEP III:

To solve for versed sine and chord of segment. In the Triangles, xop, and zop, there 2 knowns. Angle A = 35°, and side c = Radius, or c = 9.00 inches.

Formulas: Side a = c Sine A, and b = c Cos. A.

From Tables of functions: Sin 35° = 0.57358, and Cosine = 0.81915.

Side a = ½ Segment Chord: a = 9.00 x 0.57358 = 5.162 inches.

Segment Chord Length = 2 x 5.162 = 10.324 inches. (x-z).

EXAMPLE: Solving circle sectors and segments, continued 5.2.6.2

STEP IV:
For Versed Sine dimension:
Side b = 9.00 x 0.81915 = 7.372 Inches. yo = 9.00 = Radius zo.

Versed Sine = Rise in Segment arc h.
h = 9.000 − 7.372 = 1.628 Inches.

STEP V:
If area of Sector = 49.480 Sq. Inches, then the Area of the larger triangle deducted from Sector, will leave the area for Segment.
Area 2 Small triangles is same as rectangle = a x b.
Triangle Area = 5.162 x 7.372 = 38.054 Sq. inches
Area in Segment = 49.480 − 38.054 = 11.426 Sq. In.

EXAMPLE: Pipe pontoon raft 5.2.6.3

A group of Steel Pipes with 10.0 foot lengths are to be used to support a raft in fresh water. Wall thickness of cylinders is 0.172 inches and Outside diameter is 16.0 inches. Each end of pipes will be closed by welding a 3/4 inch plate. By using a fresh water displacement weight, these pontoon pipes must be capable of supporting a load of 500 Pounds each, or 50 pounds per lineal foot. Fresh water weight is assumed to be 62.50 Lbs. per cubic foot. The steel tie rods connecting the pontoons and the wood plank surface deck weights shall be neglected in computing the supporting reaction on empty hollow cylinders.

REQUIRED:

(a) Calculate the Dead weight of each pontoon unit and then determine the volume of water required to float the pipe.

(b) Locate the Water Line with respect to the cylinders' horizontal centroid when floating without other loads.

(c) Delineate the submerged segment by drawing a cross-section or end elevation and water line.

(d) Utilize the Coefficient tables to solve for segment area and versed sine. Obtain pipe weight from Steel Pipe Pile tables in Section IX.

STEP I:

From Pipe pile tables: Weight pipe with 16.0" OD and $t = 0.172$ inch wall thickness, weight per foot: 29.06 Lbs. The end area is $0.7854\,D^2$ or $A = 0.7854 \times 16.0 \times 16.0 = 201.06$ Square inches.

From Sect. VI, Steel plate weights: 3/16" Plate = 7.65 Lbs. Sq. Foot.

Weight of 2 end plates $= \dfrac{2 \times 201.06 \times 7.65}{144} = 21.36$ Pounds.

Weight of pipe 10.0 foot long: $10.0 \times 29.06 = 290.60$ Pounds.

Steel dead weight $= 21.36 + 290.60 = 311.96$# (Call it 312 Lbs.)

Water volume required to displace equal weight and float hollow cylinder $= \dfrac{312.0}{62.50} = 5.00$ cubic feet.

STEP II:

Now, consider a length of pipe section one inch long. This ring section will simplify the solution for solving for segment, arc and chord at water line.

Weight of 1.00 inch depth ring $= \dfrac{312.0}{10.0 \times 12} = 2.60$ Pounds

EXAMPLE: Pipe pontoon raft, continued 5.2.6.3

Step II Continued:

To calculate the weight of 1 Cubic inch of Water:

Wt. $1^{3''} = \frac{62.5}{12^3} = 0.0362$ Lbs. Volume in 1.0" thick ring required to support dead weight $= \frac{2.600}{0.0362} = 71.80$ Cubic inches.

Checking figures by using 10.0 Ft. Length $= \frac{5.0 \times 12^3}{10.0 \times 12} = 71.80$ Cu. In.

STEP III:

Making a drawing of Raft Cross-Section for Water Line mark:

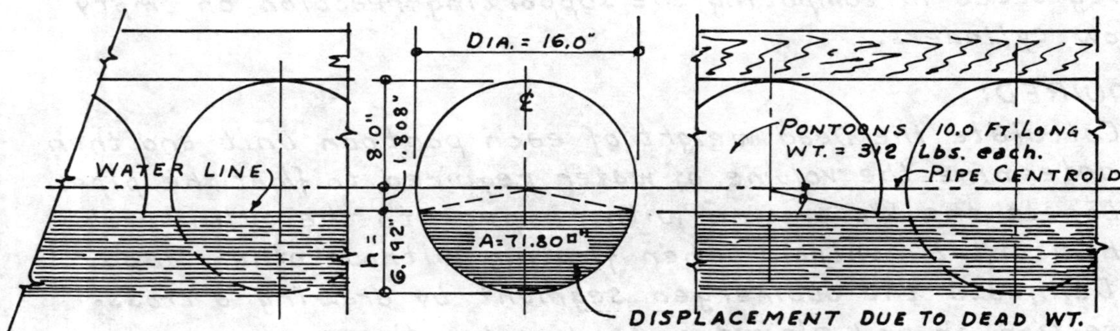

• END ELEVATION & SECTION - PONTOON SUPPORTS •

To locate the Water Line and Chord of Segment. The area of 71.80 □" only represents displacement of Water required to float empty Pontoon. It is also Area of Plane Segment. To find displaced Volume for 10.0 Foot length.

$\Delta = \frac{71.80 \times 10.0 \times 12}{12^3} = 5.00$ Cubic Ft. Wt. $= 5.00 \times 62.5 = 312$ Lbs. (OK).

STEP IV:

To solve for depth of Segment or versed Sine.

Refer to Segment Tables with respect to, D and A.

Formula to find $A = D^2 \times$ Coefficient. Since Area is known, transpose formula to find Coefficient.

Coef. $= \frac{A}{D^2}$ or $\frac{71.80}{16.0 \times 16.0} = 0.280$ From tables: When the Coefficient = 0.280, the ratio of $\frac{H}{D}$ is given as, 0.387

Then $H = 0.387 \times 16.0 = 6.192$ inches.

With respect to Centroid, dimension = 8.000 - 6.192 = 1.808 In.

EXAMPLE: Pipe pontoon raft, continued 5.2.6.3

STEP V:
Draw the Sector of the Circle as shown by dash lines extending from ℄ of Centroid, to where water line intersects the circumference. Two triangles are formed above water line and centroid. Also 2 Sides of each Right Angled Triangle are known. Radius of Circle = 8.00 inches for side c, and distance from WL., to Centroid represents side a, and equals, 1.808 inches. To check this work with another method, solve for Sector Angle in degrees.

a = 1.808″ c = 8.00 C = 90° $\text{Sin. } A = \frac{a}{c} = \frac{1.808}{8.00} = 0.2260$

From Tables of Functions:
Angle A = 13° 4′ Then Angle B = 76° 56′
The Angle of Sector = 2 × 76° 56′ = 153° 52′.

STEP VI:
The Volume above the Water Line must be weighted in order to be displaced, and therefore becomes the support for loading.
When entire Pontoon is submerged, find volume of the displaced water.

Volume of 10.0 Ft. Pipe = $\frac{D^2 \times 0.7854 \times l}{12^3}$. Then substituting:

$$\text{Displacement} = \frac{16.0 \times 16.0 \times 0.7854 \times 120}{12 \times 12 \times 12} = 13.96 \text{ Cubic Feet.}$$

Weight required to sink = 13.96 × 62.50 = 872.50 Lbs.
Applied Load, less Dead Weight = 872.50 − 312.00 = 560.50 Lbs.

STEP VII:
Check the applied Load to sink by volume above WL.
Area full Circle = 16.0 × 16.0 × 0.7854 = 201.06 Sq. Inches.
Deducting Segment Area: A = 201.06 − 71.80 = 129.26 Sq. In.

Volume above WL in 10.0 Ft. Tube = $\frac{129.26 \times 120}{1728} = 8.96$ Cubic Ft.

Weight to Sink = 8.96 × 62.5 = 560.50 Lbs. (checks OK)

STEP VIII:
The example requires that each Pontoon will be required to support 50 Pounds per lineal foot.

Pontoon will support: $\frac{560.50}{10.0} = 56.05$ Lbs. Per Foot. (OK)

EXAMPLE: Solving circle sectors 5.2.6.4

A previous design example of a Pontoon Raft unit had a pipe support with a diameter of 16.0 inches. In step V of that example, the sector angle was found to be 153° 52'. The versed sine H of the segment was 6.192 inches, and the area of segment was computed to equal 71.80 Sq. inches.

REQUIRED:
Make another drawing of same circle and show the sector, chord and segment. Now employ the tables and formulas 5.6.5 to 5.6.7 to solve for areas. If necessary, transpose the formulas to obtain the ratio of $\frac{H}{D}$ and the Coefficients. A slight variance is expected, however the area result should be close to 71.80 Sq. inches considering the end plate closures.

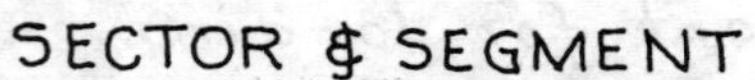

SECTOR & SEGMENT

STEP I:
Illustration is drawn as shown: Versed sine H = height of segment = 6.192 inches. Segment = XYZ. Sector angle = 153° 52'. From tables: Area of 360° circle = 201.062 Sq. inches and 180° half circle = 100.531 Sq. Inches.

STEP III:
If 180° half circle = 100.531 □" then 1 degree area can be found.

Area 1° = $\frac{100.531}{180°}$ = 0.55851 Sq. inches., and calculating the area for 1 minute = $\frac{0.55851}{60}$ = 0.0093085 Sq. inches.

Area for 153° =	0.55851 × 153	=	85.452 □"
Area for 52 minutes =	0.0093092 × 52	=	0.484 □"
	Total area of sector	=	85.936 Sq. inches.

To calculate the area of segment, deduct the area of the triangle OXZ or the area of 2 identical right triangles. Radius represents side c in triangle OXV and angle A is $\frac{180° - (153° 52')}{2}$ or 13° 4'. Side a = 8.000 − 6.192 = 1.808 inches.

$$\text{Sine } A = \frac{a}{c} \text{ or } \frac{1.808}{8.000} = 0.22600 \quad \text{and Cosine } A = \sqrt{1.000 - \text{Sin.}^2}$$

$$\text{Cosine } A = \sqrt{1.0000 - 0.2260^2} = 0.97411$$

EXAMPLE: Solving circle sectors, continued 5.2.6.4

Step III Continued:

Side b of triangles is represented as XV and VZ. Length of segment chord = 2 side b. b = C Cos. A
XV + VZ = 2 x 8.000 x 0.97411 = 15.586 inches.

STEP IV:
Computing area in 2 right angle triangles same size or 1 Triangle OXZ: Rectangle = 1.808 x 15.586 = 28.1795 Square inches.
Area in segment equals area Sector minus area in triangle.
Segment Area = $85.936 - \left(\frac{28.1795}{2}\right) = 71.846$ Sq. inches.

DESIGNERS NOTATION:
In example of Pontoon raft 5.2.6.3 the segment area was computed on the basis of weight and displacement. This factor was also used to compute the coefficient of $\frac{A}{D^2} = 0.280$ in step IV when the segment chord length was neglected.

STEP V:
Employing the tables 5.6.6 and 5.6.7

Formula: A = Chord (c) x Rise (H) x Coefficient.
C = 15.586" and H = 6.192" Ratio $\frac{H}{C} = \frac{6.192}{15.586} = 0.39728$

Table gives the coefficient for 154° of 0.7447 and the angle is 8 minutes larger. Substituting values in formula:
Area segment = 15.586 x 6.192 x 0.7447 = 71.8698 Sq. Inches. This is close enough to check previous work.

STEP VI:
Checking result when Coefficient tables the known diameter instead of chord length. Ratio becomes $\frac{H}{D}$ to obtain coefficient. Area = D^2 x Coefficient. D = 16.0"
Ratio: $\frac{6.192}{16.00} = 0.387$ Coefficient in table = 0.280669

Segment Area = $16.00^2 \times 0.280669 = 71.8513$ Sq. inches. (checks).

AUTHOR'S NOTE:
Table 5.6.7 has the advantage over similar tables since the simple formula can be transposed to solve for other values such as H and Diameter. Coefficient $C = \frac{A}{D^2}$

Graphic design systems

Graphics is defined as a system of diagrammatic line pictures which the designer uses to determine the force in members by direct scale measurements. When elevations and force diagrams are carefully drawn with an engineers scale, the accuracy is equal to the trigonometric algebraic solution; and considerable time can be saved. A force diagram or polygon can also serve as a good check for the trigonometric calculations. More than reading texts on graphic analysis it is necessary for the student to construct the geometric forms to comprehend their use and advantages.

The analysis of forces can be better understood by the use of graphic construction. The shear and moment beam diagrams in Section I *Mechanics of Beams* is an example of the use of graphics. The examples and illustrations which follow will show the practical solutions for reactions, resultants, and forces in such members as beams, derricks, joists, and trusses.

In the graphic method, the terms vector, magnitude, plane, resultant, component, funicular polygon have a distinct meaning which must be recognized and understood.

VECTOR

The lines which are used to represent a force in magnitude and direction (or sense) are called *vectors*. They do not denote a line of action. *These vectors must be drawn parallel to the direction of the structural member* which they represent. When read with an engineer's scale, each vector will provide a value for force in the member which it represents. Use care in selecting the scale to be used, and note it immediately upon the drawing. Using the correct scale to determine the magnitude of each vector is of the utmost importance for accurate readings.

RECIPROCAL FUNCTIONS

The reciprocal of any number is the quotient obtained by dividing unity by that number. Thus, the reciprocal of 2.00 is 1.00 ÷ 2.00 = ½, the reciprocal of 4.00 is 1.00 ÷ 4.00 = ¼. The reciprocal of a fraction is that fraction in inverted form. The reciprocal of ¾ is 4/3. The following trigonometric functions are reciprocals: Sin A = 1/cosec A, cos B = 1/sec B, tan C = 1/cotan C.

ABSCISSA AND ORDINATE

A horizontal axis line x–x is drawn below and another axis line y–y is drawn perpendicular and bisecting x–x. If a point P is located above x–x and to the right of y–y, we have two lines of distance. Identify the horizontal line parallel with axis x–x as line x′, and the vertical line as y′.

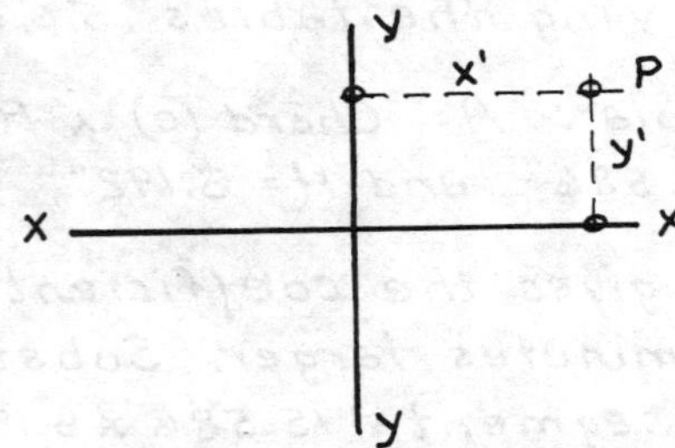

The *abscissa* of point P is the distance x′ in a horizontal direction. The *ordinate* of point P is the distance y′, measured in a vertical direction.

When the *abscissa* and *ordinate* are both referred to, they are called the *coordinates* of a given point. In large petro-chemical plants, coordinates are referenced to established base lines: N–1600.0′ x W–290.0′. The point indicated is 1600.0 feet north of the East-West base line and 290.0 feet west of the North-South base line.

MAGNITUDE

The magnitude of a stress or force represented by the scale length of a vector may be measured in pounds, tons or kips, and

Graphic design systems, continued 5.3.1

may be either tension or compression stress. Wind or moving loads may cause forces which can change stress from tension to compression.

DIRECTION PLANES

Force systems in graphics are classified by the line of action of the forces which make up the system. The action line is the plane along which a force acts.

(a) Concurrent: All force action lines have one common point.
(b) Coplaner: Force action lines lie in one plane.
(c) Non-Coplaner: Force action lines lie in different planes.
(d) Non-Concurrent: Force action lines do not intersect at a common point.
(e) Colinear: Forces have a common line of action.
(f) Non-Colinear: Forces have no common action line.
(g) Parallel: Forces which have their action lines in the same plane.

RESULTANT

A single force which will hold two or more forces in equilibrium is called a resultant force. Several forces acting in different directions may be resolved into a single force; this *resultant* becomes an equivalent of all the forces. In structural design, it is often necessary to resolve a wind load reaction with a vertical roof load reaction.

An illustration of the resultant of two forces follows. Start at the polar point O, and, using an engineers scale, measure off force F_1 of 350 pounds acting along vector OA. Next, draw line OB horizontal to represent force F_2 of 450 pounds. Now construct a parallelogram OA CB. The diagonal line from O to C is the resultant of F_1 and F_2. Scale the vector with the same scale used to construct the other forces. R = 675 pounds. The arrows show the direction of the forces.

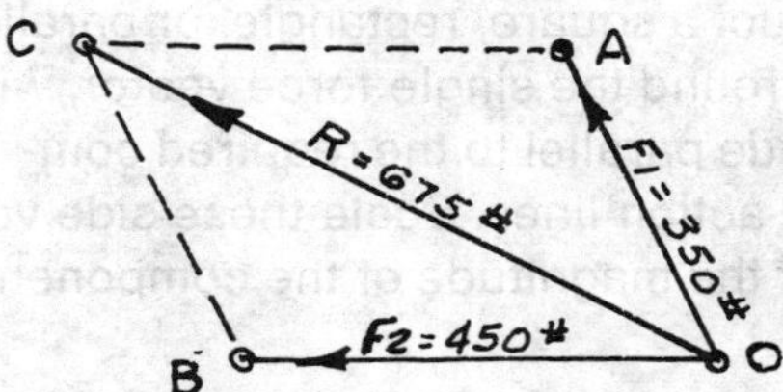

UNIT MEASUREMENT

Angles may be measured by instruments with the graphic system, and in an emergency, this may be the only solution in the field. Lines are marked off in units of ten (10) equal spaces. To measure angle A by this improvised method, extend line OB horizontally. Start at O, and mark off 10 equal spaces. Extend resultant line OC until a right angle can be made. To solve for angle A, which OC makes with OB, a right triangle is drawn.

Side a = 5.20 Units
Side b = 10.00 Units

$$\text{Tangent } A = \frac{a}{b} = \frac{5.20}{10.00} = 0.520$$

Angle A = 27° 30'

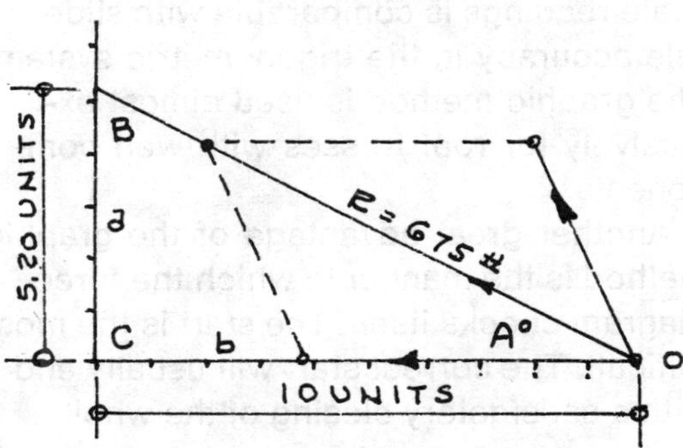

Graphic design systems, continued 5.3.1

RESOLUTION OF A FORCE

To resolve a single force into two component forces, reverse the procedure used to obtain the resultant. Lay off a vector of the given magnitude in the line of action. Construct a square, rectangle, or parallelogram around the single force vector, with each side parallel to the required component action lines. Scale these side vectors for the magnitude of the component forces.

COMPONENTS

The original vectors which are used to find the resultant force are called *components*. A force diagram or polygon is composed of two or more force components.

FUNICULAR POLYGON

A funicular polygon is a polygon which is formed by chords. Usually the funicular polygon is the method used to find the *load reactions* in beams and trusses. It should not be confused with the force diagram. To construct a funicular polygon, first draw a load line. Measure off each load, starting at the left end of the beam or truss. Point off each load in its line of action, using the proper scale. Above and to the left of the load line, draw an elevation of the truss. Extend the line of action of each load below the elevation drawing. At some convenient point to the left of the load line, place the polar point O. Connect the polar point to each load on the load line and the result is a *Ray Diagram*.

Now select a space below the elevation drawing for the funicular polygon. Start with the first load on the left and draw a line parallel to the first ray vector. Continue for each ray line until each load is included. Close the polygon with the closing string. Transfer the action line of the closing string to the load line by drawing a parallel line through the polar point and intersecting the load line. To find the reactions at each end of the truss, scale the distance from each end of the load line to the intersection of the closing string action line.

Advantages of the graphic method 5.3.1.1

Experienced designers often turn to a graphic analysis of a problem, if for no other reason than to check their own work or the work of others. The accuracy of the scale readings is comparable with slide rule accuracy in the trigonometric system. The graphic method is used almost exclusively for roof trusses with web components.

Another great advantage of the graphic method is the manner in which the force diagram checks itself. The start is the most difficult. The correct start will usually end with a satisfactory closing of the whole stress diagram. For this reason, the illustrations and examples in this section will represent the vectors for the popular truss types. Accuracy in the force magnitudes will depend upon the care and ability of the draftsman. Larger force diagrams produce more accuracy than the smaller scales. Work can also be improved by employing the newer types of drafting machines and adjustable triangles. Adjustable steel protractors are excellent instruments for transfering parallel components to corresponding vectors in the force diagram.

Bow's notation 5.3.1.2

An orderly force diagram requires systematic identification of the vectors. The notation should identify each component member to permit instant reference to the corresponding vector. When drawing the elevation of a truss, identify the chord and web members by starting at the left. Make capital letter symbols between the forces and members as shown in the illustrations. A truss member is identified by the letters in the spaces which it separates. The examples will illustrate this system by placing capital letters in a clockwise order around the truss profile and then in the web spaces. The load line is drawn parallel to the line of action of the applied loads. Point off the load line with the scaled length for each separate load. Identify the vectors with lower case letters to correspond with the capital letters on the truss elevation. Each component stress vector is drawn parallel to the corresponding member in the elevation.

This procedure of relating each truss member with a single vector in the Force Diagram is known to the engineering designers as *Bows Notation*. Initiating the first step for truss member identification will be simplified by referring to the pilot elevations and force diagrams presented in 5.3.5.

Analyzing trusses 5.3.2

A truss is a framed structure incorporating elements which provide static equilibrium under the action of the applied loads and reactions. A roof joist is a form of fabricated light truss. The truss elements must satisfy these equations of equilibrium:

(a) The sum of the vertical components of all forces = O.
(b) The sum of the horizontal components of all forces = O.
(c) The sum of the moments of all forces about any point = O.

Methods to evaluate stresses 5.3.2.1

After calculating the reactions from the applied loads and their line of action, the forces in the frame members can be found by using the three equilibrium algebraic equations or by the graphic method. In general practice, the graphic method is used, because it gives less chance for error.

When using the algebraic method, the designer must keep separate the members as the equations move to the right of reaction R_1. This method is more often used for the Howe and Warren type of truss. Computation for stresses by algebraic equations is similar to beam mechanics; both involve the method of moments.

Another useful method for the solution of truss forces involves coefficients for magnitude. This method requires the pitch of the top chord, the number of web panels, and the type of truss to have the exact values in the coefficient tables.

Indicating stresses in diagrams 5.3.2.2

Forces acting upward and to the right are positive and those forces acting downward and to the left are negative.

Wind stress diagrams 5.3.2.3

Make separate stress diagrams for loads which have different lines of action. Wind action is assumed to exert force in the horizontal plane on incline. The roof pitch will determine the wind force component normal to the top chord. A simple, three-line force diagram will solve this problem, as will be shown in the Pilot Force Diagrams. Steep roof pitches, often used for A-frame arches and scissor trusses, can have the wind loads resolved into vertical forces. In such cases, the wind load can be added to the vertical roof loads.

Pilot diagrams for trusses 5.3.2.4

To provide assistance in the initial stages of truss analysis by the graphic method, this section provides a number of Pilot Diagrams for quick reference. When the truss type has been selected, the pilot force diagram with proper Bow's Notation may be used as a guide. It is not possible to become proficient in the use of the graphic method by simply studying texts. The method must be learned by performing the actual work. Do not hesitate to apply the method to structures encountered each day, and practice setting up force diagrams.

EXAMPLE: Resolving force and deflection 5.3.4.1

A cantilever canopy beam consists of a 4×6 Southern Yellow Pine section with a 7.50 foot clear overhang. The major axis x-x is horizontal. Beam supports a concentrated load of 350 pounds at extreme end. A horizontal force of 400 pounds is also applied at end of beam. These load forces produce deflection when applied simultaneously and the resultant line of action must not exceed 3.0 inches. Modulus of Elasticity E= 1,600,000

REQUIRED:
Determine the amount of deflection in each direction by using the deflection formula for Cantilever beams. Sketch this beam, and draw force diagrams with the resultants for loads and deflection. Max. Slope Δ = 3.00 inches.

STEP I:
Drawing beam in isometric manner, the net size of beam section is 3⅝" × 5½". L = 7.50 Feet. The moment of Inertia about axis x-x = 50.25"^{4}, and I_y = 21.84"^{4} The applicable formula for deflection is: $\Delta = \frac{Pl^3}{3EI}$. L = 7.50 Ft.

STEP II:
Sketch beam in isometric.

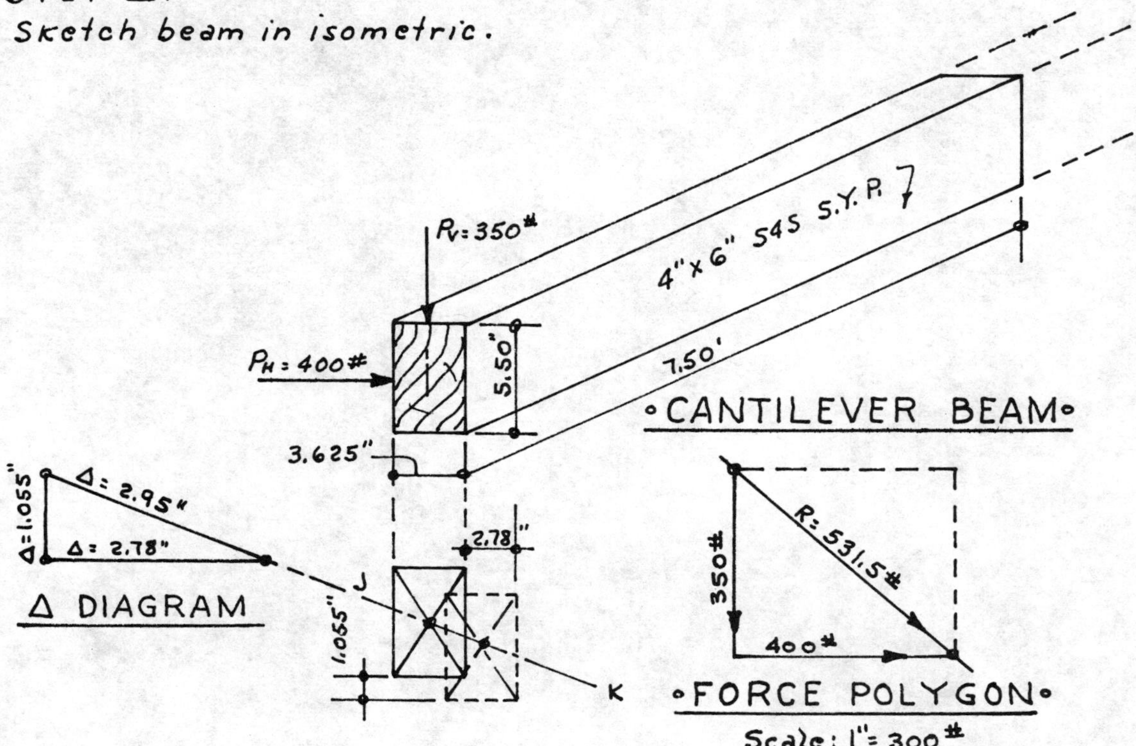

EXAMPLE: Resolving force and deflection, continued 5.3.4.1

STEP II:

Calculating amount of deflection from vertical load and about axis x-x: $l = 7.50 \times 12 = 90.0$ Inches. $I_x = 50.25''^4$

Substituting in formula:

$$\Delta_x = \frac{350 \times 90.0^3}{3 \times 1{,}600{,}000 \times 50.25} = \frac{255{,}150{,}000}{241{,}200{,}000} = 1.055 \text{ Inches}$$

STEP III:

Calculation of Lateral deflection:

$$\Delta_y = \frac{400 \times 90.0^3}{3 \times 1{,}600{,}000 \times 21.84} = \frac{291{,}600{,}000}{104{,}832{,}000} = 2.78 \text{ Inches}$$

STEP IV:

Drawing Deflection Diagram as triangle:

Side $a = 1.055''$ $b = 2.78''$ and $c = \sqrt{1.055^2 + 2.78^2} = 2.92$ In.

Resultant of Forces in like manner.

Side $a = 350^{\#}$ $b = 400^{\#}$ and $c = \sqrt{350^2 + 400^2} = 531.5$ lbs.

EXAMPLE: Solving Warren truss by moments and graphics 5.3.4.2

A Warren Truss with a 75.0 Foot span is divided into 9 Web Panels. Height of Truss is 18.0 Feet. There a 5 Roof Loads of 2500 Lbs., each placed at top on panel points.

REQUIRED:
Draw an elevation of truss to scale, identify each member, then calculate stress in members by moment method. Check the work by constructing a stress diagram.

STEP I:
Drawing Truss Elevation with member notations:

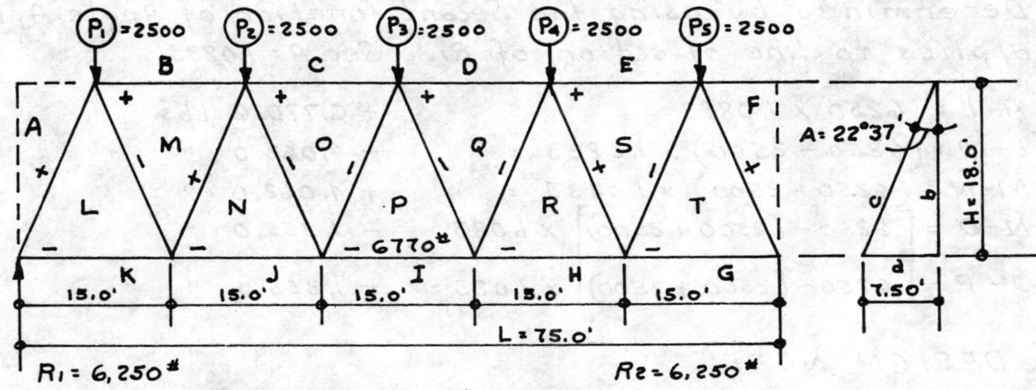

•TRUSS ELEVATION•
Scale: 1/16" = 1.0'

STEP II:
For convenience, the Stress Diagram is drawn close to elevation, and transfer vectors are better observed.

STEP III:
To determine angle A. 2 Sides known.

$\text{Tan. } A = \frac{7.50}{18.00} = 0.41666 \qquad A = 22^\circ\, 37'$

Secant $A = 1.0833$

STEP IV
STRESSES IN BOTTOM CHORD MEMBERS:

$$LK = \frac{6250 \times 7.50}{18.0} = -2604 \text{ Lbs.}$$

$$NJ = \frac{(6250 \times 22.5) - (2500 \times 15.0)}{18.0} = -5729 \text{ Lbs.}$$

$$PI = \frac{(6250 \times 37.5) - \left[(2500 \times 15.0) + (2500 \times 30.0)\right]}{18.0} = -6{,}770 \text{ Lbs.}$$

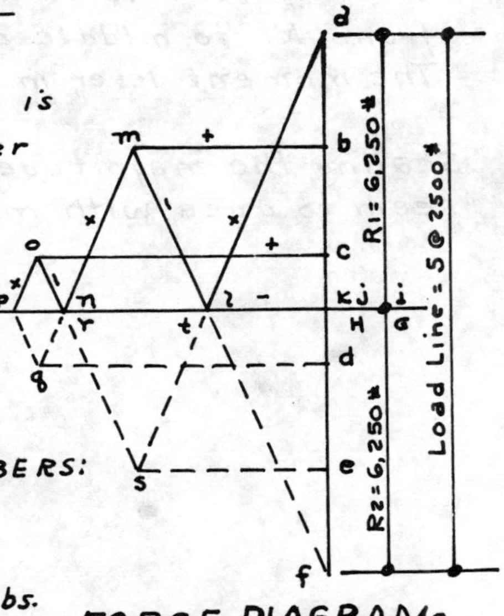

•FORCE DIAGRAM•
Scale: 1" = 4000#

EXAMPLE: Solving Warren truss by moments and graphics, continued 5.3.4.2

STEP V

STRESSES IN TOP CHORD:

$$B\text{-}M = \frac{(6250 \times 15.0) - (2500 \times 7.50)}{18.0} = +\ 4,166.66 \text{ Lbs}$$

$$C\text{-}O = \frac{(6250 \times 30.0) - [(2500 \times 7.50) + (2500 \times 22.50)]}{18.0} = +6,250 \text{ Lbs}$$

C-O = DQ and BM = ES, because of symmetrical conditions.

STEP VI:

STRESSES IN WEB MEMBERS:

Determined by using the Secant function of Angle A, which applies to line of action of R_1. Sec. A = 1.0833

$$A\text{-}L = 6250 \times 1.0833 = +\ 6,770.6 \text{ Lbs.}$$

$$L\text{-}M = (6250 - 2500) \times 1.0833 = -\ 4,062.0 \text{ "}$$

$$M\text{-}N = (6250 - 2500) \times 1.0833 = +\ 4,062.0 \text{ "}$$

$$N\text{-}O = [6250 - (2500 + 2500)] \times 1.083 = -\ 1,355.0 \text{ "}$$

$$O\text{-}P = [6250 - (2500 + 2500)] \times 1.083 = -\ 1,355.0 \text{ "}$$

DESIGN NOTE:

The moment lever in Bottom chord members is taken from R_1 to middle of chord member.
The moment lever in top chord is length to load panel.

Scaling the magnitude of vectors in force diagram, they seem to agee with moment method.

EXAMPLE: Solving Howe truss by moments and graphics 5.3.4.3

A Howe Truss with a span of 75.0 Feet. has a web member of 6 equal panels from Bottom Chord. There are 5 Loads of 2500 Lbs., each, which are suspended from bottom chord, and directly under the vertical member. Truss height = 18.0 Ft.

REQUIRED:
Prepare a drawing for elevation and identify each member. Construct a Force diagram, and check by the moment method. Denote the Tension stress by minus sign (−), and compressive stress with plus (+) sign.

STEP I:
Drawing 75.0 Ft. Truss with 5 vertical hangers. Divisions = 12.5 Ft.

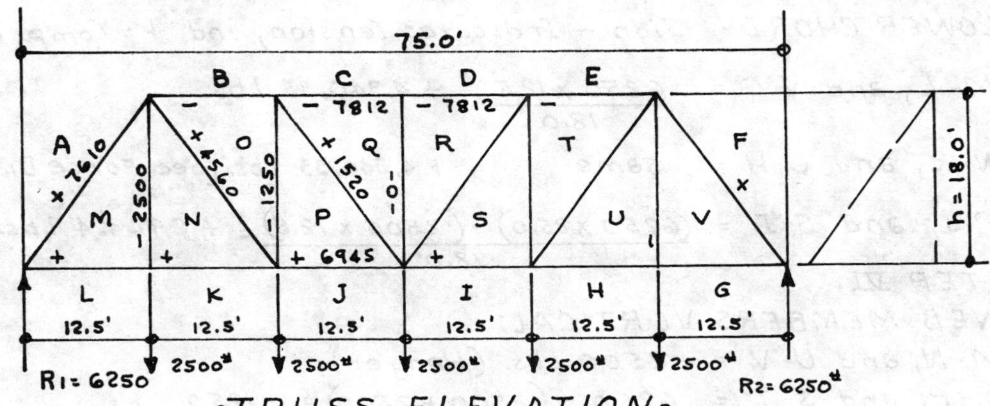

∘TRUSS ELEVATION∘
Scale: 1/16" = 1.0'

Loads = 2500 × 5 = 12,500 Lbs. $R_1 = R_2 = 6,250$ Lbs.

STEP II:
Bow's notation used to identify members.
Constructing Force Diagram to scale.
Load line action vertical = 12,500 Lbs.
Max. Top Chord: CQ = −7812 Lbs.
Max. Bot. Chord: PJ = +6945 Lbs.
Max. Web Member: AM = +7610 Lbs.
Member QR is a Redundant member, or one which is not necessary and is not stressed.

∘FORCE DIAGRAM∘
Scale: 1" = 4000#

EXAMPLE: Solving Howe truss by moments and graphics, continued 5.3.4.3

STEP III:

Determine angle A at Top Chord. Known sides: a = 12.50 Ft. and b = 18.0 Feet. $\text{Tan. } A = \frac{a}{b} = \frac{12.50}{18.00} = 0.69444$

Tables give angle A = 34° 47′ Secant A: 1.2174

STEP IV

TOP CHORD – Stresses by Moment method.

$$\text{B-O, and E-T} = \frac{(6250 \times 25.0) - (2500 \times 12.5)}{18.0} = -6{,}945 \text{ Lbs.}$$

$$\text{C-Q, and R-D} = \frac{(6250 \times 37.5) - \left[(2500 \times 12.5) + (2500 \times 25.0)\right]}{18.0'} = -7800 \text{ Lbs.}$$

STEP V:

LOWER CHORD – Sign – Indicates Tension, and + = Compression.

$$\text{M-L, and V-G} = \frac{6250 \times 12.5}{18.0} = +4{,}340.33 \text{ Lbs.}$$

N-K, and U,H = same = + 4,340.33 Lbs. (See Force Diagram)

$$\text{P-J, and S-I} = \frac{(6250 \times 25.0) - (2500 \times 12.5)}{18.0'} = +6944.4 \text{ Lbs.}$$

STEP VI:

WEB MEMBERS VERTICAL:

M-N, and U-V = −2500 Lbs. (Hanger)

O-P, and S-T = 6250 − (2500 + 2500) = −1250 Lbs.

Q-R = 6250 − (2500 + 2500 + 1250) = 0 (Redundant Member).

STEP VII:

DIAGONAL WEB MEMBERS:

Secant Angle A = 1.2174 (See Step III.)

A-M, and F-V = 6,250 × 1.2174 = + 7,610 Lbs.

N-O, and T-U = (6,250 − 2500) × 1.2174 = + 4,565 "

P-Q, and R-S = [6250 − (2500 + 2500)] × 1.2174 = + 1,522 "

A Flat Type of Truss or Joist spans 56.0 Feet and must support a uniform load of 100 Pounds per lineal foot. Height of Truss is limited to 7.0 feet and web shall be divided into 8 Panels.

EXAMPLE: Solving flat truss by moments and graphics 5.3.4.4

REQUIRED:

Draw a truss elevation or outline of a Howe design and calculate the forces in members by the algebraic method of moments. Check the work by drawing a Graphic Force Diagram. The Reactions effective on truss shall be used for both methods.

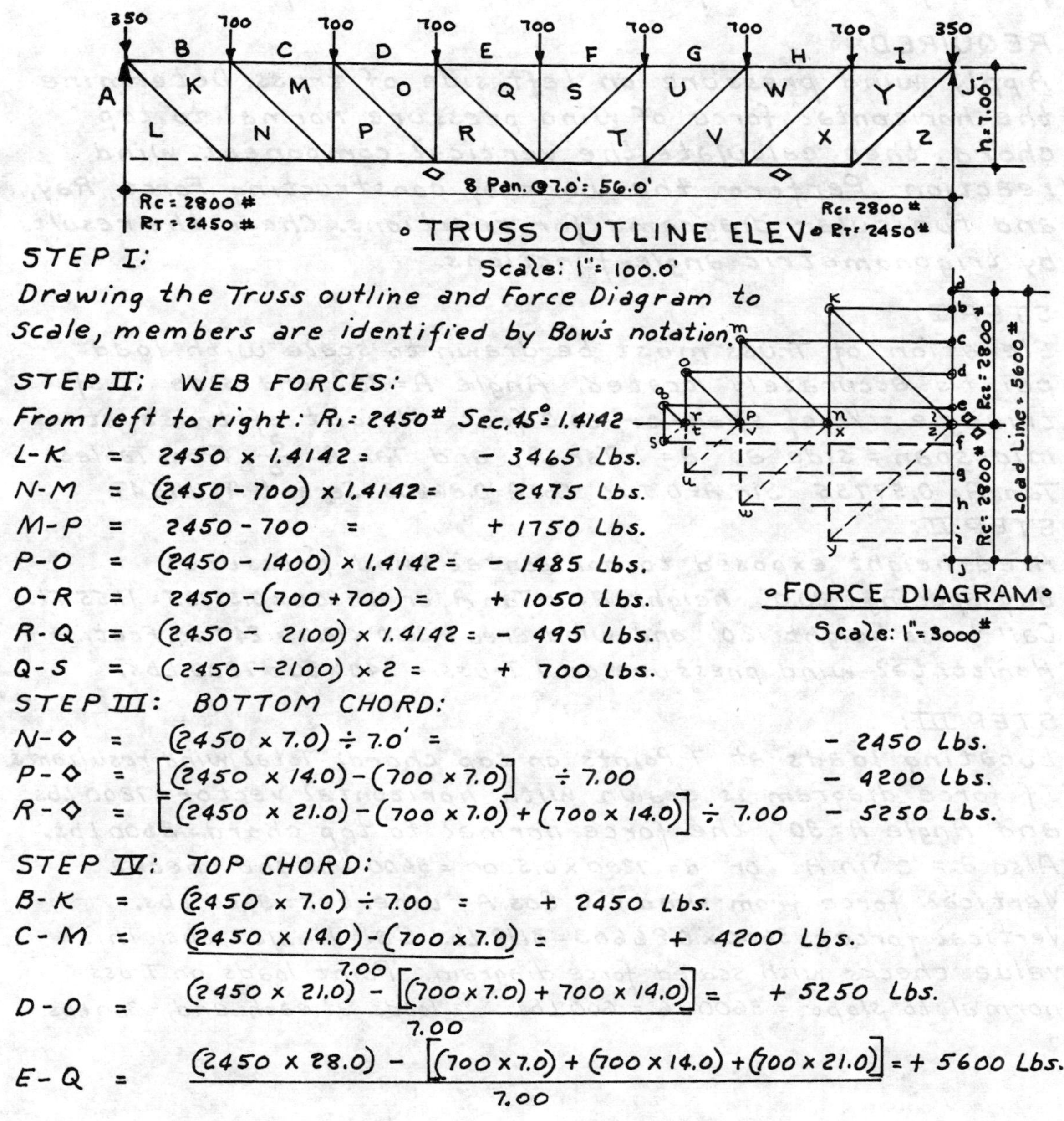

STEP I:

Drawing the Truss outline and Force Diagram to scale, members are identified by Bow's notation.

STEP II: WEB FORCES:

From left to right: $R_1 = 2450^{\#}$ Sec. $45° = 1.4142$

L-K = 2450 × 1.4142 = − 3465 Lbs.

N-M = (2450 − 700) × 1.4142 = − 2475 Lbs.

M-P = 2450 − 700 = + 1750 Lbs.

P-O = (2450 − 1400) × 1.4142 = − 1485 Lbs.

O-R = 2450 − (700 + 700) = + 1050 Lbs.

R-Q = (2450 − 2100) × 1.4142 = − 495 Lbs.

Q-S = (2450 − 2100) × 2 = + 700 Lbs.

STEP III: BOTTOM CHORD:

N-◇ = (2450 × 7.0) ÷ 7.0' = − 2450 Lbs.

P-◇ = [(2450 × 14.0) − (700 × 7.0)] ÷ 7.00 − 4200 Lbs.

R-◇ = [(2450 × 21.0) − (700 × 7.0) + (700 × 14.0)] ÷ 7.00 − 5250 Lbs.

STEP IV: TOP CHORD:

B-K = (2450 × 7.0) ÷ 7.00 = + 2450 Lbs.

$$C\text{-}M = \frac{(2450 \times 14.0) - (700 \times 7.0)}{7.00} = +\ 4200 \text{ Lbs.}$$

$$D\text{-}O = \frac{(2450 \times 21.0) - [(700 \times 7.0) + 700 \times 14.0]}{7.00} = +\ 5250 \text{ Lbs.}$$

$$E\text{-}Q = \frac{(2450 \times 28.0) - [(700 \times 7.0) + (700 \times 14.0) + (700 \times 21.0)]}{7.00} = +\ 5600 \text{ Lbs.}$$

EXAMPLE: Wind load on a sloped truss 5.3.4.5

A Truss with fixed end supports has a span of 40.0 Feet. Roof pitch of top chord = 30 degrees with horizontal plane. Top chord is divided into 6 Panels giving 7 points to apply the loads. Code requires the design wind pressure to be not less than 30 Lbs. square foot and acting in horizontal plane full height of truss. Truss bays are spaced 20.0 Ft. c-c.

REQUIRED:

Apply wind pressure on Left side of truss. Determine the horizontal force of wind pressure normal to top chord, then calculate the vertical component wind reaction. Perform the work by constructing Force, Ray, and Funicular diagrams for reactions. Check the results by trigonometric angle functions.

STEP I:

Elevation of Truss must be drawn to scale with load points accurately located. Angle A = 30° and side b of triangle = ½ of span or 20.0 feet. Height of truss at mid-span = side a. a = b Tan A, and Tan. A = $\frac{a}{b}$. From Tables:

Tan. A = 0.57735 Sin. A = 0.5000 Cos. A = 0.86603 Secant A = 1.1547

STEP II:

Area height exposed to horizontal wind pressure:

Bay spacing = 20.0' height a = b Tan A, or a = 20.0 × 0.57735 = 11.55 Ft.

Call truss height 12.0' and wind area = 20.0 × 12.0 = 240 Sq. Feet.

Horizontal wind pressure on 1 Truss = 240 × 30 = 7200 Lbs.

STEP III:

Locating loads at 7 Points on top chord: Total wind resultants. If force diagram is drawn with horizontal vector = 7200 Lbs., and Angle A = 30°, the force normal to top chord = 3600 Lbs. Also a = c Sin A. or a = 7200 × 0.5000 = 3600 Lbs. and checks. Vertical force from wind = c Cos. A, where c = 3600 Lbs. Vertical force = 3600 × 0.86603 = 3118 Lbs. Call it 3120 Lbs. This value checks with scaled force diagram. Point loads on Truss normal to slope = 3600 ÷ 6 = 600 Lbs. ½ loads at each end, = 300 Lbs.

EXAMPLE: Wind load on a sloped truss, continued 5.3.4.5

STEP IV:

Horizontal load at each point = 1200 Lbs., with 600 Lbs. at each end. Vertical load at each point = 520 Lbs., with 260 Lbs. at each end. These are the components of horizontal forces.

By drawing Ray Diagram to scale, the load line is based on the forces normal to slope and therefore the same action plane must be used. Polar point may be located at any convenient location. Closing string of Funicular Polygon is carried over to Ray Diagram to intersect load line and produce Reactions R_1 and R_2.

STEP V:

Resultant of Force 3600 Lbs., normal to slope is at the middle of Top chord slope length. This was confirmed by lines O-a and O-h.

The normal force resultant also cuts the bottom chord at 1/3 of span. The vertical Force component and its reactions are also calculated thus, with the greater reaction on the windward side which is the left. Like reactions are reversed when wind direction changes and is applied to right side.

STEP VI:

Calculating Reactions in Vertical Plane and Normal to top chord of Truss:

Span L = 40.0 Load Resultant point from left end = $\frac{40.0}{3}$ = 13.33'

Taking Moments about R_2 to solve for R_1:

Vertical Load = 3120 Lbs. Normal load = 3600 Lbs.

R_1 Vertical = $\frac{3120 \times 26.67}{40.0}$ = 2080 Lbs. R_2 = 3120 − 2080 = 1040 Lbs.

R_1 Normal = $\frac{3600 \times 26.67}{40.0}$ = 2400 Lbs. R_2 = 3600 − 2400 = 1200 Lbs.

DESIGNER'S NOTE:

Arrows in Force Diagrams indicate the direction of force.

EXAMPLE: Wind load on a sloped truss, continued 5.3.4.5

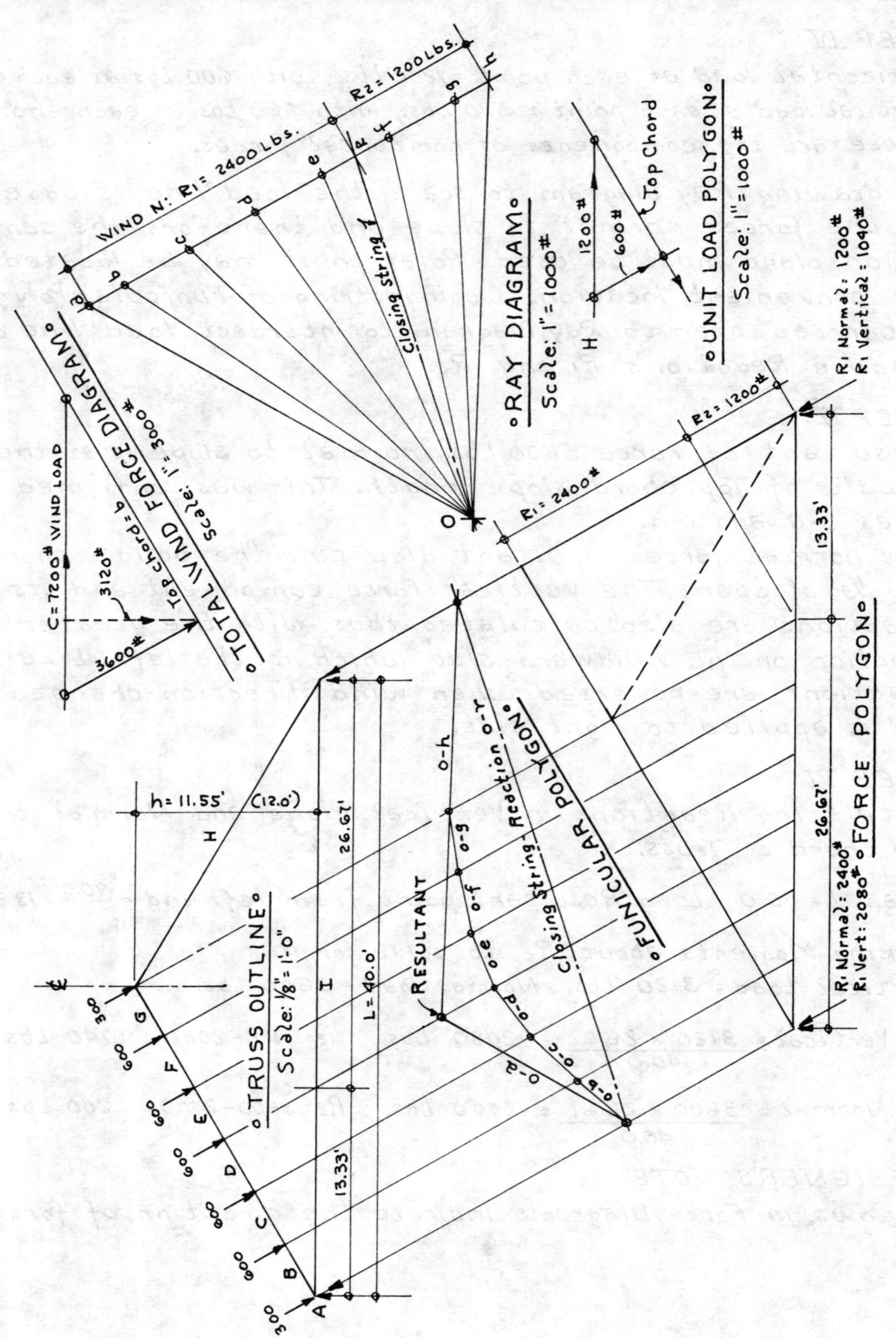

EXAMPLE: Welding machine tower analysis 5.3.4.6

A Steel Tower Rack is to support a battery of Electric Welding Machines near an outfitting dock in a shipyard. Rack is to have 4 Tiers as illustrated by the preliminary sketch shown herein. Vertical Load Reaction on each Tier is 2000 Lbs. Lateral forces from Wind and Cable Pull may come from either direction, therefore forces in diagonal braces may change from Tension to Compressive stress. When framed structures are placed in 12.0 Foot Bay spacing, the estimated lateral load is 650 Lbs. per tier. Panel heights are 9.0 Foot each in height, and Width of Tower is also 9.0 Feet.

REQUIRED:

(a) Determine Vertical Reaction resulting from horizontal loads. Calculate the Tipping Moment.

(b.) Combine all vertical Reactions and note on end elevation.

(c) Neglect the vertical loads and calculate the forces for each component resulting from lateral loads only. Use Graphic System, then check work by Trigonometry.

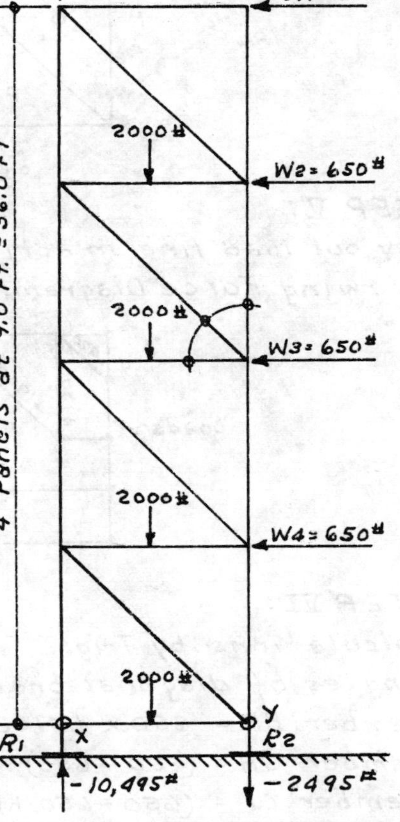

END ELEVATION

STEP I:

Vertical Loads = 8000 Lbs., and the reactions from these loads = 4000 Lbs Forces act down.

STEP II:

From horizontal forces, the vertical reaction at point y, will be up a in a counter-clock wise direction about point x. Then for R_2, take moments about R_1.

Horizontal Shear at Base:

$V = 4 \times 650 = 2600$ Lbs.

STEP III:

Horizontal Force Reactions:

$$R_2 = \frac{(650 \times 9.0) + (650 \times 18.0) + (650 \times 27.0) + (650 \times 36.0)}{9.0} = -6495 \text{ Lbs}$$

EXAMPLE: Welding machine tower analysis, continued 5.3.4.6

Step III Continued:
With Vertical Load Reacting down, and the Lateral Load forces counter-acting at Point y, the vertical Reaction,
$R_2 = {}^{+}4000 - 6495 = -2495$ Lbs.
Forces at Point x, all act downward, then:
$R_1 = {}^{+}4000 + 6495 = +10{,}495$ Lbs.

STEP IV:
To simplify the analysis by Stress Diagram, treat this structure as a Cantilever Truss. Turn Rack 90° left to a horizontal position and draw elevation. Make Bow's notation for member identification.

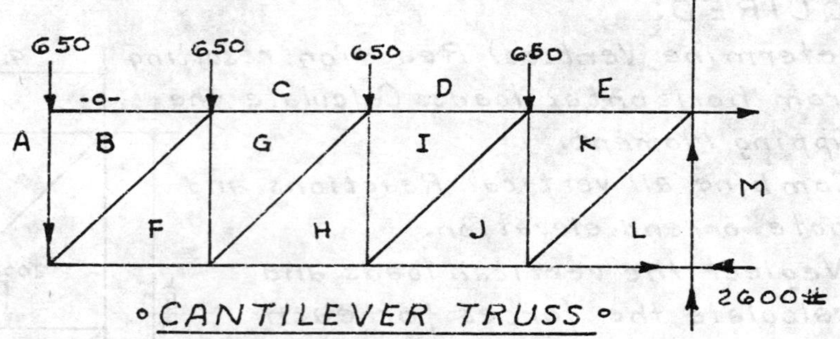

°CANTILEVER TRUSS°
Scale: 1.0" = 5.0'

STEP V:
Lay out load line in action of Loads. Length = 2600 Lbs.
Drawing Force Diagram:

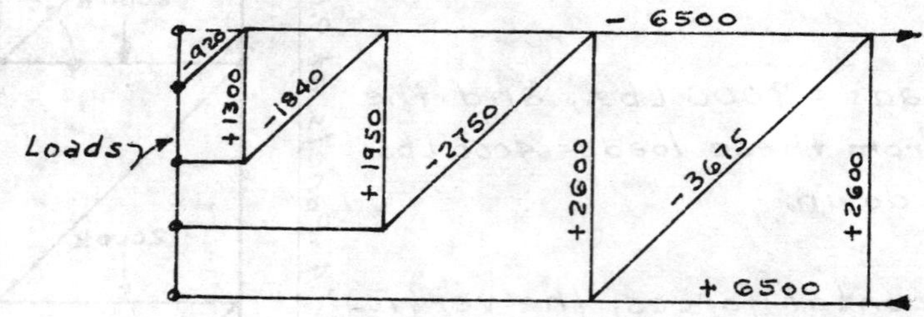

°FORCE DIAGRAM°
Scale: 1.0" = 2000#

STEP VI:
Calculations by Trig:
Angles of diagonal braces is 45°, and Secant = 1.4142

Member BF = 650 x 1.4142 = - 919.25 Lbs.
Member GH = (650 + 650) x 1.4142 = - 1838.45 "
Member IJ = (650 + 650 + 650) x 1.4142 = - 2757.70 "
Member KL = 2600 x 1.4142 = - 3676.90 "

EXAMPLE: Graphic analysis of simple span beam 5.3.4.7

A simple span beam with a length of 25.0 Feet between the supports must sustain three concentrated loads as follows:

Load P_1 = 10,000 Lbs. and located 4.833 Feet from left end.
Load P_2 = 6,000 " " " 11.667 " " " "
Load P_3 = 5,000 " " " 17.750 " " " "

Neglect weight of beam in calculations.

REQUIRED:

(a). Compute the mechanics of beam with the Algebraic method as delineated in Section I on Mechanics.

(b). Make Elevation drawing of Beam with loads, and show Shear Diagram. Construct a Ray Diagram, then make the Funicular Polygon to solve for Reactions and Bending Moments.

STEP I:

Drawing Elevation of Beam to scale with load and end line in same vertical plane of force action. Shear Diagram, and Funicular Polygon will be constructed with these lines.

STEP II:

Total Loads = 10,000 + 6,000 + 5,000 = 21,000 Lbs. L = 25.0 Ft.

$$R_1 = \frac{(5000 \times 7.25) + (6000 \times 13.33) + (10,000 \times 20.167)}{25.0'} = 12,716 \text{ Lbs.}$$

$$R_2 = \frac{(10,000 \times 4.833) + (6000 \times 11.667) + (5000 \times 17.750)}{25.0} = 8,284 \text{ Lbs.}$$

STEP III

After drawing shear diagram, Maximum bending moment appears to be under load P_2, and 11.667 Ft. from left end.

Mom. 4.833 = 12,716 × 4.833 = + 61,455 Ft. Lbs.

Mom. 11.667 = (12,716 × 11.667) − (10,000 × 6.833) = + 80,060 Ft. Lbs.

Mom. 17.750 = (12,716 × 17.750) − [(6000 × 6.083) + (10,000 × 12.916)] = +60,000 Ft. Lbs.

STEP IV:

Laying out Load Line for Ray Diagram with the given loads: Line of action is Vertical, hence, the load line is vertical. Polar Point O, will be selected at a point which will be convenient for constructing Funicular Polygon below the shear diagram. By scaling pole distance from load line, it measures 20,000 Lbs. The 3 Co-ordinate lines in the Polygon are moment arms. Pole value times arm = Moment.

EXAMPLE: Graphic analysis of simple span beam, continued 5.3.4.7

Step IV: Continued.
Co-ordinates within Funicular Polygon must be measured for length by using same scale for which Elevation was drawn.

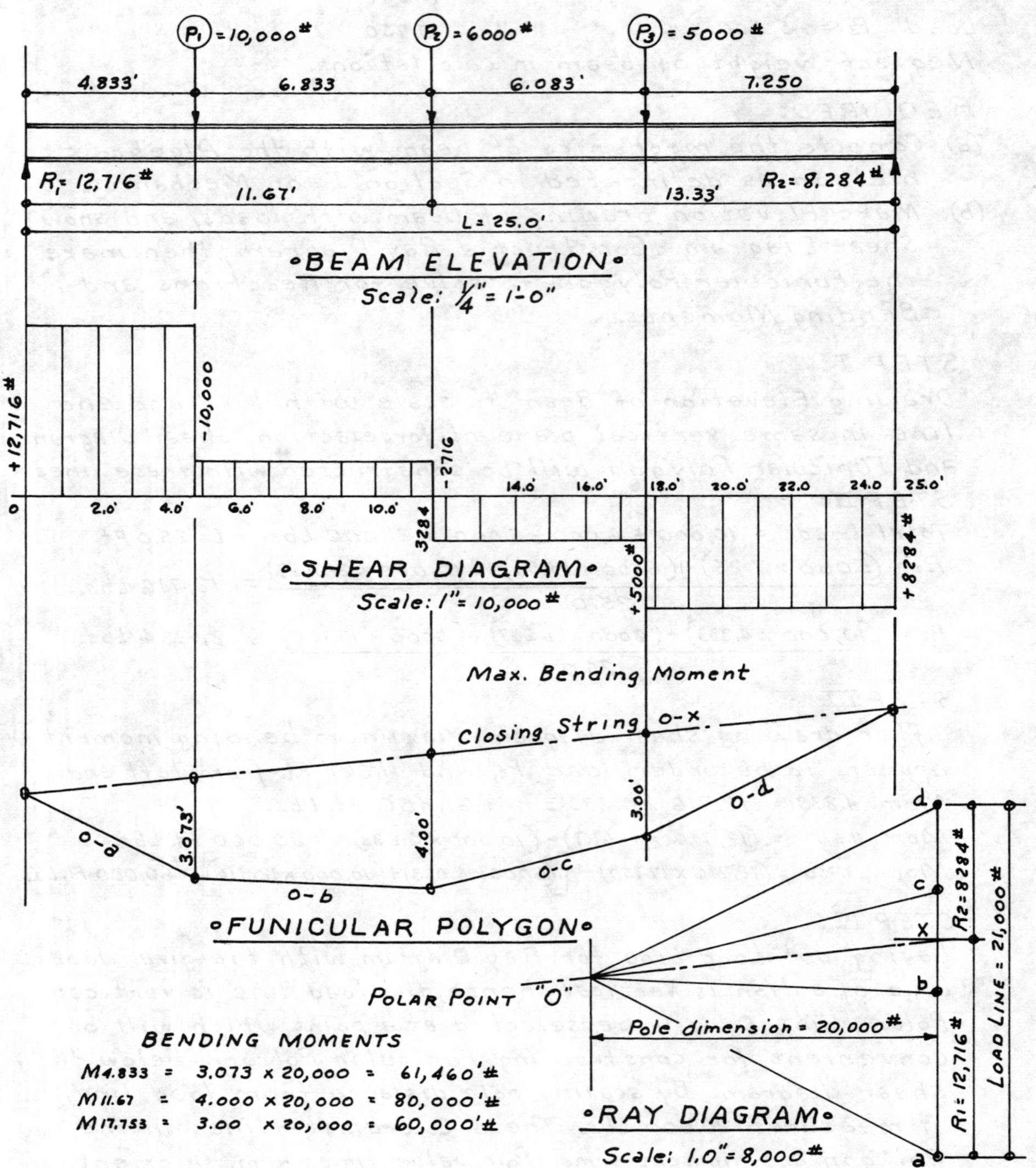

EXAMPLE: Graphic analysis of cantilever beam 5.3.4.8

A beam 22.0 feet in length with 17.0 foot spacing for supports, is cantilevered 5.0 feet at right ends. Concentrated Loads are placed as follows:
From Left ends, P_1 = 450 Lbs. at 5.0 feet. P_2 = 920 Lbs. at 12.0 feet, and P_3 = 630 Lbs., at extreme right end.

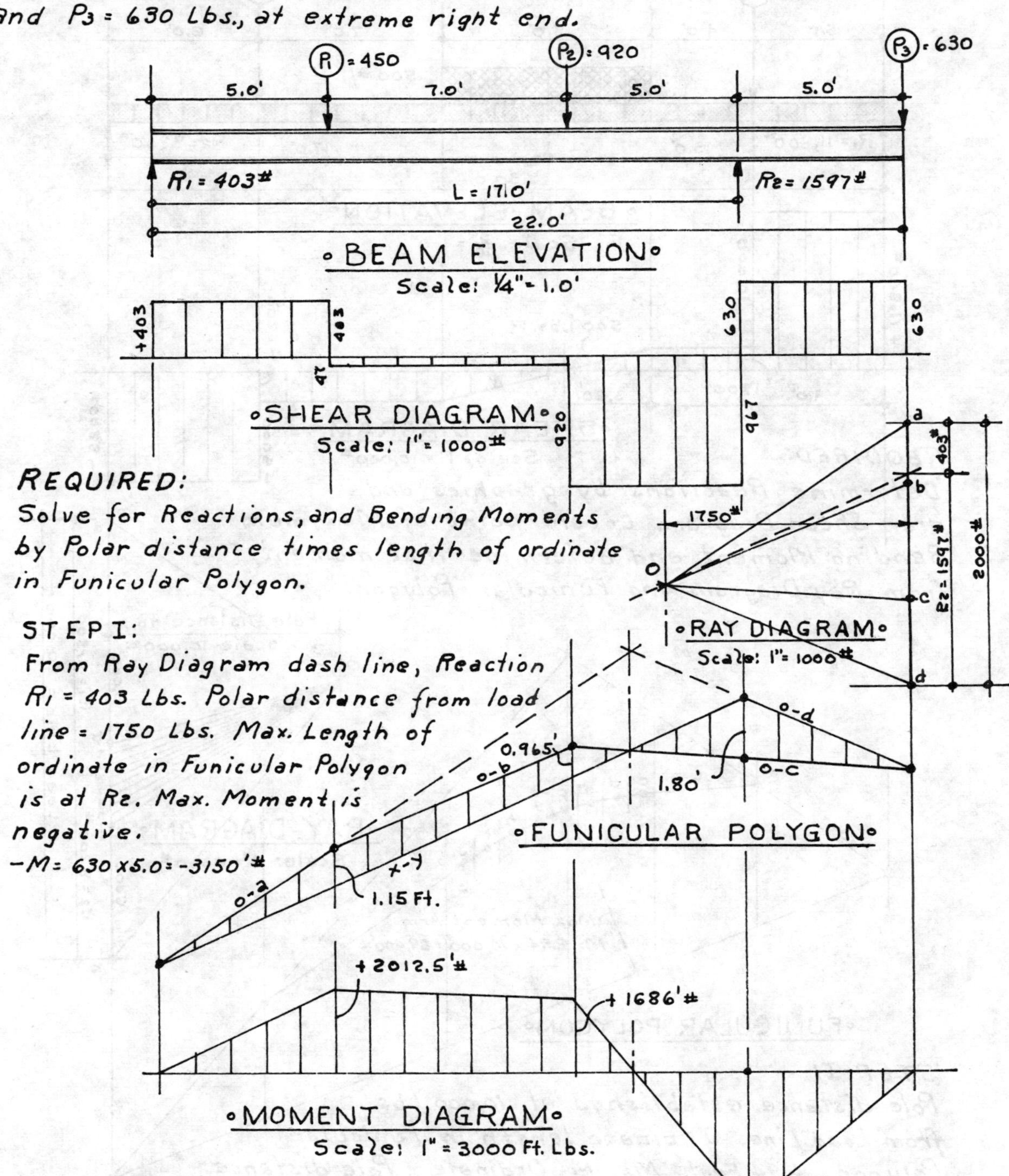

REQUIRED:
Solve for Reactions, and Bending Moments by Polar distance times length of ordinate in Funicular Polygon.

STEP I:
From Ray Diagram dash line, Reaction R_1 = 403 Lbs. Polar distance from load line = 1750 Lbs. Max. Length of ordinate in Funicular Polygon is at R_2. Max. Moment is negative.
$-M$ = 630 x 5.0 = -3150'#

EXAMPLE: Graphic analysis of beam with combined loads 5.3.4.9

A Simple 30.0 foot span beam is to support 3 Loads which total 20,000 Pounds, and are located as illustrated on elevation. Concentrated Loads total 16,000 Lbs., and Uniform load = 500 Lbs. per foot for 8.0 Feet.

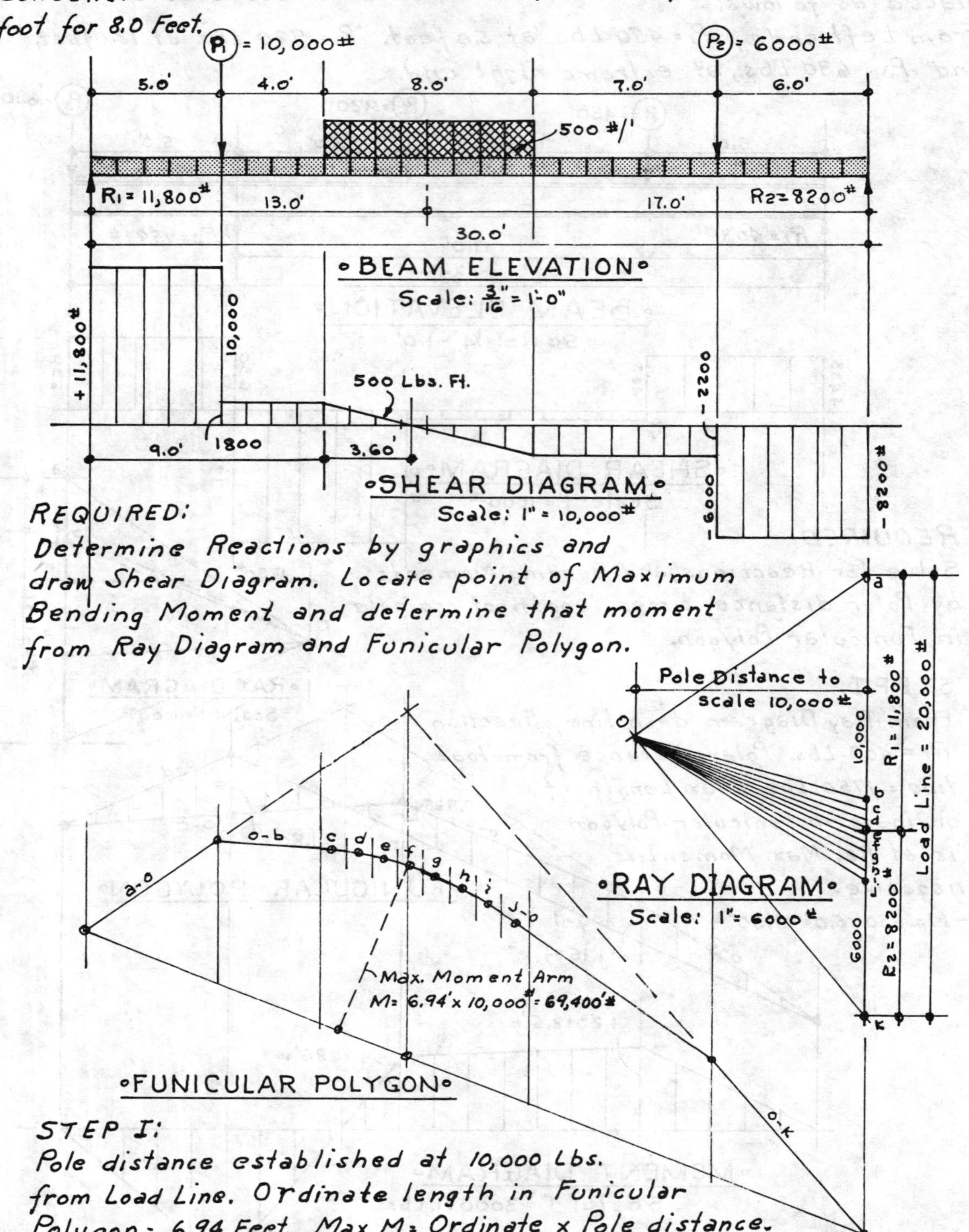

REQUIRED:
Determine Reactions by graphics and draw Shear Diagram. Locate point of Maximum Bending Moment and determine that moment from Ray Diagram and Funicular Polygon.

STEP I:
Pole distance established at 10,000 Lbs. from Load Line. Ordinate length in Funicular Polygon = 6.94 Feet. Max. M = Ordinate x Pole distance.

EXAMPLE: Analysis of timber conveyer truss 5.3.4.10

Five (5) Truss Sections of 80.0 Feet are to support a log conveyer extending 400.0 Feet on horizontal. Total Rise at end is 80.0 Feet. Width of Conveyer is 5.0 Feet, and each truss will have a depth of 8.0 Feet.
Dead Loads plus Live Timber Load is estimated to be a maximum of 1000 Lbs. per lineal foot for each side truss.

REQUIRED:

Prepare a drawing for a single 80.0 Foot Section of Conveyer after determining the slope. All Load Forces are to be in the Vertical Plane of action, except that the whole assembly will produce a force parallel to slope. Neglect for the time being, any bracing of Column supports, and calculate the force on slope at ground level for all 5 Sections.

STEP I:

To get a clear picture of Conveyer, layout the 5 Section arrangement and determine angle the truss makes with the horizontal plane.

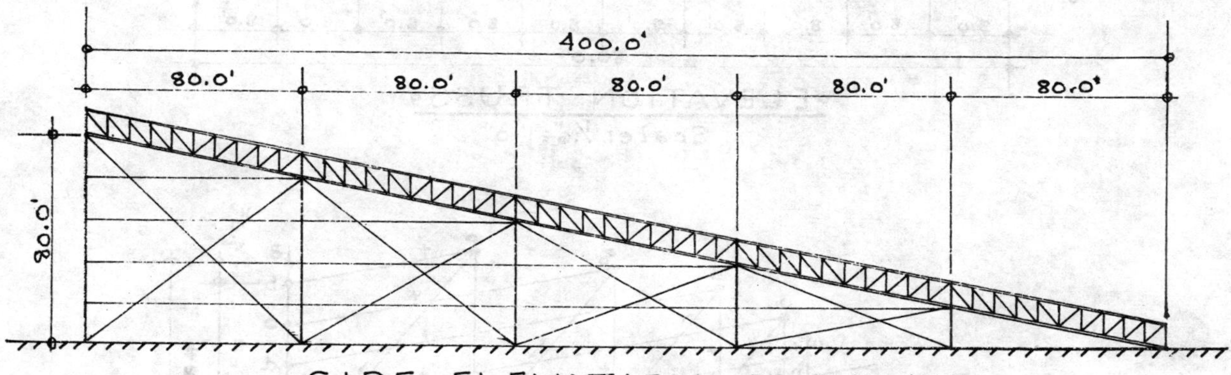

• SIDE ELEVATION CONVEYER •

Scale: 1.0" = 120.0 Ft.

STEP II:

Selecting a Howe Type Truss with 10 Panel Points for loads.
Total Load on 1 Section 80.0 Ft. long = 1000 × 80.0 = 80,000 Lbs.

End Panels: Loads =	2 @ 4000 =	8,000 Lbs
Interior Point Loads :	9 @ 8000 =	72,000 "
	Total =	80,000 Lbs.

Reaction at each end = 40,000 Lbs.

EXAMPLE: Analysis of timber conveyer truss, continued 5.3.4.10

STEP III:

Determine angle and functions Truss makes with horizontal.

Total Rise = 80.0 Ft. (a). Rise for each Section = 16.0 Ft.

$\text{Tan. } A = \dfrac{16.0}{80.0} = 0.2000$ Angle A = 11° 19' From Tables:

Sin. A = 0.19623 Cos. A = 0.98056 Sec. A = 1.0198

Total Length Slope = 1.0198 x 400 = 407.92 Ft.

Slope length of 1 Section = 1.0198 x 80.0 = 81.584 Ft. (Bottom Chord)

STEP IV:

Drawing 1 Truss Section elevation and Force Diagram:

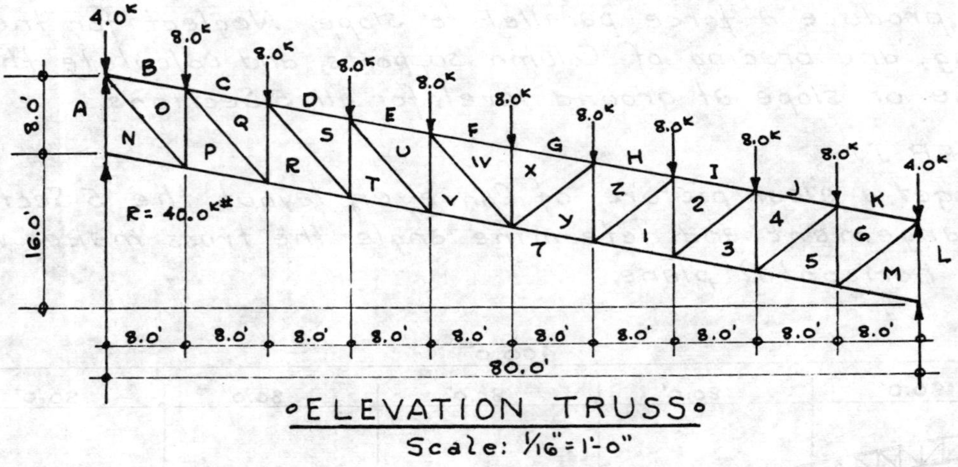

∘ELEVATION TRUSS∘

Scale: 1/16" = 1'-0"

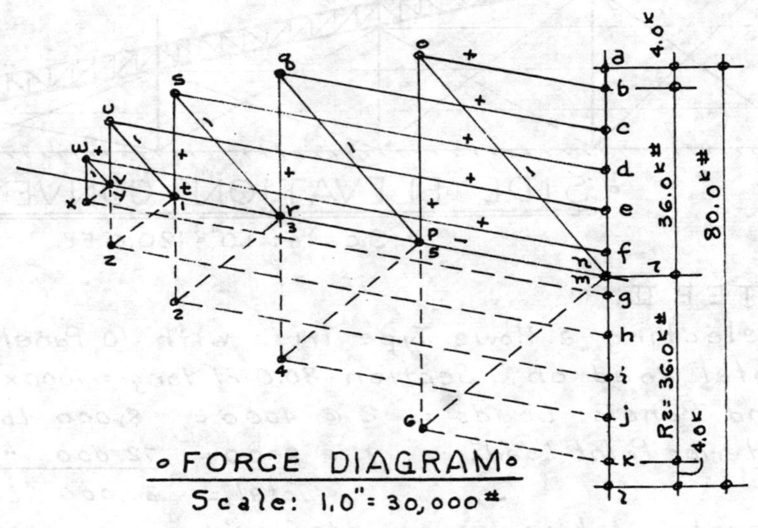

∘FORCE DIAGRAM∘

Scale: 1.0" = 30,000#

EXAMPLE: Analysis of timber conveyer truss, continued 5.3.4.10

STEP V:

Draw force Polygon to obtain force from loads on plane parallel to Truss Chords, and force horizontal which will produce bending in supporting Columns. Known force is 80,000 Lbs. in Vertical plane of action.

Drawing vertical load line of 80.0 kip Lbs. with 1" = 30,000 Lbs.

Let N = Force normal to slope, perpendicular to Chord.

Let P = Force parallel to Slope of Top Chord.

Let H = Force horizontal to Ground.

A right angle is formed at intersect of P and N, with angle at H being 11° 19'.

Then V = side c, and N = side b. b = c Cos. A

P = Side a, and a = c Sin A.

Force Normal, N = 80,000 × 0.98056 = 78,445 Lbs.

Force Parallel P = 80,000 × 0.19623 = 15,700 "

Force Horizontal, H = N Sine A.

H = 78,445 × 0.19623 = 15,390 Lbs.

Values check with diagram.

STEP VI:

Scale Force diagram and tabulate stresses for designing.

SUMMARY OF TRUSS STRESSES

MEMBER		LOCATION	STRESS	TYPE
B-O	K-6	TOP CHORD	35,000 LBS.	COMPRESSIVE
C-Q	J-4	" "	62,000 "	"
D-S	I-2	" "	81,500 "	"
E-U	H-Z	" "	92,500 "	"
F-W	G-X	" "	97,000 "	"
N-O	6-M	BOT. CHORD	55,000 "	TENSION
P-7	5-7	" "	55,000 "	"
R-7	3-7	" "	62,000 "	"
T-7	1-7	" "	81,500 "	"
V-7	Y-7	" "	92,500 "	"
P-Q	4-5	DIAGONALS	41,500 "	TENSION
R-S	2-3	"	30,000 "	"
U-T	Z-1	"	17,500 "	"
W-V	X-Y	"	11,500 "	"
O-P	5-6	VERTICALS	36,000 "	COMPRESSIVE
Q-R	3-4	"	28,000 "	"
S-T	1-2	"	20,000 "	"
U-V	Z-Y	"	12,000 "	"
W-X		"	8,000 "	"

EXAMPLE: Graphic analysis of king-post truss 5.3.4.11

A steel storage building with trusses 14.0 feet long and spaced 10.0 feet on centers is to be converted to other uses and require a 2 Ton capacity hoist. The proposed monorail is to be a W 8x13 Section welded to bottom chord at truss mid-span and run full length of building. Outline of truss shows a roof pitch of 1/5 and 4 panels wide with 3 load points. Truss is a King-Post type. Design vertical loads are 36 Lbs. per square foot and steel is A-36 with welded gusset plates 5/16 inches thick. Connections to columns at end have 6 3/4"ϕ A-307 Bolts. Columns are 4x4 M 13.0 with axis x-x unbraced a length of 9.50 feet. Wind load is to be neglected.

Members in truss are as follows.

Top Chord = 2 ⅂⅂ 2½x2x5/16 with long legs vertical.

Bot. Chord = 2 ⅃L 2½x2x5/16 with short legs vertical.

Web member at mid-span - bottom chord to ridge is: 1-L 2x2x5/16"

All other web members are 2 ⅃L 2x2x¼ welded to gussets.

REQUIRED:

Investigate truss, end connections to column to determine what changes or additions must be made to make the structure safe with the hoist load on bottom chord. Add 15% to hoist load to sustain impact force. Analyze truss by using graphic force diagrams and identify members by markings with Bow's notations. Column investigation is not required.

STEP I:

Drawing the truss elevation accurately to scale, the height is L/5 or 1/5 of 14.0 = 2.80 feet. There are 5 load points and only 3 point loads produce stress in truss. The other 2 points are plumb with column.

STEP II:

Load on 1 Truss with 10.0 foot Bay spacing. L = 14.0 Ft., and design roof load is 36 PSF. Load = 10.0 x 14.0 x 36 = 5040 Lbs.

With 4 Panels, point loads = 5040/4 = 1260 Lbs. Load plumb with Column = 630 and 3 loads on truss will be 1260 x 3 = 3780 Lbs. and length of effective load line. These loads and Bow's notation are now noted on elevation drawing.

Compressive stress is indicated by plus sign (+).

Tension stress in drawings is indicated by minus sign (-).

EXAMPLE: Graphic analysis of king-post truss, continued 5.3.4.11

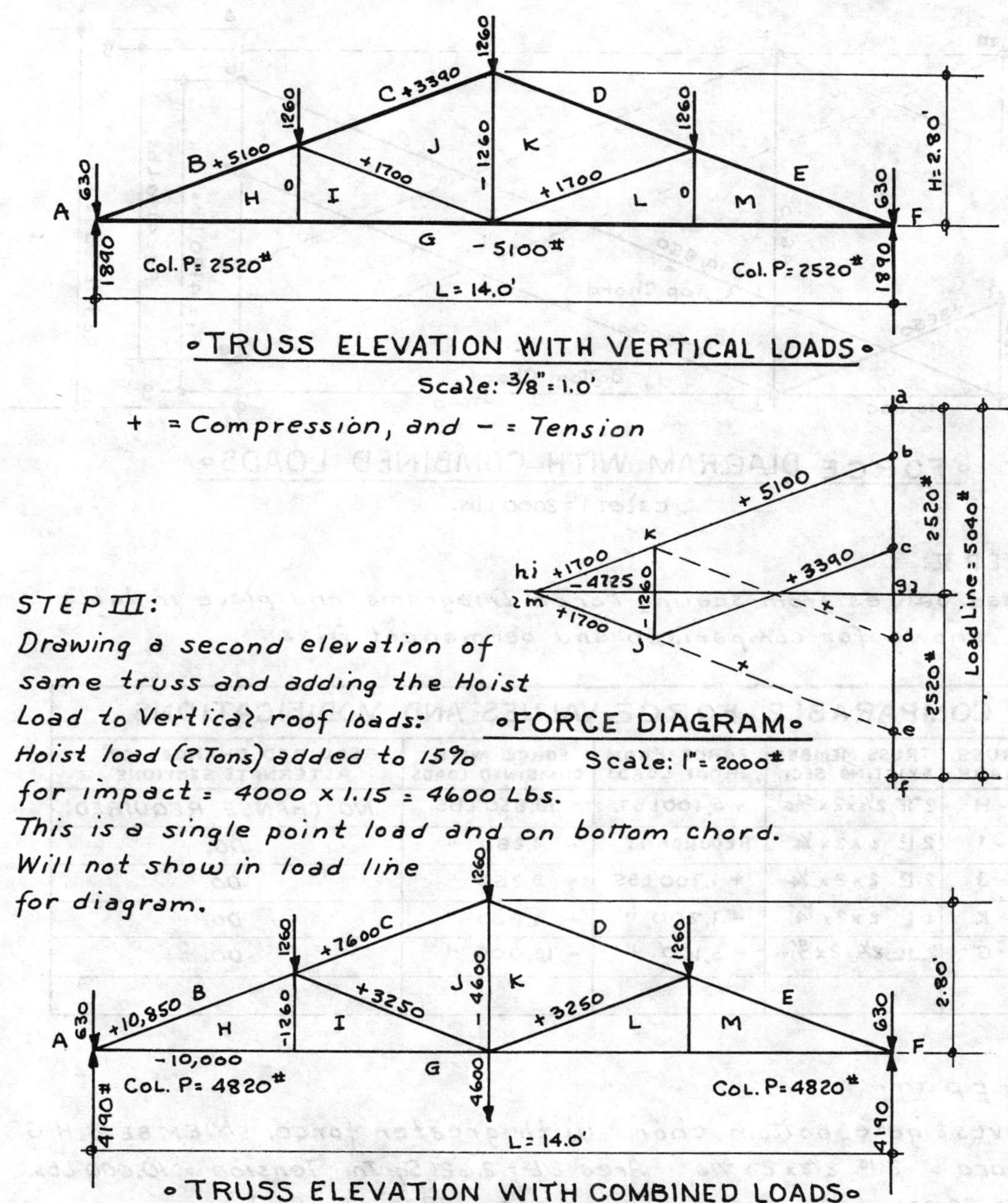

STEP III:
Drawing a second elevation of same truss and adding the Hoist Load to Vertical roof loads:
Hoist load (2 Tons) added to 15% for impact = 4000 × 1.15 = 4600 Lbs.
This is a single point load and on bottom chord.
Will not show in load line for diagram.

STEP IV:
The load line is drawn to scale similar to vertical roof loads. The reaction effective in truss is now 4190 Lbs., and position of bottom chord is 4190 Pounds from point b on load line. Stress diagram is drawn as illustrated.

EXAMPLE: Graphic analysis of king-post truss, continued 5.3.4.11

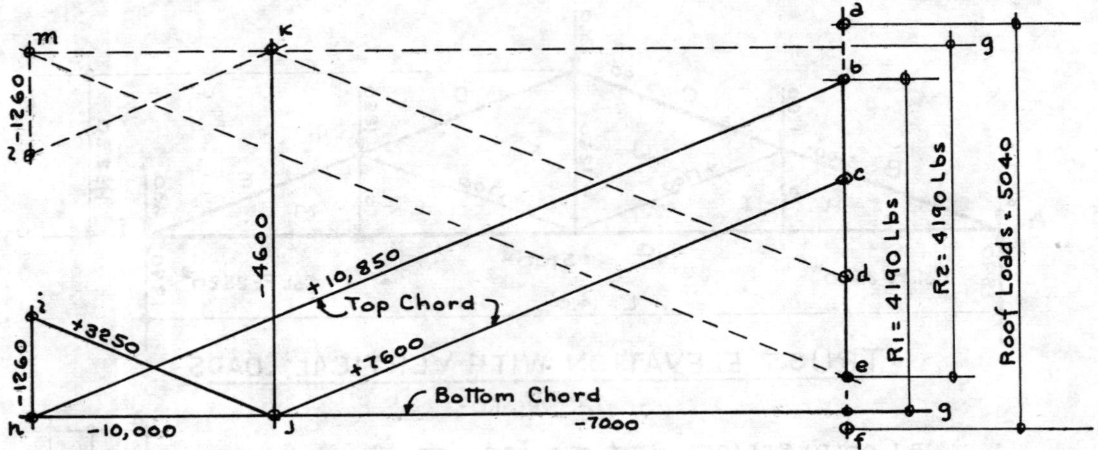

FORCE DIAGRAM WITH COMBINED LOADS

Scale: 1" = 2000 Lbs.

STEP V:

Read values from scaling Force Diagrams and place in table as shown for comparison and permanent files:

COMPARABLE FORCE VALUES AND MODIFICATIONS				
TRUSS MARK	TRUSS MEMBER EXISTING SECT.	FORCE FROM ROOF LOADS	FORCE WITH COMBINED LOADS	REQUIRED CHANGES OR ALTERNATE SECTIONS
B-H	2TT 2½ x 2 x 5/16	+ 5,100 LBS.	+ 10,850 LBS.	NO CHANGE REQUIRED
H-I	2 Ls 2 x 2 x ¼	REDUNDANT	- 1,260 "	DO.
I-J	2 Ls 2 x 2 x ¼	+ 1,700 LBS.	+ 3,250 "	DO.
J-K	1 L 2 x 2 x ¼	- 1,200 "	- 4,600 "	DO.
H-G	2 ⅃L 2½ x 2 x 5/16	- 5,100 "	- 10,000 "	DO.

STEP VI:

Investigate bottom chord with greater force: MEMBER H-G.

Chord = 2 Ls 2½ x 2 x 5/16 Area 2 Ls = 2.62 Sq. In. Tension = 10,000 Lbs.

F_t = 20,000 PSI. Actual $f_t = \frac{10,000}{2.62}$ = 3820 Lbs. Sq. In.

Bottom Chords are satisfactory with weight of hoist load added to present roof loads.

EXAMPLE: Graphic analysis of king-post truss, continued 5.3.4.11

STEP VII:
Check Top Chord Member B-H:
Section = 2 L^s $2\frac{1}{2} \times 2 \times 5/16$ with long legs back to back and 5/16 inch space between. Length unsupported as scaled from elevation = 3.75 feet. Compression force P= 10,850 Lbs. Figure as a Column. From AISC Manual of angle struts: A = 2.62 □"
r_{x-x} = 0.78 and r_y = 0.93 l = 3.75 x 12 = 45.0 inches. Slenderness ratio = l ÷ least radius of gyration (r).
Ratio = 45.0/0.78 = 57.70
Actual stress $f_a = P/A$ or $f_a = \frac{10,850}{2.62}$ = 4,150 Lbs. sq. inch.
From Table of allowable axial compressive stresses in Steel Columns Section II, the allowable F_a = 15,400 PSI. The bottom truss chord is satisfactory with hoist load added.

STEP VIII
Member H-I and L-M are redundant under roof loads. With Hoist load added, Tension = 1260 Lbs. Member H-I = 2L^s 2x2x¼, and 2 Angle Area = 1.88 □" f_t = 1260/1.88 = 670 PSI. With F_t = 20,000 PSI, the members are acceptable without change.

STEP IX:
Members I-J and K-L: Scaled length = L = 3.83' l = 3.83 x 12 = 46.0 In.
Space between L^s = 5/16. Then r_x = 0.61 and r_y = 0.96 A = 1.88 □"
Slenderness ratio: l/r = 46.0/0.61 = 75.5 Allowable F_a = 14,200 PSI.
Actual $f_a = P/A$ and P = 3250 Lbs. $f_a = \frac{3250}{1.88}$ = 1730 PSI.

Members I-J and K-L are acceptable without modification.
STEP X:
Member J-K at Mis
Tension = 4600 Lbs. Section = 1 L 2x2x¼ Area = 0.94 □" F_t = 20,000 PSI.
Actual f_t = 4600/0.94 = 4900 PSI. No change required for J-K.

STEP XI:
Check shear on Bolts at end connections.
Greatest reaction with end Purlins on top chord = 4820 Lbs.
Using 6-¾"φ Bolts in connection. Area 6-Bolts = 0.44 x 6 = 2.64 □"
In single shear, A-307 Bolt allowable F_v = 10,000 PSI.
Connection will safely sustain 2.64 x 10,000 = 26,400 Lbs. OK as is.
STEP XII:
Make note in Tabular form that no changes are necessary.

EXAMPLE: Graphic analysis of Fink truss 5.3.4.12

A Truss with a 64.0 foot span is illustrated with Roof Loads applied at panel points. Owner of building desires to attach a hoist with a 1 Ton capacity to the lower chord. Point of hoist location will be at intersection of web members and 1/3 span length from either support column.

REQUIRED:

Draw a new elevation of truss to scale with Hoist Load in location, then construct a force diagram. Design Steel web members for shop riveted fabrication. Gusset plates are to be cut from 3/8 inch plate. Run Top and Bottom chords in continuous length through entire truss. Prepare a form of tabulated forces and list selected shapes for fabrication.

STEP I:

Drawing a type of Fink Truss with applied loads and then constructing the load line and force diagram. Members are identified by using Bow's notations.

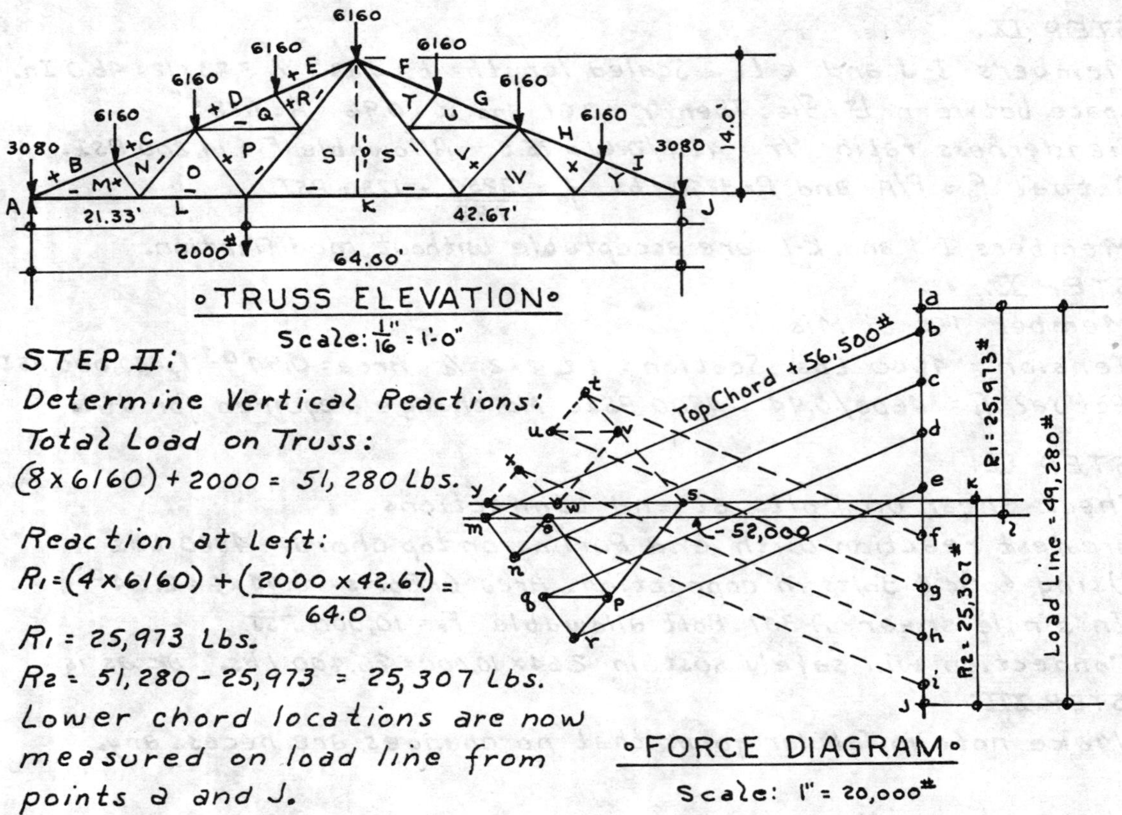

∘TRUSS ELEVATION∘
Scale: $\frac{1}{16}" = 1'\text{-}0"$

∘FORCE DIAGRAM∘
Scale: 1" = 20,000#

STEP II:

Determine Vertical Reactions:

Total Load on Truss:

$(8 \times 6160) + 2000 = 51{,}280$ Lbs.

Reaction at Left:

$R_1 = (4 \times 6160) + \dfrac{2000 \times 42.67}{64.0} =$

$R_1 = 25{,}973$ Lbs.

$R_2 = 51{,}280 - 25{,}973 = 25{,}307$ Lbs.

Lower chord locations are now measured on load line from points a and J.

EXAMPLE: Graphic analysis of Fink truss, continued 5.3.4.12

STEP III:

Force diagram is constructed to scale. Scale magnitude of force from diagram and tabulate in following format. Indicate type of stress in each member as: + = Compression and − = tension. Length of members are scaled from elevation.

FINK TRUSS - TABULATED FORCES FOR DESIGN				
MEMBER	LOCATION	LENGTH	FORCE #	FABRICATION
B − M	TOP CHORD	8.70'	+56,500	2 ⅃L 4" x 3" x 3/8"
M − L	BOT. CHORD	64.00'	−52,000	2 ⅃L 3" x 3" x 5/16"
M − N	WEB STRUT	4.60'	+ 5,750	1 L 2½ x 2½" x ¼"
N − O	Do.	8.80'	− 5,750	1 L 2½" x 2½" x ¼"
O − P	Do.	9.00'	+ 11,500	2 Ls 2" x 2" x ¼"
P − Q	Do.	10.20'	− 6,500	1 L 2½" x 2½" x ¼"
V − W	Do.	9.00'	+11,500	2 Ls 2" x 2" x ¼"
Q − R	Do.	4.60'	+ 5,750	1 L 2½" x 2½" x ¼"
R − S	Do.	8.50	− 20,000	2 Ls 2" x 2" x ¼"
P − S	Do.	9.00	− 14,000	Connect to R-S

STEP IV:

Designing Top Chord members BM and IY with greater force. Compressive Stress P = 56,500 Lbs. Length = 8.70 feet. Figure as a column with 2Ls back to back and 3/8 inch Gusset Plate space. $l = 8.70 \times 12 = 104.5$ inches. Slenderness ratio $\frac{l}{r}$ will control stress. Try 2Ls 4" x 3" x 3/8" A = 4.96□" Deduct area of 1 Rivet line with 7/8" ϕ holes. A□" hole deduct = 0.875 x 0.375 = 0.328□ Net area of 2 angles with holes = 4.96 − (2 x 0.328) = 4.304□"

From A.I.S.C. Manual - Struts with long legs vertical: $r_y = 1.31"$

Ratio $l/r = \frac{104.5}{1.31} = 79.8$ From allowable stress tables in Sect. II:

Allowable F_a = 13,900 PSI. Max. P = 4,304 x 13,900 = 59,825 Lbs. OK.

STEP V

Design for Lower Chord ML and YK: Run continuous 64.0 feet.

T = −52,000 Lbs. Allowable A36 Steel F_t = 22,000 PSI.

Net area required with 2 holes cut out = $\frac{52,000}{22,000}$ = 2.36 Sq. In.

Select 2Ls 3" x 3" x 5/16" Net A = 3.56 − 0.656 = 2.90

STEP VI:

Design Web strut MN and XY: Length = 4.50 feet. P = +5750 Lbs. Rivet hole need not be deducted for web compression members. The holes were deducted in top chord angles because of frequent bending force in lateral direction.

For struts MN and XY, select for trial a 2½ x 2½ x ¼ angle.

EXAMPLE: Graphic analysis of Fink truss, continued 5.3.4.12

From A.I.S.C. Manual: Area for $2\frac{1}{2} \times 2\frac{1}{2} \times \frac{1}{4}$ L = 1.19 Sq. In.
Area of 7/8" φ Hole cutout = 0.875 × 0.25 = 0.219 □" Call it 0.22 □"
$l = 4.50 \times 12 = 54.0$ inches. $r = 0.77$ $\frac{l}{r} = \frac{54.0}{0.77} = 70.1$
From Tables: $F_a = 14,620$ #□"
Max. P = 1.19 × 14,620 = 17,400 Lbs. Accept this single angle since fabrication is with rivets.

STEP VII:
Design of Web Struts N-O, and W-H. Tension T = −5750 Lbs.
Allowable $F_t = 22,000$ #□" Area Required = $\frac{5750}{22,000} = 0.261$ Sq. In.
Using 1 L $2\frac{1}{2} \times 2\frac{1}{2} \times \frac{1}{4}$ with rivet hole:
A = 1.19 − 0.219 = 0.971 Sq. In. Accept this angle.

STEP VIII:
Design of Web Struts, O-P and V-W. Length = 9.0 Feet.
$l = 9.0 \times 12 = 108.0$ Inches. Axial P = +11,500 Lbs.
Try 2 ⅃L $2\frac{1}{2}" \times 2\frac{1}{2}" \times \frac{1}{4}"$ Ls back to back with 3/8" spacing.
From Tables: 2 Ls have A = 2.38 □" and r = 1.19
$\frac{l}{r} = \frac{108.0}{1.19} = 90.7$ Allowable $F_a = 12,980$ #□"
Max. P = 2.38 × 12,980 = 30,900 Lbs. This is oversized, thus another trial should be made with smaller angles.
Try 2 Ls $2" \times 2" \times \frac{1}{4}"$ A = 1.88 □" $r = 0.99$ $\frac{l}{r} = \frac{108.0}{0.99} = 109.0$
Allowable Fa = 11,240 #□" Max. P = 1.88 × 11,240 = 21,100 Lbs.
Accept 2 ⅃L $2" \times 2" \times \frac{1}{4}"$.

STEP IX:
Refer to Tabulation Table: Member Q-R can be same as member M-N.
From Step VII, a $2\frac{1}{2} \times 2\frac{1}{2} \times \frac{1}{4}$ L with 7/8" φ hole, has Area = 0.971 □"
$F_t = 22,000$ #□" Then Max. T = 0.971 × 22,000 = 21,360 Lbs.
Use this for members P-S, and P-Q.

STEP X:
For lateral support of S-K, members O-P and V-W, were designed with 2 Angles. Better fabrication will be possible when members P-S and R-S, are in one length of about 18.0 Feet.
2 Angles $2" \times 2" \times \frac{1}{4}"$ have Area of 1.88 Sq. In.
Deducting for rivet holes in each angle, A = 1.88 − (2 × 0.219) = 1.42 □"
Tension is greater in R-S, = −20,000 Lbs. $F_t = 22,000$ P.S.I.
Then Max. T = 1.42 × 22,000 = 31,240 Lbs. Accept 2 Ls $2" \times 2" \times \frac{1}{4}"$

EXAMPLE: Graphic analysis of Pratt truss 5.3.4.13

This problem was submitted to applicants for State Architects Registration by the Board of Governors of Licensed Architects of Oklahoma, on November 30, 1945 as part of the structural phase of a nine part written test.

Identify the type of truss illustrated. The Roof and Wind Loads given are to be used and applicant shall work only with the data given. Truss span is 60.0 feet, and Roof Pitch is 30 degrees with horizontal.

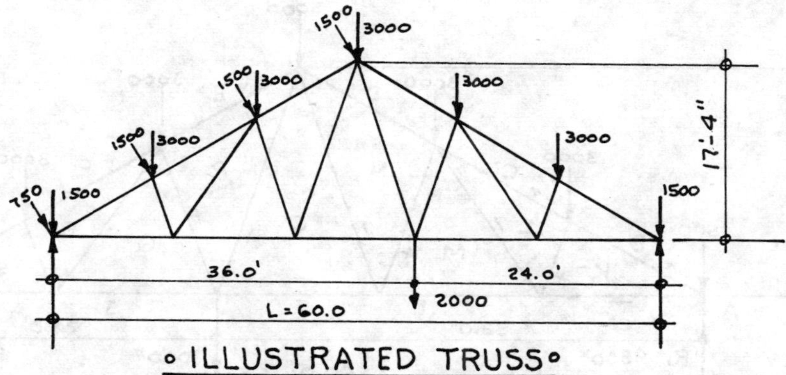

∘ ILLUSTRATED TRUSS ∘

REQUIRED:

Determine the total height of Truss and note on illustration. Construct a Force Diagram for the vertical loads, and another Force Diagram for the wind loads normal to Top Chord. Calculate the Vertical Reactions for Wind and Roof Loads separately. Show magnitude of forces and type stress on diagrams, but do not design the truss members.

STEP I:

Type of Truss a sloped Pratt design. Concentrated load is suspended from bottom chord.

STEP II:

Vertical Reactions for suspended load will be solved by the method of moments and added to uniform roof load Reactions. Roof Loads total 18,000 Lbs. $R_1 = R_2 = 9000$ Lbs. each.

For concentrated Load Reactions:

$R_1 = \frac{2000 \times 24.0}{60.0} = 800$ Lbs. $R_2 = \frac{2000 \times 36.0}{60.0} = 1200$ Lbs.

Combined Vertical Reactions:

$R_1 = 9000 + 800 = 9800$ Lbs. $R_2 = 9000 + 1200 = 10,200$ Lbs.

EXAMPLE: Graphic analysis of Pratt truss, continued 5.3.4.13

STEP III:

To determine Truss height: Side b of Triangle = ½ L or 30.0 Ft.
Angle A = 30 degrees. From Trig. Tables: Sine A = 0.5000
Tan A = 0.57735 and side a = b Tan. A. Then b = 30.0 x 0.57735 = 17.32 Ft.
or about 17'-4" as noted on illustrated drawing.

STEP IV:

Re-draw an accurate elevation to scale for use in constructing Force Diagram with Vertical Loads only.

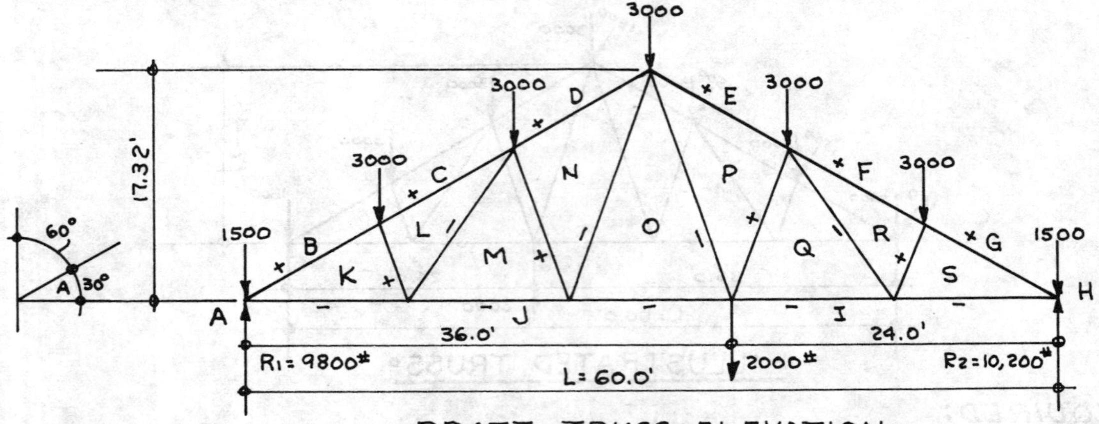

◦PRATT TRUSS ELEVATION◦
Scale: 1.0" = 12.0 Ft.

STEP V:

In constructing Force diagram, Truss members are identified with Bow's notations. Magnitude of force in each member has been determined by scaling the length of vectors.

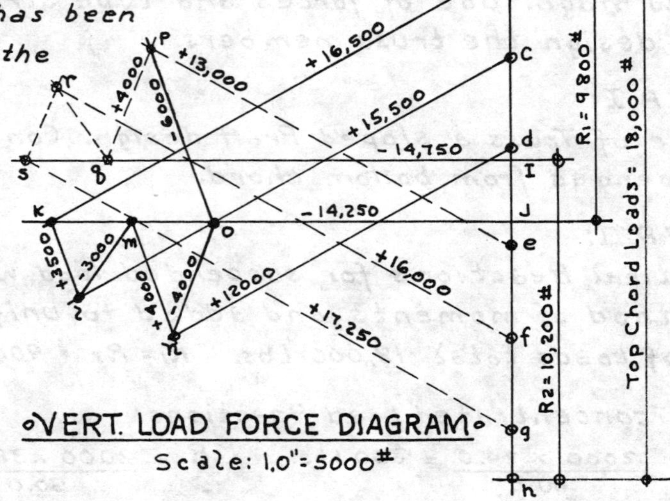

◦VERT. LOAD FORCE DIAGRAM◦
Scale: 1.0" = 5000#

EXAMPLE: Graphic analysis of Pratt truss, continued 5.3.4.13

STEP VI:

A Funicular Polygon will be constructed from a Ray Diagram and the load line will be extended to serve for Force Diagram. Vertical Reactions from wind loads normal to slope will be determined by Force triangle diagram.

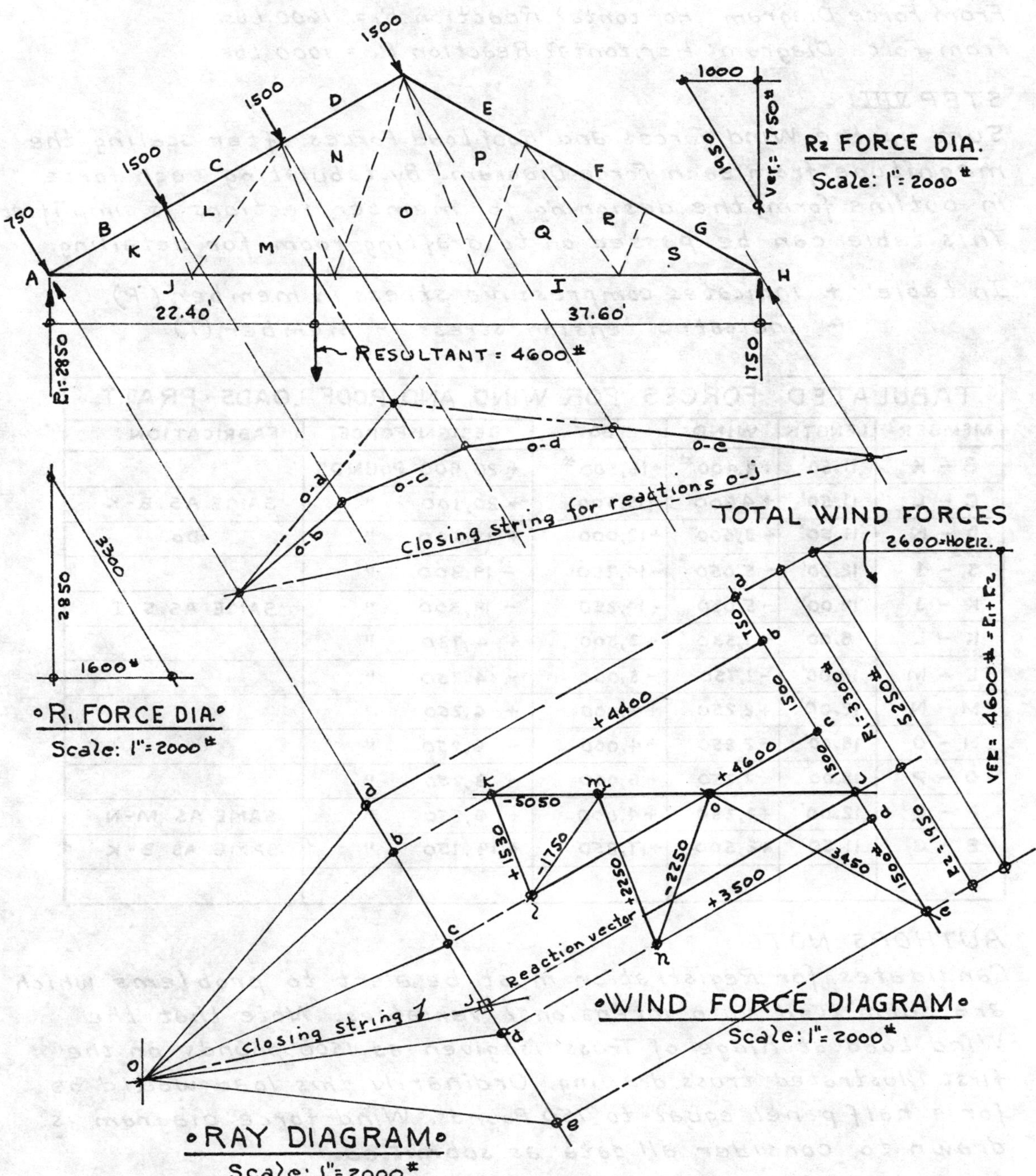

EXAMPLE: Graphic analysis of Pratt truss, continued 5.3.4.13

STEP VII:

To add up Reactions acting in Vertical plane.

Roof Loads taken from Step IV:	R_1 = 9,800 Lbs.	R_2 = 10,200 Lbs.
Wind Loads taken from Step VI:	R_1 = 2,850 "	R_2 = 1,750 "
Total Vertical Reactions:	R_1 = 12,650 Lbs.	R_2 = 11,950 Lbs.

From Force Diagram: Horizontal Reaction R_1 = 1600 Lbs.
From Force Diagram: Horizontal Reaction R_2 = 1000 Lbs.

STEP VIII:

Sum up the Wind Forces and Roof Load Forces after scaling the magnitude from each Force Diagram. By tabulating each force in outline form, the designing for member sections is simplified. This table can be passed on to drafting room for detailing.

In table: + indicates compressive stress in member, (P).
− indicates tension stress in member, (T)

TABULATED FORCES FOR WIND AND ROOF LOADS - PRATT

MEMBER	LENGTH	WIND	ROOF	DESIGN FORCE	FABRICATION
B − K	11.50'	+4,400#	+16,500#	+ 20,900 POUNDS	
C − L	11.50'	+ 4,600	+15,500	+ 20,100 "	SAME AS B-K
D − N	11.50'	+ 3,500	+12,000	+ 15,500 "	Do.
S − I	12.00'	- 5,050	-14,750	- 19,800 "	
K − J	12.00'	- 5,050	-14,250	- 19,300 "	SAME AS S-I
K − L	6.00'	+ 1,530	+ 3,500	+ 4,730 "	
L − M	14.00'	- 1,750	- 3,000	- 4,750 "	
M − N	12.00'	+ 2,250	+ 4,000	+ 6,250 "	
N − O	18.00'	- 2,250	- 4,000	- 6,250 "	
O − P	18.00'	- 2,250	- 6,000	- 8,250 "	
P − Q	12.00'	+ 2,250	+ 4,000	+ 6,250 "	SAME AS M-N
S − G	11.50'	+ 2,500	+ 17,250	+ 19,750 "	SAME AS B-K

AUTHORS NOTE:

Candidates for Registration must be alert to problems which are submitted with intentional variables. Note that the Wind Load at Ridge of Truss is given as 1500 pounds on the first illustrated truss drawing. Ordinarily this load would be for a half panel equal to 750 Pounds. Wind force diagram is drawn to consider all data as submitted.

PILOT DIAGRAMS: Warren, Howe and Fink trusses 5.3.5.1

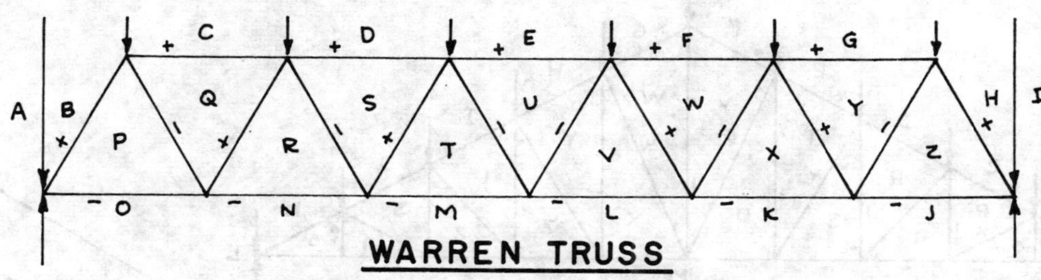

WARREN TRUSS

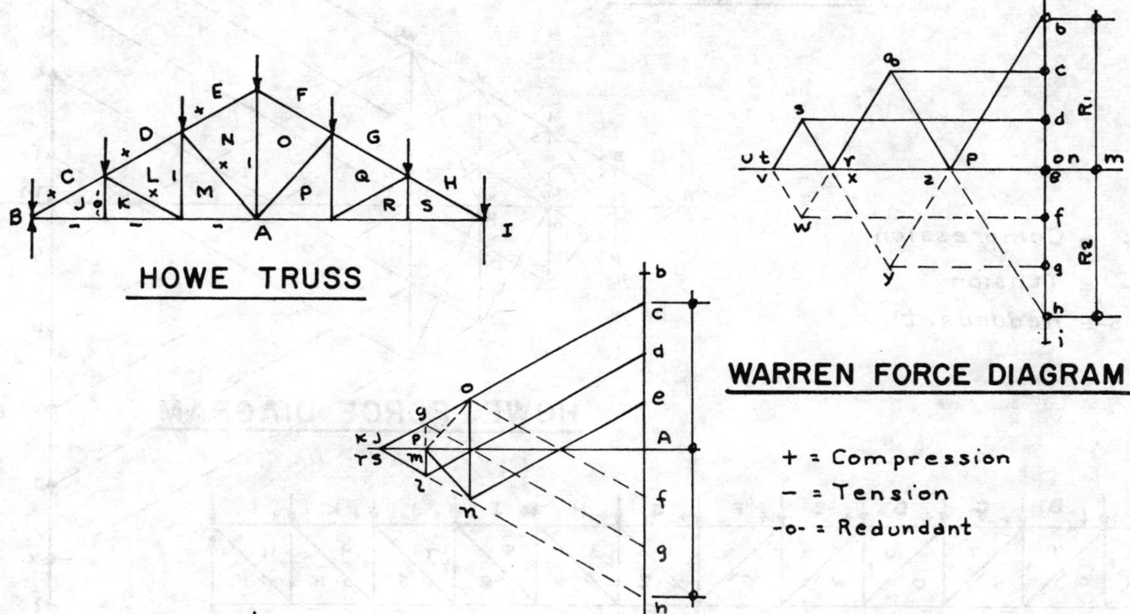

HOWE TRUSS

WARREN FORCE DIAGRAM

HOWE FORCE DIAGRAM

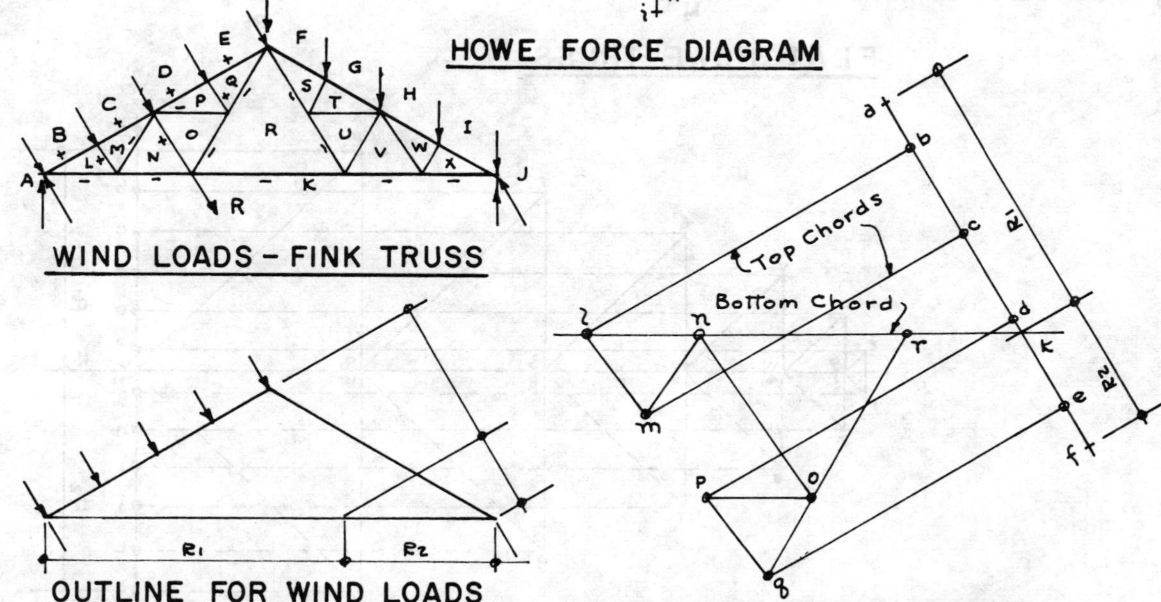

WIND LOADS – FINK TRUSS

OUTLINE FOR WIND LOADS

WIND FORCE DIAGRAM

PILOT DIAGRAMS: Sloped and flat Howe trusses 5.3.5.2

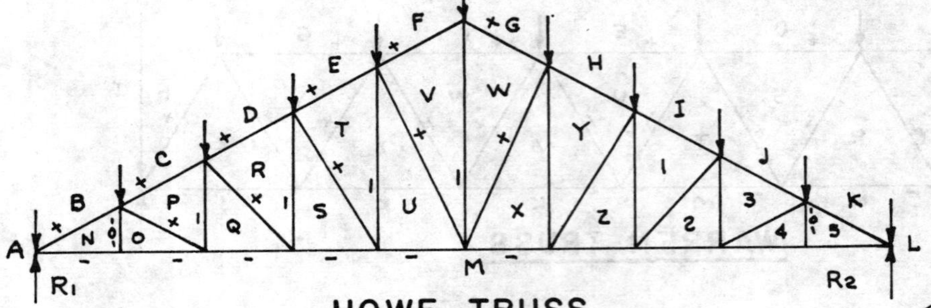

HOWE TRUSS

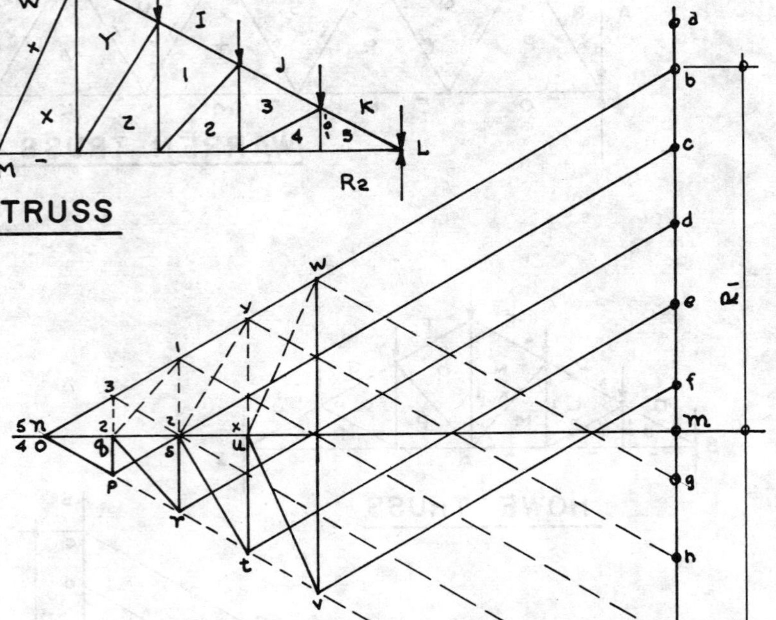

+ = Compression

− = Tension

-o- = Redundant

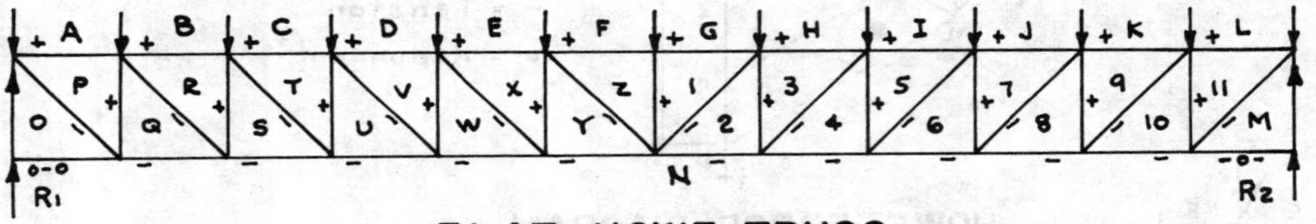

HOWE FORCE DIAGRAM

FLAT HOWE TRUSS

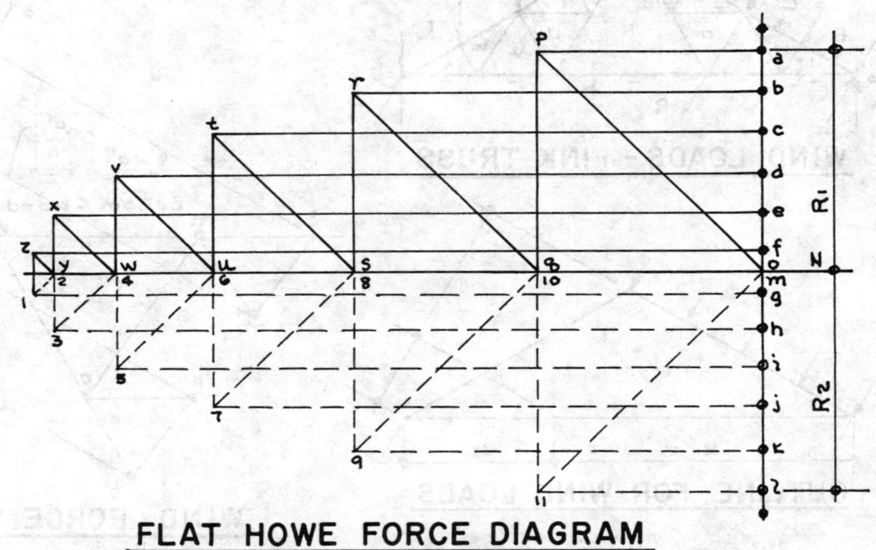

FLAT HOWE FORCE DIAGRAM

PILOT DIAGRAMS: Sloped and flat Pratt trusses 5.3.5.3

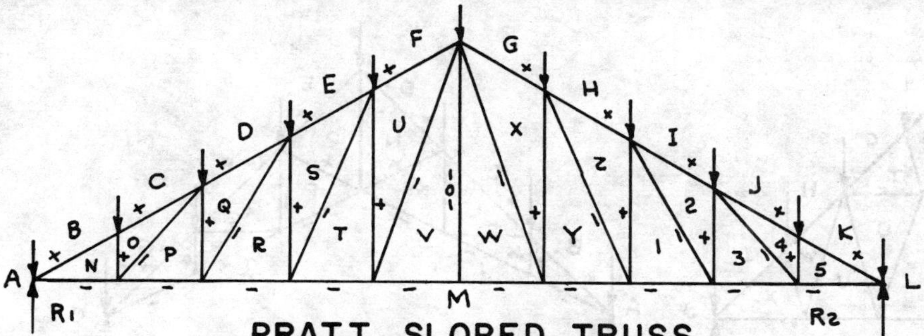

PRATT SLOPED TRUSS

+ = Compression

− = Tension

-o- = Redundant

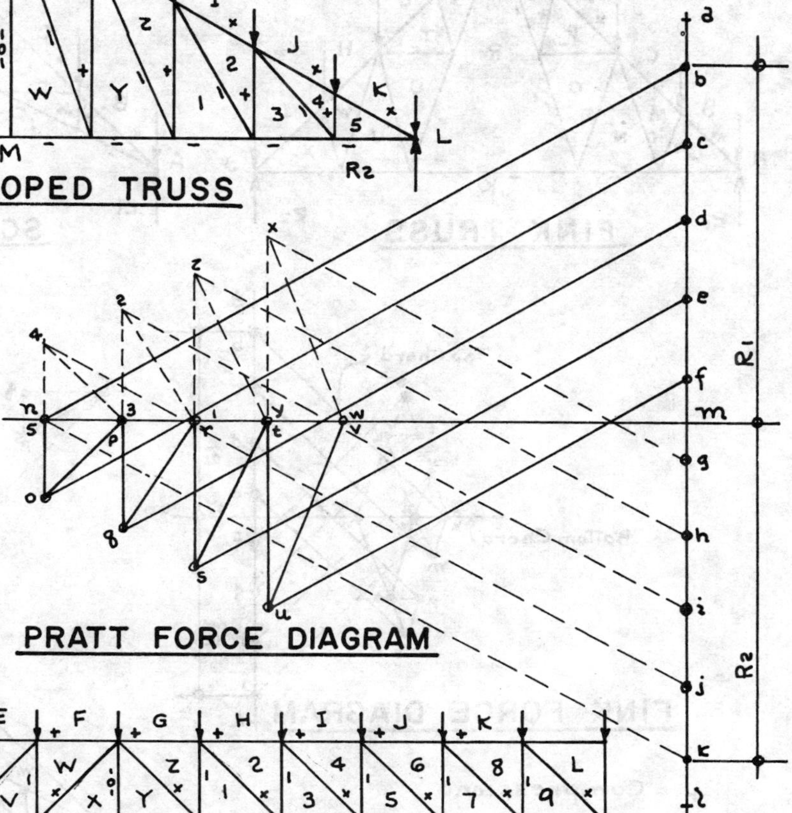

PRATT FORCE DIAGRAM

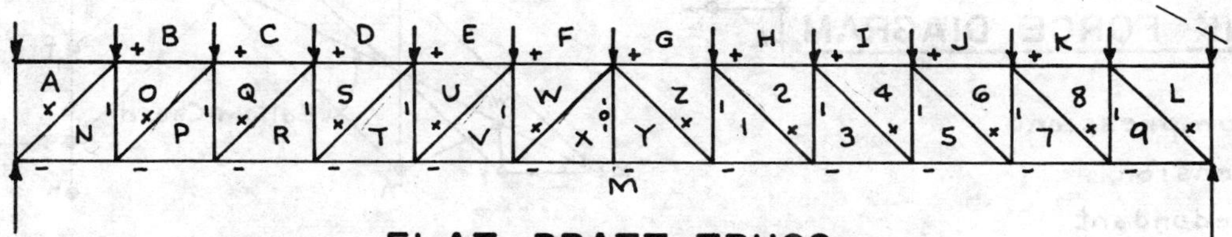

FLAT PRATT TRUSS

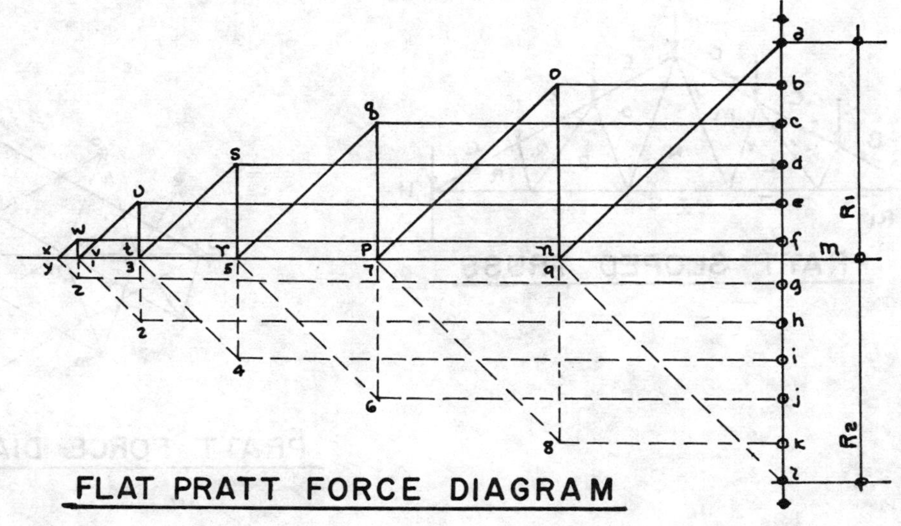

FLAT PRATT FORCE DIAGRAM

PILOT DIAGRAMS: Fink, scissors, and reverse-web Pratt trusses 5.3.5.4

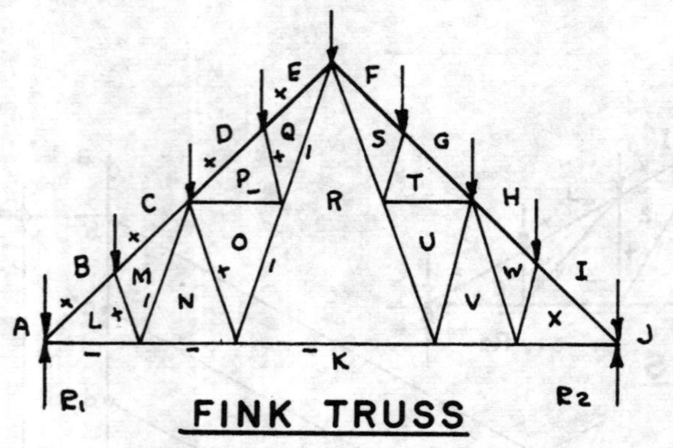

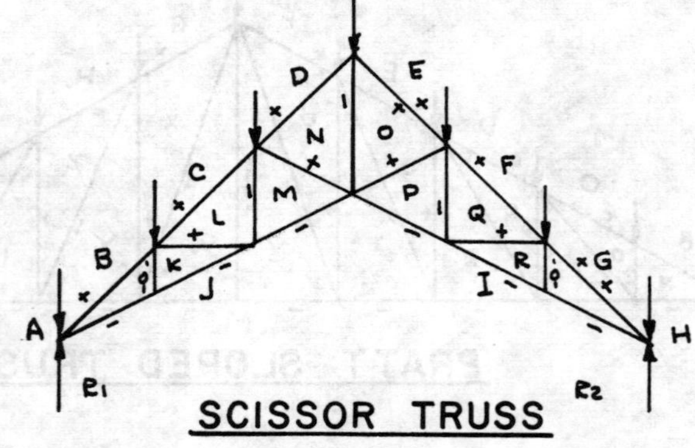

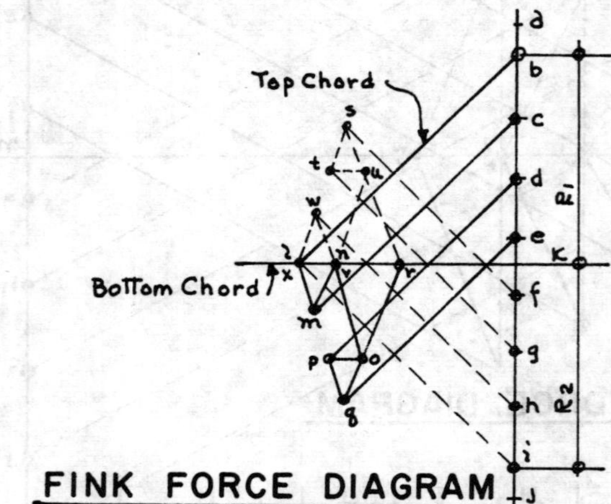

+ = Compression

– = Tension

–o– = Redundant

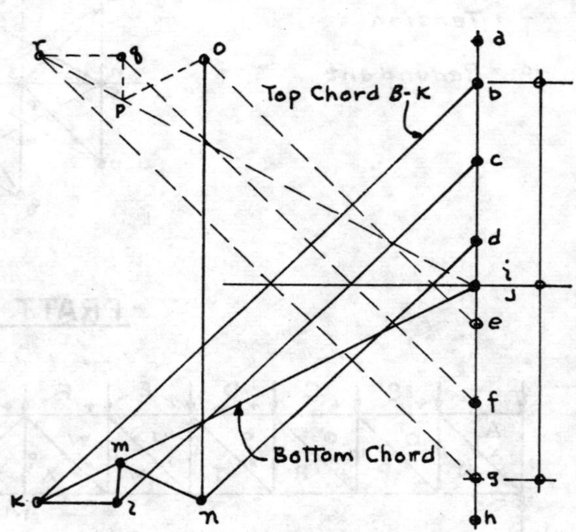

SCISSOR FORCE DIAGRAM

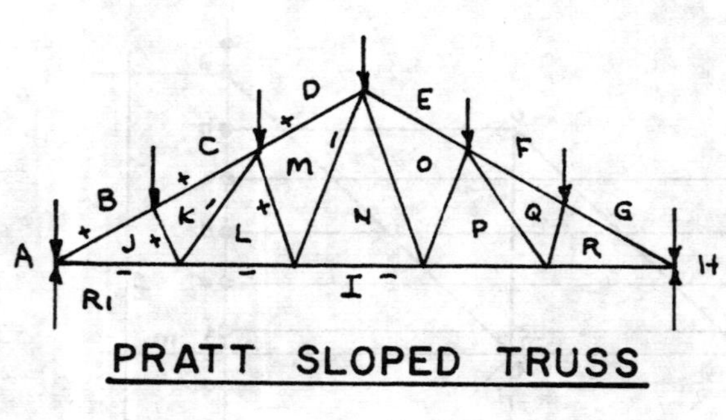

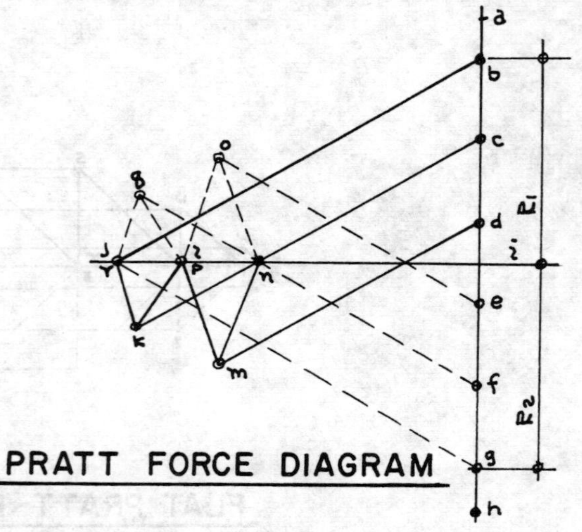

PILOT DIAGRAMS: Fink and Pratt trusses with roof loads 5.3.5.5

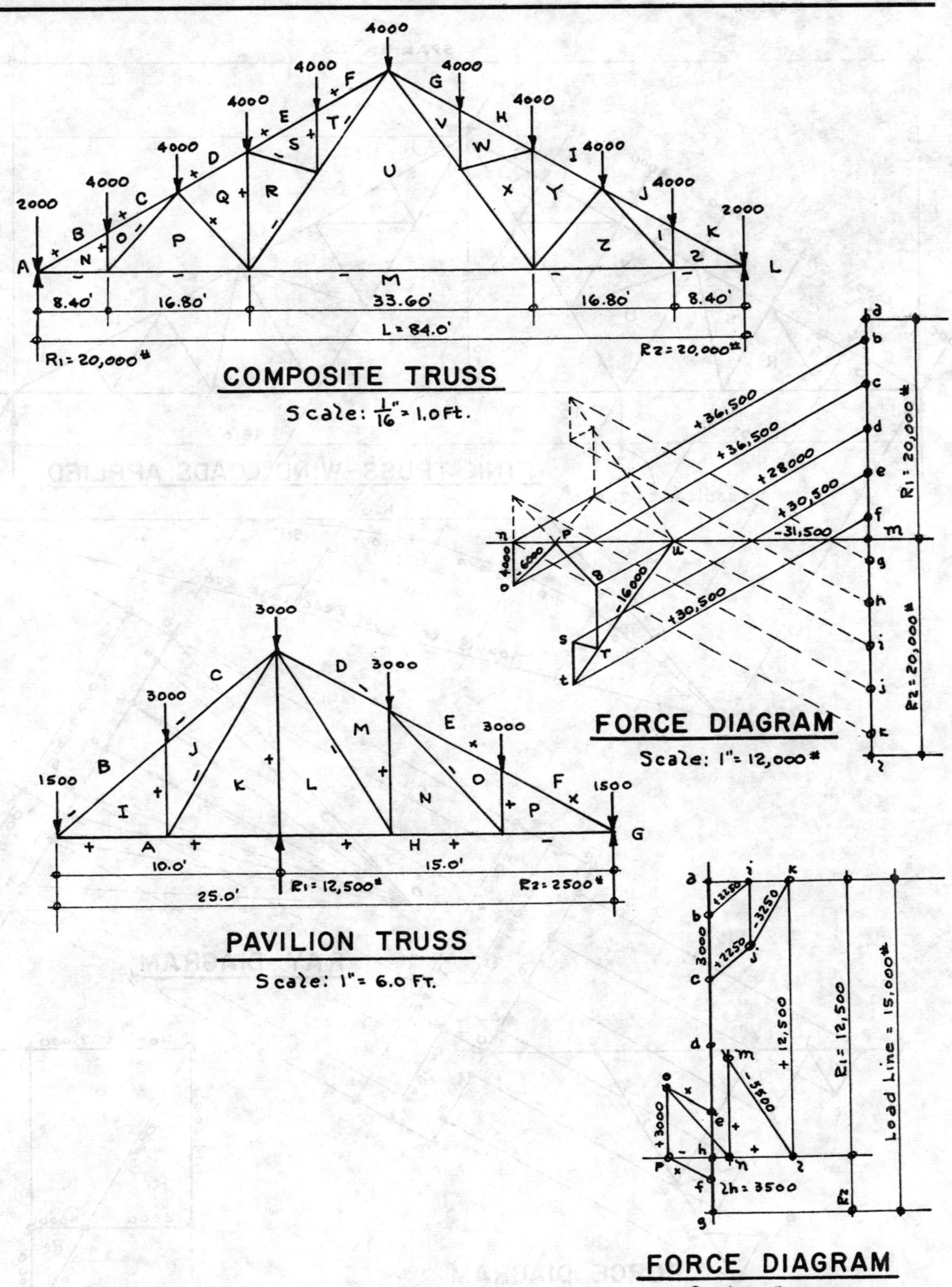

PILOT DIAGRAMS: Fink truss with Ray Diagram for wind load 5.3.5.6

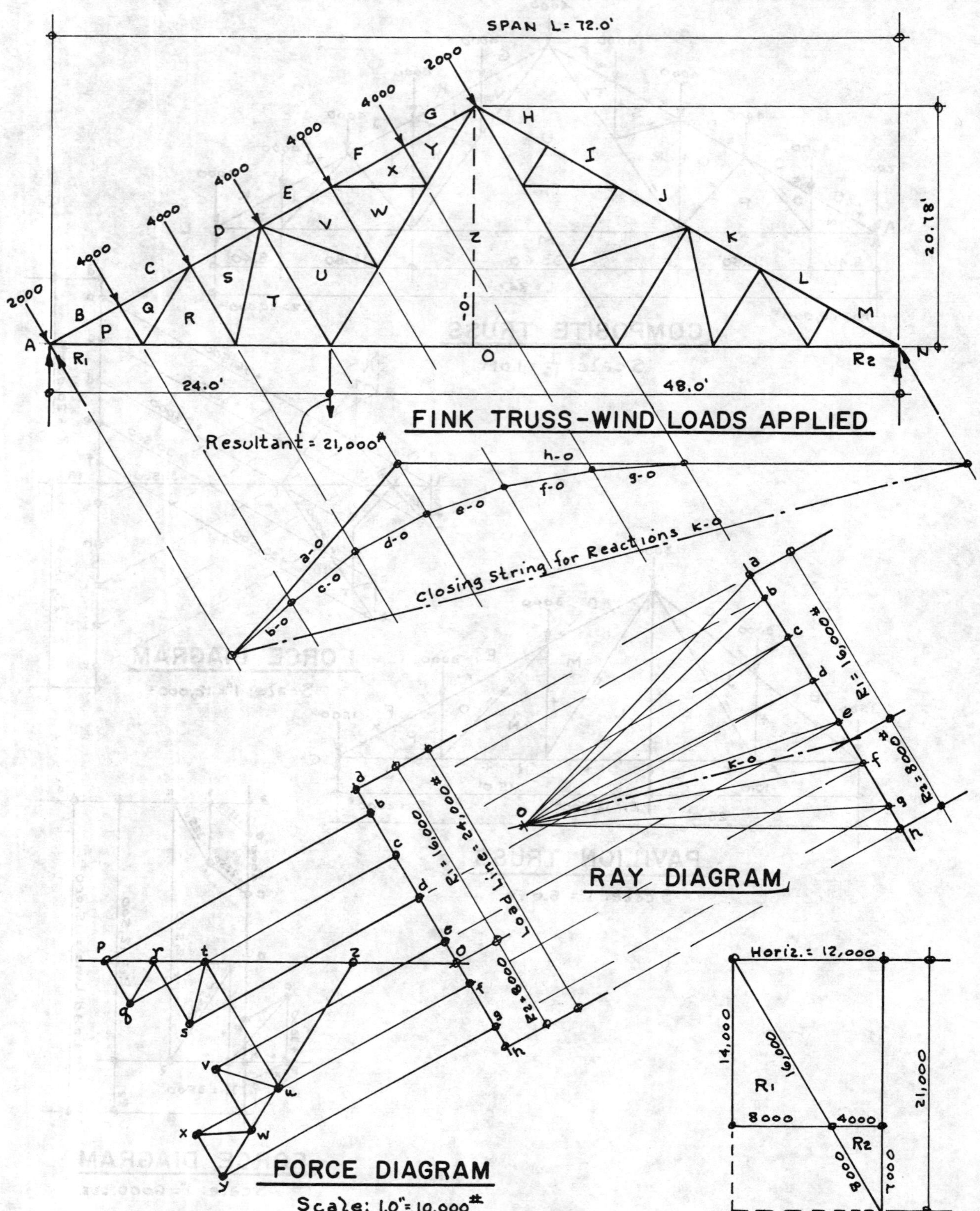

PILOT DIAGRAMS: Cantilever truss with Ray Diagram 5.3.5.7

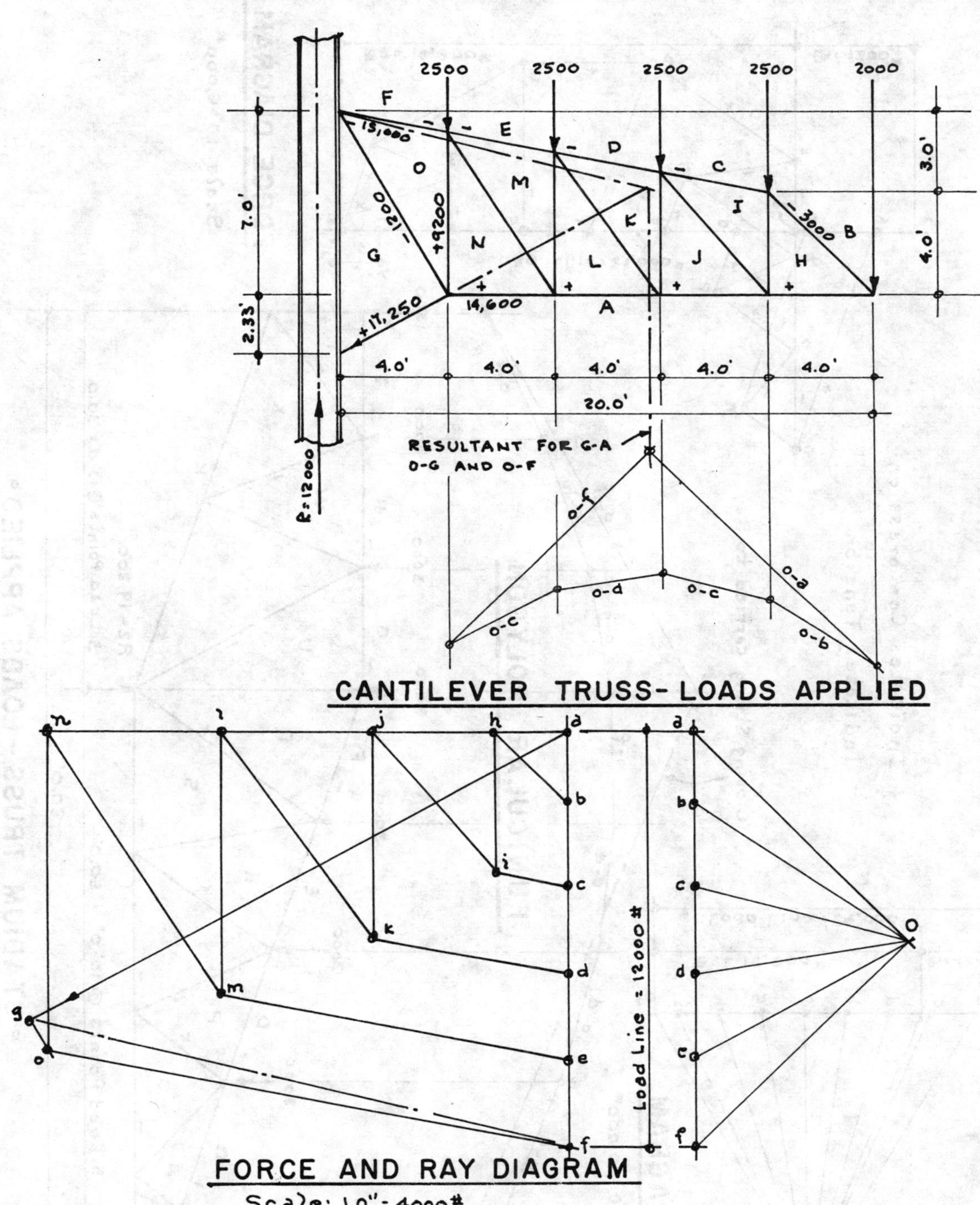

PILOT DIAGRAMS: Stadium truss with overhang 5.3.5.8

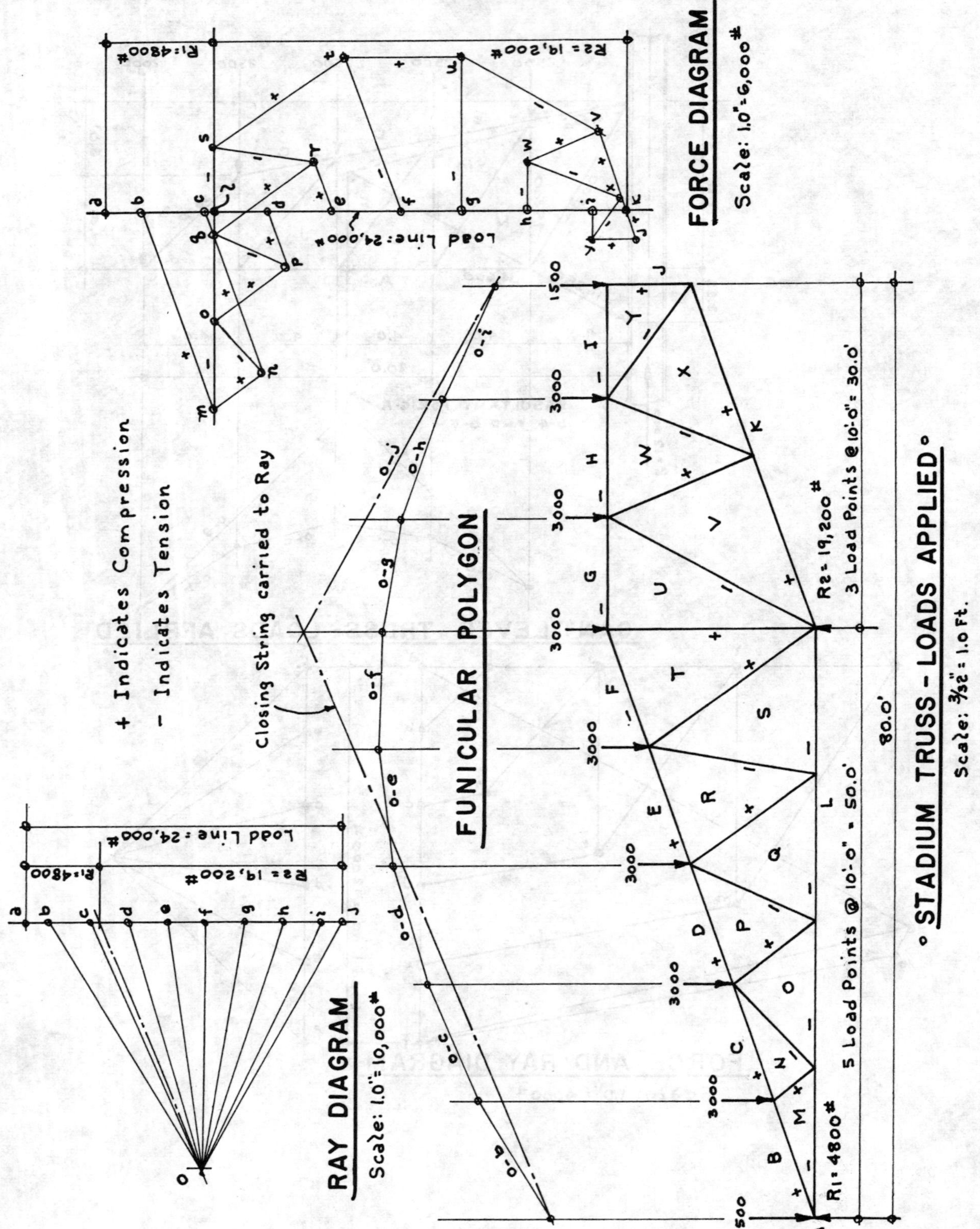

Friction 5.4

When two surfaces are compressed in contact, the force which resists sliding motion is frictional resistance or friction. Friction is a passive force, and acts only when an attempt is made to move one body over another. When a weight is resting on an incline, the force of friction is restraining the weight from sliding down the incline. This is *Static Friction.* If the contact surfaces are moving over one another as a pile moving downward into the earth under the blows of a hammer, then the resisting force on the moving body is termed *kinetic friction.*

Coefficient of friction 5.4.1

The coefficient of friction is the ratio of the force of friction (F) to the normal force between the surfaces (N): $C = \frac{F}{N}$. Remember, friction resists motion, and the coefficient of friction is dependent on the abrasive condition of the surfaces in contact.

Friction on an inclined plane 5.4.1.1

Coefficients of sliding friction for various surfaces can only be determined by tests under actual conditions. A study was made on several ship launchings at the Consolidated Steel Corporation Shipyard at Orange, Texas, to calculate the coefficient of sliding friction for the launching grease used on launching ways. This grease is placed between the timbers of the ground way and the skids of the sliding ways attached to the cradle. The weight of each vessel hull at launch was approximately the same, 680 tons, including dunnage, men, ballast and cradle. The declivity of the launching ways was 11⁄16 inches per foot or an angle of 3 degrees 17 minutes. The acceleration of the ship moving down the ways was recorded for each second of time, and the following formula applied: $C = \text{Tan}\,A - \left(\frac{a}{g\,\text{Cos}\,A}\right)$, where C = coefficient of sliding friction, A = angle of declivity, a = acceleration, and g = gravity fall of 32.20 feet per second per second. When acceleration a = 0.990 feet per second per second, the coefficient of sliding kinetic friction will be 0.02657. Due to temperature changes it was found that the grease coefficient on twenty launchings varied from 0.0257 to 0.0377.

TABLES OF COEFFICIENTS

Published coefficients of sliding friction show considerable scatter with only approximate values in text books which may vary twenty to thirty percent. The designer can avoid these extremes by giving careful consideration to potential and existing conditions. Trowelled concrete surfaces when wet have friction values reduced fifty percent from dry conditions. Table 5.4.2 lists values for coefficients of static friction. These values also will vary from wet to dry conditions. The engineer must consider the surface finish and the lubricant between surfaces.

Angles of sliding friction 5.4.1.2

There is a certain incline angle for which the gravity component force down the slope is exactly balanced by friction so that the body remains at rest on the verge of sliding. The tangent of the angle measured just before the sliding begins, or when forces P and F are equal, is the Coefficient of Static friction C.

Simple rules are: Body will slide when P is greater than F or NC. Body will remain static when F is greater than P. Where:
N = Weight normal to incline.
P = Force parallel to incline.
F = Force holding by friction.

Angles of repose 5.4.1.3

There is some similarity between sliding friction angles and the angles of repose for heaped, loose materials. In the case of retaining walls to hold the force of sliding earth, not all of the earth will slide and exert pressure on the wall. Refer to the design of retaining walls in Concrete Design, Section IV which provides data on friction angles and coefficients.

TABLE: Coefficients of static friction 5.4.2

COEFFICIENTS OF SLIDING FRICTION-STATIC	
CONTACTED SURFACES	COEFFICIENT "C"
WOOD ON CONCRETE	0.25 To 0.65
METAL ON CONCRETE	0.30 To 0.40
STONE ON CONCRETE	0.40 To 0.70
LEATHER ON CONCRETE	0.45 To 0.60
WOOD ON WOOD	0.30 To 0.65
WOOD ON METAL	0.20 To 0.60
METAL ON METAL	0.14 To 0.36
LEATHER ON METAL	0.30 To 0.58
RUBBER ON METAL	0.50 To 0.65
RUBBER ON WOOD	0.40 To 0.60
RUBBER ON CONCRETE	0.40 To 0.65
LEATHER ON WOOD	0.25 To 0.55

Values listed are for Smooth surfaces with Wet to Dry Range. For Kinetic Coefficients, reduce values 25 to 40 Percent.

EXAMPLE: Sliding friction resistance 5.4.3

A wood box containing 350 Lbs. of cargo is stored on a smooth trowelled dry concrete floor. Coefficient of Sliding friction normal to floor surface is 0.30 = (C).

REQUIRED:

Determine the force which must be applied to start the slide when a dock worker pulls the container on a direct plane parallel to floor.

Next- assume that worker will attempt to move the container by exerting a pulling pressure on a line of 30 degrees with the horizontal plane and what force is required to start action.

STEP I:

Draw a sketch for each condition:

Let W = 350 Lbs. F = Holding force of friction.

N = Force normal to surface, and

P = Pressure to start on line or value of push or pull. C = Coefficient of Sliding friction = 0.30

STEP II:

Force normal to floor = 350 Lbs., or same as W for parallel start.

If $C = \frac{F}{N}$, then $P = CN$. $P = 0.30 \times 350 = 105$ Lbs.

This is the minimum pull to start movement.

STEP III:

Condition No. 2 requires that the forces be resolved into 2 Components. 1 Force vertical and the other force horizontal.

Angle of pull: A = 30 degrees. Sine A = 0.5000 Cos. A = 0.86603

To balance the forces: $N = W - (P \sin A)$ $N = 350 - (0.5000 \times P)$

$CN = P \cos A$, or $0.86603P = 0.30(350 - 0.5000P)$

Equating for a solution, it becomes:

$0.86603P = 105 - 0.1500P$. Then: $(0.86603P) + (0.1500P) = 105$ Lbs., or $(0.86603 + 0.1500)P = 105$ Lbs. Now, $0.86603 + 0.1500 = 1.01603$

Thus: $P = \frac{105}{1.01603} = 103.343$ Lbs. for Condition 2.

EXAMPLE: Sliding friction on a ramp 5.4.4

A group of wood boxes, each weighing 250 Pounds are being unloaded from an Army Transport Plane onto a steel plate ramp surface. The Ramp is on wheels and is 30.0 Long on horizontal plane. Unloading end is raised until boxes begin a slow slide. The required rise was measured at 10.25 Feet at high end.

REQUIRED:

Calculate the Tangent of the Slope angle with horizontal plane, then determine the Coefficient of Sliding Friction between wood and steel under the conditions.

Assume that surface of ramp was greased and friction coefficient was reduced to 0.175, then at what height would ramp be raised at end.

STEP I:

Drawing a triangle with sides a, b, and c, Angle A will have Tangent $= \frac{a}{b}$. a = 10.25' b = 30.0' Tan. A $= \frac{10.25}{30.00} = 0.3417$

From Trig. Tables: Angle A = 18° 52'

Under Conditions, Coefficient of Friction = 0.3417

STEP II:

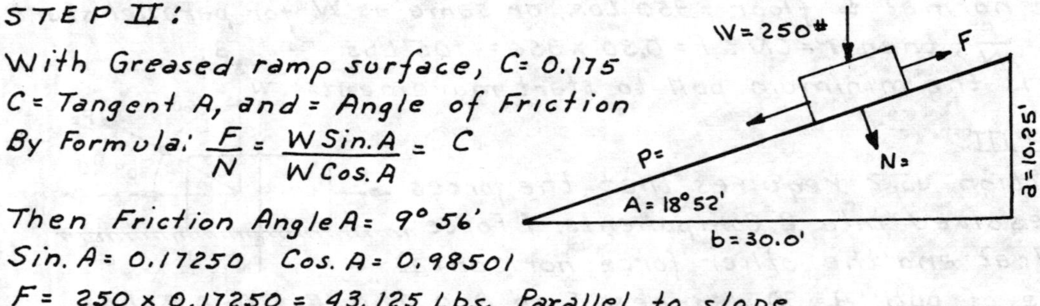

With Greased ramp surface, C = 0.175

C = Tangent A, and = Angle of Friction

By Formula: $\frac{F}{N} = \frac{W \text{ Sin. } A}{W \text{ Cos. } A} = C$

Then Friction Angle A = 9° 56'

Sin. A = 0.17250 Cos. A = 0.98501

F = 250 × 0.17250 = 43.125 Lbs. Parallel to slope

N = 250 × 0.98501 = 246.253 Lbs. Normal to slope.

$C = \frac{F}{N} = \frac{43.125}{246.253} = 0.175$ = Tangent Friction Angle.

W = 250#
F: 43.125#
P
Greased
N = 246.253#
A = 9° 56'
a = 5.25'
b = 30.0'

STEP III:

To determine Ramp height.

Solve for side a = b Tan. A.

a = 30.0 × 0.175 = 5.25 Feet

This checks out because C = Tang. A or 0.175

EXAMPLE: Friction in a ramp and cable system 5.4.5

A weight of 600 Pounds is placed on a horizontal Concrete surface. Another weight of 180 Pounds is resting on a 42 degree slope angle and held from sliding by a cable tied to the 600 Lbs. weight. Contact surfaces are Concrete and steel in both cases, with the Coefficient of Sliding Friction C = 0.400

REQUIRED:

(a) Draw a illustration of this problem, and determine whether the 2 Weights will slide.

(b) Calculate the Tension Force in connecting cable.

(c) Assuming that Weights will NOT slide, determine the force necessary to start motion to the right for both Weights.

STEP I:

Drawing illustration, and recording the Angle functions:

∠ = 42° Sine = 0.66913 Cos. = 0.74314 Tan. = 0.90040 Sec. = 1.3456

Identify 600 Lbs. Weight as W_1, and 180 Lbs. weight as W_2.

STEP II:

Calculate Forces in other lines.

$N_2 = W_2 \text{Cos.} A$ $N_2 = 180 \times 0.74314 = 133.75^{\#}$

Horizontal For $P_1 = CW_1 = 240^{\#}$

W_2 force parallel with slope.

$P_2 = W_2 \text{Sin.} A$. $P_2 = 180 \times 0.669 = 120^{\#}$

Restraining force from friction under $W_2 = F_2$ and $F_2 = N_2 C$. $F_2 = 133.7 \times 0.40 = 53.50$ Lbs

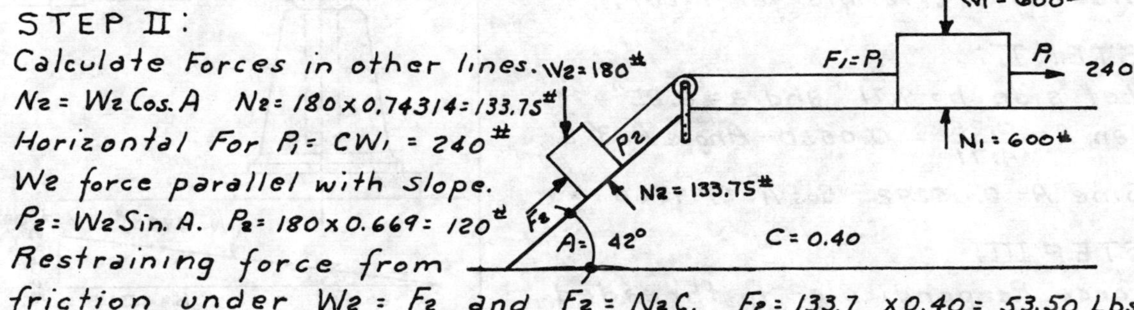

STEP III:

Review Forces: P_2 is pull force from W_1, and F_2 is a force of Friction Resistance. The force P_2 is greater than F_2, and W_2 would slide if not held by cable. The forces resisting the slide of W_2 are $F_1 + F_2$ or $F_3 = 53.50 + 240 = 293.50$ Lbs.

When connected by Cable the single pulling force is P_2 of 120 Lbs., and F_3 is greater. Bodies W_1 and W_2 will not slide.

Tension in Cable when no action is present = $P_2 - F_2 = 120 - 53.5 = 66.50^{\#}$

To start movement upward and to the right, the forces must be overcome by a single force at P_2, and call this force P_3.

$P_3 = P_1 + P_2 + F_2$. $P_3 = 240.0 + 120.0 + 53.50 = 413.50$ Lbs.

Tension in Cable when movement is started = $P_2 + F_2 = 173.50$ Lbs.

EXAMPLE: Friction in a screw jack 5.4.6

A load of 5000 Pounds is to be raised with a Screw Type Jack which has a threaded shaft diameter of 1½ inches. One complete 360 degree of Jack Lever will raise the load 0.25 Inches. Lever operating handle is 13.50 inches long. With proper lubricating grease, the Coefficient of sliding friction between steel is given as; C= 0.095

REQUIRED:

Determine the required Pressure on end of handle to start turning screw and raising load W.

If only 30 Pounds of Pressure can be applied on handle, calculate the length of Lever handle necessary.

STEP I:

If diameter of Screw is 1.50", then Circumference = πD.

C= 3.1416 × 1.50 = 4.71" The rise per turn = 0.25 inches. The angle of incline can be found by making a triangle for 1 turn.

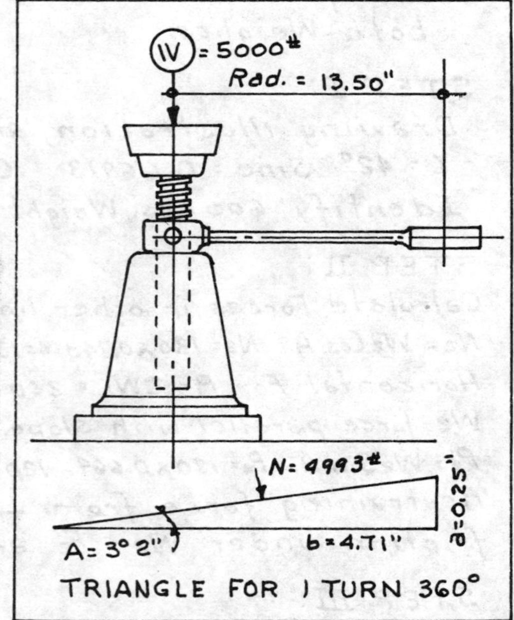

TRIANGLE FOR 1 TURN 360°

STEP II:

Let side b= 4.71" and a= 0.25"

$$\text{Tan. } A = \frac{0.25}{4.71} = 0.0530 \quad \text{Angle } A = 3^\circ 2'$$

Sine A= 0.05292 Cos. A= 0.99860

STEP III:

Force Perpendicular to threaded slope = W Cos. A = N (Normal)

N= 5000 × 0.99860 = 4993 Lbs.

Force Parallel to slope = W Sin. A. = P= F.

F= 5000 × 0.05292 = 264.60 Lbs.

STEP IV:

Normal force on Incline = 4993 Lbs. and C= 0.095

Sliding Force on incline = CF = 4993 × 0.095 = 474.3 Lbs.

Total force to turn Shaft at ℄ = 264.60 + 474.3 = 738.94# (Call it 740#)

STEP V:

Pressure required to turn with 13.5" handle: $\frac{740}{13.5}$ = 54.8 Lbs. at end.

With 30 Lbs. pressure on end of lever, the length of Lever handle required = $\frac{740}{30}$ = 24.67 Inches long.

EXAMPLE: Friction in a launching way 5.4.7

A steel Chemical Barge is 180.0 feet long with a Dead Weight of approximately 300 tons. Vessel is to be launched from a inclined wood constructed way which has an angle of 12° 30' with the horizontal plane. Sliding way and cradle are of smooth surfaced yellow pine timbers. In lieu of launching grease between sliding wood contact surfaces, bananas with skins will be used. From the data obtained from tests and experiments, the Coefficient of bananas for sliding friction is established at 0.175

REQUIRED:

(a) Barge must be held on incline preceding the launching slide, thus the force parallel with incline must be determined to construct the proper size of holding plates for release by burning between holes.

(b) Calculate the weight of barge normal to sliding incline, so that weight upon bananas may be determined.

(c) Assuming that vessel will slide freely from its own weight under the given conditions, determine the least declivity angle which will permit barge to slide alone under its parallel force.

STEP I:

An illustration of Problem will be drawn and a triangle to represent the conditions. Short tons will be used for weight as 1 Ton = 2000 Lbs. W = 300 x 2000 = 600,000 Lbs.

Coefficient C = 0.175

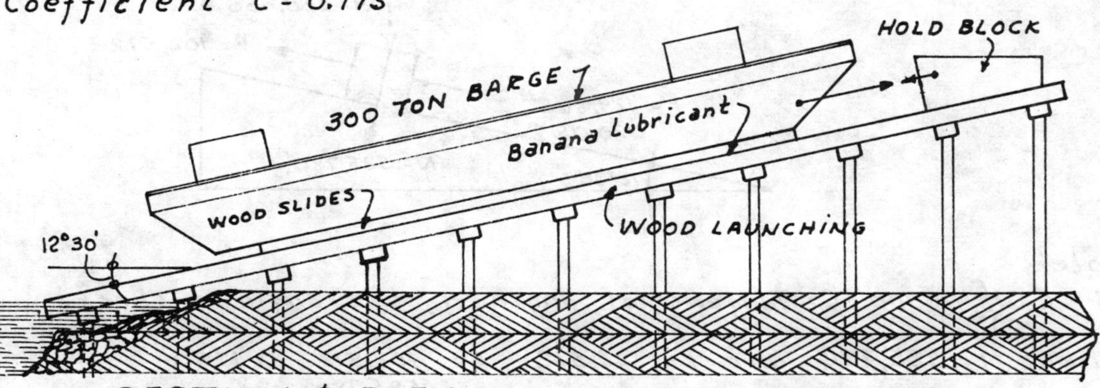

SECTION & ELEVATION LAUNCHING WAYS

EXAMPLE: Friction in a launching way, continued 5.4.7

STEP II:
Obtaining the function of angle A from tables:
For angle of 12° 30', Tan = 0.22169 Sine = 0.21644 Cos. = 0.97630
Force Parallel with incline: P = W Sin. A
Force Normal with incline: N = W Cos. A
Then:
P = 600,000 × 0.21644 = 129,865 Lbs.
N = 600,000 × 0.97630 = 585,780 Lbs.

Since Coefficient of sliding friction is 0.175 and is less than the tangent of the declivity angle A, the vessel will slide under its potential force P.

STEP III:
The force pulling vessel down incline when the friction is decreased by banana lubrication is: $\frac{F}{N} = C$, and F = NC.
Then F = 585,780 × 0.175 = 102,512 Lbs.
Again the force P, is greater than holding force F, and the holding plates must be designed to sustain full force of P.

STEP IV:
To determine the minimum angle A, or declivity of the sliding incline where Barge will just slide under its own force without any help except gravity.
When C = 0.175 or banana sliding coefficient, it is also the Tangent of the minimum angle, or A = 9° 56'. Tan = C = 0.17513
A decline of approximately 10° 30' would be ideal.

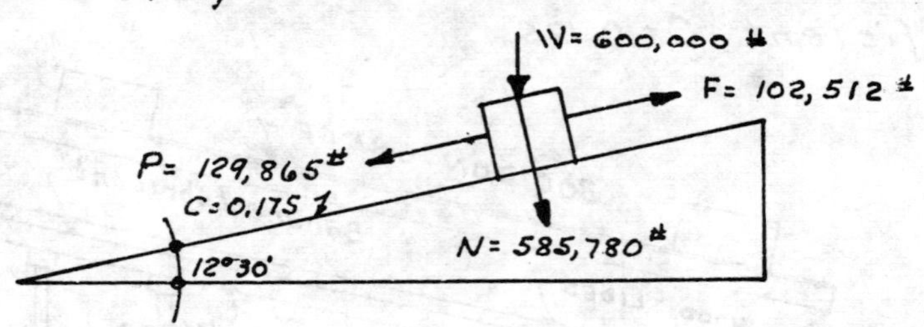

Note:
If $C = \frac{F}{N}$ and C = Tangent of angle A, the angle is checks when: N = 585,780 and F = 102,512 $C = \frac{102,512}{585,780} = 0.1750$

Solving force systems in structures 5.5

If three forces act on a body, and if these forces are in equilibrium, then their lines of action will meet at a common point. If two forces are known, the third force may be found. In the examples which follow, this principle of concurrent forces will be illustrated as a useful tool to solve for magnitude and direction of unknown forces. Three concurrent forces in equilibrium may be portrayed as a triangle of forces, and can be solved by the graphic or the algebraic method.

Forces in simple structures 5.5.1

A frame, machine, or structure which is formed of members which are hinged or pinned together so that the components sustain only tension and compressive stresses is referred to as a simple structure. Rigging for shipboard cargo-handling, pile driver leads, and crane booms are common types of tackle which the structural engineer is frequently called upon to design or analyse. Each component member is examined as a free body. The pinned joints are assumed frictionless. External forces are assumed to act only at the joints or connections. In calculating the forces, the work is often simplified by neglecting the dead weight of the members.

When examining each member of a simple structure, it is necessary to determine the direction of the deformation resulting from the applied loads or the forces from other members. If the load or force causes deformation parallel to the axis, then the force must be acting on a line thru the end pins and parallel to the member axis. Conversely, if the applied forces cause deformation in a direction other than parallel to the axis, these forces cannot be classed as co-linear. In such cases, the unknown reactions must be solved for by finding the point of intersection of the unknown forces and then taking moments about this point to find moment equilibrium.

Parallelogram of forces 5.5.1.1

If two forces which act at a point are represented in magnitude and direction by the adjacent sides of a parallelogram, their resultant will be represented in magnitude and direction by the diagonal drawn from the intersection of the two component forces.

If the resultant of three or more forces having the same point of application is required, first find the resultant of any two of the forces, and then combine this resultant with the third force. Should there be more than three forces, continue in this manner until the resultant of all the forces has been found. See Example 5.5.3.2.

Polygon of forces 5.5.1.2

When several forces in a system are applied at a point and act in a single plane, their resultant may be found more simply by using a force polygon. Construct the layout of the forces and indicate the sense by arrows. From the extreme end of the line representing the first force, draw a line representing the second force, parallel to its action line. Then from the extreme end of this line, draw a third line representing the third force, parallel to the line of action of the third force. Continue this manner until all the forces have been represented. Then draw the closing line or chord from the point of application of the forces (starting point for the first force) to the extreme end of the last force line drawn. This closing chord is the *resultant* of the force system. This method is illustrated in Example 5.5.3.3, for the same system of forces used in the parallelogram method. By comparing the force polygons for each method, one can see the relationship between the two methods.

Indeterminate systems 5.5.1.3

When using the graphic method for the analysis of a group of non-concurrent forces (which do not act through a common point), the unknowns are limited to three in number. If more than three unknown forces are involved, the problem falls into the class of force systems which are called indeterminate, meaning that they cannot be solved by either the graphic or the algebraic method, unless the deformations are considered along with the forces.

Rigging on the construction site 5.5.2

Speed and efficiency are the most important factors in modern contracting, to enable the contractor to realize a profit. Whether the project is the overhaul of a seagoing vessel in drydock or a swing derrick to transfer wet concrete to job locations, there is always the need for some type of rigging to simplify the operations. No rig should be allowed on the job unless it has been designed and built for safety, and reputable contractors should not hesitate to consult with engineers. A safe working stage is an excellent insurance policy, but if designed by a novice or careless mechanic, it can become a death trap. Rigging structures embrace a large field of structural types. Ship rigging is of a permanent nature while construction rigs serve temporarily during the construction period. Pile driving rigs and special crane booms are frequently fabricated or constructed to aid in the erection of church spires. Rigging is generally defined and understood to refer to structures with ropes, chains and tackle used to handle loads in connection with masts, booms and derricks.

EXAMPLE: Analyzing non-concurrent forces in one plane 5.5.3.1

A typical awning extends 9.0 Feet over a sidewalk or from wall to end. A tie rod supports the end and is connected to wall a distance of 5.20 feet from the horizontal member at bottom.
Concentrated loads are suspended from bottom and are as follows: Load P_1 = 1500 Lbs. located 4.0 feet from wall. Load P_2 = 2200 Lbs., and located 7.0 feet from wall.

REQUIRED:
Draw the triangle as ABC to represent this problem. Determine the stress in Tie Rod, then solve for horizontal force in bottom member at wall. Compute the vertical Reaction at wall for bottom member.

STEP I:
Drawing the Triangle and placing loads. Identifications same as Right Triangle.

STEP II:
Solve for Horizontal and Vertical Force Reactions at connection C. Take moments about point B.

$$C_H = \frac{(1500 \times 4.0) + (2200 \times 7.0)}{5.20'} = 4115^{\#}$$

For vertical C_V, take moments about point A.

$$C_V = \frac{(2200 \times 2.0) + (1500 \times 5.0)}{9.0'} = 1322^{\#}$$

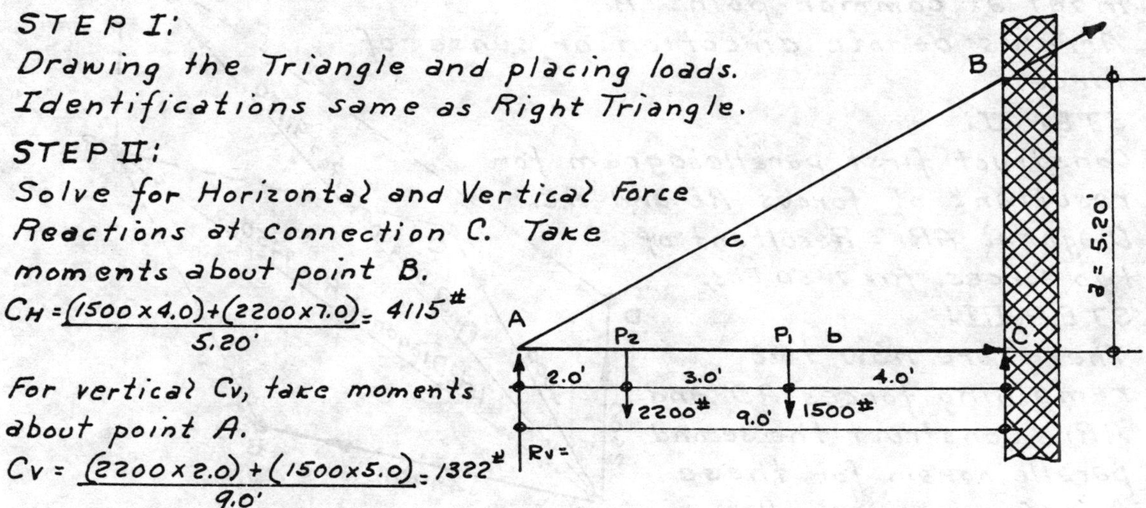

$$R_V = \frac{(1500 \times 4.0) + (2200 \times 7.0)}{9.0} = 2378^{\#}$$ Vertical Reactions total loads.

STEP III:
Force parallel with AB = side c. $\tan A = \frac{a}{b} = \frac{5.20}{9.00} = 0.5777$
Angle A = 30° and Cosine A = 0.866
Force at B = C. Take moments about C

$$B = \frac{(1500 \times 4.0) + (2200 \times 7.0)}{5.20 \times 0.866} = 4755.5 \text{ Lbs. (Force in Tie Rod).}$$

STEP IV:
To complete the reaction at point, the design joint will be the Resultant of C_V and C_H. This could be done with a Force diagram. $R = \sqrt{1322^2 + 4115^2} = 4325$ Lbs.

EXAMPLE: Parallelogram to resolve three forces 5.5.3.2

Three forces meet at a common point indicated as A.
Force AB = 4300 Lbs. and action line is 77° Right of Plumb.
Force AC = 4000 Lbs. and action line is 27° Right of Plumb.
Force AD = 2600 Lbs. and action line is 2° Left of Plumb.

REQUIRED:
Resolve the system of Forces into a single component (Resultant), show its direction and scale for the magnitude of Force. Use the method of Parallelograms.

STEP I:
Using an adjustable triangle with a protractor scale, the force system is constucted to scale of 1 inch = 4000 Lbs. Solid lines AB, AC, and AD = Forces which meet at common point A.
Arrows denote direction or sense of force.

STEP II:
Construct first parallelogram for resultant of forces AB and AC. Diagonal AR_1 = Resultant of two forces. R_1 = 7450 Lbs.

STEP III:
There are now two remaining forces AD and AR_1. Construct the second parallelogram for these two forces. Draw the closing string as the diagonal for final Resultant. R_2 will scale out to 9,150 Lbs.

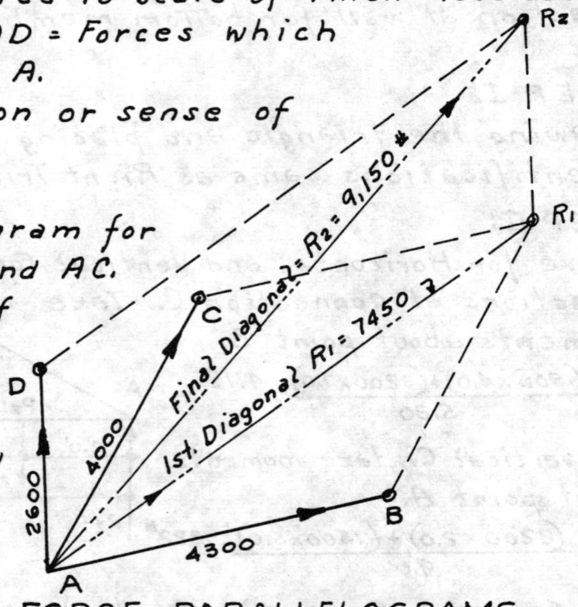

R2
R1
C
D
B
A
Final Diagonal = R2 = 9,150 #
1st. Diagonal R1 = 7450 #
4000
2600
4300

• FORCE PARALLELOGRAMS •
Scale: 1" = 4000 Lbs.

DESIGNER'S NOTE:
The unknowns found by the example are as follows:
(a) The magnitude of the Resultant for three forces.
(b) The direction line of action for the resultant.
(c) The location of Resultant, measures 39° 50' to right of vertical plumb line.

EXAMPLE: Force polygon to resolve three forces 5.5.3.3

The previous example illustrated the parallelogram method for finding the resultant of forces which met as a common point. By drawing a force polygon, the method will delete the parallelograms and resemble the original. Using the same force system thus:

Force AB = 4300 Lbs. Action line is 77° Right of Plumb.
Force AC = 4000 Lbs. Action line is 27° Right of Plumb.
Force AD = 2600 Lbs. Action line is 2° Left of Plumb.

REQUIRED:
Layout to scale 1" = 4000 Lbs., the force system. Draw the Polygon from common meeting point. Let closing string be the Resultant.

STEP I:
Adjustable triangle with protractor scale is used for reading the angles.

STEP II:
Force Polygon is started at extreme end of first force AB. From point B, bc is drawn parallel to force AC with same magnitude. From point c, vector line cd is drawn parallel to AD, and magnitude equals force of AD.

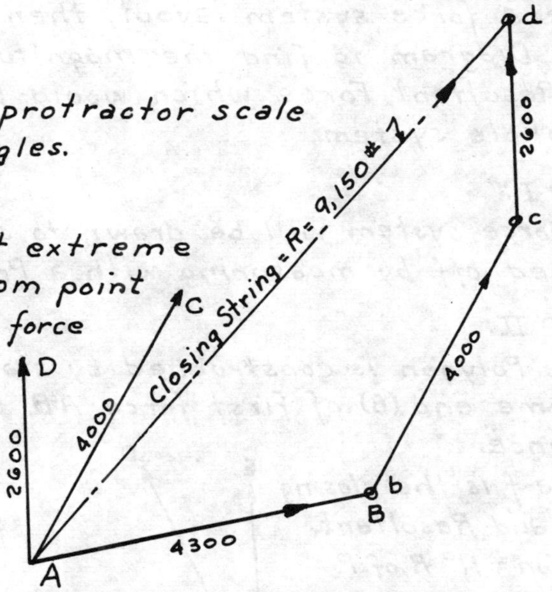

STEP III:
Close the Polygon with line connecting Ad, then scale for magnitude which is Resultant = 9,150 Lbs. Action line of Resultant is measured and reads as approximately 39° 50' to right of plumb line.

THREE FORCE POLYGON
Scale: 1" = 4000 Lbs.

EXAMPLE: Resolving five forces with force polygon 5.5.3.4

A gusset plate in a truss has 5 forces extending away from the common point indicated as A. The magnitude of each force is scaled from the stress diagram, and their action lines are given in relation to a vertical plumb line as follows:

Force AB = 11,500 Lbs. and is suspend on Plumb Line below point A.
Force AC = 7000 Lbs., acting above point A in line 60° to right of Plumb.
Force AD = 17,000 Lbs., acting above point A in line 27° to right of Plumb.
Force AE = 14,000 Lbs., acting above point A in line 3°30' to right of Plumb.
Force AF = 8000 Lbs., acting above point A in line 71° to left of Plumb.

REQUIRED:
Make a force system layout, then construct a separate Force Diagram to find the magnitude, direction, and position of a Resultant Force which would provide equilibrium to the whole system.

STEP I:
The force system will be drawn to scale, and lines of action pointed off by measuring with a Protractor.

STEP II:
Force Polygon is constructed by starting at the extreme end (B) of first force AB. Continue the sequence.
Line a-f is the closing string and Resultant.
Direction = 11° R of P.
Magnitude = 24,500 Lbs

Position = extends upward from point A.

Arrows denote the sense of each force.

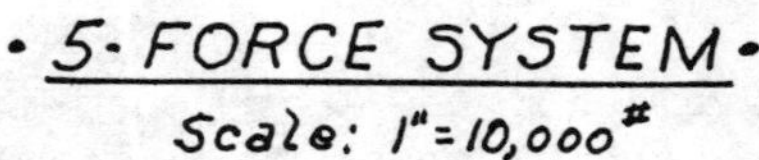
• 5-FORCE SYSTEM •
Scale: 1" = 10,000#

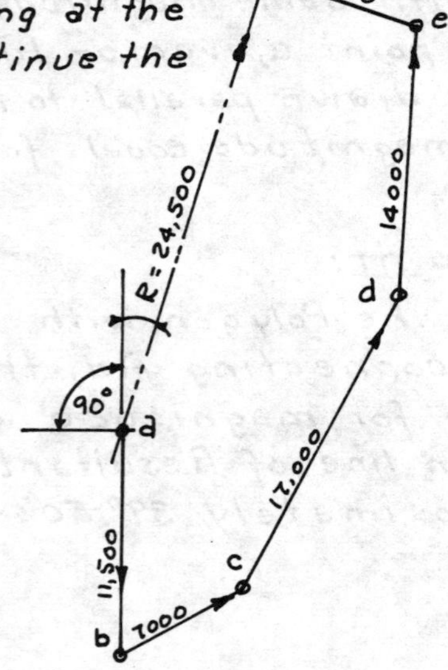

• FORCE POLYGON •
Scale: 1" = 10,000#

EXAMPLE: Analyzing a crane hoist 5.5.3.5

The illustrated elevation for a Crane Hoist was a part of the examination submitted to applicants by the Texas State Board of Registration for Professional Engineers at Austin, Texas., on March 12, 1944.

REQUIRED:
Select a standard H Column for strut member AB to support the Boom for the 10 Ton Crane Hoist shown in the illustration. Use A.I.S.C. Column specifications, and give exact length of strut.

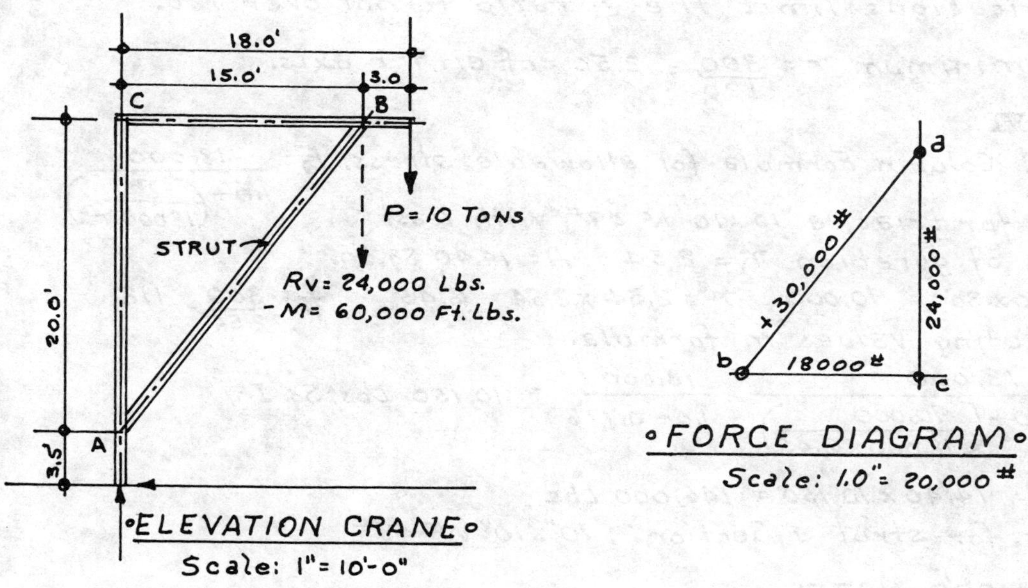

•ELEVATION CRANE•
Scale: 1" = 10'-0"

•FORCE DIAGRAM•
Scale: 1.0" = 20,000 #

STEP I:
Determine length of strut AB. Side $c = \sqrt{15.0^2 + 20.0^2} = 25.0$ Ft.

STEP II:
Move load to point B, and find vertical reaction.

$$R_V = \frac{20{,}000 \times 18.0}{15.0} = 24{,}000 \text{ Lbs.}$$

Vertical load line for force diagram = 24,000 Lbs.

STEP III:
Tipping or horizontal force BC can be checked by the method of moments.

$$A_H = \frac{24{,}000 \times 15.0}{20.0} = 18{,}000 \text{ Lbs.}$$ Checks with horizontal b-c.

EXAMPLE: Analyzing a crane hoist, continued 5.5.3.5

STEP IV:
Solving for angle A: Tan. = $\frac{15.0}{20.0}$ = 0.75000 From Tables:
Angle A = 36° 52' Sine = 0.6000 Cos. 0.8000 Secant: 1.2500
Stress in strut AB = b Sec. A. AB = 24,000 x 1.2500 = 30,000 Lbs.

STEP V:
Designing strut as a Column 25.0 feet long unsupported, except at ends, and axial load of 30,000 Lbs. compressive.

P = 30,000 l = 25.0 x 12 = 300 inches.
Classify this strut as a Main member, where AISC specifications limit the $\frac{l}{r}$ ratio to not over 120.

Then minimum $r = \frac{300}{120} = 2.50$ of either axis.

STEP VI:
A.I.S.C. Column Formula for allowable stress: $F_a = \dfrac{18{,}000}{1.0 + \left(\dfrac{l^2}{18000\, r^2}\right)}$

Select for trial, a 10 x 10 WF 49# with least radius of gyration r_y = 2.54 A = 14.40 Sq. In.
l^2 = 300 x 300 = 90,000 r^2 = 2.54 x 2.54 = 6.45 $\frac{l}{r} = \frac{300}{2.54} = 118$
Substituting values in formula:

$$F_a = \frac{18{,}000}{1.0 + \left(\dfrac{90{,}000}{18{,}000 \times 6.45}\right)} = \frac{18{,}000}{1.0 + 0.776} = 10{,}150 \text{ Lbs. Sq. In.}$$

Max. P = 14.40 x 10,150 = 146,000 Lbs.
Accept for strut a Section: 10" x 10" WF 49#

DESIGNER'S NOTE:
In a great number of the problems which are submitted to applicants for examinations to ascertain proficiency, the problem will contain circumstances which are deceiving. This deception is purposely done to test the applicant's judgement of conditions. By calling the diagonal brace a strut, one could be led to figure the brace as a secondary member where the slenderness ratio could be acceptable between 120 and 200. Obviously member AB, is as important in this structure as the mast CA.

EXAMPLE: Graphic analysis of a boom derrick 5.5.3.6

The preliminary drawing of a Hoist Derrick calls for a Mast Height of 27.0 feet, and a boom length of 23.0 Ft. Boom is hinged at bottom 4.0 feet from base. The control line for lifting load is located at ends of Mast and Boom. Winch operator for load cable is in a stationary position where the control line makes an angle of 29° 10' with vertical mast.

REQUIRED:

Draw a sketch of this problem to scale, with the Boom positioned at an angle of 52° 45' with horizontal plane. Draw force diagram to determine the stress in each member, when the vertical load on Boom end is not to exceed 2500 Lbs. Neglect weight of Mast and Boom.

STEP I:

Choosing a scale of 1.0 inch equals 10.0 feet, drawing is made with Boom in required position.

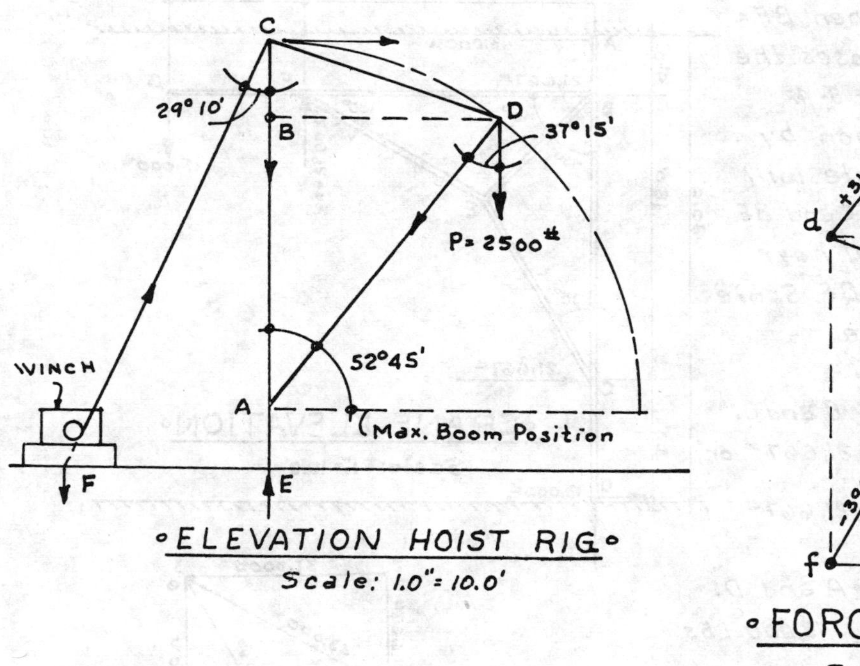

∘ELEVATION HOIST RIG∘

Scale: 1.0" = 10.0'

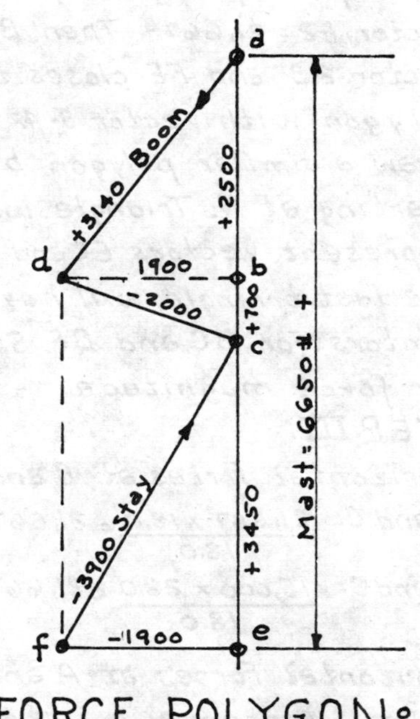

∘FORCE POLYGON∘

Scale: 1" = 2000#

EXAMPLE: Analyzing a jib crane 5.5.3.7

A Jib Crane is supported with swivel connections at floor and Ceiling for 360 degree swing. A 30.0 foot Mast height has a 26.0 foot boom connected with struts as shown. Clearance between boom and ceiling is 4.0 feet. At extreme end of boom a 7½ Ton capacity hoist is attached. Supporting struts are BE, FE and CE.

REQUIRED:

Calculate the horizontal force at Floor and Ceiling for Mast connections, then construct a force polygon to find forces in struts, mast and boom.

STEP I:

The boom will have a cantilever type bending moment which be add to force BF when designing member. M= 15,000 x 8.0= 120,000'#

Determine Vertical Reaction at F for using as known force.

$$R_v = \frac{15000 \times 26.0}{18.0} = 21{,}667 \text{ Lbs.}$$

STEP II:

Start force polygon by drawing vector 1-2 = 21,667# Then BF= vector 2-3 and FE closes the polygon with vector 3-4. Draw a similar polygon by starting at 4. Triangle will represent vectors EC and BE. The last triangle will have vectors for BC and BF. Scale for force magnitude.

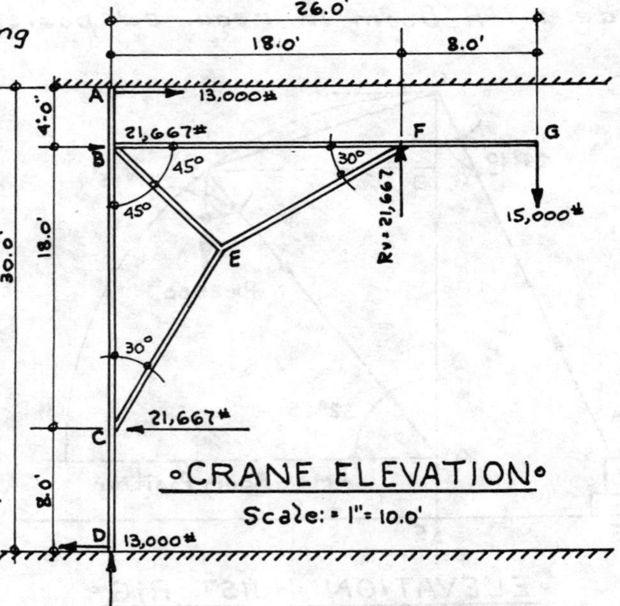

•CRANE ELEVATION•

Scale: 1"= 10.0'

STEP III:

Horizontal Forces at B and C:

$$B \text{ and } C = \frac{21{,}667 \times 18.0}{18.0} = 21{,}667\# \text{ or}$$

$$B \text{ and } C = \frac{15{,}000 \times 26.0}{18.0} = 21{,}667\#$$

Horizontal Forces at A and D:

$$A = D = \frac{15{,}000 \times 26.0}{30.0} = 13{,}000 \text{ Lbs.}$$

Vertical Force at A due to end load G:

$$B_v = A_v = \frac{15{,}000 \times 8.0}{18.0} = 6{,}667 \text{ Lbs.}$$

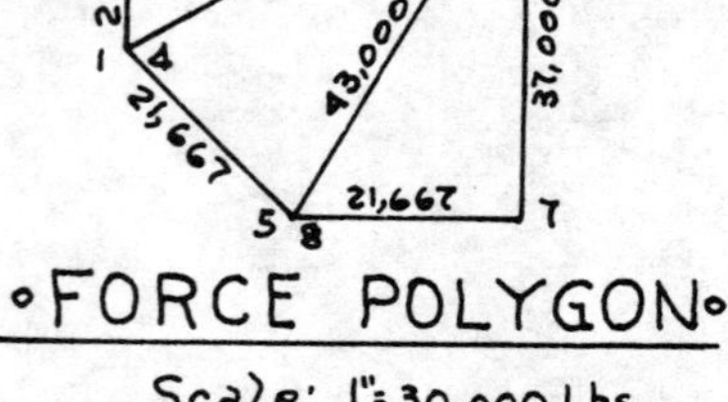

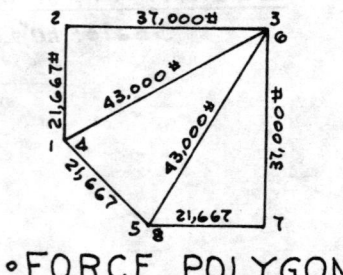

•FORCE POLYGON•

Scale: 1"= 30,000 Lbs.

EXAMPLE: Analysis of a boom crane 5.5.3.8

A manual gear operated derrick is to be constructed with a Mast height of 24.0 Feet. A cable stay extends from mast top to floor, and makes an angle of 30 degrees with vertical. Boom is 24.0 feet long and operates in one plane only or in same action line as stay.
Boom is hinged 4.0 feet from floor, and control line with pulley wheel is located 9.0 feet from end of boom, and mast control pulley is located 19.0 feet from floor. Pick up hook is placed on extreme end of boom.

REQUIRED:
Calculate stresses in derrick's boom, mast, and stay, when raising a 2200 Pound load on boom. Boom position will be on a 36 degree angle with mast, or 54° from horizontal. Consider the Dead Loads. Mast weight = 1450 Lbs. Boom = 1200.#

STEP I:
Solution will require an elevation of derrick with dimensions, and points identified. Forces are Co-planer.

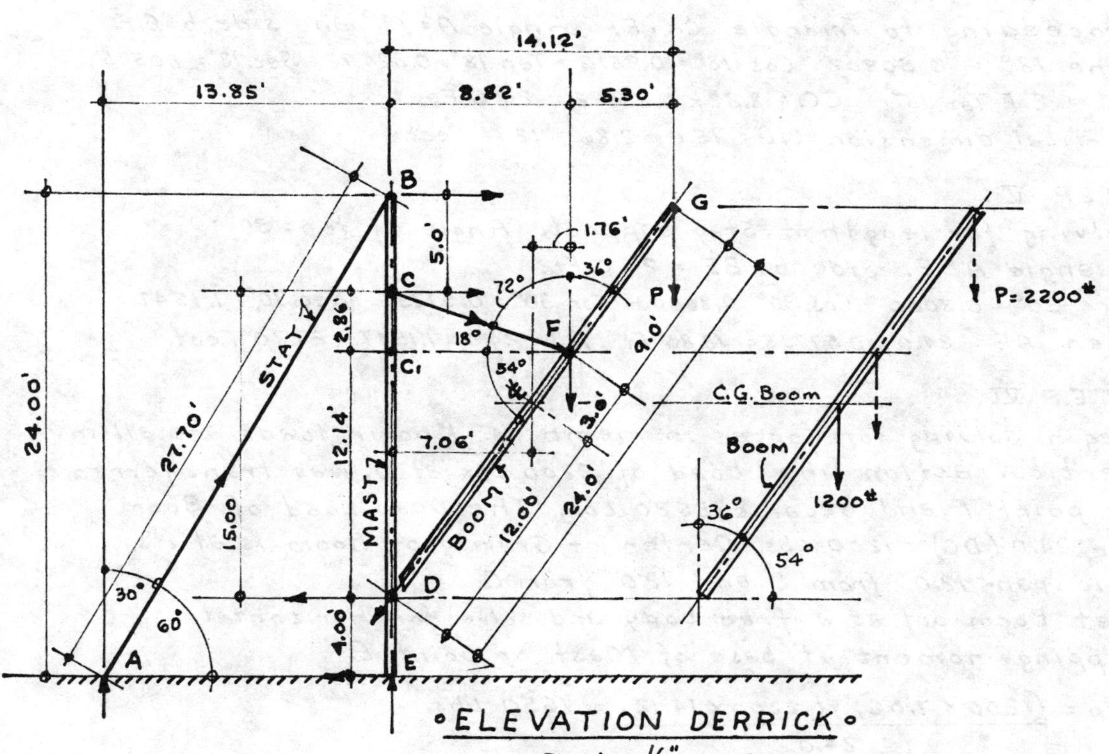

•ELEVATION DERRICK•
Scale: 1/8" = 1.0 FT.

EXAMPLE: Analysis of a boom crane, continued 5.5.3.8

STEP II:
All angles have been converted to Right Angle Triangles for convenience in solving for the dimensions.
Hoist load P= 2200 Lbs. can be relocated at point F, by the moment method.
Solve for horizontal distances points F and G are from Mast.
A = 36° Sin. A = 0.58778, Cos. A = 0.809, Tan A = 0.72654, and Sec. A = 1.2361
DF = 24.0' - 9.0' = 15.0' C_1F = Side a = C Sin A. C_1F = 15.0 x 0.58778 = 8.82 Ft.
Solve for side C_1D = b. C_1D = DF Cos. A. C_1D = 15.0 x 0.809 = 12.14 Ft.
Horizontal Distance from Mast to Point G = DG Sin. A
BG = 24.0 x 0.58778 = 14.12 Ft. Horiz. FG = 14.12 - 8.82 = 5.30 Ft.

STEP III:
Take moment about Point B to solve for Reaction at Point F.
Vertical Reaction at F = $\frac{2200 \times 14.12}{8.82}$ = 3520 Lbs.
This load will be substituted for load P at point G. There is a bending moment at F which is 2200 x 5.30 = 11,660 Ft. Lbs.
When Boom DG is in horizontal position the cantilever bending moment becomes 2200 x 9.0 = 19,800 Foot Lbs., and could be the critical factor in the design.

STEP IV:
Proceeding to Triangle CC_1F. Angle A = 18° and side b = C_1F.
Sine 18° = 0.30902 Cos. 18° = 0.9516 Tan. 18° = 0.32492 Sec. 18° = 1.0515
CC_1 = C_1F Tan. 18° CC_1 = 8.82 x 0.32492 = 2.86 Feet.
Vertical dimension C_1D = 15.0 - 2.86 = 12.14 Feet.

STEP V:
Solving for length of Stay support. Angle at top = 30°
Triangle ABE. Side b = BE = 24.0 Ft.
Sine 30° = 0.5000 Cos. 30° = 0.86603 Tan. 30° = 0.57735 Sec. 30° = 1.1547
Then AE = 24.0 x 0.57735 = 13.85 Ft. AB = 24.0 x 1.1547 = 27.70 Feet.

STEP VI:
Begin solving for forces in members. Known loads are all in vertical action line. Load of 2200 Lbs. at G, was transferred to point F, and became 3520 Lbs. The Dead Load of Boom for 24.0 (DG) = 1200 Lbs. Center of Gravity of Boom is at its midspan 12.0' from D and 12.0' from G.
Set Boom out as a free body and solve for Horizontal Tipping moment at base of Mast or point E.

$$E_H = \frac{(1200 \times 7.06) + (2200 \times 14.12)}{24.0'} = 1650 \text{ Lbs.}$$

EXAMPLE: Analysis of a boom crane, continued 5.5.3.8

STEP VII:

Solving for Stress in Stay member AB. Known is horizontal force AE = 1650 Lbs. Sine 30° = 0.5000 $AB = \frac{AE}{\text{Sine } 30°}$

Stress $AB = \frac{1650}{0.5000} = 3300$ Lbs.

STEP VIII:

Solving for Stress in BE (Mast). There are other forces in Mast in addition to force from tipping moment. Only from Triangle ABE is this force applicable. Cos. 30° = 0.86603

Force BE = AB × Cos. 30° BE = 3300 × 0.86603 = 2860 Lbs.

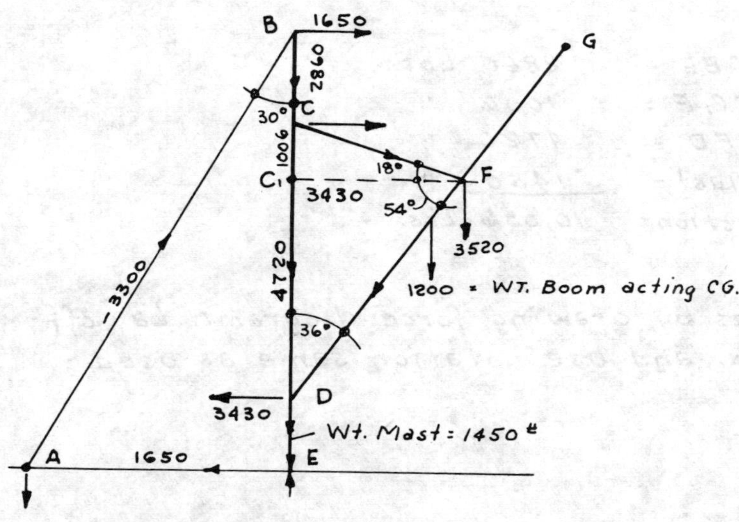

°DERRICK PROFILE°

Scale: 1/8" = 1'-0"

STEP IX:

To solve for Forces in Triangle CFD which is not a Right Angle Triangle. Drawing dash line C_1F, a triangle C_1FD is formed. Vertical Forces = 4720 Lbs. Solve for Horizontal Reaction at D.

$D_H = C_1D$ Tan. 36° $C_1F = 4720 \times 0.72654 = 3430$ Lbs.

STEP X:

Solve for Stress in Control Cable CF. Stress along $C_1F = 3430$ as Reaction D_H. Angle in Triangle $CFC_1 = 18°$ Sin. = 0.30902, Cos. = 0.951

Secant 18° = 1.0515.

Stress CF = 3430 × 1.0515 = 3610 Lbs.

STEP XI:

Additional stress in Mast member CC_1

CC_1 = CF Sin 18° $CC_1 = 3255 \times 0.30902 =$ 10,036 Lbs.

EXAMPLE: Analysis of a boom crane, continued 5.5.3.8

STEP XII:

Compressive Stress in Boom DF. (Axial) Bending Moment at F was solved in Step III to be 11,660 Foot Lbs. when in existing position.

Vertical Loads = 4720 Lbs. Angle = 36° Secant 36° = 1.2361

Cos. 36° = 0.80902

Stress in DF = 4720 × 1.2361 = 5835 Lbs. Axial.

STEP XIII:

Compressive stress. Maximum Compressive stress is at lowest point a E.

Stress from Triangle ABE =	2860	Lbs.
Stress from Triangle CC_1F =	1006	"
Stress from Triangle CFD =	4720	"
Stress from Weight of Mast =	1450	"
Total Vertical Reaction =	10,036	Lbs.

STEP XIV:

Checking stress figures by drawing force diagrams. Lay off load line of 10,036 Lbs., and use notation same as used in elevation.

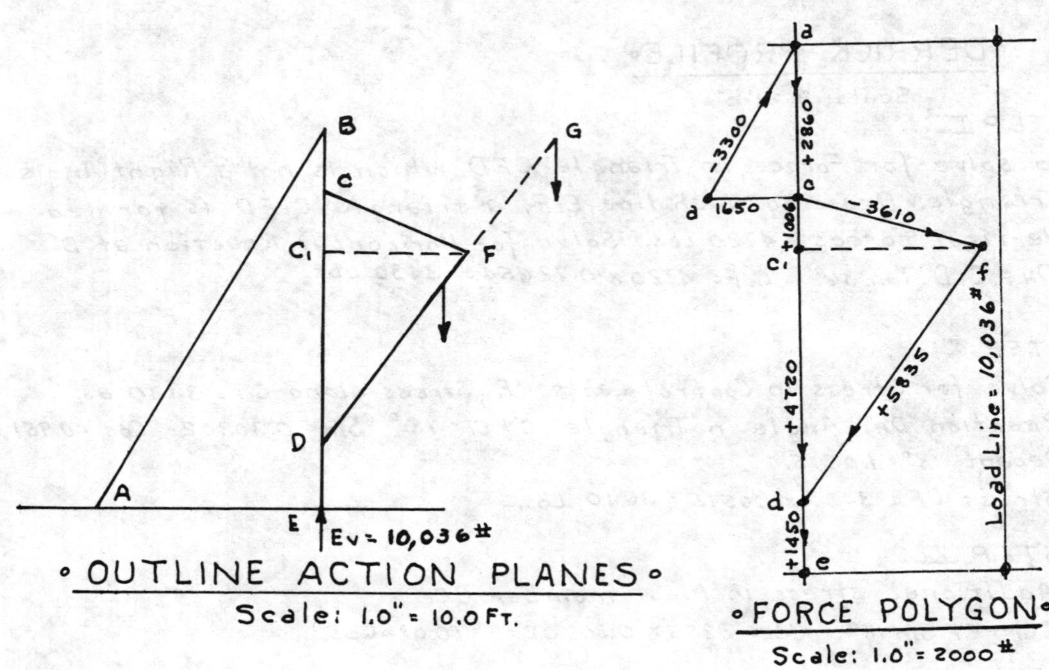

• OUTLINE ACTION PLANES •

Scale: 1.0" = 10.0 FT.

• FORCE POLYGON •

Scale: 1.0" = 2000#

EXAMPLE: Analyzing non-coplanar forces in rotating derrick 5.5.3.9

The 30.0 Foot Boom of a Derrick can swing 90° about the base of a 20.0 Foot Mast and approximately within 10° of mast in Vertical position. Supporting Mast tipping at top are two angular pipe stanchions. Each stanchion is located 45°

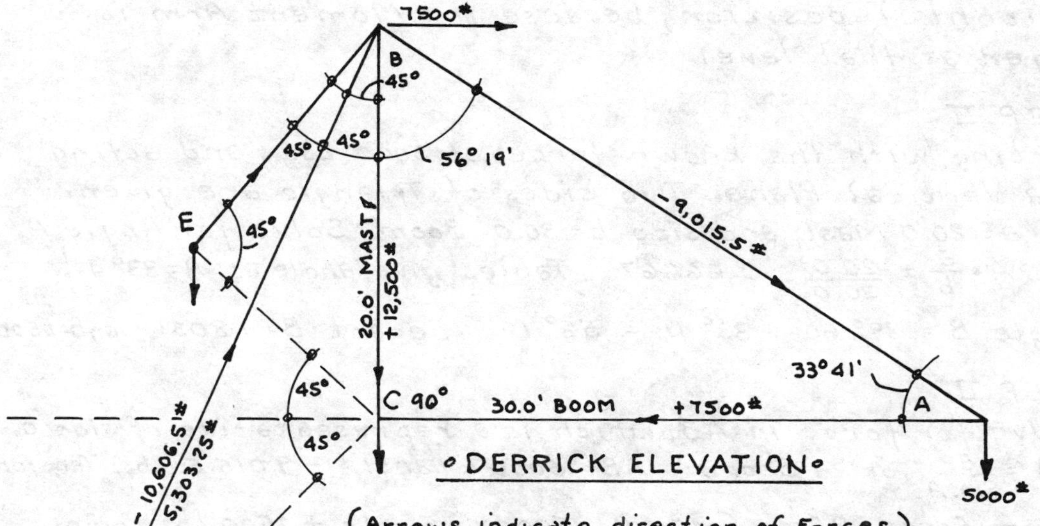

(Arrows indicate direction of Forces.)

from both the horizontal and vertical plane. Angle of 90° separates the stanchions at their bottom anchor locations. Maximum load at end of Boom is 5000 Lbs. Derrick is illustrated above.

REQUIRED:

Re-draw the elevation of Derrick to convenient scale and make a plan. Identify the members by letters placed at vertex of each angle. Calculate the maximum and minimum force in stanchions for anchorage when boom is swung in position to produce the force change.

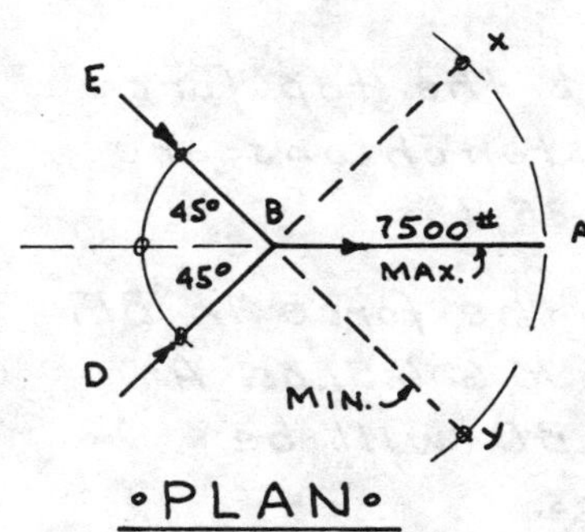

STEP I:

Derrick Members are identified thus: Boom = AC, Mast = BC and Stanchions are BE and BD. Boom support cable = AB. In plan, the maximum swing is arc X-Y. Forces are in same plane when AB is in line with BD or BE and will be maximum for on stanchion. The other stanchion will have no force when normal to the line of action of Boom AC.

EXAMPLE: Analyzing non-coplaner forces in rotating derrick, continued 5.5.3.9

Step I Continued:
Stanchions BD and BE, will have equal stress when Boom is in position A.
Maximum forces will be produced when Boom is in the horizontal position, because the Moment Arm is longer at that level.

STEP II:
Starting with the known force of 5000 Lbs., and acting in a Vertical Plane. Two sides of Triangle are given.
Let a = 20.0' Mast, and side b = 30.0' Boom. Solve for Angle A.
$\text{Tan. } A = \frac{a}{b} = \frac{20.0}{30.0} = 0.66667$ Tables give angle as: A = 33° 41'.

Angle B = 89° 60' − 33° 41' = 56° 19' Secant B = 1.8031 Tan B = 1.5004

STEP III:
Solve for force in AB, which is a representative of side c.
$AB = \frac{a}{\text{Sin. } A}$ or a Sec. B. AB = 5000 × 1.8031 = − 9,015.5 Lbs. Tension.

Boom Force AC = a Cotan. A = 5000 × 1.5004 = + 7500 Lbs. Comp.

STEP IV
The horizontal tipping force at Top of Mast is the same as in boom, equals 7500 Lbs.
Swinging Boom on Plan to plane DX, or DY, only one stanchion will be stressed, and this will be maximum.
Known force is 7500 Lbs., horizontal and is side b.
Angles A and B are 45°. Secant of 45° = 1.4142.
Then Maximum stress in BE or BD, when boom is swing to line of same action, equals 7,500 × 1.4142 = 10,606.5 Lbs.

STEP V:
Forces in stanchions BD and BE, when Boom is located with load at point A, in plan drawing.
When both stanchions equally support the top force of 7500 Lbs., acting horizontal, both stanchions are stressed. BD = BE or $\frac{10,606.50}{2}$ = 5,303.25 Lbs.
As boom is turned from point A to Y, the force in BE, will increase from 5303.25 Lbs, to Max. 10,606.5 Lbs. At the same time, the force in stanchion BD will be decreasing from 5,303.25 to zero 0.0 Lbs.

EXAMPLE: Analyzing non-coplaner forces in rotating derrick, continued 5.5.3.9

STEP VI:

Forces in Mast BC. These forces must act in a Vertical plane. They are compressive and tend to push the mast into its base.

The known force from triangle ABC = 5000 Lbs., and is acting in same plane of mast.

The force from Stanchions act at an angle of 45° with mast Force at Top is horizontal = 7500 Lbs. and when boom is at position X or Y, the diagonal C = 10,606.50 Lbs.

Solving for side a, or side b, of a 45° triangle, when side C = 10,606.50 Lbs. Sine 45° = 0.70711.

Then, a = 10,606.50 × 0.70711 = 7,500 Lbs.

Reaction at bottom of Mast = 7,500 + 5000 = 12,500 Lbs.

STEP VII:

The work can be checked by drawing scaled Force Diagrams. Draw triangle a, b, c, and vertical line bc, is drawn to scale 5000 Lbs. Close diagram and scale the magnitude of ab and ab.

Since 7500 Lbs., is a horizontal known force and 12,500 Lbs., is a vertical force, extend vertical or draw a seperate force diagram. Close diagram with 45° line, and scale magnitude of 10,606.50 Lbs. Lines be and bd, are 45° lines from load line CX.

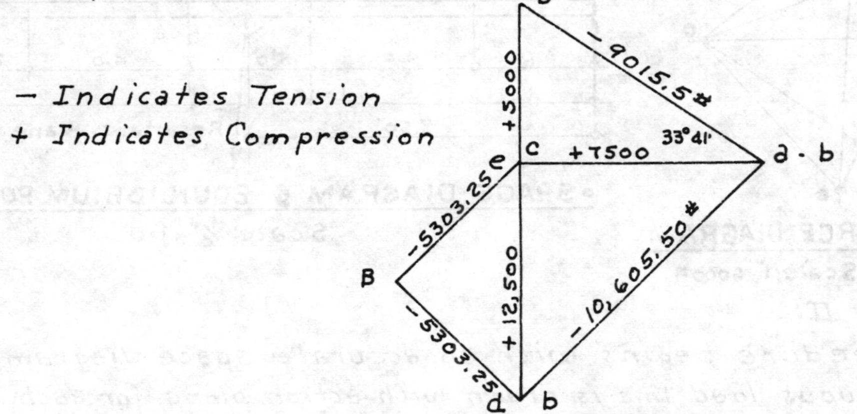

STRESS DIAGRAM
Scale: 1.0" = 5000 #

EXAMPLE: Graphic solution for non-parallel forces on a beam 5.5.3.10

A simple supported beam is 16.0 feet between end supports and sustains four (4) concentrated loads in a system of non-coplaner forces acting in different action lines as follows, from left to right:

P_1 = 2000 Lbs., Locate 3.0 feet. Action line = 20° 15' Right of Plumb
P_2 = 2600 " Locate 7.0 " Action line = 13° 30' Left of Plumb
P_3 = 4000 " Locate 10.0 " Action line: — On plumb line
P_4 = 1800 " Locate 14.0 " Action line = 22° 50' Left of Plumb.

REQUIRED:
Draw an elevation of beam with loads, then by resolution of forces, demonstrate how graphics may be used to solve for Resultant and line of action. Resolve resultant into a vertical load and calculate Reaction. Use trigonomentry for the last phase of problem. Use Bow's notations for identifying.

STEP I:
Scaled drawing will have loads placed in position by use of a protractor. Resultant magnitude and action line will be found by constructing a ray diagram and funicular polygon.

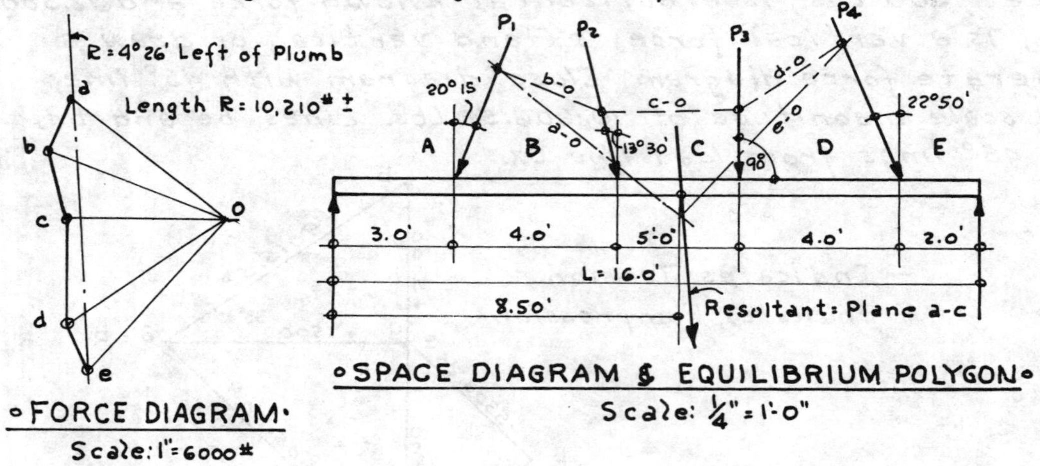

∘FORCE DIAGRAM∘
Scale: 1" = 6000#

∘SPACE DIAGRAM & EQUILIBRIUM POLYGON∘
Scale: 1/4" = 1'-0"

STEP II:
Proceedure begins with an accurate space diagram, then the continuous load line is drawn with action plane for each load.

(a) Polar point (O) for Ray diagram may be located at any point convenient but close to space diagram.

EXAMPLE: Graphic solution for non-parallel forces on a beam, continued 5.5.3.10

(b) Connect load line points to polar point O and construct the funicular polygon by transferring ray lines o-b, o-c and o-d.

(c) Extend lines o-a and o-e to locate resultant. Resultant magnitude and action line is closing string a-e at load line

STEP III:

The resultant a-e when carried over to funicular polygon intersects beam at a point 8.50 feet from left end. Magnitude of Resultant line a-e scales approximately 10,210 Lbs., and becomes a single concentrated load on beam. The line of action measured by protractor = 4° 26' to Right of Plumb line.

STEP IV:

The component single vertical force of all loads and the resultant is a force normal to beam.
Let angle A = 4° 26' and side b = vertical concentrated Load.
Resultant 10,210 Lbs. equals side c of triangle. Known values are angle A and magnitude c.
By Trig: side b = C Cos. A.
From Tables, the cosine of 4° 26' = 0.99701
Vertical concentrated force on beam = 10,210 × 0.99701 = 10,179 Lbs.

STEP V:

Calculating Reactions R_1 and R_2 by moment method:

$$R_1 = \frac{(16.0 - 8.50) \times 10{,}179}{16.0} = 4771.40 \text{ Lbs.}$$

$$R_2 = \frac{8.50 \times 10{,}179}{16.0} = 5408.60 \text{ Lbs.}$$

DESIGN NOTE:

This problem may be solved by trigonomentry by resolving each load into a vertical component normal to beam. The bending moment should be calculated under these conditions. A simple force diagram will also serve to convert a slanted load into a vertical load component force.

TABLE: Decimals of an inch 5.6.1

DECIMALS OF AN INCH
For each 64th of an inch
With Millimeter Equivalents

Fraction	1/64ths	Decimal	Millimeters (Approx.)	Fraction	1/64ths	Decimal	Millimeters (Approx.)
...	1	.015625	0.397	...	33	.515625	13.097
1/32	2	.03125	0.794	17/32	34	.53125	13.494
...	3	.046875	1.191	...	35	.546875	13.891
1/16	4	.0625	1.588	9/16	36	.5625	14.288
...	5	.078125	1.984	...	37	.578125	14.684
3/32	6	.09375	2.381	19/32	38	.59375	15.081
...	7	.109375	2.778	...	39	.609375	15.478
1/8	8	.125	3.175	5/8	40	.625	15.875
...	9	.140625	3.572	...	41	.640625	16.272
5/32	10	.15625	3.969	21/32	42	.65625	16.669
...	11	.171875	4.366	...	43	.671875	17.066
3/16	12	.1875	4.763	11/16	44	.6875	17.463
...	13	.203125	5.159	...	45	.703125	17.859
7/32	14	.21875	5.556	23/32	46	.71875	18.256
...	15	.234375	5.953	...	47	.734375	18.653
1/4	16	.250	6.350	3/4	48	.750	19.050
...	17	.265625	6.747	...	49	.765625	19.447
9/32	18	.28125	7.144	25/32	50	.78125	19.844
...	19	.296875	7.541	...	51	.796875	20.241
5/16	20	.3125	7.938	13/16	52	.8125	20.638
...	21	.328125	8.334	...	53	.828125	21.034
11/32	22	.34375	8.731	27/32	54	.84375	21.431
...	23	.359375	9.128	...	55	.859375	21.828
3/8	24	.375	9.525	7/8	56	.875	22.225
...	25	.390625	9.922	...	57	.890625	22.622
13/32	26	.40625	10.319	29/32	58	.90625	23.019
...	27	.421875	10.716	...	59	.921875	23.416
7/16	28	.4375	11.113	15/16	60	.9375	23.813
...	29	.453125	11.509	...	61	.953125	24.209
15/32	30	.46875	11.906	31/32	62	.96875	24.606
...	31	.484375	12.303	...	63	.984375	25.003
1/2	32	.500	12.700	1	64	1.000	25.400

TABLE: Decimals of a foot 5.6.2

For each 32nd of an inch

Inch	0	1	2	3	4	5
0	0	.0833	.1667	.2500	.3333	.4167
1/32	.0026	.0859	.1693	.2526	.3359	.4193
1/16	.0052	.0885	.1719	.2552	.3385	.4219
3/32	.0078	.0911	.1745	.2578	.3411	.4245
1/8	.0104	.0938	.1771	.2604	.3438	.4271
5/32	.0130	.0964	.1797	.2630	.3464	.4297
3/16	.0156	.0990	.1823	.2656	.3490	.4323
7/32	.0182	.1016	.1849	.2682	.3516	.4349
1/4	.0208	.1042	.1875	.2708	.3542	.4375
9/32	.0234	.1068	.1901	.2734	.3568	.4401
5/16	.0260	.1094	.1927	.2760	.3594	.4427
11/32	.0286	.1120	.1953	.2786	.3620	.4453
3/8	.0313	.1146	.1979	.2812	.3646	.4479
13/32	.0339	.1172	.2005	.2839	.3672	.4505
7/16	.0365	.1198	.2031	.2865	.3698	.4531
15/32	.0391	.1224	.2057	.2891	.3724	.4557
1/2	.0417	.1250	.2083	.2917	.3750	.4583
17/32	.0443	.1276	.2109	.2943	.3776	.4609
9/16	.0469	.1302	.2135	.2969	.3802	.4635
19/32	.0495	.1328	.2161	.2995	.3828	.4661
5/8	.0521	.1354	.2188	.3021	.3854	.4688
21/32	.0547	.1380	.2214	.3047	.3880	.4714
11/16	.0573	.1406	.2240	.3073	.3906	.4740
23/32	.0599	.1432	.2266	.3099	.3932	.4766
3/4	.0625	.1458	.2292	.3125	.3958	.4792
25/32	.0651	.1484	.2318	.3151	.3984	.4818
13/16	.0677	.1510	.2344	.3177	.4010	.4844
27/32	.0703	.1536	.2370	.3203	.4036	.4870
7/8	.0729	.1563	.2396	.3229	.4063	.4896
29/32	.0755	.1589	.2422	.3255	.4089	.4922
15/16	.0781	.1615	.2448	.3281	.4115	.4948
31/32	.0807	.1641	.2474	.3307	.4141	.4974

TABLE: Decimals of a foot, continued 5.6.2

For each 32nd of an inch

Inch	6	7	8	9	10	11
0	.5000	.5833	.6667	.7500	.8333	.9167
1/32	.5026	.5859	.6693	.7526	.8359	.9193
1/16	.5052	.5885	.6719	.7552	.8385	.9219
3/32	.5078	.5911	.6745	.7578	.8411	.9245
1/8	.5104	.5938	.6771	.7604	.8438	.9271
5/32	.5130	.5964	.6797	.7630	.8464	.9297
3/16	.5156	.5990	.6823	.7656	.8490	.9323
7/32	.5182	.6016	.6849	.7682	.8516	.9349
1/4	.5208	.6042	.6875	.7708	.8542	.9375
9/32	.5234	.6068	.6901	.7734	.8568	.9401
5/16	.5260	.6094	.6927	.7760	.8594	.9427
11/32	.5286	.6120	.6953	.7786	.8620	.9453
3/8	.5313	.6146	.6979	.7813	.8646	.9479
13/32	.5339	.6172	.7005	.7839	.8672	.9505
7/16	.5365	.6198	.7031	.7865	.8698	.9531
15/32	.5391	.6224	.7057	.7891	.8724	.9557
1/2	.5417	.6250	.7083	.7917	.8750	.9583
17/32	.5443	.6276	.7109	.7943	.8776	.9609
9/16	.5469	.6302	.7135	.7969	.8802	.9635
19/32	.5495	.6328	.7161	.7995	.8828	.9661
5/8	.5521	.6354	.7188	.8021	.8854	.9688
21/32	.5547	.6380	.7214	.8047	.8880	.9714
11/16	.5573	.6406	.7240	.8073	.8906	.9740
23/32	.5599	.6432	.7266	.8099	.8932	.9766
3/4	.5625	.6458	.7292	.8125	.8958	.9792
25/32	.5651	.6484	.7318	.8151	.8984	.9818
13/16	.5677	.6510	.7344	.8177	.9010	.9844
27/32	.5703	.6536	.7370	.8203	.9036	.9870
7/8	.5729	.6563	.7396	.8229	.9063	.9896
29/32	.5755	.6589	.7422	.8255	.9089	.9922
15/16	.5781	.6615	.7448	.8281	.9115	.9948
31/32	.5807	.6641	.7474	.8307	.9141	.9974

TABLE: Functions of numbers (.01 through .49) 5.6.3

No.	Square	Cube	Square Root	Cube Root	Logarithm	1000 × Reciprocal	No. = Diameter	
							Circum.	Area
.01	.0001	.000001	0.1000	0.2154	$\bar{2}$.00000	100000.000	.03142	.000079
.02	.0004	.000008	0.1414	0.2714	$\bar{2}$.30103	50000.000	.06283	.000314
.03	.0009	.000027	0.1732	0.3107	$\bar{2}$.47712	33333.333	.09425	.000707
.04	.0016	.000064	0.2000	0.3420	$\bar{2}$.60206	25000.000	.12566	.001257
.05	.0025	.000125	0.2236	0.3684	$\bar{2}$.69897	20000.000	.15708	.001964
.06	.0036	.000216	0.2449	0.3915	$\bar{2}$.77815	16666.667	.18850	.002827
.07	.0049	.000343	0.2646	0.4121	$\bar{2}$.84510	14285.714	.21991	.003849
.08	.0064	.000512	0.2828	0.4309	$\bar{2}$.90309	12500.000	.25133	.005027
.09	.0081	.000729	0.3000	0.4481	$\bar{2}$.95424	11111.111	.28274	.006362
.10	.0100	.001000	0.3162	0.4642	$\bar{1}$.00000	10000.000	.31416	.007854
.11	.0121	.001331	0.3317	0.4791	$\bar{1}$.04139	9090.909	.34558	.009503
.12	.0144	.001728	0.3464	0.4932	$\bar{1}$.07918	8333.333	.37699	.011310
.13	.0169	.002197	0.3606	0.5066	$\bar{1}$.11394	7692.308	.40841	.013273
.14	.0196	.002744	0.3742	0.5192	$\bar{1}$.14613	7142.857	.43982	.015394
.15	.0225	.003375	0.3873	0.5313	$\bar{1}$.17609	6666.667	.47124	.017672
.16	.0256	.004096	0.4000	0.5429	$\bar{1}$.20412	6250.000	.50265	.020106
.17	.0289	.004913	0.4123	0.5540	$\bar{1}$.23045	5882.353	.53407	.022698
.18	.0324	.005832	0.4243	0.5646	$\bar{1}$.25527	5555.556	.56549	.025447
.19	.0361	.006859	0.4359	0.5749	$\bar{1}$.27875	5263.158	.59690	.028353
.20	.0400	.008000	0.4472	0.5848	$\bar{1}$.30103	5000.000	.62832	.031416
.21	.0441	.009261	0.4583	0.5944	$\bar{1}$.32222	4761.905	.65973	.034636
.22	.0484	.010648	0.4690	0.6037	$\bar{1}$.34242	4545.455	.69115	.038013
.23	.0529	.012167	0.4796	0.6127	$\bar{1}$.36173	4347.826	.72257	.041548
.24	.0576	.013824	0.4899	0.6214	$\bar{1}$.38021	4166.667	.75398	.045239
.25	.0625	.015625	0.5000	0.6300	$\bar{1}$.39794	4000.000	.78540	.049087
.26	.0676	.017576	0.5099	0.6383	$\bar{1}$.41497	3846.154	.81681	.053093
.27	.0729	.019683	0.5196	0.6463	$\bar{1}$.43136	3703.704	.84823	.057256
.28	.0784	.021952	0.5292	0.6542	$\bar{1}$.44716	3571.429	.87965	.061575
.29	.0841	.024389	0.5385	0.6619	$\bar{1}$.46240	3448.276	.91106	.066052
.30	.0900	.027000	0.5477	0.6694	$\bar{1}$.47712	3333.333	.94248	.070686
.31	.0961	.029791	0.5568	0.6768	$\bar{1}$.49136	3225.807	.97389	.075477
.32	.1024	.032768	0.5657	0.6840	$\bar{1}$.50515	3125.000	1.00531	.080425
.33	.1089	.035937	0.5745	0.6910	$\bar{1}$.51851	3030.303	1.03673	.085530
.34	.1156	.039304	0.5831	0.6980	$\bar{1}$.53148	2941.177	1.06814	.090792
.35	.1225	.042875	0.5916	0.7047	$\bar{1}$.54407	2857.143	1.09956	.096211
.36	.1296	.046656	0.6000	0.7114	$\bar{1}$.55630	2777.778	1.13097	.101788
.37	.1369	.050653	0.6083	0.7179	$\bar{1}$.56820	2702.703	1.16239	.107521
.38	.1444	.054872	0.6164	0.7243	$\bar{1}$.57978	2631.579	1.19381	.113411
.39	.1521	.059319	0.6245	0.7306	$\bar{1}$.59106	2564.103	1.22522	.119459
.40	.1600	.064000	0.6325	0.7368	$\bar{1}$.60206	2500.000	1.2566	.125664
.41	.1681	.068921	0.6403	0.7429	$\bar{1}$.61278	2439.024	1.2881	.132025
.42	.1764	.074088	0.6481	0.7489	$\bar{1}$.62325	2380.952	1.3195	.138544
.43	.1849	.079507	0.6557	0.7548	$\bar{1}$.63347	2325.581	1.3509	.145220
.44	.1936	.085184	0.6633	0.7606	$\bar{1}$.64345	2272.727	1.3823	.152053
.45	.2025	.091125	0.6708	0.7663	$\bar{1}$.65321	2222.222	1.4137	.159043
.46	.2116	.097336	0.6782	0.7719	$\bar{1}$.66276	2173.913	1.4451	.166190
.47	.2209	.103823	0.6856	0.7775	$\bar{1}$.67210	2127.660	1.4765	.173494
.48	.2304	.110592	0.6928	0.7830	$\bar{1}$.68124	2083.333	1.5080	.180956
.49	.2401	.117649	0.7000	0.7884	$\bar{1}$.69020	2040.816	1.5394	.188574

TABLE: Functions of numbers, continued (.50 through .99) 5.6.3

No.	Square	Cube	Square Root	Cube Root	Logarithm	1000 × Reciprocal	No. = Diameter	
							Circum.	Area
.50	.2500	.125000	0.7071	0.7937	$\bar{1}$.69897	2000.000	1.5708	.19635
.51	.2601	.132651	0.7141	0.7990	$\bar{1}$.70757	1960.784	1.6022	.20428
.52	.2704	.140608	0.7211	0.8041	$\bar{1}$.71600	1923.077	1.6336	.21237
.53	.2809	.148877	0.7280	0.8093	$\bar{1}$.72428	1886.793	1.6650	.22062
.54	.2916	.157464	0.7348	0.8143	$\bar{1}$.73239	1851.852	1.6965	.22902
.55	.3025	.166375	0.7416	0.8193	$\bar{1}$.74036	1818.182	1.7279	.23758
.56	.3136	.175616	0.7483	0.8243	$\bar{1}$.74819	1785.714	1.7593	.24630
.57	.3249	.185193	0.7550	0.8291	$\bar{1}$.75587	1754.386	1.7907	.25518
.58	.3364	.195112	0.7616	0.8340	$\bar{1}$.76343	1724.138	1.8221	.26421
.59	.3481	.205379	0.7681	0.8387	$\bar{1}$.77085	1694.915	1.8535	.27340
.60	.3600	.216000	0.7746	0.8434	$\bar{1}$.77815	1666.667	1.8850	.28274
.61	.3721	.226981	0.7810	0.8481	$\bar{1}$.78533	1639.344	1.9164	.29225
.62	.3844	.238328	0.7874	0.8527	$\bar{1}$.79239	1612.903	1.9478	.30191
.63	.3969	.250047	0.7937	0.8573	$\bar{1}$.79934	1587.302	1.9792	.31173
.64	.4096	.262144	0.8000	0.8618	$\bar{1}$.80618	1562.500	2.0106	.32170
.65	.4225	.274625	0.8062	0.8662	$\bar{1}$.81291	1538.462	2.0420	.33183
.66	.4356	.287496	0.8124	0.8707	$\bar{1}$.81954	1515.152	2.0735	.34212
.67	.4489	.300763	0.8185	0.8750	$\bar{1}$.82607	1492.537	2.1049	.35257
.68	.4624	.314432	0.8246	0.8794	$\bar{1}$.83251	1470.588	2.1363	.36317
.69	.4761	.328509	0.8307	0.8837	$\bar{1}$.83885	1449.275	2.1677	.37393
.70	.4900	.343000	0.8367	0.8879	$\bar{1}$.84510	1428.571	2.1991	.38485
.71	.5041	.357911	0.8426	0.8921	$\bar{1}$.85126	1408.451	2.2305	.39592
.72	.5184	.373248	0.8485	0.8963	$\bar{1}$.85733	1388.889	2.2620	.40715
.73	.5329	.389017	0.8544	0.9004	$\bar{1}$.86332	1369.863	2.2934	.41854
.74	.5476	.405224	0.8602	0.9045	$\bar{1}$.86923	1351.351	2.3248	.43008
.75	.5625	.421875	0.8660	0.9086	$\bar{1}$.87506	1333.333	2.3562	.44179
.76	.5776	.438976	0.8718	0.9126	$\bar{1}$.88081	1315.790	2.3876	.45365
.77	.5929	.456533	0.8775	0.9166	$\bar{1}$.88649	1298.701	2.4190	.46566
.78	.6084	.474552	0.8832	0.9205	$\bar{1}$.89209	1282.051	2.4504	.47784
.79	.6241	.493039	0.8888	0.9244	$\bar{1}$.89763	1265.823	2.4819	.49017
.80	.6400	.512000	0.8944	0.9283	$\bar{1}$.90309	1250.000	2.5133	.50266
.81	.6561	.531441	0.9000	0.9322	$\bar{1}$.90849	1234.568	2.5447	.51530
.82	.6724	.551368	0.9055	0.9360	$\bar{1}$.91381	1219.512	2.5761	.52810
.83	.6889	.571787	0.9110	0.9398	$\bar{1}$.91908	1204.819	2.6075	.54106
.84	.7056	.592704	0.9165	0.9435	$\bar{1}$.92428	1190.476	2.6389	.55418
.85	.7225	.614125	0.9220	0.9473	$\bar{1}$.92942	1176.471	2.6704	.56745
.86	.7396	.636056	0.9274	0.9510	$\bar{1}$.93450	1162.791	2.7018	.58088
.87	.7569	.658503	0.9327	0.9546	$\bar{1}$.93952	1149.425	2.7332	.59447
.88	.7744	.681472	0.9381	0.9583	$\bar{1}$.94448	1136.364	2.7646	.60821
.89	.7921	.704969	0.9434	0.9619	$\bar{1}$.94939	1123.596	2.7960	.62211
.90	.8100	.729000	0.9487	0.9655	$\bar{1}$.95424	1111.111	2.8274	.63617
.91	.8281	.753571	0.9539	0.9691	$\bar{1}$.95904	1098.901	2.8589	.65039
.92	.8464	.778688	0.9592	0.9726	$\bar{1}$.96379	1086.957	2.8903	.66476
.93	.8649	.804357	0.9644	0.9761	$\bar{1}$.96848	1075.269	2.9217	.67929
.94	.8836	.830584	0.9695	0.9796	$\bar{1}$.97313	1063.830	2.9531	.69398
.95	.9025	.857375	0.9747	0.9830	$\bar{1}$.97772	1052.632	2.9845	.70882
.96	.9216	.884736	0.9798	0.9865	$\bar{1}$.98227	1041.667	3.0159	.72382
.97	.9409	.912673	0.9849	0.9899	$\bar{1}$.98677	1030.928	3.0473	.73898
.98	.9604	.941192	0.9899	0.9933	$\bar{1}$.99123	1020.408	3.0788	.75430
.99	.9801	.970299	0.9950	0.9967	$\bar{1}$.99564	1010.101	3.1102	.76977

TABLE: Functions of numbers, continued (1 through 49) 5.6.3

No.	Square	Cube	Square Root	Cube Root	Logarithm	1000 × Reciprocal	No. = Diameter	
							Circum.	Area
1	1	1	1.0000	1.0000	0.00000	1000.000	3.142	0.7854
2	4	8	1.4142	1.2599	0.30103	500.000	6.283	3.1416
3	9	27	1.7321	1.4422	0.47712	333.333	9.425	7.0686
4	16	64	2.0000	1.5874	0.60206	250.000	12.566	12.5664
5	25	125	2.2361	1.7100	0.69897	200.000	15.708	19.6350
6	36	216	2.4495	1.8171	0.77815	166.667	18.850	28.2743
7	49	343	2.6458	1.9129	0.84510	142.857	21.991	38.4845
8	64	512	2.8284	2.0000	0.90309	125.000	25.133	50.2655
9	81	729	3.0000	2.0801	0.95424	111.111	28.274	63.6173
10	100	1000	3.1623	2.1544	1.00000	100.000	31.416	78.5398
11	121	1331	3.3166	2.2240	1.04139	90.9091	34.558	95.0332
12	144	1728	3.4641	2.2894	1.07918	83.3333	37.699	113.097
13	169	2197	3.6056	2.3513	1.11394	76.9231	40.841	132.732
14	196	2744	3.7417	2.4101	1.14613	71.4286	43.982	153.938
15	225	3375	3.8730	2.4662	1.17609	66.6667	47.124	176.715
16	256	4096	4.0000	2.5198	1.20412	62.5000	50.265	201.062
17	289	4913	4.1231	2.5713	1.23045	58.8235	53.407	226.980
18	324	5832	4.2426	2.6207	1.25527	55.5556	56.549	254.469
19	361	6859	4.3589	2.6684	1.27875	52.6316	59.690	283.529
20	400	8000	4.4721	2.7144	1.30103	50.0000	62.832	314.159
21	441	9261	4.5826	2.7589	1.32222	47.6190	65.973	346.361
22	484	10648	4.6904	2.8020	1.34242	45.4545	69.115	380.133
23	529	12167	4.7958	2.8439	1.36173	43.4783	72.257	415.476
24	576	13824	4.8990	2.8845	1.38021	41.6667	75.398	452.389
25	625	15625	5.0000	2.9240	1.39794	40.0000	78.540	490.874
26	676	17576	5.0990	2.9625	1.41497	38.4615	81.681	530.929
27	729	19683	5.1962	3.0000	1.43136	37.0370	84.823	572.555
28	784	21952	5.2915	3.0366	1.44716	35.7143	87.965	615.752
29	841	24389	5.3852	3.0723	1.46240	34.4828	91.106	660.520
30	900	27000	5.4772	3.1072	1.47712	33.3333	94.248	706.858
31	961	29791	5.5678	3.1414	1.49136	32.2581	97.389	754.768
32	1024	32768	5.6569	3.1748	1.50515	31.2500	100.531	804.248
33	1089	35937	5.7446	3.2075	1.51851	30.3030	103.673	855.299
34	1156	39304	5.8310	3.2396	1.53148	29.4118	106.814	907.920
35	1225	42875	5.9161	3.2711	1.54407	28.5714	109.956	962.113
36	1296	46656	6.0000	3.3019	1.55630	27.7778	113.097	1017.88
37	1369	50653	6.0828	3.3322	1.56820	27.0270	116.239	1075.21
38	1444	54872	6.1644	3.3620	1.57978	26.3158	119.381	1134.11
39	1521	59319	6.2450	3.3912	1.59106	25.6410	122.522	1194.59
40	1600	64000	6.3246	3.4200	1.60206	25.0000	125.66	1256.64
41	1681	68921	6.4031	3.4482	1.61278	24.3902	128.81	1320.25
42	1764	74088	6.4807	3.4760	1.62325	23.8095	131.95	1385.44
43	1849	79507	6.5574	3.5034	1.63347	23.2558	135.09	1452.20
44	1936	85184	6.6332	3.5303	1.64345	22.7273	138.23	1520.53
45	2025	91125	6.7082	3.5569	1.65321	22.2222	141.37	1590.43
46	2116	97336	6.7823	3.5830	1.66276	21.7391	144.51	1661.90
47	2209	103823	6.8557	3.6088	1.67210	21.2766	147.65	1734.94
48	2304	110592	6.9282	3.6342	1.68124	20.8333	150.80	1809.56
49	2401	117649	7.0000	3.6593	1.69020	20.4082	153.94	1885.74

TABLE: Functions of numbers, continued (50 through 99) 5.6.3

No.	Square	Cube	Square Root	Cube Root	Logarithm	1000 × Reciprocal	No. = Diameter	
							Circum.	Area
50	2500	125000	7.0711	3.6840	1.69897	20.0000	157.08	1963.50
51	2601	132651	7.1414	3.7084	1.70757	19.6078	160.22	2042.82
52	2704	140608	7.2111	3.7325	1.71600	19.2308	163.36	2123.72
53	2809	148877	7.2801	3.7563	1.72428	18.8679	166.50	2206.18
54	2916	157464	7.3485	3.7798	1.73239	18.5185	169.65	2290.22
55	3025	166375	7.4162	3.8030	1.74036	18.1818	172.79	2375.83
56	3136	175616	7.4833	3.8259	1.74819	17.8571	175.93	2463.01
57	3249	185193	7.5498	3.8485	1.75587	17.5439	179.07	2551.76
58	3364	195112	7.6158	3.8709	1.76343	17.2414	182.21	2642.08
59	3481	205379	7.6811	3.8930	1.77085	16.9492	185.35	2733.97
60	3600	216000	7.7460	3.9149	1.77815	16.6667	188.50	2827.43
61	3721	226981	7.8102	3.9365	1.78533	16.3934	191.64	2922.47
62	3844	238328	7.8740	3.9579	1.79239	16.1290	194.78	3019.07
63	3969	250047	7.9373	3.9791	1.79934	15.8730	197.92	3117.25
64	4096	262144	8.0000	4.0000	1.80618	15.6250	201.06	3216.99
65	4225	274625	8.0623	4.0207	1.81291	15.3846	204.20	3318.31
66	4356	287496	8.1240	4.0412	1.81954	15.1515	207.35	3421.19
67	4489	300763	8.1854	4.0615	1.82607	14.9254	210.49	3525.65
68	4624	314432	8.2462	4.0817	1.83251	14.7059	213.63	3631.68
69	4761	328509	8.3066	4.1016	1.83885	14.4928	216.77	3739.28
70	4900	343000	8.3666	4.1213	1.84510	14.2857	219.91	3848.45
71	5041	357911	8.4261	4.1408	1.85126	14.0845	223.05	3959.19
72	5184	373248	8.4853	4.1602	1.85733	13.8889	226.19	4071.50
73	5329	389017	8.5440	4.1793	1.86332	13.6986	229.34	4185.39
74	5476	405224	8.6023	4.1983	1.86923	13.5135	232.48	4300.84
75	5625	421875	8.6603	4.2172	1.87506	13.3333	235.62	4417.86
76	5776	438976	8.7178	4.2358	1.88081	13.1579	238.76	4536.46
77	5929	456533	8.7750	4.2543	1.88649	12.9870	241.90	4656.63
78	6084	474552	8.8318	4.2727	1.89209	12.8205	245.04	4778.36
79	6241	493039	8.8882	4.2908	1.89763	12.6582	248.19	4901.67
80	6400	512000	8.9443	4.3089	1.90309	12.5000	251.33	5026.55
81	6561	531441	9.0000	4.3267	1.90849	12.3457	254.47	5153.00
82	6724	551368	9.0554	4.3445	1.91381	12.1951	257.61	5281.02
83	6889	571787	9.1104	4.3621	1.91908	12.0482	260.75	5410.61
84	7056	592704	9.1652	4.3795	1.92428	11.9048	263.89	5541.77
85	7225	614125	9.2195	4.3968	1.92942	11.7647	267.04	5674.50
86	7396	636056	9.2736	4.4140	1.93450	11.6279	270.18	5808.80
87	7569	658503	9.3274	4.4310	1.93952	11.4943	273.32	5944.68
88	7744	681472	9.3808	4.4480	1.94448	11.3636	276.46	6082.12
89	7921	704969	9.4340	4.4647	1.94939	11.2360	279.60	6221.14
90	8100	729000	9.4868	4.4814	1.95424	11.1111	282.74	6361.73
91	8281	753571	9.5394	4.4979	1.95904	10.9890	285.88	6503.88
92	8464	778688	9.5917	4.5144	1.96379	10.8696	289.03	6647.61
93	8649	804357	9.6437	4.5307	1.96848	10.7527	292.17	6792.91
94	8836	830584	9.6954	4.5468	1.97313	10.6383	295.31	6939.78
95	9025	857375	9.7468	4.5629	1.97772	10.5263	298.45	7088.22
96	9216	884736	9.7980	4.5789	1.98227	10.4167	301.59	7238.23
97	9409	912673	9.8489	4.5947	1.98677	10.3093	304.73	7389.81
98	9604	941192	9.8995	4.6104	1.99123	10.2041	307.88	7542.96
99	9801	970299	9.9499	4.6261	1.99564	10.1010	311.02	7697.69

TABLE: Functions of numbers, continued (100 through 149) 5.6.3

No.	Square	Cube	Square Root	Cube Root	Logarithm	1000 × Reciprocal	No. = Diameter	
							Circum.	Area
100	10000	1000000	10.0000	4.6416	2.00000	10.0000	314.16	7853.98
101	10201	1030301	10.0499	4.6570	2.00432	9.90099	317.30	8011.85
102	10404	1061208	10.0995	4.6723	2.00860	9.80392	320.44	8171.28
103	10609	1092727	10.1489	4.6875	2.01284	9.70874	323.58	8332.29
104	10816	1124864	10.1980	4.7027	2.01703	9.61538	326.73	8494.87
105	11025	1157625	10.2470	4.7177	2.02119	9.52381	329.87	8659.01
106	11236	1191016	10.2956	4.7326	2.02531	9.43396	333.01	8824.73
107	11449	1225043	10.3441	4.7475	2.02938	9.34579	336.15	8992.02
108	11664	1259712	10.3923	4.7622	2.03342	9.25926	339.29	9160.88
109	11881	1295029	10.4403	4.7769	2.03743	9.17431	342.43	9331.32
110	12100	1331000	10.4881	4.7914	2.04139	9.09091	345.58	9503.32
111	12321	1367631	10.5357	4.8059	2.04532	9.00901	348.72	9676.89
112	12544	1404928	10.5830	4.8203	2.04922	8.92857	351.86	9852.03
113	12769	1442897	10.6301	4.8346	2.05308	8.84956	355.00	10028.7
114	12996	1481544	10.6771	4.8488	2.05690	8.77193	358.14	10207.0
115	13225	1520875	10.7238	4.8629	2.06070	8.69565	361.28	10386.9
116	13456	1560896	10.7703	4.8770	2.06446	8.62069	364.42	10568.3
117	13689	1601613	10.8167	4.8910	2.06819	8.54701	367.57	10751.3
118	13924	1643032	10.8628	4.9049	2.07188	8.47458	370.71	10935.9
119	14161	1685159	10.9087	4.9187	2.07555	8.40336	373.85	11122.0
120	14400	1728000	10.9545	4.9324	2.07918	8.33333	376.99	11309.7
121	14641	1771561	11.0000	4.9461	2.08279	8.26446	380.13	11499.0
122	14884	1815848	11.0454	4.9597	2.08636	8.19672	383.27	11689.9
123	15129	1860867	11.0905	4.9732	2.08991	8.13008	386.42	11882.3
124	15376	1906624	11.1355	4.9866	2.09342	8.06452	389.56	12076.3
125	15625	1953125	11.1803	5.0000	2.09691	8.00000	392.70	12271.8
126	15876	2000376	11.2250	5.0133	2.10037	7.93651	395.84	12469.0
127	16129	2048383	11.2694	5.0265	2.10380	7.87402	398.98	12667.7
128	16384	2097152	11.3137	5.0397	2.10721	7.81250	402.12	12868.0
129	16641	2146689	11.3578	5.0528	2.11059	7.75194	405.27	13069.8
130	16900	2197000	11.4018	5.0658	2.11394	7.69231	408.41	13273.2
131	17161	2248091	11.4455	5.0788	2.11727	7.63359	411.55	13478.2
132	17424	2299968	11.4891	5.0916	2.12057	7.57576	414.69	13684.8
133	17689	2352637	11.5326	5.1045	2.12385	7.51880	417.83	13892.9
134	17956	2406104	11.5758	5.1172	2.12710	7.46269	420.97	14102.6
135	18225	2460375	11.6190	5.1299	2.13033	7.40741	424.12	14313.9
136	18496	2515456	11.6619	5.1426	2.13354	7.35294	427.26	14526.7
137	18769	2571353	11.7047	5.1551	2.13672	7.29927	430.40	14741.1
138	19044	2628072	11.7473	5.1676	2.13988	7.24638	433.54	14957.1
139	19321	2685619	11.7898	5.1801	2.14301	7.19424	436.68	15174.7
140	19600	2744000	11.8322	5.1925	2.14613	7.14286	439.82	15393.8
141	19881	2803221	11.8743	5.2048	2.14922	7.09220	442.96	15614.5
142	20164	2863288	11.9164	5.2171	2.15229	7.04225	446.11	15836.8
143	20449	2924207	11.9583	5.2293	2.15534	6.99301	449.25	16060.6
144	20736	2985984	12.0000	5.2415	2.15836	6.94444	452.39	16286.0
145	21025	3048625	12.0416	5.2536	2.16137	6.89655	455.53	16513.0
146	21316	3112136	12.0830	5.2656	2.16435	6.84932	458.67	16741.5
147	21609	3176523	12.1244	5.2776	2.16732	6.80272	461.81	16971.7
148	21904	3241792	12.1655	5.2896	2.17026	6.75676	464.96	17203.4
149	22201	3307949	12.2066	5.3015	2.17319	6.71141	468.10	17436.6

TABLE: Functions of numbers, continued (150 through 199) 5.6.3

No.	Square	Cube	Square Root	Cube Root	Logarithm	1000 × Reciprocal	No. = Diameter	
							Circum.	Area
150	22500	3375000	12.2474	5.3133	2.17609	6.66667	471.24	17671.5
151	22801	3442951	12.2882	5.3251	2.17898	6.62252	474.38	17907.9
152	23104	3511808	12.3288	5.3368	2.18184	6.57895	477.52	18145.8
153	23409	3581577	12.3693	5.3485	2.18469	6.53595	480.66	18385.4
154	23716	3652264	12.4097	5.3601	2.18752	6.49351	483.81	18626.5
155	24025	3723875	12.4499	5.3717	2.19033	6.45161	486.95	18869.2
156	24336	3796416	12.4900	5.3832	2.19312	6.41026	490.09	19113.4
157	24649	3869893	12.5300	5.3947	2.19590	6.36943	493.23	19359.3
158	24964	3944312	12.5698	5.4061	2.19866	6.32911	496.37	19606.7
159	25281	4019679	12.6095	5.4175	2.20140	6.28931	499.51	19855.7
160	25600	4096000	12.6491	5.4288	2.20412	6.25000	502.65	20106.2
161	25921	4173281	12.6886	5.4401	2.20683	6.21118	505.80	20358.3
162	26244	4251528	12.7279	5.4514	2.20952	6.17284	508.94	20612.0
163	26569	4330747	12.7671	5.4626	2.21219	6.13497	512.08	20867.2
164	26896	4410944	12.8062	5.4737	2.21484	6.09756	515.22	21124.1
165	27225	4492125	12.8452	5.4848	2.21748	6.06061	518.36	21382.5
166	27556	4574296	12.8841	5.4959	2.22011	6.02410	521.50	21642.4
167	27889	4657463	12.9228	5.5069	2.22272	5.98802	524.65	21904.0
168	28224	4741632	12.9615	5.5178	2.22531	5.95238	527.79	22167.1
169	28561	4826809	13.0000	5.5288	2.22789	5.91716	530.93	22431.8
170	28900	4913000	13.0384	5.5397	2.23045	5.88235	534.07	22698.0
171	29241	5000211	13.0767	5.5505	2.23300	5.84795	537.21	22965.8
172	29584	5088448	13.1149	5.5613	2.23553	5.81395	540.35	23235.2
173	29929	5177717	13.1529	5.5721	2.23805	5.78035	543.50	23506.2
174	30276	5268024	13.1909	5.5828	2.24055	5.74713	546.64	23778.7
175	30625	5359375	13.2288	5.5934	2.24304	5.71429	549.78	24052.8
176	30976	5451776	13.2665	5.6041	2.24551	5.68182	552.92	24328.5
177	31329	5545233	13.3041	5.6147	2.24797	5.64972	556.06	24605.7
178	31684	5639752	13.3417	5.6252	2.25042	5.61798	559.20	24884.6
179	32041	5735339	13.3791	5.6357	2.25285	5.58659	562.35	25164.9
180	32400	5832000	13.4164	5.6462	2.25527	5.55556	565.49	25446.9
181	32761	5929741	13.4536	5.6567	2.25768	5.52486	568.63	25730.4
182	33124	6028568	13.4907	5.6671	2.26007	5.49451	571.77	26015.5
183	33489	6128487	13.5277	5.6774	2.26245	5.46448	574.91	26302.2
184	33856	6229504	13.5647	5.6877	2.26482	5.43478	578.05	26590.4
185	34225	6331625	13.6015	5.6980	2.26717	5.40541	581.19	26880.3
186	34596	6434856	13.6382	5.7083	2.26951	5.37634	584.34	27171.6
187	34969	6539203	13.6748	5.7185	2.27184	5.34759	587.48	27464.6
188	35344	6644672	13.7113	5.7287	2.27416	5.31915	590.62	27759.1
189	35721	6751269	13.7477	5.7388	2.27646	5.29101	593.76	28055.2
190	36100	6859000	13.7840	5.7489	2.27875	5.26316	596.90	28352.9
191	36481	6967871	13.8203	5.7590	2.28103	5.23560	600.04	28652.1
192	36864	7077888	13.8564	5.7690	2.28330	5.20833	603.19	28952.9
193	37249	7189057	13.8924	5.7790	2.28556	5.18135	606.33	29255.3
194	37636	7301384	13.9284	5.7890	2.28780	5.15464	609.47	29559.2
195	38025	7414875	13.9642	5.7989	2.29003	5.12821	612.61	29864.8
196	38416	7529536	14.0000	5.8088	2.29226	5.10204	615.75	30171.9
197	38809	7645373	14.0357	5.8186	2.29447	5.07614	618.89	30480.5
198	39204	7762392	14.0712	5.8285	2.29667	5.05051	622.04	30790.7
199	39601	7880599	14.1067	5.8383	2.29885	5.02513	625.18	31102.6

TABLE: Functions of numbers, continued (200 through 249) 5.6.3

No.	Square	Cube	Square Root	Cube Root	Logarithm	1000 × Reciprocal	No. = Diameter	
							Circum.	Area
200	40000	8000000	14.1421	5.8480	2.30103	5.00000	628.32	31415.9
201	40401	8120601	14.1774	5.8578	2.30320	4.97512	631.46	31730.9
202	40804	8242408	14.2127	5.8675	2.30535	4.95050	634.60	32047.4
203	41209	8365427	14.2478	5.8771	2.30750	4.92611	637.74	32365.5
204	41616	8489664	14.2829	5.8868	2.30963	4.90196	640.88	32685.1
205	42025	8615125	14.3178	5.8964	2.31175	4.87805	644.03	33006.4
206	42436	8741816	14.3527	5.9059	2.31387	4.85437	647.17	33329.2
207	42849	8869743	14.3875	5.9155	2.31597	4.83092	650.31	33653.5
208	43264	8998912	14.4222	5.9250	2.31806	4.80769	653.45	33979.5
209	43681	9129329	14.4568	5.9345	2.32015	4.78469	656.59	34307.0
210	44100	9261000	14.4914	5.9439	2.32222	4.76190	659.73	34636.1
211	44521	9393931	14.5258	5.9533	2.32428	4.73934	662.88	34966.7
212	44944	9528128	14.5602	5.9627	2.32634	4.71698	666.02	35298.9
213	45369	9663597	14.5945	5.9721	2.32838	4.69484	669.16	35632.7
214	45796	9800344	14.6287	5.9814	2.33041	4.67290	672.30	35968.1
215	46225	9938375	14.6629	5.9907	2.33244	4.65116	675.44	36305.0
216	46656	10077696	14.6969	6.0000	2.33445	4.62963	678.58	36643.5
217	47089	10218313	14.7309	6.0092	2.33646	4.60829	681.73	36983.6
218	47524	10360232	14.7648	6.0185	2.33846	4.58716	684.87	37325.3
219	47961	10503459	14.7986	6.0277	2.34044	4.56621	688.01	37668.5
220	48400	10648000	14.8324	6.0368	2.34242	4.54545	691.15	38013.3
221	48841	10793861	14.8661	6.0459	2.34439	4.52489	694.29	38359.6
222	49284	10941048	14.8997	6.0550	2.34635	4.50450	697.43	38707.6
223	49729	11089567	14.9332	6.0641	2.34830	4.48430	700.58	39057.1
224	50176	11239424	14.9666	6.0732	2.35025	4.46429	703.72	39408.1
225	50625	11390625	15.0000	6.0822	2.35218	4.44444	706.86	39760.8
226	51076	11543176	15.0333	6.0912	2.35411	4.42478	710.00	40115.0
227	51529	11697083	15.0665	6.1002	2.35603	4.40529	713.14	40470.8
228	51984	11852352	15.0997	6.1091	2.35793	4.38596	716.28	40828.1
229	52441	12008989	15.1327	6.1180	2.35984	4.36681	719.42	41187.1
230	52900	12167000	15.1658	6.1269	2.36173	4.34783	722.57	41547.6
231	53361	12326391	15.1987	6.1358	2.36361	4.32900	725.71	41909.6
232	53824	12487168	15.2315	6.1446	2.36549	4.31034	728.85	42273.3
233	54289	12649337	15.2643	6.1534	2.36736	4.29185	731.99	42638.5
234	54756	12812904	15.2971	6.1622	2.36922	4.27350	735.13	43005.3
235	55225	12977875	15.3297	6.1710	2.37107	4.25532	738.27	43373.6
236	55696	13144256	15.3623	6.1797	2.37291	4.23729	741.42	43743.5
237	56169	13312053	15.3948	6.1885	2.37475	4.21941	744.56	44115.0
238	56644	13481272	15.4272	6.1972	2.37658	4.20168	747.70	44488.1
239	57121	13651919	15.4596	6.2058	2.37840	4.18410	750.84	44862.7
240	57600	13824000	15.4919	6.2145	2.38021	4.16667	753.98	45238.9
241	58081	13997521	15.5242	6.2231	2.38202	4.14938	757.12	45616.7
242	58564	14172488	15.5563	6.2317	2.38382	4.13223	760.27	45996.1
243	59049	14348907	15.5885	6.2403	2.38561	4.11523	763.41	46377.0
244	59536	14526784	15.6205	6.2488	2.38739	4.09836	766.55	46759.5
245	60025	14706125	15.6525	6.2573	2.38917	4.08163	769.69	47143.5
246	60516	14886936	15.6844	6.2658	2.39094	4.06504	772.83	47529.2
247	61009	15069223	15.7162	6.2743	2.39270	4.04858	775.97	47916.4
248	61504	15252992	15.7480	6.2828	2.39445	4.03226	779.12	48305.1
249	62001	15438249	15.7797	6.2912	2.39620	4.01606	782.26	48695.5

TABLE: Functions of numbers, continued (250 through 299) 5.6.3

No.	Square	Cube	Square Root	Cube Root	Logarithm	1000 × Reciprocal	No. = Diameter	
							Circum.	Area
250	62500	15625000	15.8114	6.2996	2.39794	4.00000	785.40	49087.4
251	63001	15813251	15.8430	6.3080	2.39967	3.98406	788.54	49480.9
252	63504	16003008	15.8745	6.3164	2.40140	3.96825	791.68	49875.9
253	64009	16194277	15.9060	6.3247	2.40312	3.95257	794.82	50272.6
254	64516	16387064	15.9374	6.3330	2.40483	3.93701	797.96	50670.7
255	65025	16581375	15.9687	6.3413	2.40654	3.92157	801.11	51070.5
256	65536	16777216	16.0000	6.3496	2.40824	3.90625	804.25	51471.9
257	66049	16974593	16.0312	6.3579	2.40993	3.89105	807.39	51874.8
258	66564	17173512	16.0624	6.3661	2.41162	3.87597	810.53	52279.2
259	67081	17373979	16.0935	6.3743	2.41330	3.86100	813.67	52685.3
260	67600	17576000	16.1245	6.3825	2.41497	3.84615	816.81	53092.9
261	68121	17779581	16.1555	6.3907	2.41664	3.83142	819.96	53502.1
262	68644	17984728	16.1864	6.3988	2.41830	3.81679	823.10	53912.9
263	69169	18191447	16.2173	6.4070	2.41996	3.80228	826.24	54325.2
264	69696	18399744	16.2481	6.4151	2.42160	3.78788	829.38	54739.1
265	70225	18609625	16.2788	6.4232	2.42325	3.77358	832.52	55154.6
266	70756	18821096	16.3095	6.4312	2.42488	3.75940	835.66	55571.6
267	71289	19034163	16.3401	6.4393	2.42651	3.74532	838.81	55990.2
268	71824	19248832	16.3707	6.4473	2.42813	3.73134	841.95	56410.4
269	72361	19465109	16.4012	6.4553	2.42975	3.71747	845.09	56832.2
270	72900	19683000	16.4317	6.4633	2.43136	3.70370	848.23	57255.5
271	73441	19902511	16.4621	6.4713	2.43297	3.69004	851.37	57680.4
272	73984	20123648	16.4924	6.4792	2.43457	3.67647	854.51	58106.9
273	74529	20346417	16.5227	6.4872	2.43616	3.66300	857.65	58534.9
274	75076	20570824	16.5529	6.4951	2.43775	3.64964	860.80	58964.6
275	75625	20796875	16.5831	6.5030	2.43933	3.63636	863.94	59395.7
276	76176	21024576	16.6132	6.5108	2.44091	3.62319	867.08	59828.5
277	76729	21253933	16.6433	6.5187	2.44248	3.61011	870.22	60262.8
278	77284	21484952	16.6733	6.5265	2.44404	3.59712	873.36	60698.7
279	77841	21717639	16.7033	6.5343	2.44560	3.58423	876.50	61136.2
280	78400	21952000	16.7332	6.5421	2.44716	3.57143	879.65	61575.2
281	78961	22188041	16.7631	6.5499	2.44871	3.55872	882.79	62015.8
282	79524	22425768	16.7929	6.5577	2.45025	3.54610	885.93	62458.0
283	80089	22665187	16.8226	6.5654	2.45179	3.53357	889.07	62901.8
284	80656	22906304	16.8523	6.5731	2.45332	3.52113	892.21	63347.1
285	81225	23149125	16.8819	6.5808	2.45484	3.50877	895.35	63794.0
286	81796	23393656	16.9115	6.5885	2.45637	3.49650	898.50	64242.4
287	82369	23639903	16.9411	6.5962	2.45788	3.48432	901.64	64692.5
288	82944	23887872	16.9706	6.6039	2.45939	3.47222	904.78	65144.1
289	83521	24137569	17.0000	6.6115	2.46090	3.46021	907.92	65597.2
290	84100	24389000	17.0294	6.6191	2.46240	3.44828	911.06	66052.0
291	84681	24642171	17.0587	6.6267	2.46389	3.43643	914.20	66508.3
292	85264	24897088	17.0880	6.6343	2.46538	3.42466	917.35	66966.2
293	85849	25153757	17.1172	6.6419	2.46687	3.41297	920.49	67425.6
294	86436	25412184	17.1464	6.6494	2.46835	3.40136	923.63	67886.7
295	87025	25672375	17.1756	6.6569	2.46982	3.38983	926.77	68349.3
296	87616	25934336	17.2047	6.6644	2.47129	3.37838	929.91	68813.4
297	88209	26198073	17.2337	6.6719	2.47276	3.36700	933.05	69279.2
298	88804	26463592	17.2627	6.6794	2.47422	3.35570	936.19	69746.5
299	89401	26730899	17.2916	6.6869	2.47567	3.34448	939.34	70215.4

TABLE: Functions of numbers, continued (300 through 349) 5.6.3

No.	Square	Cube	Square Root	Cube Root	Logarithm	1000 × Reciprocal	No. = Diameter	
							Circum.	Area
300	90000	27000000	17.3205	6.6943	2.47712	3.33333	942.48	70685.8
301	90601	27270901	17.3494	6.7018	2.47857	3.32226	945.62	71157.9
302	91204	27543608	17.3781	6.7092	2.48001	3.31126	948.76	71631.5
303	91809	27818127	17.4069	6.7166	2.48144	3.30033	951.90	72106.6
304	92416	28094464	17.4356	6.7240	2.48287	3.28947	955.04	72583.4
305	93025	28372625	17.4642	6.7313	2.48430	3.27869	958.19	73061.7
306	93636	28652616	17.4929	6.7387	2.48572	3.26797	961.33	73541.5
307	94249	28934443	17.5214	6.7460	2.48714	3.25733	964.47	74023.0
308	94864	29218112	17.5499	6.7533	2.48855	3.24675	967.61	74506.0
309	95481	29503629	17.5784	6.7606	2.48996	3.23625	970.75	74990.6
310	96100	29791000	17.6068	6.7679	2.49136	3.22581	973.89	75476.8
311	96721	30080231	17.6352	6.7752	2.49276	3.21543	977.04	75964.5
312	97344	30371328	17.6635	6.7824	2.49415	3.20513	980.18	76453.8
313	97969	30664297	17.6918	6.7897	2.49554	3.19489	983.32	76944.7
314	98596	30959144	17.7200	6.7969	2.49693	3.18471	986.46	77437.1
315	99225	31255875	17.7482	6.8041	2.49831	3.17460	989.60	77931.1
316	99856	31554496	17.7764	6.8113	2.49969	3.16456	992.74	78426.7
317	100489	31855013	17.8045	6.8185	2.50106	3.15457	995.88	78923.9
318	101124	32157432	17.8326	6.8256	2.50243	3.14465	999.03	79422.6
319	101761	32461759	17.8606	6.8328	2.50379	3.13480	1002.2	79922.9
320	102400	32768000	17.8885	6.8399	2.50515	3.12500	1005.3	80424.8
321	103041	33076161	17.9165	6.8470	2.50651	3.11526	1008.5	80928.2
322	103684	33386248	17.9444	6.8541	2.50786	3.10559	1011.6	81433.2
323	104329	33698267	17.9722	6.8612	2.50920	3.09598	1014.7	81939.8
324	104976	34012224	18.0000	6.8683	2.51055	3.08642	1017.9	82448.0
325	105625	34328125	18.0278	6.8753	2.51188	3.07692	1021.0	82957.7
326	106276	34645976	18.0555	6.8824	2.51322	3.06749	1024.2	83469.0
327	106929	34965783	18.0831	6.8894	2.51455	3.05810	1027.3	83981.8
328	107584	35287552	18.1108	6.8964	2.51587	3.04878	1030.4	84496.3
329	108241	35611289	18.1384	6.9034	2.51720	3.03951	1033.6	85012.3
330	108900	35937000	18.1659	6.9104	2.51851	3.03030	1036.7	85529.9
331	109561	36264691	18.1934	6.9174	2.51983	3.02115	1039.9	86049.0
332	110224	36594368	18.2209	6.9244	2.52114	3.01205	1043.0	86569.7
333	110889	36926037	18.2483	6.9313	2.52244	3.00300	1046.2	87092.0
334	111556	37259704	18.2757	6.9382	2.52375	2.99401	1049.3	87615.9
335	112225	37595375	18.3030	6.9451	2.52504	2.98507	1052.4	88141.3
336	112896	37933056	18.3303	6.9521	2.52634	2.97619	1055.6	88668.3
337	113569	38272753	18.3576	6.9589	2.52763	2.96736	1058.7	89196.9
338	114244	38614472	18.3848	6.9658	2.52892	2.95858	1061.9	89727.0
339	114921	38958219	18.4120	6.9727	2.53020	2.94985	1065.0	90258.7
340	115600	39304000	18.4391	6.9795	2.53148	2.94118	1068.1	90792.0
341	116281	39651821	18.4662	6.9864	2.53275	2.93255	1071.3	91326.9
342	116964	40001688	18.4932	6.9932	2.53403	2.92398	1074.4	91863.3
343	117649	40353607	18.5203	7.0000	2.53529	2.91545	1077.6	92401.3
344	118336	40707584	18.5472	7.0068	2.53656	2.90698	1080.7	92940.9
345	119025	41063625	18.5742	7.0136	2.53782	2.89855	1083.8	93482.0
346	119716	41421736	18.6011	7.0203	2.53908	2.89017	1087.0	94024.7
347	120409	41781923	18.6279	7.0271	2.54033	2.88184	1090.1	94569.0
348	121104	42144192	18.6548	7.0338	2.54158	2.87356	1093.3	95114.9
349	121801	42508549	18.6815	7.0406	2.54283	2.86533	1096.4	95662.3

TABLE: Functions of numbers, continued (350 through 399) 5.6.3

No.	Square	Cube	Square Root	Cube Root	Logarithm	1000 × Reciprocal	No. = Diameter	
							Circum.	Area
350	122500	42875000	18.7083	7.0473	2.54407	2.85714	1099.6	96211.3
351	123201	43243551	18.7350	7.0540	2.54531	2.84900	1102.7	96761.8
352	123904	43614208	18.7617	7.0607	2.54654	2.84091	1105.8	97314.0
353	124609	43986977	18.7883	7.0674	2.54777	2.83286	1109.0	97867.7
354	125316	44361864	18.8149	7.0740	2.54900	2.82486	1112.1	98423.0
355	126025	44738875	18.8414	7.0807	2.55023	2.81690	1115.3	98979.8
356	126736	45118016	18.8680	7.0873	2.55145	2.80899	1118.4	99538.2
357	127449	45499293	18.8944	7.0940	2.55267	2.80112	1121.5	100098
358	128164	45882712	18.9209	7.1006	2.55388	2.79330	1124.7	100660
359	128881	46268279	18.9473	7.1072	2.55509	2.78552	1127.8	101223
360	129600	46656000	18.9737	7.1138	2.55630	2.77778	1131.0	101788
361	130321	47045881	19.0000	7.1204	2.55751	2.77008	1134.1	102354
362	131044	47437928	19.0263	7.1269	2.55871	2.76243	1137.3	102922
363	131769	47832147	19.0526	7.1335	2.55991	2.75482	1140.4	103491
364	132496	48228544	19.0788	7.1400	2.56110	2.74725	1143.5	104062
365	133225	48627125	19.1050	7.1466	2.56229	2.73973	1146.7	104635
366	133956	49027896	19.1311	7.1531	2.56348	2.73224	1149.8	105209
367	134689	49430863	19.1572	7.1596	2.56467	2.72480	1153.0	105785
368	135424	49836032	19.1833	7.1661	2.56585	2.71739	1156.1	106362
369	136161	50243409	19.2094	7.1726	2.56703	2.71003	1159.2	106941
370	136900	50653000	19.2354	7.1791	2.56820	2.70270	1162.4	107521
371	137641	51064811	19.2614	7.1855	2.56937	2.69542	1165.5	108103
372	138384	51478848	19.2873	7.1920	2.57054	2.68817	1168.7	108687
373	139129	51895117	19.3132	7.1984	2.57171	2.68097	1171.8	109272
374	139876	52313624	19.3391	7.2048	2.57287	2.67380	1175.0	109858
375	140625	52734375	19.3649	7.2112	2.57403	2.66676	1178.1	110447
376	141376	53157376	19.3907	7.2177	2.57519	2.65957	1181.2	111036
377	142129	53582633	19.4165	7.2240	2.57634	2.65252	1184.4	111628
378	142884	54010152	19.4422	7.2304	2.57749	2.64550	1187.5	112221
379	143641	54439939	19.4679	7.2368	2.57864	2.63852	1190.7	112815
380	144400	54872000	19.4936	7.2432	2.57978	2.63158	1193.8	113411
381	145161	55306341	19.5192	7.2495	2.58093	2.62467	1196.9	114009
382	145924	55742968	19.5448	7.2558	2.58206	2.61780	1200.1	114608
383	146689	56181887	19.5704	7.2622	2.58320	2.61097	1203.2	115209
384	147456	56623104	19.5959	7.2685	2.58433	2.60417	1206.4	115812
385	148225	57066625	19.6214	7.2748	2.58546	2.59740	1209.5	116416
386	148996	57512456	19.6469	7.2811	2.58659	2.59067	1212.7	117021
387	149769	57960603	19.6723	7.2874	2.58771	2.58398	1215.8	117628
388	150544	58411072	19.6977	7.2936	2.58883	2.57732	1218.9	118237
389	151321	58863869	19.7231	7.2999	2.58995	2.57069	1222.1	118847
390	152100	59319000	19.7484	7.3061	2.59106	2.56410	1225.2	119459
391	152881	59776471	19.7737	7.3124	2.59218	2.55754	1228.4	120072
392	153664	60236288	19.7990	7.3186	2.59329	2.55102	1231.5	120687
393	154449	60698457	19.8242	7.3248	2.59439	2.54453	1234.6	121304
394	155236	61162984	19.8494	7.3310	2.59550	2.53807	1237.8	121922
395	156025	61629875	19.8746	7.3372	2.59660	2.53165	1240.9	122542
396	156816	62099136	19.8997	7.3434	2.59770	2.52525	1244.1	123163
397	157609	62570773	19.9249	7.3496	2.59879	2.51889	1247.2	123786
398	158404	63044792	19.9499	7.3558	2.59988	2.51256	1250.4	124410
399	159201	63521199	19.9750	7.3619	2.60097	2.50627	1253.5	125036

TABLE: Functions of numbers, continued (400 through 449) 5.6.3

No.	Square	Cube	Square Root	Cube Root	Logarithm	1000 × Reciprocal	No. = Diameter	
							Circum.	Area
400	160000	64000000	20.0000	7.3681	2.60206	2.50000	1256.6	125664
401	160801	64481201	20.0250	7.3742	2.60314	2.49377	1259.8	126293
402	161604	64964808	20.0499	7.3803	2.60423	2.48756	1262.9	126923
403	162409	65450827	20.0749	7.3864	2.60531	2.48139	1266.1	127556
404	163216	65939264	20.0998	7.3925	2.60638	2.47525	1269.2	128190
405	164025	66430125	20.1246	7.3986	2.60746	2.46914	1272.3	128825
406	164836	66923416	20.1494	7.4047	2.60853	2.46305	1275.5	129462
407	165649	67419143	20.1742	7.4108	2.60959	2.45700	1278.6	130100
408	166464	67917312	20.1990	7.4169	2.61066	2.45098	1281.8	130741
409	167281	68417929	20.2237	7.4229	2.61172	2.44499	1284.9	131382
410	168100	68921000	20.2485	7.4290	2.61278	2.43902	1288.1	132025
411	168921	69426531	20.2731	7.4350	2.61384	2.43309	1291.2	132670
412	169744	69934528	20.2978	7.4410	2.61490	2.42718	1294.3	133317
413	170569	70444997	20.3224	7.4470	2.61595	2.42131	1297.5	133965
414	171396	70957944	20.3470	7.4530	2.61700	2.41546	1300.6	134614
415	172225	71473375	20.3715	7.4590	2.61805	2.40964	1303.8	135265
416	173056	71991296	20.3961	7.4650	2.61909	2.40385	1306.9	135918
417	173889	72511713	20.4206	7.4710	2.62014	2.39808	1310.0	136572
418	174724	73034632	20.4450	7.4770	2.62118	2.39234	1313.2	137228
419	175561	73560059	20.4695	7.4829	2.62221	2.38663	1316.3	137885
420	176400	74088000	20.4939	7.4889	2.62325	2.38095	1319.5	138544
421	177241	74618461	20.5183	7.4948	2.62428	2.37530	1322.6	139205
422	178084	75151448	20.5426	7.5007	2.62531	2.36967	1325.8	139867
423	178929	75686967	20.5670	7.5067	2.62634	2.36407	1328.9	140531
424	179776	76225024	20.5913	7.5126	2.62737	2.35849	1332.0	141196
425	180625	76765625	20.6155	7.5185	2.62839	2.35294	1335.2	141863
426	181476	77308776	20.6398	7.5244	2.62941	2.34742	1338.3	142531
427	182329	77854483	20.6640	7.5302	2.63043	2.34192	1341.5	143201
428	183184	78402752	20.6882	7.5361	2.63144	2.33645	1344.6	143872
429	184041	78953589	20.7123	7.5420	2.63246	2.33100	1347.7	144545
430	184900	79507000	20.7364	7.5478	2.63347	2.32558	1350.9	145220
431	185761	80062991	20.7605	7.5537	2.63448	2.32019	1354.0	145896
432	186624	80621568	20.7846	7.5595	2.63548	2.31481	1357.2	146574
433	187489	81182737	20.8087	7.5654	2.63649	2.30947	1360.3	147254
434	188356	81746504	20.8327	7.5712	2.63749	2.30415	1363.5	147934
435	189225	82312875	20.8567	7.5770	2.63849	2.29885	1366.6	148617
436	190096	82881856	20.8806	7.5828	2.63949	2.29358	1369.7	149301
437	190969	83453453	20.9045	7.5886	2.64048	2.28833	1372.9	149987
438	191844	84027672	20.9284	7.5944	2.64147	2.28311	1376.0	150674
439	192721	84604519	20.9523	7.6001	2.64246	2.27790	1379.2	151363
440	193600	85184000	20.9762	7.6059	2.64345	2.27273	1382.3	152053
441	194481	85766121	21.0000	7.6117	2.64444	2.26757	1385.4	152745
442	195364	86350888	21.0238	7.6174	2.64542	2.26244	1388.6	153439
443	196249	86938307	21.0476	7.6232	2.64640	2.25734	1391.7	154134
444	197136	87528384	21.0713	7.6289	2.64738	2.25225	1394.9	154830
445	198025	88121125	21.0950	7.6346	2.64836	2.24719	1398.0	155528
446	198916	88716536	21.1187	7.6403	2.64933	2.24215	1401.2	156228
447	199809	89314623	21.1424	7.6460	2.65031	2.23714	1404.3	156930
448	200704	89915392	21.1660	7.6517	2.65128	2.23214	1407.4	157633
449	201601	90518849	21.1896	7.6574	2.65225	2.22717	1410.6	158337

TABLE: Functions of numbers, continued (450 through 499) 5.6.3

No.	Square	Cube	Square Root	Cube Root	Logarithm	1000 × Reciprocal	No. = Diameter	
							Circum.	Area
450	202500	91125000	21.2132	7.6631	2.65321	2.22222	1413.7	159043
451	203401	91733851	21.2368	7.6688	2.65418	2.21729	1416.9	159751
452	204304	92345408	21.2603	7.6744	2.65514	2.21239	1420.0	160460
453	205209	92959677	21.2838	7.6801	2.65610	2.20751	1423.1	161171
454	206116	93576664	21.3073	7.6857	2.65706	2.20264	1426.3	161883
455	207025	94196375	21.3307	7.6914	2.65801	2.19780	1429.4	162597
456	207936	94818816	21.3542	7.6970	2.65896	2.19298	1432.6	163313
457	208849	95443993	21.3776	7.7026	2.65992	2.18818	1435.7	164030
458	209764	96071912	21.4009	7.7082	2.66087	2.18341	1438.8	164748
459	210681	96702579	21.4243	7.7138	2.66181	2.17865	1442.0	165468
460	211600	97336000	21.4476	7.7194	2.66276	2.17391	1445.1	166190
461	212521	97972181	21.4709	7.7250	2.66370	2.16920	1448.3	166914
462	213444	98611128	21.4942	7.7306	2.66464	2.16450	1451.4	167639
463	214369	99252847	21.5174	7.7362	2.66558	2.15983	1454.6	168365
464	215296	99897344	21.5407	7.7418	2.66652	2.15517	1457.7	169093
465	216225	100544625	21.5639	7.7473	2.66745	2.15054	1460.8	169823
466	217156	101194696	21.5870	7.7529	2.66839	2.14592	1464.0	170554
467	218089	101847563	21.6102	7.7584	2.66932	2.14133	1467.1	171287
468	219024	102503232	21.6333	7.7639	2.67025	2.13675	1470.3	172021
469	219961	103161709	21.6564	7.7695	2.67117	2.13220	1473.4	172757
470	220900	103823000	21.6795	7.7750	2.67210	2.12766	1476.5	173494
471	221841	104487111	21.7025	7.7805	2.67302	2.12314	1479.7	174234
472	222784	105154048	21.7256	7.7860	2.67394	2.11864	1482.8	174974
473	223729	105823817	21.7486	7.7915	2.67486	2.11416	1486.0	175716
474	224676	106496424	21.7715	7.7970	2.67578	2.10970	1489.1	176460
475	225625	107171875	21.7945	7.8025	2.67669	2.10526	1492.3	177205
476	226576	107850176	21.8174	7.8079	2.67761	2.10084	1495.4	177952
477	227529	108531333	21.8403	7.8134	2.67852	2.09644	1498.5	178701
478	228484	109215352	21.8632	7.8188	2.67943	2.09205	1501.7	179451
479	229441	109902239	21.8861	7.8243	2.68034	2.08768	1504.8	180203
480	230400	110592000	21.9089	7.8297	2.68124	2.08333	1508.0	180956
481	231361	111284641	21.9317	7.8352	2.68215	2.07900	1511.1	181711
482	232324	111980168	21.9545	7.8406	2.68305	2.07469	1514.2	182467
483	233289	112678587	21.9773	7.8460	2.68395	2.07039	1517.4	183225
484	234256	113379904	22.0000	7.8514	2.68485	2.06612	1520.5	183984
485	235225	114084125	22.0227	7.8568	2.68574	2.06186	1523.7	184745
486	236196	114791256	22.0454	7.8622	2.68664	2.05761	1526.8	185508
487	237169	115501303	22.0681	7.8676	2.68753	2.05339	1530.0	186272
488	238144	116214272	22.0907	7.8730	2.68842	2.04918	1533.1	187038
489	239121	116930169	22.1133	7.8784	2.68931	2.04499	1536.2	187805
490	240100	117649000	22.1359	7.8837	2.69020	2.04082	1539.4	188574
491	241081	118370771	22.1585	7.8891	2.69108	2.03666	1542.5	189345
492	242064	119095488	22.1811	7.8944	2.69197	2.03252	1545.7	190117
493	243049	119823157	22.2036	7.8998	2.69285	2.02840	1548.8	190890
494	244036	120553784	22.2261	7.9051	2.69373	2.02429	1551.9	191665
495	245025	121287375	22.2486	7.9105	2.69461	2.02020	1555.1	192442
496	246016	122023936	22.2711	7.9158	2.69548	2.01613	1558.2	193221
497	247009	122763473	22.2935	7.9211	2.69636	2.01207	1561.4	194000
498	248004	123505992	22.3159	7.9264	2.69723	2.00803	1564.5	194782
499	249001	124251499	22.3383	7.9317	2.69810	2.00401	1567.7	195565

TABLE: Functions of numbers, continued (500 through 549) 5.6.3

No.	Square	Cube	Square Root	Cube Root	Logarithm	1000 × Reciprocal	No. = Diameter	
							Circum.	Area
500	250000	125000000	22.3607	7.9370	2.69897	2.00000	1570.8	196350
501	251001	125751501	22.3830	7.9423	2.69984	1.99601	1573.9	197136
502	252004	126506008	22.4054	7.9476	2.70070	1.99203	1577.1	197923
503	253009	127263527	22.4277	7.9528	2.70157	1.98807	1580.2	198713
504	254016	128024064	22.4499	7.9581	2.70243	1.98413	1583.4	199504
505	255025	128787625	22.4722	7.9634	2.70329	1.98020	1586.5	200296
506	256036	129554216	22.4944	7.9686	2.70415	1.97628	1589.6	201090
507	257049	130323843	22.5167	7.9739	2.70501	1.97239	1592.8	201886
508	258064	131096512	22.5389	7.9791	2.70586	1.96850	1595.9	202683
509	259081	131872229	22.5610	7.9843	2.70672	1.96464	1599.1	203482
510	260100	132651000	22.5832	7.9896	2.70757	1.96078	1602.2	204282
511	261121	133432831	22.6053	7.9948	2.70842	1.95695	1605.4	205084
512	262144	134217728	22.6274	8.0000	2.70927	1.95312	1608.5	205887
513	263169	135005697	22.6495	8.0052	2.71012	1.94932	1611.6	206692
514	264196	135796744	22.6716	8.0104	2.71096	1.94553	1614.8	207499
515	265225	136590875	22.6936	8.0156	2.71181	1.94175	1617.9	208307
516	266256	137388096	22.7156	8.0208	2.71265	1.93798	1621.1	209117
517	267289	138188413	22.7376	8.0260	2.71349	1.93424	1624.2	209928
518	268324	138991832	22.7596	8.0311	2.71433	1.93050	1627.3	210741
519	269361	139798359	22.7816	8.0363	2.71517	1.92678	1630.5	211556
520	270400	140608000	22.8035	8.0415	2.71600	1.92308	1633.6	212372
521	271441	141420761	22.8254	8.0466	2.71684	1.91939	1636.8	213189
522	272484	142236648	22.8473	8.0517	2.71767	1.91571	1639.9	214008
523	273529	143055667	22.8692	8.0569	2.71850	1.91205	1643.1	214829
524	274576	143877824	22.8910	8.0620	2.71933	1.90840	1646.2	215651
525	275625	144703125	22.9129	8.0671	2.72016	1.90476	1649.3	216475
526	276676	145531576	22.9347	8.0723	2.72099	1.90114	1652.5	217301
527	277729	146363183	22.9565	8.0774	2.72181	1.89753	1655.6	218128
528	278784	147197952	22.9783	8.0825	2.72263	1.89394	1658.8	218956
529	279841	148035889	23.0000	8.0876	2.72346	1.89036	1661.9	219787
530	280900	148877000	23.0217	8.0927	2.72428	1.88679	1665.0	220618
531	281961	149721291	23.0434	8.0978	2.72509	1.88324	1668.2	221452
532	283024	150568768	23.0651	8.1028	2.72591	1.87970	1671.3	222287
533	284089	151419437	23.0868	8.1079	2.72673	1.87617	1674.5	223123
534	285156	152273304	23.1084	8.1130	2.72754	1.87266	1677.6	223961
535	286225	153130375	23.1301	8.1180	2.72835	1.86916	1680.8	224801
536	287296	153990656	23.1517	8.1231	2.72916	1.86567	1683.9	225642
537	288369	154854153	23.1733	8.1281	2.72997	1.86220	1687.0	226484
538	289444	155720872	23.1948	8.1332	2.73078	1.85874	1690.2	227329
539	290521	156590819	23.2164	8.1382	2.73159	1.85529	1693.3	228175
540	291600	157464000	23.2379	8.1433	2.73239	1.85185	1696.5	229022
541	292681	158340421	23.2594	8.1483	2.73320	1.84843	1699.6	229871
542	293764	159220088	23.2809	8.1533	2.73400	1.84502	1702.7	230722
543	294849	160103007	23.3024	8.1583	2.73480	1.84162	1705.9	231574
544	295936	160989184	23.3238	8.1633	2.73560	1.83824	1709.0	232428
545	297025	161878625	23.3452	8.1683	2.73640	1.83486	1712.2	233283
546	298116	162771336	23.3666	8.1733	2.73719	1.83150	1715.3	234140
547	299209	163667323	23.3880	8.1783	2.73799	1.82815	1718.5	234998
548	300304	164566592	23.4094	8.1833	2.73878	1.82482	1721.6	235858
549	301401	165469149	23.4307	8.1882	2.73957	1.82149	1724.7	236720

TABLE: Functions of numbers, continued (550 through 599) 5.6.3

No.	Square	Cube	Square Root	Cube Root	Logarithm	1000 × Reciprocal	No. = Diameter	
							Circum.	Area
550	302500	166375000	23.4521	8.1932	2.74036	1.81818	1727.9	237583
551	303601	167284151	23.4734	8.1982	2.74115	1.81488	1731.0	238448
552	304704	168196608	23.4947	8.2031	2.74194	1.81159	1734.2	239314
553	305809	169112377	23.5160	8.2081	2.74273	1.80832	1737.3	240182
554	306916	170031464	23.5372	8.2130	2.74351	1.80505	1740.4	241051
555	308025	170953875	23.5584	8.2180	2.74429	1.80180	1743.6	241922
556	309136	171879616	23.5797	8.2229	2.74507	1.79856	1746.7	242795
557	310249	172808693	23.6008	8.2278	2.74586	1.79533	1749.9	243669
558	311364	173741112	23.6220	8.2327	2.74663	1.79211	1753.0	244545
559	312481	174676879	23.6432	8.2377	2.74741	1.78891	1756.2	245422
560	313600	175616000	23.6643	8.2426	2.74819	1.78571	1759.3	246301
561	314721	176558481	23.6854	8.2475	2.74896	1.78253	1762.4	247181
562	315844	177504328	23.7065	8.2524	2.74974	1.77936	1765.6	248063
563	316969	178453547	23.7276	8.2573	2.75051	1.77620	1768.7	248947
564	318096	179406144	23.7487	8.2621	2.75128	1.77305	1771.9	249832
565	319225	180362125	23.7697	8.2670	2.75205	1.76991	1775.0	250719
566	320356	181321496	23.7908	8.2719	2.75282	1.76678	1778.1	251607
567	321489	182284263	23.8118	8.2768	2.75358	1.76367	1781.3	252497
568	322624	183250432	23.8328	8.2816	2.75435	1.76056	1784.4	253388
569	323761	184220009	23.8537	8.2865	2.75511	1.75747	1787.6	254281
570	324900	185193000	23.8747	8.2913	2.75587	1.75439	1790.7	255176
571	326041	186169411	23.8956	8.2962	2.75664	1.75131	1793.8	256072
572	327184	187149248	23.9165	8.3010	2.75740	1.74825	1797.0	256970
573	328329	188132517	23.9374	8.3059	2.75815	1.74520	1800.1	257869
574	329476	189119224	23.9583	8.3107	2.75891	1.74216	1803.3	258770
575	330625	190109375	23.9792	8.3155	2.75967	1.73913	1806.4	259672
576	331776	191102976	24.0000	8.3203	2.76042	1.73611	1809.6	260576
577	332929	192100033	24.0208	8.3251	2.76118	1.73310	1812.7	261482
578	334084	193100552	24.0416	8.3300	2.76193	1.73010	1815.8	262389
579	335241	194104539	24.0624	8.3348	2.76268	1.72712	1819.0	263298
580	336400	195112000	24.0832	8.3396	2.76343	1.72414	1822.1	264208
581	337561	196122941	24.1039	8.3443	2.76418	1.72117	1825.3	265120
582	338724	197137368	24.1247	8.3491	2.76492	1.71821	1828.4	266033
583	339889	198155287	24.1454	8.3539	2.76567	1.71527	1831.6	266948
584	341056	199176704	24.1661	8.3587	2.76641	1.71233	1834.7	267865
585	342225	200201625	24.1868	8.3634	2.76716	1.70940	1837.8	268783
586	343396	201230056	24.2074	8.3682	2.76790	1.70648	1841.0	269703
587	344569	202262003	24.2281	8.3730	2.76864	1.70358	1844.1	270624
588	345744	203297472	24.2487	8.3777	2.76938	1.70068	1847.3	271547
589	346921	204336469	24.2693	8.3825	2.77012	1.69779	1850.4	272471
590	348100	205379000	24.2899	8.3872	2.77085	1.69492	1853.5	273397
591	349281	206425071	24.3105	8.3919	2.77159	1.69205	1856.7	274325
592	350464	207474688	24.3311	8.3967	2.77232	1.68919	1859.8	275254
593	351649	208527857	24.3516	8.4014	2.77305	1.68634	1863.0	276184
594	352836	209584584	24.3721	8.4061	2.77379	1.68350	1866.1	277117
595	354025	210644875	24.3926	8.4108	2.77452	1.68067	1869.2	278051
596	355216	211708736	24.4131	8.4155	2.77525	1.67785	1872.4	278986
597	356409	212776173	24.4336	8.4202	2.77597	1.67504	1875.5	279923
598	357604	213847192	24.4540	8.4249	2.77670	1.67224	1878.7	280862
599	358801	214921799	24.4745	8.4296	2.77743	1.66945	1881.8	281802

TABLE: Functions of numbers, continued (600 through 649) 5.6.3

No.	Square	Cube	Square Root	Cube Root	Logarithm	1000 × Reciprocal	No. = Diameter	
							Circum.	Area
600	360000	216000000	24.4949	8.4343	2.77815	1.66667	1885.0	282743
601	361201	217081801	24.5153	8.4390	2.77887	1.66389	1888.1	283687
602	362404	218167208	24.5357	8.4437	2.77960	1.66113	1891.2	284631
603	363609	219256227	24.5561	8.4484	2.78032	1.65837	1894.4	285578
604	364816	220348864	24.5764	8.4530	2.78104	1.65563	1897.5	286526
605	366025	221445125	24.5967	8.4577	2.78176	1.65289	1900.7	287475
606	367236	222545016	24.6171	8.4623	2.78247	1.65017	1903.8	288426
607	368449	223648543	24.6374	8.4670	2.78319	1.64745	1906.9	289379
608	369664	224755712	24.6577	8.4716	2.78390	1.64474	1910.1	290333
609	370881	225866529	24.6779	8.4763	2.78462	1.64204	1913.2	291289
610	372100	226981000	24.6982	8.4809	2.78533	1.63934	1916.4	292247
611	373321	228099131	24.7184	8.4856	2.78604	1.63666	1919.5	293206
612	374544	229220928	24.7386	8.4902	2.78675	1.63399	1922.7	294166
613	375769	230346397	24.7588	8.4948	2.78746	1.63132	1925.8	295128
614	376996	231475544	24.7790	8.4994	2.78817	1.62866	1928.9	296092
615	378225	232608375	24.7992	8.5040	2.78888	1.62602	1932.1	297057
616	379456	233744896	24.8193	8.5086	2.78958	1.62338	1935.2	298024
617	380689	234885113	24.8395	8.5132	2.79029	1.62075	1938.4	298992
618	381924	236029032	24.8596	8.5178	2.79099	1.61812	1941.5	299962
619	383161	237176659	24.8797	8.5224	2.79169	1.61551	1944.6	300934
620	384400	238328000	24.8998	8.5270	2.79239	1.61290	1947.8	301907
621	385641	239483061	24.9199	8.5316	2.79309	1.61031	1950.9	302882
622	386884	240641848	24.9399	8.5362	2.79379	1.60772	1954.1	303858
623	388129	241804367	24.9600	8.5408	2.79449	1.60514	1957.2	304836
624	389376	242970624	24.9800	8.5453	2.79518	1.60256	1960.4	305815
625	390625	244140625	25.0000	8.5499	2.79588	1.60000	1963.5	306796
626	391876	245314376	25.0200	8.5544	2.79657	1.59744	1966.6	307779
627	393129	246491883	25.0400	8.5590	2.79727	1.59490	1969.8	308763
628	394384	247673152	25.0599	8.5635	2.79796	1.59236	1972.9	309748
629	395641	248858189	25.0799	8.5681	2.79865	1.58983	1976.1	310736
630	396900	250047000	25.0998	8.5726	2.79934	1.58730	1979.2	311725
631	398161	251239591	25.1197	8.5772	2.80003	1.58479	1982.3	312715
632	399424	252435968	25.1396	8.5817	2.80072	1.58228	1985.5	313707
633	400689	253636137	25.1595	8.5862	2.80140	1.57978	1988.6	314700
634	401956	254840104	25.1794	8.5907	2.80209	1.57729	1991.8	315696
635	403225	256047875	25.1992	8.5952	2.80277	1.57480	1994.9	316692
636	404496	257259456	25.2190	8.5997	2.80346	1.57233	1998.1	317690
637	405769	258474853	25.2389	8.6043	2.80414	1.56986	2001.2	318690
638	407044	259694072	25.2587	8.6088	2.80482	1.56740	2004.3	319692
639	408321	260917119	25.2784	8.6132	2.80550	1.56495	2007.5	320695
640	409600	262144000	25.2982	8.6177	2.80618	1.56250	2010.6	321699
641	410881	263374721	25.3180	8.6222	2.80686	1.56006	2013.8	322705
642	412164	264609288	25.3377	8.6267	2.80754	1.55763	2016.9	323713
643	413449	265847707	25.3574	8.6312	2.80821	1.55521	2020.0	324722
644	414736	267089984	25.3772	8.6357	2.80889	1.55280	2023.2	325733
645	416025	268336125	25.3969	8.6401	2.80956	1.55039	2026.3	326745
646	417316	269586136	25.4165	8.6446	2.81023	1.54799	2029.5	327759
647	418609	270840023	25.4362	8.6490	2.81090	1.54560	2032.6	328775
648	419904	272097792	25.4558	8.6535	2.81158	1.54321	2035.8	329792
649	421201	273359449	25.4755	8.6579	2.81224	1.54083	2038.9	330810

TABLE: Functions of numbers, continued (650 through 699) 5.6.3

No.	Square	Cube	Square Root	Cube Root	Logarithm	1000 × Reciprocal	No. = Diameter	
							Circum.	Area
650	422500	274625000	25.4951	8.6624	2.81291	1.53846	2042.0	331831
651	423801	275894451	25.5147	8.6668	2.81358	1.53610	2045.2	332853
652	425104	277167808	25.5343	8.6713	2.81425	1.53374	2048.3	333876
653	426409	278445077	25.5539	8.6757	2.81491	1.53139	2051.5	334901
654	427716	279726264	25.5734	8.6801	2.81558	1.52905	2054.6	335927
655	429025	281011375	25.5930	8.6845	2.81624	1.52672	2057.7	336955
656	430336	282300416	25.6125	8.6890	2.81690	1.52439	2060.9	337985
657	431649	283593393	25.6320	8.6934	2.81757	1.52207	2064.0	339016
658	432964	284890312	25.6515	8.6978	2.81823	1.51976	2067.2	340049
659	434281	286191179	25.6710	8.7022	2.81889	1.51745	2070.3	341084
660	435600	287496000	25.6905	8.7066	2.81954	1.51515	2073.5	342119
661	436921	288804781	25.7099	8.7110	2.82020	1.51286	2076.6	343157
662	438244	290117528	25.7294	8.7154	2.82086	1.51057	2079.7	344196
663	439569	291434247	25.7488	8.7198	2.82151	1.50830	2082.9	345237
664	440896	292754944	25.7682	8.7241	2.82217	1.50602	2086.0	346279
665	442225	294079625	25.7876	8.7285	2.82282	1.50376	2089.2	347323
666	443556	295408296	25.8070	8.7329	2.82347	1.50150	2092.3	348368
667	444889	296740963	25.8263	8.7373	2.82413	1.49925	2095.4	349415
668	446224	298077632	25.8457	8.7416	2.82478	1.49701	2098.6	350464
669	447561	299418309	25.8650	8.7460	2.82543	1.49477	2101.7	351514
670	448900	300763000	25.8844	8.7503	2.82607	1.49254	2104.9	352565
671	450241	302111711	25.9037	8.7547	2.82672	1.49031	2108.0	353618
672	451584	303464448	25.9230	8.7590	2.82737	1.48810	2111.2	354673
673	452929	304821217	25.9422	8.7634	2.82802	1.48588	2114.3	355730
674	454276	306182024	25.9615	8.7677	2.82866	1.48368	2117.4	356788
675	455625	307546875	25.9808	8.7721	2.82930	1.48148	2120.6	357847
676	456976	308915776	26.0000	8.7764	2.82995	1.47929	2123.7	358908
677	458329	310288733	26.0192	8.7807	2.83059	1.47710	2126.9	359971
678	459684	311665752	26.0384	8.7850	2.83123	1.47493	2130.0	361035
679	461041	313046839	26.0576	8.7893	2.83187	1.47275	2133.1	362101
680	462400	314432000	26.0768	8.7937	2.83251	1.47059	2136.3	363168
681	463761	315821241	26.0960	8.7980	2.83315	1.46843	2139.4	364237
682	465124	317214568	26.1151	8.8023	2.83378	1.46628	2142.6	365308
683	466489	318611987	26.1343	8.8066	2.83442	1.46413	2145.7	366380
684	467856	320013504	26.1534	8.8109	2.83506	1.46199	2148.8	367453
685	469225	321419125	26.1725	8.8152	2.83569	1.45985	2152.0	368528
686	470596	322828856	26.1916	8.8194	2.83632	1.45773	2155.1	369605
687	471969	324242703	26.2107	8.8237	2.83696	1.45560	2158.3	370684
688	473344	325660672	26.2298	8.8280	2.83759	1.45349	2161.4	371764
689	474721	327082769	26.2488	8.8323	2.83822	1.45138	2164.6	372845
690	476100	328509000	26.2679	8.8366	2.83885	1.44928	2167.7	373928
691	477481	329939371	26.2869	8.8408	2.83948	1.44718	2170.8	375013
692	478864	331373888	26.3059	8.8451	2.84011	1.44509	2174.0	376099
693	480249	332812557	26.3249	8.8493	2.84073	1.44300	2177.1	377187
694	481636	334255384	26.3439	8.8536	2.84136	1.44092	2180.3	378276
695	483025	335702375	26.3629	8.8578	2.84198	1.43885	2183.4	379367
696	484416	337153536	26.3818	8.8621	2.84261	1.43678	2186.5	380459
697	485809	338608873	26.4008	8.8663	2.84323	1.43472	2189.7	381553
698	487204	340068392	26.4197	8.8706	2.84386	1.43266	2192.8	382649
699	488601	341532099	26.4386	8.8748	2.84448	1.43062	2196.0	383746

TABLE: Functions of numbers, continued (700 through 749) 5.6.3

No.	Square	Cube	Square Root	Cube Root	Logarithm	1000 × Reciprocal	No. = Diameter	
							Circum.	Area
700	490000	343000000	26.4575	8.8790	2.84510	1.42857	2199.1	384845
701	491401	344472101	26.4764	8.8833	2.84572	1.42653	2202.3	385945
702	492804	345948408	26.4953	8.8875	2.84634	1.42450	2205.4	387047
703	494209	347428927	26.5141	8.8917	2.84696	1.42248	2208.5	388151
704	495616	348913664	26.5330	8.8959	2.84757	1.42045	2211.7	389256
705	497025	350402625	26.5518	8.9001	2.84819	1.41844	2214.8	390363
706	498436	351895816	26.5707	8.9043	2.84880	1.41643	2218.0	391471
707	499849	353393243	26.5895	8.9085	2.84942	1.41443	2221.1	392580
708	501264	354894912	26.6083	8.9127	2.85003	1.41243	2224.2	393692
709	502681	356400829	26.6271	8.9169	2.85065	1.41044	2227.4	394805
710	504100	357911000	26.6458	8.9211	2.85126	1.40845	2230.5	395919
711	505521	359425431	26.6646	8.9253	2.85187	1.40647	2233.7	397035
712	506944	360944128	26.6833	8.9295	2.85248	1.40449	2236.8	398153
713	508369	362467097	26.7021	8.9337	2.85309	1.40252	2240.0	399272
714	509796	363994344	26.7208	8.9378	2.85370	1.40056	2243.1	400393
715	511225	365525875	26.7395	8.9420	2.85431	1.39860	2246.2	401515
716	512656	367061696	26.7582	8.9462	2.85491	1.39665	2249.4	402639
717	514089	368601813	26.7769	8.9503	2.85552	1.39470	2252.5	403765
718	515524	370146232	26.7955	8.9545	2.85612	1.39276	2255.7	404892
719	516961	371694959	26.8142	8.9587	2.85673	1.39082	2258.8	406020
720	518400	373248000	26.8328	8.9628	2.85733	1.38889	2261.9	407150
721	519841	374805361	26.8514	8.9670	2.85794	1.38696	2265.1	408282
722	521284	376367048	26.8701	8.9711	2.85854	1.38504	2268.2	409415
723	522729	377933067	26.8887	8.9752	2.85914	1.38313	2271.4	410550
724	524176	379503424	26.9072	8.9794	2.85974	1.38122	2274.5	411687
725	525625	381078125	26.9258	8.9835	2.86034	1.37931	2277.7	412825
726	527076	382657176	26.9444	8.9876	2.86094	1.37741	2280.8	413965
727	528529	384240583	26.9629	8.9918	2.86153	1.37552	2283.9	415106
728	529984	385828352	26.9815	8.9959	2.86213	1.37363	2287.1	416248
729	531441	387420489	27.0000	9.0000	2.86273	1.37174	2290.2	417393
730	532900	389017000	27.0185	9.0041	2.86332	1.36986	2293.4	418539
731	534361	390617891	27.0370	9.0082	2.86392	1.36799	2296.5	419686
732	535824	392223168	27.0555	9.0123	2.86451	1.36612	2299.6	420835
733	537289	393832837	27.0740	9.0164	2.86510	1.36426	2302.8	421986
734	538756	395446904	27.0924	9.0205	2.86570	1.36240	2305.9	423138
735	540225	397065375	27.1109	9.0246	2.86629	1.36054	2309.1	424293
736	541696	398688256	27.1293	9.0287	2.86688	1.35870	2312.2	425447
737	543169	400315553	27.1477	9.0328	2.86747	1.35685	2315.4	426604
738	544644	401947272	27.1662	9.0369	2.86806	1.35501	2318.5	427762
739	546121	403583419	27.1846	9.0410	2.86864	1.35318	2321.6	428922
740	547600	405224000	27.2029	9.0450	2.86923	1.35135	2324.8	430084
741	549081	406869021	27.2213	9.0491	2.86982	1.34953	2327.9	431247
742	550564	408518488	27.2397	9.0532	2.87040	1.34771	2331.1	432412
743	552049	410172407	27.2580	9.0572	2.87099	1.34590	2334.2	433578
744	553536	411830784	27.2764	9.0613	2.87157	1.34409	2337.3	434746
745	555025	413493625	27.2947	9.0654	2.87216	1.34228	2340.5	435916
746	556516	415160936	27.3130	9.0694	2.87274	1.34048	2343.6	437087
747	558009	416832723	27.3313	9.0735	2.87332	1.33869	2346.8	438259
748	559504	418508992	27.3496	9.0775	2.87390	1.33690	2349.9	439433
749	561001	420189749	27.3679	9.0816	2.87448	1.33511	2353.1	440609

TABLE: Functions of numbers, continued (750 through 799) 5.6.3

No.	Square	Cube	Square Root	Cube Root	Logarithm	1000 × Reciprocal	No. = Diameter	
							Circum.	Area
750	562500	421875000	27.3861	9.0856	2.87506	1.33333	2356.2	441786
751	564001	423564751	27.4044	9.0896	2.87564	1.33156	2359.3	442965
752	565504	425259008	27.4226	9.0937	2.87622	1.32979	2362.5	444146
753	567009	426957777	27.4408	9.0977	2.87680	1.32802	2365.6	445328
754	568516	428661064	27.4591	9.1017	2.87737	1.32626	2368.8	446511
755	570025	430368875	27.4773	9.1057	2.87795	1.32450	2371.9	447697
756	571536	432081216	27.4955	9.1098	2.87852	1.32275	2375.0	448883
757	573049	433798093	27.5136	9.1138	2.87910	1.32100	2378.2	450072
758	574564	435519512	27.5318	9.1178	2.87967	1.31926	2381.3	451262
759	576081	437245479	27.5500	9.1218	2.88024	1.31752	2384.5	452453
760	577600	438976000	27.5681	9.1258	2.88081	1.31579	2387.6	453646
761	579121	440711081	27.5862	9.1298	2.88138	1.31406	2390.8	454841
762	580644	442450728	27.6043	9.1338	2.88196	1.31234	2393.9	456037
763	582169	444194947	27.6225	9.1378	2.88252	1.31062	2397.0	457234
764	583696	445943744	27.6405	9.1418	2.88309	1.30890	2400.2	458434
765	585225	447697125	27.6586	9.1458	2.88366	1.30719	2403.3	459635
766	586756	449455096	27.6767	9.1498	2.88423	1.30548	2406.5	460837
767	588289	451217663	27.6948	9.1537	2.88480	1.30378	2409.6	462041
768	589824	452984832	27.7128	9.1577	2.88536	1.30208	2412.7	463247
769	591361	454756609	27.7308	9.1617	2.88593	1.30039	2415.9	464454
770	592900	456533000	27.7489	9.1657	2.88649	1.29870	2419.0	465663
771	594441	458314011	27.7669	9.1696	2.88705	1.29702	2422.2	466873
772	595984	460099648	27.7849	9.1736	2.88762	1.29534	2425.3	468085
773	597529	461889917	27.8029	9.1775	2.88818	1.29366	2428.5	469298
774	599076	463684824	27.8209	9.1815	2.88874	1.29199	2431.6	470513
775	600625	465484375	27.8388	9.1855	2.88930	1.29032	2434.7	471730
776	602176	467288576	27.8568	9.1894	2.88986	1.28866	2437.9	472948
777	603729	469097433	27.8747	9.1933	2.89042	1.28700	2441.0	474168
778	605284	470910952	27.8927	9.1973	2.89098	1.28535	2444.2	475389
779	606841	472729139	27.9106	9.2012	2.89154	1.28370	2447.3	476612
780	608400	474552000	27.9285	9.2052	2.89209	1.28205	2450.4	477836
781	609961	476379541	27.9464	9.2091	2.89265	1.28041	2453.6	479062
782	611524	478211768	27.9643	9.2130	2.89321	1.27877	2456.7	480290
783	613089	480048687	27.9821	9.2170	2.89376	1.27714	2459.9	481519
784	614656	481890304	28.0000	9.2209	2.89432	1.27551	2463.0	482750
785	616225	483736625	28.0179	9.2248	2.89487	1.27389	2466.2	483982
786	617796	485587656	28.0357	9.2287	2.89542	1.27226	2469.3	485216
787	619369	487443403	28.0535	9.2326	2.89597	1.27065	2472.4	486451
788	620944	489303872	28.0713	9.2365	2.89653	1.26904	2475.6	487688
789	622521	491169069	28.0891	9.2404	2.89708	1.26743	2478.7	488927
790	624100	493039000	28.1069	9.2443	2.89763	1.26582	2481.9	490167
791	625681	494913671	28.1247	9.2482	2.89818	1.26422	2485.0	491409
792	627264	496793088	28.1425	9.2521	2.89873	1.26263	2488.1	492652
793	628849	498677257	28.1603	9.2560	2.89927	1.26103	2491.3	493897
794	630436	500566184	28.1780	9.2599	2.89982	1.25945	2494.4	495143
795	632025	502459875	28.1957	9.2638	2.90037	1.25786	2497.6	496391
796	633616	504358336	28.2135	9.2677	2.90091	1.25628	2500.7	497641
797	635209	506261573	28.2312	9.2716	2.90146	1.25471	2503.8	498892
798	636804	508169592	28.2489	9.2754	2.90200	1.25313	2507.0	500145
799	638401	510082399	28.2666	9.2793	2.90255	1.25156	2510.1	501399

TABLE: Functions of numbers, continued (800 through 849) 5.6.3

No.	Square	Cube	Square Root	Cube Root	Logarithm	1000 × Reciprocal	No. = Diameter	
							Circum.	Area
800	640000	512000000	28.2843	9.2832	2.90309	1.25000	2513.3	502655
801	641601	513922401	28.3019	9.2870	2.90363	1.24844	2516.4	503912
802	643204	515849608	28.3196	9.2909	2.90417	1.24688	2519.6	505171
803	644809	517781627	28.3373	9.2948	2.90472	1.24533	2522.7	506432
804	646416	519718464	28.3549	9.2986	2.90526	1.24378	2525.8	507694
805	648025	521660125	28.3725	9.3025	2.90580	1.24224	2529.0	508958
806	649636	523606616	28.3901	9.3063	2.90634	1.24069	2532.1	510223
807	651249	525557943	28.4077	9.3102	2.90687	1.23916	2535.3	511490
808	652864	527514112	28.4253	9.3140	2.90741	1.23762	2538.4	512758
809	654481	529475129	28.4429	9.3179	2.90795	1.23609	2541.5	514028
810	656100	531441000	28.4605	9.3217	2.90849	1.23457	2544.7	515300
811	657721	533411731	28.4781	9.3255	2.90902	1.23305	2547.8	516573
812	659344	535387328	28.4956	9.3294	2.90956	1.23153	2551.0	517848
813	660969	537367797	28.5132	9.3332	2.91009	1.23001	2554.1	519124
814	662596	539353144	28.5307	9.3370	2.91062	1.22850	2557.3	520402
815	664225	541343375	28.5482	9.3408	2.91116	1.22699	2560.4	521681
816	665856	543338496	28.5657	9.3447	2.91169	1.22549	2563.5	522962
817	667489	545338513	28.5832	9.3485	2.91222	1.22399	2566.7	524245
818	669124	547343432	28.6007	9.3523	2.91275	1.22249	2569.8	525529
819	670761	549353259	28.6182	9.3561	2.91328	1.22100	2573.0	526814
820	672400	551368000	28.6356	9.3599	2.91381	1.21951	2576.1	528102
821	674041	553387661	28.6531	9.3637	2.91434	1.21803	2579.2	529391
822	675684	555412248	28.6705	9.3675	2.91487	1.21655	2582.4	530681
823	677329	557441767	28.6880	9.3713	2.91540	1.21507	2585.5	531973
824	678976	559476224	28.7054	9.3751	2.91593	1.21359	2588.7	533267
825	680625	561515625	28.7228	9.3789	2.91645	1.21212	2591.8	534562
826	682276	563559976	28.7402	9.3827	2.91698	1.21065	2595.0	535858
827	683929	565609283	28.7576	9.3865	2.91751	1.20919	2598.1	537157
828	685584	567663552	28.7750	9.3902	2.91803	1.20773	2601.2	538456
829	687241	569722789	28.7924	9.3940	2.91855	1.20627	2604.4	539758
830	688900	571787000	28.8097	9.3978	2.91908	1.20482	2607.5	541061
831	690561	573856191	28.8271	9.4016	2.91960	1.20337	2610.7	542365
832	692224	575930368	28.8444	9.4053	2.92012	1.20192	2613.8	543671
833	693889	578009537	28.8617	9.4091	2.92065	1.20048	2616.9	544979
834	695556	580093704	28.8791	9.4129	2.92117	1.19904	2620.1	546288
835	697225	582182875	28.8964	9.4166	2.92169	1.19760	2623.2	547599
836	698896	584277056	28.9137	9.4204	2.92221	1.19617	2626.4	548912
837	700569	586376253	28.9310	9.4241	2.92273	1.19474	2629.5	550226
838	702244	588480472	28.9482	9.4279	2.92324	1.19332	2632.7	551541
839	703921	590589719	28.9655	9.4316	2.92376	1.19190	2635.8	552858
840	705600	592704000	28.9828	9.4354	2.92428	1.19048	2638.9	554177
841	707281	594823321	29.0000	9.4391	2.92480	1.18906	2642.1	555497
842	708964	596947688	29.0172	9.4429	2.92531	1.18765	2645.2	556819
843	710649	599077107	29.0345	9.4466	2.92583	1.18624	2648.4	558142
844	712336	601211584	29.0517	9.4503	2.92634	1.18483	2651.5	559467
845	714025	603351125	29.0689	9.4541	2.92686	1.18343	2654.6	560794
846	715716	605495736	29.0861	9.4578	2.92737	1.18203	2657.8	562122
847	717409	607645423	29.1033	9.4615	2.92788	1.18064	2660.9	563452
848	719104	609800192	29.1204	9.4652	2.92840	1.17925	2664.1	564783
849	720801	611960049	29.1376	9.4690	2.92891	1.17786	2667.2	566116

TABLE: Functions of numbers, continued (850 through 899) 5.6.3

No.	Square	Cube	Square Root	Cube Root	Logarithm	1000 × Reciprocal	No. = Diameter	
							Circum.	Area
850	722500	614125000	29.1548	9.4727	2.92942	1.17647	2670.4	567450
851	724201	616295051	29.1719	9.4764	2.92993	1.17509	2673.5	568786
852	725904	618470208	29.1890	9.4801	2.93044	1.17371	2676.6	570124
853	727609	620650477	29.2062	9.4838	2.93095	1.17233	2679.8	571463
854	729316	622835864	29.2233	9.4875	2.93146	1.17096	2682.9	572803
855	731025	625026375	29.2404	9.4912	2.93197	1.16959	2686.1	574146
856	732736	627222016	29.2575	9.4949	2.93247	1.16822	2689.2	575490
857	734449	629422793	29.2746	9.4986	2.93298	1.16686	2692.3	576835
858	736164	631628712	29.2916	9.5023	2.93349	1.16550	2695.5	578182
859	737881	633839779	29.3087	9.5060	2.93399	1.16414	2698.6	579530
860	739600	636056000	29.3258	9.5097	2.93450	1.16279	2701.8	580880
861	741321	638277381	29.3428	9.5134	2.93500	1.16144	2704.9	582232
862	743044	640503928	29.3598	9.5171	2.93551	1.16009	2708.1	583585
863	744769	642735647	29.3769	9.5207	2.93601	1.15875	2711.2	584940
864	746496	644972544	29.3939	9.5244	2.93651	1.15741	2714.3	586297
865	748225	647214625	29.4109	9.5281	2.93702	1.15607	2717.5	587655
866	749956	649461896	29.4279	9.5317	2.93752	1.15473	2720.6	589014
867	751689	651714363	29.4449	9.5354	2.93802	1.15340	2723.8	590375
868	753424	653972032	29.4618	9.5391	2.93852	1.15207	2726.9	591738
869	755161	656234909	29.4788	9.5427	2.93902	1.15075	2730.0	583102
870	756900	658503000	29.4958	9.5464	2.93952	1.14943	2733.2	594468
871	758641	660776311	29.5127	9.5501	2.94002	1.14811	2736.3	595835
872	760384	663054848	29.5296	9.5537	2.94052	1.14679	2739.5	597204
873	762129	665338617	29.5466	9.5574	2.94101	1.14548	2742.6	598575
874	763876	667627624	29.5635	9.5610	2.94151	1.14416	2745.8	599947
875	765625	669921875	29.5804	9.5647	2.94201	1.14286	2748.9	601320
876	767376	672221376	29.5973	9.5683	2.94250	1.14155	2752.0	602696
877	769129	674526133	29.6142	9.5719	2.94300	1.14025	2755.2	604073
878	770884	676836152	29.6311	9.5756	2.94349	1.13895	2758.3	605451
879	772641	679151439	29.6479	9.5792	2.94399	1.13766	2761.5	606831
880	774400	681472000	29.6648	9.5828	2.94448	1.13636	2764.6	608212
881	776161	683797841	29.6816	9.5865	2.94498	1.13507	2767.7	609595
882	777924	686128968	29.6985	9.5901	2.94547	1.13379	2770.9	610980
883	779689	688465387	29.7153	9.5937	2.94596	1.13250	2774.0	612366
884	781456	690807104	29.7321	9.5973	2.94645	1.13122	2777.2	613754
885	783225	693154125	29.7489	9.6010	2.94694	1.12994	2780.3	615143
886	784996	695506456	29.7658	9.6046	2.94743	1.12867	2783.5	616534
887	786769	697864103	29.7825	9.6082	2.94792	1.12740	2786.6	617927
888	788544	700227072	29.7993	9.6118	2.94841	1.12613	2789.7	619321
889	790321	702595369	29.8161	9.6154	2.94890	1.12486	2792.9	620717
890	792100	704969000	29.8329	9.6190	2.94939	1.12360	2796.0	622114
891	793881	707347971	29.8496	9.6226	2.94988	1.12233	2799.2	623513
892	795664	709732288	29.8664	9.6262	2.95036	1.12108	2802.3	624913
893	797449	712121957	29.8831	9.6298	2.95085	1.11982	2805.4	626315
894	799236	714516984	29.8998	9.6334	2.95134	1.11857	2808.6	627718
895	801025	716917375	29.9166	9.6370	2.95182	1.11732	2811.7	629124
896	802816	719323136	29.9333	9.6406	2.95231	1.11607	2814.9	630530
897	804609	721734273	29.9500	9.6442	2.95279	1.11483	2818.0	631938
898	806404	724150792	29.9666	9.6477	2.95328	1.11359	2821.2	633348
899	808201	726572699	29.9833	9.6513	2.95376	1.11235	2824.3	634760

TABLE: Functions of numbers, continued (900 through 949) 5.6.3

No.	Square	Cube	Square Root	Cube Root	Logarithm	1000 × Reciprocal	No. = Diameter	
							Circum.	Area
900	810000	729000000	30.0000	9.6549	2.95424	1.11111	2827.4	636173
901	811801	731432701	30.0167	9.6585	2.95472	1.10988	2830.6	637587
902	813604	733870808	30.0333	9.6620	2.95521	1.10865	2833.7	639003
903	815409	736314327	30.0500	9.6656	2.95569	1.10742	2836.9	640421
904	817216	738763264	30.0666	9.6692	2.95617	1.10619	2840.0	641840
905	819025	741217625	30.0832	9.6727	2.95665	1.10497	2843.1	643261
906	820836	743677416	30.0998	9.6763	2.95713	1.10375	2846.3	644683
907	822649	746142643	30.1164	9.6799	2.95761	1.10254	2849.4	646107
908	824464	748613312	30.1330	9.6834	2.95809	1.10132	2852.6	647533
909	826281	751089429	30.1496	9.6870	2.95856	1.10011	2855.7	648960
910	828100	753571000	30.1662	9.6905	2.95904	1.09890	2858.8	650388
911	829921	756058031	30.1828	9.6941	2.95952	1.09769	2862.0	651818
912	831744	758550528	30.1993	9.6976	2.95999	1.09649	2865.1	653250
913	833569	761048497	30.2159	9.7012	2.96047	1.09529	2868.3	654684
914	835396	763551944	30.2324	9.7047	2.96095	1.09409	2871.4	656118
915	837225	766060875	30.2490	9.7082	2.96142	1.09290	2874.6	657555
916	839056	768575296	30.2655	9.7118	2.96190	1.09170	2877.7	658993
917	840889	771095213	30.2820	9.7153	2.96237	1.09051	2880.8	660433
918	842724	773620632	30.2985	9.7188	2.96284	1.08932	2884.0	661874
919	844561	776151559	30.3150	9.7224	2.96332	1.08814	2887.1	663317
920	846400	778688000	30.3315	9.7259	2.96379	1.08696	2890.3	664761
921	848241	781229961	30.3480	9.7294	2.96426	1.08578	2893.4	666207
922	850084	783777448	30.3645	9.7329	2.96473	1.08460	2896.5	667654
923	851929	786330467	30.3809	9.7364	2.96520	1.08342	2899.7	669103
924	853776	788889024	30.3974	9.7400	2.96567	1.08225	2902.8	670554
925	855625	791453125	30.4138	9.7435	2.96614	1.08108	2906.0	672006
926	857476	794022776	30.4302	9.7470	2.96661	1.07991	2909.1	673460
927	859329	796597983	30.4467	9.7505	2.96708	1.07875	2912.3	674915
928	861184	799178752	30.4631	9.7540	2.96755	1.07759	2915.4	676372
929	863041	801765089	30.4795	9.7575	2.96802	1.07643	2918.5	677831
930	864900	804357000	30.4959	9.7610	2.96848	1.07527	2921.7	679291
931	866761	806954491	30.5123	9.7645	2.96895	1.07411	2924.8	680752
932	868624	809557568	30.5287	9.7680	2.96942	1.07296	2928.0	682216
933	870489	812166237	30.5450	9.7715	2.96988	1.07181	2931.1	683680
934	872356	814780504	30.5614	9.7750	2.97035	1.07066	2934.2	685147
935	874225	817400375	30.5778	9.7785	2.97081	1.06952	2937.4	686615
936	876096	820025856	30.5941	9.7819	2.97128	1.06838	2940.5	688084
937	877969	822656953	30.6105	9.7854	2.97174	1.06724	2943.7	689555
938	879844	825293672	30.6268	9.7889	2.97220	1.06610	2946.8	691028
939	881721	827936019	30.6431	9.7924	2.97267	1.06496	2950.0	692502
940	883600	830584000	30.6594	9.7959	2.97313	1.06383	2953.1	693978
941	885481	833237621	30.6757	9.7993	2.97359	1.06270	2956.2	695455
942	887364	835896888	30.6920	9.8028	2.97405	1.06157	2959.4	696934
943	889249	838561807	30.7083	9.8063	2.97451	1.06045	2962.5	698415
944	891136	841232384	30.7246	9.8097	2.97497	1.05932	2965.7	699897
945	893025	843908625	30.7409	9.8132	2.97543	1.05820	2968.8	701380
946	894916	846590536	30.7571	9.8167	2.97589	1.05708	2971.9	702865
947	896809	849278123	30.7734	9.8201	2.97635	1.05597	2975.1	704352
948	898704	851971392	30.7896	9.8236	2.97681	1.05485	2978.2	705840
949	900601	854670349	30.8058	9.8270	2.97727	1.05374	2981.4	707330

TABLE: Functions of numbers, continued (950 through 999) 5.6.3

No.	Square	Cube	Square Root	Cube Root	Logarithm	1000 × Reciprocal	No. = Diameter	
							Circum.	Area
950	902500	857375000	30.8221	9.8305	2.97772	1.05263	2984.5	708822
951	904401	860085351	30.8383	9.8339	2.97818	1.05152	2987.7	710315
952	906304	862801408	30.8545	9.8374	2.97864	1.05042	2990.8	711809
953	908209	865523177	30.8707	9.8408	2.97909	1.04932	2993.9	713306
954	910116	868250664	30.8869	9.8443	2.97955	1.04822	2997.1	714803
955	912025	870983875	30.9031	9.8477	2.98000	1.04712	3000.2	716303
956	913936	873722816	30.9192	9.8511	2.98046	1.04603	3003.4	717804
957	915849	876467493	30.9354	9.8546	2.98091	1.04493	3006.5	719306
958	917764	879217912	30.9516	9.8580	2.98137	1.04384	3009.6	720810
959	919681	881974079	30.9677	9.8614	2.98182	1.04275	3012.8	722316
960	921600	884736000	30.9839	9.8648	2.98227	1.04167	3015.9	723823
961	923521	887503681	31.0000	9.8683	2.98272	1.04058	3019.1	725332
962	925444	890277128	31.0161	9.8717	2.98318	1.03950	3022.2	726842
963	927369	893056347	31.0322	9.8751	2.98363	1.03842	3025.4	728354
964	929296	895841344	31.0483	9.8785	2.98408	1.03734	3028.5	729867
965	931225	898632125	31.0644	9.8819	2.98453	1.03627	3031.6	731382
966	933156	901428696	31.0805	9.8854	2.98498	1.03520	3034.8	732899
967	935089	904231063	31.0966	9.8888	2.98543	1.03413	3037.9	734417
968	937024	907039232	31.1127	9.8922	2.98588	1.03306	3041.1	735937
969	938961	909853209	31.1288	9.8956	2.98632	1.03199	3044.2	737458
970	940900	912673000	31.1448	9.8990	2.98677	1.03093	3047.3	738981
971	942841	915498611	31.1609	9.9024	2.98722	1.02987	3050.5	740506
972	944784	918330048	31.1769	9.9058	2.98767	1.02881	3053.6	742032
973	946729	921167317	31.1929	9.9092	2.98811	1.02775	3056.8	743559
974	948676	924010424	31.2090	9.9126	2.98856	1.02669	3059.9	745088
975	950625	926859375	31.2250	9.9160	2.98900	1.02564	3063.1	746619
976	952576	929714176	31.2410	9.9194	2.98945	1.02459	3066.2	748151
977	954529	932574833	31.2570	9.9227	2.98989	1.02354	3069.3	749685
978	956484	935441352	31.2730	9.9261	2.99034	1.02249	3072.5	751221
979	958441	938313739	31.2890	9.9295	2.99078	1.02145	3075.6	752758
980	960400	941192000	31.3050	9.9329	2.99123	1.02041	3078.8	754296
981	962361	944076141	31.3209	9.9363	2.99167	1.01937	3081.9	755837
982	964324	946966168	31.3369	9.9396	2.99211	1.01833	3085.0	757378
983	966289	949862087	31.3528	9.9430	2.99255	1.01729	3088.2	758922
984	968256	952763904	31.3688	9.9464	2.99300	1.01626	3091.3	760466
985	970225	955671625	31.3847	9.9497	2.99344	1.01523	3094.5	762013
986	972196	958585256	31.4006	9.9531	2.99388	1.01420	3097.6	763561
987	974169	961504803	31.4166	9.9565	2.99432	1.01317	3100.8	765111
988	976144	964430272	31.4325	9.9598	2.99476	1.01215	3103.9	766662
989	978121	967361669	31.4484	9.9632	2.99520	1.01112	3107.0	768214
990	980100	970299000	31.4643	9.9666	2.99564	1.01010	3110.2	769769
991	982081	973242271	31.4802	9.9699	2.99607	1.00908	3113.3	771325
992	984064	976191488	31.4960	9.9733	2.99651	1.00806	3116.5	772882
993	986049	979146657	31.5119	9.9766	2.99695	1.00705	3119.6	774441
994	988036	982107784	31.5278	9.9800	2.99739	1.00604	3122.7	776002
995	990025	985074875	31.5436	9.9833	2.99782	1.00503	3125.9	777564
996	992016	988047936	31.5595	9.9866	2.99826	1.00402	3129.0	779128
997	994009	991026973	31.5753	9.9900	2.99870	1.00301	3132.2	780693
998	996004	994011992	31.5911	9.9933	2.99913	1.00200	3135.3	782260
999	998001	997002999	31.6070	9.9967	2.99957	1.00100	3138.5	783828

TABLE: Trigonometric formulas 5.6.4

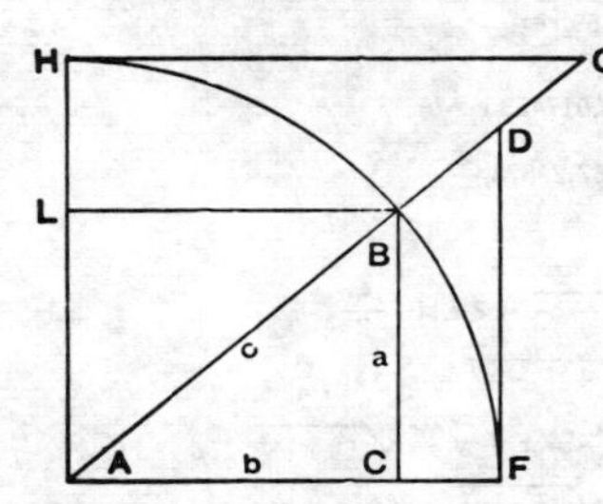

$$\text{Radius AF} = 1 = \sin^2 A + \cos^2 A = \sin A \operatorname{cosec} A = \cos A \sec A = \tan A \cot A$$

$$\text{Sine A} = \frac{\cos A}{\cot A} = \frac{1}{\operatorname{cosec} A} = \cos A \tan A = \sqrt{1 - \cos^2 A} = BC$$

$$\text{Cosine A} = \frac{\sin A}{\tan A} = \frac{1}{\sec A} = \sin A \cot A = \sqrt{1 - \sin^2 A} = AC$$

$$\text{Tangent A} = \frac{\sin A}{\cos A} = \frac{1}{\cot A} = \sin A \sec A = FD$$

$$\text{Cotangent A} = \frac{\cos A}{\sin A} = \frac{1}{\tan A} = \cos A \operatorname{cosec} A = HG$$

$$\text{Secant A} = \frac{\tan A}{\sin A} = \frac{1}{\cos A} = AD$$

$$\text{Cosecant A} = \frac{\cot A}{\cos A} = \frac{1}{\sin A} = AG$$

RIGHT ANGLED TRIANGLES

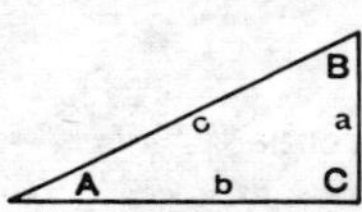

$$a^2 = c^2 - b^2$$
$$b^2 = c^2 - a^2$$
$$c^2 = a^2 + b^2$$

Known	Required					
	A	B	a	b	c	Area
a, b	$\tan A = \frac{a}{b}$	$\tan B = \frac{b}{a}$			$\sqrt{a^2 + b^2}$	$\frac{ab}{2}$
a, c	$\sin A = \frac{a}{c}$	$\cos B = \frac{a}{c}$		$\sqrt{c^2 - a^2}$		$\frac{a\sqrt{c^2 - a^2}}{2}$
A, a		$90° - A$		$a \cot A$	$\frac{a}{\sin A}$	$\frac{a^2 \cot A}{2}$
A, b		$90° - A$	$b \tan A$		$\frac{b}{\cos A}$	$\frac{b^2 \tan A}{2}$
A, c		$90° - A$	$c \sin A$	$c \cos A$		$\frac{c^2 \sin 2A}{4}$

OBLIQUE ANGLED TRIANGLES

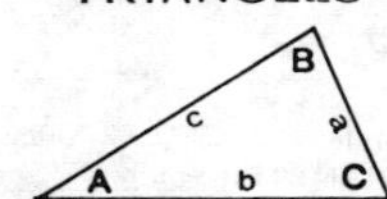

$$s = \frac{a + b + c}{2}$$

$$K = \sqrt{\frac{(s - a)(s - b)(s - c)}{s}}$$

$$a^2 = b^2 + c^2 - 2bc \cos A$$
$$b^2 = a^2 + c^2 - 2ac \cos B$$
$$c^2 = a^2 + b^2 - 2ab \cos C$$

Known	Required					
	A	B	C	b	c	Area
a, b, c	$\tan \frac{1}{2} A' = \frac{K}{s - a}$	$\tan \frac{1}{2} B = \frac{K}{s - b}$	$\tan \frac{1}{2} C = \frac{K}{s - c}$			$\sqrt{s(s-a)(s-b)(s-c)}$
a, A, B			$180° - (A + B)$	$\frac{a \sin B}{\sin A}$	$\frac{a \sin C}{\sin A}$	
a, b, A		$\sin B = \frac{b \sin A}{a}$			$\frac{b \sin C}{\sin B}$	
a, b, C	$\tan A = \frac{a \sin C}{b - a \cos C}$				$\sqrt{a^2 + b^2 - 2ab \cos C}$	$\frac{ab \sin C}{2}$

TABLE: Properties of the circle 5.6.5

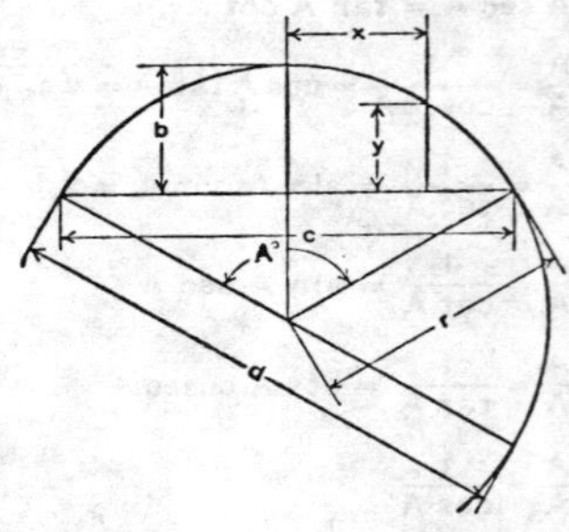

Circumference $= 6.28318\ r = 3.14159\ d$
Diameter $= 0.31831$ circumference
Area $= 3.14159\ r^2$

Arc $a = \frac{\pi r\ A°}{180°} = 0.017453\ r\ A°$

Angle $A° = \frac{180°\ a}{\pi r} = 57.29578\ \frac{a}{r}$

Radius $r = \frac{4\ b^2 + c^2}{8\ b}$

Chord $c = 2\sqrt{2\ br - b^2} = 2\ r \sin\frac{A}{2}$

Rise $b = r - \tfrac{1}{2}\sqrt{4\ r^2 - c^2} = \frac{c}{2}\tan\frac{A}{4}$

$= 2\ r \sin^2\frac{A}{4} = r + y - \sqrt{r^2 - x^2}$

$y = b - r + \sqrt{r^2 - x^2}$

$x = \sqrt{r^2 - (r + y - b)^2}$

Diameter of circle of equal periphery as square = 1.27324 side of square
Side of square of equal periphery as circle = 0.78540 diameter of circle
Diameter of circle circumscribed about square = 1.41421 side of square
Side of square inscribed in circle = 0.70711 diameter of circle

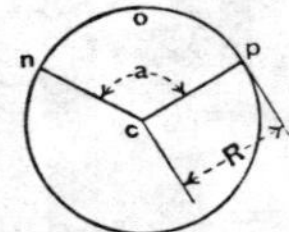

CIRCULAR SECTOR

R = radius of circle   a = angle ncp in degrees

Area of Sector ncpo = ½ (length of arc nop × R) = Area of Circle × $\frac{a}{360}$
$= 0.0087266 \times R^2 \times a$

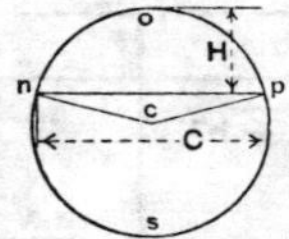

CIRCULAR SEGMENT

R = radius of circle   C = chord   H = rise

Area of Segment nop = Area of Sector ncpo – Area of triangle ncp
$= \frac{(\text{Length of arc nop} \times R) - C\ (R - H)}{2}$

Area of Segment nsp = Area of Circle – Area of Segment nop

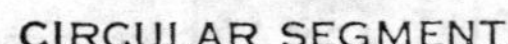

CIRCULAR SEGMENT,

Given: Rise, H, and Chord, C

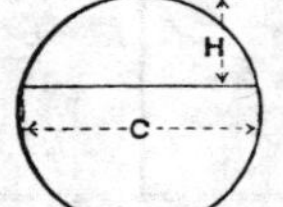

Area = Coefficient × H × C

Coefficient found opposite $\frac{H}{C}$

Interpolate for intermediate values of $\frac{H}{C}$

Example:

RISE = 1.49   CHORD = 3.52

$\frac{H}{C} = \frac{1.49}{3.52} = 0.4233$ Coeff. = 0.7542

Area = H × C × Coeff. = 1.49 × 3.52 × 0.7542 = 3.9556

CIRCULAR SEGMENT,

Given: RISE, H, and DIAMETER, D = 2R

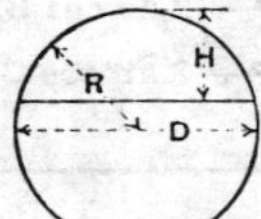

Area = Coefficient × D^2

Coefficient opposite $\frac{H}{D}$

Interpolate for intermediate values of $\frac{H}{D}$

Example:

RISE = $2\frac{7}{16}$ and DIAMETER = $5\frac{3}{32}$

$\frac{H}{D} = 2.4375 \div 5.09375 = 0.478528$

Coefficient = 0.371233

Area = Coef. × d^2 = 0.371233 × 25.94629
= 9.6321

TABLE: Areas of circular segments: ratio of rise and chord 5.6.6

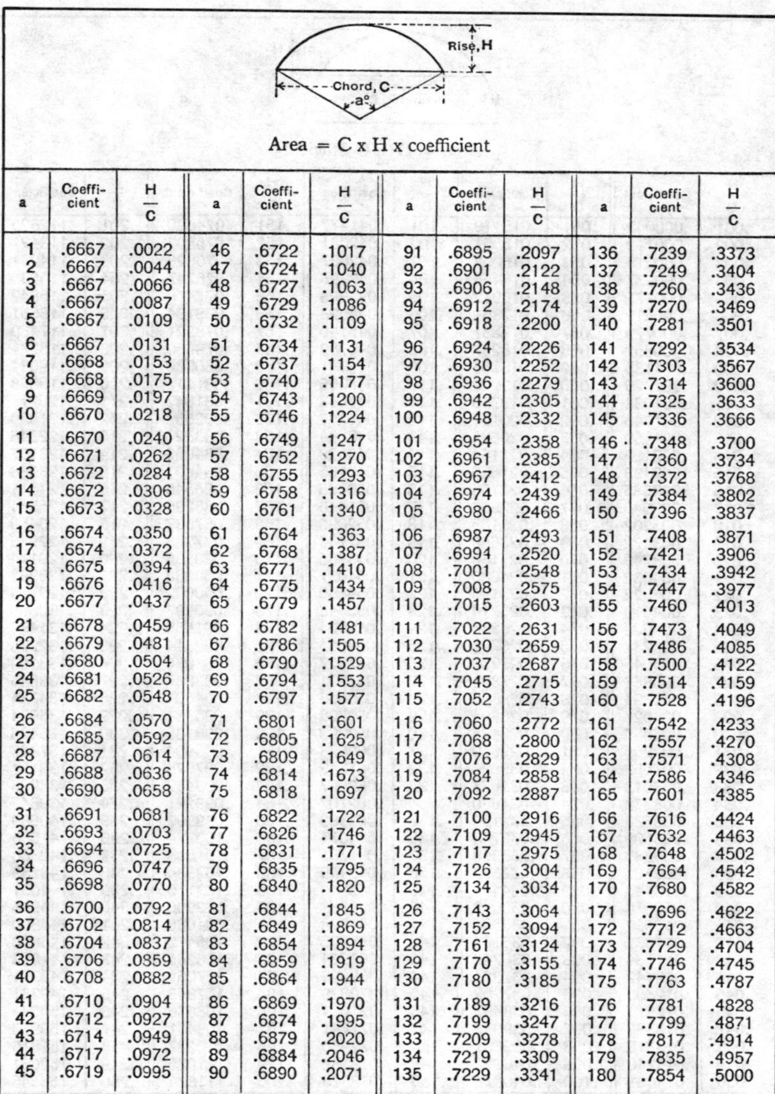

Area = C x H x coefficient

a	Coefficient	H/C	a	Coefficient	H/C	a	Coefficient	H/C	a	Coefficient	H/C
1	.6667	.0022	46	.6722	.1017	91	.6895	.2097	136	.7239	.3373
2	.6667	.0044	47	.6724	.1040	92	.6901	.2122	137	.7249	.3404
3	.6667	.0066	48	.6727	.1063	93	.6906	.2148	138	.7260	.3436
4	.6667	.0087	49	.6729	.1086	94	.6912	.2174	139	.7270	.3469
5	.6667	.0109	50	.6732	.1109	95	.6918	.2200	140	.7281	.3501
6	.6667	.0131	51	.6734	.1131	96	.6924	.2226	141	.7292	.3534
7	.6668	.0153	52	.6737	.1154	97	.6930	.2252	142	.7303	.3567
8	.6668	.0175	53	.6740	.1177	98	.6936	.2279	143	.7314	.3600
9	.6669	.0197	54	.6743	.1200	99	.6942	.2305	144	.7325	.3633
10	.6670	.0218	55	.6746	.1224	100	.6948	.2332	145	.7336	.3666
11	.6670	.0240	56	.6749	.1247	101	.6954	.2358	146	.7348	.3700
12	.6671	.0262	57	.6752	.1270	102	.6961	.2385	147	.7360	.3734
13	.6672	.0284	58	.6755	.1293	103	.6967	.2412	148	.7372	.3768
14	.6672	.0306	59	.6758	.1316	104	.6974	.2439	149	.7384	.3802
15	.6673	.0328	60	.6761	.1340	105	.6980	.2466	150	.7396	.3837
16	.6674	.0350	61	.6764	.1363	106	.6987	.2493	151	.7408	.3871
17	.6674	.0372	62	.6768	.1387	107	.6994	.2520	152	.7421	.3906
18	.6675	.0394	63	.6771	.1410	108	.7001	.2548	153	.7434	.3942
19	.6676	.0416	64	.6775	.1434	109	.7008	.2575	154	.7447	.3977
20	.6677	.0437	65	.6779	.1457	110	.7015	.2603	155	.7460	.4013
21	.6678	.0459	66	.6782	.1481	111	.7022	.2631	156	.7473	.4049
22	.6679	.0481	67	.6786	.1505	112	.7030	.2659	157	.7486	.4085
23	.6680	.0504	68	.6790	.1529	113	.7037	.2687	158	.7500	.4122
24	.6681	.0526	69	.6794	.1553	114	.7045	.2715	159	.7514	.4159
25	.6682	.0548	70	.6797	.1577	115	.7052	.2743	160	.7528	.4196
26	.6684	.0570	71	.6801	.1601	116	.7060	.2772	161	.7542	.4233
27	.6685	.0592	72	.6805	.1625	117	.7068	.2800	162	.7557	.4270
28	.6687	.0614	73	.6809	.1649	118	.7076	.2829	163	.7571	.4308
29	.6688	.0636	74	.6814	.1673	119	.7084	.2858	164	.7586	.4346
30	.6690	.0658	75	.6818	.1697	120	.7092	.2887	165	.7601	.4385
31	.6691	.0681	76	.6822	.1722	121	.7100	.2916	166	.7616	.4424
32	.6693	.0703	77	.6826	.1746	122	.7109	.2945	167	.7632	.4463
33	.6694	.0725	78	.6831	.1771	123	.7117	.2975	168	.7648	.4502
34	.6696	.0747	79	.6835	.1795	124	.7126	.3004	169	.7664	.4542
35	.6698	.0770	80	.6840	.1820	125	.7134	.3034	170	.7680	.4582
36	.6700	.0792	81	.6844	.1845	126	.7143	.3064	171	.7696	.4622
37	.6702	.0814	82	.6849	.1869	127	.7152	.3094	172	.7712	.4663
38	.6704	.0837	83	.6854	.1894	128	.7161	.3124	173	.7729	.4704
39	.6706	.0859	84	.6859	.1919	129	.7170	.3155	174	.7746	.4745
40	.6708	.0882	85	.6864	.1944	130	.7180	.3185	175	.7763	.4787
41	.6710	.0904	86	.6869	.1970	131	.7189	.3216	176	.7781	.4828
42	.6712	.0927	87	.6874	.1995	132	.7199	.3247	177	.7799	.4871
43	.6714	.0949	88	.6879	.2020	133	.7209	.3278	178	.7817	.4914
44	.6717	.0972	89	.6884	.2046	134	.7219	.3309	179	.7835	.4957
45	.6719	.0995	90	.6890	.2071	135	.7229	.3341	180	.7854	.5000

TABLE: Areas of circular segments: ratio of rise and diameter 5.6.7

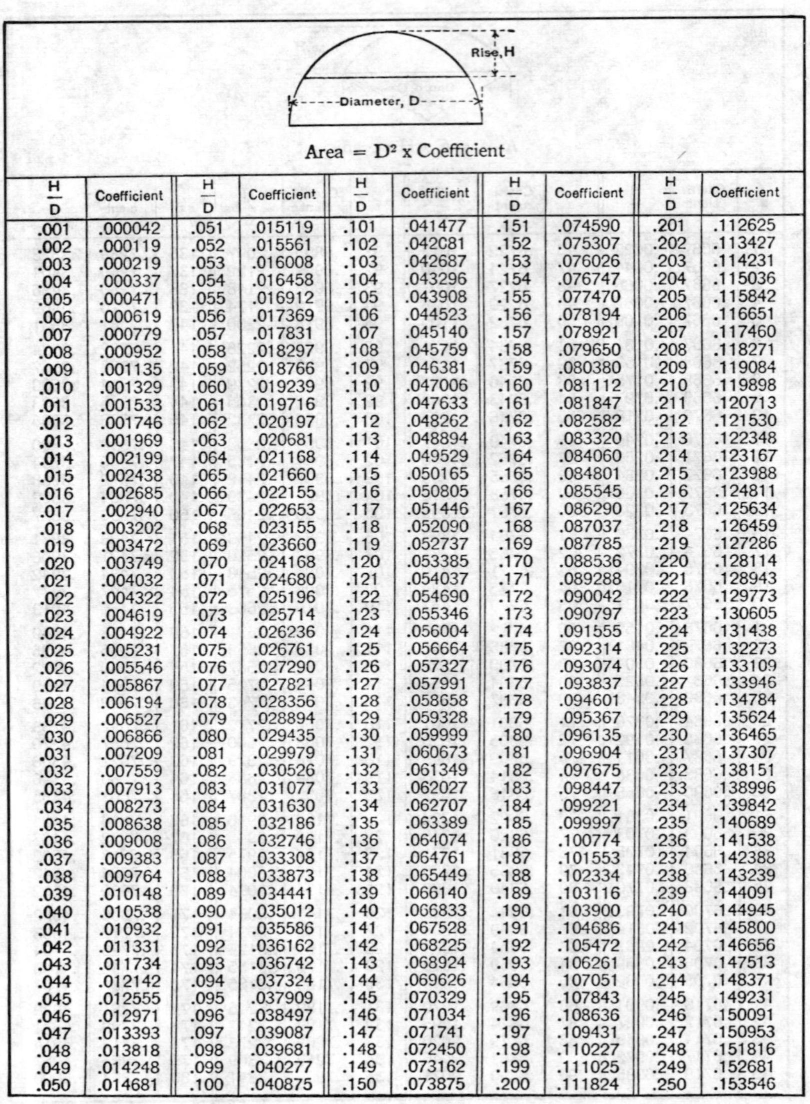

Area = D^2 x Coefficient

H/D	Coefficient	H/D	Coefficient	H/D	Coefficient	H/D	Coefficient	H/D	Coefficient
.001	.000042	.051	.015119	.101	.041477	.151	.074590	.201	.112625
.002	.000119	.052	.015561	.102	.042081	.152	.075307	.202	.113427
.003	.000219	.053	.016008	.103	.042687	.153	.076026	.203	.114231
.004	.000337	.054	.016458	.104	.043296	.154	.076747	.204	.115036
.005	.000471	.055	.016912	.105	.043908	.155	.077470	.205	.115842
.006	.000619	.056	.017369	.106	.044523	.156	.078194	.206	.116651
.007	.000779	.057	.017831	.107	.045140	.157	.078921	.207	.117460
.008	.000952	.058	.018297	.108	.045759	.158	.079650	.208	.118271
.009	.001135	.059	.018766	.109	.046381	.159	.080380	.209	.119084
.010	.001329	.060	.019239	.110	.047006	.160	.081112	.210	.119898
.011	.001533	.061	.019716	.111	.047633	.161	.081847	.211	.120713
.012	.001746	.062	.020197	.112	.048262	.162	.082582	.212	.121530
.013	.001969	.063	.020681	.113	.048894	.163	.083320	.213	.122348
.014	.002199	.064	.021168	.114	.049529	.164	.084060	.214	.123167
.015	.002438	.065	.021660	.115	.050165	.165	.084801	.215	.123988
.016	.002685	.066	.022155	.116	.050805	.166	.085545	.216	.124811
.017	.002940	.067	.022653	.117	.051446	.167	.086290	.217	.125634
.018	.003202	.068	.023155	.118	.052090	.168	.087037	.218	.126459
.019	.003472	.069	.023660	.119	.052737	.169	.087785	.219	.127286
.020	.003749	.070	.024168	.120	.053385	.170	.088536	.220	.128114
.021	.004032	.071	.024680	.121	.054037	.171	.089288	.221	.128943
.022	.004322	.072	.025196	.122	.054690	.172	.090042	.222	.129773
.023	.004619	.073	.025714	.123	.055346	.173	.090797	.223	.130605
.024	.004922	.074	.026236	.124	.056004	.174	.091555	.224	.131438
.025	.005231	.075	.026761	.125	.056664	.175	.092314	.225	.132273
.026	.005546	.076	.027290	.126	.057327	.176	.093074	.226	.133109
.027	.005867	.077	.027821	.127	.057991	.177	.093837	.227	.133946
.028	.006194	.078	.028356	.128	.058658	.178	.094601	.228	.134784
.029	.006527	.079	.028894	.129	.059328	.179	.095367	.229	.135624
.030	.006866	.080	.029435	.130	.059999	.180	.096135	.230	.136465
.031	.007209	.081	.029979	.131	.060673	.181	.096904	.231	.137307
.032	.007559	.082	.030526	.132	.061349	.182	.097675	.232	.138151
.033	.007913	.083	.031077	.133	.062027	.183	.098447	.233	.138996
.034	.008273	.084	.031630	.134	.062707	.184	.099221	.234	.139842
.035	.008638	.085	.032186	.135	.063389	.185	.099997	.235	.140689
.036	.009008	.086	.032746	.136	.064074	.186	.100774	.236	.141538
.037	.009383	.087	.033308	.137	.064761	.187	.101553	.237	.142388
.038	.009764	.088	.033873	.138	.065449	.188	.102334	.238	.143239
.039	.010148	.089	.034441	.139	.066140	.189	.103116	.239	.144091
.040	.010538	.090	.035012	.140	.066833	.190	.103900	.240	.144945
.041	.010932	.091	.035586	.141	.067528	.191	.104686	.241	.145800
.042	.011331	.092	.036162	.142	.068225	.192	.105472	.242	.146656
.043	.011734	.093	.036742	.143	.068924	.193	.106261	.243	.147513
.044	.012142	.094	.037324	.144	.069626	.194	.107051	.244	.148371
.045	.012555	.095	.037909	.145	.070329	.195	.107843	.245	.149231
.046	.012971	.096	.038497	.146	.071034	.196	.108636	.246	.150091
.047	.013393	.097	.039087	.147	.071741	.197	.109431	.247	.150953
.048	.013818	.098	.039681	.148	.072450	.198	.110227	.248	.151816
.049	.014248	.099	.040277	.149	.073162	.199	.111025	.249	.152681
.050	.014681	.100	.040875	.150	.073875	.200	.111824	.250	.153546

TABLE: Areas of cicular segments: ratio of rise and diameter, continued 5.6.7

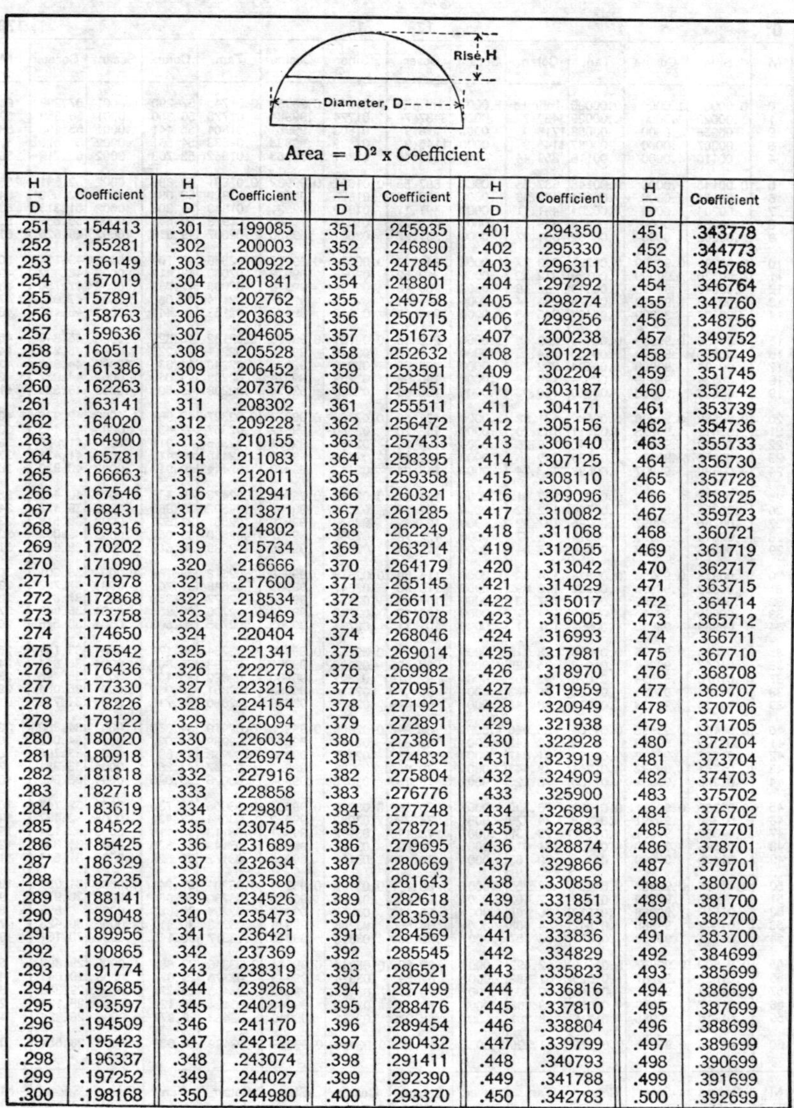

Area = D^2 x Coefficient

H/D	Coefficient	H/D	Coefficient	H/D	Coefficient	H/D	Coefficient	H/D	Coefficient
.251	.154413	.301	.199085	.351	.245935	.401	.294350	.451	.343778
.252	.155281	.302	.200003	.352	.246890	.402	.295330	.452	.344773
.253	.156149	.303	.200922	.353	.247845	.403	.296311	.453	.345768
.254	.157019	.304	.201841	.354	.248801	.404	.297292	.454	.346764
.255	.157891	.305	.202762	.355	.249758	.405	.298274	.455	.347760
.256	.158763	.306	.203683	.356	.250715	.406	.299256	.456	.348756
.257	.159636	.307	.204605	.357	.251673	.407	.300238	.457	.349752
.258	.160511	.308	.205528	.358	.252632	.408	.301221	.458	.350749
.259	.161386	.309	.206452	.359	.253591	.409	.302204	.459	.351745
.260	.162263	.310	.207376	.360	.254551	.410	.303187	.460	.352742
.261	.163141	.311	.208302	.361	.255511	.411	.304171	.461	.353739
.262	.164020	.312	.209228	.362	.256472	.412	.305156	.462	.354736
.263	.164900	.313	.210155	.363	.257433	.413	.306140	.463	.355733
.264	.165781	.314	.211083	.364	.258395	.414	.307125	.464	.356730
.265	.166663	.315	.212011	.365	.259358	.415	.308110	.465	.357728
.266	.167546	.316	.212941	.366	.260321	.416	.309096	.466	.358725
.267	.168431	.317	.213871	.367	.261285	.417	.310082	.467	.359723
.268	.169316	.318	.214802	.368	.262249	.418	.311068	.468	.360721
.269	.170202	.319	.215734	.369	.263214	.419	.312055	.469	.361719
.270	.171090	.320	.216666	.370	.264179	.420	.313042	.470	.362717
.271	.171978	.321	.217600	.371	.265145	.421	.314029	.471	.363715
.272	.172868	.322	.218534	.372	.266111	.422	.315017	.472	.364714
.273	.173758	.323	.219469	.373	.267078	.423	.316005	.473	.365712
.274	.174650	.324	.220404	.374	.268046	.424	.316993	.474	.366711
.275	.175542	.325	.221341	.375	.269014	.425	.317981	.475	.367710
.276	.176436	.326	.222278	.376	.269982	.426	.318970	.476	.368708
.277	.177330	.327	.223216	.377	.270951	.427	.319959	.477	.369707
.278	.178226	.328	.224154	.378	.271921	.428	.320949	.478	.370706
.279	.179122	.329	.225094	.379	.272891	.429	.321938	.479	.371705
.280	.180020	.330	.226034	.380	.273861	.430	.322928	.480	.372704
.281	.180918	.331	.226974	.381	.274832	.431	.323919	.481	.373704
.282	.181818	.332	.227916	.382	.275804	.432	.324909	.482	.374703
.283	.182718	.333	.228858	.383	.276776	.433	.325900	.483	.375702
.284	.183619	.334	.229801	.384	.277748	.434	.326891	.484	.376702
.285	.184522	.335	.230745	.385	.278721	.435	.327883	.485	.377701
.286	.185425	.336	.231689	.386	.279695	.436	.328874	.486	.378701
.287	.186329	.337	.232634	.387	.280669	.437	.329866	.487	.379701
.288	.187235	.338	.233580	.388	.281643	.438	.330858	.488	.380700
.289	.188141	.339	.234526	.389	.282618	.439	.331851	.489	.381700
.290	.189048	.340	.235473	.390	.283593	.440	.332843	.490	.382700
.291	.189956	.341	.236421	.391	.284569	.441	.333836	.491	.383700
.292	.190865	.342	.237369	.392	.285545	.442	.334829	.492	.384699
.293	.191774	.343	.238319	.393	.286521	.443	.335823	.493	.385699
.294	.192685	.344	.239268	.394	.287499	.444	.336816	.494	.386699
.295	.193597	.345	.240219	.395	.288476	.445	.337810	.495	.387699
.296	.194509	.346	.241170	.396	.289454	.446	.338804	.496	.388699
.297	.195423	.347	.242122	.397	.290432	.447	.339799	.497	.389699
.298	.196337	.348	.243074	.398	.291411	.448	.340793	.498	.390699
.299	.197252	.349	.244027	.399	.292390	.449	.341788	.499	.391699
.300	.198168	.350	.244980	.400	.293370	.450	.342783	.500	.392699

TABLE: Natural Trigonometric functions 5.6.8

0°						179°	1°						178°
M	Sine	Cosine	Tan.	Cotan.	Secant	Cosec.	Sine	Cosine	Tan.	Cotan.	Secant	Cosec.	M
0	0.00000	1.0000	0.00000	Infinite	1.0000	Infinite	0.01745	0.99985	0.01745	57.290	1.0001	57.299	60
1	.00029	.0000	.00029	3437.7	.0000	3437.7	.01774	.99984	.01775	56.350	.0001	56.359	59
2	.00058	.0000	.00058	1718.9	.0000	1718.9	.01803	.99984	.01804	55.441	.0001	55.450	58
3	.00087	.0000	.00087	1145.9	.0000	1145.9	.01832	.99983	.01833	54.561	.0002	54.570	57
4	.00116	.0000	.00116	859.44	.0000	859.44	.01861	.99983	.01862	53.708	.0002	53.718	56
5	0.00145	1.0000	0.00145	687.55	1.0000	687.55	0.01891	0.99982	0.01891	52.882	1.0002	52.891	55
6	.00174	.0000	.00174	572.96	.0000	572.96	.01920	.99981	.01920	52.081	.0002	52.090	54
7	.00204	.0000	.00204	491.11	.0000	491.11	.01949	.99981	.01949	51.303	.0002	51.313	53
8	.00233	.0000	.00233	429.72	.0000	429.72	.01978	.99980	.01978	50.548	.0002	50.558	52
9	.00262	.0000	.00262	381.97	.0000	381.97	.02007	.99980	.02007	49.816	.0002	49.826	51
10	0.00291	0.99999	0.00291	343.77	1.0000	343.77	0.02036	0.99979	0.02036	49.104	1.0002	49.114	50
11	.00320	.99999	.00320	312.52	.0000	312.52	.02065	.99979	.02066	48.412	.0002	48.422	49
12	.00349	.99999	.00349	286.48	.0000	286.48	.02094	.99978	.02095	47.739	.0002	47.750	48
13	.00378	.99999	.00378	264.44	.0000	264.44	.02123	.99977	.02124	47.085	.0002	47.096	47
14	.00407	.99999	.00407	245.55	.0000	245.55	.02152	.99977	.02153	46.449	.0002	46.460	46
15	0.00436	0.99999	0.00436	229.18	1.0000	229.18	0.02181	0.99976	0.02182	45.829	1.0002	45.840	45
16	.00465	.99999	.00465	214.86	.0000	214.86	.02210	.99975	.02211	45.226	.0002	45.237	44
17	.00494	.99999	.00494	202.22	.0000	202.22	.02240	.99975	.02240	44.638	.0002	44.650	43
18	.00524	.99999	.00524	190.98	.0000	190.99	.02269	.99974	.02269	44.066	.0002	44.077	42
19	.00553	.99998	.00553	180.93	.0000	180.93	.02298	.99974	.02298	43.508	.0003	43.520	41
20	0.00582	0.99998	0.00582	171.88	1.0000	171.89	0.02326	0.99973	0.02327	42.964	1.0003	42.976	40
21	.00611	.99998	.00611	163.70	.0000	163.70	.02356	.99972	.02357	42.433	.0003	42.445	39
22	.00640	.99998	.00640	156.26	.0000	156.26	.02385	.99971	.02386	41.916	.0003	41.928	38
23	.00669	.99998	.00669	149.46	.0000	149.47	.02414	.99971	.02415	41.410	.0003	41.423	37
24	.00698	.99997	.00698	143.24	.0000	143.24	.02443	.99970	.02444	40.917	.0003	40.930	36
25	0.00727	0.99997	0.00727	137.51	1.0000	137.51	0.02472	0.99969	0.02473	40.436	1.0003	40.448	35
26	.00756	.99997	.00756	132.22	.0000	132.22	.02501	.99969	.02502	39.965	.0003	39.978	34
27	.00785	.99997	.00785	127.32	.0000	127.32	.02530	.99968	.02531	39.506	.0003	39.518	33
28	.00814	.99997	.00814	122.77	.0000	122.78	.02559	.99967	.02560	39.057	.0003	39.069	32
29	.00843	.99996	.00844	118.54	.0000	118.54	.02589	.99966	.02589	38.618	.0003	38.631	31
30	0.00873	0.99996	0.00873	114.59	1.0000	114.59	0.02618	0.99966	0.02618	38.188	1.0003	38.201	30
31	.00902	.99996	.00902	110.89	.0000	110.90	.02647	.99965	.02648	37.769	.0003	37.782	29
32	.00931	.99996	.00931	107.43	.0000	107.43	.02676	.99964	.02677	37.358	.0003	37.371	28
33	.00960	.99995	.00960	104.17	.0000	104.17	.02705	.99963	.02706	36.956	.0004	36.969	27
34	.00989	.99995	.00989	101.11	.0000	101.11	.02734	.99963	.02735	36.563	.0004	36.576	26
35	0.01018	0.99995	0.01018	98.218	1.0000	98.223	0.02763	0.99962	0.02764	36.177	1.0004	36.191	25
36	.01047	.99994	.01047	95.489	.0000	95.495	.02792	.99961	.02793	35.800	.0004	35.814	24
37	.01076	.99994	.01076	92.908	.0000	92.914	.02821	.99960	.02822	35.431	.0004	35.445	23
38	.01105	.99994	.01105	90.463	.0001	90.469	.02850	.99959	.02851	35.069	.0004	35.084	22
39	.01134	.99993	.01134	88.143	.0001	88.149	.02879	.99958	.02880	34.715	.0004	34.729	21
40	0.01163	0.99993	0.01164	85.940	1.0001	85.946	0.02908	0.99958	0.02910	34.368	1.0004	34.382	20
41	.01193	.99993	.01193	83.843	.0001	83.849	.02937	.99957	.02939	34.027	.0004	34.042	19
42	.01222	.99992	.01222	81.847	.0001	81.853	.02967	.99956	.02968	33.693	.0004	33.708	18
43	.01251	.99992	.01251	79.943	.0001	79.950	.02996	.99955	.02997	33.366	.0004	33.381	17
44	.01280	.99992	.01280	78.126	.0001	78.133	.03025	.99954	.03026	33.045	.0004	33.060	16
45	0.01309	0.99991	0.01309	76.390	1.0001	76.396	0.03054	0.99953	0.03055	32.730	1.0005	32.745	15
46	.01338	.99991	.01338	74.729	.0001	74.736	.03083	.99952	.03084	32.421	.0005	32.437	14
47	.01367	.99991	.01367	73.139	.0001	73.146	.03112	.99951	.03113	32.118	.0005	32.134	13
48	.01396	.99990	.01396	71.615	.0001	71.622	.03141	.99951	.03143	31.820	.0005	31.836	12
49	.01425	.99990	.01425	70.153	.0001	70.160	.03170	.99950	.03172	31.528	.0005	31.544	11
50	0.01454	0.99989	0.01454	68.750	1.0001	68.757	0.03199	0.99949	0.03201	31.241	1.0005	31.257	10
51	.01483	.99989	.01484	67.402	.0001	67.409	.03228	.99948	.03230	30.960	.0005	30.976	9
52	.01512	.99988	.01513	66.105	.0001	66.113	.03257	.99947	.03259	30.683	.0005	30.699	8
53	.01542	.99988	.01542	64.858	.0001	64.866	.03286	.99946	.03288	30.411	.0005	30.428	7
54	.01571	.99988	.01571	63.657	.0001	63.664	.03315	.99945	.03317	30.145	.0005	30.161	6
55	0.01600	0.99987	0.01600	62.499	1.0001	62.507	0.03344	0.99944	0.03346	29.882	1.0005	29.899	5
56	.01629	.99987	.01629	61.383	.0001	61.391	.03374	.99943	.03375	29.624	.0006	29.641	4
57	.01658	.99987	.01658	60.306	.0001	60.314	.03403	.99942	.03405	29.371	.0006	29.388	3
58	.01687	.99986	.01687	59.266	.0001	59.274	.03432	.99941	.03434	29.122	.0006	29.139	2
59	.01716	.99985	.01716	58.261	.0001	58.270	.03461	.99940	.03463	28.877	.0006	28.894	1
60	0.01745	0.99985	0.01745	57.290	1.0001	57.299	0.03490	0.99939	0.03492	28.636	1.0006	28.654	0
M	Cosine	Sine	Cotan.	Tan.	Cosec.	Secant	Cosine	Sine	Cotan.	Tan.	Cosec.	Secant	M
90°						89°	91°						88°

TABLE: Natural Trigonometric functions, continued 5.6.8

2°						177°	3°					176°	
M	Sine	Cosine	Tan.	Cotan.	Secant	Cosec.	Sine	Cosine	Tan.	Cotan.	Secant	Cosec.	M
0	0.03490	0.99939	0.03492	28.636	1.0006	28.654	0.05234	0.99863	0.05241	19.081	1.0014	19.107	60
1	.03519	.99938	.03521	28.399	.0006	28.417	.05263	.99861	.05270	18.975	.0014	19.002	59
2	.03548	.99937	.03550	28.166	.0006	28.184	.05292	.99860	.05299	18.871	.0014	18.897	58
3	.03577	.99936	.03579	27.937	.0006	27.955	.05321	.99858	.05328	18.768	.0014	18.794	57
4	.03606	.99935	.03608	27.712	.0006	27.730	.05350	.99857	.05357	18.665	.0014	18.692	56
5	0.03635	0.99934	0.03638	27.490	1.0007	27.508	0.05379	0.99855	0.05387	18.564	1.0014	18.591	55
6	.03664	.99933	.03667	27.271	.0007	27.290	.05408	.99854	.05416	18.464	.0015	18.491	54
7	.03693	.99932	.03696	27.056	.0007	27.075	.05437	.99852	.05445	18.365	.0015	18.393	53
8	.03722	.99931	.03725	26.845	.0007	26.864	.05466	.99850	.05474	18.268	.0015	18.295	52
9	.03751	.99930	.03754	26.637	.0007	26.655	.05495	.99849	.05503	18.171	.0015	18.198	51
10	0.03781	0.99928	0.03783	26.432	1.0007	26.450	0.05524	0.99847	0.05532	18.075	1.0015	18.103	50
11	.03810	.99927	.03812	26.230	.0007	26.249	.05553	.99846	.05562	17.980	.0015	18.008	49
12	.03839	.99926	.03842	26.031	.0007	26.050	.05582	.99844	.05591	17.886	.0016	17.914	48
13	.03868	.99925	.03871	25.835	.0007	25.854	.05611	.99842	.05620	17.793	.0016	17.821	47
14	.03897	.99924	.03900	25.642	.0008	25.661	.05640	.99841	.05649	17.701	.0016	17.730	46
15	0.03926	0.99923	0.03929	25.452	1.0008	25.471	0.05669	0.99839	0.05678	17.610	1.0016	17.639	45
16	.03955	.99922	.03958	25.264	.0008	25.284	.05698	.99837	.05707	17.520	.0016	17.549	44
17	.03984	.99921	.03987	25.080	.0008	25.100	.05727	.99836	.05737	17.431	.0016	17.460	43
18	.04013	.99919	.04016	24.898	.0008	24.918	.05756	.99834	.05766	17.343	.0017	17.372	42
19	.04042	.99918	.04045	24.718	.0008	24.739	.05785	.99832	.05795	17.256	.0017	17.285	41
20	0.04071	0.99917	0.04075	24.542	1.0008	24.562	0.05814	0.99831	0.05824	17.169	1.0017	17.198	40
21	.04100	.99916	.04104	24.367	.0008	24.388	.05843	.99829	.05853	17.084	.0017	17.113	39
22	.04129	.99915	.04133	24.196	.0008	24.216	.05872	.99827	.05883	16.999	.0017	17.028	38
23	.04158	.99913	.04162	24.026	.0009	24.047	.05902	.99826	.05912	16.915	.0017	16.944	37
24	.04187	.99912	.04191	23.859	.0009	23.880	.05931	.99824	.05941	16.832	.0018	16.861	36
25	0.04217	0.99911	0.04220	23.694	1.0009	23.716	0.05960	0.99822	0.05970	16.750	1.0018	16.779	35
26	.04246	.99910	.04249	23.532	.0009	23.553	.05989	.99820	.05999	16.668	.0018	16.698	34
27	.04275	.99908	.04279	23.372	.0009	23.393	.06018	.99819	.06029	16.587	.0018	16.617	33
28	.04304	.99907	.04308	23.214	.0009	23.235	.06047	.99817	.06058	16.507	.0018	16.538	32
29	.04333	.99906	.04337	23.058	.0009	23.079	.06076	.99815	.06087	16.428	.0018	16.459	31
30	0.04362	0.99905	0.04366	22.904	1.0009	22.925	0.06105	0.99813	0.06116	16.350	1.0019	16.380	30
31	.04391	.99903	.04395	22.752	.0010	22.774	.06134	.99812	.06145	16.272	.0019	16.303	29
32	.04420	.99902	.04424	22.602	.0010	22.624	.06163	.99810	.06175	16.195	.0019	16.226	28
33	.04449	.99901	.04453	22.454	.0010	22.476	.06192	.99808	.06204	16.119	.0019	16.150	27
34	.04478	.99900	.04483	22.308	.0010	22.330	.06221	.99806	.06233	16.043	.0019	16.075	26
35	0.04507	0.99898	0.04512	22.164	1.0010	22.186	0.06250	0.99804	0.06262	15.969	1.0019	16.000	25
36	.04536	.99897	.04541	22.022	.0010	22.044	.06279	.99803	.06291	15.894	.0020	15.926	24
37	.04565	.99896	.04570	21.881	.0010	21.904	.06308	.99801	.06321	15.821	.0020	15.853	23
38	.04594	.99894	.04599	21.742	.0010	21.765	.06337	.99799	.06350	15.748	.0020	15.780	22
39	.04623	.99893	.04628	21.606	.0011	21.629	.06366	.99797	.06379	15.676	.0020	15.708	21
40	0.04652	0.99892	0.04657	21.470	1.0011	21.494	0.06395	0.99795	0.06408	15.605	1.0020	15.637	20
41	.04681	.99890	.04687	21.337	.0011	21.360	.06424	.99793	.06437	15.534	.0021	15.566	19
42	.04711	.99889	.04716	21.205	.0011	21.228	.06453	.99791	.06467	15.464	.0021	15.496	18
43	.04740	.99888	.04745	21.075	.0011	21.098	.06482	.99790	.06496	15.394	.0021	15.427	17
44	.04769	.99886	.04774	20.946	.0011	20.970	.06511	.99788	.06525	15.325	.0021	15.358	16
45	0.04798	0.99885	0.04803	20.819	1.0011	20.843	0.06540	0.99786	0.06554	15.257	1.0021	15.290	15
46	.04827	.99883	.04832	20.693	.0012	20.717	.06569	.99784	.06583	15.189	.0022	15.222	14
47	.04856	.99882	.04862	20.569	.0012	20.593	.06598	.99782	.06613	15.122	.0022	15.155	13
48	.04885	.99881	.04891	20.446	.0012	20.471	.06627	.99780	.06642	15.056	.0022	15.089	12
49	.04914	.99879	.04920	20.325	.0012	20.350	.06656	.99778	.06671	14.990	.0022	15.023	11
50	0.04943	0.99878	0.04949	20.205	1.0012	20.230	0.06685	0.99776	0.06700	14.924	1.0022	14.958	10
51	.04972	.99876	.04978	20.087	.0012	20.112	.06714	.99774	.06730	14.860	.0023	14.893	9
52	.05001	.99875	.05007	19.970	.0012	19.995	.06743	.99772	.06759	14.795	.0023	14.829	8
53	.05030	.99873	.05037	19.854	.0013	19.880	.06772	.99770	.06788	14.732	.0023	14.765	7
54	.05059	.99872	.05066	19.740	.0013	19.766	.06801	.99768	.06817	14.668	.0023	14.702	6
55	0.05088	0.99870	0.05095	19.627	1.0013	19.653	0.06830	0.99766	0.06846	14.606	1.0023	14.640	5
56	.05117	.99869	.05124	19.515	.0013	19.541	.06859	.99764	.06876	14.544	.0024	14.578	4
57	.05146	.99867	.05153	19.405	.0013	19.431	.06888	.99762	.06905	14.482	.0024	14.517	3
58	.05175	.99866	.05182	19.296	.0013	19.322	.06918	.99760	.06934	14.421	.0024	14.456	2
59	.05204	.99864	.05212	19.188	.0013	19.214	.06947	.99758	.06963	14.361	.0024	14.395	1
60	0.05234	0.99863	0.05241	19.081	1.0014	19.107	0.06976	0.99756	0.06993	14.301	1.0024	14.335	0
M	Cosine	Sine	Cotan.	Tan.	Cosec.	Secant	Cosine	Sine	Cotan.	Tan.	Cosec.	Secan	M
92°						87°	93°					86°	

TABLE: Natural Trigonometric functions, continued 5.6.8

4°						175°	5°						174°
M	Sine	Cosine	Tan.	Cotan.	Secant	Cosec.	Sine	Cosine	Tan.	Cotan.	Secant	Cosec.	M
0	0.06976	0.99756	0.06993	14.301	1.0024	14.335	0.08715	0.99619	0.08749	11.430	1.0038	11.474	60
1	.07005	.99754	.07022	14.241	.0025	14.276	.08744	.99617	.08778	11.392	.0038	11.436	59
2	.07034	.99752	.07051	14.182	.0025	14.217	.08773	.99614	.08807	11.354	.0039	11.398	58
3	.07063	.99750	.07080	14.123	.0025	14.159	.08802	.99612	.08837	11.316	.0039	11.360	57
4	.07092	.99748	.07110	14.065	.0025	14.101	.08831	.99609	.08866	11.279	.0039	11.323	56
5	0.07121	0.99746	0.07139	14.008	1.0025	14.043	0.08860	0.99607	0.08895	11.242	1.0039	11.286	55
6	.07150	.99744	.07168	13.951	.0026	13.986	.08889	.99604	.08925	11.205	.0040	11.249	54
7	.07179	.99742	.07197	13.894	.0026	13.930	.08918	.99601	.08954	11.168	.0040	11.213	53
8	.07208	.99740	.07226	13.838	.0026	13.874	.08947	.99599	.08983	11.132	.0040	11.176	52
9	.07237	.99738	.07256	13.782	.0026	13.818	.08976	.99596	.09013	11.095	.0040	11.140	51
10	0.07266	0.99736	0.07285	13.727	1.0026	13.763	0.09005	0.99594	0.09042	11.059	1.0041	11.104	50
11	.07295	.99733	.07314	13.672	.0027	13.708	.09034	.99591	.09071	11.024	.0041	11.069	49
12	.07324	.99731	.07343	13.617	.0027	13.654	.09063	.99588	.09101	10.988	.0041	11.033	48
13	.07353	.99729	.07373	13.563	.0027	13.600	.09092	.99586	.09130	10.953	.0041	10.998	47
14	.07382	.99727	.07402	13.510	.0027	13.547	.09121	.99583	.09159	10.918	.0042	10.963	46
15	0.07411	0.99725	0.07431	13.457	1.0027	13.494	0.09150	0.99580	0.09189	10.883	1.0042	10.929	45
16	.07440	.99723	.07460	13.404	.0028	13.441	.09179	.99578	.09218	10.848	.0042	10.894	44
17	.07469	.99721	.07490	13.351	.0028	13.389	.09208	.99575	.09247	10.814	.0043	10.860	43
18	.07498	.99718	.07519	13.299	.0028	13.337	.09237	.99572	.09277	10.780	.0043	10.826	42
19	.07527	.99716	.07548	13.248	.0028	13.286	.09266	.99570	.09306	10.746	.0043	10.792	41
20	0.07556	0.99714	0.07577	13.197	1.0029	13.235	0.09295	0.99567	0.09335	10.712	1.0043	10.758	40
21	.07585	.99712	.07607	13.146	.0029	13.184	.09324	.99564	.09365	10.678	.0044	10.725	39
22	.07614	.99710	.07636	13.096	.0029	13.134	.09353	.99562	.09394	10.645	.0044	10.692	38
23	.07643	.99707	.07665	13.046	.0029	13.084	.09382	.99559	.09423	10.612	.0044	10.659	37
24	.07672	.99705	.07694	12.996	.0029	13.034	.09411	.99556	.09453	10.579	.0044	10.626	36
25	0.07701	0.99703	0.07724	12.947	1.0030	12.985	0.09440	0.99553	0.09482	10.546	1.0045	10.593	35
26	.07730	.99701	.07753	12.898	.0030	12.937	.09469	.99551	.09511	10.514	.0045	10.561	34
27	.07759	.99698	.07782	12.849	.0030	12.888	.09498	.99548	.09541	10.481	.0045	10.529	33
28	.07788	.99696	.07812	12.801	.0030	12.840	.09527	.99545	.09570	10.449	.0046	10.497	32
29	.07817	.99694	.07841	12.754	.0031	12.793	.09556	.99542	.09599	10.417	.0046	10.465	31
30	0.07846	0.99692	0.07870	12.706	1.0031	12.745	0.09584	0.99540	0.09629	10.385	1.0046	10.433	30
31	.07875	.99689	.07899	12.659	.0031	12.698	.09613	.99537	.09658	10.354	.0046	10.402	29
32	.07904	.99687	.07929	12.612	.0031	12.652	.09642	.99534	.09688	10.322	.0047	10.371	28
33	.07933	.99685	.07958	12.566	.0032	12.606	.09671	.99531	.09717	10.291	.0047	10.340	27
34	.07962	.99682	.07987	12.520	.0032	12.560	.09700	.99528	.09746	10.260	.0047	10.309	26
35	0.07991	0.99680	0.08016	12.474	1.0032	12.514	0.09729	0.99525	0.09776	10.229	1.0048	10.278	25
36	.08020	.99678	.08046	12.429	.0032	12.469	.09758	.99523	.09805	10.199	.0048	10.248	24
37	.08049	.99675	.08075	12.384	.0032	12.424	.09787	.99520	.09834	10.168	.0048	10.217	23
38	.08078	.99673	.08104	12.339	.0033	12.379	.09816	.99517	.09864	10.138	.0048	10.187	22
39	.08107	.99671	.08134	12.295	.0033	12.335	.09845	.99514	.09893	10.108	.0049	10.157	21
40	0.08136	0.99668	0.08163	12.250	1.0033	12.291	0.09874	0.99511	0.09922	10.078	1.0049	10.127	20
41	.08165	.99666	.08192	12.207	.0033	12.248	.09903	.99508	.09952	10.048	.0049	10.098	19
42	.08194	.99664	.08221	12.163	.0034	12.204	.09932	.99505	.09981	10.019	.0050	10.068	18
43	.08223	.99661	.08251	12.120	.0034	12.161	.09961	.99503	.10011	9.9893	.0050	10.039	17
44	.08252	.99659	.08280	12.077	.0034	12.118	.09990	.99500	.10040	9.9601	.0050	10.010	16
45	0.08281	0.99656	0.08309	12.035	1.0034	12.076	0.10019	0.99497	0.10069	9.9310	1.0050	9.9812	15
46	.08310	.99654	.08339	11.992	.0035	12.034	.10048	.99494	.10099	9.9021	.0051	9.9525	14
47	.08339	.99652	.08368	11.950	.0035	11.992	.10077	.99491	.10128	9.8734	.0051	9.9239	13
48	.08368	.99649	.08397	11.909	.0035	11.950	.10106	.99488	.10158	9.8448	.0051	9.8955	12
49	.08397	.99647	.08426	11.867	.0035	11.909	.10134	.99485	.10187	9.8164	.0052	9.8672	11
50	0.08426	0.99644	0.08456	11.826	1.0036	11.868	0.10163	0.99482	0.10216	9.7882	1.0052	9.8391	10
51	.08455	.99642	.08485	11.785	.0036	11.828	.10192	.99479	.10246	9.7601	.0052	9.8112	9
52	.08484	.99639	.08514	11.745	.0036	11.787	.10221	.99476	.10275	9.7322	.0053	9.7834	8
53	.08513	.99637	.08544	11.704	.0036	11.747	.10250	.99473	.10305	9.7044	.0053	9.7558	7
54	.08542	.99634	.08573	11.664	.0037	11.707	.10279	.99470	.10334	9.6768	.0053	9.7283	6
55	0.08571	0.99632	0.08602	11.625	1.0037	11.668	0.10308	0.99467	0.10363	9.6493	1.0053	9.7010	5
56	.08600	.99629	.08632	11.585	.0037	11.628	.10337	.99464	.10393	9.6220	.0054	9.6739	4
57	.08629	.99627	.08661	11.546	.0037	11.589	.10366	.99461	.10422	9.5949	.0054	9.6469	3
58	.08658	.99624	.08690	11.507	.0038	11.550	.10395	.99458	.10452	9.5679	.0054	9.6200	2
59	.08687	.99622	.08719	11.468	.0038	11.512	.10424	.99455	.10481	9.5411	.0055	9.5933	1
60	0.08715	0.99619	0.08749	11.430	1.0038	11.474	0.10453	0.99452	0.10510	9.5144	1.0055	9.5668	0
M	Cosine	Sine	Cotan.	Tan.	Cosec.	Secant	Cosine	Sine	Cotan.	Tan.	Cosec.	Secant	M
94°						85°	95°						84°

TABLE: Natural Trigonometric functions, continued 5.6.8

6°						173°	7°					172°	
M	Sine	Cosine	Tan.	Cotan.	Secant	Cosec.	Sine	Cosine	Tan.	Cotan.	Secant	Cosec.	M
0	0.10453	0.99452	0.10510	9.5144	1.0055	9.5668	0.12187	0.99255	0.12278	8.1443	1.0075	8.2055	60
1	.10482	.99449	.10540	.4878	.0055	.5404	.12216	.99251	.12308	.1248	.0075	.1861	59
2	.10511	.99446	.10569	.4614	.0056	.5141	.12245	.99247	.12337	.1053	.0076	.1668	58
3	.10540	.99443	.10599	.4351	.0056	.4880	.12273	.99244	.12367	.0860	.0076	.1476	57
4	.10568	.99440	.10628	.4090	.0056	.4620	.12302	.99240	.12396	.0667	.0076	.1285	56
5	0.10597	0.99437	0.10657	9.3831	1.0057	9.4362	0.12331	0.99237	0.12426	8.0476	1.0077	8.1094	55
6	.10626	.99434	.10687	.3572	.0057	.4105	.12360	.99233	.12456	.0285	.0077	.0905	54
7	.10655	.99431	.10716	.3315	.0057	.3850	.12389	.99229	.12485	.0095	.0078	.0717	53
8	.10684	.99428	.10746	.3060	.0057	.3596	.12418	.99226	.12515	7.9906	.0078	.0529	52
9	.10713	.99424	.10775	.2806	.0058	.3343	.12447	.99222	.12544	.9717	.0078	.0342	51
10	0.10742	0.99421	0.10805	9.2553	1.0058	9.3092	0.12476	0.99219	0.12574	7.9530	1.0079	8.0156	50
11	.10771	.99418	.10834	.2302	.0058	.2842	.12504	.99215	.12603	.9344	.0079	7.9971	49
12	.10800	.99415	.10863	.2051	.0059	.2593	.12533	.99211	.12633	.9158	.0079	.9787	48
13	.10829	.99412	.10893	.1803	.0059	.2346	.12562	.99208	.12662	.8973	.0080	.9604	47
14	.10858	.99409	.10922	.1555	.0059	.2100	.12591	.99204	.12692	.8789	.0080	.9421	46
15	0.10887	0.99406	0.10952	9.1309	1.0060	9.1855	0.12620	0.99200	0.12722	7.8606	1.0080	7.9240	45
16	.10916	.99402	.10981	.1064	.0060	.1612	.12649	.99197	.12751	.8424	.0081	.9059	44
17	.10944	.99399	.11011	.0821	.0060	.1370	.12678	.99193	.12781	.8243	.0081	.8879	43
18	.10973	.99396	.11040	.0579	.0061	.1129	.12706	.99189	.12810	.8062	.0082	.8700	42
19	.11002	.99393	.11069	.0338	.0061	.0890	.12735	.99186	.12840	.7882	.0082	.8522	41
20	0.11031	0.99390	0.11099	9.0098	1.0061	9.0651	0.12764	0.99182	0.12869	7.7703	1.0082	7.8344	40
21	.11060	.99386	.11128	8.9860	.0062	.0414	.12793	.99178	.12899	.7525	.0083	.8168	39
22	.11089	.99383	.11158	.9623	.0062	.0179	.12822	.99174	.12928	.7348	.0083	.7992	38
23	.11118	.99380	.11187	.9387	.0062	8.9944	.12851	.99171	.12958	.7171	.0084	.7817	37
24	.11147	.99377	.11217	.9152	.0063	.9711	.12879	.99167	.12988	.6996	.0084	.7642	36
25	0.11176	0.99373	0.11246	8.8918	1.0063	8.9479	0.12908	0.99163	0.13017	7.6821	1.0084	7.7469	35
26	.11205	.99370	.11276	.8686	.0063	.9248	.12937	.99160	.13047	.6646	.0085	.7296	34
27	.11234	.99367	.11305	.8455	.0064	.9018	.12966	.99156	.13076	.6473	.0085	.7124	33
28	.11262	.99364	.11335	.8225	.0064	.8790	.12995	.99152	.13106	.6300	.0085	.6953	32
29	.11291	.99360	.11364	.7996	.0064	.8563	.13024	.99148	.13136	.6129	.0086	.6783	31
30	0.11320	0.99357	0.11393	8.7769	1.0065	8.8337	0.13053	0.99144	0.13165	7.5957	1.0086	7.6613	30
31	.11349	.99354	.11423	.7542	.0065	.8112	.13081	.99141	.13195	.5787	.0087	.6444	29
32	.11378	.99350	.11452	.7317	.0065	.7888	.13110	.99137	.13224	.5617	.0087	.6276	28
33	.11407	.99347	.11482	.7093	.0066	.7665	.13139	.99133	.13254	.5449	.0087	.6108	27
34	.11436	.99344	.11511	.6870	.0066	.7444	.13168	.99129	.13284	.5280	.0088	.5942	26
35	0.11465	0.99341	0.11541	8.6648	1.0066	8.7223	0.13197	0.99125	0.13313	7.5113	1.0088	7.5776	25
36	.11494	.99337	.11570	.6427	.0067	.7004	.13226	.99121	.13343	.4946	.0089	.5611	24
37	.11523	.99334	.11600	.6208	.0067	.6786	.13254	.99118	.13372	.4780	.0089	.5446	23
38	.11551	.99330	.11629	.5989	.0067	.6569	.13283	.99114	.13402	.4615	.0089	.5282	22
39	.11580	.99327	.11659	.5772	.0068	.6353	.13312	.99110	.13432	.4451	.0090	.5119	21
40	0.11609	0.99324	0.11688	8.5555	1.0068	8.6138	0.13341	0.99106	0.13461	7.4287	1.0090	7.4957	20
41	.11638	.99320	.11718	.5340	.0068	.5924	.13370	.99102	.13491	.4124	.0090	.4795	19
42	.11667	.99317	.11747	.5126	.0069	.5711	.13399	.99098	.13520	.3961	.0091	.4634	18
43	.11696	.99314	.11777	.4913	.0069	.5499	.13427	.99094	.13550	.3800	.0091	.4474	17
44	.11725	.99310	.11806	.4701	.0069	.5289	.13456	.99090	.13580	.3639	.0092	.4315	16
45	0.11754	0.99307	0.11836	8.4489	1.0070	8.5079	0.13485	0.99086	0.13609	7.3479	1.0092	7.4156	15
46	.11783	.99303	.11865	.4279	.0070	.4871	.13514	.99083	.13639	.3319	.0092	.3998	14
47	.11811	.99300	.11895	.4070	.0070	.4663	.13543	.99079	.13669	.3160	.0093	.3840	13
48	.11840	.99296	.11924	.3862	.0071	.4457	.13571	.99075	.13698	.3002	.0093	.3683	12
49	.11869	.99293	.11954	.3655	.0071	.4251	.13600	.99071	.13728	.2844	.0094	.3527	11
50	0.11898	0.99290	0.11983	8.3449	1.0071	8.4046	0.13629	0.99067	0.13757	7.2687	1.0094	7.3372	10
51	.11927	.99286	.12013	.3244	.0072	.3843	.13658	.99063	.13787	.2531	.0094	.3217	9
52	.11956	.99283	.12042	.3040	.0072	.3640	.13687	.99059	.13817	.2375	.0095	.3063	8
53	.11985	.99279	.12072	.2837	.0073	.3439	.13716	.99055	.13846	.2220	.0095	.2909	7
54	.12014	.99276	.12101	.2635	.0073	.3238	.13744	.99051	.13876	.2066	.0096	.2757	6
55	0.12042	0.99272	0.12131	8.2434	1.0073	8.3039	0.13773	0.99047	0.13906	7.1912	1.0096	7.2604	5
56	.12071	.99269	.12160	.2234	.0074	.2840	.13802	.99043	.13935	.1759	.0097	.2453	4
57	.12100	.99265	.12190	.2035	.0074	.2642	.13831	.99039	.13965	.1607	.0097	.2302	3
58	.12129	.99262	.12219	.1837	.0074	.2446	.13860	.99035	.13995	.1455	.0097	.2152	2
59	.12158	.99258	.12249	.1640	.0075	.2250	.13888	.99031	.14024	.1304	.0098	.2002	1
60	0.12187	0.99255	0.12278	8.1443	1.0075	8.2055	0.13917	0.99027	0.14054	7.1154	1.0098	7.1853	0
M	Cosine	Sine	Cotan.	Tan.	Cosec.	Secant	Cosine	Sine	Cotan.	Tan.	Cosec.	Secant	M
96°						83°	97°					82°	

TABLE: Natural Trigonometric functions, continued 5.6.8

8°						171°	9°						170°
M	Sine	Cosine	Tan.	Cotan.	Secant	Cosec.	Sine	Cosine	Tan.	Cotan.	Secant	Cosec.	M
0	0.13917	0.99027	0.14054	7.1154	1.0098	7.1853	0.15643	0.98769	0.15838	6.3137	1.0125	6.3924	60
1	.13946	.99023	.14084	.1004	.0099	.1704	.15672	.98764	.15868	.3019	.0125	.3807	59
2	.13975	.99019	.14113	.0854	.0099	.1557	.15701	.98760	.15898	.2901	.0125	.3690	58
3	.14004	.99015	.14143	.0706	.0099	.1409	.15730	.98755	.15928	.2783	.0126	.3574	57
4	.14032	.99010	.14173	.0558	.0100	.1263	.15758	.98750	.15958	.2665	.0126	.3458	56
5	0.14061	0.99006	0.14202	7.0410	1.0100	7.1117	0.15787	0.98746	0.15987	6.2548	1.0127	6.3343	55
6	.14090	.99002	.14232	.0264	.0101	.0972	.15816	.98741	.16017	.2432	.0127	.3228	54
7	.14119	.98998	.14262	.0117	.0101	.0827	.15844	.98737	.16047	.2316	.0128	.3113	53
8	.14148	.98994	.14291	6.9972	.0102	.0683	.15873	.98732	.16077	.2200	.0128	.2999	52
9	.14176	.98990	.14321	.9827	.0102	.0539	.15902	.98727	.16107	.2085	.0129	.2885	51
10	0.14205	0.98986	0.14351	6.9682	1.0102	7.0396	0.15931	0.98723	0.16137	6.1970	1.0129	6.2772	50
11	.14234	.98982	.14380	.9538	.0103	.0254	.15959	.98718	.16167	.1856	.0130	.2659	49
12	.14263	.98978	.14410	.9395	.0103	.0112	.15988	.98714	.16196	.1742	.0130	.2546	48
13	.14292	.98973	.14440	.9252	.0104	6.9971	.16017	.98709	.16226	.1628	.0131	.2434	47
14	.14320	.98969	.14470	.9110	.0104	.9830	.16045	.98704	.16256	.1515	.0131	.2322	46
15	0.14349	0.98965	0.14499	6.8969	1.0104	6.9690	0.16074	0.98700	0.16286	6.1402	1.0132	6.2211	45
16	.14378	.98961	.14529	.8828	.0105	.9550	.16103	.98695	.16316	.1290	.0132	.2100	44
17	.14407	.98957	.14559	.8687	.0105	.9411	.16132	.98690	.16346	.1178	.0133	.1990	43
18	.14436	.98952	.14588	.8547	.0106	.9273	.16160	.98685	.16376	.1066	.0133	.1880	42
19	.14464	.98948	.14618	.8408	.0106	.9135	.16189	.98681	.16405	.0955	.0134	.1770	41
20	0.14493	0.98944	0.14648	6.8269	1.0107	6.8998	0.16218	0.98676	0.16435	6.0844	1.0134	6.1661	40
21	.14522	.98940	.14677	.8131	.0107	.8861	.16246	.98671	.16465	.0734	.0135	.1552	39
22	.14551	.98936	.14707	.7993	.0107	.8725	.16275	.98667	.16495	.0624	.0135	.1443	38
23	.14579	.98931	.14737	.7856	.0108	.8589	.16304	.98662	.16525	.0514	.0136	.1335	37
24	.14608	.98927	.14767	.7720	.0108	.8454	.16333	.98657	.16555	.0405	.0136	.1227	36
25	0.14637	0.98923	0.14796	6.7584	1.0109	6.8320	0.16361	0.98652	0.16585	6.0296	1.0136	6.1120	35
26	.14666	.98919	.14826	.7448	.0109	.8185	.16390	.98648	.16615	.0188	.0137	.1013	34
27	.14695	.98914	.14856	.7313	.0110	.8052	.16419	.98643	.16644	.0080	.0137	.0906	33
28	.14723	.98910	.14886	.7179	.0110	.7919	.16447	.98638	.16674	5.9972	.0138	.0800	32
29	.14752	.98906	.14915	.7045	.0111	.7787	.16476	.98633	.16704	.9865	.0138	.0694	31
30	0.14781	0.98901	0.14945	6.6911	1.0111	6.7655	0.16505	0.98628	0.16734	5.9758	1.0139	6.0588	30
31	.14810	.98897	.14975	.6779	.0111	.7523	.16533	.98624	.16764	.9651	.0139	.0483	29
32	.14838	.98893	.15004	.6646	.0112	.7392	.16562	.98619	.16794	.9545	.0140	.0379	28
33	.14867	.98889	.15034	.6514	.0112	.7262	.16591	.98614	.16824	.9439	.0140	.0274	27
34	.14896	.98884	.15064	.6383	.0113	.7132	.16619	.98609	.16854	.9333	.0141	.0170	26
35	0.14925	0.98880	0.15094	6.6252	1.0113	6.7003	0.16648	0.98604	0.16884	5.9228	1.0141	6.0066	25
36	.14953	.98876	.15123	.6122	.0114	.6874	.16677	.98600	.16914	.9123	.0142	5.9963	24
37	.14982	.98871	.15153	.5992	.0114	.6745	.16705	.98595	.16944	.9019	.0142	.9860	23
38	.15011	.98867	.15183	.5863	.0115	.6617	.16734	.98590	.16973	.8915	.0143	.9758	22
39	.15040	.98862	.15213	.5734	.0115	.6490	.16763	.98585	.17003	.8811	.0143	.9655	21
40	0.15068	0.98858	0.15243	6.5605	1.0115	6.6363	0.16791	0.98580	0.17033	5.8708	1.0144	5.9554	20
41	.15097	.98854	.15272	.5478	.0116	.6237	.16820	.98575	.17063	.8605	.0144	.9452	19
42	.15126	.98849	.15302	.5350	.0116	.6111	.16849	.98570	.17093	.8502	.0145	.9351	18
43	.15155	.98845	.15332	.5223	.0117	.5985	.16878	.98565	.17123	.8400	.0145	.9250	17
44	.15183	.98840	.15362	.5097	.0117	.5860	.16906	.98560	.17153	.8298	.0146	.9150	16
45	0.15212	0.98836	0.15391	6.4971	1.0118	6.5736	0.16935	0.98556	0.17183	5.8196	1.0146	5.9049	15
46	.15241	.98832	.15421	.4845	.0118	.5612	.16964	.98551	.17213	.8095	.0147	.8950	14
47	.15270	.98827	.15451	.4720	.0119	.5488	.16992	.98546	.17243	.7994	.0147	.8850	13
48	.15298	.98823	.15481	.4596	.0119	.5365	.17021	.98541	.17273	.7894	.0148	.8751	12
49	.15328	.98818	.15511	.4472	.0119	.5243	.17050	.98536	.17303	.7794	.0148	.8652	11
50	0.15356	0.98814	0.15540	6.4348	1.0120	6.5121	0.17078	0.98531	0.17333	5.7694	1.0149	5.8554	10
51	.15385	.98809	.15570	.4225	.0120	.4999	.17107	.98526	.17363	.7594	.0150	.8456	9
52	.15413	.98805	.15600	.4103	.0121	.4878	.17136	.98521	.17393	.7495	.0150	.8358	8
53	.15442	.98800	.15630	.3980	.0121	.4757	.17164	.98516	.17423	.7396	.0151	.8261	7
54	.15471	.98796	.15659	.3859	.0122	.4637	.17193	.98511	.17453	.7297	.0151	.8163	6
55	0.15500	0.98791	0.15689	6.3737	1.0122	6.4517	0.17221	0.98506	0.17483	5.7199	1.0152	5.8067	5
56	.15528	.98787	.15719	.3616	.0123	.4398	.17250	.98501	.17513	.7101	.0152	.7970	4
57	.15557	.98782	.15749	.3496	.0123	.4279	.17279	.98496	.17543	.7004	.0153	.7874	3
58	.15586	.98778	.15779	.3376	.0124	.4160	.17307	.98491	.17573	.6906	.0153	.7778	2
59	.15615	.98773	.15809	.3257	.0124	.4042	.17336	.98486	.17603	.6809	.0154	.7683	1
60	0.15643	0.98769	0.15838	6.3137	1.0125	6.3924	0.17365	0.98481	0.17633	5.6713	1.0154	5.7588	0
M	Cosine	Sine	Cotan.	Tan.	Cosec.	Secant	Cosine	Sine	Cotan.	Tan.	Cosec.	Secant	M
98°						81°	99°						80°

TABLE: Natural Trigonometric functions, continued 5.6.8

10°						169°	11°						168°
M	Sine	Cosine	Tan.	Cotan.	Secant	Cosec.	Sine	Cosine	Tan.	Cotan.	Secant	Cosec.	M
0	0.17365	0.98481	0.17633	5.6713	1.0154	5.7588	0 19081	0.98163	0.19438	5.1445	1.0187	5.2408	60
1	.17393	.98476	.17663	.6616	.0155	.7493	.19109	.98157	.19468	.1366	.0188	.2330	59
2	.17422	.98471	.17693	.6520	.0155	.7398	.19138	.98152	.19498	.1286	.0188	.2252	58
3	.17451	.98465	.17723	.6425	.0156	.7304	.19166	.98146	.19529	.1207	.0189	.2174	57
4	.17479	.98460	.17753	.6329	.0156	.7210	.19195	.98140	.19559	.1128	.0189	.2097	56
5	0.17508	0.98455	0.17783	5.6234	1.0157	5.7117	0.19224	0.98135	0 19589	5 1049	1.0190	5.2019	55
6	.17537	-.98450	.17813	.6140	.0157	.7023	.19252	.98129	.19619	.0970	.0191	.1942	54
7	.17565	.98445	.17843	.6045	.0158	.6930	.19281	.98124	.19649	.0892	.0191	.1865	53
8	.17594	.98440	.17873	.5951	.0158	.6838	.19309	.98118	.19680	.0814	.0192	.1788	52
9	.17622	.98435	.17903	.5857	.0159	.6745	.19338	.98112	.19710	.0736	.0192	1712	51
10	0.17651	0.98430	0.17933	5.5764	1.0159	5.6653	0.19366	0.98107	0.19740	5.0658	1.0193	5.1636	50
11	.17680	.98425	.17963	.5670	.0160	.6561	.19395	.98101	.19770	.0581	.0193	.1560	49
12	.17708	.98419	.17993	.5578	.0160	.6470	.19423	.98095	.19800	.0504	.0194	.1484	48
13	.17737	.98414	.18023	.5485	.0161	.6379	.19452	.98090	.19831	.0427	.0195	.1409	47
14	.17766	.98409	.18053	.5393	.0162	.6288	.19480	.98084	.19861	.0350	.0195	.1333	46
15	0.17794	0.98404	0.18083	5.5301	1.0162	5.6197	0.19509	0.98078	0.19891	5.0273	1.0196	5.1258	45
16	.17823	.98399	.18113	.5209	.0163	.6107	.19537	.98073	.19921	.0197	.0196	.1183	44
17	.17852	.98394	.18143	.5117	.0163	.6017	.19566	.98067	.19952	.0121	.0197	.1109	43
18	.17880	.98388	.18173	.5026	.0164	.5928	.19595	.98061	.19982	.0045	.0198	.1034	42
19	.17909	.98383	.18203	.4936	.0164	.5838	.19623	.98056	.20012	4.9969	.0198	.0960	41
20	0.17937	0.98378	0.18233	5 4845	1.0165	5.5749	0.19652	0.98050	0.20042	4.9894	1.0199	5.0886	40
21	.17966	.98373	.18263	.4755	.0165	.5660	.19680	.98044	.20073	.9819	.0199	.0812	39
22	.17995	.98368	.18293	.4665	.0166	.5572	.19709	.98039	.20103	.9744	.0200	.0739	38
23	.18023	.98362	.18323	.4575	.0166	.5484	.19737	.98033	.20133	.9669	.0201	.0666	37
24	.18052	.98357	.18353	.4486	.0167	.5396	.19766	.98027	.20163	.9594	.0201	.0593	36
25	0.18080	0.98352	0.18383	5.4396	1.0167	5.5308	0.19794	0.98021	0.20194	4.9520	1.0202	5.0520	35
26	.18109	.98347	.18413	.4308	.0168	.5221	.19823	.98016	.20224	.9446	.0202	.0447	34
27	.18138	.98341	.18444	.4219	.0169	.5134	.19851	.98010	.20254	.9372	.0203	.0375	33
28	.18166	.98336	.18474	.4131	.0169	.5047	.19880	.98004	.20285	.9298	.0204	.0302	32
29	.18195	.98331	.18504	.4043	.0170	.4960	.19908	.97998	.20315	.9225	.0204	.0230	31
30	0.18223	0.98325	0.18534	5.3955	1.0170	5.4874	0.19937	0.97992	0.20345	4.9151	1.0205	5.0158	30
31	.18252	.98320	.18564	.3868	.0171	.4788	.19965	.97987	.20375	.9078	.0205	.0087	29
32	.18281	.98315	.18594	.3780	.0171	.4702	.19994	.97981	.20406	.9006	.0206	.0015	28
33	.18309	.98309	.18624	.3694	.0172	.4617	.20022	.97975	.20436	.8933	.0207	4.9944	27
34	.18338	.98304	.18654	.3607	.0172	.4532	.20051	.97969	.20466	.8860	.0207	.9873	26
35	0.18366	0.98299	0.18684	5.3521	1.0173	5.4447	0.20079	0.97963	0.20497	4.8788	1.0208	4.9802	25
36	.18395	.98293	.18714	.3434	.0174	.4362	.20108	.97957	.20527	.8716	.0208	.9732	24
37	.18424	.98288	.18745	.3349	.0174	.4278	.20136	.97952	.20557	.8644	.0209	.9661	23
38	.18452	.98283	.18775	.3263	.0175	.4194	.20165	.97946	.20588	.8573	.0210	.9591	22
39	.18481	.98277	.18805	.3178	.0175	.4110	.20193	.97940	.20618	.8501	.0210	.9521	21
40	0.18509	0.98272	0.18835	5.3093	1.0176	5.4026	0.20222	0.97934	0.20648	4.8430	1.0211	4.9452	20
41	.18538	.98267	.18865	.3008	.0176	.3943	.20250	.97928	.20679	.8359	.0211	.9382	19
42	.18567	.98261	.18895	.2923	.0177	.3860	.20279	.97922	.20709	.8288	.0212	.9313	18
43	.18595	.98256	.18925	.2839	.0177	.3777	.20307	.97916	.20739	.8217	.0213	.9243	17
44	.18624	.98250	.18955	.2755	.0178	.3695	.20336	.97910	.20770	.8147	.0213	.9175	16
45	0.18652	0.98245	0.18985	5.2671	1.0179	5.3612	0.20364	0.97904	0.20800	4.8077	1.0214	4.9106	15
46	.18681	.98240	.19016	.2588	.0179	.3530	.20393	.97899	.20830	.8007	.0215	.9037	14
47	.18709	.98234	.19046	.2505	.0180	.3449	.20421	.97893	.20861	.7937	.0215	.8969	13
48	.18738	.98229	.19076	.2422	.0180	.3367	.20450	.97887	.20891	.7867	.0216	.8901	12
49	.18767	.98223	.19106	.2339	.0181	.3286	.20478	.97881	.20921	.7798	.0216	.8833	11
50	0.18795	0.98218	0.19136	5.2257	1.0181	5.3205	0.20506	0.97875	0.20952	4.7728	1.0217	4.8765	10
51	.18824	.98212	.19166	.2174	.0182	.3124	.20535	.97869	.20982	.7659	.0218	.8697	9
52	.18852	.98207	.19197	.2092	.0182	.3044	.20563	.97863	.21012	.7591	.0218	.8630	8
53	.18881	.98201	.19227	.2011	.0183	.2963	.20592	.97857	.21043	.7522	.0219	.8563	7
54	.18909	.98196	.19257	.1929	.0184	.2883	.20620	.97851	.21073	.7453	.0220	.8496	6
55	0.18938	0.98190	0.19287	5.1848	1.0184	5.2803	0.20649	0.97845	0.21104	4.7385	1.0220	4.8429	5
56	.18967	.98185	.19317	.1767	.0185	.2724	.20677	.97839	.21134	.7317	.0221	.8362	4
57	.18995	.98179	.19347	.1686	.0185	.2645	.20706	.97833	.21164	.7249	.0221	.8296	3
58	.19024	.98174	.19378	.1606	.0186	.2566	.20734	.97827	.21195	.7181	.0222	.8229	2
59	.19052	.98168	.19408	.1525	.0186	.2487	.20763	.97821	.21225	.7114	.0223	.8163	1
60	0.19081	0.98163	0.19438	5.1445	1.0187	5.2408	0.20791	0.97815	0.21256	4.7046	1.0223	4.8097	0
M	Cosine	Sine	Cotan.	Tan.	Cosec.	Secant	Cosine	Sine	Cotan.	Tan.	Cosec.	Secant	M
100°						79°	101°						78°

TABLE: Natural Trigonometric functions, continued 5.6.8

12°						167°	13°					166°
M	Sine	Cosine	Tan.	Cotan.	Secant	Cosec.	Sine	Cosine	Tan.	Cotan.	Secant	Cosec.	M
0	0.20791	0.97815	0.21256	4.7046	1.0223	4.8097	0.22495	0.97437	0.23087	4.3315	1.0263	4.4454	60
1	.20820	.97809	.21286	.6979	.0224	.8032	.22523	.97430	.23117	.3257	.0264	.4398	59
2	.20848	.97803	.21316	.6912	.0225	.7966	.22552	.97424	.23148	.3200	.0264	.4342	58
3	.20876	.97797	.21347	.6845	.0225	.7901	.22580	.97417	.23179	.3143	.0265	.4287	57
4	.20905	.97790	.21377	.6778	.0226	.7835	.22608	.97411	.23209	.3086	.0266	.4231	56
5	0.20933	0.97784	0.21408	4.6712	1.0226	4.7770	0.22637	0.97404	0.23240	4.3029	1.0266	4.4176	55
6	.20962	.97778	.21438	.6646	.0227	.7706	.22665	.97398	.23270	.2972	.0267	.4121	54
7	.20990	.97772	.21468	.6580	.0228	.7641	.22693	.97391	.23301	.2916	.0268	.4065	53
8	.21019	.97766	.21499	.6514	.0228	.7576	.22722	.97384	.23332	.2859	.0268	.4011	52
9	.21047	.97760	.21529	.6448	.0229	.7512	.22750	.97378	.23363	.2803	.0269	.3956	51
10	0.21076	0.97754	0.21560	4.6382	1.0230	4.7448	0.22778	0.97371	0.23393	4.2747	1.0270	4.3901	50
11	.21104	.97748	.21590	.6317	.0230	.7384	.22807	.97364	.23424	.2691	.0271	.3847	49
12	.21132	.97741	.21621	.6252	.0231	.7320	.22835	.97358	.23455	.2635	.0271	.3792	48
13	.21161	.97735	.21651	.6187	.0232	.7257	.22863	.97351	.23485	.2579	.0272	.3738	47
14	.21189	.97729	.21682	.6122	.0232	.7193	.22892	.97344	.23516	.2524	.0273	.3684	46
15	0.21218	0.97723	0.21712	4.6057	1.0233	4.7130	0.22920	0.97338	0.23547	4.2468	1.0273	4.3630	45
16	.21246	.97717	.21742	.5993	.0234	.7067	.22948	.97331	.23577	.2413	.0274	.3576	44
17	.21275	.97711	.21773	.5928	.0234	.7004	.22977	.97324	.23608	.2358	.0275	.3522	43
18	.21303	.97704	.21803	.5864	.0235	.6942	.23005	.97318	.23639	.2303	.0276	.3469	42
19	.21331	.97698	.21834	.5800	.0235	.6879	.23033	.97311	.23670	.2248	.0276	.3415	41
20	0.21360	0.97692	0.21864	4.5736	1.0236	4.6817	0.23061	0.97304	0.23700	4.2193	1.0277	4.3362	40
21	.21388	.97686	.21895	.5673	.0237	.6754	.23090	.97298	.23731	.2139	.0278	.3309	39
22	.21417	.97680	.21925	.5609	.0237	.6692	.23118	.97291	.23762	.2084	.0278	.3256	38
23	.21445	.97673	.21956	.5546	.0238	.6631	.23146	.97284	.23793	.2030	.0279	.3203	37
24	.21473	.97667	.21986	.5483	.0239	.6569	.23175	.97277	.23823	.1976	.0280	.3150	36
25	0.21502	0.97661	0.22017	4.5420	1.0239	4.6507	0.23203	0.97271	0.23854	4.1921	1.0280	4.3098	35
26	.21530	.97655	.22047	.5357	.0240	.6446	.23231	.97264	.23885	.1867	.0281	.3045	34
27	.21559	.97648	.22078	.5294	.0241	.6385	.23260	.97257	.23916	.1814	.0282	.2993	33
28	.21587	.97642	.22108	.5232	.0241	.6324	.23288	.97250	.23946	.1760	.0283	.2941	32
29	.21615	.97636	.22139	.5169	.0242	.6263	.23316	.97244	.23977	.1706	.0283	.2838	31
30	0.21644	0.97630	0.22169	4.5107	1.0243	4.6201	0.23344	0.97237	0.24008	4.1653	1.0284	4.2836	30
31	.21672	.97623	.22200	.5045	.0243	.6142	.23373	.97230	.24039	.1600	.0285	.2785	29
32	.21701	.97617	.22230	.4983	.0244	.6081	.23401	.97223	.24069	.1546	.0285	.2733	28
33	.21729	.97611	.22261	.4921	.0245	.6021	.23429	.97216	.24100	.1493	.0286	.2681	27
34	.21757	.97604	.22291	.4860	.0245	.5961	.23458	.97210	.24131	.1440	.0287	.2630	26
35	0.21786	0.97598	0.22322	4.4799	1.0246	4.5901	0.23486	0.97203	0.24162	4.1388	1.0288	4.2579	25
36	.21814	.97592	.22353	.4737	.0247	.5841	.23514	.97196	.24192	.1335	.0288	.2527	24
37	.21843	.97585	.22383	.4676	.0247	.5782	.23542	.97189	.24223	.1282	.0289	.2476	23
38	.21871	.97579	.22414	.4615	.0248	.5722	.23571	.97182	.24254	.1230	.0290	.2425	22
39	.21899	.97573	.22444	.4555	.0249	.5663	.23599	.97175	.24285	.1178	.0291	.2375	21
40	0.21928	0.97566	0.22475	4.4494	1.0249	4.5604	0.23627	0.97169	0.24316	4.1126	1.0291	4.2324	20
41	.21956	.97560	.22505	.4434	.0250	.5545	.23655	.97162	.24346	.1073	.0292	.2273	19
42	.21985	.97553	.22536	.4373	.0251	.5486	.23684	.97155	.24377	.1022	.0293	.2223	18
43	.22013	.97547	.22566	.4313	.0251	.5428	.23712	.97148	.24408	.0970	.0293	.2173	17
44	.22041	.97541	.22597	.4253	.0252	.5369	.23740	.97141	.24439	.0918	.0294	.2122	16
45	0.22070	0.97534	0.22628	4.4194	1.0253	4.5311	0.23768	0.97134	0.24470	4.0867	1.0295	4.2072	15
46	.22098	.97528	.22658	.4134	.0253	.5253	.23797	.97127	.24501	.0815	.0296	.2022	14
47	.22126	.97521	.22689	.4074	.0254	.5195	.23825	.97120	.24531	.0764	.0296	.1972	13
48	.22155	.97515	.22719	.4015	.0255	.5137	.23853	.97113	.24562	.0713	.0297	.1923	12
49	.22183	.97508	.22750	.3956	.0255	.5079	.23881	.97106	.24593	.0662	.0298	.1873	11
50	0.22211	0.97502	0.22781	4.3897	1.0256	4.5021	0.23910	0.97099	0.24624	4.0611	1.0299	4.1824	10
51	.22240	.97495	.22811	.3838	.0257	.4964	.23938	.97092	.24655	.0560	.0299	.1774	9
52	.22268	.97489	.22842	.3779	.0257	.4907	.23966	.97086	.24686	.0509	.0300	.1725	8
53	.22297	.97483	.22872	.3721	.0258	.4850	.23994	.97079	.24717	.0458	.0301	.1676	7
54	.22325	.97476	.22903	.3662	.0259	.4793	.24023	.97072	.24747	.0408	.0302	.1627	6
55	0.22353	0.97470	0.22934	4.3604	1.0260	4.4736	0.24051	0.97065	0.24778	4.0358	1.0302	4.1578	5
56	.22382	.97463	.22964	.3546	.0260	.4679	.24079	.97058	.24809	.0307	.0303	.1529	4
57	.22410	.97457	.22995	.3488	.0261	.4623	.24107	.97051	.24840	.0257	.0304	.1481	3
58	.22438	.97450	.23025	.3430	.0262	.4566	.24136	.97044	.24871	.0207	.0305	.1432	2
59	.22467	.97443	.23056	.3372	.0262	.4510	.24164	.97037	.24902	.0157	.0305	.1384	1
60	0.22495	0.97437	0.23087	4.3315	1.0263	4.4454	0.24192	0.97029	0.24933	4.0108	1.0306	4.1336	0
M	Cosine	Sine	Cotan.	Tan.	Cosec.	Secant	Cosine	Sine	Cotan.	Tan.	Cosec.	Secant	M
102°						77°	103°					76°	

TABLE: Natural Trigonometric functions, continued 5.6.8

14° 165° 15° 164°

M	Sine	Cosine	Tan.	Cotan.	Secant	Cosec.	Sine	Cosine	Tan.	Cotan.	Secant	Cosec.	M
0	0.24192	0.97029	0.24933	4.0108	1.0306	4.1336	0.25882	0.96592	0.26795	3.7320	1.0353	3.8637	60
1	.24220	.97022	.24964	.0058	.0307	.1287	.25910	.96585	.26826	.7277	.0353	.8595	59
2	.24249	.97015	.24995	.0009	.0308	.1239	.25938	.96577	.26857	.7234	.0354	.8553	58
3	.24277	.97008	.25025	3.9959	.0308	.1191	.25966	.96570	.26888	.7191	.0355	.8512	57
4	.24305	.97001	.25056	.9910	.0309	.1144	.25994	.96562	.26920	.7147	.0356	.8470	56
5	0.24333	0.96994	0.25087	3.9861	1.0310	4.1096	0.26022	0.96555	0.26951	3.7104	1.0357	3.8428	55
6	.24361	.96987	.25118	.9812	.0311	.1048	.26050	.96547	.26982	.7062	.0358	.8387	54
7	.24390	.96980	.25149	.9763	.0311	.1001	.26078	.96540	.27013	.7019	.0358	.8346	53
8	.24418	.96973	.25180	.9714	.0312	.0953	.26107	.96532	.27044	.6976	.0359	.8304	52
9	.24446	.96966	.25211	.9665	.0313	.0906	.26135	.96524	.27076	.6933	.0360	.8263	51
10	0.24474	0.96959	0.25242	3.9616	1.0314	4.0859	0.26163	0.96517	0.27107	3.6891	1.0361	3.8222	50
11	.24502	.96952	.25273	.9568	.0314	.0812	.26191	.96509	.27138	.6848	.0362	.8181	49
12	.24531	.96944	.25304	.9520	.0315	.0765	.26219	.96502	.27169	.6806	.0362	.8140	48
13	.24559	.96937	.25335	.9471	.0316	.0718	.26247	.96494	.27201	.6764	.0363	.8100	47
14	.24587	.96930	.25366	.9423	.0317	.0672	.26275	.96486	.27232	.6722	.0364	.8059	46
15	0.24615	0.96923	0.25397	3.9375	1.0317	4.0625	0.26303	0.96479	0.27263	3.6679	1.0365	3.8018	45
16	.24643	.96916	.25428	.9327	.0318	.0579	.26331	.96471	.27294	.6637	.0366	.7978	44
17	.24672	.96909	.25459	.9279	.0319	.0532	.26359	.96463	.27326	.6596	.0367	.7937	43
18	.24700	.96901	.25490	.9231	.0320	.0486	.26387	.96456	.27357	.6554	.0367	.7897	42
19	.24728	.96894	.25521	.9184	.0320	.0440	.26415	.96448	.27388	.6512	.0368	.7857	41
20	0.24756	0.96887	0.25552	3.9136	1.0321	4.0394	0.26443	0.96440	0.27419	3.6470	1.0369	3.7816	40
21	.24784	.96880	.25583	.9089	.0322	.0348	.26471	.96433	.27451	.6429	.0370	.7776	39
22	.24813	.96873	.25614	.9042	.0323	.0302	.26499	.96425	.27482	.6387	.0371	.7736	38
23	.24841	.96865	.25645	.8994	.0323	.0256	.26527	.96417	.27513	.6346	.0371	.7697	37
24	.24869	.96858	.25676	.8947	.0324	.0211	.26556	.96409	.27544	.6305	.0372	.7657	36
25	0.24897	0.96851	0.25707	3.8900	1.0325	4.0165	0.26584	0.96402	0.27576	3.6263	1.0373	3.7617	35
26	.24925	.96844	.25738	.8853	.0326	.0120	.26612	.96394	.27607	.6222	.0374	.7577	34
27	.24953	.96836	.25769	.8807	.0327	.0074	.26640	.96386	.27638	.6181	.0375	.7538	33
28	.24982	.96829	.25800	.8760	.0327	.0029	.26668	.96378	.27670	.6140	.0376	.7498	32
29	.25010	.96822	.25831	.8713	.0328	3.9984	.26696	.96371	.27701	.6100	.0376	.7459	31
30	0.25038	0.96815	0.25862	3.8667	1.0329	3.9939	0.26724	0.96363	0.27732	3.6059	1.0377	3.7420	30
31	.25066	.96807	.25893	.8621	.0330	.9894	.26752	.96355	.27764	.6018	.0378	.7380	29
32	.25094	.96800	.25924	.8574	.0330	.9850	.26780	.96347	.27795	.5977	.0379	.7341	28
33	.25122	.96793	.25955	.8528	.0331	.9805	.26808	.96340	.27826	.5937	.0380	.7302	27
34	.25151	.96785	.25986	.8482	.0332	.9760	.26836	.96332	.27858	.5896	.0381	.7263	26
35	0.25179	0.96778	0.26017	3.8436	1.0333	3.9716	0.26864	0.96324	0.27889	3.5856	1.0382	3.7224	25
36	.25207	.96771	.26048	.8390	.0334	.9672	.26892	.96316	.27920	.5816	.0382	.7186	24
37	.25235	.96763	.26079	.8345	.0334	.9627	.26920	.96308	.27952	.5776	.0383	.7147	23
38	.25263	.96756	.26110	.8299	.0335	.9583	.26948	.96301	.27983	.5736	.0384	.7108	22
39	.25291	.96749	.26141	.8254	.0336	.9539	.26976	.96293	.28014	.5696	.0385	.7070	21
40	0.25319	0.96741	0.26172	3.8208	1.0337	3.9495	0.27004	0.96285	0.28046	3.5656	1.0386	3.7031	20
41	.25348	.96734	.26203	.8163	.0338	.9451	.27032	.96277	.28077	.5616	.0387	.6993	19
42	.25376	.96727	.26234	.8118	.0338	.9408	.27060	.96269	.28109	.5576	.0387	.6955	18
43	.25404	.96719	.26266	.8073	.0339	.9364	.27088	.96261	.28140	.5536	.0388	.6917	17
44	.25432	.96712	.26297	.8027	.0340	.9320	.27116	.96253	.28171	.5497	.0389	.6878	16
45	0.25460	0.96704	0.26328	3.7983	1.0341	3.9277	0.27144	0.96245	0.28203	3.5457	1.0390	3.6840	15
46	.25488	.96697	.26359	.7938	.0341	.9234	.27172	.96238	.28234	.5418	.0391	.6802	14
47	.25516	.96690	.26390	.7893	.0342	.9190	.27200	.96230	.28266	.5378	.0392	.6765	13
48	.25544	.96682	.26421	.7848	.0343	.9147	.27228	.96222	.28297	.5339	.0393	.6727	12
49	.25573	.96675	.26452	.7804	.0344	.9104	.27256	.96214	.28328	.5300	.0393	.6689	11
50	0.25601	0.96667	0.26483	3.7759	1.0345	3.9061	0.27284	0.96206	0.28360	3.5261	1.0394	3.6651	10
51	.25629	.96660	.26514	.7715	.0345	.9018	.27312	.96198	.28391	.5222	.0395	.6614	9
52	.25657	.96652	.26546	.7671	.0346	.8976	.27340	.96190	.28423	.5183	.0396	.6576	8
53	.25685	.96645	.26577	.7627	.0347	.8933	.27368	.96182	.28454	.5144	.0397	.6539	7
54	.25713	.96638	.26608	.7583	.0348	.8890	.27396	.96174	.28486	.5105	.0398	.6502	6
55	0.25741	0.96630	0.26639	3.7539	1.0349	3.8848	0.27424	0.96166	0.28517	3.5066	1.0399	3.6464	5
56	.25769	.96623	.26670	.7495	.0349	.8805	.27452	.96158	.28549	.5028	.0399	.6427	4
57	.25798	.96615	.26701	.7451	.0350	.8763	.27480	.96150	.28580	.4989	.0400	.6390	3
58	.25826	.96608	.26732	.7407	.0351	.8721	.27508	.96142	.28611	.4951	.0401	.6353	2
59	.25854	.96600	.26764	.7364	.0352	.8679	.27536	.96134	.28643	.4912	.0402	.6316	1
60	0.25882	0.96592	0.26795	3.7320	1.0353	3.8637	0.27564	0.96126	0.28674	3.4874	1.0403	3.6279	0
M	Cosine	Sine	Cotan.	Tan.	Cosec.	Secant	Cosine	Sine	Cotan.	Tan.	Cosec.	Secant	M

104° 75° 105° 74°

TABLE: Natural Trigonometric functions, continued 5.6.8

16°						163°	17°					162°	
M	Sine	Cosine	Tan.	Cotan.	Secant	Cosec.	Sine	Cosine	Tan.	Cotan.	Secant	Cosec.	M
0	0.27564	0.96126	0.28674	3.4874	1.0403	3.6279	0.29237	0.95630	0.30573	3.2708	1.0457	3.4203	60
1	.27592	.96118	.28706	.4836	.0404	.6243	.29265	.95622	.30605	.2674	.0458	.4170	59
2	.27620	.96110	.28737	.4798	.0405	.6206	.29293	.95613	.30637	.2640	.0459	.4138	58
3	.27648	.96102	.28769	.4760	.0406	.6169	.29321	.95605	.30668	.2607	.0460	.4106	57
4	.27675	.96094	.28800	.4722	.0406	.6133	.29348	.95596	.30700	.2573	.0461	.4073	56
5	0.27703	0.96086	0.28832	3.4684	1.0407	3.6096	0.29376	0.95588	0.30732	3.2539	1.0461	3.4041	55
6	.27731	.96078	.28863	.4646	.0408	.6060	.29404	.95579	.30764	.2505	.0462	.4009	54
7	.27759	.96070	.28895	.4608	.0409	.6024	.29432	.95571	.30796	.2472	.0463	.3977	53
8	.27787	.96062	.28926	.4570	.0410	.5987	.29460	.95562	.30828	.2438	.0464	.3945	52
9	.27815	.96054	.28958	.4533	.0411	.5951	.29487	.95554	.30859	.2405	.0465	.3913	51
10	0.27843	0.96045	0.28990	3.4495	1.0412	3.5915	0.29515	0.95545	0.30891	3.2371	1.0466	3.3881	50
11	.27871	.96037	.29021	.4458	.0413	.5879	.29543	.95536	.30923	.2338	.0467	.3849	49
12	.27899	.96029	.29053	.4420	.0413	.5843	.29571	.95528	.30955	.2305	.0468	.3817	48
13	.27927	.96021	.29084	.4383	.0414	.5807	.29598	.95519	.30987	.2271	.0469	.3785	47
14	.27955	.96013	.29116	.4346	.0415	.5772	.29626	.95511	.31019	.2238	.0470	.3754	46
15	0.27983	0.96005	0.29147	3.4308	1.0416	3.5736	0.29654	0.95502	0.31051	3.2205	1.0471	3.3722	45
16	.28011	.95997	.29179	.4271	.0417	.5700	.29682	.95493	.31083	.2172	.0472	.3690	44
17	.28039	.95989	.29210	.4234	.0418	.5665	.29710	.95485	.31115	.2139	.0473	.3659	43
18	.28067	.95980	.29242	.4197	.0419	.5629	.29737	.95476	.31146	.2106	.0474	.3627	42
19	.28094	.95972	.29274	.4160	.0420	.5594	.29765	.95467	.31178	.2073	.0475	.3596	41
20	0.28122	0.95964	0.29305	3.4124	1.0420	3.5559	0.29793	0.95459	0.31210	3.2041	1.0476	3.3565	40
21	.28150	.95956	.29337	.4087	.0421	.5523	.29821	.95450	.31242	.2008	.0477	.3534	39
22	.28178	.95948	.29368	.4050	.0422	.5488	.29848	.95441	.31274	.1975	.0478	.3502	38
23	.28206	.95940	.29400	.4014	.0423	.5453	.29876	.95433	.31306	.1942	.0478	.3471	37
24	.28234	.95931	.29432	.3977	.0424	.5418	.29904	.95424	.31338	.1910	.0479	.3440	36
25	0.28262	0.95923	0.29463	3.3941	1.0425	3.5383	0.29932	0.95415	0.31370	3.1877	1.0480	3.3409	35
26	.28290	.95915	.29495	.3904	.0426	.5348	.29959	.95407	.31402	.1845	.0481	.3378	34
27	.28318	.95907	.29526	.3868	.0427	.5313	.29987	.95398	.31434	.1813	.0482	.3347	33
28	.28346	.95898	.29558	.3832	.0428	.5279	.30015	.95389	.31466	.1780	.0483	.3316	32
29	.28374	.95890	.29590	.3795	.0428	.5244	.30043	.95380	.31498	.1748	.0484	.3286	31
30	0.28401	0.95882	0.29621	3.3759	1.0429	3.5209	0.30070	0.95372	0.31530	3.1716	1.0485	3.3255	30
31	.28429	.95874	.29653	.3723	.0430	.5175	.30098	.95363	.31562	.1684	.0486	.3224	29
32	.28457	.95865	.29685	.3687	.0431	.5140	.30126	.95354	.31594	.1652	.0487	.3194	28
33	.28485	.95857	.29716	.3651	.0432	.5106	.30154	.95345	.31626	.1620	.0488	.3163	27
34	.28513	.95849	.29748	.3616	.0433	.5072	.30181	.95337	.31658	.1588	.0489	.3133	26
35	0.28541	0.95840	0.29780	3.3580	1.0434	3.5037	0.30209	0.95328	0.31690	3.1556	1.0490	3.3102	25
36	.28569	.95832	.29811	.3544	.0435	.5003	.30237	.95319	.31722	.1524	.0491	.3072	24
37	.28597	.95824	.29843	.3509	.0436	.4969	.30265	.95310	.31754	.1492	.0492	.3042	23
38	.28624	.95816	.29875	.3473	.0437	.4935	.30292	.95301	.31786	.1460	.0493	.3011	22
39	.28652	.95807	.29906	.3438	.0438	.4901	.30320	.95293	.31818	.1429	.0494	.2981	21
40	0.28680	0.95799	0.29938	3.3402	1.0438	3.4867	0.30348	0.95284	0.31850	3.1397	1.0495	3.2951	20
41	.28708	.95791	.29970	.3367	.0439	.4833	.30375	.95275	.31882	.1366	.0496	.2921	19
42	.28736	.95782	.30001	.3332	.0440	.4799	.30403	.95266	.31914	.1334	.0497	.2891	18
43	.28764	.95774	.30033	.3296	.0441	.4766	.30431	.95257	.31946	.1303	.0498	.2861	17
44	.28792	.95765	.30065	.3261	.0442	.4732	.30459	.95248	.31978	.1271	.0499	.2831	16
45	0.28820	0.95757	0.30096	3.3226	1.0443	3.4698	0.30486	0.95239	0.32010	3.1240	1.0500	3.2801	15
46	.28847	.95749	.30128	.3191	.0444	.4665	.30514	.95231	.32042	.1209	.0501	.2772	14
47	.28875	.95740	.30160	.3156	.0445	.4632	.30542	.95222	.32074	.1177	.0502	.2742	13
48	.28903	.95732	.30192	.3121	.0446	.4598	.30569	.95213	.32106	.1146	.0503	.2712	12
49	.28931	.95723	.30223	.3087	.0447	.4565	.30597	.95204	.32138	.1115	.0504	.2683	11
50	0.28959	0.95715	0.30255	3.3052	1.0448	3.4532	0.30625	0.95195	0.32171	3.1084	1.0505	3.2653	10
51	.28987	.95707	.30287	.3017	.0448	.4498	.30653	.95186	.32203	.1053	.0506	.2624	9
52	.29014	.95698	.30319	.2983	.0449	.4465	.30680	.95177	.32235	.1022	.0507	.2594	8
53	.29042	.95690	.30350	.2948	.0450	.4432	.30708	.95168	.32267	.0991	.0508	.2565	7
54	.29070	.95681	.30382	.2914	.0451	.4399	.30736	.95159	.32299	.0960	.0509	.2535	6
55	0.29098	0.95673	0.30414	3.2879	1.0452	3.4366	0.30763	0.95150	0.32331	3.0930	1.0510	3.2506	5
56	.29126	.95664	.30446	.2845	.0453	.4334	.30791	.95141	.32363	.0899	.0511	.2477	4
57	.29154	.95656	.30478	.2811	.0454	.4301	.30819	.95132	.32395	.0868	.0512	.2448	3
58	.29181	.95647	.30509	.2777	.0455	.4268	.30846	.95124	.32428	.0838	.0513	.2419	2
59	.29209	.95639	.30541	.2742	.0456	.4236	.30874	.95115	.32460	.0807	.0514	.2390	1
60	0.29237	0.95630	0.30573	3.2708	1.0457	3.4203	0.30902	0.95106	0.32492	3.0777	1.0515	3.2361	0
M	Cosine	Sine	Cotan.	Tan.	Cosec.	Secant	Cosine	Sine	Cotan.	Tan.	Cosec.	Secant	M
106°						73°	107°					72°	

TABLE: Natural Trigonometric functions, continued 5.6.8

18°						161°	19°						160°
M	Sine	Cosine	Tan.	Cotan.	Secant	Cosec.	Sine	Cosine	Tan.	Cotan.	Secant	Cosec.	M
0	0.30902	0.95106	0.32492	3.0777	1.0515	3.2361	0.32557	0.94552	0.34433	2.9042	1.0576	3.0715	60
1	.30929	.95097	.32524	.0746	.0516	.2332	.32584	.94542	.34465	.9015	.0577	.0690	59
2	.30957	.95088	.32556	.0716	.0517	.2303	.32612	.94533	.34498	.8987	.0578	.0664	58
3	.30985	.95079	.32588	.0686	.0518	.2274	.32639	.94523	.34530	.8960	.0579	.0638	57
4	.31012	.95070	.32621	.0655	.0519	.2245	.32667	.94514	.34563	.8933	.0580	.0612	56
5	0.31040	0.95061	0.32653	3.0625	1.0520	3.2216	0.32694	0.94504	0.34595	2.8905	1.0581	3.0586	55
6	.31068	.95051	.32685	.0595	.0521	.2188	.32722	.94495	.34628	.8878	.0582	.0561	54
7	.31095	.95042	.32717	.0565	.0522	.2159	.32749	.94485	.34661	.8851	.0584	.0535	53
8	.31123	.95033	.32749	.0535	.0523	.2131	.32777	.94476	.34693	.8824	.0585	.0509	52
9	.31150	.95024	.32782	.0505	.0524	.2102	.32804	.94466	.34726	.8797	.0586	.0484	51
10	0.31178	0.95015	0.32814	3.0475	1.0525	3.2074	0.32832	0.94457	0.34758	2.8770	1.0587	3.0458	50
11	.31206	.95006	.32846	.0445	.0526	.2045	.32859	.94447	.34791	.8743	.0588	.0433	49
12	.31233	.94997	.32878	.0415	.0527	.2017	.32887	.94438	.34824	.8716	.0589	.0407	48
13	.31261	.94988	.32910	.0385	.0528	.1989	.32914	.94428	.34856	.8689	.0590	.0382	47
14	.31289	.94979	.32943	.0356	.0529	.1960	.32942	.94418	.34889	.8662	.0591	.0357	46
15	0.31316	0.94970	0.32975	3.0326	1.0530	3.1932	0.32969	0.94409	0.34921	2.8636	1.0592	3.0331	45
16	.31344	.94961	.33007	.0296	.0531	.1904	.32996	.94399	.34954	.8609	.0593	.0306	44
17	.31372	.94952	.33039	.0267	.0532	.1876	.33024	.94390	.34987	.8582	.0594	.0281	43
18	.31399	.94942	.33072	.0237	.0533	.1848	.33051	.94380	.35019	.8555	.0595	.0256	42
19	.31427	.94933	.33104	.0208	.0534	.1820	.33079	.94370	.35052	.8529	.0596	.0231	41
20	0.31454	0.94924	0.33136	3.0178	1.0535	3.1792	0.33106	0.94361	0.35085	2.8502	1.0598	3.0206	40
21	.31482	.94915	.33169	.0149	.0536	.1764	.33134	.94351	.35117	.8476	.0599	.0181	39
22	.31510	.94906	.33201	.0120	.0537	.1736	.33161	.94341	.35150	.8449	.0600	.0156	38
23	.31537	.94897	.33233	.0090	.0538	.1708	.33189	.94332	.35183	.8423	.0601	.0131	37
24	.31565	.94888	.33265	.0061	.0539	.1681	.33216	.94322	.35215	.8396	.0602	.0106	36
25	0.31592	0.94878	0.33298	3.0032	1.0540	3.1653	0.33243	0.94313	0.35248	2.8370	1.0603	3.0081	35
26	.31620	.94869	.33330	.0003	.0541	.1625	.33271	.94303	.35281	.8344	.0604	.0056	34
27	.31648	.94860	.33362	2.9974	.0542	.1598	.33298	.94293	.35314	.8318	.0605	.0031	33
28	.31675	.94851	.33395	.9945	.0543	.1570	.33326	.94283	.35346	.8291	.0606	.0007	32
29	.31703	.94841	.33427	.9916	.0544	.1543	.33353	.94274	.35379	.8265	.0607	2.9982	31
30	0.31730	0.94832	0.33459	2.9887	1.0545	3.1515	0.33381	0.94264	0.35412	2.8239	1.0608	2.9957	30
31	.31758	.94823	.33492	.9858	.0546	.1488	.33408	.94254	.35445	.8213	.0609	.9933	29
32	.31786	.94814	.33524	.9829	.0547	.1461	.33435	.94245	.35477	.8187	.0611	.9908	28
33	.31813	.94805	.33557	.9800	.0548	.1433	.33463	.94235	.35510	.8161	.0612	.9884	27
34	.31841	.94795	.33589	.9772	.0549	.1406	.33490	.94225	.35543	.8135	.0613	.9859	26
35	0.31868	0.94786	0.33621	2.9743	1.0550	3.1379	0.33518	0.94215	0.35576	2.8109	1.0614	2.9835	25
36	.31896	.94777	.33654	.9714	.0551	.1352	.33545	.94206	.35608	.8083	.0615	.9810	24
37	.31923	.94767	.33686	.9686	.0552	.1325	.33572	.94196	.35641	.8057	.0616	.9786	23
38	.31951	.94758	.33718	.9657	.0553	.1298	.33600	.94186	.35674	.8032	.0617	.9762	22
39	.31978	.94749	.33751	.9629	.0554	.1271	.33627	.94176	.35707	.8006	.0618	.9738	21
40	0.32006	0.94740	0.33783	2.9600	1.0555	3.1244	0.33655	0.94167	0.35739	2.7980	1.0619	2.9713	20
41	.32034	.94730	.33816	.9572	.0556	.1217	.33682	.94157	.35772	.7954	.0620	.9689	19
42	.32061	.94721	.33848	.9544	.0557	.1190	.33709	.94147	.35805	.7929	.0622	.9665	18
43	.32089	.94712	.33880	.9515	.0558	.1163	.33737	.94137	.35838	.7903	.0623	.9641	17
44	.32116	.94702	.33913	.9487	.0559	.1137	.33764	.94127	.35871	.7878	.0624	.9617	16
45	0.32144	0.94693	0.33945	2.9459	1.0560	3.1110	0.33792	0.94118	0.35904	2.7852	1.0625	2.9593	15
46	.32171	.94684	.33978	.9431	.0561	.1083	.33819	.94108	.35936	.7827	.0626	.9569	14
47	.32199	.94674	.34010	.9403	.0562	.1057	.33846	.94098	.35969	.7801	.0627	.9545	13
48	.32226	.94665	.34043	.9375	.0563	.1030	.33874	.94088	.36002	.7776	.0628	.9521	12
49	.32254	.94655	.34075	.9347	.0565	.1004	.33901	.94078	.36035	.7751	.0629	.9497	11
50	0.32282	0.94646	0.34108	2.9319	1.0566	3.0977	0.33928	0.94068	0.36068	2.7725	1.0630	2.9474	10
51	.32309	.94637	.34140	.9291	.0567	.0951	.33956	.94058	.36101	.7700	.0632	.9450	9
52	.32337	.94627	.34173	.9263	.0568	.0925	.33983	.94049	.36134	.7675	.0633	.9426	8
53	.32364	.94618	.34205	.9235	.0569	.0898	.34011	.94039	.36167	.7650	.0634	.9402	7
54	.32392	.94608	.34238	.9208	.0570	.0872	.34038	.94029	.36199	.7625	.0635	.9379	6
55	0.32419	0.94599	0.34270	2.9180	1.0571	3.0846	0.34065	0.94019	0.36232	2.7600	1.0636	2.9355	5
56	.32447	.94590	.34303	.9152	.0572	.0820	.34093	.94009	.36265	.7575	.0637	.9332	4
57	.32474	.94580	.34335	.9125	.0573	.0793	.34120	.93999	.36298	.7550	.0638	.9308	3
58	.32502	.94571	.34368	.9097	.0574	.0767	.34147	.93989	.36331	.7525	.0639	.9285	2
59	.32529	.94561	.34400	.9069	.0575	.0741	.34175	.93979	.36364	.7500	.0641	.9261	1
60	0.32557	0.94552	0.34433	2.9042	1.0576	3.0715	0.34202	0.93969	0.36397	2.7475	1.0642	2.9238	0
M	Cosine	Sine	Cotan.	Tan.	Cosec.	Secant	Cosine	Sine	Cotan.	Tan.	Cosec.	Secant	M
108°						71°	109°						70°

TABLE: Natural Trigonometric functions, continued 5.6.8

20°						159°	21°					158°	
M	Sine	Cosine	Tan.	Cotan.	Secant	Cosec.	Sine	Cosine	Tan.	Cotan.	Secant	Cosec.	M
0	0.34202	0.93969	0.36397	2.7475	1.0642	2.9238	0.35837	0.93358	0.38386	2.6051	1.0711	2.7904	60
1	.34229	.93959	.36430	.7450	.0643	.9215	.35864	.93348	.38420	.6028	.0713	.7883	59
2	.34257	.93949	.36463	.7425	.0644	.9191	.35891	.93337	.38453	.6006	.0714	.7862	58
3	.34284	.93939	.36496	.7400	.0645	.9168	.35918	.93327	.38486	.5983	.0715	.7841	57
4	.34311	.93929	.36529	.7376	.0646	.9145	.35945	.93316	.38520	.5960	.0716	.7820	56
5	0.34339	0.93919	0.36562	2.7351	1.0647	2.9122	0.35972	0.93306	0.38553	2.5938	1.0717	2.7799	55
6	.34366	.93909	.36595	.7326	.0648	.9098	.36000	.93295	.38587	.5916	.0719	.7778	54
7	.34393	.93899	.36628	.7302	.0650	.9075	.36027	.93285	.38620	.5893	.0720	.7757	53
8	.34421	.93889	.36661	.7277	.0651	.9052	.36054	.93274	.38654	.5871	.0721	.7736	52
9	.34448	.93879	.36694	.7252	.0652	.9029	.36081	.93264	.38687	.5848	.0722	.7715	51
10	0.34475	0.93869	0.36727	2.7228	1.0653	2.9006	0.36108	0.93253	0.38720	2.5826	1.0723	2.7694	50
11	.34502	.93859	.36760	.7204	.0654	.8983	.36135	.93243	.38754	.5804	.0725	.7674	49
12	.34530	.93849	.36793	.7179	.0655	.8960	.36162	.93232	.38787	.5781	.0726	.7653	48
13	.34557	.93839	.36826	.7155	.0656	.8937	.36189	.93222	.38821	.5759	.0727	.7632	47
14	.34584	.93829	.36859	.7130	.0658	.8915	.36217	.93211	.38854	.5737	.0728	.7611	46
15	0.34612	0.93819	0.36892	2.7106	1.0659	2.8892	0.36244	0.93201	0.38888	2.5715	1.0729	2.7591	45
16	.34639	.93809	.36925	.7082	.0660	.8869	.36271	.93190	.38921	.5693	.0731	.7570	44
17	.34666	.93799	.36958	.7058	.0661	.8846	.36298	.93180	.38955	.5671	.0732	.7550	43
18	.34693	.93789	.36991	.7033	.0662	.8824	.36325	.93169	.38988	.5649	.0733	.7529	42
19	.34721	.93779	.37024	.7009	.0663	.8801	.36352	.93158	.39022	.5627	.0734	.7509	41
20	0.34748	0.93769	0.37057	2.6985	1.0664	2.8778	0.36379	0.93148	0.39055	2.5605	1.0736	2.7488	40
21	.34775	.93758	.37090	.6961	.0666	.8756	.36406	.93137	.39089	.5583	.0737	.7468	39
22	.34803	.93748	.37123	.6937	.0667	.8733	.36433	.93127	.39122	.5561	.0738	.7447	38
23	.34830	.93738	.37156	.6913	.0668	.8711	.36460	.93116	.39156	.5539	.0739	.7427	37
24	.34857	.93728	.37190	.6889	.0669	.8688	.36488	.93105	.39189	.5517	.0740	.7406	36
25	0.34884	0.93718	0.37223	2.6865	1.0670	2.8666	0.36515	0.93095	0.39223	2.5495	1.0742	2.7386	35
26	.34912	.93708	.37256	.6841	.0671	.8644	.36542	.93084	.39257	.5473	.0743	.7366	34
27	.34939	.93698	.37289	.6817	.0673	.8621	.36569	.93074	.39290	.5451	.0744	.7346	33
28	.34966	.93687	.37322	.6794	.0674	.8599	.36596	.93063	.39324	.5430	.0745	.7325	32
29	.34993	.93677	.37355	.6770	.0675	.8577	.36623	.93052	.39357	.5408	.0747	.7305	31
30	0.35021	0.93667	0.37388	2.6746	1.0676	2.8554	0.36650	0.93042	0.39391	2.5386	1.0748	2.7285	30
31	.35048	.93657	.37422	.6722	.0677	.8532	.36677	.93031	.39425	.5365	.0749	.7265	29
32	.35075	.93647	.37455	.6699	.0678	.8510	.36704	.93020	.39458	.5343	.0750	.7245	28
33	.35102	.93637	.37488	.6675	.0679	.8488	.36731	.93010	.39492	.5322	.0751	.7225	27
34	.35130	.93626	.37521	.6652	.0681	.8466	.36758	.92999	.39525	.5300	.0753	.7205	26
35	0.35157	0.93616	0.37554	2.6628	1.0682	2.8444	0.36785	0.92988	0.39559	2.5278	1.0754	2.7185	25
36	.35184	.93606	.37587	.6604	.0683	.8422	.36812	.92978	.39593	.5257	.0755	.7165	24
37	.35211	.93596	.37621	.6581	.0684	.8400	.36839	.92967	.39626	.5236	.0756	.7145	23
38	.35239	.93585	.37654	.6558	.0685	.8378	.36866	.92956	.39660	.5214	.0758	.7125	22
39	.35266	.93575	.37687	.6534	.0686	.8356	.36893	.92945	.39694	.5193	.0759	.7105	21
40	0.35293	0.93565	0.37720	2.6511	1.0688	2.8334	0.36921	0.92935	0.39727	2.5171	1.0760	2.7085	20
41	.35320	.93555	.37754	.6487	.0689	.8312	.36948	.92924	.39761	.5150	.0761	.7065	19
42	.35347	.93544	.37787	.6464	.0690	.8290	.36975	.92913	.39795	.5129	.0763	.7045	18
43	.35375	.93534	.37820	.6441	.0691	.8269	.37002	.92902	.39828	.5108	.0764	.7026	17
44	.35402	.93524	.37853	.6418	.0692	.8247	.37029	.92892	.39862	.5086	.0765	.7006	16
45	0.35429	0.93513	0.37887	2.6394	1.0694	2.8225	0.37056	0.92881	0.39896	2.5065	1.0766	2.6986	15
46	.35456	.93503	.37920	.6371	.0695	.8204	.37083	.92870	.39930	.5044	.0768	.6967	14
47	.35483	.93493	.37953	.6348	.0696	.8182	.37110	.92859	.39963	.5023	.0769	.6947	13
48	.35511	.93482	.37986	.6325	.0697	.8160	.37137	.92848	.39997	.5002	.0770	.6927	12
49	.35538	.93472	.38020	.6302	.0698	.8139	.37164	.92838	.40031	.4981	.0771	.6908	11
50	0.35565	0.93462	0.38053	2.6279	1.0699	2.8117	0.37191	0.92827	0.40065	2.4960	1.0773	2.6888	10
51	.35592	.93451	.38086	.6256	.0701	.8096	.37218	.92816	.40098	.4939	.0774	.6869	9
52	.35619	.93441	.38120	.6233	.0702	.8074	.37245	.92805	.40132	.4918	.0775	.6849	8
53	.35647	.93431	.38153	.6210	.0703	.8053	.37272	.92794	.40166	.4897	.0776	.6830	7
54	.35674	.93420	.38186	.6187	.0704	.8032	.37299	.92784	.40200	.4876	.0778	.6810	6
55	0.35701	0.93410	0.38220	2.6164	1.0705	2.8010	0.37326	0.92773	0.40233	2.4855	1.0779	2.6791	5
56	.35728	.93400	.38253	.6142	.0707	.7989	.37353	.92762	.40267	.4834	.0780	.6772	4
57	.35755	.93389	.38286	.6119	.0708	.7968	.37380	.92751	.40301	.4813	.0781	.6752	3
58	.35782	.93379	.38320	.6096	.0709	.7947	.37407	.92740	.40335	.4792	.0783	.6733	2
59	.35810	.93368	.38353	.6073	.0710	.7925	.37434	.92729	.40369	.4772	.0784	.6714	1
60	0.35837	0.93358	0.38386	2.6051	1.0711	2.7904	0.37461	0.92718	0.40403	2.4751	1.0785	2.6695	0
M	Cosine	Sine	Cotan.	Tan.	Cosec.	Secant	Cosine	Sine	Cotan.	Tan.	Cosec.	Secant	M
110°						69°	111°					68°	

TABLE: Natural Trigonometric functions, continued 5.6.8

22°						157°	23°						156°
M	Sine	Cosine	Tan.	Cotan.	Secant	Cosec.	Sine	Cosine	Tan.	Cotan.	Secant	Cosec.	M
0	0.37461	0.92718	0.40403	2.4751	1.0785	2.6695	0.39073	0.92050	0.42447	2.3558	1.0864	2.5593	60
1	.37488	.92707	.40436	.4730	.0787	.6675	.39100	.92039	.42482	.3539	.0865	.5575	59
2	.37514	.92696	.40470	.4709	.0788	.6656	.39126	.92028	.42516	.3520	.0866	.5558	58
3	.37541	.92686	.40504	.4689	.0789	.6637	.39153	.92016	.42550	.3501	.0868	.5540	57
4	.37568	.92675	.40538	.4668	.0790	.6618	.39180	.92005	.42585	.3482	.0869	.5523	56
5	0.37595	0.92664	0.40572	2.4647	1.0792	2.6599	0.39207	0.91993	0.42619	2.3463	1.0870	2.5506	55
6	.37622	.92653	.40606	.4627	.0793	.6580	.39234	.91982	.42654	.3445	.0872	.5488	54
7	.37649	.92642	.40640	.4606	.0794	.6561	.39260	.91971	.42688	.3426	.0873	.5471	53
8	.37676	.92631	.40673	.4586	.0795	.6542	.39287	.91959	.42722	.3407	.0874	.5453	52
9	.37703	.92620	.40707	.4565	.0797	.6523	.39314	.91948	.42757	.3388	.0876	.5436	51
10	0.37730	0.92609	0.40741	2.4545	1.0798	2.6504	0.39341	0.91936	0.42791	2.3369	1.0877	2.5419	50
11	.37757	.92598	.40775	.4525	.0799	.6485	.39367	.91925	.42826	.3350	.0878	.5402	49
12	.37784	.92587	.40809	.4504	.0801	.6466	.39394	.91913	.42860	.3332	.0880	.5384	48
13	.37811	.92576	.40843	.4484	.0802	.6447	.39421	.91902	.42894	.3313	.0881	.5367	47
14	.37838	.92565	.40877	.4463	.0803	.6428	.39448	.91891	.42929	.3294	.0882	.5350	46
15	0.37865	0.92554	0.40911	2.4443	1.0804	2.6410	0.39474	0.91879	0.42963	2.3276	1.0884	2.5333	45
16	.37892	.92543	.40945	.4423	.0806	.6391	.39501	.91868	.42998	.3257	.0885	.5316	44
17	.37919	.92532	.40979	.4403	.0807	.6372	.39528	.91856	.43032	.3238	.0886	.5299	43
18	.37946	.92521	.41013	.4382	.0808	.6353	.39554	.91845	.43067	.3220	.0888	.5281	42
19	.37972	.92510	.41047	.4362	.0810	.6335	.39581	.91833	.43101	.3201	.0889	.5264	41
20	0.37999	0.92499	0.41081	2.4342	1.0811	2.6316	0.39608	0.91822	0.43136	2.3183	1.0891	2.5247	40
21	.38026	.92488	.41115	.4322	.0812	.6297	.39635	.91810	.43170	.3164	.0892	.5230	39
22	.38053	.92477	.41149	.4302	.0813	.6279	.39661	.91798	.43205	.3145	.0893	.5213	38
23	.38080	.92466	.41183	.4282	.0815	.6260	.39688	.91787	.43239	.3127	.0895	.5196	37
24	.38107	.92455	.41217	.4262	.0816	.6242	.39715	.91775	.43274	.3109	.0896	.5179	36
25	0.38134	0.92443	0.41251	2.4242	1.0817	2.6223	0.39741	0.91764	0.43308	2.3090	1.0897	2.5163	35
26	.38161	.92432	.41285	.4222	.0819	.6205	.39768	.91752	.43343	.3072	.0899	.5146	34
27	.38188	.92421	.41319	.4202	.0820	.6186	.39795	.91741	.43377	.3053	.0900	.5129	33
28	.38214	.92410	.41353	.4182	.0821	.6168	.39821	.91729	.43412	.3035	.0902	.5112	32
29	.38241	.92399	.41387	.4162	.0823	.6150	.39848	.91718	.43447	.3017	.0903	.5095	31
30	0.38268	0.92388	0.41421	2.4142	1.0824	2.6131	0.39875	0.91706	0.43481	2.2998	1.0904	2.5078	30
31	.38295	.92377	.41455	.4122	.0825	.6113	.39901	.91694	.43516	.2980	.0906	.5062	29
32	.38322	.92366	.41489	.4102	.0826	.6095	.39928	.91683	.43550	.2962	.0907	.5045	28
33	.38349	.92354	.41524	.4083	.0828	.6076	.39955	.91671	.43585	.2944	.0908	.5028	27
34	.38376	.92343	.41558	.4063	.0829	.6058	.39981	.91659	.43620	.2925	.0910	.5011	26
35	0.38403	0.92332	0.41592	2.4043	1.0830	2.6040	0.40008	0.91648	0.43654	2.2907	1.0911	2.4995	25
36	.38429	.92321	.41626	.4023	.0832	.6022	.40035	.91636	.43689	.2889	.0913	.4978	24
37	.38456	.92310	.41660	.4004	.0833	.6003	.40061	.91625	.43723	.2871	.0914	.4961	23
38	.38483	.92299	.41694	.3984	.0834	.5985	.40088	.91613	.43758	.2853	.0915	.4945	22
39	.38510	.92287	.41728	.3964	.0836	.5967	.40115	.91601	.43793	.2835	.0917	.4928	21
40	0.38537	0.92276	0.41762	2.3945	1.0837	2.5949	0.40141	0.91590	0.43827	2.2817	1.0918	2.4912	20
41	.38564	.92265	.41797	.3925	.0838	.5931	.40168	.91578	.43862	.2799	.0920	.4895	19
42	.38591	.92254	.41831	.3906	.0840	.5913	.40195	.91566	.43897	.2781	.0921	.4879	18
43	.38617	.92242	.41865	.3886	.0841	.5895	.40221	.91554	.43932	.2763	.0922	.4862	17
44	.38644	.92231	.41899	.3867	.0842	.5877	.40248	.91543	.43966	.2745	.0924	.4846	16
45	0.38671	0.92220	0.41933	2.3847	1.0844	2.5859	0.40275	0.91531	0.44001	2.2727	1.0925	2.4829	15
46	.38698	.92209	.41968	.3828	.0845	.5841	.40301	.91519	.44036	.2709	.0927	.4813	14
47	.38725	.92197	.42002	.3808	.0846	.5823	.40328	.91508	.44070	.2691	.0928	.4797	13
48	.38751	.92186	.42036	.3789	.0847	.5805	.40354	.91496	.44105	.2673	.0929	.4780	12
49	.38778	.92175	.42070	.3770	.0849	.5787	.40381	.91484	.44140	.2655	.0931	.4764	11
50	0.38805	0.92164	0.42105	2.3750	1.0850	2.5770	0.40408	0.91472	0.44175	2.2637	1.0932	2.4748	10
51	.38832	.92152	.42139	.3731	.0851	.5752	.40434	.91461	.44209	.2619	.0934	.4731	9
52	.38859	.92141	.42173	.3712	.0853	.5734	.40461	.91449	.44244	.2602	.0935	.4715	8
53	.38886	.92130	.42207	.3692	.0854	.5716	.40487	.91437	.44279	.2584	.0936	.4699	7
54	.38912	.92118	.42242	.3673	.0855	.5699	.40514	.91425	.44314	.2566	.0938	.4683	6
55	0.38939	0.92107	0.42276	2.3654	1.0857	2.5681	0.40541	0.91414	0.44349	2.2548	1.0939	2.4666	5
56	.38966	.92096	.42310	.3635	.0858	.5663	.40567	.91402	.44383	.2531	.0941	.4650	4
57	.38993	.92084	.42344	.3616	.0859	.5646	.40594	.91390	.44418	.2513	.0942	.4634	3
58	.39019	.92073	.42379	.3597	.0861	.5628	.40620	.91378	.44453	.2495	.0943	.4618	2
59	.39046	.92062	.42413	.3577	.0862	.5610	.40647	.91366	.44488	.2478	.0945	.4602	1
60	0.39073	0.92050	0.42447	2.3558	1.0864	2.5593	0.40674	0.91354	0.44523	2.2460	1.0946	2.4586	0
M	Cosine	Sine	Cotan.	Tan.	Cosec.	Secant	Cosine	Sine	Cotan.	Tan.	Cosec.	Secant	M
112°						67°	113°						66°

TABLE: Natural Trigonometric functions, continued 5.6.8

24°						155°	25°						154°
M	Sine	Cosine	Tan.	Cotan.	Secant	Cosec.	Sine	Cosine	Tan.	Cotan.	Secant	Cosec.	M
0	0.40674	0.91354	0.44523	2.2460	1.0946	2.4586	0.42262	0.90631	0.46631	2.1445	1.1034	2.3662	60
1	.40700	.91343	.44558	.2443	.0948	.4570	.42288	.90618	.46666	.1429	.1035	.3647	59
2	.40727	.91331	.44593	.2425	.0949	.4554	.42314	.90606	.46702	.1412	.1037	.3632	58
3	.40753	.91319	.44627	.2408	.0951	.4538	.42341	.90594	.46737	.1396	.1038	.3618	57
4	.40780	.91307	.44662	.2390	.0952	.4522	.42367	.90581	.46772	.1380	.1040	.3603	56
5	0.40806	0.91295	0.44697	2.2373	1.0953	2.4506	0.42394	0.90569	0.46808	2.1364	1.1041	2.3588	55
6	.40833	.91283	.44732	.2355	.0955	.4490	.42420	.90557	.46843	.1348	.1043	.3574	54
7	.40860	.91271	.44767	.2338	.0956	.4474	.42446	.90544	.46879	.1331	.1044	.3559	53
8	.40886	.91260	.44802	.2320	.0958	.4458	.42473	.90532	.46914	.1315	.1046	.3544	52
9	.40913	.91248	.44837	.2303	.0959	.4442	.42499	.90520	.46950	.1299	.1047	.3530	51
10	0.40939	0.91236	0.44872	2.2286	1.0961	2.4426	0.42525	0.90507	0.46985	2.1283	1.1049	2.3515	50
11	.40966	.91224	.44907	.2268	.0962	.4411	.42552	.90495	.47021	.1267	.1050	.3501	49
12	.40992	.91212	.44942	.2251	.0963	.4395	.42578	.90483	.47056	.1251	.1052	.3486	48
13	.41019	.91200	.44977	.2234	.0965	.4379	.42604	.90470	.47092	.1235	.1053	.3472	47
14	.41045	.91188	.45012	.2216	.0966	.4363	.42630	.90458	.47127	.1219	.1055	.3457	46
15	0.41072	0.91176	0.45047	2.2199	1.0968	2.4347	0.42657	0.90445	0.47163	2.1203	1.1056	2.3443	45
16	.41098	.91164	.45082	.2182	.0969	.4332	.42683	.90433	.47199	.1187	.1058	.3428	44
17	.41125	.91152	.45117	.2165	.0971	.4316	.42709	.90421	.47234	.1171	.1059	.3414	43
18	.41151	.91140	.45152	.2147	.0972	.4300	.42736	.90408	.47270	.1155	.1061	.3399	42
19	.41178	.91128	.45187	.2130	.0973	.4285	.42762	.90396	.47305	.1139	.1062	.3385	41
20	0.41204	0.91116	0.45222	2.2113	1.0975	2.4269	0.42788	0.90383	0.47341	2.1123	1.1064	2.3371	40
21	.41231	.91104	.45257	.2096	.0976	.4254	.42815	.90371	.47376	.1107	.1065	.3356	39
22	.41257	.91092	.45292	.2079	.0978	.4238	.42841	.90358	.47412	.1092	.1067	.3342	38
23	.41284	.91080	.45327	.2062	.0979	.4222	.42867	.90346	.47448	.1076	.1068	.3328	37
24	.41310	.91068	.45362	.2045	.0981	.4207	.42893	.90333	.47483	.1060	.1070	.3313	36
25	0.41337	0.91056	0.45397	2.2028	1.0982	2.4191	0.42920	0.90321	0.47519	2.1044	1.1072	2.3299	35
26	.41363	.91044	.45432	.2011	.0984	.4176	.42946	.90308	.47555	.1028	.1073	.3285	34
27	.41390	.91032	.45467	.1994	.0985	.4160	.42972	.90296	.47590	.1013	.1075	.3271	33
28	.41416	.91020	.45502	.1977	.0986	.4145	.42998	.90283	.47626	.0997	.1076	.3256	32
29	.41443	.91008	.45537	.1960	.0988	.4130	.43025	.90271	.47662	.0981	.1078	.3242	31
30	0.41469	0.90996	0.45573	2.1943	1.0989	2.4114	0.43051	0.90258	0.47697	2.0965	1.1079	2.3228	30
31	.41496	.90984	.45608	.1926	.0991	.4099	.43077	.90246	.47733	.0950	.1081	.3214	29
32	.41522	.90972	.45643	.1909	.0992	.4083	.43104	.90233	.47769	.0934	.1082	.3200	28
33	.41549	.90960	.45678	.1892	.0994	.4068	.43130	.90221	.47805	.0918	.1084	.3186	27
34	.41575	.90948	.45713	.1875	.0995	.4053	.43156	.90208	.47840	.0903	.1085	.3172	26
35	0.41602	0.90936	0.45748	2.1859	1.0997	2.4037	0.43182	0.90196	0.47876	2.0887	1.1087	2.3158	25
36	.41628	.90924	.45783	.1842	.0998	.4022	.43208	.90183	.47912	.0872	.1088	.3143	24
37	.41654	.90911	.45819	.1825	.1000	.4007	.43235	.90171	.47948	.0856	.1090	.3129	23
38	.41681	.90899	.45854	.1808	.1001	.3992	.43261	.90158	.47983	.0840	.1092	.3115	22
39	.41707	.90887	.45889	.1792	.1003	.3976	.43287	.90145	.48019	.0825	.1093	.3101	21
40	0.41734	0.90875	0.45924	2.1775	1.1004	2.3961	0.43313	0.90133	0.48055	2.0809	1.1095	2.3087	20
41	.41760	.90863	.45960	.1758	.1005	.3946	.43340	.90120	.48091	.0794	.1096	.3073	19
42	.41787	.90851	.45995	.1741	.1007	.3931	.43366	.90108	.48127	.0778	.1098	.3059	18
43	.41813	.90839	.46030	.1725	.1008	.3916	.43392	.90095	.48162	.0763	.1099	.3046	17
44	.41839	.90826	.46065	.1708	.1010	.3901	.43418	.90082	.48198	.0747	.1101	.3032	16
45	0.41866	0.90814	0.46101	2.1692	1.1011	2.3886	0.43444	0.90070	0.48234	2.0732	1.1102	2.3018	15
46	.41892	.90802	.46136	.1675	.1013	.3871	.43471	.90057	.48270	.0717	.1104	.3004	14
47	.41919	.90790	.46171	.1658	.1014	.3856	.43497	.90044	.48306	.0701	.1106	.2990	13
48	.41945	.90778	.46206	.1642	.1016	.3841	.43523	.90032	.48342	.0686	.1107	.2976	12
49	.41972	.90765	.46242	.1625	.1017	.3826	.43549	.90019	.48378	.0671	.1109	.2962	11
50	0.41998	0.90753	0.46277	2.1609	1.1019	2.3811	0.43575	0.90006	0.48414	2.0655	1.1110	2.2949	10
51	.42024	.90741	.46312	.1592	.1020	.3796	.43602	.89994	.48449	.0640	.1112	.2935	9
52	.42051	.90729	.46348	.1576	.1022	.3781	.43628	.89981	.48485	.0625	.1113	.2921	8
53	.42077	.90717	.46383	.1559	.1023	.3766	.43654	.89968	.48521	.0609	.1115	.2907	7
54	.42103	.90704	.46418	.1543	.1025	.3751	.43680	.89956	.48557	.0594	.1116	.2894	6
55	0.42130	0.90692	0.46454	2.1527	1.1026	2.3736	0.43706	0.89943	0.48593	2.0579	1.1118	2.2880	5
56	.42156	.90680	.46489	.1510	.1028	.3721	.43732	.89930	.48629	.0564	.1120	.2866	4
57	.42183	.90668	.46524	.1494	.1029	.3706	.43759	.89918	.48665	.0548	.1121	.2853	3
58	.42209	.90655	.46560	.1478	.1031	.3691	.43785	.89905	.48701	.0533	.1123	.2839	2
59	.42235	.90643	.46595	.1461	.1032	.3677	.43811	.89892	.48737	.0518	.1124	.2825	1
60	0.42262	0.90631	0.46631	2.1445	1.1034	2.3662	0.43837	0.89879	0.48773	2.0503	1.1126	2.2812	0
M	Cosine	Sine	Cotan.	Tan.	Cosec.	Secant	Cosine	Sine	Cotan.	Tan.	Cosec.	Secant	M
114°						65°	115°						64°

TABLE: Natural Trigonometric functions, continued 5.6.8

26°						153°	27°						152°	
M	Sine	Cosine	Tan.	Cotan.	Secant	Cosec.	Sine	Cosine	Tan.	Cotan.	Secant	Cosec.	M	
0	0.43837	0.89879	0.48773	2.0503	1.1126	2.2812	0.45399	0.89101	0.50952	1.9626	1.1223	2.2027	60	
1	.43863	.89867	.48809	.0488	.1127	.2798	.45425	.89087	.50989	9612	.1225	.2014	59	
2	.43889	.89854	.48845	.0473	.1129	.2784	.45451	89074	.51026	.9598	.1226	.2002	58	
3	.43915	.89841	.48881	.0458	.1131	.2771	.45477	89061	.51062	.9584	1228	.1989	57	
4	.43942	.89828	.48917	0443	.1132	.2757	.45503	.89048	.51099	.9570	.1230	.1977	56	
5	0.43968	0.89815	0.48953	2.0427	1.1134	2.2744	0.45528	0.89034	0.51136	1.9556	1.1231	2.1964	55	
6	.43994	.89803	.48989	.0412	1135	.2730	.45554	.89021	.51172	.9542	.1233	.1952	54	
7	.44020	.89790	.49025	.0397	.1137	.2717	.45580	.89008	.51209	.9528	.1235	.1939	53	
8	.44046	89777	.49062	.0382	1139	.2703	.45606	.88995	.51246	.9514	.1237	.1927	52	
9	.44072	.89764	.49098	.0367	.1140	.2690	45632	.88981	.51283	.9500	.1238	.1914	51	
10	0.44098	0.89751	0.49134	2.0352	1.1142	2.2676	0 45658	0.88968	0.51319	1.9486	1.1240	2.1902	50	
11	44124	.89739	.49170	.0338	.1143	.2663	.45684	.88955	.51356	.9472	.1242	.1889	49	
12	.44150	.89726	.49206	.0323	.1145	.2650	.45710	.88942	.51393	.9458	.1243	.1877	48	
13	.44177	.89713	.49242	.0308	.1147	.2636	.45736	.88928	.51430	.9444	.1245	.1865	47	
14	.44203	.89700	.49278	.0293	.1148	.2623	.45761	.88915	.51466	.9430	.1247	.1852	46	
15	0.44229	0.89687	0.49314	2.0278	1.1150	2.2610	0.45787	0.88902	0.51503	1.9416	1.1248	2.1840	45	
16	.44255	.89674	.49351	.0263	.1151	.2596	.45813	.88888	.51540	.9402	.1250	.1828	44	
17	.44281	.89661	.49387	.0248	.1153	.2583	.45839	.88875	.51577	.9388	.1252	.1815	43	
18	.44307	.89649	.49423	.0233	.1155	.2570	.45865	.88862	.51614	.9375	.1253	.1803	42	
19	.44333	.89636	49459	.0219	.1156	.2556	.45891	.88848	.51651	.9361	.1255	.1791	41	
20	0.44359	0.89623	0.49495	2.0204	1 1158	2.2543	0.45917	0.88835	0.51687	1.9347	1.1257	2.1778	40	
21	.44385	.89610	.49532	.0189	.1159	.2530	.45942	.88822	.51724	.9333	.1258	.1766	39	
22	.44411	.89597	.49568	.0174	.1161	.2517	.45968	.88808	.51761	.9319	.1260	.1754	38	
23	.44437	89584	.49604	.0159	.1163	.2503	.45994	.88795	.51798	.9306	.1262	.1742	37	
24	.44463	.89571	.49640	.0145	.1164	.2490	.46020	.88781	.51835	.9292	1264	.1730	36	
25	0.44489	0.89558	0.49677	2.0130	1.1166	2.2477	0.46046	0.88768	0.51872	1.9278	1.1265	2.1717	35	
26	.44516	.89545	.49713	.0115	.1167	.2464	.46072	.88755	.51909	.9264	.1267	.1705	34	
27	44542	.89532	.49749	.0101	.1169	.2451	.46097	.88741	.51946	.9251	.1269	.1693	33	
28	.44568	.89519	.49785	.0086	.1171	.2438	.46123	.88728	.51983	.9237	.1270	.1681	32	
29	.44594	.89506	.49822	.0071	.1172	.2425	.46149	.88714	.52020	.9223	.1272	.1669	31	
30	0.44620	0.89493	0 49858	2.0057	1.1174	2.2411	0.46175	0.88701	0.52057	1.9210	1.1274	2.1657	30	
31	.44646	89480	.49894	.0042	.1176	.2398	.46201	.88688	.52094	.9196	.1275	.1645	29	
32	.44672	.89467	.49931	.0028	.1177	.2385	.46226	.88674	.52131	.9182	.1277	.1633	28	
33	.44698	.89454	.49967	.0013	.1179	.2372	.46252	.88661	.52168	.9169	.1279	.1620	27	
34	.44724	.89441	50003	1.9998	.1180	.2359	.46278	.88647	.52205	.9155	.1281	.1608	26	
35	0.44750	0.89428	0 50040	1 9984	1.1182	2.2346	0.46304	0.88634	0.52242	1.9142	1.1282	2.1596	25	
36	.44776	.89415	.50076	.9969	.1184	.2333	.46330	.88620	.52279	.9128	.1284	.1584	24	
37	.44802	.89402	.50113	.9955	.1185	.2320	.46355	.88607	.52316	.9115	.1286	.1572	23	
38	.44828	.89389	.50149	.9940	.1187	.2307	.46381	.88593	.52353	.9101	.1287	.1560	22	
39	.44854	.89376	.50185	.9926	.1189	.2294	.46407	88580	.52390	.9088	.1289	.1548	21	
40	0.44880	0.89363	0.50222	1.9912	1.1190	2.2282	0.46433	0.88566	0.52427	1 9074	1.1291	2.1536	20	
41	.44906	.89350	.50258	.9897	.1192	.2269	.46458	.88553	.52464	.9061	.1293	.1525	19	
42	44932	.89337	.50295	.9883	.1193	.2256	.46484	.88539	.52501	.9047	.1294	.1513	18	
43	.44958	.89324	.50331	.9868	.1195	.2243	.46510	.88526	.52538	.9034	.1296	.1501	17	
44	.44984	.89311	.50368	.9854	.1197	.2230	46536	.88512	.52575	.9020	.1298	.1489	16	
45	0.45010	0 89298	0.50404	1 9840	1.1198	2.2217	0.46561	0.88499	0.52612	1.9007	1.1299	2.1477	15	
46	.45036	.89285	.50441	.9825	.1200	.2204	46587	.88485	.52650	8993	.1301	.1465	14	
47	.45062	.89272	.50477	.9811	.1202	.2192	.46613	.88472	.52687	.8980	.1303	.1453	13	
48	.45088	.89258	.50514	.9797	.1203	.2179	.46639	.88458	.52724	.8967	.1305	.1441	12	
49	.45114	.89245	.50550	.9782	.1205	.2166	.46664	.88444	.52761	.8953	.1306	.1430	11	
50	0.45140	0.89232	0 50587	1 9768	1.1207	2.2153	0.46690	0.88431	0.52798	1.8940	1.1308	2.1418	10	
51	.45166	.89219	.50623	.9754	.1208	.2141	46716	.88417	.52836	.8927	.1310	.1406	9	
52	.45191	.89206	.50660	.9739	.1210	.2128	.46741	.88404	.52873	.8913	.1312	.1394	8	
53	.45217	.89193	.50696	.9725	.1212	.2115	46767	.88390	.52910	.8900	.1313	.1382	7	
54	.45243	.89180	.50733	.9711	.1213	.2103	.46793	.88376	.52947	.8887	.1315	.1371	6	
55	0.45269	0.89166	0.50769	1.9697	1.1215	2.2090	0.46819	0.88363	0.52984	1.8873	1.1317	2.1359	5	
56	.45295	.89153	.50806	.9683	.1217	.2077	.46844	.88349	.53022	.8860	.1319	.1347	4	
57	.45321	.89140	.50843	.9668	.1218	.2065	.46870	.88336	.53059	.8847	.1320	.1335	3	
58	.45347	.89127	.50879	.9654	.1220	.2052	.46896	.88322	.53096	.8834	.1322	.1324	2	
59	.45373	.89114	.50916	.9640	.1222	.2039	.46921	.88308	.53134	.8820	.1324	.1312	1	
60	0.45399	0.89101	0.50952	1.9626	1.1223	2.2027	0.46947	0.88295	0.53171	1.8807	1.1326	2.1300	0	
M	Cosine	Sine	Cotan.	Tan.	Cosec.	Secant	Cosine	Sine	Cotan.	Tan.	Cosec.	Secant	M	
116°						63°	117°						62°	

TABLE: Natural Trigonometric functions, continued 5.6.8

28°						151°	29°						150°
M	Sine	Cosine	Tan.	Cotan.	Secant	Cosec.	Sine	Cosine	Tan.	Cotan.	Secant	Cosec.	M
0	0.46947	0.88295	0.53171	1.8807	1.1326	2.1300	0.48481	0.87462	0.55431	1.8040	1.1433	2.0627	60
1	.46973	.88281	.53208	.8794	.1327	.1289	.48506	.87448	.55469	.8028	.1435	.0616	59
2	.46998	.88267	.53245	.8781	.1329	.1277	.48532	.87434	.55507	.8016	.1437	.0605	58
3	.47024	.88254	.53283	.8768	.1331	.1266	.48557	.87420	.55545	.8003	.1439	.0594	57
4	.47050	.88240	.53320	.8754	.1333	.1254	.48583	.87405	.55583	.7991	.1441	.0583	56
5	0.47075	0.88226	0.53358	1.8741	1.1334	2.1242	0.48608	0.87391	0.55621	1.7979	1.1443	2.0573	55
6	.47101	.88213	.53395	.8728	.1336	.1231	.48633	.87377	.55659	.7966	.1445	.0562	54
7	.47127	.88199	.53432	.8715	.1338	.1219	.48659	.87363	.55697	.7954	.1446	.0551	53
8	.47152	.88185	.53470	.8702	.1340	.1208	.48684	.87349	.55735	.7942	.1448	.0540	52
9	.47178	.88171	.53507	.8689	.1341	.1196	.48710	.87335	.55774	.7930	.1450	.0530	51
10	0.47204	0.88158	0.53545	1.8676	1.1343	2.1185	0.48735	0.87320	0.55812	1.7917	1.1452	2.0519	50
11	.47229	.88144	.53582	.8663	.1345	.1173	.48760	.87306	.55850	.7905	.1454	.0508	49
12	.47255	.88130	.53619	.8650	.1347	.1162	.48786	.87292	.55888	.7893	.1456	.0498	48
13	.47281	.88117	.53657	.8637	.1349	.1150	.48811	.87278	.55926	.7881	.1458	.0487	47
14	.47306	.88103	.53694	.8624	.1350	.1139	.48837	.87264	.55964	.7868	.1459	.0476	46
15	0.47332	0.88089	0.53732	1.8611	1.1352	2.1127	0.48862	0.87250	0.56003	1.7856	1.1461	2.0466	45
16	.47357	.88075	.53769	.8598	.1354	.1116	.48887	.87235	.56041	.7844	.1463	.0455	44
17	.47383	.88061	.53807	.8585	.1356	.1104	.48913	.87221	.56079	.7832	.1465	.0444	43
18	.47409	.88048	.53844	.8572	.1357	.1093	.48938	.87207	.56117	.7820	.1467	.0434	42
19	.47434	.88034	.53882	.8559	.1359	.1082	.48964	.87193	.56156	.7808	.1469	.0423	41
20	0.47460	0.88020	0.53919	1.8546	1.1361	2.1070	0.48989	0.87178	0.56194	1.7795	1.1471	2.0413	40
21	.47486	.88006	.53957	.8533	.1363	.1059	.49014	.87164	.56232	.7783	.1473	.0402	39
22	.47511	.87992	.53995	.8520	.1365	.1048	.49040	.87150	.56270	.7771	.1474	.0392	38
23	.47537	.87979	.54032	.8507	.1366	.1036	.49065	.87136	.56309	.7759	.1476	.0381	37
24	.47562	.87965	.54070	.8495	.1368	.1025	.49090	.87121	.56347	.7747	.1478	.0370	36
25	0.47588	0.87951	0.54107	1.8482	1.1370	2.1014	0.49116	0.87107	0.56385	1.7735	1.1480	2.0360	35
26	.47613	.87937	.54145	.8469	.1372	.1002	.49141	.87093	.56424	.7723	.1482	.0349	34
27	.47639	.87923	.54183	.8456	.1373	.0991	.49166	.87078	.56462	.7711	.1484	.0339	33
28	.47665	.87909	.54220	.8443	.1375	.0980	.49192	.87064	.56500	.7699	.1486	.0329	32
29	.47690	.87895	.54258	.8430	.1377	.0969	.49217	.87050	.56539	.7687	.1488	.0318	31
30	0.47716	0.87882	0.54295	1.8418	1.1379	2.0957	0.49242	0.87035	0.56577	1.7675	1.1489	2.0308	30
31	.47741	.87868	.54333	.8405	.1381	.0946	.49268	.87021	.56616	.7663	.1491	.0297	29
32	.47767	.87854	.54371	.8392	.1382	.0935	.49293	.87007	.56654	.7651	.1493	.0287	28
33	.47792	.87840	.54409	.8379	.1384	.0924	.49318	.86992	.56692	.7639	.1495	.0276	27
34	.47818	.87826	.54446	.8367	.1386	.0912	.49343	.86978	.56731	.7627	.1497	.0266	26
35	0.47844	0.87812	0.54484	1.8354	1.1388	2.0901	0.49369	0.86964	0.56769	1.7615	1.1499	2.0256	25
36	.47869	.87798	.54522	.8341	.1390	.0890	.49394	.86949	.56808	.7603	.1501	.0245	24
37	.47895	.87784	.54559	.8329	.1391	.0879	.49419	.86935	.56846	.7591	.1503	.0235	23
38	.47920	.87770	.54597	.8316	.1393	.0868	.49445	.86921	.56885	.7579	.1505	.0224	22
39	.47946	.87756	.54635	.8303	.1395	.0857	.49470	.86906	.56923	.7567	.1507	.0214	21
40	0.47971	0.87742	0.54673	1.8291	1.1397	2.0846	0.49495	0.86892	0.56962	1.7555	1.1508	2.0204	20
41	.47997	.87728	.54711	.8278	.1399	.0835	.49521	.86877	.57000	.7544	.1510	.0194	19
42	.48022	.87715	.54748	.8265	.1401	.0824	.49546	.86863	.57039	.7532	.1512	.0183	18
43	.48048	.87701	.54786	.8253	.1402	.0812	.49571	.86849	.57077	.7520	.1514	.0173	17
44	.48073	.87687	.54824	.8240	.1404	.0801	.49596	.86834	.57116	.7508	.1516	.0163	16
45	0.48099	0.87673	0.54862	1.8227	1.1406	2.0790	0.49622	0.86820	0.57155	1.7496	1.1518	2.0152	15
46	.48124	.87659	.54900	.8215	.1408	.0779	.49647	.86805	.57193	.7484	.1520	.0142	14
47	.48150	.87645	.54937	.8202	.1410	.0768	.49672	.86791	.57232	.7473	.1522	.0132	13
48	.48175	.87631	.54975	.8190	.1411	.0757	.49697	.86776	.57270	.7461	.1524	.0122	12
49	.48201	.87617	.55013	.8177	.1413	.0746	.49723	.86762	.57309	.7449	.1526	.0111	11
50	0.48226	0.87603	0.55051	1.8165	1.1415	2.0735	0.49748	0.86748	0.57348	1.7437	1.1528	2.0101	10
51	.48252	.87588	.55089	.8152	.1417	.0725	.49773	.86733	.57386	.7426	.1530	.0091	9
52	.48277	.87574	.55127	.8140	.1419	.0714	.49798	.86719	.57425	.7414	.1531	.0081	8
53	.48303	.87560	.55165	.8127	.1421	.0703	.49823	.86704	.57464	.7402	.1533	.0071	7
54	.48328	.87546	.55203	.8115	.1422	.0692	.49849	.86690	.57502	.7390	.1535	.0061	6
55	0.48354	0.87532	0.55241	1.8102	1.1424	2.0681	0.49874	0.86675	0.57541	1.7379	1.1537	2.0050	5
56	.48379	.87518	.55279	.8090	.1426	.0670	.49899	.86661	.57580	.7367	.1539	.0040	4
57	.48405	.87504	.55317	.8078	.1428	.0659	.49924	.86646	.57619	.7355	.1541	.0030	3
58	.48430	.87490	.55355	.8065	.1430	.0648	.49950	.86632	.57657	.7344	.1543	.0020	2
59	.48455	.87476	.55393	.8053	.1432	.0637	.49975	.86617	.57696	.7332	.1545	.0010	1
60	0.48481	0.87462	0.55431	1.8040	1.1433	2.0627	0.50000	0.86603	0.57735	1.7320	1.1547	2.0000	0
M	Cosine	Sine	Cotan.	Tan.	Cosec.	Secant	Cosine	Sine	Cotan.	Tan.	Cosec.	Secant	M
118°						61°	119°						60°

TABLE: Natural Trigonometric functions, continued 5.6.8

30°						149°	31°						148°
M	Sine	Cosine	Tan.	Cotan.	Secant	Cosec.	Sine	Cosine	Tan.	Cotan.	Secant	Cosec.	M
0	0.50000	0.86603	0.57735	1.7320	1.1547	2.0000	0.51504	0.85717	0.60086	1.6643	1.1666	1.9416	60
1	.50025	.86588	.57774	.7309	.1549	1.9990	.51529	.85702	.60126	.6632	.1668	.9407	59
2	.50050	.86573	.57813	.7297	.1551	.9980	.51554	.85687	.60165	.6621	.1670	.9397	58
3	.50075	.86559	.57851	.7286	.1553	.9970	.51578	.85672	.60205	.6610	.1672	.9388	57
4	.50101	.86544	.57890	.7274	.1555	.9960	.51603	.85657	.60244	.6599	.1674	.9378	56
5	0.50126	0.86530	0.57929	1.7262	1.1557	1.9950	0.51628	0.85642	0.60284	1.6588	1.1676	1.9369	55
6	.50151	.86515	.57968	.7251	.1559	.9940	.51653	.85627	.60324	.6577	.1678	.9360	54
7	.50176	.86500	.58007	.7239	.1561	.9930	.51678	.85612	.60363	.6566	.1681	.9350	53
8	.50201	.86486	.58046	.7228	.1562	.9920	.51703	.85597	.60403	.6555	.1683	.9341	52
9	.50226	.86471	.58085	.7216	.1564	.9910	.51728	.85582	.60443	.6544	.1685	.9332	51
10	0.50252	0.86457	0.58123	1.7205	1.1566	1.9900	0.51753	0.85566	0.60483	1.6534	1.1687	1.9322	50
11	.50277	.86442	.58162	.7193	.1568	.9890	.51778	.85551	.60522	.6523	.1689	.9313	49
12	.50302	.86427	.58201	.7182	.1570	.9880	.51803	.85536	.60562	.6512	.1691	.9304	48
13	.50327	.86413	.58240	.7170	.1572	.9870	.51827	.85521	.60602	.6501	.1693	.9295	47
14	.50352	.86398	.58279	.7159	.1574	.9860	.51852	.85506	.60642	.6490	.1695	.9285	46
15	0.50377	0.86383	0.58318	1.7147	1.1576	1.9850	0.51877	0.85491	0.60681	1.6479	1.1697	1.9276	45
16	.50402	.86369	.58357	.7136	.1578	.9840	.51902	.85476	.60721	.6469	.1699	.9267	44
17	.50428	.86354	.58396	.7124	.1580	.9830	.51927	.85461	.60761	.6458	.1701	.9258	43
18	.50453	.86339	.58435	.7113	.1582	.9820	.51952	.85446	.60801	.6447	.1703	.9248	42
19	.50478	.86325	.58474	.7101	.1584	.9811	.51977	.85431	.60841	.6436	.1705	.9239	41
20	0.50503	0.86310	0.58513	1.7090	1.1586	1.9801	0.52002	0.85416	0.60881	1.6425	1.1707	1.9230	40
21	.50528	.86295	.58552	.7079	.1588	.9791	.52026	.85400	.60920	.6415	.1709	.9221	39
22	.50553	.86281	.58591	.7067	.1590	.9781	.52051	.85385	.60960	.6404	.1712	.9212	38
23	.50578	.86266	.58630	.7056	.1592	.9771	.52076	.85370	.61000	.6393	.1714	.9203	37
24	.50603	.86251	.58670	.7044	.1594	.9761	.52101	.85355	.61040	.6383	.1716	.9193	36
25	0.50628	0.86237	0.58709	1.7033	1.1596	1.9752	0.52126	0.85340	0.61080	1.6372	1.1718	1.9184	35
26	.50653	.86222	.58748	.7022	.1598	.9742	.52151	.85325	.61120	.6361	.1720	.9175	34
27	.50679	.86207	.58787	.7010	.1600	.9732	.52175	.85309	.61160	.6350	.1722	.9166	33
28	.50704	.86192	.58826	.6999	.1602	.9722	.52200	.85294	.61200	.6340	.1724	.9157	32
29	.50729	.86178	.58865	.6988	.1604	.9713	.52225	.85279	.61240	.6329	.1726	.9148	31
30	0.50754	0.86163	0.58904	1.6977	1.1606	1.9703	0.52250	0.85264	0.61280	1.6318	1.1728	1.9139	30
31	.50779	.86148	.58944	.6965	.1608	.9693	.52275	.85249	.61320	.6308	.1730	.9130	29
32	.50804	.86133	.58983	.6954	.1610	.9683	.52299	.85234	.61360	.6297	.1732	.9121	28
33	.50829	.86118	.59022	.6943	.1612	.9674	.52324	.85218	.61400	.6286	.1734	.9112	27
34	.50854	.86104	.59061	.6931	.1614	.9664	.52349	.85203	.61440	.6276	.1737	.9102	26
35	0.50879	0.86089	0.59100	1.6920	1.1616	1.9654	0.52374	0.85188	0.61480	1.6265	1.1739	1.9093	25
36	.50904	.86074	.59140	.6909	.1618	.9645	.52398	.85173	.61520	.6255	.1741	.9084	24
37	.50929	.86059	.59179	.6898	.1620	.9635	.52423	.85157	.61560	.6244	.1743	.9075	23
38	.50954	.86044	.59218	.6887	.1622	.9625	.52448	.85142	.61601	.6233	.1745	.9066	22
39	.50979	.86030	.59258	.6875	.1624	.9616	.52473	.85127	.61641	.6223	.1747	.9057	21
40	0.51004	0.86015	0.59297	1.6864	1.1626	1.9606	0.52498	0.85112	0.61681	1.6212	1.1749	1.9048	20
41	.51029	.86000	.59336	.6853	.1628	.9596	.52522	.85096	.61721	.6202	.1751	.9039	19
42	.51054	.85985	.59376	.6842	.1630	.9587	.52547	.85081	.61761	.6191	.1753	.9030	18
43	.51079	.85970	.59415	.6831	.1632	.9577	.52572	.85066	.61801	.6181	.1756	.9021	17
44	.51104	.85955	.59454	.6820	.1634	.9568	.52597	.85050	.61842	.6170	.1758	.9013	16
45	0.51129	0.85941	0.59494	1.6808	1.1636	1.9558	0.52621	0.85035	0.61882	1.6160	1.1760	1.9004	15
46	.51154	.85926	.59533	.6797	.1638	.9549	.52646	.85020	.61922	.6149	.1762	.8995	14
47	.51179	.85911	.59572	.6786	.1640	.9539	.52671	.85004	.61962	.6139	.1764	.8986	13
48	.51204	.85896	.59612	.6775	.1642	.9530	.52695	.84989	.62003	.6128	.1766	.8977	12
49	.51229	.85881	.59651	.6764	.1644	.9520	.52720	.84974	.62043	.6118	.1768	.8968	11
50	0.51254	0.85866	0.59691	1.6753	1.1646	1.9510	0.52745	0.84959	0.62083	1.6107	1.1770	1.8959	10
51	.51279	.85851	.59730	.6742	.1648	.9501	.52770	.84943	.62123	.6097	.1772	.8950	9
52	.51304	.85836	.59770	.6731	.1650	.9491	.52794	.84928	.62164	.6086	.1775	.8941	8
53	.51329	.85821	.59809	.6720	.1652	.9482	.52819	.84912	.62204	.6076	.1777	.8932	7
54	.51354	.85806	.59849	.6709	.1654	.9473	.52844	.84897	.62244	.6066	.1779	.8924	6
55	0.51379	0.85791	0.59888	1.6698	1.1656	1.9463	0.52868	0.84882	0.62285	1.6055	1.1781	1.8915	5
56	.51404	.85777	.59928	.6687	.1658	.9454	.52893	.84866	.62325	.6045	.1783	.8906	4
57	.51429	.85762	.59967	.6676	.1660	.9444	.52918	.84851	.62366	.6034	.1785	.8897	3
58	.51454	.85747	.60007	.6665	.1662	.9435	.52942	.84836	.62406	.6024	.1787	.8888	2
59	.51479	.85732	.60046	.6654	.1664	.9425	.52967	.84820	.62446	.6014	.1790	.8879	1
60	0.51504	0.85717	0.60086	1.6643	1.1666	1.9416	0.52992	0.84805	0.62487	1.6003	1.1792	1.8871	0
M	Cosine	Sine	Cotan.	Tan.	Cosec.	Secant	Cosine	Sine	Cotan.	Tan.	Cosec.	Secant	M
120°						59°	121°						58°

TABLE: Natural Trigonometric functions, continued 5.6.8

32° 147° 33° 146°

M	Sine	Cosine	Tan.	Cotan.	Secant	Cosec.	Sine	Cosine	Tan.	Cotan.	Secant	Cosec.	M
0	0.52992	0.84805	0.62487	1.6003	1.1792	1.8871	0.54464	0.83867	0.64941	1.5399	1.1924	1.8361	60
1	.53016	.84789	.62527	.5993	.1794	.8862	.54488	.83851	.64982	.5389	.1926	.8352	59
2	.53041	.84774	.62568	.5983	.1796	.8853	.54513	.83835	.65023	.5379	.1928	.8344	58
3	.53066	.84758	.62608	.5972	.1798	.8844	.54537	.83819	.65065	.5369	.1930	.8336	57
4	.53090	.84743	.62649	.5962	.1800	.8836	.54561	.83804	.65106	.5359	.1933	.8328	56
5	0.53115	0.84728	0.62689	1.5952	1.1802	1.8827	0.54586	0.83788	0.65148	1.5350	1.1935	1.8320	55
6	.53140	.84712	.62730	.5941	.1805	.8818	.54610	.83772	.65189	.5340	.1937	.8311	54
7	.53164	.84697	.62770	.5931	.1807	.8809	.54634	.83756	.65231	.5330	.1939	.8303	53
8	.53189	.84681	.62811	.5921	.1809	.8801	.54659	.83740	.65272	.5320	.1942	.8295	52
9	.53214	.84666	.62851	.5910	.1811	.8792	.54683	.83724	.65314	.5311	.1944	.8287	51
10	0.53238	0.84650	0.62892	1.5900	1.1813	1.8783	0.54708	0.83708	0.65355	1.5301	1.1946	1.8279	50
11	.53263	.84635	.62933	.5890	.1815	.8775	.54732	.83692	.65397	.5291	.1948	.8271	49
12	.53288	.84619	.62973	.5880	.1818	.8766	.54756	.83676	.65438	.5282	.1951	.8263	48
13	.53312	.84604	.63014	.5869	.1820	.8757	.54781	.83660	.65480	.5272	.1953	.8255	47
14	.53337	.84588	.63055	.5859	.1822	.8749	.54805	.83644	.65521	.5262	.1955	.8246	46
15	0.53361	0.84573	0.63095	1.5849	1.1824	1.8740	0.54829	0.83629	0.65563	1.5252	1.1958	1.8238	45
16	.53386	.84557	.63136	.5839	.1826	.8731	.54854	.83613	.65604	.5243	.1960	.8230	44
17	.53411	.84542	.63177	.5829	.1828	.8723	.54878	.83597	.65646	.5233	.1962	.8222	43
18	.53435	.84526	.63217	.5818	.1831	.8714	.54902	.83581	.65688	.5223	.1964	.8214	42
19	.53460	.84511	.63258	.5808	.1833	.8706	.54926	.83565	.65729	.5214	.1967	.8206	41
20	0.53484	0.84495	0.63299	1.5798	1.1835	1.8697	0.54951	0.83549	0.65771	1.5204	1.1969	1.8198	40
21	.53509	.84479	.63339	.5788	.1837	.8688	.54975	.83533	.65813	.5195	.1971	.8190	39
22	.53533	.84464	.63380	.5778	.1839	.8680	.54999	.83517	.65854	.5185	.1974	.8182	38
23	.53558	.84448	.63421	.5768	.1841	.8671	.55024	.83501	.65896	.5175	.1976	.8174	37
24	.53583	.84433	.63462	.5757	.1844	.8663	.55048	.83485	.65938	.5166	.1978	.8166	36
25	0.53607	0.84417	0.63503	1.5747	1.1846	1.8654	0.55072	0.83469	0.65980	1.5156	1.1980	1.8158	35
26	.53632	.84402	.63543	.5737	.1848	.8646	.55097	.83453	.66021	.5147	.1983	.8150	34
27	.53656	.84386	.63584	.5727	.1850	.8637	.55121	.83437	.66063	.5137	.1985	.8142	33
28	.53681	.84370	.63625	.5717	.1852	.8629	.55145	.83421	.66105	.5127	.1987	.8134	32
29	.53705	.84355	.63666	.5707	.1855	.8520	.55169	.83405	.66147	.5118	.1990	.8126	31
30	0.53730	0.84339	0.63707	1.5697	1.1857	1.8611	0.55194	0.83388	0.66188	1.5108	1.1992	1.8118	30
31	.53754	.84323	.63748	.5687	.1859	.8603	.55218	.83372	.66230	.5099	.1994	.8110	29
32	.53779	.84308	.63789	.5677	.1861	.8595	.55242	.83356	.66272	.5089	.1997	.8102	28
33	.53803	.84292	.63830	.5667	.1863	.8586	.55266	.83340	.66314	.5080	.1999	.8094	27
34	.53828	.84276	.63871	.5657	.1866	.8578	.55291	.83324	.66356	.5070	.2001	.8086	26
35	0.53852	0.84261	0.63912	1.5646	1.1868	1.8569	0.55315	0.83308	0.66398	1.5061	1.2004	1.8078	25
36	.53877	.84245	.63953	.5636	.1870	.8561	.55339	.83292	.66440	.5051	.2006	.8070	24
37	.53901	.84229	.63994	.5626	.1872	.8552	.55363	.83276	.66482	.5042	.2008	.8062	23
38	.53926	.84214	.64035	.5616	.1874	.8544	.55388	.83260	.66524	.5032	.2010	.8054	22
39	.53950	.84198	.64076	.5606	.1877	.8535	.55412	.83244	.66566	.5023	.2013	.8047	21
40	0.53975	0.84182	0.64117	1.5596	1.1879	1.8527	0.55436	0.83228	0.66608	1.5013	1.2015	1.8039	20
41	.53999	.84167	.64158	.5586	.1881	.8519	.55460	.83211	.66650	.5004	.2017	.8031	19
42	.54024	.84151	.64199	.5577	.1883	.8510	.55484	.83195	.66692	.4994	.2020	.8023	18
43	.54048	.84135	.64240	.5567	.1886	.8502	.55509	.83179	.66734	.4985	.2022	.8015	17
44	.54073	.84120	.64281	.5557	.1888	.8493	.55533	.83163	.66776	.4975	.2024	.8007	16
45	0.54097	0.84104	0.64322	1.5547	1.1890	1.8485	0.55557	0.83147	0.66818	1.4966	1.2027	1.7999	15
46	.54122	.84088	.64363	.5537	.1892	.8477	.55581	.83131	.66860	.4957	.2029	.7992	14
47	.54146	.84072	.64404	.5527	.1894	.8468	.55605	.83115	.66902	.4947	.2031	.7984	13
48	.54171	.84057	.64446	.5517	.1897	.8460	.55629	.83098	.66944	.4938	.2034	.7976	12
49	.54195	.84041	.64487	.5507	.1899	.8452	.55654	.83082	.66986	.4928	.2036	.7968	11
50	0.54220	0.84025	0.64528	1.5497	1.1901	1.8443	0.55678	0.83066	0.67028	1.4919	1.2039	1.7960	10
51	.54244	.84009	.64569	.5487	.1903	.8435	.55702	.83050	.67071	.4910	.2041	.7953	9
52	.54268	.83993	.64610	.5477	.1906	.8427	.55726	.83034	.67113	.4900	.2043	.7945	8
53	.54293	.83978	.64652	.5467	.1908	.8418	.55750	.83017	.67155	.4891	.2046	.7937	7
54	.54317	.83962	.64693	.5458	.1910	.8410	.55774	.83001	.67197	.4881	.2048	.7929	6
55	0.54342	0.83946	0.64734	1.5448	1.1912	1.8402	0.55799	0.82985	0.67239	1.4872	1.2050	1.7921	5
56	.54366	.83930	.64775	.5438	.1915	.8394	.55823	.82969	.67282	.4863	.2053	.7914	4
57	.54391	.83914	.64817	.5428	.1917	.8385	.55847	.82952	.67324	.4853	.2055	.7906	3
58	.54415	.83899	.64858	.5418	.1919	.8377	.55871	.82936	.67366	.4844	.2057	.7898	2
59	.54439	.83883	.64899	.5408	.1921	.8369	.55895	.82920	.67408	.4835	.2060	.7891	1
60	0.54464	0.83867	0.64941	1.5399	1.1922	1.8361	0.55919	0.82904	0.67451	1.4826	1.2062	1.7883	0
M	Cosine	Sine	Cotan.	Tan.	Cosec.	Secant	Cosine	Sine	Cotan.	Tan.	Cosec.	Secant	M

122° 57° 123° 56°

TABLE: Natural Trigonometric functions, continued 5.6.8

34°						145°	35°						144°
M	Sine	Cosine	Tan.	Cotan.	Secant	Cosec.	Sine	Cosine	Tan.	Cotan.	Secant	Cosec.	M
0	0.55919	0.82904	0.67451	1.4826	1.2062	1.7883	0.57358	0.81915	0.70021	1.4281	1.2208	1.7434	60
1	.55943	.82887	.67493	.4816	.2064	.7875	.57381	.81898	.70064	.4273	.2210	.7427	59
2	.55967	.82871	.67535	.4807	.2067	.7867	.57405	.81882	.70107	.4264	.2213	.7420	58
3	.55992	.82855	.67578	.4798	.2069	.7860	.57429	.81865	.70151	.4255	.2215	.7413	57
4	.56016	.82839	.67620	.4788	.2072	.7852	.57453	.81848	.70194	.4246	.2218	.7405	56
5	0.56040	0.82822	0.67663	1.4779	1.2074	1.7844	0.57477	0.81832	0.70238	1.4237	1.2220	1.7398	55
6	.56064	.82806	.67705	.4770	.2076	.7837	.57500	.81815	.70281	.4228	.2223	.7391	54
7	.56088	.82790	.67747	.4761	.2079	.7829	.57524	.81798	.70325	.4220	.2225	.7384	53
8	.56112	.82773	.67790	.4751	.2081	.7821	.57548	.81781	.70368	.4211	.2228	.7377	52
9	.56136	.82757	.67832	.4742	.2083	.7814	.57572	.81765	.70412	.4202	.2230	.7369	51
10	0.56160	0.82741	0.67875	1.4733	1.2086	1.7806	0.57596	0.81748	0.70455	1.4193	1.2233	1.7362	50
11	.56184	.82724	.67917	.4724	.2088	.7798	.57619	.81731	.70499	.4185	.2235	.7355	49
12	.56208	.82708	.67960	.4714	.2091	.7791	.57643	.81714	.70542	.4176	.2238	.7348	48
13	.56232	.82692	.68002	.4705	.2093	.7783	.57667	.81698	.70586	.4167	.2240	.7341	47
14	.56256	.82675	.68045	.4696	.2095	.7776	.57691	.81681	.70629	.4158	.2243	.7334	46
15	0.56280	0.82659	0.68087	1.4687	1.2098	1.7768	0.57714	0.81664	0.70673	1.4150	1.2245	1.7327	45
16	.56304	.82643	.68130	.4678	.2100	.7760	.57738	.81647	.70717	.4141	.2248	.7319	44
17	.56328	.82626	.68173	.4669	.2103	.7753	.57762	.81630	.70760	.4132	.2250	.7312	43
18	.56353	.82610	.68215	.4659	.2105	.7745	.57786	.81614	.70804	.4123	.2253	.7305	42
19	.56377	.82593	.68258	.4650	.2107	.7738	.57809	.81597	.70848	.4115	.2255	.7298	41
20	0.56401	0.82577	0.68301	1.4641	1.2110	1.7730	0.57833	0.81580	0.70891	1.4106	1.2258	1.7291	40
21	.56425	.82561	.68343	.4632	.2112	.7723	.57857	.81563	.70935	.4097	.2260	.7284	39
22	.56449	.82544	.68386	.4623	.2115	.7715	.57881	.81546	.70979	.4089	.2263	.7277	38
23	.56473	.82528	.68429	.4614	.2117	.7708	.57904	.81530	.71022	.4080	.2265	.7270	37
24	.56497	.82511	.68471	.4605	.2119	.7700	.57928	.81513	.71066	.4071	.2268	.7263	36
25	0.56521	0.82495	0.68514	1.4595	1.2122	1.7693	0.57952	0.81496	0.71110	1.4063	1.2270	1.7256	35
26	.56545	.82478	.68557	.4586	.2124	.7685	.57975	.81479	.71154	.4054	.2273	.7249	34
27	.56569	.82462	.68600	.4577	.2127	.7678	.57999	.81462	.71198	.4045	.2276	.7242	33
28	.56593	.82445	.68642	.4568	.2129	.7670	.58023	.81445	.71241	.4037	.2278	.7234	32
29	.56617	.82429	.68685	.4559	.2132	.7663	.58047	.81428	.71285	.4028	.2281	.7227	31
30	0.56641	0.82413	0.68728	1.4550	1.2134	1.7655	0.58070	0.81411	0.71329	1.4019	1.2283	1.7220	30
31	.56664	.82396	.68771	.4541	.2136	.7648	.58094	.81395	.71373	.4011	.2286	.7213	29
32	.56688	.82380	.68814	.4532	.2139	.7640	.58118	.81378	.71417	.4002	.2288	.7206	28
33	.56712	.82363	.68857	.4523	.2141	.7633	.58141	.81361	.71461	.3994	.2291	.7199	27
34	.56736	.82347	.68899	.4514	.2144	.7625	.58165	.81344	.71505	.3985	.2293	.7192	26
35	0.56760	0.82330	0.68942	1.4505	1.2146	1.7618	0.58189	0.81327	0.71549	1.3976	1.2296	1.7185	25
36	.56784	.82314	.68985	.4496	.2149	.7610	.58212	.81310	.71593	.3968	.2298	.7178	24
37	.56808	.82297	.69028	.4487	.2151	.7603	.58236	.81293	.71637	.3959	.2301	.7171	23
38	.56832	.82280	.69071	.4478	.2153	.7596	.58259	.81276	.71681	.3951	.2304	.7164	22
39	.56856	.82264	.69114	.4469	.2156	.7588	.58283	.81259	.71725	.3942	.2306	.7157	21
40	0.56880	0.82247	0.69157	1.4460	1.2158	1.7581	0.58307	0.81242	0.71769	1.3933	1.2309	1.7151	20
41	.56904	.82231	.69200	.4451	.2161	.7573	.58330	.81225	.71813	.3925	.2311	.7144	19
42	.56928	.82214	.69243	.4442	.2163	.7566	.58354	.81208	.71857	.3916	.2314	.7137	18
43	.56952	.82198	.69286	.4433	.2166	.7559	.58378	.81191	.71901	.3908	.2316	.7130	17
44	.56976	.82181	.69329	.4424	.2168	.7551	.58401	.81174	.71945	.3899	.2319	.7123	16
45	0.57000	0.82165	0.69372	1.4415	1.2171	1.7544	0.58425	0.81157	0.71990	1.3891	1.2322	1.7116	15
46	.57023	.82148	.69415	.4406	.2173	.7537	.58448	.81140	.72034	.3882	.2324	.7109	14
47	.57047	.82131	.69459	.4397	.2175	.7529	.58472	.81123	.72078	.3874	.2327	.7102	13
48	.57071	.82115	.69502	.4388	.2178	.7522	.58496	.81106	.72122	.3865	.2329	.7095	12
49	.57095	.82098	.69545	.4379	.2180	.7514	.58519	.81089	.72166	.3857	.2332	.7088	11
50	0.57119	0.82082	0.69588	1.4370	1.2183	1.7507	0.58543	0.81072	0.72211	1.3848	1.2335	1.7081	10
51	.57143	.82065	.69631	.4361	.2185	.7500	.58566	.81055	.72255	.3840	.2337	.7075	9
52	.57167	.82048	.69674	.4352	.2188	.7493	.58590	.81038	.72299	.3831	.2340	.7068	8
53	.57191	.82032	.69718	.4343	.2190	.7485	.58614	.81021	.72344	.3823	.2342	.7061	7
54	.57214	.82015	.69761	.4335	.2193	.7478	.58637	.81004	.72388	.3814	.2345	.7054	6
55	0.57238	0.81998	0.69804	1.4326	1.2195	1.7471	0.58661	0.80987	0.72432	1.3806	1.2348	1.7047	5
56	.57262	.81982	.69847	.4317	.2198	.7463	.58684	.80970	.72477	.3797	.2350	.7040	4
57	.57286	.81965	.69891	.4308	.2200	.7456	.58708	.80953	.72521	.3789	.2353	.7033	3
58	.57310	.81948	.69934	.4299	.2203	.7449	.58731	.80936	.72565	.3781	.2355	.7027	2
59	.57334	.81932	.69977	.4290	.2205	.7442	.58755	.80919	.72610	.3772	.2358	.7020	1
60	0.57358	0.81915	0.70021	1.4281	1.2208	1.7434	0.58778	0.80902	0.72654	1.3764	1.2361	1.7013	0
M	Cosine	Sine	Cotan.	Tan.	Cosec.	Secant	Cosine	Sine	Cotan.	Tan.	Cosec.	Secant	M
124°						55°	125°						54°

TABLE: Natural Trigonometric functions, continued 5.6.8

36°						143°	37°					142°	
M	Sine	Cosine	Tan.	Cotan.	Secant	Cosec.	Sine	Cosine	Tan.	Cotan.	Secant	Cosec.	M
0	0.58778	0.80902	0.72654	1.3764	1.2361	1.7013	0.60181	0.79863	0.75355	1.3270	1.2521	1.6616	60
1	.58802	.80885	.72699	.3755	.2363	.7006	.60205	.79846	.75401	.3262	.2524	.6610	59
2	.58825	.80867	.72743	.3747	.2366	.6999	.60228	.79828	.75447	.3254	.2527	.6603	58
3	.58849	.80850	.72788	.3738	.2368	.6993	.60251	.79811	.75492	.3246	.2530	.6597	57
4	.58873	.80833	.72832	.3730	.2371	.6986	.60274	.79793	.75538	.3238	.2532	.6591	56
5	0.58896	0.80816	0.72877	1.3722	1.2374	1.6979	0.60298	0.79776	0.75584	1.3230	1.2535	1.6584	55
6	.58920	.30799	.72921	.3713	.2376	.6972	.60320	.79758	.75629	.3222	.2538	.6578	54
7	.58943	.80782	.72966	.3705	.2379	.6965	.60344	.79741	.75675	.3214	.2541	.6572	53
8	.58967	.80765	.73010	.3697	.2382	.6959	.60367	.79723	.75721	.3206	.2543	.6565	52
9	.58990	.80747	.73055	.3688	.2384	.6952	.60390	.79706	.75767	.3198	.2546	.6559	51
10	0.59014	0.80730	0.73100	1.3680	1.2387	1.6945	0.60413	0.79688	0.75812	1.3190	1.2549	1.6552	50
11	.59037	.80713	.73144	.3672	.2389	.6938	.60437	.79670	.75858	.3182	.2552	.6546	49
12	.59060	.80696	.73189	.3663	.2392	.6932	.60460	.79653	.75904	.3174	.2554	.6540	48
13	.59084	.80679	.73234	.3655	.2395	.6925	.60483	.79635	.75950	.3166	.2557	.6533	47
14	.59107	.80662	.73278	.3647	.2397	.6918	.60506	.79618	.75996	.3159	.2560	.6527	46
15	0.59131	0.80644	0.73323	1.3638	1.2400	1.6912	0.60529	0.79600	0.76042	1.3151	1.2563	1.6521	45
16	.59154	.80627	.73368	.3630	.2403	.6905	.60552	.79582	.76088	.3143	.2565	.6514	44
17	.59178	.80610	.73412	.3622	.2405	.6898	.60576	.79565	.76134	.3135	.2568	.6508	43
18	.59201	.80593	.73457	.3613	.2408	.6891	.60599	.79547	.76179	.3127	.2571	.6502	42
19	.59225	.80576	.73502	.3605	.2411	.6885	.60622	.79530	.76225	.3119	.2574	.6496	41
20	0.59248	0.80558	0.73547	1.3597	1.2413	1.6878	0.60645	0.79512	0.76271	1.3111	1.2577	1.6489	40
21	.59272	.80541	.73592	.3588	.2416	.6871	.60668	.79494	.76317	.3103	.2579	.6483	39
22	.59295	.80524	.73637	.3580	.2419	.6865	.60691	.79477	.76364	.3095	.2582	.6477	38
23	.59318	.80507	.73681	.3572	.2421	.6858	.60714	.79459	.76410	.3087	.2585	.6470	37
24	.59342	.80489	.73726	.3564	.2424	.6851	.60737	.79441	.76456	.3079	.2588	.6464	36
25	0.59365	0.80472	0.73771	1.3555	1.2427	1.6845	0.60761	0.79424	0.76502	1.3071	1.2591	1.6458	35
26	.59389	.80455	.73816	.3547	.2429	.6838	.60784	.79406	.76548	.3064	.2593	.6452	34
27	.59412	.80437	.73861	.3539	.2432	.6831	.60807	.79388	.76594	.3056	.2596	.6445	33
28	.59435	.80420	.73906	.3531	.2435	.6825	.60830	.79371	.76640	.3048	.2599	.6439	32
29	.59459	.80403	.73951	.3522	.2437	.6818	.60853	.79353	.76686	.3040	.2602	.6433	31
30	0.59482	0.80386	0.73996	1.3514	1.2440	1.6812	0.60876	0.79335	0.76733	1.3032	1.2605	1.6427	30
31	.59506	.80368	.74041	.3506	.2443	.6805	.60899	.79318	.76779	.3024	.2607	.6420	29
32	.59529	.80351	.74086	.3498	.2445	.6798	.60922	.79300	.76825	.3016	.2610	.6414	28
33	.59552	.80334	.74131	.3489	.2448	.6792	.60945	.79282	.76871	.3009	.2613	.6408	27
34	.59576	.80316	.74176	.3481	.2451	.6785	.60968	.79264	.76918	.3001	.2616	.6402	26
35	0.59599	0.80299	0.74221	1.3473	1.2453	1.6779	0.60991	0.79247	0.76964	1.2993	1.2619	1.6396	25
36	.59622	.80282	.74266	.3465	.2456	.6772	.61014	.79229	.77010	.2985	.2622	.6389	24
37	.59646	.80264	.74312	.3457	.2459	.6766	.61037	.79211	.77057	.2977	.2624	.6383	23
38	.59669	.80247	.74357	.3449	.2461	.6759	.61061	.79193	.77103	.2970	.2627	.6377	22
39	.59692	.80230	.74402	.3440	.2464	.6752	.61084	.79176	.77149	.2962	.2630	.6371	21
40	0.59716	0.80212	0.74447	1.3432	1.2467	1.6746	0.61107	0.79158	0.77196	1.2954	1.2633	1.6365	20
41	.59739	.80195	.74492	.3424	.2470	.6739	.61130	.79140	.77242	.2946	.2636	.6359	19
42	.59762	.80177	.74538	.3416	.2472	.6733	.61153	.79122	.77289	.2938	.2639	.6352	18
43	.59786	.80160	.74583	.3408	.2475	.6726	.61176	.79104	.77335	.2931	.2641	.6346	17
44	.59809	.80143	.74628	.3400	.2478	.6720	.61199	.79087	.77382	.2923	.2644	.6340	16
45	0.59832	0.80125	0.74673	1.3392	1.2480	1.6713	0.61222	0.79069	0.77428	1.2915	1.2647	1.6334	15
46	.59856	.80108	.74719	.3383	.2483	.6707	.61245	.79051	.77475	.2907	.2650	.6328	14
47	.59879	.80090	.74764	.3375	.2486	.6700	.61268	.79033	.77521	.2900	.2653	.6322	13
48	.59902	.80073	.74809	.3367	.2488	.6694	.61290	.79015	.77568	.2892	.2656	.6316	12
49	.59926	.80056	.74855	.3359	.2491	.6687	.61314	.78998	.77614	.2884	.2659	.6309	11
50	0.59949	0.80038	0.74900	1.3351	1.2494	1.6681	0.61337	0.78980	0.77661	1.2876	1.2661	1.6303	10
51	.59972	.80021	.74946	.3343	.2497	.6674	.61360	.78962	.77708	.2869	.2664	.6297	9
52	.59995	.80003	.74991	.3335	.2499	.6668	.61383	.78944	.77754	.2861	.2667	.6291	8
53	.60019	.79986	.75037	.3327	.2502	.6661	.61405	.78926	.77801	.2853	.2670	.6285	7
54	.60042	.79968	.75082	.3319	.2505	.6655	.61428	.78908	.77848	.2845	.2673	.6279	6
55	0.60065	0.79951	0.75128	1.3311	1.2508	1.6648	0.61451	0.78890	0.77895	1.2838	1.2676	1.6273	5
56	.60088	.79933	.75173	.3303	.2510	.6642	.61474	.78873	.77941	.2830	.2679	.6267	4
57	.60112	.79916	.75219	.3294	.2513	.6636	.61497	.78855	.77988	.2822	.2681	.6261	3
58	.60135	.79898	.75264	.3286	.2516	.6629	.61520	.78837	.78035	.2815	.2684	.6255	2
59	.60158	.79881	.75310	.3278	.2519	.6623	.61543	.78819	.78082	.2807	.2687	.6249	1
60	0.60181	0.79863	0.75355	1.3270	1.2521	1.6616	0.61566	0.78801	0.78128	1.2799	1.2690	1.6243	0
M	Cosine	Sine	Cotan.	Tan.	Cosec.	Secant	Cosine	Sine	Cotan.	Tan.	Cosec.	Secant	M
126°						53°	127°					52°	

TABLE: Natural Trigonometric functions, continued 5.6.8

38°						141°	39°						140°
M	Sine	Cosine	Tan.	Cotan.	Secant	Cosec.	Sine	Cosine	Tan.	Cotan.	Secant	Cosec.	M
0	0.61566	0.78801	0.78128	1.2799	1.2690	1.6243	0.62932	0.77715	0.80978	1.2349	1.2867	1.5890	60
1	.61589	.78783	.78175	.2792	.2693	.6237	.62955	.77696	.81026	.2342	.2871	.5884	59
2	.61612	.78765	.78222	.2784	.2696	.6231	.62977	.77678	.81075	.2334	.2874	.5879	58
3	.61635	.78747	.78269	.2776	.2699	.6224	.63000	.77660	.81123	.2327	.2877	.5873	57
4	.61658	.78729	.78316	.2769	.2702	.6218	.63022	.77641	.81171	.2320	.2880	.5867	56
5	0.61681	0.78711	0.78363	1.2761	1.2705	1.6212	0.63045	0.77623	0.81219	1.2312	1.2883	1.5862	55
6	.61703	.78693	.78410	.2753	.2707	.6206	.63067	.77605	.81268	.2305	.2886	.5856	54
7	.61726	.78675	.78457	.2746	.2710	.6200	.63090	.77586	.81316	.2297	.2889	.5850	53
8	.61749	.78657	.78504	.2738	.2713	.6194	.63113	.77568	.81364	.2290	.2892	.5845	52
9	.61772	.78640	.78551	.2730	.2716	.6188	.63135	.77549	.81413	.2283	.2895	.5839	51
10	0.61795	0.78622	0.78598	1.2723	1.2719	1.6182	0.63158	0.77531	0.81461	1.2276	1.2898	1.5833	50
11	.61818	.78604	.78645	.2715	.2722	.6176	.63180	.77513	.81509	.2268	.2901	.5828	49
12	.61841	.78586	.78692	.2708	.2725	.6170	.63203	.77494	.81558	.2261	.2904	.5822	48
13	.61864	.78568	.78739	.2700	.2728	.6164	.63225	.77476	.81606	.2254	.2907	.5816	47
14	.61886	.78550	.78786	.2692	.2731	.6159	.63248	.77458	.81655	.2247	.2910	.5811	46
15	0.61909	0.78532	0.78834	1.2685	1.2734	1.6153	0.63270	0.77439	0.81703	1.2239	1.2913	1.5805	45
16	.61932	.78514	.78881	.2677	.2737	.6147	.63293	.77421	.81752	.2232	.2916	.5799	44
17	.61955	.78496	.78928	.2670	.2739	.6141	.63315	.77402	.81800	.2225	.2919	.5794	43
18	.61978	.78478	.78975	.2662	.2742	.6135	.63338	.77384	.81849	.2218	.2922	.5788	42
19	.62001	.78460	.79022	.2655	.2745	.6129	.63360	.77365	.81898	.2210	.2926	.5783	41
20	0.62023	0.78441	0.79070	1.2647	1.2748	1.6123	0.63383	0.77347	0.81946	1.2203	1.2929	1.5777	40
21	.62046	.78423	.79117	.2639	.2751	.6117	.63405	.77329	.81995	.2196	.2932	.5771	39
22	.62069	.78405	.79164	.2632	.2754	.6111	.63428	.77310	.82043	.2189	.2935	.5766	38
23	.62092	.78387	.79212	.2624	.2757	.6105	.63450	.77292	.82092	.2181	.2938	.5760	37
24	.62115	.78369	.79259	.2617	.2760	.6099	.63473	.77273	.82141	.2174	.2941	.5755	36
25	0.62137	0.78351	0.79306	1.2609	1.2763	1.6093	0.63495	0.77255	0.82190	1.2167	1.2944	1.5749	35
26	.62160	.78333	.79354	.2602	.2766	.6087	.63518	.77236	.82238	.2160	.2947	.5743	34
27	.62183	.78315	.79401	.2594	.2769	.6081	.63540	.77218	.82287	.2152	.2950	.5738	33
28	.62206	.78297	.79449	.2587	.2772	.6077	.63563	.77199	.82336	.2145	.2953	.5732	32
29	.62229	.78279	.79496	.2579	.2775	.6070	.63585	.77181	.82385	.2138	.2956	.5727	31
30	0.62251	0.78261	0.79543	1.2572	1.2778	1.6064	0.63608	0.77162	0.82434	1.2131	1.2960	1.5721	30
31	.62274	.78243	.79591	.2564	.2781	.6058	.63630	.77144	.82482	.2124	.2963	.5716	29
32	.62297	.78224	.79639	.2557	.2784	.6052	.63653	.77125	.82531	.2117	.2966	.5710	28
33	.62320	.78206	.79686	.2549	.2787	.6046	.63675	.77107	.82580	.2109	.2969	.5705	27
34	.62342	.78188	.79734	.2542	.2790	.6040	.63697	.77088	.82629	.2102	.2972	.5699	26
35	0.62365	0.78170	0.79781	1.2534	1.2793	1.6034	0.63720	0.77070	0.82678	1.2095	1.2975	1.5694	25
36	.62388	.78152	.79829	.2527	.2795	.6029	.63742	.77051	.82727	.2088	.2978	.5688	24
37	.62411	.78134	.79876	.2519	.2798	.6023	.63765	.77033	.82776	.2081	.2981	.5683	23
38	.62433	.78116	.79924	.2512	.2801	.6017	.63787	.77014	.82825	.2074	.2985	.5677	22
39	.62456	.78097	.79972	.2504	.2804	.6011	.63810	.76996	.82874	.2066	.2988	.5672	21
40	0.62479	0.78079	0.80020	1.2497	1.2807	1.6005	0.63832	0.76977	0.82923	1.2059	1.2991	1.5666	20
41	.62501	.78061	.80067	.2489	.2810	.6000	.63854	.76958	.82972	.2052	.2994	.5661	19
42	.62524	.78043	.80115	.2482	.2813	.5994	.63877	.76940	.83022	.2045	.2997	.5655	18
43	.62547	.78025	.80163	.2475	.2816	.5988	.63899	.76921	.83071	.2038	.3000	.5650	17
44	.62570	.78007	.80211	.2467	.2819	.5982	.63921	.76903	.83120	.2031	.3003	.5644	16
45	0.62592	0.77988	0.80258	1.2460	1.2822	1.5976	0.63944	0.76884	0.83169	1.2024	1.3006	1.5639	15
46	.62615	.77970	.80306	.2452	.2825	.5971	.63966	.76865	.83218	.2016	.3010	.5633	14
47	.62638	.77952	.80354	.2445	.2828	.5965	.63989	.76847	.83267	.2009	.3013	.5628	13
48	.62660	.77934	.80402	.2437	.2831	.5959	.64011	.76828	.83317	.2002	.3016	.5622	12
49	.62683	.77915	.80450	.2430	.2834	.5953	.64033	.76810	.83366	.1995	.3019	.5617	11
50	0.62706	0.77897	0.80498	1.2423	1.2837	1.5947	0.64056	0.76791	0.83415	1.1988	1.3022	1.5611	10
51	.62728	.77879	.80546	.2415	.2840	.5942	.64078	.76772	.83465	.1981	.3025	.5606	9
52	.62751	.77861	.80594	.2408	.2843	.5936	.64100	.76754	.83514	.1974	.3029	.5600	8
53	.62774	.77842	.80642	.2400	.2846	.5930	.64123	.76735	.83563	.1967	.3032	.5595	7
54	.62796	.77824	.80690	.2393	.2849	.5924	.64145	.76716	.83613	.1960	.3035	.5590	6
55	0.62819	0.77806	0.80738	1.2386	1.2852	1.5919	0.64167	0.76698	0.83662	1.1953	1.3038	1.5584	5
56	.62841	.77788	.80786	.2378	.2855	.5913	.64189	.76679	.83712	.1946	.3041	.5579	4
57	.62864	.77769	.80834	.2371	.2858	.5907	.64212	.76660	.83761	.1939	.3044	.5573	3
58	.62887	.77751	.80882	.2364	.2861	.5901	.64234	.76642	.83811	.1932	.3048	.5568	2
59	.62909	.77733	.80930	.2356	.2864	.5896	.64256	.76623	.83860	.1924	.3051	.5563	1
60	0.62932	0.77715	0.80978	1.2349	1.2867	1.5890	0.64279	0.76604	0.83910	1.1917	1.3054	1.5557	0
M	Cosine	Sine	Cotan.	Tan.	Cosec.	Secant	Cosine	Sine	Cotan.	Tan.	Cosec.	Secant	M
128°						51°	129°						50°

TABLE: Natural Trigonometric functions, continued 5.6.8

40°						139°	41°					138°	
M	Sine	Cosine	Tan.	Cotan.	Secant	Cosec.	Sine	Cosine	Tan.	Cotan.	Secant	Cosec.	M
0	0.64279	0.76604	0.83910	1.1917	1.3054	1.5557	0.65606	0.75471	0.86929	1.1504	1.3250	1.5242	60
1	.64301	.76586	.83959	.1910	.3057	.5552	.65628	.75452	.86980	.1497	.3253	.5237	59
2	.64323	.76567	.84009	.1903	.3060	.5546	.65650	.75433	.87031	.1490	.3257	.5232	58
3	.64345	.76548	.84059	.1896	.3064	.5541	.65672	.75414	.87082	.1483	.3260	.5227	57
4	.64368	.76530	.84108	.1889	.3067	.5536	.65694	.75394	.87133	.1477	.3263	.5222	56
5	0.64390	0.76511	0.84158	1.1882	1.3070	1.5530	0.65716	0.75375	0.87184	1.1470	1.3267	1.5217	55
6	.64412	.76492	.84208	.1875	.3073	.5525	.65737	.75356	.87235	.1463	.3270	.5212	54
7	.64435	.76473	.84257	.1868	.3076	.5520	.65759	.75337	.87287	.1456	.3274	.5207	53
8	.64457	.76455	.84307	.1861	.3080	.5514	.65781	.75318	.87338	.1450	.3277	.5202	52
9	.64479	.76436	.84357	.1854	.3083	.5509	.65803	.75299	.87389	.1443	.3280	.5197	51
10	0.64501	0.76417	0.84407	1.1847	1.3086	1.5503	0.65825	0.75280	0.87441	1.1436	1.3284	1.5192	50
11	.64523	.76398	.84457	.1840	.3089	.5498	.65847	.75261	.87492	.1430	.3287	.5187	49
12	.64546	.76380	.84506	.1833	.3092	.5493	.65869	.75241	.87543	.1423	.3290	.5182	48
13	.64568	.76361	.84556	.1826	.3096	.5487	.65891	.75222	.87595	.1416	.3294	.5177	47
14	.64590	.76342	.84606	.1819	.3099	.5482	.65913	.75203	.87646	.1409	.3297	.5171	46
15	0.64612	0.76323	0.84656	1.1812	1.3102	1.5477	0.65934	0.75184	0.87698	1.1403	1.3301	1.5166	45
16	.64635	.76304	.84706	.1805	.3105	.5471	.65956	.75165	.87749	.1396	.3304	.5161	44
17	.64657	.76286	.84756	.1798	.3109	.5466	.65978	.75146	.87801	.1389	.3307	.5156	43
18	.64679	.76267	.84806	.1791	.3112	.5461	.66000	.75126	.87852	.1383	.3311	.5151	42
19	.64701	.76248	.84856	.1785	.3115	.5456	.66022	.75107	.87904	.1376	.3314	.5146	41
20	0.64723	0.76229	0.84906	1.1778	1.3118	1.5450	0.66044	0.75088	0.87955	1.1369	1.3318	1.5141	40
21	.64745	.76210	.84956	.1771	.3121	.5445	.66066	.75069	.88007	.1363	.3321	.5136	39
22	.64768	.76191	.85006	.1764	.3125	.5440	.66087	.75049	.88058	.1356	.3324	.5131	38
23	.64790	.76173	.85056	.1757	.3128	.5434	.66109	.75030	.88110	.1349	.3328	.5126	37
24	.64812	.76154	.85107	.1750	.3131	.5429	.66131	.75011	.88162	.1343	.3331	.5121	36
25	0.64834	0.76135	0.85157	1.1743	1.3134	1.5424	0.66153	0.74992	0.88213	1.1336	1.3335	1.5116	35
26	.64856	.76116	.85207	.1736	.3138	.5419	.66175	.74973	.88265	.1329	.3338	.5111	34
27	.64878	.76097	.85257	.1729	.3141	.5413	.66197	.74953	.88317	.1323	.3342	.5106	33
28	.64900	.76078	.85307	.1722	.3144	.5408	.66218	.74934	.88369	.1316	.3345	.5101	32
29	.64923	.76059	.85358	.1715	.3148	.5403	.66240	.74915	.88421	.1309	.3348	.5096	31
30	0.64945	0.76041	0.85408	1.1708	1.3151	1.5398	0.66262	0.74895	0.88472	1.1303	1.3352	1.5092	30
31	.64967	.76022	.85458	.1702	.3154	.5392	.66284	.74876	.88524	.1296	.3355	.5087	29
32	.64989	.76003	.85509	.1695	.3157	.5387	.66305	.74857	.88576	.1290	.3359	.5082	28
33	.65011	.75984	.85559	.1688	.3161	.5382	.66327	.74838	.88628	.1283	.3362	.5077	27
34	.65033	.75965	.85609	.1681	.3164	.5377	.66349	.74818	.88680	.1276	.3366	.5072	26
35	0.65055	0.75946	0.85660	1.1674	1.3167	1.5371	0.66371	0.74799	0.88732	1.1270	1.3369	1.5067	25
36	.65077	.75927	.85710	.1667	.3170	.5366	.66393	.74780	.88784	.1263	.3372	.5062	24
37	.65100	.75908	.85761	.1660	.3174	.5361	.66414	.74760	.88836	.1257	.3376	.5057	23
38	.65121	.75889	.85811	.1653	.3177	.5356	.66436	.74741	.88888	.1250	.3379	.5052	22
39	.65144	.75870	.85862	.1647	.3180	.5351	.66458	.74722	.88940	.1243	.3383	.5047	21
40	0.65166	0.75851	0.85912	1.1640	1.3184	1.5345	0.66479	0.74702	0.88992	1.1237	1.3386	1.5042	20
41	.65188	.75832	.85963	.1633	.3187	.5340	.66501	.74683	.89044	.1230	.3390	.5037	19
42	.65210	.75813	.86013	.1626	.3190	.5335	.66523	.74664	.89097	.1224	.3393	.5032	18
43	.65232	.75794	.86064	.1619	.3193	.5330	.66545	.74644	.89149	.1217	.3397	.5027	17
44	.65254	.75775	.86115	.1612	.3197	.5325	.66566	.74625	.89201	.1211	.3400	.5022	16
45	0.65276	0.75756	0.86165	1.1605	1.3200	1.5319	0.66588	0.74606	0.89253	1.1204	1.3404	1.5018	15
46	.65298	.75737	.86216	.1599	.3203	.5314	.66610	.74586	.89306	.1197	.3407	.5013	14
47	.65320	.75718	.86267	.1592	.3207	.5309	.66631	.74567	.89358	.1191	.3411	.5008	13
48	.65342	.75700	.86318	.1585	.3210	.5304	.66653	.74548	.89410	.1184	.3414	.5003	12
49	.65364	.75680	.86368	.1578	.3213	.5299	.66675	.74528	.89463	.1178	.3418	.4998	11
50	0.65386	0.75661	0.86419	1.1571	1.3217	1.5294	0.66697	0.74509	0.89515	1.1171	1.3421	1.4993	10
51	.65408	.75642	.86470	.1565	.3220	.5289	.66718	.74489	.89567	.1165	.3425	.4988	9
52	.65430	.75623	.86521	.1558	.3223	.5283	.66740	.74470	.89620	.1158	.3428	.4983	8
53	.65452	.75604	.86572	.1551	.3227	.5278	.66762	.74450	.89672	.1152	.3432	.4979	7
54	.65474	.75585	.86623	.1544	.3230	.5273	.66783	.74431	.89725	.1145	.3435	.4974	6
55	0.65496	0.75566	0.86674	1.1537	1.3233	1.5268	0.66805	0.74412	0.89777	1.1139	1.3439	1.4969	5
56	.65518	.75547	.86725	.1531	.3237	.5263	.66826	.74392	.89830	.1132	.3442	.4964	4
57	.65540	.75528	.86775	.1524	.3240	.5258	.66848	.74373	.89882	.1126	.3446	.4959	3
58	.65562	.75509	.86826	.1517	.3243	.5253	.66870	.74353	.89935	.1119	.3449	.4954	2
59	.65584	.75490	.86878	.1510	.3247	.5248	.66891	.74334	.89988	.1113	.3453	.4949	1
60	0.65606	0.75471	0.86929	1.1504	1.3250	1.5242	0.66913	0.74314	0.90040	1.1106	1.3456	1.4945	0
M	Cosine	Sine	Cotan.	Tan.	Cosec.	Secant	Cosine	Sine	Cotan.	Tan.	Cosec.	Secant	M
130°						49°	131°					48°	

TABLE: Natural Trigonometric functions, continued 5.6.8

42°						137°	43°						136°
M	Sine	Cosine	Tan.	Cotan.	Secant	Cosec.	Sine	Cosine	Tan.	Cotan.	Secant	Cosec.	M
0	0.66913	0.74314	0.90040	1.1106	1.3456	1.4945	0.68200	0.73135	0.93251	1.0724	1.3673	1.4663	60
1	.66935	.74295	.90093	.1100	.3460	.4940	.68221	.73115	.93306	.0717	.3677	.4658	59
2	.66956	.74276	.90146	.1093	.3463	.4935	.68242	.73096	.93360	.0711	.3681	.4654	58
3	.66978	.74256	.90198	.1086	.3467	.4930	.68264	.73076	.93415	.0705	.3684	.4649	57
4	.66999	.74236	.90251	.1080	.3470	.4925	.68285	.73056	.93469	.0699	.3688	.4644	56
5	0.67021	0.74217	0.90304	1.1074	1.3474	1.4921	0.68306	0.73036	0.93524	1.0692	1.3692	1.4640	55
6	.67043	.74197	.90357	.1067	.3477	.4916	.68327	.73016	.93578	.0686	.3695	.4635	54
7	.67064	.74178	.90410	.1061	.3481	.4911	.68349	.72996	.93633	.0680	.3699	.4631	53
8	.67086	.74158	.90463	.1054	.3485	.4906	.68370	.72976	.93687	.0674	.3703	.4626	52
9	.67107	.74139	.90515	.1048	.3488	.4901	.68391	.72956	.93742	.0667	.3707	.4622	51
10	0.67129	0.74119	0.90568	1.1041	1.3492	1.4897	0.68412	0.72937	0.93797	1.0661	1.3710	1.4617	50
11	.67150	.74100	.90621	.1035	.3495	.4892	.68433	.72917	.93851	.0655	.3714	.4613	49
12	.67172	.74080	.90674	.1028	.3499	.4887	.68455	.72897	.93906	.0649	.3718	.4608	48
13	.67194	.74061	.90727	.1022	.3502	.4882	.68476	.72877	.93961	.0643	.3722	.4604	47
14	.67215	.74041	.90780	.1015	.3506	.4877	.68497	.72857	.94016	.0636	.3725	.4599	46
15	0.67237	0.74022	0.90834	1.1009	1.3509	1.4873	0.68518	0.72837	0.94071	1.0630	1.3729	1.4595	45
16	.67258	.74002	.90887	.1003	.3513	.4868	.68539	.72817	.94125	.0624	.3733	.4590	44
17	.67280	.73983	.90940	.0996	.3517	.4863	.68561	.72797	.94180	.0618	.3737	.4586	43
18	.67301	.73963	.90993	.0990	.3520	.4858	.68582	.72777	.94235	.0612	.3740	.4581	42
19	.67323	.73943	.91046	.0983	.3524	.4854	.68603	.72757	.94290	.0605	.3744	.4577	41
20	0.67344	0.73924	0.91099	1.0977	1.3527	1.4849	0.68624	0.72737	0.94345	1.0599	1.3748	1.4572	40
21	.67366	.73904	.91153	.0971	.3531	.4844	.68645	.72717	.94400	.0593	.3752	.4568	39
22	.67387	.73885	.91206	.0964	.3534	.4839	.68666	.72697	.94455	.0587	.3756	.4563	38
23	.67409	.73865	.91259	.0958	.3538	.4835	.68688	.72677	.94510	.0581	.3759	.4559	37
24	.67430	.73845	.91312	.0951	.3542	.4830	.68709	.72657	.94565	.0575	.3763	.4554	36
25	0.67452	0.73826	0.91366	1.0945	1.3545	1.4825	0.68730	0.72637	0.94620	1.0568	1.3767	1.4550	35
26	.67473	.73806	.91419	.0939	.3549	.4821	.68751	.72617	.94675	.0562	.3771	.4545	34
27	.67495	.73787	.91473	.0932	.3552	.4816	.68772	.72597	.94731	.0556	.3774	.4541	33
28	.67516	.73767	.91526	.0926	.3556	.4811	.68793	.72577	.94786	.0550	.3778	.4536	32
29	.67537	.73747	.91580	.0919	.3560	.4806	.68814	.72557	.94841	.0544	.3782	.4532	31
30	0.67559	0.73728	0.91633	1.0913	1.3563	1.4802	0.68835	0.72537	0.94896	1.0538	1.3786	1.4527	30
31	.67580	.73708	.91687	.0907	.3567	.4797	.68856	.72517	.94952	.0532	.3790	.4523	29
32	.67602	.73688	.91740	.0900	.3571	.4792	.68878	.72497	.95007	.0525	.3794	.4518	28
33	.67623	.73669	.91794	.0894	.3574	.4788	.68899	.72477	.95062	.0519	.3797	.4514	27
34	.67645	.73649	.91847	.0888	.3578	.4783	.68920	.72457	.95118	.0513	.3801	.4510	26
35	0.67666	0.73629	0.91901	1.0881	1.3581	1.4778	0.68941	0.72437	0.95173	1.0507	1.3805	1.4505	25
36	.67688	.73610	.91955	.0875	.3585	.4774	.68962	.72417	.95229	.0501	.3809	.4501	24
37	.67709	.73590	.92008	.0868	.3589	.4769	.68983	.72397	.95284	.0495	.3813	.4496	23
38	.67730	.73570	.92062	.0862	.3592	.4764	.69004	.72377	.95340	.0489	.3816	.4492	22
39	.67752	.73551	.92116	.0856	.3596	.4760	.69025	.72357	.95395	.0483	.3820	.4487	21
40	0.67773	0.73531	0.92170	1.0849	1.3600	1.4755	0.69046	0.72337	0.95451	1.0476	1.3824	1.4483	20
41	.67794	.73511	.92223	.0843	.3603	.4750	.69067	.72317	.95506	.0470	.3828	.4479	19
42	.67816	.73491	.92277	.0837	.3607	.4746	.69088	.72297	.95562	.0464	.3832	.4474	18
43	.67837	.73472	.92331	.0830	.3611	.4741	.69109	.72277	.95618	.0458	.3836	.4470	17
44	.67859	.73452	.92385	.0824	.3614	.4736	.69130	.72256	.95673	.0452	.3839	.4465	16
45	0.67880	0.73432	0.92439	1.0818	1.3618	1.4732	0.69151	0.72236	0.95729	1.0446	1.3843	1.4461	15
46	.67901	.73412	.92493	.0812	.3622	.4727	.69172	.72216	.95785	.0440	.3847	.4457	14
47	.67923	.73393	.92547	.0805	.3625	.4723	.69193	.72196	.95841	.0434	.3851	.4452	13
48	.67944	.73373	.92601	.0799	.3629	.4718	.69214	.72176	.95896	.0428	.3855	.4448	12
49	.67965	.73353	.92655	.0793	.3633	.4713	.69235	.72156	.95952	.0422	.3859	.4443	11
50	0.67987	0.73333	0.92709	1.0786	1.3636	1.4709	0.69256	0.72136	0.96008	1.0416	1.3863	1.4439	10
51	.68008	.73314	.92763	.0780	.3640	.4704	.69277	.72115	.96064	.0410	.3867	.4435	9
52	.68029	.73294	.92817	.0774	.3644	.4699	.69298	.72095	.96120	.0404	.3870	.4430	8
53	.68051	.73274	.92871	.0767	.3647	.4695	.69319	.72075	.96176	.0397	.3874	.4426	7
54	.68072	.73254	.92926	.0761	.3651	.4690	.69340	.72055	.96232	.0391	.3878	.4422	6
55	0.68093	0.73234	0.92980	1.0755	1.3655	1.4686	0.69361	0.72035	0.96288	1.0385	1.3882	1.4417	5
56	.68115	.73215	.93034	.0749	.3658	.4681	.69382	.72015	.96344	.0379	.3886	.4413	4
57	.68136	.73195	.93088	.0742	.3662	.4676	.69403	.71994	.96400	.0373	.3890	.4408	3
58	.68157	.73175	.93143	.0736	.3666	.4672	.69424	.71974	.96456	.0367	.3894	.4404	2
59	.68178	.73155	.93197	.0730	.3669	.4667	.69445	.71954	.96513	.0361	.3898	.4400	1
60	0.68200	0.73135	0.93251	1.0724	1.3673	1.4663	0.69466	0.71934	0.96569	1.0355	1.3902	1.4395	0
M	Cosine	Sine	Cotan.	Tan.	Cosec.	Secant	Cosine	Sine	Cotan.	Tan.	Cosec.	Secant	M
132°						47°	133°						46°

TABLE: Natural Trigonometric functions, continued 5.6.8

44° **135°**

M	Sine	Cosine	Tangent	Cotangent	Secant	Cosecant	M
0	0.69466	0.71934	0.96569	1.0355	1.3902	1.4395	60
1	.69487	.71914	.96625	.0349	.3905	.4391	59
2	.69508	.71893	.96681	.0343	.3909	.4387	58
3	.69528	.71873	.96738	.0337	.3913	.4382	57
4	.69549	.71853	.96794	.0331	.3917	.4378	56
5	0.69570	0.71833	0.96850	1.0325	1.3921	1.4374	55
6	.69591	.71813	.96907	.0319	.3925	.4370	54
7	.69612	.71792	.96963	.0313	.3929	.4365	53
8	.69633	.71772	.97020	.0307	.3933	.4361	52
9	.69654	.71752	.97076	.0301	.3937	.4357	51
10	0.69675	0.71732	0.97133	1.0295	1.3941	1.4352	50
11	.69696	.71711	.97189	.0289	.3945	.4348	49
12	.69716	.71691	.97246	.0283	.3949	.4344	48
13	.69737	.71671	.97302	.0277	.3953	.4339	47
14	.69758	.71650	.97359	.0271	.3957	.4335	46
15	0.69779	0.71630	0.97416	1.0265	1.3960	1.4331	45
16	.69800	.71610	.97472	.0259	.3964	.4327	44
17	.69821	.71589	.97529	.0253	.3968	.4322	43
18	.69841	.71569	.97586	.0247	.3972	.4318	42
19	.69862	.71549	.97643	.0241	.3976	.4314	41
20	0.69883	0.71529	0.97700	1.0235	1.3980	1.4310	40
21	.69904	.71508	.97756	.0229	.3984	.4305	39
22	.69925	.71488	.97813	.0223	.3988	.4301	38
23	.69945	.71468	.97870	.0218	.3992	.4297	37
24	.69966	.71447	.97927	.0212	.3996	.4292	36
25	0.69987	0.71427	0.97984	1.0206	1.4000	1.4288	35
26	.70008	.71406	.98041	.0200	.4004	.4284	34
27	.70029	.71386	.98098	.0194	.4008	.4280	33
28	.70049	.71366	.98155	.0188	.4012	.4276	32
29	.70070	.71345	.98212	.0182	.4016	.4271	31
30	0.70091	0.71325	0.98270	1.0176	1.4020	1.4267	30
31	.70112	.71305	.98327	.0170	.4024	.4263	29
32	.70132	.71284	.98384	.0164	.4028	.4259	28
33	.70153	.71264	.98441	.0158	.4032	.4254	27
34	.70174	.71243	.98499	.0152	.4036	.4250	26
35	0.70194	0.71223	0.98556	1.0146	1.4040	1.4246	25
36	.70215	.71203	.98613	.0141	.4044	.4242	24
37	.70236	.71182	.98671	.0135	.4048	.4238	23
38	.70257	.71162	.98728	.0129	.4052	.4233	22
39	.70277	.71141	.98786	.0123	.4056	.4229	21
40	0.70298	0.71121	0.98843	1.0117	1.4060	1.4225	20
41	.70319	.71100	.98901	.0111	.4065	.4221	19
42	.70339	.71080	.98958	.0105	.4069	.4217	18
43	.70360	.71059	.99016	.0099	.4073	.4212	17
44	.70381	.71039	.99073	.0093	.4077	.4208	16
45	0.70401	0.71018	0.99131	1.0088	1.4081	1.4204	15
46	.70422	.70998	.99189	.0082	.4085	.4200	14
47	.70443	.70977	.99246	.0076	.4089	.4196	13
48	.70463	.70957	.99304	.0070	.4093	.4192	12
49	.70484	.70936	.99362	.0064	.4097	.4188	11
50	0.70505	0.70916	0.99420	1.0058	1.4101	1.4183	10
51	.70525	.70895	.99478	.0052	.4105	.4179	9
52	.70546	.70875	.99536	.0047	.4109	.4175	8
53	.70566	.70854	.99593	.0041	.4113	.4171	7
54	.70587	.70834	.99651	.0035	.4117	.4167	6
55	0.70608	0.70813	0.99709	1.0029	1.4122	1.4163	5
56	.70628	.70793	.99767	.0023	.4126	.4159	4
57	.70649	.70772	.99826	.0017	.4130	.4154	3
58	.70669	.70752	.99884	.0012	.4134	.4150	2
59	.70690	.70731	.99942	.0006	.4138	.4146	1
60	0.70711	0.70711	1.00000	1.0000	1.4142	1.4142	0
M	Cosine	Sine	Cotangent	Tangent	Cosecant	Secant	M

134° **45°**

VI—PROPERTIES OF SECTIONS

VI

PROPERTIES OF SECTIONS

PROPERTIES OF SECTIONS

Contents

Properties of sections 6.1

In Section I *Mechanics of Beams* the bending moments and reactions were calculated for beams with various spans and load types. The algebraic moment method was used to obtain the results from the external forces. A beam must now be selected which will have an adequate size, shape, and be of a material which will sustain the calculated bending moment. The resisting moment is the internal capacity of the beam to support equilibrium and must be equal to or greater than the maximum imposed bending moment. (Also, as discussed in Sections II, III and IV, the resisting shear capacity of the beam must exceed the maximum reaction and horizontal shear.) In order to calculate the resisting moment capacity of any beam or joist, the properties of the cross-section must be calculated.

Plane surfaces 6.1.1

Geometry defines a plane as a flat or level surface. A plane figure is a part of a plane surface bounded by straight or curved lines or by a combination of both. Suppose that a 2 x 4 wood joist is sawn into two pieces. The cut end will correspond to a rectangular plane section with a cross-sectional area of 8.0 square inches. The area is obtained by multiplying the breadth (b) times the depth (d). Handbooks illustrade the end profiles of steel beams, channels, angles and pipes, and also list their weights per foot, cross-sectional areas, and other properties. Each cross-section has certain mathematical elements such as: Area, Moment of Inertia and Center of Gravity. From these fundamental elements, other properties can be derived.

Neutral axis 6.1.2

The neutral axis is defined as a plane where there is neither tensile or compressive stress when the beam section is subjected to bending. It can be shown that the neutral axis of any cross section passes thru the center of gravity. Thus, if we locate the section center of gravity, we may draw the neutral axis as an imaginary line passing through the center of gravity.

The neutral axis should not be confused with terms such as centroid or axis. With respect to symmetrical sections such as

Neutral axis, continued 6.1.2

rectangles and squares, the location of the center of gravity can be obtained by observation. The neutral axis is simply located at the center, or one half the depth. In most cross sections representing steel shapes, the tables will show the location of two neutral axes. The major axis will be indicated as x–x. The minor axis will be noted as y–y. The minor axis is more significant in the design of steel columns. Compound sections are built up by an assembly of simple component shapes, and the overall outline of the plane figure is usually irregular and unsymmetrical. For such sections, the method of moments is employed to find the center of gravity point from which to draw the neutral axis.

Format for calculations 6.1.3

The following method for calculating the properties of sections, as used in examples, is a time-honored and well-accepted system which encourages accuracy. Many naval architects and bridge designers require their employees to follow this established system, since it offers continuity from designer to designer and a good file record. In some offices, the calculation outline is furnished as a printed form.

Begin the work by drawing the cross section to be studied, broken into several simple shapes. Establish an arbitrary base line from which to take moments for each part of the section. Calculate the area of each part. Add the moments and divide by the total area. The result will be the distance from the base line to center of gravity. A line drawn through the CG, parallel to the base line, will be the neutral axis. The other neutral axis (perpendicular to the first) can be calculated in the same manner. A simplified version of this calculation is:

$$\text{Neutral axis} = \frac{\text{Sum of moments about the base line}}{\text{Sum of the areas}} = \text{distance from base line.}$$

Center of gravity 6.1.4

Weight and volume are usually associated with the term *center of gravity*, and to be precise, a body must have weight to have a center of gravity. Nevertheless, it has long been the custom of engineers to apply this term to plane shapes. Thus the CG of a circular pipe is at its center. The CG of a triangle is at the intersection of the medians (bisectors of each angle). A square or rectangle will have its CG at a point where diagonals cross. When a bolt hole is punch out of a cross section, the cross-section is reduced so that the gravity axis is changed from the original location. The cut-out, or void, must be considered as a component part with a negative contribution and its moment and area must be deducted from the summation of the positive moments and areas. The examples will show how this deduction is best accomplished.

Extreme fiber distance 6.1.5

The distance from the neutral axis to the outermost fiber in a section is given the symbol (c). This distance is most important because it is critical to solving for the property of Section Modulus (S). With respect to symmetrical shapes, the distance c is one half the depth, or $c = \frac{d}{2}$. Note in the examples that this dimension may come either above or below the neutral axis. Bending stress increases in a section directly as c increases.

Moment of Inertia 6.1.6

Moment of Inertia (I) is the sum of the products obtained by multiplying each of the elementary areas of which the section is composed times the square of its perpendicular distance from the neutral axis of the section. By using calculus, mathematicians have developed many formulas which shorten the work for finding the value of I. Moment of inertia is expressed in biquadratic inches or inches to the fourth power I''^4).

Before the other sections properties can be obtained, the value of I must be calculated, because it is the factor from which the Section Modulus (S) and Radius of Gyration (r) is calculated. A convenient bending factor property, used in the design of eccentrically loaded columns, is also related to the value of I.

In order to calculate the Moment of Inertia for a complete section, we use the the formula:

$I = \text{Sum of } (Io + A\,l^2)$ for each component part.

where Io = the Moment of Inertia of each component part about its own center of gravity axis

A = the area of each component part in square inches

l = the perpendicular lever arm in inches from the neutral axis of the whole section to the center of gravity axis for each component part

TABLE 6.2.9 will give formulas for the value of Io for various simple shapes, into which a complete section may be broken. Be sure to chose the value of Io for the correct axis : the axis which is parallel to the neutral axis of the whole section. This may become confusing when designing angle sections with unequal legs. In dividing a section with unequal legs into component parts for calculations, it is best to sketch the relative orientation of each part as it enters the calculations.

Section modulus 6.1.7

When the moment of inertia (I) is divided by the distance from neutral axis to extreme fibers of section (c), the resulting value is the Section Modulus, which is denoted as S with dimensions of inches to the third power. By formula: $S = \frac{I}{c}$. Should there be any question concerning the applicable neutral axis, all doubts can removed by using subscript thus: $S_x = \frac{I_x}{c_x}$.

It is this property (S) that is most often used

Section modulus, continued 6.1.7

in beam design and bending formulas.

Using the section modulus in conjunction with the allowable unit stress of a material (F_b), the resisting moment is found as: RM = SF_b. To preserve equilibrium, RM must be equal to or greater than the imposed bending moment (M). To find the actual stress a section is sustaining under external loads, the formula is written as:

$f_b = \frac{M}{S}$, *with the bending moment in inch pounds.*

Radius of gyration 6.1.8

The standard symbol for the *radius of gyration* is the small letter r, and it is the distance from the neutral axis of a plane area to an imaginary point where the whole area could be concentrated while the moment of inertia is left unchanged. Stated as a formula: $I = Ar^2$, or by transposing the formula, $r = \sqrt{\frac{I}{A}}$. Also $r_x = \frac{I_x}{A}$.
There is a radius of gyration property related to axis x–x, and another for axis y-y. Denote them as r_x or r_y. In the design of columns, the axis with the least value of r will be the governing factor for finding the allowable unit compressive stress. The other factor in the design of columns is the unbraced length; the slenderness ratio of a column is $\frac{l}{r}$. Remember, use the least radius of gyration value for this ratio, and the length of column *l* must always be given in inches. Handbooks containing tables of rolled steel sections will provide the value of both r_x and r_y. This property of a section will only be important for axial or concentric loads which develop compressive stress.

Bending factor 6.1.9

The bending factors about the major and minor axes are indicated as B and B_y. These properties are derived from the value of I or S. The bending factor for axis x–x is found by dividing the Section Area by the Section modulus for axis x–x. Stated in formula: $B_x = \frac{A}{S_x}$, or $B_y = \frac{A}{S_y}$.
These bending factors are considered properties of sections, but have often not been given adequate consideration. This property is a convenient aid in the design of columns with eccentric loads which develop a bending moment in the cross section. Certain designers feel that a small amount of eccentricity and bending moment can be neglected; however, it may be important in locations exposed to fire or extreme heat. The design examples for eccentric loaded steel columns included in Section II of this manual will illustrate the advantage of using the bending factor.

Calculating rectangular section properties 6.2.1

Timber sections used on construction projects are usually square or rectangular shapes, and the properties may be obtained by fundamental formulas, or taken from Table 6.4.3. Remember, the actual size of a dressed timber is not the nominal size.

In using the tables, remember to select the proper axis to calculate properties. When plate girders are built up into a compound section, as will be illustrated in the examples, the value of Io will be taken from these tables.

Calculating hollow tube section properties 6.2.2

Symmetrical sections such as circular pipe and square hollow tubes may have their properties computed by several methods. The shortest method is perhaps the most accurate. Simply assume the section to be a solid plane figure and calculate the value of I. Since the hollow portion is on the same neutral axis, calculate the moment of inertia of the void, then deduct the result from the value of I for a solid section. For a square tube section, the formula is: $I = \frac{bd^3 - b_1d_1^3}{12}$. The area of solid would be $A = bd - b_1d_1$, where b and d are the outer dimensions of the tube and b_1 and d_1 are the dimensions of the void.

Calculating compound section properties 6.2.3

It is not uncommon for bridge designers to build up compound sections of plate and angle girders which have overall depths of 60 inches or more. Extremely long spans require great depth in order to build up the moment of inertia. Another common use for built-up compound sections is for the lintels over wide window openings or groups of main entrance doors. Component parts may consist of flat plate for flanges and web, with unequal leg angles for connections and to give lateral support.

When designing a compound section we use the same formula as in calculating the Moment of Inertia for a complex single shape. However, in this case each component part is actually a separate structural shape which we will fabricate into the compound section. Repeating the formula in short form:

$I = \Sigma(Io + Ao\,l^2)$ where

- Io = component Moment of Inertia
- Ao = component area
- l = perpendicular lever distance from the compound section neutral axis to the component neutral axis

Transferring Moments of Inertia 6.2.4

Instructions presented in the preceding paragraph for the Moment of Inertia (I) in a compound section involves transferring the Moment of Inertia to another parallel axis. Let this term be thoroughly understood as the subject may be submitted to applicants seeking state registration. In performing calculations for the Moment of Inertia (I) of a compound section it is necessary to transfer the value of I of a component member to another axis not passing through the center of gravity of the component area. This axis must be parallel to the component gravity axis.

I represents the Moment of Inertia about the neutral axis of the compound section. Io is the Moment of Inertia of a component section about its own gravity axis. Aol^2 equals the area of the component section multiplied by the lever distance squared from axis x–x of the compound section. Follow the examples, and note that the components are tabulated separately with their individual properties readily available for use. As a general rule, the Moment of Inertia may be calculated about any axis parallel to the neutral axis of a section in like manner.

Calculating composite section properties 6.2.5

A *composite* section is a section which is composed of two or more materials of different characteristics. Concrete offers little resistance to bending and tension stress, but excels in sustaining compressive forces. Steel and concrete composite beam and girder sections eliminate the need for the costly form material and labor that is required for solid concrete monolithic tee beams. Composite construction permits a concrete slab floor to be placed upon steel ribbed decking over steel beams, with shear connectors providing the horizontal shear resistance between the steel and the slab.

When calculating the properties of a composite section, only a portion of the slab is considered in compression. This effective concrete area is determined by the modulus of elasticity (E) of the two materials. The neutral axis of the section will be near the top flange of the steel component. In isolated cases, the neutral axis may fall in the lower part of slab leaving the steel component to resist the total forces of tension. An example is provided to illustrate the merit of employing the form system in calculating the properties. To make the composite section effective, the sag in the steel must be shored up until the concrete attains the necessary strength to sustain compression. The need for shoring is described in Section IV which discusses concrete design and composite construction.

Tables for Rectangular sections 6.2.6

TABLE 6.4.3 includes Moment of Inertia values for rectangular shapes, and may be used to compute the Moments of Inertia of plate girders, columns, and other compound sections in which plates are used. To obtain the Moment of Inertia of any rectangle, multiply the tabulated value for its depth by its width in inches. To illustrate the use of the tables: An alternate size Southern Pine 2 x 4 S4S, has a net size of 1½ x 3½ inches. The value of $I = \frac{bd}{12}$ or 5.36″. Turning to the tables, it is shown that a rectangle 1″ x 3½″ has an I value of 3.573″, and a ½″ x 3½″ size lists I = 1.787″. Adding the two values together, I = 3.573 + 1.787 = 5.36″. The areas may be combined in a like manner.

Calculating rivet hole deductions 6.2.7

The area of a rivet or bolt hole is the diameter times the plate thickness. All holes are to be considered as ⅛ inch larger than the rivet diameter. Accuracy is improved when the holes are figured separately, and the sum of the moments deducted from the solid components. In general practice, rivets are placed in a pattern symmetrical to the neutral axis, and the hole areas can be easily totaled. This method is used in the examples. In observing voids in a section, keep in mind that the holes are to be treated as rectangles and each void has its own gravity axis, area, and I values. The lever (*l*) distance is taken from the gravity axis of the void to the neutral axis of the section.

Calculating laced compound section properties 6.2.8

Many bridge designers have used angle and channel shapes for struts and columns in designs of the older spans. With the sections separated by a lacing of short, diagonal, flat bars on the outside, the area of the section (and weight) remains low, while the value of I is raised considerably. It follows that the radius of gyration is also increased considerably. Laced sections are not likely to become obsolete because the welding lacing improves the section property values by eliminating the rivet holes. Refer to EXAMPLE 6.3.16 of a laced section, and note the value of r is calculated to be 3.14″ for a column eight inches square. With such a large value, the slenderness ratio would permit the column to be extremely long, and still not require any intermediate bracing support. This advantage and the reduction in weight is best illustrated in the lacing of long crane booms in current use.

TABLE: Section property formulas 6.2.9

PLANE SHAPE OR SECTION	DISTANCE TO OUTER FIBER "C"	MOMENT OF INERTIA I_o "4 ABOUT AXIS X-X	SECTION MODULUS S "3 ABOUT AXIS X-X	RADIUS OF GYRATION r ABOUT AXIS X-X
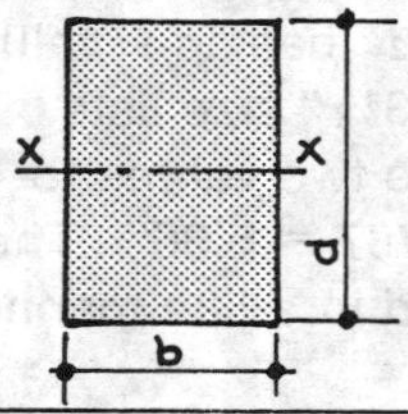 x, x, d, b	$\frac{d}{2}$	$\frac{bd^3}{12}$	$\frac{bd^2}{6}$	$\frac{d}{\sqrt{12}} = 0.289\,d$
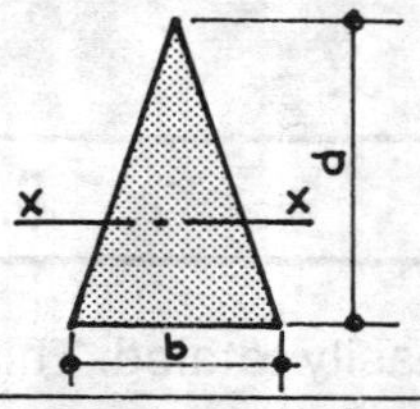 x, x, d, b	$\frac{2d}{3}$	$\frac{bd^3}{36}$	MIN. $\frac{bd^2}{24}$	$\frac{d}{\sqrt{18}} = 0.236\,d$
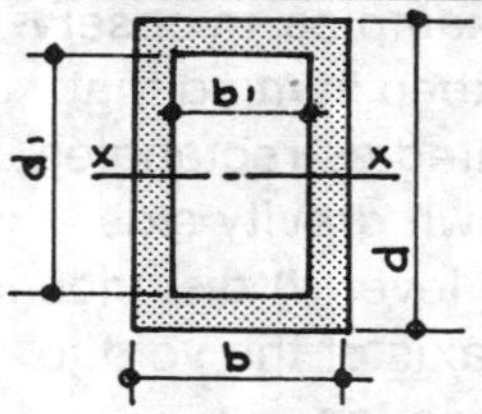 b_1, d_1, x, x, d, b	$\frac{d}{2}$	$\frac{bd^3 - b_1 d_1^3}{12}$	$\frac{bd^3 - b_1 d_1^3}{6d}$	$\sqrt{\frac{bd^3 - b_1 d_1^3}{12(bd - b_1 d_1)}}$
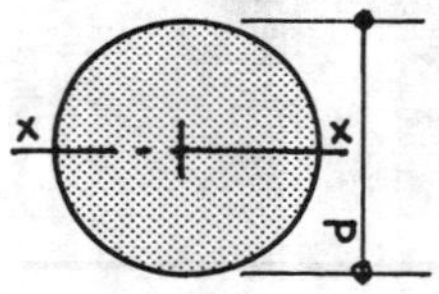 x, x, d	$\frac{d}{2}$	$\frac{\pi d^4}{64} = 0.0491\,d^4$	$\frac{\pi d^3}{32} = 0.0982\,d^3$	$\frac{d}{4}$
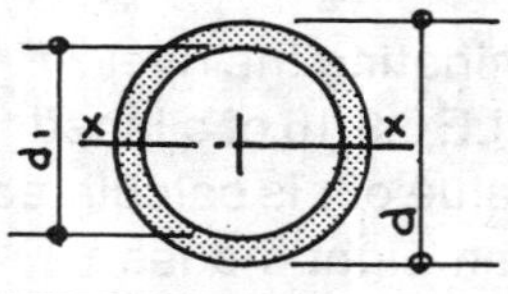 d_1, x, x, d	$\frac{d}{2}$	$\frac{\pi(d^4 - d_1^4)}{64} = 0.0491\,(d^4 - d_1^4)$	$\frac{\pi(d^4 - d_1^4)}{32d} = 0.0982\,\frac{(d^4 - d_1^4)}{d}$	$\frac{\sqrt{d^2 + d_1^2}}{4}$
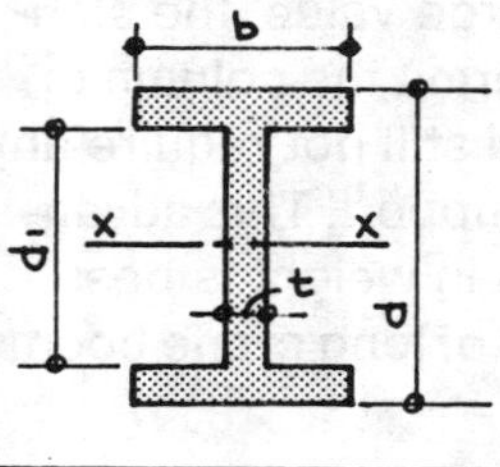 b, d_1, x, x, t, d	$\frac{d}{2}$	$\frac{bd^3 - d_1^3(b-t)}{12}$	$\frac{bd^3 - d_1^3(b-t)}{6d}$	$\sqrt{\frac{bd^3 - d_1^3(b-t)}{12[bd - d_1(b-t)]}}$
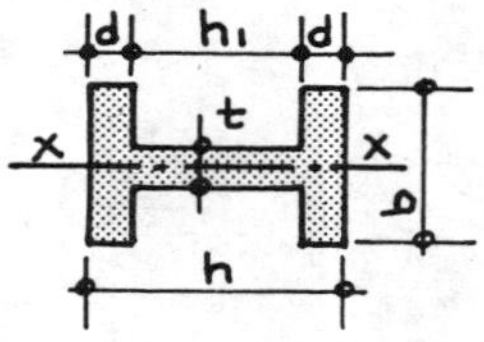 d, h_1, d, x, t, x, b, h	$\frac{b}{2}$	$\frac{2db^3 + h_1 t^3}{12}$	$\frac{2db^3 + h_1 t^3}{6b}$	$\sqrt{\frac{2db^3 + h_1 t^3}{12[bh - h_1(b-t)]}}$

EXAMPLE I: Locate centroid: moment method 6.3.1

A cross section is composed of four (4) rectangular parts as shown in illustration. Each part is to be an integral unit of the full plane, and are considered to be acting in unison.

REQUIRED:
Layout the figure as illustrated with each part given a mark of identification. Take the necessary steps to calculate the Horizontal Neutral Axis and its distance from Base line at bottom. From NA, identify the distance from Axis to extreme fibers or known as dimension c.

STEP I:
Moment arms from base line to center of gravity axis for each part is noted in outline form as distance "d."

Moment = Area × distance.

STEP II:
Summations:

$$\frac{\Sigma M}{\Sigma A} = \frac{230.0}{66.00} = 3.48 \text{ Inches.}$$

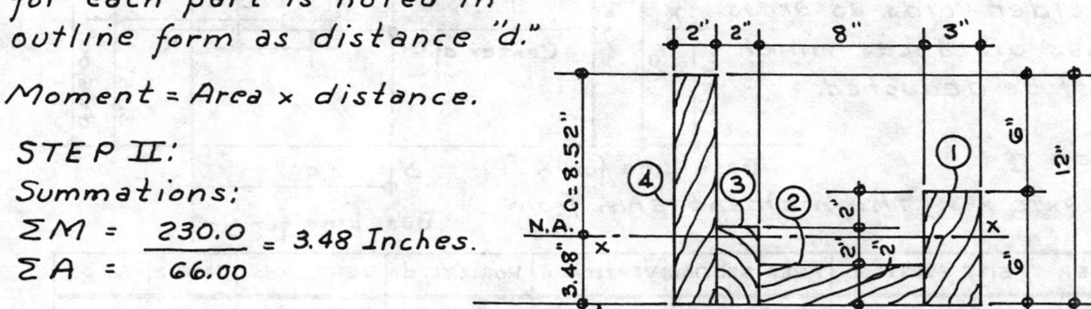

Location NA from Base line = 3.48 Inches.

SECTION	SIZE	AREA A□"	DISTANCE "d"	Ad=MOM.
1	3" × 6"	18.00	3.00"	54.0
2	8" × 2"	16.00	1.00"	16.0
3	2" × 4"	8.00	2.00"	16.0
4	2" × 12"	24.00	6.00"	144.0
	ΣA =	66.00□"		ΣM=230.0

STEP III:
Distance from Neutral Axis to extreme fibers:
Total depth section involved in calculations was 12.0 Inches.
C = 12.00 − 3.48 = 8.52 Inches

EXAMPLE II: Locate centroid: moment method 6.3.2

Assume a plane surface 10.0" x 6.00" represents a steel plate. Thickness of plate is of no consequence. There are two cutout voids. 1 Square void, and 1 hole are cut and placed in location shown in drawing.

REQUIRED:

Take moments about base to locate horizontal centroid x-x, then take moments about right side to locate vertical Axis y-y. Note on drawing where the exact Center of Gravity is located.

STEP I:

Make an outlined form to designate sizes, areas, moment arms (d) and consider voids as areas times distances which must be deducted.

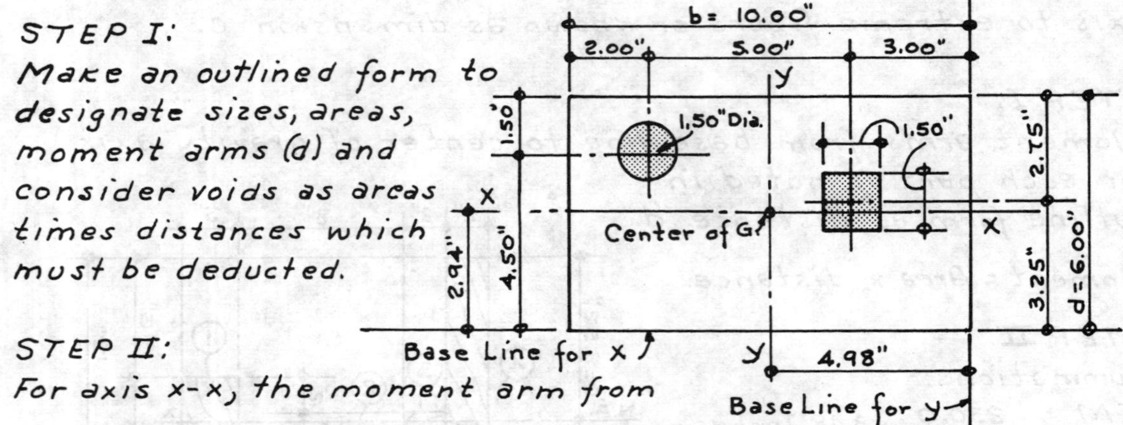

STEP II:

For axis x-x, the moment arm from

FOR AREA	SIZE SECTION	AREA SECTION	VERTICAL d	HORIZ'TL. d	VERT. Ad.	HORIZ. Ad
bd	10.0" x 6.0"	+ 60.00 □"	3.00"	5.00"	+ 180.00	+ 300.00
0.7854 D²	1.50" DIAMETER	− 1.77	4.50"	8.00"	− 7.97	− 14.16
bd	1.50" x 1.50"	− 2.25	3.25"	3.00"	− 7.31	− 6.75
		+ 55.98			+ 164.72	+ 279.09

base line to center of gravity of each section is noted as the vertical moment arm. The lever for axis y-y is horizontal.

STEP III

Summation of Vertical Moments = 164.72 $''^2$

Summation of all Areas = 55.98 □"

Distance from Base Line to Centroid x-x = $\frac{164.72}{55.98}$ = 2.94 Inches.

For vertical axis y-y:

Horizontal Moments: $\Sigma M = \frac{279.09}{55.98}$ = 4.98 Inches from Base line.

Three Sections net. ΣA = 55.98

Center of Gravity of plane surface is located at intersecting point of x-x and y-y.

EXAMPLE: Properties of rectangular section 6.3.3

The net size of a surfaced laminated wood section is 4.0 inches for breadth (b), and 8.0 inches for depth (d). Plane figure is a solid rectangle.

REQUIRED:

Draw the section to scale and designate the major and minor axes. Use formulas to calculate the properties of; A, I and S.

STEP I:

Axis x-x is major axis because it is normal to greater dimension and contains greater value of I.

About axis x-x.

$A = bd$ or $A = 4.0 \times 8.0 = 32.0 \square''$ $\quad S_x = \frac{bd^2}{6}$ or $S_x = \frac{4.0 \times 8.0 \times 8.0}{6} = 42.67''^3$

$I_x = \frac{bd^3}{12}$ and $I_x = \frac{4.0 \times 8.0 \times 8.0 \times 8.0}{12} = 170.67''^4$ $\quad c = \frac{d}{2}$ or $4.00''$

Also $S_x = \frac{I_x}{c}$ $\quad S_x = \frac{170.67}{4} = 42.67''^3$

STEP II:

About minor axis y-y. In formula, b becomes 8.00" and d, becomes 4.0".

$I_y = \frac{8.0 \times 4.0 \times 4.0 \times 4.0}{12} = 42.67''^4$ $\quad c = \frac{4.00}{2} = 2.00''$

$S_y = \frac{42.67}{2.0} = 21.33''^3$ or $S_y = \frac{8.0 \times 4.0 \times 4.0}{6} = 21.33''^3$

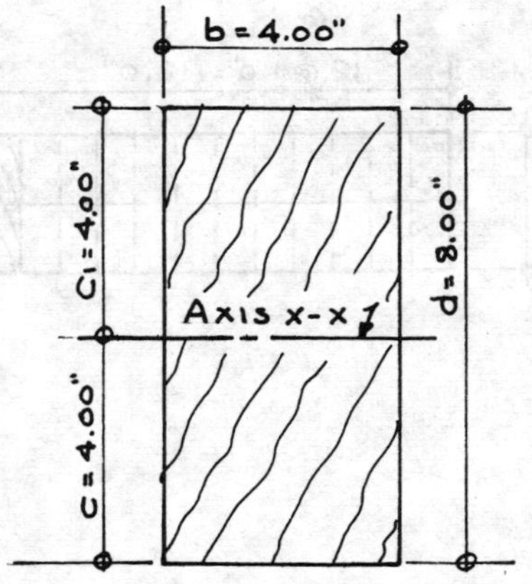

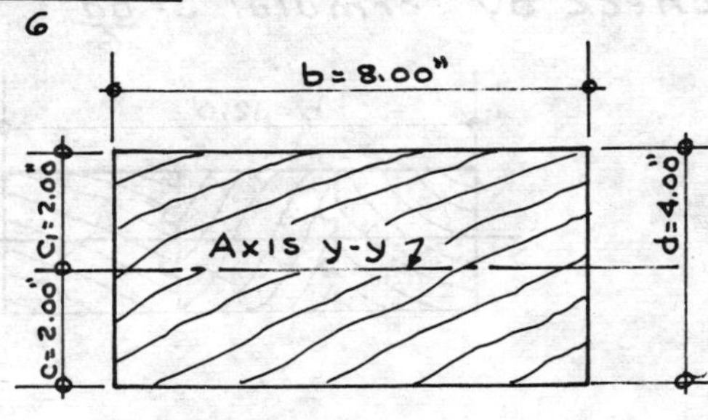

EXAMPLE: Properties of strip plank decking 6.3.4

A tongue and grooved wood deck material is 3⅝ inches in depth and consists of random widths. When designers desire to space joists or segment arches, the design is based upon the combined dead and live load as required by the applicable building code. The figures given for design loads are given in; Pounds per square foot, and therefore a strip load is 1.0 foot wide.

REQUIRED:
Calculate the Section Modulus of a strip load section of wood deck 3⅝ inches in depth. Use tables for I.

STEP I:
This will be a symmetrical rectangular shape 12.0"x3.625", and concerns the minor axis, however the values of I in tables must be taken from x-x axis. The strip load rectangle section is: b= 12.0" and d= 3.625"

STEP II:
Cut the 12 inch width into twelve 1"x3.625" components.
From tables: A 1.0"x3.625" has this value: $I_x = 3.970''^4$
For 12" width, $I_y = 3.970 \times 12 = 47.64''^4$ $C = \frac{3.625}{2} = 1.8125''$

For strip section: $S = \frac{47.64}{1.8125} = 26.3''^3$

To check by Formula: $S = \frac{bd^2}{6}$ $S = \frac{12.0 \times 3.625 \times 3.625}{6} = 26.3''^3$

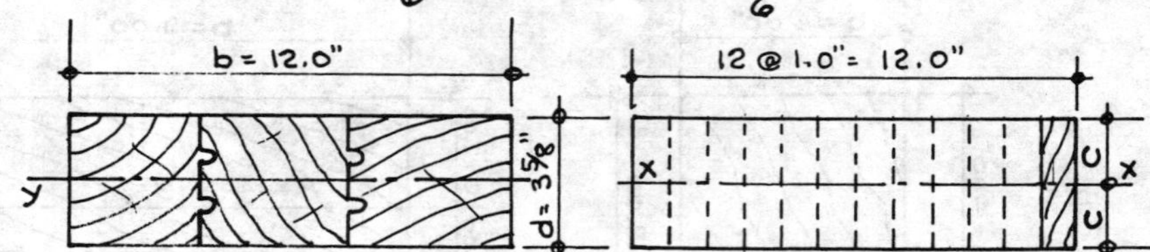

EXAMPLE: Properties of ribbed decking 6.3.5

A flat 20 Gauge steel sheet is rolled into a ribbed type section for use as a deck or form for concrete. Sheets are produced in various width of 24, 30, and 36 inches, and depths up to 8.0 inches. The listed properties provided by the manufacturer will generally refer a deck width of 12 inches, which is comparable to strip load designing.

REQUIRED:
Illustration represents a 12 inch width of a simple "A" type of ribbed deck. Calculate the values of the following: Weight in steel, Cross section area, I_x, and S_x. Treat the section as having symmetry about axis x-x, which is not the case in most decks.

STEP I:
Divide the section into vertical and horizontal pieces.

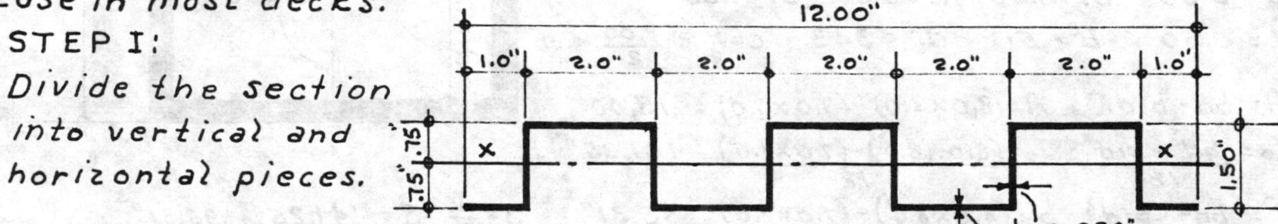

Area of Vertical Ribs = 6 x 1.50 x 0.036 = 0.324 □"
Area of Horizontal Parts = 6 x 2.00 x 0.036 = 0.432 □"
Sum of A = 0.756 □"

Weight per square foot in steel = 0.756 x 3.40 = 2.57 Lbs.

STEP II:
Total width of Vertical Ribs = 6 x 0.032 = 0.216" (Becomes b)
d = 1.50" and $I_o = \frac{bd^3}{12}$ $I_x = \frac{0.216 \times 1.50^3}{12} = 0.06075''^4$
The I_o of horizontal components will be of too little value to be a consideration.

STEP III
Lever distance from axis x-x to gravity axis of horizontal components is: $l = 0.750 - \frac{0.36}{2} = 0.732''$
Horizontals: A = 0.432 □"
Then $Al^2 = 0.432 \times 0.732^2 = 0.23191$
Total value $I_x = Al^2 + I_o = 0.06075 + 0.23191 = 0.29266''^4$

STEP IV:
For Section Modulus of 12 inch strip width, $S_x = \frac{I_x}{c}$

$S_x = \frac{0.29266}{0.75} = 0.3902''^3$

EXAMPLE: Hollow sections: square and round 6.3.6

The square and circular plane figures illustrated represent structural steel sections. Each tube has the same wall thickness, with side dimensions of square tube being the same as round tube.

REQUIRED:
Calculate the following properties for comparison: A, I_o, S, and r. Use the formulas applicable to each section, then calculate the net weight with, steel = 3.40 Pounds per square inch per lineal foot.

b = 8.00"
b1 = 7.00"
t = 0.50"
d = 8.00"
d1 = 7.00"
Axes

STEP I (Square Tube Properties)

$b = 8.00''$ $b_1 = 7.00''$ $d = 8.00''$ $d_1 = 7.00''$

$t = 0.50''$ $b^3 = 512$ $d_1^3 = 343$ $c = \frac{d}{2} = \frac{8.00}{2} = 4.00''$

$A = bd - b_1 d_1.$ $A = (8.0\times8.0) - (7.0\times7.0) = 15.00 \,\square''$

$I_o = \frac{bd^3 - b_1 d_1^3}{12}.$ $I_o = \frac{(8.0\times8.0^3) - (7.0\times7.0^3)}{12} = 141.25''^4$

$S = \frac{bd^3 - b_1 d_1^3}{6d}.$ $S = \frac{(8.0\times8.0^3) - (7.0\times7.0^3)}{6\times8.00} = 35.31''^3$ $S = \frac{I}{C}$ $S = \frac{141.25}{4.00} = 35.31''^3$

$r = \sqrt{\frac{bd^3 - b_1 d_1^3}{12A}}.$ $r = \sqrt{\frac{(8.0\times8.0^3) - (7.0\times7.0^3)}{12\times15.00}} = 3.07''$ or $r = \frac{I}{A}$ $r = \sqrt{\frac{141.25}{15.00}} = 3.07''$

Weight per lineal foot = 15.00 x 3.40 = 5.10# (In steel).

STEP II: (Circular Tube Properties)

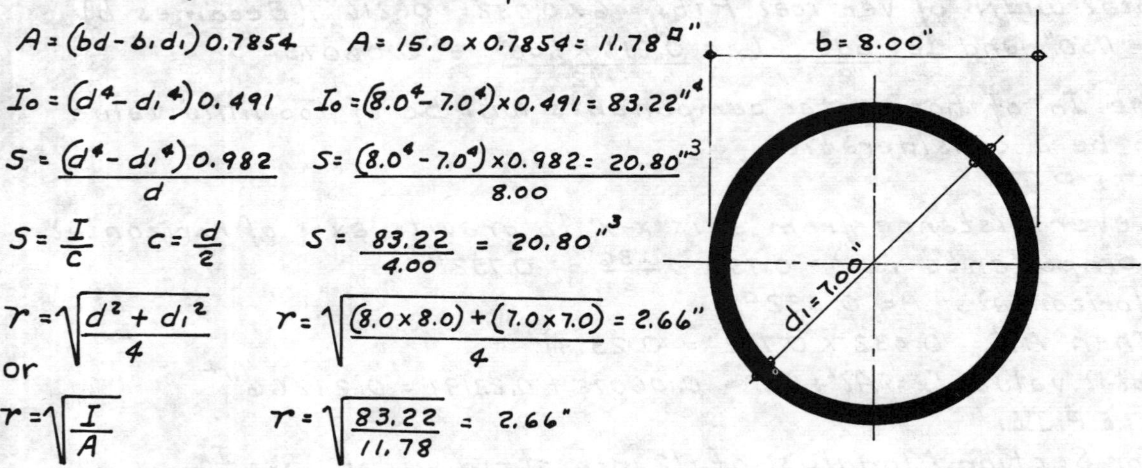

$A = (bd - b_1 d_1)\,0.7854$ $A = 15.0 \times 0.7854 = 11.78 \,\square''$

$I_o = (d^4 - d_1^4)\,0.491$ $I_o = (8.0^4 - 7.0^4)\times0.491 = 83.22''^4$

$S = \frac{(d^4 - d_1^4)\,0.982}{d}$ $S = \frac{(8.0^4 - 7.0^4)\times0.982}{8.00} = 20.80''^3$

$S = \frac{I}{C}$ $c = \frac{d}{2}$ $S = \frac{83.22}{4.00} = 20.80''^3$

$r = \sqrt{\frac{d^2 + d_1^2}{4}}$ $r = \sqrt{\frac{(8.0\times8.0) + (7.0\times7.0)}{4}} = 2.66''$

or

$r = \sqrt{\frac{I}{A}}$ $r = \sqrt{\frac{83.22}{11.78}} = 2.66''$

Weight per lineal foot = 11.78 x 3.40 = 4.00 Lbs. (In steel)

EXAMPLE: Locate centroid for compound section 6.3.7

A compound Section is built up of 4 Rectangular parts as shown in illustration. Whole plane is symmetrical about axis y-y, but axis x-x is not known.

REQUIRED:

Prepare a neat form to tabulate areas, distances and the moments required to find axis x-x. Extend the form to include moment arms from each section to axis x-x, then calculate the moment of Inertia of whole plane. Denote the distances from axis x-x to centers of gravity as lever (λ).

Note that the moment of Inertia is thus:

$I_{x-x} = A\lambda^2 + I_o$

Take the value of I_o from the tables or calculate it by formula $I = \frac{bd^3}{12}$.

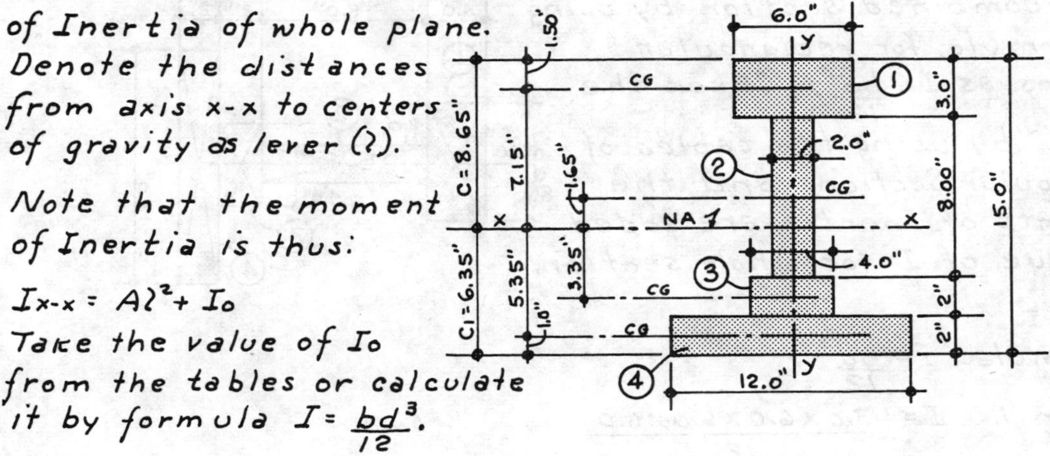

STEP I:

Preparing form and all distances will be noted on sections.
Sect. 1: $A\lambda^2 = 18.00 \times 7.15 \times 7.15 = 920.20$ I_{xo} will be found in tables of Rectangular Section Properties in Table 6.4.3.

FOR CENTROID SOLUTIONS					FOR MOMENTS OF INERTIA			
SECT.	SIZE"	AREA□"	d"	Ad	λ_{x-x}	$A\lambda^2$	I_o	$A\lambda^2 + I_o$
1	6.0×3.0	18.00	13.50	243.00	7.15"	920.20	13.50	933.71
2	2.0×8.0	16.00	8.00	128.00	1.65"	16.33	85.33	101.66
3	4.0×2.0	8.00	3.00	24.00	3.35	89.78	2.67	92.45
4	12.0×2.0	24.00	1.00	24.00	5.35	686.94	8.00	694.94
		Σ= 66.00		ΣM=419.00				Σ=1822.76"⁴

Neutral Axis x-x $= \frac{419.00}{66.00} = 6.35"$ from bottom base line.

Distance to extreme fiber: $C = 15.00 - 6.35 = 8.65$ inches.

Section Modulus: $S_x = \frac{I_x}{C} = \frac{1822.76''^4}{8.65''} = 210.72''^3$

EXAMPLE: Properties of symmetrical shapes 6.3.8

The same cross-section as illustrated in previous example. The whole section is symmetrical about the minor axis y-y which is the centroid for each of the four (4) sectional parts. The combined section is now turned 90 degrees counter-clockwise so that the breadth (b) dimension is horizontal.

REQUIRED:

Calculate the Moment of Inertia of the combined section by using the formula for rectangular sections as $I = \frac{bd^3}{12}$. Check the results by using the tables of Rectangular Sections and the Moments of Inertia are added for value of I for whole section.

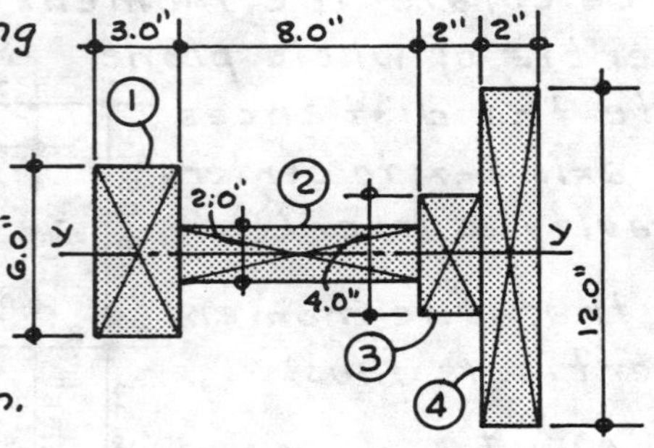

STEP I:

By Formula: $I = \frac{bd^3}{12}$

Section 1: $I = \frac{3.0 \times 6.0 \times 6.0 \times 6.0}{12} =$ 54.00

Section 2: $I = \frac{8.0 \times 2.0 \times 2.0 \times 2.0}{12} =$ 5.33

Section 3: $I = \frac{2.0 \times 4.0 \times 4.0 \times 4.0}{12} =$ 10.67

Section 4: $I = \frac{2.0 \times 12.0 \times 12.0 \times 12.0}{12} =$ 288.00

$\Sigma I_o = 358.00''^4$

STEP II:

By using Rectangular shape tables: 1 inch is greatest b.

Section 1. I_x = 3 × 18.00 = 54.00
Section 2. I_y = 8 × 0.66 = 5.33
Section 3. I_x = 2 × 5.333 = 10.67
Section 4. I_x = 2 × 144.00 = 288.00

$\Sigma I_o = 358.00''^4$

NOTE:

The I_o for section 2 was obtained by cutting section into 8 pieces 1.0"×2.0" and taking I about x-x axis. I_x for a section 1.0"×2.0" = 0.666. 8 pieces = 5.33"^{4}

EXAMPLE: Rolled section: slopes and fillets 6.3.9

Prior to 1971, the United States Steel Corporation advertised they had rolled a section with a weight of 730 Pounds per lineal foot. They referred to this section as a 14 WF 730, although it is not listed as being of 14 inch width. The listed properties of this 14 WF 730 Section were given as follows:
$A = 214.65^{\square''}$ $S_x = 1280.6''^3$ $I_x = 14{,}371.4''^4$ $I_y = 4{,}716.8''^4$ and $r_y = 4.69$
The physical dimensions are: Flanges 18.0" x 5.0", and Web = 3.07" x 12.50". Depth d = 22.50"

REQUIRED:

Neglect the flange slopes and web fillets in drawing the cross section and make up with 3 rectangular components. Calculate the properties about major and minor axes to determine the deduction of area and value of I caused by roller bevel shaped rims.

STEP I:

Section is drawn and dimensions noted from neutral axis to gravity axis of each part 1, 2, and 3. Distances = ʅ.
I_o is calculated as $\frac{bd^3}{12}$ or taken from tables of Rectangular shape properties.

STEP II:

Section is symmetrical and c = 11.25"

$$S_x = \frac{I_x}{c} = \frac{14{,}655.91}{11.25} = 1300.0''^3$$

$$r_x = \sqrt{\frac{I_x}{A}} = \sqrt{\frac{14{,}655.91}{218.38}} = 8.21$$

STEP III:

About axis y-y, an equation is thus:

$$I_y = \left[\frac{2 \times (5.0 \times 18.0^3)}{12}\right] + \left(\frac{12.5 \times 3.07^3}{12}\right) =$$

$$I_y = 4890.18''^4$$

SECT.	SIZE	A□"	ʅ"	Aʅ²	Io x-x	Io + Aʅ²
1	18.0" x 5.0"	90.00	8.75	6890.62	187.50	7078.12
2	3.07" x 12.50"	38.38	-0-	-0-	499.67	499.67
3	18.0" x 5.0"	90.00	8.75	6890.62	187.50	7078.12
		Σ = 218.38			Ix = Σ =	14,655.91"⁴

STEP IV:

The value of I contained in slopes and fillets:

For I_x: $14{,}655.91 - 14{,}371.4 = 284.51''^4$

For I_y: $4{,}890.18 - 4{,}716.8 = 173.38''^4$

EXAMPLE: Cover plates to reinforce a beam 6.3.10

A beam section must have a minimum Section Modulus of $31.8''^3$ to meet job requirements, and available are a number of 8 WF 17 Sections, light channels and some 3/8 inch thick plates. Emergency conditions prevail and time is important. Cutting torch and welding equipment is available on site.

REQUIRED:

Determine the width of 2 cover plates necessary to raise the Section Modulus to $31.80''^3$. Use 3/8" ℞. Check work by inserting figures in tabular form used in office.

STEP I:

Collect property data on 8 WF 17 Section:

$A = 5.00^{\square''}$ $S_x = 14.10''^3$ $c = 4.00''$ and $I_x = Sc$ or $I_x = 14.10 \times 4.00 = 56.40''^4$

STEP II:

With 3/8" cover plates, $c = 4.000 + 0.375 = 4.375''$

Required $I_x = 31.80 \times 4.375 = 139.22''^4$ (with cover plates)

Required for 2 ℞, $I_x = 139.22 - 56.40 = 82.82''^4$

For 1 plate, $I_x = \frac{82.82}{2} = 41.41''^4$

STEP III

Gravity axis of 3/8" ℞ $= \frac{0.375}{2} = 0.1875''$

Then lever $l = 4.000 + 0.1875 = 4.1875''$

Neglecting I_o for ℞, use $Al^2 = I$

$l^2 = 4.1875 \times 4.1875 = 17.48$ $A = \frac{I}{l^2} = \frac{41.41}{17.48} = 2.37^{\square''}$

Width of Plate $= \frac{2.37}{0.375} = 6.32''$ (use $6\frac{3}{8}''$)

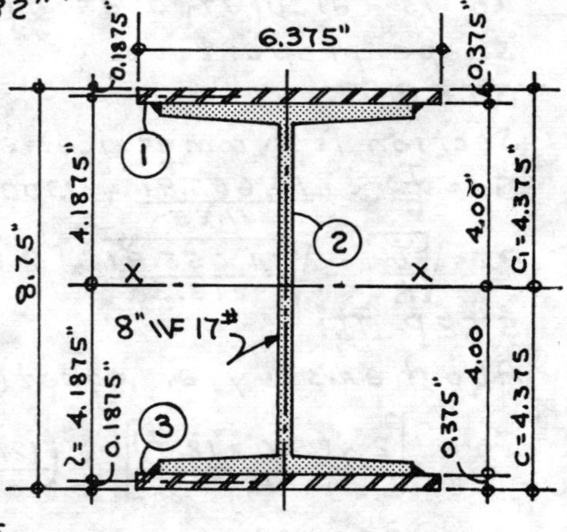

STEP IV:

Check results in tabular form for files.

COVER PLATES ADDED TO ROLLED SHAPE.						INCREASE $I_x = 82.87''^4$		
M'K.	SECTION	A□"	d"	Ad	l"	Al^2	I_o	$Al^2 + I_o$
1	℞ 6.375 x 0.375	2.37			4.1875	41.41	0.028	41.438
2	8" WF 17#	5.00			-0-	—	56.400	56.400
3	℞ 6.375 x 0.375	2.37			4.1875	41.41	0.028	41.438
								$\Sigma = 139.276''^4$

$S_x = \frac{I_x}{c}$ $S_x = \frac{139.276}{4.375} = 31.84''^3$ Meets requirements.

Use 2 Cover plates $\frac{3}{8}'' \times 6\frac{3}{8}''$ and weld.

EXAMPLE: Rectangular structural tube 6.3.11

A rectangular structural tube has outside dimensions of 3.0" x 8.0", and inside dimensions of 2.0" x 7.0". The wall thickness is therefore ½ inches, and section is symmetrical.

REQUIRED:

Use the formulas to calculate the properties of: A, I, S, and r about the major and minor axes. Neglect the rounded corners as shown in A.I.S.C. Manual which will reduce the values to a small extent.

STEP I:

About major axis x-x. Outside $b = 3.0''$, and inside $b_1 = 2.0''$. $d = 8.0''$ and $d_1 = 7.0''$. For area: $A = bd - b_1 d_1$.

$A = (3.0 \times 8.0) - (2.0 \times 7.0) = 10.0$ Sq. In.

$$I_x = \frac{bd^3 - b_1 d_1^3}{12}. \quad I_x = \frac{(3.0 \times 8.0^3) - (2.0 \times 7.0^3)}{12} = 70.83''^4$$

$$S_x = \frac{bd^3 - b_1 d_1^3}{6d}. \quad S_x = \frac{(3.0 \times 8.0^3) - (2.0 \times 7.0^3)}{6 \times 8.0} = 17.71''^3$$

$$r_x \sqrt{\frac{bd^3 - b_1 d_1^3}{12A}}. \quad r_x = \sqrt{\frac{(3.0 \times 8.0^3) - (2.0 \times 7.0^3)}{12 \times 10.0}} = 2.65$$

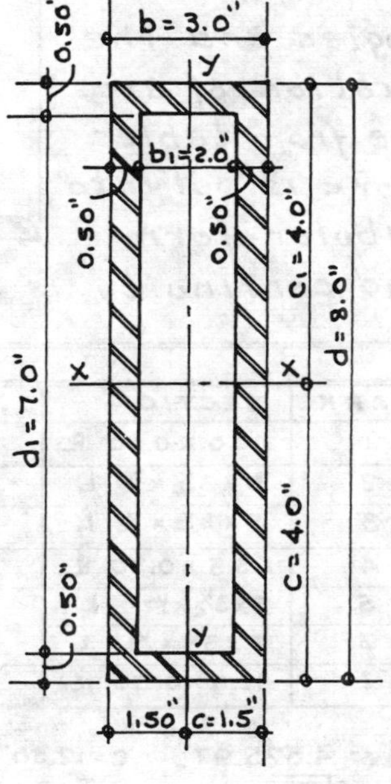

STEP II:

Computing about minor axis y-y. dimensions for b and d are reversed.

$$I_y = \frac{(8.0 \times 3.0 \times 3.0 \times 3.0) - (7.0 \times 2.0 \times 2.0 \times 2.0)}{12} = 13.33''^4$$

$$S_y = \frac{(8.0 \times 3.0 \times 3.0 \times 3.0) - (7.0 \times 2.0 \times 2.0 \times 2.0)}{6 \times 3.0} = 8.88''^3$$

$$r_y = \sqrt{\frac{I_y}{A}}. \quad r_y = \sqrt{\frac{13.33}{10.0}} = 1.17$$

DESIGNERS NOTE:

When comparing the results obtained in this example with the properties given in the AISC Handbook, it will emphasize this fact: If the area contained in the round corners is lost, and these small area were multiplied times their (2) distances squared, then deducted from I_x as found, the final rusults of values would be the same as given by A.I.S.C. tables.

EXAMPLE: Symmetrical compound section 6.3.12

A compound section is to be constructed with ½ plate. Flange is 12.0" wide. Total depth: 25.0". Connecting plate are 4 angles 5"x 3½"x ½" with long legs horizontal. Welding to be used.

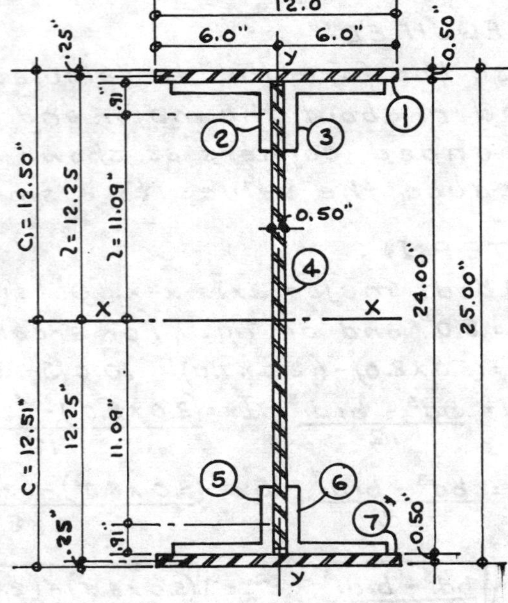

REQUIRED:

Draw the Cross section assembly. Calculate the properties about both axes. The moments of Inertia of angles are to be tranferred to NA of compound section.

STEP I:

Properties of angles and the location of axes are from tables. Work is put into tabular form and continuity is shown.

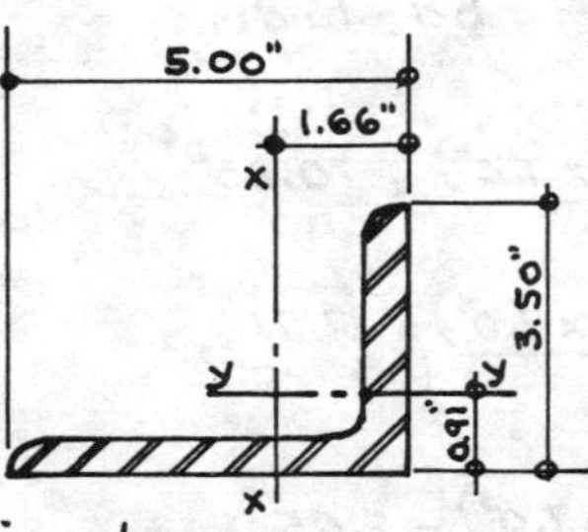

MAJOR AXIS X-X						
MARK	SECTION	A□"	ℓ"	Aℓ²	Io	Aℓ² + Io
1	12.0 x 0.50 ℞	6.00	12.25	900.36	0.125	900.485
2	5 x 3½ x ½ L	4.00	11.09	491.95	4.100	496.050
3	5 x 3½ x ½ L	4.00	11.09	491.95	4.100	496.050
4	23.5 x 0.50 ℞	11.75	-0-	-0-	540.800	540.800
5	5x3½x½ L	4.00	11.09	491.95	4.100	496.050
6	5 x 3½ x ½ L	4.00	11.09	491.95	4.100	496.050
7	12.0 x 0.50 ℞	6.00	12.25	900.36	0.125	900.485
		Σ= 39.75 □"				Σ= 4,325.970"⁴

$I_x = 4{,}325.97''^4$ $\quad c = 12.50''$ $\quad S_x = \frac{4325.97}{12.50} = 346.0''^3$

$r_x = \sqrt{\frac{I_x}{A}}$ $\quad S_x = \frac{I_x}{c}$ $\quad A = 39.75\,□''$ $\quad r_x = \sqrt{\frac{4325.97}{39.75}} = 10.44$

MINOR AXIS Y-Y						
MARK	SECTION	A□"	ℓ"	Aℓ²	Io	Aℓ² + Io
1	0.50"x 12.00" ℞	6.00	-0-	-0-	72.00	72.000
2	5"x3½"x ½ L×	4.00	1.91	14.592	10.00	24.592
3	5"x 3½"x ½" L×	4.00	1.91	14.592	10.00	24.592
4	23.5 x 0.50 ℞	11.75	-0-	-0-	0.25	0.250
5	5 x 3½ x ½ L×	4.00	1.91	14.592	10.00	24.592
6	5 x 3½ x ½ L×	4.00	1.91	14.592	10.00	24.592
7	0.50 x 12.00 ℞	6.00	-0-	-0-	72.00	72.000
		Σ= 39.75 □"				Σ= 242.618"⁴

$S_y = \frac{242.618}{6.00} = 40.50''^3$ $\quad c = \frac{12.0}{2} = 6.00''$ $\quad r_y = \sqrt{\frac{242.618}{39.75}} = 2.46$

EXAMPLE: Designing a special lintel 6.3.13

A wide window opening in a 12 inch masonary wall must be a special type for window installation. An 11.0" x 0.25" steel plate is proposed to serve for end bearing. Section modulus required is minimum of 5.00"³. Architects detail designates vertical plates to separate brick courses.

REQUIRED:
Check the Architects detail to ascertain the property of S on the horizontal Neutral Axis.

STEP I:
Detail will be revised to delete masonry, plaster and sash.

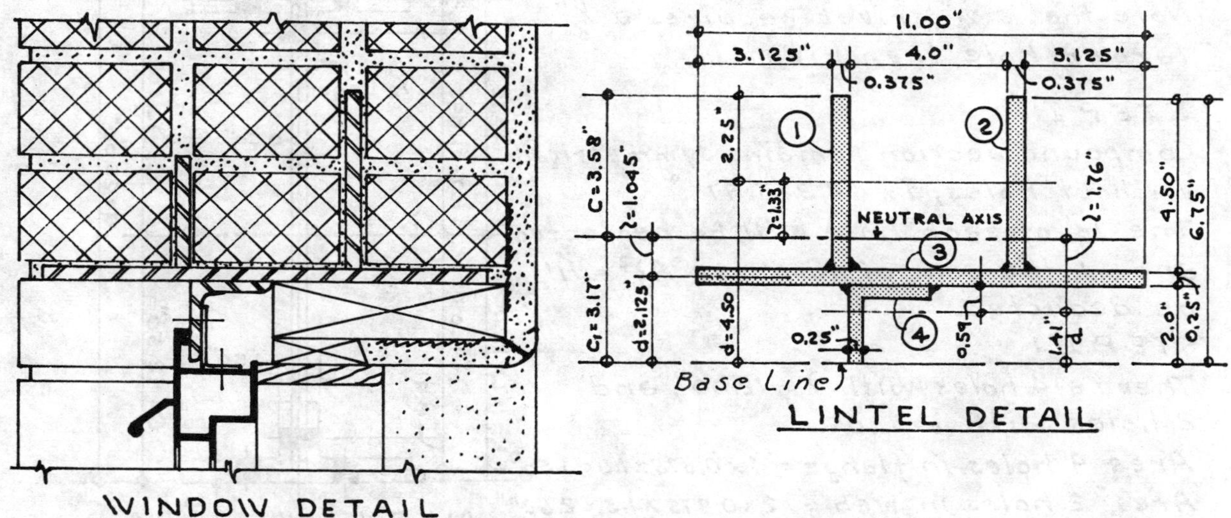

STEP II:
Elements in tabluar form for continuity:

PROPERTIES OF WINDOW LINTEL "A" IN 12" MASONRY WALL								
M'K.	SECTION	A□"	d"	Ad	ℓ"	Aℓ²	I₀	Aℓ² + I₀
1	0.375 x 4.50 ℞	1.688	4.500	7.596	1.330	2.986	2.848	5.834
2	0.375 x 4.50 ℞	1.688	4.500	7.596	1.330	2.986	2.848	5.834
3	11.0 x 0.25 ℞	2.750	2.125	5.844	1.045	3.003	0.350	3.353
4	2x2x¼ L	0.940	1.410	1.325	1.760	2.912	0.014	2.926
	ΣA=	7.066 □"		ΣM: 22.361			IΣ =	17.947"⁴

Neutral Axis distance from Base Line $= \frac{\Sigma M}{\Sigma A} = \frac{22.361}{7.066} = 3.17''$

$I_{NA} = 17.947''^4$ $c = 3.58''$ $S_{NA} = \frac{I}{C}$ $S = \frac{17.947}{3.58} = 5.03''^3$ (OK Accept lintel)

Weight per foot in steel = 3.40A Wt = 3.40 x 7.066 = 24.04#/'

EXAMPLE: Deducting for rivet hole areas 6.3.14

The illustrated Compound Cross Section will be recognized as a previous example with its properties calculated on the basis of welded assembly. The Moment of Inertia, I_x, was found to be $4{,}325.970''^4$, without any voids in area.

REQUIRED:

Assume that the compound section is to be shop fabricated with 3/4 inch diameter hot rivets power driven. Calculate the Moments of Inertia about axis x-x, and the value of S_x. Designate on drawing the lever distances from gravity axes of rivet holes to main Neutral Axis. Note that a $\frac{3}{4}''\phi$ rivet requires a $\frac{1}{8}''$ larger hole than rivet.

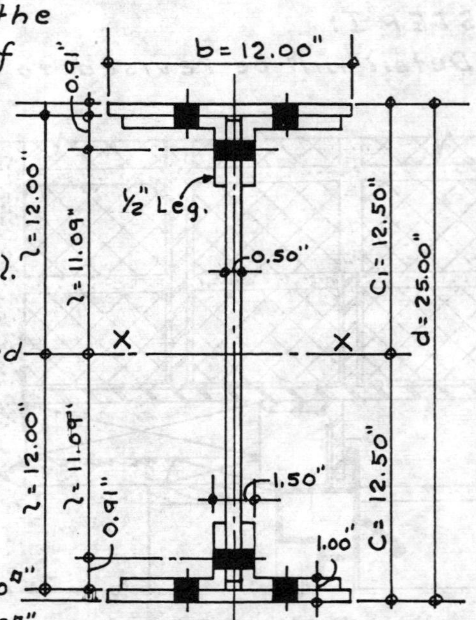

ALL RIVET HOLES $\frac{7}{8}''$ DIA.

STEP I:

Compound Section remains symmetrical. Without holes, $I_x = 4{,}325.97''^4$.

The I_o of each hole will be neglected and only the sum of moments $A\ell^2$ will be deducted.

STEP II:

There a 4 holes with $\ell = 12.00''$, and 2 holes with $\ell = 11.09$.

Area 4 holes in flange $= 4 \times 0.875 \times 1.0 = 3.50\square''$

Area 2 holes in web $= 2 \times 0.875 \times 1.5 = 2.62\square''$

STEP III:

Moments for rivet holes: $(A\ell^2)$.

In flange: $4 \times 12.0 \times 12.0 = 576.00$

In web: $2 \times 11.09 \times 11.09 = 245.97$

$\Sigma A\ell^2 = 821.97$

STEP IV:

Net value of I_x with holes $= 4{,}325.97 - 821.97 = 3504.00''^4$

$S_x = \dfrac{I_x}{c}$ $\quad S_x = \dfrac{3504.00}{12.50} = 280.2''^3$

For welded section, $S_x = 346.0''^3$ Rivet assembly reduces the Section Modulus $65.80''^3$ and internal resisting moment would be reduced considerably. For a steel beam with A36 Steel, $F_b = 22{,}000$ PSI. $RM = S_x F_b$.

Lost Resisting Moment $= 65.80 \times 22{,}000 = 1{,}447{,}600$ Inch Lbs.

EXAMPLE: Locate centroid: complex shape 6.3.15

The cross section illustration represents a compound section with 4 Component parts. Dimensions for each member is given, and the gravity axis for each component is noted. The illustration is to be treated as a plane figure without substance.

REQUIRED:
By method of moments, use bottom as a base line and find the horizontal Neutral Axis. Redraw the cross section and show the moment levers (λ) which are the distances from each gravity axis to the Neutral Axis.

FOR LOCATING NEUTRAL AXIS					FOR MOMENTS OF INERTIA			
M'K.	SECTION	A□"	d"	Ad"	λ"	$A\lambda^2$	I_o	$A\lambda^2 + I_o$
1	4.0"x2.0" ◺	4.000	1.67	6.667	0.46	0.846	0.888	1.734
2	2.0" Dia. ϕ	3.142	2.00	6.284	0.79	1.960	0.785	2.745
3	2.0" Sq. ⊡	1.750	2.00	3.500	0.79	1.092	0.911	2.003
4	8.0"x1.0" PL	8.000	0.50	4.000	0.71	4.033	0.667	4.700
	ΣA= 16,897□"		ΣM=20.451				Σ =	11.182"⁴

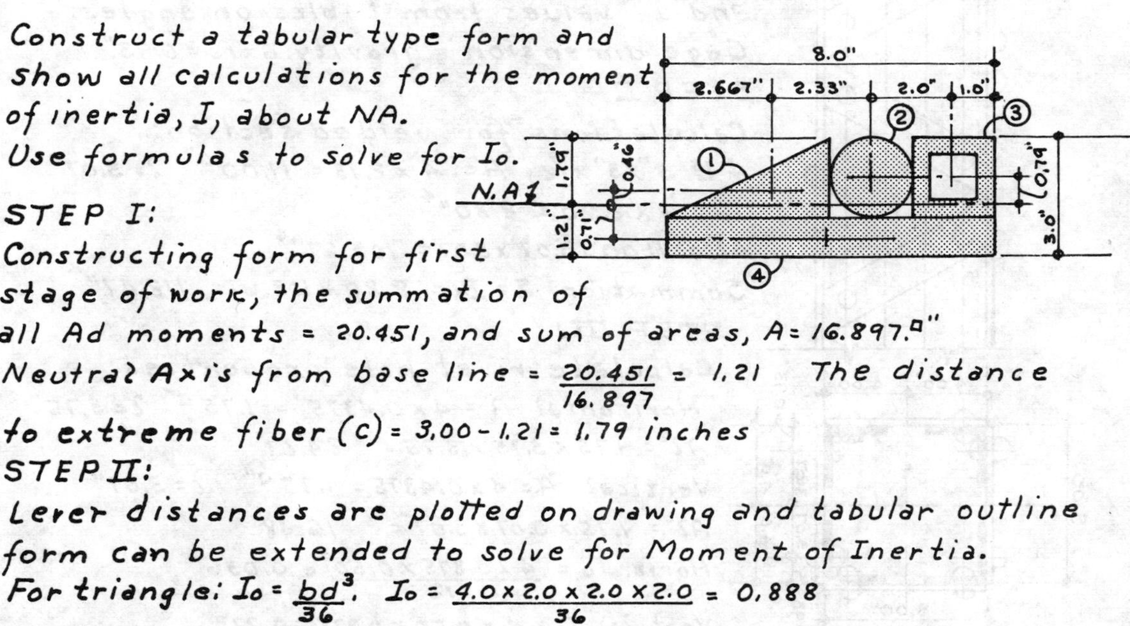

Construct a tabular type form and show all calculations for the moment of inertia, I, about NA.
Use formulas to solve for I_o.

STEP I:
Constructing form for first stage of work, the summation of all Ad moments = 20.451, and sum of areas, A = 16.897.□"

Neutral Axis from base line = $\frac{20.451}{16.897} = 1.21$ The distance to extreme fiber (c) = 3.00 − 1.21 = 1.79 inches

STEP II:
Lever distances are plotted on drawing and tabular outline form can be extended to solve for Moment of Inertia.

For triangle: $I_o = \frac{bd^3}{36}$. $I_o = \frac{4.0 \times 2.0 \times 2.0 \times 2.0}{36} = 0.888$

For Circle: $I = \frac{\pi d^4}{64}$ $I_o = \frac{3.1416 \times 2.0 \times 2.0 \times 2.0 \times 2.0}{64} = 0.7854$

For Hollow Square: $I_o = \frac{bd^3 - b_1 d_1^3}{12}$ $I_o = \frac{(2.0 \times 2.0^3) - (1.50 \times 1.50^3)}{12} = 0.911$

EXAMPLE: Properties of laced sections 6.3.16

A column section 8.00 inches square is composed of 4 equal leg angles 3"x3"x½. Fabrication is to be with ¾" φ Rivets placed in same plane on each angles gravity axis. Lacing to consist of ¼"x2" Flat bars placed on 4 sides.

REQUIRED:

Draw a partial elevation of column lacing, and symmetrical cross section. Calculate the Radius of Gyration and net Area when rivets are used for assembly.

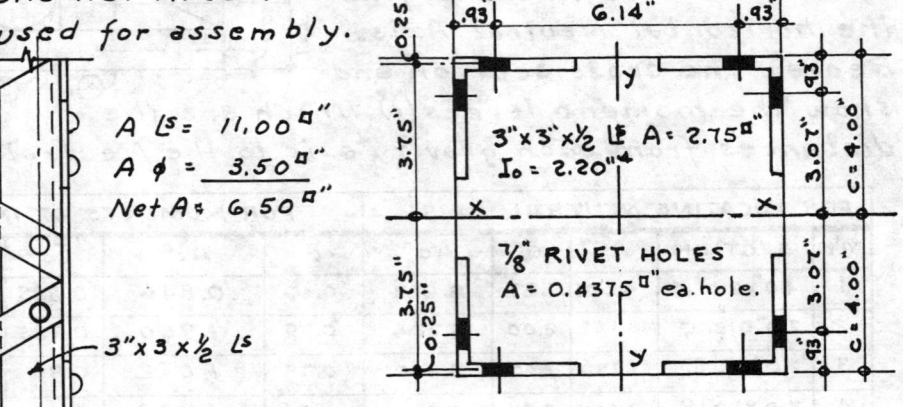

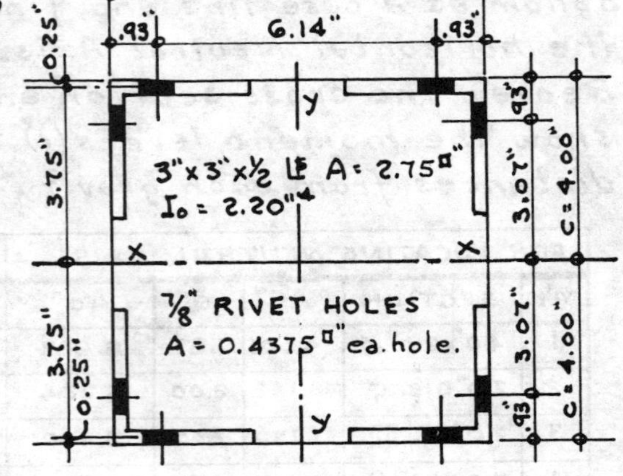

A L^s = 11.00□"
A φ = 3.50□"
Net A = 6.50□"

3"x3x½ L^s

4.00" 4.00"
8.00"
0.93"
3.07
4.0"

STEP I:

Draw sections to scale, and obtain A, and I_o values from tables on angles. Gage dimension = gravity axis = 0.93".

STEP II:

Calculations for welded section:

4 L^s 3"x3"x½, A = 4 x 2.75 = 11.00□" l = 3.07"

$I_o = 4 \times 2.20 = 8.80''^4$

$Al^2 = 11.00 \times 3.07 \times 3.07 = 103.67''^4$

Summation $\Sigma = I_x = 8.80 + 103.67 = 112.47''^4$

STEP III:

Calculate rivet hole properties:

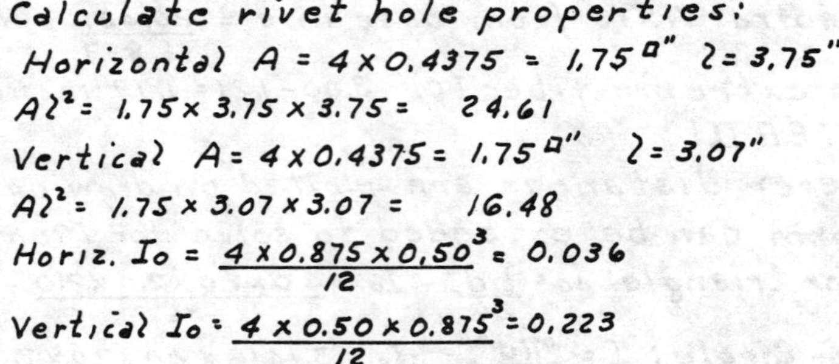

Horizontal A = 4 x 0.4375 = 1.75□" l = 3.75"

$Al^2 = 1.75 \times 3.75 \times 3.75 = 24.61$

Vertical A = 4 x 0.4375 = 1.75□" l = 3.07"

$Al^2 = 1.75 \times 3.07 \times 3.07 = 16.48$

Horiz. $I_o = \frac{4 \times 0.875 \times 0.50^3}{12} = 0.036$

Vertical $I_o = \frac{4 \times 0.50 \times 0.875^3}{12} = 0.223$

$-\Sigma = 24.61 + 16.48 + 0.036 + 0.223 = 41.349$

$I_x = 112.470 - 41.439 = 71.121''^4$ $r = \sqrt{\frac{71.121}{6.50}}$ 3.14"

EXAMPLE: Truss cord: properties of two angles 6.3.17

A 2 Angle truss chord in compression requires a section with a gross area of 3.25 Sq. In. Least radius of gyration cannot be less than 0.90." Gusset plates of ½ inch separate the angles. Proposed are 2 Ls 4x3x¼ with welded joining.

REQUIRED:
Calculate the value of r for angles short legs are back to back, then when long legs are back to back. Put all figures in format as shown in Section VI.

STEP I:
Drawings will be made for each arrangement of angles. From Table of rolled shapes, the properties of one angle are:

A = 1.69□" gauge y = 1.24" r_x = 1.28 I_x = 2.80"^{4} (Major axis).
gauge x = 0.74" r_y = 0.90 I_y = 1.40"^{4} (Minor axis).

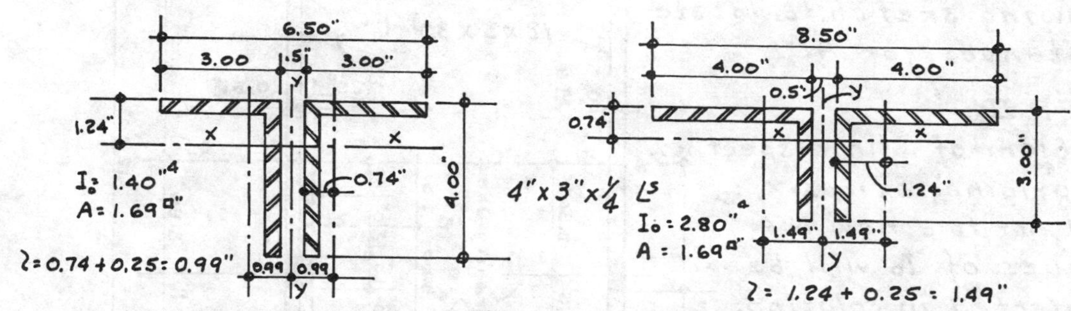

LONG LEGS BACK TO BACK SHORT LEGS BACK TO BACK

STEP II:

TRUSS TOP CHORD - ABOUT Y-Y.					LONG LEGS BACK TO BACK			
M'K	SECTION	A□"	d"	Ad	l"	Al²	Io	Al² + Io
1	4 x 3 x ¼ L	1.69			0.99	1.67	1.40	3.07
2	Do.	1.69			0.99	1.67	1.40	3.07
	ΣA: 3.38						Σ:	6.14"^{4}

On Vertical Axis y-y: $r_y = \sqrt{\frac{I}{A}}$ $r_y = \sqrt{\frac{6.14}{3.38}} = 1.33"$ $r_x = 1.28"$

TRUSS TOP CHORD - ABOUT Y-Y.					SHORT LEGS BACK TO BACK			
M'K.	SECTION	A□"	d"	Ad	l"	Al²	Io	Al² + Io
1	4 x 3 x ¼	1.69			1.49	3.75	2.80	6.55
2	Do.	1.69			1.49	3.75	2.80	6.55
	Σ: 3.38							Σ: 13.10"^{4}

On Vertical Axis y-y: $r_y = \sqrt{\frac{13.10}{3.38}} = 1.96"$ $r_x = 0.90"$

CONCLUSION: Use Long legs back to back

EXAMPLE: Scab plate to restore section properties 6.3.18

A 12"x3"x30# Channel is subjected to bending stress in the fibers below axis x-x. A bolt hole is to be drilled in tension fibers 2½ inches from bottom. Hole diameter is 1.0 inch.

REQUIRED;
Calculate the moment of Inertia (I) of Channel Section with hole cutout about it original axis x-x. Restore the lost value of I, and area with a scab plate in which the hole will be drilled through both web and scab.

STEP I:
From tables on Channel Sections, obtain the data thus:
$I_x = 161.20''^4$ $A = 8.79^{\square''}$
Web thickness, t = 0.51"
Drawing sketch to locate distances for $A\ell^2$.

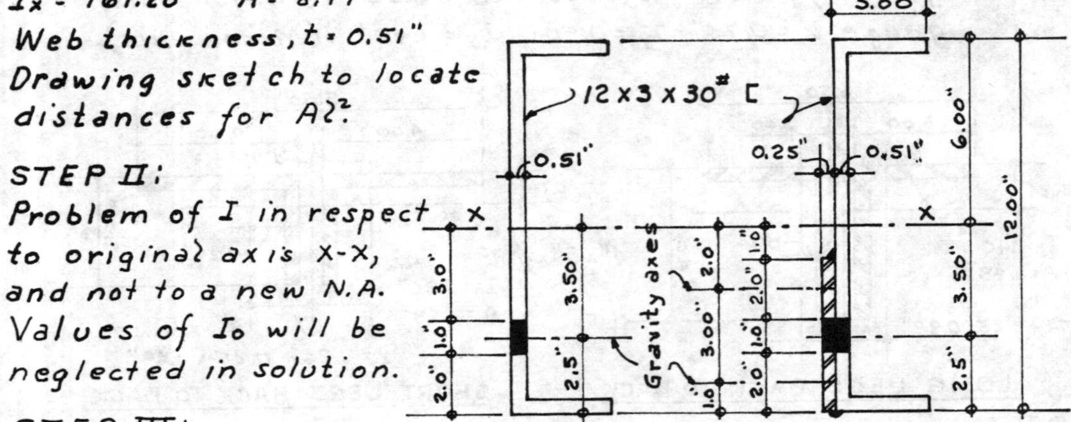

STEP II:
Problem of I in respect to original axis X-X, and not to a new N.A. Values of I_o will be neglected in solution.

STEP III:
Net area of hole = 1.00 x 0.51 = $0.51^{\square''}$
Net area [with hole = 8.79 − 0.51 = $8.28^{\square''}$
Distance of hole from x-x. ℓ = 3.50"
$A\ell^2$ = 0.51 x 3.50 x 3.50 = $6.25''^4$ (lost I_x)
Reduced Sections I_x = 161.20 − 6.25 = $154.75''^4$

STEP IV:
By adding a scab plate ¼"x5.00", two components remain on each side of hole to restore lost value of I_x.
Area scab above: A = 0.25 x 2.00 = $0.50^{\square''}$ 2 Pieces = $1.00^{\square''}$
Top part; $A\ell^2 = 0.50 \times 2.0^2 =$ $2.00''^4$
Bot. part; $A\ell^2 = 0.50 \times 5.0^2 =$ $12.50''^4$
With hole [Section I_x = $154.75''^4$
Total of Restored Sect. I_x $168.25''^4$ with scab plate.
Area restored Section = 8.28 + 1.00 = $9.28^{\square''}$
Section will be satisfactory with hole when treated thus.

EXAMPLE: Restore properties in deck girder 6.3.19

It is required that a flat oval ventilating duct pass through a transverse ship girder where top flange is of 3/4" plate with continuous deck. Web plate is of 1/2" plate with 22.50 inch depth. Bottom flange is 12.00"×3/4" plate. Size of cutout for duct collar has an outside dimension of 20.5×9.0 inches. All work shall be welded and provision made for watertight connection.

REQUIRED:

Assume top flange is also 12.0 inches in width for section, and before cutout the NA is symmetrical about flanges. Leave enough room at top for welding and locate collar above NA if possible.

Calculate the Moment of Inertia and NA for the 3 Sections by using standard form thus:

(a) Original symmetrical Section.
(b) Section with rectangular oval void in web.
(c) Section with collar and restoring section value of I''^4
(d) Draw cross-sections and elevation as necessary for accuracy.

STEP I:

Draw 3 Cross-sections and identify sections as A, B and C. The section noted C, will be for trial and altered if required.

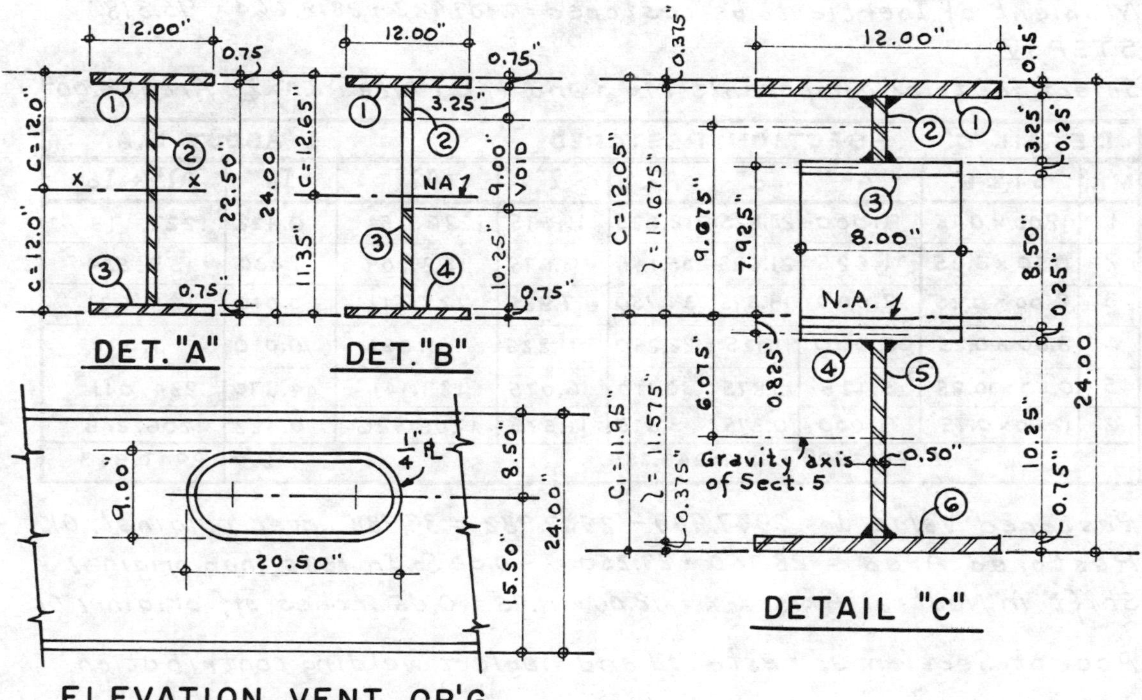

EXAMPLE: Restore properties in deck girder, continued 6.3.19

STEP II:

Original Section symmetrical and axis X-X is gravity axis or NA.

DETAIL "A"	ORIGINAL SECTION					I_x ABOUT AXIS X-X		
M'K.	SIZE"	A□"	d"	Ad	ℓ"	Aℓ²	I_o	Aℓ² + I_o
1	12.00 × 0.75	9.00	23.625	212.625	11.625	1216.266	0.422	1216.688
2	0.50 × 22.50	11.25	12.000	135.000	-0-	-0-	474.606	474.606
3	12.00 × 0.75	9.00	0.375	3.375	11.625	1216.266	0.422	1216.688
	ΣA: 29.25□"		ΣM=351.000				I_x =	2907.982

STEP III:

Calculate location of NA and value of I''^4 to be restored.

DETAIL "B"	SECT. WITH WEB CUT-OUT					I_x ABOUT N.A.		
M'K.	SIZE	A□"	d"	Ad	ℓ"	Aℓ²	I_o	Aℓ² + I_o
1	12.0 × 0.75	9.000	23.375	210.375	12.352	1373.147	0.422	1373.569
2	0.50 × 3.25	1.625	21.625	35.141	10.352	174.141	1.430	175.571
3	0.50 × 10.25	5.125	5.875	30.109	5.398	149.334	44.870	194.204
4	12.0 × 0.75	9.000	0.375	3.375	10.898	1068.898	0.422	1069.320
	A = 24.750□"		ΣM=279.000				Σ I_x =	2812.664

Location of NA from Base Line = $\frac{279.000}{24.75}$ = 11.35" or 5/8" off original X-X

Area lost by void = 29.25 − 24.75 = 4.50 square inches

Moment of Inertia to be restored = 2907.982 − 2812.664 = 95.318"⁴

STEP IV:

Select for trial components Mk 3 and 4, 2 Plates 8.0"× 1/4": Area: 4.00□"

DETAIL "C"	SECTION RESTORED					I ABOUT N.A.		
M'K.	SIZE	A□"	d"	Ad	ℓ"	Aℓ²	I_o	Aℓ² + I_o
1	12.00 × 0.75	9.000	23.625	212.625	11.675	1226.751	0.422	1227.173
2	0.50 × 3.25	1.625	21.625	35.141	9.675	152.109	1.430	153.539
3	8.00 × 0.25	2.000	19.875	39.750	7.925	125.611	0.010	125.621
4	8.00 × 0.25	2.000	11.125	22.250	0.825	1.361	0.010	1.371
5	0.50 × 10.25	5.125	5.875	30.110	6.075	189.141	44.870	234.011
6	12.00 × 0.75	9.000	0.375	3.375	11.575	1205.826	0.422	1206.248
	A: 28.750□"		ΣM=343.251				I_x =	2947.963

Restored Value I = 2947.963 − 2907.982 = 39.981"⁴ over original. OK.

Restored Area = 28.750 − 29.250 = −0.50 Sq.In. less than original

Shift in Neutral Axis X-X = 12.00 − 11.95 = 0.05 inches off original. OK.

Accept Section as restored and neglect welding contribution.

EXAMPLE: Split beam to increase section modulus 6.3.20

Select a standard rolled steel section such as a 12 x 6½ WF 27# to be used under emergency circumstances. Cut the section as shown in illustration, then weld the section as shown.

REQUIRED:

The welded section with the increased depth must now be investigated through the plane of least area. Divide section into 4 components and calculate the increase or decrease in values of, I_x, S_x and r_x.

STEP I:

From tables, the data on 12 x 6½ WF 27 Section is thus:

$I_x = 201.40''^4$, $S_x = 34.10''^3$, $r_x = 5.06''$, $A = 7.97^{\square''}$ $d = 12.0''$ $b = 6.50''$ $c = 6.00''$

Flange thickness = 0.40" Web thickness = 0.24"

No consideration will be given to fillets and rounded edges.

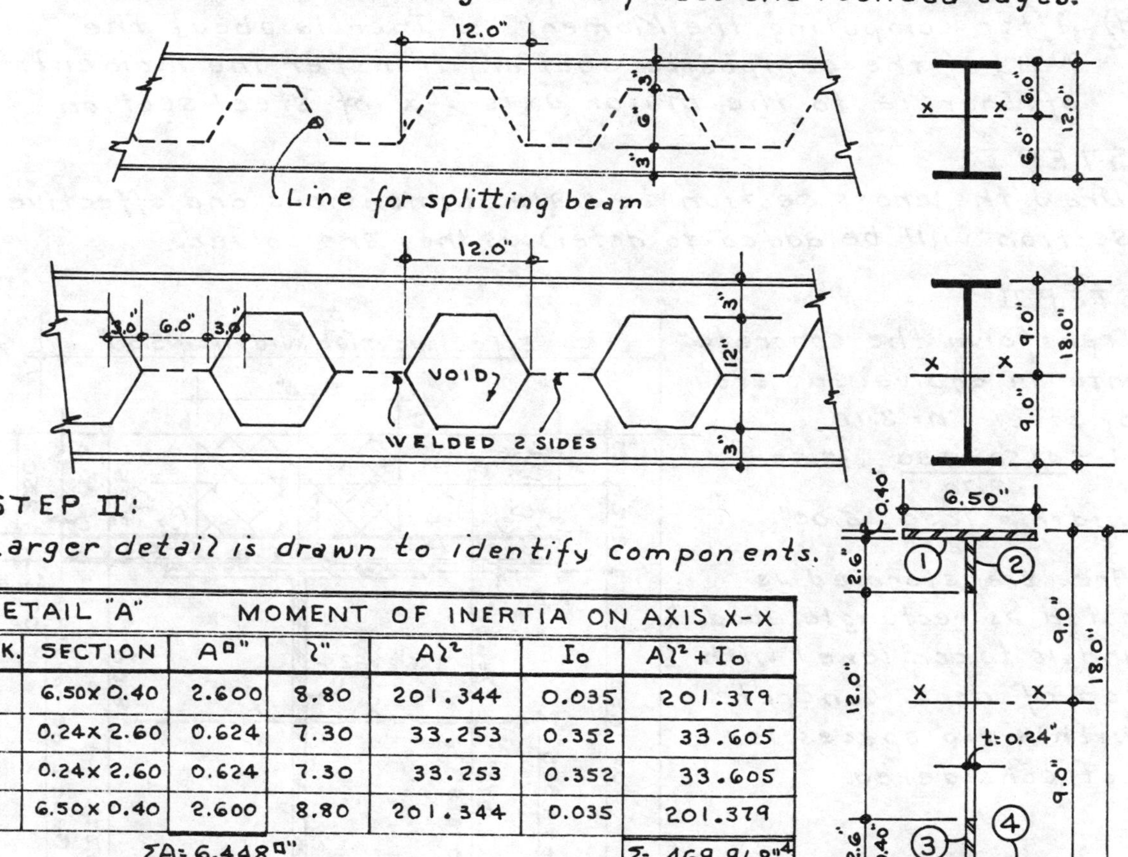

STEP II:

Larger detail is drawn to identify components.

DETAIL "A" MOMENT OF INERTIA ON AXIS X-X

M'K.	SECTION	$A^{\square''}$	r''	Ar^2	I_o	$Ar^2 + I_o$
1	6.50 x 0.40	2.600	8.80	201.344	0.035	201.379
2	0.24 x 2.60	0.624	7.30	33.253	0.352	33.605
3	0.24 x 2.60	0.624	7.30	33.253	0.352	33.605
4	6.50 x 0.40	2.600	8.80	201.344	0.035	201.379
		$\Sigma A = 6.448^{\square''}$				$\Sigma = 469.968''^4$

For $S_x = \dfrac{469.968}{9.00} = 52.20''^3$ An increase of: 52.6%

For $r_x \sqrt{\dfrac{469.968}{7.97}} = 7.66''$ An increase of 51.4%

EXAMPLE: Composite section: steel with concrete 6.3.21

The effective width of a concrete slab placed upon a metal deck form is 43.5 inches. Depth of slab is 3.25 inches. Ratio of Modulus of Elasticity between steel and concrete is: $n = \frac{E_s}{E_c}$, and $n = 8.70$. Deck depth = 3/4 inches.

Steel supports for slab consist of 8 x 6½ WF 24# Steel beams. Connection of concrete slab to steel beam is accomplished with stud type welded shear connectors.

REQUIRED:

(a) Determine the effective area and size of concrete portion which is equivalent to steel when, $n = 8.70$.

(b) Locate the Neutral Axis of the Composite section.

(c) Calculate the size of a steel cover plate on bottom flange which will be equivalent of concrete effective area.

(d) After computing the Moments of Inertia about the N.A., of the composite section, transfer the moments of Inertia to the major axis x-x of steel section.

STEP I:

Draw the cross section to scale. Dimensions and effective section will be added to detail as they are solved.

STEP II

Transform the concrete into an equivalent area of steel. $n = 8.70$

$$A_c = \frac{43.50 \times 2.50}{8.70} = 12.50\ \square''$$

$$\text{width} = \frac{12.50}{2.50} = 5.00''$$

Area transformed is noted as rectangle abcd, and is taken level with top of deck. Concrete within rib spaces is not considered

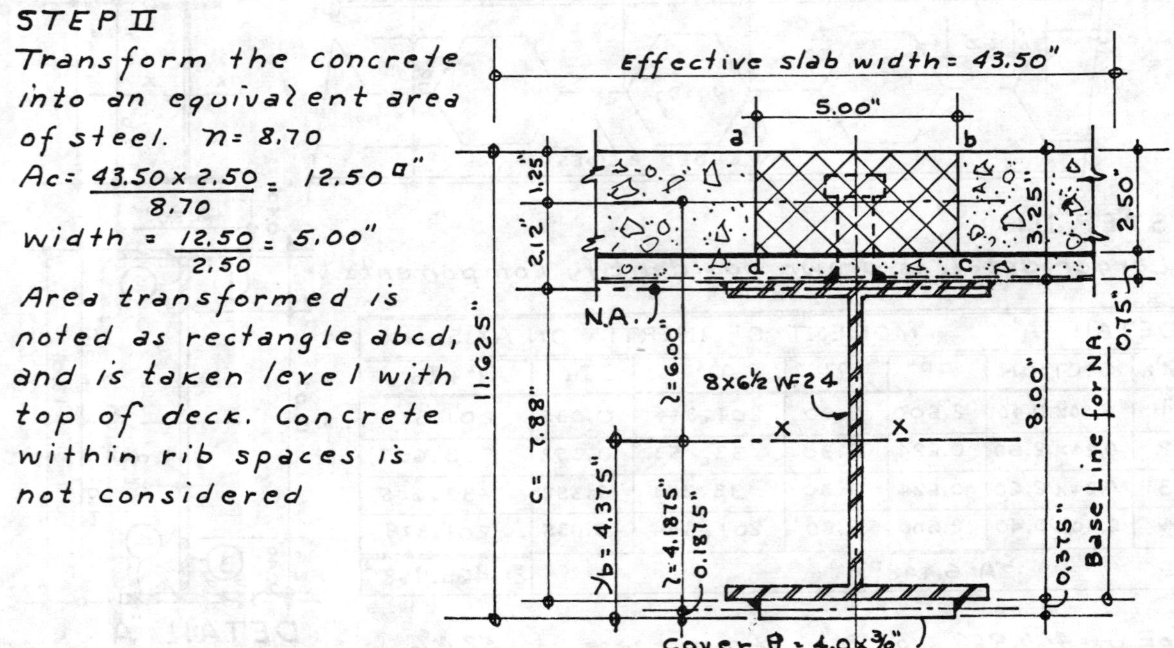

To calculate Neutral Axis, take properties of 8 WF 27 from tables. Base line for taking moments will be at bottom of steel beam. Again, the deck area will be neglected. Compiling the figures in tabular form, the value of I will be computed about the Neutral Axis. I_o for abcd = $\frac{bd^3}{12}$.

COMPOSITE SECTION-LESS COVER PLATE- ABOUT N.A.								
M'K	SIZE	A□"	d"	Ad	λ"	Aλ²	Io	Aλ² + Io
1	5.00 x 2.50	12.50	10.00	125.00	2.12	56.20	6.51	62.71
2	8 WF 24	7.06	4.00	28.24	3.88	106.28	83.80	190.08
		ΣA = 19.56□"		ΣM = 153.24				ΣI = 252.79"⁴

$C = 7.88"$ $S = \frac{I}{C}$ $S_{NA} = \frac{252.79}{7.88} = 32.08"^3$

STEP III:

From the above, with steel beam having a depth of 8.00", the location of NA is, 8.00 - 7.88 = 0.12" from top of 8 WF 24. To find area and size of cover plate equivalent to transformed concrete with n = 8.70. A_c = 12.50□"

$A_s = \frac{12.50}{8.70} = 1.437$□" Choosing a plate thickness of 3/8".

Width of Cover Pl = $\frac{1.437}{0.375} = 3.84"$ Use plate width of 4.00 inches.

With steel cover plate, the neutral axis is assumed to return to axis x-x of steel section. Area Pl = 4.0 x 0.375 = 1.50□"

STEP IV:

Calculating the value of Ix by transferring the moments from component abcd and cover plate

COMPOSITE SECTION WITH COVER PLATE- ABOUT AXIS X-X								
M'K.	SECTION	A□"	d"	Ad	λ"	Aλ²	Io	Aλ² + Io
1	5.00x2.50	12.50			6.00	450.00	6.51	456.51
2	8x6½WF 24	7.06			-0-	—	83.80	83.80
3	4.00x0.375	1.50			4.1875	26.30	0.18	26.48
								Σ = 566.79"⁴

Distance to extreme fiber is from axis x-x to top slab, (C = 4.00 + 3.25 = 7.25"), however in Composite Sections, use y_b.

$y_b = 4.375"$ $S_x = \frac{566.79}{4.375} = 129.55"^3$

Increase = 129.55 - 83.80 = 45.75"³

EXAMPLE: Exam problem: compound section properties 6.3.22

Find the Section Modulus for a beam built up with the following shapes:
2 Cover plates ½" x 20.0"
4 Angles 5x3x½ with long legs against cover plates.
1 Web plate ½" x 36.0"
Cross section and properties of angle given the detail at right. Applicant to draw a sketch of assembly.

Calculate the safe uniform load the section will support on a simple 40.0 foot span, when the unit F_b = 20,000 P.S.I.

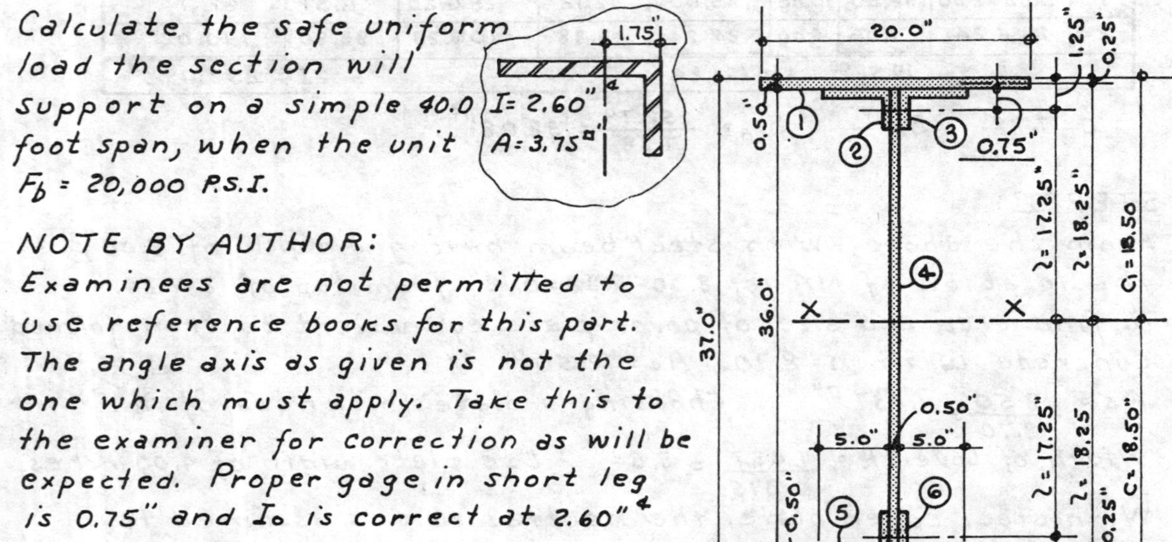

NOTE BY AUTHOR:
Examinees are not permitted to use reference books for this part. The angle axis as given is not the one which must apply. Take this to the examiner for correction as will be expected. Proper gage in short leg is 0.75" and I_o is correct at 2.60"⁴

STEP I:
Draw cross section and note that fabrication is welded. For accuracy the moment levers, ℓ, may be added to detail. Put in tabular form as follows:

COMPOUND SECTION-WELDED-MOMENTS OF INERTIA ABOUT X-X.						
M'K.	SECTION	A□"	λ"	Aλ²	Io	Aλ²+ Io
1	20.0 × 0.50 ℞	10.00	18.25	3330.625	0.208	3330.833
2	5 × 3 × ½ L	3.75	17.25	1115.859	2.600	1118.459
3	Do.	3.75	17.25	1115.859	2.600	1118.459
4	0.50 × 36.0 ℞	18.00	-0-	0.0	1944.000	1944.000
5	5 × 3 × ½ L	3.75	17.25	1115.859	2.600	1118.459
6	Do.	3.75	17.25	1115.859	2.600	1118.459
7	20.0 × 0.50 ℞	10.00	18.25	3330.625	0.208	3330.833
	ΣA= 53.00□"				ΣI=	13,079.502"⁴

STEP II:

Distance from x-x to extreme fiber = C = 18.50"

$S = \frac{I}{C}$ $S_x = \frac{13{,}079.502}{18.50} = 707.0''^3$

STEP III:

To calculate safe uniform load on simple span. L = 40.0 Ft.

Weight of Steel Beam = 3.40 A. Wt. = 3.40 × 53.0 = 180.2 Lbs. Foot.

Formula for Moment = $\frac{WL}{8}$ or $\frac{wL^2}{8}$ Transposed: $w = \frac{8M}{L^2}$

Resisting Moment of Compound Section = $S \times F_b$, and = M.

$RM = \frac{707.0 \times 20{,000}}{12} = 1{,}178{,}333$ Foot Lbs.

Including Beam Wt. $w = \frac{8 \times 1{,}178{,}333}{\frac{40.0}{5} \times 40.0} = 5891.67$ Lbs. per foot.

Superimposed Load = 5891.67 − 180.20 = 5711.47 Lbs. per foot

Total load W = 5711.47 × 40.0 = 228,458.80 Lbs.

TABLE: Weight of steel plate 6.4.1

THEORETICAL WEIGHT CARBON STEEL					
FRACTION THICKNESS IN INCHES	DECIMAL THICKNESS IN INCHES	WEIGHT IN POUNDS PER SQUARE FT.	FRACTION THICKNESS IN INCHES	DECIMAL THICKNESS IN INCHES	WEIGHT IN POUNDS PER SQUARE FT.
1/8	.1250	5.10	1 3/32	1.09375	44.63
5/32	.15625	6.38	1 1/8	1.1250	45.90
3/16	.1875	7.65	1 5/32	1.15625	47.18
7/32	.21875	8.93	1 3/16	1.1875	48.45
1/4	.2500	10.20	1 7/32	1.21875	49.73
9/32	.28125	11.48	1 1/4	1.2500	51.00
5/16	.3125	12.75	1 9/32	1.28125	52.28
11/32	.34375	14.03	1 5/16	1.3125	53.55
3/8	.3750	15.30	1 11/32	1.34376	54.83
13/32	.40625	16.58	1 3/8	1.3750	56.10
7/16	.4375	17.85	1 13/32	1.40625	57.38
15/32	.46875	19.13	1 7/16	1.4375	58.65
1/2	.5000	20.40	1 15/32	1.46875	59.93
17/32	.53125	21.68	1 1/2	1.5000	61.20
9/16	.5625	22.95	1 17/32	1.53125	62.48
19/32	.59375	24.23	1 9/16	1.5625	63.75
5/8	.6250	25.50	1 19/32	1.59375	65.03
21/32	.65625	26.78	1 5/8	1.6250	66.30
11/16	.6875	28.05	1 21/32	1.65625	67.58
23/32	.71875	29.33	1 11/16	1.6875	68.85
3/4	.7500	30.60	1 23/32	1.71875	70.13
25/32	.78125	31.88	1 3/4	1.7500	71.40
13/16	.8125	33.15	1 25/32	1.78125	72.68
27/32	.84375	34.43	1 13/16	1.8125	73.95
7/8	.8750	35.70	1 27/32	1.84375	75.23
29/32	.90625	36.98	1 7/8	1.8750	76.50
15/16	.9375	38.25	1 29/32	1.90625	77.78
31/32	.96875	39.53	1 15/16	1.9375	79.05
1.	1.0000	40.80	1 31/32	1.969	80.33
1 1/32	1.03125	42.08	2	2.0000	81.60
1 1/16	1.0625	43.35	2 1/4	2.2500	91.80

TABLE: Sheet metal gage and weight 6.4.2

CARBON AND STAINLESS STEEL-SHEET METAL GAGES & WEIGHTS

STANDARD U.S. GAGE NUMBER	DECIMAL THICKNESS IN INCHES	WEIGHT STEEL MILL FINISH LBS. SQ. FOOT	WEIGHT STEEL GALVANIZED LBS. SQ. FOOT	STAINLESS STEELS THICKNESS	STRAIGHT CHROME FIN. LBS. SQ. FOOT	CHROME NICKEL LBS. SQ. FOOT
THICKNESS 0.25 IN. AND OVER ARE PLATES						
7	.1793	7.500				
8	.1644	6.875				
9	.1494	6.250				
10	.1345	5.625	5.7812	.140625	5.7937	5.9062
11	.1196	5.000	5.1562	.125000	5.1500	5.2500
12	.1046	4.375	4.5312	.109375	4.5063	4.5937
13	.0897	3.750	3.9062	.093750	3.8625	3.9375
14	.0747	3.125	3.2812	.078125	3.2187	3.2812
15	.0673	2.812	2.9687	.070312	2.8968	2.9531
16	.0598	2.500	2.6562	.062500	2.5750	2.6250
17	.0538	2.250	2.4062	.056250	2.3175	2.3625
18	.0478	2.000	2.1562	.050000	2.0600	2.1000
19	.0418	1.750	1.9062	.043750	1.8025	1.8375
20	.0359	1.500	1.6562	.037500	1.5450	1.5750
21	.0329	1.375	1.5312	.034375	1.4160	1.4437
22	.0299	1.250	1.4062	.031250	1.2875	1.3125
23	.0269	1.125	1.2812	.028125	1.1587	1.1813
24	.0239	1.000	1.1562	.025000	1.0300	1.0500
25	.0209	.875	1.0312	.021875	.9013	.9187
26	.0179	.750	.9062	.018750	.7725	.7875
27	.0164	.6875	.8437	.017187	.7081	.7218
28	.0149	.625	.7812	.015625	.6438	.6562
29	.0135	.5625	.7187	.014062	.5794	.5906
30	.0120	.500	.6562	.012500	.5150	.5250

THE U.S. STANDARD GAGE FOR STEEL SHEETS WAS ESTABLISHED BY ACT OF CONGRESS IN 1893. IT SPECIFIED THAT WEIGHTS PER SQUARE FOOT SHALL BE INDICATED BY GAUGE NUMBER. THE DETERMINING FACTOR IN ORDERING OR SPECIFYING STEEL SHEETS IS THE WEIGHT, RATHER THAN THICKNESS.
THE WEIGHTS GIVEN FOR GALVANIZED SHEETS, IS BASED ON THE U.S. STANDARD GAGE WITH THE CUSTOMARY 2.5 OUNCE PER SQ. FOOT OF ZINC COATING, REGARDLESS OF OTHER COATING WEIGHTS.

PROPERTIES OF RECTANGULAR SHAPES

$r_{x-x} = 0.289\,d$

$r_{y-y} = 0.289\,t$

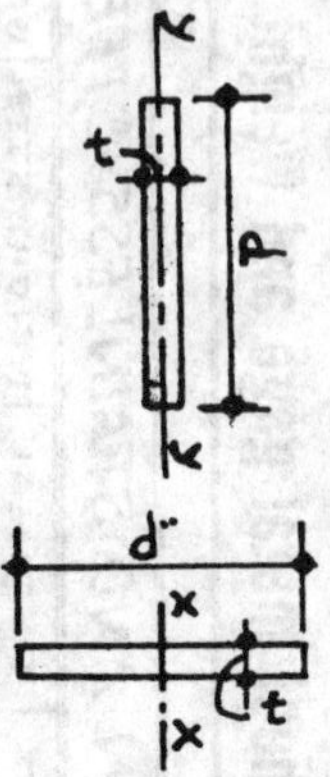

TABLE: Properties of rectangular shapes, 1/8" x 1" to 1" x 38"

6.4.3

DEPTH d"	THICKNESS "t" DECIMAL"	0.1250	0.1875	0.2500	0.3125	0.3750	0.4375	0.5000	0.5625	0.6250	0.7500	0.8750	1.0000
	FRACTION"	1/8	3/16	1/4	5/16	3/8	7/16	1/2	9/16	5/8	3/4	7/8	1.0
1	AREA □"	0.125	0.188	0.250	0.313	0.375	0.438	0.500	0.563	0.625	0.750	0.875	1.000
	I_{x-x}	0.010	0.016	0.021	0.026	0.031	0.037	0.042	0.047	0.052	0.063	0.073	0.083
	I_{y-y}	0.000	0.001	0.001	0.003	0.004	0.007	0.010	0.015	0.020	0.035	0.056	0.083
1 1/8	AREA	0.141	0.211	0.281	0.352	0.422	0.492	0.563	0.633	0.703	0.844	0.984	1.125
	I_{x-x}	0.015	0.022	0.030	0.037	0.045	0.052	0.059	0.067	0.074	0.089	0.104	0.119
	I_{y-y}	0.000	0.001	0.002	0.003	0.005	0.009	0.013	0.019	0.025	0.044	0.070	0.094
1 1/4	AREA	0.156	0.234	0.313	0.391	0.469	0.547	0.625	0.703	0.781	0.938	1.094	1.250
	I_{x-x}	0.020	0.031	0.041	0.051	0.061	0.071	0.081	0.092	0.102	0.122	0.142	0.163
	I_{y-y}	0.000	0.001	0.002	0.003	0.005	0.009	0.013	0.019	0.025	0.044	0.070	0.104
1 3/8	AREA	0.172	0.258	0.344	0.430	0.516	0.602	0.688	0.773	0.859	1.031	1.203	1.375
	I_{x-x}	0.027	0.041	0.054	0.068	0.081	0.095	0.108	0.122	0.135	0.163	0.190	0.217
	I_{y-y}	0.000	0.001	0.002	0.003	0.006	0.010	0.014	0.020	0.028	0.048	0.077	0.115
1 1/2	AREA	0.188	0.281	0.375	0.469	0.563	0.656	0.750	0.844	0.938	1.125	1.313	1.500
	I_{x-x}	0.035	0.053	0.070	0.088	0.106	0.123	0.141	0.158	0.176	0.211	0.246	0.281
	I_{y-y}	0.000	0.001	0.002	0.004	0.007	0.010	0.016	0.022	0.031	0.053	0.084	0.125
1 5/8	AREA	0.203	0.305	0.406	0.508	0.609	0.711	0.813	0.914	1.016	1.219	1.422	1.625
	I_{x-x}	0.045	0.067	0.089	0.112	0.134	0.156	0.179	0.201	0.224	0.268	0.313	0.358
	I_{y-y}	0.000	0.001	0.002	0.004	0.007	0.011	0.017	0.024	0.033	0.057	0.091	0.135
1 3/4	AREA	0.219	0.328	0.438	0.547	0.656	0.766	0.875	0.984	1.094	1.313	1.531	1.750
	I_{x-x}	0.056	0.084	0.112	0.140	0.168	0.195	0.223	0.251	0.279	0.335	0.391	0.447
	I_{y-y}	0.000	0.001	0.002	0.004	0.008	0.012	0.018	0.026	0.036	0.062	0.098	0.146
1 7/8	AREA	0.234	0.352	0.469	0.586	0.703	0.820	0.938	1.055	1.172	1.406	1.641	1.875
	I_{x-x}	0.069	0.103	0.137	0.172	0.206	0.240	0.275	0.309	0.343	0.412	0.481	0.549
	I_{y-y}	0.000	0.001	0.002	0.005	0.008	0.013	0.020	0.028	0.038	0.066	0.105	0.156
2	AREA	0.250	0.375	0.500	0.625	0.750	0.875	1.000	1.125	1.250	1.500	1.750	2.000
	I_{x-x}	0.083	0.125	0.167	0.208	0.250	0.292	0.333	0.375	0.417	0.500	0.583	0.667
	I_{y-y}	0.000	0.001	0.003	0.005	0.009	0.014	0.021	0.030	0.041	0.070	0.112	0.167

PROPERTIES OF RECTANGULAR SHAPES

$r_{x-x} = 0.289\ d.$

$r_{y-y} = 0.289\ t$

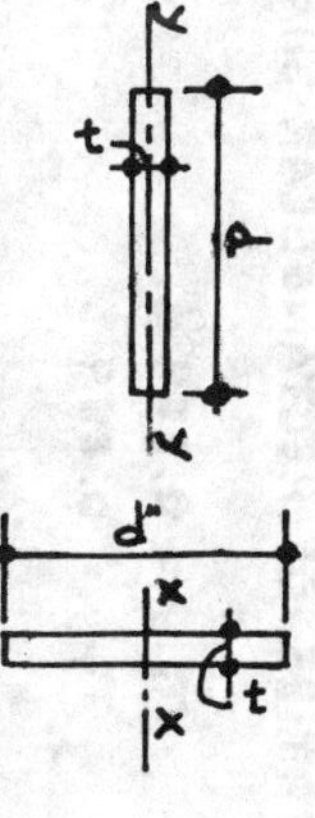

DEPTH d:		THICKNESS "t"											
	DECIMAL"	0.1250	0.1875	0.2500	0.3125	0.3750	0.4375	0.5000	0.5625	0.6250	0.7500	0.8750	1.0000
	FRACTION"	1/8	3/16	1/4	5/16	3/8	7/16	1/2	9/16	5/8	3/4	7/8	1.0
2 1/8	AREA □"	0.266	0.398	0.531	0.664	0.794	0.930	1.063	1.195	1.328	1.594	1.859	2.125
	I_{x-x}	0.100	0.150	0.200	0.250	0.300	0.350	0.400	0.450	.500	.600	.700	0.800
	I_{y-y}	0.000	0.001	0.003	0.005	0.009	0.015	0.022	.032	.043	.075	.119	0.177
2 1/4	AREA	0.281	0.422	0.563	0.703	0.844	0.984	1.125	1.266	1.406	1.688	1.969	2.250
	I_{x-x}	0.119	0.178	0.237	0.297	0.356	0.415	0.475	0.534	0.593	0.712	0.831	0.949
	I_{y-y}	0.000	0.001	0.003	0.006	0.010	0.016	0.023	0.033	0.046	0.079	0.126	0.187
2 3/8	AREA	0.297	0.445	0.594	0.742	0.891	1.039	1.188	1.336	1.484	1.781	2.078	2.375
	I_{x-x}	0.140	0.209	0.279	0.349	0.419	0.488	0.558	0.628	0.698	0.837	0.977	1.116
	I_{y-y}	0.000	0.001	0.003	0.006	0.010	0.017	0.025	0.035	0.048	0.083	0.133	0.198
2 1/2	AREA	0.313	0.469	0.625	0.781	0.938	1.094	1.250	1.406	1.563	1.875	2.188	2.500
	I_{x-x}	0.163	0.244	0.326	0.407	0.488	0.570	0.651	0.732	0.814	.977	1.139	1.302
	I_{y-y}	0.000	0.001	0.003	0.006	0.011	0.017	0.026	0.037	0.051	.088	0.140	0.208
2 5/8	AREA	0.328	0.492	0.656	0.820	0.984	1.148	1.313	1.477	1.641	1.969	2.297	2.625
	I_{x-x}	0.188	0.283	0.377	0.471	0.565	0.660	0.754	0.848	0.942	1.131	1.319	1.507
	I_{y-y}	0.000	0.001	0.003	0.007	0.012	0.018	0.027	0.039	0.053	0.092	0.147	0.219
2 3/4	AREA	0.344	0.516	0.688	0.859	1.031	1.203	1.375	1.547	1.719	2.063	2.406	2.750
	I_{x-x}	0.217	0.325	0.433	0.542	0.650	0.758	0.867	0.975	1.083	1.300	1.517	1.733
	I_{y-y}	0.000	0.002	0.004	0.007	0.012	0.019	0.029	0.041	0.056	0.097	0.154	0.229
2 7/8	AREA	0.359	0.539	0.719	0.898	1.078	1.258	1.438	1.617	1.797	2.156	2.516	2.875
	I_{x-x}	0.248	0.371	0.495	0.619	0.743	0.866	0.990	1.114	1.238	1.485	1.733	1.980
	I_{y-y}	0.000	0.002	0.004	0.007	0.013	0.020	0.030	0.043	0.058	0.101	0.161	0.240
3	AREA	0.375	0.563	0.750	0.938	1.125	1.313	1.500	1.688	1.875	2.250	2.625	3.000
	I_{x-x}	0.281	0.422	0.563	0.703	0.844	0.984	1.125	1.266	1.406	1.688	1.969	2.250
	I_{y-y}	0.000	0.002	0.004	0.008	0.013	0.021	0.031	0.044	0.061	0.105	0.167	0.250
3 1/8	AREA	0.391	0.586	0.781	0.977	1.172	1.367	1.563	1.758	1.953	2.344	2.734	3.125
	I_{x-x}	0.318	0.477	0.636	0.795	0.954	1.113	1.272	1.431	1.590	1.907	2.225	2.543
	I_{y-y}	0.001	0.002	0.004	0.008	0.014	0.022	0.033	0.046	0.064	0.110	0.174	0.260

PROPERTIES OF RECTANGULAR SHAPES

$r_{x-x} = 0.289\,d.$

$r_{y-y} = 0.289\,t$

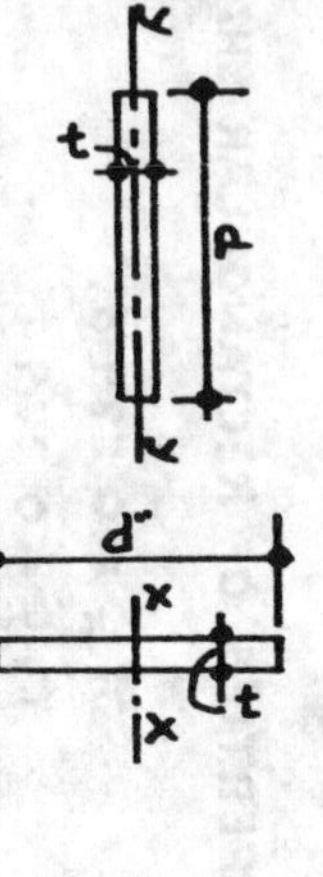

DEPTH d"	DECIMAL"	0.1250	0.1875	0.2500	0.3125	0.3750	0.4375	0.5000	0.5625	0.6250	0.7500	0.8750	1.0000
	FRACTION"	1/8	3/16	1/4	5/16	3/8	7/16	1/2	9/16	5/8	3/4	7/8	1.0
3 1/4	AREA □"	0.406	0.609	0.813	1.016	1.219	1.422	1.625	1.828	2.031	2.438	2.844	3.250
	I_{x-x}	0.358	0.536	0.715	0.894	1.073	1.252	1.430	1.609	1.788	2.146	2.503	2.861
	I_{y-y}	0.001	0.002	0.004	0.008	0.014	0.023	0.034	0.048	0.066	0.114	0.181	0.271
3 3/8	AREA	0.422	0.633	0.844	1.055	1.266	1.477	1.688	1.898	2.109	2.531	2.953	3.375
	I_{x-x}	0.401	0.601	0.801	1.001	1.201	1.402	1.602	1.802	2.002	2.403	2.803	3.204
	I_{y-y}	0.001	0.002	0.004	0.009	0.015	.024	.035	.050	0.069	0.119	0.188	0.281
3 1/2	AREA	0.438	0.656	0.875	1.094	1.313	1.531	1.750	1.969	2.188	2.625	3.063	3.500
	I_{x-x}	0.447	0.670	0.893	1.117	1.340	1.563	1.787	2.010	2.233	2.680	3.126	3.573
	I_{y-y}	0.001	0.002	0.005	0.009	0.015	0.024	0.036	0.052	0.071	0.123	0.195	0.292
3 5/8	AREA	0.453	0.680	0.906	1.133	1.359	1.586	1.813	2.039	2.266	2.719	3.172	3.625
	I_{x-x}	0.496	0.744	0.992	1.241	1.489	1.787	1.985	2.233	2.481	2.977	3.473	3.970
	I_{y-y}	0.001	0.002	0.005	0.009	0.016	0.025	0.038	0.054	0.074	0.127	0.202	0.302
3 3/4	AREA	0.469	0.703	0.938	1.172	1.406	1.641	1.875	2.109	2.344	2.813	3.281	3.750
	I_{x-x}	0.549	0.824	1.099	1.373	1.648	1.923	2.197	2.472	2.747	3.296	3.845	4.395
	I_{y-y}	0.001	0.002	0.005	0.010	0.016	0.026	0.039	0.056	0.076	0.132	0.209	0.312
3 7/8	AREA	0.484	0.727	0.969	1.211	1.453	1.695	1.938	2.180	2.422	2.906	3.391	3.875
	I_{x-x}	0.606	0.909	1.212	1.515	1.818	2.121	2.424	2.728	3.031	3.637	4.243	4.849
	I_{y-y}	0.001	0.002	0.005	0.010	0.017	0.027	0.040	0.057	0.079	0.136	0.216	0.323
4	AREA	0.500	0.750	1.000	1.250	1.500	1.750	2.000	2.250	2.500	3.000	3.500	4.000
	I_{x-x}	0.667	1.000	1.333	1.667	2.000	2.333	2.667	3.000	3.333	4.000	4.667	5.333
	I_{y-y}	0.001	0.002	0.005	0.010	0.018	0.028	0.042	0.059	0.081	0.141	0.223	0.333
4 1/8	AREA	0.516	0.773	1.031	1.289	1.547	1.805	2.063	2.320	2.578	3.094	3.609	4.125
	I_{x-x}	0.731	1.097	1.462	1.828	2.193	2.559	2.925	3.290	3.656	4.387	5.118	5.849
	I_{y-y}	0.001	0.002	0.005	0.010	0.018	0.029	0.043	0.061	0.084	0.145	0.230	0.344
4 1/4	AREA	0.531	0.797	1.063	1.328	1.594	1.859	2.125	2.391	2.656	3.188	3.719	4.250
	I_{x-x}	0.800	1.200	1.599	1.999	2.399	2.799	3.199	3.598	3.998	4.798	5.598	6.397
	I_{y-y}	0.001	0.002	0.006	0.011	0.019	0.030	0.044	0.063	0.086	0.149	0.237	0.354

THICKNESS "t"

PROPERTIES OF RECTANGULAR SHAPES

$r_{x-x} = 0.289\ d.$

$r_{y-y} = 0.289\ t$

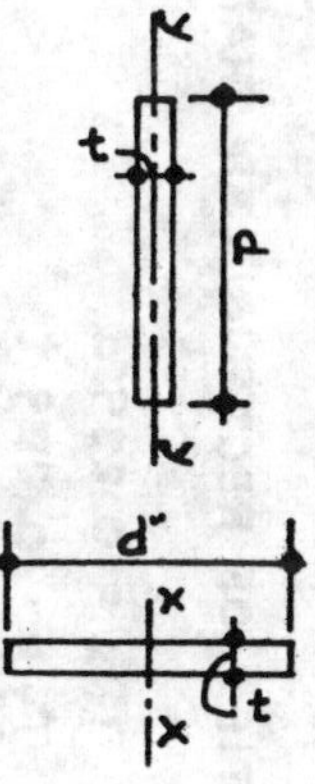

DEPTH d"	DECIMAL"	0.1250	0.1875	0.2500	0.3125	0.3750	0.4375	0.5000	0.5625	0.6250	0.7500	0.8750	1.0000
	FRACTION"	1/8	3/16	1/4	5/16	3/8	7/16	1/2	9/16	5/8	3/4	7/8	1.0
4 3/8	AREA □"	0.547	0.820	1.094	1.367	1.641	1.914	2.188	2.461	2.734	3.281	3.828	4.375
	I_{x-x}	0.872	1.308	1.745	2.181	2.617	3.053	3.489	3.925	4.362	5.234	6.106	6.987
	I_{y-y}	0.001	0.002	0.006	0.011	0.019	0.031	0.046	0.065	0.089	0.154	0.244	0.365
4 1/2	AREA	0.563	0.844	1.125	1.406	1.688	1.969	2.250	2.531	2.813	3.375	3.938	4.500
	I_{x-x}	0.949	1.424	1.898	2.373	2.848	3.322	3.797	4.272	4.746	5.695	6.645	7.594
	I_{y-y}	0.001	0.002	0.006	0.011	0.020	0.031	0.047	0.067	0.092	0.158	0.251	0.375
4 5/8	AREA	0.578	0.867	1.156	1.445	1.734	2.023	2.313	2.602	2.891	3.469	4.047	4.625
	I_{x-x}	1.031	1.546	2.061	2.576	3.092	3.607	4.122	4.637	5.153	6.183	7.214	8.244
	I_{y-y}	0.001	0.003	0.006	0.012	0.020	0.032	0.048	0.069	0.094	0.163	0.258	0.385
4 3/4	AREA	0.594	0.891	1.188	1.484	1.781	2.078	2.375	2.672	2.969	3.563	4.156	4.750
	I_{x-x}	1.116	1.675	2.233	2.791	3.349	3.907	4.466	5.024	5.582	6.698	7.815	8.931
	I_{y-y}	0.001	0.003	0.006	0.012	0.021	0.033	0.049	0.070	0.097	0.167	0.265	0.396
4 7/8	AREA	0.609	0.914	1.219	1.523	1.828	2.133	2.438	2.742	3.047	3.656	4.266	4.875
	I_{x-x}	1.207	1.810	2.414	3.017	3.621	4.224	4.827	5.431	6.034	7.241	8.448	9.655
	I_{y-y}	0.001	0.003	0.006	0.012	0.021	0.034	0.051	0.072	0.099	0.171	0.272	0.406
5	AREA	0.625	0.938	1.250	1.563	1.875	2.186	2.500	2.813	3.125	3.750	4.375	5.000
	I_{x-x}	1.302	1.953	2.604	3.255	3.906	4.557	5.208	5.859	6.510	7.813	9.115	10.420
	I_{y-y}	0.001	0.003	0.007	0.013	0.022	0.035	0.052	0.074	0.102	0.176	0.279	0.417
5 1/8	AREA	0.641	0.961	1.281	1.602	1.922	2.242	2.563	2.883	3.203	3.844	4.484	5.125
	I_{x-x}	1.402	2.103	2.804	3.506	4.207	4.908	5.609	6.310	7.011	8.413	9.815	11.220
	I_{y-y}	0.001	0.003	0.007	0.013	0.023	0.036	0.053	0.076	0.104	0.180	0.286	0.427
5 1/4	AREA	0.656	0.984	1.313	1.641	1.969	2.297	2.625	2.953	3.281	3.938	4.594	5.250
	I_{x-x}	1.507	2.261	3.015	3.768	4.522	5.276	6.029	6.783	7.537	9.044	10.550	12.060
	I_{y-y}	0.001	0.003	0.007	0.013	0.023	0.037	0.055	0.078	0.107	0.185	0.293	0.437
5 3/8	AREA	0.672	1.008	1.344	1.680	2.016	2.352	2.688	3.023	3.359	4.031	4.703	5.375
	I_{x-x}	1.618	2.426	3.235	4.044	4.858	5.662	6.470	7.279	8.088	9.705	11.320	12.940
	I_{y-y}	0.001	0.003	0.007	0.014	0.024	0.038	0.056	0.080	0.109	0.189	0.300	0.448

PROPERTIES OF RECTANGULAR SHAPES

$r_{x-x} = 0.289\,d.$

$r_{y-y} = 0.289\,t$

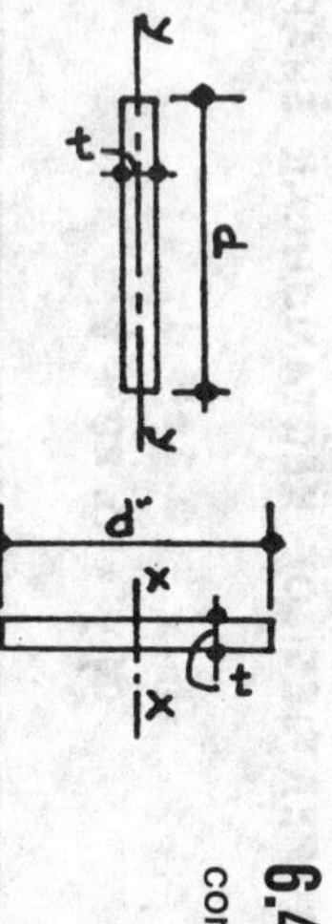

DEPTH d"		THICKNESS "t"											
	DECIMAL"	0.1250	0.1875	0.2500	0.3125	0.3750	0.4375	0.5000	0.5625	0.6250	0.7500	0.8750	1.0000
	FRACTION"	1/8	3/16	1/4	5/16	3/8	7/16	1/2	9/16	5/8	3/4	7/8	1.0
5 1/2	AREA □"	0.688	1.031	1.375	1.719	2.063	2.406	2.750	3.094	3.438	4.125	4.813	5.500
	I_{x-x}	1.733	2.600	3.466	4.333	5.199	6.066	6.932	7.799	8.655	10.400	12.130	13.860
	I_{y-y}	0.001	0.003	0.007	0.014	0.024	0.038	0.057	0.082	0.112	0.193	0.307	0.458
5 5/8	AREA	0.703	1.055	1.406	1.758	2.109	2.461	2.813	3.164	3.516	4.219	4.922	5.625
	I_{x-x}	1.854	2.781	3.708	4.635	5.562	6.489	7.416	8.343	9.270	11.120	12.980	14.830
	I_{y-y}	0.001	0.003	0.007	0.014	0.025	0.039	0.059	0.083	0.114	0.198	0.314	0.469
5 3/4	AREA	0.719	1.078	1.438	1.793	2.156	2.516	2.875	3.234	3.594	4.313	5.031	5.750
	I_{x-x}	1.980	2.971	3.961	4.951	5.941	6.931	7.921	8.911	9.902	11.880	13.860	15.840
	I_{y-y}	0.001	0.003	0.007	0.015	0.025	0.040	0.060	0.085	0.117	0.202	0.321	0.479
5 7/8	AREA	0.734	1.102	1.469	1.836	2.203	2.570	2.937	3.305	3.672	4.406	5.141	5.875
	I_{x-x}	2.112	3.168	4.225	5.281	6.337	7.393	8.449	9.505	10.560	12.670	14.790	16.900
	I_{y-y}	0.001	0.003	0.008	0.015	0.026	0.041	0.061	0.087	0.120	0.207	0.328	0.490
6	AREA	0.750	1.125	1.500	1.875	2.250	2.625	3.000	3.375	3.750	4.500	5.250	6.000
	I_{x-x}	2.250	3.375	4.500	5.625	6.750	7.875	9.000	10.130	11.250	13.500	15.750	18.000
	I_{y-y}	0.001	0.003	0.008	0.015	0.026	0.042	0.063	0.089	0.122	0.211	0.335	0.500
6 1/8	AREA	0.766	1.148	1.531	1.914	2.297	2.680	3.063	3.445	3.828	4.594	5.359	6.125
	I_{x-x}	2.394	3.590	4.787	5.984	7.181	8.378	9.574	10.770	11.970	14.360	16.760	19.150
	I_{y-y}	0.001	0.003	0.008	0.016	0.027	0.043	0.064	0.091	0.125	0.215	0.342	0.510
6 1/4	AREA	0.781	1.172	1.563	1.953	2.344	2.734	3.125	3.516	3.906	4.688	5.469	6.250
	I_{x-x}	2.543	3.815	5.086	6.358	7.629	8.901	10.170	11.440	12.720	15.260	17.800	20.35
	I_{y-y}	0.001	0.003	0.008	0.016	0.027	0.044	0.065	0.093	0.127	0.220	0.349	0.521
6 3/8	AREA	0.797	1.195	1.594	1.992	2.391	2.789	3.188	3.586	3.984	4.781	5.578	6.375
	I_{x-x}	2.699	4.048	5.398	6.747	8.096	9.466	10.800	12.140	13.490	16.190	18.890	21.59
	I_{y-y}	0.001	0.004	0.008	0.016	0.028	0.044	0.066	0.095	0.130	0.224	0.356	0.531
6 1/2	AREA	0.813	1.219	1.625	2.031	2.438	2.844	3.250	3.656	4.063	4.875	5.688	6.500
	I_{x-x}	2.861	4.291	5.721	7.152	8.582	10.010	11.440	12.870	14.300	17.160	20.020	22.89
	I_{y-y}	0.001	0.004	0.008	0.017	0.029	0.045	0.068	0.096	0.132	0.229	0.363	0.542

PROPERTIES OF RECTANGULAR SHAPES

$r_{x-x} = 0.289\,d$

$r_{y-y} = 0.289\,t$

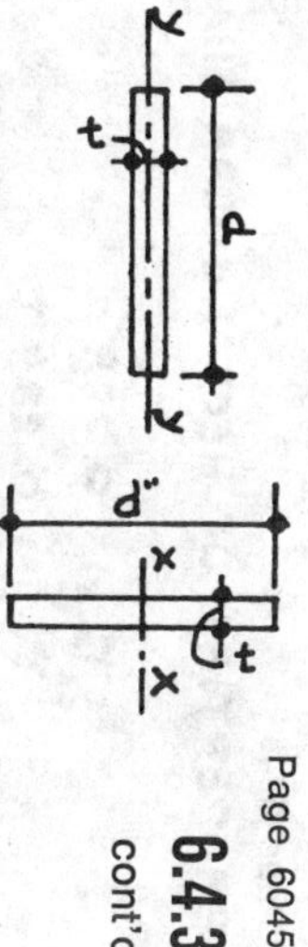

DEPTH d"		THICKNESS "t"											
	DECIMAL"	0.1250	0.1875	0.2500	0.3125	0.3750	0.4375	0.5000	0.5625	0.6250	0.7500	0.8750	1.0000
	FRACTION"	1/8	3/16	1/4	5/16	3/8	7/16	1/2	9/16	5/8	3/4	7/8	1.0
6 5/8	AREA □"	0.828	1.242	1.656	2.070	2.484	2.898	3.313	3.727	4.141	4.969	5.797	6.625
	I_{x-x}	3.029	4.543	6.058	7.572	9.087	10.60	12.12	13.63	15.14	18.17	21.20	24.23
	I_{y-y}	0.001	0.004	0.009	0.017	0.029	0.046	0.069	0.098	0.135	0.233	0.370	0.522
6 3/4	AREA	0.844	1.266	1.688	2.109	2.531	2.953	3.375	3.797	4.219	5.063	5.906	6.750
	I_{x-x}	3.204	4.805	6.407	8.009	9.611	11.21	12.81	14.42	16.02	19.22	22.43	25.63
	I_{y-y}	0.001	0.004	0.009	0.017	0.030	0.047	0.070	0.100	0.137	0.237	0.377	0.562
6 7/8	AREA	0.859	1.289	1.719	2.148	2.578	3.008	3.438	3.867	4.297	5.156	6.016	6.875
	I_{x-x}	3.385	5.077	6.770	8.462	10.15	11.85	13.54	15.23	16.92	20.31	23.69	27.08
	I_{y-y}	0.001	0.004	0.009	0.017	0.030	0.048	0.072	0.102	0.140	0.242	0.384	0.573
7	AREA	0.875	1.313	1.750	2.188	2.625	3.063	3.500	3.938	4.375	5.250	6.125	7.000
	I_{x-x}	3.573	5.359	7.146	8.932	10.72	12.51	14.29	16.08	17.86	21.44	25.01	28.58
	I_{y-y}	0.001	0.004	0.009	0.018	0.031	0.049	0.073	0.104	0.142	0.246	0.391	0.583
7 1/8	AREA	0.891	1.336	1.781	2.227	2.672	3.117	3.563	4.008	4.453	5.344	6.234	7.125
	I_{x-x}	3.768	5.652	7.536	9.419	11.30	13.19	15.07	16.95	18.84	22.61	26.37	30.14
	I_{y-y}	0.001	0.004	0.009	0.018	0.031	0.050	0.074	0.106	0.145	0.250	0.398	0.594
7 1/4	AREA	0.906	1.359	1.813	2.266	2.719	3.172	3.625	4.078	4.531	5.438	6.344	7.250
	I_{x-x}	3.970	5.954	7.930	9.924	11.91	13.89	15.88	17.86	19.85	23.82	27.79	31.76
	I_{y-y}	0.001	0.004	0.009	0.018	0.032	0.051	0.076	0.108	0.148	0.255	0.405	0.604
7 3/8	AREA	0.922	1.383	1.844	2.305	2.766	3.227	3.688	4.148	4.609	5.531	6,453	7.375
	I_{x-x}	4.178	6.268	8.357	10.45	12.54	14.62	16.71	18.80	20.89	25.07	29.25	33.43
	I_{y-y}	0.001	0.004	0.010	0.019	0.032	0.051	0.077	0.109	0.150	0.259	0.412	0.615
7 1/2	AREA	0.938	1.406	1.875	2.344	2.814	3.281	3.750	4.219	4.688	5.625	6.565	7.500
	I_{x-x}	4.395	6.592	8.789	10.99	13.18	15.38	17.58	19.78	21.97	26.37	30.76	35.16
	I_{y-y}	0.001	0.004	0.010	0.019	0.033	0.052	0.078	0.111	0.153	0.264	0.419	0.625
7 5/8	AREA	0.953	1.430	1.906	2.383	2.859	3.336	3.813	4.289	4.766	5.719	6.672	7.625
	I_{x-x}	4.618	6.927	9.236	11.54	13.85	16.16	18.47	20.78	23.00	27.71	32.33	36.94
	I_{y-y}	0.001	0.004	0.010	0.019	0.034	0.053	0.079	0.113	0.155	0.268	0.426	0.635

PROPERTIES OF RECTANGULAR SHAPES

$r_{x-x} = 0.289\,d$

$r_{y-y} = 0.289\,t$

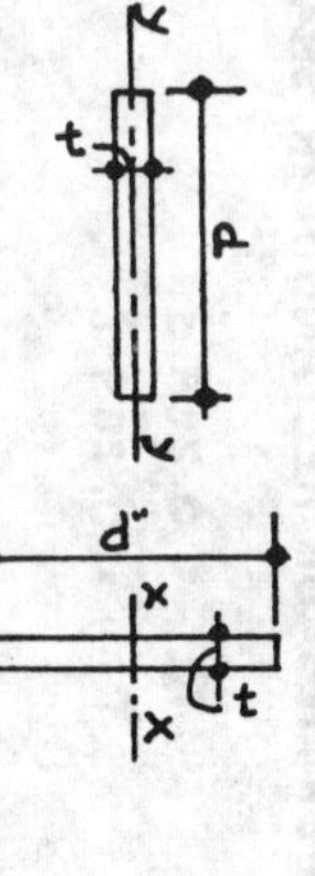

DEPTH d"		THICKNESS "t"											
	DECIMAL"	0.1250	0.1875	0.2500	0.3125	0.3750	0.4375	0.5000	0.5625	0.6250	0.7500	0.8750	1.0000
	FRACTION"	1/8	3/16	1/4	5/16	3/8	7/16	1/2	9/16	5/8	3/4	7/8	1.0
7 3/4	AREA □"	0.969	1.453	1.938	2.422	2.906	3.391	3.875	4.359	4.844	5.813	6.781	7.750
	I_{x-x}	4.849	7.273	9.698	12.12	14.55	16.97	19.40	21.82	24.24	29.09	33.94	38.79
	I_{y-y}	0.001	0.004	0.010	0.020	0.034	0.054	0.081	0.115	0.158	0.272	0.433	0.646
7 7/8	AREA	0.984	1.477	1.969	2.461	2.953	3.445	3.938	4.430	4.922	5.906	6.891	7.875
	I_{x-x}	5.087	7.631	10.17	12.72	15.26	17.81	20.35	22.89	25.44	30.52	35.61	40.70
	I_{y-y}	0.001	0.004	0.010	0.020	0.035	0.055	0.082	0.117	0.160	0.277	0.440	0.656
8	AREA	1.000	1.500	2.000	2.500	3.000	3.500	4.000	4.500	5.000	6.000	7.000	8.000
	I_{x-x}	5.333	8.000	10.67	13.33	16.00	18.67	21.33	24.00	26.67	32.00	37.33	42.67
	I_{y-y}	0.001	0.004	0.010	0.020	0.035	0.056	0.083	0.119	0.163	0.281	0.447	0.667
8 1/8	AREA	1.016	1.523	2.031	2.539	3.047	3.555	4.063	4.570	5.078	6.094	7.109	8.125
	I_{x-x}	5.587	8.381	11.17	13.97	16.76	19.56	22.35	25.14	27.94	33.54	39.11	44.70
	I_{y-y}	0.001	0.004	0.011	0.021	0.036	0.057	0.085	0.124	0.165	0.286	0.454	0.677
8 1/4	AREA	1.031	1.547	2.063	2.578	3.094	3.609	4.125	4.641	5.156	6.188	7.219	8.250
	I_{x-x}	5.849	8.774	11.70	14.62	17.55	20.47	23.40	26.32	29.25	35.09	40.94	46.79
	I_{y-y}	0.001	0.005	0.011	0.021	0.036	0.058	0.086	0.122	0.168	0.290	0.461	0.687
8 3/8	AREA	1.047	1.570	2.094	2.617	3.142	3.664	4.188	4.711	5.234	6.281	7.328	8.375
	I_{x-x}	6.119	9.179	12.24	15.30	18.36	21.42	24.48	27.54	30.60	36.71	42.83	48.95
	I_{y-y}	0.001	0.005	0.011	0.021	0.037	0.058	0.087	0.124	0.170	0.294	0.468	0.698
8 1/2	AREA	1.063	1.594	2.125	2.656	3.188	3.719	4.250	4.781	5.313	6.375	7.438	8.500
	I_{x-x}	6.397	9.596	12.79	15.99	19.19	22.39	25.59	28.79	31.99	38.38	44.78	51.18
	I_{y-y}	0.001	0.005	0.011	0.022	0.037	0.059	0.089	0.126	0.173	0.299	0.475	0.708
8 5/8	AREA	1.078	1.617	2.156	2.695	3.234	3.773	4.313	4.852	5.391	6.469	7.547	8.625
	I_{x-x}	6.684	10.03	13.37	16.71	20.05	23.39	26.73	30.08	33.42	40.10	46.78	53.47
	I_{y-y}	0.001	0.005	0.011	0.022	0.038	0.060	0.090	0.128	0.175	0.303	0.482	0.719
8 3/4	AREA	1.094	1.641	2.188	2.734	3.281	3.828	4.375	4.922	5.469	6.563	7.656	8.750
	I_{x-x}	6.978	10.47	13.96	17.45	20.94	24.42	27.91	31.40	34.89	41.87	48.85	55.83
	I_{y-y}	0.001	0.005	0.011	0.022	0.038	0.061	0.091	0.130	0.178	0.308	0.488	0.729

PROPERTIES OF RECTANGULAR SHAPES

$r_{x-x} = 0.289\,d.$

$r_{y-y} = 0.289\,t$

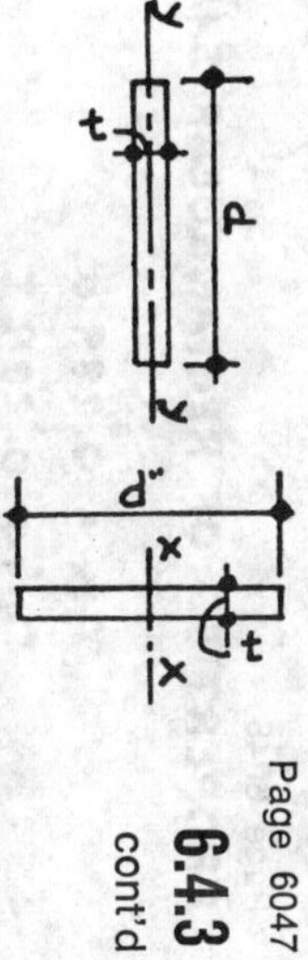

DEPTH "d" =	THICKNESS "t"												
	DECIMAL"	0.1250	0.1875	0.2500	0.3125	0.3750	0.4375	0.5000	0.5625	0.6250	0.7500	0.8750	1.0000
	FRACTION"	1/8	3/16	1/4	5/16	3/8	7/16	1/2	9/16	5/8	3/4	7/8	1.0
8 7/8	AREA □"	1.109	1.664	2.219	2.773	3.328	3.883	4.438	4.992	5.547	6.656	7.766	8.875
	I_{x-x}	7.282	10.92	14.56	18.20	21.85	25.49	29.13	32.77	36.41	43.69	50.97	58.25
	I_{y-y}	0.001	0.005	0.012	0.023	0.039	0.062	0.092	0.132	0.181	0.312	0.495	0.740
9	AREA	1.125	1.688	2.250	2.813	3.375	3.938	4.500	5.063	5.625	6.750	7.875	9.000
	I_{x-x}	7.594	11.39	15.19	18.98	22.78	26.58	30.38	34.17	37.97	45.56	53.16	60.75
	I_{y-y}	0.001	0.005	0.012	0.023	0.040	0.063	0.094	0.133	0.183	0.316	0.502	0.750
9 1/8	AREA	1.141	1.711	2.281	2.852	3.422	3.992	4.563	5.135	5.703	6.844	7.984	9.125
	I_{x-x}	7.915	11.87	15.83	19.79	23.74	27.70	31.66	35.62	39.57	47.49	55.40	63.32
	I_{y-y}	0.001	0.005	0.012	0.023	0.040	0.064	0.095	0.135	0.186	0.321	0.509	0.760
9 1/4	AREA	1.156	1.734	2.313	2.891	3.469	4.047	4.625	5.203	5.781	6.938	8.094	9.250
	I_{x-x}	8.244	12.37	16.49	20.61	24.73	28.80	32.96	37.10	41.22	49.47	57.71	65.95
	I_{y-y}	0.002	0.005	0.012	0.024	0.041	0.065	0.096	0.137	0.188	0.325	0.516	0.771
9 3/8	AREA	1.172	1.758	2.344	2.930	3.516	4.102	4.688	5.273	5.850	7.031	8.203	9.375
	I_{x-x}	8.583	12.87	17.17	21.46	25.75	30.04	34.33	38.62	42.92	51.50	60.08	68.66
	I_{y-y}	0.002	0.005	0.012	0.024	0.041	0.065	0.098	0.139	0.191	0.330	0.523	0.781
9 1/2	AREA	1.188	1.781	2.375	2.969	3.563	4.156	4.750	5.344	5.938	7.125	8.313	9.500
	I_{x-x}	8.931	13.40	17.86	22.33	26.79	31.26	35.72	40.19	44.66	53.59	62.52	71.45
	I_{y-y}	0.002	0.005	0.012	0.024	0.042	0.066	0.099	0.141	0.193	0.334	0.530	0.792
9 5/8	AREA	1.203	1.805	2.406	3.008	3.609	4.211	4.813	5.414	6.016	7.219	8.422	9.625
	I_{x-x}	9.288	13.93	18.58	23.22	27.86	32.51	37.16	41.80	46.44	55.73	65.02	74.31
	I_{y-y}	0.002	0.005	0.013	0.024	0.042	0.067	0.100	0.143	0.196	0.338	0.537	0.802
9 3/4	AREA	1.219	1.828	2.438	3.047	3.656	4.266	4.875	5.484	6.094	7.313	8.531	9.750
	I_{x-x}	9.655	14.48	19.31	24.14	28.96	33.79	38.62	43.45	48.27	57.93	67.58	77.24
	I_{y-y}	0.002	0.005	0.013	0.025	0.043	0.068	0.102	0.145	0.198	0.343	0.544	0.812
9 7/8	AREA	1.234	1.852	2.469	3.086	3.703	4.320	4.938	5.555	6.172	7.406	8.641	9.875
	I_{x-x}	10.03	15.05	20.06	25.08	30.09	35.11	40.12	45.14	50.15	60.19	70.22	80.25
	I_{y-y}	0.022	0.005	0.013	0.025	0.043	0.069	0.103	0.146	0.201	0.347	0.551	0.823

PROPERTIES OF RECTANGULAR SHAPES

$r_{x-x} = 0.289\,d$

$r_{y-y} = 0.289\,t$

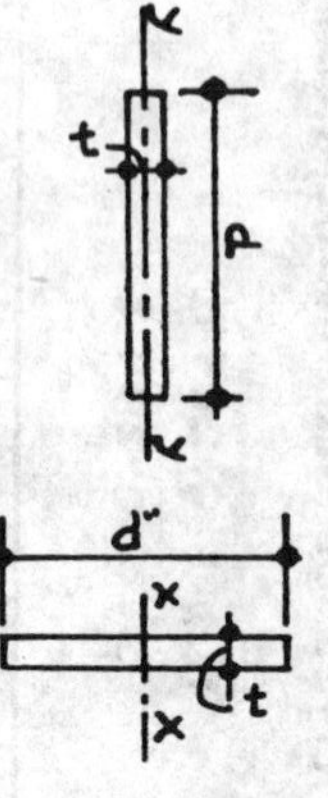

DEPTH d"		THICKNESS "t"											
	DECIMAL"	0.1250	0.1875	0.2500	0.3125	0.3750	0.4375	0.5000	0.5625	0.6250	0.7500	0.8750	1.0000
	FRACTION"	1/8	3/16	1/4	5/16	3/8	7/16	1/2	9/16	5/8	3/4	7/8	1.0
10	AREA □"	1.250	1.875	2.500	3.125	3.750	4.375	5.000	5.625	6.250	7.500	8.750	10.00
	I_{x-x}	10.42	15.63	20.83	26.04	31.25	36.46	41.67	46.88	52.08	62.50	72.92	83.33
	I_{y-y}	0.002	0.005	0.013	0.025	0.044	0.070	0.104	0.148	0.203	0.352	0.558	0.833
10 1/8	AREA	1.266	1.898	2.531	3.164	3.797	4.430	5.063	5.695	6.328	7.594	8.859	10.13
	I_{x-x}	10.81	16.22	21.62	27.03	32.44	37.84	43.25	48.65	54.06	64.87	75.69	86.50
	I_{y-y}	0.002	0.006	0.013	0.026	0.044	0.071	0.105	0.150	0.206	0.356	0.565	0.844
10 1/4	AREA	1.281	1.922	2.563	3.203	3.844	4.484	5.125	5.766	6.406	7.688	8.969	10.25
	I_{x-x}	11.22	16.83	22.44	28.04	33.65	39.26	44.87	50.48	56.09	67.31	78.52	89.74
	I_{y-y}	0.002	0.006	0.013	0.026	0.045	0.072	0.107	0.152	0.209	0.360	0.572	0.854
10 3/8	AREA	1.297	1.945	2.594	3.242	3.891	4.539	5.188	5.836	6.484	7.781	9.078	10.38
	I_{x-x}	11.63	17.45	23.27	29.08	34.90	40.72	46.53	52.35	58.17	69.80	81.43	93.06
	I_{y-y}	0.002	0.006	0.014	0.026	0.046	0.072	0.108	0.154	0.211	0.365	0.579	0.865
10 1/2	AREA	1.313	1.969	2.625	3.281	3.938	4.594	5.250	5.906	6.563	7.875	9.188	10.50
	I_{x-x}	12.06	18.09	24.12	30.15	36.18	42.21	48.23	54.26	60.29	72.35	84.41	96.47
	I_{y-y}	0.002	0.006	0.014	0.027	0.046	0.073	0.109	0.156	0.214	0.369	0.586	0.875
10 5/8	AREA	1.328	1.992	2.656	3.320	3.984	4.648	5.313	5.977	6.641	7.969	9.297	10.63
	I_{x-x}	12.49	18.74	24.99	31.24	37.48	43.73	49.98	56.22	62.47	74.97	87.46	99.95
	I_{y-y}	0.002	0.006	0.014	0.027	0.047	0.074	0.111	0.158	0.216	0.374	0.593	0.885
10 3/4	AREA	1.344	2.016	2.688	3.359	4.031	4.703	5.375	6.047	6.719	8.063	9.406	10.75
	I_{x-x}	12.94	19.41	25.88	32.35	38.82	45.29	51.76	58.23	64.70	77.64	90.58	103.5
	I_{y-y}	0.002	0.006	0.014	0.027	0.047	0.075	0.112	0.159	0.219	0.378	0.600	0.896
10 7/8	AREA	1.359	2.039	2.719	3.398	4.078	4.758	5.438	6.117	6.797	8.156	9.516	10.88
	I_{x-x}	13.40	20.10	26.79	33.49	40.19	46.89	53.59	60.29	66.99	80.38	93.78	107.2
	I_{y-y}	0.002	0.006	0.014	0.028	0.048	0.076	0.113	0.161	0.221	0.382	0.607	0.906
11	AREA	1.375	2.063	2.750	3.438	4.125	4.813	5.500	6.188	6.875	8.250	9.625	11.000
	I_{x-x}	13.86	20.80	27.73	34.63	41.59	48.53	55.40	62.39	69.32	83.19	97.05	110.9
	I_{y-y}	0.002	0.006	0.014	0.028	0.048	0.077	0.115	0.163	0.224	0.387	0.614	0.917

PROPERTIES OF RECTANGULAR SHAPES

$r_{x-x} = 0.289\,d.$

$r_{y-y} = 0.289\,t$

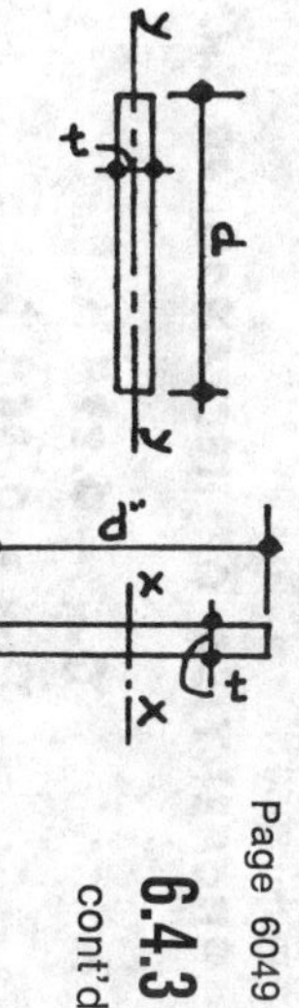

DEPTH d"		THICKNESS "t"											
	DECIMAL"	0.1250	0.1875	0.2500	0.3125	0.3750	0.4375	0.5000	0.5625	0.6250	0.7500	0.8750	1.0000
	FRACTION"	1/8	3/16	1/4	5/16	3/8	7/16	1/2	9/16	5/8	3/4	7/8	1.0
11 1/8	AREA □"	1.391	2.086	2.781	3.477	4.172	4.867	5.563	6.258	6.953	8.344	9.734	11.13
	I_{x-x}	14.34	21.51	28.69	35.86	43.03	50.20	57.37	64.54	71.71	86.06	100.4	114.7
	I_{y-y}	0.002	0.006	0.014	0.028	0.049	0.078	0.116	0.165	0.226	0.391	0.621	0.927
11 1/4	AREA	1.406	2.109	2.813	3.516	4.219	4.922	5.625	6.328	7.031	8.438	9.844	11.25
	I_{x-x}	14.83	22.25	29.66	37.08	44.49	51.91	59.33	66.74	74.16	88.99	103.8	118.7
	I_{y-y}	0.002	0.006	0.015	0.029	0.049	0.079	0.117	0.167	0.229	0.396	0.628	0.937
11 3/8	AREA	1.422	2.133	2.844	3.555	4.266	4.977	5.688	6.398	7.109	8.531	9.953	11.38
	I_{x-x}	15.33	23.00	30.66	38.33	45.99	53.67	61.33	68.99	76.66	91.99	107.3	122.7
	I_{y-y}	0.002	0.006	0.015	0.029	0.050	0.079	0.118	0.169	0.231	0.400	0.635	0.948
11 1/2	AREA	1.438	2.156	2.875	3.594	4.313	5.031	5.750	6.469	7.188	8.625	10.06	11.50
	I_{x-x}	15.84	23.76	31.69	39.61	47.53	55.45	63.37	71.29	79.21	95.06	110.9	126.7
	I_{y-y}	0.002	0.006	0.015	0.029	0.051	0.080	0.120	0.171	0.234	0.404	0.642	0.958
11 5/8	AREA	1.453	2.180	2.906	3.633	4.359	5.086	5.813	6.539	7.266	8.719	10.17	11.63
	I_{x-x}	16.36	24.55	32.73	40.91	49.09	57.28	65.46	73.64	81.82	98.19	114.6	130.9
	I_{y-y}	0.002	0.006	0.015	0.030	0.051	0.081	0.121	0.172	0.237	0.409	0.649	0.969
11 3/4	AREA	1.469	2.203	2.938	3.672	4.406	5.141	5.875	6.609	7.344	8.813	10.28	11.75
	I_{x-x}	16.90	25.35	33.80	42.25	50.69	59.14	67.59	76.04	84.49	101.4	118.3	135.2
	I_{y-y}	0.002	0.006	0.015	0.030	0.052	0.082	0.122	0.174	0.239	0.413	0.656	0.979
11 7/8	AREA	1.484	2.227	2.969	3.711	4.453	5.195	5.938	6.680	7.422	8.906	10.39	11.88
	I_{x-x}	17.44	26.17	34.89	43.61	52.33	61.05	69.77	78.49	87.22	104.7	122.1	139.5
	I_{y-y}	0.002	0.007	0.015	0.030	0.052	0.083	0.124	0.176	0.242	0.417	0.663	0.990
12	AREA	1.500	2.250	3.000	3.750	4.500	5.250	6.000	6.750	7.500	9.000	10.50	12.00
	I_{x-x}	18.00	27.00	36.00	45.00	54.00	63.00	72.00	81.00	90.00	108.0	126.0	144.0
	I_{y-y}	0.002	0.007	0.016	0.031	0.053	0.084	0.125	0.178	0.244	0.422	0.670	1.000
12 1/4	AREA	1.531	2.297	3.063	3.828	4.594	5.359	6.125	6.891	7.656	9.188	10.72	12.25
	I_{x-x}	19.15	28.72	38.30	47.87	57.45	67.02	76.59	86.17	95.74	114.9	134.0	153.2
	I_{y-y}	0.002	0.007	0.016	0.031	0.054	0.085	0.128	0.182	0.249	0.431	0.684	1.021

PROPERTIES OF RECTANGULAR SHAPES

$r_{x-x} = 0.289\,d$

$r_{y-y} = 0.289\,t$

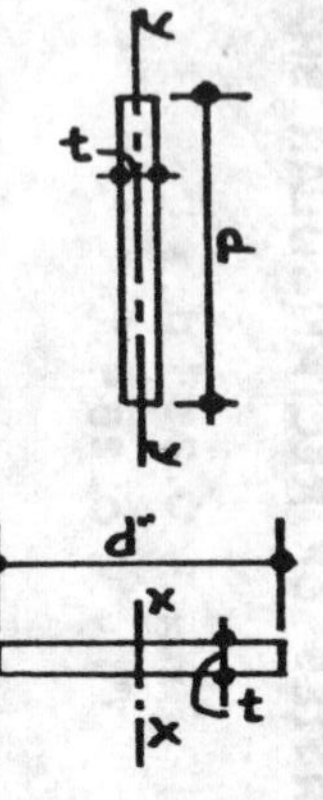

DEPTH d"		THICKNESS "t"											
	DECIMAL"	0.1250	0.1875	0.2500	0.3125	0.3750	0.4375	0.5000	0.5625	0.6250	0.7500	0.8750	1.0000
	FRACTION"	1/8	3/16	1/4	5/16	3/8	7/16	1/2	9/16	5/8	3/4	7/8	1.0
12½	AREA □"	1.563	2.344	3.125	3.906	4.688	5.469	6.250	7.031	7.813	9.375	10.94	12.50
	I_{x-x}	20.35	30.52	40.69	50.86	61.04	71.21	81.38	91.55	101.7	122.1	142.4	162.8
	I_{y-y}	0.002	0.007	0.016	0.032	0.055	0.087	0.130	0.185	0.254	0.439	0.698	1.042
12¾	AREA	1.594	2.391	3.188	3.984	4.781	5.578	6.375	7.172	7.969	9.563	11.16	12.75
	I_{x-x}	21.59	32.39	43.18	53.98	64.77	75.57	86.36	97.16	108.0	129.5	151.1	172.7
	I_{y-y}	0.002	0.007	0.017	0.032	0.056	0.089	0.133	0.189	0.259	0.448	0.712	1.062
13	AREA	1.625	2.438	3.250	4.063	4.875	5.688	6.500	7.313	8.125	9.750	11.38	13.00
	I_{x-x}	22.89	34.33	45.77	57.21	68.66	80.10	91.54	103.0	114.4	137.3	160.2	183.1
	I_{y-y}	0.002	0.007	0.017	0.033	0.057	0.091	0.135	0.193	0.264	0.457	0.726	1.083
13¼	AREA	1.656	2.484	3.313	4.141	4.969	5.797	6.625	7.453	8.281	9.938	11.59	13.25
	I_{x-x}	24.23	36.35	48.46	60.58	72.69	84.81	96.93	109.0	121.2	145.4	169.6	193.9
	I_{y-y}	0.002	0.007	0.017	0.034	0.058	0.092	0.138	0.197	0.270	0.466	0.740	1.104
13½	AREA	1.688	2.531	3.375	4.219	5.063	5.906	6.750	7.594	8.438	10.13	11.81	13.51
	I_{x-x}	25.68	38.44	51.26	64.07	76.89	89.70	102.5	115.3	128.1	153.8	179.4	205.0
	I_{y-y}	0.002	0.007	0.018	0.034	0.059	0.094	0.141	0.200	0.275	0.475	0.754	
13¾	AREA	1.719	2.578	3.438	4.297	5.156	6.016	6.875	7.734	8.594	10.31	12.03	13.75
	I_{x-x}	27.08	40.62	54.16	67.70	81.24	94.78	108.3	121.9	135.4	162.5	189.6	216.6
	I_{y-y}	0.002	0.008	0.018	0.035	0.060	0.096	0.143	0.204	0.280	0.483	0.768	1.146
14	AREA	1.750	2.625	3.500	4.375	5.250	6.125	7.000	7.875	8.750	10.50	12.25	14.00
	I_{x-x}	28.58	42.88	57.17	71.46	85.75	100.0	114.3	128.6	142.9	171.5	200.1	228.7
	I_{y-y}	0.002	0.008	0.018	0.036	0.062	0.098	0.146	0.208	0.285	0.492	0.782	1.167
14¼	AREA	1.781	2.672	3.563	4.453	5.344	6.234	7.125	8.016	8.906	10.69	12.47	14.25
	I_{x-x}	30.14	45.21	60.28	75.36	90.43	105.5	120.6	135.6	150.7	180.9	211.0	241.1
	I_{y-y}	0.002	0.008	0.019	0.036	0.063	0.099	0.148	0.211	0.290	0.501	0.796	1.187
14½	AREA	1.813	2.719	3.625	4.531	5.438	6.344	7.250	8.156	9.063	10.88	12.69	14.50
	I_{x-x}	31.76	47.63	63.51	79.39	95.27	111.1	127.0	142.9	158.8	190.5	222.3	254.1
	I_{y-y}	0.002	0.008	0.019	0.037	0.064	0.101	0.151	0.215	0.295	0.510	0.809	1.208

PROPERTIES OF RECTANGULAR SHAPES

$r_{x-x} = 0.289\ d.$
$r_{y-y} = 0.289\ t$

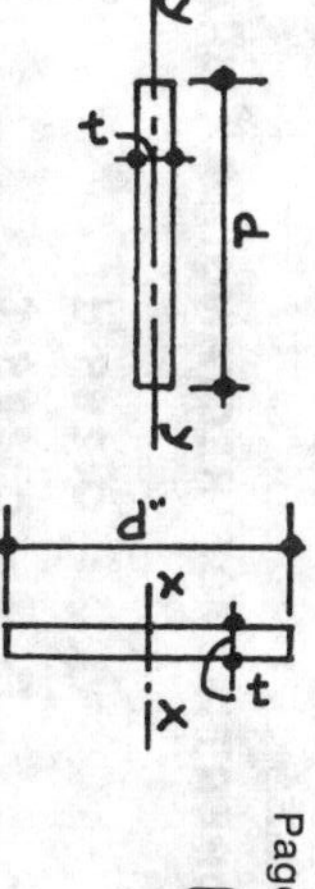

DEPTH d"	THICKNESS "t"												
	DECIMAL"	0.1250	0.1875	0.2500	0.3125	0.3750	0.4375	0.5000	0.5625	0.6250	0.7500	0.8750	1.0000
	FRACTION"	1/8	3/16	1/4	5/16	3/8	7/16	1/2	9/16	5/8	3/4	7/8	1.0
14 3/4	AREA □"	1.813	2.719	3.625	4.531	5.438	6.344	7.250	8.156	9.063	10.88	12.69	14.50
	I_{x-x}	33.43	50.14	66.86	83.57	100.3	117.0	133.7	150.4	167.1	200.6	222.3	254.1
	I_{y-y}	0.002	0.008	0.019	0.038	0.065	0.103	0.154	0.219	0.300	0.519	0.809	1.208
15	AREA	1.875	2.813	3.750	4.688	5.625	6.563	7.500	8.438	9.375	11.25	13.13	15.00
	I_{x-x}	35.16	52.73	70.31	87.89	105.5	123.0	140.6	158.2	175.8	210.9	246.1	281.3
	I_{y-y}	0.002	0.008	0.020	0.038	0.066	0.105	0.156	0.222	0.305	0.527	0.837	1.250
15 1/4	AREA	1.906	2.859	3.813	4.766	5.719	6.672	7.625	8.578	9.531	11.44	13.34	15.25
	I_{x-x}	36.94	55.42	73.89	92.36	110.8	129.3	147.8	166.2	184.7	221.7	258.6	295.5
	I_{y-y}	0.002	0.008	0.020	0.039	0.067	0.106	0.159	0.226	0.310	0.536	0.851	1.271
15 1/2	AREA	1.938	2.906	3.875	4.844	5.813	6.781	7.750	8.719	9.688	11.63	13.56	15.50
	I_{x-x}	38.79	58.19	77.58	96.98	116.4	135.8	155.2	174.6	194.0	232.7	271.5	310.3
	I_{y-y}	0.003	0.009	0.020	0.039	0.068	0.108	0.161	0.230	0.315	0.545	0.865	1.292
15 3/4	AREA	1.969	2.953	3.938	4.922	5.906	6.891	7.875	8.859	9.844	11.81	13.78	15.75
	I_{x-x}	40.70	61.05	81.40	101.7	122.1	142.4	162.8	183.1	203.5	244.2	284.9	325.6
	I_{y-y}	0.003	0.009	0.021	0.040	0.069	0.110	0.164	0.234	0.320	0.554	0.879	1.312
16	AREA	2.000	3.000	4.000	5.000	6.000	7.000	8.000	9.000	10.00	12.00	14.00	16.00
	I_{x-x}	42.67	64.00	85.33	106.7	128.0	149.3	170.7	192.0	213.3	256.0	298.7	341.3
	I_{y-y}	0.003	0.009	0.021	0.041	0.070	0.112	0.167	0.237	0.326	0.562	0.893	1.333
16 1/4	AREA	2.031	3.047	4.063	5.078	6.094	7.109	8.125	9.141	10.16	12.19	14.22	16.25
	I_{x-x}	44.70	67.05	89.40	111.7	134.1	156.4	178.8	201.1	223.5	268.2	312.9	357.6
	I_{y-y}	0.003	0.009	0.021	0.041	0.071	0.113	0.169	0.241	0.331	0.571	0.907	1.354
16 1/2	AREA	2.063	3.094	4.125	5.156	6.188	7.219	8.250	9.281	10.31	12.38	14.44	16.50
	I_{x-x}	46.79	70.19	93.59	117.0	140.4	163.8	187.2	210.6	234.1	280.8	327.6	374.3
	I_{y-y}	0.003	0.009	0.021	0.042	0.073	0.115	0.172	0.245	0.336	0.580	0.921	1.375
16 3/4	AREA	2.094	3.141	4.188	5.234	6.281	7.328	8.375	9.422	10.47	12.56	14.66	16.75
	I_{x-x}	48.95	73.43	97.90	122.4	146.9	171.3	195.8	220.3	244.8	293.7	342.7	391.6
	I_{y-y}	0.003	0.009	0.022	0.043	0.074	0.117	0.174	0.248	0.341	0.589	0.935	1.396

PROPERTIES OF RECTANGULAR SHAPES

$r_{x-x} = 0.289\,d.$

$r_{y-y} = 0.289\,t$

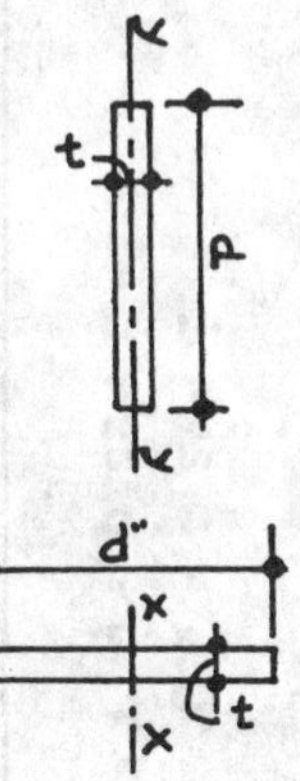

DEPTH d"		THICKNESS "t"											
	DECIMAL "	0.1250	0.1875	0.2500	0.3125	0.3750	0.4375	0.5000	0.5625	0.6250	0.7500	0.8750	1.0000
	FRACTION"	1/8	3/16	1/4	5/16	3/8	7/16	1/2	9/16	5/8	3/4	7/8	1.0
17	AREA □"	2.125	3.188	4.250	5.313	6.375	7.438	8.500	9.563	10.63	12.75	14.88	17.00
	I_{x-x}	51.18	76.77	102.4	127.9	153.5	179.1	204.7	230.3	255.9	307.1	358.2	409.4
	I_{y-y}	0.003	0.009	0.022	0.043	0.075	0.119	0.177	0.252	0.346	0.598	0.949	1.417
17¼	AREA	2.156	3.234	4.313	5.391	6.469	7.547	8.625	9.703	10.78	12.94	15.09	17.25
	I_{x-x}	53.47	80.20	106.9	133.7	160.4	187.1	213.9	240.6	267.3	320.8	374.3	487.7
	I_{y-y}	0.003	0.009	0.022	0.044	0.076	0.120	0.180	0.256	0.351	0.606	0.963	1.437
17½	AREA	2.188	3.281	4.375	5.469	6.563	7.656	8.750	9.844	10.94	13.13	15.31	17.50
	I_{x-x}	55.83	83.74	111.7	139.6	167.5	195.4	223.3	251.2	279.1	335.0	390.8	446.6
	I_{y-y}	0.003	0.010	0.023	0.045	0.077	0.122	0.182	0.260	0.356	0.615	0.977	1.458
17¾	AREA	2.219	3.328	4.438	5.547	6.656	7.766	8.875	9.984	11.09	13.31	15.53	17.75
	I_{x-x}	58.29	87.38	116.5	145.6	174.8	203.9	233.0	262.1	291.3	349.5	407.8	466.0
	I_{y-y}	0.003	0.010	0.023	0.045	0.078	0.124	0.185	0.263	0.361	0.624	0.991	1.479
18	AREA	2.250	3.375	4.500	5.625	6.750	7.875	9.000	10.13	11.25	13.50	15.75	18.00
	I_{x-x}	60.75	91.13	121.5	151.9	182.3	212.6	243.0	273.4	303.8	364.5	425.3	486.0
	I_{y-y}	0.003	0.010	0.023	0.046	0.079	0.126	0.188	0.267	0.366	0.633	1.005	1.500
18¼	AREA	2.281	3.422	4.563	5.703	6.844	7.984	9.125	10.27	11.41	13.69	15.97	18.25
	I_{x-x}	63.32	94.97	126.6	158.3	189.9	221.6	253.3	284.9	316.6	379.9	442.3	506.5
	I_{y-y}	0.003	0.010	0.024	0.046	0.080	0.127	0.190	0.271	0.371	0.642	1.019	1.521
18½	AREA	2.313	3.469	4.625	5.781	6.938	8.094	9.250	10.41	11.56	13.88	16.19	18.50
	I_{x-x}	65.95	98.93	131.9	164.9	197.9	230.8	263.8	296.8	329.8	397.5	461.7	527.6
	I_{y-y}	0.003	0.010	0.024	0.047	0.081	0.129	0.193	0.274	0.376	0.650	1.033	1.542
18¾	AREA	2.344	3.516	4.688	5.859	7.031	8.203	9.375	10.55	11.72	14.06	16.41	18.75
	I_{x-x}	68.66	103.0	137.3	171.7	206.0	240.3	274.7	309.0	343.3	412.0	480.7	549.3
	I_{y-y}	0.003	0.010	0.024	0.048	0.082	0.131	0.195	0.278	0.381	0.659	1.047	1.562
19	AREA	2.375	3.563	4.750	5.938	7.125	8.313	9.500	10.69	11.88	14.25	16.63	19.00
	I_{x-x}	71.45	107.2	142.9	178.6	214.3	250.1	285.8	321.5	357.2	428.7	500.1	571.6
	I_{y-y}	0.003	0.010	0.025	0.048	0.083	0.133	0.198	0.282	0.387	0.668	1.061	1.583

PROPERTIES OF RECTANGULAR SHAPES

$r_{x-x} = 0.289\,d.$

$r_{y-y} = 0.289\,t$

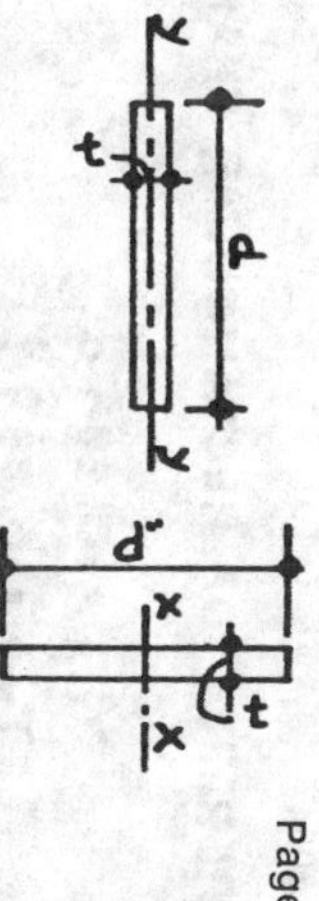

DEPTH d =		THICKNESS "t"											
	DECIMAL"	0.1250	0.1875	0.2500	0.3125	0.3750	0.4375	0.5000	0.5625	0.6250	0.7500	0.8750	1.0000
	FRACTION"	1/8	3/16	1/4	5/16	3/8	7/16	1/2	9/16	5/8	3/4	7/8	1.0
19¼	AREA □"	2.406	3,609	4.813	6.016	7.219	8.422	9.625	10.83	12.03	14.44	16.84	19.25
	I_{x-x}	74.31	111.5	148.6	185.8	222.9	260.1	297.2	334.4	371.5	445.8	520.1	594.4
	I_{y-y}	0.003	0.011	0.025	0.049	0.085	0.134	0.201	0.286	0.392	0.677	1.075	1.604
19½	AREA	2.438	3.656	4.875	6.094	7.313	8.531	9.750	10.97	12.19	14.63	17.06	19.50
	I_{x-x}	77.24	115.9	154.5	193.1	231.7	270.3	309.0	347.6	386.2	463.4	540.7	617.9
	I_{y-y}	0.003	0.011	0.025	0.050	0.086	0.136	0.203	0.289	0.397	0.686	1.089	1.625
19¾	AREA	2.469	3.703	4.938	6.172	7.406	8.641	9.875	11.11	12.34	14.81	17.28	19.75
	I_{x-x}	80.25	120.4	160.5	200.6	240.7	280.9	321.0	361.1	401.2	481.5	561.7	642.0
	I_{y-y}	0.003	0.011	0.026	0.050	0.087	0.138	0.206	0.293	0.402	0.694	1.103	1.646
20	AREA	2.500	3.750	5.000	6.250	7.500	8.750	10.00	11.25	12.50	15.00	17.50	20.00
	I_{x-x}	83.33	125.0	166.7	208.3	250.0	291.7	333.3	375.0	416.7	500.0	583.3	666.7
	I_{y-y}	0.003	0.011	0.026	0.051	0.088	0.141	0.208	0.297	0.407	0.703	1.117	1.667
20¼	AREA	2.531	3.797	5.063	6.328	7.594	8.859	10.13	11.39	12.66	15.19	17.72	20.25
	I_{x-x}	86.50	129.7	173.0	216.2	259.5	302.7	346.0	389.2	432.5	519.0	605.5	692.0
	I_{y-y}	0.003	0.011	0.026	0.051	0.089	0.141	0.211	0.300	0.412	0.712	1.130	1.687
20½	AREA	2.563	3.844	5.125	6.406	7.688	8.969	10.25	11.53	12.81	15.38	17.94	20.50
	I_{x-x}	89.74	134.6	179.5	224.4	269.2	314.1	359.0	403.8	448.7	538.4	628.2	717.9
	I_{y-y}	0.003	0.011	0.027	0.052	0.090	0.143	0.214	0.304	0.417	0.721	1.144	1.708
20¾	AREA	2.594	3.891	5.188	6.484	7.781	9.078	10.38	11.67	12.97	15.56	18.16	20.75
	I_{x-x}	93.06	139.6	186.1	232.7	279.2	325.7	372.3	418.8	465.3	558.4	651.5	744.5
	I_{y-y}	0.003	0.011	0.027	0.053	0.091	0.145	0.216	0.308	0.422	0.729	1.158	1.729
21	AREA	2.625	3.938	5.250	6.563	7.875	9.188	10.50	11.81	13.13	15.75	18.38	21.00
	I_{x-x}	96.47	144.7	192.9	241.2	289.4	337.6	385.9	431.1	482.3	578.8	675.3	771.8
	I_{y-y}	0.003	0.012	0.027	0.053	0.092	0.147	0.219	0.311	0.427	0.738	1.172	1.750
21¼	AREA	2.656	3.984	5.313	6.641	7.969	9.297	10.63	11.95	13.28	15.94	18.59	21.25
	I_{x-x}	99.96	149.9	199.9	249.9	299.9	349.8	399.8	449.8	499.8	599.7	699.7	799.6
	I_{y-y}	0.003	0.012	0.028	0.054	0.093	0.148	0.221	0.315	0.432	0.747	1.186	1.771

PROPERTIES OF RECTANGULAR SHAPES

$r_{x-x} = 0.289\,d.$

$r_{y-y} = 0.289\,t$

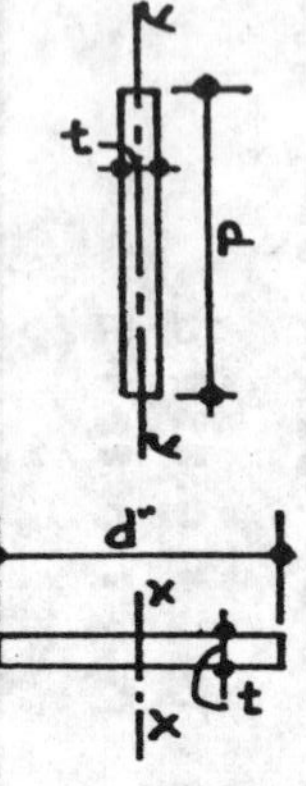

DEPTH d"	DECIMAL "	0.1250	0.1875	0.2500	0.3125	0.3750	0.4375	0.5000	0.5625	0.6250	0.7500	0.8750	1.0000
	FRACTION"	1/8	3/16	1/4	5/16	3/8	7/16	1/2	9/16	5/8	3/4	7/8	1.0
21 1/2	AREA □"	2.688	4.031	5.375	6.719	8.063	9.406	10.75	12.09	13.44	16.13	18.81	21.50
	I_{x-x}	103.5	155.3	207.1	258.8	310.6	362.3	414.1	465.9	517.6	621.1	724.7	828.2
	I_{y-y}	0.003	0.012	0.028	0.055	0.094	0.150	0.224	0.319	0.437	0.756	1.200	1.792
21 3/4	AREA	2.719	4.078	5.438	6.797	8.156	9.516	10.88	12.23	13.59	16.31	19.03	21.75
	I_{x-x}	107.2	160.8	214.4	267.9	321.5	375.1	428.7	482.3	535.9	643.1	750.2	857.4
	I_{y-y}	0.004	0.012	0.028	0.055	0.096	0.152	0.227	0.323	0.443	0.765	1.214	1.812
22	AREA	2.750	4.125	5.500	6.875	8.250	9.625	11.00	12.38	13.75	16.50	19.25	22.00
	I_{x-x}	110.9	166.4	221.8	277.3	332.8	388.2	443.7	499.1	554.6	665.5	776.4	887.3
	I_{y-y}	0.004	0.012	0.029	0.056	0.097	0.154	0.229	0.326	0.448	0.773	1.228	1.833
22 1/4	AREA	2.781	4.172	5.563	6.953	8.344	9.734	11.13	12.52	13.91	16.69	19.47	22.25
	I_{x-x}	114.7	172.1	229.5	286.8	344.2	401.6	459.0	516.3	573.7	688.4	803.2	917.9
	I_{y-y}	0.004	0.012	0.029	0.057	0.098	0.155	0.232	0.330	0.453	0.782	1.242	1.854
22 1/2	AREA	2.813	4.219	5.625	7.031	8.438	9.844	11.25	12.66	14.06	16.88	19.69	22.50
	I_{x-x}	118.7	178.0	237.3	296.6	356.0	415.3	474.6	534.0	593.3	711.9	830.6	949.3
	I_{y-y}	0.004	0.012	0.029	0.057	0.099	0.157	0.234	0.334	0.458	0.791	1.256	1.875
22 3/4	AREA	2.844	4.266	5.688	7.109	8.531	9.953	11.38	12.80	14.22	17.06	19.91	22.75
	I_{x-x}	122.7	184.0	245.3	306.6	368.0	429.3	490.6	552.0	613.3	735.9	858.6	981.3
	I_{y-y}	0.004	0.012	0.030	0.058	0.100	0.159	0.237	0.337	0.463	0.800	1.270	1.896
23	AREA	2.875	4.313	5.750	7.188	8.625	10.06	11.50	12.94	14.38	17.25	20.13	23.00
	I_{x-x}	126.7	190.1	253.5	316.8	380.2	443.6	507.0	570.3	633.7	760.4	887.2	1014.0
	I_{y-y}	0.004	0.013	0.030	0.058	0.101	0.161	0.240	0.341	0.468	0.809	1.284	1.917
23 1/4	AREA	2.906	4.359	5.813	7.266	8.719	10.17	11.63	13.08	14.53	17.44	20.34	23.25
	I_{x-x}	130.9	196.4	261.8	327.3	392.8	458.2	523.7	589.1	654.6	785.5	916.4	1047.0
	I_{y-y}	0.004	0.013	0.030	0.059	0.102	0.162	0.242	0.345	0.473	0.817	1.298	1.937
23 1/2	AREA	2.938	4.406	5.875	7.344	8.813	10.28	11.75	13.22	14.69	17.63	20.56	23.50
	I_{x-x}	135.2	202.8	270.4	338.0	405.6	473.2	540.8	608.3	675.9	811.1	946.3	1082.0
	I_{y-y}	0.004	0.013	0.031	0.060	0.103	0.164	0.245	0.349	0.078	0.826	1.312	1.958

PROPERTIES OF RECTANGULAR SHAPES

$r_{x-x} = 0.289\,d.$

$r_{y-y} = 0.289\,t$

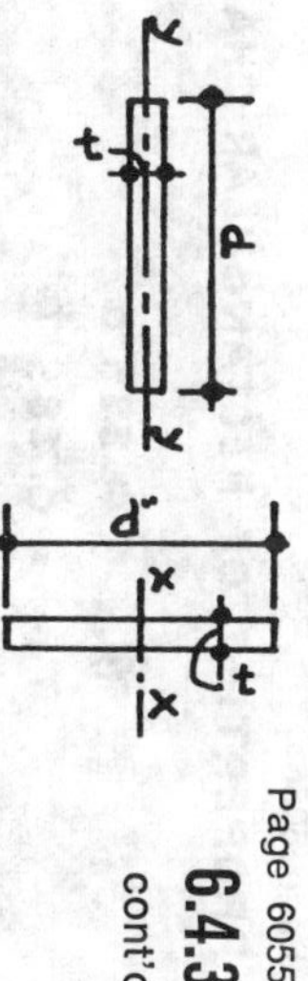

DEPTH d=		THICKNESS "t"											
	DECIMAL"	0.1250	0.1875	0.2500	0.3125	0.3750	0.4375	0.5000	0.5625	0.6250	0.7500	0.8750	1.0000
	FRACTION"	1/8	3/16	1/4	5/16	3/8	7/16	1/2	9/16	5/8	3/4	7/8	1.0
23 3/4	AREA □"	2.906	4.359	5.813	7.266	8.719	10.17	11.63	13.08	14.53	17.44	20.34	23.25
	I_{x-x}	139.5	209.3	279.1	348.9	418.6	488.4	558.2	627.9	697.7	837.3	976.8	1047.0
	I_{y-y}	0.004	0.013	0.031	0.060	0.104	0.166	0.247	0.352	0.483	0.835	1.326	1.937
24	AREA	3.000	4.500	6.000	7.500	9.000	10.50	12.00	13.50	15.00	18.00	21.00	24.00
	I_{x-x}	144.0	216.0	288.0	360.0	432.0	504.0	576.0	648.0	720.0	864.0	1008.0	1152.0
	I_{y-y}	0.004	0.013	0.031	0.061	0.105	0.167	0.250	0.356	0.488	0.844	1.340	2.000
24 1/2	AREA	3.063	4.594	6.125	7.656	9.188	10.72	12.25	13.78	15.31	18.38	21.44	24.50
	I_{x-x}	153.2	229.8	306.4	383.0	459.6	536.2	612.8	689.3	765.9	919.1	1072.0	1226.0
	I_{y-y}	0.004	0.013	0.032	0.062	0.108	0.171	0.255	0.363	0.498	0.861	1.368	2.042
25	AREA	3.125	4.688	6.250	7.813	9.375	10.94	12.50	14.06	15.63	18.75	21.88	25.00
	I_{x-x}	162.8	244.1	325.5	406.9	488.3	569.7	651.0	732.4	813.8	976.6	1139.0	1302.0
	I_{y-y}	0.004	0.014	0.033	0.064	0.110	0.174	0.260	0.371	0.509	0.879	1.396	2.083
25 1/2	AREA	3.188	4.781	6.375	7.969	9.563	11.16	12.75	14.34	15.94	19.13	22.31	25.50
	I_{x-x}	172.7	259.1	345.1	431.8	518.2	604.5	690.9	777.2	863.6	1036.0	1209.0	1382.0
	I_{y-y}	0.004	0.014	0.033	0.065	0.112	0.178	0.266	0.378	0.519	0.896	1.424	2.125
26	AREA	3.250	4.875	6.500	8.125	9.750	11.38	13.00	14.63	16.25	19.50	22.75	26.00
	I_{x-x}	183.1	274.6	366.2	457.7	549.3	640.8	732.3	823.9	915.4	1099.0	1282.0	1465.0
	I_{y-y}	0.004	0.014	0.034	0.066	0.114	0.181	0.272	0.386	0.529	0.914	1.452	2.167
26 1/2	AREA	3.313	4.969	6.625	8.281	9.938	11.59	13.25	14.91	16.56	19.88	23.19	26.50
	I_{x-x}	193.9	290.8	387.7	484.6	581.6	678.5	775.4	872.3	969.3	1163.0	1357.0	1551.0
	I_{y-y}	0.004	0.015	0.035	0.067	0.116	0.185	0.276	0.393	0.539	0.932	1.479	2.208
27	AREA	3.375	5.063	6.750	8.438	10.13	11.81	13.50	15.19	16.88	20.25	23.63	27.00
	I_{x-x}	205.0	307.5	410.1	512.6	615.1	717.6	820.1	922.6	1025.0	1230.0	1435.0	1640.0
	I_{y-y}	0.004	0.015	0.035	0.069	0.119	0.188	0.281	0.400	0.549	0.949	1.507	2.250
27 1/2	AREA	3.438	5.156	6.875	8.594	10.31	12.03	13.75	15.47	17.19	20.63	24.00	27.50
	I_{x-x}	216.6	325.0	433.3	541.6	649.9	758.2	866.5	974.9	1083.0	1300.0	1516.0	1733.0
	I_{y-y}	0.004	0.015	0.036	0.070	0.121	0.192	0.286	0.408	0.559	0.967	1.535	2.292

PROPERTIES OF RECTANGULAR SHAPES

$r_{x-x} = 0.289\,d.$

$r_{y-y} = 0.289\,t$

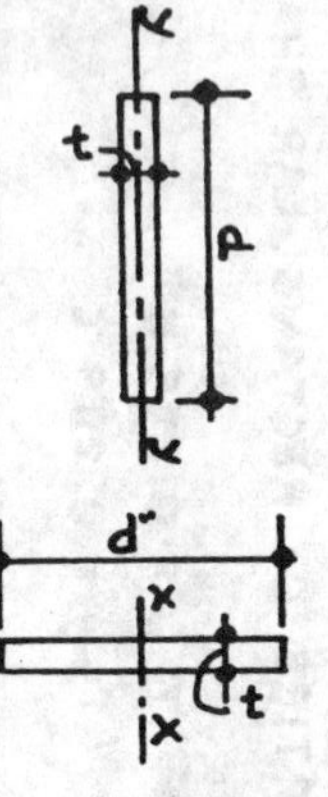

DEPTH d"		THICKNESS "t"											
	DECIMAL"	0.1250	0.1875	0.2500	0.3125	0.3750	0.4375	0.5000	0.5625	0.6250	0.7500	0.8750	1.0000
	FRACTION"	1/8	3/16	1/4	5/16	3/8	7/16	1/2	9/16	5/8	3/4	7/8	1.0
28	AREA □"	3.500	5.250	7.000	8.750	10.50	12.25	14.00	15.75	17.50	21.00	24.50	28.00
	I_{x-x}	228.7	343.0	457.3	571.7	686.0	800.3	914.7	1029.0	1143.0	1372.0	1601.0	1829.0
	I_{y-y}	0.005	0.015	0.036	0.071	0.123	0.195	0.292	0.415	0.570	0.984	1.563	2.333
28½	AREA	3.563	5.344	7.125	8.906	10.69	12.47	14.25	16.03	17.81	21.38	24.94	28.50
	I_{x-x}	241.1	361.7	482.3	602.8	723.4	844.0	964.5	1085.0	1206.0	1447.0	1688.0	1929.0
	I_{y-y}	0.005	0.016	0.037	0.072	0.125	0.199	0.297	0.423	0.580	1.002	1.591	2.375
29	AREA	3.625	5.438	7.250	9.063	10.88	12.69	14.50	16.31	18.13	21.75	25.38	29.00
	I_{x-x}	254.1	381.1	508.1	635.1	762.2	889.2	1016.0	1143.0	1270.0	1524.0	1778.0	2032.0
	I_{y-y}	0.005	0.016	0.038	0.074	0.127	0.202	0.302	0.430	0.590	1.020	1.619	2.417
29½	AREA	3.688	5.531	7.375	9.219	11.06	12.91	14.75	16.59	18.44	22.13	25.81	29.50
	I_{x-x}	267.4	401.1	534.8	668.5	802.3	936.0	1070.0	1203.0	1337.0	1605.0	1872.0	2139.0
	I_{y-y}	0.005	0.016	0.038	0.075	0.130	0.206	0.307	0.438	0.600	1.037	1.647	2.458
30	AREA	3.750	5.625	7.500	9.375	11.25	13.13	15.00	16.88	18.75	22.50	26.25	30.00
	I_{x-x}	281.3	421.9	562.5	703.1	843.8	984.4	1125.0	1266.0	1406.0	1688.0	1969.0	2250.0
	I_{y-y}	0.005	0.016	0.039	0.076	0.132	0.209	0.313	0.445	0.610	1.055	1.675	2.500
30½	AREA	3.813	5.719	7.625	9.531	11.44	13.34	15.25	17.16	19.06	22.88	26.69	30.50
	I_{x-x}	295.6	443.3	591.1	738.9	886.7	1034.0	1182.0	1330.0	1478.0	1773.0	2069.0	2364.0
	I_{y-y}	0.005	0.017	0.040	0.078	0.134	0.213	0.318	0.452	0.621	1.072	1.703	2.542
31	AREA	3.875	5.813	7.750	9.688	11.63	13.56	15.50	17.44	19.38	23.25	27.13	31.00
	I_{x-x}	310.3	465.5	620.6	775.8	931.0	1086.0	1241.0	1396.0	1552.0	1862.0	2172.0	2483.0
	I_{y-y}	0.005	0.017	0.040	0.079	0.136	0.216	0.323	0.460	0.631	1.090	1.731	2.583
31½	AREA	3.938	5.906	7.875	9.844	11.81	13.78	15.75	17.72	19.69	23.63	27.56	31.50
	I_{x-x}	325.6	488.4	651.2	814.0	976.8	1140.0	1302.0	1465.0	1628.0	1954.0	2279.0	2605.0
	I_{y-y}	0.005	0.017	0.041	0.080	0.138	0.220	0.328	0.467	0.641	1.107	1.759	2.625
32	AREA	4.000	6.000	8.000	10.00	12.00	14.00	16.00	18.00	20.00	24.00	28.00	32.00
	I_{x-x}	341.3	512.0	682.7	853.3	1024.0	1195.0	1365.0	1536.0	1707.0	2048.0	2389.0	2731.0
	I_{y-y}	0.005	0.018	0.042	0.081	0.141	0.223	0.333	0.475	0.651	1.125	1.786	2.667

PROPERTIES OF RECTANGULAR SHAPES

$r_{x-x} = 0.289\,d$

$r_{y-y} = 0.289\,t$

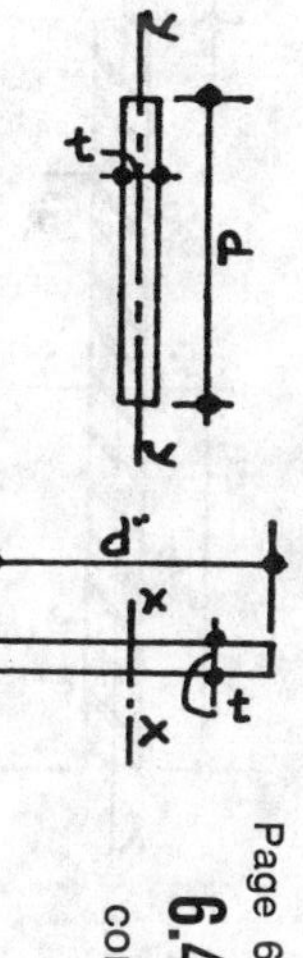

DEPTH d"	DECIMAL"	0.1250	0.1875	0.2500	0.3125	0.3750	0.4375	0.5000	0.5625	0.6250	0.7500	0.8750	1.0000
	FRACTION"	1/8	3/16	1/4	5/16	3/8	7/16	1/2	9/16	5/8	3/4	7/8	1.0
32½	AREA □"	4.063	6.094	8.125	10.16	12.19	14.22	16.25	18.28	20.31	24.38	28.44	32.50
	I_{x-x}	357.6	536.4	715.2	894.0	1073.0	1252.0	1430.0	1609.0	1788.0	2146.0	2503.0	2861.0
	I_{y-y}	0.005	0.018	0.042	0.083	0.143	0.227	0.339	0.482	0.661	1.143	1.814	2.708
33	AREA	4.125	6.188	8.250	10.31	12.38	14.44	16.50	18.56	20.63	24.75	28.88	33.00
	I_{x-x}	374.3	561.5	748.7	935.9	1123.0	1310.0	1497.0	1685.0	1872.0	2246.0	2620.0	2995.0
	I_{y-y}	0.005	0.018	0.043	0.084	0.145	0.230	0.344	0.489	0.671	1.160	1.842	2.750
33½	AREA	4.188	6.281	8.375	10.47	12.56	14.66	16.75	18.84	20.94	25.13	29.31	33.50
	I_{x-x}	391.6	587.4	783.2	979.0	1175.0	1371.0	1566.0	1762.0	1958.0	2350.0	2741.0	3133.0
	I_{y-y}	0.005	0.018	0.044	0.085	0.147	0.234	0.349	0.497	0.682	1.178	1.870	2.792
34	AREA	4.250	6.375	8.500	10.63	12.75	14.88	17.00	19.13	21.25	25.50	29.75	34.00
	I_{x-x}	409.4	614.1	818.8	1024.0	1228.0	1433.0	1638.0	1842.0	2047.0	2457.0	2866.0	3275.0
	I_{y-y}	0.006	0.019	0.044	0.086	0.149	0.237	0.354	0.504	0.692	1.195	1.898	2.833
34½	AREA	4.313	6.469	8.625	10.78	12.94	15.09	17.25	19.41	21.56	25.88	30.19	34.50
	I_{x-x}	427.8	641.6	855.5	1069.0	1283.0	1497.0	1711.0	1925.0	2139.0	2567.0	2994.0	3422.0
	I_{y-y}	0.006	0.019	0.045	0.088	0.152	0.241	0.359	0.512	0.702	1.213	1.926	2.875
35	AREA	4.375	6.563	8.750	10.94	13.13	15.31	17.50	19.69	21.88	26.25	30.63	35.00
	I_{x-x}	446.6	669.9	893.2	1117.0	1340.0	1563.0	1786.0	2010.0	2233.0	2680.0	3126.0	3573.0
	I_{y-y}	0.006	0.019	0.046	0.089	0.154	0.244	0.365	0.519	0.712	1.230	1.954	2.917
35½	AREA	4.438	6.656	8.875	11.09	13.31	15.53	17.75	19.97	22.19	26.63	31.06	35.50
	I_{x-x}	466.0	699.0	932.1	1165.0	1398.0	1631.0	1864.0	2097.0	2330.0	2796.0	3262.0	3728.0
	I_{y-y}	0.006	0.020	0.046	0.090	0.156	0.248	0.378	0.527	0.722	1.248	1.982	2.958
36	AREA	4.500	6.750	9.000	11.25	13.50	15.75	18.00	20.25	22.50	27.00	31.50	36.00
	I_{x-x}	486.0	729.0	972.0	1215.0	1458.0	1701.0	1944.0	2187.0	2430.0	2916.0	3402.0	3888.0
	I_{y-y}	0.006	0.020	0.047	0.092	0.158	0.251	0.375	0.534	0.732	1.266	2.010	3.000
38	AREA	4.750	7.125	9.500	11.88	14.25	16.63	19.00	21.38	23.75	28.50	33.25	38.00
	I_{x-x}	571.6	857.4	1143.0	1429.0	1715.0	2001.0	2286.0	2572.0	2858.0	3430.0	4001.0	4573.0
	I_{y-y}	0.006	0.021	0.049	0.097	0.167	0.265	0.396	0.564	0.773	1.336	2.121	3.167

$$I_x = \frac{td^3}{12}$$

$$S_x = \frac{td^2}{6}$$

TABLE: Properties of steel web plates, 1/8" x 40" to 1" x 72" 6.4.4

PROPERTIES OF SECTIONS

$r_x = 0.289\,d$ $r_y = 0.289\,t$

DEPTH "d"	THICKNESS "t"												
	FRACTIONS	1/8	3/16	1/4	5/16	3/8	7/16	1/2	9/16	5/8	3/4	7/8	1.0
	DECIMALS	0.125	0.1875	0.250	0.3125	0.375	0.4375	0.500	0.5625	0.625	0.750	0.875	1.00
40	WEIGHT	17.00	25.50	34.00	42.50	51.00	59.50	68.00	76.50	85.00	102.00	119.00	136.00
	AREA □"	5.000	7.500	10.000	12.500	15.000	17.500	20.000	22.500	25.000	30.000	35.000	40.000
	I_{x-x}	666.7	1000.0	1333.3	1666.7	2000.0	2333.3	2666.7	3000.0	3333.3	4000.0	4666.7	5333.0
	I_{y-y}	0.007	0.022	0.052	0.102	0.176	0.279	0.417	0.593	0.814	1.406	2.233	3.333
42	WEIGHT	17.85	26.78	35.70	44.63	53.55	62.48	71.40	80.33	89.25	107.10	125.00	142.80
	AREA	5.250	7.875	10.500	13.133	15.750	18.380	21.000	23.633	26.250	31.500	36.750	42.000
	I_{x-x}	771.80	1158.0	1544.0	1929.0	2315.0	2701.0	3087.0	3473.0	3859.0	4631.0	5402.0	6174.0
	I_{y-y}	0.007	0.023	0.055	0.107	0.185	0.293	0.438	0.623	0.854	1.477	2.345	3.500
48	WEIGHT	20.40	30.60	40.80	51.00	61.20	71.40	81.60	91.80	102.00	122.40	142.80	163.20
	AREA	6.000	9.000	12.000	15.000	18.000	21.000	24.000	27.000	30.000	36.000	42.000	48.000
	I_{x-x}	1152.0	1728.0	2304.0	2880.0	3456.0	4032.0	4608.0	5184.0	5760.0	6912.0	8064.0	9216.0
	I_{y-y}	0.008	0.026	0.063	0.122	0.211	0.335	0.500	0.712	0.977	1.687	2.680	4.000
54	WEIGHT	22.95	34.43	45.90	57.38	68.85	80.33	91.80	103.30	114.80	137.70	160.70	183.60
	AREA	6.750	10.130	13.500	16.880	20.250	23.630	27.000	30.380	33.750	40.500	47.250	54.000
	I_{x-x}	1640.0	2460.0	3281.0	4101.0	4921.0	5741.0	6561.0	7381.0	8201.0	9842.0	11,482.0	13,122.0
	I_{y-y}	0.009	0.030	0.070	0.137	0.237	0.377	0.563	0.801	1.099	1.898	3.015	4.500
60	WEIGHT	25.50	38.25	51.00	63.75	76.50	89.25	102.00	114.80	127.50	153.00	178.50	204.00
	AREA	7.500	11.250	15.000	18.750	22.500	26.250	30.000	33.750	37.500	45.000	52.500	60.000
	I_{x-x}	2250.0	3375.0	4500.0	5625.0	6750.0	7875.0	9000.0	10,125.0	11,250.0	11,250.0	15,750.0	18,000.0
	I_{y-y}	0.010	0.033	0.078	0.153	0.264	0.419	0.625	0.890	1.221	2.109	3.350	5.000
66	WEIGHT	28.05	42.08	56.10	70.13	84.15	98.18	112.20	126.20	140.30	168.30	196.40	224.40
	AREA	8.250	12.380	16.500	20.630	24.750	28.880	33.000	37.130	41.250	49.500	57.750	66.000
	I_{x-x}	2995.0	4492.0	5990.0	7487.0	8984.0	10,482.0	11,979.0	13,476.0	14,974.0	17,969.0	20,963.0	23,958.0
	I_{y-y}	0.011	0.036	0.086	0.168	0.298	0.461	0.690	0.979	1.343	2.320	3.685	5.500
72	WEIGHT	30.60	45.90	61.20	76.50	91.80	107.10	122.40	137.70	153.00	183.60	214.20	244.80
	AREA	9.000	13.500	18.000	22.500	27.000	31.500	36.000	40.500	45.000	54.000	63.000	72.000
	I_{x-x}	3890.0	5832.0	7776.0	9720.0	11,664.0	13,608.0	15,552.0	17,496.0	19,440.0	23,328.0	27,216.0	31,104.0
	I_{y-y}	0.012	0.040	0.094	0.183	0.316	0.502	0.750	1.068	1.465	2.531	4.020	6.000

VII—RIGID FRAME DESIGN

VII

RIGID FRAME DESIGN

RIGID FRAME DESIGN

Contents

Rigid frame structures 7.0

The amazing growth in the popularity of pre-engineered Rigid Steel Arch buildings can be attributed to several factors: package selling, economy, short erection time, free planning and colorful appearance. Developments during World War II gave steel fabricators improved electric arc welding techniques and a large surplus of skilled metalworkers. The pre-war introduction of the rigid arch, with light gage components, had appeared in Berlin in 1937, where exhibition halls were erected with large span lengths. During the war, Germany erected many airplane and munitions plants using the light-weight rigid arch. These plants proved extremely vulnerable to percussion forces; Allied bombing resulted in almost complete devastation of these buildings.

After the war, in this country, the rigid arch structure was an attractive product for steel fabricators, because no expensive tooling or jigs were required. Indeed an over-abundance of small welding shops sprouted along main highways. They offered design, fabrication and erection of pipe and plate trusses and columns. Unfortunately, many shops did not offer satisfactory products which were competently engineered or which complied with local Building Code requirements.

METAL BUILDING MANUFACTURERS ASSOCIATION

Prior to 1963, each fabricator of metal buildings made his own decisions on design theory for engineering purposes. This was reflected in the sales brochures and claims of each producer. It became apparent that a set of standards was necessary. Accordingly, the M.B.M.A. was formally organized. It was composed of a group of fabricators whose main objectives were to devise production standards, adopt a flexible engineering approach to design, and establish a number of types and sizes best suited for the packaged building market.

The 1963 Edition of the M.B.M.A. "Recommended Design Practices Manual" presents a minimum set of standards for package designs. A more flexible and conservative approach is presented in the AISC recommended specifications. The Southern Standard Building Code, 1965 Edition, Chapter XV on Steel Construction, has not yet been revised to include the Design of Rigid Arches or approve the M.B.M.A. specifications.

Where there are hazards, such as explosion, hurricane winds, or close proximity to overhead high voltage lines, the AISC specifications and rules should be used. Also, buildings designed for petro-chemical and refining plants should follow the AISC system, which we will use exclusively in our examples.

Rigid frame design 7.1

In the design of the Rigid Arch, it is to be assumed that the rafter and column are considered a single member. With this approach in mind, the designer should experience no difficulty in comprehending the behavior of the wind and vertical forces. The simple formulas should be studied until the action of forces is clear and the nomenclature used in the formulas becomes familiar. Accuracy is improved when the applicable formula is noted for each operation, as we will illustrate in the examples to follow.

Rigid frame theory 7.1.1

The rigid frame arch when used for a single span, single story is considered as a single member with hinged column bases. It is referred to in some texts as a single member with free ends. It is better to approach the design of rigid arches by assuming that there is a combination of two statically determinate frames. The vertical forces from roof loads produce a vertical reaction at the columns and at the ends of the rafter. Bending stress is also present in the rafter, and this produces a horizontal force which is greatest at base of the column.

The horizontal wind force applied against the side wall also produces a horizontal reaction which has the greatest value at the column base. Wind horizontal reactions also produce a bending moment in the complete frame, and must be considered in each design. By simply drawing an outline, it is not difficult to visualize the action of these forces and to understand that the knee section is the critical part of the design. Assuming that the horizontal force of the base of each column is known, and that the column base is a hinged joint, the support therefore is at the knee in the fashion of a cantilever. Thus if P = He, the horizontal reaction, and h equals column height, then the moment at top of column is simply Heh. Since the roof loads tend to spread the column bases, and the forces of Ha and He must be tied together to obtain equilibrium, it can be seen that the tension stress is in the outside flange of the column.

POSITIVE AND NEGATIVE MOMENTS

Negative bending moments are indicated by a minus sign (−) when tension occurs on the outside flange of the frame arch. Positive moments (+) are indicated when tension is produced on the inside flange. This is the usual condition when the roof vertical loads act downward, and the horizontal forces applied to the left side of the frame act toward the right. It is necessary that all design work be carried on from left to right, otherwise the signs would be reversed, and the moment diagrams, charts and formulas available for the design procedure would be of no value.

Horizontal forces 7.1.2

The dimensions of a building are usually determined in advance by the client, however the roof slope angle may be set by the designer. The dimension from base of column vertically to the roof ridge will govern the horizontal forces Ha and He at the base of each column. The height of the building on the Center Line of the arch is denoted as h + f (eave height plus total roof pitch).

WIND LOADS

Industrial manufacturers or refiners of petroleum products are concerned with safety and will require steel buildings to be designed with a capability to resist wind loads considerably above code requirements. High velocity winds are capable of collapsing rigid frame structures, and the danger of light gauge steel sheets coming into contact with high voltage plant electric lines is a great hazard. The Southern Building Code requires higher wind loads be used for Gulf Coast Areas: from 25 to 35 pounds per square foot of wall surface. The devastation to the city of Corpus Christi, Texas, from hurricane Celia in 1970 has raised the possibility that the present wind loads will be raised to require designing for wind velocities of 160 miles per hour.

Vertical forces 7.1.3

Live loads on roof should be not less than the building code requirements, and snow loads must also be considered. After the structure has been completed and in use for a period of time, there is always the chance that a hoist will be suspended from the rafters, creating a dangerous condition. It is also a common practice of many owners to look upon the open area in the roof gables as a storage room, and to construct storage racks suspended from the roof structure. The amount which should be added to roof live load for supporting a hoist or other concentrated load on a rafter can be solved in a manner similar to solving for a beam. Convert the concentrated load to an equivalent load by using the formula: $W = \frac{8\,Pab.}{l}$ An illustration of concentrated load design is given on the Pilot Diagram III (7.3.3.).

Eccentric forces 7.1.4

A close study of the moment diagrams in the examples will reveal that the columns are constantly sustaining bending and compressive stress. The vertical reaction is the axial force, and the tie rod reaction is the eccentric bending force. By consulting the separate wind and roof load moment diagrams, each moment can be determined and the moments combined to give the maximum stress locations. By

Eccentric forces, continued 7.1.4

close examination of the problem and examples given, it will be understood why each condition should be investigated separately. In some cases a member may show a maximum stress and moment when subjected to only one type of load, rather than to a combination of loads. In each design, it can be stated that the axial force of R and horizontal force of Ha which produce bending are always present at the same time and operate together. The design therefore is governed by this interaction and will comply with Rule III.

Secondary considerations 7.1.5

In none of the great number of rigid frame steel structures in use under widely varying conditions, has the stress behavior in the arch with respect to temperature, deflection and tie rod elongation ever presented any problem. Experienced designers agree that any attempt to make an exact analysis on their effect is time wasted. The approximate amount of deflection which would occur in the horizontal direction may be predicted by assuming that one column is free to move horizontally. The total horizontal deflection is the sum of the following:

Δ due to temperature change = etl

Δ due to direct stress = $\frac{Hl}{AE}$

Δ due to elongation of ties = $\frac{HL}{A_t E}$

Where the nomenclature for formula is:

e = Coefficient of expansion. (See Table Sect. I. .0000067 per degree)

t = Temperature range – degrees Fahr.

l = Length of Arch. (Span in feet).

Δ = Deflection in inches

H = Maximum Horizontal Reaction under all loads.

A = Average or mean cross-sectional area of girder.

A_t = Cross-sectional summation of areas in tie rods in square inches.

E = Modulus of Elasticity of material used.

Pilot diagrams for the rigid arch 7.2

In using the Pilot Diagrams which follow, choose the diagram which corresponds to the load conditions. Lay out the design to a convenient scale and note the magnitude of loads. Show all dimensions necessary for finding the values of Q and k. The grapic charts will provide the coefficient of C_1, C_2 and C_6. The proper pair of pilot diagram and coefficient chart must be used for each load condition. Do not attempt to combine the load reactions of roof and wind pressure until each condition has been examined, and the proper coefficients have been found. To avoid errors, it is suggested that the pilot diagrams be used by substituting values in the formulas as shown on each diagram.

Moment diagrams and design formulas 7.2.1

The action of roof and wind loads applied to a rigid frame arch is best illustrated when the actual moments are computed and plotted on a scale drawing. Examples which follow will provide the formula for calculating the bending moments on rafter and columns. Since the moment diagrams usually serve as a check on stress requirements, they also present a pattern which will, to some extent, confirm the theory of design. All work in the calculations is best accomplished by working from the left side and proceeding to the right. With respect to concentrated loads on a rafter, remember to use the proper formula when calculating the bending moment.

FORMULAS: Rigid frame moment diagrams 7.2.1.1

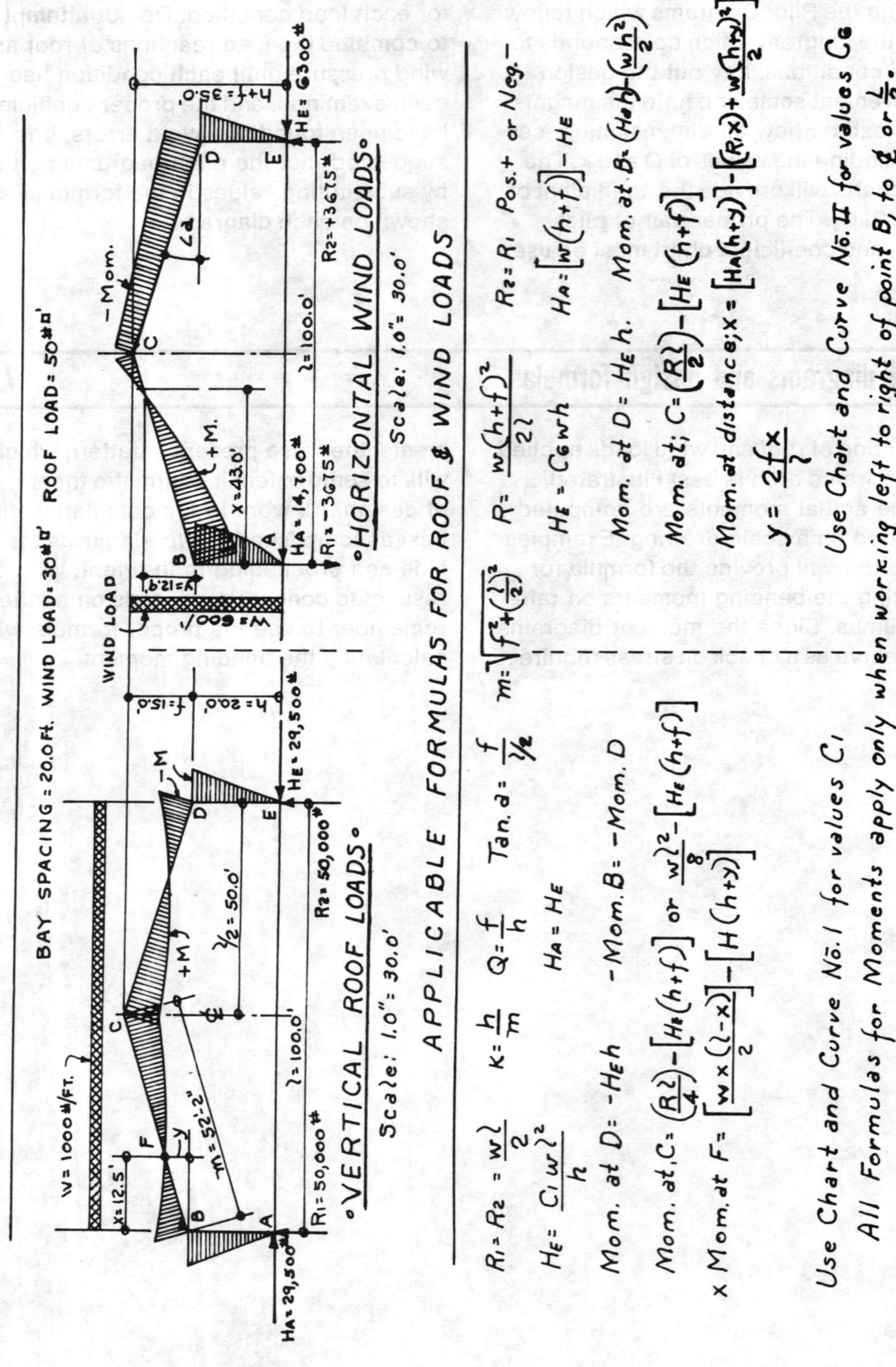

Knee area 7.2.2

The critical section in any rigid arch structure is the connection of column to rafter. This area is called the knee. The moment diagrams will usually show the greatest bending moment at the knee. In most designs, the compressive stress in the inner flange will be of great intensity, and require stiffeners and considerable lateral bracing. Refer to the design rules when making computations, because the allowable unit stress could very well be the limiting factor in the knee area. Rule V particularly limits the web shear stress to 13,000 PSI.

Erection and splices 7.2.3

The bending moment diagrams will show the logical locations for splices. Shipping size limits must be considered when using rail or truck transportation. When splice joints are necessary they should only be made at the less critical points on the rafter and column. A column may be bolted at the knee joint as shown in the examples and illustrations. By assuming a rotating condition at the connection of column to rafter, the number of bolts required in the tension side is easily determined. After all bolts are tightened and seat is level and plumbed, the scab plate lapping the outside flanges of knee and column should be welded to insure a truly rigid connection.

With respect to longer spans of over 100 feet, the rafter splices should be close to the inflection point or where the moment diagrams indicate a lesser magnitude of bending. Structures which are to receive light gauge ribbed steel panels should be completely wind braced, plumb, with all girts levelled and held by tight sag rods, before any wall or roof panels are installed.

The selection of fasteners for ribbed wall panels should be set forth in the specifications with additional notes put on the drawings.

Tie rods at base 7.2.4

Another critical component in a rigid arch structure is the connecting tie rod between the column bases. Since the vertical loads tend to spread the columns outward at their bases, a suitable tie rod or permanent anchorage is required. It is not a good policy to depend upon the bolt anchors in the base plates to serve the purpose of tie rods, nor to rely upon the concrete slab to sustain the tension forces. The rod ties must be protected from corrosion for the life of the structure. Rods which are to be placed under slabs in soils which may attack the steel should be protected. Analyze the soil for the presence of acids, salinity or electrolysis. When electrical conduits are placed adjacent to tie rods under the slab, the tie rods should be coated with a heavy covering of coal tar epoxy protective paint. Use a safety factor of three to determine the rod size and number. On long spans, the rods may be installed with turnbuckles which are drawn tight during erection. If these turnbuckles are overtightened, they will exert additional stress in the knee or haunch. Tie rods which must pass through concrete beams or haunches can be installed through transite (asbestos) pipe sleeves. This will protect the floor slab and permit the rod forces to act independently of any other influences.

PILOT DIAGRAM I: Uniform roof load on full span 7.3.1

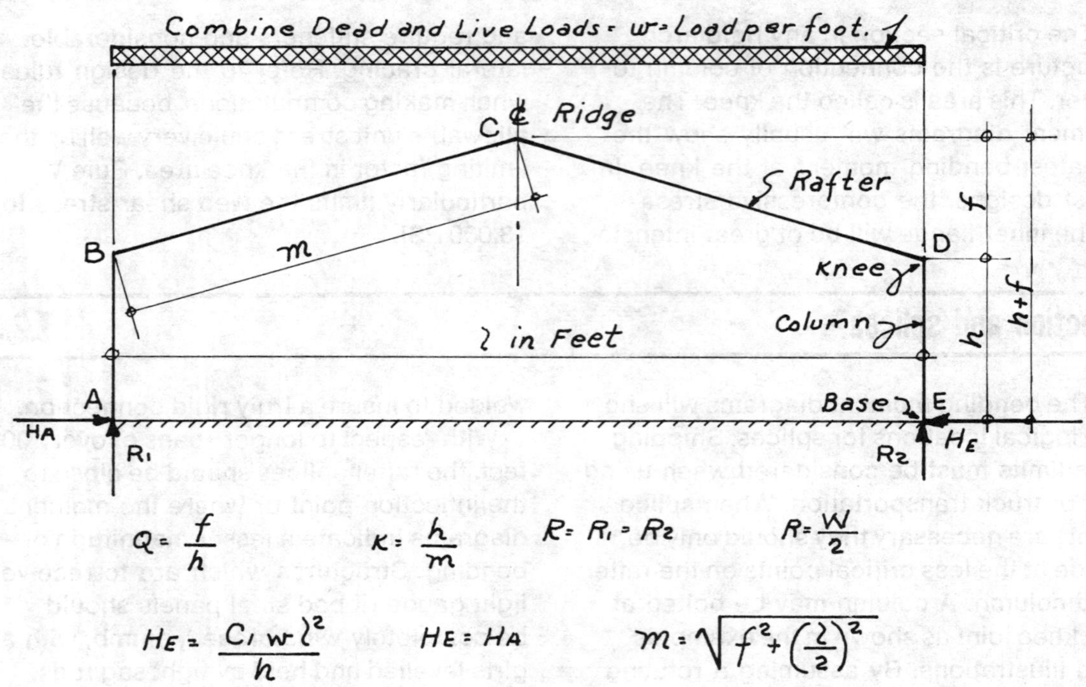

$Q = \dfrac{f}{h}$ $\quad K = \dfrac{h}{m}$ $\quad R = R_1 = R_2$ $\quad R = \dfrac{W}{2}$

$H_E = \dfrac{C_1 w l^2}{h}$ $\quad H_E = H_A$ $\quad m = \sqrt{f^2 + \left(\dfrac{l}{2}\right)^2}$

ILLUSTRATION:

Assume Arch spacing in Bays = 20.0 Ft. C.C. Uniform Load = 50 Lbs Sq. Foot

$w = 20.0 \times 50 = 1000$ Lbs. per foot lin. $h = 20.0'$ $f = 15.0'$ Span $l = 100.0$ Feet.

$m = \sqrt{15.0^2 + 50.0^2} = 52.30$ Feet $\quad K = \dfrac{20.0}{52.30} = 0.383$ $\quad Q = \dfrac{15.0}{20.0} = 0.75$

Refer to Chart I: With $K = 0.383$ and $Q = 0.75$ Coefficient $C_1 = 0.0545$

Substituting the values in formula to find Horizontal reaction H_E:

$H_E = \dfrac{0.0545 \times 1000 \times 100.0^2}{20.0} = 27{,}250$ Lbs. $\quad H_A$ = same as H_E

VERTICAL REACTIONS:

Area of Roof supported by 1 Arch = $20.0 \times 100.0 = 2000$ Square Feet.

Design Load = 50 P.S.F. $W = 2000 \times 50 = 100{,}000$ Lbs.

Reactions: $R_1 = \dfrac{100{,}000}{2} = 50{,}000$ Lbs.

Lineal foot roof load $w = 50 \times 20.0 = 1000$ Lbs. per foot.

MOMENT DIAGRAM I: Uniform roof load on full span 7.3.1

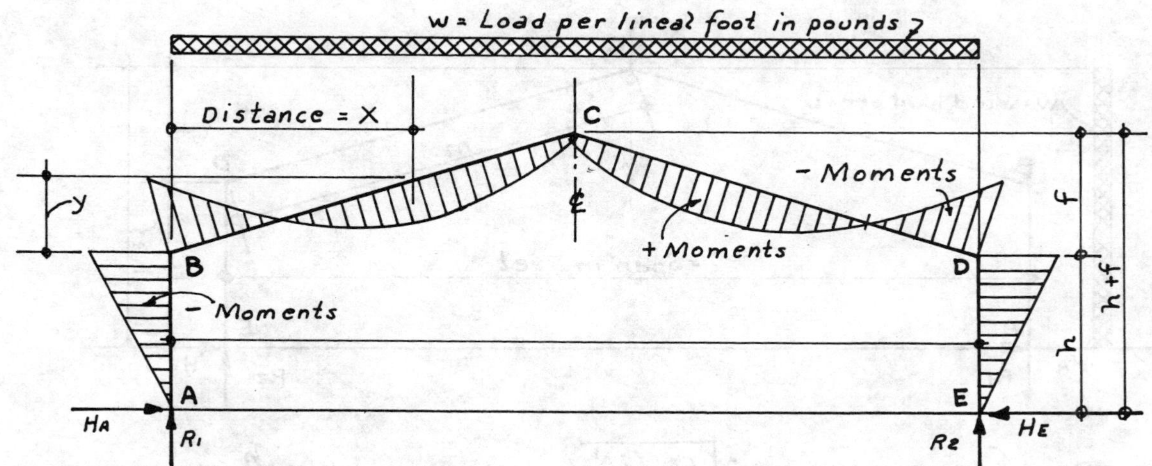

Moments at Knee points B and D are same when symmetrical.

$M_B = H_A h$. Moment at ℄ Ridge: $M_C = \left(\frac{Rl}{4}\right) - \left[H(h+f)\right]$.

Moment at point X on Rafter is left of ℄.

$M_X = \left[\frac{wX(l-X)}{2}\right] - \left[H(h+y)\right]$. Dimension y = Rise from horizontal point B to point on Rafter designated X.

ILLUSTRATION:

Assumed: $l = 100.0$ Feet. $w = 1000$#/'. $h = 20.0$ Feet and $f = 15.0$ Feet.

$H = 27{,}250$ Lbs. $R = 50{,}000$ Lbs. Substitute the values in formulas:

$M_B = 27{,}250 \times 20.0 = -545{,}000$ Foot Lbs. $M_B = M_D$.

$M_C = \left(\frac{50{,}000 \times 100.0}{4}\right) - \left[27{,}250(20.0 + 15.0)\right] = +296{,}250$ Foot Lbs.

Assume point $X = 30.0$ Ft. from B. When left of ℄: $y = \frac{2fX}{l}$.

$y = \frac{2 \times 15.0 \times 30.0}{100.0} = 9.0$ Ft. Placing values in formula for $M_{30.0}$:

$M_{30.0} = \left[\frac{1000 \times 30.0 \times (100.0 - 30.0)}{2}\right] - \left[27{,}250 \times (20.0 + 9.0)\right] = +259{,}750$ Ft. Lbs.

When distance X is to right of ℄: Rise $y = \frac{2f(l-X)}{l}$.

PILOT DIAGRAM II: Horizontal wind load on full height 7.3.2

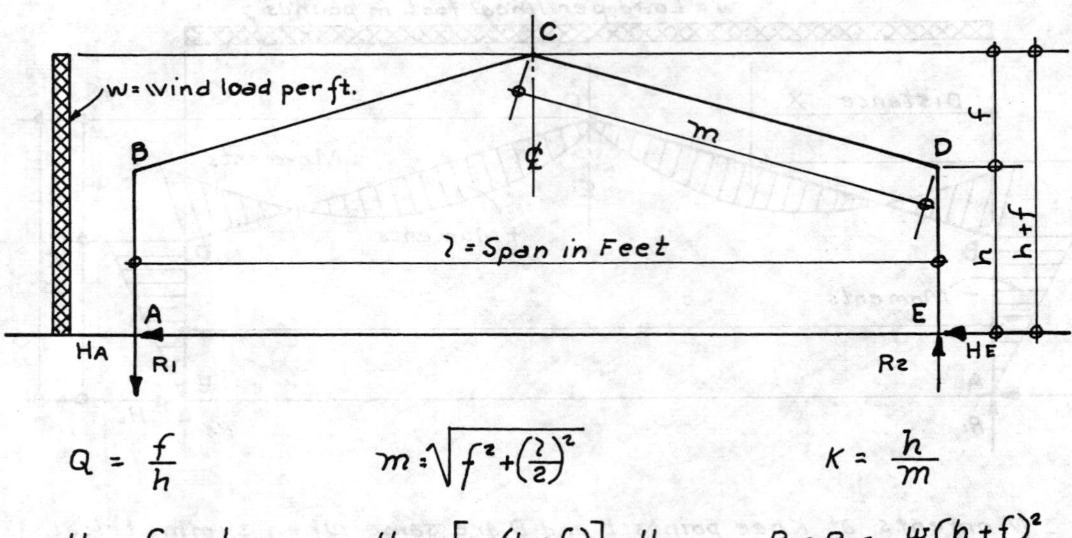

$Q = \frac{f}{h}$ $\quad m = \sqrt{f^2 + \left(\frac{l}{2}\right)^2}$ $\quad K = \frac{h}{m}$

$H_E = C_6 wh.$ $\quad H_A = \left[w(h+f)\right] - H_E$ $\quad R_1 = R_2 = \frac{w(h+f)^2}{2l}$

ILLUSTRATION:

Assume Arches are spaced in bents at 20.0 Foot Centers.

Wind Pressure on left = 30 Lbs. Sq. Foot for full height building.

Eave height, h = 20.0′ Roof rise, f = 15.0 Ft. Span l = 100.0 Feet.

$m = \sqrt{15.0^2 + 50.0^2} = 52.3$ Feet. $h + f = 35.0'$

Wind Load on one Arch bay = 30 x 20.0 = 600 Lbs. per foot vertical.

Wind pressure tends to tip Arch upward at R_1, down at R_2.

For Vertical Reactions:

$R = \frac{600 \times (20.0 + 15.0)^2}{2 \times 100.0} = 3675$ Lbs. Force R_1 is up in action.

For Horizontal Reactions H_A and H_E with Wind from left.

$Q = \frac{15.0}{20.0} = 0.75$ $\quad K = \frac{20.0}{52.3} = 0.383$ Refer to Chart II for $C_6 = 0.53$

$H_E = 0.53 \times 600 \times 20.0 = 6,360$ Lbs.

$H_A = \left[600 \times (20.0 + 15.0)\right] - 6,360 = 14,640$ Lbs.

MOMENT DIAGRAM II: Horizontal wind load on full height 7.3.2

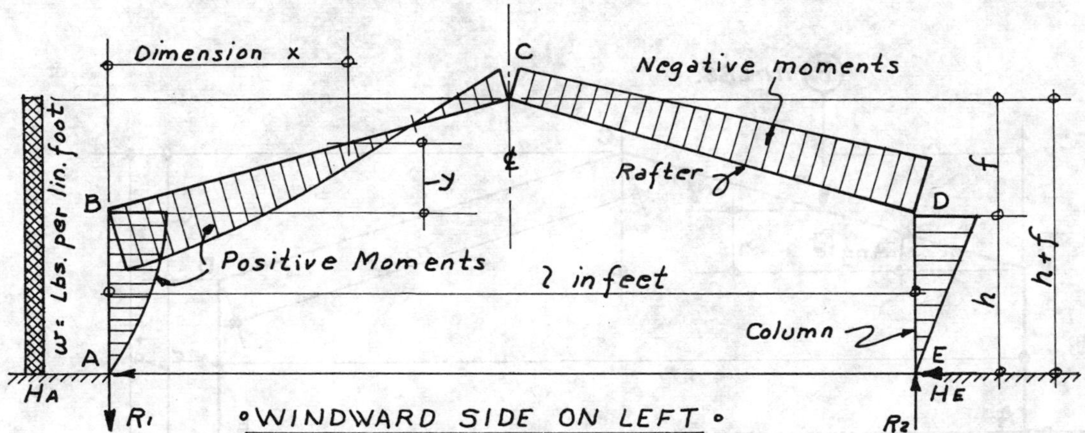

°WINDWARD SIDE ON LEFT°

Wind moment at Ridge C: $M_C = \left(\frac{R_1 l}{2}\right) - \left[H_E(h+f)\right]$ $M_D = H_E h$

Moment at knee, B. $M_B = (H_A h) - \left(\frac{wh^2}{2}\right)$. When dim. x is left of ℄: $y = \frac{2fx}{l}$.

Moment on Rafter left of ℄: $M_x = \left[H_A(h+y)\right] - \left[(R_1 x) + w\frac{(h+y)^2}{2}\right]$.

When dim. x is right of ℄: $y = \frac{2f(l-x)}{l}$.

Rafter moments at right of ℄: $M_x \left[R(l-x)\right] - \left[H_E(h+y)\right]$.

ILLUSTRATION:

Assume Span $l = 100.0$ Ft. $h = 20.0'$ $f = 15.0'$ and $w = 600$ Lbs. per foot.

From Pilot 7.3.2: $R = 3675$ Lbs. $H_E = 6360$ Lbs., and $H_A = 14{,}640$ Lbs.

Distance y at Ridge point C = 15.0 feet, or same as f.

Moment at C: $M_C = \left(\frac{3675 \times 100.0}{2}\right) - \left[6360 \times (20.0 + 15.0)\right] = -38{,}850$ Ft. Lbs.

Moment at Knee B: $M_B = (14{,}640 \times 20.0) - \left(\frac{600 \times 20.0^2}{2}\right) = +172{,}800$ Ft. Lbs.

Moment at knee D: $M_D = 6360 \times 20.0 = -127{,}200$ Foot Lbs.

Rafter moment left of ℄ when dimension $x = 25.0$ feet:

Dimension $y = \frac{2 \times 15.0 \times 25.0}{100.0} = 7.50$ Feet

Substituting known values in formula:

$$M_{25.0'} = \left[14{,}640 \times (20.0 + 7.50)\right] - \left[(3675 \times 25.0) + \left(600 \times \frac{27.50^2}{2}\right)\right] =$$

$$M_{25.0'} = (+402{,}600) - \left[(91{,}875 + 226{,}875)\right] = +83{,}850 \text{ Ft. Lbs.}$$

Moment 25.0 feet to right of ℄. $y = \frac{2 \times 15.0(100.0 - 75.0)}{100.0} = 7.50$ Feet.

$$M_{75.0'} = \left[3675(100.0 - 75.0)\right] - \left[6360 \times (20.0 + 7.50)\right] = -83{,}025 \text{ Ft. Lbs.}$$

PILOT DIAGRAM III: Concentrated load on roof rafter 7.3.3

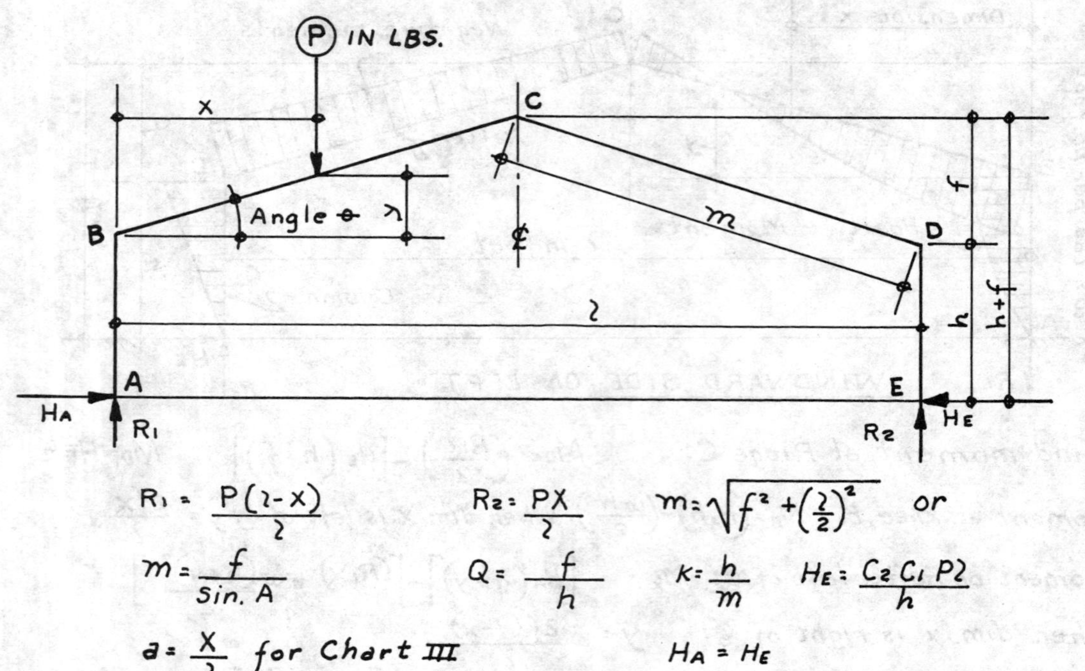

$R_1 = \frac{P(l-x)}{l}$ $R_2 = \frac{PX}{l}$ $m = \sqrt{f^2 + \left(\frac{l}{2}\right)^2}$ or

$m = \frac{f}{\sin. A}$ $Q = \frac{f}{h}$ $K = \frac{h}{m}$ $H_E = \frac{C_2 C_1 P l}{h}$

$a = \frac{X}{l}$ for Chart III $H_A = H_E$

When P is at left of ℄: $y = \frac{2fX}{l}$

When P is to right of ℄: $y = \frac{2f(l-X)}{l}$

ILLUSTRATION:

Assumed: Span $l = 80.0$ Feet. $h = 20.0$ Feet. $f = 10.0$ Feet.

Load $P = 20{,}000$ Lbs. $X = 20.0$ Feet.

$R_1 = \frac{20{,}000 \times (80.0 - 20.0)}{80.0} = 15{,}000$ Lbs. $R_2 = 20{,}000 - 15{,}000 = 5000$ Lbs.

Calculating the Horizontal forces H_A and H_E:

$Q = \frac{10.0}{20.0} = 0.50$ $m = \sqrt{40.0^2 + 10.0^2} = 41.2'$ $K = \frac{20.0}{41.2} = 0.485$

From Chart I for C_1: $C_1 = 0.0628$

For value of a: $a = \frac{20.0}{80.0} = 0.25$

From Chart III: With $a = 0.25$ and $Q = 0.50$ $C_2 = 1.10$

Substituting values in formula:

$H_E = \frac{1.10 \times 0.0628 \times 20{,}000 \times 80.0}{20.0} = 5526.5$ Lbs. $H_A = H_E$

MOMENT DIAGRAM III: Concentrated load on roof rafter 7.3.3

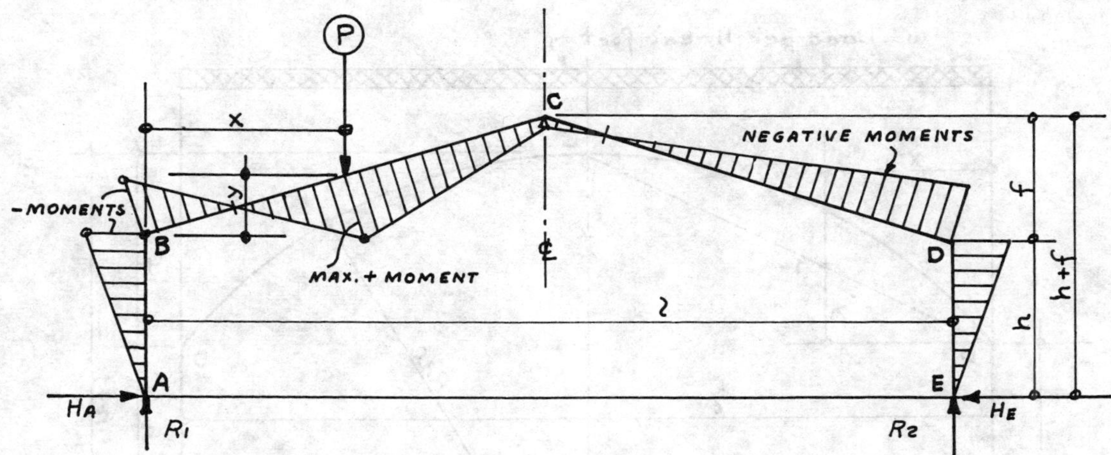

Bending Moment at B or D: $M_B = H_A h.$ $M_C = H_E h.$

Moment at ℄ Ridge C: $M_C = \left(\frac{R_2 l}{2}\right) - \left[H\,(h+f)\right]$

Rafter Moments to left of load P: $M_x = (R_1 X) - \left[H(h+y)\right]$

Rafter Moments at right of load P: $M_x = R_2\,(l-x) - \left[H(h+y)\right]$

ILLUSTRATION:

Assumed: Span $l = 80.0$ Ft. $h = 20.0'$ $f = 10.0'$ $P = 20{,}000$ Lbs.

Locate load P at 20.0 Feet to right of B. $x = 20.0'$ $y = 5.0$ Ft.

From Pilot Diagram III: Use the following reactions for R and H.

$R_1 = 15000\#$ $R_2 = 5000\#$ Horizontal $H_A = 5780\#$ $H_E = 5780\#$

Moment at C: $M_C = \left(\frac{5000 \times 80}{2}\right) - \left[5780(20.0+10.0)\right] = +26{,}600$ Ft. Lbs.

Moment at B: $M_B = 5780 \times 20.0 = 115{,}600$ Ft. Lbs.

Moment under P: $M_{20.0} = (15{,}000 \times 20.0) - \left[5780(20.0+5.0)\right] = +155{,}500'\#$

At right of ℄ 10.0' $M_{50.0} = \left[5000 \times (80.0-50)\right] - \left[5780 + (20.0+7.5)\right] = -8{,}950'\#$

TO DETERMINE DIMENSION y AT ANY POINT:

When dimension x is left of ℄: $y = \frac{2fx}{l}.$

When dimension x is right of ℄: $y = \frac{2f(l-x)}{l}.$

PILOT DIAGRAM IV: Uniform roof load on full segment arch 7.3.4

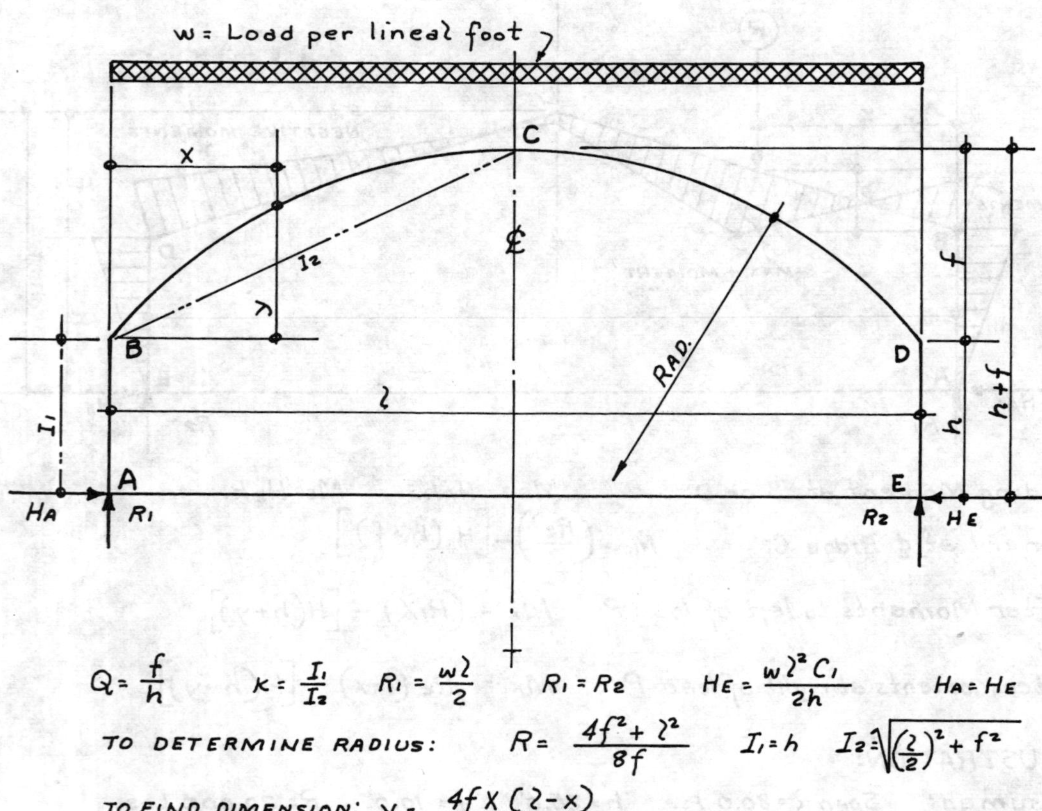

$Q = \frac{f}{h}$ $K = \frac{I_1}{I_2}$ $R_1 = \frac{wl}{2}$ $R_1 = R_2$ $H_E = \frac{wl^2 C_1}{2h}$ $H_A = H_E$

TO DETERMINE RADIUS: $R = \frac{4f^2 + l^2}{8f}$ $I_1 = h$ $I_2 = \sqrt{\left(\frac{l}{2}\right)^2 + f^2}$

TO FIND DIMENSION: $y = \frac{4f\,x\,(l - x)}{l^2}$

ILLUSTRATION:

Assume: $f = 26.0$ feet. $h = 15.0'$ $w = 630$ #/' Span $l = 100.0$ feet

$Q = \frac{26.0}{15.0} = 1.73$ $I_1 = h.$ $I_2 = \sqrt{50.0^2 + 26.0^2} = 56.8$ Ft.

$K = \frac{15.0}{56.8} = 0.264$ Use Chart I for coefficient C_1. $C_1 = 0.044$

VERTICAL REACTIONS:

$w = 630$ #' $W = 630 \times 100.0 = 63{,}000$ Lbs. R_1 and $R_2 = \frac{63{,}000}{2} = 31{,}500$ Lbs.

HORIZONTAL REACTIONS:

$H_E = \frac{630 \times 100.0 \times 100.0 \times 0.044}{2 \times 15.0} = 9{,}240$ Lbs. $H_A = H_E$

DETERMINE RADIUS OF SEGMENT:

Radius $R = \frac{(4 \times 26.0^2) + 100.0^2}{8 \times 26.0} = 61.08$ Feet.

DIMENSION FOR y WHEN X = 20.0 FEET:

$y = \frac{4 \times 26.0 \times 20.0\,(100.0 - 20.0)}{100.0 \times 100.0} = 16.64$ Feet.

MOMENT DIAGRAM IV: Uniform roof load on full segment arch 7.3.4

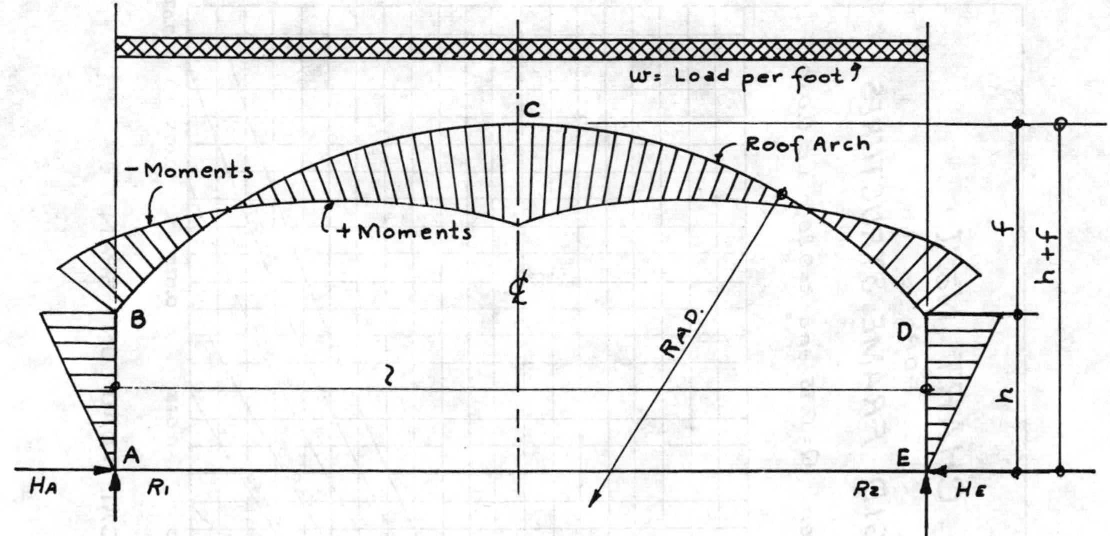

Moment at ℄, point C: $M_C = \left(\frac{Rl}{4}\right) - \left[H_A(h+f)\right]$. $M_B = M_D = Hh$.

Rafter Moments at left of ℄: $M_x = \left[\frac{wx(l-x)}{2}\right] - \left[H_A(h+y)\right]$

ILLUSTRATION:

Assumed: Span l = 100.0 Ft. h = 15.0' f = 26.0' w = 630 Lbs. Lin. Foot.
R_1 = 31,500 Lbs. H_A and H_E = 9,240 Lbs. See Pilot Diagram IV.

Moment at Knee: B & C: M_B = 9240 × 15.0 = − 138,600 Foot Lbs.

Rafter moments taken at certain point on segment:

$$M_{4.0} = \left[\frac{630 \times 4.0\,(100.0-4.0)}{2}\right] - \left[9{,}240\ (25.0+3.99)\right] = -\ 146{,}907 \text{ Ft. Lbs.}$$

$$M_{10.0} = \left[\frac{630 \times 10.0\,(100.0-10.0)}{2}\right] - \left[9{,}240\ (25.0+9.36)\right] = -\ 33{,}986 \text{ Ft. Lbs.}$$

$$M_{20.0} = \left[\frac{630 \times 20.0\,(100.0-20.0)}{2}\right] - \left[9{,}240\ (25.0+16.64)\right] = +\ 119{,}245 \text{ Ft. Lbs.}$$

$$M_{30.0} = \left[\frac{630 \times 30.0\,(100.0-30.0)}{2}\right] - \left[9{,}240\ (25.0+21.84)\right] = +\ 128{,}700 \text{ Ft. Lbs.}$$

$$M_{40.0} = \left[\frac{630 \times 40.0\,(100.0-40.0)}{2}\right] - \left[9{,}240\ (25.0+24.96)\right] = +\ 294{,}370 \text{ Ft. Lbs.}$$

$$M_{C.} = \left[\frac{(31{,}500 \times 100.0)}{4}\right] - \left[9{,}240\ (25.0+26.0)\right] = +\ 316\ 260 \text{ Ft. Lbs.}$$

CHART I: Rigid frame design, to find C_1 7.3.5

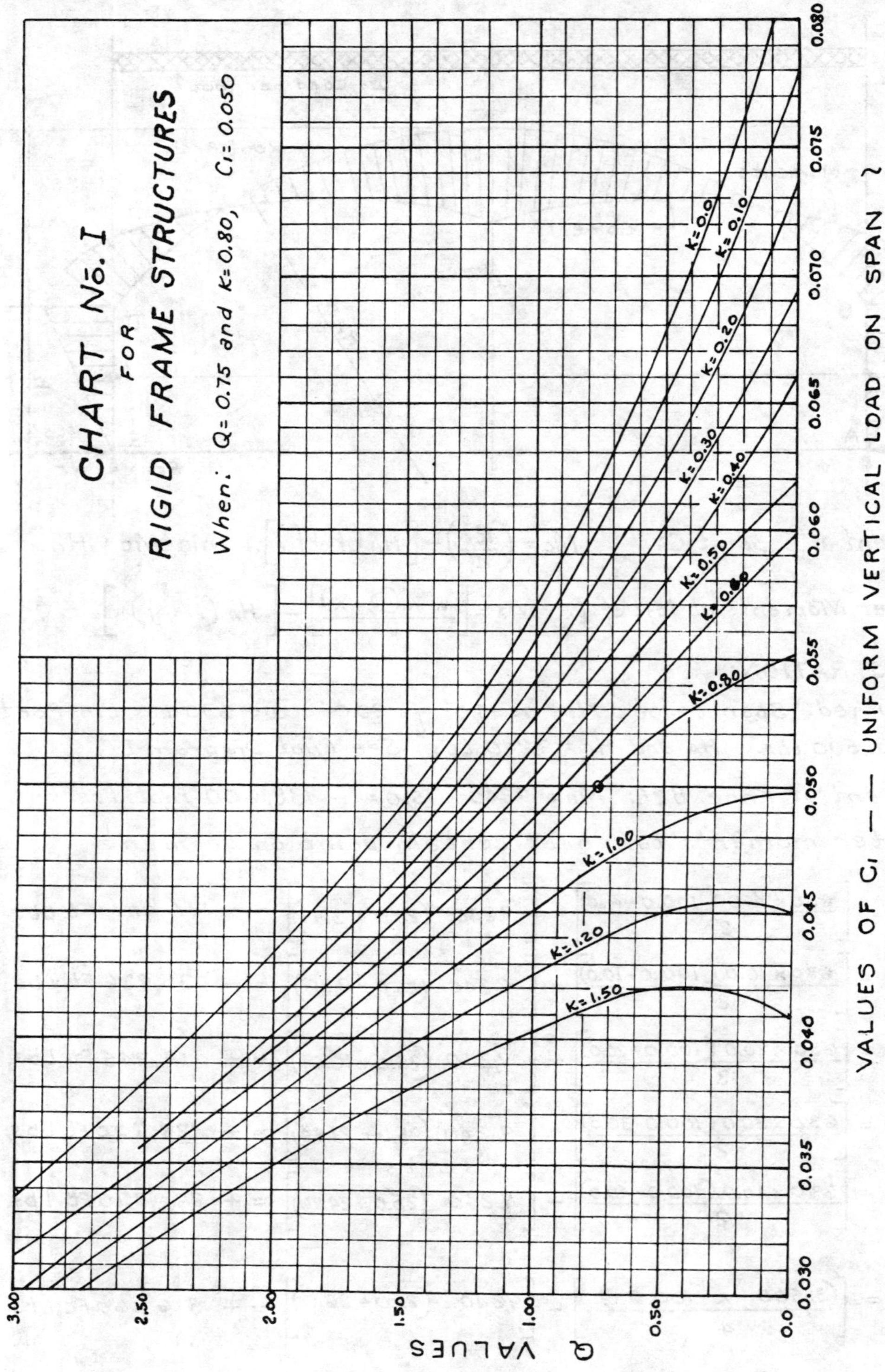

CHART II: Rigid frame design, to find C_6 7.3.6

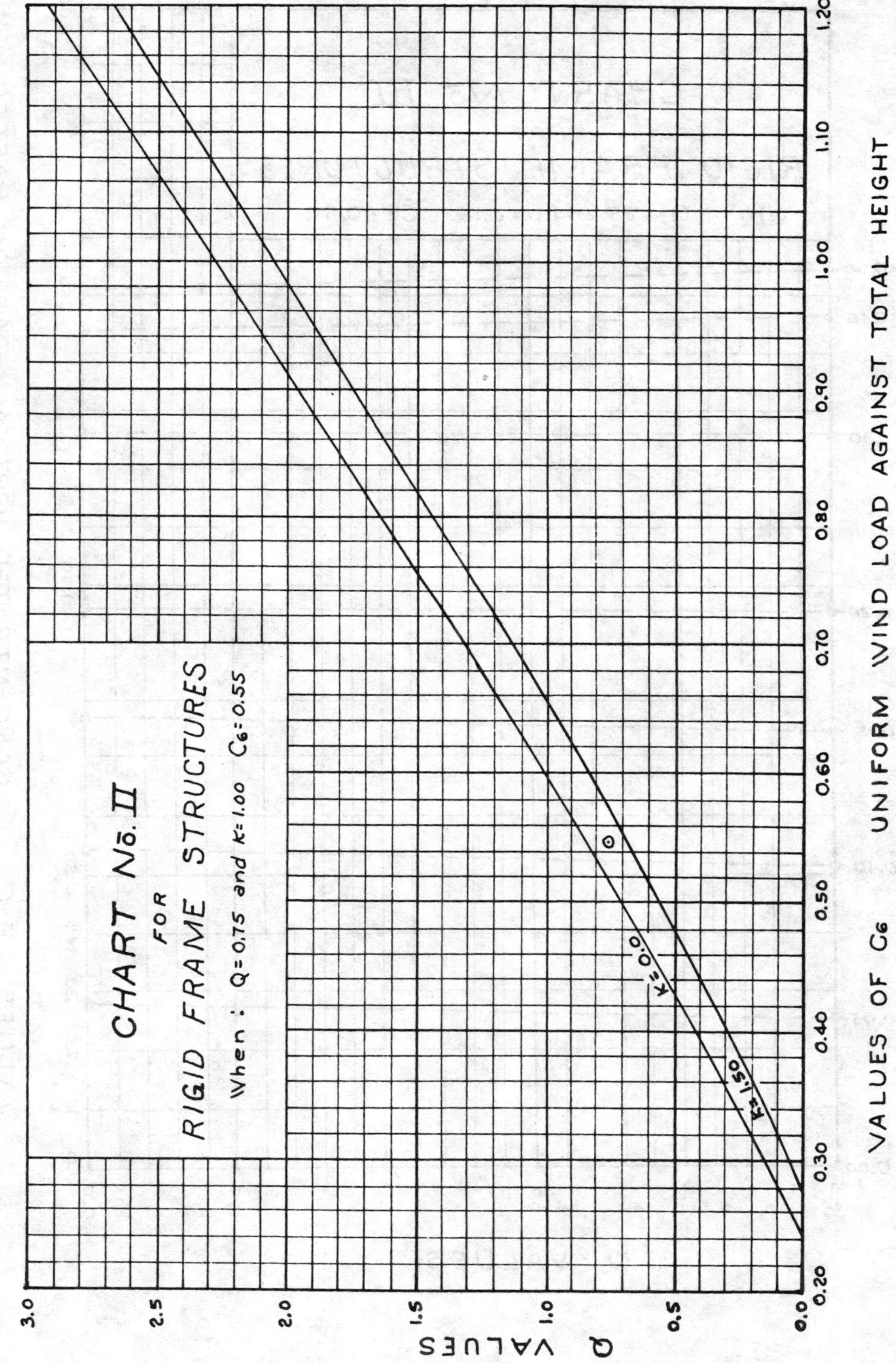

CHART III: Rigid frame design, to find C_2 7.3.7

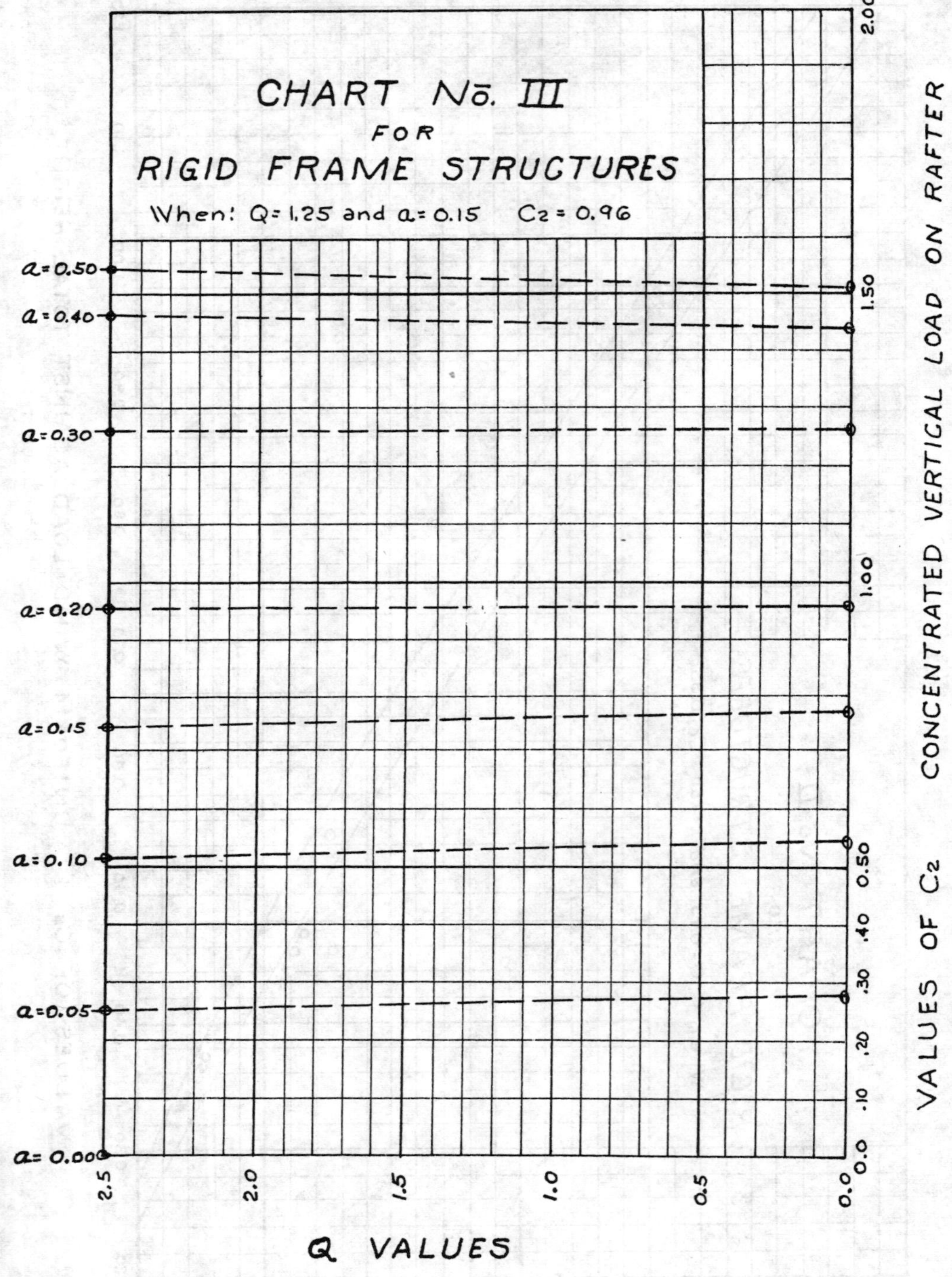

Recommended AISC Rules for rigid frame design 7.4

To comply with the AISC Specifications for the design of Rigid Frame Arch Structures, the critical sections or location points for the design shall be based on these fundamental rules:

RULE I

Critical design of Knee Section shall be taken thus:

(a) At the inside face of column and bottom of rafter for a straight knee.

(b) At the tangent points for a circular haunched knee.

(c) At the extremities and common intersection point for haunch made up of tapered or multiple side knee.

RULE II:

The following formula shall be employed to limit the allowable unit stress where maximum $F_b = 20{,}000$ PSI, or modified thus:

$F_b = \dfrac{12{,}000{,}000}{\left(\frac{ld}{bt}\right)}$. Hence: $\dfrac{ld}{bt}$ is limited to 600 for maximum F_b.

That allowable axial compressive stress be limited by the formula: $F_a = 17{,}000 - \left(\dfrac{0.485 l^2}{r^2}\right)$ and the maximum combined stress be limited to the provisions of unity or less: $u = \left(\dfrac{f_a}{F_a}\right) + \left(\dfrac{f_b}{F_b}\right)$.

In the examples to follow, the formula for unity may be rewritten in a compact equation thus: $u = \dfrac{M}{S_x F_b} + \dfrac{P}{A F_a}$.

RULE III:

That the maximum combined stress be determined by the conventional formula: $f = \dfrac{P}{A} + \dfrac{Mc}{I}$.

RULE IV

Adequate provision shall be provided to resist lateral sway movement of the inside knee compression flange. Bracing shall be installed.

RULE V:

The average shear stress on haunch web or critical points as designated in Rule I shall be limited to $F_v = 13{,}000$ P.S.I.

RULE VI:

In the case of a curved haunch, the ratio of Radius (R) to depth (d), shall be not less than that determined by the curve in the following curve given in chart.

Recommended AISC Rules for rigid frame design, continued 7.4

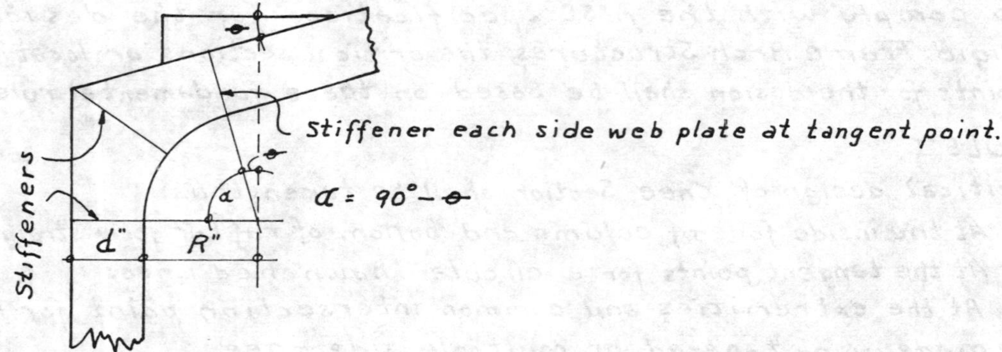

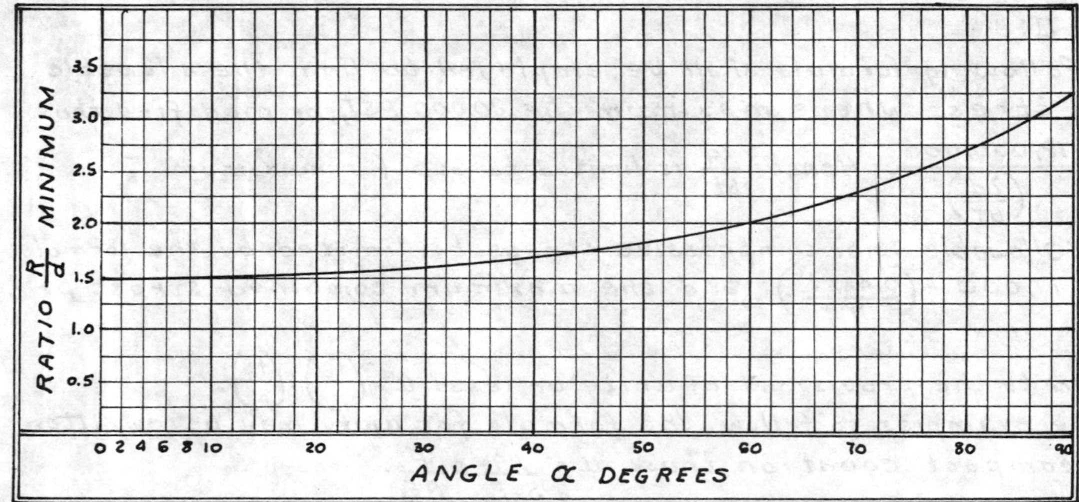

RULE VII:
With respect to a Curved haunch, the relationship of $\frac{b^2}{Rt}$ shall be not more than 2, where R is radius of inner flange curvature.

RULE VIII:
Stiffeners shall be provided in a curved or straight haunched knee at mid-point, and at extremities of knee in line with inner flanges of rafter and column.

NOMENCLATURE FOR RULES:

α = Angle for tangent points in knee section, in degrees.
θ = Angle of rafter and roof pitch, in degrees. ($\alpha = 90° - \theta$.)
d = Depth of Cross-section of rafter and column, in inches.
l = Length unsupported laterally for rafter or column, in inches.
t = Thickness of web or flange plate, in inches.
b = Width of flange plate, in inches.
R = Radius of curvature of inner flange plate, in inches.

EXAMPLE: Rigid frame with concentrated rafter load 7.5.1

A Rigid Arch is required to be designed which will support an additional concentrated hoist load at middle of rafter on one side only. Span $l = 80.0'$ Eave $h = 20.0'$ $f = 10.0'$

REQUIRED:

(a) Draw section through structure and place load with dimensions.

(b) Calculate Vertical Reactions R_1 and R_2.

(c) Calculate Horizontal Reactions H_A and H_E.

(d) Calculate moments and construct Moment Diagram.

STEP I:

Scale drawing of Arch profile = 1.0" = 20.0 Feet.

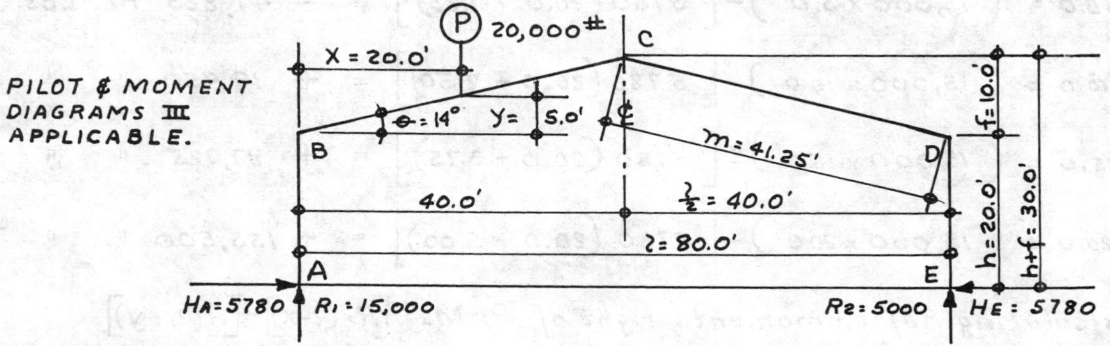

Vertical Reactions: Calculate by moments same as a horizontal beam.

$R_1 = \dfrac{20{,}000 \times 60.0}{80.0'} = 15{,}000$ Lbs. $\quad R_2 = \dfrac{20{,}000 \times 20.0}{80.0} = 5000$ Lbs.

STEP II:

Horizontal Reactions: Requires use of Charts for C_1 and C_2.

$m = \sqrt{10.0^2 + 40.0^2} = 41.25'$ $\quad Q = \dfrac{f}{h} = \dfrac{10.0}{20.0} = 0.50$ $\quad k = \dfrac{h}{m} = \dfrac{20.0}{41.25} = 0.485$

From Chart I: With $k = 0.4$ and $Q = 0.50$ point is: $C_1 = 0.0628$

From Chart III: With $a = \dfrac{X}{l} = \dfrac{20.0}{80.0} = 0.25$, $Q = 0.50$. $\quad C_2 = 1.15$

From Pilot Diagram: Formula for H_A and $H_E = \dfrac{C_1 C_2 P l}{h}$

Substituting values in formula:

$H = \dfrac{0.0628 \times 1.15 \times 20{,}000 \times 80.0}{20.0} = 5777.6$ $\quad$ Call it 5780 Lbs.

EXAMPLE: Rigid frame with concentrated rafter load, continued 7.5.1

STEP III: (Computing moments).

From Pilot Diagram: $M_D = Hh$. $M_D = 5780 \times 20.0 = 115,600$ Ft. Lbs. (Negative)

At Ridge ℄: $M_C = \left(\frac{R_2 l}{2}\right) - \left[H(h+f)\right]$. Substitute values in formula:

$$M_C = \left(\frac{5000 \times 80.0}{2}\right) - \left[5780(20.0 + 10.0)\right] = +26,600 \text{ Foot Lbs.}$$

When dimension $X = 20.0'$, $y = 5.0'$ Tangent $\theta = \frac{5.0}{20.0} = 0.25$ Angle $= 14°$

Calculating Rafter moments left of P: $M_x = (R_1 X) - \left[H(h+y)\right]$.

$M_{5.0} = (15,000 \times 5.0) - \left[5780(20.0 + 1.25)\right] = -47,825$ Ft. Lbs.

$M_{10.0} = (15,000 \times 10.0) - \left[5780(20.0 + 2.50)\right] = +19,950$ " "

$M_{15.0} = (15,000 \times 15.0) - \left[5780(20.0 + 3.75)\right] = +87,725$ " "

$M_{20.0} = (15,000 \times 20.0) - \left[5780(20.0 + 5.00)\right] = +155,500$ " "

Calculating Rafter moments right of, P: $M_x = \left[R_2(l-x)\right] - \left[H(h+y)\right]$.

$M_{25.0} = \left[5000(80.0-25.0)\right] \left[5780(20.0+6.25)\right] = +123,275$ Ft. Lbs.

$M_{30.0} = \left[5000(80.0-30.0)\right] \left[5780(20.0+7.50)\right] = +91,050$ " "

$M_{35.0} = \left[5000(80.0-35.0)\right] \left[5780(20.0+8.75)\right] = +58,825$ " "

$M_{40.0} = \left[5000(80.0-35.0)\right] \left[5780(20.0+10.0)\right] = +26,600$ " "

EXAMPLE: Rigid frame with concentrated rafter load, continued 7.5.1

STEP IV:

Calculate Compressive Axial Force in Rafter resulting from the concentrated load P_R: Angle $\theta = 14.0°$ Sine $\theta = 0.242$ Cos. $\theta = 0.970$

Axial force, $P = (H_A \text{ Cos.}\theta) + (R_1 \text{ Sin.}\theta)$. Take the greater values of H_A and R_1, and substitute in formula:

$P_R = (5780 \times 0.970) + (15{,}000 \times 0.242) = 9{,}248$ Pounds.

STEP V:

From the Moment values obtained in Step III the plotted points result in a straight line. At right of ℄ producing a line as from the positive moment of 26,600'# to the point at knee indicated D. The moment at D = Heh or same value as at B.

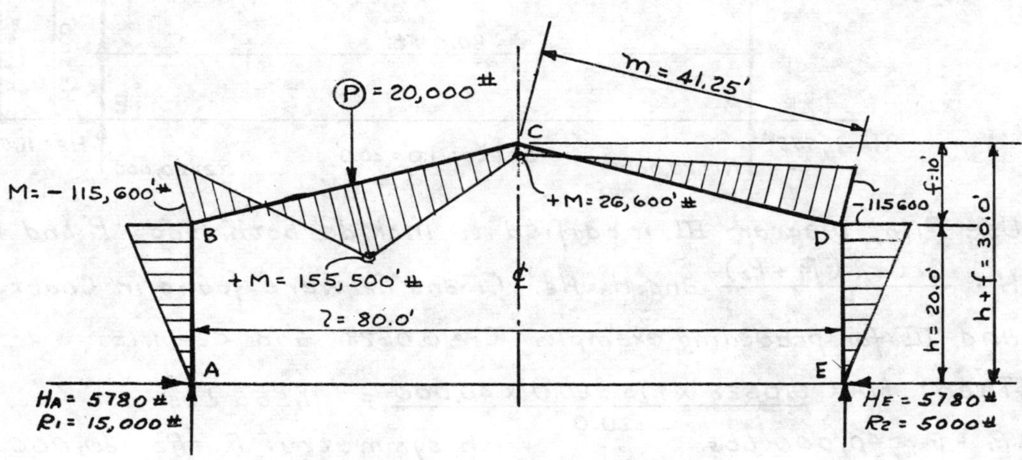

MOMENT DIAGRAM

SCALE: 1" = 400,000 Ft. LBS.

EXAMPLE: Rigid frame with symmetrical concentrated loads 7.5.2

Assume the identical Arch as used in preceding example. Place an additional 20,000 Pound Load on rafter 20.0 feet to right of Center Line (℄). Loads are to be symmetrical about ℄. Span l = 80.0 ft. Eave h = 20.0 feet. Rise f = 10.0 feet. P_1 = 20,000 Lbs. P_2 = 20,000 Lbs.

REQUIRED:

Calculate Vertical and Horizontal reactions R and H. Draw to scale the section and construct a moment diagram.

STEP I:

Drawing:

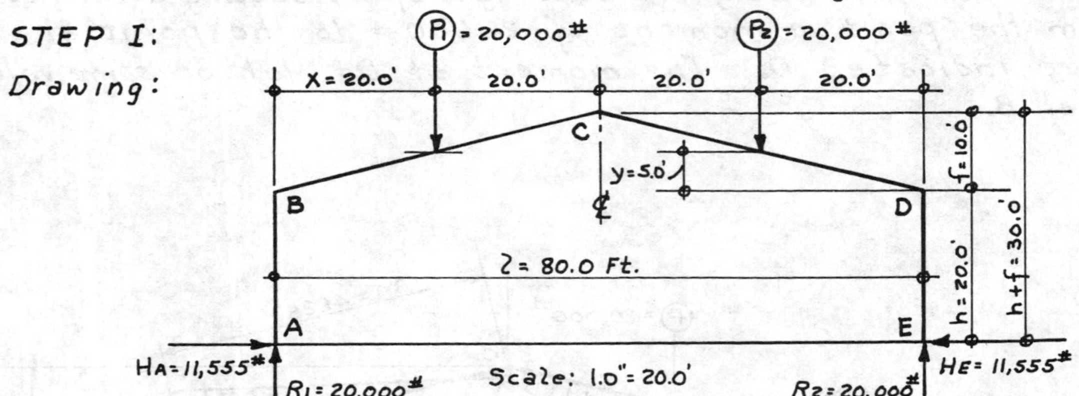

Use Pilot Diagram III modified to include both loads P_1 and P_2.

$H = \dfrac{C_1 C_2 l (P_1 + P_2)}{h}$ and $H_A = H_E$. C_1 and C_2 were found in Charts I and III for preceding example. $C_1 = 0.0628$ and $C_2 = 1.15$

Then: $H_A = \dfrac{0.0628 \times 1.15 \times 80.0 \times 40{,}000}{20.0'} =$ 11,555 Lbs.

$P_1 + P_2$ = 40,000 Lbs. With symmetry: $R_1 = R_2$ = 20,000 Lbs.

STEP II:

With 2 loads of equal value and placed same distance from ℄, Moment under P_1 same under P_2. $M_x = (R_1 X) - [H(h+y)]$.

Working from left at point B: X = 20.0 Ft. and y = 5.0 Feet.

Moment at Knee: $M_B = H_A h$, or M_B = 11,555 x 20.0 = − 231,100 Ft. Lbs.

Other moments along Rafters:

$M_{5.0} = (20{,}000 \times 5.0) - [11{,}555(20.0 + 1.25)] = -$ 144,545 Foot Lbs.

$M_{10.0} = (20{,}000 \times 10.0) - [11{,}555(20.0 + 2.50)] = -$ 60,000 Foot Lbs.

EXAMPLE: Rigid frame with symmetrical concentrated loads 7.5.2

$M_{15.0'} = (20{,}000 \times 15.0) - [11{,}555(20.0 + 3.75)] = +25{,}570$ Foot Lbs.

$M_{20.0} = (20{,}000 \times 20.0) - [11{,}555(20.0 + 5.00)] = +111{,}125$ Foot Lbs.

Moments at points B and D were computed in Step II previously.

At some point on Rafter between $M_{10.0}$ and $M_{15.0}$ will be inflection.

STEP III

Change formula when computing moments to right side of load P_1.

$M_{25.0} = (20{,}000 \times 25.0) - [11{,}555\ (20.0 + 6.25) + (20{,}000 \times 5.0)] = +96{,}680$ Ft. Lbs.

$M_{30.0} = (20{,}000 \times 30.0) - [11{,}555\ (20.0 + 7.50) + (20{,}000 \times 10.0)] = +82{,}238$ Ft. Lbs.

$M_{35.0} = (20{,}000 \times 35.0) - [11{,}555\ (20.0 + 8.75) + (20{,}000 \times 15.0)] = +67{,}800$ Ft. Lbs.

$M_{40.0} = (20{,}000 \times 40.0) - [11{,}555\ (20.0 + 10.0) + (20{,}000 \times 20.0)] = +53{,}350$ Ft. Lbs.

STEP IV

A moment $M_{65.0}$ is at right of both loads P_1 and P_2. Because of symmetrical placement, the moment point corresponds with $M_{15.0'}$ and should be the same. Check this moment. $y = 20.0' + 3.75'$.

$M_{65.0} = (20{,}000 \times 65.0) - [(20{,}000 \times 45.0) + (20{,}000 \times 5.0) + (11{,}555 \times 23.75)] =$

or $M_{65.0} = 1{,}300{,}000 - (900{,}000 + 100{,}000 + 274{,}430) = +25{,}570$ Ft. Lbs.

STEP V

In plotting the diagram with value points taken from the above calculations, the product will be a straight chord and any error will immediately show an off line plotted point.

If a tapered column or rafter is desired, take other moments to determine the Section Modulus at several critical points.

Treat the Column as a cantilever beam with H_A or H_E becoming a concentrated load thus: Column moment at 10.0 feet above the base is, $11{,}555 \times 10.0 = 115{,}550$ Foot pounds.

EXAMPLE: Rigid frame with symmetrical concentrated loads, continued 7.5.2

Moment diagram is plotted on ordinates spaced 2.0 feet vertical and drawn normal to slope of rafter.
Moment changes from negative at Knee B to positive at a point 13.7 feet, and is greatest under loads.

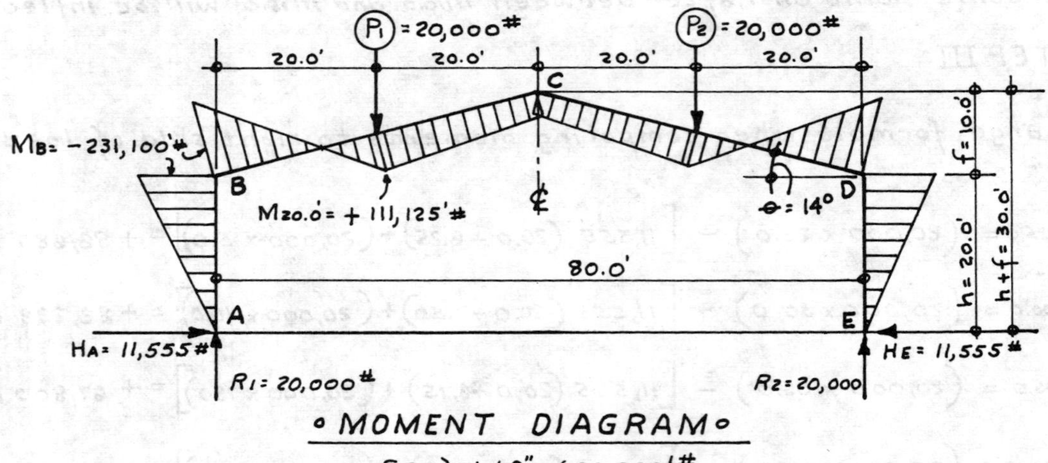

∘ MOMENT DIAGRAM ∘

Scale: 1.0" = 600,000'#

STEP VI:

Calculate the Axial Force in Rafter: Axial stress $f = \frac{\text{Force}}{\text{Area}}$

Tangent Rafter slope = $\frac{10.0'}{40.0'} = 0.25$ From Trigonometric tables in Section V, Angle θ = 14 degrees. Cos. θ = 0.970 and Sine θ = 0.242

Formula for force: $P_R = (H_A \text{ Cos. } \theta) + (R \text{ Sin. } \theta)$. Substituting values:

$P_R = (11{,}555 \times 0.970) + (20{,}000 \times 0.242) =$ 16,048.35 Pounds.

AUTHOR'S NOTATION:

This example - like the preceding example with a single concentrated load on rafter, considers the loads only as shown. The results obtained in this example must be added to results obtained from roof and wind loads. Refer to Pilot and Moment Diagrams I and II for guide and design formulas.

EXAMPLE I: Complete design of rigid frame 7.5.3

Warehouse project: 120'-0" x 100'-0"
Plans call for the layout as follows:
Total of 6 Bays spaced 20.0 feet on centers.
Clear Arch span to outside columns = 100.0 feet.
Eave height from column base plate = 20.0 feet.
Ridge height on Center Line above eave level = 15.0 feet.

LOADINGS:
Roof with combined Dead and Live loads = 50 Lbs. Sq. Foot on level.
Wind Load applied full height of structure = 30 Lbs. Sq. Foot. V=

REQUIRED:
Complete structural design for the following:

(a) Rigid Arch with haunched knee and tie rods at base columns.
(b) Cross Section for rafter and Column at critical points.
(c) Spacing for lateral bracing for Girts, Rafter, Purlins. Use sag rods. Design to comply with A.I.S.C. Rules. Straight or curved knee haunch may be at option of designer.
(d) Wind and Roof load moments should be calculated separately and Moment Diagram constructed for each.
(e) Draw a trial knee component and analyze for Radius, axial force and moments at tangent points of curvature.
(f) A tapered Column and Rafter is desirable, however if more economical a straight rolled section should be used.
(g) In each case use the greater moment value which may be with or without wind or a combination of both moments.

STEP I:
Drawing an illustration of arch with loading system:

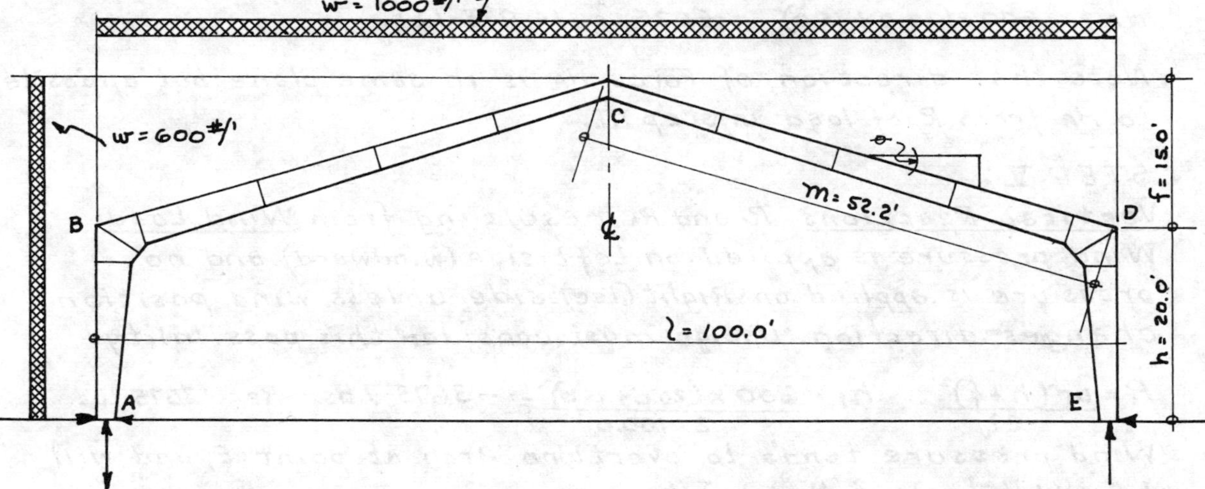

EXAMPLE I: Complete design of rigid frame, continued 7.5.3

STEP II:

To determine loads for 1 bay when arch spacing is 20.0 feet on Centers: Roof loads acting vertical are 50 Lbs. square foot.

$w = 50 \times 20.0 = 1000$ Lbs. Lineal foot.

Wind load acting horizontal full height = 30 Lbs per square foot.

Wind $w = 30 \times 20.0 = 600$ Lbs per Lineal foot on windward side only.

Vertical Reactions of Roof Load: $l = 100.0$ Ft. $W = 1000 \times 100.0 = 100{,}000^{\#}$

$R = \frac{Wl}{2}$ or $R_1 = \frac{100{,}000}{2} = 50{,}000$ Lbs. $R_2 = 50{,}000$ Lbs.

STEP III:

To calculate <u>Horizontal reactions</u> at base of Columns. See Pilot Diagram I for H_A and H_E resulting from Roof Loads. Find values of Q and k then use Chart I to obtain coefficient C_1.

$Q = \frac{f}{h}$ and $k = \frac{h}{m}$ $H_E = \frac{C_1 w l^2}{h}$ $Q = \frac{15.0}{20.0} = 0.75$ $k = \frac{20.0}{52.2} = 0.383$

From chart curve $C_1 = 0.055$ Formula: $H_A = H_E$ for Roof Loads.

$H_E = \frac{0.055 \times 1000 \times 100.0^2}{20.0} = 27{,}500$ Lbs. $H_A = 27{,}500$ Lbs.

Roof Loads acting in vertical plane tend to spread the Columns at Base and should be tied. This force produces a Tension stress in outer flange and bending moment is negative.

STEP IV:

Calculate Horizontal Reactions from Wind Pressure:

From Pilot Diagram II: $H_E = C_6 wh$, and $H_A = \left[w(h+f)\right] - H_E$.

From Chart Coefficient $C_6 = 0.502$. With values:

$H_E = 0.502 \times 600 \times 20.0 = 6024$ Lbs. (Call it 6025 Lbs.)

$H_A = 600 \times (20.0 + 15.0) - 6025 = 14{,}975$ Lbs.

Note that direction of Force H_A is in same plane but opposite to H_A from Roof load in step III.

STEP V:

<u>Vertical Reactions</u> R_1 and R_2 resulting from <u>Wind Load:</u>

Wind pressure is applied on Left side (windward) and no pressure is applied on Right (lee) side unless wind position changes direction. Design must consider this possibility.

$R = \frac{w(h+f)^2}{2l}$ $R_1 = \frac{600 \times (20.0 + 15.0)^2}{2 \times 100.0} = -3675$ Lbs. $R_2 = +3675$ Lbs.

Wind pressure tends to overturn Arch at point E, and will be additional at that point.

STEP VI:

To calculate bending moments in knee, columns and rafter, tabulate the reactions thus:

R_1, Roof Load =	+ 50,000 Lbs.	R_2, Roof Load =	+ 50,000 Lbs.
R_1, Wind Load =	− 3,675 "	R_2, Wind Load =	+ 3,675 "
Combined R_1 =	+ 46,375 Lbs.	Combined R_2 =	+ 53,675 Lbs.
H_A, Roof Load =	− 27,500 Lbs.	H_E, Roof Load:	− 27,500 Lbs.
H_A, Wind Load =	+ 14,975 "	H_E, Wind Load:	− 6,025 "
Combined H_A =	− 12,525 Lbs.	Combined H_E =	− 33,525 Lbs.

STEP VII:

To design knee, rafter and column draw a working sketch as shown on next page #7034. A fair idea of required depth at knee may be obtained by referring to table 7.7.2 on page 7073. Assume a 32.0 inch depth for rafter and column at knee. Center line of column and rafter intersect at a point 19.01 Feet from base. This is the exact moment arm for rafter. Column moment arm is 17.62 feet, however most designers will use the dimension h for both rafter and column due to safety. Moment arm of vertical force R_2 at point D is 1.33 feet.

$$M_D = (33,525 \times 19.01) + (53,675 \times 1.33) = -708,698 \text{ Foot Lbs.}$$

Probable value of Section Modulus at knee will be thus:

$S_x = \dfrac{708,698 \times 12}{20,000} = 425.22''^3$ Refer to AISC Rule II, page 7023 to ascertain the allowable stresses for tension and compression in knee. Width of flange b is important in section selection. In checking the 32.0 inch depth, a 30WF 172 with 15 inch flange has a $S_x = 530.0''^3$. A flange width of 10.0 inches is desired and a section welded from plate will be tried in order to get a tapered profile on column and rafter.

STEP VIII:

To calculate rafter moments, the rise dimension y will be required for each equation. Ascertain the angle of roof pitch and record the required angle functions:

Tangent $A = \dfrac{a}{b}$ or $\dfrac{f}{l/2}$. Then Tan $\theta = \dfrac{15.0}{50.0} = 0.3000$ $\theta = 16° 42'$

Sine θ = 0.28736 Cos. θ = 0.95782 Secant θ = 1.0440

EXAMPLE I: Complete design of rigid frame, continued 7.5.3

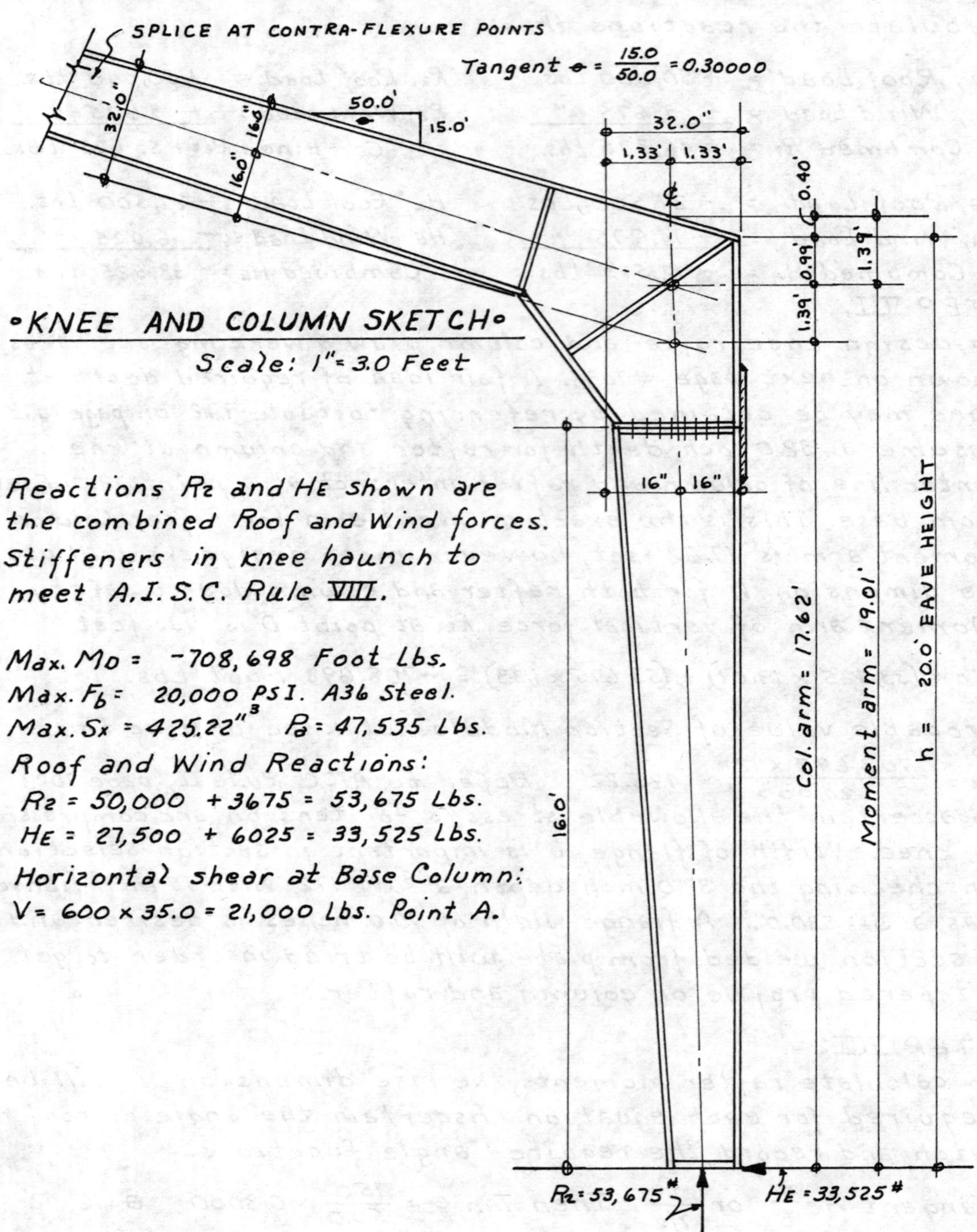

Reactions R_Z and H_E shown are the combined Roof and Wind forces. Stiffeners in knee haunch to meet A.I.S.C. Rule VIII.

Max. M_D = −708,698 Foot Lbs.
Max. F_b = 20,000 PSI. A36 Steel.
Max. S_x = 425.22"^{3} P_a = 47,535 Lbs.

Roof and Wind Reactions:
R_Z = 50,000 + 3675 = 53,675 Lbs.
H_E = 27,500 + 6025 = 33,525 Lbs.

Horizontal shear at Base Column:
V = 600 × 35.0 = 21,000 Lbs. Point A.

STEP IX:

Since a section 32.0 inches in depth appeared to be the size for knee joint, calculate a satisfactory welded plate section which must have the property of $S_x = 425.22''^3$ or slightly over that value. See Section VI properties of Sections and follow format given on page 6024. Use table 6.4.3 for properties of rectangular shapes.

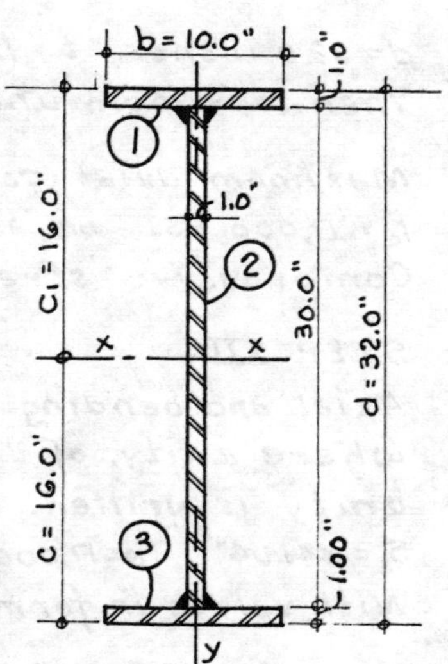

RAFTER PROPERTIES ABOUT AXIS X-X						
M'K.	SECTION	A□"	ʅ"	Aʅ²	Io	Aʅ² + Io
1	10.0 x 1.00 ℞	10.00	15.50	2402.50	0.833	2403.333
2	1.00 x 30.0 ℞	30.00	0	0	2250.00	2250.000
3	10.0 x 1.00 ℞	10.00	15.50	2402.50	0.833	2403.333
	A = 50.00 □"				ΣM = I = 7056.666	

$S_x = \frac{I}{C}$ $S_x = \frac{7056.666}{16.0} = 441.04''^3$ $r_x = \sqrt{\frac{7056.666}{50.00}}$

RAFTER PROPERTIES ABOUT AXIS Y-Y						
M'K.	SECTION	A□"	ʅ"	Aʅ²	Io	Aʅ² + Io
1	1.00 x 10.0 ℞	10.00	0	0	83.33	83.333
2	30.0 x 1.00 ℞	30.00	0	0	2.50	2.500
3	1.00 x 10.0 ℞	10.00	0	0	83.33	83.333
		50.00			ΣM = I = 169.166	

$r_y = \sqrt{\frac{I_y}{A}}$ $r_y = \sqrt{\frac{169.16}{50.0}} = 1.84$

Section Modulus S_x is greater than $425.22''^3$ required and section is acceptable. Resisting Moment = $\frac{441.04 \times 20{,}000}{12} = 735{,}067$ Foot Lbs.

STEP X:

Columns and Rafters are subjected to a combination of bending and axial compressive stress. Unit stresses may require a reduced allowable or modified as required in Rule II page 7023.

To obtain the axial force in Rafter use the formula thus: $P_a = (H_E \cos\theta) + (R_2 \sin\theta)$. Using combined wind and roof load reaction values in formula: Angle functions in Step VIII.

$P_a = (33{,}525 \times 0.95782) + (53{,}675 \times 0.28736) = 47{,}535$ Lbs. Show on detail.

Area knee section: A = 50.0 Sq. In. Stress = $\frac{P_a}{A}$ or $f = \frac{47{,}535}{50.00} = 950.7$ P.S.I

EXAMPLE I: Complete design of rigid frame, continued 7.5.3

STEP XI

AISC Rule II requires that $\frac{l d}{b t}$ shall be limited to 600 where l is the length between lateral bracing and given in inches. By transposing formula, the maximum bracing length may be determined thus: $l = \frac{600 bt}{d}$. Known values of Section are:

$d = 32.0$ inches, $t = 1.00$ inch and flange $b = 10.0$ inches.

Then maximum unbraced $l = \frac{600 \times 10.0 \times 1.00}{32.0} = 187.5$ inches or 15.62 Ft.

Maximum axial compressive stress is limited by Rule II to unit $F_a = 17{,}000$ PSI or as modified by the formula given.

Compressive stress was found in Step X to be: $f_a = 950.7$ PSI.

STEP XII

Axial and bending stresses must be proportioned as per Rule III, where unity of 1.00 or less is obtained. This equation for unity is written: $u = \frac{M}{S_x F_b} + \frac{P_a}{A F_a}$. Largest $-M = 708{,}698$ Foot Lbs.

$S_x = 441.04''^3$ $F_a = 17{,}000$ PSI, Area $A = 50.0$ Sq. In. $P_a = 47{,}535$ Lbs. $F_b = 20{,}000$ PSI.

With values in formula: $u = \left(\frac{708{,}698 \times 12}{441.04 \times 20{,}000}\right) + \left(\frac{47{,}535}{50.0 \times 17{,}000}\right) = 1.018$ (This is close enough since other bending moments are less in value.)

STEP XIII:

Lateral support bracing for Rafter and Column: Max. $l = 187.5$ inches.

Brace Column at or near mid-height or 9.50 ft. from base. $l = 114.0$ In.

Brace Rafters with purlins spaced $\frac{52.2}{5} = 10.44$ Ft. or $\pm$ 125.0 inches.

STEP XIV:

When Rafters are to be tapered from knee joint to crown ridge, take bending moments at several uniform distances X and calculate the required value of Sx and the point of Contra-Flexure. All these moments are necessary to construct the three moment diagrams which will follow.

Calculate the moments for Wind Pressure and Roof Loads each in a separate group for tabulating. Work from left side only from point B on work diagram.

Dimension X equals the distance in feet from point B to the location of moment and distance measured on the horizontal.

Check the formulas for Wind load moments carefully as the equations differ for moment on each side of Ridge Center Line.

Use the Tangent of pitch angle θ to calculate the rise y at each point. Tan $\theta = 0.300$ At $M_{25.0}$ $y = 25.0 \times 0.300 = 7.50$ feet.

Rafter Moments resulting from ROOF LOADS and taken at x.

Roof load moments are same value on right side ℄ for Uniform load From Moment Diagram 7.3.1 Page 7013, the formula is written:

$$M_x = \left[\frac{wx(l-x)}{2}\right] - \left[H(h+y)\right].$$ Values: $w = 1000\#/'$. $l = 100.0'$ $h = 20.0'$

y = Tangent θ times x, or $0.300x$.

$$M_{10.0} = \left[\frac{(1000 \times 10.0) \times (100.0 - 10.0)}{2}\right] - \left[27500\,(20.0 + 3.00)\right] = -182,500'\#$$

$$M_{18.0} = \left[\frac{(1000 \times 18.0) \times (100.0 - 18.0)}{2}\right] - \left[27,500\,(20.0 + 5.40)\right] = +39,500'\#$$

$$M_{18.5} = \left[\frac{(1000 \times 18.5) \times (100.0 - 18.5)}{2}\right] - \left[27,500\,(20.0 + 5.55)\right] = +51,250'\#$$

$$M_{20.0} = \left[\frac{(1000 \times 20.0) \times (100.0 - 20.0)}{2}\right] - \left[27,500\,(20.0 + 6.00)\right] = +85,000'\#$$

$$M_{30.0} = \left[\frac{(1000 \times 30.0) \times (100.0 - 30.0)}{2}\right] - \left[27,500\,(20.0 + 9.00)\right] = +252,500'\#$$

$$M_{40.0} = \left[\frac{(1000 \times 40.0) \times (100.0 - 40.0)}{2}\right] - \left[27,500\,(20.0 + 12.0)\right] = +320,000'\#$$

$$M_{50.0} = \left[\frac{(1000 \times 50.0) \times (100.0 - 50.0)}{2}\right] - \left[27,500\,(20.0 + 15.0)\right] = +287,500'\#$$

Moment at Ridge ℄: $M_c = \left(\frac{wl^2}{8}\right) - \left[H_E(h+f)\right]$ or $M_c = \left(\frac{Rl}{4}\right) - \left[H(h+f)\right]$.

$$M_c = \left(\frac{50,000 \times 100.0}{4}\right) - \left[27,500 \times (20.0 + 15.0)\right] = +287,500 \text{ Foot Lbs.}$$

STEP XV

Rafter Moments resulting from WIND PRESSURE and taken at x.

$$M_x = \left[H_A\,(h+y)\right] - \left[(R_1 x) + w\,\frac{(h+y)^2}{2}\right].$$ H_A = 14,975 Lbs. R_1 = 3675 Lbs.

$$M_{10.0'} = \left[14,975 \times (20.0 + 3.0)\right] - \left[(3675 \times 10.0) + \left(\frac{600 \times 23.0^2}{2}\right)\right] = +148,975'\#$$

$$M_{18.0'} = \left[14,975 \times (20.0 + 5.4)\right] - \left[(3675 \times 18.0) + \frac{(600 \times 25.4^2)}{2}\right] = +120,667'\#$$

$$M_{20.0'} = \left[14,975 \times (20.0 + 6.0)\right] - \left[(3675 \times 20.0) + \left(\frac{600 \times 26.0^2}{2}\right)\right] = +113,050'\#$$

$$M_{30.0'} = \left[14,975 \times (20.0 + 9.0)\right] - \left[(3675 \times 30.0) + \left(\frac{600 \times 29.0^2}{2}\right)\right] = +71,725'\#$$

EXAMPLE I: Complete design of rigid frame, continued 7.5.3

$$M_{40.0'} = \left[14{,}975 \times (20.0 + 12.0)\right] - \left[(3675 \times 40.0) + \left(\frac{600 \times 32.0^2}{2}\right)\right] = +25{,}000\text{'\#}$$

$$M_{43.0'} = \left[14{,}975 \times (20.0 + 12.9)\right] - \left[(3675 \times 43.0) + \left(\frac{600 \times 32.9^2}{2}\right)\right] = +9{,}929.5\text{'\#}$$

$$M_{50.0'} = \left[14{,}975 \times (20.0 + 15.0)\right] - \left[(3675 \times 50.0) + \left(\frac{600 \times 35.0^2}{2}\right)\right] = -27{,}125\text{'\#}$$

$M_{50.0}$ is on ℄ at Ridge. Check value with Alternate Formula thus:

$$M_c \quad \left(\frac{Rl}{2}\right) - \left[H_E(h+f)\right] \qquad M_c = \left(\frac{3675 \times 100.0}{2}\right) - \left[6025(20.0 + 15.0)\right] = -27{,}125\text{'\#}$$

STEP XVI

Tabulate Rafter Moments resulting from Horizontal and Vertical forces separately, then determine the maximum M. Negative moments indicated with minus sign (—) and Positive moment with plus sign (+). All moments are left side of ridge ℄.

TABULATED RAFTER MOMENTS

LOCATION	VER. ROOF LOAD	WIND LOAD	DESIGN MOMENT
KNEE	−550,000	−120,500	−708,698 Ft. Lbs.
$M_{10.0'}$	−182,500 '#	+148,975 '#	−182,500 " "
$M_{18.0'}$	+39,500	+120,667	+160,167 " "
$M_{18.5'}$	+51,250	+118,783	+170,033 " "
$M_{20.0'}$	+85,000	+113,050	+198,050 " "
$M_{30.0'}$	+252,500	+71,725	+324,225 " "
$M_{40.0'}$	+320,000	+25,000	+345,000 " "
$M_{43.0'}$	+320,750	+9,929.5	+330,679.5 " "
$M_{50.0'}$	+287,500	−27,125	+287,500 " "

To construct Moment diagrams, other moments will be required thus:

WIND $M_B = H_A h - \frac{wh^2}{2}$, and $M_D = -H_E h$. ROOF moments are same on each side of ℄ due to symmetrical uniform loading.

$$\text{Wind } M_B = (+14{,}975 \times 20.0) - \left(\frac{600 \times 20.0^2}{2}\right) = +179{,}500 \text{ Foot Lbs.}$$

$$\text{Wind } M_D = -6025 \times 20.0 = -120{,}500 \text{ Foot Lbs.}$$

Combined wind and roof moments in Knee were computed in step VII where lever arms were less. In many cases a designer will calculate knee moment simply thus:

For Roof: $M_B = -27{,}500 \times 20.0 = -550{,}000$ Foot Lbs.

For Wind: $M_D = -6{,}025 \times 20.0 = -120{,}500$ " "

Total at M_B and $M_D = -670{,}500$ Foot Lbs.

This method is not recommended for knee moment as it does not consider eccentricity due to axial stress. See Step VII.

STEP XVII:

Constructing Moment Diagrams for Wind, Roof and Combined loading. Ordinate values obtained from Table in step XVI.

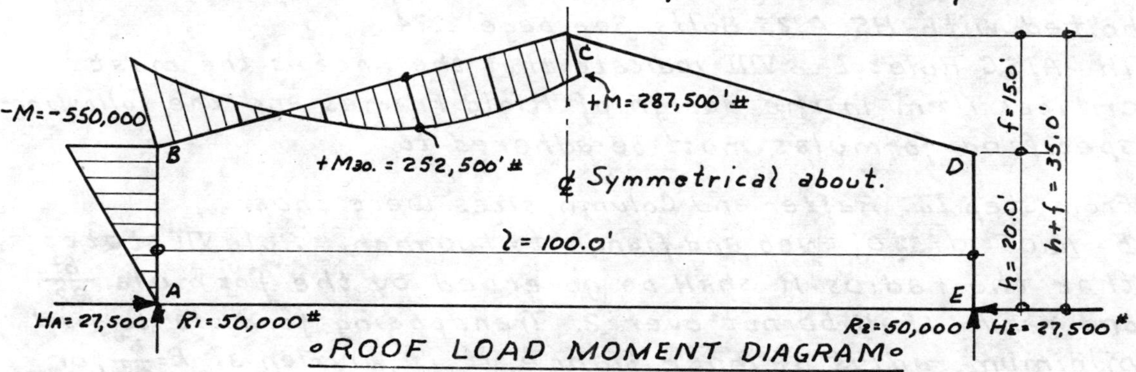

∘ROOF LOAD MOMENT DIAGRAM∘

Scale: 1.0" = 1,000,000'#

-M = 27,125'#
C
-M = 120,500'#
B
-M = 120,500'#
D
+M = 179,500'#
A
E
R1 = -3675
HA = +14,975#
R2 = +3675#
HE = -6025#

∘WIND LOAD MOMENT DIAGRAM∘

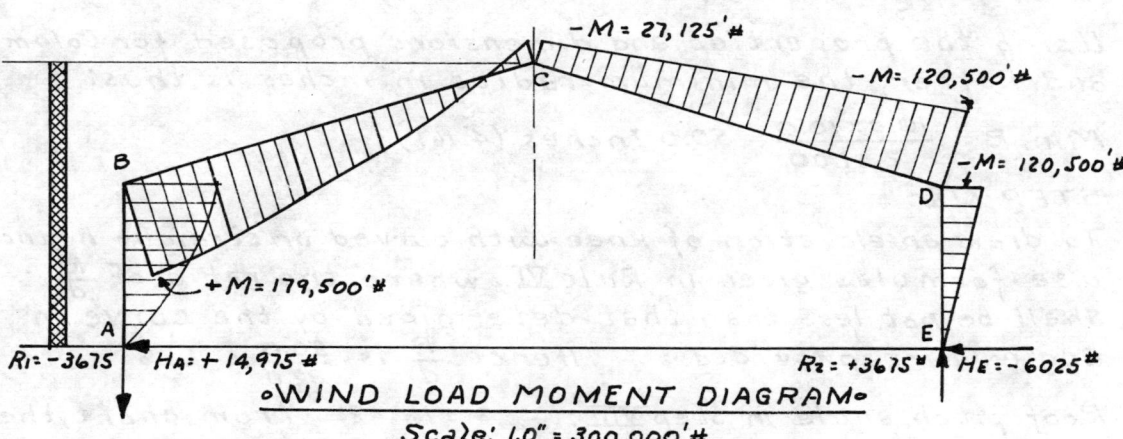

Scale: 1.0" = 300,000'#

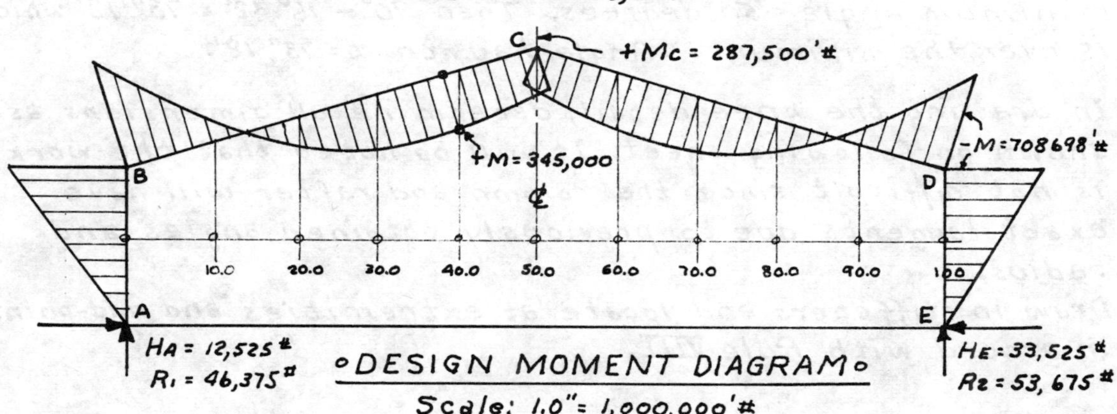

∘DESIGN MOMENT DIAGRAM∘

Scale: 1.0" = 1,000,000'#

EXAMPLE I: Complete design of rigid frame, continued 7.5.3

STEP XVIII :

Designing and detailing the knee:

In scaling the preceding Design Moment Diagram, take note of the point of contra-flexure where stress is minimum and moments change from positive to negative. Splices in Rafters must be made near these points. This example will assume that knee will be welded to rafter and knee will be bolted with HS A325 Bolts. See page 2224.

The AISC Rules I to VIII indicate that the knee is the most critical item in the design of Rigid Frames and the following specified formulas must be adhered to.

From step IX: Rafter and Column sizes were thus: $b = 10.0''$ $d = 32.0''$ web and flange $t = 1.00$ inch. Rule VII states that the radius R shall be governed by the formula $\frac{b^2}{R2}$, and ratio shall be not over 2. Transposing formula to find minimum radius of inner flange, it is written as: $R = \frac{b^2}{2t}$, or again, $t = \frac{b^2}{2R}$, and $b = \sqrt{2Rt}$.

Using the properties and dimensions proposed for Column and Rafter, the minimum radius in inches is thus:

$$\text{Min. } R = \frac{10.0 \times 10.0}{2 \times 1.00} = 50.0 \text{ Inches } (4.167').$$

STEP XIX:

To draw an elevation of knee with curved or straight haunch, use formulas given in Rule VI, where the ratio of $\frac{R}{d}$ shall be not less than that determined by the curve in angle Chart provided. Hence $\frac{R}{d}$ is $\frac{50.0}{32.0} = 1.56$

Roof pitch angle in step VIII: $\theta = 16° 42'$ From chart, the minimum angle = 50 degrees. Then $90° - 16° 42' = 73° 18'$ which is over the minimum. Angle haunch $\alpha = 73° 18'$.

In drawing the knee detail, determine all dimensions as shown on following sheet. It will be noted that the work is not difficult since the column and rafter will have exact tangents due to previously obtained angles and radius.

Draw in stiffeners and locate at extremities and mid-point in accord with Rule VIII.

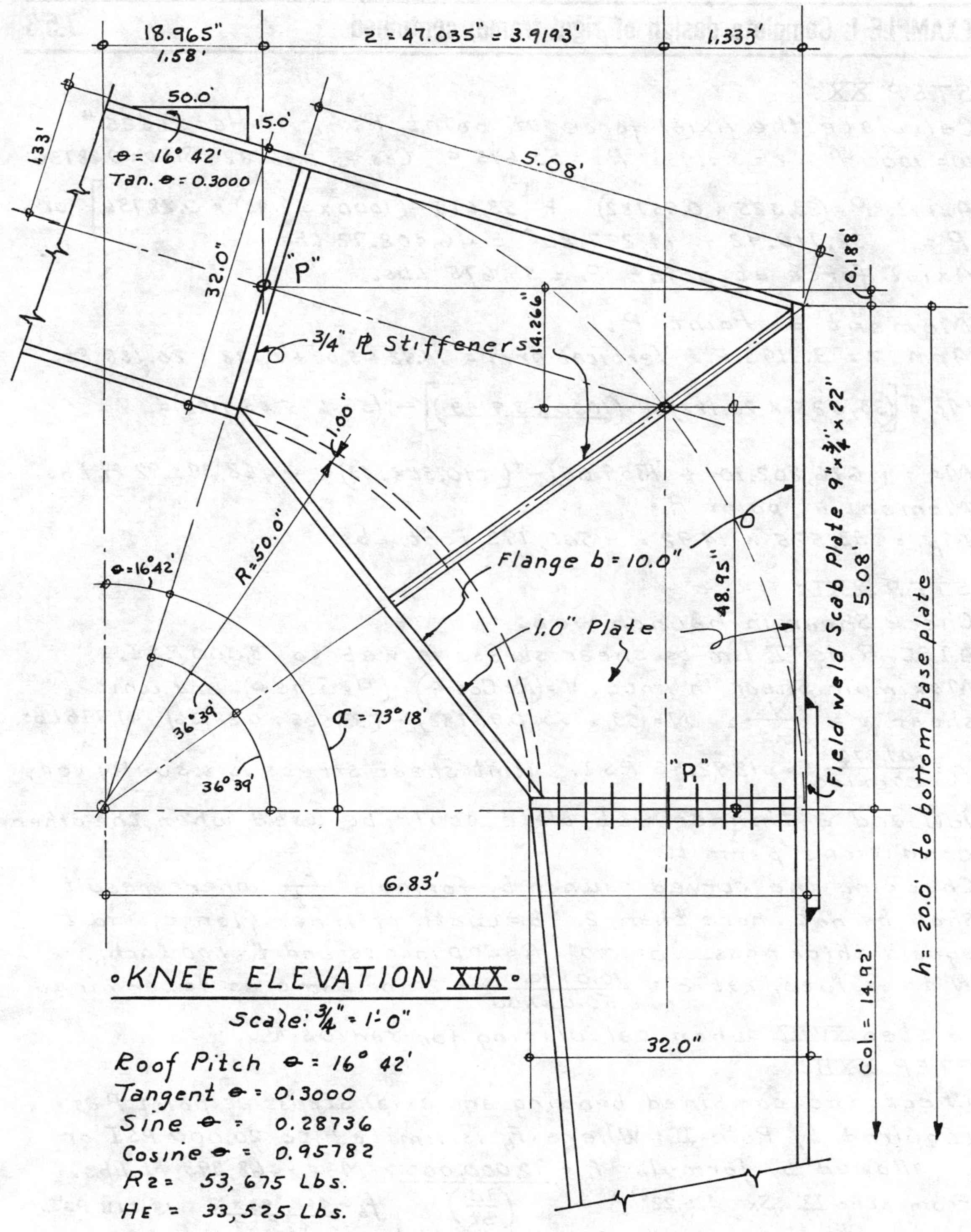

KNEE ELEVATION XIX

Scale: 3/4" = 1'-0"

Roof Pitch θ = 16° 42'

Tangent θ = 0.3000

Sine θ = 0.28736

Cosine θ = 0.95782

R_2 = 53,675 Lbs.

H_E = 33,525 Lbs.

EXAMPLE I: Complete design of rigid frame, continued 7.5.3

STEP XX:

Calculate the Axial force at point P: $H_E = 33{,}525^{\#}$

$w = 1000^{\#/'}$ $z = 3.9193$ $R_2 = 53{,}675^{\#}$ $Cos.\theta = 0.95782$ $Sin\theta = 0.28736$

Axial $P = (33{,}525 \times 0.95782) + \left[53{,}675 - (1000 \times 3.9193) \times 0.28736\right]$ or

$P = 32{,}110.92 + 14{,}297.80 = 46{,}408.72$ Lbs.

Axial force at point $P_1 = 53{,}675$ Lbs.

Moment at Point P:

Arm $z = 3.9193$ Ft. Vertical arm $= 14.92 + 5.08 + 0.188 = 20.188$ Ft.

$M_P = {}^{-}\left[(33{,}525 \times 20.188) + \left(\frac{1000 \times 3.9193}{2}\right)\right] -{}^{+}(53{,}675 \times 3.9193) =$

$M_P = {}^{-}(676{,}802.70 + 1959.65) -{}^{+}(210{,}368.43) = -468{,}393.92$ Ft. Lbs.

Moment at point P_1:

$M_{P'} = 33{,}525 \times 14.92 = -500{,}193$ Foot Lbs.

STEP XXI:

Check Shear in haunch web:

A.I.S.C. Rule V limits shear stress in web to 13,000 P.S.I.

Maximum Shear in knee: $V = (R_2 \cos\theta) - (H_E \sin\theta)$ and unit shear $v = \frac{V}{d \times t}$. $V = (53{,}675 \times 0.95782) - (33{,}525 \times 0.28736) = 41{,}776$ Lbs.

$v = \frac{41{,}776}{30.0 \times 1.00} = 1392.56$ P.S.I. Unit shear stress is usually very low and a thinner web plate could be used when the other conditions permit.

Checking the curved haunch by formula $\frac{b^2}{Rt}$ where result shall be not more than 2. b = width of inner flange, and t equals thickness. b = 10.0" R = 50.0 inches and t = 1.00 Inch.

With values, ratio $= \frac{10.0 \times 10.0}{50.0 \times 1.00} = 2$ or same as determined in step XVIII when calculating for radius R.

STEP XXII:

Check the combined bending and axial stress at point P as required by Rule II: Where F_b is limited to 20,000 PSI or as allowed by formula: $F_b = \frac{12{,}000{,}000}{\left(\frac{ld}{bt}\right)}$. $M_P = -468{,}393$ Ft Lbs.

From step IX: $S_x = 425.22''^3$ $f_b = \frac{468{,}393 \times 12}{425.22} = 13{,}218$ PSI.

l = Length determined according to bracing. t = 1.00 inch.

Assume lateral bracing at 16.67 feet = 200.0 inches and put values in formula: Allowable for rafter at point P is thus:

$F_b = \frac{12{,}000{,}000}{\left(\frac{200 \times 32.0}{10.0 \times 1.00}\right)} = 18{,}750$ PSI. By closer bracing for lateral

support, the allowable will be raised where 20,000 PSI can be used for rafter design.
Since a tapered rafter is desired, it would be advantageous to continue knee from point P to the point of contra-flexure as shown on moment diagram. This distance is approximately 12.5 feet. $l = 12.5 \times 12 = 150.0$ inches.

Rule III states that maximum combined unit stresses for axial compression and bending be determined by the usual formula thus: $f = \frac{P}{A} + \frac{M}{S_x}$. Allowable $F_b = 20{,}000$ PSI. and Rule II gives the allowable compressive be governed by the formula:

$F_a = 17{,}000 - \frac{0.485\, l^2}{r^2}$. $l^2 = 150.0 \times 150.0 = 22{,}500$. Least value of r from Step IX is: $r_y = 1.84$ Substituting values:

$$F_a = 17{,}000 - \left(\frac{0.485 \times 22{,}500}{1.84 \times 1.84}\right) = 13{,}777 \text{ PSI.}$$

Combined Unit stress:
Axial at point P for rafter, $P_a = 46{,}408.72$ Lbs. Bending at P: $M_P = 468{,}393.92$ Ft. Lbs. Area = 50.0 Sq. In. $S_x = 441.04''^3$
Combined stress: $f = \frac{P}{A} + \frac{M}{S_x}$. With values substituted:

$$\text{Combined } f = \left(\frac{46{,}408.72}{50.00}\right) + \left(\frac{468{,}393 \times 12}{441.04}\right) = 928.17 + 12{,}744.25 = 13{,}672.4\ \#\square''$$

Checking ratio: Combined stress for unity of 1.00 or less, the common formula is written: $U = \frac{f_a}{F_a} + \frac{f_b}{F_b}$. With values:

$$U = \frac{928.17}{13{,}777} + \frac{12{,}744.25}{20{,}000} = 0.7046 \quad \text{(Less than unity 1.00 and OK.)}$$

STEP XXIII:
Column design for base shear, bending and axial load:
Height = 14.92 feet. Assume lateral bracing at mid-height, or $l = \frac{14.92 \times 12}{2} = 89.52$ inches. $R_2 = P_E = 53{,}675$ Lbs. $M_P' = -500{,}193$ Ft. Lbs

Reqd. $S_x = \frac{M}{F_b}$ or $S_x = \frac{500{,}193 \times 12}{20{,}000} = 300.116''^3$

At mid-height: $M = 33{,}525 \times 89.52 = 3{,}001{,}158$ Inch Lbs.
At mid-height: $S_x = \frac{3{,}001{,}158}{20{,}000} = 150.06''^3$
At top $S_x = 300.116''^3$ and a Wide Flange section such as a 30WF 108# with flange b = 10½". $S_x = 300.0''^3$, could be used. Cut off inner flange to provide taper and replace with plate.
Shear area of web plate at bottom must have, $A = \frac{33{,}525}{13{,}000} = 2.58$ Sq. In.
Web t = 0.548 width web = $\frac{2.58}{0.548} = 4.71$ Sq. inches. (Low).

EXAMPLE I: Complete design of rigid frame, continued 7.5.3

Depth of column at base points A and E will be governed by space necessary to accomodate anchor bolts. When H_E = 33,525 Lbs. to be resisted by bolts which have a unit shear stress F_v = 10,000 PSI for A-307 Bolts. See table 2.10.5.2 on page 2225 for single shear. Assume 4 ϕ bolts are desired. Required shear value / Bolt: $\frac{33,525}{4}$ = 8381.25 Lbs. Accept 4ϕ 1⅛" Bolts. When columns are tied together at the base with rods or flat bar, bolts may be reduced in number and ties will be in Tension stress.

STEP XXIV:

Rafter design from ridge to point of contra-flexure:
Referring to Design Moment Diagram in Step XVII:
$M_{40.0}$ = + 345,000 Foot Lbs. (Maximum bending in rafter with wind and roof loads combined.)
Axial force at P = 46,408.72 Lbs. (Step XX) Assume that rafter will join knee at this point or knee can be extended on slope to have same depth as selected section for a compact rafter. See Rule IV for lateral bracing and length ℓ.

$$S_x = \frac{M}{F_b} \quad S_x = \frac{345,000 \times 12}{20,000} = 207.0''^3$$

From Tables 2.3.3.1, select for trial a 27 WF 94 with these properties: S_x = 243.0"^{3} Flange b = 10.0", d = 26.99", Flange t = 0.747", Web t = 0.490", r_y = 2.12, and A = 27.7 Sq. Inches.
Apply combined stress formula from Rule III:

$$f = \frac{46,408.72}{27.70} + \frac{345,000 \times 12}{243.0} = 1675.405 + 17,037.037 = 18,712.442 \text{ PSI.}$$

Stresses are below allowables and lateral bracing of the top compression flange must be adequate.

STEP XXV:

Rule II permits full allowable stress only when $\frac{\ell d}{bt}$ is less than 600 ratio. Then to transpose for maximum bracing lengths ℓ:
Max. $\ell = \frac{600\,bt}{d}$ or $\ell = \frac{600 \times 10.0 \times 0.747}{27.0}$ = 166.0 inches. From Ridge to eave rafter length m = 52.20 feet. Bracing at 5 equal points = $\frac{52.20 \times 12}{4}$ = 156.6 inches, and acceptable. Check Rule II formula for stress in compression flange: $F_a = 17000 - \left(\frac{.485\,\ell^2}{r_y^2}\right)$.

$$F_a = 17,000 - \left(\frac{0.485 \times 156.6^2}{2.12^2}\right) = 14,353.61 \text{ PSI.}$$ For bending allowable unit stress: $F_b = \frac{12,000,000}{\left(\frac{156.6 \times 27.0}{10.0 \times 0.747}\right)}$ = 21,200.6 PSI. Section is acceptable

since neither F_a or F_b has been exceeded. Then, when checking combined stresses for unity of 1.00 or less with formula $u = \frac{f_a}{F_a} + \frac{f_b}{F_b}$, it is: $u = \frac{1675.405}{14,353.61} + \frac{17,037.037}{20,000} = 0.969-$. If a lighter Section than the 27WF94 with lesser flange thickness and S_x property had been selected, unity would be over maximum 1.00.

STEP XXVI:

Designing the tapered column. Column height to point $P_1 = 14.92'$ feet and base dimensions must accomodate room for anchor bolts resisting shear. $V = H_E$ and critical with combined loads. $V = 33,525$ Lbs. Bolts A-305 allowable in single shear $F_v = 10,000$ PSI. From Table 2.10.5.2: Assume 4 ϕ bolts in number, and value of 1 bolt to resist $= \frac{33,525}{4} = 8381.25$ Lbs. Will require 4- $1\frac{1}{8}$ ϕ bolts, therefore, make base of column 12.0 inches deep.

Bending moment at top is: $M_p = 33,525 \times 14.92 = 500,193$ Ft. Lbs. Required $S_x = \frac{500,193 \times 12}{20,000} = 300.16''^3$ Ascertain from Table 2.3.3.1 that a 30WF108 has a $S_x = 300.0$ and flange $t = 0.76''$ and web thickness $t = 0.548$ inches. Try fabricating a column section with ½ inch plate for web and flanges of ¾ plate. Width of flanges same as at haunch or 10.0 inches. Greater depth of 32.0 inches should give adequate moment of inertia. Using the format given in Section VI for properties of Sections:

COLUMN SECTION AT P_1

M'K.	SIZE	A□"	ℓ"	Aℓ²	Io	Aℓ² + Io
1	10.0 × 0.75	7.500	15.625	1831.055	0.352	1831.407
2	10.0 × 0.75	7.500	15.625	1831.055	0.352	1831.407
3	0.50 × 30.50	15.250	0	—	1182.193	1182.193
		30.250				Σo = 4845.007

$$S_x = \frac{I}{C} = \frac{4845.007}{16.0} = 302.813''^3$$

10.0" 12.0" x x c=16.0" d=32.0" ① ② ③

Web plate at Column base = 10.5 × 0.50 inches. $A_w = 5.25$ Sq.In. Shear is $H_E = V$ and $f_v = \frac{V}{A_w}$. Thus $f_v = \frac{33,525}{5.25} = 6385.71$ P.S.I. Dimensions at base are also acceptable.

EXAMPLE I: Complete design of rigid frame, continued 7.5.3

STEP XXVII:

Knee with haunch will be tapered to join 27WF94 rafter, which is rolled section. Scaling the moment diagram the apparent point of contra-flexure is located at 16.39 feet from point B. Check this by calculating the moment at 16.39 feet. M value should low or zero.

$$M_{16.39} = \left[\frac{1000 \times 16.39(100.0-16.39)}{2}\right] - \left[27{,}500\left(20.0 + 4.917\right)\right] = -33.55 \text{ Ft. Lbs.}$$

Inflection point is close enough to make connection to rafter at 16.40 feet. Show this on sketch and detail on final plans.

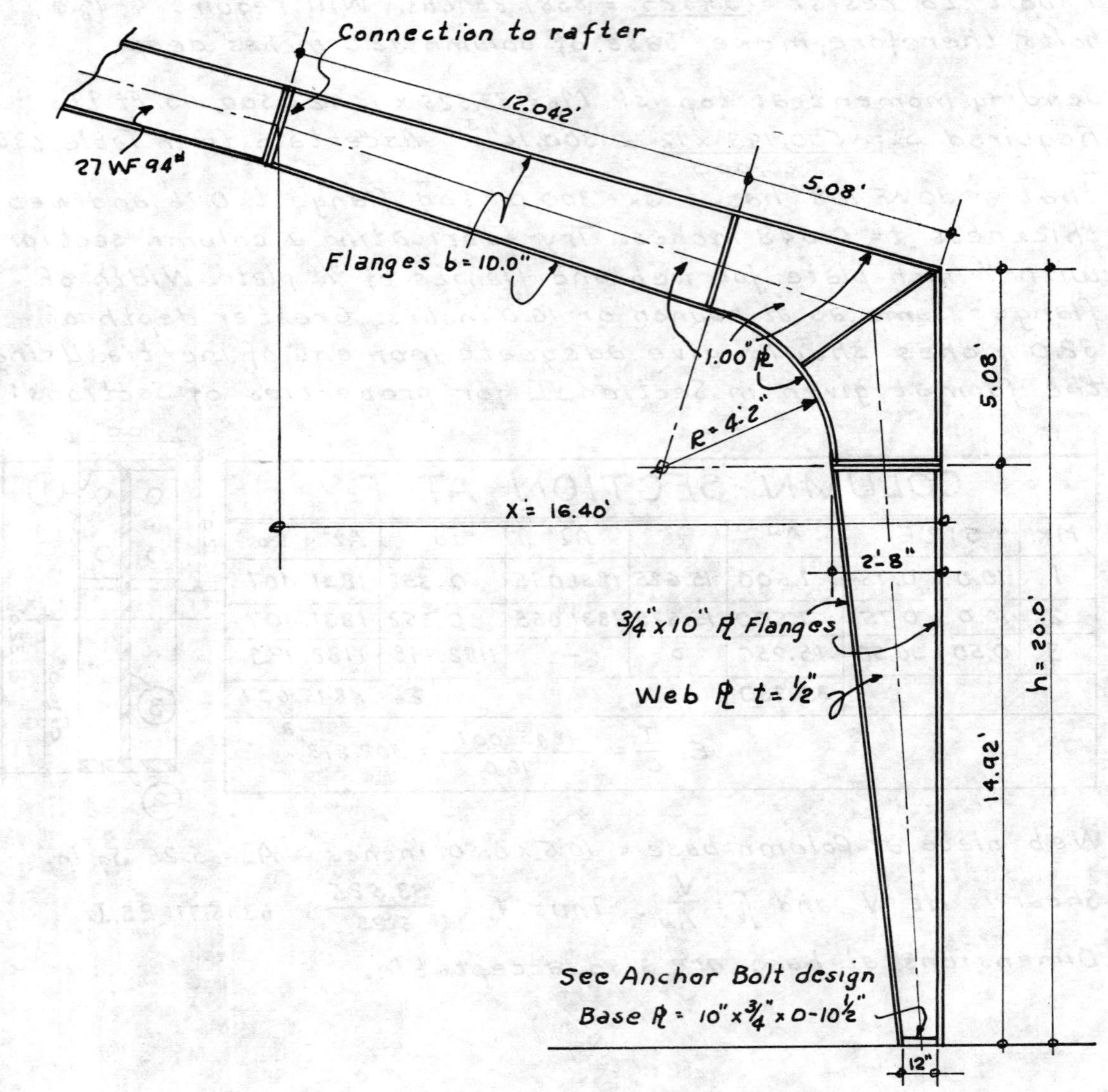

STEP XXVIII:

Wall Girts and Roof Purlin design:

The design data calls for wall wind load pressure of 30 PSF and Roof load of 50 PSF. Assuming that roof and wall siding will be of 26 Gauge metal ribbed sheets, light gauge girts of A242 type steel can be used for both girts and purlins. Select a Z Section 823.7 listed in table 7.7.4 with these properties: Flange width with lip $b = 3.00$ inches. $S_x = 2.50''^3$. $F_y = 50{,}000$ PSI and max $F_b = 30{,}000$ PSI. When girts are placed inside columns abutting web of Column, the moment, $M = \frac{WL}{8}$ and $L = 20.0$ feet. Then $l = 20.0 \times 12 = 240$ inches. The ratio of $\frac{l}{b}$ is therefore $\frac{240}{3.0} = 80$ and too high to permit use of full allowable $F_b = 30{,}000$ PSI when maximum ratio is limited to 15 similar to table 2.4.4.1. Lateral support for girts is necessary and by reverting to a similar stress reducing formula to use for the higher strength steels, modify the formula thus:

$$F_b = \frac{30{,}000}{1.00 + \left(\frac{l^2}{3000\,b^2}\right)}$$

Use sag rods for this lateral support and with 2 Sag rods per bay l becomes $\frac{l}{3}$ or $l = \frac{240.0}{3} = 80.0$ inches.

$$\text{Max. } F_b = \frac{30{,}000}{1.00 + \left(\frac{80.0^2}{3000 \times 3.00^2}\right)} = 24{,}252 \text{ PSI.}$$

Determine the Resisting Moment of 1 Girt thus: $RM = S_x F_b$ or $RM = \frac{2.50 \times 24{,}252}{12} = 5052.50$ Ft. Lbs.

Convert RM to total load: $W = \frac{8M}{L}$ and $W = \frac{8 \times 5052.50}{20.0} = 2021$ Lbs.

A strip 1.0' wide has a load as: $W = 30 \times 20.0 = 600$ Lbs. Spacing of Girts: $s = \frac{2021}{600} = 3.368$ Feet. Max. spacing is therefore 3'-4⅜" when 2 sag rods are used to divide span into 3 equal spaces.

STEP XXIX:

Designing purlins to rest on top of rafter, the end span $M = \frac{WL}{10}$ and from RM load becomes $W = \frac{10 \times 5052.50}{20.0} = 2526.25$ Lbs.

Strip load $W = 50 \times 20.0 = 1000$ Lbs. Spacing of purlins with roof load greater than wind load, $s = \frac{2526.25}{1000} = 2.526$ Ft. Max. (2'-6 5/16").

DESIGN NOTE:

With 823.70 Sections used for girts and purlins, 2 sag rods must be used in each bent. When windows or skylites are used the sag rod should be replaced with a suitable section used for framing around sash or similar openings.

EXAMPLE I: Complete design of rigid frame, continued 7.5.3

STEP XXX:

Tie rods connecting column bases:

Referring to Moment diagram for Roof loads in Step XVII the vertical forces tend to spread the columns and must be tied securely together for the life of structure. Placing tie rods under slab may result in corrosion and be weakened by rust. Use sufficient rod material and coat surfaces to retard decay.

Standard steel bolt material has F_t = 15000 PSI and H_E = 33,525 Lbs.

Area required = $\frac{33{,}525}{15{,}000}$ = 2.235 Sq. In. Use 3 - 1.00"ϕ Rod stud bolts which will give an A = 2.356 Sq. In. Do not rely on anchors to replace ties.

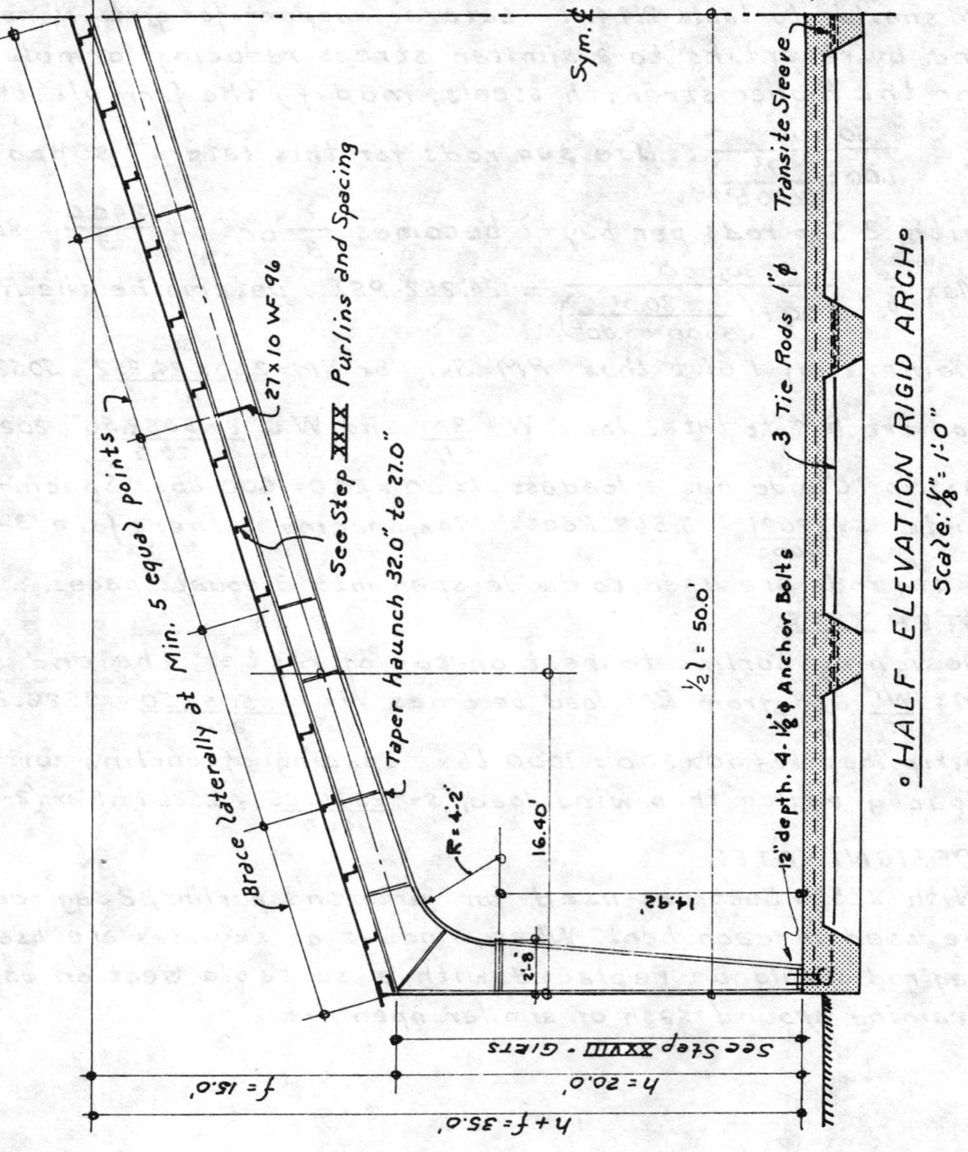

EXAMPLE II: Complete design of rigid frame 7.5.4

Available Data:

Span out to out of Columns, l = 160.0 Feet.

Spacing of bents = 20.0 Ft. on Centers. Number Bents = 15

Combined Dead and Live Roof Loads = 35 Lbs. Sq. Foot.

Wind Load Full Height = 125 Miles Per Hour.

Roof Pitch to be 4.0 Inches per foot

Eave height to top of Column = 14.0 Feet.

All structural Steel to be A36. Use AISC specifications.

Max. Allowable, F_b = 22,000 PSI.

For wind pressure use Formula: $P = 0.00256 V^2$

Identify this project as: AFE-66-28

REQUIRED:

(a) Calculate Horizontal and Vertical Load Reactions separately.

(b) Draw Rafter Moment Diagram by calculating moment on intervals of 10.0 feet. Also determine wind moments on right side of ℄.

(c) Combine and tabulate moments to determine critical points.

(d) Draw a half elevation moment diagram with design moments.

(e) Calculate Axial stress in rafter and shear forces combined.

(f) Draw an elevation of the probable knee section. Limit flange width to 10 inches and check Profile tables to determine the preliminary dimension for depth.

(g) Do not design rafter or column, but calculate the value of the moment of inertia for the critical points at knee.

STEP I:

Draw a scale elevation of arch to work from. Show loads and dimensions. After calculating reactions, note the values on sketch.

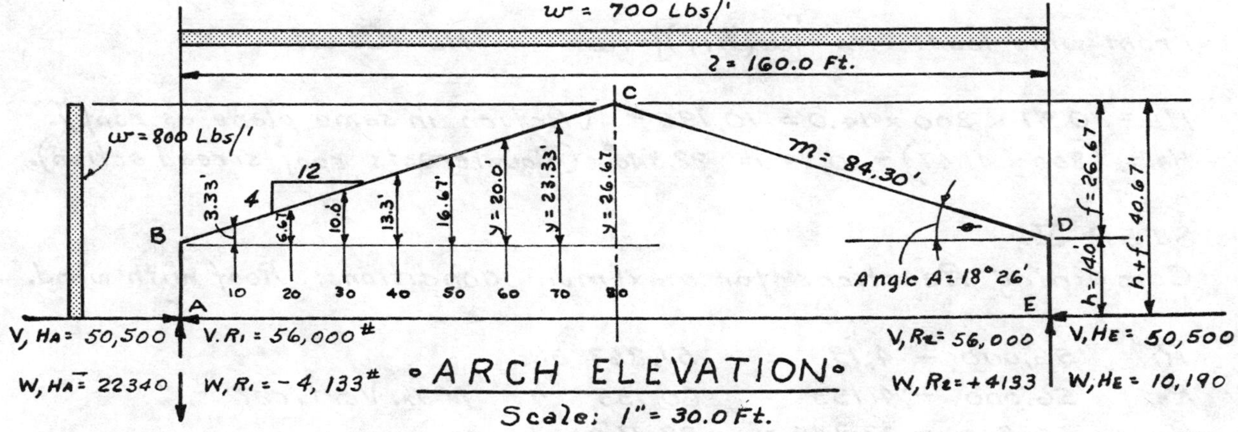

ARCH ELEVATION

Scale: 1" = 30.0 Ft.

EXAMPLE II: Complete design of rigid frame, continued 7.5.4

STEP II

Solve for dimensions and ratios:

Angle of slope: $\text{Tan } A = \dfrac{26.67}{80.0} = 0.33337$ Angle $A = 18° 26'$

Sine $A = 0.31620$ Cos. $A = 0.94869$ Sec. $A = 1.0541$

$m = b \text{ Sec.} A = 1.0541 \times 80.0 = 84.328$ Ft. Call $m = 84.30$ Ft.

From Pilot Diagrams:

$Q = \dfrac{f}{h}$ $Q = \dfrac{26.67}{14.0} = 1.90$ $k = \dfrac{h}{m}$ $k = \dfrac{14.0}{84.30} = 0.166$

Use Chart I, $C_1 = 0.043$

Use Chart II, $C_6 = 0.91$

STEP III:

Reactions from Roof Load. $W = 700 \times 160.0 = 112{,}000$ Lbs.

$R_1 = R_2$ and $R = \dfrac{112{,}000}{2} = 56{,}000$ Lbs.

Wind Load Reactions acting on vertical plane at column bases.

$W = 800 \times 40.67 = 32{,}530\#$ Wind $R = \dfrac{\omega(h+f)^2}{2l}$. Values in formula,

$R_1 = \dfrac{800 \times 40.67^2}{2 \times 160.0} = -4{,}133$ Lbs. $R_2 = +4{,}133$ Lbs.

With combined Loads: $R_1 = 56{,}000 - 4133 = 51{,}867$ Lbs

$R_2 = 56{,}000 + 4133 = 60{,}133$ "

STEP IV:

Calculating Horizontal Reactions:

From roof load: $H_A = H_E$ $H = \dfrac{C_1 \omega l^2}{h}$ $H = \dfrac{0.043 \times 700 \times 160.0^2}{14.0} = 50{,}500$ Lbs.

From wind load: $H_A = [\omega(h+f)] - H_E$. $H_E = C_6 \omega h$

$H_E = 0.91 \times 800 \times 14.0 = 10{,}190\#$ (Action in same plane as roof).

$H_A = (800 \times 41.67) - 10{,}190 = 22{,}340\#$ (Counteracts roof spread action).

STEP V:

Combining Reactions for maximum conditions: Roof with wind.

$R_1 = 56{,}000 - 4{,}133 = 51{,}867$ Lbs.

$R_2 = 56{,}000 + 4{,}133 = 60{,}133$ " Max. Vertical

$H_A = 50{,}500 - 22{,}340 = 28{,}160$ "

$H_E = 50{,}500 + 10{,}190 = 60{,}690$ " Max. Horizontal

STEP VI:
Computing Bending Moments at critical points.
Roof Load at Ridge C. $M_C = \left(\frac{R_1 l}{4}\right) - \left[H_E(h+f)\right]$

$$M_C = \left(\frac{56,000 \times 160.0}{4}\right) - \left(50,500 \times 40.67\right) = +186,333 \text{ Ft. Lbs.}$$

Roof Load at Points B and D. $M = H_E h$

$$M_B = 50,500 \times 14.0 = -707,000 \text{ Ft. Lbs.}$$

STEP VII
Same as step VI, but with wind load.
$M_C = \frac{R_1 l}{2} - \left[H_E(h+f)\right]$ and at D. $M_D = H_E h$.

$$M_C = \left(\frac{4,133 \times 160.0}{2}\right) - \left(10,190 \times 40.67\right) = -83,755 \text{ Ft. Lbs.}$$

$$M_D = 10,190 \times 14.0 = -142,660 \text{ Ft. Lbs.}$$

STEP VIII:
At this point of design, determine if critical points have the greater moment with a single wind moment, or a single roof load moment, or a combination of the two.

At point B and D: Max. $-M_D = 707,000 + 142,600 = -849,660$ Ft. Lbs.
At point C on ℄: Use roof load $+M_C = +186,333$ Ft. Lbs.

$-M$, denotes tension in outside flange, and $+M$, denotes tension in inner flange.

STEP IX:
Computing bending moments at points along rafter to make a moment diagram.
Working from left side: For Vertical Roof Loads, the formula:

$M_x = \frac{\omega x(l-x)}{2} - \left[H(h+y)\right]$ where x = distance from the point B, and y = rise of rafter from horizontal eave line where point of moment x is being taken. $y = \frac{2fx}{l}$ when moment is between point B and ℄.

EXAMPLE II: Complete design of rigid frame, continued 7.5.4

STEP IX CONTINUED.

VERTICAL ROOF LOAD MOMENTS

$M_B = Hah.$ $M_{0.0} = 50{,}500 \times 14.0 = -707{,}000$ Ft. Lbs

$$M_{10.0'} = \left[\frac{(700 \times 10.0) \times (160.0-10.0)}{2}\right] - \left[50{,}500(14.0+3.33)\right] = -350{,}333 \text{ Ft. Lbs.}$$

$$M_{20.0} = \left[\frac{(700 \times 20.0) \times (160.0-20.0)}{2}\right] - \left[50{,}500(14.0+6.66)\right] = -63{,}666 \text{ '\#}$$

$$M_{30.0} = \left[\frac{(700 \times 30.0) \times (160.0-30.0)}{2}\right] - \left[50{,}500(14.0+10.0)\right] = +153{,}000 \text{ '\#}$$

$$M_{35.0} = \left[\frac{(700 \times 35.0) \times (160.0-35.0)}{2}\right] - \left[50{,}500(14.0+11.67)\right] = +235{,}085 \text{ '\#}$$

$$M_{40.0} = \left[\frac{(700 \times 40.0) \times (160.0-40.0)}{2}\right] - \left[50{,}500(14.0+13.33)\right] = +299{,}667 \text{ '\#}$$

$$M_{50.0} = \left[\frac{(700 \times 50.0) \times (160.0-50.0)}{2}\right] - \left[50{,}500(14.0+16.67)\right] = +376{,}333 \text{ '\#}$$

$$M_{55.0} = \left[\frac{(700 \times 55.0) \times (160.0-55.0)}{2}\right] - \left[50{,}500(14.0+18.33)\right] = +388{,}420 \text{ '\#}$$

$$M_{60.0} = \left[\frac{(700 \times 60.0) \times (160.0-60.0)}{2}\right] - \left[50{,}500(14.0+20.0)\right] = +383{,}000 \text{ '\#}$$

$$M_{65.0} = \left[\frac{(700 \times 65.0) \times (160.0-65.0)}{2}\right] - \left[50{,}500(14.0+21.67)\right] = +345{,}833 \text{ '\#}$$

$$M_{70.0} = \left[\frac{(700 \times 70.0) \times (160.0-70.0)}{2}\right] - \left[50{,}500(14.0+23.33)\right] = +319{,}667 \text{ '\#}$$

$$M_{75.0} = \left[\frac{(700 \times 75.0) \times (160.0-75.0)}{2}\right] - \left[50{,}500(14.0+25.0)\right] = +261{,}750 \text{ '\#}$$

$$M_{80.0} = \left[\frac{(700 \times 80.0) \times (160.0-80.0)}{2}\right] - \left[50{,}500(14.0+26.67)\right] = +186{,}333 \text{ '\#}$$

STEP X

BENDING MOMENTS ON RAFTER LEFT OF ℄

WIND LOADS MOMENTS

Formula: $M = \left[H_A(h+y)\right] - \left[(R_1X) + \dfrac{w(h+y)^2}{2}\right]$ $y = X \tan A$

$M_{0.0'}$ Same as Moment at B $M_B = 22{,}340 \times 14.0 = +312{,}760'\#$

Pos. + NEG. −

$M_{10.0} = \left[22{,}340\ (14.0+3.33)\right] - \left[(4133 \times 10.0) + 800\ \dfrac{(14.0+3.33)^2}{2}\right] = +225{,}765'\#$

$M_{20.0} = \left[22{,}340\ (14.0+6.66\right] - \left[(4133 \times 20.0) + 800\ \dfrac{(14.0+6.66)^2}{2}\right] = +208{,}295'\#$

$M_{30.0} = \left[22{,}340\ (14.0+10.00\right] - \left[(4133 \times 30.0) + 800\ \dfrac{(14.0+10.00)^2}{2}\right] = +181{,}770'\#$

$M_{35.0} = \left[22{,}340\ (14.0+11.67\right] - \left[(4133 \times 35.0) + 800\ \dfrac{(14.0+11.67)^2}{2}\right] = +166{,}070'\#$

$M_{40.0} = \left[22{,}340\ (14.0+13.33\right] - \left[(4133 \times 40.0) + 800\ \dfrac{(14.0+13.33)^2}{2}\right] = +147{,}172'\#$

$M_{50.0} = \left[22{,}340\ (14.0+16.67\right] - \left[(4133 \times 50.0) + 800\ \dfrac{(14.0+16.67)^2}{2}\right] = +102{,}200'\#$

$M_{55.0} = \left[22{,}340\ (14.0+18.33\right] - \left[(4133 \times 55.0) + 800\ \dfrac{(14.0+18.33)^2}{2}\right] = +78{,}700'\#$

$M_{60.0} = \left[22{,}340\ (14.0+20.00\right] - \left[(4133 \times 60.0) + 800\ \dfrac{(14.0+20.00)^2}{2}\right] = +49{,}180'\#$

$M_{65.0} = \left[22{,}340\ (14.0+21.67\right] - \left[(4133 \times 65.0) + 800\ \dfrac{(14.0+21.67)^2}{2}\right] = +19{,}495'\#$

$M_{70.0} = \left[22{,}340\ (14.0+23.33\right] - \left[(4133 \times 70.0) + 800\ \dfrac{(14.0+23.33)^2}{2}\right] = -12{,}680'\#$

$M_{75.0} = \left[22{,}340\ (14.0+25.00\right] - \left[(4133 \times 75.0) + 800\ \dfrac{(14.0+25.00)^2}{2}\right] = -47{,}115'\#$

$M_{80.0} = \left[22{,}340\ (14.0+26.67\right] - \left[(4133 \times 80.0) + 800\ \dfrac{(14.0+26.67)^2}{2}\right] = -83{,}755'\#$

Continue with wind load moments right of ℄

EXAMPLE II: Complete design of rigid frame, continued 7.5.4

STEP X CONT'D.

BENDING MOMENTS ON RAFTER - RIGHT SIDE OF ℄

WIND LOAD MOMENTS

From Step III $R_1 = 4{,}133^{\#}$ From Step IV $H_E = 10{,}190\#$

$M_x = \left[R_1(l-x)\right] - \left[H_E(h+y)\right]$ = Formula when Dimension "X" is Right ℄

$$M_{80.0'} = \left[4{,}133\,(160.0-80.0)\right] - \left[10{,}190\,(14.0+26.67)\right] = -83{,}755'\#$$

$$M_{85.0'} = \left[4{,}133\,(160.0-85.0)\right] - \left[10{,}190\,(14.0+25.00)\right] = -87{,}410'\#$$

$$M_{90.0'} = \left[4{,}133\,(160.0-90.0)\right] - \left[10{,}190\,(14.0+23.33)\right] = -91{,}100'\#$$

$$M_{95.0'} = \left[4{,}133\,(160.0-95.0)\right] - \left[10{,}190\,(14.0+21.67)\right] = -92{,}766'\#$$

$$M_{100.0'} = \left[4{,}133\,(160.0-100.0)\right] - \left[10{,}190\,(14.0+20.00)\right] = -98{,}460'\#$$

$$M_{105.0'} = \left[4{,}133\,(160.0-105.0)\right] - \left[10{,}190\,(14.0+18.33)\right] = -102{,}144'\#$$

$$M_{110.0'} = \left[4{,}133\,(160.0-110.0)\right] - \left[10{,}190\,(14.0+16.67)\right] = -103{,}827'\#$$

$$M_{120.0'} = \left[4{,}133\,(160.0-120.0)\right] - \left[10{,}190\,(14.0+13.33)\right] = -113{,}194'\#$$

$$M_{125.0'} = \left[4{,}133\,(160.0-125.0)\right] - \left[10{,}190\,(14.0+11.67)\right] = -114{,}877'\#$$

$$M_{130.0'} = \left[4{,}133\,(160.0-130.0)\right] - \left[10{,}190\,(14.0+10.00)\right] = -120{,}560'\#$$

$$M_{140.0'} = \left[4{,}133\,(160.0-140.0)\right] - \left[10{,}190\,(14.0+6.66)\right] = -125{,}927'\#$$

$$M_{150.0'} = \left[4{,}133\,(160.0-150.0)\right] - \left[10{,}190\,(14.0+3.33)\right] = -135{,}295'\#$$

$$M_{160.0'} = \left[4.133\,(160.0-160.0)\right] - \left[10{,}190\,(14.0+0.00)\right] = -142{,}666'\#$$

STEP XI:

Although not always a requirement, it is most desirable to make a moment diagram on the rafters for each type of loading. Comparison may be better obtained for determining the critical points to be used as maximum design points.

WIND AND ROOF LOAD MOMENT DIAGRAMS:

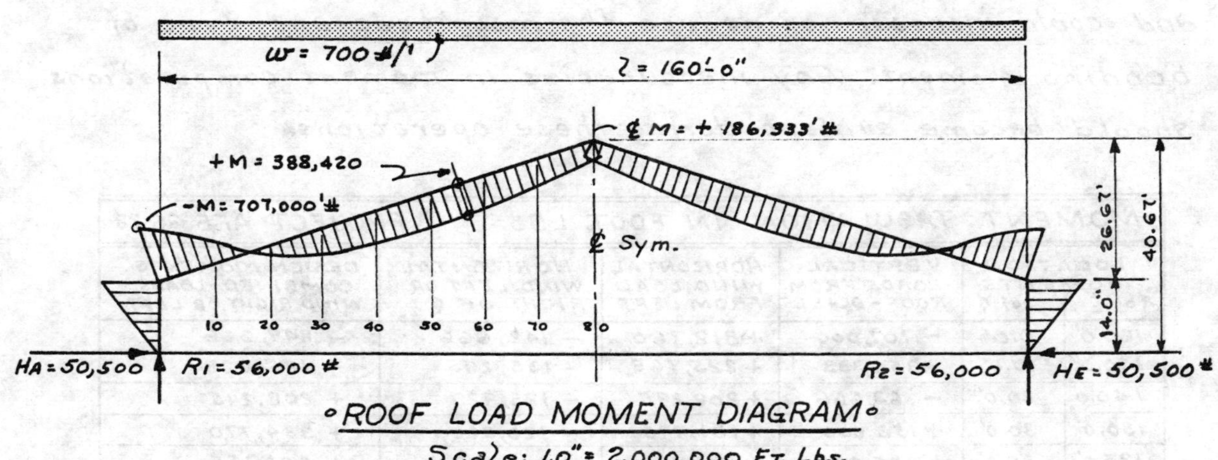

∘ ROOF LOAD MOMENT DIAGRAM ∘

Scale: 1.0" = 2,000,000 Ft. Lbs.

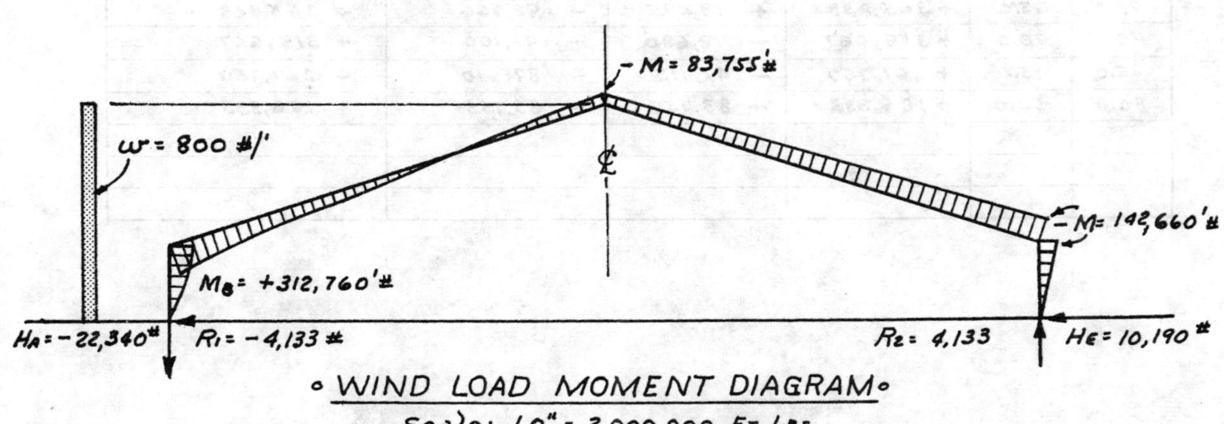

∘ WIND LOAD MOMENT DIAGRAM ∘

Scale: 1.0" = 2,000,000 Ft. Lbs.

EXAMPLE II: Complete design of rigid frame, continued 7.5.4

STEP XII:

By using the results obtained in steps IX and X an accurate method should be used for final tabulation and to determine the critical design moment. Check with the diagrams to ascertain whether moment is positive or negative. Consider the fact that the wind direction is also from the right side and could increase or reduce the magnitude and type of bending moment. Any inaccuracies in moment computations should become evident during these operations.

MOMENT TABULATION IN FOOT LBS.					PROJECT AFE 66-28
LOCATION MOMENTS R of ℄	L of ℄	VERTICAL LOADS FROM ROOF - DL+LL	HORIZONTAL WIND LOAD FROM LEFT	HORIZONTAL WIND LEFT OR RIGHT OF ℄	DESIGN MOMENTS COMBINED LOADS WIND RIGHT OR LEFT
160.0'	0.0'	− 707,000	+ 312,760	− 142,666	− 849,666
150.0'	10.0'	− 350,333	+ 225,765	− 135,295	− 485,628
140.0'	20.0'	− 63,666	+ 208,295	− 125,927	+ 208,295
130.0'	30.0'	+ 153,000	+ 181,770	− 120,560	+ 334,770
125.0'	35.0'	+ 235,085	+ 166,070	− 114,877	+ 401,155
120.0'	40.0'	+ 299,667	+ 147,172	− 113,195	+ 446,838
110.0'	50.0'	+ 376,333	+ 102,200	− 103,827	+ 478,533
105.0'	55.0'	+ 388,420	+ 78,700	− 102,144	+ 467,120
100.0'	60.0'	+ 383,000	+ 49,180	− 98,460	+ 432,180
95.0'	65.0'	+ 345,833	+ 19,495	− 92,766	+ 365,328
90.0'	70.0'	+ 319,667	− 12,680	− 91,100	+ 319,667
85.0'	75.0'	+ 261,750	− 47,115	− 87,410	+ 261,750
80.0'	80.0	+ 186,333	− 83,755	− 83,755	+ 186,333

STEP XIII:

Using the moment values in last table column in the tabulation, construct a final moment diagram by pointing off the locations on rafter and use the design moment. Only half of arch frame need be shown, as conditions will be symmetrical when wind direction comes from right. This diagram meets requirement (d) which is necessary to file with Code authorities. On this diagram drawing, include all information on loads and reactions. Check the inflection point on rafter, which should be approximately the location of the splice for knee to rafter. Greatest depth of rafter will be same as ordinate on points 50.0 and 110.0 feet from outside left column.

EXAMPLE II: Complete design of rigid frame, continued 7.5.4

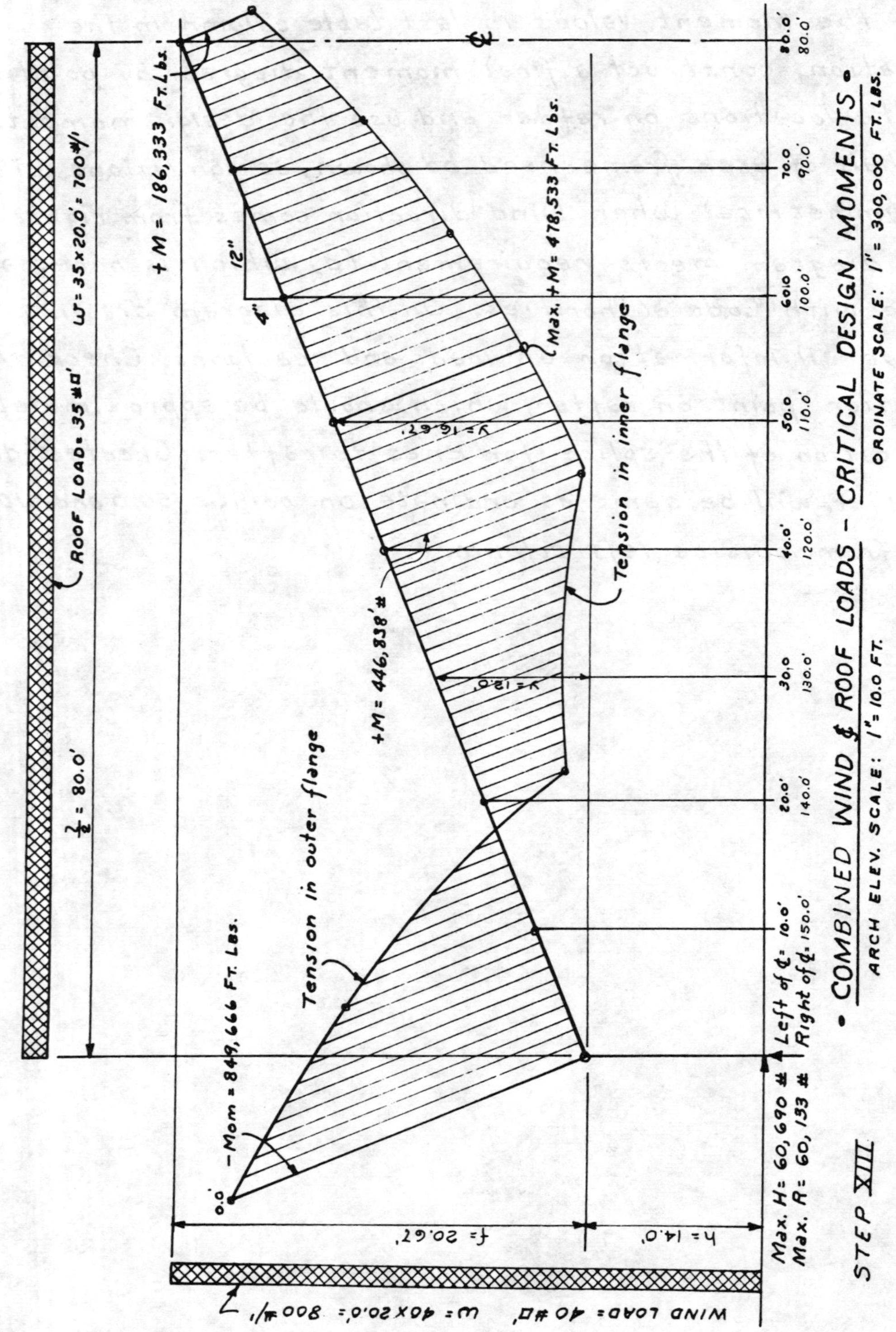

STEP XIV:

Axial forces in rafter and knee at points C and D. Functions of pitch angle A of 18° 26'. From Trig. Tables: Tan. A = 0.333 Sine A = 0.3162 Cos. A = 0.9487

Max. H = 60,690 Lbs. Max. R = 60,133 Lbs. Written a formula for axial force: $P_a = (H \text{ Cos.} A) + (R \text{ sine } A.)$. substituting values in formula:

$$P_a = (60{,}690 \times 0.9487) + (60{,}133 \times 0.3162) = 76{,}590 \text{ Lbs.}$$

Since rafter length was computed in step II, and is 84.30 feet in length, the slenderness ratio of $\frac{l}{r}$ will govern the spacing for lateral bracing.

STEP XV:

Depth of knee section.
By screening the arch profile tables, a fair estimate may be obtained for most economical depth to use for a trial investigation. Assume that depth will be 44.0 inches with 10.0 inch flange widths. Dimension c = 22.0 inches.

Probable Section Modulus at knee: $S = \frac{M}{F_b}$. when $F_b = 22{,}000$ psi.
Moment at knee = −849,666 Ft. Lbs.

Required $S = \frac{849{,}666 \times 12}{22{,}000} = 464.0''^3$. A section must be built up to have a moment of Inertia of: $464.0 \times 22.0 = 10{,}208.''^4$

STEP XVI:

To check the assumed 44.0 inch depth of knee section before drawing an elevation of column with a portion of rafter, compute the value of I_o. Assume 10.0" x 0.75" for flanges, and web plate of 42.5" x 0.75".

Draw cross section and use convential form system as guide.

EXAMPLE II: Complete design of rigid frame, continued 7.5.4

KNEE SECTION PROPERTIES - AXIS X-X

M'K	SECTION	A□"	l	Al²	Io	Al² + Io
1	10.0 x 0.75	7.50	21.625	3547.20	0.352	3547.552
2	0.625 x 42.5	25.31	0	—	3460.000	3460.000
3	10.0 x 0.75	7.50	21.625	3547.20	0.352	3547.552
	A = 40.31 □"					10,555.104

$S_x = \frac{10,555}{21.625} = 487.0''^3$ $r_x = \sqrt{\frac{10,555}{40.31}} = 51.3''$

KNEE SECTION PROPERTIES - AXIS Y-Y

M'K	SECTION	A□"	l	Al²	Io	Al² + Io
1	0.75 x 10.0	7.50	0	—	62.50	62.50
2	42.5 x 0.625	25.31	0	—	0.82	0.82
3	0.75 x 10.0	7.50	0	—	62.50	62.50
						125.82"⁴

$r_y = \sqrt{\frac{125.82}{40.31}} = 1.77''$

The value of I_x, is unusually close to the 10,208 required and adjustments may be made, however values should be on the side of safety when allowable bending stress is to be determined by Rule II.

STEP XVII:

The axial force of 76,590 pounds computed in step XIV will be allowed only when the lateral bracing meets the requirement of Rule II, paragraph (a) and (b).

The ratio of $\frac{ld}{bt}$ must be limited to 600 or less. Then by transposing formula, the maximum length $l = \frac{600\,bt}{d}$

Max. $l = \frac{600 \times 10.0 \times 0.75}{44.0} = 102.25$ inches. (8'-6½ spacing).

For allowable axial compressive stress use: $F_a = 17000 - \left(.485 \frac{l^2}{r^2}\right)$.

STEP XVIII:

Drawing elevation of Column and Knee.

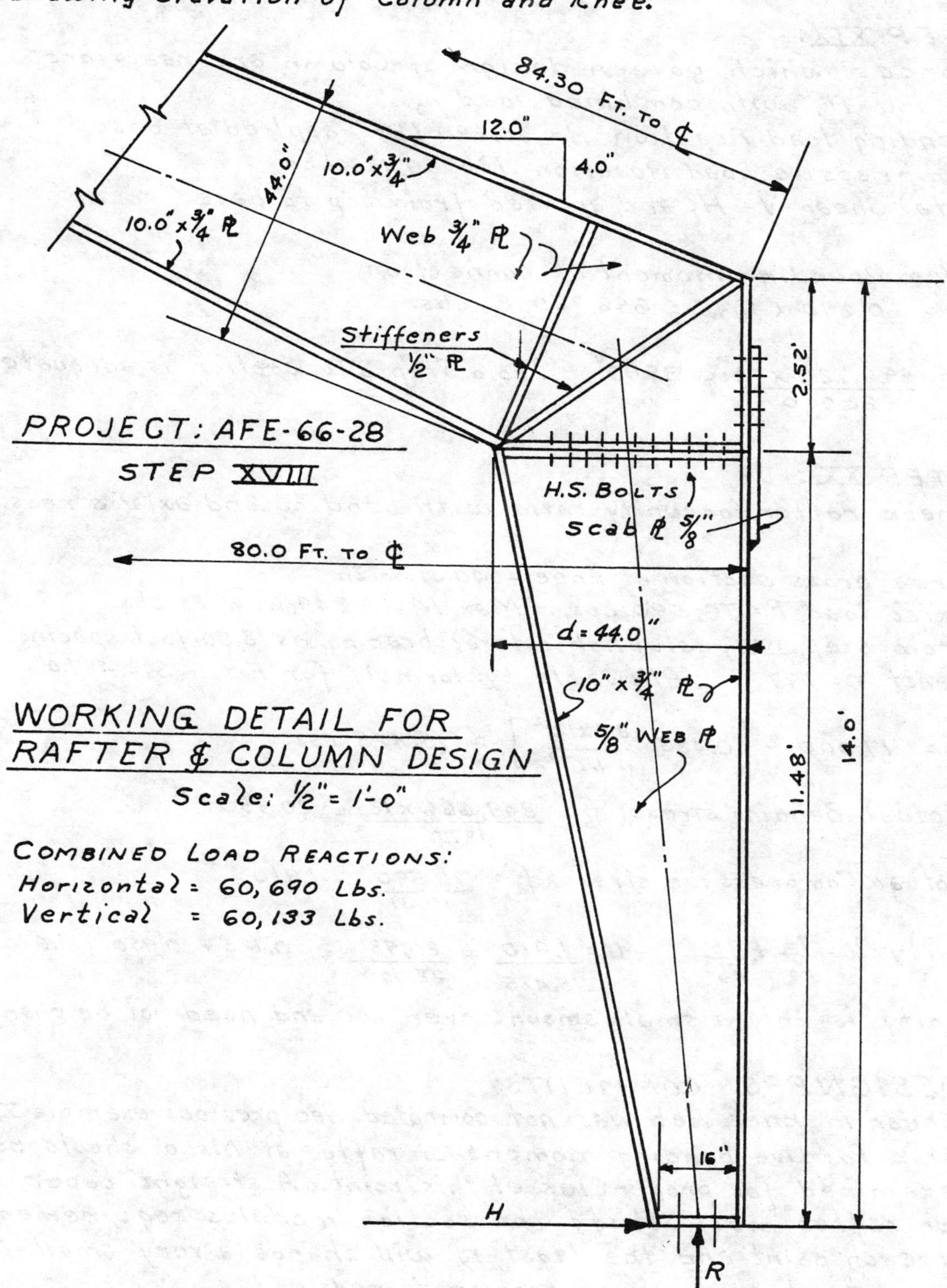

EXAMPLE II: Complete design of rigid frame, continued 7.5.4

STEP XIX:

Forces which govern design of Column are reactions R and H, with combined loading.
Bending load Reaction H = 60,690 Lbs., applied at base.
Compressive Load Reaction R = 60,133 Lbs.
Total Shear V = H, and applied from top to base.

Column bending moment at connection:
M = 60,690 x 11.48 = 696,720 Ft. Lbs.

$$S = \frac{696,720 \times 12}{22,000} = 381.0''^3$$ From Step XVI, section is adequate.

STEP XX:

Check rafter for unity ratio with bending and axial stress.

Area cross section at knee = 40.31 Sq. In.
Axial load P = 76,590 Lbs. Max. M = − 849,666 Ft. Lbs.
From step XVII, establish lateral bracing as 8.50 foot spacing
Least r = 1.77 Allowable by formula for Fa. $S_x = 487.0''^3$

$$F_a = 17,000 - \left[0.485 \frac{(8.5 \times 12)^2}{1.77^2}\right] = 15,425 \text{ PSI.}$$

Actual Bending stress: $f_b = \frac{849,666 \times 12}{487.0} = 20,930 \#\square''$

Actual Compressive stress: $f_a = \frac{76,590}{40.31} = 1,910 \#\square''$

Unity $u = \frac{f_a}{F_a} + \frac{f_b}{F_b}$ $u = \frac{1,910}{15,425} + \frac{20,930}{22,000} = 0.125 + 0.950 = 1.075$

Unity is only a small amount over 1.00 and need not be changed.

DESIGNERS COMMENTS:

Shear in knee web was not computed. See previous example XXI. Max. Positive bending moment on rafter at M50.0' should be examined for properties at this point. A straight taper for rafter must satisfy the section modulus requirements at each point and the least r_y will change a very small amount when depth of section is reduced.

Laminated glued wood 7.6.1

Glued laminated rigid arches are often selected by architects for structures such as churches and field houses. The exposed wood rafters and columns, combined with masonry walls, lend an air of sanctity to the nave and better the acoustics. Stained wood radiates a warm feeling, not associated with steel.

The segment arch with buttressed supports has been used with spans of 250 feet; the parabolic arch can be employed to give height to an interior vault. Either of these two geometric profiles is ideal from an engineering viewpoint, because a uniformly distributed load across the span induces mainly longitudinal and compressive stresses.

Straight beams and girders of greater length than the standard lengths of solid sawn timbers are available, which can be designed for better appearance and higher unit allowable stresses.

Wood species most common in fabricating glued members are Southern Yellow Pine and West Coast Douglas Fir. Both contain the long fibers necessary for strength and have the ability to absorb the glue and bond the laminations together in a safe unit.

GRADES

The American Institute of Timber Construction (AITC) and Inspection Bureaus adopted a recommendation on October 26, 1961, that the grading for laminated timber be as follows:

(a) Industrial appearance with sound knots etc., for painting.

(b) Architectural appearance for staining or paint.

(c) Premium appearance, minimum imperfections for staining.

The three grades apply to the surface of the wood and the growth characteristics, inserts, wood fillers and planing or sanding scars may be included. The surface appearances do not change or modify the design stresses nor change the method of fabrication.

GLUE ADHESIVES

Laminated wood members which are to be used for dry areas, or interior use, may be fabricated with a casein glue which has some degree of moisture resistance and is identified as: Federal Specification, MMM–A125 II. The wood should not be subjected to a humidity reading which would cause the wood to absorb moisture to exceed 16% over prolonged periods of service. Laminated wood members exposed to weather or locations where the moisture content may exceed 16 percent for long durations, such as covers for swimming pools or exposed column legs of arches should be fabricated with a waterproof phenol rescorcinol or melamine resin glue which will meet Federal Specification MIL–A–397 B.

It is beyond the scope of this work to elaborate on the requirements of satisfactory glued laminated members. Designers and specification writers should provide their offices with a copy of the SPIB Glued Lumber Standards for Southern Pine. This sixteen page booklet was adopted on January 1, 1965, and may be obtained by writing the Southern Pine Inspection Bureau at New Orleans, Louisiana.

ERECTION METHODS

Careful consideration must be given to the conditions under which glued laminated arches are to be transported from the fabricating plant to job site. Transportation from plant to site is made difficult due to large sizes. Fabricators should carefully wrap their products with water repellent and padded coverings for protection during shipment. These covers should not be re-

Laminated glued wood, continued 7.6.1

moved until after unloading and erection. All slings for lifting should be of rope; steel slings will cause damage to finish. Heavy flat belting is recommended for slings when steel rope lines are employed with cranes. After placement, the edges of the wood can be protected by nailing flat boards near the edges. These stipulations for handling should be the concern of the architect and written into the specifications for erectors to follow.

Laminated arch design stress 7.6.2

Laminations used in glued members must be kiln dried, before gluing, to a lower moisture content than is normal for commercial drying of regular lumber yard dimension grades. The wood strength is materially increased by this drying, and higher allowable unit stresses are possible. The many combinations and circumstances in which the members are used will alter the stresses to the extent that a table which listed all the variables would be too complicated to be useful. Using the members under dry conditions, the following design stresses are considered safe and allowable for glued laminated sections when the loading is perpendicular to the wide face of the laminations.

(a) Extreme fiber in bending $F_b = 2400$ PSI
(b) Tension, parallel to grain, $F_t = 1800$ "
(c) Compression, parallel to grain, $F_c = 1400$ "
(d) Horizontal shear, $F_v = 195$ "
(e) Compression, Perpendicular to grain, $\perp F_c = 400$ "
(f) Modulus of Elasticity E = 1,820,000

FOR WET CONDITION OF USE

(a) Extreme fiber in bending, 80% or $F_b = 1920$ PSI
(b) Tension, parallel to grain, 80% or $F_t = 1400$ "
(c) Compression, parallel to grain, 73% or $F_c = 1025$ "
(d) Horizontal shear, 89% or $F_v = 185$ "
(e) Compression, perpendicular to grain, 67% or $\perp F_c = 270$ "
(f) Modulus of Elasticity, 83% or E = 1,510,000

Laminated arch design

Fabricators of glued laminated arches use design methods which are in some instances similar to the Steel method as proposed by the AISC. The basic formula for determining the horizontal spreading force from roof loads is written: Ha = He or $H = \frac{wl^2}{8(h+f)}$. For computing the reactions from wind loads, several conditions need to be considered. The wind pressure could possibly be against only the roof slope with the columns protected by masonry walls. The methods used for calculating wind load forces are characterised by a variety of constants of uncertain origin. A graphic force diagram is used by a number of fabricators to compute the bending moment at the knee. In each case, the several methods appear to be conservative and on the side of safety.

Comparing the calculations for steel and wood arches, it will be seen that the value derived for horizontal thrust from roof loads is 30 or 35 percent greater for the wood system. Investigations for wind loads causing horizontal reactions are likewise more conservative for the wood system of design. Since the static moments produced by the roof loads are symmetrical about the center line of the arch, the formula for computing He was derived as summations of $\frac{My\Delta}{I}$, and $\frac{y^2\Delta}{I}$. These values may be computed on the basis of using half of the arch frame. For static moments produced by the horizontal wind loads, the full reaction is placed only on the windward side at base of column. The lee side of the arch frame is treated as if disconnected and free to move or deflect horizontally. The static moments resulting from wind loads are not symmetrical about center line, and thus the calculations must be based upon the full frame. With respect to point A, the horizontal displacement of point E, may be studied by using the expression of $\sum = \frac{y^2\Delta}{EI}$.

DESIGN PROCEDURE

The design of laminated glued arches is dependent upon the initial calculations for the vertical and horizontal reactions at base of columns. The span length, eave height, roof slope and spacing are selected in advance by the architect. The structural designer then prepares a drawing to scale, and notes all dimensions and load conditions. The reactions are calculated, and the bending moment determined at knee points B and D. Horizontal and vertical reactions are best understood when computed separately. As work progresses through the various design steps, it will be seen that a vertical load moment will have positive bending stress, while, at the same point, the wind load moment will produce a negative bending stress. These two moments will then counterbalance each other on the left side of center. At the same point on the rafter on right side of center, the moments can both become either positive or negative. To insure accuracy and permit the computations to be checked without difficulty, the designer should follow the same steps and format as used in the example.

EXAMPLE: Laminated wood rigid arch 7.6.4

Preliminary plans call for a arch with 80 foot span.
Spacing = 18.0 feet on centers.
Eave height = 20.0 feet. Roof pitch is 4½ inches per foot.
Arch is enclosed inside walls and wind load is neglected.
Dead Loads plus Live Roof Load = 45 Lbs. Square foot.
Radius at knee shown by Architect scales 7.0 feet inside.
Laminations to be of Southern Yellow Pine width of the laminations not to exceed 8 inches.

REQUIRED:
Design of Arch and Columns, with b = 8.0 inches. Use following allowable unit stresses:
F_b = 2200 #□" F_v = 200 #□" F_c = 2000 #□" F_t = 1800 #□"
Make a moment diagram to submit to Code Authorities. Locate moments on rafter at approximate 5.0 foot intervals.

STEP I:
Draw a working sketch of arch elevation with dimensions.
Calculate Vertical Reactions and Horizontal spread forces.

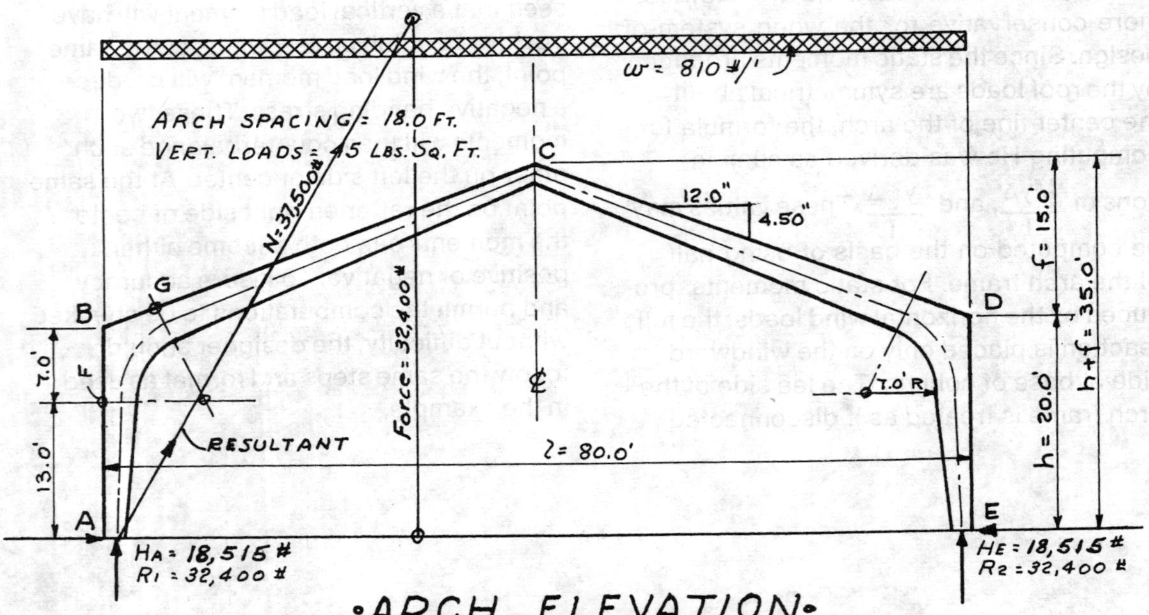

•ARCH ELEVATION•

Scale: 1" = 15.0 Ft.

FORCE DIAGRAM SCALE: 1" = 10,000 Lbs.

Lineal load on rafter, $w = 45 \times 18.0' = 810$ Lbs. Foot

$R_1 = \dfrac{810 \times 80.0}{2} = 32{,}400$ Lbs.

Horizontal Reactions: $H = \dfrac{w\ell^2}{8(h+f)}$ $H = \dfrac{810 \times 6400}{8 \times 35.0} = 18{,}515$ Lbs.

STEP II:

H_A = Shear magnitude at A. Allowable $F_v = 200$ PSI.

$b = 8.00$ inches (Desirable only) $d = \dfrac{3H}{2bF_v}$

Area required at base and depth.

$d = \dfrac{3 \times 18{,}515}{2 \times 8 \times 200} = 17.36$ in. (Call it 17.50 inches)

STEP III:

Calculate for depth at knee, points B and D, also points C and F.

$M_B = H_A h$. $M_B = 18{,}515 \times 20.0' = -370{,}300$ Ft. Lbs.

Moment at Ridge C = $\left(\dfrac{w\ell^2}{8}\right) - \left[H_E(h+f)\right]$

$M_C = \left(\dfrac{810 \times 6400}{8}\right) - (18{,}515 \times 35.0) = -25$ Ft. Lbs. (will have to determine depth here to satisfy architectural appearance.)

Depth required at knee: Allowable $F_b = 2200$ PSI. $S = \dfrac{M}{F_b}$

$S = \dfrac{370{,}300 \times 12}{2200} = 2020.0''^3$ Also: $S = \dfrac{bd^2}{6}$ and $b = 8.0''$

solving for $d = \sqrt{\dfrac{6S}{b}}$ $d = \sqrt{\dfrac{6 \times 2020.0}{8}} = 39.0$ inches.

At point F on column: $M_F = H_A F$.

$M_F = 18{,}515 \times 13.0 = -240{,}695$ Ft. Lbs. Req'd. $S = \dfrac{240{,}695 \times 12}{2200} = 1313.0''^3$

$d^2 = \dfrac{6 \times 1313.0}{8} = 985$, $d = \sqrt{985} = 31.4$ inches.

EXAMPLE: Laminated wood rigid arch, continued 7.6.4

STEP IV

An accurate elevation of arch can now be drawn and rafter moments computed to determine taper. Use the dimensions thus:

Depth at Ridge C = Min. 12 inches.
Depth at Knee B = " 39 "
Depth at Tangent F = " 32 "
Depth at Column A and E = " 17.5 "

STEP V

Compute Rafter moments and plot for moment diagram.

Formula: $M_x = \left[\frac{w x (l-x)}{2}\right] - \left[H_A(h+y)\right]$. $y = 0.375x$.

$$M_{5.0'} = \left[\frac{(810 \times 5.0) \times (80.0-5.0)}{2}\right] - \left[18{,}515\,(20.0 + 1.875)\right] = -253{,}140 \text{ FT. LBS.}$$

$$M_{10.0} = \left[\frac{(810 \times 10.0) \times (80.0-10.0)}{2}\right] - \left[18{,}515\,(20.0 + 3.750)\right] = -156{,}230 \text{ ''}$$

$$M_{15.0} = \left[\frac{(810 \times 15.0) \times (80.0-15.0)}{2}\right] - \left[18{,}515\,(20.0 + 5.625)\right] = -81{,}125 \text{ ''}$$

$$M_{20.0} = \left[\frac{(810 \times 20.0) \times (80.0-20.0)}{2}\right] - \left[18{,}515\,(20.0 + 7.500)\right] = -24{,}000 \text{ ''}$$

$$M_{25.0} = \left[\frac{(810 \times 25.0) \times (80.0-25.0)}{2}\right] - \left[18{,}515\,(20.0 + 9.375)\right] = +13{,}000 \text{ ''}$$

$$M_{30.0} = \left[\frac{(810 \times 30.0) \times (80.0-30.0)}{2}\right] - \left[18{,}515\,(20.0 + 11.250)\right] = +28{,}900 \text{ ''}$$

$$M_{35.0} = \left[\frac{(810 \times 35.0) \times (80.0-35.0)}{2}\right] - \left[18{,}515\,(20.0 + 13.125)\right] = +24{,}875 \text{ ''}$$

$$M_{40.0} = \left[\frac{(810 \times 40.0) \times (80.0-40.0)}{2}\right] - \left[18{,}515\,(20.0 + 15.000)\right] = -25 \text{ ''}$$

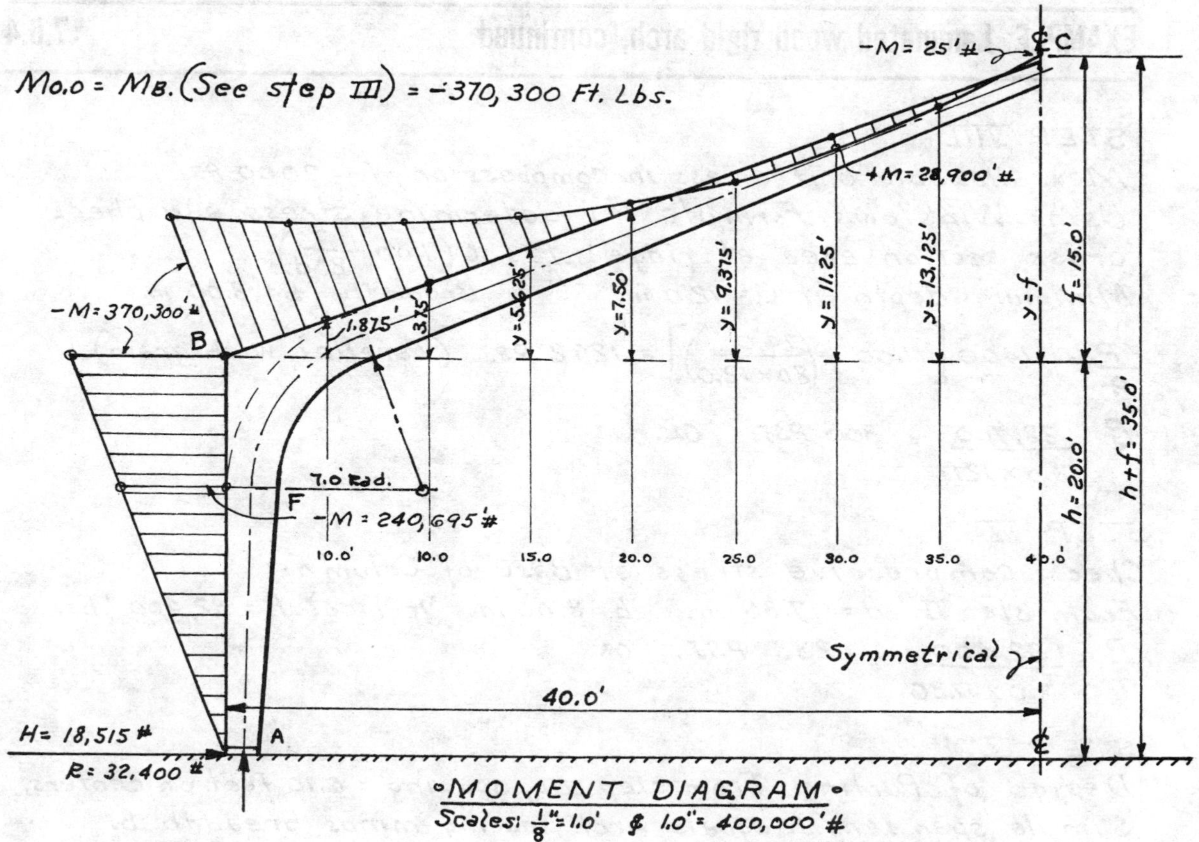

MOMENT DIAGRAM

Scales: $\frac{1}{8}'' = 1.0'$ & $1.0'' = 400{,}000'\#$

STEP VI:

Compute Axial Force in Rafter:

By formula: $P_a = (H_E \text{ Cos. } A) + (R_2 \text{ Sin. } A)$. Find slope angle A.

$\text{Tan. } A = \frac{f}{(l/2)} = \frac{15.0}{40.0} = 0.375$ From Trig Tables, $A = 20° 34'$

Cosine A = 0.93625 Sine A = 0.3513

Point G on rafter is 7.0 foot from B and $y = 7.0 \times 0.375 = 2.625$ Ft.

$P_a = (18{,}515 \times 0.93625) + (32{,}400 \times 0.3513) = 28{,}700$ Lbs.

STEP VII:

Dimension along rafter, points B to C $= \frac{f}{\text{Sin. } A}$

$BC = \frac{15.0}{0.3513} = 42.70$ Feet. Purlins to provide lateral bracing and installed inside rafter, flush with top. Assume purlins are spaced between 6.0 and 7.0 feet on centers. Using 7 spaces equal, purlin centers $= \frac{42.70}{7} = 6.10$ Feet, (call it 74 in.).

EXAMPLE: Laminated wood rigid arch, continued 7.6.4

STEP VIII:

Max. allowable unit stress in compression, $F_c = 2000$ PSI.

Using Winslow's Formula to determine stress and check cross section area at ridge: $\frac{P}{A} = C\left(1.00 - \frac{l}{80d}\right)$

Minimum depth at C = 12.0 in. Breadth b = 8.00 in.

$$\frac{P}{A} = 1400\left[1.00 - \left(\frac{74.0}{80 \times 12.0}\right)\right] = 1292 \text{ PSI. (Parallel with grain).}$$

$$\frac{P}{A} = \frac{28,700}{8.0 \times 12.0} = 300 \text{ PSI. OK.}$$

STEP IX

Check compressive stress at Base of Columns:

From step II, d = 17.36 in. b = 8.00 in. Vertical R = 32,400 Lbs.

$$\frac{P}{A} = \frac{32,400}{8.0 \times 17.36} = 233 \text{ PSI. OK}$$

STEP X:

Design of Purlins. From step VII, spacing = 6.10 feet on centers.

Simple span length equals Arch spacing minus breadth b.

L = 18.0 − 0.67 = 17.33 Feet. Support area = 6.10 X 17.33 = 105.75 Sq. Ft.

Roof Loads = 45 Lbs. Sq. Ft. $M = \frac{WL}{8}$ W = 45 X 105.75 = 4,760 Lbs.

$$M = \frac{4760 \times 17.33}{8} = 10,315 \text{ Ft. Lbs.} \quad S = \frac{M}{F_b} \quad S = \frac{10,315 \times 12}{2200} = 56.27''^3$$

Select from the Tables giving the properties of Glued Laminated Sections, the purlin desired.

A size of $5\frac{1}{4}'' \times 8\frac{1}{8}''$ net, has a $S = 57.8''^3$ and is acceptable.

DESIGNERS NOTE:

The tension stress in the outer face of knee and the compressive stress in the inside face are not always consistent with the depths of sections previously computed. When the two (2) forces R and H are working together, there is a resultant force which will equal those forces only when acting in another direction. Call this resultant force N, and obtain its direction and magnitude by drawing a force polygon. Lay outwith engineers scale on the elevation drawn in step I: The closing string = N, and is 37,500 Pounds when measured to same scale:

A formula can be written which will produce the results for the actual unit stress for both tension and compression. Thus: f_t or $f_c = \left(\frac{-N}{bd}\right) \pm \left(\frac{6M}{bd^2}\right)$. The following values are known: Allowable $F_t = 1800$ #□" and $F_c = 2000$ #□"

From step III, $d = 39.0$ inches. $b = 8.0$ inches, and $M_B = -370{,}300$ Ft. Lbs.

Resultant $N = 37{,}500$ Lbs.

Inserting the known values in the formula:

At outer face, $f_t = \frac{-37{,}500}{8.0 \times 39.0} + \frac{6 \times 370{,}300 \times 12}{8.0 \times 39.0 \times 39.0} = -120 + 2200 = 2080$ PSI.

Compression at inner face, $f_c = -120$ and $-2200 = 2320$ Lbs. Sq. Inch.

Both compressive and tension stresses are over the allowable and the dimensions of b or d, must be increased. By using a slide rule, increased values for divisors b and d will provide the necessary dimensions for cross section at knee.

Use for final acceptance, $b = 8.50"$ and $d = 40.0"$

$$f_t = \frac{-37{,}500}{8.50 \times 40.0} + \frac{6 \times 370{,}300 \times 12}{8.50 \times 40.0 \times 40.0} = -110 + 1965 = 1855 \text{ P.S.I. (OK)}$$

$f_c = -110 - 1965 = 2075$ PSI. Accept these dimension for the arch design.

TABLE: Quick reference for low-profile steel rigid frames 7.7.1

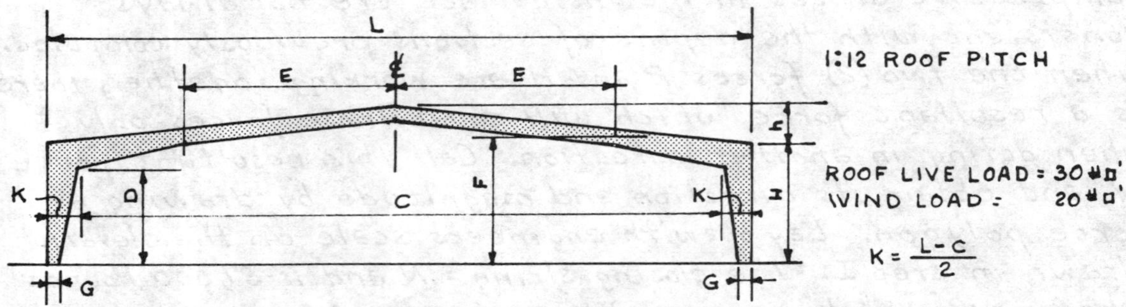

LOW ROOF PROFILE - RIGID STEEL FRAMES

WIDTH	H	L	C	D	E	F	G	WIDTH	H	L	C	D	E	F	G
50'	10'-0"	48-8	44-4	7-7			0'-6"	100'	14'-0"	98'-8"	90'-6"	9'-8"	24'-0"	13'-6"	1'-2"
	12'-0"	48-8	44-4	9'-7"			0'-6"		16'-0"	98'-8"	90'-6"	11'-8"	24'-0"	15'-6	1'-0"
	14'-0"	48-8	44-4	11'-7"			0'-6"		20'-0"	98'-8"	90'-4"	15'-8"	24'-0"	19'-6"	1'-0"
	16'-0"	48-8	44-4	13'-7"			0'-6"		24'-0"	98'-8"	90'-4"	19'-8"	24'-0"	23'-6"	1'-0"
	20'-0"	48-8	44-4	17'-9"			0'-6"	110'	14'-0"	108'-8"	100'-8"	9'-9"	27'-0"	13'-4"	1'-6"
	24'-0"	48-8	44-8	2:'-9"			0'-6"		16'-0"	108'-8"	100'-8"	11'-9"	27'-0"	15'-4"	1'-2"
60'	10'-0"	58-8	53-4	6'-11"	15'-6"	9'-3"	0'-10"		20'-0"	108'-8"	100'-6"	15'-3"	27'-0"	19'-6"	1'-0"
	12'-0"	58-8	53-4	8'-11"	15'-6"	11'-3"	0'-8"		24'-0"	108'-8"	100'-6"	19'-3"	27'-0"	23'-6"	1'-0"
	14'-0"	58-8	53-4	10'-11"	15'-6"	13'-3"	0'-8"	120'	14'-0"	118-8	110-8	9-4	31'-0	13'-5"	1-4
	16'-0"	58-8	53-0	12'-11"	15'-6"	15'-3"	0'-8"		16'-0"	118 8	110-8	11-4	31-0	15'-7"	1-4
	20'-0"	58-8	53-0	16'-11"	15'-6"	19'-3"	0'-8"		20'-0"	118 8	110-0	15-3	31-0	19'-7"	1'-4
	24'-0"	58-8	53-0	20'-11"	15'-6"	23'-3"	0'-8"		24'-0"	118 8	110-0	19-3	31-0	23-7"	1'-4
70'	12'-0"	68-8	62-4	8-7	12-0	11-11	1-0	125'	14'-0"	123-8	119-8	9'-4"	33-6	13'-4"	1'-6
	14'-0"	68-8	62-4	10-7	12-0	13-11	0-10		16'-0"	123-8	119-8	11'-4"	33-6	13-5	1'-6
	16'-0"	68-8	62-4	12-7	12-0	15-11	0-10		20'-0"	123-8	119-0	15'-4"	33-6	19'-5"	1'-6
	20'-0"	68-8	62-4	16-7	12-0	19-11	0-10		24'-0"	123-8	119-0	19'-3"	33-6	23'-5"	1'-6
	24'-0"	68-8	62-4	20-7	12-0	23-11	0-10	130'	14'-0"	128-8	120-0	9-4	36-0	13'-3"	1-6
80'	12'-0"	78-8	72-4	8-9	15-6	11-9	0-10		16'-0"	128-8	120-0	11-4	36-0	15'-3"	1-6
	14'-0"	78-8	72-4	10-9	15-6	13-9	0-10		20'-0"	128-8	120-0	15-4	36-0	19'-3"	1-8
	16'-0"	78-8	72-0	12-4	15-6	15-9	0-10		24'-0"	128-8	119-6	19-3	36-0	23'-3"	1'-8
	20'-0"	78-8	72-0	16-4	15-6	19-9	0-10	140'	14'-0"	138-8	129-6	9-0	36-0	13'-2"	1-8
	24'-0"	78-8	72-0	20-4	15-6	23-9	0-10		16'-0"	138-8	129'-6"	11-0	36-0	15'-2"	1'-8
90'	12'-0"	88-8	81-8	8-0	24-0	11'-5"	1-0		20'-0"	138-8	129'-0"	14-10	36-0	19'-2"	1'-8
	14'-0"	88-8	81-8	10-0	24-0	13'-5"	1-0		24'-0"	138-8	129'-0"	18-10	36-0	23'-2"	1'-8
	16'-0"	88-8	81-8	12-0	24-0	15'-5"	0-10	150'	14'-0"	148-8	139-4	8-10	36-0	12'-10"	2-0
	20'-0"	88-8	81-8	16-0	17-3	20'-0"	0-10		16'-0"	148-8	139-4	10-10	36-0	14'-10"	2-0
	24'-0"	88-8	81-8	20-0	17-3	24'-0"	0-10		20'-0"	148-8	139-0	14-9	36-0	19'-7"	2-0
									24'-0"	148-8	139-0	18-9	36-0	23'-7"	2-0

POINT CONTRAFLEXURE AT INTERSECTION OF POINTS E AND F

TABLE: Quick reference for high-profile steel rigid frames 7.7.2

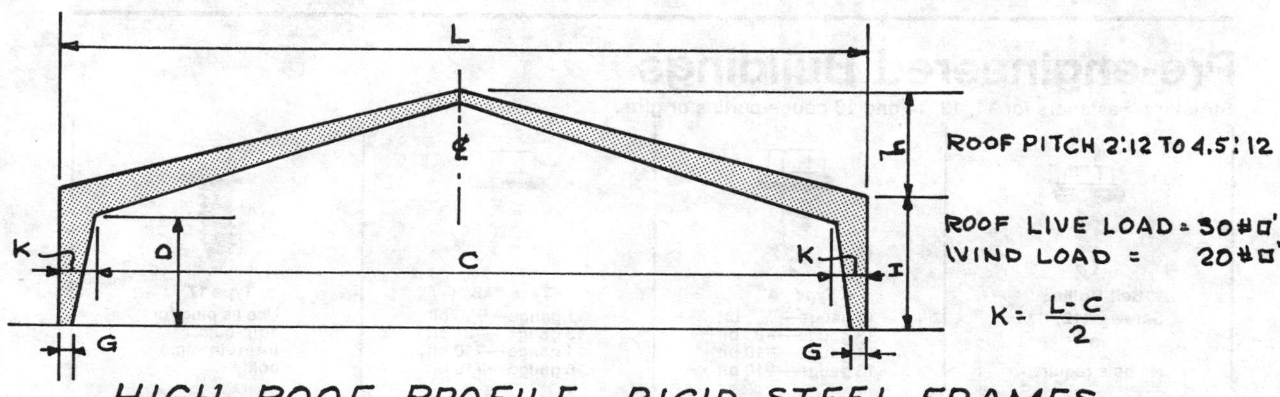

HIGH ROOF PROFILE - RIGID STEEL FRAMES

WIDTH	H	L	C	D	G	WIDTH	H	L	C	D	G
40'	10'-0"	38'-8"	35'-4"	8'-4"	0'-6"	80'	12'-0"	78'-8"	72'-8"	9'-3"	0'-10"
	12'-0"	38'-8"	35'-4"	10'-4"	0'-6"		14'-0"	78'-8"	72'-8"	11'-3"	0'-10"
	14'-0"	38'-8"	35'-4"	12'-4"	0'-6"		16'-0"	78'-8"	72'-2"	13'-1"	0'-10"
	16'-0"	38'-8"	35'-4"	14-4	0'-6"		20'-0"	78'-8"	72'-2"	17'-1"	0'-10"
	20'-0"	38'-8"	35'-4"	18-4"	0'-6"		24'-0"	78'-8"	72'-2"	21'-1"	0'-10"
	24'-0"	38'-8"	35'-4"	22'-4"	0'-6"	90'	12'-0"	88'-8"	82'-4"	9'-2"	0'-10"
50'	10'-0"	48'-8"	44'-8"	8'-1"	0'-8"		14'-0"	88'-8"	82'-4"	11'-2"	0'-10"
	12'-0"	48'-8"	44'-8"	10'-1"	0'-8"		16'-0"	88'-8"	82'-4"	13'-2"	0'-10"
	14'-0"	48'-8"	44'-8"	12'-1"	0'-8"		20'-0"	88'-8"	81'-8"	17'-1"	0'-10"
	16'-0"	48'-8"	44'-8"	14'-1"	0'-8"		24'-0"	88'-8"	81'-8"	21'-1"	0'-10"
	20'-0"	48'-8"	44'-8"	18'-1"	0'-8"	100'	14'-0"	98'-8"	92'-0"	11'-1"	0'-10"
	24'-0"	48'-8"	44'-8"	22'-1"	0'-8"		16'-0"	98'-8"	92'-0"	13'-1"	0'-10"
60'	10'-0"	58-8"	53'-8"	7'-9"	0'-8"		20'-0"	98'-8"	91'-8"	17'-1"	0'-10"
	12'-0"	58'-8"	53'-8"	9'-9	0'-8"		24'-0"	98'-8"	91'-8"	21'-1"	0'-10"
	14'-0"	58'-8"	53'-8"	11'-9"	0'-8"	110'	14'-0"	108'-8"	102'-0"	11'-1"	1'-0"
	16'-0"	58'-8"	53'-8"	13'-9"	0'-8"		16'-0"	108'-8"	102'-0"	13'-1"	1'-0"
	20'-0"	58'-8"	53'-8"	17'-8"	0'-8"		20'-0"	108'-8"	101'-8"	16'-10"	1'-0"
	24'-0"	58'-8"	53'-10"	21'-8"	0'-8"		24'-0"	108'-8"	101'-8"	20'-10"	1'-0"
70'	12'-0"	68'-8"	63'-0"	9'-6"	0'-8"	120	14'-0"	118'-8"	111'-8"	10'-10"	1'-0"
	14'-0"	68'-8"	63'-0"	11'-6"	0'-8"		16'-0"	118'-8"	111'-8"	12'-10"	1'-0"
	16'-0"	68'-8"	63'-0"	13'-6"	0'-8"		20'-0"	118'-8"	110'-8"	16'-7"	1'-0"
	20'-0"	68'-8"	62'-4"	17'-3"	0'-8"		24'-0"	118'-8"	110'-8"	20'-7"	1'-0"
	24'-0"	68'-8"	62'-4"	21'-3"	0'-8"						

PILOT DIMENSIONS FOR TRIAL ANALYSIS
MASONRY WALLS OR STEEL GIRTS ASSUMED PLACED OUTSIDE COLUMNS

TABLE: Standard fasteners for steel buildings 7.7.3

Pre-engineered Buildings

Structural Fasteners for 11, 13, 14 and 16 gauge purlins or girts.

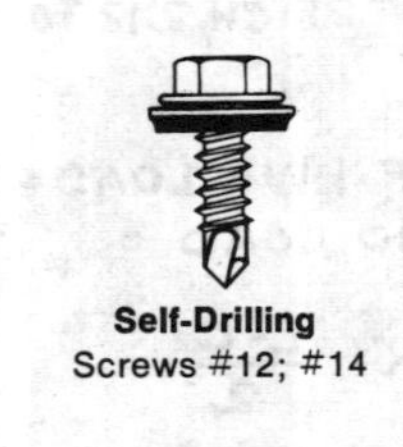

Self-Drilling
Screws #12; #14

No hole required.

Type "A"
11 gauge—13/64" bit
13 gauge—#8 bit
14 gauge—#10 bit
16 gauge—#10 bit

Type "AB"
11 gauge—13/64" bit
13 gauge—#8 bit
14 gauge—#10 bit
16 gauge—#10 bit

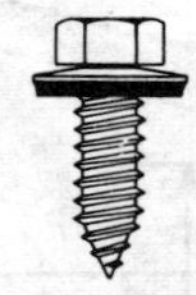

Type 17
Use as plug for strip-outs. Insert in same hole.

Sidelaps of 24 and 26 gauge sheets.

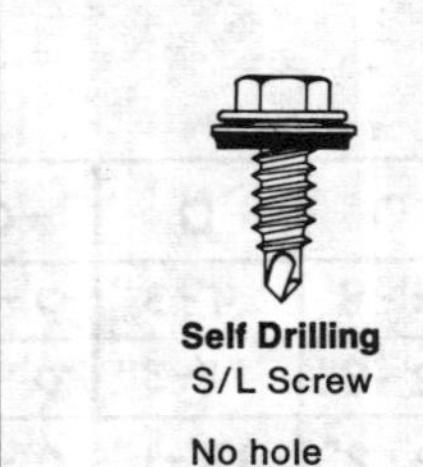

Self Drilling
S/L Screw

No hole required.

Type "A"
24 gauge, 26 gauge: 1/8" bit

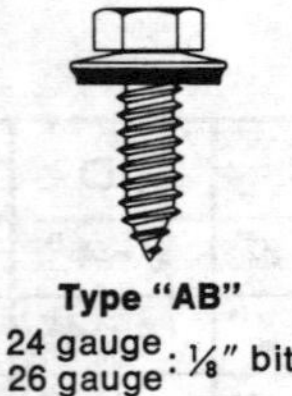

Type "AB"
24 gauge, 26 gauge: 1/8" bit

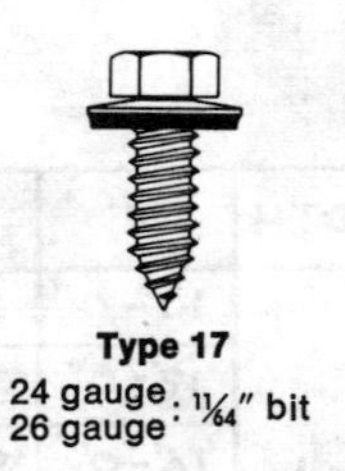

Type 17
24 gauge, 26 gauge: 11/64" bit

Heavy Construction

Steels to ASTM A-7, A-36.

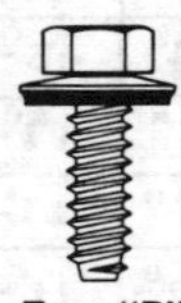

Type "B"

Carbon steel screws.
Valley: Use 1" fastener
Crest: Height of corr. plus 7/8"
Use #1 drill bit in 3/16" or heavier steel.

Type "BP"

H-3 stainless steel for aluminum, asbestos, stainless, protected metal, or painted sheets.
Valley: Use 1" fastener
Crest: Height of corr. plus 7/8"
Use #1 bit up to 3/16" thick steel. Heaviest steel use .231" drill bit.

Sidelaps of 18 or 20 gauge steel sheets; .040" or .032" aluminum sheets.

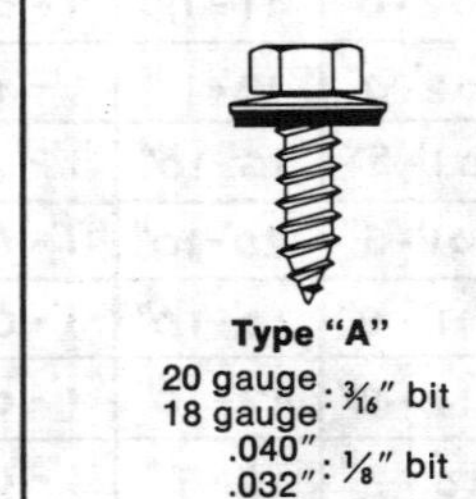

Type "A"
20 gauge, 18 gauge: 3/16" bit
.040", .032": 1/8" bit

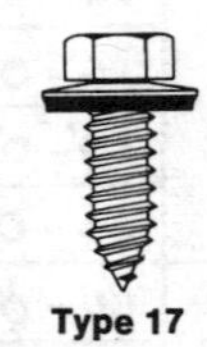

Type 17
20 gauge, 18 gauge: #1 bit

Weath-R-Lok
5/16" hole throughout

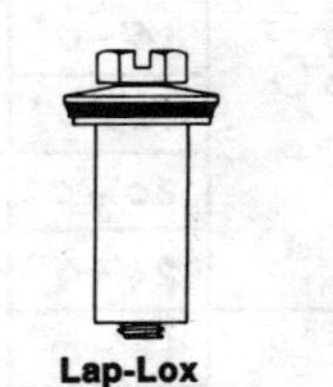

Lap-Lox
3/8" hole throughout for translucent plastic sheets

Note: Hole sizes shown above for the various gauges are applicable under most conditions. For conditions other than "normal", consult us for alternate sizes.

TABLE: Properties for light gauge steel sections 7.7.4

SECTION	D (IN)	B (IN)	L (IN)	THK. (IN)	R (IN)	A (IN²)	WT/FT. (#/FT.)	AXIS X-X Ix (IN⁴)	AXIS X-X Sx (IN³)	AXIS X-X rx (IN)	AXIS Y-Y Iy (IN⁴)	AXIS Y-Y ry (IN)	$\bar{X}$ (IN)	Q	Fy (KSI)	Fb (KSI)
8 C 4.1	8.000	3.250	1.000	.075	.188	1.19	4.10	11.85	2.86	3.17	1.75	1.22	1.02	0.71	40.00	24.24
8 C 5.7	8.000	3.250	1.000	.105	.188	1.65	5.70	16.33	4.08	3.15	2.38	1.20	1.02	0.81	40.00	24.24
8 C 6.8	8.000	3.250	1.000	.125	.188	1.95	6.75	19.20	4.80	3.14	2.77	1.19	1.02	0.84	42.00	25.45
8 C 8.0	8.000	3.250	1.000	.150	.188	2.32	8.01	22.65	5.65	3.13	3.23	1.18	1.02	0.89	42.00	25.45
8 C 9.9	8.000	3.250	1.000	.188	.188	2.86	9.86	27.63	6.91	3.11	3.87	1.16	1.02	0 94	42.00	25.45
8 C 4.5	8.000	4.000	1.000	.075	.188	1.30	4.46	13.30	3.08	3.24	2.88	1.49	1.32	0.69	40.00	24.24
8 C 6.4	8.000	4.000	1.188	.105	.188	1.84	6.37	19.13	4.68	3.23	4.19	1.50	1.38	0.81	40.00	24.24
9 C 4.8	9.000	4.000	1.000	.075	.188	1.37	4.75	17.44	3.60	3.61	3.00	1.48	1.25	0.65	40.00	24.24
9 C 6.6	9.000	4.000	1.000	.105	.188	1.91	6.61	24.64	5.37	3.59	4.09	1.46	1.25	0.77	40.00	24.24
9 C 7.8	9.000	4.000	1.000	.125	.188	2.26	7.84	29.00	6.44	3.58	4.77	1.45	1.25	0.82	42.00	25.45
11 C 5.3	11.375	4.000	0.875	.075	.188	1.53	5.31	29.63	4.87	4.45	3.07	1.42	1.08	0.58	40.00	24.24
11 C 7.7	11.375	4.000	1.312	.105	.188	2.22	7.70	43.30	7.57	4.45	4.91	1.49	1.20	0.70	40.00	24.24

SECTION	D (IN)	B (IN)		THK. (IN)	R (IN)	A (IN²)	WT/FT (#/FT.)	AXIS X-X Ix (IN⁴)	AXIS X-X Sx (IN³)	AXIS X-X rx (IN)	AXIS Y-Y Iy (IN⁴)	AXIS Y-Y ry (IN)	$\bar{X}$ (IN)	Q	Fy (KSI)	Fc (KSI)
4 CH 1.3	4.000	1.125	--	.060	.188	0.35	1.25	0.77	0.39	1.47	0.04	0.33	0.23	0.72	40.00	20.76
4 CH 1.6	4.000	1.125	--	.075	.188	0.44	1.56	0.95	0.47	1.46	0.05	0.32	0.24	0.85	40.00	23.10
4 CH 1.8	4.125	1.563	--	.075	.188	0.52	1.82	1.29	0.63	1.58	0.12	0.48	0.38	0.71	40.00	18.72
4 CH 2.1	4.000	1.125	--	.105	.188	0.61	2.05	1.29	0.64	1.45	0.06	0.32	0.25	0.98	40.00	24.24
6 CH 1.9	6.250	1.625	--	.060	.188	0.55	1.91	2.92	0.93	2.30	0.12	0.47	0.31	0.44	40.00	14.46
6 CH 2.4	6.250	1.625	--	.075	.188	0.69	2.38	3.60	1.15	2.29	0.15	0.47	0.31	0.57	40.00	18.06
6 CH 3.3	6.250	1.625	--	.105	.188	0.95	3.30	4.95	1.58	2.28	0.20	0.46	0.33	0.77	40.00	22.20
7 CH 1.9	6.500	1.500	--	.060	.188	0.55	1.91	3.04	0.94	2.35	0.10	0.42	0.27	0.46	40.00	16.03
7 CH 2.4	6.500	1.500	--	.075	.188	0.69	2.38	3.77	1.16	2.34	0.12	0.41	0.27	0.59	40.00	19.35
7 CH 3.3	6.500	1.500	--	.105	.188	0.95	3.30	5.17	1.59	2.33	0.16	0.41	0.28	0.78	40.00	23.10
7 CH 3.9	6.500	1.500	--	.125	.188	1.13	3.92	6.07	1.87	2.32	0.19	0.41	0.28	0.87	42.00	25.45
7 CH 5.9	6.750	1.500	--	.188	.188	1.71	5.87	9.60	2.84	2.37	0.27	0.39	0.31	0.97	42.00	25.45
7 CH 7.9	6.875	1.625	--	.250	.250	2.33	7.86	13.35	3.89	2.39	0.42	0.43	0.37	1.00	42.00	25.45
8 CH 3.0	8.250	1.875	--	.075	.188	0.87	3.03	7.77	1.88	2.98	0.24	0.52	0.33	0.44	40.00	15.59
8 CH 4.3	8.312	2.000	--	.105	.188	1.25	4.33	11.35	2.73	3.01	0.39	0.56	0.39	0.61	40.00	19.53
8 CH 10.8	8.625	2.500	--	.250	.250	3.20	10.84	31.61	7.33	3.14	1.60	0.71	0.57	0.98	42.00	25.45
12 CH 12.4	11.875	3.938	--	.188	.188	3.59	12.42	72.60	12.23	4.50	4.95	1.18	0.87	0.57	42.00	17.96

SECTION	L (IN)	THK. (IN)	R (IN)	A (IN²)	WT/FT. (#/FT.)	AXIS X-X Ix (IN⁴)	AXIS X-X Sx (IN³)	AXIS X-X rx (IN)	AXIS Y-Y Iy (IN⁴)	AXIS Y-Y ry (IN)	AXIS Z-Z α (°)	AXIS Z-Z rz (IN)	Ixy (IN⁴)	Q	Fy (KSI)	Fb (KSI)
8 Z 2.9	0.688	0.060	1.56	0.84	2.92	8.09	1.95	3.10	1.16	1.18	16.36	0.78	2.22	0.63	40.00	24.24
8 Z 3.3	0.750	0.067	.156	0.95	3.28	9.09	2.20	3.10	1.36	1.20	16.69	0.79	2.55	0.64	50.00	30.30
8 Z 3.7	0.750	0.075	.156	1.06	3.65	10.13	2.50	3.10	1.51	1.20	16.70	0.79	2.84	0.67	50.00	30.30

NOTES:
1. SECTION PROPERTIES AND ALLOWABLE STRESSES ARE CALCULATED IN ACCORDANCE WITH THE 1962 EDITION OF THE AISI SPECIFICATIONS. Ix IS FOR DEFLECTION DETERMINATION; Sx IS FOR LOAD DETERMINATION.

TYPICAL DETAILS: Rigid frame plan layout and framing 7.7.5

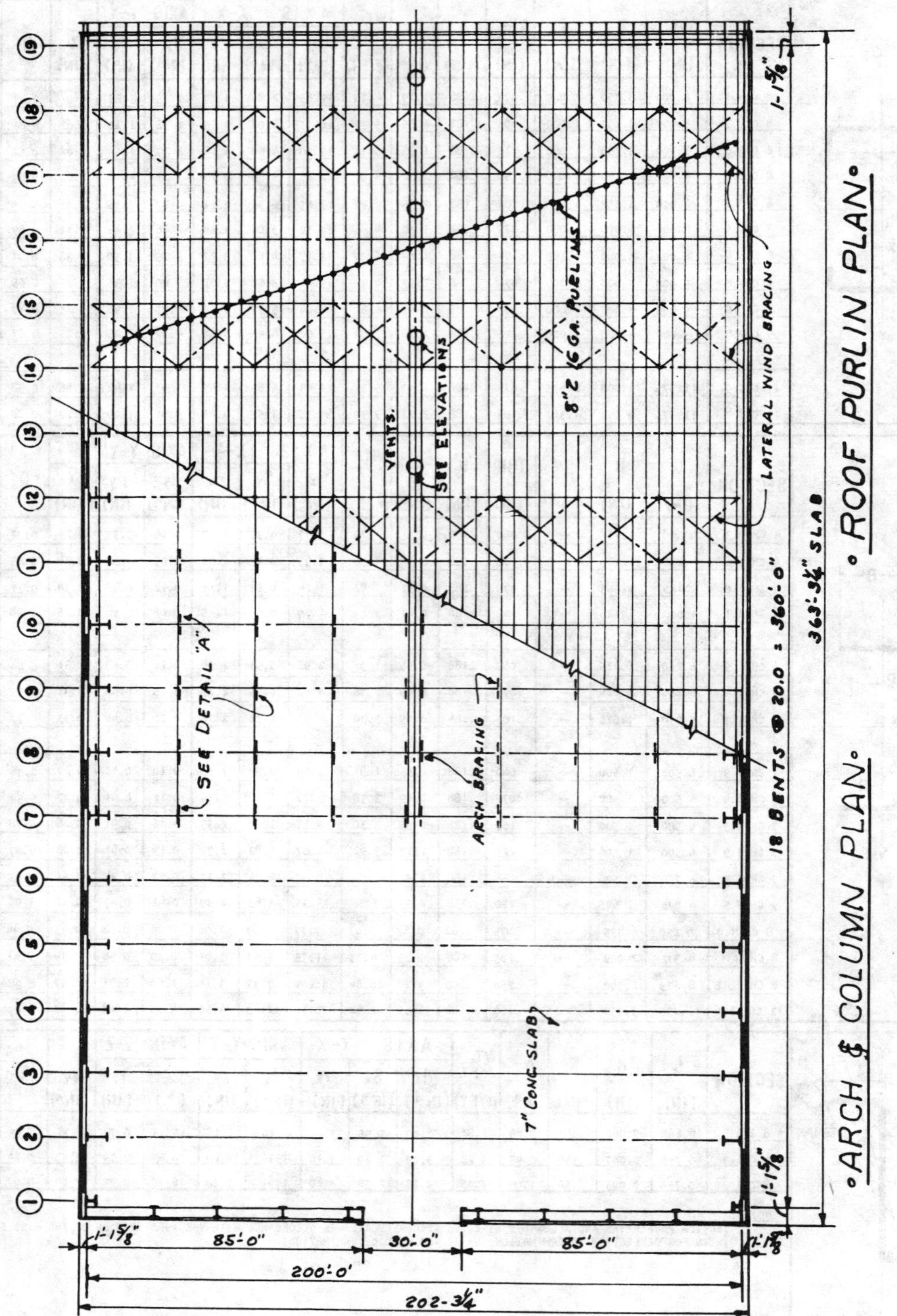

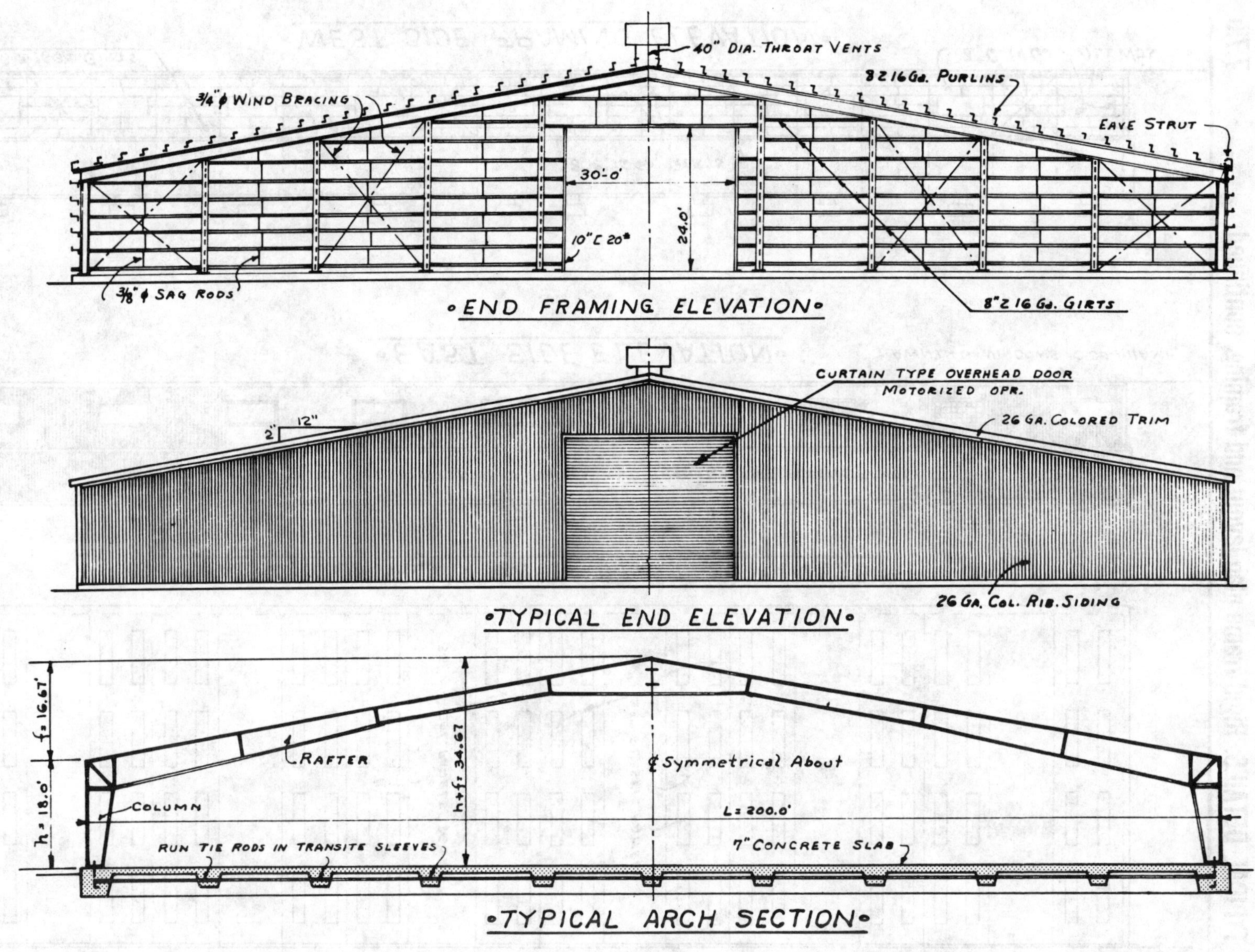
40" DIA. THROAT VENTS
3/4"⌀ WIND BRACING
8Z 16 GA. PURLINS
EAVE STRUT
30'-0"
24.0'
10"⊏ 20#
3/8"⌀ SAG RODS
8"Z 16 GA. GIRTS
• END FRAMING ELEVATION •
CURTAIN TYPE OVERHEAD DOOR
MOTORIZED OPR.
12"
2"
26 GA. COLORED TRIM
26 GA. COL. RIB. SIDING
• TYPICAL END ELEVATION •
f=16.67'
h=18.0'
RAFTER
h+f=34.67
℄ Symmetrical About
COLUMN
L=200.0'
RUN TIE RODS IN TRANSITE SLEEVES
7" CONCRETE SLAB
• TYPICAL ARCH SECTION •

TYPICAL DETAILS: Rigid frame plan layout and framing, continued 7.7.5

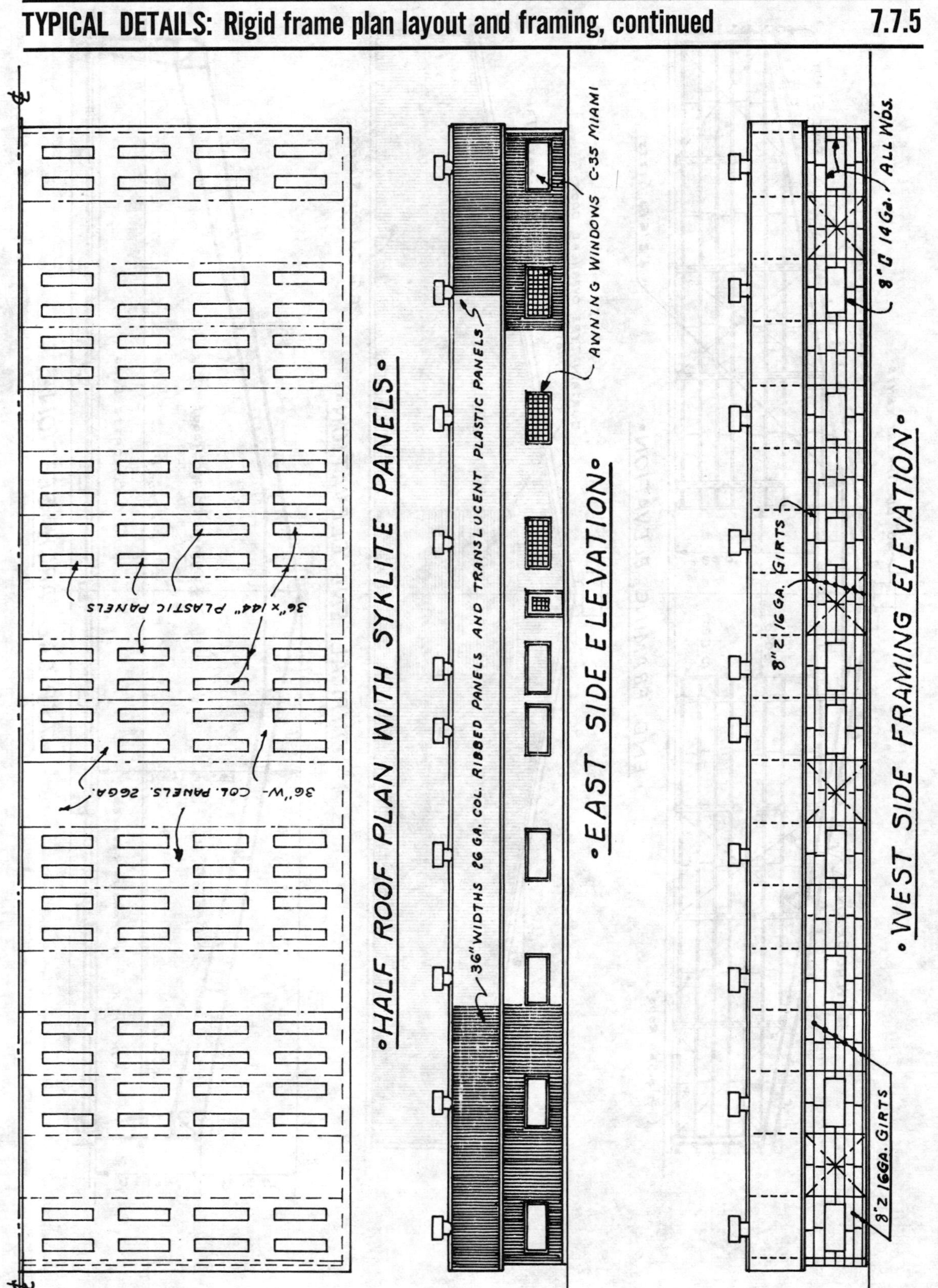

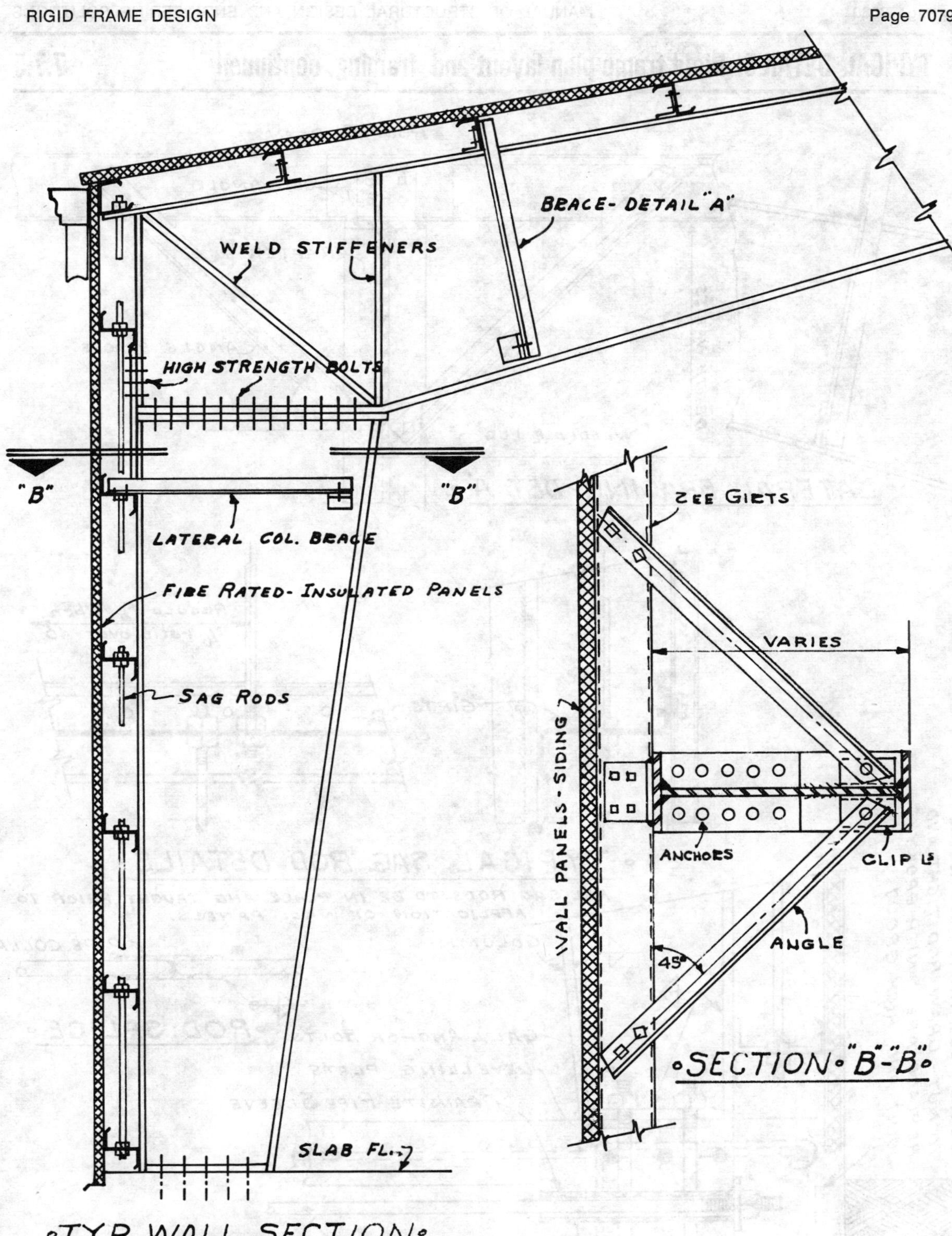
BRACE- DETAIL "A"
WELD STIFFENERS
HIGH STRENGTH BOLTS
"B"
"B"
LATERAL COL. BRACE
FIBE RATED- INSULATED PANELS
SAG RODS
SLAB FL.
•TYP. WALL SECTION•
ZEE GIRTS
VARIES
WALL PANELS - SIDING
ANCHORS
CLIP Ls
ANGLE
45°
•SECTION "B"-"B"•

TYPICAL DETAILS: Rigid frame plan layout and framing, continued 7.7.5

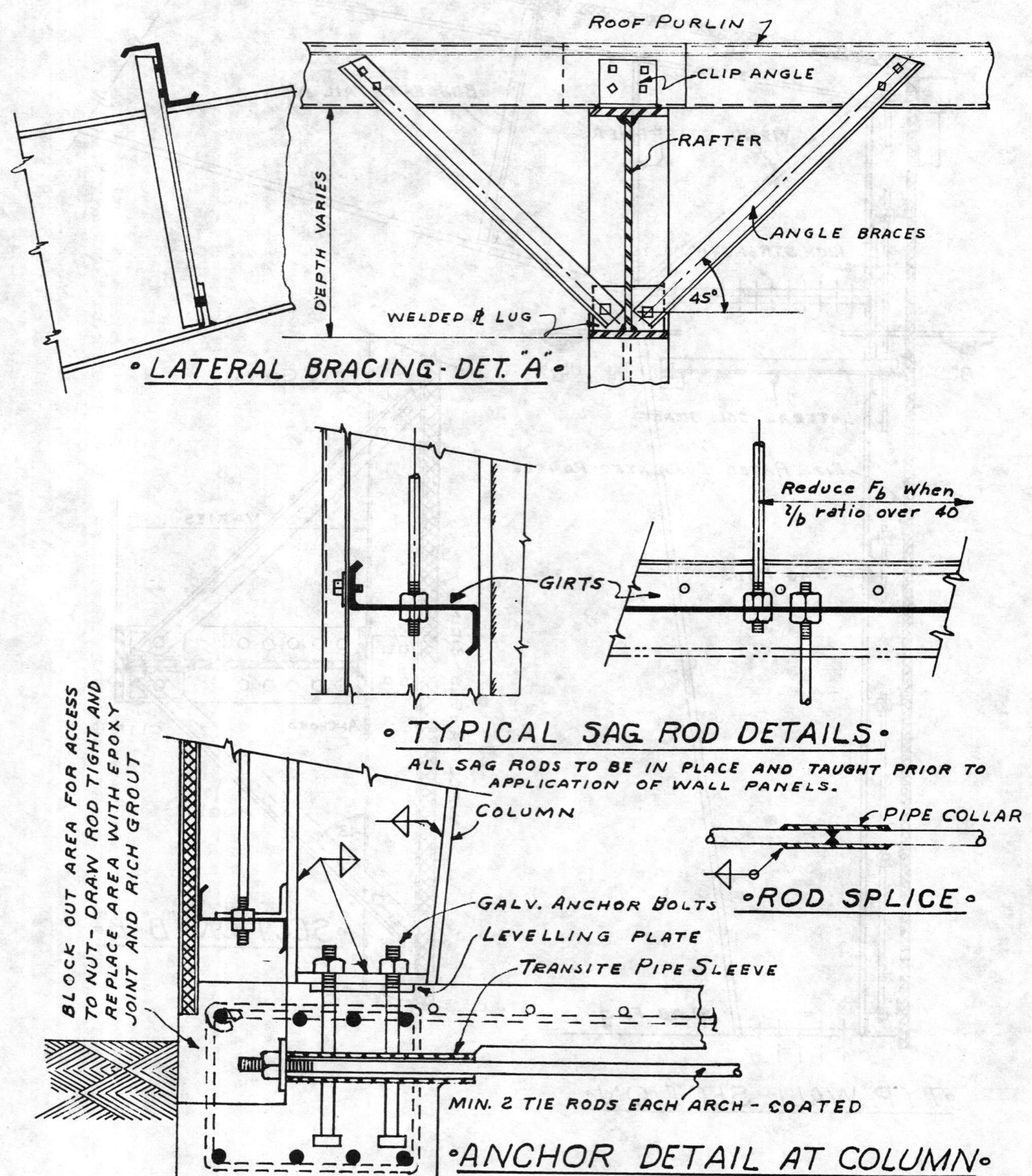

VIII—HIGH RISE DESIGN

VIII

HIGH RISE DESIGN

HIGH RISE DESIGN

Contents

Evolution of the high rise 8.1

Throughout history, man has had to make use of the available building materials. The spans which timber and stone could bridge, either as beams, lintels or arches were limited. However, the whole style and sense of Classic, Gothic and Renaissance architecture was established with those basic materials. Yet, timber and stone have basic limitations. The Cathedral at Beauvais, France will be remembered primarily because of its over-ambitious builders. During construction (1247–1500) several attempts to erect the highest vault ended in failure. Finally, in 1320, the roof ridge was completed and rose 154 feet above the ground. In 1500, the transepts and towering 500 foot spire were started. But these crashed to earth before the end of the century.

In the early 1890's, an American architect, Louis Sullivan, became the creator of the modern skyscraper. Sullivan realized that an office building could be erected using totally different sources for materials. He chose steel for the Guaranty Building of Buffalo, and thus gave the clearest expression to the architectural trend toward height. Others followed, such as the Woolworth Building by Architect Cass Gilbert. Europeans joined the high rise trend, led by Otto Wagner with his Postal Savings Bank in Vienna. Wagner demonstrated to other architects that tall, multi-story structures must be developed incorporating an understanding of sound engineering for the basic foundation and structural components.

The demand for tall buildings increased. Large corporations recognized the advertising and publicity advantages of connecting their name with an imposing hi-rise office building, even though their operations might have required only one floor. The other floors were leased out to eager business tenants.

World War I temporarily slowed the building of tall structures. The United States got the trend underway again at the end of hostilities. America, still considered a new country, had provided the Allied countries with the fighting men, material, money and resources to terminate the struggle with a convincing Allied victory. But at home, economic problems and inflated cost appeared. The political leaders, flushed with the pride of victory, proceeded to make postwar plans. Exciting speeches and emotional appeals promised continued prosperity and cities with an entirely new concept.

Competition for the leading metropolis, as judged by tall building skyline, developed between Chicago and New York. The people of these cities showed a readiness to elect to high offices men who were progressive, flambuoyant and not known to be conservative spenders. The cities embarked upon a contest of height, with the Empire State Building winning tallest honors.

This period was to become known as the "Roaring Twenties." The automobile industry was putting people on wheels, factories were working multiple shifts and living was luxurious. The boom ended

Evolution of the high rise, continued 8.1

abruptly with the stock market crash on Wall Street in July 1929. The effects of the economic collapse were felt in all countries, particularly the defeated countries of the recent war. The Great Depression followed. Chaos developed in many countries. Governments crumbled, dictators took over and promised relief. Hitler converted factories from producing consumer goods to the production of war materials. Workers were conscripted for industry and the military. Other countries reacted to the Depression in a like manner. The result of such preparations led to World War II.

Since the war, the predicted population increase has become a reality, and a problem to the large cities. The coming years of the 1970's will see the completion and occupancy of thousands of hi-rise buildings throughout this country and the world. These structures will rise to heights thought impossible a few short years ago. The late Frank Lloyd Wright, this century's most illustrious architect predicted that high rise structures one-half mile high could be constructed. Years back, most people thought it absurd that a building over twenty stories could be constructed. The same doubts were expressed over the ability of man ever to place foot upon the moon. Yet, on July 20, 1969, two Americans, Neil Armstrong and Edwin Aldrin, Jr., landed a lunar module on the moon, spent nearly two hours walking upon its surface and returned safely to earth. This project was a result of eight short years of preparation, prompted by competition with the Soviet Union.

Man's competitive planning will continue to create taller, more ambitious structures. Within the last few years, many major cities have had imaginative new shapes thrusting above their skyline. The World Trade Center Towers in New York has become the world's tallest at 1350 feet. The Lake Point Towers Apartments in Chicago are triform in shape, and afford the unique prospect of a home on the seventieth floor. Many other new "hi-rises" will use a plan shape other than rectangular, to better cope with wind pressure. A 913 foot circular office tower and hotel will soon be constructed in Houston. Dealing with the forces of nature can lead to surprising departures from accepted designs.

Designing for wind loading 8.2

In the examination of all building codes, it will be noted that specific instructions will be given for the application of wind loads to most buildings over thirty feet in height. As the building height becomes greater, the wind pressure will increase. For example, the Southern Standard Building Code, 1965 Edition, requires that the minimum design wind loads shall be as follows:

	Design wind load in PSF.	
Height in Feet	*Inland Region*	*Coastal Region*
30 or Less	10	25
31 to 50	20	35
51 to 99	24	45
100 to 199	28	50
200 to 299	30	50
300 to 399	32	50
400 and Over	40	50

These loads shall be applied as acting horizontally from any direction against the exterior walls, either inward or outward. In this case, the coastal area is an area lying within 125 miles of the coast line, subject to hurricanes and tropical disturbances. Those high velocity winds have been recorded as exceeding 150 miles per hour. Using the U.S. Navy Bureau of Yards and Docks formula, $P = 0.00256\ V^2$, for wind pressure on a flat surface perpendicular to wind direction, a 150 MPH wind would have a 57.60 pound per square foot pressure application. Other formulas such as $P = 0.004V^2$ will exceed the U.S. Navy formula, and give a result of 90 pounds per square foot.

The fact that the meteorologists do not agree on which of the two formulas should be used, has caused the Code writing authorities to establish the requirement in pounds per square foot, not in wind velocity. The wind loads are to be added to the gravity loads acting vertically.

Wind load direction 8.2.1

The assumption that wind pressure acts uniformly horizontally is not exactly correct. Wind direction may be at an incline to the horizontal, due to obstructions or ground profile. The maximum resisting reaction is produced when the wind strikes the walls squarely against the sides. Hence, the code will require the direction of the force produced by wind to be considered in a horizontal direction.

Tipping moment 8.2.2

Wind loads acting against the side of a multi-story building from one direction only will produce a force which could overturn a slender and light weight structure. This action may be referred to as "overturning moment" or "tipping moment." The only resistance a multi-story has against tipping over is its dead weight. The Houston, Texas Building Code requires that the tipping moment produced by wind loads shall not exceed ⅔ of the moment of stability of the building structure as determined by dead loads only.

Before starting the more complex labors of designing the structural members, it is better to get definite decisions from the Architect on type of materials for exterior walls, floor slabs, elevators, stairs, and anything else which will determine the dead loads to be supported by the skeleton framing.

Methods for wind load design 8.2.3

Engineering judgment requires that the greatest portion of wind pressure on walls must be assumed to be resisted by the rigid structural framework and particularly the connections. Past investigations made on many of the older buildings considered as tall structures, have shown that the floors, partitions and exterior walls have shared much of this load. In the initial stage of design, had the section properties of these components been known, an exact method for design could have been developed. Rarely, if ever, will the structural designer receive preliminary plans from the Architect which show the precise location and direction of partition walls, mechanical equipment supports, and desired joist spacing. (Enclosures for elevators and stair wells are usually located early in design because of the Code requirements.) While it is agreed that these additional walls will offer some resistance to wind pressure, calculating their values would be time-consuming and become more of a burden than an economy. Therefore a method must be used which will be simple, accurately productive, and incorporate an acceptable safety factor.

Cantilever wind load calculations 8.2.4

A method of High Rise Design commonly called for in building codes will require that the lateral forces on exterior walls shall vary from top to bottom, and the whole structure shall be treated as a cantilever. This system assumes that the frame acts as a cantilever beam with the axial stress in the columns varying directly with the moment distance from the centroid of the area which is formed by column bays. In short, there is a point of inflection at the mid-span of each beam and mid-height of each column. The action of lateral forces on columns and girders can be better understood by referring to Plate 6 of Example 8.6. All wind loads are presumed to act horizontally against the wall surface. Building Codes will stipulate the wind pressure in pounds per square foot on the wall, and require higher pressure values for the higher stories.

Wind pressure loads are applied at each floor level as indicated by W'_1, W'_7, etc. Load applied at roof level is identified as W'_r. The total shear at each floor level is computed from the top and represented by W'_7, W'_{10}, and W'_b. The total horizontal shear at any floor is the sum of all loads applied to the stories above. For a five story structure the summation becomes: $W'_b = W'_r + W'_5 + W'_4 + W'_3 + W'_2 + W'_1$. When the shear magnitude at each story has been computed for a single bay, the distribution among columns and panels to resist this force is reduced to simple rules and formulas which are part of the recommended design system.

Seismic design 8.3

A rigid type of structure is preferred to resist forces resulting from earthquakes. This is a requirement in certain localities which have experienced quakes in past years. The entire structure must be so braced and rigidly connected, that it will move as a unit. The structural members cannot be loaded to ultimate unit stresses, and connections are of greater concern. Diagonal cross-frame bracing and knee brackets should be employed. See Section VI on Steel Design for more on Moment Connectors.

High rise design systems 8.4

In the design of multi-story structures the assumption is made that the calculations shall represent only the skeletal framework which consists of the columns, girders, spandrel beams, and, most important, the connections of columns to girders. It is not difficult to calculate the gravity loads which act in a vertical direction, and are supported by the girders or exterior spandrels. Exterior masonry walls are supported by the spandrel girders. A part of the floor area is also supported on the spandrels. In most cases, the interior girders support the loads from elevators and stairs, in addition to the thick fire walls which enclose these main facilities. The machinery for freight and passenger elevators is enclosed in a pent house area above the roof, although with the underslung electric type elevator, the penthouse can be omitted.

In any system of design for tall structures, the columns must transmit the loads to the foundation. Because wind loads must be considered and calculated in a separate category, and combined with the gravity loads later, a method should be adopted which will, in a simple manner, relate the two kinds of load stresses, and make the work less laborious. The Architect's preliminary plans should be carefully studied, and the designer should make up layouts for sections and elevations of the skeleton structure. Provide separate drawings for each column group which differs in loading. These column elevations are illustrated on Plates 1, 2 and 3, of Example 8.6. Start at the Roof Top and continue down by posting the results in regular sequence until the total load is transmitted to footing. The vertically acting, gravity loads on each column are determined before considering the lateral wind loads. Record load on each column at each floor on the diagram, and identify columns by number. The Live Loads and Dead Loads will have to be separated in order to calculate the Resisting Moment against overturning. Only the Dead Loads may be used to determine the resistance to overturning.

It is always considered good practice to make the designer's work sheets in a short and neat arrangement so that errors may most easily be detected as the work progresses. Building Code Officials most often ask the Engineer for multi-story structures to submit the design data for examination before a building permit will be issued. In the offices of established consulting engineers, each designer is expected to follow the same system, so that the continuity of the work may be checked by others without spending unnecessary time explaining and debating the methods.

For computing the stresses in the rectangular framing of the skeletal structure resulting from lateral wind loads, a system was devised by Henry J. Burt, a structural engineer for the Architectural firm of Holabird and Roche of Chicago, Illinois. The system is not an exact method for stress computations, but rather an approximation, because no system can be exact when the action of wind on a tall structure is the most indeterminate of all forces. The Burt system is sound and has found wide acceptance. The theory in this system considers the structure to act as a cantilever beam with its support at ground level. The distribution of stresses produced by wind force is dependent upon the number of panels or portals in the framing, and the analysis may extend to any number of stories and any number of portals.

In the rectangular type of framing, the

High rise design systems, continued 8.4

maximum bending moments will occur at the connections of girders to the columns, and could require considerably larger beams or knee brackets on the lower floors. By using diagonal cross bracing this problem is reduced because such bracing transforms the stresses produced by wind force into direct axial stress in the columns and girders. The connections in such cases may then be designed primarily for the gravity loads.

Wind analysis symbols 8.4.1

There are many advantages to be gained by using a standard set of symbols for each force or action which can be immediately interpreted and translated into common usage. The following letter symbols with subscripts are accepted nomenclature for this system of wind stress analysis:

N = Number of panels or portals in considered elevation.
L = Span distance between columns ℄ each panel, in feet.
H = Height of column in feet. Taken from floor to floor.
H_7 = Denotes column at 7th. floor or between 6th. and 7th. floor.
M_7 = Indicates bending moment in column H_7.
M_a = Designates moment above floor to determine mean M.
M_b = Designates moment below floor to determine mean M.
W_R = Wind force at Roof line and starting point.
W_7 = Single panel wind pressure acting at point of 7th. floor.
W_7' = Sum of wind forces above down to 7th. mid-height of column H_7 and equals horizontal shear. See 8.6.4. Plate 4.
W_a' = Panel wind force above. W_b'= below. Used in formulas.
M_7 = Column bending moment at 7th. floor.
M_a = Bending moment at girder
G_7 = Girder designation for 7th. floor and column connection

Column bending moments 8.4.2

The bending moment resulting from wind forces in a column is tabulated separately from axial gravity loads. The bending moment is treated as an eccentric load on the column. Refer to either Section II or IV for a complete explanation of designing for eccentric loads on steel or concrete columns.

An eccentric moment for a steel column can be converted to a close enough equivalent axial load by the formula thus:

$W_w' = \frac{W'ec}{r^2}$. where: W_w' = Equivalent axial load.

W' = Horizontal shear on column applied at mid-height of column or at point of assumed contra-flexure. See locations for W' on Plate 6 (8.6.6.).

e = Eccentric moment arm distance in inches.

c = Extreme fiber distance from axis on compression side.

r = Radius of gyration of column section in effective plane or direction

Let these values be assumed to illustrate formula:

$W' = 22{,}500\#$ $e = 26.0$ inches. Column $c = 6.75''$ and $r_x = 4.12''$

Then: $W_w' = \frac{22{,}500 \times 26.0 \times 6.75}{4.12 \times 4.12} = 232{,}745$ Pounds

When the equivalent axial load derived from the wind bending moment is less than half of the gravity axial load on a column, it is not likely that the column section will need to be increased on account of wind forces. Check out all corner columns first as this approximation will not always apply to these columns.

Steel columns required to support axial plus eccentric loads must be designed in accordance with the AISC., or where the equivalent axial load equals moment times bending factor (B) of the steel section. Bending factors B_x and B_y are found in column tables or can be determined by dividing the area by the section modulus as: $B_x = A/S_x$.

The combined axial and eccentric equivalent load or steel columns is governed by the formula: Unity $= \frac{f_a}{F_a} + \frac{f_b}{F_b}$.

Unit is 1.0 or less. Examples in Section II will provide a full explanation for using bending factors.

Column moment distribution 8.4.3

The distribution of moments for columns can be accomplished in several ways; the work can be shortened if the total moment is computed and then proportioned, based upon the number (N) panels or portals and columns. The final distribution of column bending moments must agree with the assumption that the exterior column bending moment is one half that for an interior or intermediate column.

The total column moment formula is: $M = \frac{W'H}{2}$, *or it may be written as;* $M_2 = W' \tfrac{1}{2}H$.
When the formula is written: $M = \frac{W'H}{2N}$, *it provides the value of moment for an interior column. For an outside or exterior column, the formula becomes:* $M = \frac{W'H}{4N}$.

Illustration:

Assume horizontal shear from wind loads above the 2nd. floor is W'_2 *= 180,375 Lbs., from east direction on 1 bay only. Column height* H_2 *= 18.0 feet. Panels number 3 or N=3, making 2 columns interior and 2 exterior. Then with values:*

Exterior Columns: $M_2 = \frac{180{,}375 \times 18.0}{4 \times 3} = 270{,}562.5$ *Foot Lbs.*

Interior Columns: $M_2 = \frac{180{,}375 \times 18.0}{2 \times 3} = 541{,}125$ *Foot Lbs.*

Total $\Sigma M = (270{,}565{,}5 \times 2) + (541{,}125 \times 2) = 1{,}623{,}375$ *Foot Lbs.*

Girder bending moments and connectors 8.4.4

The bending moment in a girder connection at an exterior column is the same as that at each connection for the interior columns. Therefore the distribution of moment will depend upon the number of panels (N). Each end connection carries an equal moment and the connector can be of typical design for each side of column.

The total bending moment in all the girder connections is the mean value of the bending moment in column above the girder and the bending moment in the column below the girder. The formula for distributed girder moments is written:

$$M_g = \tfrac{1}{2}\left[\left(\frac{W_a'H_a}{2N}\right) + \left(\frac{W_b'H_b}{2N}\right)\right]$$

, or the mean moment divided by the number of panels. This formula can be revised to a simplified equation as follows:

$$M_g = \frac{M_a' + M_b'}{2N}$$

, where M_a = Moment above and M_b = Moment below. The column moments can be calculated as: $M_a = W_a' \times \tfrac{1}{2} H$. This will be the method used in examples to follow for finding bending moments in columns.

Vertical shear in girders 8.4.5

The bending moment in girders resulting from wind forces also produces vertical shear at the connection of column to girder. The formula for calculating vertical shear is expressed thus: $V = \frac{(W_a'H_a) + (W_b'H_b)}{2NL}$. As in the previous paragraph, the formula can be simplified since the bending moments in columns will have been previously calculated by the formula: $M_a = \frac{W_a'H_a}{2}$. The formula for shear is now rewritten in this equation: $V = \frac{M_a + M_b}{NL}$, where L is the length of span, in feet. M_a and M_b are the column bending moments above and below the girder. N is the number of spans or panels. To these results for vertical shear there must be added the shear values from static loads. See example.

Combining gravity and wind loads 8.4.6

The column bending moment resulting from wind pressure is usually treated as an eccentric load, and transformed into an axial load equivalent. The application of wind loads is the same for steel or concrete framing.

The girder wind bending moments are added to the static load moments and should be recorded in tabular form. The vertical shear from wind loads is combined with the shear from gravity and live loads at the connections of girders to columns.

The critical point for design of the connections is most often located at the end of girder, and may require some extensive investigation.

When architectural considerations permit, it is advisable to use brackets rather than going to a deeper section. The brackets can be fabricated from angles or plate in triangular shape. For design of steel moment connectors, see examples in Steel Design Section II.

EXAMPLE: High rise overturning stability 8.5

An eight (8) story structure is 48.0 x 100.0 feet in plan and 120.0 in height. Dead Loads of floors are approximately 50 Pounds per square foot. The dead load of the exterior brick and tile walls is estimated at 75 Pounds per square foot. The building code stipulates the following wind pressures shall be applied to vertical wall surface to calculate the moment on cantilever:

Lower 30.0 feet of height, wind pressure = 20 Lbs. Sq. Foot.
Next 30.0 feet of height, wind pressure = 30 Lbs. Sq. Foot.
Next 60.0 feet to top, wind pressure = 40 Lbs. sq. Foot.

EXAMPLE: High rise overturning stability, continued 8.5

East and West elevations are identical: 4 bays wide @ 25.0 Ft., and North-South elevations have 3 bays at 16.0 Ft.

REQUIRED:

Calculate the weight of structure above grade and use the dead loads only. Apply the wind pressure loads on east elevation and calculate the resultant overturning moment from wind. Calculate the resisting moment to tipping, then determine the percentage of stability. Code requires that wind moment shall not exceed 2/3 of resisting moment. Make single line drawings to aid calculations.

STEP I:

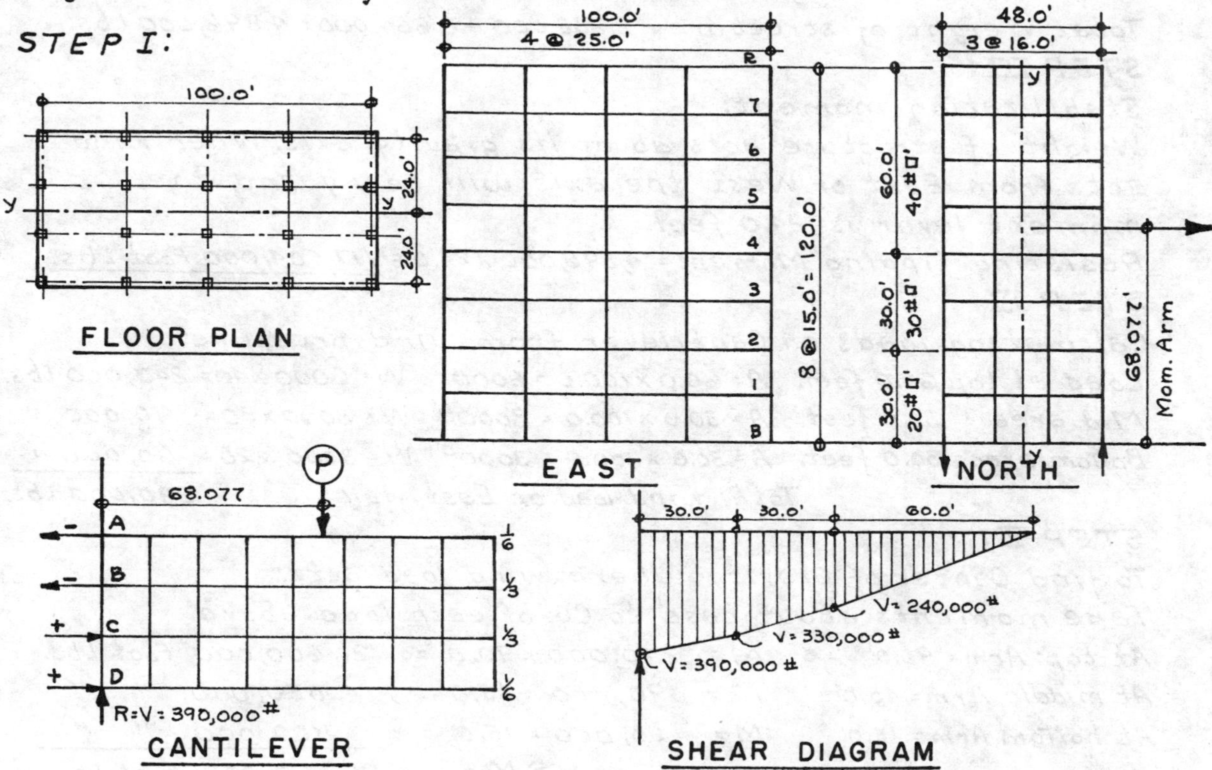

STEP II:

Calculating Weight of entire structure using dead load of roof, floors and exterior masonry walls.
Roof slab weight will be taken at 45 PSF and ground floor (B) will be taken as 70 PSF. All other floors at 50 PSF.

EXAMPLE: High rise overturning stability, continued 8.5

Area each floor and roof is 100.0 x 48.0 = 4800 Sq. Feet.

Roof DL Weight = 4800 x 45 =		216,000 Lbs.
Floor (B) Weight = 4800 x 70 =		336,000 "
Floors 1 to 7 incl. = 4800 x 50 x 7. =		1,680,000 "
	Total Weight Roof and Floors =	2,232,000 Lbs.

Exterior masonry wall areas and weight:

North and South walls:	A = 2 x 48.0 x 120.0 =	11,520 Sq. Ft.
East and West walls:	A = 2 x 100.0 x 120.0 =	24,000 " "
	Total exterior wall area =	35,520 Sq. Ft.

Total Dead Load weight ext. walls = 35,520 x 75 = 2,664,000 Lbs.

Total weight of structure = 2,232,000 + 2,664,000 = 4,896,000 Lbs.

STEP III:

Stabilization moment:

Weight of structure acts about its gravity axis. When wind acts from East or West the axis will be y-y and the moment lever is 24.0 feet.

Resisting tipping Moment = 4,896,000 x 24.0 = 117,504,000 Foot Lbs.

STEP IV:

Calculating loads on Cantilever from wind pressure:

Load at top 60.0 feet.	A = 60.0 x 100.0 = 6000□'	W = 6000 x 40 =	240,000 lbs.
Mid area = 30.0 feet.	A = 30.0 x 100.0 = 3000□'	W = 3000 x 30 =	90,000 "
Bottom area = 30.0 feet.	A = 30.0 x 100.0 = 3000□'	W = 3000 x 20 =	60,000 "
	Total wind load on East wall =		390,000 Lbs.

STEP V:

To find Center of Gravity where wind load acts:

Take moments about base to CG of each load above.

At top: Arm = 90.0'	M = 240,000 x 90.0 =	21,600,000 Foot Lbs.
At middle, Arm = 45.0'	M = 90,000 x 45.0 =	4,050,000 " "
At bottom Arm = 15.0'	M = 60,000 x 15.0 =	900,000 " "
	ΣM =	26,550,000 Foot Lbs.

$$\text{Location of Resultant} = \frac{26,550,000}{390,000} = 68.077 \text{ Feet from base.}$$

STEP VI:

Resisting moment from step III is 117,504,000 Foot Pounds and tipping moment from wind pressure = 26,550,000 Ft. Lbs. Tipping moment must not exceed 2/3 of resisting moment.

EXAMPLE: High rise overturning stability, continued 8.5

Percent of RM = $\frac{26,550,000}{117,504,000}$ = 22.4 percent which is less than 66.67%.

Percentage above Code requirements = 66.67 − 22.40 = 44.27 %.

P = 390,000 Lbs. Tipping moment arm = 68.077 Feet. and wind tipping moment = 390,000 x 68.077 = 26,550,000 Foot Lbs.

STEP VII:

Column Load distribution as shown on cantilever drawing. Since interior columns assume twice the load of the exterior columns, the distribution should be noted on all drawings. Column are indicated as A, B, C and D. Also tension is noted by minus (−) sign, and compression by plus (+) sign.

The total wind load can be converted to a single 25.0 foot bay while the moment arm of 68.077 feet will remain the same. For a single bay, P = $\frac{390,000}{4}$ = 97,500 Lbs.

STEP VIII:

Horizontal shear in bottom columns: The shear acting on a horizontal plane at base for each bay = 97,500 Lbs., and is distributed among 4 colums. The shear value of each column at its base times ½ the column will equal the bending moment in column. Then V for columns A and D = 1/6 of 97,500 Lbs., or 16,250 Lbs. each. Colums B and C have shear values of 32,500 Lbs. each.

STEP IX:

Bending moments in bottom columns: Height = 15.0 feet, and moment lever = ½ 15.0 or 7.50 feet.

M for Columns A and D = 16,250 x 7.50 = 121,875 Foot Lbs.

M for Columns B and C = 32,500 x 7.50 = 243,750 Foot Lbs.

STEP X:

The shear and bending moment distribution for columns when wind pressure is applied to north or south wall elevation which is shared by 5 column is thus:

3 Interior columns ¼ of total V for a single bent.

2 Exterior columns ⅛ of total V for a single bent.

EXAMPLE: Design criteria for high rise gravity and wind stresses 8.6

Design criteria and specifications: Applicable to Plates I to VI.

Project: Ten (10) floor apartment building. Ground floor to contain stores, salons, storage and maintenance facilities, AC and Heating, Elevator equipment. Eleventh floor to have play rooms, social club, etc.

Owner: Bayport Developement Corp.

Location: Hempsted Community, North Houston, Texas.

Architect: Jay Carroll & Associates, Houston, Texas.

Mech. Engr.: Michael Barr, PE., Beaumont, Texas.

Str. Engr.: Milo V. Walmer, PE., Beaumont-Galveston-Houston, Tex.

Contractor: Not selected

Dimensions: East-West elevations = 102.0' South-North = 62.0'

Floor Area: Approximately 6000 Sq. Feet per story

Code Appl.: Southern Standard and Houston municipal code.

Soil Engrs: McClelland Engr's. Houston, Tex.

ARCHITECTURAL REQUIREMENTS

Structural: Class A36 Steel. Columns, girders, beams, open web Joists, metal deck, concrete slabs and roof.

Code Group: Type H, Section 1301. Fire resisting.

Stair Wells: 12 inch masonary, concrete treads and risers.

Elevator: Electric, Shaft of 12 inch masonry. 3000 Lb. Cap.

Int. Walls: Dry-wall with steel studs-gypsum panels.

Ext. Walls: Exposed precast aggregate on 8 in. tile back-up.

Bracing: No wind diagonals. Rigid moment connections.

LOADS - CODE REQUIREMENTS

Roof Str. :	Live Load = 40#□'	Dead Load = 65#□'	Total = 105#□'
Ground Fl. :	Live Load = 100#□'	Dead Load = 180#□'	Total = 280#□'
Apartments :	Live Load = 40#□'	Dead Load = 58#□'	Total = 98#□'
Corridors :	Live Load = 100#□'	Dead Load = 65#□'	Total = 165#□'
Stairs :	Live Load = 100#□'	Dead Load = 90#□'	Total = 190#□'
Exterior :	Masonry walls 12"	Dead Load = 85#□'	
Sash etc. :	Steel or Alum.	Dead Load = 10#□'	

Live loads on apartments living quarters may be reduced 10% at top with same reduction for successive floors. In no case shall any LL be reduced more than 30%.

WIND LOADS AND PRESSURES
(Coastal Region)

Using Southern Standard Building Code: The pressures given have been adjusted to bring the pressure allowed per square foot to a point which will bring change on a level with lowest story floor line as follows:

1st. to 3rd. floor = 53.0 feet. Applicable W = 35 Lbs. Sq. Foot.
3rd. to 6th. floor = 45.0 feet. Applicable w = 45 Lbs. Sq. Foot.
6th. to Roof top = 87.0 feet. Applicable w = 50 Lbs. Sq. Foot.

Over-turning moment cannot exceed 66⅔ percent of the Resisting moment as calculated in any direction for wind and based on Dead Loads only. Wind pressure will be calculated by assuming critical wind direction coming from EAST. All bays in East Elevation are for 25.0 Ft. span. girder spans. Axis y-y for Resisting moment = 30.0 feet.

DESIGN STEP I:

From preliminary plans furnished by Architect, layout necessary elevations and plan for structural framing. Assume for time being that gravity loads on girders will be calculated later and the concern at this point is to determine the column loads. Gravity loads consist of Live loads plus Dead loads. Only the Dead Loads can determine the structures stability to resist the overtuning moment from wind pressure.

Calculations shall begin by computing the gravity loads transmitted to columns. The method of tabulating the column loads will be illustrated on Plate I.

By computing the roof or floor load as area x Dead load plus live load and noting the product on section, it can be very readily checked. In computing wall loads for corner colums, above the 11th. floor the east portion contains windows while the south elevation contains solid masonry. This

EXAMPLE: Gravity and wind stresses, continued 8.6

means that front and end elevations are needed for a quick reference as provided in Plates 2 and 3. The wall load is calculated and noted on section on inside as 16.23^k for Columns 6 and 9. Adding roof load 26.25^k to wall load 16.23^k the column load at 11th. floor level equals 42.38 kips.

Since Building Code allows a reduced live load for upper floors, because the assumption is made that no area in multi-story buildings is occupied simultaneously, and thus the reduced Live load is justified. These reduced loads are recorded is tabular form as shown.

• GRAVITY LOADS • ESTIMATED UNITS FOR COLUMN DATA •							
LOCATED FLOOR	DEAD LOAD PSF	LIVE LOAD PSF	REDUCED %	LOCATED FLOOR	DEAD LOAD PSF	LIVE LOAD PSF	REDUCED %
ROOF	65	40	NONE	4th.	58	32	20
11th.	58	32	20	3rd.	58	32	20
10th.	58	32	20	2nd.	58	36	10
9th.	58	32	20	1st.	58	40	NONE
8th.	58	32	20	BAS.	180	100	NONE
7th.	58	32	20				
6th.	53	32	20				
5th.	58	32	20				
EXTERIOR WALL MASONRY = 85 PSF				GLAZED SASH = 10 PSF.			

CORNER COLUMNS 1-2-3-4 - STATIC LOADS

AT ROOF: (Plot on Plate 1).

Live Load = 40#□' D.L = 65#□' Total: 105# Area on Col. = 10.0 x 12.5 = 125.0□'

Flat slab load = 125.0 x 105 = 13,130 Lbs. (Plotted as 13.13^k)

For exterior wall D.L. Height = 12.0' Area wall = (10.0 + 12.5) x 12.0' = 270.0□'

Neglect Glazing. Wall weight = 270.0 x 85 = 22,950 Lbs. (Plot as 22.45 kips)

For weight on Corner Cols. above level of 11th. Floor: Wt. = 13.13 + 22.45 = 36.08^k

AT FLOORS 11th. DOWN TO 2nd:

L.L. + DL. = 32 + 58 = 90#□' Area = 125.0□' Floor weight = 125.0 x 90 = 11,250 Lbs.

Exterior wall Heights are now 15.0 feet down to 2nd. floor level.

Area wall = (10.0 + 12.5) x 15.0' = 337.5□' Wall weight = 337.5 x 85 = 28,688# (28.69^k).

Weight on Corner Column at 10th. Fl. = 36.08 + 11.25 + 28.69 = 76.02 kips (Plot.

Now floor loads down to 2nd. Floor = 11.25^k and walls = 28.69^k These will be plotted on section Plate 1.

Column Loads on Corners:	9th. Fl. = 76.02 + 11.25 + 28.69 = 115.96^K
" " " "	8th. Fl. = 115.96 + 11.25 + 28.69 = 155.90^K
" " " "	7th. Fl. = 155.90 + 11.25 + 28.69 = 195.84^K
" " " "	6th. Fl. = 195.84 + 11.25 + 28.69 = 235.78^K
" " " "	5th. Fl. = 235.78 + 11.25 + 28.69 = 275.72^K
" " " "	4th. Fl. = 275.72 + 11.25 + 28.69 = 315.66^K
" " " "	3rd. Fl. = 315.66 + 11.25 + 28.69 = 355.60^K
" " " "	2nd. Fl = 355.60 + 11.25 + 28.69 = 395.54^K

At Floor level of 2nd. floor slab:
Loads on column to this point total 395.44 KIPS. (395,540 Lbs.)
Load reduction for Live Load = 10% or LL = 40#□' x 90% = 36#□'
DL. = 58#□' Combined Floor loads = 36 + 58 = 94#□' A = 125.0 Sq. Ft.
Floor load on column from 2nd. Floor = 125.0 x 94 = 11,750# (Post 11.75^K)
Wall height is increased to 18.0' Wall area = 22.5 x 18.0 = 405.0 Sq. Ft.-
Wall load supported by corner columns = 405.0 x 85 = 34,425# (34.43^K) (post.)
Column Load at 1st. Fl. level = 395.54 + 11.75 + 34.43 = 441.42 kips. (Posted)

At Ground Floor Level B: (Plate 2 Index 8.6.2)
East wall has plate glass with masonry bulkhead. Floor to floor height = 20.0 feet. Future tenants of building may replace glass with masonry. Therefore use 85#□' for weight of wall. No reduction in live load from 40 PSF. A = 125.0□' Total DL + LL = 98#□'
Floor weight on Corner columns = 125.0 x 98 = 12,250 Lbs. (Post 12.25^K plate 1)
Area wall = (12.5 + 10.0) x 20.0 = 450 Sq. Ft. Wt. wall = 450 x 85 = 38,250 Lbs.
Load on column at ground floor level = 441.42 + 12.25 + 38.25 = 492.22 kips.

At Basement below Ground Floor:
Floor assumed to be framed into columns. Live load = 100 PSF.
Dead load = 180 PSF. Total DL + LL = 280 PSF. Area = 125.0 Sq. Ft.
Load to each corner column from ground floor = 125.0 x 280 = 35,000 Lbs.
Column load below floor B = 492.22 + 35.00 = 527.22^K (Post on Plate 1).

REMOVING LIVE LOADS FROM COLUMNS.

In the previous table for loads, add up the column of live loads from Roof down to floor B. Total for 12 floors plus Roof = 504 Lbs. Sq. foot. Area floor at corner = 125.0 Sq. Ft.
Column deduction for LL = 125.0 x 504 = 63,000 Lbs. (63.0^K)
Dead Load on each corner column = 527.22 - 63.00 = 464.22 kips.

STEP II:
STATIC COLUMN LOADS ON EXTERIOR COLS. 6 and 9. (See Plate 1)
Area column supports = 25.0 x 10.0 = 250 Sq. Ft. (Hatched in plan.)

EXAMPLE: Gravity and wind stresses, continued 8.6

At ROOF: Combined DL + LL = 105 #□' Col. Load = 250 × 105 = 26.25 kips.
Wall exterior at East elevation has length of 25.0 feet with a glased sash height of 5.0 feet. Balance = 7.0' masonry and conc.
Glazed unit wt. = 10 Lbs. Sq. Ft. Mas. unit wt = 85 #□'.
Weight glazed wall on Column = 25.0 × 5.0 × 10 = 1250 Lbs.
Weight masonry on column = 25.0 × 7.0 × 85 = 14,875 Lbs.
Total weight wall = 16,125 Lbs.

Total weight on column 6 at 11th. floor level = 26.25 + 16.13 = 42.38 kips.

Floor loads + Wall loads on Column 6 from 11th to 2nd Floor level.
Reduced LL = 32 #□' D.L = 58 #□' Combined LL + DL = 90 #□' Floor area on Column 6 = 25.0 × 10.0 = 250 □' Then 11th. Floor load = 250 × 90 = 22.50 k.
Post these values on Plate 1 for column 6 and 9.
Height walls = 15.0 ft. Sash height = 9.0' and masonry height = 6.0 ft.
Wall weight on col. 6 = (25.0 × 9.0 × 10) + (25.0 × 6.0 × 85) = 15.00 kips.
Post these values on section shown on Plate 1 down to 2nd. fl.
Compiling loads on Column 6 from floor and exterior walls.

Column	6	Load	at	11th.	floor	level	= 42.38 kips.	
"	6	"	"	10th.	"	"	= 42.38 + 22.50 + 15.00 =	79.88 k.
"	6	"	"	9th.	"	"	= 79.88 + 22.50 + 15.00 =	117.38 k.
"	6	"	"	8th.	"	"	= 117.38 + 22.50 + 15.00 =	154.88 k.
"	6	"	"	7th.	"	"	= 154.88 + 22.50 + 15.00 =	192.38 k.
"	6	"	"	6th.	"	"	= 192.38 + 22.50 + 15.00 =	229.88 k.
"	6	"	"	5th.	"	"	= 229.88 + 22.50 + 15.00 =	267.38 k.
"	6	"	"	4th.	"	"	= 267.38 + 22.50 + 15.00 =	304.88 k.
"	6	"	"	3rd.	"	"	= 304.88 + 22.50 + 15.00 =	342.38 k.
"	6	"	"	2nd.	"	"	= 342.38 + 22.50 + 15.00 =	379.88 k.

Load on Floor and column at 1st floor level:
Live load on 2nd Floor = 36 PSF DL = 58 PSF. Combined Loads = 94 PSF.
Floor Load = 25.0 × 10.0 × 94 = 23.50 kips. (Plotted on section plate I)
Wall height Fl. 1 to 2 = 18.0' Sash height = 11.0' Mas. height = 7.0'
Weight of wall = (25.0 × 11.0 × 10) + (25.0 × 7.0 × 85) = 17.63 kips. (Plot)
Column 6 Load at 1st floor level = 379.88 + 23.50 + 17.63 = 421.01 kips.

Load on Colum 6 at ground level:
First floor DL + LL = 58 + 40 = 98 PSF. Load = 98 × 250 = 24.50 kips.
Height wall = 20.0 Ft. Figure as all masonry, same as corners.

Wt. Wall between ground floor and 1st. = 25.0 x 20.0 x 85 = 42.50 K.
Column 6 Load at ground floor level = 421.01 + 24.50 + 42.50 = 488.01 KIPS.

Adding Ground Floor to Column 6:
Live Load = 100 PSF DL = 180 PSF Fl. Wt = 10.0 x 25.0 x 280 = 70.00 KIPS.
Load on Column 6 below grade = 488.01 + 70.00 = 558,000 Lbs.

Removing all Live Loads from Column 6. Floor area = 250 Sq. Ft.
Live Load PSF Roof to below grade = 504 PSF.

Total Live Load on Column 6 = 504 x 250.0 = 126,000 Lbs.
Then Dead Load on Column 6 = $558.0^K - 126.0^K$ = 432.0 KIPS.

STEP III:
Estimate the Dead Load weight of structure for stability and resistance to overturning from wind pressure:
From Plate 2, Determine weight and areas of exterior walls.
East Elevation: (West elevation is same).

2 End bays of masonry	=	2 x 25.0 x 153.0 x 85 =	650,250 Lbs.
2 End bays Glazed	=	2 x 25.0 x 20.0 x 10 =	10,000 "
2 End Top bays Glazed	=	2 x 25.0 x 12.0 x 10 =	6,000 "
		Total Wt. =	748,750 Lbs.
		Add for identical West Elev. =	748,750 "
		Total weight East & West Ext. Walls =	1,497,500 Lbs.

North Elevation: (South elevation is same)
Neglect glazed Panels and figure all masonry.
Area North & South Elevations: = 2 x 60.0 x 185.0 = 22,200 Sq. Feet.
Weight Exterior walls North & South = 22,200 x 85 = 1,887,000 Lbs.
Total Weight of ALL Exterior Walls = 3,384,500 Pounds

Permanent Interior Walls: Plate 1 Plan.
Walls of Stair wells and Elevator shaft are 12" tile - extend Top to Bottom.
Periphery of 2 Stair wells and 1 Shaft: 150 Feet each floor.
Weight of tile walls = 150.0 x 185.0 x 55 = 1,526,250 Lbs.

Reactions and structure for 2-3000# Cap. Elevators with 185.0 height taken from Elevator Catalog. DL. Wt = 95,080 Lbs.

Dead Load Weight Roof and Floor structure: (Plate 1)
Use Table in Step I: Total PSF D.L for full height = 883 Lbs. Sq. Ft.
Area in Plan = 100.0 x 60.0 = 6000 Sq. Feet.
D.L. Weight Roof and 12 Floors = 6000 x 883 = 5,298,000 Lbs.

EXAMPLE: Gravity and wind stresses, continued 8.6

Total Dead Load Weight of Structure Top to Ground:
Weight = 1,497,500 + 3,384,500 + 1,526,250 + 95,080 + 5,298,000 =
Total Wt = 11,801,330 Lbs.

STEP IV:
Resisting Moment for over-turning from Wind Pressure:
Critical axis occurs when wind from East or West.
On Plan in Plate 1, Moment arm is axis y-y to where the Wind Pressure is applied, ie, East or West Wall. If wind is against South or North wall, the moment arm is axis x-x.
Critical = least moment arm for Resisting Moment.

About Axis Y-Y: RM = 11,801,330# x 30.0' = 354,039,900 Foot Pounds.

STEP V:
Tipping Moment from Horizontal East Wind Pressure:
Plate 4 will now be drawn to tabulate wind loads with cantilever indicated to locate Resultant of wind forces. Calculations can be taken based upon a single bay width of 25.0, however to compare tipping moment with result of Resisting Moment above, all four bays must be involved.

At 3 Lower stories, pressure = 35 PSF. Height = 53.0' Moment arm = 26.50'
At 3 next stories up, pressure = 45 PSF. Height = 45.0' Moment arm = 75.50'
At 6 Upper stories, pressure = 50 PSF. Height = 87.0' Moment arm = 141.50'

STEP VI:
Total Wind Pressure on East Wall (4 bays) and Moments:
Upper 6 Stories: Wind Pressure = 100.0 x 87.0 x 50 = 435,000 Lbs.
Next 3 Stories: " " 100.0 x 45.0 x 45 = 202,500 "
Lower 3 Stories: " " 100.0 x 53.0 x 35 = 185,500 "
Total Wind Load on Cantilever = 823,000 Lbs.

Take Moment about Base to find Center of Gravity:
Upper 6 Stories: M = 435,000 x 141.50 = 61,552,500 Foot Lbs.
Next 3 Stories: M = 202,500 x 75.50 = 15,288,750 " "
Lower 3 Stories: M = 185,500 x 26.50 = 4,915,750 " "
Summation (Σ) Moments = 81,757,000 Foot Lbs

Location of CG and distance from Base = Moment lever. Then:
Moment Arm = $\frac{81,757,000}{823,000}$ = 99.34 Feet. The tipping moment is: M = 823,000 x 99.34 = 81,757,000 '# (or same as ΣM).

STEP VI:

Comparing Resisting Moment to Wind overturning Moment:
Max. allowed Tipping moment = 2/3 RM or 66.67 Percent.
2/3 of RM = 354,039,900 x 66.67 = 236,026,600 Foot Lbs. (Stable).

$$\text{Percentage of TM of RM} = \frac{81,757,000}{354,039,900} = 23.1 \text{ Percent (ok).}$$

STEP VII:

Wind Loads at each floor must be determined. These values will be figured and inserted in tabular form befor they are plotted on North Elevation drawn as Plate 4. Only one (1) panel width of 25.0 feet will be used in making calculations. Only 1 Column is involed on East elevation where wind load is to be applied, however the distribution of wind forces for all 4 Columns in same plane will come later. The following for tabulating Wind pressure W and horizontal shear W' is suggested.

LOAD AND SHEAR TABULATION		•WIND FROM EAST•	
W_R	25.0' x 6.0' x 50 = 7500	W'_R	= 7,500 LBS.
W_{11}	(25.0 x 6.0 x 50) + (25.0 x 7.5 x 50) = 16,875	W'_{11}	16,875 + 7500 = 24,375
W_{10}	25.0 x 15.0 x 50 = 18,750	W'_{10}	24,375 + 18,750 = 43,125
W_9	25.0 x 15.0 x 50 = 18,750	W'_9	43,125 + 18,750 = 61,875
W_8	25.0 x 15.0 x 50 = 18,750	W'_8	61,875 + 18,750 = 80,625
W_7	25.0 x 15.0 x 50 = 18,750	W'_7	80,625 + 18,750 = 99,375
W_6	(25.0 x 7.5 x 50) + (25.0 x 7.5 x 45) = 17,812.5	W'_6	99,375 + 17,812.5 = 117,187.50
W_5	25.0 x 15.0 x 45 = 16,875	W'_5	117,187.5 + 16,875 = 134,062.50
W_4	25.0 x 15.0 x 45 = 16,875	W'_4	134,062.5 + 16,875 = 150,937.50
W_3	(25.0 x 7.5 x 45) + (25.0 x 7.5 x 35) = 15,000	W'_3	150,937.5 + 15,000 = 165,937.50
W_2	(25.0 x 7.5 x 35) + (25.0 x 9.0 x 35) = 14,437.5	W'_2	165,937.5 + 14,375.5 = 180,375
W_1	(25.0 x 9.0 x 35) + (25.0 x 10.0 x 35) = 16,625	W'_1	180,375 + 16,625 = 197,000
W_B	25.0 x 10.0 x 35 = 8,750	W'_B	197,000 + 8,750 = 205,750

Use Plate 4 for computing Wind Load in left column in above:
Start a Roof: Bay width = 25.0' Height = 12.0' Wind Pressure = 50#□'
Horizontal force at Roof = 25.0 x ½H or 25.0 x 6.0 x 50 = 7500.#
The Horizontal Force at 11th. Floor is on height ½ of 12.0 + ½ of 15.0 = 13.50'
W_{11} = 25.0 x 13.5 x 50 = 16,875# Shear from above is equal to sum of all loads above. The W'_{11} = 7,500 + 16,875 = 24,375 Lbs., and is placed in right side column in table. These values will again be checked when they are plotted on North Elevation of Plate-4-. Finally: Refer back to Step VI where the total wind load on East Elevation was found to be 823,000 Lbs. This is total horizontal shear at ground

EXAMPLE: Gravity and wind stresses, continued 8.6

level applicable to 4 Bays. Then the last figure for W'_B in table should check and equal to: $W'_B = \frac{823,000}{4} = 205,750$ Lbs. (checks OK).

STEP VIII:

Wind Pressure applied to South wall:

In the event Steel framing is to be the choice, all steel columns will be turned with axis x-x parallel to longest wall dimension. The moment connectors for column to girder will in such case be less and perpendicular to column axis y-y.

LOAD AND SHEAR TABULATION • WIND FROM SOUTH •					
W_R	20.0 x 6.0 x 50 =	6,000 LBS.	W'_R	=	6,000 LBS.
W_{11}	(20.0 x 6.0 x 50) + (20.0 x 7.5 x 50) =	13,500 "	W'_{11}	6,000 + 13,500 =	19,500 "
W_{10}	20.0 x 15.0 x 50 =	15,000 "	W'_{10}	19,500 + 15,000 =	34,500 "
W_9	20.0 x 15.0 x 50 =	15,000 "	W'_9	34,500 + 15,000 =	49,500 "
W_8	20.0 x 15.0 x 50 =	15,000 "	W'_8	49,500 + 15,000 =	64,500 "
W_7	20.0 x 15.0 x 50 =	15,000 "	W'_7	64,500 + 15,000 =	79,500 "
W_6	(20.0 x 7.5 x 50) + (20.0 x 7.5 x 45) =	14,250 "	W'_6	79,500 + 14,250 =	93,750 "
W_5	20.0 x 15.0 x 45 =	13,500 "	W'_5	93,750 + 13,500 =	107,250 "
W_4	20.0 x 15.0 x 45 =	13,500 "	W'_4	107,250 + 13,500 =	120,750 "
W_3	(20.0 x 7.5 x 45) + (20.0 x 7.5 x 35) =	12,000 "	W'_3	120,750 + 12,000 =	132,750 "
W_2	(20.0 x 7.5 x 35) + (20.0 x 9.0 x 35) =	11,550 "	W'_2	132,750 + 11,550 =	144,300 "
W_1	(20.0 x 9.0 x 35) + (20.0 x 10.0 x 35) =	13,300 "	W'_1	144,300 + 13,300 =	157,600 "
W_B	20.0 x 10.0 x 35 =	7,000 "	W'_B	157,600 + 7,000 =	164,600 "

Plate 5 will now be drawn to aid in calculations for values of W and W' with respect to wind from south or north. The moment lever for cantilever will be the same or 99.34 ft. from base. South Elevation has 20.0 foot bays and width structure = 60.0 feet. Calculate Wind pressure on South Elevation! (With 3 Bents).

Upper 6 Stories - Pressure = 60.0 x 87.0 x 50 = 261,000 Lbs.
Middle 3 Stories - Pressure = 60.0 x 45.0 x 45 = 121,500 "
Lower 3 Stories - Pressure = 60.0 x 53.0 x 35 = 111,300 "
Total Wind Pressure on South Wall = 493,800 Lbs.

Tipping Moment about base = 493,800 x 99.34 = 49,054,092 Foot Lbs.
The moment arm for Resisting Moment is 50.0 Ft. or axis x-x.
Total Wt. from Step III = 11,801,330 Lbs. RM_x = 11,801,330 x 50.0 = 590,066,500 '#
Percentage of Resisting Moment = $\frac{49,054,092}{590,066,500}$ = 8.32 Percent.

Checking last figure for W'_B in table above: $W'_B = \frac{493,800}{3} = 164,600$#

STEP IX:

CALCULATING BENDING MOMENT IN COLUMNS- (WIND FROM EAST)

The Column bending moments resulting from wind forces is to be divided in such a manner that the Interior Columns will have twice the Bending Moment of the Exterior Columns. This distribution will be noted on Plate 4. Exterior columns will take 1/6 of the total bending moment and the interior columns each take 1/3 of the Total Bending Moment. This distribution is conveniently accomplished with two simple formulas.

Exterior Column $BM = \frac{W'H}{4N}$. Interior Column $M = \frac{W'H}{2N}$.

Where: W' = Horizontal Shear value at a given story and taken from Table compiled is Step VII (East) or Step VIII (South wind). The total moment for all Columns can be calculated as W'½H, where H equals heigth of Column floor to floor. The Total Moment can then be dirstributed according to the number of panels. This Total Moment shall be recorded in a tabular form since it will be required for later computations. The tabular form recommended is shown with column moments given for total, exterior and interior columns.

• COLUMN WIND BENDING MOMENTS •			WIND FROM EAST •		
COL'S. MARK	SHEAR W' #	LEVER ½H'	TOTAL MOMENT W' × ½ H = M	EXTERIOR COL'S. 1/6 M	INTERIOR COLS 1/3 M
H_R	7,500 LBS.	6.0'	45,000 FT. LBS	7,500 FT. LBS.	15,000 FT. LBS
H_{11}	24,375 "	7.5'	182,813 " "	30,470 "	60,940 "
H_{10}	43,125 "	7.5'	323,437 " "	53,906 "	107,812 "
H_9	61,875 "	7.5'	464,062 " "	77,346 "	154,692 "
H_8	80,625 "	7.5'	604,687 " "	100,781 "	201,562 "
H_7	99,375 "	7.5'	745,313 " "	124,219 "	248,438 "
H_6	117,187.5 "	7.5'	878,906 " "	146,484 "	292,968 "
H_5	134,062.5"	7.5'	1,005,465 " "	167,577 "	335,154 "
H_4	150,937.5"	7.5'	1,132,045 " "	188,674 "	377,348 "
H_3	165,937.5"	7.5'	1,244,535 " "	207,428 "	414,856 "
H_2	180,375 "	9.0'	1,623,375 " "	270,562 "	541,124 "
H_1	197,000 "	10.0'	1,970,000 " "	328,333 "	656,667 "
H_B	205,750 "				

EXAMPLE: Gravity and wind stresses, continued 8.6

The quantities in Table will be reduced to kips and plotted on North Elevation shown as Plate 4. Since building is symmetrical about ℄ or axis y-y, the plotting may be on 1 side only. Space on skelton frame will be required to plot values of bending moments and vertical shear will now be calculated.

STEP X:

BENDING MOMENTS IN GIRDERS & CONNECTIONS:

The bending moment at each connection for a girder to column is the same for all columns in the same plane. Therefore the connection is the same design for both exterior and interior columns. At an interior column the bending moment is the mean value of two adjecent columns. The bending moment in the column above and below the girder are added together and divided by 2. The distribution of moment is made according to the number of panels (N). Written into a formula, the bending moment for each girder connection is thus:

$M = \frac{1}{2}\left(\frac{W_a' H_a}{2N}\right) + \left(\frac{W_b' H_b}{2N}\right)$. Where W_a' = Horizontal shear as shown on plate 4. H_a = Column height above and H_b = Column height below.

When Table of Moments as prepared in step VIII is used, the work is shortened. For instance; the Total Moment in column H_8 = 604,687'# and M for H_7 = 745,313'#. Adding $H_8 + H_7$ together and dividing by 2, the Mean moment = $\frac{1,350,000}{2}$ = 675,000 Ft. Lbs.

Number of panels: N = 3

For each girder connection: $M_7 = \frac{675,000}{3}$ = 225,000 Ft. Lbs.

The formula is simplified thus:

$M_g = \frac{M_a + M_b}{2N}$. Where M_a = Total Moments above and M_b = Total Column bending moments below.

Girder moments must be recorded and the suggested form given below will assist the designer in calculations and permit easy checking:

• GIRDER BENDING MOMENTS • WIND FROM EAST • PLATE-4-						
GIRDER MARK	TOTAL COLUMN BENDING MOMENTS			MEAN MOMENT	PANELS	GIRDER MOMENT
	M_a = Ft. Lbs.	M_b = Ft. Lbs.	$M_a + M_b$	$\frac{M_a + M_b}{2}$	N	$\frac{M_a + M_b}{2N}$ = '#.
G_R	0	45,000	45,000	45,000	3	15,000
G_{11}	45,000	182,813	227,813	113,906	3	37,968
G_{10}	182,813	323,437	505,750	252,875	3	84,292
G_9	323,437	464,062	787,499	393,750	3	131,250
G_8	464,062	604,687	1,068,749	534,375	3	178,125
G_7	604,687	745,313	1,350,000	675,000	3	225,000
G_6	745,313	878,906	1,624,219	812,109	3	270,703
G_5	878,906	1,005,465	1,884,371	942,186	3	314,062
G_4	1,005,465	1,132,045	2,137,510	1,068,755	3	356,252
G_3	1,132,045	1,244,535	2,376,580	1,188,290	3	396,097
G_2	1,244,535	1,623,375	2,867,910	1,433,955	3	477,985
G_1	1,623,375	1,970,000	3,593,375	1,796,688	3	588,896
G_B	1,970,000	1,646,000	3,616,000	1,808,000	3	602,667

Moment for M_b below ground floor was assumed that Column height is 16.0 Ft. $M = 205,750 \times 8.0 = 1,646,000$ Ft. Lbs.

Since the bending moment in all girders from wind forces is the same value for each girder on same floor, the plotting on elevation Plate 4 only requires a single entry. These moments are to be added to gravity load bending moments.

STEP XI:

VERTICAL SHEAR IN GIRDERS RESULT OF WIND FORCE:

The Vertical Shear in a girder produced by wind pressure is calculated similar to the bending moments performed in the preceding step X. Like bending moment, the vertical shear in the girder is a function of the horizontal shears above and below the girder, of the column heights, and of the span or panel lengths (L). Distribution of shear is the same as used for bending moments. The established formula for distributed vertical is thus: $V = \frac{(W_a' H_a) + (W_b' H_b)}{2NL}$. *Where: L = Span or Panel length is feet. On close examination, much of this formula has previously been equated, because;* $\frac{(W_a' H_a) + (W_b' H_b)}{2}$ *is the Column bending moment above and below.*

EXAMPLE: Gravity and wind stresses, continued 8.6

The equation amounts to the Total Moments as: $M_a + M_b$. These values will have been recorded in preceding table for step X. Revised, the formula is: $V = \frac{M_a + M_b}{NL}$. Let the comparisons be illustrated for checking:

From Plate 4 the shear values will be taken above and below the girder at 8th. Floor. $W_9' = 61{,}875^{\#}$ and $W_8' = 80{,}625^{\#}$. H_9 and H_8 are same height at 15.0 feet. $N = 3$ Panels and $L = 20.0$ feet. Then:

$$V = \frac{(61{,}875 \times 15.0) + (80{,}625 \times 15.0)}{2 \times 3 \times 20.0} = \frac{2{,}137{,}150}{120} = 17{,}812.5 \text{ Lbs.}$$

Also, the value in table for $M_9 + M_8 = 464{,}062 + 604{,}687 = 1{,}068{,}749$ Ft. Lbs., and

$$V = \frac{1{,}068{,}749}{3 \times 20.0'} = 17{,}812.5 \text{ Lbs.}$$ (Check same in value)

The fact that the Wind direction may change from East to West, requires that all bending moments in each girder on a single floor shall be of the same magnitude. Under like circumstances the vertical shear value (V) must be the same at each column connection.

When the number of panels correspond to the number used in preceding tabulation and the panel lengths (L) are of the same length, the vertical shear (V) will be equal to the Girder Bending Moment divided by ½ of length L. This method is not advised for calculating shear and the tabulated system should be used as follows:

•VERTICAL SHEAR IN GIRDERS• WIND FROM EAST• LBS.						
GIRDER MARK	•COLUMN BENDING MOMENTS• IN FT. LBS.			PANELS	LENGTH	$V = \frac{M_A + M_B}{NL}$
	M Above G	M below G	$M_A + M_B$	N	L = Ft.	
G_R	------	45,000	45,000	3	20.0	750 #
G_{11}	45,000	182,813	227,813	3	20.0	3,797
G_{10}	182,813	323,437	506,250	3	20.0	8,404
G_9	323,437	464,062	787,499	3	20.0	13,125
G_8	464,062	604,687	1,068,749	3	20.0	17,812.5
G_7	604,687	745,313	1,350,000	3	20.0	22,500
G_6	745,313	878,906	1,624,219	3	20.0	27,070
G_5	878,906	1,005,465	1,884,371	3	20.0	31,406
G_4	1,005,465	1,132,045	2,137,510	3	20.0	35,625
G_3	1,132,045	1,244,535	2,376,580	3	20.0	39,610
G_2	1,244,535	1,623,375	2,867,910	3	20.0	47,800
G_1	1,623,375	1,970,000	3,593,375	3	20.0	59,890
G_B	1,970,000	1,646,000	3,616,000	3	20.0	60,267

STEP XII:

The value of vertical shear in girders from Wind Load is to be combined with vertical shear from static floor loads. The value in right hand column in above table is the vertical shear at each girder connection to column. Refer to plate 4 and note the method for plotting vertical shear values. By referring to plan and Column sections in Plate 1, it will be seen that the floor loads on girders extending in East to West direction have only a small amount of shear from floor loads. The main shear loads from floor joists is transmitted to girders running in the North-South direction. Plate 5 is provided to aid the student designer in the work of extending the example to include the results of wind from the south. This work may be accomplished by starting with Step VII. By preparing the suggested form for tabulating horizontal shears W and W', the work will have proper sequence and continuity which is desired for accuracy.

PLATE 1: Gravity dead and live column loads 8.6.1

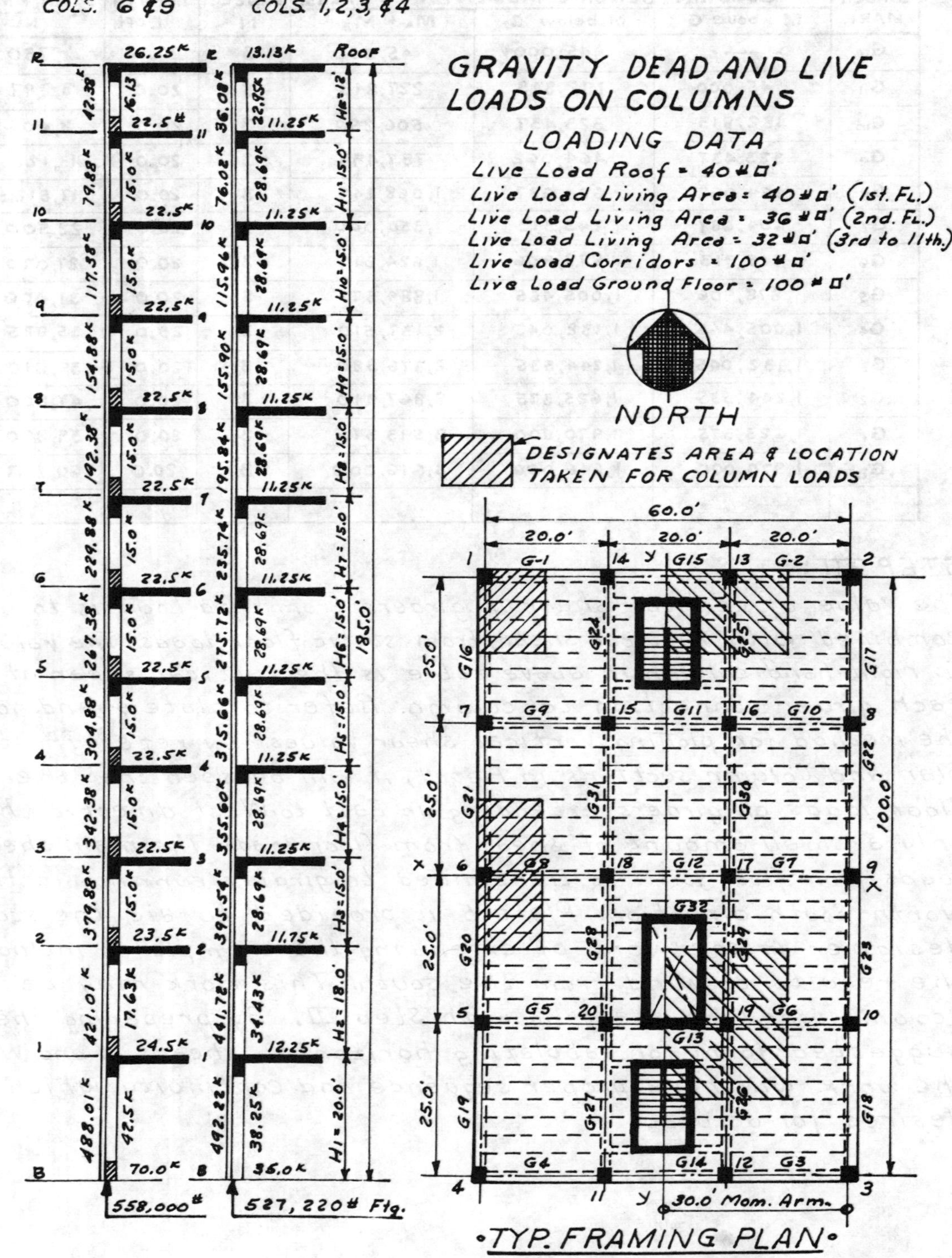

•TYP. FRAMING PLAN•

SCALE: 1" = 20.0'

PLATE 2: East and West elevation 8.6.2

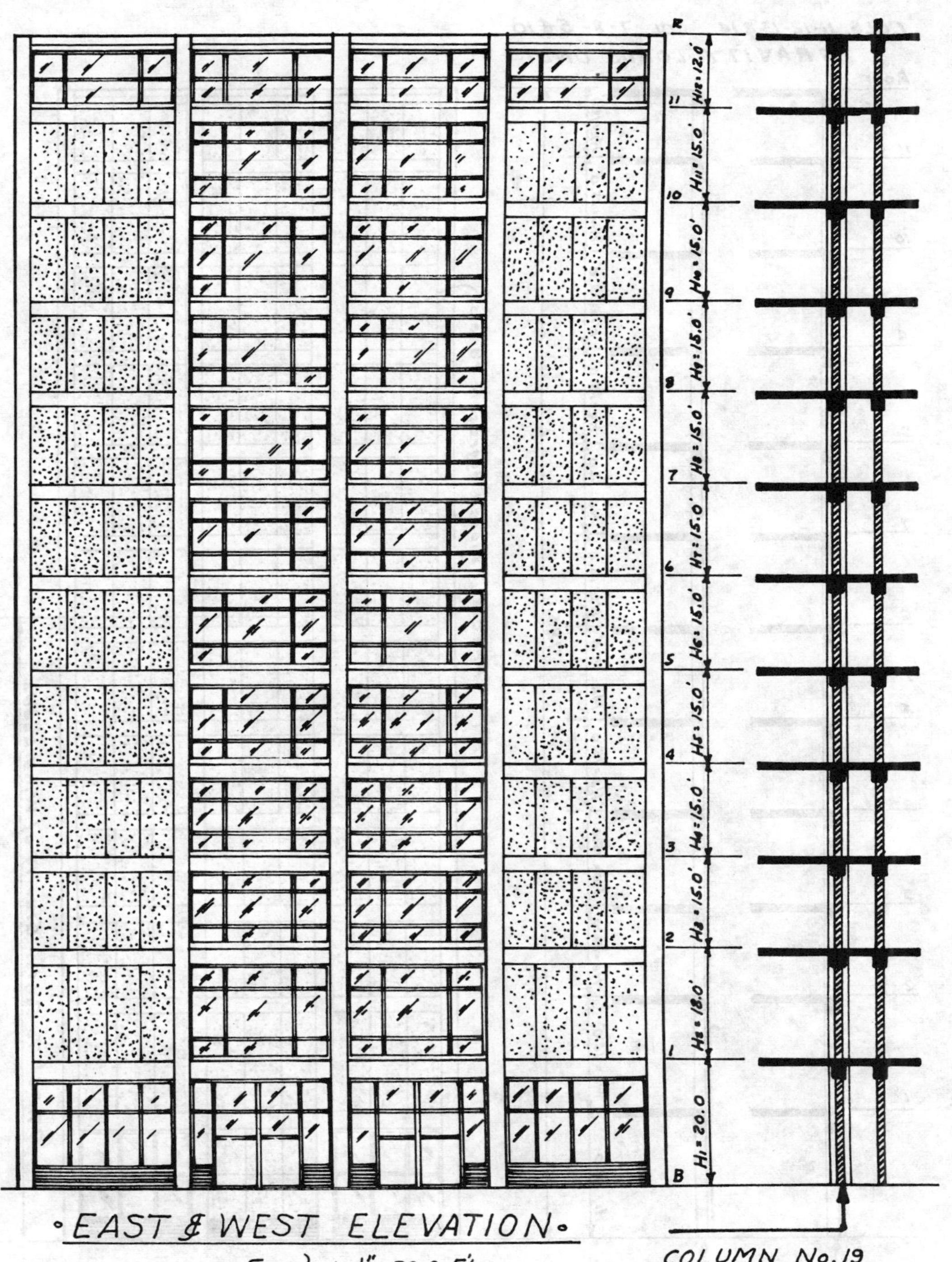

EAST & WEST ELEVATION

Scale: 1"=20.0 Ft.

PLATE 3: North and South elevation 8.6.3

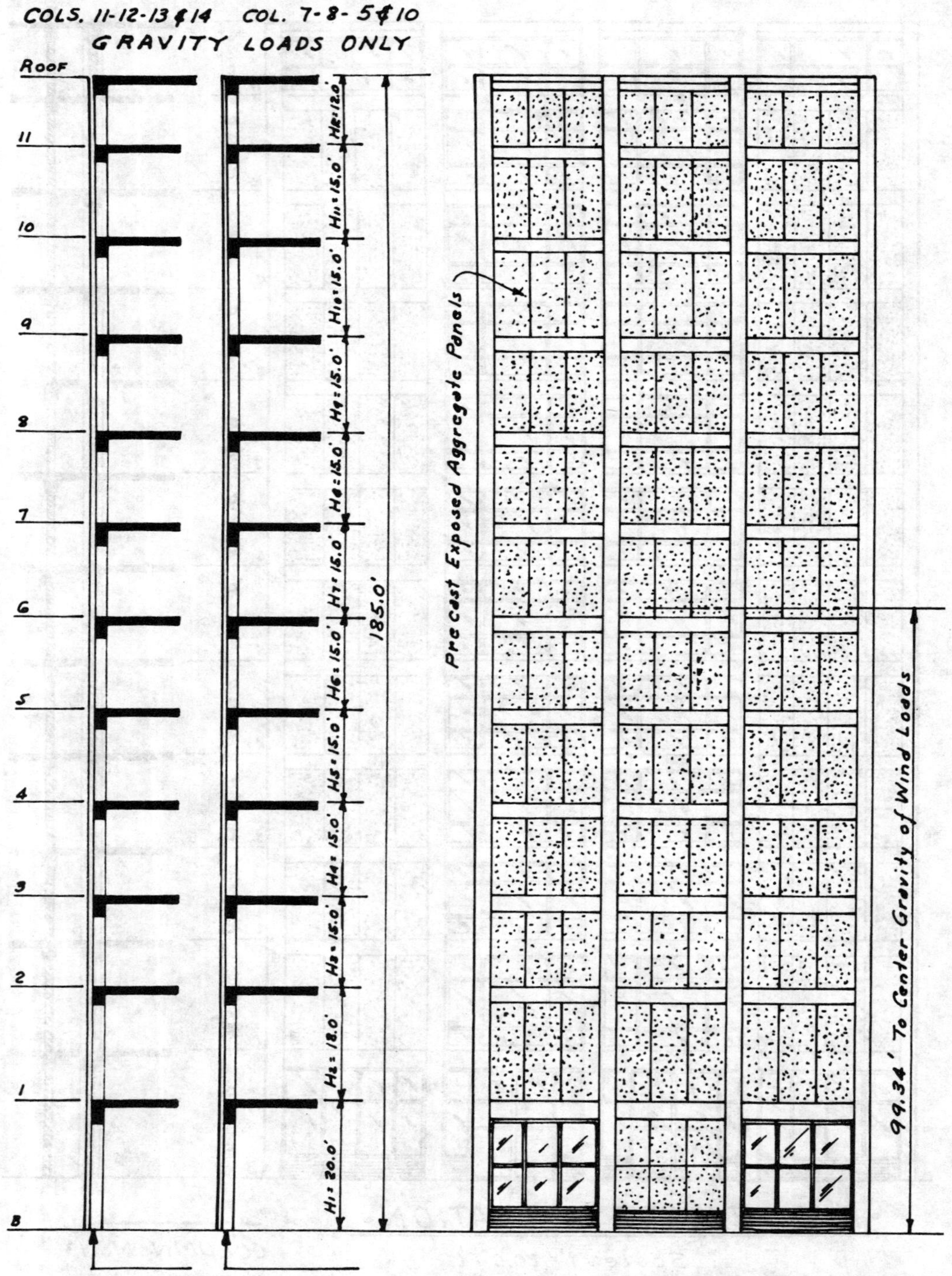

NORTH & SOUTH ELEVATIONS

Scale: 1" = 20.0 Ft.

PLATE 4: Wind loading from East, with cantilever 8.6.4

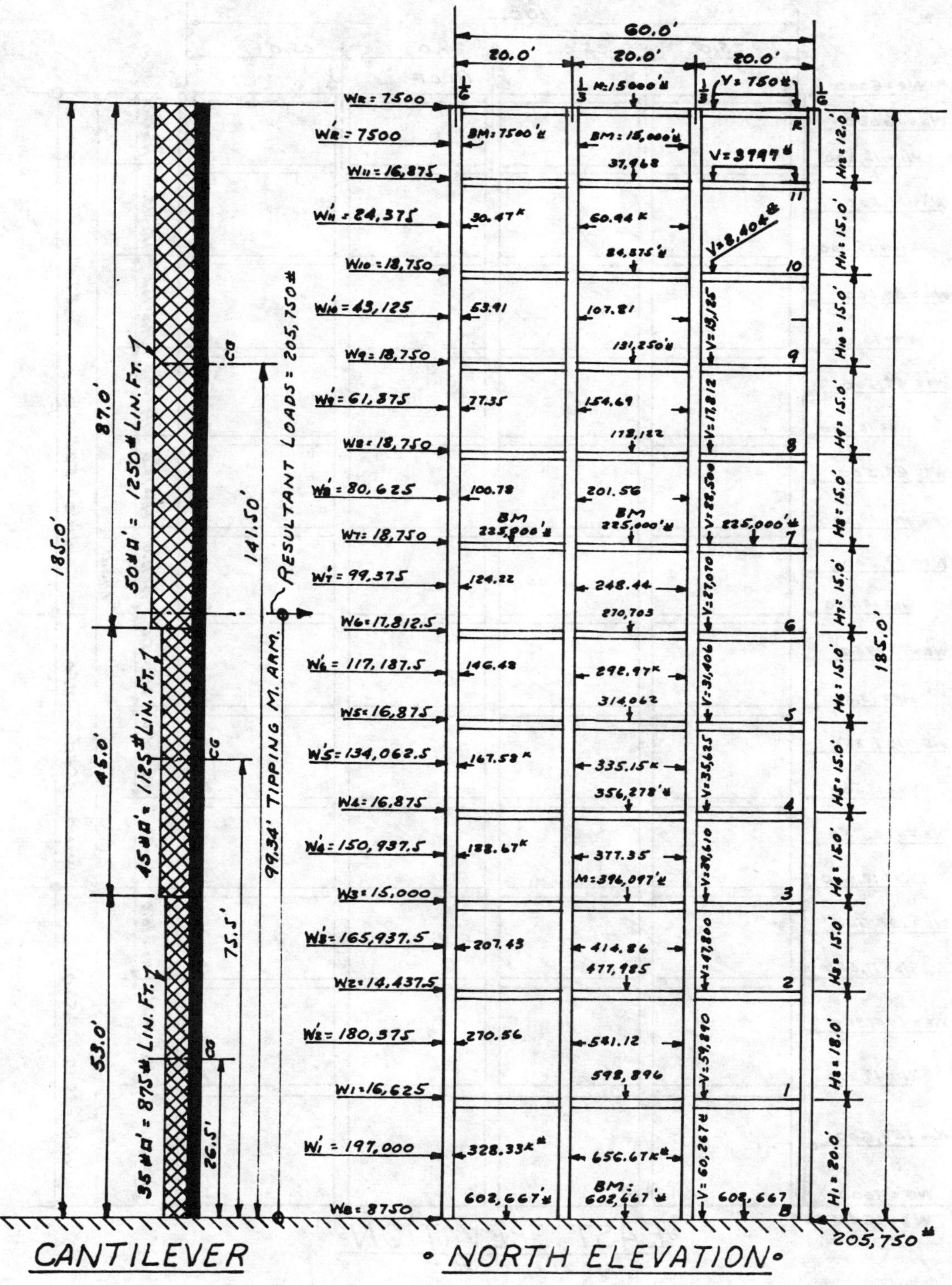

PLATE 5: Wind loading from South, shear values 8.6.5

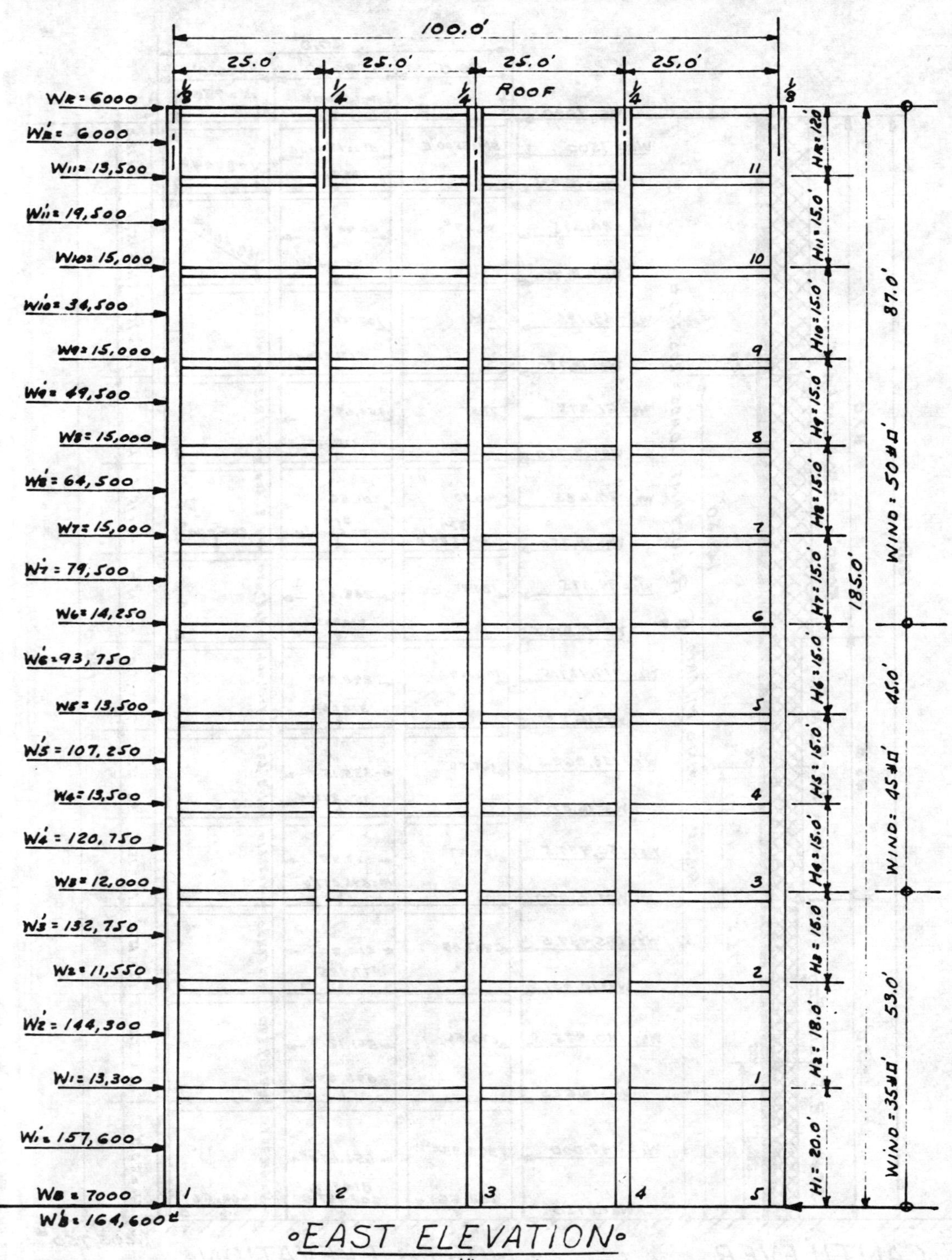

EAST ELEVATION

SCALE: 1" = 20.0'

PLATE 6: Moment and shear diagram, with South wind 8.6.6

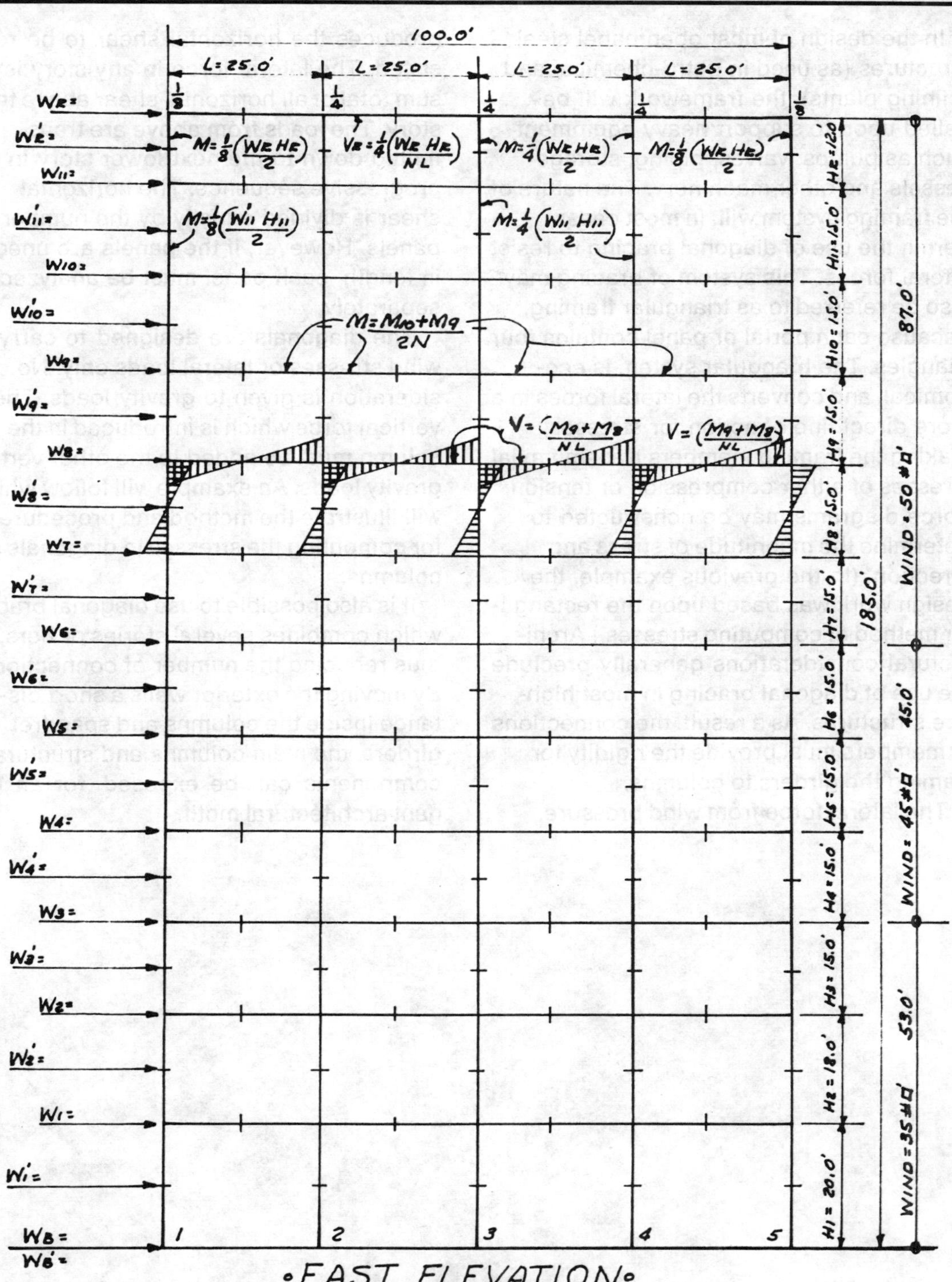

Diagonal bracing 8.7

In the design of most open panel steel structures (as used in petro-chemical and refining plants), the framework will be called upon to support heavy equipment such as pumps, valves, piping, storage vessels and other machinery. The nature of the framing system will, in most cases, permit the use of diagonal bracing to resist lateral forces. This system of bracing may also be referred to as triangular framing, because each portal or panel contains four triangles. The triangular system is economical, and converts the lateral forces in a more direct line of action for stresses, making the framing members transmit axial stresses of either compression or tension. Force diagrams may be constructed to determine the magnitude of stress and direction. (In the previous example, the design work was based upon the rectangular method of computing stresses.) Architectural considerations generally preclude the use of diagonal bracing in most high-rise structures. As a result, the connections of members must provide the rigidity for framing the girders to columns.

The lateral force from wind pressure produces the horizontal shear to be resisted. The lateral force in any story is the sum total of all horizontal shear above that story. The loads from above are transmitted down to the next lower story in a progressive sequence. The horizontal shear is divided equally by the number of panels. However, if the panels are unequal in length, each panel must be analyzed separately.

The diagonals are designed to carry wind stresses or lateral loads only. No consideration is given to gravity loads. The vertical force which is introduced in the column must be added to the other vertical gravity loads. An example will follow which will illustrate the method and procedure for computing the stresses in diagonals and columns.

It is also possible to use diagonal bracing which combines several stories or tiers, thus reducing the number of connections. By moving the exterior walls a short distance inside the columns and spandrel girders, the main columns and structural components can be exposed, for an elegant architectural motif.

EXAMPLE: Diagonal high rise wind bracing 8.7.1

Take design data from Plate 7 and example for 11 Story Hi-Rise.
W = Wind pressure on 1 Panel per floor height.
W' = Horizontal shear load from tiers above.
Assume that framed structure will permit diagonal wind bracing in lieu of rigid rectangular connected framing.

REQUIRED:
Calculate the axial loads in girders, columns and bracing when Wind is acting in either East or West direction. Restrict the design to the following:
(a) Forces in Roof Columns H_R. Make a force Diagram for panel.
(b) Forces in panels, 7th, to 8th. " " " " "
(c) Forces in panels, 1st. to 2nd. " " " " "

STEP I:
Wind Load at Roof = W' = 7500 Lbs. For each Panel = 7500/3 = 2500 Lbs.
Let diagonal brace form Triangle ABC with opposite sides abc.
H_R = Side a = 12.0' L = b = 20.0' c = Diagonal brace. Tan A = $\frac{12.0}{20.0}$ = 0.6000
From Trig. Tables: Angle A = 31° Secant A = 1.1666 c = b Secant A.
Length diagonal c = 20.0 × 1.1666 = 23.33' (23'4"). Force b = 2500 Lbs.
Force in brace c = 2500 × 1.1666 = 2,916.5 Lbs. Side a = b Tan A, and
Force in H_R = a = 2500 × 0.6000 = 1500 Lbs.
These forces are now plotted on Plate 7. Stresses in members will depend on wind direction. When from East direction, brace c will be in tension as indicated by minus (−) sign.

STEP II:
At 7th. Floor: H_8 = 15.0' L = 20.0' Tangent A = 15.0/20.0 = 0.7500 and angle A = 36°52' Secant A = 1.2500 Length diagonal C = 20.0 × 1.2500 = 25.0 Ft.
Horizontal Force at W_8' = 80,625 Each panel = 80,625/3 = 26,875 Lbs. (L).
Tension force in diagonal = L Sec. A or 26,875 × 1.25 = 33,594 Lbs.
Force in Column H_8 = L Tan A, or 26,875 × 0.7500 = 20,156 Lbs.

STEP III:
At 2nd. Floor: H_2 = 18.0' L = 20.0 W_2' = 180,375 Lbs. Number Panels N = 3
Force in each Panel = $\frac{180,375}{3}$ = 60,125 Lbs (L). Tan. A = $\frac{18.0}{20.0}$ = 0.9000
Angle A = 32° Secant A = 1.3456
Force in Diagonal = 60,125 × 1.3456 = 80,900 Lbs. Length c = 20.0 × 1.3456 = 26.91 Ft.
Force in Columns (Axial) = 60,125 × 0.9000 = 54,115 Lbs.

STEP IV:
Force diagrams are drawn on Plate 7 and results check with above.

PLATE 7: Diagonal wind bracing, North elevation 8.7.2

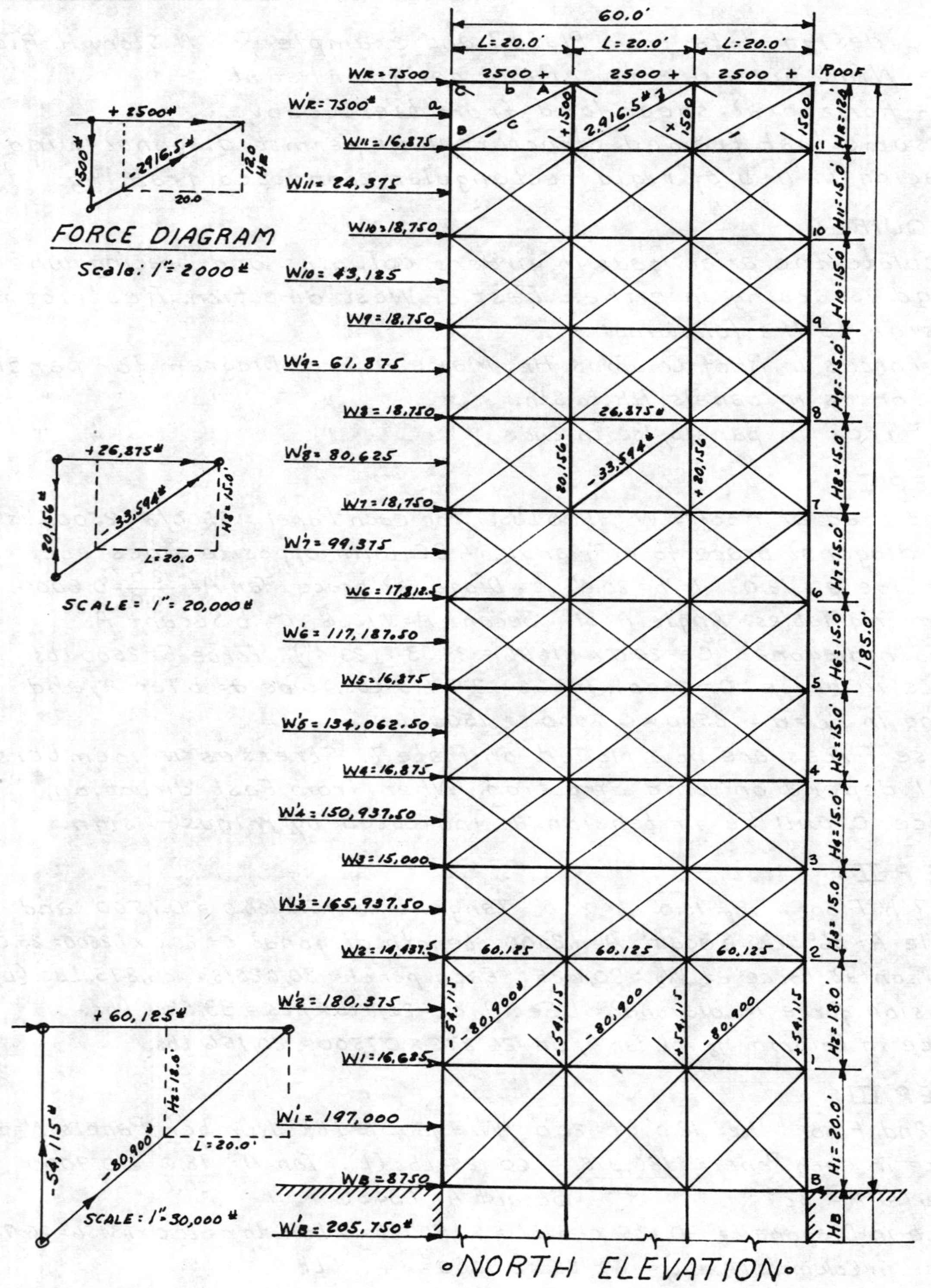

TABLE: Wind velocity and pressure formulas 8.8

The table below lists three formulas for comparison and tends to show the absence of formality in existing codes.

Formula No. 1: Established by the U.S. Navy Bureau of Yards and Docks. Extensively adopted by the M.B.M.A for design of Pre-Engineered and packaged structures.

Formula No. 2: Basic conversion equation for many building codes.

Formula No. 3: In general use for design of bridges, grain elevators and petro-chemical plant structures.

HORIZONTAL WIND PRESSURES ON WALL SURFACE			
V= WIND VELOCITY IN MILES PER HOUR	FORMULA -1- $P = 0.00236\ V^2$ P= LBS. SQ. FOOT	FORMULA -2- $P = 0.00320\ V^2$ P= LBS. SQ. FOOT	FORMULA -3- $P = 0.0040\ V^2$ P= LBS. SQ. FOOT
30	2.30	2.88	3.60
35	3.14	3.92	4.90
40	4.10	5.12	6.40
45	5.18	6.48	8.10
50	6.40	8.00	10.00
55	7.74	9.68	12.10
60	9.22	11.52	14.40
65	10.82	13.52	16.90
70	12.54	15.68	19.60
75	14.40	18.00	22.50
80	16.38	20.48	25.60
85	18.50	23.12	28.90
90	20.74	25.92	32.40
100	23.60	32.00	40.00
105	28.22	35.28	44.10
110	30.98	38.72	48.40
115	33.85	42.32	52.90
120	36.86	46.08	57.60
125	40.00	50.00	62.50
130	43.26	54.08	67.60
140	50.18	62.72	78.40
150	57.60	72.00	90.00
160	65.54	81.92	102.40
175	78.40	98.00	122.50
200	102.40	128.00	160.00

IX—PILE DRIVING AND DOCK FENDERING

IX—PILE DRIVING
AND DOCK FENDERING

IX

PILE DRIVING AND DOCK FENDERING

PILE DRIVING AND DOCK FENDERING

Contents

Early use of piling 9.1.1

Primitive man discovered that driven poles would serve as supports for his hut in the swamp or marsh. The earliest evidence of a pile supported structure was left by the lake dwellers in Switzerland of about 4000 B.C. The first pile supported structure of written record was a wood bridge spanning the Tiber river, built by the Romans about 1620 B.C. Ancient historians state that this bridge was repaired and lasted for over a thousand years. Julius Caesar (100–44 B.C.), the Roman general, led his legions across the Rhine on a wood bridge which he claimed was built in the short period of only ten days. The Carthaginian general Hannibal (247–183 B.C.) transported his army through the Alps and over rivers on wood bridges. The early seafarers on their sailing expeditions to the west coast of Africa discovered that the natives in the tropical zones were living in thatched dwellings supported by poles driven into the earth. These huts were constructed in groups, and extended over water for protection from roving animals. The greatest use of structural piling occurred in the city of Venice, Italy. This old city was constructed on the mud flats of the north Adriatic sea. Canals were used for transportation. Old Venice is slowly sinking into the sea; the ground floor of many old buildings is below water level.

Sheet piling 9.1.2

Sheet piling was first employed by the Romans, when they decided to build larger bridges. Cofferdams were constructed of wood sheet piles, and made watertight with clay mixtures. The water could then be lifted out giving greater access to the work of placing huge stones in the river bed. Stones were hewn and served as piers; many are still in use, although new superstructures have replaced the original spans.

Wood sheet piling was produced in the United States beginning in 1830. A groove in each side of pile timber was provided, and a spline was driven into this groove. When the piles and splines became well saturated with water, swelling would make the joints watertight. As the United States grew, the railroads carried the expansion to the West. Bridges were necessary to make this growth possible, and the use of wood piles for rail trestles was the moving force which started the wood pile industry. As passenger traffic increased, the railroads were forced to construct huge terminals and central stations for the convenience of travelers. Strangely, without exception, the architects designing these buildings used the Roman baths as a model. Illustrations are the Grand Central Station in New York, the LaSalle Street Station in Chicago, the Municipal Station in

Steel piling, continued 9.1.2

Kansas City, and many others.

Sheet piling made of cast iron was introduced in England about 1835, and a small amount of this type was shipped to the Americas for use along the coast. In 1902 an ironmaster, Luther Friestedt, patented a type of sheet with interlocking joints. It remained for the Carnegie Steel Company to design and develop the rolling equipment for production. The interlocks consisted of a bulb shape on one edge of the sheet and a cylindrical slot on the other side. Since the introduction of the steam powered pile hammer, steel sheet piling has been used for temporary and permanent construction in retaining walls, bulkheads, dolphins, cofferdams and foundation supports.

Pile driving contracts 9.1.3

The equipment necessary for a contractor to engage in pile driving operations on a continuous basis requires considerable capital investment. The size of this investment in large equipment prevents the greater majority of general contractors from engaging in this specialized field of work. To carry the large investment in driving rig, crew, and hammers, the expert in these operations must move from job to job all over the country. Driven piling for foundations has become a science. Regrettably, it is not included in the curriculum of many colleges and universities. Architects and design engineers should consult with experienced pile driving contractors prior to writing specifications and issuing plans for bidding. Such firms as Raymond, International; L. B. Foster Company; and Western Foundation have sales offices in the larger cities. With well organized personnel to supervise and operate their equipment, they are able to perform work in the shortest possible time and with reliable results.

The newer developments in Diesel and steam hammers have simplified the work of driving piles. Virtually every type of hammer is available to the contractor on a lease or rental basis. Such an arrangement enables the contractor in heavy construction to undertake the task of driving his own piles. In some cases, this is desirable since it can possibly reduce the cost and waiting period. Designers must make certain that the operating personnel have adequate experience. Specifications on this part of the work must be concise, rigid and informative.

Designing pile foundations 9.2.1

It remains the responsibility of the Structural Engineer to choose and design a foundation after receiving the reports from test borings and soil laboratories. The general goal is to transfer the weight of the structure through a poor soil strata to one which will sustain satisfactory bearing. When considering which type of pile is best at any given location, the Engineer must be familiar with the many types, and know exactly what the pile is expected to do. Economy enters into these investigations, and one must remember that the optimum type of pile will depend on the various soil and water conditions at the site.

Soil strata may be composed of solid rock similar to hard-pan, or it may be capable of improvement by the introduction of outside material and compaction. The load that a pile will support depends upon the type of soil into which it is driven, and its resistance to penetration.

Test piles 9.2.2

There is only one dependable method to determine the safe load capacity of a pile driven to a stipulated soil depth. Although a pile may be driven down to a strata which seems to have sufficient resistance to penetration under the impact of hammer blows, there is no guarantee that the pile will sustain a heavy *static* load. In making a static load test, a pile is driven with a hammer of known energy rating to a specified depth as recommended from soil borings. The hammer blows are counted to ascertain the amount of penetration under each blow of the hammer. This penetration per blow is called the "set per blow." To illustrate: A count of hammer blows at a 40.0 foot depth is recorded by an observer. The pile penetrated 12 inches under the last 80 blows. The set per blow is computed as $\frac{12}{80}$ = 0.15 inches. This figure will be centered into a formula later in order to equate the hammer's kinetic energy to a static load.

ILLUSTRATION: Static pile load test 9.2.2.1

Test loads are customarily placed upon piles in a sequence of increasing loads. Loads are added at intervals, and a settlement reading taken by instrument before additional load is applied. The final loading will bring the test load up to approximately 150 to 200 percent of the desired working static load. The test load should remain on the pile for several days, and the amount of settlement recorded. The Code requirements usually require that pile settlement shall not exceed 0.01 inch per ton of applied test load. When the static load test has met the requirements, it is reasonable to predict that other piles driven to the same set per blow will have a corresponding capacity and safety factor.

ILLUSTRATION: Static pile load test, continued 9.2.2.1

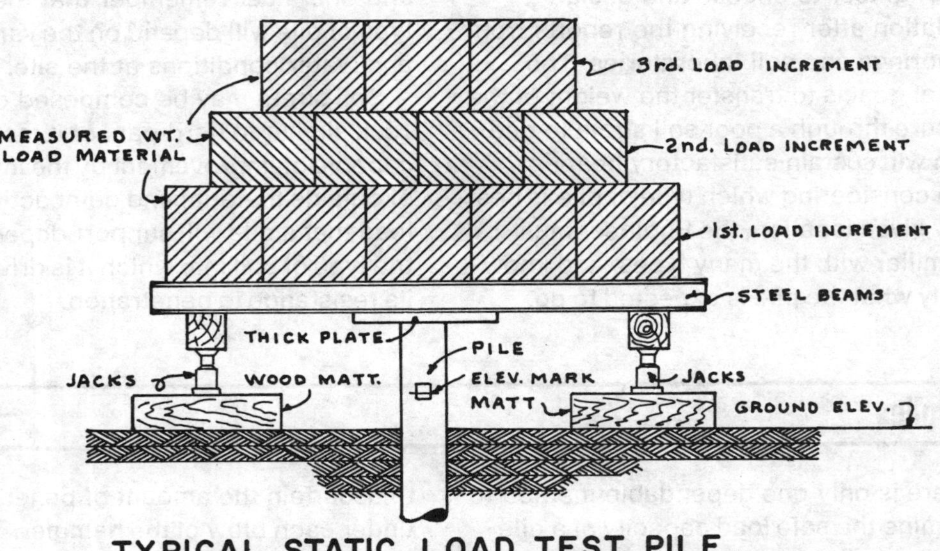

TYPICAL STATIC LOAD TEST PILE

Choosing a pile type 9.2.3

Modern construction requirements have resulted in the development of many types of piles and new hammers to drive them to greater depths. Manufacturers have developed steel piles which are intended to derive their support strength from skin friction as well as tip bearing. Each type has advantages and limitations. Some are only suitable for light loads. Others are limited in length, or cannot stand up under hard driving. In deep water, piles must be designed as long columns, and the pile cross-section must contain the necessary structural properties to support the axial load and resist bending action. Many materials lack resistance to chemicals, electrolysis, salt water or sulphur.

As the complexity and variety of foundation requirements increase and the loads from modern projects tend to grow larger, the selection of a pile type to meet the design requirements is of greater significance. A few of the following questions must be answered:

(a) How large are the loads and how are they concentrated?
(b) At what depth will firm bearing or friction be adequate?
(c) What will be the bearing capacity at the established depth of penetration?
(d) What resistance does the overlaying material offer as skin friction during the course of driving?
(e) What are the load capacity limits of the various types of piles available?
(f) White type offers the most practical installation method?
(g) Will drilled pilot holes or jetting be necessary?
(h) Which type offers the most in economy?
(i) Which type offers the best resistance to the elements?
(j) Finally the site conditions: Will driving be required to be performed from solid earth, or from a floating barge? Are there any high voltage overhead power lines in vicinity of driving area? What equipment is required to transport pile driving equipment over streets to the site, and what are the restrictions and costs of permits? What are labor conditions and what is the potential quake damage to nearby structures?

Soil investigation 9.2.4

The investigation of the strata below the proposed site is generally the work of a specialized firm. The boring samples are analyzed to determine the characteristics of each layer. The depth and location of each strata are charted. Since the same strata will vary in depth over a large site, it is well to require that several test holes be bored so that the profile charts can be compared. The soil mechanics engineer will furnish a full description of the formations. With this data on the profile charts, he will make recommendations as to bearing pressure, skin friction coefficients, and probable length of piles.

A test pile may be driven near the test hole, and the hammer blows for each foot of penetration recorded. Then a profile of pile and strata may be drawn (similar to Chart 9.2.4.1). After confirming the ability of the pile to support the test load, the hammer formula becomes a reliable guide for calculating other pile loads.

CHART: Profile of soil strata 9.2.4.1

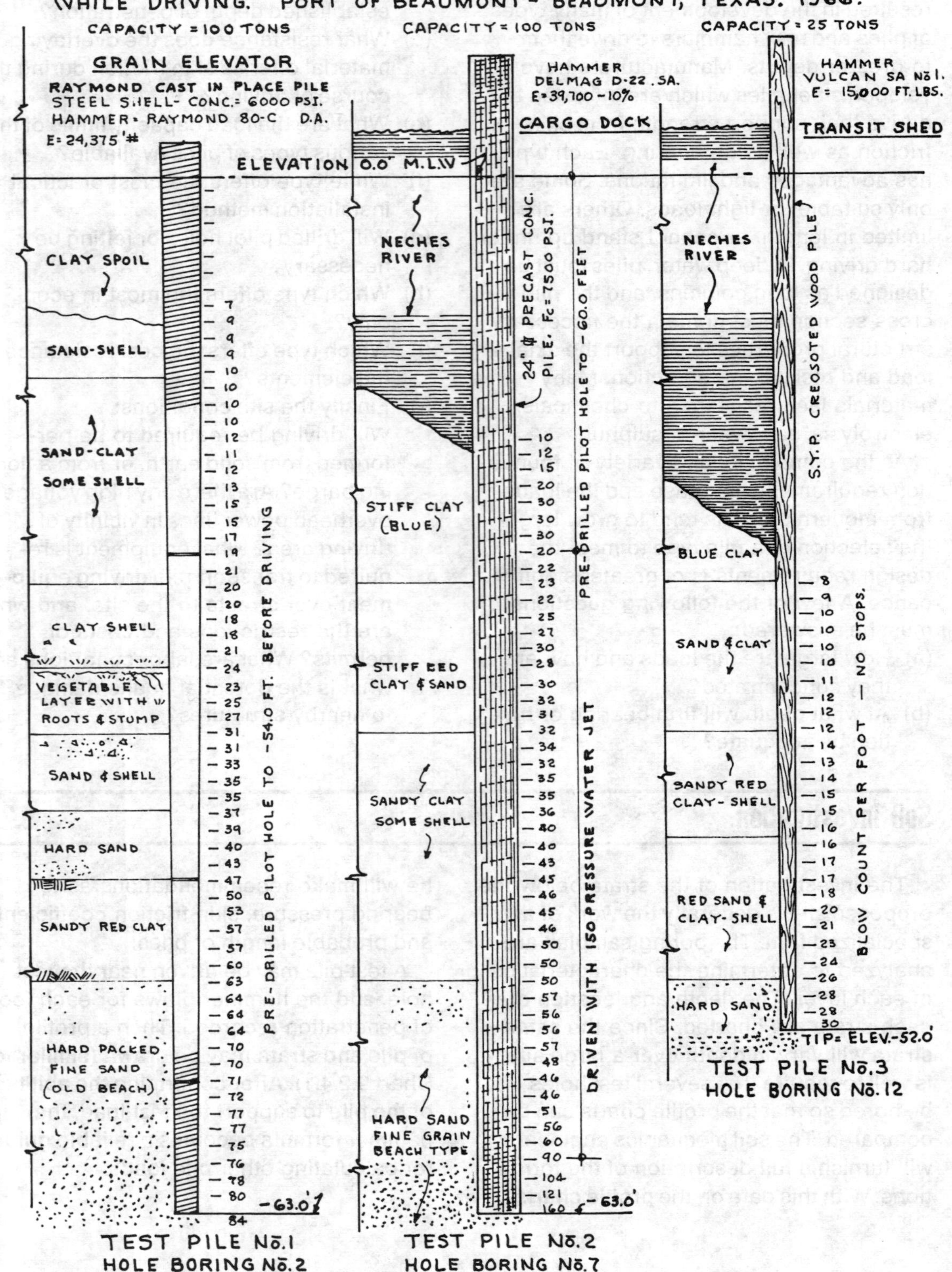

Alluvial formations 9.2.4.2

A good soil formation which lies over a compressible stratum will need longer piles which extend through the soft, intermediate layers, until a suitable stratum is found to sustain tip bearing. The use of a shortened pile, which can simply be punched through several questionable layers and with the tip end bearing on a compressible stratum, is worse than no pile at all.

Alluvial soil areas are often found in the coastal regions, because much of the present land surface was formed by silt deposits and a beach sand and shell mixture. Solid earth deposits extend to depths ranging from six to twenty feet, and usually cover a silt vein four to ten feet in depth. After a pile is driven through the layer of top soil, it will drop through the silt layer with no additional hammer blows. The top soil layer will remain stable as long as the silt is contained. If, perhaps, some river dredging operations release this silt vein, the top soil may become unstable and sink. This can be observed in many tank farms, where large storage tanks are supported by piles, and the earth surface under the tank has subsided to a level below the concrete slab. In scattered regions where alluvial formations exist, there will be layers of fine sand containing fresh or salt water. These water veins are not artesian (water will not spout up through a drilled hole) but great quantities may be pumped to the surface. The practice of relying on these fresh water wells for municipal and industrial consumption has produced a noticeable subsidence in several areas, including the city of Houston, Texas. The U.S. Army Corp of Engineers refers to this earth sinkage as being caused by the lowering of the "water table," and huge lakes and dams are being constructed in order to obtain another source for fresh water.

Pile driving operations 9.2.5

When piles are to be driven to a specified load bearing capacity by an impact type hammer, the design engineering firm is obligated to provide an observer to count the hammer blows. The observer should be an experienced individual who will be able to ascertain when the pile has come into contact with underground obstructions such as old stumps, abandoned piles, rock bolders, and buried pipes. Tree logs and stumps lie in "vegetable layers." As will be observed from the profile chart of test piles, the blow count will reveal the particular stratum into which the pile is penetrating. Any sudden change in the number of blows per foot of penetration will indicate a change in the soil compactness and the amount of frictional resistance. During the driving of a pile, there should be no stops. The penetration should be steady with a uniform increase in the number of blows per foot. A sudden increase in blow count will indicate that an obstruction has been hit or a soil stratum encountered which is well compacted. A pile which refuses to penetrate and bounces under each impact is usually against an obstruction. Wood piles, in particular, are of a resilient material and will bounce when in contact with old logs and stumps. In such cases, a probe should be inserted into the hole to find the obstruction, or another pile driven close to the hole. In the alluvial formations bordering the Caribbean Coast,

Pile driving operations, continued 9.2.5

it is not unusual to note that during driving, a pile will stiffen up at less than the specified elevation. Test borings have revealed this stratum to contain a layer of hard beach sand with shell content. Continued driving will penetrate this sand layer which may be only two or three feet in thickness. After punching through this sand, there will be a drop in blow count until the tip is bearing upon a deeper sand stratum at the desired elevation.

Driving equipment required 9.2.5.1

In addition to employing the proper hammer for driving, the crane or rig should have sufficient capacity to lift several tons in excess of the combined weight of the pile, hammer and mandrel. Since most refusals to penetrate result from obstructions at less than half the pile length, this extra crane strength will be necessary to pull the pile. Prompt action should be taken when the observer orders a pile to be pulled for investigation. Removing the hammer and permitting a delay will result in the penetrated soil returning to its compacted state. In such cases, the adhesion of the soil to the pile surface will increase to such an extent that the resistance to pulling cannot be overcome by the crane. This adhesion of the soil to the pile surface is referred to as "skin friction," and will be investigated later. A representative of the design engineer should maintain a constant vigil over each pile driven. Otherwise, many driving crews may claim that certain piles fall into the refusal class, and may cut off piles short without authority or poper records.

Refusal to penetration 9.2.5.2

Even though a site may appear on the surface to be an ideal location for driving piles, there is the possibility that somewhere beneath the surface, certain conditions will be encountered in which the penetration will diminish or cease entirely. The blow count for final set will be exhausted; yet the pile will not attain sufficient depth. When a pile refuses to move down under repeated hammer blows, it is referred to as "refusal to penetration." Refusal is only properly applied to a pile that has stopped before the necessary or required depth has been reached. The circumstances in such cases will be open to question, and a difference of opinion between contractor and observer may occur. To preclude the possibility of any misunderstanding, the pile specifications must stipulate what constitutes refusal.
It may be required in the specifications, that the piles must in no event be driven to a depth less than the given tip end elevation. As an alternate stipulation, it may be required that continuous hammer blows shall reach a certain count before refusal is claimed.

When wood piles are driven through a silt stratum or an old vegetable layer, stumps, roots and logs are frequently encountered which could prevent penetration. Such conditions are readily recognized by experienced observers. In the majority of cases, with continuous pounding, the tip

Refusal to penetration, continued 9.2.5.2

will break through and penetration will resume to a stratum where the blow count will actually reflect tip bearing. Surface soil layers built up from spoil deposits resulting from dredging operations will often contain water pockets under vegetable layers. When these water-filled cavities are under pressure, a capillary action will tend to repel the pile after each blow from the hammer. Again, the pile cannot be considered in the category of refusal. Withdrawing the pile and pointing the end, or attaching a cast steel driving point, will usually provide the means for penetration.

Wood piles subjected to continuous pounding have a tendency to show the effects of hammer blows at the top and bottom ends. This separation of the wood fibers is called "brooming." This damage to the pile should be avoided, as the resilience of the broomed area will absorb the hammer energy, and the bearing requirements will not meet the load capacity.

The driving of steel pipe or BP sections does not usually encounter the problem of refusal, except in rock stratums. Precast concrete and large pipe piles are driven into a predrilled hole. In sand strata and water front structures, the use of pressure jets is usually necessary to assist the hammer in penetrating to the required depth.

Steel BP (H sections) are primarily adapted to piles which obtain their capacity to support loads by skin friction. Driving is less difficult, especially with hammers which do not require a wood cushion block. Under corresponding conditions, these piles will be more capable of penetrating rock formations, and refusal will result only at hardpan or solid rock. In the driving of steel BP and open end pipe piles, "refusal to penetration" may be assumed to apply to either or both of the following conditions:

(a) If, after 50 continuous hammer blows, the pile has not penetrated in excess of 1.0 inch.

(b) If, after 15 minutes of continuous driving, or with a minimum of 250 hammer blows, no appreciable penetration is noted.

The specifications for pile driving should state that the determination of the refusal of piles to penetrate under continuous driving, shall remain the exclusive responsibility of the Engineer Observer.

Hammer quake 9.2.5.3

Under rapid and continuous blows from a pile hammer, the soil surrounding the pile has a tendency to loose adhesion. This disturbance to the soil and its subsequent loosening effect is referred to as "quake," or as used in the Hiley Hammer Formula, "ground quake."

When comparing the hammers listed in the tables, it will be noted that each manufacturer has provided pertinent information for each type of hammer. The greater the number of blows per minute, the greater will be the quake, and the greater the quake the less the skin resistance on driving. The pile penetration is more rapid under the faster double-acting hammers. The vibrating hammer was invented to take further advantage of hammer quake to increase driving speed.

Soil pressure 9.2.5.4

Any type of pile driven to a considerable depth will have displaced an equal volume of compacted soil. Unless a pilot hole is drilled before driving, the pile will compact the surrounding soil, and this pressure will increase the skin friction. This may be observed quite clearly when a cluster of piles is driven in close proximity. Indeed, when driving tapered piles, the soil pressure is occasionally built up to such a degree that adjacent piles are forced up. The compacted soil in many instances has caused the collapse of metal, thin-wall piles, before the concrete core was installed. To guard against the possibility of any rise in a pile cluster, the piles should be cut off at frequent intervals as driving proceeds. Before the driving rig leaves the site, it may be found that several piles may require "tapping down." Elevations at cut-off are easily and rapidly checked with a level or transit. Building Codes usually stipulate that piles shall not be driven closer than 2½ feet on centers. Even at greater spacings, the soil pressure from compaction can be enough to cause surrounding piles to rise after the hammer has been removed.

Estimating pile lengths 9.2.5.5

Several methods may be employed to find the lengths of piles before placing the pile order for an entire project. A single pile may be obtained and driven at the site. Then the load capacity may be calculated by using the hammer formulas. Regardless of the method used for estimating the lengths, a test pile should be used for final verification of load capacity.

In the event a pile project extends over a large area, additional test piles would be necessary. The cost would be prohibitive. Employing a soil consulting firm to take soil borings is more economical. A pattern of bored test holes could be analyzed to determine the contour of the existing strata. The soil engineer will test each stratum for compactness and adhesion, and draw a cross-sectional chart for each test hole. These charts are called "stratiforms." With such information available, each pile type can be calculated for probable resistance to penetration, which is proportional to the safe ultimate load capacity.

An older method for soil exploration was used by road builders to determine soil bearing and the existence of rock and cavities. The "sounding rod" consisting of an inexpensive device on wheels was used. The sounding rod was made of a 1 to 2 inch diameter tempered steel rod with pointed tip. Extension lengths were attached with pin and sleeve joints. The "anvil" consisted of a larger solid steel rod, tempered to a hardness of a sledge hammer. When a manually struck blow was applied to the anvil, a distinct "ping" would result as each stratum was penetrated. When the sounding rod entered water and sand, the sound and vibrations would be damped down. The sounding rod can furnish information about penetration resistance, but cannot provide any characteristics on the soil strata compositions.

Jetting piles 9.2.5.6

A common practice used as an aid in the driving of large concrete and closed end pipe piles is called "jetting." This operation consists of inserting water pressure around the tip and sides of pile during driving. Starting about 1960, state highway engineers began to realize that the use of jets provided economical bids, and if properly controlled, it did not diminish the pile capacity.

Specifications now permit the use of jetting for certain types of piles after thorough investigations. The main concern over jetted piles is their tip bearing capacity. The observer must concentrate on the blow count during the final three or four feet of pile penetration. The jet stream must be withdrawn before final cut-off, and the timing of pulling the jet is a critical decision for even the most experienced pile observers.

Normal jetting operations can be accomplished satisfactorily with water pressure between 75 to 175 pounds. Higher pressures will force the pile out of position, and alignment will be difficult, because excessive soil is removed from the compacted surroundings. As the wet soil returns to surround the pile surface, compaction from natural causes will return the skin friction resistance to a normal value.

Pile driving record form 9.2.5.7

PILE DRIVING RECORD FORM

CLIENT: RAYMOND R. RAPP & ASSOCIATES-ARCHITECT. Galveston, Tex.

PROJECT-LOCATION: SERVICE CENTER- HOUSTON POWER & LIGHT Co.

CONTRACTOR: P.G. BELL COMPANY, HOUSTON, TEXAS

HAMMER-TYPE: VULCAN IRON WORKS- No. 1 SA. RATED ENERGY: 15,000'#

CUT-OFF ELEVATION: - 57'-0" MAX. DRIVE CUSHION: 2" Oak

PILE LENGTH: 60'-0" PILE MAT'L. SO. YEL. PINE TIP ELEVATION: -51'-5"

PILE IDENTIFICATION Bent 3, Line C

FEET	COUNT	FEET	COUNT	FEET	COUNT	FEET	COUNT
1.0		29.0	10	57.0		85.0	
2.0		30.0	11	58.0		86.0	
3.0		31.0	11	59.0		87.0	
4.0		32.0	12	60.0		88.0	
5.0		33.0	13	61.0		89.0	
6.0		34.0	13	62.0		90.0	
7.0		35.0	15	63.0		91.0	
8.0		36.0	14	64.0		92.0	
9.0		37.0	15	65.0		93.0	
10.0		38.0	17	66.0		94.0	
11.0		39.0	19	67.0		95.0	
12.0		40.0	18	68.0		96.0	
13.0		41.0	20	69.0		97.0	
14.0		42.0	21	70.0		98.0	
15.0	Count Std	43.0	21	71.0		99.0	
16.0	6	44.0	21	72.0		100.0	
17.0	6	45.0	22	73.0		101.0	
18.0	6	46.0	22	74.0		102.0	
19.0	7	47.0	22	75.0		103.0	
20.0	6	48.0	24	76.0		104.0	
21.0	8	49.0	25	77.0		105.0	
22.0	8	50.0	25	78.0		106.0	
23.0	9	51.0	30	79.0		107.0	
24.0	9	52.0	32 CO	80.0		108.0	
25.0	10	53.0		81.0		109.0	
26.0	9	54.0		82.0		110.0	
27.0	10	55.0		83.0		111.0	
28.0	10	56.0		84.0		112.0	

OBSERVER: Tom McKenna FINAL SET PER BLOW: 0.375

REMARKS: $L = \frac{2 \times 15{,}000}{0.10 + .375}$ $\frac{63{,}250}{2000} = 31.6$ Tons.

Types of piles 9.2.6

Various types of piles have been developed with particular characteristics. Piles may be classified as friction type or tip bearing; however, in most cases, the supporting strength is derived from a combination of friction and end bearing. Wood piles are limited in length. The tapered sides and small tip ends give a combination of both friction and end bearing. The thin-shell, corrugated cast-in-place pile is tapered, and the corrugations provide the wall friction when the soil returns to its compacted state after driving operations are completed.

Precast concrete piles can be made in a length which is limited only by their pick up strength. They are cast on a horizontal bed, and the reinforcing must be capable of resisting the bending due to their own length and weight. Tips may be tapered and the cross sections may be circular, square or octagonal. Concrete piles may be driven either for end bearing in deep water or for friction support in soil strata. For extremely heavy loads, pipe piles are used to greater advantage. These seamless tubes are produced in several standard diameters with a choice of wall thickness. The pipe pile may be driven with the end closed or open. When the end is left open during the driving operation, it is necessary to clean out the interior before filling the void with concrete. This is accomplished by blowing out the core with water and air pressure. Pipe piles are available in long lengths, and are provided with splicing sleeves when extreme length is required.

Steel, rolled BP Sections are listed in most steel catalogs, and are used in foundations as friction and end bearing piles. Their use is particularly advantageous where hard driving conditions are encountered, or where longer pile lengths are required. The BP piles are commonly called H-Piles since their cross section is a true H.

Skin friction 9.2.6.1

Load bearing piles driven into the earth derive their capacity to support loads from the end bearing at the tip end and the adhesion of the soil to the surface of the pile. This adhesion of soil to pile is called skin friction. Piles are classified as bearing piles, friction piles, or a combination of both.

The earth's surface is built up of layers which vary in composition and characteristics. These layers of unlike soils are called strata, and a single layer is referred to as a stratum. The qualities of adhesion to wood, concrete and steel surfaces vary with each stratum. Tapered piles such as wood and thin shell cast-in-place piles develop their greatest support resistance from skin friction. Tip end load bearing piles are popular in rock formations and where driving is difficult. Steel H-Piles and hollow pipe piles are the better selection for rocky regions, since they can absorb the heavy pounding until the tip comes into contact with hardpan or solid rock.

It is beyond the scope of this manual to enter into a study of the characteristics of soil strata. The study of the earth's layers and formations is a course of study for geologists and students entering the specialized field of soil mechanics. It is from these people, that the design engineer will obtain the reports and data on soil borings.

Skin friction, continued 9.2.6.1

EMPIRICAL VALUES FOR SKIN FRICTION			
TYPE SOIL IN STRATUM - DETERMINE BY BORINGS	SKIN FRICTION - P.S.Ft.		
COMPACTED COARSE SAND AND GRAVEL	1,000	TO	2000
COMPACTED SAND AND SHELL. PUG MILL MIXES	750	TO	1500
FINE TO COARSE SAND - MOIST.	500	TO	1000
SAND, SHELL AND CLAY MIXED.	400	TO	800
SANDY CLAY - STIFF BLUE AND HARD.	350	TO	900
SANDY AND CLAYEY - STIFF RED AND MOIST.	300	TO	800
SHALE, SAND AND CLAYEY - MEDIUM MOIST.	300	TO	750
DREDGED RIVER SPOIL WITH STIFF MED. CLAY.	250	TO	500
LOOSE SPOIL - CLAY AND SANDY - DAMP OR WET.	150	TO	300
COMPACTED FINE BEACH SAND - WELL CONTAINED.	750	TO	1000
FROM TEST OBSERVATIONS IN AREAS OF: GALVESTON, NEW ORLEANS AND MOBILE.			

Cast-in-place piles 9.2.6.2

A popular type of pile to support the foundations of grain elevators and hi-rise structures is the step-taper, thin-shell, corrugated-wall, ringed pile as installed by Raymond International. The outside shell is driven into a pre-bored hole by internal methods. The use of a steel mandrel inserted into the shell permits hard driving, and the driving energy is transmitted directly to the tip. Each pile can be internally inspected for its full length after having been driven and before concreting begins. The shells are assembled and joined together by using the ring corrugations as screw threads. Joints are made watertight with picked oakum and mastic. Any length can be obtained by adding an additional length section, to continue driving until the blow count reaches the required number. Transportation and unloading the pile sections is simplified. Tapered sections are approximately twenty feet in length and may be shipped by truck or freight car.

The concrete which is placed in the longer pile lengths should consist of rough aggregate, not over ¾ inch to preclude the possibility of any voids in the completed pile. This type of pile can be driven in closely-spaced patterns to form a cluster that is capable of supporting high capacity loads. Since the cast-in-place piles do not require longitudinal rod reinforcing when the full length is enclosed by soil strata, their use is limited when lateral support is lacking.

Wood piles 9.2.6.3

The primary source of wood piling is Southern Yellow Pine and West Coast Douglas Fir. Pine and fir have long been the choice, due to their long length and ability to withstand the weather elements. In 1968, over 800,000 pine trees were felled in the state of Mississippi and used for utility poles and piles. The abundant supply reflects the advantages of forest conservation, re-planting and harvesting methods. The wood fibers of the yellow pine and fir tend to absorb and retain the impregnation of chemicals, which makes the wood resistant to decay, insects, and other destructive conditions.

There are over 160 listed chemical agents which can be used for treating wood products; however, nearly eight out of ten pressure-treating plants will be equipped to use creosote, coal tar and petroleum. Others will treat with such chemical preservatives as pentachlorophenal (Penta), Wolman Salts, Tanalith, Woodlife, and others. In writing specifications for creosoted treated piles, they should stipulate that all pressure treatment shall comply with the American Wood Preservers Association Specification C–12, Edition 1951. The net retention of preservative should be 12 pounds per cubic foot of content after pressure treatment. This amount of retention will be adequate, and to require a higher ratio would be harmful to wood fibers. The increased pressure necessary to raise the retention would break down the fibers, until their original strength would fall below the requirements for sound timber.

All wood timber which is to be treated and used for piles should be cut from sound, close-grained, live trees cut not over twelve months prior to treating. The taper should be uniform from tip end to butt. Before treating, the piles should be peeled of both outer and inner layers of bark. The pile should be straight: a line drawn from center of tip to center of butt shall not fall outside center of pile at any point more than one percent of the length of pile. The tip diameters of wood piles should be not less than the following:

Pile Length	*Tip Diameter*
40.0 feet long or less	8.0 Inches
40.0 feet to 60.0 feet	7.0 "
60.0 feet or over	6.0 "

Butt diameters should be not less than thirteen inches nor more than twenty inches in diameter. These dimensions apply to piles after being peeled.

During the driving operations, the lead enclosure for maintaining plumb should have a guide cage on three sides, with loading to be done from the front side. Long piles occasionally are difficult to keep in proper alignment and frequently break under hard driving when penetration is first started. When piles are cut off at the desired elevation, the saw cuts should be clean and square. The exposed cut must be given a coating of a hot sealing mixture composed of 60 percent creosote oil and 40 percent waterproof coal tar pitch. The popular size of hammer for driving wood piles will have an energy rating of approximately 15,000 foot pounds. All hammer types will perform satisfactorily.

Precast concrete piles 9.2.6.4

The method used for designing concrete piles is similar to the design of spiral-hooped and vertical rod reinforced concrete columns. The entire forming and placing of steel is carried on above ground on a casting bed. The bottom and sides of the casting form may be long enough to form from five to six 100-foot piles. With this type of operation, production is rapid and the cost is reduced. In many instances the straight reinforcing will consist of high strength, stranded steel wire rope. The stranded reinforcing is twisted in shape which provides a good bond, even though the wire strands have a smooth surface. The number of ropes will vary according to size of the pile cross section. Inside the form, these ropes are stretched taut by ratchet rigs, and may be arranged in circles, squares, or other profiles to suit the shape of the pile. Spiral type hooping is wrapped around the stranded steel, and in many cases welding is used to insure uniformity for the longer pile lengths. The tip end can be cast with a taper for a short length; however this practice is of little advantage since this type of pile must be driven into a pre-drilled hole, and may require a jet to assist the hammer during driving. Damage to tip and cap end by continuous driving is avoided by including additional hoop wrapping. In order to direct the hammer energy to the outer sides or perimeter of the pile and into the reinforcing, a circular void may be provided by inserting a fiber tube exactly into the center of the pile form before concrete is placed. This hole may extend from the top to approximately seven feet above tip end. Forms for casting piles should be lined with smooth sheet metal or similar lining to give the finished pile a neat and smooth surface. Square piles should have their corners chamfered. This is accomplished by adding a bevel slat into the form.

Concrete mixes for piles must be above average strength to resist shattering under hammer blows. Rich cement mixes are used which have strengths up to 8000 PSI at 28 days. Constant care must attend the curing period, and steam curing is necessary for good control. Precast concrete piles are susceptible to cracking through their cross section. When in a horizontal position, they should be supported at several points. They must never be rolled or slid down an incline or ramp when being unloaded from a truck trailer. The engineer at the casting bed will provide each pile with a strong pick-up loop. This pick-up will be located at a point where the driving crew can attach the crane lead line and raise the pile to a vertical position. There will be minimum danger to pile if the pick-up loop is used properly.

During driving, a good oak cushion between hammer and pile is necessary. Cushions are placed inside the heavy iron helmet which slips over the top of pile. The effect of the hammer impact on a concrete pile is somewhat delayed, when compared to a steel pile. The pile material must contain sufficient resilience to recover after each blow. This action is called "restitution after impact." This term will appear when investigating the Hiley formula for pile hammers (see 9.6.4.6). As a result of the restitution in this type of pile, the slower speed Diesels and single-acting hammers seem to minimize damage to precast piles.

Steel bearing piles 9.2.6.5

The AISC Steel Construction Manual lists H-bearing piles in the column tables. They are produced in A-36 steel which has a yield stress (F_y) = 36,000 PSI. They are also furnished in A242, A440 and A441 steels. The yield stress for the latter specifications is F_y = 50,000 PSI. Usually a pile functions structurally as a short column; further investigation is called for if the pile functions as a long column. When a pile passes through a considerable depth of water or silt, the design capacity must be verified.

For all steel piles, whether they be H-piles or pipe sections with wall thickness over 0.10 inches, the maximum unit working stress is based upon 35 percent of the minimum yield stress specified for steel. This is less than ⅔ the maximum stress allowed for steel building columns. These recommended regulations were provided by the American Iron and Steel Institute in 1963 to assist municipalities in bringing their Building Codes up to date. Unfortunately, only a very few have done so.

For designing steel H-Piles and circular tube piles, an older code required that a ¹⁄₁₆ inch thickness deduction be made at the outer periphery of a steel pile section when computing the load bearing capacity of the pile. This deduction was intended to allow for corrosion, but in the opinion of many designers, the practice is unnecessary. Modern designers prefer to add protective pile coatings when unfavorable soil conditions or salt water exists.

H-Piles may be used as friction piles or tip bearing piles when soil conditions are suitable. To calculate the exterior pile surface exposed to the soil for adherence and friction, the cross section is converted to a square. The effective perimeter is less than the apparent cross-section perimeter (see cross sections in 9.6.4.6). To illustrate: A section of H bearing pile listed as BP 12, 53# has a flange width of 12 inches and a depth of 11¾ inches. For 1 lineal foot of pile, the area for skin friction is (12 x 12 x 2) + (11.75 x 12 x 2) = 570 Sq. Inches = 3.96 Sq. Ft.

Steel pipe piles 9.2.6.6

Two types of steel pipe piles are in common use. Pipe piles driven with open ends are used when soil tests reveal the existence of rock formations, and when load capacities are relatively high. The open end pipe pile is normally used to a depth of 40 to 70 feet. Closed end piles are recommended when soil borings indicate the absence of rock formations or when the rock is much deeper than the pile depth. These piles may be driven to the desired resistance by the results of test loads and pile-hammer formulas. In other instances, closed end piles are driven to refusal, after penetration of the tip end has reached a minimum depth of fifty feet. Refusal must be carefully defined in the specifications. To avoid the possibility of the pile being cut-off short after hitting an obstruction, the specifications should require that the tip ends penetrate to a given elevation.

Before concreting can begin, open end piles must be cleaned out. The soil material accumulated inside the pipe during driving can be blown out by compressed air. A small pipe is connected to a compressor by a gooseneck and hose, and then lowered into the pile at intervals. Pile interiors are

Steel pipe piles, continued 9.2.3.6

inspected by lowering an electric lamp into the hollow tube. After open end piles are blown out, they are sometimes sealed with cement grout containing metal filings. This type of grout will prevent entry of seep water, and the concrete placement can be deferred for a time. Splicing of steel pipe piles should not be done without sleeves. For 10 to 20 inch diameter pipe, the sleeves are available in cast steel. For larger sizes, a short length of pipe is split to reduce the perimeter. Compressing the ring with a chain and coffing jack will permit the sleeve to be slipped inside the tube and welded. The pile extension can then be slipped over the sleeve and welded.

Designing for load bearing 9.3

Open and closed end steel pipe piles filled with concrete are designed for load capacity by several methods. When the compressive strength of the steel tube and concrete core are known, the computations become relatively simple. For these pile types the calculations ignore the value of skin friction, and the design is on the basis of a braced column. When the piles extend above ground such as in a viaduct, and the exposed length constitutes a part of the superstructure, the design is governed by the unbraced length above ground. When driven into exceptionally deep water, these considerations apply to the total unbraced length above and below water.

Standard specifications for Welded and Seamless Steel Pipe Piles (ASTM A252) indicate that the material for pipes is produced as Grade 2 or Grade 3. The choice of grade must be kept in mind, and firmly stated in the design computations, because the formulas are based on the yield stresses obtained by testing each grade. The following data should be used in design:

Yield point minimum = F_{syp}.
Grade 2; F_{syp}= 35,000 PSI.
Grade 3; F_{syp}= 45,000 PSI.

For unbraced pipe pile columns the allowable unit stress on the steel area of pipe is:

$$F_s = \frac{F_{syp}}{45{,}000}\left[18{,}000 - \left(70\frac{l}{r}\right)\right].$$

It should be noted, that the AISC column formulas given in Section II for steel columns are developed for the design of members made of Grade 1 carbon steels. The nomenclature used in the formulas, remarks, tables and examples are as follows:

Designing for load bearing, continued 9.3

A_c = Area of concrete core, in square inches.
A_s = Area of steel in pipe ring, in square inches.
F_s = Allowable unit steel stress from formula, in PSI.
F_{syp} = Strength of yield point of steel pipe, in PSI.
F_c' = Compressive strength of concrete at 28 days, PSI.
l = Unbraced length of pile, in inches.
P = Safe design load, in pounds or tons.
P_u = Ultimate load, in pounds.
r = Radius of gyration of steel pipe.
n = Ratio of Modulus of Elasticity of steel to the Modulus of Elasticity of concrete. $n = \frac{E_s}{E_c}$.

Ultimate load method 9.3.1

The ultimate strength of a pipe pile with a concrete core is equal to 85 percent of the 28 day concrete strength times its area, plus the area of steel times its yield point strength. Written into a formula, it becomes:

$$P_u = (0.85\, F_c' A_c) + (A_s F_{syp}).$$

To obtain the allowable working load, divide the value of P_u by the desired factor of safety. A safety factor of 2.5 is generally considered adequate to satisfy code authorities. Then $P = \frac{P_u}{2.50}$. At the tip end of the pile, the steel and concrete are supporting the superimposed loads in addition to the dead weight of pile. The weight of the pile must be deducted from the calculated safe design load to find the allowable load for each pile.

Proportionate method 9.3.2

Several building codes which lack modern revisions are still in use, and were patterned after the old New York City Code. Such codes stipulate that the outer 1/16 inch of steel pipe is to be deducted in computing the area of steel. As discussed earlier, it was believed that corrosion should be allowed for in the design. The allowable concrete stress was 500 PSI and n = 15. The allowable working stress was therefore $F_s = F_c n = 500 \times 15 = 7500$ PSI.

It was finally recognized that concrete encased in a steel pipe is capable of considerably greater loading. The modern formula permits an increase in the allowable unit stress, and is written:

$$P = 1.20 \times 0.225\ F_c' \left[A_c + (n A_s) \right].$$

This formula permits the concrete mix to govern the stress in the steel, and is rather conservative as will be illustrated in the following examples.

USING THE ALLOWABLE LOAD TABLES:

When referring to table as a possible guide for selecting a diameter and wall thickness of pipe, several conditions are to be observed as follows;

(a) The calculated loads were computed by using the Joint Committe Formula.

(b) Grade 2 steel pipe is designated with F_{syp} = 35,000 PSI.

(c) Concrete strength at age of 28 days, F_c' = 3000 PSI.

(d) The area of steel is based upon reducing wall thickness 1/16 inch all around, or diameter minus 0.125 inches.

Joint Committee formula 9.3.3

A 1940 report of the Joint Committee Specifications for Reinforced Concrete required that safe allowable load capacities for concrete filled pipe piles should be determined by the following formula:

$$P = \left[(0.225\ F_c' A_c) + (0.40\ F_{syp}\ A_s) \right].$$

In this formula, the committee called for a deduction of 1/8 inch from the outside diameter of the steel pipe when calculating the area of steel. This reduction amounts to the 1/16 inch corrosion allowance as mentioned in the previous formula.

Some design engineers have taken exception to the rule concerning the steel area deduction for corrosion. In salt water and soils which contain corrosive chemicals the piles should be cleaned and coated with coal tar epoxy prior to driving. Then they would use the gross area in the design. An account of the recommendations of the Joint Committee is contained in Section IV.

TABLES: Allowable loads—concrete core pipe piles 9.3.3.1

CONCRETE FILLED STEEL PIPE PILES - OPEN END TYPE					
PIPE			ALLOWABLE LOAD IN LBS.		
OUTSIDE DIAMETER IN INCHES	WALL THICKNESS IN INCHES	WEIGHT PER FOOT IN LBS.	CONCRETE F'_c = 3000 PSI.	STEEL F_{yp} = 35,000 PSI.	COMBINED STEEL + CONC.
10.75	0.307	34.24	54,400	111,600	166,000
	.365	40.48	53,200	137,400	190,600
	.438	48.19	51,600	162,200	220,800
	.500	54.74	50,400	196,000	246,400
12.75	0.312	41.51	78,000	135,800	213,800
	.330	43.77	77,400	145,400	222,800
	.375	49.56	76,400	169,200	245,600
	.438	57.53	74,800	202,400	277,200
	.500	65.42	73,200	234,600	307,800
14.00	0.375	54.57	93,000	186,400	279,400
	.438	63.37	91,400	223,000	314,400
	.500	72.09	89,600	258,600	348,200
16.00	0.375	62.58	123,200	214,000	337,200
	.438	72.72	121,200	256,000	377,200
	.500	82.77	119,200	297,000	416,200
18.00	0.375	70.59	157,800	241,400	399,200
	.438	82.06	155,400	289,000	444,400
	.500	93.45	153,200	335,600	488,800
20.00	0.375	78.60	196,400	268,800	465,200
	.438	91.41	193,800	322,000	515,800
	.500	104.13	191,400	374,000	565,400
22.00	0.375	86.61	239,400	296,400	535,800
	.438	100.75	236,600	355,000	591,600
	.500	114.81	233,800	412,600	646,400
24.00	0.500	125.49	280,400	451,000	731,400

Using the allowable load tables 9.3.3.2

When referring to the tables as a possible guide for selecting a diameter and wall thickness of pipe, several conditions are to be observed as follows:

(a) The calculated loads were computed by using the Joint Committee Formula.

(b) Grade 2 steel pipe is used with F_{syp} = 35,000 PSI.

(c) Concrete strength at age of 28 days with F_c = 3000 PSI.

(d) The area of steel is based upon reducing wall thickness 1/16 inch all around, or diameter minus 0.125 inches.

Closed-end pipe piles 9.3.3.3

When driving closed end pipe piles to a definite load bearing resistance, as determined by hammer formulas, it is the practice of most designers to reduce the design load capacity to 60 to 70 percent of the open type. This safety factor allows for the fact that closed end driving reduces the skin friction value. The load tables are based upon open end driving with the 1/16 inch steel reduction for corrosion. The applicable Building Code should be studied to ascertain whether a distinction is made between open and closed end piles.

EXAMPLE: Designing concrete filled pipe pile 9.3.4

A seamless steel pipe with a 16.0" outside diameter is 50.0 foot long, and has a wall thickness of 0.375 inches. Grade of Steel is #2 with F_{syp} = 35,000 PSI. Pile is to have open ends during driving, cleaned out inside and filled with concrete. F_c' = 3000 PSI at age of 28 days.

REQUIRED:
Use the Joint Committee Formula to calculate the Safe Allowable Load Capacity. Deduct dead weight of Steel and Concrete from load calculations, and allow 1/16 inch of pipe wall circumference for corrosion.
Use the Ultimate Load Method to determine the safety factor as it applies to the JC allowable load formula.

STEP I:
Allowing 1/16 inch for corrosion, the OD becomes 15.875", and ID = 15.250". From Tables for Circles:
Steel Area in ring = 197.933 − 182.655 = 15.278 □"

STEP II:
Allowable load calculated by JC Formula:
$P = (0.225\ F_c' A_c) + (0.40\ F_{syp} A_s)$. Substituting values in formula:

$P = (0.225 \times 3000 \times 182.655) + (0.40 \times 35{,}000 \times 15.278) = 337{,}185$ Lbs.
Check with table and converting to Tons: P = 168.59 Tons.

STEP III
Deducting weight of pile for safe load:
Wt. of Steel: 62.58 Lbs. Lineal Foot.
Wt. of Concrete = 1.268 × 150 = 190.20 Lbs. Lineal Foot.
Total Weight of Pile = 50.0 × (62.58 + 190.20) = 12,639 Lbs.
Safe applied load P = 337,185 − 12,639 = <u>324,546 Lbs.</u>

STEP IV:
Ultimate Load Formula: $P_u = (0.85\ F_c' A_c) + (F_{syp} A_s)$.

$P_u = (0.85 \times 3000 \times 182.655) + (35{,}000 \times 15.278) = 1{,}000{,}500$ Lbs.
Converting Ultimate to Tons = 500.25 Tons.
Safety Factor of JC Formula: $SF = \frac{500.25}{168.59} = 2.96$

Weight of Pile = $\frac{12{,}639}{2000} = 6.32$ Tons

EXAMPLE: Designing with the proportionate formula 9.3.5

Concrete filled steel pipe pile is 50.0 ft. in length, and 16.0 inches in outside diameter. Wall thickness of pipe is 3/8 inches. Concrete is 3000 PSI at age of 28 days. Steel Pipe is of Grade 2 with Fsyp = 35,000 PSI.

REQUIRED:
Note that this pile is same as used in the preceding example and shall be designed by using the Proportionate method Formula as: $P = 1.20 \times 0.225\ F_c' \left[A_c + (n\ A_s)\right]$.
Deduct 1/16 inch for corrosion loss.

STEP I:
Outside Diameter = 16.00" Corrossion deduction = 0.125 inches.
Net outside diameter = 16.000 − 0.125 = 15.875" (3/8" = 0.375")
Inside diameter = 16.000 − 0.750 = 15.250 in.
Area circle with 15.875" OD = 197.933□" (use table 9.3.8)
Area core inside diameter of 15.250" = 182.655□" " " "
Net area wall steel: A_s = 15.278 Square in.

STEP II:
Concrete; F_c' = 3000 #□" and n = 10
Substituting values in formula:

$$P = 1.20 \times 0.225 \times 3000 \quad 182.655 + \left[(10 \times 15.278)\right] = 271{,}700 \text{ Lbs.}$$

Converting to U.S. Tons: $P = \dfrac{271{,}700}{2000} = 135.85$ Tons.

STEP III
From previous example, the ultimate load P_u = 500.25 Tons.
The safety factor for the Proportionate Method will be as follows: $SF = \dfrac{500.25}{135.85} = 3.69$

Safety Factor with dead weight of Pile deducted:
Wt. Pile = 6.32 Tons. Max. Allowable P = 135.85 − 6.32 = 129.53 Tons
$SF = \dfrac{500.25}{129.53} = 3.86$ This is too conservative and the Joint Committee formula is acceptable in previous example.

EXAMPLE: Concrete core pile as a column 9.3.6

A Steel Pipe Pile with OD of 10.75 inches, and length of 65.0 feet is shown in place at final driving. Wall thickness of steel is 0.50 inches. The unbraced length in water is 38.0 feet. Core concrete will test 3000 PSI at 28 days. Pile was driven with open end and bearing penetration was computed from hammer blows to sustain a load of 75 Tons.

REQUIRED:
Use the Joint Committee's pile column formula to calculate the maximum load on steel area. Check with J.C. Formula for pile design load. Deduct dead weight, and neglect the deduction for corrosion. Use Grade 2 steel.

STEP I:
Column Formula for $F_s = \frac{F_{syp}}{45{,}000}\left[18{,}000 - \left(70\frac{l}{r}\right)\right]$

From Tables:
$A_s = 16.10\,\square''$ $A_c = 74.66\,\square''$ $F_{syp} = 35{,}000$ PSI. $F'_c = 3000$ PSI
$r = 3.63''$ Unbraced $L = 38.0'$ $l = 38.0 \times 12 = 456$ inches.

STEP II:
Put values in formula:
$F_s = \frac{35{,}000}{45{,}000} \times \left[18000 - \left(70 \times \frac{456}{3.63}\right)\right] = 7160\,\#\square''$ $P_s = 16.10 \times 7160 = 115{,}275\#$

STEP III:
Using the J.C. Pile Formula for both Concrete and Steel:
$P_c + P_s = P = (0.225\,F'_c A_c) + (0.40\,F_{syp}\,A_s)$
Compute P_s to compare with result of Step II
$P_s = 0.40 \times 35{,}000 \times 16.10 = 225{,}400$ Lbs.
$P_c = 0.225 \times 3000 \times 74.66 = 50{,}395$ Lbs.
Column Formula will govern steel design.

STEP IV
$P_g = 115{,}275 + 50{,}395 = 165{,}670$ Lbs.
Dead Weight of Steel Pile $= 65.0 \times 54.7 =$ 3555 Lbs
Dead Weight of Conc. Core $= \left(\frac{74.66}{144}\right) \times 150 \times 65.0 =$ 5050 "
Pile Wt. = 8605 Lbs.

Safe superimposed Load $P = \frac{165{,}670 - 8605}{2000} = 78.53$ Tons.

Pile will be satisfactory for 75 Tons.
The unsupported length in most cases should have been about 4.0 feet longer. Bottom surface soil is often not reliable.

TABLE: Standard pipe piles 9.3.7

OUTSIDE DIA. IN INCHES	WALL THICK'S. IN INCHES	WEIGHT PER FT. IN LBS.	I''^4	r''	S''^3	AREA OUTSIDE SURFACE PER LIN. FT., IN SQ. FT.	AREA STEEL IN WALL SECTION IN SQ. INCHES	AREA VOID IN SECTION INT'R. IN SQ. INCHES	CONC. VOLUME PER LIN. FOOT IN CUBIC FEET
8.625	0.322	28.55	72.49	2.94	16.81	2.26	8.40	50.03	0.347
10.750	0.307	34.24	137.42	3.69	25.57	2.82	10.07	80.69	0.560
	.365	40.48	160.73	3.67	29.90	2.82	11.91	78.85	.548
	.438	48.19	188.95	3.65	35.15	2.82	14.19	76.57	.532
	.500	54.74	211.95	3.63	39.43	2.82	16.10	74.66	.518
12.750	0.312	41.51	235.91	4.40	37.00	3.34	12.19	115.49	0.802
	.330	43.77	248.38	4.39	38.96	3.34	12.88	114.80	.797
	.375	49.56	279.34	4.38	43.82	3.34	14.58	113.10	.785
	.438	57.53	321.42	4.36	50.42	3.34	16.94	110.74	.769
	.500	65.42	361.54	4.34	56.71	3.34	19.24	108.43	.753

TABLE: Standard pipe piles, continued 9.3.7

OUTSIDE DIA. IN INCHES	WALL THK'S. IN INCHES	WEIGHT PER FT. IN LBS.	I"4	r"	S"3	AREA OUTSIDE SURFACE PER LIN.FT.-IN SQ.FT.	AREA STEEL IN WALL SECTION IN SQ.INCHES	AREA VOID IN SECTION INT. IN SQ. INCHES	CONC. VOLUME PER LIN. FOOT. IN CUBIC YDS.
10.0	0.141	14.81	52.90	3.49	10.58	2.62	4.367	74.17	0.0191
	.172	18.04	64.10	3.48	12.82	2.62	5.311	73.23	.0189
	.188	19.70	69.60	3.47	13.92	2.62	5.795	72.74	.0187
	.219	22.88	80.45	3.46	16.09	2.62	6.730	71.81	.0185
	.250	26.03	91.05	3.45	18.21	2.62	7.658	70.88	.0182
10.75	0.141	15.93	65.95	3.75	12.27	2.81	4.699	86.06	0.0221
	.172	19.42	79.93	3.74	14.87	2.81	5.716	85.05	.0219
	.188	21.15	86.81	3.74	16.15	2.81	6.238	84.52	.0217
	.219	24.60	100.41	3.72	18.68	2.81	7.245	83.52	.0215
	.250	28.04	113.74	3.71	21.16	2.81	8.245	82.52	.0212
	.279	31.20	126.85	3.70	23.60	2.81	9.242	81.52	.0210
12.0	0.141	17.81	92.16	4.19	15.36	3.14	5.253	107.84	0.0277
	.172	21.71	111.72	4.18	18.62	3.14	6.391	106.71	.0274
	.188	23.72	121.38	4.18	20.23	3.14	6.976	106.12	.0273
	.219	27.56	140.52	4.17	23.42	3.14	8.105	105.00	.0270
	.250	31.37	159.36	4.16	26.56	3.14	9.228	103.87	.0267
	.281	35.17	177.90	4.14	29.65	3.14	10.345	102.75	.0264
	.312	38.95	195.78	4.13	32.63	3.14	11.456	101.64	.0261
12.75	0.141	18.94	110.73	4.46	17.37	3.34	5.585	122.09	0.0314
	.172	23.09	134.39	4.45	21.08	3.34	6.797	120.88	.0311
	.188	25.16	146.05	4.44	22.91	3.34	7.419	120.26	.0309
	.219	29.28	169.13	4.43	26.53	3.34	8.621	119.06	.0306
	.250	33.38	191.82	4.42	30.09	3.34	9.818	117.86	.0303
	.281	37.45	214.26	4.41	33.61	3.34	11.008	116.67	.0300
	.312	41.51	236.26	4.40	37.06	3.34	12.191	115.49	.0297
14.0	0.141	20.82	147.00	4.90	21.00	3.67	6.139	147.80	0.0380
	.172	25.38	178.50	4.89	25.50	3.67	7.472	146.47	.0377
	.188	27.66	194.04	4.88	27.72	3.67	8.158	145.78	.0375
	.219	32.20	224.84	4.87	32.13	3.67	9.482	144.46	.0372

TABLE: Standard pipe piles, continued 9.3.7

OUTSIDE DIA. IN INCHES	WALL THK'S. IN INCHES	WEIGHT PER FT. IN LBS.	I''^4	r''	S''^3	AREA OUTSIDE SURFACE PER LIN.FT. IN SQ.FT.	AREA STEEL IN WALL SECTION IN SQ. IN.	AREA VOID IN SECTION INT. IN SQ. IN.	CONC. VOLUME PER LIN. FOOT. IN CUBIC YDS.
14.0	0.250	36.71	255.36	4.86	36.48	3.67	10.800	143.14	0.0368
	.281	41.21	283.08	4.85	40.44	3.67	12.110	141.83	.0365
	.312	45.68	314.86	4.84	44.98	3.67	13.417	140.52	.0362
16.0	0.172	29.06	267.68	5.60	33.46	4.19	8.553	192.51	0.0495
	.188	31.66	291.12	5.59	36.39	4.19	9.339	191.72	.0493
	.219	36.87	337.76	5.58	42.22	4.19	10.858	190.20	.0489
	.250	42.05	383.68	5.57	47.96	4.19	12.370	188.69	.0485
	.281	47.22	429.12	5.56	53.64	4.19	13.877	187.19	.0482
	.312	52.36	473.92	5.55	59.24	4.19	15.377	185.69	.0477
	.375	62.58	562.08	5.53	70.26	4.19	18.408	182.65	.0470
18.0	0.219	41.54	483.03	6.29	53.67	4.71	12.234	242.24	0.0630
	.250	47.39	549.09	6.28	61.01	4.71	13.941	240.53	.0619
	.281	53.22	610.52	6.27	68.28	4.71	15.642	238.83	.0614
	.312	59.03	679.23	6.26	75.47	4.71	17.337	237.13	.0610
	.375	70.59	806.58	6.23	89.62	4.71	20.764	233.71	.0601
20.0	0.250	52.73	756.50	6.98	75.65	5.24	15.512	298.65	0.0768
	.281	59.23	847.10	6.97	84.71	5.24	17.408	296.75	.0763
	.312	65.71	936.70	6.96	93.67	5.24	19.298	294.86	.0758
	.375	78.60	1113.50	6.94	111.35	5.24	23.120	291.04	.0749
24.0	0.250	63.71	1315.44	8.40	109.62	6.28	18.653	433.74	0.1115
	.281	71.25	1474.20	8.39	122.85	6.28	20.939	431.45	.1110
	.312	79.06	1631.52	8.38	135.96	6.28	23.218	429.17	.1104
	.375	94.62	1942.44	8.35	161.87	6.28	27.833	424.56	.1092
	.500	125.49	2549.64	8.31	212.47	6.28	36.914	415.48	.1069
30.0	0.312	99.08	3211.80	10.50	214.12	7.85	29.100	677.60	0.1743
	.375	118.05	3829.95	10.48	255.33	7.85	34.901	672.00	.1728
	.500	157.53	5043.00	10.43	336.20	7.85	46.339	660.52	.1699
36.0	0.312	119.11	5578.02	12.62	309.89	9.42	34.981	982.90	0.2528
	.375	142.68	6658.74	12.60	369.93	9.42	41.970	975.91	.2510
	.500	189.57	8785.98	12.55	488.11	9.42	55.763	962.12	.2475

TABLE: Areas and circumferences of circles, 1/16 to 19 7/8 9.3.8

(π = 3.1416)

Diameter	Area	Circumference	Diameter	Area	Circumference	Diameter	Area	Circumference	Diameter	Area	Circumference
1/16	.0031	.1963	5	19.6350	15.7080	10	78.540	31.4160	15	176.715	47.1240
1/8	.0123	.3927	1/8	20.6290	16.1007	1/8	80.516	31.8087	1/8	179.673	47.5167
1/4	.0491	.7854	1/4	21.6476	16.4934	1/4	82.516	32.2014	1/4	182.655	47.9094
3/8	.1104	1.1781	3/8	22.6907	16.8861	3/8	84.541	32.5941	3/8	185.661	48.3021
1/2	.1963	1.5708	1/2	23.7583	17.2788	1/2	86.590	32.9868	1/2	188.692	48.6948
5/8	.3068	1.9635	5/8	24.8505	17.6715	5/8	88.664	33.3795	5/8	191.748	49.0875
3/4	.4418	2.3562	3/4	25.9673	18.0642	3/4	90.763	33.7722	3/4	194.828	49.4802
7/8	.6013	2.7489	7/8	27.1086	18.4569	7/8	92.886	34.1649	7/8	197.933	49.8729
1	.7854	3.1416	6	28.2744	18.8496	11	95.033	34.5576	16	201.062	50.2656
1/8	.9940	3.5343	1/8	29.4648	19.2423	1/8	97.205	34.9503	1/8	204.216	50.6583
1/4	1.2272	3.9270	1/4	30.6797	19.6350	1/4	99.402	35.3430	1/4	207.395	51.0510
3/8	1.4849	4.3197	3/8	31.9191	20.0277	3/8	101.623	35.7357	3/8	210.598	51.4437
1/2	1.7671	4.7124	1/2	33.1831	20.4204	1/2	103.869	36.1284	1/2	213.825	51.8364
5/8	2.0739	5.1051	5/8	34.4717	20.8131	5/8	106.139	36.5211	5/8	217.077	52.2291
3/4	2.4053	5.4978	3/4	35.7848	21.2058	3/4	108.434	36.9138	3/4	220.354	52.6218
7/8	2.7612	5.8905	7/8	37.1224	21.5985	7/8	110.754	37.3065	7/8	223.655	53.0145
2	3.1416	6.2832	7	38.4846	21.9912	12	113.098	37.6992	17	226.981	53.4072
1/8	3.5466	6.6759	1/8	39.8713	22.3839	1/8	115.466	38.0919	1/8	230.331	53.7999
1/4	3.9761	7.0686	1/4	41.2826	22.7766	1/4	117.859	38.4846	1/4	233.706	54.1926
3/8	4.4301	7.4613	3/8	42.7184	23.1693	3/8	120.277	38.8773	3/8	237.105	54.5853
1/2	4.9087	7.8540	1/2	44.1787	23.5620	1/2	122.719	39.2700	1/2	240.529	54.9780
5/8	5.4119	8.2467	5/8	45.6636	23.9547	5/8	125.185	39.6627	5/8	243.977	55.3707
3/4	5.9396	8.6394	3/4	47.1731	24.3474	3/4	127.677	40.0554	3/4	247.450	55.7634
7/8	6.4918	9.0321	7/8	48.7071	24.7401	7/8	130.192	40.4481	7/8	250.948	56.1561
3	7.0686	9.4248	8	50.2656	25.1328	13	132.733	40.8408	18	254.470	56.5488
1/8	7.6699	9.8175	1/8	51.8487	25.5255	1/8	135.297	41.2335	1/8	258.016	56.9415
1/4	8.2958	10.2102	1/4	53.4563	25.9182	1/4	137.887	41.6262	1/4	261.587	57.3342
3/8	8.9462	10.6029	3/8	55.0884	26.3109	3/8	140.501	42.0189	3/8	265.183	57.7269
1/2	9.6211	10.9956	1/2	56.7451	26.7036	1/2	143.139	42.4116	1/2	268.803	58.1196
5/8	10.3206	11.3883	5/8	58.4264	27.0963	5/8	145.802	42.8043	5/8	272.448	58.5123
3/4	11.0447	11.7810	3/4	60.1322	27.4890	3/4	148.490	43.1970	3/4	276.117	58.9050
7/8	11.7933	12.1737	7/8	61.8625	27.8817	7/8	151.202	43.5897	7/8	279.811	59.2977
4	12.5664	12.5664	9	63.6174	28.2744	14	153.938	43.9824	19	283.529	59.6904
1/8	13.3641	12.9591	1/8	65.3968	28.6671	1/8	156.700	44.3751	1/8	287.272	60.0831
1/4	14.1863	13.3518	1/4	67.2008	29.0598	1/4	159.485	44.7678	1/4	291.040	60.4758
3/8	15.0330	13.7445	3/8	69.0293	29.4525	3/8	162.296	45.1605	3/8	294.832	60.8685
1/2	15.9043	14.1372	1/2	70.8823	29.8452	1/2	165.130	45.5532	1/2	298.648	61.2612
5/8	16.8002	14.5299	5/8	72.7599	30.2379	5/8	167.990	45.9459	5/8	302.489	61.6539
3/4	17.7206	14.9226	3/4	74.6621	30.6306	3/4	170.874	46.3386	3/4	306.355	62.0466
7/8	18.6655	15.3153	7/8	76.5889	31.0233	7/8	173.782	46.7313	7/8	310.245	62.4393

TABLE: Areas and circumferences of circles, 20 to 39 7/8 9.3.8

($\pi = 3.1416$)

Diameter	Area	Circumference	Diameter	Area	Circumference	Diameter	Area	Circumference	Diameter	Area	Circumference
20	314.160	62.8320	25	490.875	78.5400	30	706.860	94.248	35	962.115	109.956
⅛	318.099	63.2247	⅛	495.796	78.9327	⅛	712.763	94.641	⅛	969.000	110.349
¼	322.063	63.6174	¼	500.742	79.3254	¼	718.690	95.033	¼	975.909	110.741
⅜	326.051	64.0101	⅜	505.712	79.7181	⅜	724.642	95.426	⅜	982.842	111.134
½	330.064	64.4028	½	510.706	80.1108	½	730.618	95.819	½	989.800	111.527
⅝	334.102	64.7955	⅝	515.726	80.5035	⅝	736.619	96.212	⅝	996.783	111.919
¾	338.164	65.1828	¾	520.769	80.8962	¾	742.645	96.604	¾	1003.790	112.312
⅞	342.250	65.5809	⅞	525.838	81.2889	⅞	748.695	96.997	⅞	1010.822	112.705
21	346.361	65.9736	26	530.930	81.6816	31	754.769	97.390	36	1017.878	113.098
⅛	350.497	66.3663	⅛	536.048	82.0743	⅛	760.869	97.782	⅛	1024.960	113.490
¼	354.657	66.7590	¼	541.190	82.4670	¼	766.992	98.175	¼	1032.065	113.883
⅜	358.842	67.1517	⅜	546.356	82.8597	⅜	773.140	98.568	⅜	1039.195	114.276
½	363.051	67.5444	½	551.547	83.2524	½	779.313	98.960	½	1046.349	114.668
⅝	367.285	67.9371	⅝	556.763	83.6451	⅝	785.510	99.353	⅝	1053.528	115.061
¾	371.543	68.3298	¾	562.003	84.0378	¾	791.732	99.746	¾	1060.732	115.454
⅞	375.826	68.7225	⅞	567.267	84.4305	⅞	797.979	100.138	⅞	1067.960	115.846
22	380.134	69.1152	27	572.557	84.8232	32	804.250	100.531	37	1075.213	116.239
⅛	384.466	69.5079	⅛	577.870	85.2159	⅛	810.545	100.924	⅛	1082.490	116.632
¼	388.822	69.9006	¼	583.209	85.6086	¼	816.865	101.317	¼	1089.792	117.025
⅜	393.203	70.2933	⅜	588.571	86.0013	⅜	823.210	101.709	⅜	1097.118	117.417
½	397.609	70.6860	½	593.959	86.3940	½	829.579	102.102	½	1104.469	117.810
⅝	402.038	71.0787	⅝	599.371	86.7867	⅝	835.972	102.495	⅝	1111.844	118.203
¾	406.494	71.4714	¾	604.807	87.1794	¾	842.391	102.887	¾	1119.244	118.595
⅞	410.973	71.8641	⅞	610.268	87.5721	⅞	848.833	103.280	⅞	1126.669	118.988
23	415.477	72.2568	28	615.754	87.9648	33	855.301	103.673	38	1134.118	119.381
⅛	420.004	72.6495	⅛	621.264	88.3575	⅛	861.792	104.065	⅛	1141.591	119.773
¼	424.558	73.0422	¼	626.798	88.7502	¼	868.309	104.458	¼	1149.089	120.166
⅜	429.135	73.4349	⅜	632.357	89.1429	⅜	874.850	104.851	⅜	1156.612	120.559
½	433.737	73.8276	½	637.941	89.5356	½	881.415	105.244	½	1164.159	120.952
⅝	438.364	74.2203	⅝	643.549	89.9283	⅝	888.005	105.636	⅝	1171.731	121.344
¾	443.015	74.6130	¾	649.182	90.3210	¾	894.620	106.029	¾	1179.327	121.737
⅞	447.690	75.0057	⅞	654.840	90.7137	⅞	901.259	106.422	⅞	1186.948	122.130
24	452.390	75.3984	29	660.521	91.1064	34	907.922	106.814	39	1194.593	122.522
⅛	457.115	75.7911	⅛	666.228	91.4991	⅛	914.611	107.207	⅛	1202.263	122.915
¼	461.864	76.1838	¼	671.959	91.8918	¼	921.323	107.600	¼	1209.958	123.308
⅜	466.638	76.5765	⅜	677.714	92.2845	⅜	928.061	107.992	⅜	1217.677	123.700
½	471.436	76.9692	½	683.494	92.6772	½	934.822	108.385	½	1225.420	124.093
⅝	476.259	77.3619	⅝	689.299	93.0699	⅝	941.609	108.778	⅝	1233.188	124.486
¾	481.107	77.7546	¾	695.128	93.4626	¾	948.420	109.171	¾	1240.981	124.879
⅞	485.979	78.1473	⅞	700.982	93.8553	⅞	955.255	109.563	⅞	1248.798	125.271

TABLE: Areas and circumferences of circles, 40 to 59 7/8 9.3.8

(π = 3.1416)

Diameter	Area	Circumference	Diameter	Area	Circumference	Diameter	Area	Circumference	Diameter	Area	Circumference
40	1256.64	125.664	45	1590.43	141.372	50	1963.50	157.080	55	2375.83	172.788
⅛	1264.51	126.057	⅛	1599.28	141.765	⅛	1973.33	157.473	⅛	2386.65	173.181
¼	1272.40	126.449	¼	1608.16	142.157	¼	1983.18	157.865	¼	2397.48	173.573
⅜	1280.31	126.842	⅜	1617.05	142.550	⅜	1993.06	158.258	⅜	2408.34	173.966
½	1288.25	127.235	½	1625.97	142.943	½	2002.97	158.651	½	2419.23	174.359
⅝	1296.22	127.627	⅝	1634.92	143.335	⅝	2012.89	159.043	⅝	2430.14	174.751
¾	1304.21	128.020	¾	1643.89	143.728	¾	2022.85	159.436	¾	2441.07	175.144
⅞	1312.22	128.413	⅞	1652.89	144.121	⅞	2032.82	159.829	⅞	2452.03	175.537
41	1320.26	128.806	46	1661.91	144.514	51	2042.83	160.222	56	2463.01	175.930
⅛	1328.32	129.198	⅛	1670.95	144.906	⅛	2052.85	160.614	⅛	2474.02	176.322
¼	1336.41	129.591	¼	1680.02	145.299	¼	2062.90	161.007	¼	2485.05	176.715
⅜	1344.52	129.984	⅜	1689.11	145.692	⅜	2072.98	161.400	⅜	2496.11	177.108
½	1352.66	130.376	½	1698.23	146.084	½	2083.08	161.792	½	2507.19	177.500
⅝	1360.82	130.769	⅝	1707.37	146.477	⅝	2093.20	162.185	⅝	2518.30	177.893
¾	1369.00	131.162	¾	1716.54	146.870	¾	2103.35	162.578	¾	2529.43	178.286
⅞	1377.21	131.554	⅞	1725.73	147.262	⅞	2113.52	162.970	⅞	2540.58	178.678
42	1385.45	131.947	47	1734.95	147.655	52	2123.72	163.363	57	2551.76	179.071
⅛	1393.70	132.340	⅛	1744.19	148.048	⅛	2133.94	163.756	⅛	2562.97	179.464
¼	1401.99	132.733	¼	1753.45	148.441	¼	2144.19	164.149	¼	2574.20	179.857
⅜	1410.30	133.125	⅜	1762.74	148.833	⅜	2154.46	164.541	⅜	2585.45	180.249
½	1418.63	133.518	½	1772.06	149.226	½	2164.76	164.934	½	2596.73	180.642
⅝	1426.99	133.911	⅝	1781.40	149.619	⅝	2175.08	165.327	⅝	2608.03	181.035
¾	1435.37	134.303	¾	1790.76	150.011	¾	2185.42	165.719	¾	2619.36	181.427
⅞	1443.77	134.696	⅞	1800.15	150.404	⅞	2195.79	166.112	⅞	2630.71	181.820
43	1452.20	135.089	48	1809.56	150.797	53	2206.19	166.505	58	2642.09	182.213
⅛	1460.66	135.481	⅛	1819.00	151.189	⅛	2216.61	166.897	⅛	2653.49	182.605
¼	1469.14	135.874	¼	1828.46	151.582	¼	2227.05	167.290	¼	2664.91	182.998
⅜	1477.64	136.267	⅜	1837.95	151.975	⅜	2237.52	167.683	⅜	2676.36	183.391
½	1486.17	136.660	½	1847.46	152.368	½	2248.01	168.076	½	2687.84	183.784
⅝	1494.73	137.052	⅝	1856.99	152.760	⅝	2258.53	168.468	⅝	2699.33	184.176
¾	1503.30	137.445	¾	1866.55	153.153	¾	2269.07	168.861	¾	2710.86	184.569
⅞	1511.91	137.838	⅞	1876.14	153.546	⅞	2279.64	169.254	⅞	2722.41	184.962
44	1520.53	138.230	49	1885.75	153.938	54	2290.23	169.646	59	2733.98	185.354
⅛	1529.19	138.623	⅛	1895.38	154.331	⅛	2300.84	170.039	⅛	2745.57	185.747
¼	1537.86	139.016	¼	1905.04	154.724	¼	2311.48	170.432	¼	2757.20	186.140
⅜	1546.56	139.408	⅜	1914.72	155.116	⅜	2322.15	170.824	⅜	2768.84	186.532
½	1555.29	139.801	½	1924.43	155.509	½	2332.83	171.217	½	2780.51	186.925
⅝	1564.04	140.194	⅝	1934.16	155.902	⅝	2343.55	171.610	⅝	2792.21	187.318
¾	1572.81	140.587	¾	1943.91	156.295	¾	2354.29	172.003	¾	2803.93	187.711
⅞	1581.61	140.979	⅞	1953.69	156.687	⅞	2365.05	172.395	⅞	2815.67	188.103

TABLE: Areas and circumferences of circles, 60 to 79 7/8 9.3.8

($\pi = 3.1416$)

Diameter	Area	Circumference	Diameter	Area	Circumference	Diameter	Area	Circumference	Diameter	Area	Circumference
60	2827.44	188.496	65	3318.31	204.204	70	3848.46	219.912	75	4417.87	235.620
1/8	2839.23	188.889	1/8	3331.09	204.597	1/8	3862.22	220.305	1/8	4432.61	236.013
1/4	2851.05	189.281	1/4	3343.89	204.989	1/4	3876.00	220.697	1/4	4447.38	236.405
3/8	2862.89	189.674	3/8	3356.71	205.382	3/8	3889.80	221.090	3/8	4462.16	236.798
1/2	2874.76	190.067	1/2	3369.56	205.775	1/2	3903.63	221.483	1/2	4476.98	237.191
5/8	2886.65	190.459	5/8	3382.44	206.167	5/8	3917.49	221.875	5/8	4491.81	237.583
3/4	2898.57	190.852	3/4	3395.33	206.560	3/4	3931.37	222.268	3/4	4506.67	237.976
7/8	2910.51	191.245	7/8	3408.26	206.953	7/8	3945.27	222.661	7/8	4521.56	238.369
61	2922.47	191.638	66	3421.20	207.346	71	3959.20	223.054	76	4536.47	238.762
1/8	2934.46	192.030	1/8	3434.17	207.738	1/8	3973.15	223.446	1/8	4551.41	239.154
1/4	2946.48	192.423	1/4	3447.17	208.131	1/4	3987.13	223.839	1/4	4566.36	239.547
3/8	2958.52	192.816	3/8	3460.19	208.524	3/8	4001.13	224.232	3/8	4581.35	239.940
1/2	2970.58	193.208	1/2	3473.24	208.916	1/2	4015.16	224.624	1/2	4596.36	240.332
5/8	2982.67	193.601	5/8	3486.30	209.309	5/8	4029.21	225.017	5/8	4611.39	240.725
3/4	2994.78	193.994	3/4	3499.40	209.702	3/4	4043.29	225.410	3/4	4626.45	241.118
7/8	3006.92	194.386	7/8	3512.52	210.094	7/8	4057.39	225.802	7/8	4641.53	241.510
62	3019.08	194.779	67	3525.66	210.487	72	4071.51	226.195	77	4656.64	241.903
1/8	3031.26	195.172	1/8	3538.83	210.880	1/8	4085.66	226.588	1/8	4671.77	242.296
1/4	3043.47	195.565	1/4	3552.02	211.273	1/4	4099.84	226.981	1/4	4686.92	242.689
3/8	3055.71	195.957	3/8	3565.24	211.665	3/8	4114.04	227.373	3/8	4702.10	243.081
1/2	3067.97	196.350	1/2	3578.48	212.058	1/2	4128.26	227.766	1/2	4717.31	243.474
5/8	3080.25	196.743	5/8	3591.74	212.451	5/8	4142.51	228.159	5/8	4732.54	243.867
3/4	3092.56	197.135	3/4	3605.04	212.843	3/4	4156.78	228.551	3/4	4747.79	244.259
7/8	3104.89	197.528	7/8	3618.35	213.236	7/8	4171.08	228.944	7/8	4763.07	244.652
63	3117.25	197.921	68	3631.69	213.629	73	4185.40	229.337	78	4778.37	245.045
1/8	3129.64	198.313	1/8	3645.05	214.021	1/8	4199.74	229.729	1/8	4793.70	245.437
1/4	3142.04	198.706	1/4	3658.44	214.414	1/4	4214.11	230.122	1/4	4809.05	245.830
3/8	3154.47	199.099	3/8	3671.86	214.807	3/8	4228.51	230.515	3/8	4824.43	246.223
1/2	3166.93	199.492	1/2	3685.29	215.200	1/2	4242.93	230.908	1/2	4839.83	246.616
5/8	3179.41	199.884	5/8	3698.76	215.592	5/8	4257.37	231.300	5/8	4855.26	247.008
3/4	3191.91	200.277	3/4	3712.24	215.985	3/4	4271.84	231.693	3/4	4870.71	247.401
7/8	3204.44	200.670	7/8	3725.75	216.378	7/8	4286.33	232.086	7/8	4886.18	247.794
64	3217.00	201.062	69	3739.29	216.770	74	4300.85	232.478	79	4901.68	248.186
1/8	3229.58	201.455	1/8	3752.85	217.163	1/8	4315.39	232.871	1/8	4917.21	248.579
1/4	3242.18	201.848	1/4	3766.43	217.556	1/4	4329.96	233.264	1/4	4932.75	248.972
3/8	3254.81	202.240	3/8	3780.04	217.948	3/8	4344.55	233.656	3/8	4948.33	249.364
1/2	3267.46	202.633	1/2	3793.68	218.341	1/2	4359.17	234.049	1/2	4963.92	249.757
5/8	3280.14	203.026	5/8	3807.34	218.734	5/8	4373.81	234.442	5/8	4979.55	250.150
3/4	3292.84	203.419	3/4	3821.02	219.127	3/4	4388.47	234.835	3/4	4995.19	250.543
7/8	3305.56	203.811	7/8	3834.73	219.519	7/8	4403.16	235.227	7/8	5010.86	250.935

TABLE: Areas and circumferences of circles, 80 to 100 9.3.8

($\pi = 3.1416$)

Diameter	Area	Circumference	Diameter	Area	Circumference	Diameter	Area	Circumference	Diameter	Area	Circumference
80	5026.56	251.328	85	5674.51	267.036	90	6361.74	282.744	95	7088.24	298.452
⅛	5042.28	251.721	⅛	5691.22	267.429	⅛	6379.42	283.137	⅛	7106.90	298.845
¼	5058.03	252.113	¼	5707.94	267.821	¼	6397.13	283.529	¼	7125.59	299.237
⅜	5073.79	252.506	⅜	5724.69	268.214	⅜	6414.86	283.922	⅜	7144.31	299.630
½	5089.59	252.899	½	5741.47	268.607	½	6432.62	284.315	½	7163.04	300.023
⅝	5105.41	253.291	⅝	5758.27	268.999	⅝	6450.40	284.707	⅝	7181.81	300.415
¾	5121.25	253.684	¾	5775.10	269.392	¾	6468.21	285.100	¾	7200.60	300.808
⅞	5137.12	254.077	⅞	5791.94	269.785	⅞	6486.04	285.493	⅞	7219.41	301.201
81	5153.01	254.470	86	5808.82	270.178	91	6503.90	285.886	96	7238.25	301.594
⅛	5168.93	254.862	⅛	5825.72	270.570	⅛	6521.78	286.278	⅛	7257.11	301.986
¼	5184.87	255.255	¼	5842.64	270.963	¼	6539.68	286.671	¼	7275.99	302.379
⅜	5200.83	255.648	⅜	5859.59	271.356	⅜	6557.61	287.064	⅜	7294.91	302.772
½	5216.82	256.040	½	5876.56	271.748	½	6575.56	287.456	½	7313.84	303.164
⅝	5232.84	256.433	⅝	5893.55	272.141	⅝	6593.54	287.849	⅝	7332.80	303.557
¾	5248.88	256.826	¾	5910.58	272.534	¾	6611.55	288.242	¾	7351.79	303.950
⅞	5264.94	257.218	⅞	5927.62	272.926	⅞	6629.57	288.634	⅞	7370.79	304.342
82	5281.03	257.611	87	5944.69	273.319	92	6647.63	289.027	97	7389.83	304.735
⅛	5297.14	258.004	⅛	5961.79	273.712	⅛	6665.70	289.420	⅛	7408.89	305.128
¼	5313.28	258.397	¼	5978.91	274.105	¼	6683.80	289.813	¼	7427.97	305.521
⅜	5329.44	258.789	⅜	5996.05	274.497	⅜	6701.93	290.205	⅜	7447.08	305.913
½	5345.63	259.182	½	6013.22	274.890	½	6720.08	290.598	½	7466.21	306.306
⅝	5361.84	259.575	⅝	6030.41	275.283	⅝	6738.25	290.991	⅝	7485.37	306.699
¾	5378.08	259.967	¾	6047.63	275.675	¾	6756.45	291.383	¾	7504.55	307.091
⅞	5394.34	260.360	⅞	6064.87	276.068	⅞	6774.68	291.776	⅞	7523.75	307.484
83	5410.62	260.753	88	6082.14	276.461	93	6792.92	292.169	98	7542.98	307.877
⅛	5426.93	261.145	⅛	6099.43	276.853	⅛	6811.20	292.562	⅛	7562.24	308.270
¼	5443.26	261.538	¼	6116.74	277.246	¼	6829.49	292.954	¼	7581.52	308.662
⅜	5459.62	261.931	⅜	6134.08	277.638	⅜	6847.82	293.347	⅜	7600.82	309.055
½	5476.01	262.324	½	6151.45	278.032	½	6866.16	293.740	½	7620.15	309.448
⅝	5492.41	262.716	⅝	6168.84	278.424	⅝	6884.53	294.132	⅝	7639.50	309.840
¾	5508.84	263.109	¾	6186.25	278.817	¾	6902.93	294.525	¾	7658.88	310.233
⅞	5525.30	263.502	⅞	6203.69	279.210	⅞	6921.35	294.918	⅞	7678.28	310.626
84	5541.78	263.894	89	6221.15	279.602	94	6939.79	295.310	99	7697.71	311.018
⅛	5558.29	264.287	⅛	6238.64	279.995	⅛	6958.26	295.703	⅛	7717.16	311.411
¼	5574.82	264.680	¼	6256.15	280.388	¼	6976.76	296.096	¼	7736.63	311.804
⅜	5591.37	265.072	⅜	6273.69	280.780	⅜	6995.28	296.488	⅜	7756.13	312.196
½	5607.95	265.465	½	6291.25	281.173	½	7013.82	296.881	½	7775.66	312.589
⅝	5624.56	265.858	⅝	6308.84	281.566	⅝	7032.39	297.274	⅝	7795.21	312.982
¾	5641.18	266.251	¾	6326.45	281.959	¾	7050.98	297.667	¾	7814.78	313.375
⅞	5657.84	266.643	⅞	6344.08	282.351	⅞	7069.59	298.059	⅞	7834.38	313.767
									100	7854.00	314.160

Pile hammers 9.4

Prior to 1900, piles were driven by manually operated drop hammers. This type of hammer consists of a heavy weight, which is raised to a pre-determined height, then released, allowing it to drop and impact upon the pile. Blows are as rapid as the weight can be raised and released. The hammer is raised by hand cranking a rope on a winch. At a certain hammer height, a trigger automatically releases the winch. Larger drop hammers used two horses hitched on each side of the winch. Later, the horses were replaced by a gasoline engine. The drop hammer is still in wide use, especially on very small projects, where a more mechanized hammer would be too costly.

In order to drive a slanted batter pile, the older rigs were equipped with an inclined trough to guide the drop hammer. These guides were greased to reduce friction and permit the hammer to develop greater energy at impact. Various experiments were made, such as putting wheels or rollers on the hammer, to eliminate the need for grease.

The railroad builders, constructing routes to open up the West, built drop hammer equipment on flat cars. The raising winch was operated with steam pressure from the locomotive. These antiquated methods made possible the construction of many trestles, over which the trains made their regular runs for many years.

Single acting hammers 9.4.1

The ironmasters soon provided the pile driving trade with a hammer built as a compact unit. The striking ram was made heavier and attached to a piston. The piston raised the ram when high pressure steam was injected into the cylinder. At full rise the pressure was released through an exhaust outlet. Upon release of steam pressure, the heavy ram fell by gravity to impact upon the pile. Single action means that the force for moving the piston is in a single direction. By referring to Table 9.4.6 which lists the manufacturers' data on various hammers, it may be seen that the single acting steam types will have striking speeds up to sixty blows per minute. When the weight of the ram is reduced below 5000 pounds, and the stroke shortened, the single acting (SA) hammer may exceed the sixty blow per minute speed.

Energy for SA hammers is calculated by the formula: $E = WH$. Where E = Kinetic Energy in foot pounds, W = Weight of ram, and H = Height of fall. A hammer with a ram weight of 3000 pounds, and a stroke height of 20 inches, would have

$$\text{energy } E = \frac{3000 \times 20}{12} = 5000 \text{ Foot Pounds.}$$

Double acting hammers 9.4.2

Steam pressure acting in an upward direction to raise the striking ram may also be applied to start the piston back down. In mechanics, this action is called reciprocating motion. The action is produced by arranging the port holes in the cylinder

Double acting hammers, continued 9.4.2

in such a manner that the injection of steam pressure at one end of the piston will alternate with the injection at the opposite end. This action in pile hammer is identified as double action, or differential action. If the exhaust outlets are led through a manifold to a flexible hose, the double acting hammer can operate under water. Only the exhaust hose end needs to be above water.

The reciprocating motion increases the speed count of blows per minute, which produces faster penetration. The skin friction resistance is decreased as greater vibrations are set up in the pile during the driving process.

To determine the kinetic energy of a double acting hammer in a vertical position, the force of steam pressure acting downward during the stroke is added to the single acting energy formula. The force of steam pressure is the area of piston surface times the steam pressure applied against piston. For double acting hammers, the formula for kinetic energy is written as: $E = WH + Ap\ Sp$, where Ap = Area of piston in square inches, and Sp = Steam pressure in pounds per square inch.

In comparing the rated energy listed in Table 9.4.6, it must be pointed out that manufacturers also consider other factors to establish the value of rated energy. Pile driving observers counting the blows should make frequent note of the steam pressure gauge reading at the boiler. A lower pressure may continue to operate the hammer, but will result in lower hammer energy. The friction in the hose from boiler to hammer will also reduce the steam pressure at the hammer; therefore extreme distance from boiler to hammer is to be avoided.

EXAMPLE: Computing hammer energy 9.4.2.1

A double acting hammer has a stroke height of 1.5833 feet, and a piston diameter of 10.0 inches.
Weight of Ram with striking parts is given at 3150 Lbs.
Net gross weight of hammer is 10,950 Lbs.

REQUIRED:
Determine the kinetic energy of hammer. Assume that steam pressure at hammer is 100 PSI, and neglect the weight contribution of hammer.

STEP I:
Area of piston = $0.7854 D^2$ or $Ap = 10.0^2 \times 0.7854 = 78.54$ Sq. In.
$W = 3150$ Lbs. $H = 1.5833$ Ft. Steam pressure = 100 PSI.
Formula for energy, $E = WH + (Ap\ Sp)$

EXAMPLE: Computing hammer energy, continued 9.4.2.1

STEP II:
Substituting values in Formula:

$E = (3150 \times 1.5833) + (78.54 \times 100) = 12{,}840$ Foot Pounds.

Design Note:
When steam pressure is reduced to 90 PSI, resulting from a long supply hose, the energy, $E = 12{,}055$ Foot Lbs.

Diesel hammers 9.4.3

The Diesel pile hammer works on the principle of compression ignition of the fuel. The first hammers of this type were developed in Western Germany shortly after World War II. Basically the operation is single acting, with the piston being a part of the striking ram. Early models allowed the height of ram stroke to be observed, and on hard driving, it would be seen quite clearly that the ram was delivering greater energy. A reference to Table 9.4.6 reveals that the stroke of the ram-piston has a maximum length of eight feet. The developed energy is computed WH plus the impact explosive force of injected fuel.

After a Diesel hammer is placed upon a pile, operation is initiated by raising the free ram-piston to top position by a separate crane hoist line. As the piston moves up, diesel fuel is drawn into the lower combustion chamber through intake ports. When the trigger mechanism releases the heavy ram, it falls downward by gravity. During the downward stroke, the ram closes the intake and exhaust ports, and compresses the fuel and air trapped in the cylinder between the lower end of the ram and the anvil. The high pressure results in heat sufficient to explode the combustible mixture. The explosive force acts in two directions. It adds to the gravity force of ram upon the anvil, which tends to push the pile downward; and simultaneously the explosion pushes the ram upward into position for another blow. The ram, on its return upward, uncovers the exhaust ports, allowing the burned gases to escape. The hammer will continue to operate in this manner. The speed count will vary. Heavy resistance of the pile to penetration will usually cause the ram to rebound its full length which makes for a slower speed but greater impact. The Diesel hammer is stopped by manually pulling on a rope attached to a lever which opens a port in the combustion chamber. After opening this stopping port, the ram falls and there is no explosive force to return it upward.

A few refinements have been added to the original open top hammer. The stroke has been made shorter, and the top part of the cylinder closed. The ram top is equipped with compression rings which permit air to be compressed in the upper chamber and tanks. This top chamber is called the bounce cylinder. Faster operation is possible with the shorter ram because the compressed air in top chamber accelerates the ram downward as it begins its next stroke.

Diesel hammers are favored by field crews for driving steel sheet piles. Their smaller diameter allows the pile to be plumbed with less effort, and the hammer

Diesel hammers, continued 9.4.3

can be attached to a single steel section for control guide. The advantages of the Diesel over the steam hammer is that it eliminates the need for a boiler, supply hose, water tanks and fireman. In most cases the driving crew is reduced by a minimum of two men.

Vibrating hammers 9.4.4

The "Vibro" or vibrating type of pile driver is a device which has evolved from the double acting pile extractor. It has long been known that a pile could be pulled if lifting force was applied in combination with oscillations. These oscillations tend to shake the pile and break the adhesion of earth to the pile surface previously described as skin friction. By referring to the table on pile hammer data, it will be noted that the McKiernan-Terry Model 3 Extractor is listed as a double action hammer with a stroke of 5¾ inches and a speed of 400 blows per minute. Converting this speed to 6⅔ strokes per second, with an energy rating of only 385 foot pounds, it is seen that the Model 3 is a vibrating device rather than a pile hammer.

The vibrating driver reverses the extraction action by applying a heavily constructed, motorized vibrator to the top of the pile. The integral vibrator is clamped to the pile in such a position that oscillations will be produced over the entire pile length. The vibrating drivers are powered by electric motors, although earlier models were successful using steam and compressed air. The electric models are provided with portable generators which enable the driving operations to be carried on in isolated regions. When the motor is started and the vibrating mechanism clutch engaged, the crane hoist releases the weight to bear upon the pile. In ordinary soils the vibro-driver will install a long pile in a matter of minutes. This factor leads to economic advantages, especially on large projects where the bulky equipment can be most efficiently employed.

Hammer selection 9.4.5

For normal pile driving operations, an efficient hammer may be selected from Table 9.6.6, safe allowable loads determined by blow count. Hammers which deliver insufficient energy to drive piles to the higher capacity loads will require a large number of blows. A high blow count with less penetration per blow may give deceptive and disappointing results. The ratio of pile weight to ram weight should be given a thorough analysis to avoid wasting hammer energy. Conversely, a hammer with too high an impact energy may cause damage to the pile, and also provide results which are inconclusive. Before a hammer selection is made final the Hiley or Terzaghi formula should be applied to the conditions of pile and hammer characteristics, in order to obtain the correct energy rating.

TABLE: Pile hammer data and characteristics 9.4.6

MANUFACTURER OR TRADE NAME	NUMBER OR MODEL	TYPE OF ACTION	RATED ENERGY IN FOOT POUNDS	RAM SPEED BLOWS PER MIN.	WEIGHT OF RAM IN LBS.	LINE MARK
	020	SINGLE	60,000	60	20,000	1
	200-C	DOUBLE	50,000	98	20,000	2
	014	SINGLE	42,000	60	14,000	3
	140-C	DOUBLE	36,000	103	14,000	4
	010	SINGLE	32,500	50	10,000	5
	08	SINGLE	26,000	50	8,000	6
	80-C	DOUBLE	24,450	111	8,000	7
VULCAN IRON WORKS	06	SINGLE	19,500	60	6,500	8
	65-C	DOUBLE	19,200	117	6,500	9
	50-C	DOUBLE	15,100	120	5,000	10
	1	SINGLE	15,000	60	5,000	11
	30-C	DOUBLE	7,260	133	3,000	12
	2	SINGLE	7,260	70	3,000	13
	DGH-900	DOUBLE	4,000	238	900	14
	S-20	SINGLE	60,000	60	20,000	15
	S-14	SINGLE	37,500	60	14,000	16
	DA-35	SINGLE	35,500	48	2,800	17
	DA-35	DOUBLE	21,000	82	2,800	18
	S-10	SINGLE	32,500	55	10,000	19
	DE-40	DIESEL	32,000	50	4,000	20
	S-8	SINGLE	26,000	55	8,000	21
	C-8	DOUBLE	26,000	78	8,000	22
	DE-30	DIESEL	22,400	50	2,800	23
	11-B-3	DOUBLE	19,150	95	5,000	24
	S-5	SINGLE	16,250	60	5,000	25
McKIERNAN-TERRY	DE-20	SINGLE	16,000	50	2,000	26
	C-5	DOUBLE	16,000	100	5,000	27
	10-B-3	DOUBLE	13,100	105	3,000	28
	C-3	DOUBLE	9,000	135	3,000	29
	S-3	SINGLE	9,000	65	3,000	30
	DE-10	DIESEL	8,800	50	1,100	31
	9-B-3	DOUBLE	8,750	145	1,600	32
	7	DOUBLE	4,150	225	800	33
	6.5	DOUBLE	3,200	280	600	34
	6	DOUBLE	2,500	275	400	35
	5	DOUBLE	1,000	300	200	36
	3	DOUBLE	385	400	68	37
	520	DIESEL	30,000	82	5,070	38
	440	DIESEL	18,200	90	4,000	39
LINK-BELT CORP.	312	DIESEL	18,000	100	3,855	40
	180	DIESEL	8,100	95	1,725	41
	105	DIESEL	7,500	95	1,445	42
	D-30	DIESEL	54,250	45		43
DELMAG-GERMANY	D-22	DIESEL	39,700	48	4,850	44
	D-12	DIESEL	22,500	51	2,750	45
	D-5	DIESEL	9,100	56	1,100	46
	MB-70	DIESEL	155,507	60	15,840	47
	MB-40	DIESEL	91,135	60	9,039	48
	MB-22	DIESEL	42,674	60	4,840	49
MITSUBISHI-JAPAN	M-43	DIESEL	92,580	60	9,460	50
	M-33	DIESEL	61,840	60	7,260	51
	M-23	DIESEL	44,000	60	5,060	52
	M-145	DIESEL	26,000	60	2,970	53

TABLE: Pile hammer data and characteristics, continued 9.4.6

MATCH LINES	RAM STROKE HEIGHT IN FT. OR IN.	CAPACITY BOILER H.P.	OPERATING PRESSURE P.S.I.	GROSS W'T. HAMMER IN LBS.	HOSE DIA. INLET IN INCHES	TOTAL LENGTH OF HAMMER FEET + INCHES	REMARKS OR GENERAL USE	MATCH LINES
1	36.0"	278	120	39,000	3.00	15'-0"		1
2	15.5"	260	142	39,050	4.00	13'-2"	DIFFERENTIAL	2
3	36.0"	200	110	27,500	3.00	14'-6"		3
4	15.5'	210	140	27,985	3.00	12'-3"	DIFFERENTIAL	4
5	39.0"	157	105	18,750	2.50	15'-0"		5
6	39.0"	127	83	16,750	2.50	15'-0"		6
7	16.5"	180	120	17,885	2.50	12'-2"	DIFFERENTIAL	7
8	36.0"	94	100	11,200	2.00	13'-0"		8
9	15.5"	152	150	14,885	2.00	12'-2"	DIFFERENTIAL	9
10	15.5"	125	120	11,780	2.00	11'-0"	DIFFERENTIAL	10
11	36.0"	85	80	9,700	2.00	13'-0"		11
12	12.5"	70	120	7,050	1.50	9'-8"	DIFFERENTIAL	12
13	29.0"	50	80	6,700	1.50	11'-6"		13
14	10.0"	75	78	5,000	1.50	6'-9"	EXTRACTOR	14
15	36.0"	280	150	38,650	3.00	15'-5"		15
16	32.0"	155	100	31,100	3.00	13'-7"		16
17	8'-0"	---	---	10,000	---	17'-0"		17
18	5'-10"	---	---	10,000	---	17'-0"		18
19	39.0"	130	80	22,380	2.50	13'-0"		19
20	8'-0"	---	---	9,900	---	14'-2"		20
21	39.0"	120	80	18,300	2.50	13'-3"		21
22	20.0"	110	100	18,750	2.50	9'-9"		22
23	8'-0"	---	---	8,125	---	14'-0"		23
24	19.0"	126	100	14,000	2.50	11'-1½"	NO CUSHIONS	24
25	39.0"	85	80	12,460	2.00	12'-2"		25
26	8'-0"	---	---	5,500	---	12'-2"	DIESEL	26
27	18.0"	80	100	11,880	2.50	8'-9"		27
28	19.0"	104	100	10,850	2.50	9'-2"	NO CUSHIONS	28
29	16.0"	60	100	8,500	2.00	7'-9½		29
30	36.0"	57	80	9,030	1.50	11'-4"		30
31	8'-0"	---	---	2,900	---	11'-3"		31
32	17.0"	85	100	7,000	2.00	8'-4"	NO CUSHIONS	32
33	9.5"	65	100	5,000	1.50	6'-1"	EXTRACTIONS	33
34	8.375"	65	100	4,550	1.50	6'-2"	EXTRACTIONS	34
35	8.75"	45	100	2,900	1.25	5'-3"	EXTRACTIONS	35
36	7.0"	35	100	1,500	1.25	4'-9"	EXTRACTIONS	36
37	5.75"	25	100	675	1.00	4'-10"	EXTRACTIONS	37
38	43.17"	---	---	12,545	---	13'-6"	RECIPROCATING	38
39	48.55"	---	---	10,300	---	14'-6"	RECIPROCATING	39
40	30.89"	---	---	10,375	---	10'-9"	RECIPROCATING	40
41	37.60"	---	---	4,550	---	11'-3"	RECIPROCATING	41
42	35.23"	---	---	3,885	---	10'-3"	RECIPROCATING	42
43	NOT LISTED	---	---	NOT AVAIL.	---	NOT AVAILABLE		43
44	VISIBLE	---	---	10,055	---	12'-10½"		44
45	VISIBLE	---	---	5,440	---	12'-7½		45
46	VISIBLE	---	---	2,400	---	11'-2½"		46
47	9'-6"	---	---	40,700	---	18'-8⅛"	FOR STEEL PILES	47
48	8'-2"	---	---	24,030	---	18'-5¼"	FOR STEEL PILES	48
49	8'-2"	---	---	11,660	---	16-0¼"		49
50	NOT LISTED	---	---	22,660	---	15'-4⅜"	FOR STEEL PILES	50
51	NOT LISTED	---	---	16,940	---	14'-8⅞"		51
52	NOT LISTED	---	---	11,220	---	13'-3¼"		52
53	NOT LISTED	---	---	7,260	---	13'-4"		53

Work and energy 9.5.1

Work is defined as a product of a moving force (W) times the distance (H) through which it moves. A unit of work is the foot pound ('#). When work is stored, it becomes energy (E), which also has units of foot pounds.

The earlier formulas used to determine the load bearing capacity of a pile did not take into account all of the elements involved in the driving operation. They were simply a statement of the basic work equation: $WH = Rs$ or $R = \frac{WH}{s}$. Where R equals the resistance to penetration and s equals the set per blow given in inches. Set means the amount of penetration after each blow of the hammer upon the pile. This early formula did not give realistic results when compared with control test piles. This led to the search for a formula which would provide a reliable safety factor and predict the test load values. The changes made in the basic formula will be illustrated in 9.6.2 on the Engineering-News formula.

Power 9.5.2

Power equals work divided by time. Written in a formula: $\text{Power} = \frac{\text{Work}}{t}$. Power has units of Horse Power, named by the British engineer and inventor, James Watt (1736–1819). Credit is given to this man for the first successful steam engine. Horse Power is defined: One horse power is equal to 33,000 foot pounds of work done in one minute. $HP = \frac{\text{Work}}{33{,}000\,t}$ where Work is in foot pounds and Time (t) is in minutes. An example to follow will illustrate the practical method for calculating the horse power and operating time.

Momentum 9.5.3

Momentum = Mass times Velocity. Momentum must not be confused with such terms as work, power, or mass times distance (WH).

EXAMPLE: Hammer velocity and energy 9.5.4

Assume that a Diesel Hammer contains a Ram with a weight of 5000 Pounds. The maximum stroke is given as 8.00 feet, and the number of blows is 48 per minute.

REQUIRED:
Calculate the Velocity of falling Ram at bottom of its stroke. Neglect any cylinder friction or compressed air in combustion chamber being a factor in the free fall. If 60 blows per minute are struck, determine the time required to raise the Ram after each stroke.

STEP I:
The fall will be less than 32.174′(g) and therefore will be less than 1 second.

Formula for $KE = \left(\frac{W}{2g}\right) V^2$, and Basic Work Formula: $KE = WH$.

STEP II:
Transpose first formula thus: $V = \sqrt{2gH}$, and put in values.

$V = \sqrt{2 \times 32.174 \times 8.00} = 22.7$ Feet per second.

Then $KE = \left(\frac{5000}{2 \times 32.174}\right) \times 22.7^2 = 40{,}000$ Foot Pounds.

STEP III:
Using Basic Work Formula: $KE = 5000 \times 8.00 = 40{,}000$ Ft. Lbs.

Time required for fall $= \frac{22.70}{32.174} = 0.706$ seconds.

STEP IV:
At a speed of 48 blows per minute, each blow with the return of ram $= \frac{60}{48} = 1.25$ seconds.

Time used to return ram $= 1.250 - 0.706 = 0.544$ seconds.

EXAMPLE: Computing required horsepower 9.5.5

An Elevator to raise men & materials to a High-Rise structure is to be installed with a capacity of 2500 Pounds. Desired speed is to be between 60 and 75 feet per minute. A standard size electric motor will be employed with manual operation for stopping.

REQUIRED:
Use maximum speed data to calculate the necessary Horse Power to operate the carrier.

STEP I:
At maximum speed the work required is thus:
2500 x 75 = 187,500 Foot Lbs. Time (t) = 1.0 minute.
By Formula: $HP = \frac{Work}{33{,}000\,t}$ or $H.P. = \frac{187{,}500}{33{,}000 \times 1.0} = 5.68$

STEP I:
A 5.68 Horse Power Motor is not standard. A 5.0 HP is close enough to use. Calculating speed with this size is computed thus: Capacity of lift remains as 2500 Lbs.
Work maximum for 5.0 H.P. = 33,000 x 5.0 = 165,000 '# per minute.
$Speed = \frac{165{,}000}{2500} = 66$ feet per minute.

The importance of the test pile 9.6.1

It must be emphasized that when the design load for any pile is in doubt, a control-test pile should be tested in the area, under increasing increments of constant load (see 9.2.2.1). The allowable load permitted by Code authorities is generally not more than one-half of the test load that caused a settlement of one-half inch.

Assuming that a record of pile penetration and blow count was made when driving the control-pile, the information may be used in subsequent driving of the balance of the same type of piles. When the conditions of driving correspond, the other piles may be assumed to have a supporting capacity equal to the control-pile. In any event, to apply a pile hammer formula, the rate of penetration must be equal to or less than that of the control-pile tested. The set penetration per blow should show similar values through a comparable driving distance.

Allowable safe load capacities computed from hammer formulas are not a substitute for the test pile. The data obtained when driving the test pile is evaluated and used in formula for design records.

Test pile exemptions 9.6.1.1

Petro-chemical and processing plants located in tideland regions may not require a test-pile, when similar projects have been completed in the area. With pile records available for reference, it would be a waste of time and money to construct another control test. In such circumstances, the pile hammer formulas may be depended upon to provide the ultimate and allowable safe load capacity for new piling.

Engineering - News hammer formula 9.6.2

Probably the most extensive set of hammer formulas used in America are the equations developed in 1920 by the engineer A. M. Wellington. These formulas came to be known as the Engineering-News Formula: $L = \frac{2WH}{s + c}$. Wellington introduced the constant (c) in the denominator of the formula. The E-N Formula is applied to three types of hammers thus:

For Drop Hammer: $L = \frac{2WH}{s + 1.0}$

For Single Acting Hammers: $L = \frac{2WH}{s + 0.10}$

For Double Acting Hammers: $L = \frac{2E}{s + 0.10}$

Where L = Safe load bearing.

W = Weight of ram or striking parts, in pounds.

H = Height of stroke, in feet.

s = Set penetration per blow, in inches.

E = Rated energy as listed by manufacture of hammer, in foot pounds.

In using the E-N Formulas to calculate the load bearing capacity of a pile, it is recommended that the set penetration (s) be an average from the last ten blows.

Modified Engineering - News formula 9.6.2.1

It must be realized that the E-N Formula is only a guide to determine load capacities based upon the control test pile observations. The majority of highway and municipal Building Codes call for the equations in the preceding paragraph to be used, while others will call for the same formula with modifications. The 1966 Building Code for the city of Houston, Texas, specified that the pile formulas be modified and written as follows: $L = \frac{2WH}{s + (c \times \frac{P}{W})}$; Where P = weight of pile in pounds, and W = weight of ram in pounds. As can be seen, the modified E-N Formula is identical to the original when pile and ram are equal in weight.

Safety factors in the formula 9.6.2.2

When using the Engineering-News formulas, the results apply to the safe allowable load capacity and contain a safety factor. When the formula was initially developed, it was thought that the safety factor was about six. Seldom, if ever, do the control test pile loads confirm this arbitrary figure. The more complex formulas developed by such engineers as Terzaghi, Redtenbacher, and Hiley, are based on the ultimate resistance to penetration, and should be divided by a safety factor to establish the safe load. The succeeding paragraphs and examples will serve to illustrate the complexity of the mechanics of the driving operation. It may be interesting to note that a recent survey revealed a particular formula preference among foreign engineers. British engineers prefer the Hiley formula, while Japanese engineers use a modified version of the Terzaghi formula called the Karl-Terzaghi Equation. Many other hammer formulas which are based on ultimate resistance to penetration are used, such as formulas developed by Meyerhaff, Dorr, Dunham, Caquot and Karl.

Terzaghi hammer formula 9.6.3

An eminent authority on soil mechanics and heavy foundations, Dr. Charles Terzaghi, in 1929, proposed a formula which equates the available hammer energy to the pile's resistance to penetrate. To this result is added the loss of energy in temporary compression at impact. The efficiency (e) of the hammer blow on the pile will depend upon the restitution coefficient (N) of the type of material under impact, the cushion and the hammer weight. For inelastic materials the coefficient N = 0.0, and for elastic materials, the coefficient is equal to 1.0. With respect to the restitution or recovery coefficient, it is the ratio of the velocity of one body after impact with another body, to the relative velocity before impact was made.

The coefficient of restitution (N) may be derived with the formula: $N = \frac{V_p' - V_h'}{V_h - V_p}$.

Where symbols denote the following:

V_p' = Velocity of Pile after impact blow.

V_p = Velocity of Pile before impact. (Equals 0.0)

V_h' = Velocity of Hammer Ram after impact blow.

V_h = Velocity of Hammer Ram

Hammer blow efficiency (e) upon impact is: $e = \frac{W + PN^2}{W + P}$. Equating the energy in the hammer to the energy required to drive down the pile plus the energy absorbed in the pile, the formula may be written:

$$WHe = \left[RS + \left(\frac{R^2 l}{2AE}\right)\right].$$

Terzaghi hammer formula 9.6.3

Where nomenclature is as follows:

WH = Weight of hammer ram times height of fall. In double-acting hammers, WH = E, given in foot pounds.
P = Weight of pile, given in pounds.
S = Set penetration per blow, given in inches.
L = Length of pile, in feet, or l = length of pile, in inches.
A = Area of piles cross-section, given in square inches.
R = Resistance to penetration, given in pounds.
E = Young's modulus of elasticity of pile material.
e = Efficiency of blow upon impact, given in foot pounds, and converted to inch pounds in formula for R.

This formula for the energy transferred from the hammer to the pile is derived as follows:

RS is the energy absorbed by the pile as it moves downward.

$\frac{R}{2} \times \frac{Rl}{AE} = \frac{R^2 l}{2AE}$ is the energy absorbed by the deformation of the pile. Using Hooke's Law for the elastic deformation of a solid: $\Delta = \frac{Rl}{AE}$, and using $\frac{R}{2}$ as the average value of energy lost and the force acting on this deformation.

Using the formula for hammer efficiency, the energy formula may be rewritten:

$$\frac{W + PN^2}{W + P} \times WH = RS + \frac{R^2 l}{2AE}.$$

Solving this formula for R, the Ultimate Resistance to Penetration:

$$R_u = \frac{-S \pm \sqrt{S^2 + \frac{2l}{AE} \times \left(\frac{W + PN^2}{W + P}\right) \times WH}}{\frac{l}{AE}}$$

If the hammer ram weight (W) is equal or greater than the weight of the pile times the restitution coefficient (PN), the blow efficiency must be obtained by the first formula as: $e = \frac{W + PN^2}{W + P}$. Conversely, when the ram weight is less than PN, the bounce of the ram will reduce the effective energy transferred to the pile, and the blow efficiency is obtained with the formula: $e = \left(\frac{W + PN^2}{W + P}\right) - \left(\frac{W - PN^2}{W + P}\right)$. Table 9.6.3.1 presents the coefficient e for various values of N.

Designers employing this type of hammer formula should obtain information from the hammer manufacturer relative to the use of wood cushion blocks between the pile and the hammer anvil. A number of hammers do not require cushions. When these hammers are driving steel piles, the impact of steel upon steel will produce a greater blow efficiency. In such a circumstance, the Engineering-News Formula fails to describe driving performance variations for different pile materials. Since the Terzaghi Formula involves many terms it is not possible to evaluate the complete formula with any degree of haste. When the terms are evaluated separately and then combined, the problem becomes less difficult and more accurate. The succeeding examples will illustrate the simplified procedure.

TABLE: Terzaghi hammer efficiency 9.6.3.1

HAMMER BLOW EFFICIENCY COEFFICIENT= "e" WITH VARIABLE VALUES OF "N." •TERZAGHI•

WEIGHT OF PILE TO RAM RATIO P/W	NEW-FRESH OAK CUSHION BLOCK N=0.25	MEDIUM USED OAK CUSHION BLOCK N=0.40	COMPACTED OAK CUSHION BLOCK N=0.50	NO CUSHION STEEL ON STEEL N=0.55
0.50	0.690	0.720	0.750	0.770
0.75	0.610	0.650	0.690	0.710
1.00	0.530	0.580	0.630	0.650
1.25	0.485	0.540	0.590	0.615
1.50	0.440	0.500	0.550	0.580
1.75	0.405	0.470	0.525	0.560
2.00	0.370	0.440	0.500	0.540
2.25	0.350	0.420	0.475	0.525
2.50	0.330	0.400	0.450	0.510
2.75	0.315	0.380	0.435	0.495
3.00	0.300	0.360	0.420	0.480
3.50	0.275	0.340	0.390	0.460
4.00	0.250	0.320	0.360	0.440
4.50	0.230	0.295	0.335	0.430
5.00	0.210	0.270	0.310	0.420
5.50	0.200	0.255	0.290	0.410
6.00	0.190	0.240	0.270	0.400

Courtesy of: Mitsubishi Heavy Industries L'td., Tokyo.

EXAMPLE: Ultimate load by Terzaghi formula 9.6.3.2

A steel 12 BP 74 Pile is 72.0 feet long and is to be driven to cut-off elevation. At cut-off, penetration into clay-sand soil will have been 48.0 feet. A double acting 10-B-3 steam hammer of McKiernan-Terry manufacturer was used to drive this pile, and final blows produced a set penetration of 0.10 inches per blow.

REQUIRED:

Calculate the Ultimate Resistance to Penetration (Ru) by using the Terzaghi hammer formula. State maximum load capacity in pounds and U.S. Tons.

STEP I:

Gather all data required to put in formula. This hammer uses no cushion and impact will be steel on steel.

Weight Pile $P = 72.0 \times 74 = 5328$ Lbs. $l' = 72.0 \times 12 = 864$ inches.

Effective length $l = 48.0 \times 12 = 576$ inches. $E = 29,000,000$

From Pile Tables: Area cross-section = 21.75 Square inches.

STEP II:

From Table on hammers: McK-T 10-B-3 has Ram W = 3000 Lbs. Stroke height, H = 1.583 Ft. Rated E = 13,100 Foot Lbs. Reducing for possible low steam, effective WH = 13,100 x 0.90 = 11,790 Ft. Lbs. Used 90% of E for effective operation.

Ratio of $P/W = \frac{5328}{3000} = 1.776$ From table: N = 0.55 and e = 0.560

STEP III:

Check efficiency of blow at impact by formula previously given. $e = \frac{W + PN^2}{W + P}$. (Pile has greater weight than Ram)

substituting values in formula:

$$e = \frac{3000 + (5328 \times 0.55^2)}{3000 + 5328} = \frac{4612}{8328} = 0.553 \text{ (Use for e value)}$$

STEP IV:

Complete Terzaghi Formula: $Ru = \dfrac{-S \pm \sqrt{S^2 + \frac{2l}{AE} \times \left(\frac{W + PN^2}{W + P}\right) \times WH}}{\frac{l}{AE}}$

Equate formula by sections:

$S^2 = 0.10 \times 0.10 = 0.010$ WH = 11,790'#

$$\frac{2l}{AE} = \frac{2 \times 576}{21.75 \times 29,000,000} = \frac{1152}{630,750,000} = 0.00000182$$

$$\frac{l}{AE} = \frac{576}{630,750,000} = 0.000000912 \text{ (Take half of above)}$$

EXAMPLE: Ultimate load by Terzaghi formula (continued) 9.6.3.2

STEP V:

Substituting values obtained in complete formula:

$$R_u = \frac{-0.10 \pm \sqrt{0.010 + (0.00000182 \times 0.553 \times 11{,}790 \times 12)}}{0.000000912} =$$

$$R_u = \frac{-0.10 \pm \sqrt{0.010 + 0.142394}}{0.000000912} = \frac{-0.10 + \sqrt{0.1324}}{0.000000912} = \frac{0.264}{0.000000912} =$$

R_u = 289,500 Lbs. In tons R_u = 144.75

Using a safety factor of 2, Safe Load Capacity = 72⅜ Tons.

DESIGN NOTATION:

(a) The Engineering-News's Formula for the M-K 10-B-3 Hammer is: $L = \frac{2E}{S + 0.10}$. With manufacturer rated energy, E = 13,100 Foot Lbs., and S = 0.10 inches per blow.

Safe Allowable Load = $\frac{2 \times 13{,}100}{0.10 + 0.10}$ = 131,000 Lbs.

With the Terzaghi Formula serving as a guide for calculating the ultimate Resistance to penetration, the E-N Safety Factor becomes $\frac{289{,}500}{131{,}000}$ = 2.21

Several comparisons similar to the above have been made on different pile types and hammer model. In general terms, the comparison will result in the Terzaghi Formula Ultimate being approximately double the E-N Formula's Allowable.

(b) When the Terzaghi Formula is assembled together with the several components transposed, it results into a quadratic equation as given previously. Contemplating that the values in the example were to be substituted in the quadratic form, the formula would be evaluated accordingly:

$$R_u = \frac{-0.10 \pm \sqrt{\left(0.10^2 + \frac{2 \times 576}{21.75 \times 29{,}000{,}000}\right) \times \left[\frac{3000 + (5328 \times 0.55^2)}{3000 + 5328}\right] \times (13{,}100 \times 12 \times 0.90)}}{\frac{576}{21.75 \times 29{,}000{,}000}}$$

Hiley hammer formula 9.6.4

When the composition of the soil offers a steady driving resistance, the Hiley Formula will provide a better estimate of the ultimate bearing capacity of a pile. Because this equation is based upon the laws which deal with the impacts of elastic bodies, it is considered more rational than the others. The Hiley formula is used almost exclusively by British and European pile engineers.

Hiley has taken into consideration all of the loss factors of the Terzaghi and Redtenbacher formulas. In addition, he accounts for the energy lost during driving as a result of the temporary compression of the soil and pile material.

Conditions of driving must fall within certain limits for the Hiley formula to produce reliable results for estimating load capacities. It cannot be used for driving into soft sands, silts, or soft clays. It can be used when piles are driven into compacted sand strata, hard clay, shell or gravel formations, or any soil with permeable characteristics.

As a guide for determining if a dynamic hammer formula may be used to make a reliable estimate, try making a "sleep test" during the driving operation. This test is conducted as follows: Start driving the pile into the soil for a partial distance, then stop and remove the hammer. Allow the pile to rest a period of twelve hours or more. Resume the driving operation after the rest period, and drive the pile another comparable distance into soil strata. These stops and starts may require two to three days, but are well worth the labor when soil conditions are in doubt. If, upon resuming the driving after each stop and rest period, *the set per blow is greater* on redrive than before, the Hiley formula cannot be used. In the event *the set per blow is less or equal* on the redrive, the formula can be used to good advantage.

Hiley formula nomenclature 9.6.4.1

HILEY FORMULA NOMENCLATURE			
MARK	UNIT	EXPLANATION	SOURCE
R_u	TON	ULTIMATE DRIVING RESISTANCE	FORMULA
R_w	TON	ALLOWED WORKING LOAD PILE CAPACITY	
F		FACTOR OF SAFETY. $F = R_u \div R_w$	CODES
K		COEFFICIENT APPLIED TO HAMMER ACTION	TABLE
W	TON	WEIGHT OF RAM OR STRIKING PARTS	TABLE
H	INCHES	EFFECTIVE HEIGHT OF FALL - (K x Table)	TABLE
η		EFFICIENCY OF BLOW - RESTITUTION - P/W	
S	INCHES	SET PENETRATION OF PILE FROM IMPACT	COUNT
P	TON	PILE WEIGHT, + ANVIL, CUSHION OR HELMET	
A	SQ. INCH'S.	OVERALL CROSS SECTION OF PILE AREA	SECTIONS
e		COEFFICIENT OF RESTITUTION OF PILE	TABLE
A_p	SQ. INCH'S.	NET AREA OF PILE CROSS SECTION OR MATERIAL SUSTAINING BLOW	
C	INCHES	SUM TOTAL OF TEMPORARY COMPRESSION. EQUALS: $\Sigma = C_c + C_p + C_q$	
C_c	INCH	TEMPORARY COMPRESSION OF PILES TOP PORTION, CUSHION, DOLLY, PACKET, HELMET OR ANVIL	CHART C_c
C_p	INCH	TEMPORARY COMPRESSION OF PILE: 1. STEEL PILES 2. CONCRETE PILES 3. WOOD PILES	CHART C_p
C_q	INCH	TEMPORARY COMPRESSION OF SOIL, OR QUAKE OF GROUND AROUND PILE	CHART C_q
m		A COEFFICIENT IN $R_u = mc$. EXPLAINED IN HILEY ALTERNATE FORMULA, $m = R_u / C$	SEE EXAMPLE
A_g	Sq. In.	Gross Area overall driving. See 9.6.4.6	Sections

NOTE: CONVERT TON UNITS FROM 2240 POUNDS TO U.S. TONS OF 2000 POUNDS WHEN USING FORMULA FOR R_u.

Hiley formula coefficients 9.6.4.2

HILEY FORMULA:- COEFFICIENT VALUES FOR "K"			
CHARACTERISTICS OF HAMMER			
ACTION TYPE	RELEASE	OPERATING FORCE	K
DROP WEIGHT	MANUAL	COILED WIRE ROPE: CABLE ATTACHED	0.80
DROP WEIGHT	TRIGGER	FREE FALL- GRAVITY FORCE	1.00
SINGLE ACTING	VALVE	STEAM RAISE RAM- FREE GRAVITY FALL	0.90
DOUBLE ACTING	PORT	STEAM PRESSURE PLUS GRAVITY	1.00
DIFFERENTIAL	PORT	STEAM PRESSURE PLUS GRAVITY	1.00
DIESEL- S.A.	SLEEVE	GRAVITY PLUS EXPLOSIVE AID	1.00

HILEY FORMULA:- RESTITUTION COEFFICIENT VALUES FOR "e"			
PILE - TOP CONDITIONS - HAMMER TYPES			
TYPE OF PILE MATERIAL	CONDITION AT JUNCTURE OF PILE AND HAMMER	DROP HAMMER OR SINGLE ACT.- DIESEL	DOUBLE ACTING DIFFERENTIAL
HARD STONE OR PRECAST CONCRETE WITH STEEL REIN.	WOOD CUSHION - ANVIL PACKING- HELMET OR FOLLOWER ON PILE CAP	0.40	0.50
SAME	FRESH CUSHION ON PILE OR WITH FOLLOWER	0.25	0.40
SAME	HAMMER DIRECTLY SET ON PILE CUSHION - HELMET	0.40	0.50
STEEL	DRIVING CAP OR WITH ANVIL AND CUSHION	0.50	0.50
STEEL	DRIVING CAP WITH WOOD FOLLOWER	0.30	0.30
STEEL	HAMMER STEEL ANVIL DIRECTLY ON PILE	—	0.50
WOOD	HAMMER SET DIRECTLY ON PILE CAP	0.25	0.40

Hiley formula graphic charts 9.6.4.3

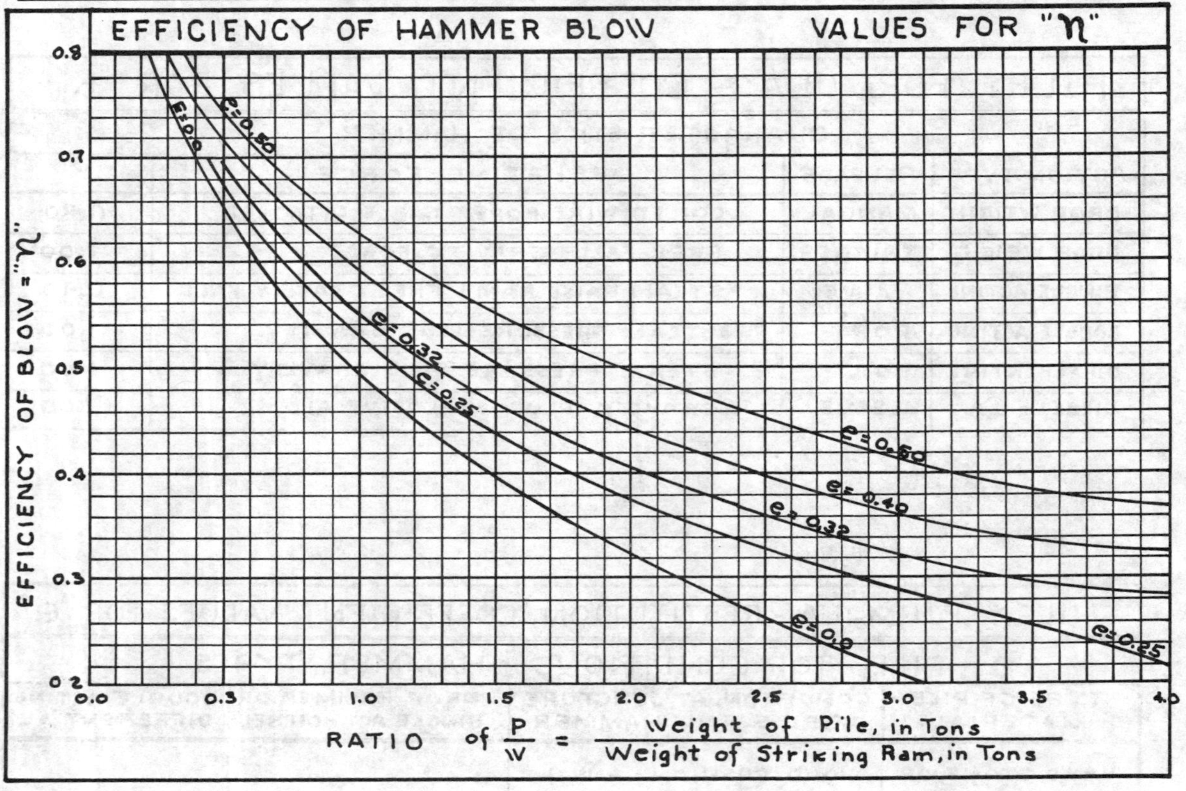

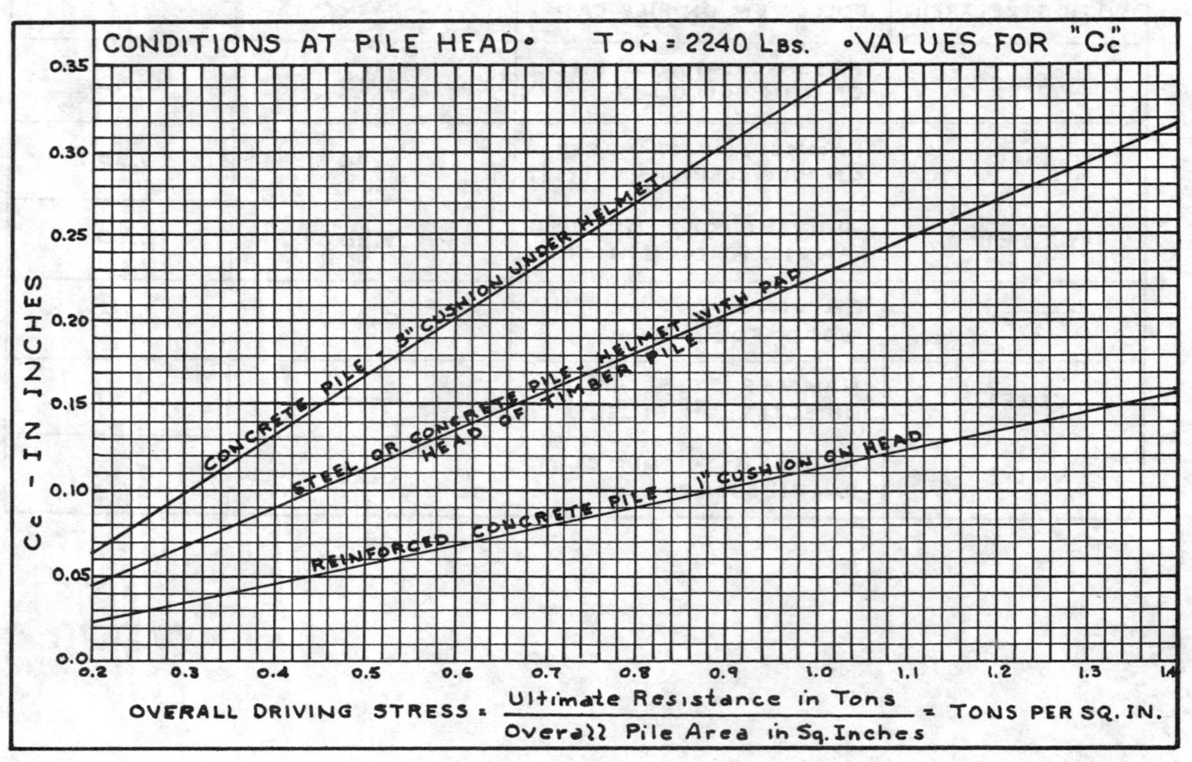

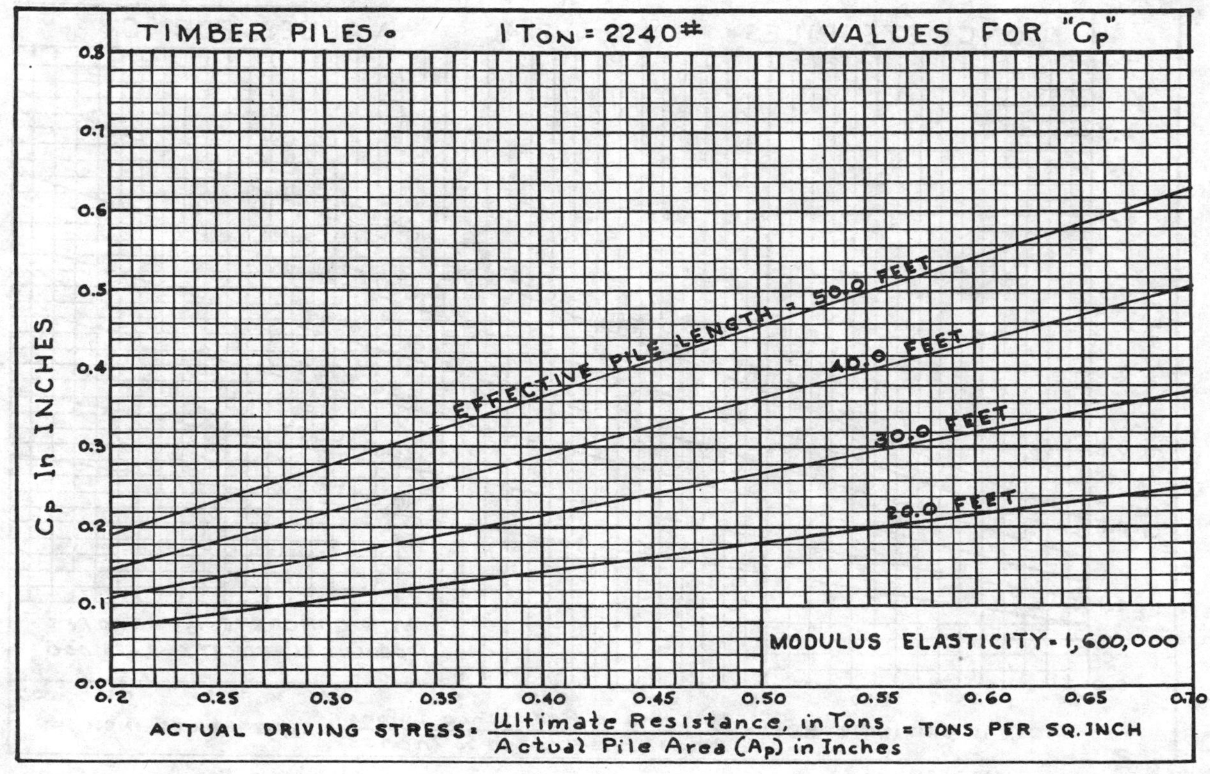
TIMBER PILES • 1 TON = 2240# VALUES FOR "C_P"
C_P In INCHES
0.8
0.7
0.6
0.5
0.4
0.3
0.2
0.1
0.0
EFFECTIVE PILE LENGTH = 50.0 FEET
40.0 FEET
30.0 FEET
20.0 FEET
MODULUS ELASTICITY - 1,600,000
0.2
0.25
0.30
0.35
0.40
0.45
0.50
0.55
0.60
0.65
0.70
ACTUAL DRIVING STRESS = Ultimate Resistance, in Tons / Actual Pile Area (A_p) in Inches = TONS PER SQ. INCH

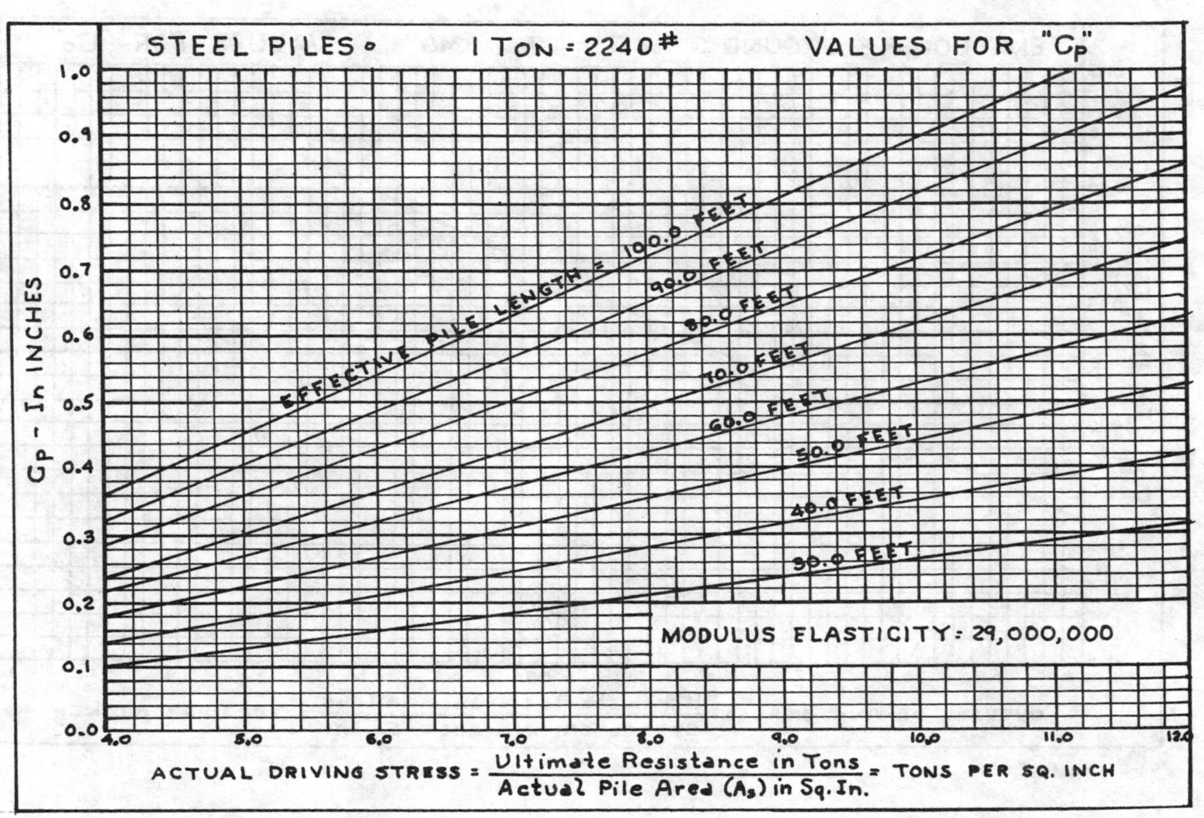
STEEL PILES • 1 TON = 2240# VALUES FOR "C_P"
C_P - In INCHES
1.0
0.9
0.8
0.7
0.6
0.5
0.4
0.3
0.2
0.1
0.0
EFFECTIVE PILE LENGTH = 100.0 FEET
90.0 FEET
80.0 FEET
70.0 FEET
60.0 FEET
50.0 FEET
40.0 FEET
30.0 FEET
MODULUS ELASTICITY = 29,000,000
4.0
5.0
6.0
7.0
8.0
9.0
10.0
11.0
12.0
ACTUAL DRIVING STRESS = Ultimate Resistance in Tons / Actual Pile Area (A_s) in Sq. In. = TONS PER SQ. INCH

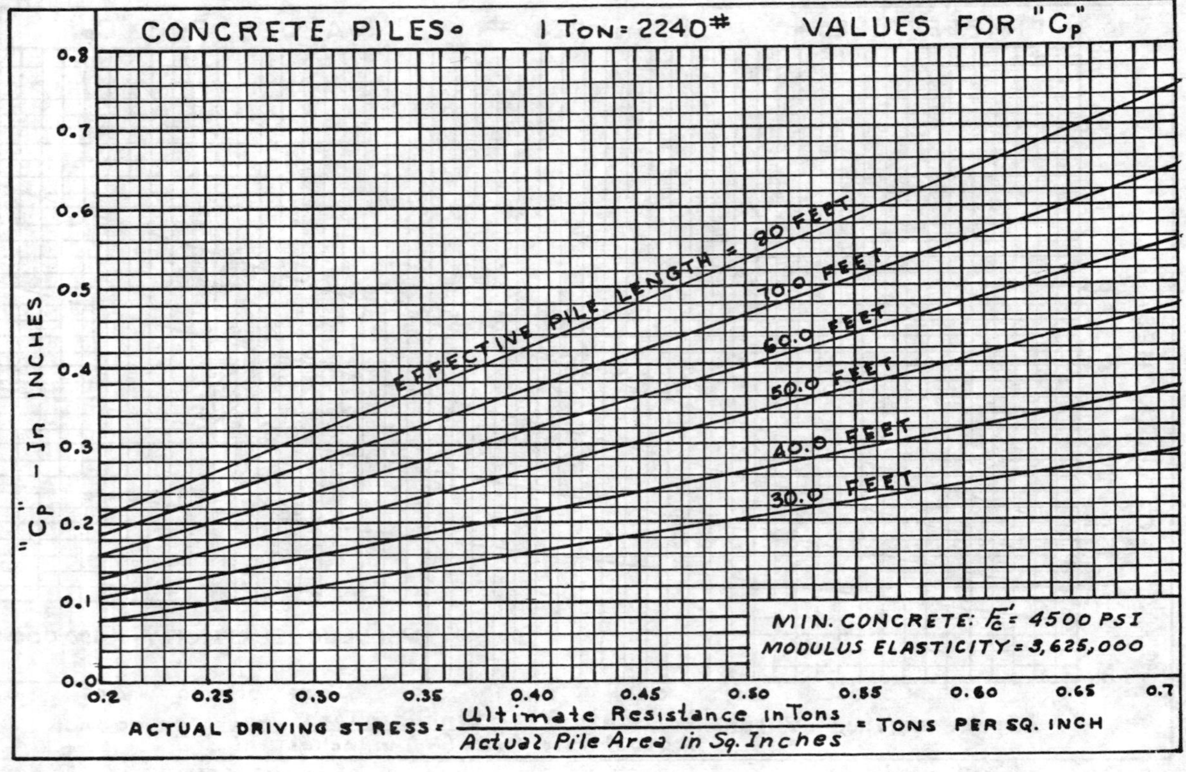
CONCRETE PILES. 1 TON = 2240# VALUES FOR "C_p"
"C_p" - In INCHES
EFFECTIVE PILE LENGTH = 80 FEET
70.0 FEET
60.0 FEET
50.0 FEET
40.0 FEET
30.0 FEET
MIN. CONCRETE: F_c' = 4500 PSI
MODULUS ELASTICITY = 3,625,000
0.0
0.1
0.2
0.3
0.4
0.5
0.6
0.7
0.8
0.2
0.25
0.30
0.35
0.40
0.45
0.50
0.55
0.60
0.65
0.7
ACTUAL DRIVING STRESS = Ultimate Resistance in Tons / Actual Pile Area in Sq. Inches = TONS PER SQ. INCH

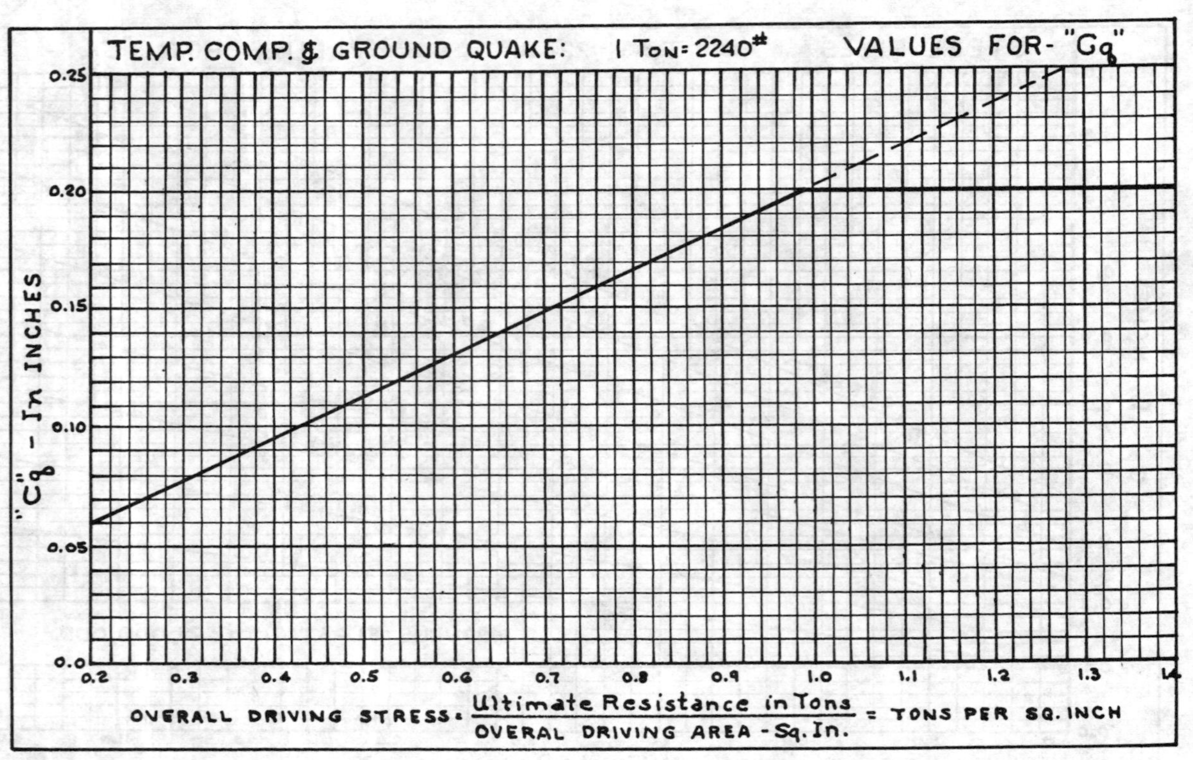
TEMP. COMP. & GROUND QUAKE: 1 TON = 2240# VALUES FOR - "C_q"
"C_q" - In INCHES
0.0
0.05
0.10
0.15
0.20
0.25
0.2
0.3
0.4
0.5
0.6
0.7
0.8
0.9
1.0
1.1
1.2
1.3
1.4
OVERALL DRIVING STRESS = Ultimate Resistance in Tons / OVERAL DRIVING AREA - Sq. In. = TONS PER SQ. INCH

Energy transfer by Hiley formula 9.6.4.4

An equation for the transfer of energy in the Hiley formula is written as: $WH\eta = SR + \frac{CR}{2}$. The following coefficients must be added together to obtain the value of C in the equation.

C_c = Temporary compression in cushion material, top of pile or other resilient parts.
C_p = Temporary compression of pile.
C_q = Temporary compression of soil and ground quake.

These coefficients are evaluated in the following graphic charts and have units of inches. Unlike other formulas, the efficiency of the hammer blow is given the symbol η, and the coefficient of restitution is indicated as e. The hammer characteristics are considered in the coefficient k. The British version of the Hiley Formula will be used with the long ton of 2240 pounds. The following tables should be carefully studied before analyzing the complete solutions of the equation in the examples.

In using the Hiley Formula for Ultimate Resistance to Penetration, it is written as: $R_u = \frac{WH\eta}{S + \frac{C}{2}}$. It is necessary to make an assumption for ultimate value of R, so that approximate values may be obtained for the coefficient values of C_c, C_p and C_q. The summation of the coefficients (C) is then entered in the formula.

Hiley alternate formula 9.6.4.5

When making an early assumption for R_u, and then finding the temporary compression, $C = C_c + C_p + C_q$, the Hiley formula may be re-written in an alternate form as follows:

$R_u = \sqrt{[(2m\ WH\eta) + (ms)^2]} - ms$. Where: $m = \frac{R_u}{C}$.

The practical examples which follow will illustrate the application of the Hiley Formula and the method used to evaluate its various terms.

Hiley temporary compression factors 9.6.4.6

Energy loss by temporary compression upon hammer impact occurs in the pile cap, pile length and penetrated soil. These three loss factors are combined in the design factor C. Temporary compression in the pile can be calculated when the modulus of elasticity is known. The graphic charts, which are essential to using the Hiley formula, are the results of many tests. Designers and students can compare the given values by conducting their own experiments for elastic compression in the pile and soil while driving is in progress. The illustration of the elastic profile shown here was obtained from a wood pile while being driven. A section of the pile surface was painted with white paint. A stationary rest was constructed in the shape of a tripod. This rest was provided with a horizontal wood straight edge. As blows were struck by the hammer, a pencil was placed on painted surface to record

Hiley temporary compression factors, continued 9.6.4.6

the pile movement and rebound. A slight movement of the pencil to the left on the straight edge produced the profile. Temporary compression due to impact is designated as C, and net amount of penetration is noted as set s.

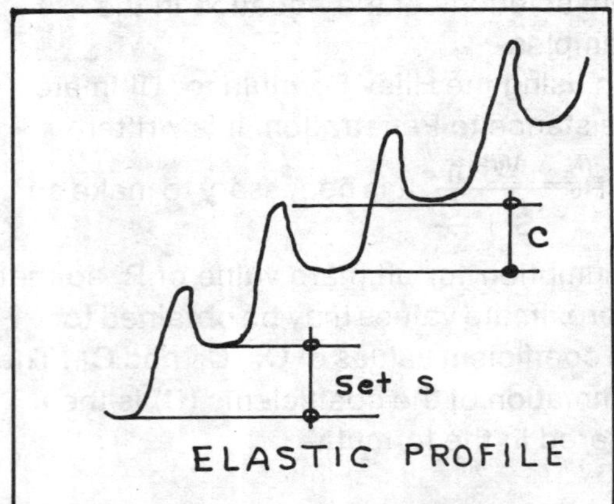

Referring to the chart values of Cq for soil compression and ground quake, Hiley makes little distinction between piles composed of wood, steel or concrete. Since soil is displaced with all piles this fact is realistic. Hiley has prepared an approximate coefficient to lay out the curve for temporary soil compression: C_q: $= \dfrac{0.20\ R_u}{A_g}$.
Where A_g is the gross soil area involved, which is used to obtain the overall driving stress.

The constant C_c assumes that the driving cap is of well-compacted material and the impact compression will be constant. The formula given to solve for temporary compression in cap and cushion is:

$$C_c = \frac{0.08 \times R_u}{A_g}.$$

For temporary compression in the pile (C_p), Hiley assumed that the effective length of pile (L) should average ⅔ of the total pile length in feet. As piles depend upon tip bearing and skin friction for bearing, the resistance to penetration is based upon the length which displaces soil, and the net area of pile cross section. Temporary compression in pile is:

$$C_p = \frac{0.0008\ R_u L}{A_p}$$

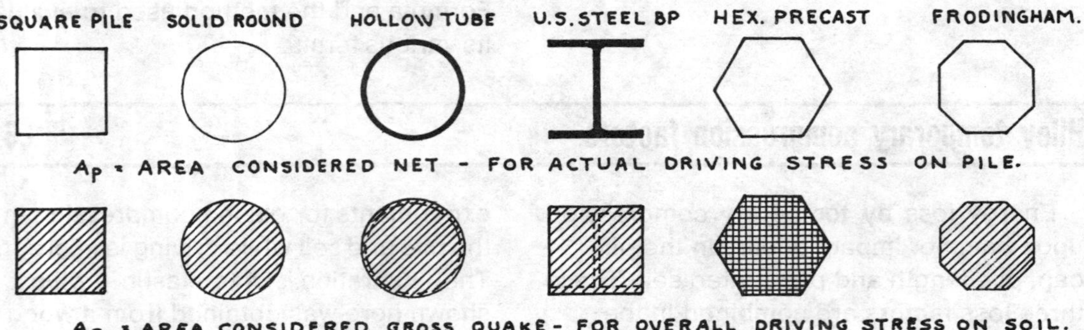

Hiley formula solution 9.6.4.7

The temporary compression factor to be used in the Hiley Formula for Ultimate Resistance to Penetration is written thus: $C = C_c + C_p + C_q$, or if the formulas for approximation without charts are used, it is written:

$$C = \frac{0.08\,Ru}{Ag} + \frac{0.0008\,Ru\,L}{Ap} + \frac{0.20\,Ru}{Ag}.$$

In order to compare the Hiley and Terzaghi Formulas it is necessary to resolve each into a quadratic equation. For the Hiley Formula the process is as follows:

have: $\frac{0.08\,Ru + 0.20\,Ru}{Ag} = \frac{0.28\,Ru}{Ag}$. From Formula: $Ru = \frac{eWH}{S + \frac{C}{2}}$, then

$$Ru = \frac{eWH}{S + \frac{Ru\left[\frac{0.0008L}{Ap} + \frac{0.28}{Ag}\right]}{2}}$$, thus in final equation, it is:

$$Ru = \frac{-S + \sqrt{S^2 + \left[\frac{0.0016L}{Ap} + \frac{0.56}{Ag}\right] eWH}}{\frac{0.0008\,L}{Ap} + \frac{0.28}{Ag}}$$

CHART: Pile load capacity 9.6.5

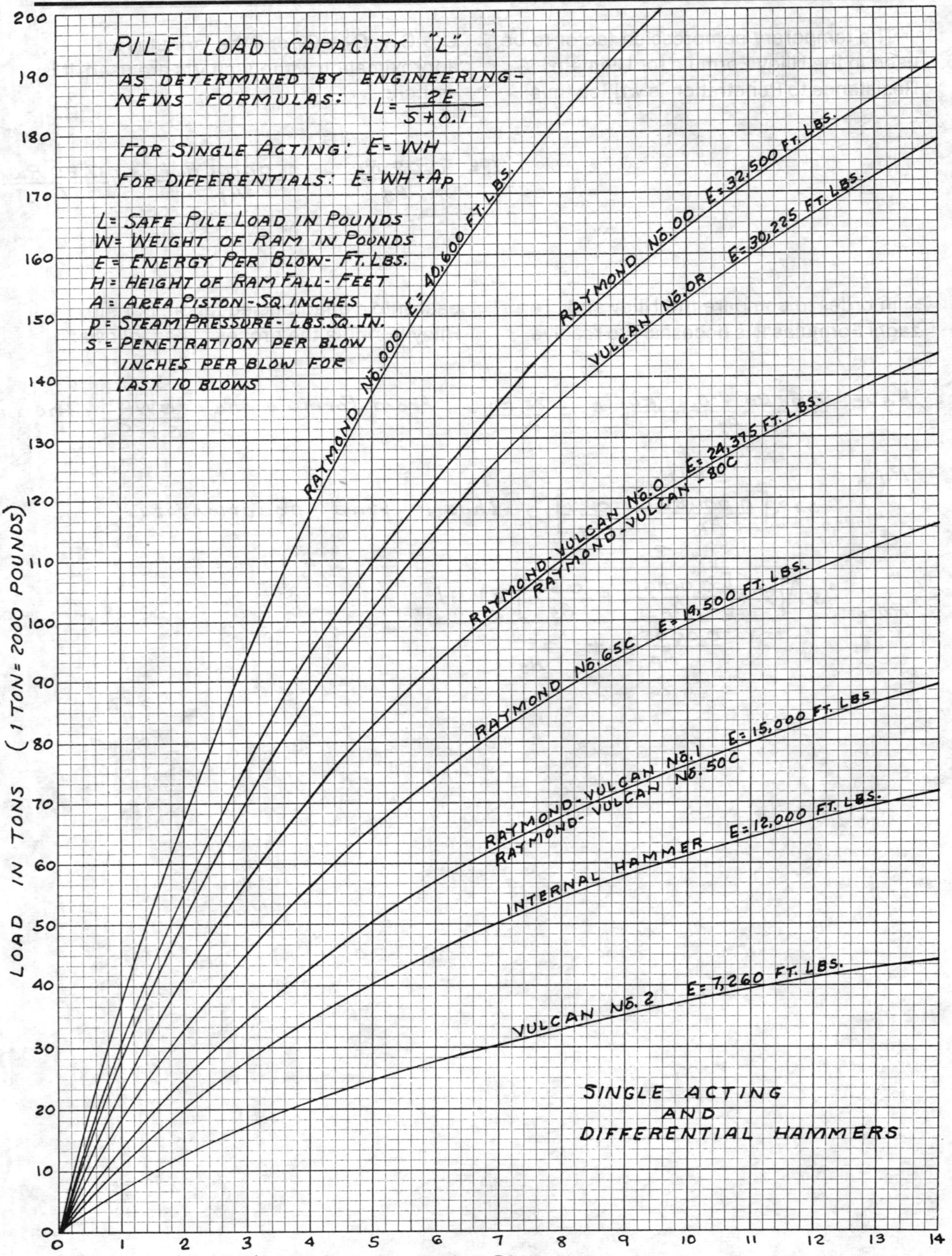

TABLE: Hammer blows per foot for safe pile load 9.6.6

MANUFACTURER OF HAMMER OR TRADE NAME	TYPE OF ACTION	HAMMER MODEL OR NŌ.	BLOWS PER MINUTE	RATED ENERGY IN FT. LBS.	LOAD IN TONS					MATCH LINES
					5	10	15	20	25	
VULCAN IRON WORKS	SINGLE	18-C	150	3,600	19.5	46.2	85.7			1
VULCAN IRON WORKS	DOUBLE	NŌ. 3	80	3,600	19.5	46.2	85.7			2
UNION IRON WORKS	DOUBLE	NŌ. 3	160	3,660	19.0	45.2	83.3	145.0		3
UNION IRON WORKS	DOUBLE	NŌ. 3A	150	4,390	15.5	35.4	62.3	100.0	160.0	4
UNION IRON WORKS	DOUBLE	NŌ. 2	145	5,755	11.5	25.3	42.3	64.0	92.5	5
VULCAN IRON WORKS	DOUBLE	30-C	135	7,260	8.9	19.2	31.3	45.7	63.3	6
VULCAN IRON WORKS	SINGLE	NŌ. 2	70	7,260	8.9	19.2	31.3	45.7	63.3	7
UNION IRON WORKS	DOUBLE	1½ A	125	8,680	7.4	15.6	25.0	36.0	48.5	8
McKIERNAN-TERRY	DOUBLE	9-B-3	145	8,750	7.3	15.5	24.8	35.5	48.0	9
McKIERNAN-TERRY	SINGLE	S-3	65	9,000	7.0	15.0	24.0	34.3	46.0	10
BROWN INDUSTRIAL	DOUBLE	1	110	9,650		14.0	22.0	31.5	42.0	11
UNION IRON WORKS	DOUBLE	1A	120	10,020		13.3	21.0	30.0	40.0	12
UNION IRON WORKS	DOUBLE	NŌ. 1	130	13,100		10.0	15.5	21.6	28.3	13
McKIERNAN-TERRY	DOUBLE	10-B-3	105	13,100		10.0	15.5	21.6	28.3	14
VULCAN IRON WORKS	SINGLE	NŌ. 1	60	15,000		10.0	13.4	18.5	24.0	15
VULCAN IRON WORKS	DOUBLE	50-C	120	15,000			13.2	18.4	23.8	16
McKIERNAN-TERRY	SINGLE	S-5	60	16,250			11.0	16.8	22.0	17
McKIERNAN-TERRY	DOUBLE	11-B-3	95	19,150			10.2	14.0	18.0	18
UNION IRON WORKS	DOUBLE	OA	90	22,050				12.0	15.4	19
VULCAN IRON WORKS	SINGLE	O	50	24,375				11.0	14.0	20
VULCAN IRON WORKS	DOUBLE	80-C	111	24,450				11.0	13.6	21
McKIERNAN-TERRY	SINGLE	S-8	55	26,000				10.0	12.8	22
VULCAN IRON WORKS	SINGLE	OR	50	30,225					10.8	23
McKIERNAN-TERRY	SINGLE	S-10	55	32,500					10.0	24
RAYMOND INTERNAT'L.		OO	50	32,500					10.0	25
VULCAN IRON WORKS	DOUBLE	140-C	103	36,000						26
McKIERNAN-TERRY	SINGLE	S-14	60	37,500						27
RAYMOND INTERNAT'L.		OOO		40,600						28
VULCAN IRON WORKS	DOUBLE	200-C	98	50,200						29
UNION IRON WORKS	DOUBLE	OO	85	54,900						30
McKIERNAN-TERRY	DIESEL	DE-20	48-52	16,000		10.9	17.1	20.0	31.6	31
McKIERNAN-TERRY	DIESEL	DE-30	48-52	22,400		7.2	11.8	16.2	20.8	32
McKIERNAN-TERRY	DIESEL	DE-40	48-52	32,000						33
DELMAG-GERMANY	DIESEL	D-5	56	9,100	9.5	20.6	34.0	50.0	70.0	34
DELMAG-GERMANY	DIESEL	D-12	51	22,500			12.0	16.3	21.0	35
DELMAG-GERMANY	DIESEL	D-22	48	39,700				8.7	11.0	36
DELMAG-GERMANY	DIESEL	D-30	45	54,250						37
MITSUBISHI-JAPAN	DIESEL	M-145	60	26,000			10.0	13.7	17.7	38
MITSUBISHI-JAPAN	DIESEL	M-23	60	44,000					12.0	39
MITSUBISHI-JAPAN	DIESEL	M-33	60	61,840						40
MITSUBISHI-JAPAN	DIESEL	M-43	60	92,580						41
MITSUBISHI-JAPAN	DIESEL	MB-22	60	42,674						42
MITSUBISHI-JAPAN	DIESEL	MB-40	60	91,135						43
MITSUBISHI-JAPAN	DIESEL	MB-70	60	155,507						44

TABLE: Hammer blows per foot for safe pile load, continued 9.6.6

MATCH LINES	SAFE LOAD CAPACITY IN U.S. TONS																	
	30	35	40	45	50	55	60	65	70	75	80	85	90	95	100	105	110	115
1																		
2																		
3																		
4																		
5	131.0																	
6	84.6	112.0	147.0															
7	84.6	112.0	147.0															
8	63.5	81.0																
9	62.6	80.0	101.0	128.0														
10	60.0	76.2	96.0	120.0	150.0													
11	54.0	68.2	85.0	105.0	129.0	159.0												
12	51.3	64.4	79.6	98.0	120.0	146.0	180.0											
13	35.7	48.8	52.8	62.8	74.0	87.0	101.5											
14	35.7	48.8	52.8	62.8	74.0	87.0	101.5	118.0	138.0									
15	30.0	36.5	43.6	51.5	60.0	69.5	80.0	92.0	105.0	120.0								
16	30.0	36.2	43.3	50.0	59.5	68.8	79.2	91.0	104.0	118.5	136.5	155.5	177.0	203.0				
17	27.2	33.0	39.2	46.0	53.4	61.4	70.0	80.0	91.0	103.0	116.5	131.5	149.0	169.0	192.0			
18	22.3	26.8	31.7	37.0	42.5	48.5	55.0	62.0	69.2	74.2	87.0	96.0	106.5	118.2	131.0	145.5	162.0	180.0
19	19.0	22.6	26.6	30.8	35.2	40.0	44.8	50.0	56.0	62.0	68.3	75.4	82.8	91.0	99.5	104.5	120.0	131.0
20	16.8	20.0	23.6	27.3	31.2	35.0	39.3	43.7	48.5	53.5	58.8	64.4	70.5	76.7	83.6	91.6	100.0	107.5
21	16.7	20.0	23.5	27.7	30.9	35.2	39.1	43.4	48.3	53.3	58.4	64.0	69.9	76.3	83.0	90.4	98.2	106.5
22	15.7	18.7	21.8	25.8	28.6	32.3	36.0	40.0	44.3	48.7	53.4	58.4	65.5	69.0	75.0	81.4	88.0	95.3
23	13.2	16.0	18.3	21.0	23.8	26.8	30.0	33.0	36.2	39.6	43.3	47.0	51.0	55.0	59.5	64.0	68.6	73.6
24	12.2	14.5	16.9	19.3	21.9	24.5	27.2	30.0	32.6	36.0	39.2	42.5	46.5	49.6	53.4	57.3	61.4	65.6
25	12.2	14.5	16.9	19.3	21.9	24.5	27.2	30.0	32.6	36.0	39.2	42.5	46.5	49.6	53.4	57.3	61.4	65.6
26	11.0	13.0	15.0	17.2	19.4	21.4	24.0	26.4	29.0	31.6	34.3	37.1	40.0	43.0	46.2	49.4	52.8	56.2
27		12.4	14.3	16.4	18.5	20.6	23.0	25.2	27.6	30.0	32.6	35.2	37.9	40.8	43.6	46.7	50.0	53.2
28		11.3	13.1	15.0	16.9	18.9	21.0	22.9	25.0	27.2	29.4	31.8	34.2	36.7	39.2	42.0	44.6	47.4
29			10.4	11.9	13.3	14.8	16.3	17.8	19.5	21.3	22.8	24.5	26.2	27.0	29.9	32.0	33.8	35.6
30			10.0	10.7	12.0	13.4	14.7	16.1	17.6	19.0	20.5	22.2	23.6	25.1	26.8	28.4	30.3	31.8
31	40.0	49.4	60.0	72.1	85.7	101.5	120.0											
32	26.1	32.3	37.5	43.8	50.8	58.7	66.7	75.7	85.7	96.7	109.0	122.8	138.5	156.0	176.5			
33	15.0	20.5	24.0	27.7	31.6	35.7	40.0	44.6	49.5	54.3	60.0	65.7	72.0	78.6	85.7	93.3	101.5	110.5
34	95.0	126.4	170.0															
35	26.0	32.0	37.4	46.0	51.0	58.2	66.5	75.4	85.0	96.0	108.0	122.0						
36	13.5	16.0	18.6	21.6	24.2	27.0	30.3	33.4	38.0	40.5	44.3	48.5	52.0	56.2	61.0	65.5	70.5	75.5
37					16.8	19.0	21.0	22.8	25.0	27.0	29.4	32.0	34.0	36.5	39.0	42.0	44.5	47.3
38	21.8	26.3	31.0	36.0	41.4	47.2	53.3	60.0	67.0	75.0	83.3	92.7	102.5	114.0	126.0	140.0	155.0	172.5
39	12.0	14.5	16.5	18.9	21.1	24.0	26.7	29.5	32.3	35.5	38.4	41.6	45.0	48.5	52.2	56.0	60.0	64.2
40	8.3	9.8	11.3	12.9	14.5	16.2	17.8	19.5	21.3	23.2	25.0	26.9	28.9	30.9	33.0	35.1	37.3	39.6
41								12.4	13.4	14.6	15.5	16.7	17.8	19.0	20.8	21.4	22.6	23.2
42	12.4	14.7	17.2	19.7	22.2	25.4	27.7	30.6	33.6	36.8	40.0	43.4	47.0	50.6	54.6	58.6	62.8	67.3
43										14.8	15.9	17.0	18.2	19.4	20.6	21.8	23.9	24.3
44															11.3	11.8	12.5	13.1

SAFE LOAD CAPACITY IN U.S. TONS

MATCH LINES	120	125	130	135	140	145	150	155	160	165	170	175	180	185	190	195	200	MATCH LINES
1																		1
2																		2
3																		3
4																		4
5																		5
6																		6
7																		7
8																		8
9																		9
10																		10
11																		11
12																		12
13																		13
14																		14
15																		15
16																		16
17																		17
18	202.0																	18
19	143.4	157.0	175.0	189.0														19
20	116.5	125.5	137.5	149.3	162.0	176.0	192.0											20
21	115.5	125.4	136.4	148.0	160.5	175.0	190.5	207.8	227.0									21
22	103.3	111.0	120.0	129.5	140.0	151.3	163.5	177.0	192.0	208.0								22
23	79.0	84.5	90.8	97.0	103.6	110.5	118.0	126.3	135.0	144.0	155.0	165.0	176.5	190.0				23
24	70.3	75.0	80.0	85.3	90.8	96.6	102.6	109.3	116.4	124.0	132.0	140.0	149.0	159.0	170.0	180.0	192.0	24
25	70.3	75.0	80.0	85.3	90.8	96.6	102.6	109.3	116.4	124.0	132.0	140.0	149.0	159.0	170.0	180.0	192.0	25
26	60.0	63.8	67.8	72.0	76.3	80.8	85.7	90.8	96.0	101.5	107.5	113.5	120.0	126.5	134.5	141.5	150.0	26
27	56.5	60.0	63.7	67.5	71.5	75.6	80.0	84.5	88.3	94.3	99.5	105.0	111.0	116.6	123.5	130.0	137.0	27
28	50.3	53.3	56.5	60.0	64.2	66.6	70.3	74.0	78.0	82.0	86.5	91.0	95.6	100.0	105.5	110.8	116.5	28
29	37.9	39.8	42.0	44.2	46.4	48.8	50.1	53.6	56.3	58.8	61.8	64.3	67.3	70.0	73.3	76.3	79.5	29
30	33.8	35.4	37.2	39.0	41.1	43.3	45.0	47.2	49.3	51.0	54.0	56.2	58.5	61.0	63.6	66.2	68.8	30
31																		31
32																		32
33	120.0	130.5	141.7	154.0	168.0	183.0	200.0											33
34																		34
35																		35
36	81.2	87.0	93.2	100.0	107.0	114.0	122.0	130.5	139.5	145.0	160.0	171.3	183.6	197.0	212.0	229.0	246.0	36
37	50.2	53.2	56.5	60.0	64.0	66.5	70.2	74.0	77.8	82.0	86.5	91	95.5	100.0	105.3	111.0	116.2	37
38	192.0	214.0																38
39	68.5	73.2	78.0	83.0	88.5	94.0	100.0	106.2	113.0	120.0	127.5	135.5	144.0	153.0	163.0	173.5	184.5	39
40	41.8	44.3	46.8	49.2	51.9	54.6	57.4	60.3	63.2	66.3	69.5	72.7	76.2	79.6	86.5	87.3	91.0	40
41	24.2	26.3	27.6	28.9	29.4	31.6	33.0	34.4	35.9	37.4	38.8	40.3	42.0	43.5	45.2	46.8	48.5	41
42	72.0	77.0	82.2	87.5	93.2	99.4	105.8	115.5	120.0	127.7	136.0	145.0	154.5	164.5	175.5	187.0	200.0	42
43	25.5	26.9	28.2	29.5	30.9	32.3	33.7	35.2	36.7	38.1	39.8	41.3	43.2	44.5	46.2	47.8	49.6	43
44	13.7	14.4	15.0	15.7	16.4	17.0	17.7	18.4	19.1	19.7	20.5	21.1	21.4	22.6	23.3	24.1	25.0	44

COMPUTATIONS BASED ON ENGINEERING-NEWS RECORD WORK

FORMULAS: $L = \dfrac{2E}{S + 0.10}$ AND $L = \dfrac{(2WH) + (AP)}{S + 0.10}$

NUMBER OF BLOWS PER FOOT $= \dfrac{120\,L}{(20E) - L}$ = AVERAGE 10 FINAL.

NUMBER OF BLOWS PER INCH $= \dfrac{n}{12}$ AND SET $= S = \dfrac{1.00}{n}$

1 U.S. TON = 2000 LBS. 1 BRITISH TON = 2240 LBS.

L = SAFE LOAD CAPACITY IN POUNDS.
E = MANUFACTURES RATED HAMMER ENERGY, IN FOOT LBS.
W = WEIGHT OF STRIKING RAM, IN POUNDS.
H = HEIGHT OF RAM'S FALL, IN FEET.
A = AREA OF PISTON, IN SQUARE INCHES. $A = 0.7854\,D^2$
P = STEAM PRESSURE AT HAMMER, IN LBS. SQ. IN. (P.S.I.)
S = SET PENETRATION PER BLOW, IN INCHES.

NOTE:
HAMMER BLOWS FOR DIESEL HAMMER LISTED ARE BASED ON 75 PERCENT OF RATED ENERGY.

EXAMPLE: Ultimate capacity by Hiley formula 9.7.1

A Steel 12BP74 Pile is 72.0 Feet long. Soil penetration is to be 48.0 feet to cut-off elevation. A Double-Acting McK-T Hammer without cushion was used for driving. Hammer was Model 10-B-3, and set penetration average for last 12 blows was 0.10 inches per blow.

REQUIRED:
This problem is identical with example used in the Terzaghi Formula: Employ the Hiley Formula to calculate the Ultimate (Ru) Resistance to Penetration, then compare the Safety Factor of the E-N Formula with final results. Calculate the equations with British Tons and convert to U.S. Tons after obtaining final figures for Ru.

STEP I:
An assumption must be made for preliminary Ru, in order to ascertain values of Cc, Cp, and Cq. The constant C equals these three added. Assume Ru = 150 Long Tons. Convert hammer Energy, pile and Ram to long tons also.

STEP II:
Weight of Pile: $P = \frac{74.0 \times 72}{2240} = 2.380$ Tons.

Estimate weight of Anvil at 0.335 Tons.

Then Total $P = 2.380 + 0.335 = 2.715$ Long Tons.

Weight of Ram $= \frac{5000}{2240} = 2.23$ Tons. From Table: E = WH or

$WH = \frac{19,150}{2240} = 8.55$ Ft. Tons, and $WH = 8.55 \times 12 = 102.60$ Inch T.

STEP III:
From Hiley Tables: Restitution Coefficient $e = 0.50$ as hammer is Double Action and uses no cushion.

Coefficient $K = 1.00$ for same condition.

Net Area in Steel 12-BP 74 Pile Cross Section = 21.75 Sq. In.

Overall Driving Area of cross-section is $12.0 \times 12.0 = 144.0$ □"

STEP IV:
With assumed Ru = 150.0 Tons.

Actual driving stress (Compressive) on steel $= \frac{150.0}{21.75} = 6.91$ Tons □"

Overall compressive stress on soil $= \frac{150.0}{144} = 1.04$ Tons Sq. Inch.

STEP V:
Ratio of Pile weight to Ram: $\frac{P}{W} = \frac{2.715}{2.23} = 1.215$

With $\frac{P}{W} = 1.215$ Use chart for hammer blow efficiency.

$\eta = 0.597$ (e = 0.50 on curve) (K = 1.0 and will be deleted)

STEP VI:

Refer to Charts and Graphs for Coefficients which make up Constant C.

For Steel Pile effective pile length is 48.0 Feet, therefore use curve for 50.0' in chart. $C_p = 0.300$

$C_c = 0.235$ and for quake, $C_q = 0.200$

Then $C = \Sigma\ C_p + C_c + C_q$. $C = 0.300 + 0.235 + 0.200 = 0.735''$

STEP VII:

The Hiley Formula: $R_u = \frac{WH\eta}{S + \frac{C}{2}}$. $\frac{C}{2} = 0.735 \times 0.5 = 0.3675''$

Substituting values:

$$R_u = \frac{102.60 \times 0.597}{0.10 + 0.3675} = \frac{61.252}{0.4675} = 131.0 \text{ Tons. (Long Tons)}$$

In U.S. Tons: $R_u = \frac{131.0 \times 2240}{2000} = 146.72$ Tons. (293,440 Lbs.)

STEP VIII:

The Terzaghi Formula for $R_u = 144.75$ Tons (289,500 Lbs.)

The Safe Allowable Load calculated by the Engineering-News Formula = 131,000 Pounds, or 65.5 Tons.

The Safety Factor of E-N Formula when based on Ultimate computed by Hiley Formula $= \frac{146.72}{65.5} = 2.24$

The Safety Factor of E-N Formula based on Terzaghi Formula is therefore: $\frac{289{,}500}{131{,}000} = 2.21$

NOTE BY AUTHOR:

Under identical conditions, the Ultimate value for the Resistance to Penetration (R_u) will be more conservative with the Terzaghi Formula than when using the above Hiley Formula.

In the absence of a control test pile to substantiate the load capacities, it is recommended that the Hiley Formula govern the allowable load capacities.

In certain circumstances, as using a hammer without cushioning of any kind for driving steel BP or Pipe piles, the Coefficient C_c is reduced to 0.0 (zero).

This practice is not recommended for any condition.

EXAMPLE: Ultimate capacity by alternate Hiley formula 9.7.2

Assume a steel pile 12B74 is 72.0 feet long, and is driven into soil a distance of 48.0 feet. Hammer used is a Vulcan DA Model 50-C. Observer counted 120 blows for last 10.0 inches of pile penetration. Steam pressure at Boiler during driving read 146 PSI.

REQUIRED:

Calculate the Ultimate Resistance to Penetration (R_u) by using the Alternate Hiley Formula: $R_u = \sqrt{2mWH\eta + (ms)^2} - ms$.

STEP I:

Assume for trial that R_u = 175.0 Long tons.

Manufacturer's Data on Vulcan 50-C: From Tables:

Hammer uses wood cushion, assume it well compacted.

Rated Energy E = 15,100'# equal to WH. Double Acting.

Weight of Ram W = 5000 # or $W = \frac{5000}{2240} = 2.235$ Tons

Weight of Pile, $P = \frac{72.0 \times 74}{2240} = 2.380$ Tons.

Estimated Weight of Anvil and Cap = 0.335 Tons.

Pile Ratio to Ram: $\frac{P}{W} = \frac{2.380 + 0.335}{2.235} = 1.215$

STEP II:

From Graph Chart; Restitution coefficient: e = 0.50

" " " Coefficient value of, K = 1.00

" " " Effective blow Coefficient, η = 0.590

Net steel area of Pile sustaining blow = A_p = 21.75 Sq. In.

Gross soil and steel area sustaining blow = A_g : 12.0 × 12.0 = 144.0 Sq. In.

STEP III:

Overall driving stress on soil = $\frac{R_u}{A_g} = \frac{175.0}{144.0} = 1.215$ Tons Sq. Inch.

Actual driving stress on steel = $\frac{R_u}{A_s} = \frac{175.0}{21.75} = 8.06$ Tons Sq. In.

STEP IV:

From Chart-Graphs, Coefficient C_c = 0.280"

" " " " C_p = 0.340"

" " " " C_q = 0.200"

For Coefficient C — Σ = 0.820"

Convert $WH\eta$ to inch Tons to use in formula.

$$WH\eta = \frac{15,100 \times 12 \times 0.590}{2240} = 47.70 \text{ Inch Tons.}$$

Since Coefficient $K = 1.00$, it need not be considered.

STEP V:

To obtain the value of m, refer to text. $m = \frac{Ru}{C}$

Then: $m = \frac{175.0}{0.820} = 213.0$ Reminder: Ru in I.

Penetration was 120 blows in 10.0 inches, then set per blow was thus: $S = \frac{10.0}{120} = 0.0833$" per blow.

STEP VI

Reconcile the values to insert in formula:

$m = 213.0$ $WH\eta = 47.70$ Inch tons. $S = 0.0833"$

Formula is shortened thus:

$$Ru = \sqrt{\left[(2 \times 213.0 \times 47.70) + (213.0 \times 0.0833)^2\right]} - (213.0 \times 0.0833)$$

$$Ru = \sqrt{20,320.20 + 315.06} - 17.75 = 126.0 \text{ Long Tons.}$$

In U.S. Tons, Ultimate $R_u = \frac{126.0 \times 2240}{2000} = 141.12$ Tons.

With Safety Factor of 100%, $R = \frac{141.12}{2} = 70.56$ Tons.

DESIGN NOTE:

The assumed Ru used at beginning was 175.0 Tons and although it was off of the final result by 49.0 tons, it remains conservative. Using a higher ultimate load assumption, with 100% Safety Factor in mind, will raise the driving stresses shown in Step III. This is on the safe side since the subsequent coefficient values will remain in line with actual driving conditions.

EXAMPLE: Ultimate load for timber pile 9.7.3

A Southern Yellow Pine Pile is 60.0 Ft. Long and is to be driven with the tip elevation at −20.0 feet. Top soil layer elevation is +17.0 feet. Effective length: 37.0 feet. With a safety factor of between 1½ and 2, this pile is to sustain a working load capacity of 30 U.S. Tons. A Vulcan No. 1 Hammer is available for the driving operation.

REQUIRED:
Use the Engineers-News Formula to determine set per blow for last 10 blows, then use Hiley Formula to solve for Ultimate load and safety factor.

STEP I:
From Hammer Table Data on Vulcan No. 1: Rated WH = 15,000'#
Single Action, Cushion Block and Ram Weight = 5000 Lbs.
H = 3.0' E-N Formula applicable: $R = \frac{2WH}{S+0.10}$. With formula based upon 10 blow average for set per blow the formula for <u>Number of blows per foot</u> is as follows.
Load R = 30 × 2000 = 60,000 Lbs. Transposing and 10 blows:
$\text{Number} = \frac{10 \times 12 \times L}{(10 \times 2WH) - L}$. Then $\text{No.} = \frac{120 \times 60{,}000}{(20 \times 15{,}000) - 60{,}000} = 30$ Per foot.

set per blow, $s = \frac{1.00}{30} = 0.0333$' blow, or 0.333 × 12 = 0.40 inches per blow.

STEP II:
Convert certain values to long tons to use Hiley formula:
$R = \frac{60{,}000}{2240} = 26.8$ $W = \frac{5000}{2240} = 2.23$ Tons. $WH = \frac{15{,}000}{2240} = 6.80$ Ft. Tons.
Average SYP Treated Wood Pile weighs approx. 1300 Pounds.
$P = \frac{1300}{2240} = 0.58$ Tons. Converting WH = 6.80 × 12 = 81.60 inch Tons.
Ratio of Pile weight to Ram weight: $\frac{P}{W} = \frac{0.58}{2.23} = 0.26$

STEP III:

From Tables: Efficiency of hammer; $K = 0.90$
" " Restitution Coefficient; $e = 0.25$
" " Efficiency of Blow; $\eta = 0.80$

STEP IV:

With a Safety Factor of 2, Ultimate: $R_u = 2 \times 26.8 = 53.6$ Tons.
Tip diameter of wood pile approximately = 6.00 Inches.
Area pile tip; $A_p = 6.0 \times 6.0 \times 0.7854 = 28.30$ Sq. Inches
Driving stress on Pile = $\frac{53.60}{28.30} = 1.90$ Tons per Sq. Inch.

STEP V:

Pile gross area is same at net area: $A_p = A_g$.
From Graphic Chart, Coefficient: $C_c = 0.32$ (Used limit)
" " " " $C_p = 0.80$ (Used maximum)
" " " " $C_q = 0.20$ " "
Coefficient for $C = \Sigma_c = 1.32$

STEP VI:

The Hiley formula for ultimate: $R_u = \dfrac{WHK\eta}{s + \frac{C}{2}}$

From step I: $s = 0.40$ inches per blow.
Substituting values in formula:

$$R_u = \frac{81.60 \times 0.90 \times 0.80}{0.40 + \left(\frac{1.32}{2}\right)} = 52.3 \text{ Long tons.}$$

Converting: $R_u = \dfrac{52.3 \times 2240}{2000} = 58.4$ U.S. Tons (Ultimate)

STEP VII:

For safety factor: $L = 30$ Ton load.
$SF = \dfrac{R_u}{L}$. Safety factor = $\dfrac{58.4}{30} = 1.95$ (Close to 100%)

Pile will be satisfactory. Accept driving with Vulcan No. 1 hammer with a blow count of 30 blows per foot.

EXAMPLE: Calculating blows per foot for safe load 9.7.4

A pile is required to sustain a safe load of 100 Tons. For driving purposes it is assumed that a Vulcan Model 80-C will perform satisfactory. Also on hand is a Diesel Delmag Model D-22, which could be used.

REQUIRED:
Calculate the required number of Blows per foot, and final set penetration per blow for each hammer. Use the Engineering-News Formula applicable to each action type.

STEP I:
For Vulcan Model 80-C: The rated energy given in tables is: E = 24,450 Foot Lbs. Hammer is Double Acting, and E-N Formula is: Load, $L = \frac{2E}{S+0.10}$.

STEP II:
Load $L = 100 \times 2000 = 200{,}000$ Pounds (This is a Safe Load)
Transpose Formula to solve for set (s).
$s + 0.10 = \frac{2E}{L}$, and $s = \frac{2E}{L} - 0.10$. Substituting values:

$$s = \frac{2 \times 24{,}450}{200{,}000} - 0.10 = 0.244 - 0.10 = 0.144 \text{ inches per blow.}$$

Number of blows required per inch: $n'' = \frac{1.00}{0.144} = 6.94$

Number of blows required per foot: $n' = \frac{12.00}{0.144} = 83.4$

Use either 7 blows per inch on last 10 blows of hammer or 84 blows for 1.0 foot of penetration.

STEP III:
For Delmag Hammer D-22. This is a single acting with explosive aid. Formula is: $L = \frac{2E}{S+0.10}$ $E = 39{,}700$ Ft. Lbs.

set: $s = \frac{2 \times 39{,}700}{200{,}000} - 0.10 = 0.297$ inches per blow.

Number of blows required per inch: $n = \frac{1.000}{0.297} = 3.39$

Number of blows required per foot: $n = 3.39 \times 12 = 40.68$

Diesels are seldom used on basis of rated energy. Reductions up to 75% of E is generally used to solve for number of blows. See table for blows required.

EXAMPLE: E-N formula required set for three hammers 9.7.5

Assume that a pile is to be driven to sustain a safe working load of 75 Tons. ($L = 150{,}000$ Lbs.).
Three hammers are available in this order:

1. Vulcan, Model 80-C, Double Action, with rated $E = 24{,}450'\#$.
2. McK-T_0, Model S-8, Single Action, with rated $E = 26{,}000'\#$.
3. Mitsubshi, Model M-145, Single Action, with rated $E = 26{,}000'\#$

REQUIRED:
Determine the required number of blows per foot of penetration for each of the hammers listed. Use the Engineering-News Formulas with an average set per blow on the last 10 blows from hammer.

STEP I:
The Engineering-News Formula applying to Single Acting Hammers is: $L = \frac{2WH}{S+0.10}$. For Double Acting Hammers: $L = \frac{2E}{S+0.10}$.
In each case, $WH = E$ = Rated Energy listed in Tables.

STEP II:
To use the Formula in transposed form, and based upon set average of final 10 blows of hammer, it is re-written thus: Number of Blows, $n = \frac{10 \times 12 \times L}{(10 \times 2E) - L}$. set $= \frac{12}{n}$, in inches.

STEP III:

1. For Vulcan Hammer 80-C; $n = \frac{10 \times 12 \times 150{,}000}{(10 \times 2 \times 24{,}450) - 150{,}000} = 53.2$ per Ft.
2. For McK-T. Hammer S-8; $n \quad \frac{120 \times 150{,}000}{(20 \times 26{,}000) - 150{,}000} = 48.6$ per Ft.
3. Mitsubshi Hammer M-145 is same as No. 2.

STEP IV:

1. For Vulcan 80-C; set per blow: $s = \frac{12.0}{53.2} = 0.226$ inches per blow.
2. For McK-T. S-8; set per blow: $s = \frac{12.0}{48.6} = 0.247$ inches per blow.

STEP V:
To check the Alternate Formula, use the set 0.226 from Vulcan 80-C in basic E-N Formula: Solve for safe load L.

$$L = \frac{2E}{S+0.10} \quad \text{or} \quad L = \frac{2 \times 24{,}450}{0.226 + 0.10} = 150{,}000 \text{ Lbs. (OK 75 Tons)}$$

EXAMPLE: Safe load by modified E-N formula 9.7.6

A Pre-cast Concrete was driven to a final set penetration of 0.15 inches per blow. A Vulcan Hammer Type S-10 was used. Energy rating for Hammer is 32,500 Foot Pounds. Pile is 18.0 inches square, solid, and 60.0 Feet long.

REQUIRED:
Use the Engineering-News Modified Formula to determine the safe allowable load. Compare result with Basic Formula.

STEP I:
Modified formula relates the weight of pile (P) to ram's weight (W). Formula: $L = \frac{2E}{S + (c\frac{P}{W})}$. From Table: W = 10,000 Lbs.

STEP II:
Weight of Pile: P = 1.50 x 1.50 x 60.0 x 150 = 20,250 Pounds.
S = 0.15" per blow. C = 0.10 $\frac{P}{W} = \frac{20,250}{10,000} = 2.025$

STEP III:
Substitute values in Modified E-N Formula:

$$L = \frac{2 \times 32,500}{0.15 + (0.10 \times 2.025)} = \frac{65,000}{0.352} = 184,500 \text{ Lbs. } (92.25 \text{ Tons})$$

STEP IV:
Check with Basic E-N Formula: $L = \frac{2WH}{S + 0.10}$. H = 3.25 Feet.
S-10 is a Single Acting Hammer.

$$L = \frac{2 \times 10,000 \times 3.25}{0.15 + 0.10} = \frac{65,000}{0.25} = 260,000 \text{ Lbs. } (130.0 \text{ Tons})$$

STEP V:
Load difference = 130.00 − 92.25 = 37.75 Tons.
Modified E-N Formula is the more conservative when the Pile Weight is greater than weight of Ram.

EXAMPLE: Hammer selection by E-N formula 9.7.7

A group of piles are to be driven to a safe load of 100 Tons each. Several hammers are available however Contractor desire a Double Acting Hammer which will accomplish 100 Ton bearing with 60 to 70 blows per foot. The Engineering-News Formula is to govern design.

REQUIRED:
Transpose the E-N Formula for Double Acting Hammer and solve for Energy Rating E. Use 65 blows per foot to determine set per blow. Refer to Table for selection of at least 3 satisfactory hammers.

STEP I:
At 65 Blows per foot, set $S = \frac{12.0}{65} = 0.184$ inches per blow.
Double Acting E-N Formula: Must consider constant.
$L = \frac{2E}{S+0.10}$ and transposed: $S+0.10 = \frac{2E}{L}$, therefore
actual $S = \left(\frac{2E}{L}\right) - 0.10$ and $2E = L(S+C)$ with $E = \frac{2E}{2}$.

STEP II:
Substituting values to solve for Energy:
$E = \frac{200{,}000 \times (0.184 + 0.10)}{2} = 28{,}400$ Foot Pounds.
Check this with original formula: $L = \frac{2E}{S+0.10}$
$L = \frac{2 \times 28{,}400}{0.184 + 0.10} = \frac{56{,}800}{0.284} = 200{,}000$ Lbs.

STEP III:
From Hammer Tables, the choice may be either of these:
Vulcan, 80-C, DA with E = 24,450 Ft. Lbs.
McKiernan-Terry, C-8, DA with E = 26,000 Ft. Lbs. (S-8 is Single Act.)
Vulcan, 140-C, DA with 36,000 Ft. Lbs.

STEP IV:
Referring to Table for Hammer Blows for Safe Loads, the blows per foot are as follows: (100 Ton Capacity).
Vulcan 80-C, requires 83 blows per foot.
McKiernan-Terry C-8, requires 75 blows per foot.
Vulcan 140-C, requires 46.2 blows per foot.

Sheet piling 9.8

Steel sheet pilings are produced in several shapes which can interlock to form retaining walls, abuttments, caissons and many other structures. Construction may be either permanent or for temporary use. It is not to be expected that the interlocks will be absolutely water-tight, but the sections will retain sufficient liquid to permit pumping out with oversized pumps.

Table of cross-sections 9.8.3 is provided for the standard shapes. Sections are rolled to position the centroid where the section will have the highest moment of inertia. Zee shaped sections have the interlocks located in the flanges. The straight web sections are used only where interlocking is desired, and little consideration is given to lateral bending stress. New sheet pile sections can be milled in lengths up to 85 feet, however truck transportation facilities will limit the usual maximum length to 60 feet or less.

A generous tolerance is provided in the interlocking edges to permit the sections to be driven on angles or in circular forms within certain radius limits. Sections can be driven to provide caissons, well digging, storage bins, dolphins and other protective shafts. Special corner sections are rolled or fabricated to interlock form shapes with straight walls.

Strength of interlocks 9.8.1

The interlock strength against pulling apart will vary according to the pile grouping. The tension on interlocks must be considered separately for each type structure. For design purposes, the amount of tension is to be taken at 8000 or 12,000 pounds per lineal inch of interlock. One of these figures should be made plain in the specifications and in the purchase order.

When interlocking pile sections are to be driven and used for retaining walls, a safety factor of two should be used. The safety factor in such cases is based on the maximum yield stress instead of the ultimate stress. In mild steel where F_y = 36,000 P.S.I., the working stress is 18,000 P.S.I. A medium tensile steel section is available from foreign sources and a number of domestic producers which will allow a working unit stress of 22,500 P.S.I.

Driving sheet pile 9.8.2

Straight walls of sheet piles of any type will show a tendency to lean forward in the direction of driving, and soon depart from plumb. This tendency can be controlled by using the system known as panel driving. It is necessary to correct the leaning as soon as it becomes apparent. If nothing is done it soon becomes very difficult to correct. The only satisfactory method for keeping the interlocks plumb is to drive in panels, and the specifications should stipulate the panel method as a requirement.

Panel driving is begun by setting and driving to part penetration a pair of pile sections, taking every precaution to ensure that they are indeed plumb in interlock and vertical in lateral set. About six to ten pairs of piles are then set and interlocked in position against template or wale guides. After being partly driven, another panel is set and partly driven. Driving in alternate stages for each direction will tend to counterbalance the leaning action; thus the piles will remain plumb.

During driving, the top sheet pile sections will have a tendency to deform by rolling and bending. A certain amount of deformation is desirable but should be kept above the cut-off line. The curled top should not be cut away until driving has progressed to a safe distance or entirely finished. Occasionally a large amount of friction will develop between interlocking piles, and the adjacent piles will be drawn down with the pile being driven. Raising a drawn down pile by jacks or an extractor is not an effective method for permanent structures. It is far better and much simpler to weld an additional length to the drawn down pile.

TABLE: Standard sheet piling sections 9.8.3

STANDARD INTERLOCKING SECTIONS

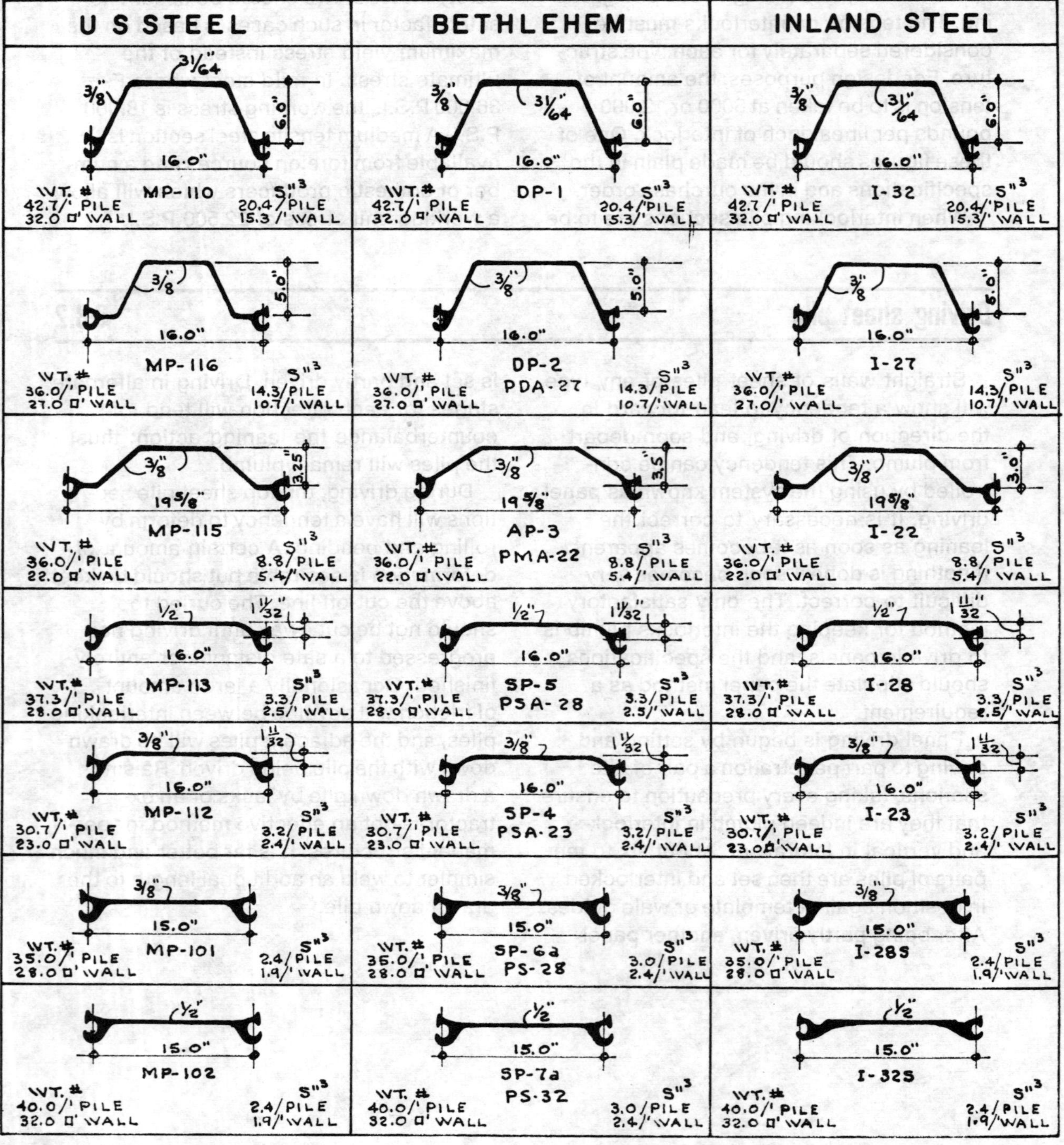

U S STEEL	BETHLEHEM	INLAND STEEL
MP-110 3/8" — 31/64" — 6.0" — 16.0" WT. # 42.7/' PILE, 32.0 □' WALL S"3 20.4/PILE, 15.3/' WALL	DP-1 3/8" — 31/64" — 6.0" — 16.0" WT. # 42.7/' PILE, 32.0 □' WALL S"3 20.4/PILE, 15.3/' WALL	I-32 3/8" — 31/64" — 6.0" — 16.0" WT. # 42.7/' PILE, 32.0 □' WALL S"3 20.4/PILE, 15.3/' WALL
MP-116 3/8" — 5.0" — 16.0" WT. # 36.0/' PILE, 27.0 □' WALL S"3 14.3/PILE, 10.7/' WALL	DP-2 PDA-27 3/8" — 5.0" — 16.0" WT. # 36.0/' PILE, 27.0 □' WALL S"3 14.3/PILE, 10.7/' WALL	I-27 3/8" — 5.0" — 16.0" WT. # 36.0/' PILE, 27.0 □' WALL S"3 14.3/PILE, 10.7/' WALL
MP-115 3/8" — 3.25" — 19 5/8" WT. # 36.0/' PILE, 22.0 □' WALL S"3 8.8/PILE, 5.4/' WALL	AP-3 PMA-22 3/8" — 3.5" — 19 5/8" WT. # 36.0/' PILE, 22.0 □' WALL S"3 8.8/PILE, 5.4/' WALL	I-22 3/8" — 3.0" — 19 5/8" WT. # 36.0/' PILE, 22.0 □' WALL S"3 8.8/PILE, 5.4/' WALL
MP-113 1/2" — 1 1/32" — 16.0" WT. # 37.3/' PILE, 28.0 □' WALL S"3 3.3/PILE, 2.5/' WALL	SP-5 PSA-28 1/2" — 1 1/32" — 16.0" WT. # 37.3/' PILE, 28.0 □' WALL S"3 3.3/PILE, 2.5/' WALL	I-28 1/2" — 1 1/32" — 16.0" WT. # 37.3/' PILE, 28.0 □' WALL S"3 3.3/PILE, 2.5/' WALL
MP-112 3/8" — 1 11/32" — 16.0" WT. # 30.7/' PILE, 23.0 □' WALL S"3 3.2/PILE, 2.4/' WALL	SP-4 PSA-23 3/8" — 1 11/32" — 16.0" WT. # 30.7/' PILE, 23.0 □' WALL S"3 3.2/PILE, 2.4/' WALL	I-23 3/8" — 1 11/32" — 16.0" WT. # 30.7/' PILE, 23.0 □' WALL S"3 3.2/PILE, 2.4/' WALL
MP-101 3/8" — 15.0" WT. # 35.0/' PILE, 28.0 □' WALL S"3 2.4/PILE, 1.9/' WALL	SP-6a PS-28 3/8" — 15.0" WT. # 35.0/' PILE, 28.0 □' WALL S"3 3.0/PILE, 2.4/' WALL	I-28S 3/8" — 15.0" WT. # 35.0/' PILE, 28.0 □' WALL S"3 2.4/PILE, 1.9/' WALL
MP-102 1/2" — 15.0" WT. # 40.0/' PILE, 32.0 □' WALL S"3 2.4/PILE, 1.9/' WALL	SP-7a PS-32 1/2" — 15.0" WT. # 40.0/' PILE, 32.0 □' WALL S"3 3.0/PILE, 2.4/' WALL	I-32S 1/2" — 15.0" WT. # 40.0/' PILE, 32.0 □' WALL S"3 2.4/PILE, 1.9/' WALL

STANDARD SHEET PILING—INTERLOCKING "Z"

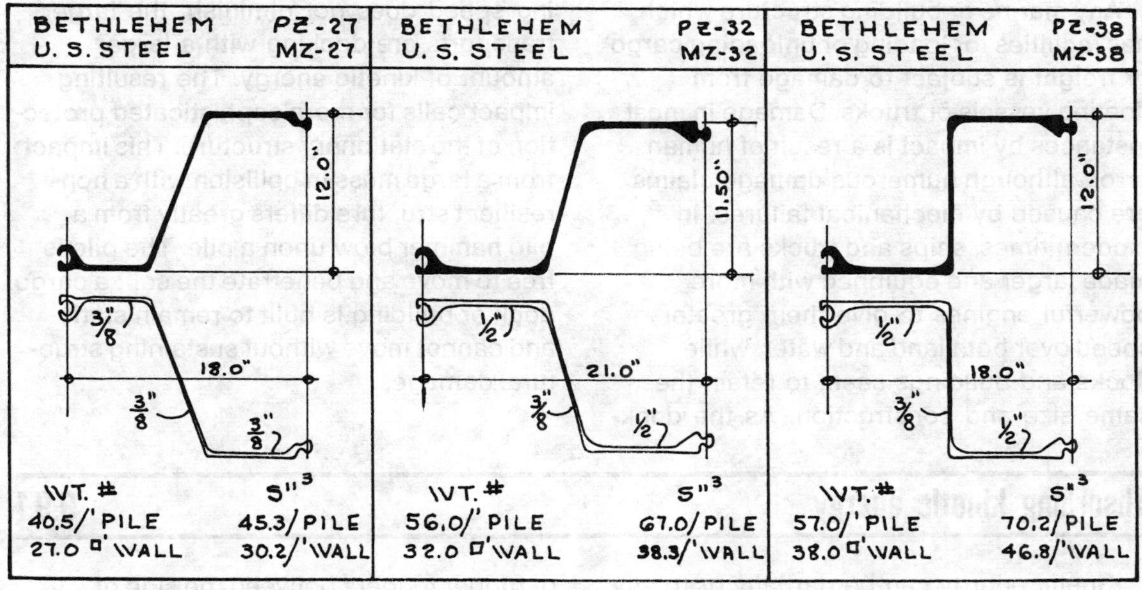

STANDARD BENT TYPE CONNECTORS

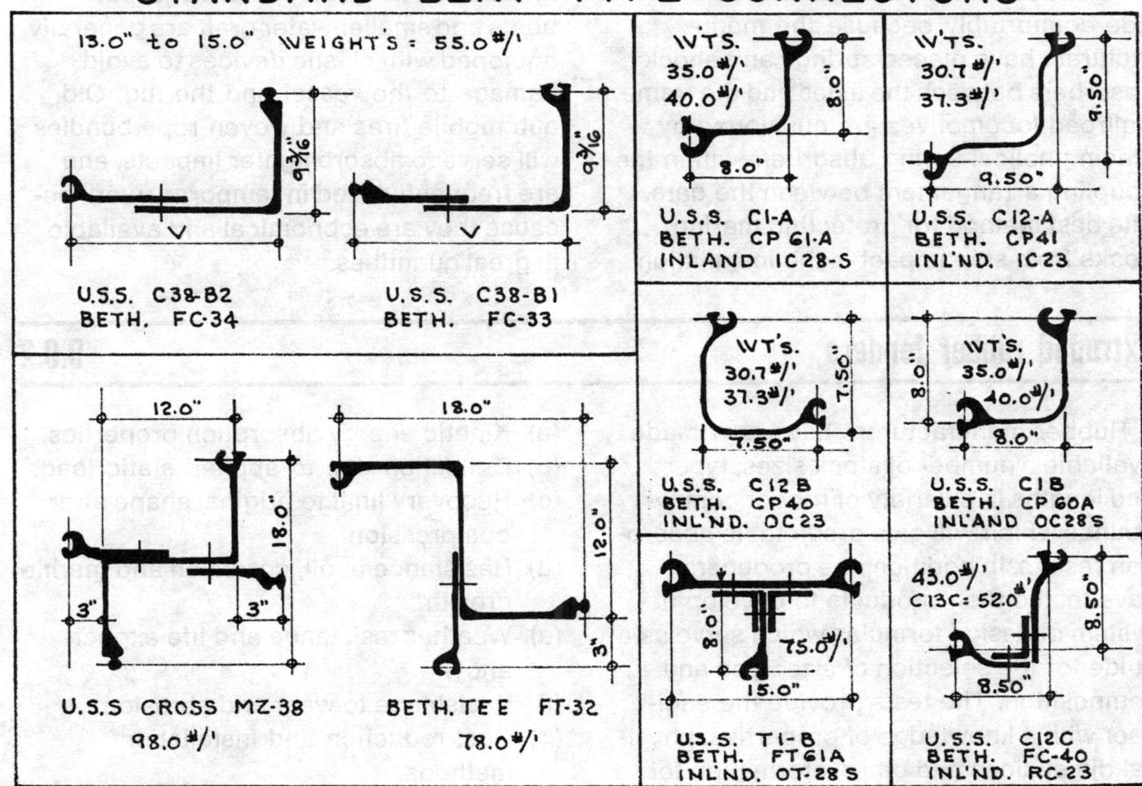

Marine and dock fendering 9.9

Any marine or building structure which has facilities for loading or unloading cargo or freight is subject to damage from floating vessels or trucks. Damage in most instances by impact is a result of human error, although numerous damage claims are caused by mechanical failures. In modern times, ships and trucks are being made larger and equipped with more powerful engines to give them greater speed over both land and water, while docks and buildings seem to retain the same size and construction. As the docking speed does not diminish, the larger transports are docking with a larger amount of kinetic energy. The resulting impact calls for more sophisticated protection of the stationary structure. This impact from a large mass in collision with a non-resilient structure differs greatly from a pile hammer blow upon a pile. The pile is free to move and penetrate the soil; a cargo dock or building is built to remain static and cannot move without sustaining structural damage.

Absorbing kinetic energy 9.9.1

Kinetic energy can be partially or totally absorbed by an elastic cushion placed between the moving mass and the immovable structure. Automobiles are able to ride comfortably because the manufacturers have placed springs and shock absorbers between the axles and the frame. Railroad locomotives are cushioned by placing helical spring absorbers within the coupling arrangement between the cars. The best method for protecting marine docks from ship impact is the installation of rubber fenders between the side of the ship and the wharf surface. Fendering is also a good insurance for buildings when installed as truck bumpers. Tugs, push boats and smaller water craft are generally equipped with elastic devices to avoid damage to the vessel and the tug. Old automobile tires and woven rope bundles will serve to absorb lighter impacts, and are frequently used in temporary work because they are economical and available in great quantities.

Extruded rubber fenders 9.9.2

Rubber manufacturers have now made available a number of stock sizes, types and lengths in a variety of rubber compounds which will give predictable absorption results. In addition, the producers have tested their products to develop a system of design formulas which serve as a guide for the selection of size, type and composition. The tests provide the engineer with a knowledge of properties, physical dimensions and other information for specifications as follows:

(a) Kinetic energy absorption properties.
(b) Deflection due to applied static loads.
(c) Recovery limit to original shape after compression.
(d) Resistance to oil, corrosion and marine growth.
(e) Weather resistance and life expectancy.
(f) Resistance to wear and abrasion.
(g) Cost reduction and installation methods.

Fender specification standards 9.9.2.1

The rubber fenders produced by domestic manufacturers are formed from extrusion molds. When the finished material is tested to ASTM Standards, it must meet the classification requirements R–725B, C, F, J, and L. These specifications are outlined fully in ASTM pamphlet D–735–56. Rubber fender application is mainly concerned with the compressive stress rather than the tension properties. The tensile strength of the rubber compound will have little bearing upon the service life of the fender. The following properties are general:

Tensile Strength;—minimum 2500 PSI.
Elongation; " 300%
Durometer; 70 ± 5
Compression set; maximum 25%
Flexibility range; To −40 Degrees Fahr.
Tolerances; 3% Outside Diameter; 8% Inside Diameter.
Wall and Length Tolerances; 4% Wall thickness; ½ inch on length.

Fender compositions 9.9.2.2

A natural rubber and synthetic rubber are compounded substances. The base properties of the compound are determined by the type of rubber used. Other properties are modified or supplied by the amount and type of carbon-black, which is used to reinforce the rubber. Other additives which are added are: antioxidants; antiozonants; oils; waxes; and butyl. All of these additives provide the agents for curing, vulcanizing, protection, and physical properties of the compound.

NATURAL RUBBER

Natural rubber is produced by extracting the sap from the rubber tree. As a pure-gum compound, a strong and highly wear-resistant product is obtained when curing agents are added and the mixture is solidified by vulcanizing.

Natural rubber has good resistance to water and normal temperatures. It is not satisfactory for use in the oil and chemical industries. Natural rubber was used in the production of some fenders until World War II when the supply of sap was halted. Synthetic rubber SBR was introduced, and the need for importing the basic gums was eliminated. The new synthetic compounds have been developed to the extent that natural rubber accounts for only a very small fraction of total rubber production.

SYNTHETIC RUBBER SBR

The standard fender compound is SBR Rubber. This product derives its name from the emulsion polymerization of Styrene and Butadiene. This product accounts for the largest part of synthetic rubber in today's market. SBR was first produced in government-owned plants during the war, and it was known as GRS: Government Rubber Styrene. SBR rubber is produced in several varieties, with each kind having its own characteristics, properties and cures. SBR is used in great quantities for the production of automobile tires, because its friction and abrasion resistance is superior to natural rubber.

BUTYL RUBBER

The manufacturers have given Butyl rubber the official designation of I.I.R: Isobutylene-Isoprene-Rubber. When compared to SBR or natural rubber, Butyl has a lower resilience and recovery property.

Fender compositions, continued 9.9.2.2

It has excellent heat-resistance and good weathering qualities for longer life. At lower temperatures it is less resistant and elastic. Butyl is favored by many engineers for use in marine dock fenders. It compares very well with SBR, except in the qualities of abrasion resistance and resilience.

CHEMIGUM NBR

There are at present several types of NBR, which is a co-polymer of Butadiene and Acrylonitrile. It is made primarily for maximum resistance to oils, fats, and strong solvents associated with the production of paints and chemical coatings. At the present time NRB is not manufactured into fenders. Research is continuing on this product. The cost of each type of NBR is considerably higher than SBR or Butyl. Until the compound is fully developed, it is recommended that designers specify the older, more economical synthetic rubbers.

Docking speed and length of impact 9.9.3

Throughout the world all major ports subscribe to certain safety rules. It is mandatory that each incoming vessel be placed under the control of a skilled pilot, who is knowledgable of channel depths, docking speeds and ship traffic. These pilots are available from an organization of former ship masters and men who have spent many years at sea. At a certain point in the sea lane, docking tugs will be attached to the ship and slowly guide her into position for docking. By radio communication between pilot and tug operators, the vessel is under complete control for safe docking.

To protect both vessel and stationary dock, the pilot will estimate the docking speed according to the vessel's weight, direction of wind and type of fendering. For large cargo vessels, the docking speed will vary from one inch to nine inches per second. These speeds are converted to feet per second and entered in the energy formula as velocity (V). A coefficient is used for normal docking operations to compensate for water movement and tug braking energy. Several tankers were observed while being docked at the Gulf Oil Refinery and Texaco Docks in Port Arthur, Texas. Their velocity at impact, with tugs attached was measured as from two to four inches per second.

Unlike barges and large tankers, the usual cargo and naval ships will have a slight curvature on the sides. The length of impact surface will therefore be less than if two parallel flat surfaces were meeting. When the dock bents are spaced on twenty foot centers, it is safe to assume that impact will occur on this length. Smaller vessels with greater hull curvature will have considerably less impact length, but the weight of the vessel will be much less, and there will be much less energy.

A Docking Chart (9.9.6) is provided to estimate the value of kinetic energy based upon various values for displacements, velocities and conditions. This chart can serve as a guide for dock and fender design, or for designing fenders for existing marine structures.

Vessel displacement 9.9.4

The quantity or volume of water necessary to keep a ship afloat is referred to as the displacement. That volume of the hull which is submerged below the water line is displacing a weight of water which is equal to the total weight of the vessel, including cargo, men, machinery and dunnage. Because the trim and profile of the hull is irregular, naval architects use the calculus in solving for the displacement volume. Draft markings on a vessel indicate the depth of water to the keel. Amidships on each side of the hull will be a marking consisting of a disc with horizontal lines called the Plimsoll or Freeboard mark. The ship cannot be legally loaded deeper than this mark.

When all of the ship's calculations are compiled, the data is assembled on a large graph plan called the "Curves of Form." Using this graph, the docking pilot can determine the values of buoyancy, area of water planes and other information he will need to dock the ship safely. In calculating the kinetic energy of a moving vessel, the displacement is equal to the weight (W), as used in the basic energy formula. In the design of docks and wharves, the engineer is obligated to visit the proposed site and take notes on the size and weight of vessels to be docked and berthed for loading.

Fender design 9.9.5

The manufacturers of standard rubber fenders have provided engineers and architects with data and properties which are the result of rigid tests. The designer must establish a criteria, based upon his observations and experience with moving bodies, which will provide the basis for his calculations for the amount of kinetic energy to be absorbed. Maximum absorption values for each size and type of standard fender are given in the tables. Properties listed are applicable to one lineal foot of fender, except where end loading is to be used.

Designers will also find the charts with deflection curves a convenient method for fender selection. In using the KE deflection chart, it should be noted that the curves end when the deflection reaches the point for complete closure. In order to provide the fender a longer life expectancy, the maximum deflection should be limited to approximately 50 percent. Excessive deflection destroys the elastic property which is necessary for recovery to original shape.

Fender design formulas 9.9.5.1

The type of vessel or truck, and conditions which reduce velocity before impact, will influence the values which are used in the design formulas. With respect to large tankers and cargo ships, docking is seldom attempted without the assistance of two or more tugs. The tug boats are attached to the sides of the vessel and push the ship into position at a very slow speed. An instant before impact, the pilot will signal the tugs to reverse their propellers which will reduce the kinetic energy. Contact between the side of the ship and the fenders will be from ten to thirty feet in length.

For calculating the ship's kinetic energy, and the length of fender required for absorption, assuming that tugs are used, the formula is written as follows: For one lineal foot of fender, $KE = \frac{0.50\left(\frac{W}{2G}\right)V^2}{L}$,

where formula nomenclature is as follows:

KE = Kinetic Energy in foot pounds.

W = Weight of vessel in pounds. (Displacement tons x 2240).

G = Gravity acceleration: Equals 32.174 feet per second.

V = Velocity of ship at instant of impact, in feet per second.

L = Length of fender subjected to simultaneous impact and given in lineal feet.

The coefficient of 0.50 is used most frequently for large, deep draft vessels fully loaded. The damping force of the water between the ship and the dock and the braking force from the tugs' reversing propellers before impact, will dissipate a considerable amount of kinetic energy. For similar vessels being docked against a weak structure or dolphin, the coefficient should be increased to 0.75.

Wind can become a contributing factor to kinetic energy during docking operations: ships bare of cargo and floating with a shallow draft tend to move laterally with strong winds and swells from passing ships. Records disclose that the most damage to water structures results from improper mooring and tide changes. In any dock or terminal constructed for barge loading, the wind and swell factor should be of primary importance in the design. Barges are fabricated with flat bottoms and straight sides; their draft depth is very shallow when empty. For fender design on barge loading terminals, the recommended formula is as follows: $KE = \frac{0.75 \times \left(\frac{W}{2G}\right)V^2}{L}$.

Truck fender design 9.9.5.2

As a truck rolls toward a loading dock, there are no forces to reduce the kinetic energy at impact. The fender must be designed to absorb the total energy. For most platforms the length of fender is governed by truck width or door opening. With no braking forces the formula for truck bumpers is written as: $KE = \frac{WV^2}{L}$. Engineers frequently find it convenient to transpose the formula when the fender has already been selected and they must solve for length, L. Then the formula becomes $L = \frac{WV^2}{F_a}$. Where W = weight of truck with tractor, and F_a = kinetic energy absorption per lineal foot for the fender under consideration.

End loaded fender design 9.9.5.3

Only cylindrical-type fenders should be used for end load design except in structures where the lengths are very short. The length of cylindrical types is limited to 1½ times the outside diameter. With this rule, the bulge at impact will be uniform, and deflection should not exceed 50 percent of length for full recovery. End loaded fendering provides excellent resilience and energy absorption qualities when installed between a buffer curtain wall and main dock surface. Where heavy ship traffic creates water swells, or in barge terminals where strong winds are a factor, the end loaded design will provide satisfactory protection.

End load stress limit 9.9.5.4

When the end loaded Deflection Curves are examined, it will be noted that the kinetic energy curve is based upon a deflection limit of approximately 50 percent of length. This limit also applies to a static load which compresses the length for a longer time period. When static load tests are conducted and the cross-section area is divided into load value for the 50 percent deflection, the unit stress will be found to be approximately 490 pounds per square inch. With this information, we can convert an amount of kinetic energy into a corresponding deflection under static load. Also, the problem can be reversed to solve for value of energy for a given deflection. The formulas for energy and load conversions are:

$E = 0.50 \times \frac{P\Delta}{12}$, and $P = \frac{12E}{0.50\Delta}$. Where symbols indicate:

Δ = Deflection in inches = 0.50 x Length. Length = 1.50 x O.D. Fender.

P = Static Load compressing fender, given in pounds.

E = Kinetic energy, in foot pounds.

Load P may also be obtained as F_cA. (Unit stress x area).

A deflection of 50 percent is on the safe side to obtain fully recovery. Actually recovery will result if, for short time periods and warm temperatures, the deflection is limited to 60 percent. In no event should the static load exceed a unit stress of 600 PSI on the fender cross-section. An example which follows will illustrate the design of end loaded fenders and the conversion of kinetic energy to static load.

DOCKING CHART: Displacement, velocity, energy 9.9.6

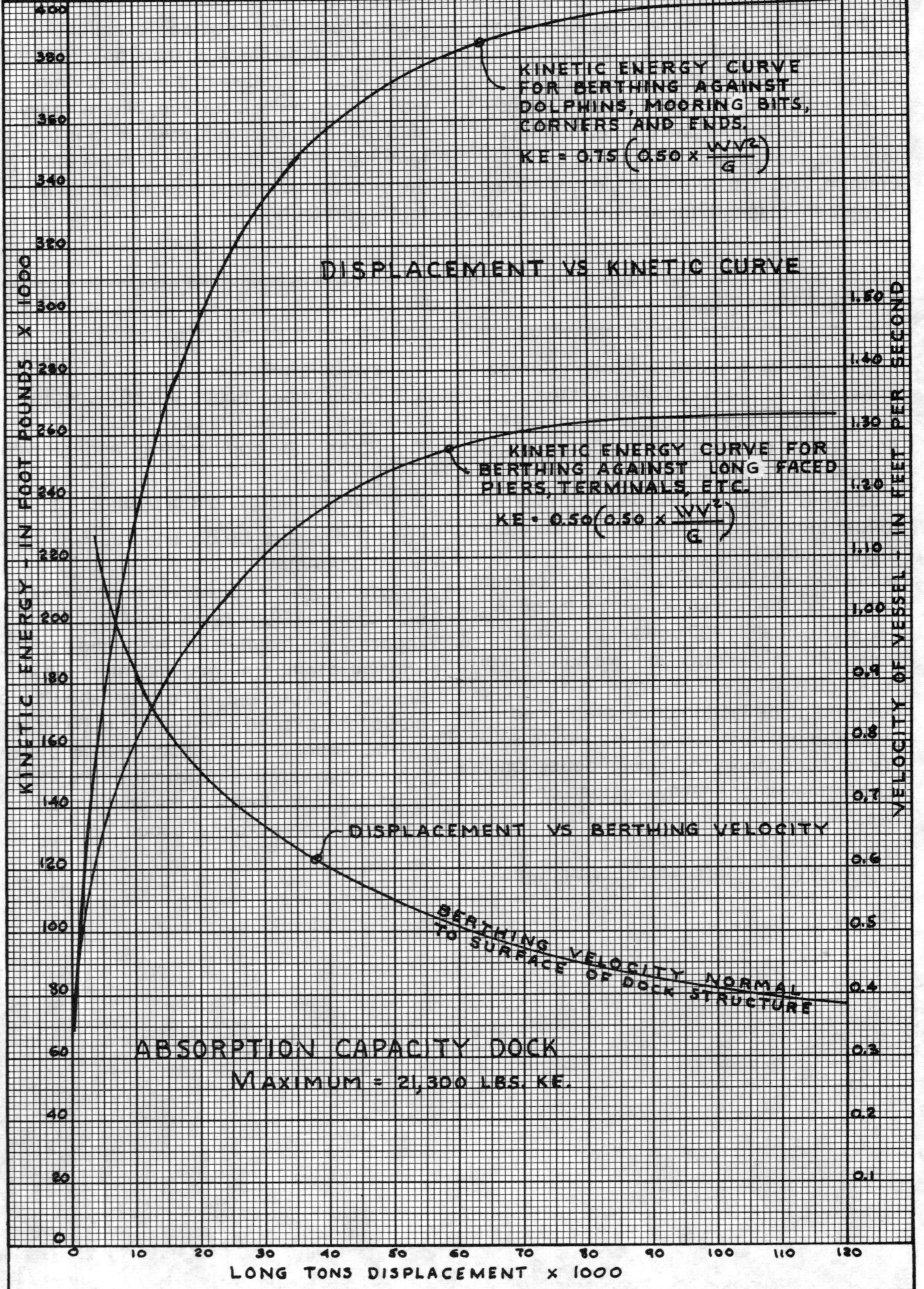

TABLE: Standard rubber dock fenders 9.9.7

TYPES – DIMENSIONS – DEFLECTIONS – WEIGHTS

STANDARD MAXIMUM LENGTHS ALL TYPES IS 19'-0"
LONGER LENGTHS AVAILABLE ON SPECIAL ORDER

CYLINDRICAL TYPE

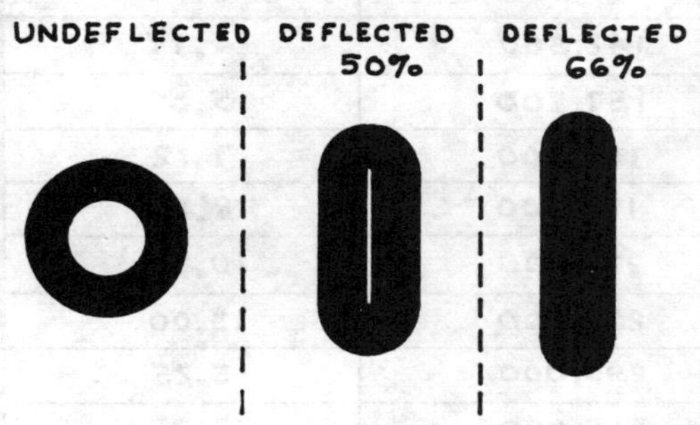

OUTSIDE DIAMETER	BORE INCHES	AREA SQ. INCH	WEIGHT PER FT.	WALL THICKN'S.
3.0	1.5	5.301	3.2	0.75
5.0	2.5	14.726	7.7	1.25
7.0	3.5	28.864	16.0	1.75
8.0	4.0	37.702	21.0	2.00
10.0	5.0	58.906	30.5	2.50
12.0	6.0	84.820	44.0	3.00
15.0	7.5	132.544	69.0	3.75
17.0	8.5	170.242	90.0	4.25
18.0	9.0	190.850	102.0	4.50

RECTANGULAR TYPE

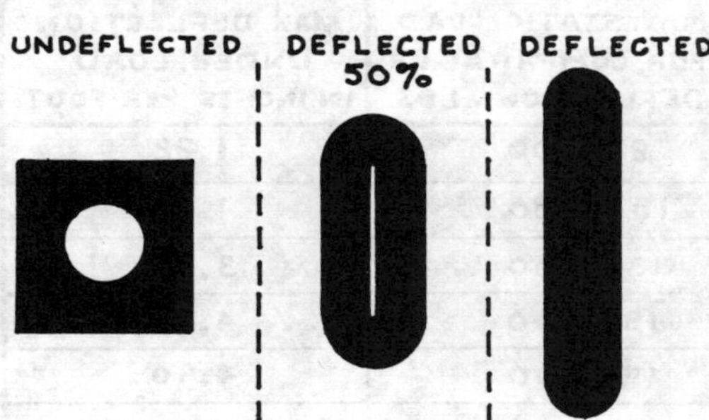

WIDTH INCHES	DEPTH INCHES	BORE INCHES	AREA SQ. INCH	WEIGHT PER FT.
2.0	4.0	NONE	8.00	4.0
3.5	4.5	1.0	14.97	7.5
5.0	6.0	2.5	25.09	14.2
8.0	8.0	3.0	56.93	30.0
8.0	10.0	3.0	72.93	38.0
10.00	12.0	4.0	107.43	56.0
12.0	12.0	5.0	124.37	65.0
7.0	10.0	3.0	62.93	32.8

WING TYPE

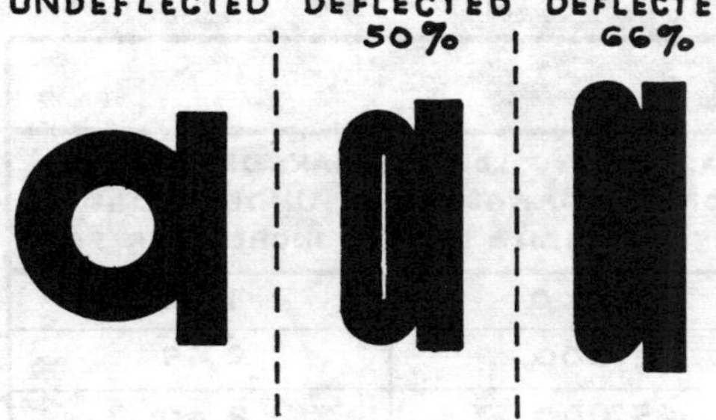

HEIGHT INCHES	WIDTH INCHES	BORE INCHES	THICKN'S. INCHES	WEIGHT PER FT.
2.5	4.0	1.0	0.50	3.0
3.0	6.0	1.0	0.75	5.0
4.0	6.5	2.0	1.00	9.0
4.0	6.5	1.0	1.00	9.5
6.0	9.0	3.0	1.50	16.5
12.0	16.0	4.0	2.50	50.0
6.0	9.5	2.0	1.50	
10.0	16.0	3.0	2.50	

TABLE: Fender resilient properties 9.9.8

CYLINDRICAL TYPE FENDER

SIZE OF FENDER IN INCHES	MAX. ABSORPTION OF KINETIC ENERGY IN FT. LBS., PER LIN. FOOT.	MAX. DEFLECTION RESULT OF KINETIC ENERGY. IN INCHES	MAX. STATIC LOAD FOR COMPARABLE DEFLECTION-LBS.	MAX. DEFLECTION UNDER LOAD IN INCHES PER FT.
3 x 1 3/8	2,000	2.18	84,000	2.10
5 x 2½	2,800	3.50	80,000	3.52
7 x 3	7,300	4.85	142,500	4.77
8 x 3½	9,000	5.50	157,500	5.52
10 x 5	10,300	7.25	157,500	7.12
12 x 6	14,200	8.52	180,000	8.67
15 x 7½	21,300	10.77	220,000	10.80
18 x 9	31,000	13.11	252,500	13.00
21 x 10½	45,700	15.23	290,000	15.25
24 x 12	51,000	17.47	315,000	17.25

RECTANGULAR TYPE FENDER

SIZE OF FENDER IN INCHES	MAX. ABSORPTION OF KINETIC ENERGY IN FT. LBS., PER LIN. FOOT.	MAX. DEFLECTION RESULT OF KINETIC ENERGY. IN INCHES	MAX. STATIC LOAD FOR COMPARABLE DEFLECTION- LBS.	MAX. DEFLECTION UNDER LOAD IN INCHES PER FOOT
2 x 4 x 0	2,000	1.20	80,000	1.22
3½ x 4½ x 1	5,500	2.10	135,000	1.95
5 x 6½ x 2½	7,300	3.13	130,000	3.04
7 x 10 x 3	16,600	4.20	192,500	4.22
8 x 8 x 3	17,000	4.91	197,500	4.90
8 x 10 x 3	21,000	5.21	230,000	5.20
10 x 10 x 4	29,800	6.33	290,000	6.33
10 x 12 x 4	34,500	6.70	320,000	6.72
12 x 12 x 5	44,250	8.09	340,000	8.07

WING TYPE FENDER

SIZE OF FENDER IN INCHES	MAX. ABSORPTION OF KINETIC ENERGY IN FT. LBS., PER LIN. FOOT.	MAX. DEFLECTION RESULT OF KINETIC ENERGY. IN INCHES.	MAX. STATIC LOAD FOR COMPARABLE DEFLECTION - LBS.	MAX. DEFLECTION UNDER LOAD IN INCHES PER FOOT
3 x 1 x 6 x ¾	1,350	1.80	36,000	1.80
4 x 2 x 6½ x 1	2,300	2.47	72,000	2.49
4 x 1 x 6½ x 1	2,850	2.40	65,000	2.40
6 x 2 x 9½ x 1½	6,600	3.60	107,000	3.60
10 x 3 x 16 x 2½	17,150	6.00	180,000	6.00

DEFLECTION CURVES for rubber fenders 9.9.9

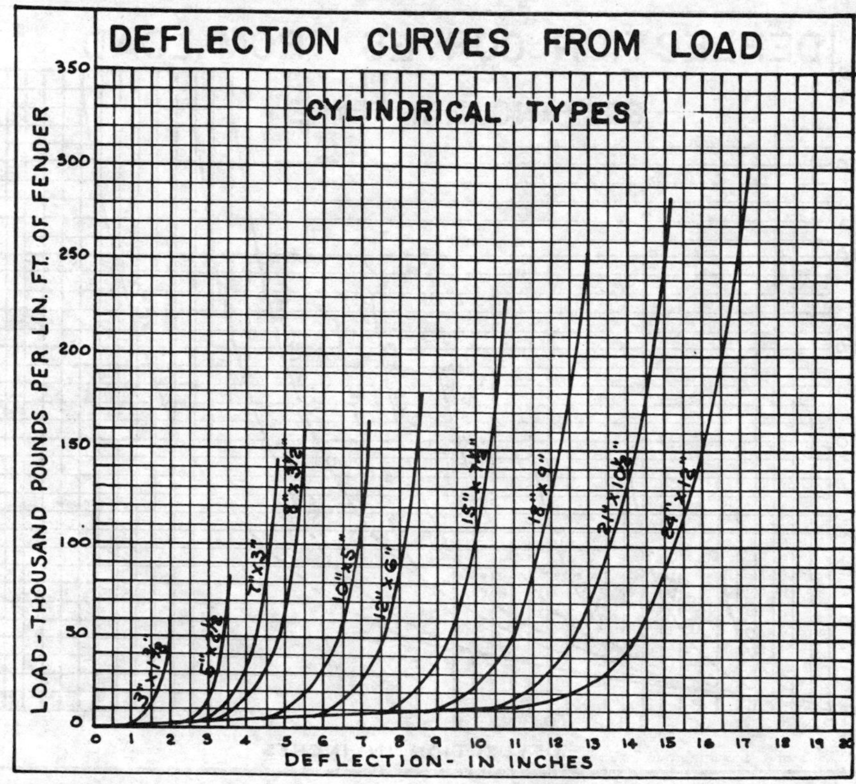

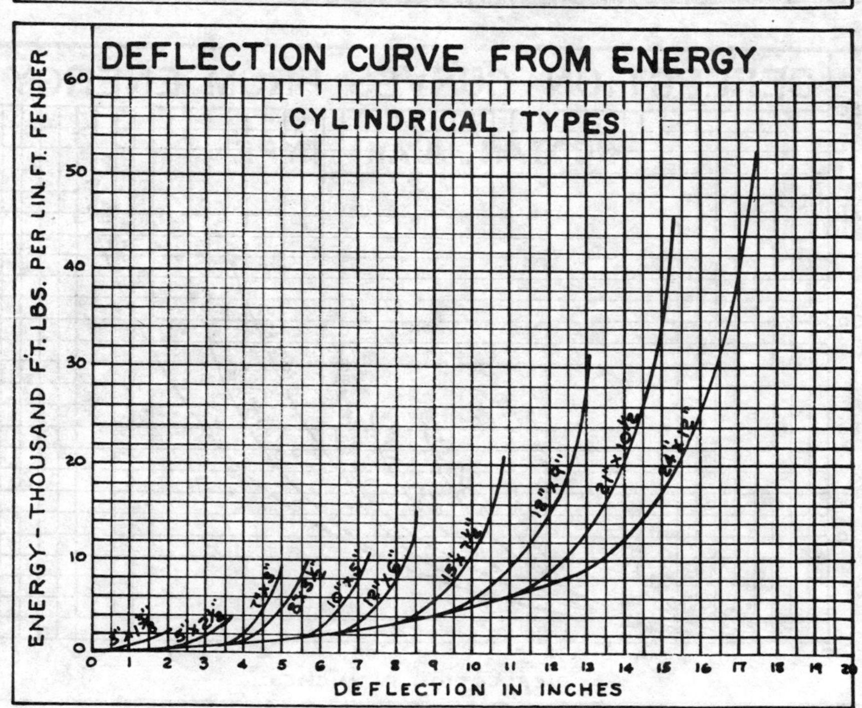

DEFLECTION CURVES for rubber fenders, continued 9.9.9

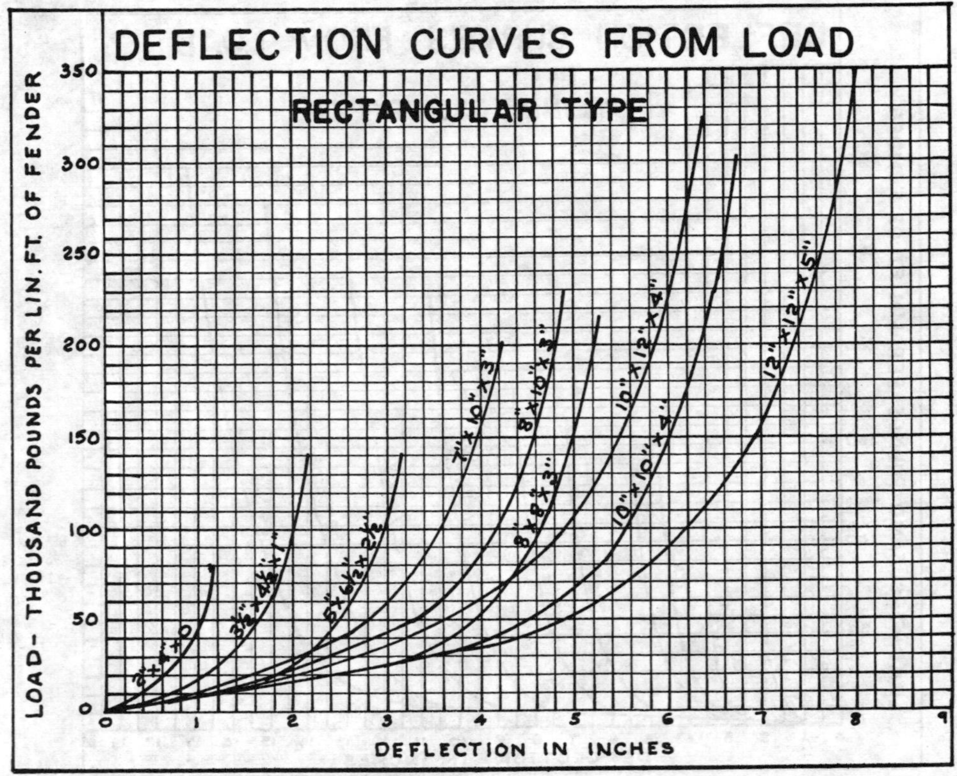

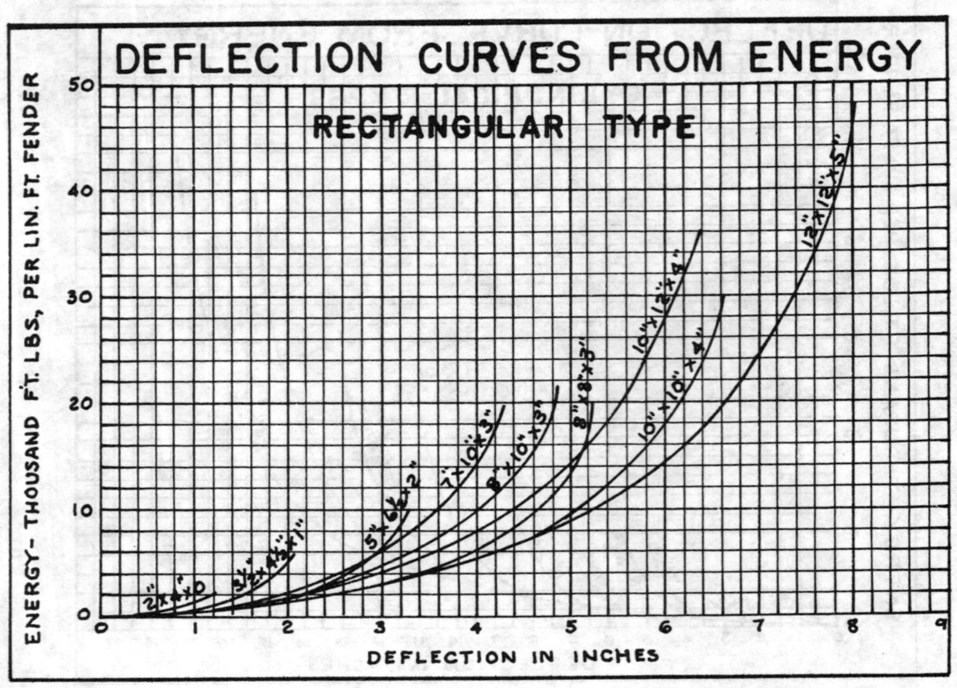

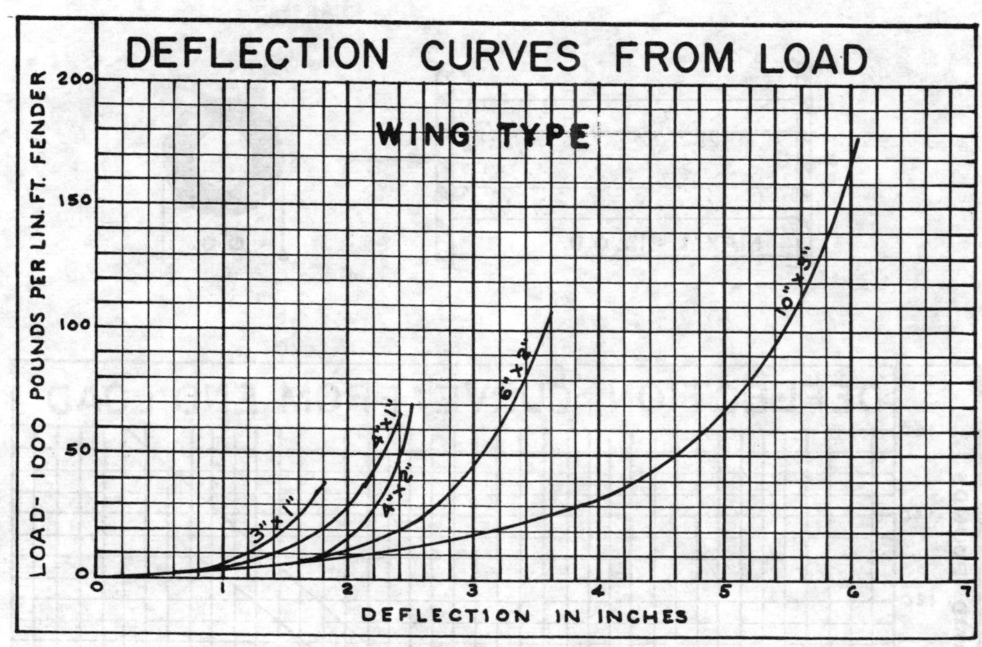
DEFLECTION CURVES FROM LOAD
WING TYPE
LOAD - 1000 POUNDS PER LIN. FT. FENDER
0
50
100
150
200
DEFLECTION IN INCHES
0
1
2
3
4
5
6
7
3"x1"
4"x1"
4"x2"
6"x2"
10"x3"

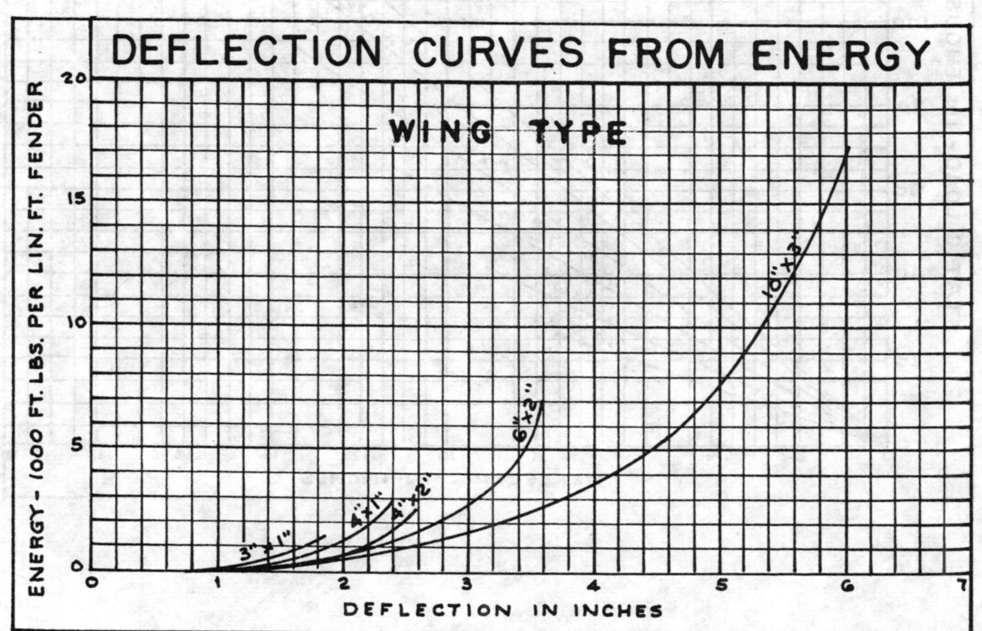
DEFLECTION CURVES FROM ENERGY
WING TYPE
ENERGY - 1000 FT. LBS. PER LIN. FT. FENDER
0
5
10
15
20
DEFLECTION IN INCHES
0
1
2
3
4
5
6
7
3"x1"
4"x1"
4"x2"
6"x2"
10"x3"

DEFLECTION CURVES for rubber fenders, continued 9.9.9

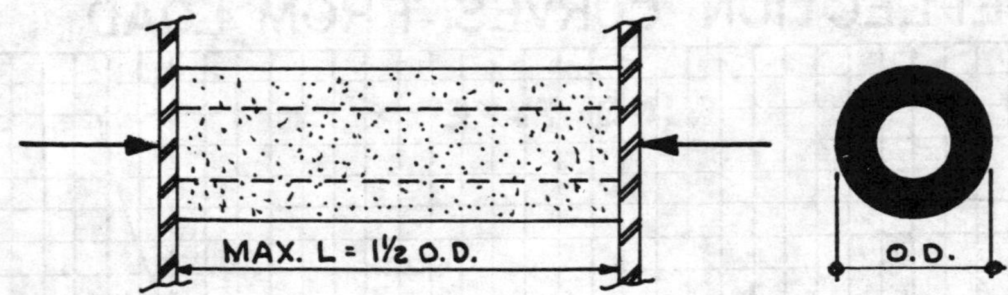

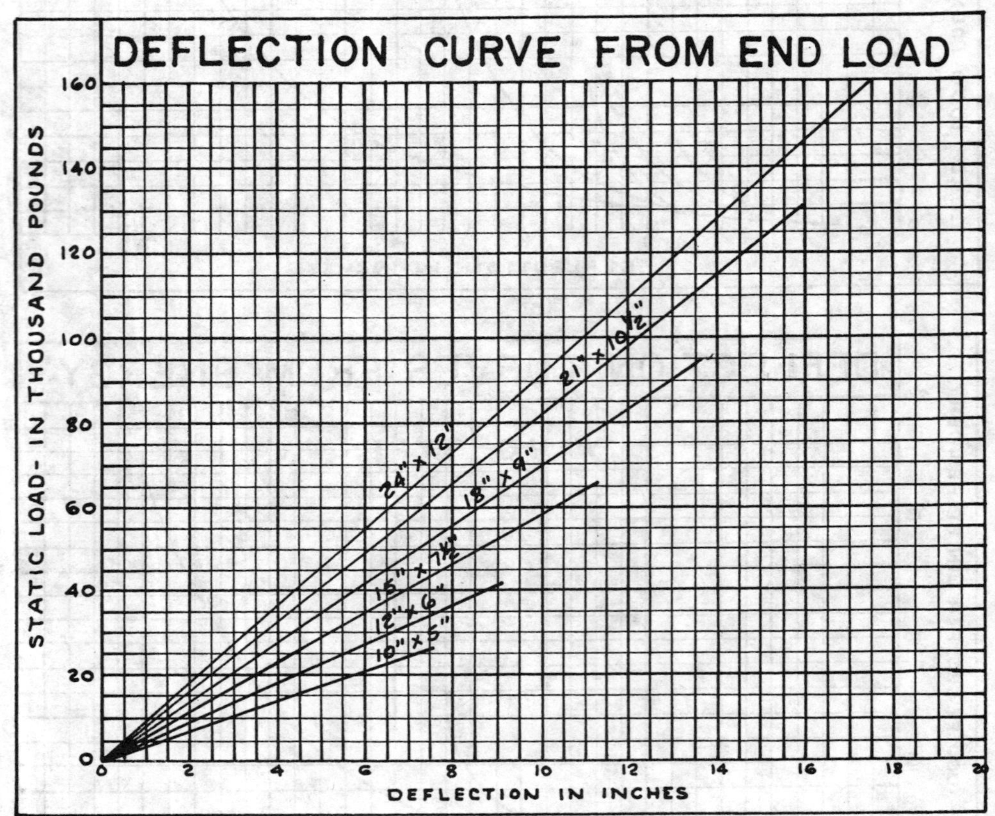

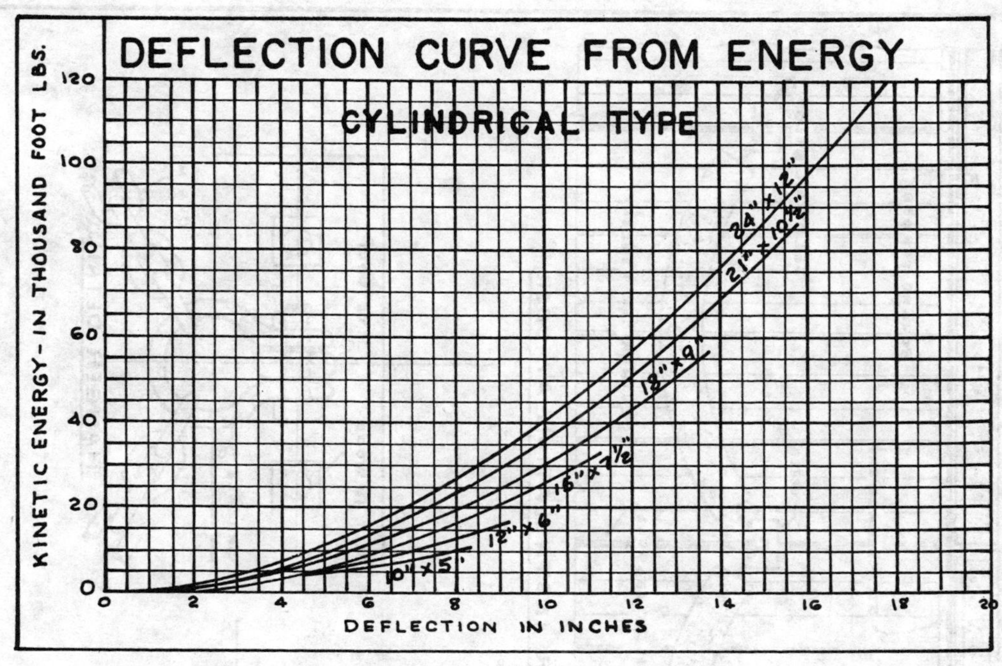
DEFLECTION CURVE FROM ENERGY
CYLINDRICAL TYPE
KINETIC ENERGY - IN THOUSAND FOOT LBS.
0
20
40
60
80
100
120
DEFLECTION IN INCHES
0
2
4
6
8
10
12
14
16
18
20
24" x 12"
21" x 10½"
18" x 9"
15" x 7½"
12" x 6"
10" x 5"

ILLUSTRATIONS: Typical fender details 9.9.10

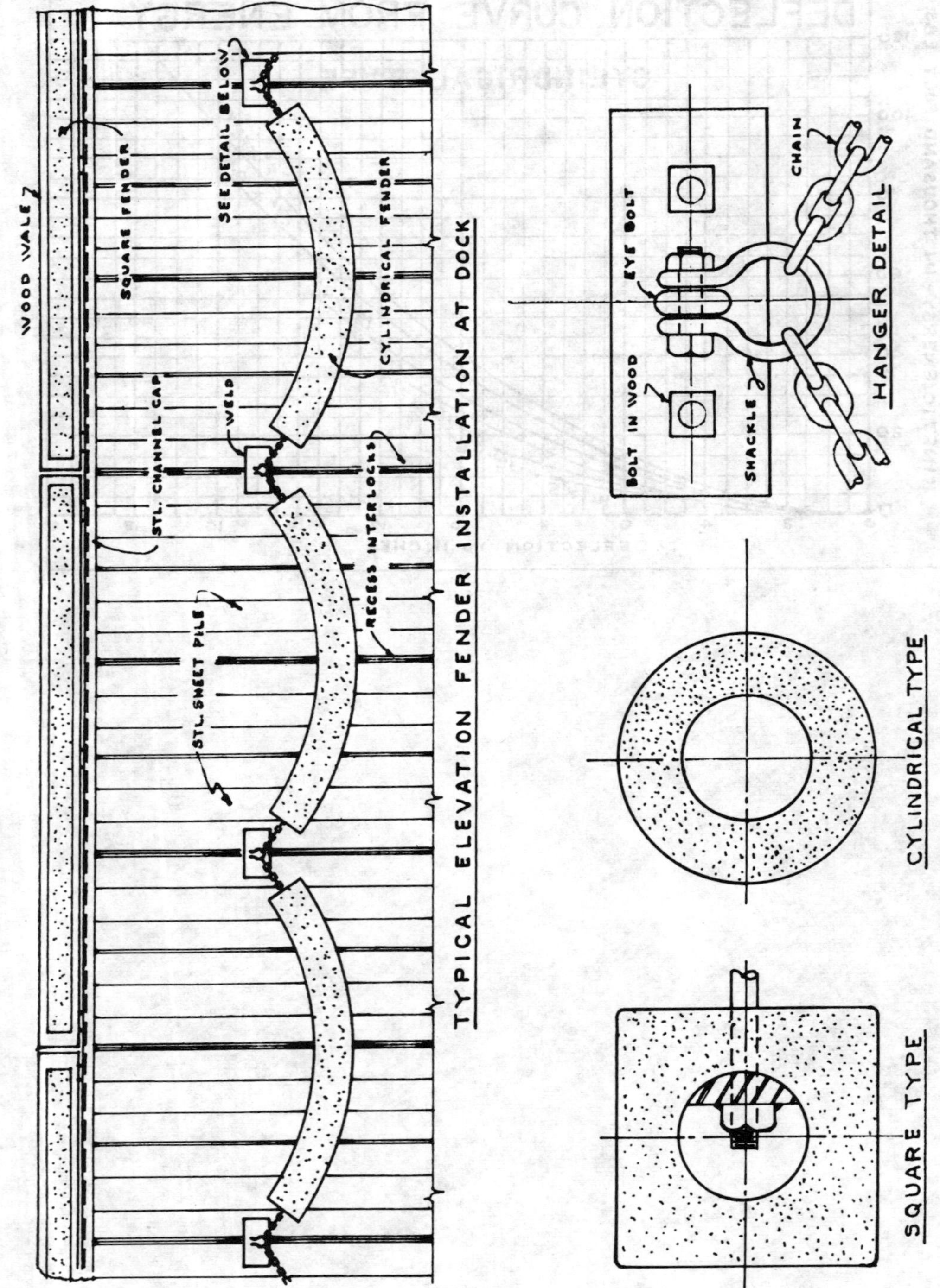

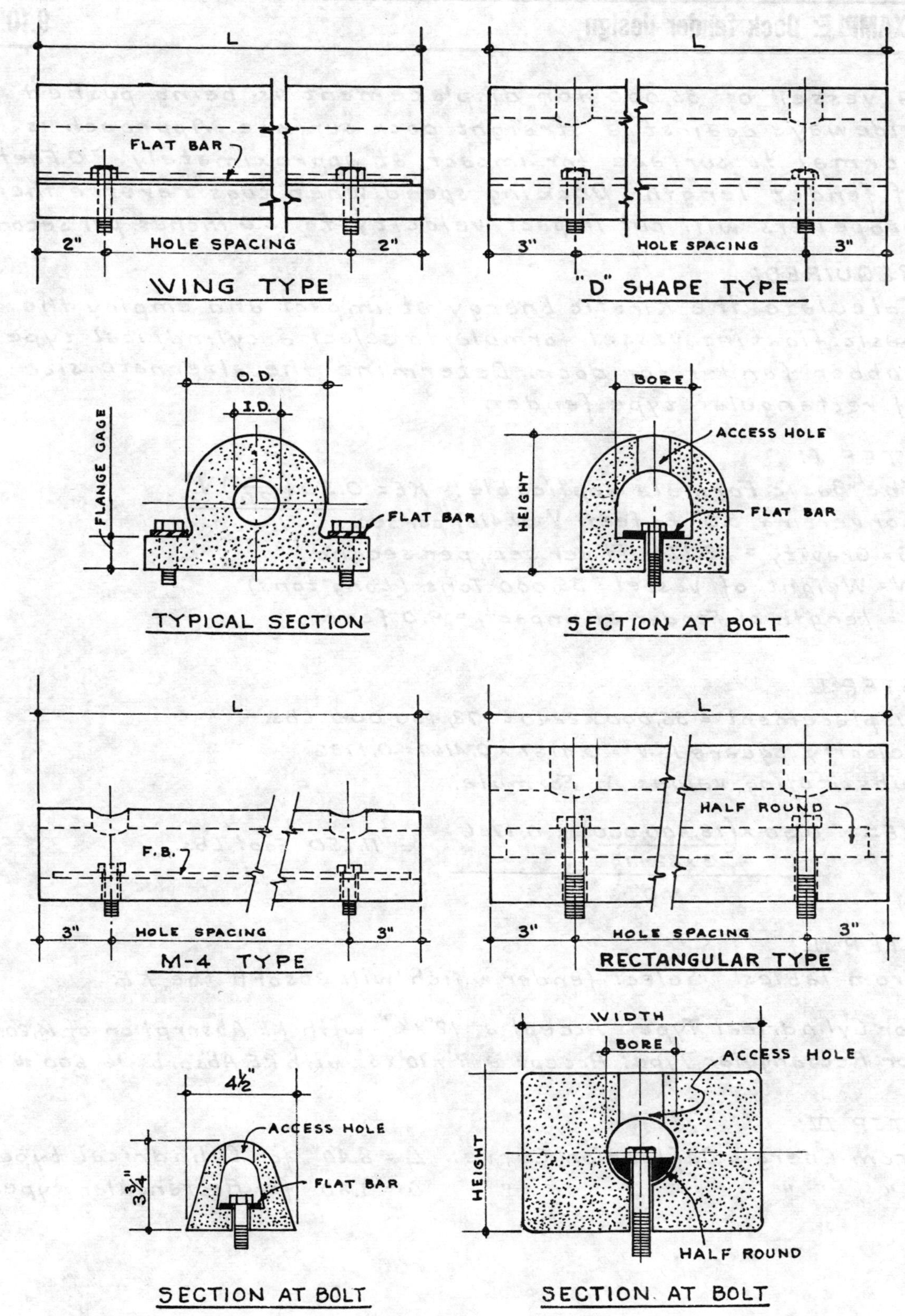
L
FLAT BAR
2"
HOLE SPACING
2"
WING TYPE
L
3"
HOLE SPACING
3"
"D" SHAPE TYPE
O.D.
I.D.
FLANGE GAGE
FLAT BAR
TYPICAL SECTION
BORE
ACCESS HOLE
HEIGHT
FLAT BAR
SECTION AT BOLT
L
F.B.
3"
HOLE SPACING
3"
M-4 TYPE
L
HALF ROUND
3"
HOLE SPACING
3"
RECTANGULAR TYPE
4½"
ACCESS HOLE
3¾"
FLAT BAR
SECTION AT BOLT
WIDTH
BORE
ACCESS HOLE
HEIGHT
HALF ROUND
SECTION AT BOLT

EXAMPLE: Dock fender design 9.10.1

A vessel of 35,000 ton displacement is being pushed sideways against a straight dock surface. Approach is normal to surface for impact at approximately 9.0 Feet of fender length. Docking speed when tugs reverse their propellers will cut impact velocity to 5.0 inches per second.

REQUIRED:
Calculate the Kinetic Energy at impact and employ the basic floating vessel formula to select a cylindrical type rubber fender for dock. Determine the alternate size of rectangular type fender.

STEP I:
The Basic Formula applicable: $KE = \dfrac{0.50\left(\frac{W}{2G}\right)V^2}{L}$.
Converting 5.0" to feet: V = 0.4167' per sec.
G = Gravity = 32.174 Ft. per sec., per second.
W = Weight of Vessel = 35,000 Tons. (Long tons.)
L = Length of Fender at impact, = 9.0 feet.

STEP II
Displacement = 35,000 x 2240 = 78,400,000 Lbs.
Velocity squared: $V^2 = 0.4167 \times 0.4167 = 0.1736$
Substituting values in Formula:

$$KE = \frac{0.50 \times \left(\frac{78,400,000}{2 \times 32.174}\right) \times 0.1736}{9.0'} = 11,750 \text{ Foot Lbs:}$$

STEP III:
From Tables: Select fender which will absorb the KE:

For Cylindrical Type: Accept a 12"x 6" with KE Absorption of 14,200'#
For Rectangular Type: Accept a 7"x 10"x 3" with KE Absorb = 16,600'#

STEP IV:
From Energy Deflection Curve: Δ = 8.40" for Cylindrical type.
" " " " Δ = 3.80" for Rectangular type.

EXAMPLE: End loaded fender design 9.10.2

A Buffer wall to receive impact is constructed parallel to main dock structure. Between main dock and buffer wall, cylindrical rubber fenders of 15" x 7½" are used in end loading position.

REQUIRED:
Determine Maximum length of fender. Limit the amount of deflection to 50 percent of length, then by using unit stress allowable, calculate the amount of Kinetic Energy, and Static Load which produces this deflection. Refer to Chart Curve for end loading for check on KE and P.

STEP I:
Max. Length of fender = 1.5 O.D. Outside Diameter = 15.0 inches.
L = 1.50 x 15.0 = 22.50 inches.
Limited Δ = 0.50 x 22.50 = 11.25 inches.
From Table: 15" x 7½ Cylindrical Type has area, A = 132.544 Sq. In.

STEP II:
Deflection curves for end loads are based on Max. of 50%
To solve for a unit stress under a static load, refer to the curve applying to an 18" x 9." Δ = 13.5" P = 94,800 Lbs. A = 190.85□"

$$\text{Max. unit stress} = \frac{P}{A} = \frac{94,800}{190.85} = 495\ \text{PSI}$$

STEP III:
Applying unit stress of 495 PSI to 15" x 7½ Fender:
Static load P = 132.544 x 495 = 65,600 Lbs. (Checks with curve)
Fender will compress 50% and absorb an energy impact as computed by formula: $KE = 0.50 \times \left(\frac{P\Delta}{12}\right)$
Substituting in formula:

$$KE = 0.50 \times \left(\frac{65,600 \times 11.25}{12}\right) = 30,800\ \text{Ft. Lbs.}$$ (Also checks with curve).

EXAMPLE: Wing fender for truck bumper 9.10.3

An overhead type of Warehouse door is 8.0 feet wide. Door sill is to be protected full width with a wing type or rectangular rubber fender.
State Highway Code limits trucks of tractor-trailer types to 36 tons with full load. Maximum velocity at impact is to be limited to 9.0 inches per second.

REQUIRED:
Design fender for size with maximum deflection limited to 60%. Provide a sketch of cross-section for draftsman. Assume that only 6.0 feet of bumper will absorb impact.

STEP I:
The gravity factor in formula will be deleted for this design, a basic energy formula will apply thus:

$KE = \frac{WV^2}{L}$. $W = 36 \times 2000 = 72{,}000$ Lbs. Convert approaching speed from inches to feet: 9" = 0.75 feet. L = 6.0 feet of bumper to absorb impact.

STEP II:
Substitute values in formula:

$$KE = \frac{72{,}000 \times 0.75 \times 0.75}{6.0} = 6{,}750 \text{ Foot Lbs. per lineal foot.}$$

STEP III:
From Fender Properties Table, a Wing Type Fender 6x2x9½x1½ will absorb 6600 Foot Lbs., of kinetic energy and produce a 50% deflection of 3.60 inches. A rectangular section 5 x 6½ x 2½ will take 7300'# and $\Delta = 3.13$ inches.

STEP IV:
Refer to deflection curve for wing type 6"x2"
At 6750'# of KE, $\Delta = 3.48$ inches. (Less than 50%).
For rectangle fender 5"x 6½" x 2½":
At 6750'# of KE, $\Delta = 3.10$ inches.

The rectangular type is of less weight and probably more economical, however either type will serve the purpose.

EXAMPLE: Using Docking Curve for berthing 9.10.4

A sea-going ship enters a harbor to await the berthing by tug boats. Draught marks on ship reveal a displacement of 60,000 tons. Dock fendering consists of 15.0" x 7.5" Cylindrical Rubber fendering in continuous arrangement.

REQUIRED:

Calculate the safe velocity limit and length of fendering required to absorb Kinetic Energy. Docking operations are to be perpendicular to pier surface. Use the Docking Chart for calculations.

STEP I:

The Absorbtion for 15.0"x 7.5" Cylindrical Fender = 21,300 Foot Pounds per lineal foot.

STEP II:

Begin at bottom of Chart and locate a Displacement equal to, or greater than 60,000 long tons. Follow upward on this line until KE curve is reached. Read on line to left and note that KE = 256,000 Ft. Lbs.

Length of fender to absorb impact is: $\frac{256,000}{21,300} = 12.0$ Feet.

STEP III:

For maximum velocity, read on bottom berthing curve at right a velocity of 0.51 feet per second. (About 6 inches.).

DESIGNERS NOTE:

The Docking Chart takes into account the breaking power from tugs reversing their propellers and reduce the velocity at impact.

To determine the amount of tug's reaction, and water assistance, proceed thus:

Let W = 60,000 Lbs. Distance H = 0.51 Ft. KE = WH or

KE = 60,000 x 0.51 = 306,000 Foot Lbs.

Tugs breaking energy = 306,000 - 256,000 = 50,000 Foot Lbs.

Actual velocity at impact = $\frac{256,000}{60,000}$ = 0.426 Ft. per second.

EXAMPLE: End loaded fenders for buffer wall 9.10.5

A curtain buffer wall is to be constructed parallel to an existing dock for damage protection. Clear space between both surfaces shall not exceed 1.50 feet. Cylindrical fenders are to be placed between structures to absorb impact The largest vessel displacement to dock is rated to be approximately 50,000 long tons. Velocity at docking is limited to seven (7.0) inches per second. Dock bents are on 20.0 foot cc spacing. Impact is not to be absorbed with less than 2 dock bents or 40.0 feet.

REQUIRED:
Calculate the Kinetic Energy developed at impact, then distribute the energy into an end load design which will meet the conditions. Provide a preliminary sketch which could be developed into a possible design and plan.

STEP I:
Determine KE at docking: Displacement = 50,000 tons, and velocity = 7.00 inches per sec. Convert V to feet per second.
$V = \frac{7.00}{12} = 0.5833$ W = 50,000 x 2240 = 112,000,000 Lbs. G = 32.174

STEP II:
Formula for ship: $KE = 0.50\left(\frac{W}{2G}\right)V^2$. Substituting values:

$$KE = \frac{0.50 \times 112{,}000{,}000 \times 0.5833^2}{2 \times 32.174} = \frac{19{,}400{,}000}{64.348} = 301{,}000 \text{ Ft. Lbs.}$$

Maximum distribution is over 2 bents or 40.0 feet.
KE per lineal foot = $\frac{301{,}000}{40.0}$ = 7260 Ft. Lbs. per foot of dock.

STEP III:
Length of end loaded fender, L = 1.50' or 18.0" clear space.
At 50% deflection, Δ = 0.50 x 18.0 = 9.00 inches.
Length cannot exceed 1½ OD of fender, then minimum outside diameter, OD = 18.0/1½ = 12.0 inches.

STEP IV:
Select for trial a 12.0" x 6.0" Clyndrical type. A = 84.82 Sq.In.
At 50% deflection, unit compressive stress F_r = 490 #□".
Max. Static Load, $P = F_r A$ or P = 490 x 84.82 = 41,560 Lbs.

STEP V:
With Static Load P known and deflection Δ = 9.0 inches, the KE Formula can be used.

Continued from Step V;

Formula for KE when P and Δ are known is: $E = \frac{0.50 \times P\Delta}{12}$

Substituting values: $KE = \frac{0.50 \times 41,560 \times 9.0}{12} = 15,600$'#.

STEP VI:

Required number of fenders = $\frac{301,000}{15,600} = 19.3$ (Call it 20).

Layout of fender contacts may be spaced on 4.0 centers and 2 fenders used at each point. With 10 points there will be 9 spaces and absorption will extend 36.0 feet.

STEP VI:

Refer to end loaded deflection curves for checking.

From static load curve: $\Delta = 9.00''$ for 12"x 6" cylindrical section, and P = 42,000 Lbs. Total P = 42,000 x 20 = 840,000 Lbs.

From KE curve: KE = 16,000'#, and total KE = 16,000 x 20 = 320,000 Ft. Lbs.

STEP VII:

Preliminary drawing for design draftsman:

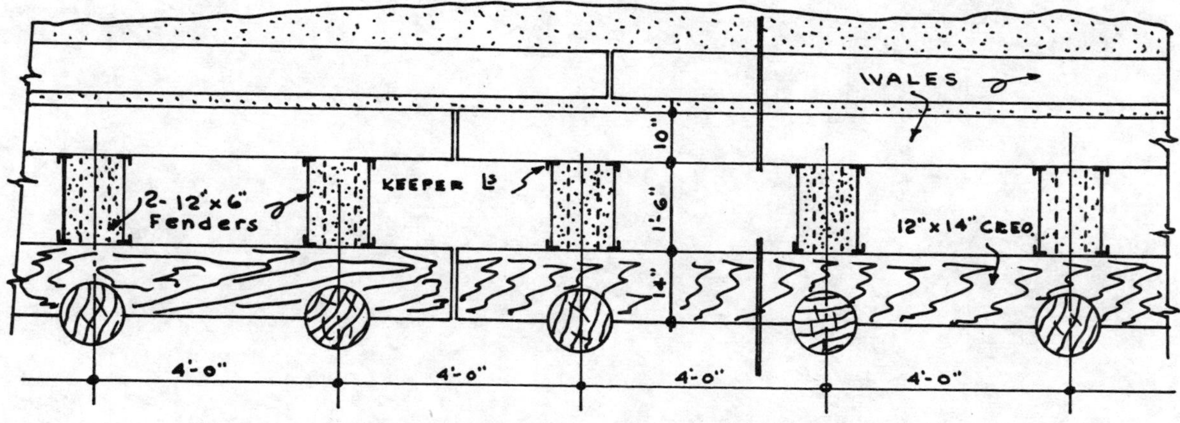

•TYPICAL PLAN-LENGTH OF DOCK•

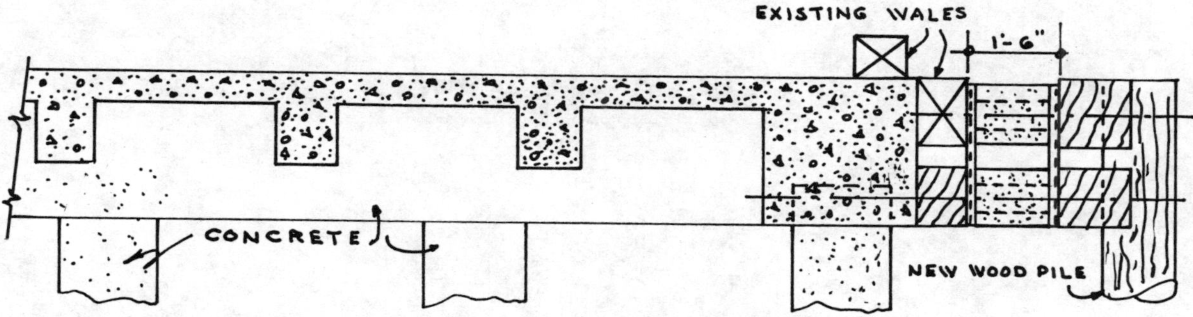

•TYPICAL SECTION THROUGH DOCK•